REVIEW COPY
We would appreciate your sending two copies of your review.

MACROPHAGEPATHOGEN INTERACTIONS
(Immunology Series/60)

Bruce S. Zwilling, Toby K. Eisenstein

September 1993
656 pages, bound, illustrated
ISBN# 0-8247-9124-X

$185.00
(price subject to change without notice)

marcel dekker, inc.
270 MADISON AVENUE • NEW YORK, NY 10016 • 212-696-9000

Macrophage–Pathogen Interactions

IMMUNOLOGY SERIES

Editor-in-chief

NOEL R. ROSE

Professor and Chairman
Department of Immunology and Infectious Diseases
The Johns Hopkins University
School of Hygiene and Public Health
Baltimore, Maryland

European Editor

ZDENEK TRNKA

Basel Institute for Immunology
Basel, Switzerland

1. Mechanisms in Allergy: Reagin-Mediated Hypersensitivity, *edited by Lawrence Goodfriend, Alec Sehon and Robert P. Orange*
2. Immuopathology: Methods and Techniques, *edited by Theodore P. Zacharia and Sidney S. Breese, Jr.*
3. Immunity and Cancer in Man: An Introduction, *edited by Arnold E. Reif*
4. *Bordetella pertussis*: Immunological and Other Biological Activities, *J. J. Munoz and R. K. Bergman*
5. The Lymphocyte: Structure and Function (in two parts), *edited by John J. Marchalonis*
6. Immunology of Receptors, *edited by B. Cinader*
7. Immediate Hypersensitivity: Modern Concepts and Development, *edited by Michael K. Bach*
8. Theoretical Immunology, *edited by George I. Bell, Alan S. Perelson, and George H. Pimbley, Jr.*
9. Immunodiagnosis of Cancer (in two parts), *edited by Ronald B. Herberman and K. Robert McIntire*
10. Immunologically Mediated Renal Diseases: Criteria for Diagnosis and Treatment, *edited by Robert T. McCluskey and Guiseppe A. Andres*
11. Clinical Immunotherapy, *edited by Albert F. LoBuglio*
12. Mechanisms of Immunity to Virus-Induced Tumors, *edited by John W. Blasecki*
13. Manual of Macrophage Methodology: Collection, Characterization, and Function, *edited by Herbert B. Herscowitz, Howard T. Holden, Joseph A. Bellanti, and Abdul Ghaffar*
14. Supressor Cells in Human Disease, *edited by James S. Goodwin*
15. Immunological Aspects of Aging, *edited by Diego Segre and Lester Smith*
16. Cellular and Molecular Mechanisms of Immunologic Tolerance, *edited by Tomáš Hraba and Milan Hašek*
17. Immune Regulation: Evolution and Biological Significance, *edited by Laurens N. Ruben and M. Eric Gershwin*
18. Tumor Immunity in Prognosis: The Role of Mononuclear Cell Infiltration, *edited by Stephen Haskill*
19. Immunopharmacology and the Regulation of Leukocyte Function, *edited by David R. Webb*

20. Pathogenesis and Immunology of Treponemal Infection, *edited by Ronald F. Schell and Daniel M. Musher*
21. Macrophage-Mediated Antibody-Dependent Cellular Cytotoxicity, *edited by Hillel S. Koren*
22. Molecular Immunology: A Textbook, *edited by M. Zouhair Atassi, Carel J. van Oss, and Darryl R. Absolom*
23. Monoclonal Antibodies and Cancer, *edited by George L. Wright, Jr.*
24. Stress, Immunity, and Aging, *edited by Edwin L. Cooper*
25. Immune Modulation Agents and Their Mechanisms, *edited by Richard L. Fenichel and Michael A. Chirigos*
26. Mononuclear Phagocyte Biology, *edited by Alvin Volkman*
27. The Lactoperoxidase System: Chemistry and Biological Significance, *edited by Kenneth M. Pruitt and Jorma O. Tenovuo*
28. Introduction to Medical Immunology, *edited by Gabriel Virella, Jean-Michel Goust, H. Hugh Fudenberg, and Christian C. Patrick*
29. Handbook of Food Allergies, *edited by James C. Breneman*
30. Human Hybridomas: Diagnostic and Therapeutic Applications, *edited by Anthony J. Strelkauskas*
31. Aging and the Immune Response: Cellular and Humoral Aspects, *edited by Edmond A. Goidl*
32. Complications of Organ Transplantation, *edited by Luis H. Toledo-Pereyra*
33. Monoclonal Antibody Production Techniques and Applications, *edited by Lawrence B. Schook*
34. Fundamentals of Receptor Molecular Biology, *Donald F. H. Wallach*
35. Recombinant Lymphokines and Their Receptors, *edited by Steven Gillis*
36. Immunology of the Male Reproductive Organs, *edited by Pierluigi E. Bigazzi*
37. The Lymphocyte: Structure and Function, *edited by John J. Marchalonis*
38. Differentiation Antigens in Lymphohemopoietic Tissues, *edited by Masayuki Miyasaka and Zdenek Trnka*
39. Cancer Diagnosis In Vitro Using Monoclonal Antibodies, *edited by Herbert Z. Kupchick*
40. Biological Response Modifiers and Cancer Therapy, *edited by J. W. Chiao*
41. Cellular Oncogene Activation, *edited by George Klein*
42. Interferon and Nonviral Pathogens, *edited by Gerald I. Byrne and Jenifer Turco*
43. Human Immunogenetics: Basic Principles and Clinical Relevance, *edited by Stephen D. Litwin, with David W. Scott, Lorraine Flaherty, Ralph A. Reisfeld, and Donald M. Marcus*
44. AIDS: Pathogenesis and Treatment, *edited by Jay A. Levy*
45. Cell Surface Antigen Thy-1: Immunology, Neurology, and Therapeutic Applications, *edited by Arnold E. Reif and Michael Schlesinger*
46. Immune Mechanisms in Cutaneous Disease, *edited by David A. Norris*
47. Immunology of Fungal Diseases, *edited by Edouard Kurstak*
48. Adoptive Cellular Immunotherapy of Cancer, *edited by Henry C. Stevenson*
49. Colony-Stimulating Factors: Molecular and Cellular Biology, *edited by T. Michael Dexter, John M. Garland, and Nydia G. Testa*
50. Introduction to Medical Immunology: Second Edition, *edited by Gabriel Virella, Jean-Michel Goust, and H. Hugh Fudenberg*
51. Tumor Suppressor Genes, *edited by George Klein*
52. Organ-Specific Autoimmunity, *edited by Pierluigi E. Bigazzi, Georg Wick, and Konrad Wicher*
53. Immunodiagnosis of Cancer: Second Edition, *edited by Ronald B. Herberman and Donald W. Mercer*

54. Systemic Autoimmunity, *edited by Pierluigi E. Bigazzi and Morris Reichlin*
55. Molecular Immunobiology of Self-Reactivity, *edited by Constantin A. Bona and Azad K. Kaushik*
56. Tumor Necrosis Factors: Structure, Function, and Mechanism of Action, *edited by Bharat B. Aggarwal and Jan Vilček*
57. Granulocyte Responses to Cytokines: Basic and Clinical Research, *edited by Ronald G. Coffey*
58. Introduction to Medical Immunology: Third Edition, *edited by Gabriel Virella*
59. Monoclonal Antibodies and Peptide Therapy in Autoimmune Diseases, *edited by Jean-François Bach*
60. Macrophage–Pathogen Interactions, *edited by Bruce S. Zwilling and Toby K. Eisenstein*

ADDITIONAL VOLUMES IN PREPARATION

Tumor Immunology and Cancer Therapy, *edited by Ronald H. Goldfarb and Theresa L. Whiteside*

Macrophage–Pathogen Interactions

edited by

Bruce S. Zwilling
The Ohio State University
Columbus, Ohio

Toby K. Eisenstein
Temple University School of Medicine
Philadelphia, Pennsylvania

Marcel Dekker, Inc. New York • Basel • Hong Kong

Library of Congress Cataloging-in-Publication Data

Macrophage–pathogen interactions / edited by Bruce S. Zwilling, Toby K. Eisenstein.
p. cm. -- (Immunology series ; v. 60)
Includes bibliographical references and index.
ISBN 0-8247-9124-X (alk. paper)
1. Macrophages. 2. Antibody–dependent cell cytotoxicity.
I. Zwilling, Bruce S. II. Eisenstein, Toby K.
III. Series.
[DNLM: 1. Macrophages--immunology. 2. Macrophage Activation--immunology. W1 IM53K v. 60 1993 / WH 650 M17385 1994]
QR185.8.M3M2947 1994
616.07'9--dc20
DNLM/DLC
for Library of Congress 93-27941
CIP

The publisher offers discounts on this book when ordered in bulk quantities. For more information, write to Special Sales/Professional Marketing at the address below.

This book is printed on acid-free paper.

Marcel Dekker, Inc.
270 Madison Avenue, New York, New York 10016

Current printing (last digit):
10 9 8 7 6 5 4 3 2 1

PRINTED IN THE UNITED STATES OF AMERICA

Series Introduction

The decade of the 1980s is often termed the era of the lymphocyte. During that period, the structure of the T-cell receptor was revealed and the details of the genetic basis of immunoglobulin and T-cell diversity worked out. As we moved into the 1990s, the molecular structure of the major histocompatibility antigens was solved, and the interactions between antigen-presenting cells and T cells have been illuminated. Thus the stage is set to advance the study of macrophage functions to the subcellular and molecular levels. The decade of the macrophage has dawned.

Clearly, the macrophage and its antigen-presenting cousins are critical elements in initiating the immune response. The full extent of the macrophage as an effector cell has more recently been perceived. As a generator of diverse cytokines, it participates in immunological protection through classical pathways of phagocytosis and through novel mechanisms involving oxygen-free radicals, reactive nitrogen intermediates, nitric oxide, and other mediators.

By bringing together the widely scattered literature relating to macrophage–pathogen interactions, the contributors to this volume have focused attention on the common mechanisms of macrophage effector functions. They foreshadow a trail leading to new and even more provocative discoveries in this decade of the macrophage.

Noel R. Rose

Foreword

The mononuclear phagocyte system (MPS) was first postulated more than 30 years ago [1]. Today the MPS has been widely accepted as a replacement for the reticulo-endothelial system (RES) [2] and is included in many textbooks of immunology, hematology, and pathology. The power of the MPS lies in the concept that cells of common origin form a cell line and thus can be distinguished from cells of another origin. The origin of the cells of the RES was unknown at the time the system was introduced, but at present the difference in origin between macrophages, endothelial cells, and reticulum cells is well known. The cells of the RES do not form a system of cells sharing origin, morphology, and function [2].

Another cell often thought to be related to the macrophage is the dendritic cell, which also originates from a precursor cell in bone marrow, the identity of which has not yet been established. These precursor cells circulate in the peripheral blood and migrate to various organs, e.g., the skin (called Langerhans cells), spleen, and lymph nodes, where they differentiate into dendritic cells. Thus the dendritic cells form a cell line separate from the MPS.

I thought it would be relevant to summarize schematically the origin and kinetics of cells that form the MPS (Figure 1) [1,3]. The precursors of monocytes are monoblasts and promonocytes, cells that divide only once in the bone marrow. After a short period of maturation, monocytes leave the bone marrow and migrate to the bloodstream, where they circulate for 1 to 3 days. Under normal steady-state conditions monocytes migrate randomly to various organs and body cavities where they differentiate into macrophages. Macrophages remain at these sites for 1–3 months and then die, locally or in draining lymph nodes. During a local infection or inflammation, chemotactic factors and other mediators of inflammation, such as cytokines and eicosanoids, stimulate monocytes to migrate to the relevant site where they differentiate into exudate macrophages. Leukocyte adhesion molecules on monocytes and their counterstructures on endothelial cells are involved in the adherence of monocytes to endothelial cells and their subsequent transendothelial migration during normal steady-state conditions and inflammation [4]. According to their localization in tissues and body cavities, macrophages vary in their morphological and functional characteristics and therefore have different names, e.g., Kupffer cells in the liver, pulmonary and alveolar macrophages in the lungs, and microglial cells in the central nervous system. Macrophages also have several characteristics in common, such as the presence of various receptors (e.g., Fcγ receptors, $\beta 2$ integrins, mannose receptors), antigens (e.g., F4/80, CD68), and certain functions (e.g., phagocytosis of opsonized particles, production of cytokines, such as TNF, IL-1, IL-6).

During the past few decades, our insight into the physiology and pathophysiology of monocytes and macrophages has increased enormously: from 1965 to 1990 the number of relevant publications increased from about 80 to 1675 [5]. The present volume is a reflection of this increase in knowledge which, as can be seen from the Table of Contents, is rather diversified.

Our current knowledge of the cellular immune response in experimental animals, especially mice, is quite extensive and we expect that similar mechanisms apply in humans. To be more explicit, many details of the macrophage at various locations in the mouse are known, but studies

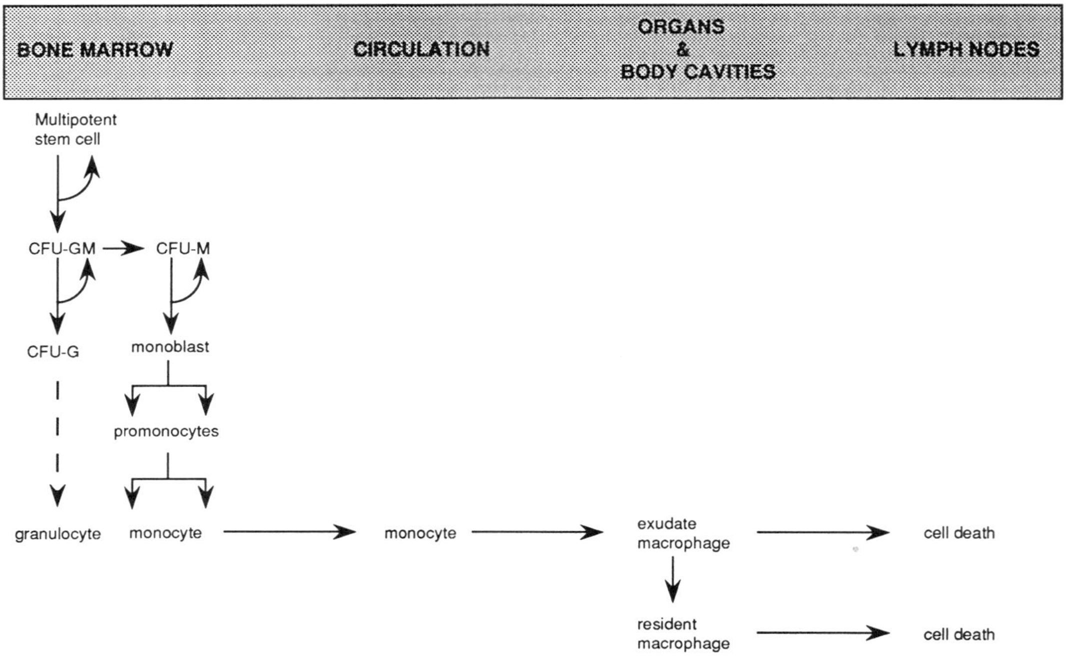

FIGURE 1 Outline of the life cycle of mononuclear phagocytes. Information on the origin of monocytes in bone marrow, their transport via the circulation, their migration to organs and body cavities, and their death in lymph nodes has been obtained mainly from studies of mice.

of the human macrophage are scarce, partly due to the difficulty or impossibility of collecting these cells.

Most studies in mice are performed with resident or exudate peritoneal macrophages. The latter population of cells, which are obtained after injection of an inflammatory stimulus, is actually a mixture of resident macrophages and exudate macrophages, i.e., recently recruited monocytes. Since these two kinds of cells differ in many aspects, the results obtained with exudate macrophages should be interpreted with caution.

Murine resident peritoneal macrophages are often considered as the representative example of tissue macrophages. However, in recent years I have begun to question the truth of this view. I have started to consider these macrophages as "burned-out" cells which differ from cells located "where the fire is", that is, at the site of infection or inflammation. The functions of resident macrophages can be enhanced by all kinds of stimuli, but this often results in a return to a previous function, i.e., as a monocyte. It is more likely that in real life exudate macrophages and not resident macrophages are stimulated at the site of infection or inflammation. Furthermore, to assume that the results obtained with murine macrophages are the same as with human cells is not justified. Murine and human monocytes and macrophages differ in many respects. The latest, intriguing distinction between murine and human macrophages is in the production of reactive nitrogen intermediates [6,7]. Since human macrophages are difficult to collect—peritoneal macrophages obtained during laparoscopy and alveolar macrophages obtained during broncho-alveolar lavage being the exception—fresh and cultured monocytes are frequently used. The latter cells are considered as monocytes that have differentiated in vitro into macrophages and are therefore called monocyte-derived macrophages. Although very often such cells are the closest possible approximation of human macrophages, especially for studies on cells from patients, the experimental results achieved after modulation of these cells in vitro may be of limited value in relation to the actual situation in vivo. Similar objections can be applied to the use of murine or human macrophage cell lines. The results of experiments with these cells can be exciting, but in vivo conditions will be different. Thus, a monocyte is not a macrophage and a murine macrophage is not a human cell.

The above remarks are only words of caution which are not intended to discourage research involving these cells. Why is research based on these in vitro cell models or experimental animals, usually mice, still valuable? Because it provides insight into the various mechanisms by which monocytes and macrophages conquer microorganisms or eliminate other deleterious materials or cells. Knowledge of these physiological and pathophysiological mechanisms can lead to the possibility of manipulating these cells with drugs or biological factors, thus increasing their functional efficacy.

The present volume, which deals with the interaction between macrophages and pathogens, discusses these aspects in a clear and stimulating way. I am particularly impressed by the first part, which covers macrophage biology and provides an excellent review of the state of the art, including new aspects, such as suppressor macrophages and antimicrobial polypeptides. The other parts include the interactions between macrophages and bacteria, protozoa, fungi, and viruses. It is of great interest to realize how much we know abut the interactions between macrophages and the various facultative intracellular microorganisms, which are all discussed in Parts II and III, and how little we know about the obligate intracellular microorganisms, the viruses. However, in the defense of viral infections natural killer cells and cytotoxic T lymphocytes are the most important effector cells. Together this monograph demonstrates that in vitro models and animal studies are indispensable for an understanding of host defense mechanisms. Although most of the insights into the interactions of macrophages with microorganisms are not yet directly applicable to the patient with infection and have not yet resulted in new medicines, novel remedies lie on the horizon.

Ralph van Furth

Professor of Internal Medicine
and Infectious Diseases
University Hospital
Leiden, The Netherlands

REFERENCES

1. van Furth R, Cohn AZ, Hirsch JG, Humphrey JH, Spector WG, Langevoort HL. The mononuclear phagocyte system: a new classification of macrophages, monocytes, and their precursor cells. Bull. Wld. Hlth. Org. 1972; 46:845–52.
2. Aschoff L. Das Reticulo-endotheliale System. Ergebn. inn. Med. Kinderheilk. 1924; 26:1–118.
3. van Furth R. Production and migration of monocytes and kinetics of macrophages. In: van Furth R, ed. Mononuclear phagocytes. Biology of monocytes and macrophages. Kluwer Academic Publishers, Dordrecht, Boston, London, 1992; 3–12.
4. Beekhuizen H, Blokland I, Corsel-van Tilburg AJ, van Furth R. Human monocytes exhibit different mechanisms of adhesion to non-stimulated or cytokine-stimulated human endothelial cells. In: van Furth R, ed. Mononuclear phagocytes. Biology of monocytes and macrophages. Kluwer Academic Publishers, Dordrecht, Boston, London, 1992; 105–16.
5. van Furth R. Preface. In: van Furth R, ed. Mononuclear phagocytes. Biology of monocytes and macrophages. Kluwer Academic Publishers, Dordrecht, Boston, London, 1992; xi–xii.
6. Stuehr J, Marletta MA. Induction of nitrate/nitrate synthesis in murine macrophages by BCG infection, lymphokines, or interferon-γ. J. Immunol. 1987; 139:518–25.
7. Murray HW, Teitelbaum RF. L-arginine-dependent reactive nitrogen intermediates and the antimicrobial effect of activated human mononuclear phagocytes. J. Infect. Dis. 1992; 162:513–17.

Preface

Macrophages are central to host defense against microbes. They are involved in both the afferent and efferent arms of the immune system, the former in antigen processing and presentation, and the latter as the major effector cells that contain and kill intracellular pathogens of macrophages. The motivating factor behind assembling information in this book was the appreciation that investigators who use particular organisms belong to different professional societies, attend different scientific meetings, and even within societies are usually grouped by organisms. Yet it is quite clear that similar kinds of strategies are used by microbes of different phylogenic groups, and similar types of host responses are mounted to these pathogens. Frequently, investigators working with one organism are not aware of advances or peculiarities observed with other organisms with respect to interactions with macrophages. We thought that a book devoted to macrophage–pathogen interactions would allow investigators access in a single volume to reviews of how a variety of obligate and facultative intracellular pathogens of macrophages interact with host cells. In addition we decided it would be useful to preface these chapters with some general chapters on macrophage cell biology.

The book is divided into five parts. Part I provides the reader with background information on macrophage function. The introduction, by Carleton Stewart and colleagues, provides information on macrophage differentiation from bone marrow stem cells. The second chapter, by Robert Crawford and Monte Meltzer et al., provides a historical review of macrophage activation together with a comparison of activation of human monocytes and rodent macrophages. This review is followed by Donna Paulnock's chapter describing the molecular biology of macrophage activation. Carl Manthey and Stefanie Vogel then discuss interactions of macrophages with lipopolysaccharides, followed by Thomas Hamilton and colleagues on the signaling events that follow LPS stimulation of macrophages. The interaction of macrophages with antibody and complement as well as other receptors is discussed by David Mosser.

One of the most important roles of the macrophage is as an antigen-presenting cell, and in Chapter 6 Susan Safley and Kirk Ziegler review work on processing and presentation of microbial antigens. The role of macrophages in inflammatory processes is reviewed by Bent Windelborg Neilsen with Kouji Matsushima et al. Richard Mortensen reviews the interactions of macrophages with acute phase proteins. The next three chapters discuss some of the antimicrobial mechanisms of macrophages. John Robinson and John Badwey review reactive oxygen metabolites, Marion Woods and John Hibbs et al. discuss their work on reactive nitrogen intermediates, and Pieter Hiemstra and Ralph van Furth discuss antimicrobial peptides. The last three chapters in Part I discuss regulatory aspects of macrophage function. Toby Eisenstein reviews work on suppressor macrophages, Julie Djeu discusses NK cell–macrophage interactions, and Bruce Zwilling and Mary Hilburger review the role of macrophage resistance genes *Bcg/Ity/Lsh* in controlling macrophage function.

Part II begins our discussion of macrophage–pathogen interactions and focuses on bacterial interactions. The first three chapters discuss macrophage–mycobacterial interactions. Graham

Rook writes about macrophages and *Mycobacterium tuberculosis*; John Johnson and Jerry Ellner and colleague review work on macrophage *Mycobacterium avium* interactions, and Jim Krahenbuhl and Linda Adams discuss the interaction of *Mycobacterium leprae* and macrophages. These reviews are followed by a discussion of salmonella infection by Nancy Dunlap with David Briles et al. Priscilla Campbell then provides us with her special insights into macrophage–*Listeria* interactions. *Legionella* interactions with macrophages are discussed by Yoshimasa Yamamoto with Herman Friedman and colleagues, and Anne Fortier with Carol Nacy et al. review work on *Francisella tularensis* and the macrophage. The last three chapters in this part consist of a review by Cynthia Baldwin and Alex Winter on macrophage–*Brucella* interactions; a review by David La Verda and Gerald Byrne on *Chlamydia* and macrophages, and Herbert Winkler and Jenifer Turco discuss macrophage–*Rickettsia* interactions.

Part III reviews macrophage–parasite interactions. Henry Koziel with Richard Rose et al. provide a very timely review of macrophage *Pneumocystis carinii* interactions. This is followed by Richard Titus's review of the interactions of *Leishmania major* and macrophages. Stephanie James and Dov Boros review schistosomiasis while Miles Beaman with Jack Remington and colleagues discuss macrophage–*Toxoplasma* interactions. Part III is concluded by Raymond Kuhn's review of *Trypanosoma cruzi* and macrophages.

Part IV reviews macrophage–fungal interactions. Richard Calderone and Joy Sturtevant review macrophage–*Candida* interactions. The interaction of *Histoplasma capsulatum* with macrophages is discussed by Simon Newman and Ward Bullock, and Stuart Levitz reviews macrophage cryptococcal interactions. Part IV is concluded by reviews by Andreas Schaffner on macrophage–*Aspergillus* interactions and by Robert Bradsher on the interaction of *Blastomyces dermatitides* with macrophages.

The last part provides a review of macrophage–viral interactions. Wen-Zhe Ho with Steven Douglas et al. discuss the interaction of HIV with macrophages. The interaction of macrophages with other lentiviruses is then discussed by Janice Clements. Finally, Jill Hendrzak and Page Morahan review macrophage–herpes simplex interactions.

We hope this volume will prove helpful to the research community and enable us to advance our knowledge at a more rapid rate.

Bruce S. Zwilling
Toby K. Eisenstein

Contents

Contributors

Linda B. Adams, Ph.D. G.W. Long Hansen's Disease Center, Laboratory Research Branch at Louisiana State University, United States Public Health Service, Baton Rouge, Louisiana

John A. Badwey, Ph.D. Department of Cell and Molecular Biology, Boston Biomedical Research Institute, and Department of Biological Chemistry and Molecular Pharmacology, Harvard Medical School, Boston, Massachusetts

Cynthia L. Baldwin, Ph.D. Department of Microbiology, The Ohio State University, Columbus, Ohio

Miles H. Beaman, M.B.B.S.* Division of Infectious Diseases, Department of Medicine, Stanford University School of Medicine, and Department of Immunology and Infectious Diseases, Palo Alto Medical Foundation, Palo Alto, California

William H. Benjamin, Jr., Ph.D. Departments of Pathology and Microbiology, University of Alabama at Birmingham, Birmingham, Alabama

D. Kay Blanchard, Ph.D. Departments of Medical Microbiology and Biochemistry, University of South Florida College of Medicine, and Immunology Program, H. Lee Moffitt Cancer Center and Research Institute, Tampa, Florida

Dov L. Boros, Ph.D. Department of Immunology and Microbiology, Wayne State University School of Medicine, Detroit, Michigan

Robert W. Bradsher, M.D. Division of Infectious Diseases, Department of Medicine, University of Arkansas for Medical Sciences, Little Rock, Arkansas

David E. Briles, Ph.D. Departments of Microbiology and Pediatrics, University of Alabama at Birmingham, Birmingham, Alabama

Ward E. Bullock, M.D. Division of Infectious Diseases, Department of Medicine, University of Cincinnati College of Medicine, Cincinnati, Ohio

Gerald I. Byrne, Ph.D. Department of Medical Microbiology and Immunology, University of Wisconsin—Madison, Madison, Wisconsin

Richard Calderone, Ph.D. Department of Microbiology, Georgetown University School of Medicine, Washington, D.C.

**Current affiliation*: Department of Medicine, Fremantle Hospital, University of Western Australia, Perth, Australia

Priscilla A. Campbell, Ph.D. Division of Basic Immunology, Department of Medicine, National Jewish Center for Immunology and Respiratory Medicine, and the Departments of Pathology and Microbiology/Immunology, The Cancer Center, University of Colorado Health Sciences Center, Denver, Colorado

Ravi Cherukuri Division of Allergy, Immunology, and Infectious Diseases, and Department of Pediatrics, The Joseph Stokes, Jr. Research Institute, The Children's Hospital of Philadelphia, University of Pennsylvania School of Medicine, Philadelphia, Pennsylvania

Janice E. Clements, Ph.D. Department of Comparative Medicine, The Johns Hopkins University School of Medicine, Baltimore, Maryland

Robert M. Crawford, Ph.D. Department of Cellular Immunology, Walter Reed Army Institute of Research, Washington, D.C.

Julie Y. Djeu, Ph.D. Departments of Medical Microbiology and Biochemistry, University of South Florida College of Medicine, and Immunology Program, H. Lee Moffitt Cancer Center and Research Institute, Tampa, Florida

Steven D. Douglas, M.D. Division of Allergy, Immunology, and Infectious Diseases, and Department of Pediatrics, The Joseph Stokes, Jr. Research Institute, The Children's Hospital of Philadelphia, University of Pennsylvania School of Medicine, Philadelphia, Pennsylvania

Nancy E. Dunlap, M.D., Ph.D. Departments of Microbiology and Medicine, University of Alabama at Birmingham, Birmingham, Alabama

Toby K. Eisenstein, Ph.D. Department of Microbiology and Immunology, Temple University School of Medicine, Philadelphia, Pennsylvania

Karen L. Elkins, Ph.D. Department of Cellular Immunology, Walter Reed Army Institute of Research, Washington, D.C.

Jerrold J. Ellner, M.D. Division of Infectious Diseases, Department of Medicine, Case Western Reserve University and University Hospitals of Cleveland, Cleveland, Ohio

Thomas G. Evans, M.D. Division of Infectious Diseases, The University of Utah School of Medicine, Salt Lake City, Utah

Anne H. Fortier, Ph.D. Department of Cellular Immunology, Walter Reed Army Institute of Research, Washington, D.C.

Herman Friedman, Ph.D. Department of Medical Microbiology and Immunology, University of South Florida College of Medicine, Tampa, Florida

Dana H. Gabuzda, M.D. Division of Human Retrovirology, Dana Farber Cancer Institute, Boston, Massachusetts

Shawn J. Green, Ph.D. Department of Cellular Immunology, Walter Reed Army Institute of Research, Washington, D.C.

Laurie R. Hall Department of Tropical Public Health, Harvard School of Public Health, Boston, Massachusetts

Thomas A. Hamilton, Ph.D. Department of Immunology, Cleveland Clinic Foundation, Cleveland, Ohio

Jill A. Hendrzak, Ph.D. Department of Microbiology and Immunology, The Medical College of Pennsylvania, Philadelphia, Pennsylvania

John B. Hibbs, Jr., M.D. Veterans Affairs Medical Center, and Division of Infectious Diseases, University of Utah School of Medicine, Salt Lake City, Utah

Pieter S. Hiemstra, Ph.D. Department of Pulmonology, University Hospital, Leiden, The Netherlands

Mary E. Hilburger Department of Microbiology, The Ohio State University, Columbus, Ohio

Wen-Zhe Ho, M.D. Division of Allergy, Immunology, and Infectious Diseases, and Department of Pediatrics, The Joseph Stokes, Jr. Research Institute, The Children's Hospital of Philadelphia, University of Pennsylvania School of Medicine, Philadelphia, Pennsylvania

Stephanie L. James, Ph.D. Laboratory of Parasitic Diseases, Immunology and Cell Biology Section, National Institutes of Health, Bethesda, Maryland

John L. Johnson, M.D. Division of Pulmonary and Critical Care Medicine, Department of Medicine, Case Western Reserve University and University Hospitals of Cleveland, Cleveland, Ohio

Tonya R. Jones Department of Cellular Immunology, Walter Reed Army Institute of Research, Washington, D.C.

Tadashi Kasahara, Ph.D. Department of Medical Biology and Parasitology, Jichi Medical School, Tochigi, Japan

Thomas W. Klein, Ph.D. Department of Medical Microbiology and Immunology, University of South Florida College of Medicine, Tampa, Florida

Henry Koziel, M.D. Division of Pulmonary and Critical Care Medicine, Department of Medicine, New England Deaconess Hospital and Harvard Medical School, Boston, Massachusetts

James L. Krahenbuhl, Ph.D. G.W. Long Hansen's Disease Center, Laboratory Research Branch at Louisiana State University, United States Public Health Service, Baton Rouge, Louisiana

Raymond E. Kuhn, Ph.D. Department of Biology, Wake Forest University, Winston-Salem, North Carolina

David La Verda Department of Medical Microbiology and Immunology, University of Wisconsin—Madison, Madison, Wisconsin

David A. Leiby, Ph.D. Department of Cellular Immunology, Walter Reed Army Institute of Research, Washington, D.C.

Stuart M. Levitz, M.D. Division of Infectious Diseases, Department of Medicine, Boston University School of Medicine, The Boston University Hospital, Boston, Massachusetts

Carl L. Manthey, Ph.D. Department of Microbiology, Uniformed Services University of the Health Sciences, Bethesda, Maryland

Kouji Matsushima, M.D., Ph.D. Department of Pharmacology, Cancer Research Institute, Kanazawa University, Kanazawa, Japan

Jeanmarie Mayer, M.D. Division of Infectious Diseases, The University of Utah School of Medicine, Salt Lake City, Utah

Monte S. Meltzer, M.D. Department of Cellular Immunology, Walter Reed Army Institute of Research, Washington, D.C.

M. Beatriz Michelini-Norris, Ph.D. Departments of Medical Microbiology and Biochemistry, University of South Florida College of Medicine, and Immunology Program, H. Lee Moffitt Cancer Center and Research Institute, Tampa, Florida

Page S. Morahan, Ph.D. Department of Microbiology and Immunology, The Medical College of Pennsylvania, Philadelphia, Pennsylvania

Richard F. Mortensen, Ph.D. Department of Microbiology, The Ohio State University, Columbus, Ohio

David M. Mosser, Ph.D. Department of Microbiology and Immunology, Temple University School of Medicine, Philadelphia, Pennsylvania

Naofumi Mukaida, M.D. Department of Pharmacology, Cancer Research Institute, Kanazawa University, Kanazawa, Japan

Carol A. Nacy, Ph.D. Department of Cellular Immunology, Walter Reed Army Institute of Research, Washington, D.C.

Opendra Narayan, D.V.M., Ph.D. Department of Comparative Medicine, The Johns Hopkins University School of Medicine, Baltimore, Maryland

Shosaku Narumi, M.D.* Department of Immunology, Cleveland Clinic Foundation, Cleveland, Ohio

Simon L. Newman, Ph.D. Division of Infectious Diseases, Department of Medicine, University of Cincinnati College of Medicine, Cincinnati, Ohio

Bent Windelborg Nielsen, M.D.† Department of Pharmacology, Cancer Research Institute, Kanazawa University, Kanazawa, Japan

Yoshihiro Ohmori, D.D.S., Ph.D. Department of Immunology, Cleveland Clinic Foundation, Cleveland, Ohio

Deirdre O'Riordan, M.D.‡ Division of Pulmonary and Critical Care Medicine, Department of Medicine, New England Deaconess Hospital and Harvard Medical School, Boston, Massachusetts

Donna M. Paulnock, Ph.D. Department of Medical Microbiology and Immunology, University of Wisconsin—Madison, Madison, Wisconsin

Tammy Polsinelli, M.S. Department of Cellular Immunology, Walter Reed Army Institute of Research, Washington, D.C.

Jack S. Remington, M.D. Division of Infectious Diseases, Department of Medicine, Stanford University School of Medicine, and Department of Immunology and Infectious Diseases, Palo Alto Medical Foundation, Palo Alto, California

Mary C. Riedy, M.S. Laboratory of Flow Cytometry, Roswell Park Cancer Institute, Buffalo, New York

John M. Robinson, Ph.D. Department of Cell Biology, Neurobiology and Anatomy, The Ohio State University, Columbus, Ohio

Graham A. W. Rook, M.D. Department of Medical Microbiology, University College London Medical School, London, England

Richard M. Rose, M.D.§ Division of Pulmonary and Critical Care Medicine, New England Deaconess Hospital and Harvard Medical School, Boston, Massachusetts

---

Current affiliations:
*Pharmaceutical Basic Research Laboratories, Japan Tobacco Inc., Kanagawa, Japan
†Department of Pediatrics, Aarhus Kommunehospital, Aarhus, Denmark
‡Thoracic Diseases Research Unit, Mayo Clinic, Rochester, Minnesota
§Cytel Corporation, San Diego, California

Susan A. Safley, M.S. Department of Microbiology and Immunology, Emory University School of Medicine, Atlanta, Georgia

Andreas Schaffner, M.D., Ph.D. Laboratory for Clinical Mycology, Department of Medicine, University Hospital, Zurich, Switzerland

Anuraj H. Shankar, D.Sc. Department of Tropical Public Health, Harvard School of Public Health, Boston, Massachusetts

Hiroe Shiratsuchi, Ph.D. Division of Infectious Diseases, Department of Medicine, Case Western Reserve University, Cleveland, Ohio

Carleton C. Stewart, Ph.D. Laboratory of Flow Cytometry, Roswell Park Cancer Institute, Buffalo, New York

Sigrid J. Stewart Laboratory of Flow Cytometry, Roswell Park Cancer Institute, Buffalo, New York

Joy Sturtevant, Ph.D. Department of Microbiology, Georgetown University School of Medicine, Washington, D.C.

Carlos S. Subauste, M.D. Division of Infectious Diseases, Department of Medicine, Stanford University School of Medicine, and Department of Immunology and Infectious Diseases, Palo Alto Medical Foundation, Palo Alto, California

Charles S. Tannenbaum, Ph.D. Department of Immunology, Cleveland Clinic Foundation, Cleveland, Ohio

Julie M. Tebo, Ph.D. Department of Immunology, Cleveland Clinic Foundation, Cleveland, Ohio

Cynthia M. Theodos, Ph.D. Department of Tropical Public Health, Harvard School of Public Health, Boston, Massachusetts

Richard G. Titus, Ph.D. Department of Tropical Public Health, Harvard School of Public Health, Boston, Massachusetts

Jenifer Turco, Ph.D. Laboratory of Molecular Biology, Department of Microbiology and Immunology, University of South Alabama College of Medicine, Mobile, Alabama

Ralph van Furth, M.D., Ph.D. Department of Infectious Diseases, University Hospital, Leiden, The Netherlands

Stefanie N. Vogel, Ph.D. Department of Microbiology, Uniformed Services University of the Health Sciences, Bethesda, Maryland

Angeline Warner, D.V.M. D.Sc. The Respiratory Biology Program, Harvard School of Public Health, Boston, Massachusetts

Herbert H. Winkler, Ph.D. Laboratory of Molecular Biology, Department of Microbiology and Immunology, University of South Alabama College of Medicine, Mobile, Alabama

Alexander J. Winter, D.V.M., Ph.D. Department of Microbiology, Immunology and Parasitology, Cornell University, Ithaca, New York

Sin-Yew Wong, M.D. Division of Infectious Diseases, Department of Medicine, Stanford University School of Medicine, and Department of Immunology and Infectious Diseases, Palo Alto Medical Foundation, Palo Alto, California

Marion L. Woods II, M.D., M.P.H. Division of Infectious Diseases, The University of Utah School of Medicine, Salt Lake City, Utah

Yoshimasa Yamamoto, Ph.D. Department of Medical Microbiology and Immunology, University of South Florida College of Medicine, Tampa, Florida

H. Kirk Ziegler, Ph.D. Department of Microbiology and Immunology, Emory University School of Medicine, Atlanta, Georgia

M. Christine Zink, D.V.M., Ph.D. Department of Comparative Medicine, The Johns Hopkins University School of Medicine, Baltimore, Maryland

Bruce S. Zwilling, Ph.D. Department of Microbiology, The Ohio State University, Columbus, Ohio

Macrophage–Pathogen Interactions

I

MACROPHAGE BIOLOGY

Introduction: The Proliferation and Differentiation of Macrophages

Carleton C. Stewart, Mary C. Riedy, and Sigrid J. Stewart

Roswell Park Cancer Institute, Buffalo, New York

I INTRODUCTION

Blood cells are produced in the bone marrow and are released into the circulation as granulocytes, erythrocytes, platelets, lymphocytes, and monocytes. Since these cells have a limited life span, multipotent stem cells, which are capable of self-renewal and which differentiate into lineage-specific progenitor cells, are required to maintain an adequate supply of these end cells. We will focus our attention on the proliferation and differentiation of the mononuclear phagocyte system. The classical differentiation sequence of morphologically recognizable cells proceeds from promonoblasts to monoblasts, promonocytes, monocytes, and mature macrophages [1]. Whereas all of these morphological types are found in the bone marrow, monocytes are released into the bloodstream, and from there they disseminate to all the tissues. Upon entering the tissue, monocytes differentiate into macrophages to perform important tissue-maintenance and regulatory functions such as phagocytosis, microbicidal and tumoricidal activity, antigen presentation, and secretion of a myriad of substances including enzymes, complement components, tissue clotting factors, growth factors and inhibitors, and immunoregulatory substances.

In the current concept of hematopoiesis, pluripotent stem cells differentiate into committed lineage-specific progenitor cells [2–9]. As these cells differentiate into the various lineages, they acquire new proteins while losing other proteins. Although we do not yet know the function of most proteins, each subset along any given differentiation pathway expresses a unique repertoire of surface proteins that, by combining several antibodies to them in a single "cocktail," can provide us with the opportunity to precisely identify the subsets and to sort them for further study.

While the mature elements of most lineages of hematopoietic cells have lost their proliferative capacity after leaving the bone marrow, both lymphocytes [10] and mononuclear phagocytes [11–13] retain it. Proliferation of these cells is dependent on specific hematopoietic growth factors, which have been called colony-stimulating factors (CSFs) because of their ability to stimulate colony formation. The CSFs may have other activities besides promoting cell proliferation. As cells differentiate they lose some growth factor receptors and acquire more differentiation-related receptors. A possible action of growth factors is to control differentiation [14–16]. Whether the factor acts to commit a cell to a differentiation pathway or whether maturation is linked to proliferation by other factors or genetic programs is not known.

Because most tissues have cells in them that produce CSFs at physiologically stimulatory levels, inhibitors of cell proliferation may play a key role in regulating proliferation [17]. This mechanism would also account for the requirement that the growth factor be present for cell survival, differentiation, and function. Thus, a growth factor has several functions, and the particular cellular activity at any time depends on how the combination of all factors, known or unknown, is presented to the cell at any particular moment. The delicate balance among growth factor production, consumption, and inhibition in the regulation of myelomonocytic proliferation is a complicated phenomenon. It is our intent to provide some insights that will help to clarify the intricacies of this balance.

II. COLONY FORMATION BY PERITONEAL MACROPHAGES

Using a method patterned after the agar culture system developed by Metcalf for bone marrow [18], in 1973 Lin and Stewart described a new class of slow-growing colony-forming cells that give rise to clones of mononuclear phagocytes in vitro [12, 19]. These cells were elicited by the injection of thioglycolate medium into the peritoneal cavity of mice. It was discovered that thioglycolate-elicited peritoneal exudate cells (PECs) would form colonies in soft agar at a plating efficiency of approximately 5% [12, 19] after 3 weeks of culture. Unlike bone marrow–derived colonies, which were mixtures of granulocytes and macrophages, these colony-forming cells (CFCs) produced colonies that were composed exclusively of macrophages [20, 21].

In 1967 Virolainen and Defendi [22] had reported that PECs, specifically peritoneal macrophages, would proliferate in mass culture in a medium containing serum and a substance present in conditioned medium from L-929 cells (LCM) or mouse fibroblasts. Accordingly, we wondered if these were indeed the same cells, and we plated a range of PECs in αMEM supplemented with 10% fetal bovine serum (FBS), 5% horse serum, and 10% LCM, without agar. (It has been found that horse serum is not necessary, and the growth medium now consists of αMEM supplemented with 10% FBS and 10% LCM). With this method, unlike the agar system, colonies of more than 50 cells were found after only 10 days. The plating efficiency was three times greater (15%) than for the same cells cultured in agar medium (5%).

Since lavaged peritoneal cells produced few colonies, the ability of several phlogogenic agents to elicit CFCs was tested. As shown in Table 1, the injection of different kinds of sterile phlogogenic agent produces differing yields of CFCs. As will be discussed later, given a cell cycle time of 18 hours, and no lag period, the resident CFCs could produce an eightfold increase in cells over 3 days. Yet by day 3, there was an increase in CFCs of more than three orders of magnitude

TABLE 1 Colony Formation by Peritoneal Cells Elicited with Different Phlogogenic Agents from C_3H Mice

Phlogogenic agent	Cells/mouse (millions)	Colonies/10^3 cells ± SEM)	Total CFC (thousands)
None	2	1 ± 1	2
IP needle stick	2	4 ± 2	8
Saline	3	8 ± 3	24
Pyran	7	20 ± 5	140
BCG	6	70 ± 8	420
Starch	6	80 ± 7	480
Peptone	12	120 ± 15	1440
Sodium percodate	15	150 ± 20	2250
Thioglycolate medium	16	150 ± 18	2400

SEM = standard error of the mean, CFC = colony-forming cells; IP = intraperitoneal; BCG = Bacillus Calmette-Guérin.

TABLE 2 Frequency of Macrophage Colony-Forming Cells from Different Murine Tissues

Source	Colonies/10^3 cells	
	Day 7	Day 14
Bone marrow	8	31
Spleen	0.5	14
Blood	0.1	12
Lung	0	53

when thioglycolate medium was used (Table 1). While some local proliferation may have occurred, these data clearly established that the major source of macrophages in an inflammatory site was from outside the site [19, 23–25].

As shown in Table 2, CFCs can also be obtained from spleen, blood, bone marrow, and lung, and no phlogogenic agent is required. Furthermore, virtually all monocytes are capable of forming colonies. These results clearly show that macrophages, like lymphocytes, are not end cells. However, human macrophages derived from any of the sources described above, except bone marrow, do not form colonies with a high plating efficiency (unpublished data). Indeed, the CFCs found in human blood are not in the monocyte fraction; they are instead $CD34^+CD14^-$ cells [26]. This perplexing difference between humans and other mammals has yet to be resolved. No one, to our knowledge, has successfully obtained with human mononuclear phagocytes the frequency of colonies or the degree of proliferation that can be obtained with cells from other mammals.

When murine mononuclear phagocytes from peripheral tissue or bone marrow are cultured in the presence of FBS and LCM in liquid culture medium, they will proliferate. If they are cultured at sufficiently low density they will form discrete colonies. We have previously shown that this proliferation depends on an adequate concentration of CSF-1 in the culture medium. Stanley and Heard have shown that these cells have receptors for CSF-1 [27]. After binding, the cells internalize the CSF-1, degrade it, and return non–trichloroacetic acid–precipitable products to the medium without biological activity.

In order to determine the extent of proliferation and better define the proliferation kinetics, were cultured PECs at various concentrations. As shown in Figure 1, no matter how many cells were initially cultured, no more than 10^5 cells/ml of growth medium (having 10% FBS and 10% LCM) could be achieved; only the time for cells to reach this stationary phase varied with the number initially plated. This time was consistent with a mean population-doubling time of 42 hours. Thus, if 340,000 cells were initially cultured in 6 ml of growth medium, they doubled only once, but if 2700 cells were cultured, 13 days were required to reach a plateau of 300,000 to 700,000 cells.

Because cells always stopped proliferating when they reached about 10^5/ml, we investigated whether removal of LCM caused the cells to stop proliferating [25]. As shown in Figure 2, removal of LCM caused cells to leave cell cycle, and its readdition resulted in their reentry into cell cycle. Using time-lapse movies, we found that when LCM was removed, cells that were no more than 9 hours old would not divide again, while cells that were 12 hours or more old always divided one more time. This clearly established that the cells' commitment point for cell cycle progression was near the G1/S boundary between 9 and 12 hours.

The first cell divisions were observed 31 hours after readdition of LCM. All cells divided over the next 10 hours. These reentry kinetics were also found when the fraction of H^3TdR-labeled cells was followed hourly after the readdition of LCM. These studies clearly established the obligatory

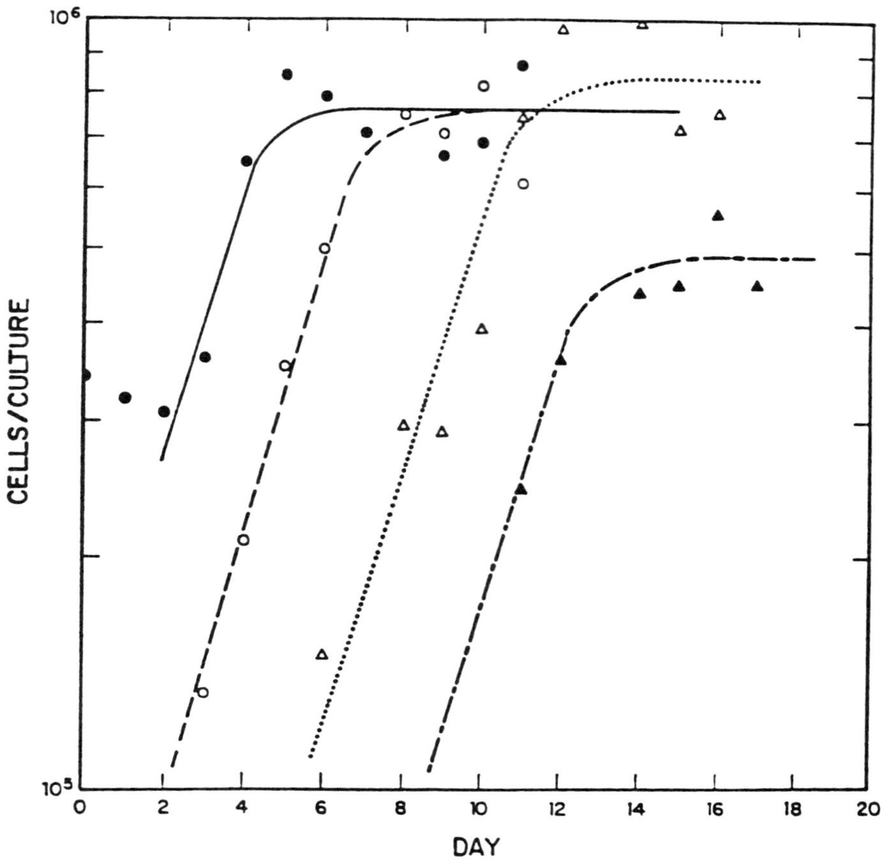

FIGURE 1 Proliferation of peritoneal exudate cells. Peritoneal cells were adjusted to 340,000 (●), 68,000 (○), 13,000 (△), or 2700 (▲) cells in 6 ml of growth medium in 60-mm culture dishes. The number of cells per dish was determined daily by removing cells with the detergent cetrimide. From Ref. 10.

role of CSF-1 in sustaining macrophage proliferation. These results are consistent with the molecular studies of Sherr et al. [28], who, using a similar culture system, showed evidence for a regulatable restriction point in late G1.

In 1977, Stanley et al. [27] successfully purified and subsequently characterized the macrophage growth factor from LCM and called in CSF-1. Since they had an immunoassay for CSF-1, we collaborated with Drs. Stanley and Guilbert to measure the depletion of CSF-1 by macrophages. The results for PECs and for pulmonary alveolar macrophages (PAMs) are shown in Figure 3. The consumption rate was found to be 0.18 units CSF-1/cell/h, and when alveolar macrophages reached 10^5 cells/ml the CSF-1 level was reduced to less than 250 units/ml, a level too low to support macrophage proliferation. The same initial kinetics were also found for PECs, but, much to our surprise, an increase in CSF-1 levels followed the cessation of macrophage proliferation, so that by the end of the observation period there was more CSF-1 in the culture medium than we initially added. Further experiments showed that the source of CSF-1 was contaminating fibroblasts found in the cultures of PECs but not in those of PAMs because there are no fibroblasts in lung avleoli. This also underscores the need for using highly purified cell populations when studying cytokines, because the presence of endogenously produced unknown

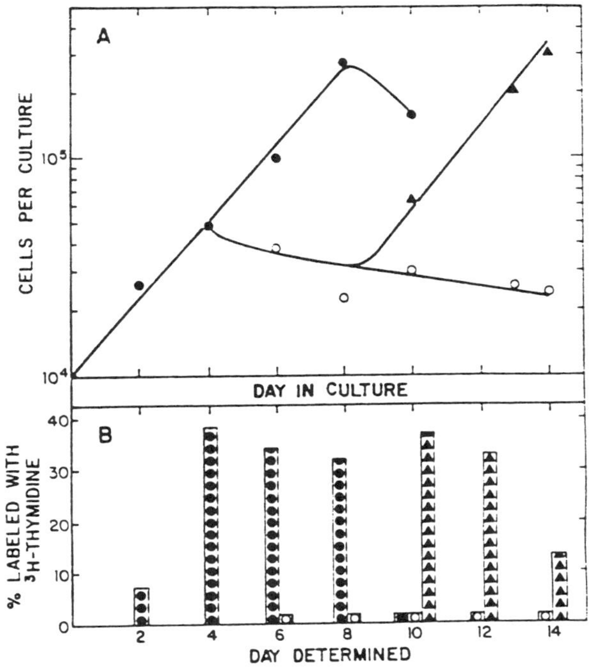

FIGURE 2 Dependence of macrophages on the continuous presence of LCM for cell division and DNA analysis. (A) Macrophages were grown in the continuous presence of LCM (●). On day 4, the medium of some plates was removed, and the cells were washed and refed with medium lacking LCM (○). On day 8, fresh medium or medium with LCM was added (▲). (B) Parallel plates were pulse labeled with [^{3}H] thymidine: the fraction of the labeled cells was then determined after autoradiography. Phagocytosis on other parallel plates was done on days 4, 8, and 12. In all cases more than 99% of the adherent cells phagocytized yeast particles. From Ref. 10.

cytokine cell contaminants can influence the measured response. Since the PECs stopped proliferating but CSF-1 was accumulating in the medium, either the cells were becoming senescent, another factor was also required for macrophages to proliferate and it too was depleted, or an inhibitor accumulated in the culture medium.

To determine if the cells were becoming senescent, we subcultured macrophages that had reached stationary phase, and they resumed their proliferative response [29] (Figure 4). Indeed, by continuing to subculture these cells, we found that the 2.4×10^6 clonogenic cells (Table 1) elicited by thioglycolate medium from a single mouse would produce by the 24th day 4.2×10^{10} progeny or 42 ml of packed macrophages. Each CFC was capable of 16 divisions, producing 26,000 progeny, before truly senescing. It appears senescence is not the reason macrophages stop proliferating in these primary cultures. Furthermore, that their proliferative potential is not achieved during an in vivo inflammatory response attests to the in vivo control of proliferation that is lost in vitro [25].

The second possibility is the depletion of a second obligatory factor that is restored by adding fresh growth medium. As shown in Figure 5, if FBS was replaced with bovine serum, macrophages stopped proliferating, and when FBS was readded, macrophages resumed proliferating. The kinetics of this response is similar to that shown in Figure 2 for CSF-1. Thus, a factor in FBS was also necessary for macrophages to proliferate. We have not been able to identify this essential factor. Because the fractionation of serum to identify the active component is complex, we sought a different approach. Several serum-free medium supplements are available, and we tested them for serum replacement activity. It was found that the supplement HB101, from Hana Biologics, Inc. (Alameda, California), completely replaced the need for FBS, suggesting that it contained the unknown factor. Further work is in progress to identify and characterize this factor. Thus, another reason why macrophages stop proliferating, besides CSF-1 depletion, is that they deplete a second factor.

The accumulation of an inhibitor that is removed when cells are subcultured into fresh growth medium can also account for the observed cessation of proliferation. In an attempt at replacing the serum requirement for macrophage cell proliferation in culture, it was found that, in the presence of FBS and CSF-1, macrophage proliferation was inhibited by interleukin-6 (IL-6) (Figure 6). This inhibition could be demonstrated when the optimal amount of IL-6 (100 U/ml) was added before the initiation of cell cycle or after macrophages had entered exponential growth. The inhibition, facilitated by 100 U/ml IL-6, is prevented by the addition of a monoclonal antibody specific for IL-6. If IL-6 was removed from macrophage cultures after preventing macrophage growth for 1 week, these macrophages could again initiate a proliferative response (Figure 6). Thus, the inhibitory activity of IL-6 occurs when all the components needed for macrophage proliferation are present, and it is reversible. These initial data indicate that IL-6 is specifically involved in the regulation of macrophage proliferation [30].

(A)

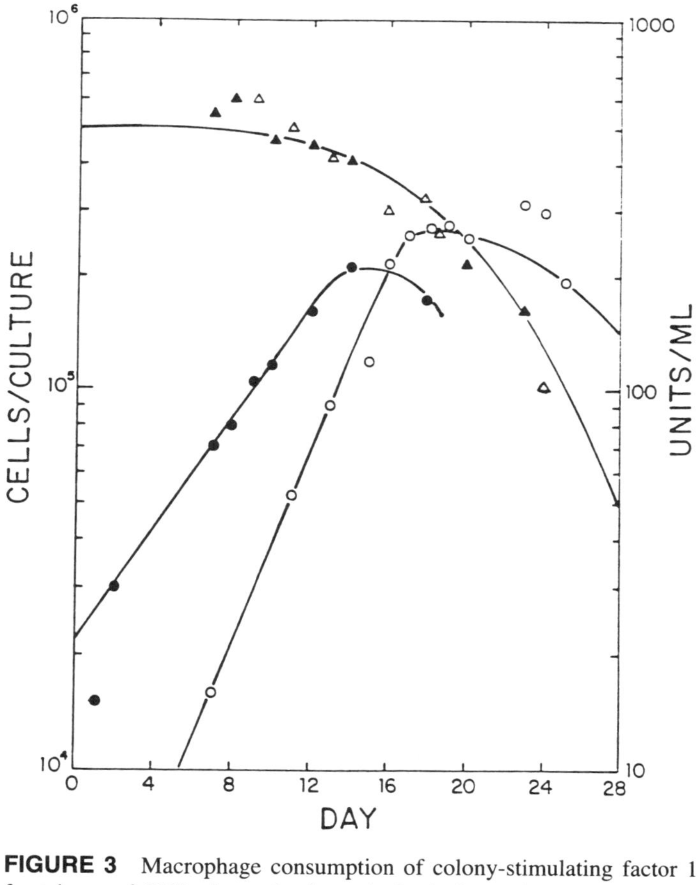

FIGURE 3 Macrophage consumption of colony-stimulating factor 1. In A, cells were lavaged from lungs of C3H mice and cultures in 6 ml of growth medium at 3500 cells/ml (●) or 700 cells/ml (○). Beginning on day 5, medium was removed and cells were counted. In B, peritoneal exudate cells were cultured at 40 × 10^3 cells/ml in 6 ml growth medium. Beginning on day 0, medium was

III. KINETICS OF MACROPHAGE PROLIFERATION

There are several methods for evaluating cell cycle parameters, but no other is as direct as time-lapse cinemicrography. Using this method, it is possible to retrospectively determine the lag time to the first mitosis, the generation time, the mitotic time, and the growth fraction at any point in time. Since each cell is observed, heterogeneity among them in each of these parameters can be directly measured. We have applied this approach to the measurement of these parameters on several thousand thyoglycolated medium–elicited PECs.

We first quantified the number of divisions each dividing cell underwent. Figure 7 is a frequency distribution of the percentage of cells that underwent mitosis over a 14-day observation period [23]. Only 19% of cells underwent no divisions at all. Thirty-three percent of cells underwent five or six divisions, forming clusters of fewer than 32 cells. The number of cells in a cluster does not necessarily increase geometrically, because not all daughters will divide a second time. Fifteen percent of cells underwent 10 divisions and formed colonies defined as clusters with 50 or more cells (there was an average of 331 cells/colony). These results show that the great

(B)

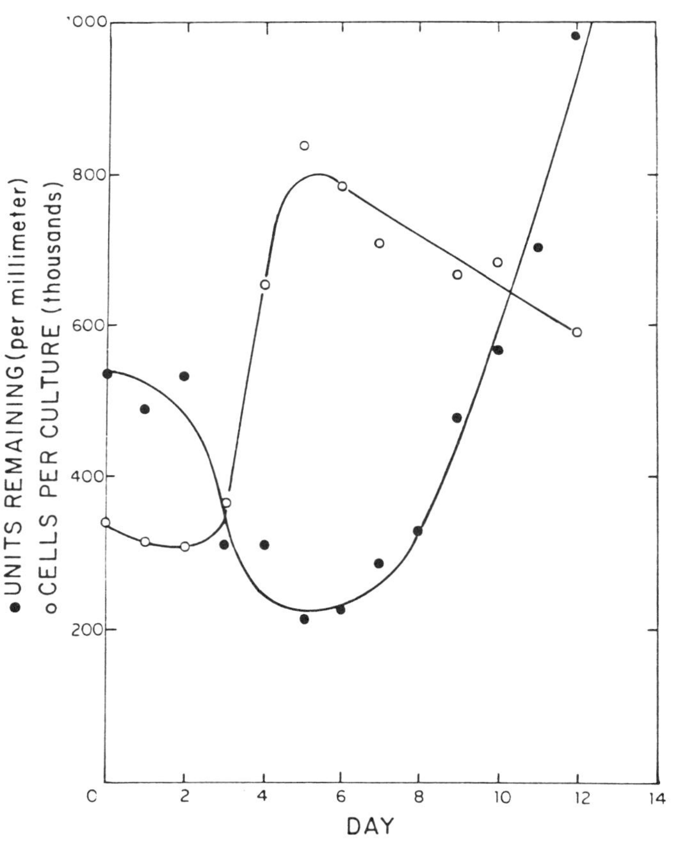

removed and cells were counted (○). The medium was analyzed by Drs. Stanley and Guilbert to determine the colony-stimulating factor 1 (CSF-1) content of each sample. In A, the units of CSF-1/ml medium is shown for cells cultured at 3500 (▲) or 700 (△) cells/ml, and in B for the PEC (●).

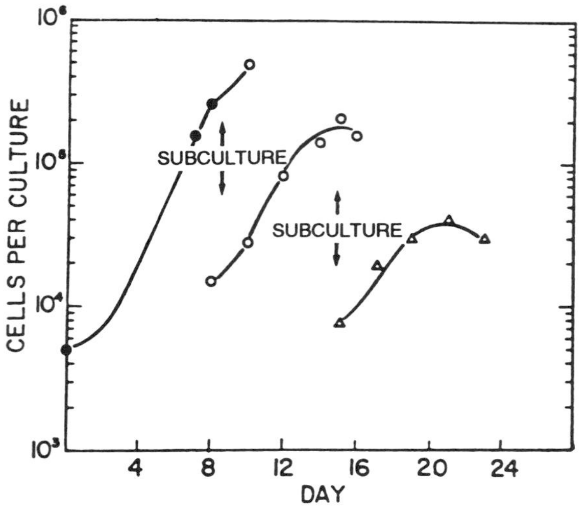

FIGURE 4 Macrophage proliferative capacity. Peritoneal exudate cells were cultured in 10 ml growth medium at an initial concentration of 500 cells/ml in 100-mm nonwettable petri dishes in αMEM containing 10% FBS and 10% LCM. On day 8 and 16 cells were removed from dishes by lifting the cells off by directing a jet stream of medium from a 10 ml syringe with a 25-gauge needle attached. This method effectively disperses the cells during the removal process (do not repeat with medium containing cells, as this passage through the needle will damage them). Cells were counted as previously described [31].

majority (85%) of exudate cells have a limited proliferative capability and become senescent after only five or six divisions.

We also determined the lag time for all cells that divided. The lag time is defined as the time between plating cells (t=0) and the first division. The results are summarized in Figure 8. Nine percent of cells divided within the first 24 hours of plating, but none of these cells formed colonies. Not until day 4, when 21% of cells divided for the first time (90–120 hours), did any of them begin to form colonies. Between day 4 and day 7, when all cells that were going to divide had finally divided, an increasing percentage of cells formed colonies. Thus, early in culture, only cells with a limited proliferative capacity divide. Cells with a greater proliferative capability can form discrete colonies and have a 4- to 7-day lag period. This is apparently why no colonies are seen before day 7 when peripheral CFCs are cultured. Only bone marrow–derived CFCs form colonies before day 7 (see below).

We also continuously observed the formation of several cells that produced very large colonies, and a family tree for one of them is shown in Figure 9. All of the cells capable of extensive proliferation to form colonies had a common epithelioid morphology. This is in contrast to the more prominent ameboid morphology of macrophages. In the family tree, one daughter created by the first division failed to undergo cytokinesis on both its second and its third divisions (lineage B). The fourth division, however, produced a mononuclear daughter and a multinucleated daughter that did not divide again. The second daughter of the first division divided normally the second time, but one of its daughters failed to undergo cytokinesis on its subsequent division. The progeny of the other daughter produced lineage A, which resulted in the effective production of a colony.

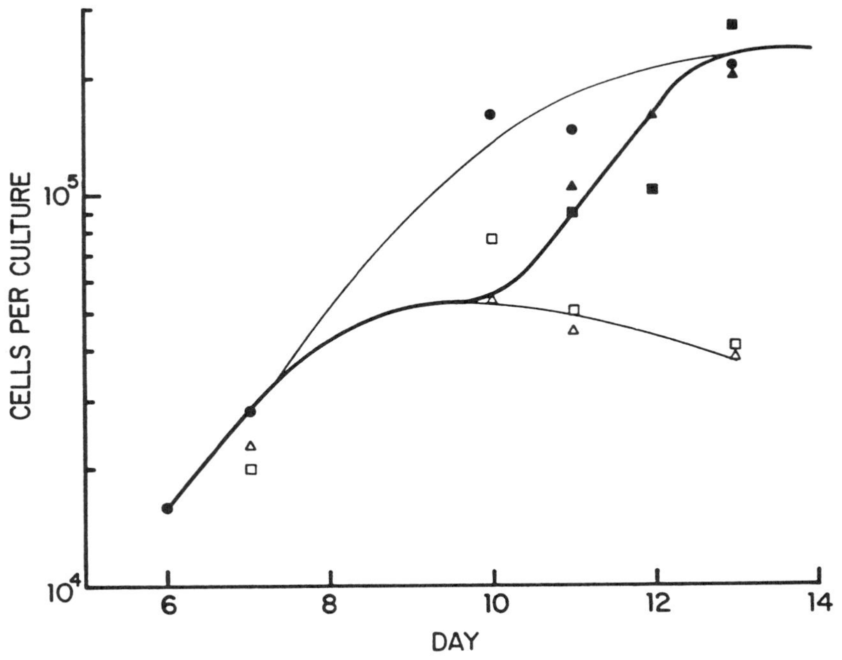

FIGURE 5 Effect of fetal bovine serum on macrophage proliferation. Thioglycolate medium–elicited peritoneal cells were cultured at 10^3 cells in 3 ml αMEM containing 10% LCM and 10% FBS, and for some cultures the medium was not changed (●). After 7 days, the medium in some cultures was changed to αMEM containing 10% LCM and 10% bovine serum (△). After 3 days, the medium was again changed on some cultures to αMEM containing 10% LCM and 10% FBS (▲). Cells were counted daily.

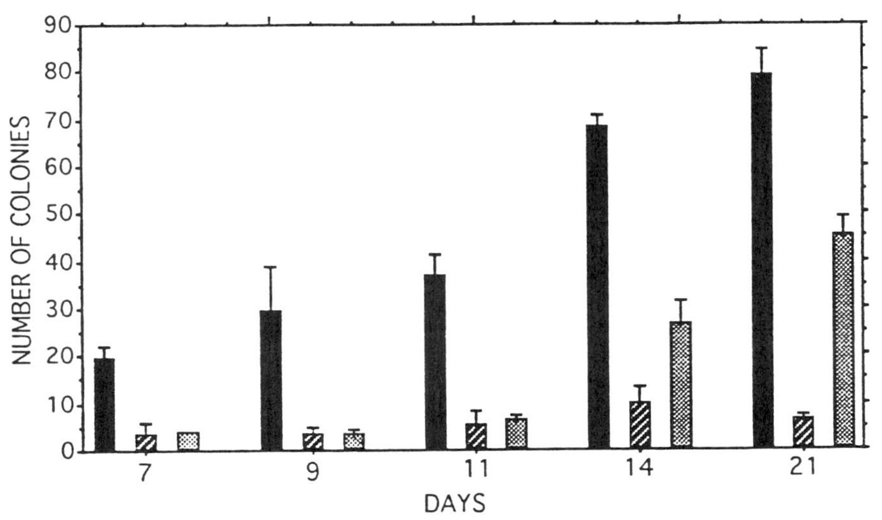

FIGURE 6 Macrophage colony counts in the presence and absence of interleukin-6. Bone marrow macrophages grown in αMEM in the presence of 20% FBS and 20% LCM. Cells were plated at 3.3 × 10^2 cells/ml in 35-mm tissue culture dishes without interleukin-6 (IL-6) (black bar), with 100 U/ml IL-6 (striped bar) and 100 U/ml IL-6 removed from the cultures on day 7 (hatched bar).

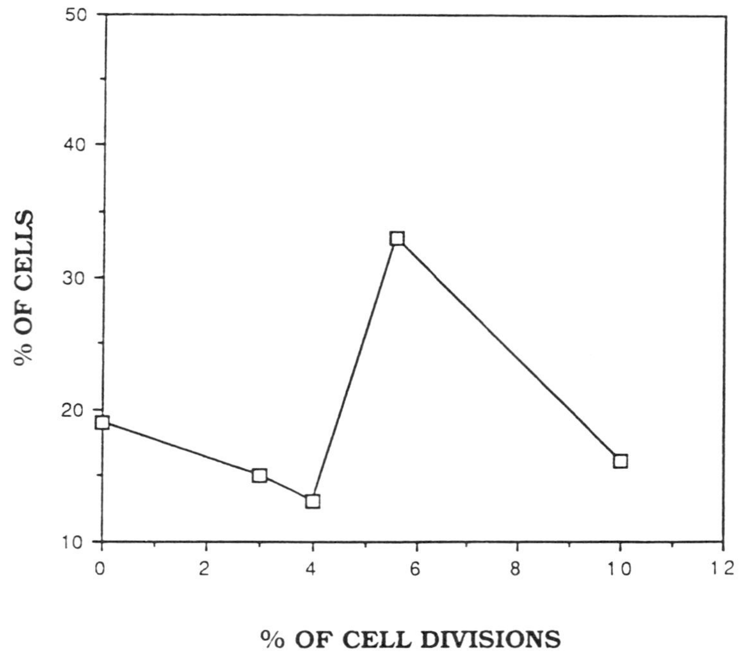

FIGURE 7 Extent of elicited macrophage proliferation. Time-lapse movies were used to photograph single cells and determine the number of divisions all progeny underwent during a 14-day observation period. Usually 10^3 or 10^4 peritoneal exudate cells were plated in growth medium on 35-mm culture dishes. Two Nikon inverted microscopes were equipped with time-lapse sequencing electronics and Bolex cameras or a JVC model BR-9000U time-lapse video camcorder with COHU high-resolution video camera. The equipment was located in a warm room providing a stable environment. A specially built chamber enclosed the culture dish in an airtight environment that was constantly purged with a humidified atmosphere of 7% CO_2 in air.

The failure of some cells to undergo cytokinesis was a common finding among all CFCs we have filmed, but the basis for the phenomenon is not known. While most of these failures occurred during the first few rounds of cell division, some of them, as appears in lineage C, can occur after several normal rounds of division. These multinucleated epitheloid macrophages are commonly found in granulomas and have previously been thought to arise by cell fusion [31]. While this may be so, clearly they also form by a failure to cytokinese.

During the period of observation, most of the daughters lost their epitheloid morphology and progressively acquired an ameboid morphology. All cells that failed to undergo cytokinesis, however, did retain their epitheloid morphology, as did some of the cells that were still dividing at the end of the observation period. These results strongly suggest that the macrophage progenitor cell has an epithelioid morphology and produces ameboid progeny that ultimately differentiate into nonproliferating end cells. From these data, it is also possible to precisely determine the generation (cell cycle) time of macrophages. This time was found to be 17 ± 4 hours (mean ± 1 standard deviation).

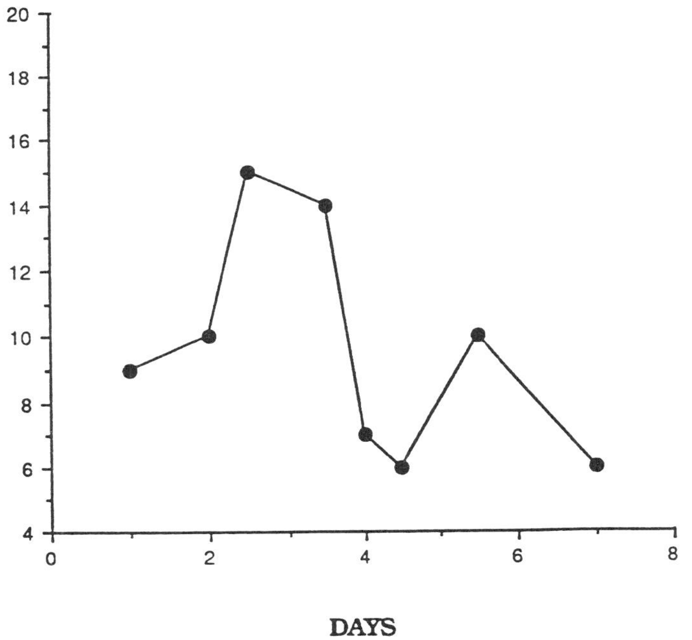

FIGURE 8 Time to first mitosis. From the films it was possible to determine the time to the first mitosis. The majority of cells require 2.5 days to divide for the first time. These cells underwent no more than six divisions. A second group of cells, most of which formed colonies of more than 50 cells, initiate their first mitosis on day 5.5.

IV. CELL GROWTH KINETICS OF BONE MARROW MACROPHAGES

Cell lines have been a major source of cellular products because they are uniform and convenient. However, they may not always be representative of normal cells. Primary cultures of bone marrow macrophages derived after CSF-1 stimulation offer a convenient and uniform source of cells to bridge the gap between normal cells and cell lines. We have used them extensively in this paradigm.

Murine bone marrow cells were cultured in αMEM containing 10% FBS and 10% LCM. While it might be argued that recombinant CSF-1 would be better, its use simply to obtain mass cultures of macrophages does not justify its cost. To measure growth kinetics, the number of nonadherent and adherent cells was determined daily. While the use of thymidine labeling as a method for assessing proliferation is popular, we do not recommend this approach unless direct cell counting is also used to attest to the actual degree of cell proliferation that has occurred. As shown in Figure 10, nonadherent cells initially decreased at the same rate as did cells in cultures with no CSF-1 added. After day 4, the number of nonadherent cells in cultures containing CSF-1 increased. First detected on day 2, adherent cells increased exponentially two orders of magnitude over the subsequent 6-day culture period. Cells in cultures without growth factor decreased continuously over the 8-day period. We thus confirmed that bone marrow cells die in the absence of growth factor, as shown by Stanley et al. (27). When CSF-1 is present, however, macrophage progenitor cells survive and proliferate.

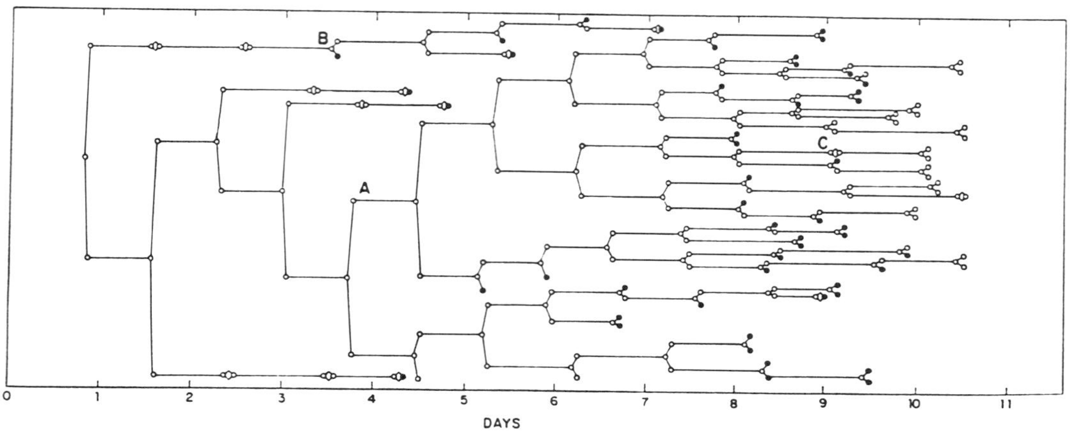

FIGURE 9 The family tree of a colony-forming cell. Peritoneal exudate cells were established in 35-mm culture dishes containing 300 cells in growth medium and continuously monitored by time-lapse cinemicrography. One cell divided for the first time after 4 days (day 0 on figure) and produced a colony of at least 77 cells in the next 10 days. Filled in circles represent cells that did not divide again while open circles are dividing cells. Two cells left the field of view and their fate could no longer be followed.

The changes with culture time in the cell populations, based on cell size, and in cell proliferation as reflected by DNA content are shown in Figure 11. On day 0, two distinct populations according to cell size are resolved, the second one trailing into a heterogeneous distribution of large cells. If sorted, the first peak is composed primarily of lymphocytes and the second primarily of granulocytes, while the larger cells are blast cells and macrophages [32]. By day 2, about 50% of cells have died (Figure 10), and the disappearance of the first peak shows that these were specifically the lymphocytes. By day 4, with the marked decrease in the second peak, most mature granulocytes have also disappeared from culture. As shown by the DNA histogram, this is the day of maximum proliferation, but a significant fraction of cells are in cell cycle. On day 6, two populations of cells are found, a population of small cells (5% of total) and a population of very large cells. There are virtually no cycling cells because the macrophages have consumed all the CSF-1 and serum factors needed to sustain growth. When sorted, the small cells that were nonadherent when cultured on wettable plastic surfaces had myelomonocytic blast cell properties, while the large cells were all macrophages. Thus, from day 0 to day 6, the culture becomes composed almost exclusively of macrophages by means of a proliferative expansion of precursor cells and by the death of the other formed elements.

V. LONG-TERM CULTURES

Dexter et al. [33] introduced a system for the long-term culture of bone marrow utilizing a stromal lawn upon which progenitor cells can proliferate. The stromal cells will presumably provide the appropriate microenvironment and factors necessary for hematopoietic progenitor cell growth and differentiation. This system has been one of the important foundations for studying hematopoiesis. Over the years, the microenvironment that includes the stromal cells has been discussed as critical to normal hematopoiesis, yet it remains the least studied and least understood area. More research emphasis is clearly warranted.

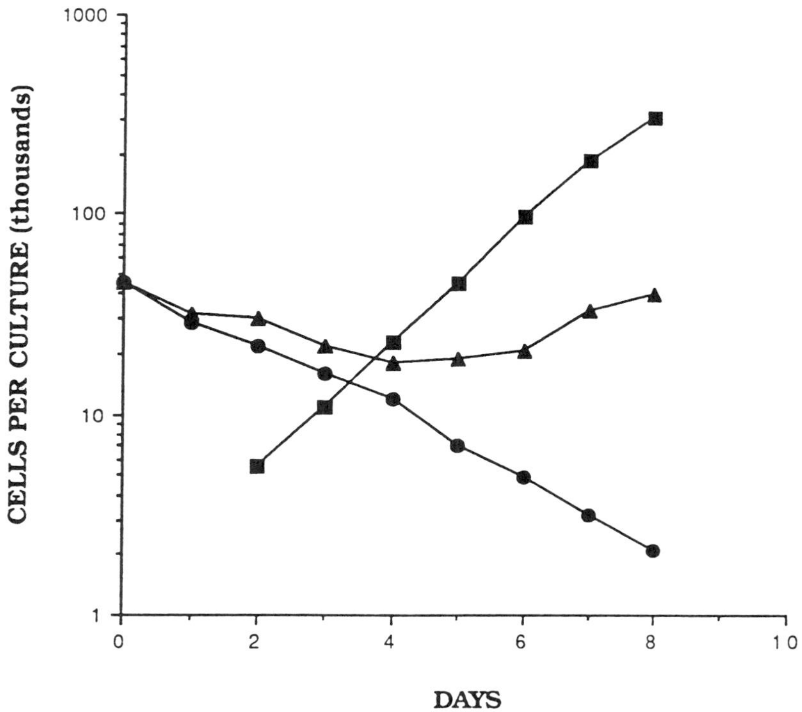

FIGURE 10 Cell growth kinetics of bone marrow macrophages. Bone marrow cells from the femurs of C3H mice were adjusted to 50×10^3 cells in 3 ml of growth medium consisting of αMEM with 20% FBS and 20% LCM and plated in 35-mm culture dishes. One set of cultures (●) contained no LCM. The nonadherent (▲) cells were resuspended in medium and transferred to 10 ml certrimide counting solution. The adherent cells (■) were obtained by directly pipetting cetrimide on the plate and rinsing. Counts were performed using an electronic particle counter.

These cells are an important source of cytokines that regulate both proliferation and differentiation of hematopoietic cells. Presumably, if one had a purified population of precursor cells and knew all the cytokines that were necessary for their regulation, each lineage could be completely reproduced in vitro in large numbers, just as we have done for macrophages. Indeed, great progress has been made in accomplishing this goal.

The culture systems used today, however, are complex, and while the cytokines may be pure, the cells used are not. Therefore, the culture represents a balance between the known exogenous cytokines that have been added and the endogenously produced unknown cytokines. The proliferative response of the cells, therefore, may not represent their true capacity, and our interpretation is tainted to the extent that unknown endogenously produced cytokines have modified the response measured.

We performed a simple but long-term experiment to evaluate what a "long-term" culture might actually represent with regard to proliferation of macrophages. It is clear from our above discussion that macrophages consume their growth factors, and that when these are reduced to a low enough level, proliferation stops. It is also clear that contaminating fibroblasts produce copious amounts of CSF-1 and, by reducing the cell concentration and resupplying adequate levels of cytokines, allow cells to resume growth. Our question then is whether the proliferation clock—i.e., the time a cell is in cell cycle through its lifetime—is the same as the elapsed real time. One model for long-term

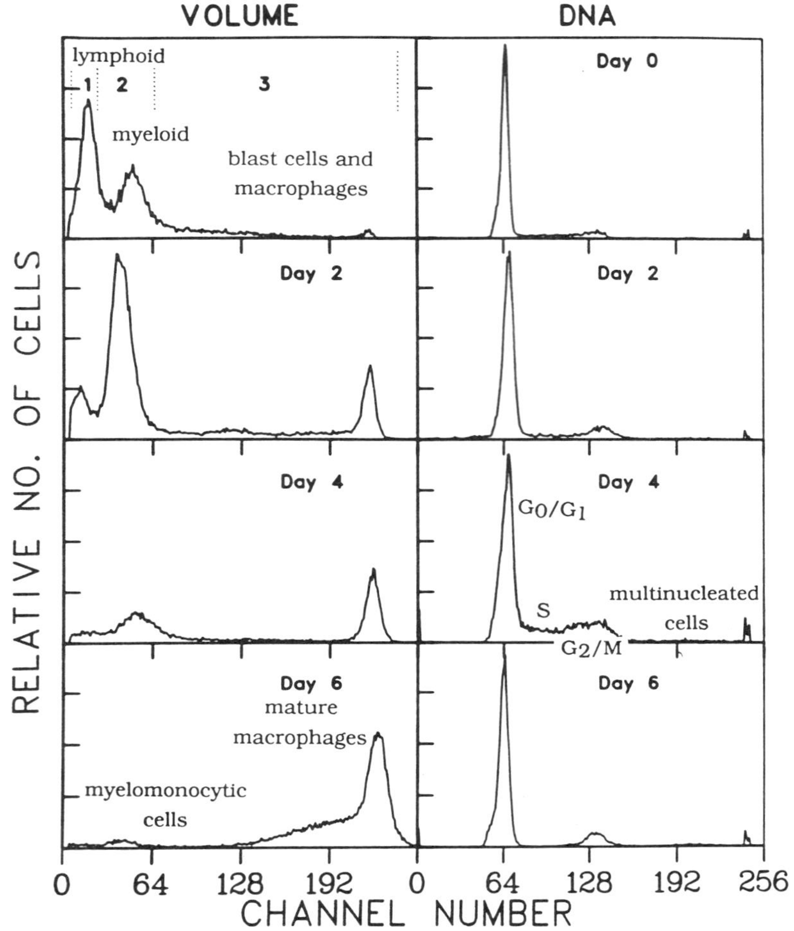

FIGURE 11 Changes in bone marrow cell populations and DNA content. Bone marrow cells prepared as described in Figure 10 were also cultured on 100-mm nonwettable petri dishes containing 15 ml of cells in growth medium (600×10^3). Cells were removed from the dish with a jet stream of medium from a 10-ml syringe with a 25-gauge needle attached. After harvesting, the cells were incubated at 37° C for 45 minutes with 3 μg Hoechst 33342 at pH 7.1. Cells were analyzed by flow cytometry unfixed, and propidium iodide was added to exclude dead cells from the analysis. Dead cells accounted for less than 15% of cells in all samples.

cultures, then, is that stromal cells and surviving hematopoietic cells produce cytokines that cause both proliferation and differentiation to occur, and that the time to which a particular response of interest is measured depends on the extent to which the endogenous factors are produced and utilized to support it. A high concentration of cells will result in rapid consumption and a long quiescence time, because not enough cells are present in the support layer to produce amounts of needed cytokines adequate to maintain the desired response.

To evaluate this possibility, we continuously subcultured bone marrow–derived macrophages in growth medium every 3 or 4 days to force them to remain in a proliferative state, or we cultured them over a long period by charging the medium with CSF-1 at monthly intervals. When the cells were subcultured continuously, they underwent 2^{27} divisions before completely senescing. This period lasts about 30 days, 6 days longer than for PECs (Figure 4). If, however, cells were refed monthly with CSF-1, they still underwent 2^{27} divisions, but required more than 6 months (10). Thus, the cell cycle clock runs only when cells are in cycle, and that occurs when there is a sufficient supply of cytokines known and unknown. Since the growth factor is at maximum levels for only 6 days, the cells remain quiescent for the next 24 hours. These data show that no more cells are produced in a long-term culture than in a short-term culture—it only takes longer to produce them when conditions are suboptimal. As soon as the limiting factors or nutrients are depleted, the cells' response is no longer to the "test" substances but rather to the culture conditions. It is the identification and characterization of these truths that must be dealt with if true advances in our understanding of hematopoiesis are to occur. Current research trends do not appear to be in this direction.

VI. CHARACTERIZATION OF MACROPHAGES

While murine bone marrow has been extensively used to study hematopoiesis, the primary emphasis has been on the influence of various hematopoietic growth factors and cytokines on colony formation of hematopoietic progenitor cells. There is a paucity of data on the precise characterization of the cells as they proliferate and differentiate in these systems.

While immunophenotyping with monoclonal antibodies to cell surface proteins has been extensively used to investigate the differentiation of murine lymphocytes, limited data have been obtained with this technique concerning macrophage differentiation and heterogeneity. Walker et al. [34–37] have studied the immunophenotype of individual splenic-derived macrophage colonies and macroglia. Colonies of cells were all positive for Mac-1 and Mac-3 but showed heterogeneity in Mac-2 and expression. By expanding individual colonies, this group has provided strong evidence that macrophage heterogeneity is based on individual functional macrophage populations that share some functions but differ in others that are both constitutive and inducible. Whether these diverse functional populations can be resolved by determining their unique repertoire of cell surface markers in a manner similar to that in which functional lymphocyte subsets are resolved remains to be determined, and this is an exciting area for further research.

We now realize that each cell type expresses a unique repertoire of surface proteins. As cells differentiate, they lose specific proteins while they gain new ones. Thus, the repertoire changes as the cells differentiate. While we currently may not know the function of each protein, a particular protein can still be used for identification providing a good monoclonal antibody specific for it is available. Bender et al. have applied this logic in combination with flow cytometry to characterize the cells that emerge in bulk cultures of bone marrow containing growth factors [38]. They have shown that murine bone marrow cells in liquid culture with IL-3 will develop into nonadherent granulocytes, mast cells, and macrophages [38]. Recently (1990) Bender et al. found that 13-*cis*-retinoic acid (13cRA) enhances the production of macrophages in murine bone marrow cultures that have been stimulated with IL-3 [39]. This enhancement was illustrated using four-color flow-cytometric techniques and the following monoclonal antibodies: Mac-1 (granulocytes and macrophages), F4/80 (macrophages), B54.2 (mast cells), and H12 (Thy-1.2:myeloid precursors). There was a twofold increase in MAC-1 and F4/80-positive cells (macrophages). Their data suggest that 13cRA increases macrophage production by modulating the commitment of IL3-expanded progenitor cells to the macrophage lineage (39).

It has recently been discovered that the antibody F4/80, which had previously demonstrated binding specificity only for cells of the monocyte/macrophage lineage, also has an affinity for eosinophils (40). This antibody in combination with Mac-1 will bind eosinophils as well as binding to monocyte/macrophages. The two cell types can be distinguished based on the fluorescence intensity of the labeling. Eosinophils will display with a less intense fluorescence for both Mac-1 and F4/80 as compared with monocytes/macrophages. Neutrophilic granulocytes, which also bind

Mac-1, will not bind F4/80 [40]. This information may demonstrate a closer lineage relationship between monocytes/macrophages and eosinophils.

Further work is being done to phenotype myeloid progenitor cells found within the peripheral blood and bone marrow of humans. Using a battery of monoclonal antibodies and four-color flow cytometry, a more definitive characterization of myeloid precursors is made possible (26). The monoclonal antibody CD34, which is known to identify hematopoietic precursor cells from human bone marrow, was used in combination with CD33, HLA-DR, CD45, CD19, CD7, CD71, CD10, CD3, CD14, CD20 and CD11b. Bender et al. found that a higher proportion of peripheral blood CD34+ cells express the CD33 antigen. This suggest that these cells are committed to the myeloid lineage. These same cells are also found in the bone marrow (26). The demonstration of CD34+ cells that possess lymphoid antigens as well as myeloid antigens supports the potential use of peripheral blood cells as an alternative source of stem cells for the study of hematopoiesis. These cells form colonies in soft agar or in liquid culture medium when supplemented with specific CSFs.

We have used this approach to define more precisely the differentation pathway of murine macrophages. Although an exhaustive list is not given, some important antibodies to murine macrophages are shown in Table 3. None of these antibodies are specific for macrophages, as they bind to other cells as well. It is the appropriate combination of them that leads to the precise identification of the various differentiation stages of the macrophage.

We screened these antibodies to determine the best combination to use for resolving the differentiation of mononuclear phagocytes (32). Three of them, HK1.4, F4/80, and 30E2, showed interesting characteristics; the results are shown in Figure 12. HK1.4 stains a fraction of the smallest nonadherent cells but not the intermediate-size or the large cells (the HK1.4 was kindly provided by Drs. Wendy Havern and Frank Fitch). The small HK1.4+ cell population can be further subdivided into bright, dim, and negative cells. When sorted, the bright HK1.4+ cells had a blast cell morpholoby characteristic of early myeloid cells. The dim and negative cells had characteristics associated with early myelomonocytic cells. The F4/80- HK1.4+ cells may be the earliest precursors capable of both myelomonocytic and granulocytic differentiation (see below). The largest cells (right panel) do not express HK1.4.

Table 3 Antibodies that Label Murine Mononuclear Phagocytes

Hybridoma	Common name	Primary target cell	ATCC	Reference
30-H12	Thy-1	T cell	TIB107	62,63
30-E2	—	B lymphomas	—	62,63
30-H11	—	T and B lymphomas	—	62,63
30-G2	—	Myeloid cells T and B lymphomas	—	62,63
MI/70	Mac-1	Granulocytes and macrophages	TIB128	59,64,65
M3/38	Mac-2	Macrophages	TIB166	60,64,65
M3/84	Mac-3	Macrophages	TIB168	61,64,65
F4/80	—	Macrophages and eosinophils	HB198	34,35,40
HK1.4	ly6	Granulocytes	—	36,41

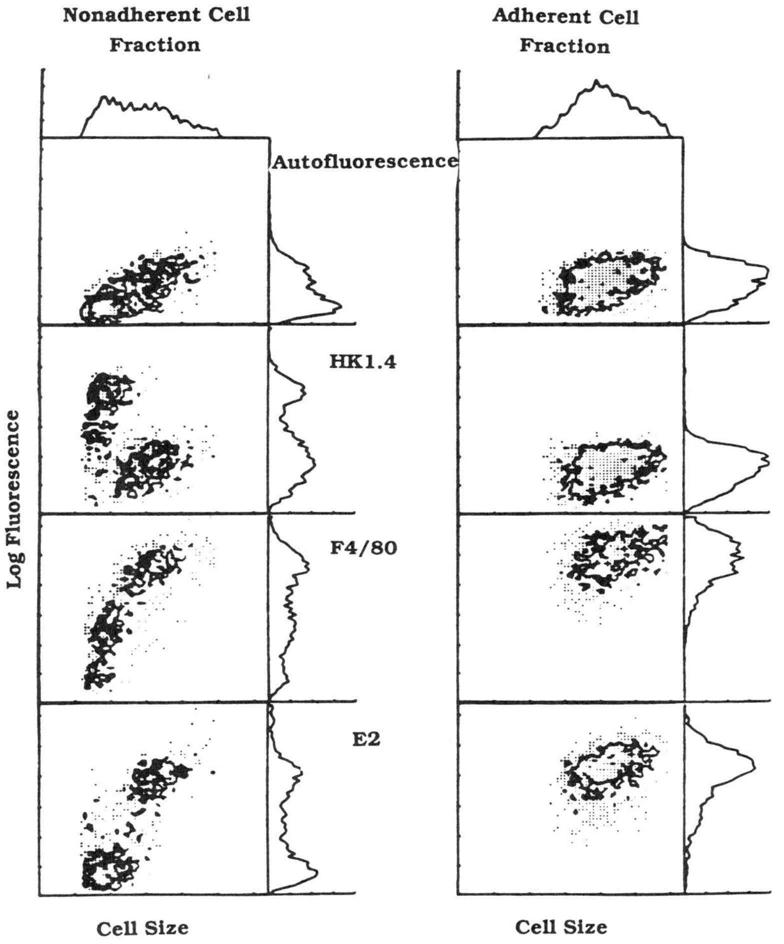

FIGURE 12 Important epitopes for resolving macrophage subsets. Bone marrow cells were cultured in growth medium at 2.5 × 104 cells/ml for 6 or 7 days in T-75 flasks. To obtain all the cells, they were cultured on nonwettable 100-mm petri dishes. Nonadherent cells were obtained from T-75 flasks by hitting the flasks along the side to jar them loose. An average recovery of 5×10^5 nonadherent cells was obtained. Since adherent cells cannot be removed from T-75 flasks with good viability, they were obtained from 100-mm petri dishes as described in Figure 11. The average recovery was 6×10^6 cells including the nonadherent fraction. The flow cytometer settings were adjusted to optimize the resolution of each cell population. For the nonadherent cells, the amplifier gain for forward scatter has been increased over that used in Figure 11. In the panels on the right, the small cells have been removed from analysis by electronic gating. For staining, the cells were centrifuged and the pellet resuspended in phosphate-buffered saline containing 0.1% bovine albumin (PAB), to a concentration of 10×10^6 cells/ml. All procedures were done using an ice bath. Aliquots of 50 μl were incubated for 10 minutes with 10 μg of normal mouse immunoglobulin G (IgG) to block Fc receptors. Without washing, 1 μg (in 10 μl) of purified antibody (each is a rat monoclonal antibody) was added to appropriate tubes and incubated 15 minutes. After washing cells with 3.5 ml PAB, 1 μ of FITC-goat anti-rat IgG. F(ab)'2 with no demonstrable anti-mouse IgG activity, was added for 15 min. The cells were excluded from analysis by adding 1 μg propidium iodide to the samples 5 minutes before data acquisition. From Ref. 58.

F4/80 produces a complementary pattern of increasing epitope density as cells become larger. While HK1.4 labels all granulocytes, F4/80 labels only macrophages and eosinophils [40]. 30E2 stains virtually none of the smallest nonadherent cells, but all the intermediate-size nonadherent and adherent cells are positive for this antibody. HK1.4 is a surface membrane protein in the Ly6 family [41]. The F4/80 antigen has a molecular weight of 150,000, but its function is not known [42, 43]. 30E2 antigen is found on macrophages and murine lymphoma cells [44].

Since the nonadherent fraction derived from the T flasks was enriched for HK1.4+ cells, they were sorted to determine their ability to form macrophage colonies. It was found that HK1.4+ cells form large colonies of macrophages when cultured in growth medium for 7 days, but, as with peritoneal exudate cells, no colonies were found on the culture dishes containing sorted cells that were HK1.4–. By day 14, however, the cultures containing the HK1.4– cells had formed distinct colonies of macrophages. These colonies were composed of both epithelioid and ameboid macrophages indentical to those formed by PECs.

As shown in Table 2, there are two kinds of macrophage CFC: cells with little or no lag period that form large colonies after only 7 days' incubation in growth medium and express HK1.4, and cells such as those found in the peritoneal cavity, that have a longer lag period before entering cell cycle and form colonies later than day 7. The loss of HK1.4 appears to be coincident with differentiation to this latter stage of macrophage development, and even though the basis for the increased lag period is not yet known, we believe IL-6 plays an important role.

VII. TWO DISTINCT MACROPHAGE LINEAGES

Once thought to be limited to T cells and some cells in the brain, possibly glial cells, Thy-1 has been discovered on several different cell types. Cells outside the hematopoietic system that apparently express Thy-1 are epidermal cells, myoepithelial cells, and fibroblasts in the breast [41]. Within the hematopoietic lineage, uncommitted stem cells and undifferentiated progenitor cells have also been shown to express Thy-1 [45]. Unlike thymocytes, which express high amounts, and peripheral T cells, which express somewhat less Thy-1, uncommitted stem cells and undifferentiated progenitor cells express low levels of Thy-1 [46]. The Thy-1 antigen may be membrane-associated adenosine deaminase (47), but this report has apparently not yet been confirmed. This enzyme may function in macrophages to produce superoxide in the catabolism of purines [48].

In studying the infiltration of leukocytes into subcutaneously implanted tumors [49, 50], we found that a high fraction of macrophages expressed Thy-1 on their surface. Thy-1 expression on various sources of macrophages was evaluated, and the data are shown in Table 4. Of recovered viable cells from the EMT6 tumor, 27% were Thy-1+ MAC-1+; this represents 61% of Mac-1+ cells. Mac-1 (CD11b) has specificity for the CR3bi receptor found on granulocytes and macrophages. It cannot be determined from these data if there are two distinct populations of cells that

TABLE 4 Distribution of Thy-1–Positive Macrophages

Source	Percent positive cells			
	T-M–	T-M+	T+M+	T+M–
EMT6 tumor	45	17	27	11
Sponge	18	31	37	14
Peritoneal cells	59	34	2	5
Peritoneal exudate cells	34	56	3	7
Pulmonary alveolar macrophages	85	6	4	8

overlap or if a single population is acquiring or losing Thy-1. Also present in the cell suspension is a population of Thy-1+ cells that do not express Mac-1 and a population of cells that are negative for both markers.

When either peritoneal cells or peritoneal exudate cells are evaluated, major populations that expressed either Mac-1, Thy-1, or neither epitope were found. A very small distinct population of Thy-1+ Mac-1+ cells was found in the peritoneal cell suspension. When pulmonary alveolar cells were used, the majority of cells were very autofluorescent.

From this data, only the tumor exhibited a high frequency of Thy-1+ MAC-1+ cells. In order to determine if the tumor environmment was responsible for Thy-1+ Mac-1+ cell infiltration, gel foam sponges were implanted subcutaneously in mice and removed 4 days later. As shown in Table 4, 37% of cells were Thy-1+ Mac-1+. This represents 54% of Mac-1+ cells. We tentatively conclude that Thy-1+ macrophages specifically home to regenerating tissues but not sites of acute inflammation.

Since blocking Fc receptors with both mouse immunoglobulin and rat immunoglobulin before labeling with Thy-1 would rule out the possibility that the antibody was binding to Fc receptors, it was possible that macrophages might have Thy-1 absorbed on their membranes. To rule this out and determine whether the Thy-1+ macrophages were a distinct subpopulation, we cultured 10^3 sponge-derived cells in 5 ml of growth medium on 60-mm nonwettable plastic petri dishes. Ten days later, 3–5 $\times$ 10^5 cells were recovered (a 300-fold increase). Cells in suspension were stained and analyzed by flow cytometry.

The bivariate distribution of freshly isolated cells from sponges is shown in Figure 14. After culturing for 10 days (right panel of Figure 14), cells were considerably brighter in their Mac-1 staining but expressed about the same amount of Thy-1 as did the initial suspension. The bright Thy-1+, dim Mac-1+ cells shown in this figure have not yet been identified but may be T lymphoblasts that are somewhat autofluorescent and have actually not bound any PE-Mac-1. These results clearly show Thy-1 is a stable marker for this distinct subpopulation of macrophages.

VIII. MACROPHAGE DIFFERENTIATION

Using this approach, we have developed a model that describes the orderly progression of murine hematopoietic lineages on the basis of their repertoire of cell surface markers. By combining the four MAbs F4/80, Mac-1, Thy-1, and HK1.4, virtually all leukocytes within murine bone marrow can be resolved. A summary of these subsets is shown in Figure 13. The earliest progenitor cell (probably the CFU-S) expresses HK1.4 and Thy-1. As this cell differentiates into the common myeloid progenitor cell, it acquires the Mac-1 antigen (CD11b). If this precursor loses Thy-1, its differentation is toward neutrophils, basophils, and mast cells. If this progenitor cell acquires F4/80, commitment to the mononuclear phagocyte series occurs. This includes both macrophages and eosinophils [38]. A loss of HK1.4 and further up-regulation of F4/80 signals the macrophage differentiation pathway, while eosinophils lose THY-1 but retain HK1.4 as they mature. Continuing to differentiate, two distinct populations of macrophages arise, one that retains Thy-1 and functions in consolidating tissues and one that loses Thy-1 and functions in acute inflammatory processes. While there are certainly many more surface proteins that can be used for stratifying myeloid cell differentiation and function, these four appear to be most useful in identifying the major subsets of differentiating myeloid cells in murine bone marrow.

IX. ROLE OF COMPETENCE AND PROGRESSION FACTORS

The mechanism for the regulation of commitment and proliferation of the stem cell and its progenitors is postulated to involve factors unique to each hematophoietic lineage [50]. These unique factors will control a hematopoietic cell's traversal of the cell cycle. These growth factors

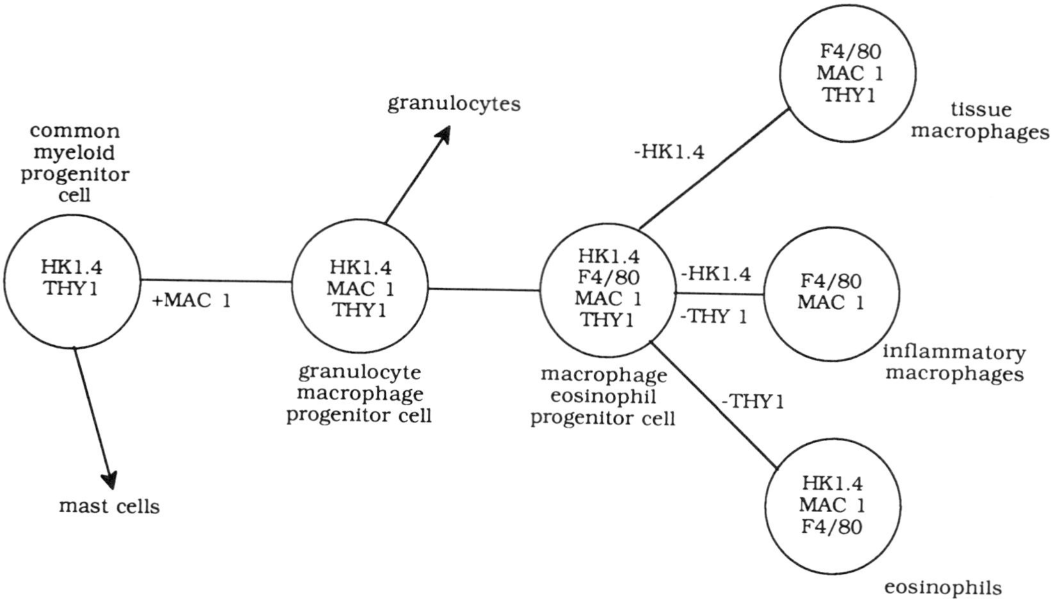

FIGURE 13 Immunophenotypic differentiation of monunuclear phagocytes. The differentiation of myelomonocytic cell subsets can be determined by the cell surface immunophenotype. When the four antibodies are combined the common myeloid progenitor cell can be uniquely identified as a Thy-1 + HK1.4 + cell. As this cell differentiates into a granulocyte-macrophage progenitor cell (GM-CFC), it acquires Mac-1 expression. The acquisition of F4/80 expression signifies differentiation to the macrophage/eosinophil progenitor cell. Loss of HK1.4 results in the tissue macrophage phenotype, while loss of both HK1.4 and Thy-1 results in the inflammatory macrophage. Loss of Thy-1 results in the eosinophil phenotype.

can be classified as competence or progression factors, depending on the phase of the cell cycle in which they are required.

Competence and progression factors may work together in a lineage-specific manner. As a cell differentiates it loses its responsiveness to one specific factor and gains the ability to respond to another. At transition there is most likely a time during differentiation at which receptors for both growth factors are present on the same cell [50]. In this postulate, competence factors act on progenitor cells to stimulate the expression of a receptor for a progression factor, thereby allowing the cell to proliferate (50). This postulate is supported by studies of Bartelmez et al. [51, 52] in which some of the factors acting upon early progenitors (i.e., interleukin-3 [IL-3] or HP-1) may behave as competence factors by stimulating the expression of progression factor receptors. It is possible that differentiated cells lose their proliferative capability because they lose responsiveness to progression factors necessary for traversing the cell cycle. Further understanding of the relationship of these factors is necessary to provide insight into the mechanism for regulating proliferation and differentation.

X. *c-FGR* EXPRESSION BY MACROPHAGES

Within the realm of macrophage proliferation, the action of cellular proto-oncogenes cannot be ignored. Proto-oncogenes, which are highly homologous to retroviral DNA, are normal genes associated with proliferation-related events [53]. The transcription of proto-oncogenes in the

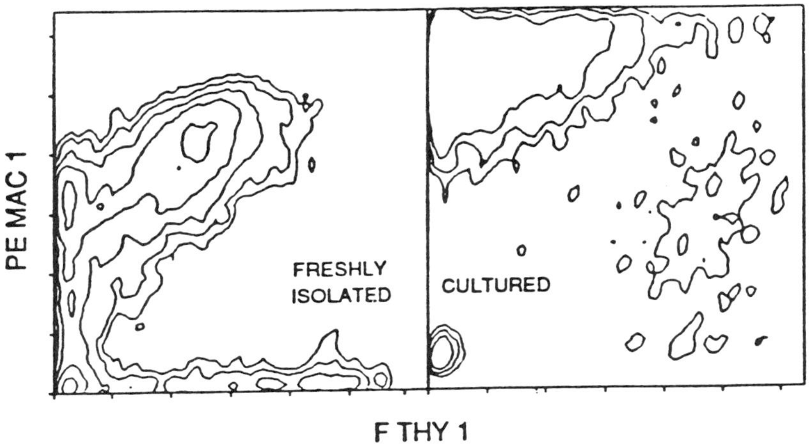

FIGURE 14 Stability of Thy-1 on cultured macrophages. As previously described [58], sponges were implanted subcutaneously on the backs of C3H mice. After 3 days, the sponge was recovered and digested in collagenase to produce a monodispersed suspension of infiltrating cells. Harvested cells were adjusted to 10×10^6 cells/ml in PAB, and 50-μl aliquots were incubated 10 minutes with 10 μg (in 10 μl) mouse immunoglobulin G to block Fc receptors. Without washing, 1 μg each (in 10μl) of directly fluoresceinated Thy-1 (Becton Dickinson) and directly phycoerythrinated Mac-1 (prepared by us and conjugated with PE by Molecular Probes) was added. After 15 minutes the cells were washed with 3.5 ml PAB, and after adding PI, the data were acquired by flow cytometry. Freshly isolated cells or cells cultured for 10 days in growth medium are shown.

normal cell is highly regulated to maintain homeostasis. If this regulation breaks down, as in neoplasia, inappropriate expression occurs. The Src family of proto-oncogenes encodes for tyrosine kinases [54], and members of this family share a highly conserved carboxy-terminal region in which is found the catalytic domain for tyrosine kinase activity. However, they show divergence in their amino-terminal domains [54]. One member of the this family, *c-fgr*, has been found to be expressed in a lineage-specific manner in differentiating myelomonocytic cells [55].

Recent results suggest that *c-fgr* expression is at a maximum at the G0/G1 boundary of the cell cycle, indicating a possible role in regulating proliferation. The potential role of *c-fgr* in regulating the cell cycle of the monocyte/macrophage lineage remains unknown. Colony-stimulating factor-1, which appears to modulate *c-fgr* expression [56], in combination with other growth factors may be necessary for competence and progression of macrophage proliferation. Interleukin-6 may act to disrupt this. Willman and Yi have elucidated the mRNA sequence of murine macrophage *c-fgr* [57] and developed specific antibodies to murine *c-fgr*; they have shown that two separated products may be active in regulating macrophage proliferation.

Our studies have centered on the relationship of *c-fgr* expression to the initiation and progression of macrophages through cell cycle. In the experiments of Willman et al. [56], it was found that *c-fgr* transcription occurred continuously, but mRNA levels were not detectable until 4 hours after CSF-1 addition. The message reached a maximum at 8 hours and was at basal levels by 12 hours. These results suggested that CSF-1 modulates c-fgr expression through a posttranscriptional mechanism. Our preliminary quantitative PCR results indicate that c-fgr expression is detectable at 2 hours and remains elevated up to 16 hours later (Figure 15). Since our study did not include time points beyond 16 hours, *c-fgr* expression modulation needs further evaluation. While these studies need to be extended to cover later time periods, it is clear that *c-fgr* expression is very low in quiescent cells. Since FBS is present in the 24-hour synchronizing medium, it seems likely

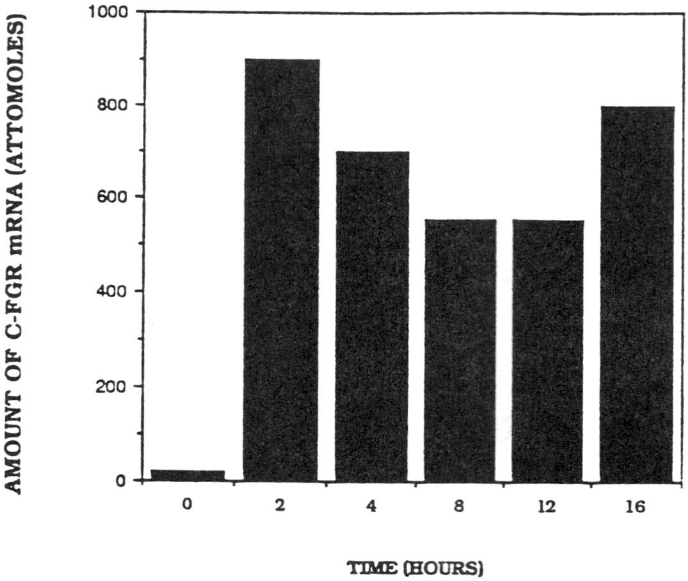

FIGURE 15 *c-fgr* mRNA expression in bone marrow macrophages. Bone marrow macrophages are cultured for 6 days in growth medium. On day 6, the growth medium is exchanged for αMEM with 20% FBS (LCM has been removed). After 24 hours, 20% LCM is given back to the cultures. The cells are harvested at the indicated time points. RNA is extracted and *c-fgr* expression is quantitated using quantitative PCR. The results of these quantitations are plotted against the individual time points.

that CSF-1 itself may not be solely inducing this oncogene. Further work is warranted on the role of this important oncogene in regulating macrophage proliferation.

ACKNOWLEDGMENTS

The authors wish to recognize the many individuals who have contributed to this work: Drs. Cheryl Adles, James Bender, Frank Fitch, Larry Guilbert, Wendy Havern, Richard Stanley, Thomas Tomasi, Ben van der Zeist, Dennis van Epps, Edwin Walker, and Cheryl Willman.

REFERENCES

1. Van Furth R, ed. Mononuclear phagocytes: characteristics, physiology and function. Dordrecht, the Netherlands: Martinus Nijhoff, 1985.
2. Till JE, McCulloch EA. A direct measure of the radiation sensitivity of normal mouse bone marrow cells. Radiat Res 1961; 14, 213–22.
3. Johnson GR, Nicola NA. Characterization of two populations of CFU-S fractionated from mouse fetal liver by fluorescence-activated cell sorting. J Cell Physiol 1984; 118:45–52.
4. Harris RA, Hogarth PM, Wadeson LJ, Collins P, McKenzie IFC, Pennington DG. An antigenic difference between cells forming early and late haematopoietic spleen colonies (CFU-S). Nature 1983; 307:638–41.
5. Miller BA, Lipton JM, Linch DC, Burakoff SJ, Nathan DG. Thy1 is a differentiation antigen that characterizes immature erythroid and myeloid hematopoietic progenitors. J Cell Physiol 1985; 123:25–32.

6. Till JE, McCulloch EA. Hemopoietic stem cell differentiation. Biochem Biophys ACTA 1980; 605:431–59.
7. Korn AP, Henkelman RM, Ottensmyer FP, Till JE. Investigations of a stochastic model of haemopoiesis. Exp Hematol 1973; 1:362–75.
8. Nakahata T, Ogawa M. Clonal origin of murine haemopoietic colonies with apparent restriction of granulocyte-macrophage-megakaryocyte (GMM) differentiation. J Cell Physiol 1982; 111:239–46.
9. Fitchen JH, Foon KA, Cline MJ. The antigenic characteristics of hematopoietic stem cells. N Engl J Med 1983; 305:17–24.
10. Stewart CC, Cramer SF, Stewart PG. The response of human peripheral blood lymphocytes to phytohemagglutinin: determination of cell numbers. Cell Immunol 1975; 16:237–50.
11. Stewart CC. Murine mononuclear phagocytes from bone marrow. In: Adams D, Edelson PJ, Koren HS, eds. Methods for studying mononuclear phagocytes. New York: Academic Press. 1981:5–20.
12. Lin H, Stewart CC. Colony formation by mouse peritoneal exudate cells in vitro. Nature 1973; 243: 176.
13. Stewart CC. Formation of colonies by mononuclear phagocytes outside the bone marrow. In: Mononuclear phagocytes: functional aspects. Dordrecht, the Netherlands: Martinus Nijhoff, 1980.
14. Newburger PE, Chovaniec ME, Greenberger JS, Cohen HJ. Functional changes in human leukemic cell line HL-60. A model for myeloid differentiation. J Cell Biol 1979; 82:315–22.
15. Stendahl O, Dahlgren C, Hed J. Physiochemical and functional changes in human leukemic cell line HL-60. J Cell Physiol 1982; 112:217–21.
16. Glaser L. Leukemic cells as probes for sequential functional differentiation of the human granulocyte. Am J Clin Pathol 1983; 79:45–51.
17. Stewart CC, Lin H. Macrophage growth factor and its relationship to colony stimulating factor. J Reticuloendothelial Soc 1978; 23:269.
18. Stewart CC, Walker EB, Johnson C, Little R. Clonal analysis of bone marrow and macrophage cultures. In: Van Furth R, ed. Mononuclear phagocytes characteristics, physiology and function. Dordrecht, the Netherlands: Martinus Nijhoff, 1985:255.
19. Lin H, Stewart CC. Peritoneal exudate cells. I. Growth requirement of cells capable of forming colonies in soft agar. J Cell Physiol 1974; 83:369–78.
20. Stewart CC, Lin HS, Adles C. Proliferation and colony-forming ability of peritoneal exudate cells in liquid culture. J Exp Med 1975; 141:1114–32.
21. Stewart CC, Adles C, Hibbs JR. Cloned macrophages can be induced to kill tumor cells. J Reticuloendothelial Soc 1978; 24:107–13.
22. Virolainen M, Defendi V. Dependence of macrophage growth in vitro upon interaction with other cell types. In: Defendi V, Stoker M, eds. Growth regulating substances for animal cells in culture. Wiistar Institute Symposium Monograph 7: 1967:67–85.
23. Van Furth R, Cohn ZA. The origin and kinetics of mononuclear phagocytes. J Exp Med 128:415–35.
24. Van Furth R, Hirsch JG, Fedorko ME. Morphology and peroxidase cytochemistry of mouse promonocytes, monocytes and macrophages. J Exp Med 1970; 132:794–812.
25. Van der Zeijst BAM, Stewart CC, Schlesinger C. Proliferative capacity of mouse peritoneal macrophages in vitro. J Exp Med 1978; 147:1253.
26. Bender JG, Unverzagt KL, Walker DE, Lee W, Van Epps DE, Smith DH, Stewart CC, Bilto L. Identification and comparison of CD34-positive cells and their subpopulations from normal peripheral blood and bone marrow using multicolor flow cytometry. Blood 1991; 77:259–6.
27. Stanley ER, Heard PM. Factors regulating macrophage production and growth. Purification and some properties of the colony stimulating factor from medium conditioned by mouse L cells. Chem 1977; 252:4305–12.
28. Matsushime H, Roussel MF, Ashmun RA, Sherr CJ. Colony-stimulating factor 1 regulates novel cyclins during the G1 phase of the cell cycle. Cell 1991; 65:701–13.
29. Lin H. Peritoneral exudate cells. II. Kinetics of appearance of colony-forming cells. J Cell Physiol 1974; 84:159.
30. Riedy MC, Stewart CC. Inhibitory role of interleukin-6 in macrophage proliferation. J Leukocyte Biol 1992; 52:125–7.
31. Weinberg, JB. Macrophage polykaryon formation in vitro by peritoneal cells from mice given injections of sodium periodate. Am J Path. Vol 110(2):182–92, 1983.

32. Walker EB, Warner NL, Stewart CC. Characterization of subsets of bone marrow–derived macrophages by flow cytometry analysis. J Leukocyte Biol 1985; 37:121–36.
33. Spooncer E, Heyworth CM, Dunn AS, Dexter TM. Self renewal and differentiation of interleukin-3–dependent multipotent stem cells are modulated by stromal cells and serum factors. Differentiation 1986; 31:111–8.
34. Walker WS. Origins of macrophage diversity: functional and phenotypic analysis of cloned populations of mouse splenic macrophages. Cell Immunol 1987; 107:417–32.
35. Walker WS. Differential antigen presentation by cloned populations of mouse splenic macrophages. J Immunol 1989; 143:2142–5.
36. McCormack JM, Sun D, Walder WS. A subset of mouse splenic macrophages can constitutively present alloantigen directly to CD8+ T cells. J Immunol 1991; 147:421–7.
37. Moore SC, McCormack JM, Armendariz E, Gatewood J, Walker WS. Phenotypes and alloantigen-presenting activity of individual clones of microglia derived from the mouse brain. J Neuroimmun 1992; 1–12.
38. Bender JG, Van Epps DE, Stewart CC. Characterization of granulocytes and mast cells in cultures of mouse bone marrow stimulated with interleukin-3. J Cell Physiol 1988; 135:71–8.
39. Bender JG, Stewart CC, Van Epps DE, Waller DE. 13-cis retinoic acid augments the production of macrophages in mouse bone marrow cultures stimulated with IL-3. Exp Hematol 1990; 18:990–4.
40. McGarry MP, Stewart CC. Murine eosinophil granulocytes bind the murine macrophage-monocyte specific monoclonal antibody F4/80. J Leukocyte Biol 1991; 50:471–8.
41. Havran WL, Lancki DW, Moldwin RL, Dialynas DP, Fitch FW. Characterization of an Anti-Ly-6 monoclonal antibody which defines and activates cytolytic T lymphocytes. J Immunol 1988; 140:1034–42.
42. Starkey PM, Turley L, Gordon S. The mouse macrophage-specific glycoprotgein. Defined by monoclonal antibody F4/80: characterization, biosynthesis and demonstration of a rat analogue. Immunology 1987; 60:117–22.
43. Austyn JM, Gordon S. F4/80, a monoclonal antibody directed specifically against the mouse macrophage. Eur J Immunol 1981; 11:805–15.
44. Ledbeller JA, Herzenberg CA. Xenogenic monoclonal antibodies to mouse lymphoid differentiation antigens. Immunol Rev 1979; 47:64.
45. Miller B, Lipton JM, Linch DC, Burakoff SJ, Nathan DG. J Cell Physiol 1985; 123:25–32.
46. Spangurde GJ, Muller-Sieburg CE, Heimfeld S, Weissman IL. Two rare populations of mouse Thy-1^{10} bone marrow cells repopulate the thymus. J Exp Med 1988; 167:1671–83.
47. Chechik BE, Schrader WP, Daddona PE. Identification of human thymus-leukemia-associated antigen as a low-molecular-weight form of adenosine deaminase. JNCI 1980; 64:1077–81.
48. Tritsch GL, Paolini NS, Bielat K. Adenosine deaminase activity associated with phagocytic vacuoles. Histochemistry 1985; 82:281–5.
49. Stevenson AP, Martin JC, Stewart CC. Simultaneous measurements of macrophage-induced cytostasis and cytotoxicity of EMT6 cells by flow cytometry. Cancer Res 1986; 46:99–105.
50. Stewart CC, Stevenson AP, Hibbs J. Effector mechanisms for macrophage-induced cytostasis and cytolysis of tumor cells. In: Heppner G, Fulton A, eds. Macrophages and cancer. CRC Press, 1988:39–59.
51. Bartelmez SH, Sacca R, Stanley ER. Lineage-specific receptors used to identify a growth factor for developmentally early hemopoietic cells: assay of hemopoietin-2. J Cell Physiol 1985; 122:362.
52. Bartelmez SH, Stanley ER. Syngergism between hemopoietic growth factors (HGFs) detected by their effects on cells bearing receptors for a lineage-specific HGF: assay of hemopoietin-1. J Cell Physiol 1985; 122:370.
53. Walker EB, Akporiaje ET, Warner NL, Stewart CC. Characterization of subsets of bone marrow-derived macrophages by flow cytometry analysis. J Leukocyte Biol 1985; 37:121–36.
54. Stewart CC. Flow cytometric analysis of oncogene expression in human neoplasias. Arch Pathol Lab Med 1989; 113:634–40.
55. Hunter T, Cooper JA. Protein-tyrosine kinases. Annu Rev Biochem 1985; 54:897–930.
56. Willman CL, Stewart CC, Griffith JK, Stewart SJ, Tomasi TB. Differential expression and regulation of

the c-SRC and c-FGR protooncogenes in myelomonocytic cells. Proc Natl Acad Sci USA, 1987; 84:4480–4.
57. Willman CL, Yi TL. Cloning of the murine c-FGR proto-oncogene cDNA and induction of C-FGR expression by proliferation and activation factors in normal bone marrow–derived monocytic cells. Oncogene 1989; 4:1081–7.
58. Stewart CC. Methods for studying the ontogeny of mononuclear phagocytes. In: Weir DM, ed. Cellular immunology. Blackwell Scientific Publications, 1986; 44.
59. Ho MK, Springer TA. MAC-1 antigen: quantitative expression in macrophage populations and tissues, and immunofluorescence localization in spleen. J Immunol 1982; 128:2281.
60. Ho MK, Springer TA. MAC-2, a novel 32,000 Mr mouse macrophage subpopulation-specific antigen defined by monoclonal antibodies. J Immunol 1982; 128:1221.
61. Ho MK, Springer TA. Tissue distribution, structural characterization, and biosynthesis of MAC-3, a macrophage surface glycoprotein exhibiting molecular weight heterogeneity. J Biol Chem 1983; 258:636.
62. Lanier LL, Warner NL, Ledbetter JA, Herzenberg LA. Quantitative immunofluorescent analysis of surface phenotypes of murine B cell lymphomas and plasmacytomas with monoclonal antibodies. J Immunol 1981; 127:1691.
63. Ledbetter JA, Herzenberg LA. Xenogenic monoclonal antibodies to mouse lymphoid differentiation antigens. Immunol Rev 1979; 47:64.
64. Ralph P, Ho MK, Litcofsky PB, Springer TA. Expression and induction in vitro of macrophage differentiation antigens on murine cell lines. J Immunol 1983; 130:108.
65. Springer TA. Murine macrophage differentiation antigens identified by monoclonal antibodies. In: Foster O, Landry M, eds. Heterogeneity of mononuclear phagocytes. New York: Academic Press, 1981:37.

1

Macrophage Activation: A Riddle of Immunological Resistance

Robert M. Crawford, David A. Leiby, Shawn J. Green, Carol A. Nacy, Anne H. Fortier, and Monte S. Meltzer

Walter Reed Army Institute of Research, Washington, D.C.

I. MACROPHAGE ACTIVATION: A RIDDLE OF IMMUNE RESISTANCE: INTRODUCTION

A. INTRODUCTION, CHRONOLOGY, AND DEFINITIONS

It is difficult to say when man first noted a connection between recovery from disease and resistance to reinfection (acquired immunity). Recordings of such observations date back to Greco-Roman times. Widespread, successful, and practical application of these observations to disease prophylaxis (vaccination) was established by the work of Jenner, Pasteur, and Koch in the eighteenth and nineteenth centuries. But even with these impressive results in protective immunization, there was no clear and cohesive theory that explained the nature and mechanisms of acquired immunity. Indeed, bodily components and physiological reactions that coalesce to protect man against disease are still only partially identified. The fact that man is not always the victor in his perpetual clash with infectious pathogens emphasizes the complexity and unpredictability of such immune reactions.

The concept that acquired immunity is a bodily function that requires previous exposure to an infectious agent eluded early scientists even after broad experience with preventive immunizations. This is not so surprising, of course, given that a formal link between human coitus and conception was not established until the fourteenth century. With time, the cellular and humoral components that comprise the "immune system" and produce the protective "immune response" were delineated. A mechanistic basis for an immune response was first revealed in 1890, when von Behring demonstrated in sera of toxin-immunized animals humoral factors (antibodies) that neutralized the toxin and could be transferred to control animals to prevent toxin-specific disease [1]. Coincident with these studies, Metchnikov described a "cellular" basis for elimination or "phagocytosis" of foreign bodies and infectious agents [2]. However, not until the discovery of allergic or "sensitivity" reactions, in which previously infected animals reacted to infectious challenge in a manner very different from that of uninfected animals, did the model of acquired immunity become clear. In 1921, Zinsser recognized that tuberculin sensitivity differed from other forms of allergy [3]. Noting that similar states of "delayed" hypersensitivity were common to other infectious diseases (typhoidin from *Salmonella typhi*, abortin from *Brucella abortus*, mallein from *Pfeifferella mallei*), could not be easily transferred with serum, and tended to parallel the general level of host resistance, Zinsser and Mueller suggested that ". . . these substances upon which [bacterial] allergy depends may possess protective functions different from and based on a different mechanism from those possessed by antibodies."

Since the pioneering studies of Metchnikov on the central role of macrophages in host resistance to infection, other cells (T cells and their subsets, B cells, NK cells) have been added to (some say supplanted) the macrophage as critical elements in the immune system. However, the macrophage retains a pivotal position in the development of all immune responses. Metchnikov's discovery of an adaptive state of macrophage function in response to foreign bodies was fully appreciated in the context of acquired resistance to infection. In the 1960s Mackaness and others developed the concept of "macrophage activation," its cellular basis, and the relationship of delayed hypersensitivity to acquired resistance. Such activated macrophages underlie the enhanced but antigenically nonspecific resistance to infection with intracellular bacteria observed in animals recovering from infection with pathogens such as *Listeria monocytogenes* [4]. Animals that recover from infection with one microorganism acquire the ability to resist infection not only by the same pathogen but also by a wide spectrum of completely unrelated microorganisms. Enhanced nonspecific resistance in the infected animal correlates directly with macrophage-mediated killing of a wide range of infectious pathogens in culture. Such antimicrobial activity defined the term "macrophage activation" and is the defining function that we use in this discussion.

The cardinal features and "immunological" bases of acquired immunity to reinfection were established by adoptive cell transfer experiments [5]. Lymphocytes are specially stimulated by infectious challenge, a mononuclear cell infiltrate of monocytes or macrophages occurs at the site of inflammation, and macrophages are activated by soluble products of lymphocytes to kill the infectious agent and resolve the infection. A dual requirement exists for expression of acquired resistance to infection: antigen-specific T cells to initiate the reaction and nonspecific "activated" macrophage accessory cells to effect the response.

Mackaness's analysis of acquired resistance to facultative and obligate intracellular parasites implied some form of interaction between different types of immune cells. Although fundamental tenets of immunology today, the notion of cell-cell interaction and the conception of how this could be achieved were controversial for many years. Do cells interact through physical contact (as originally proposed) or through secreted mediators? The first description of a soluble factor released into culture fluids of specific antigen-stimulated immune lymphocytes (termed migration inhibition factor [MIF]) prompted speculation that this factor might also activate macrophages for enhanced microbicidal activity [6,7]. The term "macrophage activation factor" (MAF) was first used to describe a lymphokine (LK) activity that induced increased macrophage spreading on glass [8]. Other MAF activities in LK induced macrophage cytotoxic activity against tumor cell and microbial targets [9]. Soon literally hundreds of LK activities were described in crude culture fluids of antigen- or mitogen-stimulated lymphocytes. In 1982, Roberts and Vasil showed that mouse recombinant interferon-γ (IFN-γ) and MAF were functionally and physicochemically identical [10]. For most investigators, IFN-γ was not only the prototypic MAF but the primary and perhaps only MAF for induction of macrophage-mediated microbicidal activity. As purification of LK evolved, so did recognition of the number of the individual cytokines, their complex interactions, and the varied induced activities. More and more factors emerged from crude lymphocyte culture fluids (lymphodreck) in purified form. It became obvious that many different cytokines existed that were able to induce potent effects on macrophage function.

While we have focused on the "activated state" of macrophages that effect nonspecific cytotoxicity, a broad range of other "altered" or "activated" states is ascribed to macrophages. Generalists view all changes in macrophage biology (morphological, phenotypic, functional, or metabolic) as "activated." Macrophages change in response to their environment in a variety of ways, only some of which may be integral to the development of cytotoxic effector activity. For example, literally scores of secretory products of macrophages are identified, but only a few appear relevant to "classical" macrophage activation and cytotoxicity [11]. This review will examine the nature of the responsive macrophage, the array of regulatory signals both endogenous (cytokines) and exogenous that regulate macrophage function, and the repertoire of effector mechanisms crucial to orchestrating the mysterious but powerful "life-protecting" function called macrophage activation.

II. EXPERIMENTAL ANALYSIS: THE VARIABLES

No discussion of macrophage activation would be complete without comment on the experimental variables involved in the study of activated cytotoxic macrophages: the responsive cell, the activation signals, the susceptible target of cytotoxicity and the assay used to model this process. Heterogeneity within and between even steady-state mononuclear phagocyte populations at all levels of analysis is well recognized [12]. What remains obscure is the nature of the driving force behind this heterogeneity. There is no evidence for predetermined subpopulations of blood monocytes for different tissues. The distinguishing property of the various resident tissue macrophages occurs subsequent to emigration of monocytes from the blood into tissue and is directly controlled by local tissue factors. Not all macrophages respond to all activation signals. For example, the level of microbicidal activity against *Rickettsia tsutsugamushi* induced by IFN-γ treatment of inflammatory macrophages is equal to that of identically treated resident peritoneal cells. With these same cells, antimicrobial activity against *Leishmania major* is 10 times lower in inflammatory macrophages than in resident peritoneal cells [13]. Other differences in cell response to a single defined stimulus are described for macrophages from different species of animals, from different strains within the same species, and from different tissues within the same strain.

Pleotropism and redundancy are the key descriptors for cytokines that regulate macrophage function. The first reports of IFN-γ as an MAF in both human and murine systems raised expectations that the regulation of macrophage activation would be simple [14,15]. Instead, the level of complexity in signaling networks during macrophage activation increased with the availability of new cytokines. The initial scattered observations of MAF activity in various cell culture fluids distinct from IFN-γ [16–18] were subsequently bolstered by a barrage of reports using purified cytokines such as granulocyte-macrophage colony-stimulating factor (GM-CSF) and interleukin-4 (IL-4) [19,20]. Instead of wondering whether multiple MAFs exist, investigators asked, "Why other MAFs?" Compounding the confusion created by the discovery of multiple MAFs is the not-so-surprising finding that these MAFs interact with each other as well as other cytokines to regulate the activation of macrophages. The best-documented example of such interaction is the reciprocal activities of IFN-γ and IL-4. Each by itself or in the presence of another stimulus activates a cytotoxic activity in macrophages, yet each is also capable of acting as an antagonist for the other [21].

In addition to responsive cell and effective signal, a third variable is that of susceptible targets. Indeed, macrophage activation for cytotoxic activity must be defined in terms of the target cell used to measure this activity. Thus, the same macrophage classified as "activated" by virtue of its ability to kill one organism would also be "not activated" with respect to its inability to kill a different organism. One example of this is the differential susceptibility of the various species and stages of *Leishmania* parasites to "activated" macrophages. Human monocytes activated to kill *L. donovani* promastigotes do not kill amastigotes of the same species [22,23]. Activated macrophages with enhanced antimicrobial activity against certain targets (promastigotes) produce increased amounts of toxic oxygen metabolites. Microorganisms that survive within these activated cells (amastigotes) fail to stimulate production of such metabolites or are resistant to their toxic effects.

Finally, small changes in the actual mechanics of cytotoxicity assays, keeping all other variables constant, can profoundly affect macrophage effector reactions; such factors include length of time macrophages are in culture before exposure to cytokines or targets, different types of serum or medium, various culture substrates, purity of macrophage populations, and many others. It is usually not clear which combinations of experimental conditions most closely resemble the physiologically relevant reactions. Such culture models may define an event that occurs in vivo most of the time, or only rarely.

The preceding issues paint a picture of macrophage activation that seems hopelessly complex: a multitude of signals (both primary and secondary, endogenous or exogenous) that amplify or suppress cell function within heterogeneous populations of cells able to adapt to small changes in environmental conditions to produce varied effects on the growth and viability of intracellular and extracellular microorganisms. Is this an advance from the fundamental observation that infection with certain microorganisms evokes a high degree of resistance to infection by other unrelated organisms? Early observations of a highly ordered "all-or-none" activation process contrast with evidence that shows this process to be a jumble of mismatched puzzle pieces. Whether these data still support the existence of an ultimate and universal macrophage-dependent resistance to infection, or instead reflect the existence of several types or stages of activation, selectively or sequentially determined and maybe even species-specific, remains a quandary.

III. MACROPHAGE ACTIVATION FACTORS

The discovery of cytokines (soluble nonimmunoglobulin proteins released by cells that usually act at close range on other cells to mediate immune as well as nonimmune functions) initially led many researchers to propose a simple one cytokine–one activity theory. Indeed, names first given to these factors were acronyms for the specific activity they performed: TCGF for T cell growth factor, BCGF for B cell growth factor, MIF for migration inhibition factor, and MAF for macrophage activation factor. When the proteins mediating each activity were purified, it was realized that we had only scratched the surface of the complexity of immunoregulation. Most cytokines arise from more than one type of cell and exert an array of activities, also performed by other cytokines, on an extraordinary number of responding cell types. We first discuss the factors that influence macrophage activation for induction of antimicrobial immunity in mouse systems: 1) cytokines that act as the principal stimulus for induction of macrophage-mediated cytotoxicity (primary signals) and 2) cytokines and certain exogenous factors that also affect macrophage function but do not necessarily induce cytotoxic reactions by themselves (secondary signals).

A. PRIMARY SIGNALS

In the 1970s and early 1980s, tumoricidal assays proved to be most convenient and reproducible for studies of macrophage effector function. Involvement of IFN-α/β in induction of macrophage cytotoxicity was first suggested by observations that various IFN inducers also activate macrophages to kill tumor cells [24]. Indeed, such cytotoxicity was completely inhibited by antibodies against type I interferon [25]. In 1965, Wheelock described a novel IFN released from specific antigen- or mitogen-stimulated lymphocytes [26,27]. Others showed that such IFN-γ–rich LK altered macrophage function and induced enhanced antimicrobial and antitumor effector reactions [28–30]. Not long afterward, physiochemical and functional identity between immune interferon (IFN-γ) and MAF was established [10]. Interferon-γ reigned supreme as the primary, if not the only, MAF. Recombinant IFN-γ induces all activities (increased class II major histocompatibility complex [MHC] expression, production of toxic oxygen intermediates, and cytotoxic activity against numerous targets) ascribed to macrophages activated in vivo [31]. More important, IFN-γ and its associated effects were found with both mouse and human mononuclear phagocytes [14,32].

The contention of a few investigators that MAFs distinct from IFN-γ exist remained alive despite the absence of conclusive corroborative evidence [16]. Such studies analyzed MAF from sources other than conventional LK: T cell hybridomas, T cell clones, or T cell lines. Interleukin-4 (or B cell stimulating factor 1), derived from the EL-4 T cell line, was first described as an MAF in 1987 [20]. Like IFN-γ, IL-4 induces macrophage cytotoxic activity against tumor targets, enhances antigen presentation by macrophages, and up-regulates expression of class II MHC [20,33]. Interleukin-4 also up-regulates respiratory burst activity by macrophages, an important event in cytotoxicity against certain microbial and neoplastic targets [21].

Colony-stimulating factors (CSFs), cytokines that stimulate clonal expansion of individual progenitor cells in bone marrow, also function as MAFs. In 1979, Handman and Burgess first reported antimicrobial effects in mouse macrophages induced by GM-CSF [35]. Mouse macrophages treated with GM-CSF or macrophage colony-stimulating factor (M-CSF) develop tumoricidal activity [36]. Reed et al. showed that GM-CSF activates mouse macrophages to inhibit replication of *Trypanosoma cruzi* and to produce H_2O_2 [37].

B. SECONDARY SIGNALS

Since its discovery by Carswell et al. as a bacterial lipopolysaccharide (LPS)-induced serum factor that causes necrosis of tumors, considerable attention has been focused on tumor necrosis factor (TNF) in regulation of macrophage effector reactions [38]. Despite extensive studies, the exact role of TNF during infection remains shrouded. Depending on experimental conditions or disease model, the effect of TNF may be either beneficial or harmful [39,40]. Tumor necrosis factor is a prime mediator of shock [41]. It is also involved in numerous proinflammatory activities as well as in resistance to various infections [42,43]. In early studies, serum that contained TNF was found to inactivate certain species of malarial parasites [44,45]. More recent studies with recombinant TNF, however, show no direct effect on the viability of several infectious pathogens [46]. Tumor necrosis factor is found in serum after injection of LPS into animals infected with any of a variety of microorganisms [48]. Interestingly, the highest levels of TNF in vitro and in vivo are detected coincidently with IFN-γ [49]. The effect of TNF alone on macrophage-mediated antimicrobial activity is arguable [50,51]. Tumor necrosis factor is reported to kill or inhibit intracellular multiplication of *Mycobacterium avium* and *T. cruzi* but not of *Toxoplasma gondii* in macrophages. In these same studies, TNF did not inhibit *T. cruzi* growth in fibroblasts. In addition to TNF, several other cytokines also have profound effects on macrophages. Cytokines such as transforming growth factor-β (TGF-β), IL-1, IL-6, IL-7, IL-8, IL-10, IFN-α, IFN-β, MIF, and several members of the chemokine family that affect macrophage effector function are discussed in other sections.

It seems eminently reasonable that infectious agents participate in regulation of the immune responses generated to combat them. Therefore, the induction process of macrophage activation has to be studied not only in relation to the other components of the immune system, but also in relation to the microorganisms that initiate the inflammatory insult. The immune system evolved in close and constant association with microorganisms that are always present in the digestive and respiratory tracts. Evolutionary forces favor protective adaptations on the part of both host and parasite to ensure mutual survival. Do immunostimulants of bacterial origin, known to activate macrophages in vivo, act directly on macrophages? Two components that have received the most attention are the bacterial cell wall components LPS and synthetic muramyl peptides (MDPs), copies of bacterial cell wall peptidoglycans. Both components activate macrophages by morphological, biochemical, or functional parameters [52]. As mentioned previously, macrophages are factories for autocrine secretory products and are exquisitely responsive to LPS for production of TNF. Further, gram-negative bacteria are not the only microorganisms that induce macrophages to produce TNF, suggesting the existence of other molecules with LPS-like activity [53]. But TNF is just one of several potentially important mediators produced by LPS-stimulated macrophages that act as autocrine regulators of macrophage function. Indeed, most cytokines that affect macrophage function are also produced by macrophages.

C. SIGNAL INTERACTIONS

The first indication that multiple signals and signal interactions might be necessary for macrophage activation occurred during the 1970s and early 1980s with the use of tumoricidal assays. Tumoricidal activity was the ultimate result of a progression of several different activation signals. Lipopolysaccharide could almost always provide the terminal signal for this process [54]. Thus, even though most strains of mice develop tumoricidal macrophages after Bacillus Calmette-Guérin (BCG) infection or after in vitro treatment with LK, cytotoxicity ultimately depends on a progression of activation signals delivered in a defined sequence. Neither IFN-γ nor LPS alone was sufficient to induce the cytotoxic state [55]. Added together or in the right sequence (IFN-γ then LPS), these two signals provided all the necessary stimuli to generate tumor cytotoxicity. Lymphokine must contain factors distinct from IFN-γ and presumably unrelated to LPS that were required for induction of macrophage cytotoxic activity. This two-signal sequence was termed priming and triggering. Although the effects of microorganisms and their components on macrophage function are readily acknowledged, the actual contribution of any of these to the generation of antimicrobial activity by macrophages is not clear or is presumed to be counterproductive and to contribute to pathogen survival. The recent discovery and quantitation of the L-arginine–dependent nitric oxide (NO) pathway provided a powerful a tool to analyze signal requirements for macrophage activation. Destruction of the protozoan parasite *L. major* in cytokine-activated macrophages is mediated by nitrogen oxides derived from L-arginine [56], the same mechanism shown by Hibbs et al. to effect potent tumoricidal activity [57]. *L. major* provides the second or trigger signal for this killing mechanism in IFN-γ–treated macrophages through induction of TNF [53]. Others have reported similar findings in a wide range of models for intracellular and even extracellular destruction of microbial targets by activated macrophages [58–61]. Thus, IFN-γ and TNF induce L-arginine–dependent cytotoxic effector mechanisms in murine macrophages that destroy neoplastic targets and act in a similar manner to destroy microbial targets. Was NO the penultimate result of the immunologically based macrophage activation process as originally defined by Mackaness, or just another facet or an assorted arsenal of antimicrobial macrophage defensive strategies?

The finding that IFN-γ does not work alone as an MAF, but only in conjunction with at least one other cytokine such as TNF to activate the NO pathway, suggests that other reported MAFs, (M-CSF, GM-CSF, etc.) might also require accessory molecules. Since many authors have in recent years eloquently expounded the physiologic significance of multiple redundant cytokines and the relevance and intricacies of cytokine networks, we will not add to that debate [62,63]. Particularly fascinating are studies on the production and role of TNF in infectious disease. That infectious agents can induce TNF, which acts as the crucial second signal for macrophage activation, implies participation of the pathogen in its own destruction. There are no substantiated claims that TNF has a direct toxic effect on any microorganisms. Yet a few authors report that macrophages treated only with TNF can kill leishmanias, trypanosomes, and mycobacteria [50,51]. The possibilities that the macrophages used were already primed in vivo or that IFN-γ was produced in culture were not addressed. Although the prevailing opinion is that IL-4 is an anti-inflammatory cytokine that acts as an antagonist of IFN-γ, two reports from a single laboratory document synergy between IFN-γ and IL-4 for production of TNF. This TNF then interacts with IFN-γ for the killing of intracellular leishmanias through the NO pathway [64]. If confirmed, these findings will magnify the complexity and confusion of cytokine networks, but they again highlight the central regulatory role of TNF in the induction of macrophage antimicrobial activities.

The need for negative cytokine signals seems clear: activation of antimicrobial processes is dangerous not only to the invading pathogen but also to the host. Destructive mechanisms potent enough to eradicate one living form may not be highly selective and therefore pose a threat to the host. Some type of control to disarm the activated macrophage and return it to steady state is indispensable. From the standpoint of the pathogen, negative regulatory factors could be exploited to ensure their survival. Interferon-γ and IL-4, which alone induce similar activities in macrophages, also interfere with each other's effects [21,65,66]. What has made this finding even more compelling are the complementary discoveries 1) that these two cytokines can be used to divide CD4+ T cell clones into two subsets, T_{H1} and T_{H2}, and 2) that different functions can be attributed to these T cell subsets, presumably through the different actions of these two cytokines [67]. That these cytokines positively or negatively influence the outcome of microbial infection is vividly demonstrated in mouse resistance to *L. major* infection. Through a variety of methods, including the quantitation of cytokine production in the tissues of infected animals and the neutralization of cytokines in vivo using anticytokine antibodies, production of IFN-γ was found to be crucial for the establishment of protective immunity; production of IL-4 was found to interfere with development [68,69]. Recently, the interaction between these two cytokines has been analyzed in vitro in an effort to understand the conditions and mechanisms of the reciprocal effects of IFN-γ and IL-4 on macrophage activation. Treatment of resident peritoneal macrophages with IL-4 four to 24 hours before IFN-γ treatment inhibits macrophage NO_2^- production and intracellular killing [65]. However, addition of IL-4 after IFN-γ treatment is ineffective for such inhibition and leads to enhanced macrophage NO synthesis and antimicrobial effects. Interleukin-4 suppresses cytokine gene expression in macrophages induced by IFN-γ or IL-2 [66]. Interestingly, expression of TNF mRNA was markedly suppressed by IL-4, whereas LPS-induced TNF mRNA was unaffected. It should be pointed out that even though neutralization of IL-4 at the time of infection with *L. major* allows susceptible mice to heal, in vivo administration of IL-4 had little effect on the course of the infection in resistant animals [70].

Another negative regulator of macrophage activation is TGF-β. Transforming growth factor-β is produced by macrophages and acts as an autocrine suppressor factor [71]. It blocks the release of reactive oxygen intermediates (ROIs) and inhibits intracellular killing of *L. major* amastigotes as well as NO production by IFN-γ–activated macrophages [72–74]. Data suggest that TGF-β acts by inhibiting TNF production, the second signal for induction of the L-arginine–dependent NO killing mechanism [73]. Interleukin-2 also has profound effects on macrophage function [75,76]. In the mouse, IFN-γ and IL-2 together induce tumoricidal activity, an antimicrobial phenomenon called resistance to infection, and TNF gene expression. Yet for intracellular killing of *L. major*, the combination of LK and IL-2 suppresses rather than enhances microbicidal activity and NO production [77]. This suppression is mediated through IL-2–stimulated secretion of autocrine TGF-β.

Interleukin-10 down-regulates IFN-γ–dependent immunity at both the T cell and the macrophage level [78,79]. It inhibits microbicidal activity of IFN-γ–treated macrophages; this suppression correlates with inhibition of NO production [80]. Maximal suppression of macrophage function requires continuous exposure to IL-10 and may be mediated indirectly through suppression of macrophage production of TNF, IL-1, and IL-6 or the release of ROIs. Much like IL-4, IL-10 is ineffective on cells previously exposed to IFN-γ. Distinct from the preceding cytokines, macrophage deactivataing factor is functionally similar in that all of these factors inhibit macrophage NO synthesis and microbicidal function [74,81].

Finally, we must consider the role of microorganisms themselves in evading, combating, or subverting defensive measures of the macrophage. Since their survival may hinge on ability to prevent or withstand macrophage activation, parasites could take an active role in their survival and possess the means to down-regulate activation of macrophages. Indeed, an important strategy might be the untimely induction of innate negative regulatory mechanisms. Induction of TNF is crucial to completion of the activation cascade for macrophage cytotoxicity. Lipopolysaccharide induces both TNF mRNA and protein secretion within 4 hours after treatment. In most instances the first signal received by the macrophage is the trigger signal in the form of the target microorganism. In 1987, Ding et al. found that exposure of mouse macrophages to extremely low concentrations of LPS for short intervals blocks subsequent activation of macrophage ROIs by either IFN-γ or TNF [82]. These results suggest that during certain infections, the same substance that contributes to release of TNF and other cytokines (such as LPS) may be associated with obstruction of another important component of the macrophage activation process, induction of nitrogen oxidation. Exposure of macrophages to low doses of LPS selectively inhibits expression of TNF, but not of IL-1 [83]. Treatment of macrophages for 4 to 24 hours with LPS or infection with *L. major* significantly decreases killing and L-arginine–dependent nitrogen oxidation induced by subsequent IFN-γ treatment.

IV. MECHANISMS OF MACROPHAGE MICROBICIDAL ACTIVITY

Just as scientists over the past 20 years searched for the one signal that elevates macrophages to cytotoxic capability, so they also searched for the one effector molecule that mediates nonspecific cytotoxicity. Although the various secretory or phagolysosomal products of the steady-state or inflammatory macrophage are usually sufficient to neutralize, degrade, and eliminate most foreign bodies, certain infectious pathogens resist this first-line defense and actually thrive within the phagolysosome. Among the myriad chemically reactive secretory products of macrophages, two most prominent in immunologically activated macrophages are ROIs and reactive nitrogen intermediates (RNIs). The production of ROIs by macrophages and ROI-associated cytotoxicity were discovered years ago [84]; production of RNIs by mammalian cells was discovered recently [85]. Macrophage killing closely correlates with capacity to mount an oxidative burst [84]. The use of molecular scavengers, such as superoxide dismutase and catalase, to eliminate oxygen radicals from culture fluids reduces macrophage effector function. Many of the microorganisms studied in models of acquired cell-mediated resistance are exquisitely susceptible to oxygen metabolites [86]. In fact, many authors have expressed their belief that ROIs mediate much of the antimicrobial function of activated macrophages. Nonetheless, solid evidence supports oxygen-independent mechanisms [87]. Cells from patients with chronic granulomatous disease have impaired respiratory burst capacity, but they also achieve antimicrobial activity following immune activation. In addition, a number of microorganisms exhibit low susceptibility to ROIs, some being able to completely withstand high levels of ROIs. Analysis of the metabolic pathway for nitrogen oxidation in activated macrophages showed an enzymatic pathway involving oxidation of the guanido nitrogens of L-arginine for production of various reactive nitrogen species [88,89]. Study of this pathway was greatly facilitated by the availability of an analogue of L-arginine, N^G-monomethyl L-arginine (N^GMMLA), that specifically inhibited this metabolic process [57]. Recent studies from several laboratories show that NO production is the principal inducible effector mechanism in many systems of macrophage-mediated microbial killing [90]: 1) IFN-γ is the only one of 12 cytokines tested capable of inducing both RNIs and ROIs; 2) the pathways for secretion of ROIs and RNIs are independent; and 3) macrophages treated with IFN-γ and TNF or IFN-γ and LPS produce high levels of only RNIs [91]. While these findings tend to obscure the role of ROIs in macrophage activation for cytotoxic effector function, there is now evidence that ROIs indirectly influence RNI synthesis. The same oxygen scavengers used to validate the role of ROIs in macrophage microbicidal activity affect RNIs. Nitric oxide synthesis and antimicrobial functions of macrophages are enhanced by superoxide dismutase and inhibited by catalase, presumably through the oxidation/reduction of a critical cofactor required for NO synthesis [92].

V. ANIMAL MODEL SYSTEMS OF MACROPHAGE ACTIVATION

Availability of purified recombinant cytokines makes it possible to administer selected cytokines to determine their effect in vivo. This is a powerful tool in the study of physiologic cytokine actions and often leads to immediate clinical applications. An equally powerful tool is the in vivo administration of monoclonal antibodies (mAb) that neutralize cytokines in response to infection. Two cytokines shown to be critical for induction of macrophage microbicidal activity, IFN-γ and TNF, are also essential to the generation and expression of acquired resistance to infection. Buchmeier and Schreiber showed a requirement for IFN-γ in resolution of *L. monocytogenes* infection [93]. Other reports using the same or similar antibodies demonstrate a protective role for IFN-γ in *L. major*, *Francisella tularensis*, *T. gondii*, and other infections [68,94,95]. Evidence for a crucial role for TNF in protecting animals against various infectious agents was obtained in much the same way and in many of the same experimental systems [46,96,98]. Finally, administration of N^GMMLA to animals and measurement of urinary NO_3^- during different stages of infection or following various treatments enabled analysis of the L-arginine–dependent NO pathway [99,100]. Recently, we reexamined, both in vitro and in vivo, various aspects of macrophage activation and nonspecific acquired resistance in the prototypic system of activated macrophage effector reactions: mice infected with *Mycobacterium bovis*, strain BCG. Mice infected with BCG resist a broad range of unrelated infectious pathogens; macrophages (the proximate effector cell in these systems) from these animals often kill the challenge agent in vitro. Stuehr and Marletta first observed that BCG infection of mice induces high levels of urinary NO_3^- [88]. Peritoneal macrophages from these mice kill microbial and neoplastic targets through activation of the L-arginine–dependent NO pathway: macrophages from BCG mice constitutively produce NO_2^-. In culture, NO_2^- levels from BCG-activated macrophages are increased following exposure to *F. tularensis*; replication of this unrelated bacterium is strongly inhibited. Addition of N^GMMLA to these cultures blocked both NO_2^- and antimicrobial activity in a dose-dependent fashion [101]. Interferon-γ and TNF play important roles in regulation of this macrophage function, since the addition of mAbs to either or both cytokines also reduces NO_2^- production and antimicrobial activity. Interferon-γ TNF, and L-arginine–dependent NO production also play a role in vivo in the same model system [102]. Anti-IFN-γ and anti-TNF administered alone or together at the time of BCG infection completely abolished BCG-mediated protection. Whereas animals infected with BCG produced levels of urinary NO_3^- six to 10 times greater than those found in normal mice, mice treated with BCG and mAbs to either cytokine failed to excrete urinary NO_3^- above background. Nitrates excreted by BCG-infected mice were derived from L-arginine, since administering N^GMMLA orally from the onset of BCG infection was as effective in blocking urinary NO_3^- excretion as the anticytokine mAb.

VI. OF MICE AND MEN

The practice in writing reviews on macrophage activation has been to indiscriminately combine work done in animal and human systems. Indeed, unless one actually read the reference, one would be oblivious to the particular system used in any study. Research in both animal and human models emphasized the critical role of IFN-γ as activator of potent antimicrobial effects and oxidative antimicrobial mechanisms. The now much-repeated failure by numerous laboratories to demonstrate in human monocytes/macrophages the L-arginine–dependent generation of RNIs mandated a critical reappraisal of human macrophage activation. Do activated macrophages play similar roles in mouse and man? Are activated macrophages from mouse and man equivalent? Whether the highly activated cytotoxic macrophages studied in BCG- or *Listeria*-infected mice have a similar counterpart in humans is not clear.

A. MACROPHAGE ACTIVATION FACTORS

Identification of IFN-γ as the factor in LK that activates macrophages for microbicidal effector function was first achieved with human cells [103,104]. All MAF activity in LK from mouse or human T cells is neutralized with anti-IFN γ antibodies. Recombinant murine or human IFN-γ

activated macrophages to kill certain intracellular pathogens, and this was closely correlated with increased capacity to secrete ROIs. The first report documenting the induction of antimicrobial effects in human macrophages by recombinant GM-CSF appeared simultaneously with a report showing similar effects with mouse cells: GM-CSF activated mouse macrophages to inhibit *T. cruzi* and release hydrogen peroxide; GM-CSF also induced intracellular killing of *L. donovani* by human monocyte-derived macrophages [37,105]. In 1990, Ho et al. demonstrated an antimicrobial effect of GM-CSF and M-CSF for killing of *L. mexicana amazonensis* [106]. Interestingly, in this system, the CSFs and IFN-γ all had similar activities, even though their effects were additive. Wang et al. described enhancement of human monocyte killing of *Candida albicans* by IL-3 [107]. We have not found a role for any of the CSFs in activating human monocytes/macrophages to kill or inhibit the growth of either *L. major* or *F. tularensis* (unpublished data).

Interleukin-2 receptors are expressed on activated murine and human macrophages, and IL-2 induces production of various inflammatory mediators (IL-1, TNF, and ROIs) in these cells [108–110]. There are very few reports of IL-2–mediated enhancement of microbicidal activity by human monocytes/macrophages. Indeed, recent evidence with mouse cells points to an inhibitory role for IL-2 in regulation of macrophage function [77]. There is overwhelming evidence that TNF is a major cofactor in the induction of macrophage cytotoxic function in the mouse. In humans, such evidence is almost nonexistent. Association of TNF with macrophage killing or antimicrobial effects in human systems has been in the absence of IFN-γ [112]. Attempts to reverse antimicrobial effects of IFN-γ on *F. tularensis* replication in human monocytes with mAb TNF show little or no effect, even though the combination of IFN-γ and *F. tularensis* induces monocytes to produce significant amounts of TNF. These results stand in stark contrast to results in mouse systems, in which the induction of TNF by the combination of IFN-γ and *F. tularensis* is necessary for the production of NO and bacterial killing. Indeed, anti-TNF mAb inhibits both macrophage-mediated activities. Thus, even though mouse and man possess comparable priming and triggering systems for certain macrophage proinflammatory products such as TNF, the overall effects of these cytokine signals are quite different.

In 1989, Weiss et al. documented inhibitory effects of IL-4 on IL-1 and TNF production in monocytes in the presence of low doses of LPS [112]. Donnelly et al. also showed that IL-4 needed to be added before or simultaneously with LPS for this effect. Further, IL-4 suppressed IL-6 mRNA and neutralized the potentiating effect of IFN-γ on IL-1 production, suggesting that IL-4 may also be an effective antagonist of certain IFN-γ–induced effects in human macrophages [113]. More pertinent to this discussion, IL-4 inhibited IFN-γ–dependent H_2O_2 production by monocytes as well as IFN-γ–induced antileishmanial activity in a system where microbicidal activity closely correlated with oxygen-dependent mechanisms [114]. Interleukin-4 inhibits not only IFN-γ–induced but also TNF-, GM-CSF– and IL-3–induced antileishmanial and oxidative burst capacities in human monocytes [115]. Interleukin-6, IL-10, and TGF-β, each active for down-regulating macrophage responses in the mouse, inhibit human macrophage production of proinflammatory cytokines, such as TNF [116,117].

Microoganisms or their components exert dramatic influences on human cells, just as they do on mouse macrophages. Macrophages from both mouse and man are uniquely responsive to LPS for production of TNF, IL-1, and IL-6 [118]. Intact microorganisms (bacteria, protozoa, etc.) also activate the oxidative burst in human monocytes [119]. Suppression of macrophage activation by endogenous factors such as TGF-β, and also by certain extrinsic factors linked to the infectious pathogen itself that improve its chances for survival, is of some interest. Engelhorn et al. found that a soluble factor produced by infection of human monocytes with *L. donovani* promastigotes suppresses IFN-γ–dependent monocyte activation [120]. Whereas cultured monocytes treated with IFN-γ increase their antileishmanial capacity, monocytes inoculated with *L. donovani* before IFN-γ treatment exhibit a marked suppression of this response.

B. ANTIMICROBIAL MECHANISMS OF HUMAN MONONUCLEAR PHAGOCYTES

Mononuclear phagocytes of both mouse and man use an oxygen-dependent mechanism to kill or inhibit the replication of a diverse group of intracellular pathogens and provide a biochemical basis for acquired nonspecific cellular resistance and macrophage activation. Although L-arginine–dependent nitrogen oxidation occurs in human cells and plays a major role in certain physiological processes (endothelium relaxation and neurotransmission), several reports now document no role for NO in human macrophage-mediated antimicrobial effects [122–123].

Human mononuclear phagocytes utilize both respiratory burst–dependent and independent antimicrobial mechanisms; IFN-γ enhances both pathways [22]. Antimicrobial pathways may exist in human macrophages independent of the oxidative burst: macrophages from patients with chronic granulomatous disease deficient in oxidative activity still exhibit potent antimicrobial capacity. Two IFN-γ–inducible antimicrobial mechanisms distinct from nitrogen oxidation represent examples of oxygen-independent pathways. Interferon-γ stimulates a variety of human cells, including monocytes, for increased activity of indoleamine 2,3-dioxygenase and increased degradation of extracellular tryptophan [124]. Since some antimicrobial activity induced by IFN-γ can be reversed by exogenous tryptophan, induction of this pathway might contribute to oxygen-independent antimicrobial effects of activated human macrophages. However, different effects of tryptophan supplementation in a variety of experimental models of human macrophage activation suggest that this mechanism is not relevant to all human cells or all targets. Further, IFN-α and IFN-β treatment of macrophages also depletes tryptophan from the culture, but this is not sufficient to achieve effective antimicrobial activity.

A second oxygen-independent antimicrobial effect of cytokine-activated human macrophages is modulation of iron metabolism [125]. Such iron depletion mimics the action of L-arginine–dependent NO production in mouse systems: the removal of essential nonheme iron from enzymes involved in vital metabolic pathways. Iron depletion in human cells is mediated through reduction in net uptake of iron via IFN-γ–induced down-regulation of transferrin receptor expression. This mechanism has been carefully explored by Byrd and Horowitz in the intracellular biology of *Legionella pneumophila* in human monocytes [126]. *L. pneumophila* has a defined metabolic requirement for iron. Addition of iron chelators, like deferoxamine, prevents bacterial multiplication in growth medium and within human monocytes. Interferon-γ–activated human monocytes also inhibit intracellular multiplication of *L. pneumophila*. This antibacterial effect is reversed by iron-saturated transferrin, suggesting that growth inhibition is achieved by limiting availability of iron. This conclusion is supported by the finding that transferrin receptors are markedly down-regulated on activated monocytes. Recently, we observed a similar phenomenon in the interaction of human monocytes with *F. tularensis*. In contrast to results in the mouse, in which IFN-γ–induced L-arginine–dependent nitrogen oxidation plays the predominant role in restricting bacterial growth in macrophages, IFN-γ–activated human monocytes depend on iron depletion to inhibit the growth of intracellular bacteria. Just as in the *Legionella* model, iron-saturated transferrin reversed macrophage-mediated inhibition of *F. tularensis*, whereas inhibitors of other pathways (N^GMMLA, catalase, superoxide desmutase, tryptophan) did not affect IFN-γ–induced antimicrobial activity. Studies by Murray et al. do not delineate a role for this mechanism in several other systems of human monocyte-mediated antimicrobial activity [127].

VII. SUMMARY

Various lines of defense against infection are present in all living creatures. The balance between symbiosis and parasitism is determined by the mechanisms through which the host resists infection and by the extent of injury induced by the parasite: both factors contribute to disease. Lines of host defense can be arbitrarily divided into three components: 1) barrier functions of skin and mucous membranes and their innate physical and secretory antimicrobial components; 2) elements of host defense that do not necessarily require prior exposure to an infectious agent or immunologic memory (mast cells, granulocytes, macrophages, NK cells, γ/δ T cells); and 3) immune responses directed against specific epitopes on the infectious agent induced by prior exposure and immunologic memory (α/β T cells, B cells). Analysis of such host defense mechanisms repeatedly documents tremendous redundancy and overlap between these lines of defense. Further, there is open communication, so that a change at any one level ripples throughout the system.

Acquired nonspecific resistance to infection is an example of such a ripple. Host response to one infection alerts the immune system, so that the general level of resistance to other infectious agents is increased. This response is initiated by an immune response (third line of defense) but effected by nonspecific elements (second line of defense). The survival value of such responses is obvious. There are numerous examples in both mouse and man of the operation of these systems in response to infection. Further, the menus of antimicrobial components available to both mouse and man for resistance to infection are very similar, but not identical. Indeed, it is said that the genetic basis for differences between mice and man revolve around a difference of less than 10% in DNA sequences. But there are differences! Mouse macrophages produce IFN-β in response to infection, human cells produce IFN-α. Mouse macrophages effect antimicrobial activity principally through induction of NO synthase and the generation of toxic nitrogen oxides. This pathway has yet to be described with human macrophages.

In both man and mouse, *F. tularensis* is an obligate intracellular parasite of macrophages that requires an essential component provided by the cell for its replication. That mouse and man are not so different is well illustrated by the effector mechanisms induced by IFN-γ for antimicrobial activity against *F. tularensis*. In the mouse, IFN-γ induced NO synthase to produce toxic nitrogen oxides, which then interact with essential iron groups in certain enzymes vital to bacterial replication. Such antimicrobial activity is inhibited by high levels of iron in the medium. With human monocytes, IFN-γ also induces potent antimicrobial activity against *F. tularensis*. Nitrogen oxides are not detected in cultures of IFN-γ–treated human monocytes, nor is monocyte antimicrobial activity affected by agents that inhibit the L-arginine–dependent NO pathway. Pointedly, addition of exogenous iron to such IFN-γ–activated human monocytes also inhibits antimicrobial activity. Thus IFN-γ–activated macrophages from man or mouse develop antimicrobial mechanisms that inhibit replication of *F. tularensis* through pathways that limit available iron. The mechanisms are different, but the effector responses are directed against the same bacterial Achilles' heel.

REFERENCES

1. Silverstein AM. The history of immunology. In: Paul WE, ed. Fundamental immunology. 2nd ed. New York: Raven Press, 1989:21–38.
2. Metchnikoff E. Immunity in the infectious diseases. New York: Macmillan, 1905.
3. Zinsser H. Studies on the tuberculin reaction and on specific hypersensitiveness in bacterial infection. J Exp Med 1921; 34:495–522.
4. Mackaness GB. Cellular resistance to infection. J Exp Med 1962; 116:381–406.
5. Mackaness GB. The immunological basis of acquired cellular resistance. J Exp Med 1964; 120:105–20.
6. David JR. Delayed hypersensitivity *in vitro:* its mediation by cell-free substances formed by lymphoid cell-antigen interaction. Proc Natl Acad Sci USA 1966; 56:72–7.
7. Bloom BR, Bennett B. Mechanism of a reaction *in vitro* associated with delayed-type hypersensitivity. Science 1966; 153:80–2.
8. Mooney JJ, Waksman BH. Activation of normal rabbit macrophage monolayers by supernatants of antigen-stimulated lymphocytes. J Immunol 1970; 105:1138–42.

9. Evans R, Alexander P. Cooperation of immune lymphoid cells with macrophages in tumor immunity. Nature 1972; 228:620–2.
10. Roberts WK, Vasil A. Evidence for the identity of murine gamma interferon and macrophage activating factor. J Interferon Res 1982; 2:519–32.
11. Nathan CF. Secretory products of macrophages. J Clin Invest 1987; 79:319–26.
12. Forster O, Landy L. Heterogeneity of mononuclear phagocytes. London: Academic Press, 1981.
13. Nacy CA, Oster DN, James SL, Meltzer MS. Microbicidal effector reactions of activated macrophages against intracellular and extracellular parasites. Cont Topics Immunobiol 1984; 14:147–70.
14. Nathan CF, Murray HW, Wiege ME, Rubin BY. Identification of interferon-γ as the lymphokine that activates human macrophage oxidative metabolism and antimicrobial activity. J Exp Med 1983; 158:670–89.
15. Schultz RM, Kleinschmidt WJ. Functional identity between murine gamma interferon and macrophage activating factor. Nature 1983; 305:239–40.
16. Meltzer MS, Benjamin WR, Farrar JJ. Macrophage activation for tumor cytotoxicity: induction of macrophage tumoricidal activity by lymphokines from EL-4, a continuous T cell line. J Immunol 1982; 129:2802–10.
17. Nacy CA, Fortier AH, Meltzer MS, Buchmeier NA, Schreiber RD. Macrophage activation to kill *Leishmania major*: activation of macrophages for intracellular destruction of amastigotes can be induced by both recombinant interferon-gamma and non-interferon lymphokines. J Immunol 1986; 135:3505–12.
18. Hoover, DL, Finbloom DS, Crawford RM, Nacy CA, Gilbreath MG, Meltzer MS. A lymphokine distinct from interferon-gamma that activates human monocytes to kill *Leishmania donavani* in vitro. J Immunol 1986; 136:1329–35.
19. Grabstein KH, Urdal DL, Tushinshi FJ, Mochizuki DY, Price VL, Cantrell MA, Gillis S, Gonlon PJ. Induction of macrophage tumoricidal activity by granulocyte-macrophage colony stimulating factor. Science 1986; 232:506–8.
20. Crawford RM, Finbloom DS, Ohara J, Paul WE, Meltzer JS. B-cell stimulatory factor-1: a macrophage activation factor. J Immunol 1987; 139:135–40.
21. Phillips WA, Croatto M, Hamilton JA. Priming the macrophage respiratory burst with IL-4: enhancement with TNFα but inhibition by IFNγ. Immunology 1990; 70:498–503.
22. Murray HW, Cartelli DM. Killing of intracellular *Leishmania donovani* by human mononuclear phagocytes. J Clin Invest 1983; 72:32–44.
23. Hoover DL, Nacy DA, Meltzer MS. Human monocyte activation for cytotoxicity against intracellular *Leishmania donovani* amastigotes: induction of microbicidal activity by interferon-γ. Cell Immunol 1985; 94:500–511.
24. Alexander P, Evans R. Endotoxin and double-stranded RNA render macrophages cytotoxic. Nature 1971; 232:76–78.
25. Schultz, RM, Chirigos MA. Selective neutralization by anti-interferon globulin of macrophage activation by L-cell interferon, *Brucella abortus* ether extract, *Salmonella typhimurium* LPS, and polyanions. Cell Immunol 1979; 48:52–7.
26. Wheelock EF. Interferon-like virus-inhibitor induced in human leukocytes by phytohemagglutinin. Science 1965; 149:310–1.
27. Dumonde DC, Wolstencroft RA, Panayi GS, Mathew M, Morley J, Howson WT. Lymphokines: non-antibody mediators of cellular immunity generated by lymphocyte activation. Nature 1969; 224:38–42.
28. Simon HB, Sheagren JN. Enhancement of macrophage bactericidal capacity by antigenically stimulated immune lymphocytes. Cell Immunol 1972; 4:163–174.
29. Fowles RE, Fajardo IM, Leibowitch JL, David JR. The enhancement of macrophage bacteriostasis by products of activated lymphocytes. J Exp Med 1973; 138:952–64.
30. Churchill WH, Piessens WF, Sulis CA, David JR. Macrophages activated as suspension cultures with lymphocyte mediators devoid of antigen become cytotoxic for tumor cells. J Immunol 1975; 115:781–5.
31. Svedersky LP, Benton CV, Berger WH, Rinderknecht E, Harkins RN, Palladino MA. Biological and antigenic similarities of murine interferon-γ and macrophage activating factor. J Exp Med 1984; 159:812–20.
32. Pace JL, Russell SW, Torres BA, Johnson HM, Gray PW. Recombinant mouse interferon-γ induces the priming step in macrophage activation for tumor cell killing. J Immunol 1983; 130:2011–5.

33. Zlotnick A, Fischer M, Roehm N, Zipori D. Evidence for effects of interleukin-4 (B cell stimulatory factor-1) on macrophages: enhancement of antigen presenting ability of bone marrow–derived macrophages. J Immunol 1987; 138:4275–81.
34. Cao H, Wolff RG, Meltzer MS, Crawford RM. Differential regulation of class II MHC determinants on macrophages by IFNγ and IL-4. J Immunol 1989; 143:3525–31.
35. Handman E, Burgess AW. Stimulation by granulocyte-macrophage colony-stimulating factor of *Leishmania tropica* killing by macrophages. J Immunol 1979; 122:1134–9.
36. Ralph P, Nakoinz I. Stimulation of macrophage tumoricidal activity by the growth and differentiation factor CSF-1. Cell Immunol 1987; 105:270–9.
37. Reed SG, Nathan CF, Pihl CL, Rodricks P, Stanebeck K, Conlon PJ, Grabstein KH. Recombinant granulocyte-macrophage colony stimulating factor activates macrophages to inhibit *Trypanosoma cruzi* and release hydrogen peroxide. Comparison with interferon-γ. J Exp Med 1987; 166:1734–46.
38. Carswell EA, Old LJ, Kassel RL, Green S, Fiore N, and Williamson B. An endotoxin-induced serum factor that causes necrosis of tumors. Proc Natl Acad Sci USA 1975; 72:3666–8.
39. Havell EA. Production of tumor necrosis factor during murine listerosis. J Immunol 1987; 139:4225–30.
40. Grau G, Fajardo LF, Piguet PF, Allet B, Lambert PH, Vassalli P. Tumor necrosis factor (cachectin) as an essential mediator in murine cerebral malaria. Science 1987; 237:1210–2.
41. Tracey KF, Fong Y, Hesse DG, Manogue KR, Lee AT, Kuo GC, Lowry SF, Cerami A. Anti-cachectin/TNF monoclonal antibodies prevent septic shock during lethal bacteraemia. Nature 1987; 330:662–4.
42. Beutler B, Cerami A. Tumor necrosis, cachexia, shock and inflammation: a common mediator. Annu Rev Biochem 1988; 57:505–30.
43. Kindler V, Sappino AP, Grau G, Piguet PF, Vassalli P. The inducing role of tumor necrosis factor in the development of bacterial granulomas during BCG infection. Cell 1989; 56:731–40.
44. Taverne J, Dockrell HM, Playfair JHL. Endotoxin-induced serum factor kills malarial parasites in vitro. Infect Immun 1981; 33:83–8.
45. Parent M. Antimicrobial resistance enhancing activity of tumor necrosis serum factor induced by endotoxin in BCG-treated mice. Recent Results Cancer Res 1980; 75:213–8.
46. Titus RG, Sherry B, Cerami A. Tumor necrosis factor plays a protective role in experimental murine cutaneous leishmaniasis. J Exp Med 1989; 170:2097–104.
47. Hoffmann P, Heinle S, Schade MF, Loppnow H, Ulmer HJ, Flad HD, Jung G, Bessler WG. Stimulation of human and murine adherent cells by bacterial lipoprotein and synthetic lipopeptide analogues. Immunobiol 1988; 177:158–70.
48. Nakane A, Minagawa T, Kato K. Endogenous tumor necrosis factor (cachectin) is essential to host resistance against *Listeria monocytogenes* infection. Infect Immun 1988; 56:2563–9.
49. Nakane A, Minagawa T, Kohanawa M, Chen Y, Sato H, Moriyama M, Tsuruoka N. Interactions between endogenous gamma interferon and tumor necrosis factor in host resistance against primary and secondary *Listeria monocytogenes* infections. Infect Immun 1989; 57:3331–7.
50. Titto EH, Catterall JR, Remington JS. Activity of recombinant tumor necrosis factor on *Toxoplasma gondii* and *Trypanosoma cruzi*. J Immunol 1986; 137:1342–5.
51. Bermudez LEM, Young LS. Tumor necrosis factor, alone or in combination with IL-2, but not IFN-g, is associated with macrophage killing of *Mycobacterium avium* complex. J Immunol 1988; 140:3006–13.
52. Adams DO, Hamilton TA. The cell biology of macrophage activation. Annu Rev Immunol 1984; 2:283–308.
53. Green SJ, Crawford RM, Hockmeyer JT, Meltzer MS, Nacy CA. *Leishmania major* amastigotes initiate the L-arginine–dependent killing mechanism in IFN-γ–stimulated macrophages by induction of tumor necrosis factor-α. J Immunol 1990; 145:4290–7.
54. Ruco LP, Meltzer MS. Macrophage activation for tumor cytotoxicity: development of macrophage cytotoxic activity requires completion of a sequence of short-lived intermediary reactions. J Immunol 1978; 121:2035–42.
55. Pace JL, Russell SW, Torres BA, Johnson HM, Gray PW. Recombinant mouse gamma interferon induces the priming step in macrophage activation for tumor cell killing. J Immunol 1983; 130:2011–7.
56. Green SJ, Meltzer MS, Hibbs JB, Nacy CA. Activated macrophages destroy intracellular *Leishmania major* amastigotes by an L-arginine–dependent killing mechanism. J Immunol 1990; 144:278–83.
57. Hibbs JB, Taintor RR, Vavrin Z. Macrophage cytotoxicity: role for L-arginine deiminase and imino nitrogen oxidation of nitrite. Science 1987; 235:473–476.

58. James SL, Glaven J. Macrophage cytotoxicity against schistosomula of *Schistosoma mansoni* involves arginine-dependent production of reactive nitrogen intermediates. J Immunol 1989; 143:4208–12.
59. Crawford RM, Green SJ, Leiby DA, Nacy CA, Fortier AH. L-arginine dependent killing mechanism involved in the destruction of *Francisella tularensis* J Leukocyte Biol 1990; (suppl 1):481.
60. Adams, LB, Hibbs JB, Taintor RR, Krahenbuhl JL. Microbistatic effects of murine-activated macrophages for *Toxoplasma gondii*: role for synthesis of inorganic nitrogen oxides from L-arginine. J Immunol 1990; 144:2725–9.
61. Chan J, Xing Y, Magliozzo RS, Bloom BR. Killing of virulent *Mycobacterium tuberculosis* by reactive nitrogen intermediates produced by activated murine macrophages. J Exp Med 1992; 175:1111–22.
62. Nacy CA, Green SJ, Leiby DL, Nelson BA, Crawford RM, Fortier AH, Hoover DL, Meltzer MS. Intercellular communication: macrophages and cytokines. In: Kiyono H, Jirillo E, DeSimone C, eds. Molecular aspects of immune response and infectious diseases. New York: Raven Press, 1990:47–53.
63. Nathan CF. Coordinate actions of growth factors in monocytes/macrophages. In Roberts AB, Sporn, eds. Handbook of experimental pharmacology. 1990:427–462.
64. Stenger S, Solbach W, Rollinghoff M, Bogdan C. Cytokine interactions in experimental cutaneous leishmaniasis. II. Endogenous tumor necrosis factor-α production by macrophages is induced by the synergistic action of interferon (IFN)-γ and interleukin (IL)-4 and accounts for the antiparasitic effect mediated by IFN-γ and IL-4. Eur J Immunol 1991; 21:1669–75.
65. Liew FY, Millott S, Li Y, Lelchuk R, Chan WL, Ziltener H. Macrophage activation by interferon-γ from host-protective T cells is inhibited by interleukin IL-3 and IL-4 produced by disease-promoting T cells in leishmaniasis. Eur J Immunol 1989; 19:1227–32.
66. Gautam S, Tebo JM, Hamilton TA. IL-4 suppresses cytokine gene expression induced by IFN-γ and/or IL-2 in murine peritoneal macrophages. J Immunol 1992; 148:1725–30.
67. Mosmann TR, Cherwinski H, Bond MW, Giedlin MA, Coffman RL. Two types of murine helper T cell clone. 1. Definition according to profiles of lymphokine activities and secreted proteins. J Immunol 1986; 136:2348–54.
68. Belosevic M, Finbloom DS, Van Der Meide PH, Slayter M, Nacy CA. Administration of monoclonal anti-IFN-g antibodies in vivo abrogates natural resistance of C3H/HeN mice to infection with *Leishmania major*. J Immunol 1989; 143:266–74.
69. Sadick MD, Heinzel FFP, Holaday BJ, Pu RT, Dawkins RS, Locksley RM. Cure of murine leishmaniasis with anti-interleukin 4 monoclonal antibody. Evidence for a T cell–dependent, interferon-γ–independent mechanism. J Exp Med 1990; 171:115–27.
70. Sadick MD, Street N, Mosmann TR, Locksley RM. Cytokine regulation of murine leishmaniasis: Interleukin 4 is not sufficient to mediate progressive disease in resistant C57BL/6 mice. Infect Immun 1991; 59:4710–4.
71. Tsunawaki S, Sporn M, Ding A, Nathan CF. Deactivation of macrophages by transforming growth factor-β. Nature 1988; 344:260–2.
72. Tsunawaki S, Sporn M, Nathan CF. Comparison of transforming growth factor-B and a macrophage-deactivating polypeptide from tumor cells. J Immunol 1989; 142:3462–7.
73. Nelson BJ, Ralph P, Green SJ, Nacy CA. Differential susceptibility of activated macrophage cytotoxic effector reactions to the suppressive effects of transforming growth factor-β1. J Immunol 1991; 146:1849–57.
74. Ding A, Nathan CF, Graycar J, Derynck R, Stuehr DJ, Srimal S. Macrophage deactivating factor and transforming growth factors-β1, β2, and β3 inhibit induction of macrophage nitrogen oxide synthesis by IFN-γ. J Immunol 1990; 145:940–4.
75. Belosevic M, Finbloom DS, Meltzer MS, Nacy CA. Interleukin-2: a cofactor for induction of macrophage resistance to infection. J Immunol 1990; 145:831–9.
76. Cox GW, Mathieson BJ, Giardina SL, Vareiso L. Characterization of IL-2 receptor expression and function on murine macrophages. J Immunol 1990; 145:1710–24.
77. Nelson BJ, Belosevic M, Green SJ, Turpin J, Nacy CA. Interleukin-2 and the regulation of activated macrophage cytotoxic activities. In: Friedman H, Klein TW, eds. Microbial infections: role of biological response modifiers. New York: Plenum Press, 1991:
78. Mosmann TR, Moore K. The role of IL-10 in crossregulation of Th-1 and Th-2 responses. Immunoparasitol Today 1991; 2:49–57.

79. Fiorentino DF, Iotnik A, Vieira P, Mossman TR, Hoard M, Moore K, O'Garra A. IL-10 acts on the antigen-presenting cell to inhibit cytokine production by Th1 cells. J Immunol 1991; 146:3444–9.
80. Gazzinelli RT, Oswald IP, James SL, Sher A. IL-10 inhibits parasite killing and nitrogen oxide production by IFN-γ-activated macrophages. J Immunol 1992; 148:1792–6.
81. Srimal S, Nathan CF. Purification of macrophage deactivating factor. J Exp Med 1990; 171:1347–59.
82. Ding AH, Nathan CF. Trace levels of bacterial lipopolysaccharide prevent interferon-γ or tumor necrosis factor-α from enhancing mouse peritoneal macrophage respiratory burst capacity. J Immunol 1987; 139:1971–7.
83. Takasuka T, Tokunaga T. Akagawa KS. Preexposure of macrophages to low doses of lipopolysaccharide inhibits the expression of tumor necrosis factor-α mRNA but not of IL-1β mRNA. J Immunol 1991; 146:3824–30.
84. Murray HW, Cohn ZA. Macrophage oxygen-dependent antimicrobial activity. III. Enhanced oxidative metabolism as an expression of macrophage activation. J Exp Med 1980; 152:1596–1610.
85. Stuehr DJ, Marletta MA. Mammalian nitrate biosynthesis: mouse macrophages produce nitrite and nitrate in response to *Escherichia coli* lipopolysaccharide. Proc Natl Acad Sci USA 1985; 82:7738–42.
86. Nathan CF. Mechanisms of macrophage antimicrobial activity. Trans R Soc Trop Med Hyg. 1983; 77:620–30.
87. Murray HW, Cartelli DM. Killing of intracellular *Leishmania donovani* by human mononuclear phagocytes. J Clin Invest 1983; 72:32–44.
88. Stuehr DJ, Marletta MA. Induction of nitrite/nitrate synthesis in murine macrophages by BCG infection, lymphokines or interferonγ. J Immunol 1987; 139:518–23.
89. Hibbs JB, Vavrin Z, Taintor R. L-arginine is required for expression of the activated macrophage effector mechanism causing selective metabolic inhibition in target cells. J Immunol 1987; 138:550–6.
90. Green SJ, Nacy CA, Meltzer MS. Cytokine-induced synthesis of nitrogen oxides in macrophages: a protective host response to *Leishmania* and other intracellular pathogens. J Leukocyte Biol 1991; 50:93–103.
91. Ding AH, Nathan CF, Stuehr DJ. Release of reactive nitrogen intermediates and reactive oxygen intermediates from mouse peritoneal macrophages. J Immunol 1988; 141:2407–12.
92. Liew FY. Role of cytokines in killing of intracellular pathogens. Immunol Letters 1991; 30:193–8.
93. Buchmeier NA, Schreiber RC. Requirement of endogenous interferon-γ production for resolution of *Listeria monocytogenes* infection. Proc Natl Acad Sci 1985; 82:7404–8.
94. Anthony LSD, Ghadirian E, Nestel FP, Kongshaven PAL. The requirement for gamma interferon in resistance of mice to experimental tularemia. Microb Pathog 1989; 7:421–8.
95. Suzuki Y, Orellana MA, Schreiber RD, Remington JS. Interferon-γ: the major mediator of resistance against *Toxoplasma gondii*. Science 1988; 240:516–8.
96. Havell EA. Evidence that tumor necrosis factor has an important role in antibacterial resistance. J Immunol 1989; 143:2894–9.
97. Johnson LL. A protective role for endogenous tumor necrosis factor in *Toxoplasma gondii* infection. Infect Immun 1992; 60:1979–83.
98. Leiby DA, Fortier AH, Crawford RM, Schreiber RD, Nacy CA. In vivo modulation of the murine immune response to *Francisella tularensis* LVS by administration of anticytokine antibodies. Infect Immun 1992; 60:84–9.
99. Granger DL, Hibbs JB, Broadnax LM. Urinary nitrate excretion in relation to murine macrophage activation: influence of dietary L-arginine and oral N^G-mono-methyl-arginine. J Immunol 1991; 146:1294–9.
100. Liew FY, Millott S, Parkinson C, Palmer RMJ, Moncada S. Macrophage killing of *Leishmania* parasite in vivo is mediated by nitric oxide from L-arginine. J Immunol 1990; 144:4794–7.
101. Fortier AH, Polsinelli T, Green SJ, Nacy CA. Activation of macrophages for destruction of *Francisella tularensis:* identification of cytokines, effector cells, and effector molecules. Infect Immun 1992; 60:817–25.
102. Green, SJ, Nacy CA, Schreiber RD, Granger DL, Crawford RM, Meltzer MS, Fortier AH. Neutralization of IFNγ and TNFα in vivo blocks production of reactive nitrogen intermediates and protection against *Francisella tularensis* infection in BCG-treated mice (submitted for publication).

103. Nathan CF, Murray HW, Wieve ME, Rubin CY. Identification of interferon-γ as the lymphokine that activates human macrophage oxidative metabolism and antimicrobial activity. J Exp Med 1983; 158:670–89.
104. Nathan CF, Prendergast TJ, Wiebe ME, Stanley ER, Platzer E, Remold H, Welte K, Rubin BY, Murray HW. Activation of human macrophages. J Exp Med 1984; 160:600–5.
105. Weiser WY, Van Niel A, Clark SC, David JR, Remold HG. Recombinant human granulocyte/macrophage colony stimulating factor activates intracellular killing of *Leishmania donovani* by human monocyte-derived macrophages. J Exp Med 1987; 166:1436–46.
106. Ho JL, Reed SG, Wick EA, Giordano M. Granulocyte macrophage and macrophage colony stimulating factors activate intramacrophage killing of *Leishmania amazonensis*. J Infect Dis 1990; 157:925–30.
107. Wang M, Friedman H, Djeu JY. Enhancement of human monocyte function against *Candida albicans* by colony stimulating factors (CSF): IL-3, granulocyte-macrophage-CSF, and macrophage-CSF. J Immunol 1989; 143:671–7.
108. Malkovsky M, Loveland B, North M, Asherson GL, Gao L, Ward P, Friers W. Recombinant interleukin-2 directly augments the cytotoxicity of human monocytes. Nature 1987; 325:262–3.
109. Wahl SM, McCartney-Francis N, Hunt DA, Smith PD, Wahl LM, Katona IM. Monocyte interleukin-2 receptor gene expression and interleukin-2 augmentation of microbicidal activity. J Immunol 1988; 139:1332–7.
110. Hermann F, Cannistra SA, Lindemann A, Blohm D, Rambaldi A, Mertlesmann RH, Griffen JD. Functional consequences of monocyte IL-2 receptor expression. Induction of IL-1β secretion by IFN and IL-2. J Immunol 1989; 142:139–45.
111. Denis M. Growth of *Mycobacterium avium* in human monocytes: identification of cytokines which reduce and enhance intracellular microbial growth. Eur J Immunol 1991; 21:391–5.
112. Weiss L, Haeffner-Cavaillon N, Laude M, Cavaillon JM, Kazatchkine MD. Human T cells and interleukin 4 inhibit the release of interleukin 1 induced by lipopolysaccharide in serum-free cultures of autologous monocytes. Eur J Immunol 1989; 19:1347–50.
113. Donnely RP, Fenton MJ, Finbloom DS, Gerrard TL. Differential regulation of IL-1 production in human monocytes by IFN-γ and IL-4. J Immunol 1990; 145:569–75.
114. Lehn M, Weiser WY, Engelhorn S, Gillis S, Remold HG. IL-4 inhibits H_2O_2 production and antileishmanial capacity of human cultured monocytes mediated by IFN-γ. J Immunol 1989; 143:3020–4.
115. Ho JL, He SH, Rios MJC, Wick EA. Interleukin-4 inhibits human macrophage activation by tumor necrosis factor, granulocyte-macrophage colony stimulating factor, and interleukin-3 for antileishmanial activity and oxidative burst capacity. J Infect Dis 1992; 165:344–51.
116. Aderka D, Le J, Vilcek J. IL-6 inhibits lipopolysaccharide-induced tumor necrosis factor production in cultured human monocytes, U937 cells, and in mice. J Immunol 1989; 143:3517–23.
117. Malefyt RW, Abrams J, Bennett B, Figdor C, deVries JE. IL-10 inhibits cytokine synthesis by human monocytes: an autoregulatory role of IL-10 produced by monocytes. J Exp Med 1991; 174:1209–20.
118. Mangan D, Welch GR, Wahl SM. Lipopolysaccharide, TNFα, and IL-1β prevent programmed cell death (apoptosis) in human peripheral blood monocytes. J Immunol 1991; 146:1541–6.
119. Murray HW. Interaction of *Leishmania* with a macrophage cell line. Correlation between intracellular killing and the generation of oxygen intermediates. J Exp Med 1981; 153:1690–5.
120. Engelhorn S, Bruckner A, Remold HG. A soluble factor produced by inoculation of human monocytes with *Leishmania donovani* promastigotes suppresses IFN-γ-dependent monocyte activation. J Immunol 1990; 145:2662–8.
121. Denis M. Interferon-gamma treated murine macrophages inhibit growth of tubercle bacilli via the generation of reactive nitrogen intermediates. Cell Immunol 1991; 132:150–6.
122. Moncada S, Palmer RMJ, Higgs EA. Biosynthesis of nitric oxide from L-arginine. A pathway for the regulation of cell function and communication. Biochem Pharmacol 1989; 38:1709–15.
123. Murray HW, Teitelbaum RF. L-arginine–dependent reactive nitrogen intermediates and the antimicrobial effect of activated human mononuclear phagocytes. J Infect Dis 1992; 165:513–17.
124. Murray HW, Szuro-Sudol A, Wellner D, Oca MJ, Granger AM, Libby DM, Rothermel CD, Rubin BY. Role of tryptophan degradation in respiratory burst–indepedent antimicrobial activity of gamma interferon–stimulated human macrophages. Infect Immun 1989; 57:845–9.

125. Weinberg ED. Iron depletion: a defense against intracellular infection and neoplasia. Life Sciences 1992; 50:1289–97.
126. Byrd TF, Horwitz MA. Interferon gamma–activated human monocytes downregulate transferrin receptors and inhibit the intracellular multiplication of *Legionella pneumophila* by limiting the availability of iron. J Clin Invest; 83:1457–65.
127. Murray HW, Granger AM, Teitelbaum RF. Gamma interferon–activated human macrophages and *Toxoplasma gondii*, *Chlamydia psittaci* and *Leishmania donovani*: Antimicrobial role of limiting intracellular iron. Infect Immun 1991; 59:4684–6.

2

The Molecular Biology of Macrophage Activation

Donna M. Paulnock

University of Wisconsin—Madison, Madison, Wisconsin

I. INTRODUCTION

The concept of macrophage activation has long been recognized as having particular relevance to the development of immunity to pathogenic agents. Whether such microbes are intracellular or extracellular organisms, the macrophage often plays a pivotal role in the antimicrobial immune response as an antigen-presenting cell for the induction of T cell immunity, as a direct effector cell in the control or elimination of the infecting organism, or as a requisite cellular component of the host inflammatory response. In each of these cases, optimal macrophage function is dependent on "activation" of the cells by a variety of stimuli, most prominently either T cell–derived lymphokines (for example, interferon-γ [IFN-γ]) or microbial products (for example, bacterial lipopolysaccharide [LPS]). Thus, "activation" is a general term that describes the induction or enhancement by endogenous or extracellular stimuli of various macrophage functions required for the generation of effective antimicrobial immunity. The potential for autocrine activation occurring by an intracellular stimulatory route also exists and may play a role in the global activation process.

As a result of the recognized central role of this process of activation in determining the ultimate outcome of host-pathogen interactions, substantial effort has been devoted to characterizing the dynamics of this process. Of necessity, most of the past work in this area has focused on a detailed description at the cellular level of changes in macrophage function that accompany activation. Such studies have focused on assessing activation-induced changes in antigen processing and presentation, on characterization of a variety of induced antimicrobial effector mechanisms, and on description of the nature and activities of cell-associated and secreted proteins following activation. More recently, the techniques of molecular biology have begun to be applied to characterization of the activation response, in an effort to extend our basic understanding of the nature and regulation in the activation process to the molecular level. These molecular explorations have focused on selected aspects of the macrophage activation response, and they largely have emphasized mechanisms of control of gene expression in macrophages in response to the prototypic activating stimuli IFN-γ and LPS. In this review, selected examples of molecular genetic changes associated with IFN-γ and LPS stimulation of macrophages will be highlighted, and they will be discussed in the context of cellular biological studies that form the extensive foundation of our understanding of the activation process. However, rather than assembling an exhaustive survey of genes that are induced during the activation process, this review will examine aspects of gene regulation of macrophages that may have particular relevance to the outcome of macrophage-pathogen interactions. Although the overall breadth and scope of the molecular studies accomplished to date are limited in comparison with the tremendous archive of information available on cellular aspects of the activation process, information generated in these efforts has revealed an unexpected complexity with respect to the control of gene expression during the activation process. While thus providing exciting initial clues to molecular aspects of the activation process, the work published to date represents only the beginning of what will become a substantial area of future investigation.

II. REGULATION OF MACROPHAGE PROTO-ONCOGENE EXPRESSION

The principal emphasis in initial molecular studies of activation has been placed on analysis of the regulation of expression of two categories of genes, the diverse family of proto-oncogenes and the broad family of cytokine genes, that represent early responses of macrophages to a variety of activating stimuli [1]. The rapid appearance of the mRNA products of these genes following activation has suggested they may be among the earliest detectable molecular regulation events in this process. These studies on regulation of proto-oncogene and cytokine expression in activated macrophages have their roots in earlier analyses of cellular regulation by factors such as platelet-derived growth factor (PDGF), in which the category of early, or "competence," genes was first described [2]. In this model of activation, early events in the molecular pathways of growth regulation following the rapid intracellular signaling events stimulated by the interaction of PDGF with its cell surface receptor are dependent on gene transcription and mRNA production. The result is the appearance of newly synthesized proteins within or secreted by the cell [2]. Thus, investigators have reasoned that an understanding of the regulation of such "early" genes also should illuminate aspects of the initial macrophage activation response, through characterization of genes that may function by regulating the function of other genes in the stimulated cell.

Prominent members of this category of early genes are various proto-oncogenes, a group of genes believed to be the normal cellular counterparts of the transforming oncogenes and thought to play a major role in the regulation of normal cellular function [3]. Proto-oncogenes identified in macrophages and associated with the activation process include *c-fos*, *c-myc*, and *c-fms* [4–12]. Although other cellular oncogenes have been screened for association with activated macrophages [7], only modulation of *c-fos*, *c-myc*, and *c-fms* has reproducibly been observed during macrophage activation. A broad survey of available studies, including analysis of both LPS- and IFN-γ–stimulated macrophage cell lines and primary macrophage populations, reveals that all three genes are constitutively expressed, albeit at varying levels, in differentiated macrophages before activation. Lipopolysaccharide generally stimulates at least a transient increase in *c-fos* and *c-myc* expression [4,6,7], a process that appears to involve stimulation of protein kinase C [4,6]. In contrast, IFN-γ alone has little effect on levels of *c-fos* and *c-myc* expression [4,5], strengthening the proposal that this mediator initiates a series of biochemical events distinct from that stimulated by LPS to achieve activation [1]. Stimulation of cells with the two mediators in combination results in distinct patterns of proto-oncogene regulation, with both increases and decreases in *c-fos* expression observed [6,7]. Expression of *c-fms* appears closely related to the stage of macrophage development, and varying levels are observed in macrophages at various stages of differentiation or activation. Expression is high in blood monocytes, with decreased expression observed in terminally differentiated and LPS-activated cells [8–11].

The theme that emerges from these studies of cellular oncogene expression following macrophage activation is that constitutive expression of such genes in macrophages can be either enhanced or decreased following activation, depending on both the type of stimulus and the nature of the target cell. Changes in proto-oncogene expression may provide the first indication of activation in macrophages and may be useful for analyzing the uniformity of responses in potentially heterogeneous macrophage populations. However, no evidence has been provided to date that directly links proto-oncogene expression with the induction of specific activation-associated macrophage functions. One exception may be found in changes in *c-fos* expression. The

AP-1 family of enhancer binding factors includes a complex composed of the protein products of the *c-fos* and *c-jun* oncogenes [13,14]. As noted below, AP-1 may be involved in regulation of the transcriptional activity of a number of genes involved in the macrophage activation response. Nonetheless, it remains unclear whether any of the observed changes in proto-oncogene expression have a requisite role in the activation process.

III. REGULATION OF MACROPHAGE CYTOKINE GENE EXPRESSION

The second category of genes associated with macrophage activation that have been amenable to molecular analysis has been the broad set of induced cytokine genes. Although perusal of the literature uncovers a surprisingly large number of macrophage-derived cytokines induced by either IFN-γ or LPS treatment whose regulation has been explored to some extent at the molecular level, complete analysis of the regulatory aspects of only a relatively limited number of cytokines has been achieved to date. Prominent among those characterized are the molecules tumor necrosis factor-α (TNF-α) and interleukin-1 (IL-1), and several members of a large group cytokines with proinflammatory and chemotactic activities [15]. The emphasis on these molecules appropriately highlights the critical role of activated macrophages in development of both acute and chronic inflammatory responses, as studies begin to examine the basic molecular mechanisms of regulation of the inflammatory response. The majority of these studies have analyzed murine-derived products; the observation that many of these factors from different animal species share similar structural elements as well as biological activities suggests they have functionally homologous roles [15]. Although these molecules provide an admittedly limited view of gene regulation, they have served as initial models for understanding the broad effects of activation on macrophage gene expression.

In considering the nature of the studies on macrophage gene regulation to date, it is perhaps useful to review briefly the sequence of cellular responses that are thought to govern modulation of gene expression cells in response to an extracellular stimulus. Initial interaction of a stimulus (such as IFN-γ or LPS) with a cell surface receptor or molecule initiates a sequence of events that ultimately results in modulation of gene expression. These effects are governed by cascade of seconday and tertiary messengers that act in sequence following stimulation to bring about changes in gene expression. First, regulatory molecules such as inositol triphosphate, cyclic nucleotides, and activated protein kinases are generated in response to the interaction of activating stimuli with the macrophage cell surface. These second messengers then effect the transcription of a variety of activation-associated genes. The ultimate production of such gene products is achieved, at least in part, by up-regulation of gene transcription, through the interaction of *cis*-acting sequences or structure motifs and *trans*-acting proteins binding to such response elements. The production of the regulatory proteins in turn is stimulated through the expression of transcription factor genes, which encode those *trans*-acting molecules regulating the transcription of genes leading to the cellular response. Thus, an understanding of the molecular biology of the activation process can be approached either through an exploration of the changes in second messenger molecules that follow cellular stimulation, or through direct analysis of the structure and regulation of transcription factor genes and the ultimate activation-associated genes they regulate. Both areas have received some attention with respect to macrophage activation. In this discussion, we will focus on examination of the information available concerning aspects of cytokine genes and those *trans*-acting factors that may influence their regulation. Unlike the situation described above for proto-oncogenes, the previous identification of a cellular role for many cytokines in paracrine and autocrine activation provides a strong link to functional components of the activation process.

The genes encoding many inflammatory cytokines have been identified over the past several years, yet surprisingly little is known about the molecular mechanisms governing their regulation. In general, these genes are not constitutively expressed by macrophages, but rather exhibit inducible expression following stimulation by one or more mediators, including IFN-γ and/or LPS. The observations of inducible expression suggested that at least one mechanism of control of gene expression for such molecules would be at the transcriptional level, as has been observed for numerous other IFN- and LPS-induced genes [16,17]. Thus, initial analyses of these genes at the molecular level attempted to identify regulatory regions in the 5'-flanking regions of the gene promoters, with emphasis on identification of sequences or regulatory elements thought to bind described transcription factors and that occur in common between genes coordinately induced by activating stimuli. Theoretically, such approaches are relatively straightforward in design and interpretation, but several aspects of macrophage biology have added complexity to these analyses. First, although some aspects of activation are initiated in macrophages by single molecules such as IFN-γ, optimal induction of many functional activities, including antimicrobial and antitumor cytotoxicity, is best achieved by stimulation of cells with multiple mediators, usually IFN-γ in combination with LPS or other lymphokines [1]. Thus, stimulus-specific aspects of activation probably play an important role in analysis of regulatory sequences required for induction of activation-associated genes. Second, many of the genes that are induced in macrophages during the activation process appear to be expressed in a tissue-specific manner. Thus, in addition to a variety of transcription factors involved in the inducible expression of a broad spectrum of genes, including, for example, NF-κB, AP-1, and AP-2, tissue-specific *trans*-acting transcription factors may be of some importance in gene regulation in these cells. Finally, superimposed on these characteristics is the well-documented functional heterogeneity of macrophage populations in vivo, both within individual tissues and between cells found at different tissue sites [18]. Such heterogeneity complicates molecular analysis of the activation process, as it may be difficult to discern at the population level whether distinct mechanisms are operating in individual cells. Despite these constraints on interpretation, current studies of cytokine gene regulation have attempted to analyze the effects of multiple mediators on several aspects of gene expression in different cell populations. The results of such experiments provide a somewhat discursive series of observations. Nonetheless, in context, the studies have revealed information concerning general mechanisms of control of gene expression in macrophages, and also have suggested some means by which the tissue-specific and activation-associated aspects of gene regulation may be accomplished by cells of this lineage.

A widely studied model of molecular regulation has come from analysis of regulation of TNF-α gene expression in macrophages in response to LPS. Early cellular studies clearly associated this molecule both with macrophage effector function and with a variety of regulatory effects on macrophage function [19], thereby implicating TNF-α in both the development and the execution of immune responses. The early cloning of this gene [20] provided key molecular tools for exploration of how the expression of a molecule with multiple potential roles in the immune response might be controlled at the molecular level. The resultant studies have revealed several aspects of interest concerning gene regulation in macrophages.

While numerous results suggested that expression of the TNF-α gene was controlled at multiple levels, including transcriptional, post-transcriptional, and post-translational [21,23], the principal interest in the area of macrophage biology has been focused on exploration of the mechanisms of transcriptional control of this gene in response to LPS. Sequences extending more than 450 bp upstream of the transcriptional initiation site were shown to be necessary for full induction of TNF-α mRNA [24]. Detailed examination of this region for sites capable of binding transcription factors revealed a role for the family of NF-κB/rel transcription factors in TNF-α gene activation that is relatively complex. First, a constitutive form of NF-κB was identified that appeared to regulate basal levels of TNF-α gene expression [25]. Although basal expression of TNF-α previously had been observed in some unstimulated macrophage populations, it was unclear whether the varying levels observed represented constitutive expression or could be attributed to unintentional gene induction by LPS [26–30]. These more recent molecular results suggest that detectable constitutive TNF-α production has a molecular basis distinct from that of induced expression. Treatment of peritoneal macrophages with IFN-γ resulted in varying patterns of gene induction. Although some studies observed that transcription and message levels, as well as secretion, of TNF-α were unaffected by IFN-γ alone [26], others demonstrated that induction of both transcriptional activity and secreted product could be identified in at least some macrophages [27]. In agreement with the latter observations is the recent detection, following IFN-γ activation, of a new NF-κB subunit that can be detected via the appearance of protein-DNA complexes of differing mobilities in gel shift analyses [25]. The inducible subunit, alone and in combination with the constitutive subunit, appears to bind with differing affinities to four κB-like elements in the TNF promoter [25]. Finally, analysis of TNF-α transcription in response to IFN-γ plus LPS suggested that IFN-γ could enhance transcription of the TNF-α gene that was stimulated by LPS [31]. In addition to a role for NF-κB factors in controlling transcription of the TNF-α gene, other regulatory elements have been implicated in transcriptional regulation of both the human and the murine TNF-α genes [24,32]. Recent analyses suggest that many, if not all, of these sequences are capable of binding NF-κB [33]. This suggests that there is some redundancy in the regulatory elements that respond to the transcriptional factors induced by an activating signal such as LPS. Ultimately, such flexibility in regulation may influence the overall level of the transcriptional response of TNF-α. Although it has been demonstrated that substantial loss of inducibility of the TNF-α gene by LPS correlates with the deletion of sites capable of binding the NF-κB factor [24,33], further studies, including DNA footprinting analysis and in vitro mutagenesis of proposed regulatory elements, will be required to determine precisely the role of diverse controlling elements in TNF-α gene regulation.

Multiple *cis*-acting regulatory sites therefore appear to control constitutive and induced TNF-α gene expression. These experiments show that the TNF-α gene provides a useful model for exploring the multiple levels of gene regulation in macrophages, and several conclusions can be drawn from the current results. First, it appears that TNF-α can be constitutively expressed, as least at the mRNA level, and that constitutive gene expression may be mediated by molecular mechanisms distinct from those controlling induced expression. This concept may apply to multiple activation-associated genes, as several additional examples of such differential regulation, including the α_1-antitrypsin gene [34] and genes involved in hemopoiesis [35], already have been identified. The contribution of macrophages at distinct differentiation or developmental stages to such different mechanisms of basal and induced TNF regulation remains to be established, but it has been explored in some studies [27,36]. Second, detailed promoter analysis suggests that both macrophage-specific and developmentally regulated expression of TNF-α involves interaction between diverse DNA sequence elements and multiple forms of the NF-κB transcription factor. Other regulatory factors also may be involved in TNF regulation. This complex response is controlled in part by the nature of the stimulus received by the macrophage. Overall, these observations demonstrate that information concerning both tissue-specific and activation state–specific regulatory mechanisms can be elucidated through analysis of even a single gene.

A second category of genes also has received attention with respect to molecular aspects of macrophage activation, namely those genes belonging to the chemotactic cytokine gene family. This superfamily of related genes includes a variety of recently identified secreted products that encode molecules with proinflammatory properties [37]. These genes can be divided into two broad subfamilies, alpha and beta, on the basis of certain similarities in their primary amino acid sequences. Genes for members of each of these subfamilies are clustered together, but the two groups are located at distinct chromosomal locations [15]. Although members of this family are produced by a wide variety of cell types, those produced by macrophages act principally to elicit the accumulation of immunological and inflammatory cells in vivo [37]. Some of the factors produced by macrophages/monocytes (for example, IL-8 and MIP-2) act principally to recruit other cell types, such as neutrophils, to an inflammatory site. Others macrophage-derived members of this family have monocyte/macrophage chemotactic activity (for example, MIP-1α, IP-10, and JE). Functions for other putative members of this large family remain to be defined [15]. Thus, via production of these molecules, macrophages can regulate the recruitment and activation of various cells, including macrophages themselves, to sites of inflammation and immunological response.

Many experimental approaches to evaluating the regulation of chemotactic cytokine expression during macrophage activation have concentrated on determination of the stimulus-specificity of gene induction and on analysis of potential sequences involved in the transcriptional control of selected genes. The emphasis on transcriptional control mechanisms was prompted by the initial identification of many genes in this superfamily as mRNA species induced in various activated cells, following stimulation with inflammatory or mitogenic stimuli [15,37]. The analysis of the molecular regulation of these family members have revealed both differences and similarities in gene regulation among the members of this cytokine family when compared with what is currently known concerning TNF-α regulation.

It is clear that, as for TNF-α, up-regulation of most of these genes occurs at least partly at the transcriptional level, and IFN-γ and LPS appear to provide the principal stimuli for elaboration of many of these factors by macrophages [38–41]. Although the 5'-flanking regions of most members of this gene family do not show substantial overall sequence homology with the same regions of other family members or other cytokines, numerous potential transcription-regulatory elements have been found in or near the transcriptional start sites. However, both the number and the nature of these sequences varies in each case. For example, putative transcription factor–binding sites for AP-1, AP-2, and NF-κB, among others, are present on the IL-8 gene, along with a site for binding of interferon regulatory factor-1 [42]. An AP-1 binding site similarly has been implicated in basal expression of the rat JE gene; however, additional factors appear to control inducible expression by phorbol esters [43]. The MIP-1α gene exhibits several 5'-promoter region consensus binding sequences [38], including CK-1 and CK-2, which were originally characterized as "cytokine-related" by virtue of their presence in the human or mouse granulocyte-macrophage colony-stimulating factor (GM-CSF), granulocyte colony-stimulating factor (G-CSF), and IL-3 genes [44–46]. At least one of these elements also is observed in the JE gene [43], which appears susceptible to induction by LPS, but not by IFN-γ, in macrophages [47]. Finally, at least one new putative member of this gene family that is regulated specifically by IFN-γ, named *mig* or CRG-10, contains a sequence in the 5'-regulatory region that has homology to the IFN-stimulated response element (ISRE) [39,48]. CRG-10 also contains a region with similarity to the NF-κB binding site and an AP-2 binding site [39].

Thus, the potential regulatory sequences observed within members of this family appear to include a spectrum of recognized transcription factor–binding sites. However, the role and relative importance of each of these sites in gene expression is not yet understood. The potential importance of at least some sites in regulating transcription is suggested by virtue of their frequent appearance in induced genes. For example, genomic regulatory sequences that bind interferon-specific transcription factors are found in the promoter regions of several genes whose transcription is activated by IFN-α/β or IFN-γ [17] and may be responsible for the IFN-γ–specific transcription of genes such as CRG-10 that show particular responsiveness to interferons. Similarly, it has been shown that AP-1 and NF-κB binding sites occur upstream of a large number of immune response–related genes, including those encoding many cytokines, including IL-2 and IL-2 receptor, IFN-β, and TNF-α [13,16]. However, even given these comparisons, it is clear that no potential regulatory sequences are universally expressed even within a class of genes that are coordinately regulated during the immune response. Thus, it still remains unclear exactly how the available regulatory sequences act to control either the cell-specific expression of genes able to be induced by the same stimulus or the transcriptional activity of multiple genes in response to a single mediator. It will be of importance to establish the minimal sequences necessary for conferring responsiveness, in order to understand the potential relationship of and requirements for such potential regulatory *cis*-acting factors in effective regulation of these structurally unrelated, but coordinately induced, genes.

One approach to such analyses has recently been demonstrated in studies of the regulatory pathways controlling the lineage-specific and developmentally regulated expression of the CD11b gene in cells of the myeloid lineage [49]. CD11b encodes a component of the constitutively expressed type 3 complement receptor, which is a cell surface molecule susceptible to modulation and involved in antimicrobial immunity. Investigators have identified that region of the promoter required for directing the tissue- and cell-specific expression of CD11b, through comparison of the promoter region of the CD11b gene with that of the panleukocytic CD18 antigen induced by the same stimuli [49]. However, even within this localized region potential binding sites remain for a number of transcriptional factors important in regulation of hemopoiesis and activation [49]. Thus, a combination of transcription factors of various types again may be of critical importance in achieving tissue-specific expression. These studies suggest a potentially important role for lineage-specific transcription factors in regulating inducible gene expression. While many DNA-binding proteins, such as NF-κB, are ubiquitous in their cellular expression, suggesting they play a general role in gene transcription, others are restricted in their expression to one or a few cell types. Although the numbers of macrophage-specific regulatory factors identified to date have been relatively small, several transcription factors expressed in macrophages have been described that may control either cell-specific or differentiation stage–specific expression of a variety of genes. These include a differentiation-specific nuclear factor potentially involved in regulation of IL-6, IL-1, TNF-α, and G-CSF genes [50], macrophage- and B cell–specific transcription factor ets-1 [51], and the product of the *egr-1* oncogene induced by GM-CSF [52]. The characterization of such tissue- or cell type–specific activators of gene expression, or of the developmentally controlled expression of the availability of requisite *cis*-acting factors [53], will be critical for achieving a complete understanding of gene regulation in macrophages.

Investigation of the mechanisms of regulation of cytokine expression has emphasized elucidation of transcriptional control. It is clear, however, that the inducible gene expression could be the result of multiple additional mechanisms, including effects on the stability or processing of mRNA. Although post-transcriptional control mechanisms have not yet been as thoroughly explored, numerous examples suggest these mechanisms can play a role in controlling the activation response in macrophages, particularly with respect to cytokine expression. For example, LPS-induced production of mRNA for tissue factor [54], for IL-1β [55, 56], for GM-CSF [57], and for TNF-α [24] is in all these cases regulated to some extent at the post-transcriptional level. Such post-transcriptional effects are thought to reflect changes in the stability of induced mRNAs for cytokines, which normally turn over rapidly within the cell as a result of the influence of structural motifs within the 3'-untranslated region [58,59]. These mechanisms of regulation may provide additional opportunities for different modulation of cytokine production during the activation process.

Analysis of even a limited number of the multiple cytokine genes expressed by activated macrophages thus has yielded a complex picture concerning the molecular mechanisms of regulation of gene expression in these cells at the molecular level. However, much work still remains to be accomplished, and several additional aspects of regulation remain to be explored. First, although IFN-γ– and LPS-responsive elements have been identified in individual cytokine genes, our understanding of the regulation of these genes by additional factors that also have activating capability (for example, IL-2 and IL-4, alone or in combination with IFN-γ) currently is limited to a few studies [60–63]. In addition, although the action of repressors has been noted in some cases of macrophage gene expression [30], little information currently is available concerning mechanisms that may suppress expression of many cytokine genes in macrophages and concerning how such negative regulation may be induced or controlled. The majority of molecular studies to date have concentrated on analysis of promoter elements essentially directly upstream of the genes of interest; however, it was recently proposed that regulatory elements distinct from promoters and enhancers, and capable of acting over some distance from the target gene, may be involved in the correct tissue- and stage-specific expression of some genes [64]. The potential role of such regions in macrophage-specific gene expression has not yet been explored. Finally, observations that activation of several of the genes encoding transcription factors involved in cytokine regulation can be activated by the factors themselves (for example, the induction of *c-jun* and NF-κB expression by TNF-α [65,66]) suggest a potential molecular basis for autocrine regulatory loops in the regulation of macrophage function that remains to be fully characterized. Thus, it is clear that the complete picture of the complex regulatory pattern of macrophage gene expression remains to be elucidated.

IV. REGULATION OF MACROPHAGE EFFECTOR FUNCTION–ASSOCIATED GENES

Analysis of genes associated with other activation–related functions of macrophages also has provided information of interest concerning molecular aspects of the activation process. Particular focus in these studies again has been placed on genes with the IFN-γ–mediated macrophage activation process. The most prominent of the genes that fall into this category are those encoding MHC class II (la) molecules. It has been clear for some time that these antigens are critical cell surface determinants involved in the antigen-presenting function of macrophages [67]; however, the nature of the molecular mechanisms associated with IFN-γ–mediated induction of class II expression has only recently been examined in detail in macrophages. The availability of a large background of literature accumulated concerning the IFN-induced transcriptional regulation of class II genes in other cell types [68] provided a good foundation for the work in macrophages, which has asked whether Ia regulation in these cells is similarly controlled. Initial aspects of this work demonstrated that IFN-γ mediates coordinate induction of mRNA for the I-region loci and the I-region–associated invariant chain [69,70]. As in other cell types, this mRNA production is regulated at the transcriptional level and is dependent on sequences in the 5'-flanking regions of the class II genes that are similar to those previously described in other cell types [71,72]. However, a number of observations suggest that macrophage-specific regulation of la expression has additional complexity at the molecular level. For example, macrophages appear to express both global and cell-specific nuclear proteins involved in regulation of class II expression [72]. In addition, class II induction by lymphokines other than IFN-γ, including TNF-α, and IL-4, may be achieved by distinct mechanisms, including the use of alternate sequence elements for effective induction [73,74]. Thus, stimulus-specific aspects of regulation of class II expression also are observed in macrophages and may depend on unique *cis*-acting elements. In addition, similar to the situation for TNF-α expression, it remains unclear whether basal class II expression and induced class II expression are regulated by the same mechanism. Recent comparative analysis of B cell Ia expression and macrophage Ia expression, as prototypes for developmentally controlled and induced Ia antigen expression, has suggested that potential differences in the mechanisms of regulation can be detected [75,76]. Finally, information is needed concerning the molecular mechanisms of down-regulation of Ia antigen expression. It has been previously established that the inhibition of Ia expression by IFN-α/β or dexamethasone is regulated by changes in Ia gene transcription [71]. Lipopolysaccharide also has been shown to down-regulate Ia expression via changes in accumulation of Ia mRNA, although the basis of this effect on RNA levels was not established [77]. The more recent observation that inhibition of class II expression observed in macrophages infected by *Leishmania donovani* occurs at the level of gene transcription [78,79] suggests that sequences susceptible to negative regulatory factors are likely to be important control points for regulation of expression of class II genes during antimicrobial immune responses.

Supplementing analyses of the regulation of expression of Ia antigens, recent work has focused on identification and characterization of other MHC-encoded genes that appear to function in antigen processing and presentation. Several genes found recently within the MHC encode proteins with possible roles in the supply of peptides to class I and class II molecules [80]. Although these studies are at present primarily at the level of gene identification and characterization, at least some of these putative "peptide transporters" are transcribed in macrophages in response to IFN-γ activation [81,82]. Thus, control of the expression of these genes also represents an attractive and critical target for molecular regulation during an activation response.

In addition to genes involved in the afferent, antigen-presenting cell role of the macrophage, recent studies have analyzed the molecular basis of genes contributing to macrophage effector function during acute and chronic infections. Although a wide variety of mediators have been implicated in direct macrophage microbicidal activity, including reactive oxygen intermediates, IL-1, and TNF-α, current interest is focused on the role of nitric oxide (NO) in macrophage effector functions. This radical is generated during arginine metabolism [83] and has been implicated as an intercellular messenger in a wide variety of biological systems [84]. Most pertinent to this discussion, NO has been shown to play a role in regulation of inflammation and lymphocyte proliferation [85,86] and in the cytotoxic action of murine macrophages against tumor cells and microbes [87]. Although endothelial cells, neurons, and other cell types constitutively express NO, NO production in macrophages is observed only after exposure of the cells to cytokines or bacterial products [88]. Thus, up-regulation of inducible NO also represents a response associated with the activation process. Although this short-lived radical originally was associated only with the activation of murine macrophages, recent evidence suggests this molecule also may be produced in at least some cases of human macrophage activation in vivo [89].

Since NO exhibits multiple roles in the development and execution of antimicrobial and antitumor immune responses, there has been considerable interest in understanding the regulation of the biochemical pathway that gives rise to both constitutive and inducible NO production. Initially, constitutively expressed NO synthase (cNOS) from brain was cloned and used to map the portions of the molecule that serve as recognition sites for the two flavins and NADPH, which are believed to serve as cofactors for enzyme activity [90]. Most recently, several reports have documented the cloning of the gene for the inducible NO synthase enzyme, which controls the production of NO in macrophages [91,92]. The inducible form of the enzyme (iNOS) differs substantially from NO synthase purified from the brain, with only 51% identity at the amino acid level compared to cNOS; however, conserved areas of high homology remain in areas of the molecule that are thought to contribute to the active site of the enzyme [91,92]. In a manner similar to other activation-associated changes in gene expression, iNOS also is regulated at the transcriptional level in macrophages, and effective induction requires activation by both IFN-γ and LPS [92]. This combination of mediators also is effective at stimulating synergistic levels of microbicidal and tumoricidal activity in macrophages [1]. However, numerous studies at the cellular level have demonstrated that IFN-γ also synergizes effectively with TNF-α in the induction of NO production, suggesting that multiple mediators can act on the iNOS gene to achieve inducible NO synthesis [93,94]. These induction requirements suggest that the activation of this enzyme is strictly controlled, and it will be of considerable interest to determine how these multiple activating factors work in concert to control transcriptional regulation of the iNOS gene.

V. IDENTIFICATION AND CHARACTERIZATION OF NOVEL GENES ASSOCIATED WITH MACROPHAGE ACTIVATION

In addition to the characterization of genes associated, through cellular studies, with the activation process, some investigators recently have begun to explore more thoroughly the spectrum of genes induced in macrophages by activating stimuli. The studies accomplished to date again have concentrated on identification of genes induced by IFN-γ and LPS, and are based on observations that multiple cellular proteins are induced by these mediators in concert with the acquisition of functional capability [95–98]. These studies represent attempts to extend these earlier observations to include a complete description of the activation response at the molecular level, in a manner similar to that accomplished for T cells and other cell types [99,100], with the additional goal of identifying and characterizing previously undescribed genes.

The majority of the studies exploring the nature of changes in macrophage gene expression at the mRNA level have used the technical approach of screening cDNA libraries, prepared from IFN-γ– or LPS-treated murine macrophages, by differential hybridization with probes from untreated versus activated cells. These efforts thus identify mRNA species induced de novo in macrophages following stimulation by either mediator. Each of the studies, although essentially identical conceptually, has been somewhat different in emphasis and tailored to address changes in gene expression associated with distinct stages of the activation process. Thus, aspects of IFN-γ–induced priming for cytolytic activity [101], aspects of LPS-induced macrophage functional competence [102], and the broader mechanisms of IFN-γ–mediated regulation of gene expression [103] all have been targeted for analysis. Both murine macrophage cell lines and primary peritoneal macrophages have been used for construction of the cDNA libraries, and to date, both have yielded positive results. A restricted number of differentially expressed genes also have been identified by this approach in human monocytes/macrophages [104,105].

These approaches have detected and characterized a number of induced genes expressed following macrophage activation. The characterization studies have concentrated on determining whether the induced genes represent novel or previously identified inducible species, and on characterization of the kinetics and regulation of their expression following macrophage activation with both the original mediators and other agents with activating capacity. The results of these studies have closely echoed the information available as a result of cell biological analyses. For example, most of the genes identified in initial studies are induced relatively early following IFN-γ or LPS stimulation [102,103]. As discussed earlier, such genes may be of importance in regulating both intracellular and intercellular aspects of the activation response. Many of the identified inducible species are preferentially or selectively expressed in macrophages [102,103,106]. This again reiterates the concept that these cells have a high degree of tissue-specific control influencing the nature of the genetic response initiated by mediators such as IFN-γ and LPS that can act on other cell types. In some cases, the genes are responsive to multiple activating stimuli, while a number of induced species demonstrate strict stimulus-specificity for induction [61,101,103,107]. Finally, several of these initial studies suggest that transcriptional control mechanisms again appear in many cases to control both the cell-specificity and stimulus-specificity of gene induction that are observed [39,101,103,107]. However, inducible expression of at least some genes appears to vary according to the developmental stage of the target macrophage [101,108], again suggesting that developmentally regulated control mechanisms play a role in expression of activation-induced genes. Interestingly, sequence analyses have revealed that many of the genes identified to date by this approach fall within categories of inducible genes that had been previously identified at the protein level in macrophages (or other cells) activated by IFN-γ or LPS. For example, many of the genes identified represent new or previously identified members of the chemotactic cytokine family discussed in Section I above [101,103,109]. Additional identifiable genes correspond to members of a previously identified cluster of interferon-inducible genes on chromosome 1 [103,110], while others encode potential regulatory proteins identified initially in IFN-treated cells on the basis of their nucleotide binding capacity [101,111]. Finally, genes encoding several known transcriptional regulatory factors have been detected, including IRF-1 [103].

Collectively, these results have demonstrated that the conceptual approach of these studies is a valid one, and in many cases the molecular probes developed have provided improved tools for analysis of functionally relevant genes. The net result of these studies so far is a first glimpse of the mosaic of genetic changes associated with the activation process, with emphasis on those genes that appear to be expressed at relatively high levels at the mRNA level following activation. Continuing studies of the 5'-flanking sequences of genes in this category should provide important information concerning aspects of the transcriptional control mechanisms that are involved in selective activation of these genes in macrophages. The current challenge in this area, since it is clear that induced genes can be identified by this approach, is twofold. First, it is likely that additional genes await identification at the mRNA level, genes that have not been described in previous studies of activation. Application of methods for the identification of more rare mRNA species to these systems may allow the detection of induced messages that may be of critical importance in regulation of the complex series of events required to achieve full activation. Such novel probes may provide the best mechanisms for identifying specific aspects of the immune response to infectious organisms or tumors, including control of the basis of genetically determined, non-MHC-linked disease resistance or susceptibility, or the nature of the control mechanisms responsible for controlling the balance of positive versus negative immune responses to tumors or microbes. A second clear challenge to these studies is how to develop appropriate systems for linking such expressed genes directly to the processes of inflammation and/or effective effector cell activation and function. Such work probably will involve the use of mutant cells expressing defective or selective activation profiles in experiments designed to replace defective or missing genes with their normal counterparts in the mutant cells. A variety of available macrophage cell lines may be particularly useful components of experimental systems designed to achieve these goals [112–114].

VI. CONCLUSION

Recent studies have begun to characterize the molecular biology of macrophage activation. Analysis of the regulation of expression of a variety of genes has revealed a complex pattern of tissue-specific and stimulus-specific control mechanisms that underlie the significant changes in gene expression grouped under the general term "activation." These studies highlight what one might consider the contemporary theme of macrophage molecular regulation, namely that multiple mechanisms generally are utilized in control of expression of a single genetic event in these cells, both developmentally and in response to acute activation. Such complex regulatory mechanisms probably represent critical control points in the development of the immune response to pathogens that involve the macrophage. A clear understanding of the molecular basis of the activation process thus ultimately may provide the most useful means for enhancement or modulation of these biological responses.

REFERENCES

1. Adams DO, Hamilton TA. Molecular transductional mechanisms by which IFN gamma and other signals regulate macrophage development. Immunol Rev 1987; 97:5–27.
2. Cochran BH. The molecular biology of action of platelet-derived growth factor. Adv Cancer Res 1992; 45:183–225.
3. Muller R, Verma IM. Expression of cellular oncogenes. Curr Top Microbiol Immunol 1984; 112:73–116.
4. Introna M, Hamilton TA, Kaufman RE, Adams DO, Bast RCJ. Treatment of murine peritoneal macrophages with bacterial lipopolysaccharide alters expression of c-fos and c-myc oncogenes. J Immunol 1986; 137:2711–5.
5. Sariban E, Mitchell T, Griffin J, Kufe DW. Effects of interferon-gamma on proto-oncogene expression during induction of human monocytic differentiation. J Immunol 1987; 138:1954–8.
6. Radzioch D, Bottazzi B, Varesio L. Augmentation of c-fos mRNA expression by activators of protein kinase C in fresh, terminally differentiated resting macrophages. Mol Cell Biol 1987; 7:595–9.

7. Higuchi Y, Setoguchi M, Yoshida S, Akizuki S, Yamamoto S. Enhancement of c-fos expression is associated with activated macrophages. Oncogene 1988; 2:515–21.
8. Sariban E, Mitchell T, Kufe D. Expression of the c-fms proto-oncogene during human monocytic differentiation. Nature 1985; 316:64–6.
9. Radzun HJ, Kreipe H, Heidorn K, Parwaresch MR. Modulation of c-fms proto-oncogene expression in human blood monocytes and macrophages. J Leukocyte Biol 1988; 44:198–204.
10. Gusella GL, Ayroldi E. Espinoza-Delgado I, Varesio L. Lipopolysaccharide, but not IFN-gamma, down-regulates c-fms mRNA proto-oncogene expression in murine macrophages. J Immunol 1990; 144:3574–80.
11. Sariban E, Imamura K, Sherman M, Rothwell V, Pantazis P, Kufe D. Downregulation of c-fms gene expression in human monocytes treated with phorbol esters and colony-stimulating factor 1. Blood 1989; 74:123–9.
12. Mitchell RL, Zokas L, Schreiber RD, Verma IM. Rapid induction of the expression of proto-oncogene fos during human monocytic differentiation. Cell 1985; 40:209–17.
13. Turner R, Tijan R. Leucine repeats and an adjacent DNA binding domain mediate the formation of functional cfos-cjun heterodimers. Science 1989; 243:1689–94.
14. Adunyah AE, Unlap TM, Wagner F, Kraft AS. Regulation of c-jun expression and AP-1 enhancer activity by granulocyte-macrophage colony-stimulating factor. J Biol Chem 1991; 266:5670–5.
15. Wolpe SD, Cerami A. Macrophage inflammatory proteins 1 and 2: members of a novel superfamily of cytokines. FASEB J 1989; 3:2565–73.
16. Baeuerle PA. The inducible transcription activator NF-κB: regulation by distinct protein subunits. Biochim Biophys Acta 1991; 1072:63–80.
17. Williams BRG. Transcriptional regulation of interferon-stimulated genes. Eur J Biochem 1991; 200:1–11.
18. Dougherty GJ, McBride WH. Macrophage heterogeneity. J Clin Lab Immunol 1984; 14:1–26.
19. Beutler B. Tumor necrosis, cachexia, shock and inflammation: a common mediator. Annu Rev Biochem 1988; 57:505–47.
20. Semon D, Kawashima E, Jongeneel CV, Shakhov AN, Nedospasov SA. Nucleotide sequence of the TNF locus, including the TNF-α (tumor necrosis factor) and TNF-β (lymphotoxin) genes. Nucleic Acids Res 1987; 15:908–12.
21. Han JH, Beutler B, Huez G. Complex regulation of tumor necrosis factor mRNA turnover in lipopolysaccharide-activated macrophages. Biochim Biophys Acta 1991; 1090:22–8.
22. Jongeneel CV, Shakhov AN, Nedospasov SA, Cerottini JC. Molecular control of tissue-specific expression at the mouse TNF locus. Eur J Immunol 1989; 19:549–52.
23. Beutler B, Krochin N, Milsark IW, Luedke C, Cerami A. Control of cachectin (tumor necrosis factor) synthesis: mechanisms of endotoxin resistance. Science 1986; 232:977–80.
24. Shakhov AN, Collart MA, Vassalli P, Nedospasov SA, Jongeneel CV. Kappa B-type enhancers are involved in lipopolysaccharide-mediated transcriptional activation of the tumor necrosis factor alpha gene in primary macrophages. J Exp Med 1990; 171:35–47.
25. Collart MA, Baeuerle P, Vassalli P. Regulation of tumor necrosis factor alpha transcription in macrophages: involvement of four kappa B–like motifs and of constitutive and inducible forms of NF-kappa B. Mol Cell Biol 1990; 10:1498–1506.
26. Koerner TJ, Adams DO, Hamilton TA. Regulation of tumor necrosis factor (TNF) expression: interferon-gamma enhances the accumulation of mRNA for TNF induced by lipopolysaccharide in murine peritoneal macrophages. Cell Immunol 1987; 109:437–43.
27. Wynn TA, Freund YR, Paulnock DM. Tumor necrosis factor-alpha differentially regulates Ia antigen expression and tumoricidal activity in two murine macrophage cell lines. Cell Immunol 1992; 140:184–196.
28. Gifford GE, Lohmann-Matthes ML. Gamma interferon priming of mouse and human macrophages for induction of tumor necrosis factor production by bacterial lipopolysaccharide. J Natl Cancer Inst 1987; 78:121–7.
29. Beutler B, Tkacenko V, Milsark I, Krochin N, Cerami A. Effect of γ interferon on cachectin expression by mononuclear phagocytes. Reversal of the Ips^d (endotoxin resistance) phenotype. J Exp Med 1986; 164:1791–8.
30. Collart MA, Belin D, Vassalli JD, DeKossodo S, Vassalli P. γ interferon enhances macrophage

transcription of the tumor necrosis factor/cachectin, interleukin 1, and urokinase genes, which are controlled by short-lived repressors. J Exp Med 1986; 164:2113–5.
31. Yu SF, Koerner TJ, Adams DO. Gene regulation in macrophage activation: differential regulation of genes encoding for tumor necrosis factor, interleukin-1, JE, and KC by interferon-gamma and lipopolysaccharide. J Leukocyte Biol 1990; 48:412–9.
32. Goldfeld AE, Doyle C, Maniatis T. Human tumor necrosis factor α gene regulation by virus and lipopolysaccharide. Proc Natl Acad Sci USA 1991; 87:9769–73.
33. Drouet C, Shakhov AN, Jongeneel CV. Enhancers and transcription factors controlling the inducibility of the tumor necrosis factor-α promoter in primary macrophages. J Immunol 1991; 147:1694–1700.
34. Hafeez W, Ciliberto G, Perlmutter DH. Constitutive and modulated expression of the human alpha 1 antitrypsin gene. Different transcriptional initiation sites used in three different cell types. J Clin Invest 1992; 89:1214–22.
35. Miyajima A, Miyatake S, Schreurs J, DeVries, J, Arai, N, Yokota, T, Arai, K. Coordinate regulation of immune and inflammatory responses by T cell–derived lymphokines. FASEB J 1988; 2:2462–73.
36. Watanabe Y, Jacob CO. Regulation of MHC class II antigen expression. Opposing effects of tumor necrosis factor-α on IFN-γ-induced HLA-DR and Ia expression depends on the maturation and differentiation stage of the cell. J Immunol 1991; 146:899–905.
37. Oppenheim JJ, Zachariae COC, Mukaida N, Matsushima K. Properties of the novel proinflammatory supergene "intercrine" cytokine family. Annu Rev Immunol 1991; 9:617–48.
38. Widmer U, Yang Z, van Deventer S, Manogue KR, Sherry B, Cerami A. Genomic structure of murine macrophage inflammatory protein-1 alpha and conservation of potential regulatory sequences with a human homolog, LD78. J Immunol 1991; 146:4031–40.
39. Wright TM, Farber JM. 5' regulatory region of a novel cytokine gene mediates selective activation by interferon gamma. J Exp Med 1991; 173:417–22.
40. Orlofsky A, Berger MS, Prystowsky MB. Novel expression pattern of a new member of the MIP-1 family of cytokine-like genes. Cell Regul 1991; 2:403–12.
41. Furutani Y, Nomura H, Notake M, Oyamada, Y, Fukui, T, Yamada, N, Larsen, CG. Cloning and sequencing of the cDNA for human monocyte chemotactic and activating factor (MCAF). Biochem Biophys Res Commun 1989; 159:249–255.
42. Mukaida N, Shiroo M, Matsushima K. Genomic structure of the human monocyte-derived neutrophil chemotactic factor IL-8. J Immunol 1989; 143:1366–71.
43. Timmers HT, Pronk GJ, Bos JL, van der Eb AJ. Analysis of the rat JE gene promoter identifies an AP-1 binding site essential for basal expression but not for TPA induction. Nucleic Acids Res 1990; 18:23–34.
44. Nimer SD, Morita EA, Martis MJ, Wachsman W, Gasson JC. Characterization of the human granulocyte/macrophage colony stimulating factor promoter region by genetic analysis: correlation with DNase I footprinting. Mol Cell Biol 1988; 8:1979–85.
45. Tsuchiya M, Kaziro Y, Nagata S. The chromosomal gene structure for murine granulocyte colony-stimulating factor. Eur J Biochem 1987; 165:7–12.
46. Stanley E, Metcalf D, Sobieszczuk P, Gough NM, Dunn AR. The structure and expression of the murine gene encoding granulocyte-macrophage colony stimulating factor: Evidence for utilisation of alternative promoters. EMBO J 1985; 4:2569–75.
47. Introna M, Bast RC Jr, Tannenbaum CS, Hamilton TA, Adams DO. The effect of LPS on expression of the early "competence" genes JE and KC in murine peritoneal macrophages. J Immunol 1987; 138:3891–96.
48. Friedman RL, Stark GR. α-interferon–induced transcription of HLA and metallothionein genes containing homologous upstream sequences. Nature 1985; 314:637–9.
49. Shelley CS, Arnaout MA. The promoter of the CD11b gene directs myeloid-specific and developmentally regulated expression. Proc Natl Acad Sci USA 1991; 88:10525–9.
50. Natsuka S, Akira S, Nishio Y, Hashimoto S, Sugita T, Ishiki H, Kishimoto T. Macrophage differentiation-specific expression of NF-IL6, a transcription factor for interleukin-6. Blood 1992; 79:460–466.
51. Klemsz MJ, McKercher SR, Celada A, Van Beveren C, Maki RA. The microphage and B cell–specific transcription factor PU.1 is related to the ets oncogene. Cell 1990; 61:113–24.
52. Liu JW, Lacy J, Sukhatme VP, Coleman DL. Granulocyte-macrophage colony-stimulating factor induces transcriptional activation of Egr-1 in murine peritoneal macrophages. J Biol Chem 1991; 266:5929–33.

53. Harada H, Willison K, Sakakibara J, Miyamoto M, Fujita T, Taniguchi T. Absence of the type I IFN system in EC cells: transcriptional activator (IRF-1) and repressor (IRF-2) genes are developmentally regulated. Cell 1990; 63:303–12.
54. Brand K, Fowler BJ, Edgington TS, Mackman N. Tissue factor mRNA in THP-1 monocytic cells is regulated at both transcriptional and posttranscriptional levels in response to lipopolysaccharide. Mol Cell Biol 1991; 11:4732–8.
55. Burchett SK, Weaver WM, Westall JA, Larsen A, Kronheim S, Wilson CB. Regulation of tumor necrosis factor/cachectin and IL-1 secretion in human mononuclear phagocytes. J Immunol 1988; 140:3473–81.
56. Arend WP, Gordon DF, Wood WM, Janson RW, Joslin FG, Jameel S. IL-1β production in cultured human monocytes is regulated at multiple levels. J Immunol 1989; 143:118–26.
57. Schuler GD, Cole MD. GM-CSF and oncogene mRNA stabilities are independently regulated in trans in a mouse monocytic tumor. Cell 1988; 55:1115–22.
58. Caput D, Beutler B, Hartog K, Thayer R, Brown-Shimer S, Cerami A. Identification of a common nucleotide sequence in the 3'-untranslated region of mRNA molecules specifying inflammatory mediators. Proc Natl Acad Sci USA 1986; 83:1670–4.
59. Shaw G, Kamen R. A conserved AU sequence from the 3'-untranslated region of GM-CSF mRNA mediates selective mRNA degradation. Cell 1986; 46:659–67.
60. Gautam S, Tebo JM, Hamilton TA. Il-4 suppresses cytokine gene expression induced by IFN-gamma and/or IL-2 in murine peritoneal macrophages. J Immunol 1992; 148:1725–30.
61. Narumi S, Finke JH, Hamilton TA. Interferon gamma and interleukin 2 synergize to induce selective monokine expression in murine peritoneal macrophages. J Biol Chem 1990; 265:7036–41.
62. Donnelly RP, Fenton MJ, Finbloom DS, Gerrard TL. Differential regulation of IL-1 production in human monocytes by IFN-gamma and IL-4. J Immunol 1990; 145:569–75.
63. Cox GW, Chattopadhyay U, Oppenheim JJ, Varesio L. IL-4 inhibits the costimulatory activity of IL-2 or picoline acid but not of lipopolysaccharide on IFN-gamma-treated macrophages. J Immunol 1991; 147:3809–14.
64. Felsenfield G. Chromatin as an essential part of the transcriptional machinery. Nature 1992; 355:219–24.
65. Brenner DA, O'Hara M, Angel P, Chojkin M, Karin M. Prolonged activation of jun and colleganase genes by tumor necrosis factor-alpha. Nature 1989; 337:661–3.
66. Osborn L, Kunkel S, Nabel GJ. Tumor necrosis factor alpha and interleukin 1 stimulate the human immunodeficiency virus enhancer by the activation of the nuclear factor kappa B. Proc Natl Acad Sci USA 1989; 86:2336–40.
67. Unanue ER, Allen PM. The basis for the immunoregulatory role of macrophages and other accessory cells. Science 1987; 236:551–7.
68. Kara CJ, Glimcher LH. Regulation of MHC class II gene transcription. Curr Opin Immunol 1991; 3:16–21.
69. Fertsch D, Schoenberg DR, Germain RN, Tou JY, Vogel SN. Induction of macrophage Ia antigen expression by rIFN-gamma and down-regulation by IFN-alpha/beta and dexamethasone are mediated by changes in steady-state levels of Ia mRNA. J Immunol 1987; 139:244–9.
70. Paulnock-King DM, Sizer, K, Freund YR, Jones PP, Parnes J. Coordinate induction of Ia α, β, and Ii mRNA in a macrophage cell line. J Immunol 1985; 135:632–6.
71. Fertsch-Ruggio D, Schoenberg DR, Vogel SN. Induction of macrophage Ia antigen expression by rIFN-γ and down-regulation by IFN-α/β and dexamethasone are regulated transcriptionally. J Immunol 1988; 141:1582–9.
72. Dedrick RL, Jones PP. Sequence elements required for activity of a murine major histocompatibility complex class II promoter bind common and cell-type-specific nuclear factors. Mol Cell Biol 1990; 10:593–604.
73. Freund YR, Dedrich RL, Jones PP. Cis-acting sequences required for Class II gene regulation by interferon-gamma and tumor necrosis factor-alpha in a murine macrophage cell line. J Exp Med 1990; 171:1283–99.
74. Cao H, Wolff RG, Meltzer MS, Crawford RM. Differential regulation of class II MHC determinants on macrophages by IFN-gamma and IL-4. J Immunol 1989.. 143:3524–31.
75. Sartoris S, Scupoli MT, Scarpellino L, Paiola F, Jotterand Bellomo M, Tridente G, Accolla RS.

Inducible and constitutive MHC class II gene expression. Distinct tissue-specific genetic controls. J Immunol 1990; 145:1960–7.

76. Maffei A, Scarpellino L, Bernard M, Carra G, Jotterand Bellomo M, Guardiola J, Accolla RS. Distinct mechanisms regulate MHC class II gene expression in B cells and macrophages. J Immunol 1987; 139:942–8.
77. Koerner TJ, Hamilton TA, Adams DO. Suppressed expression of surface Ia on macrophages by lipopolysaccharide: evidence for regulation at the level of accumulation of mRNA. J Immunol 1987; 139:239–43.
78. Kwan WC, McMaster WR, Wong N, Reiner NE. Inhibition of expression of major histocompatibility complex class II molecules in macrophages infected with *Leishmania donovani* occurs at the level of gene transcription via a cyclic AMP–independent mechanism. Infect Immun 1992; 60:2115–20.
79. Reiner NE, Ng W, Ma T, McMaster WR. Kinetics of gamma interferon binding and induction of major histocompatibility complex class II mRNA in Leishmania-infected macrophages. Proc Natl Acad Sci USA 1988; 85:4330–4.
80. DeMars R, Spies T. New genes in the MHC that encode proteins for antigen processing. Trends Cell Biol 1992; 2:81–6.
81. Martinez CK, Monaco JJ. Homology of proteasome subunits to a major histocompatibility complex–linked LMP gene. Nature 1991; 353:664–7.
82. Cho S, Attaya M, Brown MG, Monaco JJ. A cluster of transcribed sequences between the Pb and Ob genes of the murine major histocompatibility complex. Proc Natl Acad Sci USA 1991; 88:5197–201.
83. Drapier JC, Hibbs JBJ. Differentiation of murine macrophages to express nonspecific cytotoxicity for tumor cells results in L-arginine–dependent inhibition of mitochondrial iron-sulfur enzymes in the macrophage effector cells. J Immunol 1988; 140:2829–38.
84. Marletta MA, Tayeh MA, Hevel JM. Unraveling the biological significance of nitric oxide. Biofactors 1990; 2:219–25.
85. Albina JE, Abate JA, Henry WL Jr. Nitric oxide production is required for murine resident peritoneal macrophages to suppress mitogen-stimulated T cell proliferation: role of IFN-gamma in the induction of the nitric oxide–synthesizing pathway. J Immunol 1991; 147:144–48.
86. Mills CD. Molecular basis of "suppressor macrophages": arginine metabolism via the nitric oxide synthetase pathway. J Immunol 1991; 146:2719–23.
87. Nathan CF, Hibbs JB Jr. Role of nitric oxide synthesis in macrophage antimicrobial activity. Curr Opin Immunol 1991; 3:65–70.
88. Green SJ, Nacy CA, Meltzer MA. Cytokine-induced synthesis of nitric oxides in macrophages: A protective response to *Leishmania donovani* and other intracellular pathogens. J Leukocyte Biol 1991; 50:93–103.
89. Hibbs JB Jr, Westenfelder C, Taintor R, Vavrin Z, Kablitz C, Baranowski RL, Ward JH. Evidence for cytokine-inducible nitric oxide synthesis from L-arginine in patients receiving interleukin-2 therapy. J Clin Invest 1992; 89:867–77.
90. Bredt DS, Hwang PM, Glatt CE, Lowenstein C, Reed RR, Snyder SH. Cloned and expressed nitric oxide synthase structurally resembles cytochrome P-450 reductase. Nature 1991; 351:714–8.
91. Lyons CR, Orloff GJ, Cunningham JM. Molecular cloning and functional expression of an inducible nitric oxide synthase from a murine macrophage cell line. J Biol Chem 1992; 267:6370–4.
92. Xie Q, Cho HJ, Calaycay J, Mumford RA, Swiderek KM, Lee TD, Ding A. Cloning and characterization of inducible nitric oxide synthase from mouse macrophages. Science 1992; 256:225–8.
93. Nacy CA, Meierovics AI, Belosevic M, Green SJ. Tumor necrosis factor-alpha: central regulatory cytokine in the induction of macrophage antimicrobial activities. Pathobiology 1991: 59:182–4.
94. Liew FY, Li Y, Millot S. Tumor necrosis factor-a synergizes with IFN-γ in mediating killing of Leishmania major through the induction of nitric oxide. J Immunol 1990; 145:4306–4317.
95. Johnston PA, Somers SD, Hamilton TA. Expression of a 120,000 dalton protein during tumoricidal activation in murine peritoneal macrophages. J Immunol 1987; 138:2739–44.
96. Paulnock DM, Lambert LE. Identification and characterization of monoclonal antibodies specific for macrophages at intermediate stages in the tumoricidal activation pathway. J Immunol 1990; 144:765–3.
97. Pinson DM, Phillips TA, Pace JL, MacKay RJ, Russell SW. Activation-associated marker proteins: peptide mapping and their expression in macrophage cell lines [published erratum appears in Biochem Biophys Res Commun 1991; 177:895]. Biochem Biophys Res Commun 1991; 176:882–6.

98. Lokesh BR, Wrann M. Preliminary studies on the inflammatory stimulus induced proteins in mouse resident peritoneal macrophages. Agents Actions 1987; 21:145–8.
99. Zipfel PF, Irving SG, Kelly K, Siebenlist U. Complexity of the primary genetic response to mitogenic activation of human T cells. Mol Cell Biol 1989; 9:1041–8.
100. Burd PR, Freeman GJ, Wilson SD, Berman M, DeKruyff R, Billings PR, Dorf ME. Cloning and characterization of a novel T cell activation gene. J Immunol 1987; 139:3126–31.
101. Wynn TA, Nicolet CM, Paulnock DM. Identification and characterization of a new gene family induced during macrophage activation. J Immunol 1991; 147:4384–92.
102. Tannenbaum CS, Koerner TJ, Jansen MM, Hamilton TA. Characterization of lipopolysaccharide-induced macrophage gene expression. J Immunol 1988; 140:3640–5.
103. Farber JM. A collection of mRNA species that are inducible in the RAW 264.7 mouse macrophage cell line by gamma interferon and other agents. Mol Cell Biol 1992; 12:1535–45.
104. Fan XD, Stark GR, Bloom BR. Molecular cloning of a gene selectively induced by gamma interferon from human macrophage cell line U937. Mol Cell Biol 1989; 9:1922–8.
105. Vermeulen MW, David JR, Remold HG. Differential mRNA responses in human macrophages activated by interferon-gamma and muramyl dipeptide. J Immunol 1987; 139:7–9.
106. Collins S, Altman JD, Marletta MA. Development of an affinity chromatography resin for the purification of carcinogen binding proteins from mouse liver. Biochem Biophys Res Commun 1985; 129:155–62.
107. Hamilton TA, Bredon N, Ohmori Y, Tannenbaum CS. IFN-gamma and IFN-beta independently stimulate the expression of lipopolysaccharide-inducible genes in murine peritoneal macrophages. J Immunol 1989; 142:2325–31.
108. Narumi S, Hamilton TA. Inducible expression of murine IP-10 mRNA varies with the state of macrophage inflammatory activity. J Immunol 1991; 146:3038–44.
109. Ohmori Y, Hamilton TA. A macrophage LPS-inducible early gene encodes the murine homologue of IP-10. Biochem Biophys Res Commun 1990; 168:1261–7.
110. Opdenakker G, Snoddy J, Choubey D, Toniato E, Prautcheva DD, Seldin MF, Ruddle FH. Interferons as gene activators: a cluster of six interferon-activatable genes is linked to the erythroid alpha-spectrin locus on murine chromosome 1. Virology 1989; 171:568–78.
111. Cheng Y-SE, Becker-Manley MF, Chow TP, Horan DC. Affinity purification of an interferon-induced human guanylate-binding protein and its characterization. J Biol Chem 1985; 260:15834–9.
112. Ricciardi-Castagnoli P, Paglia P. New tools for investigating macrophage differentiation. Res Immunol 1992; 143:101–6.
113. Lambert LE, Paulnock DM. Differential induction of activation markers in macrophage cell lines by interferon-famma. Cell Immunol 1989; 120:401–18.
114. Blasi E, Varesio L, Wiltrout RH. Tumor formation by a murine macrophage cell line immortalized in vitro by v-raf and v-myc oncogenes. Cancer Immunol Immunother 1988; 27:109–13.

3

Interactions of Lipopolysaccharide With Macrophages

Carl L. Manthey and Stefanie N. Vogel

Uniformed Services University of the Health Sciences, Bethesda, Maryland

I. INTRODUCTION TO THE BIOMEDICAL IMPORTANCE OF THE LIPOPOLYSACCHARIDE–MACROPHAGE INTERACTION

The incidence of gram-negative sepsis in the United States has recently been estimated at 170,000 cases per year, with an associated mortality of more than 70,000 per year [1]. This staggering statistic represents an upward trend, and has been attributed to an increase in the incidence of invasive surgical procedures and a larger population of persons immunocompromised by burns, myeloblitерative cancer chemotherapy and radiation therapies, and immunosuppressive diseases such as AIDS. Unlike the pathogenesis associated with organisms that secrete classical "exotoxins," the pathophysiology seen in gram-negative sepsis has been strongly associated with "endotoxin," which consists of the major structural components of the bacterial outer membrane, namely, lipopolysaccharide (LPS) and associated proteins. During the course of a gram-negative infection, endotoxin is released from the outer membranes of dead and dying bacteria, a process that appears to be exacerbated by appropriate antibiotic therapy, so that final circulating levels of endotoxin can approach nanograms per milliliter [2]. For many years, it has been recognized that experimental administration of purified LPS mimics the symptoms elicited during gram-negative sepsis, such as fever; cardiovascular shock; accumulation of destructive inflammatory exudates in the microvasculature of the lung, intestines, and other organs; multiorgan failure, and death [reviewed in 3–5].

The structure of LPS (Figure 1a) is amphipathic, in that it is composed of a hydrophilic "polysaccharide" region covalently linked to the hydrophobic "lipid A" region (Figure 1b) [6]. In the bacterium, the lipid A region is integrated into the outer membrane, whereas the polysaccharide extends out from the surface of the organism. Thus, it is not surprising that the portion of the polysaccharide most distal from the lipid A region, i.e., the repeating oligosaccharides that comprise the O antigen, are immunodominant and confer serological specificity to the particular gram-negative bacterium. The O antigen is, in turn, covalently linked to a substituted, branched hetero-oligosaccharide, referred to as the "core region," which is more conserved than the "O antigen" within species of bacteria. Gram-negative bacteria that possess mutations in specific glucosyltransferase enzymes possess LPS molecules that lack O antigen but retain intact lipid A and, to varying degrees of completeness, core structures. These LPS molecules have been characterized as specific "R chemotypes." The core, in turn, is covalently linked to the hydrophobic lipid A region, which is highly conserved within families of bacteria. Within the *Enterobacteriaceae*, the lipid A structure consists of a diphosphorylated glucosamine disaccharide that is acylated with characteristic hydroxylated and nonhydroxylated fatty acids. Maximal toxicity is associated with lipid A structures that are hexaacylated, and the number and length of the individual acyl groups are characteristic of a particular bacterial species. Careful experimental

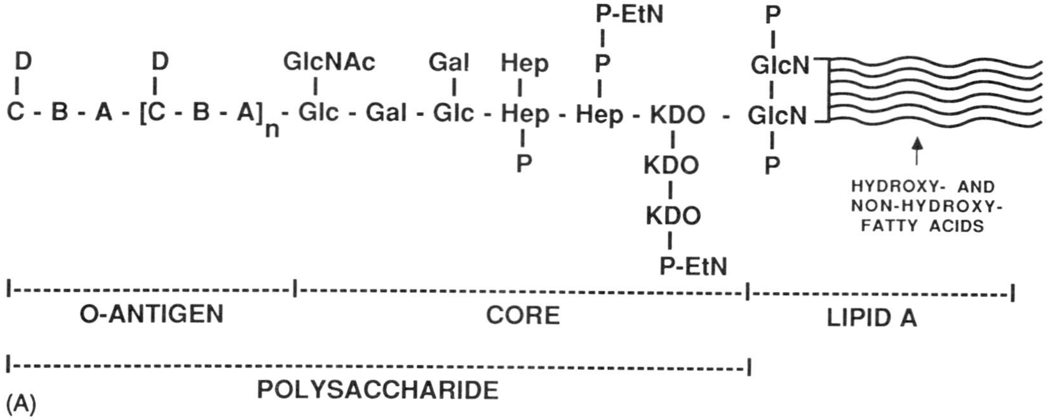

(A)

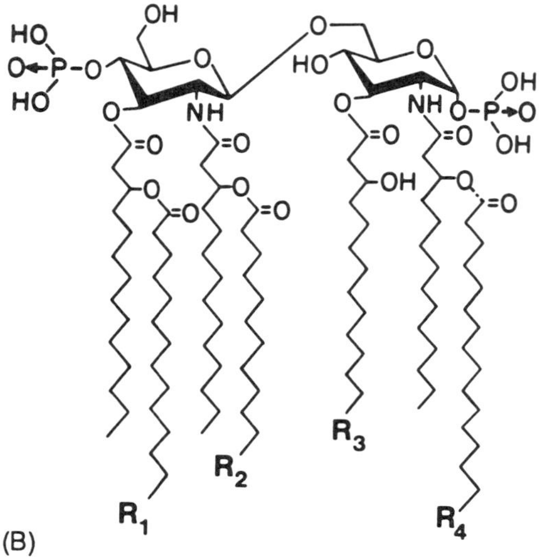

(B)

FIGURE 1 General structure of a lipopolysaccharide molecule (A) and lipid A (B) derived from *Salmonella*. A–D = sugar residues; Glc = D-glucose; Gal = D-galactose; GlcN = D-glucosamine; GlcNac = *N*-acetyl-D-glucosamine; Hep= L-glycero-D-mannoheptose; KDO = 2-keto-3-deoxy-D-manno-octonate; P = phosphate; EtN = ethanolamine. The structure of the lipid A molecule was kindly provided by Dr. Nilofer Qureshi.

evaluation of synthetic lipid A and other LPS partial structures has indicated that the lipid A region of LPS is the "endotoxic center" of the molecule.

Another feature of LPS that is in sharp contrast to the action of classical exotoxins is that toxic symptoms are not due to the direct action of LPS on the cells of the target organ. Rather, they are elicited in large part by cytokines synthesized de novo by macrophages in response to LPS [reviewed in 3–5]. For example, the induction of fever is a hallmark of LPS-induced toxicity. However, LPS induces fever indirectly, through its action on macrophages, which in turn secrete several distinct cytokines—e.g. interleukin-1 (IL-1), tumor necrosis factor-α (TNF-α), and interferon-β (IFN-β)—that act distally on the hypothalamus to invoke fevers that are brisk, monophasic, and prostaglandin-mediated. The experimental basis for the role of macrophages and their products as mediators of endotoxicity originated in a model of LPS *hyporesponsiveness*. From a spontaneous mutation in the C3H mouse colony at Jackson Laboratories a mouse strain resulted, called C3H/HeJ, that was highly resistant to virtually all effects of endotoxin [reviewed in 3]. Genetic analysis indicated that LPS responsiveness is controlled by a single autosomal gene, called *Lps*, and that the inheritance of LPS sensitivity is expressed in all cells studied. To date, the nature of the gene product has not been elucidated. In studies utilizing C3H/HeJ mice and their syngeneic LPS-responsive counterparts, Michalek et al. [7] first implicated bone marrow–derived lymphoreticular cells as central in the mediation of LPS-induced toxicity. Reconstitution of irradiated C3H/HeJ mice with the bone marrow of C3H/HeN mice rendered C3H/HeJ mice sensitive to LPS. Macrophages were more specifically implicated when adoptive transfer of C3H/HeN peritoneal macrophages into the peritoneal cavity of C3H/HeJ mice rendered the latter sensitive to the lethal effects of LPS [8]. In addition, macrophages derived from C3H/HeJ mice failed to respond to LPS or lipid A in vitro to produce the soluble factors that were then suspected to mediate LPS-induced toxicity. A large body of literature ensued that showed that LPS-inducible monokines such as TNF-α, IL-1, and IL-6 could reproduce many of the toxic actions of LPS and that the toxicity induced by LPS administration could be blocked by reagents that antagonized the actions of these same monokines [3–5]. Thus, a major approach now in clinical trials focuses on the administration of cytokine antagonists to patients with gram-negative sepsis [5,9–11].

Thus far, this introduction has focused on the role of LPS in the pathogenesis of gram-negative sepsis. However, it would be remiss to cast LPS solely in a pathogenic role, as there is strong evidence to indicate that normal exposure to LPS, released from gut-associated bacteria or as a result of low-level infection, may be very important in the development of lymphoid organs, protection against otherwise lethal irradiation, and innate resistance to infection and malignancy [reviewed in 4]. Furthermore, the local responses to LPS that occur during a superficial gram-negative infection initiate diverse host defense mechanisms that sequester and kill the bacteria. As in full-blown sepsis, these favorable responses are initiated by the interaction of LPS/lipid A with the macrophage; however, it appears that in the absence of an overwhelming exposure to this molecule, the cytokine cascade that is initiated favors the host, rather than escalating into an uncontrolled inflammatory disease.

II. THE PHYSICAL INTERACTIONS OF LIPOPOLYSACCHARIDE WITH MACROPHAGES

A. Role of Plasma Proteins in Lipopolysaccharide Binding

The cellular responses of macrophages to LPS are legion; virtually every aspect of macrophage physiology is in some way influenced by LPS [reviewed in 12]. Lipopolysaccharide induces increases in glucose and oxygen consumption, cell size, and membrane ruffling. The synthesis of an impressive number of secreted and cellular proteins and lipids is increased, while others are decreased. The cells acquire heightened functional capacities, such as the ability to kill microbes and tumor cells, but their motility is diminished. The macrophage has developed a finely tuned mechanism for recognizing LPS in its environment, and as a consequence, LPS is among the most potent of macrophage stimulators. In many cases, as little as 1 pg/ml($\sim$100 fM) elicits a measurable response. To understand better how LPS initiates changes in macrophage functions, a great deal of effort has been made to characterize the physical interactions that occur between macrophages and LPS. Lipopolysaccharide, consistent with its amphipathic structure, forms macromolecular complexes in aqueous solution analogous to micellar formation by phospholipids [13]. These LPS complexes can insert directly into cell plasma membrane bilayers and alter membrane fluidity, and at high concentrations can lead to dissolution of the cell membrane. As a result, in part, of this propensity for direct insertion, and because the macromolecular structure of LPS complexes does not lend itself to classical ligand-receptor interactions, it has been argued that

LPS-macrophage interactions may not be mediated by specific "receptor" proteins, but rather by insertional perturbation of macrophage membranes. However, it has been useful to consider that LPS encountered by macrophages under physiologic conditions differs in significant ways from purified LPS in aqueous solutions. Cell-free LPS in plasma or interstitial fluid, or LPS bound on the surface of live bacteria, is acted on by sundry plasma proteins, and several of these represent evolutionary mechanisms to facilitate bacterial/LPS clearance and LPS-induced activation of phagocytes. Characterization of the complexes formed between LPS and plasma proteins has proven invaluable for understanding LPS-macrophage binding mechanisms. An emerging consensus is that macrophages bind free LPS relatively poorly but bind LPS–plasma protein complexes efficiently via specific "receptors."

A partial list of plasma proteins that bind LPS to form complexes includes O antigen–specific antibodies, complement, high-density lipoproteins (HDL), albumin, LPS-binding protein (LBP), and septin (for a more exhaustive review of proteins that bind LPS, see Morrison [14]). In some cases, binding of LPS with a particular protein facilitates macrophage stimulation by LPS. This is true for LBP and septin, which increase the potency of LPS approximately 100-fold [15,16]. In other cases, binding modulates macrophage-mediated LPS clearance or may even block macrophage stimulation by LPS. For example, HDL binds about half the LPS injected intravenously into experimental animals, and HDL-LPS complexes, compared with other forms of LPS, are cleared inefficiently by macrophages, exhibit reduced toxicity, and are poor stimulators of macrophages in vitro [17,18]. On the other hand, O antigen–specific antibodies provide a very effective mechanism by which bacteria or free LPS can be bound and cleared by macrophages. Binding of immunoglobulin G (IgG)-LPS to macrophages is mediated by Fcγ receptors and is increased by complement factors that associate with IgG-LPS immune complexes and bind cell surface C3b receptors. Although antibodies and complement enhance LPS delivery and clearance, LPS–immune complex binding to macrophages does not appear to induce TNF expression, and in fact, O antigen–specific antibodies may even block the ability of LPS to stimulate macrophages [19]. Thus, the ability of LPS to stimulate macrophages is a function of its relative binding to one or more of these various plasma proteins. Because the plasma levels of several of these proteins, i.e., LBP, HDL, and anti–O antigen IgG, are subject to regulation, it follows that the relative plasma levels of these proteins and the complex mass action kinetics of their binding to LPS might be major determinants of host responsiveness to LPS.

Of the two proteins that facilitate LPS-stimulation of macrophages, i.e., LBP and septin, LBP is better characterized. Lipopolysaccharide-binding protein was discovered during studies into the regulation of LPS binding to HDL. Lipopolysaccharide mixed with rabbit plasma had been shown to associate nearly quantitatively with HDL, but Tobias and Ulevitch [20] noted that the rate of association of LPS with HDL was markedly retarded in acute-phase plasma. The reduced rate of association correlated with the presence of a 60-kD protein that was rendered cold-insoluble by LPS. The protein precipitates contained large amounts of LPS, and the 60-kD protein was called LPS-binding protein. Rabbit LBP was subsequently purified [21], and the first functional role for LBP was reported by Wright et al. [22], who showed that LBP bound the surfaces of gram-negative bacteria and LPS-coated erythrocytes (LPS-E) and opsonized these particles for dramatically increased binding by human macrophages. Schumann et al. [15] subsequently reported that LBP markedly increased the potency of soluble LPS as a stimulus for macrophages. While macrophages cultured in LBP-free media required at least 100 ng/ml LPS to induce maximal TNF secretion, comparable TNF secretion was achieved when macrophages were cultured with purified LBP and as little as 1 ng/ml LPS. Thus, LBP was shown to function as both a bacterial opsonin and a cofactor that increased the efficiency of LPS stimulation of macrophages. Investigation of LBP opsonic activity led to the finding that macrophage binding of LBP-treated LPS-E could be completely blocked with monoclonal antibodies (mAbs) specific for the cellular differentiation antigen CD14 [23]. Binding was not blocked by mAbs specific for other highly expressed surface proteins. Anti-CD14 monoclonal antibodies also inhibited LPS stimulation of monocytes, which led to the hypothesis that CD14 mediates macrophage binding of LBP-opsonized bacteria and is the "receptor" for soluble LBP-LPS complexes. In this model, when LPS is present in low concentrations ($\leq$10 ng/ml), macrophages are stimulated by the sequential binding of LBP to LPS and then of the LBP-LPS complexes to CD14.

Both human and rabbit LBPs have been cloned, expressed, and sequenced, and they exhibit a 69% amino acid identity [15]. Human LBP is 44% homologous to human bactericidal/permeability-increasing protein (BPI). The latter is expressed in primary granules of neutrophils and binds and disrupts the membranes of internalized bacteria [24]. Although purified BPI *inhibits* LPS-mediated neutrophil activation [25], significant levels of BPI, have not, to date, been detected in normal or acute-phase plasma. Lipopolysaccharide-binding protein and BPI bind the lipid A region of LPS specifically [25,26]. Competition binding studies have established that LBP binds smooth LPS, rough forms of LPS, and lipid A with a K_d of approximately 1 nM [26]. The minimal structure that can be bound with high affinity is the lipid IVA precursor of lipid A, and mass action considerations have implied that LBP and lipid A bind with a 1:1 stoichiometry. O antigen and core oligosaccharides do not appear to contribute greatly to LBP binding. Nonetheless, the quaternary structure of LBP-LPS complexes remains unclear. It is yet to be determined experimentally if circulating LBP-LPS complexes are monomers or multimers. It has been reported that immunoprecipitation of LBP can coprecipitate an eightfold molar excess of LPS, suggesting that LBP can bind LPS nonstoichiometrically. Also, since LBP can induce LPS-E agglutination [22], LBP multimers may exist. Clarification of the structure of LBP-LPS complexes will help to distinguish between a one ligand–one receptor model and a multivalent crosslinking model for LPS stimulation of macrophages.

Lipopolysaccharide-binding protein is made by hepatocytes [27], and average values present in normal and acute-phase human plasma have been reported to be approximately 7 and approximately 220 μg/ml, respectively [28]. Thus, even though it is an acute-phase reactant, LBP is arguably present in sufficient concentrations in normal plasma to function as an early defense mechanism during infection. The best evidence to date of a role for LBP in normal plasma is that TNF production induced with up to 10 ng/ml LPS in heparinized whole blood is inhibited when endogenous LBP is first removed by immunoprecipitation [15]. The function of the acute-phase increase in LBP may be to facilitate distribution of opsonin to extravascular sites of infection, where concentrations of LPS may be very high.

Wright et al. [16] recently reported on a novel, incompletely characterized group of plasma proteins called septins that, like LBP, appear to bind LPS and present it to phagocytes. The existence of septins was suggested by the impressive level of opsonic activity in normal human plasma, which could not be attributed to endogenous LBP, since opsonization was not significantly reduced in the presence of a neutralizing anti-LBP antibody. Chromatographic separation of plasma resulted in a loss of opsonic activity unless selected fractions were recombined. Opsonic activity in plasma could be blocked with serine protease inhibitors, but only in the presence of LPS. Therefore, a two-protein model was proposed in which deposition of "septin 1" and "septin 2" on the surface of LPS-coated particles activates a latent protease activity required for opsonization. It was suggested that septins might be analogous to the LPS-activated serine protease clotting cascade of *Limulus* amebocytes. Human plasma was also remarkably potent (half-maximal dilution was 1/3000) at rendering macrophages and neutrophils responsive to low concentrations of LPS. Because this plasma function could also be inhibited by serine protease inhibitors, septins may play a role in the presentation of "free" LPS to macrophages. Anti-CD14 monoclonal antibody blocked plasma opsonic activity, suggesting that macrophage binding of septin-LPS complexes, analogous to LBP-LPS complexes, is mediated by CD14.

The relative contributions of LBP and septins as LPS-binding proteins in normal and acute-phase plasma remains to be delineated, and the possibility that binding of septins and LBP is cooperative has not yet been explored. As noted earlier, other serum proteins also bind LPS, and each uniquely affects the manner in which LPS is presented to macrophages. Thus, the bioavailability of LPS in plasma is a function of several potentially competing processes. In this regard it is also important to consider that LPS (released from bacteria) exists in tight association with bacterial outer membrane proteins [29]. Certain of these proteins are themselves potent macrophage stimuli. It remains to be determined what role these proteins may play in modulating LPS interactions with LBP, CD14, and other plasma or macrophage proteins.

B. Role of CD14 as a Macrophage Receptor for Lipopolysaccharide-Binding Protein Complexes

CD14 was originally identified as a surface antigen highly expressed on monocytes and macrophages and weakly expressed on granulocytes [30]. Human and mouse CD14 have been cloned and sequenced and do not exhibit large regions of homology with other known proteins [31]. CD14 lacks a transmembrane region and cytoplasmic tail; it is attached to the extracellular membrane surface via a glycosylphosphatidylinositol linkage [32]. Soluble CD14 can be secreted from cells and can be released from cell surfaces through the action of a specific phospholipase.

Before the finding that anti-CD14 mAb blocks macrophage binding of opsonized LPS-E (see above), no function for CD14 was established. A variety of approaches have provided more direct evidence that CD14 mediates LPS binding. Saturable and displaceable binding of ^{3}H-LPS or FITC-LPS to human monocytes in the presence of serum or purified LBP can be completely inhibited by anti-CD14 mAb [33,34]. Monocytes enzymatically stripped of CD14 by treatment with glycosylphosphatidylinositol-specific phospholipase C no longer bound LPS [34], and a subclone of J774 macrophages has been derived that is hyporesponsive to LPS, is specifically deficient in CD14 expression, and fails to bind LPS [35]. Finally, LPS-crosslinking studies have provided direct evidence that CD14 binds LBP-LPS complexes. A bioactive, radioiodinated, photoreactive derivative of LPS (^{125}I-ASD-LPS) can be used to crosslink proteins that bind LPS [36]. When a low concentration of ^{125}I-ASD-LPS (5 ng/ml) was incubated with LBP and the THP-1 monocytic cell line, LBP and CD14 were the only major proteins crosslinked as determined by SDS-PAGE of cell lysates and autoradiography [37]. ^{125}I-ASD-LPS did not label CD14 efficiently in the absence of LBP, providing evidence that free LPS is, at best, a poor ligand for CD14. Free LPS probably does bind CD14 (albeit weakly), since monocyte stimulation by high concentrations of LPS in the absence of plasma or LBP can be inhibited by anti-CD14 [34]. A functional role for CD14-mediated LPS binding, originally suggested by the ability of anti-CD14 mAb to inhibit LPS-induced TNF synthesis, has been elegantly demonstrated by Lee et al. [38]. The murine 70/3Z B cell line expresses no detectable CD14, yet it activates NF-κB and IgG secretion when cultured in the presence of high concentrations ($\geq$100 ng/ml) of LPS. Transfection of 70/3Z cells with human CD14 resulted in a 1000-fold increase in their sensitivity to LPS. The full increase in sensitivity of the transfected cells required LBP and could be reversed with anti-CD14 mAb. Thus binding of LBP-LPS to CD14 represents a major pathway that culminates in LPS signaling.

C. Other Candidate Lipopolysaccharide "Receptors"

Because CD14 lacks transmembrane and cytoplasmic regions, it has been argued that CD14 may not be the true signaling receptor for LPS, but that CD14 may focus delivery of LPS to an as yet unidentified signaling receptor. Alternatively, there has been speculation that CD14 may be one subunit of a multisubunit LPS receptor complex analogous to the IL-2 receptor, where two receptor subunits contribute to a single high-affinity binding site, or analogous to the IL-6 receptor, where the binding of IL-6 to the IL-6 receptor binding subunit activates a signaling subunit [38]. The most compelling evidence for LPS signaling receptors distinct from CD14 is that many cell types that do not express CD14, including lymphocytes, NK cells, fibroblasts, and endothelial cells, are nonetheless responsive to relatively high concentrations of LPS. Noteworthy is a recent report that soluble CD14, present in normal human serum, was essential for endothelial cell responsiveness to LPS. This finding seems to obviate an absolute requirement for target cells to express CD14 [39].

A number of candidate LPS receptor proteins have been identified when target cells are treated with ^{125}I-ASD-LPS in the absence of plasma or serum; several membrane proteins are labeled with ^{125}I-ASD-LPS in a saturable and lipid A–displaceable manner. An approximately 73-kDa protein described by Lei and Morrison [40] is the major protein labeled on the surfaces of murine [40] and human [41] lymphoid and myeloid cells. Several other lower-molecular-weight proteins are also labeled to a lesser extent. Using a similar approach, Kirkland et al. [42] identified an 18-kDa LPS-binding protein on the surface of 70Z/3 murine B cells. A mAb against the 73-kDa LPS-binding protein has been characterized that inhibits LPS binding to the 73-kDa protein [43]. Binding of this antibody to LPS-responsive, but not to LPS-hyporesponsive (C3H/HeJ), macrophages delivers an LPS-like signal as assessed by induction of TNF production, tumoricidal activity [44], and NO synthesis [45], although the expression of this putative LPS receptor appears to be equivalent for the two mouse strains [46]. Thus, the 73-kDa protein appears to bind LPS and to transduce an LPS-like signal. Finally, this same mAb was shown to protect animals in vivo from lethal LPS challenge [47]. Definitive proof of LPS receptor function awaits cloning and transfection studies.

In addition to CD14 and the 73-kDa protein, other LPS-binding mechanisms exist that may not contribute to LPS-signaling but may facilitate bacterial and LPS clearance. *Escherichia coli* are bound weakly by macrophages in the absence of plasma by interactions between LPS and adhesion-promoting receptors (CR3, LFA-1, and p150,95) that share CD18 as a common β-subunit [48]. The scavenger receptor has been implicated in the binding, internalization, and detoxification of lipid A [49]. Nonetheless, neither scavenger receptor blockade nor genetic deficiency in CD18 expression affects the ability of monocytes to respond to LPS with TNF production. Kupffer cells and other macrophages express a variety of surface proteins that bind specific sugars, e.g., mannose and heptose receptors, and these structures may contribute to macrophage binding of O antigen and core oligosaccharides. A 48-kDa inner core–specific receptor that binds heptose-KDO oligosaccharides has been described [50]. A role for oligosaccharide receptors in LPS signaling has been viewed as unlikely because of the poor stimulatory activity of purified oligosaccharides and the appreciation that most, if not all, of the activity of LPS resides in the lipid A moiety. Nonetheless, oligosaccharide receptors may contribute to the superior potency of smooth and rough chemotypes of LPS relative to lipid A by stabilizing the overall interaction of oligosaccharide-containing LPS structures with macrophages.

Perhaps one of the most intriguing observations relating to the capacity of LPS to bind to macrophages via specific receptors resides in the discovery of an unusual lipid A, derived from *Rhodopseudomonas* (*Rhodobacter*) *sphaeroides* [51]. This lipid A species is pentaacylated, with relatively shorter hydroxydecanoic acids, rather than the typical hydroxytetradecanoic acids seen in lipid A derived from *Enterobacteriaceae*, in addition to several other structural differences. This unusual lipid A appears to bind to LPS receptors on macrophages and has recently been shown to act as a biologically inert antagonist of LPS- and lipid-A–stimulated macrophage TNF secretion [51,52]. In addition to the lipid A of *R. sphaeroides*, the lipid A precursors, lipid IVA and lipid X, and several lipid X analogues have been shown to block LPS-mediate effects in vitro and in vivo [53,54]. Thus, one approach to mitigating LPS-induced toxicity is to develop structural analogues of lipid A that are able to bind, but not signal, through lipid A–specific receptors.

III. LIPOPOLYSACCHARIDE-INDUCED SIGNAL TRANSDUCTION AND REGULATION OF GENE EXPRESSION

Following the interaction of LPS-protein complexes or free LPS with putative receptors in the macrophage membrane, intracellular signaling pathways that lead to gene expression and activated macrophage function are induced. These pathways are still relatively ill defined. There are some data to support involvement of pertussis toxin–sensitive G proteins in LPS stimulation of 70/Z3 pre-B cells and monocyte-like tumor lines [55] and for phosphatidylinositol turnover and release of Ca^{2+} from intracellular stores as early events in LPS stimulation of primary macrophages [56,57]. In any event, the currently available data are consistent with multiple divergent, but crosstalking, pathways. For example, Ohmori and Hamilton [58] have recently reported that Ca^{2+} and calmodulin antagonists differentially modulate three LPS-inducible genes. These agents strongly suppress the ability of LPS to induce IL-1β mRNA, weakly suppress LPS-induced IL-1α mRNA, and increase LPS-induced IP-10 mRNA. These data are consistent with the hypothesis that at least three signaling pathways leading to gene induction are induced by LPS, i.e., pathways that are Ca^{2+}-dependent, relatively Ca^{2+}-independent, and negatively regulated by Ca^{2+}.

A. Lipopolysaccharide-Induced Protein Phosphorylation

Perhaps easier to document than second messenger and G-protein induction are the post-translational protein modifications that rapidly follow LPS stimulation. There are reports of LPS-induced protein myristoylation [59], methylation [60], and phosphorylation [61–65]. When cells are pulsed with ^{32}P-ATP and stimulated briefly with LPS, several proteins incorporate increased levels of phosphate as determined by SDS-PAGE of lysates and autoradiography. The identity and function of these substrates are not known. One substrate, a 65-kDa protein that represents as much as 0.2% of the protein in macrophage lysates, has been purified and appears to be novel [62]. Lipopolysaccharide induces phosphorylation of this protein on a serine residue in genetically LPS-responsive macrophages only, and sequencing and identification of its function will provide additional insights into mechanisms of LPS-induced macrophage activation. Weinstein et al. [63] recently used Western blot analysis to show that LPS induces a rapid (5-minute) tyrosine phosphorylation of at least three proteins between 41 and 44 kDa. One protein (41 kDa) has been identified as mitogen-activated protein kinase (MAP kinase), a serine/threonine protein kinase that undergoes phosphorylation in a variety of cell types following mitogenic stimulation [64]. Mitogen-activated protein kinase is activated by tyrosine phosphorylation and is believed to play a role in cell proliferation [66], but its role in macrophage activation is unknown. Herbimycin inhibited LPS-induced tyrosine phosphorylation and arachidonic acid release, suggesting that the action of MAP kinase (or perhaps one of the other tyrosine phosphoproteins) may be required for LPS-induced arachidonic acid metabolism, possibly through an interaction of the kinase cascade with the operative phospholipases. Recent data obtained with taxol, an anticancer drug that acts by stabilizing microtubules, has supported a role for tyrosine phosphorylation of MAP kinase in converging signaling pathways leading to LPS-like stimulation of macrophages. Ding et al. [67] originally reported that taxol could induce macrophage TNF expression and suggested that taxol and LPS may share a common signaling intermediate because taxol was unable to stimulate C3H/HeJ macrophages. Subsequently, we have shown that the LPS-mimetic effects of taxol extend to the ability of this drug to induce expression of LPS-inducible genes as well as tyrosine phosphorylation of MAP kinases and other macrophage proteins [65].

LPS stimulation of macrophages most likely involves the serial activation of a complex, interactive network of kinases. The identity of the LPS-activated tyrosine kinase that phosphorylates MAP kinase is not known. In other cell systems, MAP kinase can be phosphorylated by members of the sarc family of kinases, which in turn may be regulated by receptor tyrosine kinases or protein kinase C (PKC) [68,69]. A major role for PKC in LPS macrophage activation remains to be clearly shown. Protein kinase C activation by phorbol myristic acid (PMA) induces an overlapping but discrete set of phosphoproteins when compared to phosphoproteins induced by LPS [61], and although LPS-induced TNF-α gene expression can be inhibited by PKC inhibitors [70], other studies have shown that LPS and PMA have somewhat divergent actions on macrophages. For example, PMA, but not LPS, induces a large macrophage respiratory burst [71]. Moreover, LPS-induced MAP kinase phosphorylation is not blocked by PKC inhibitors [64], and delineation of the kinase cascade(s) initiated by LPS will be crucial to unraveling the mechanisms by which LPS mediates its many effects on macrophages.

B. Gene Regulation

Among the most notable manifestations of LPS stimulation of macrophages is the induced expression of genes that encode for an impressive variety of structural and regulatory proteins. Best understood is LPS-induction of TNF-α. As discussed by Han et al. [72], LPS induces a 10,000-fold increase in TNF-α protein synthesis while inducing a relatively modest increase in the rate of transcription of the TNF-α gene (threefold) and in steady-state levels of TNF-α mRNA (100-fold). These findings indicate that LPS-induced TNF-α secretion is only partially controlled at the level of gene transcription and mRNA stability. Beutler and colleagues [72] have shown that in unstimulated macrophages, translation of TNF-α mRNA is repressed by an unknown mechanism dependent on the TTATTTAT element present in the 3'-untranslated region of TNF-α mRNA. Treatment of macrophages with LPS somehow releases the mRNA from this repressor activity, and the mRNA is translated with markedly greater efficiency. The frequency with which the TTATTTAT element is found in the 3' region of other LPS-induced genes suggests that translational "derepression" may represent a more general mechanism, not limited to the control of TNF synthesis. In addition to its repressor function, the TTATTTAT element also appears to target mRNA for degradation [73], and LPS appears to increase the steady-state mRNA levels of select genes, in part, by transiently increasing mRNA stability [74]. More work is required to identify proteins involved in LPS-induced derepression and mRNA stability, and the possible influence of phosphorylation, or other post-translational protein modifications.

Most LPS-inducible genes, including TNF-α, are in part induced by transcriptional activation, and subcloning of the 5'-regulatory region has led to identification of LPS-responsive elements [75,76]. Distributed within the 5'-untranslated region of the TNF-α gene are 4 NK-κB or NK-κB–like enhancer elements and a Y box [75]. Subcloning of the Y box or two or more of the NF-κB enhancers onto a CAT reporter gene renders the reporter gene LPS-responsive in transfected bone marrow–derived macrophages. Furthermore, treatment of the macrophages with LPS results in nuclear protein binding to the NF-κB enhancers and Y box by activation of an NF-κB–like protein or proteins and an uncharacterized protein or proteins, respectively. The mechanism of LPS activation of NF-κB is not known. NF-κB exists in macrophages and other cell types as a heterodimer (50 and 65 kDa) that is bound by the cytosolic protein, IκB [77]. Upon cell activation, IκB is phosphorylated, and the NF-κB heterodimer dissociates from IκB and translocates to the nucleus, where it binds upstream of specific genes. IκB can be phosphorylated by PKC [77], but the putative protein kinase operative in LPS-treated cells is not known. Recently, a role for reactive oxygen intermediates (ROIs) as mediators of NF-κB activation was proposed because H_2O_2 was able to activate latent NF-κB, and ROI scavengers were able to inhibit NF-κB activation in 70Z/3 murine B cells exposed to a variety of stimulants including LPS [78].

Although a growing number of LPS-inducible genes (IL-1β, IFN-β, IL-6, tissue factor) have been shown to encode 5'-untranslated NF-κB enhancers and to require these enhancer regions for LPS-inducibility, single NF-κB enhancers are insufficient to confer LPS inducibility, and other enhancer elements have been shown to contribute to LPS inducibility. In particular, the 5'-regulatory region of the tissue factor gene encodes two AP-1 enhancer regions as well as two NF-κB enhancers, and the coordinate activity of both enhancer types is required for optimal induction by LPS [76]. AP-1 enhancers are bound by homodimers of the protein encoded by the *c-jun* proto-oncogene [reviewed in 79]. Additionally, Jun can be bound by Fos and the heterodimer binds AP-1 sites with increased affinity. The *c-fos* gene is itself induced markedly in macrophages treated with LPS for as little as 15 minutes [80]. DNA binding of the Jun/Fos heterodimer is regulated by the cytoplasmic and/or nuclear protein IP-1 [81]. In resting cells, IP-1 is not phosphorylated, and it appears to bind either Jun or Fos protein and prevent their binding to DNA.

Upon treatment of cells with PMA, both IP-1 and the heterodimer are phosphorylated. Phosphorylated IP-1 is no longer able to inhibit DNA binding, and phosphorylation of the free heterodimer facilitates its binding to the AP-1 enhancer region. Interestingly, PKA, but not PKC, can phosphorylate IP-1 in vitro; the kinases involved in PMA- or LPS-induced AP-1 activation are not known.

By mechanisms that are currently unclear, LPS induces the production of prostaglandin E_2 (PGE_2), which activates its own receptor on the surface of macrophages and induces a slow rise in intracellular cAMP [82]. Genes such as TNF-α and IL-1β, whose induction is in part mediated by NF-κB or Jun/Fos, are induced rapidly and transiently. Expression of these genes can be regulated bidirectionally by cAMP, e.g., LPS-induced TNF-α and IL-1β mRNA are down- and up-regulated by cAMP, respectively [83,84]. Other genes or gene products are induced more slowly by LPS. A distinct set of these genes appear to be induced by cAMP-*dependent* mechanisms. For example, Wahl and colleagues [reviewed in 85] have reported that LPS induction of macrophage collagenase and ornithine decarboxylase (ODC) can be inhibited by indomethacin, an inhibitor of cyclooxygenase, a key enzyme in the synthesis of PGE_2. Collagenase production by indomethacin-treated cells can be restored with PGE_2 or dibutyryl cAMP in combination with LPS. In contrast, ODC expression by macrophages can be induced directly with PGE_2 or dibutyryl cAMP alone. Interestingly, collagenase synthesis can be blocked by specific inhibitors of ODC, and this inhibition can be overridden by putrescine, an enzymatic product of ODC thought to function as a regulator of RNA polymerase 1 activity. Thus, LPS-induced collagenase synthesis is mediated by the coordinate activity of an undefined LPS-induced pathway and a pathway involving the tandem induction of PGE_2, cAMP, ornithine decarboxylase, and putrescine.

IV. LIPOPOLYSACCHARIDE-INDUCED CHANGES IN MACROPHAGE FUNCTIONAL STATES: RELEVANCE TO HOST DEFENSE AGAINST BACTERIAL INFECTION

There is good evidence based on murine genetics and studies of germ-free animals that macrophage recognition of LPS is critical to mounting an efficient defense against gram-negative bacterial infections. For example, genetically LPS-hyporesponsive C3H/HeJ (*Lps*^{d}) mice, relative to their normally LPS-responsive, syngeneic C3H (*Lps*^{n}) counterparts, are unable to control adequately infection with *Salmonella typhimurium* or other gram-negative organisms, but are able to control gram-positive infection normally [86]. Macrophages mediate the primary response to LPS, and macrophages encountering LPS acquire new functional capabilities that act in a coordinated fashion to destroy infectious agents by two broad mechanisms: 1) sequestration of the infectious agent and 2) direct killing.

A. Lipopolysaccharide-Stimulated Macrophages Secrete Inflammatory Cytokines That Aid in the Sequestration of Infection

LPS stimulates macrophages to secrete a large array of inflammatory cytokines, e.g., TNF-α, IL-1, IL-6, IL-8, IL-10, IFN-α/β, TGF-β, MIP-1, MIP-2, CSF, and others. These peptides, by communicating with cells in the infected tissue, induce an inflammatory response that ultimately leads to the sequestration and destruction of bacteria. Excellent reviews on how cytokines mediate tissue inflammation are available [87]. In brief, certain cytokines recruit phagocytes into the infected site by inducing intercellular and matrix adhesion-promoting molecules on the surfaces of neutrophils, monocytes, and vascular endothelial cells. Circulating neutrophils and monocytes adhere to the endothelium of capillaries that irrigate the site of infection, and a host of chemotactic factors diffuse from the infected foci and induce the phagocytes to diapedese through the capillary endothelium and matrix material to the bacteria in the tissue. Several chemotactic factors, e.g., MIP-1, MIP-2, IL-8, and TGF-β are produced directly by LPS-stimulated macrophages or, in some cases, indirectly by TNFα/IL-1–stimulated endothelial cells. Migration of monocytes/macrophages through basement membrane and extracellular matrix is facilitated by plasminogen activator and neutral proteases, such as collagenase [85], whose secretion is induced by LPS and cytokines. Recruited neutrophils and monocytes engulf and kill the bacteria (see below). Moreover, LPS activates CR3 function on neutrophils, resulting in increased capacity for uptake of opsonized organisms [88]. Sequestration of the infection is greatly aided by LPS- and cytokine-induced expression of tissue factor on the surface of monocytes/macrophages and endothelial cells [76]. Tissue factor activates the tissue clotting cascade, and microthrombi occlude the surrounding vasculature, limiting dissemination of the infectious agent.

B. Lipopolysaccharide-Stimulated Macrophages Efficiently Kill Bacteria

Macrophages and neutrophils are specially equipped to engulf and kill bacteria. However, it has long been recognized that macrophages isolated from healthy, pathogen-free animals are "quiescent" and kill microbes relatively poorly, whereas macrophages from animals preadministered a sublethal dose of pathogen vigorously kill the same or even unrelated microbes in vitro [89]. The latter macrophages are said to have acquired an "activated" phenotype. Although activation of cytotoxic function can be induced by lymphokines, purified LPS can also substitute for whole bacteria as an inducer of this macrophage phenotype in vivo and in vitro [71]. More recently, it has been demonstrated that LPS-induced monokines can induce macrophage antimicrobial activity [90]. In fact, several macrophage-derived cytokines (e.g., TNF-α, IL-1, IFN-β, and GM-CSF) have been implicated as mediators of increased resistance to infection [5].

The biochemical determinants of the increased ability of LPS-stimulated macrophages to kill microbes are still poorly understood, but they clearly relate to the readiness with which macrophages can be triggered by a second stimulus to synthesize ROIs [89]. The second stimulus for ROI release is generally phagocytosis of opsonized bacteria, but bacterial peptides or phorbol esters provide convenient second stimuli for in vitro studies. Reactive oxygen intermediates are a principal set of mediators by which macrophages kill bacteria and tumor cells (see Chapter 13). Synthesis of ROIs is initiated when NADPH oxidase catalyzes the reduction of molecular oxygen to superoxide anion (O_2^-) and the concomitant oxidation of NADPH. Superoxide anion is subsequently converted to hydrogen peroxide, which can be converted to hydroxyl ions. The latter two products are potent toxins for many microbes, but mammalian cells are protected by glutathione and the expression of peroxidase. Quiescent macrophages treated in vitro with FMLP or PMA secrete only small amounts of superoxide anion. Macrophages harvested from animals after bacterial or LPS challenge, or quiescent macrophages treated in vitro with LPS, secrete copius amounts of O_2^- when stimulated with FMLP or PMA [71]. Lipopolysaccharide-stimulated macrophages secrete negligible amounts of O_2^- if the second stimulus is not presented. Thus, LPS-stimulated macrophages are said to be "primed" for O_2^- release. Priming by LPS is relatively rapid; enhancement of O_2^- release can be measured after only 30 minutes of exposure. The primed state is induced maximally by 4 to 12 hours and can be maintained for several days in the continuous presence of LPS [71]. Neutrophils are also subject to LPS-induced priming for O_2^- release. Lipopolysaccharide primes macrophages, in part, by increasing the affinity of the NADPH oxidase for NADPH [91]. Further studies into the ability of LPS to regulate the synthesis or function of the various components of the oxidase may provide insights into important mechanisms of priming for O_2^- and for antimicrobial activity.

Potentially relevant are studies reported by Aderem and coworkers [59,92]. They have shown that the synthesis and myristoylation of a 68-kDa protein called myristoylated, alanine-rich C kinase substrate (MARCKS) is induced within minutes of stimulation of either macrophages or neutrophils with LPS. Myristoylation targets MARCKS to the membrane, where it is phosphorylated more rapidly and to a much greater extent upon treatment of the cells with FMLP/PMA.

These events occur over the same time frame as priming, and it has been proposed that LPS induction of MARCKS synthesis and myristoylation may mediate the ability of LPS to increase responsiveness to agonists whose actions are mediated through PKC. MARCKS is normally associated with the cellular membrane, but upon phosphorylation it is released to the cytosol, where it binds actin [92]. It is believed that the interchangeable phosphorylated states of MARCKS mediate dynamic interactions between actin and the cell surface.

Complementing the ability of LPS to prime macrophages for O_2^- is the ability of LPS to serve as a *second* signal for the synthesis of a second bacteriacidal factor, nitric oxide (NO). Nitric oxide is the product of nitric oxide synthase, an FAD- and FMN-binding enzyme that catalyzes the reductive release of NO from arginine [93]. Nitric oxide is highly reactive with sulfhydryl groups and together with O_2^- mediates much of the bacterial (and tumor cell) killing by macrophages (see Chapter 14). Interferon-γ is the only known factor that singly induces the production of NO by macrophages. Lipopolysaccharide induces NO only poorly but markedly potentiates the ability of IFN-γ–treated macrophages to synthesize NO [94]. Tumor necrosis factors-α/β is also able to potentiate NO production, and LPS induced TNF-α apparently contributes to LPS induced NO synthesis, since anti-TNF antibodies partially block LPS-induced NO [45]. Nitric oxide release by IFN-γ plus LPS–stimulated macrophages requires several hours before detectable levels are induced, and NO release remains linear for several days provided the viability of the culture is maintained. Unstimulated macrophages contain very little nitric oxide synthase, and NO synthesis is regulated by the transcriptional induction of the nitric oxide synthase gene and the subsequent de novo synthesis of the enzyme [93].

Thus, one consequence of LPS-macrophage interaction is the ability to convert a resting macrophage to one with greatly augmented capacity for destruction of invading bacteria via the production of various cytokines (which act in a paracrine and autocrine manner to increase resistance to infection), and to arm the macrophage with enzymes and ROIs and NO, capable of acting directly on or in conjunction with other enzyme systems (e.g., myeloperoxidase) to limit bacterial growth.

C. Lipopolysaccharide Alters Macrophage Responsiveness to Inflammatory Mediators and to Lipopolysaccharide

Once a macrophage has encountered LPS, its ability to respond to additional stimuli is often altered. Lipopolysaccharide-induced changes in responsiveness can be mediated at the level of receptor expression or by feedback loops that affect proteins involved in signaling or effector function. Changes in responsiveness may in some cases ensure appropriate resolution of an inflammatory response and in other cases render macrophages more competent to respond to additional stimuli within the inflamed tissue. For example, as discussed above, LPS primes macrophages for enhanced responsiveness to agents such as bacterial peptides and opsonized bacteria that activate PKC, and this priming may be mediated by the synthesis and myristoylation of MARCKS, a major PKC substrate. In another example, LPS enhances responsiveness to the lymphokine IL-2. Lipopolysaccharide induces a marked increase in IL-2 receptors on monocytes and macrophages, and IL-2 can in turn increase macrophage microbial and tumor cell killing [95]. Lipopolysaccharide modulation of receptor expression is, however, bidirectional. Lipopolysaccharide induces the rapid internalization or down-regulation of surface receptors for TNF-α [67], TGF-β [96], M-CSF [97], and C5a [98], and renders macrophages correspondingly less responsive to these agonists.

Lipopolysaccharide modulates LPS-responsiveness in a bidirectional manner that is time- and concentration-dependent. We have consistently observed [54] that murine elicited macrophages cultured for 19 hours with low concentrations of LPS ($\leq$100 pg/ml) are rendered hyperresponsive to a subsequent challenge of 10 ng/ml LPS assessed by TNF-α mRNA induction. In contrast, we and others [99] have found that higher concentrations of LPS ($\geq$1 ng/ml) render macrophages unable to induce TNF-α mRNA fully to subsequent challenge with LPS. This refractory state is specific, i.e., LPS-refractory cells respond normally to other inducers of TNF-α, such as heat-killed *Staphylococcus aureus*, but the mechanism underlying acquired hyporesponsiveness is unknown. Surface expression of the 73-kDa putative LPS receptor is not reduced by exposure to LPS [46], although about half of cell surface CD14 is shed from monocytes within 3 hours of LPS exposure [100]. It remains to be determined if CD14 shedding contributes significantly to acquired hyporesponsiveness. It is noteworthy that LPS administered to animals induces an acute tolerance to subsequent injections of LPS, and this tolerance has been correlated with an inability of macrophages to respond to the second dose of LPS ex vivo [reviewed in 101]. Host responsiveness to LPS can be also modulated bidirectionally by cytokines. Limited infection or IFN-γ generally sensitizes an animal to the effects of administered LPS [102], whereas TNF-α and IL-1α synergize to induce acquired tolerance [103].

V. THERAPEUTIC INTERVENTION WITH LIPOPOLYSACCHARIDE MIMETICS

Because many of the biological actions of low doses of LPS are favorable, e.g., increased resistance to infection and malignancy, and to radiation, and because humans are especially susceptible to the toxic actions of LPS, analogues and mimetics of lipid A that induce the beneficial effects of LPS without its toxicity have been sought. In 1984, Ribi [104] demonstrated that the removal from Lipid A of the 1'-phosphoryl group on the reducing terminal glucosamine resulted in an analogue called monophosphoryl lipid A (MPL) that is significantly less toxic than LPS or lipid A, yet retains the ability to act as an adjuvant, to increase nonspecific resistance to infection and malignancy, and to protect against lethal irradiation. In mice, pretreatment with MPL decreased toxicity and increased survival of mice from intraperitoneal bacterial challenge with 1–2 $\times$ 10^8 *E. coli* [105]. Similarly, Stütz et al. [54] have synthesized and tested a large number of mono- and disaccharide lipid A partial structures, as potential candidates for therapeutically useful, nontoxic LPS mimetics. Most favorable was a triacylated monosaccharide analogue of the reducing sugar of lipid A called SDZ MRL 953. When given to mice intraperitoneally, SDZ MRL 953 induced resistance to infection to a variety of microorganisms. It was effective at doses $\leq$2 mg/kg, was not lethal to galactosamine-sensitized mice even at a dose of 125 mg/kg, and exhibited a therapeutic index as much as 10^4 times greater than that of LPS. At therapeutic doses, MPL and SDZ MRL 953 readily induce serum CSF activity but are poor inducers of IL-1 and TNFα [106–108]. MPL and SDZ MRL 953 are currently in clinical trials as stimulators of myelopoeisis in patients undergoing chemotherapy [54].

VI. THERAPEUTIC INTERVENTIONS TO GRAM-NEGATIVE SEPTIC SHOCK

As mentioned in the Introduction, the interaction of LPS with macrophages can lead to septic shock, a condition that represents the most extreme result of this primary interaction. If the invading gram-negative pathogen is not contained early in infection, massive amounts of LPS are released by dead and dying bacteria. The result is that sufficient LPS and/or bacteria circulate in the blood to induce a systemic cytokine cascade that spirals out of control unless blocked in some manner. Pathologic manifestations of this cascade include cytokine-mediated loss of vascular tension and inappropriate and destructive inflammatory exudation of peripheral tissues. Several therapeutic approaches are being considered to control these life-threatening sequelae. One successful approach recently reported in septic patients has been the administration of an IL-1 receptor antagonist to block the cascade and the resultant inflammatory avalanche [11,12]. At first this may seem surprising, since IL-1 is only one of several cytokines implicated in LPS-induced shock and inflammation. However, the ability of an antagonist for a single cytokine, e.g., IL-1, to reduce mortality underscores two fundamental aspects of the LPS-induced cytokine cascade: 1) full expression of the multicytokine cascade is dependent on the presence of several key cytokines such as IL-1 and TNF that support their own induction and the induction of other cytokines, and 2) full expression of toxic effects requires the synergistic activity of several cytokines. Indeed, blockade of the actions of single cytokines such as TNF-α, IL-1, IL-6, or IFN-γ with specific antibodies, antireceptor antibodies, or antagonists has been sufficient in animal models to defuse the full toxic effects of the LPS-induced cytokine cascade [reviewed in 7], and further development of anticytokine reagents for use in septic shock is ongoing.

Three other therapeutic approaches target either LPS itself or cellular responsiveness to LPS. As noted above, host responsiveness to LPS can be attenuated by prior administration of LPS. Lipopolysaccharide tolerance can also be induced with the nontoxic LPS mimetics described above, MPL and SDZ MRL 953. Clinical trials in which MPL is being used to mitigate the effects of gram-negative sepsis are currently under way [109]. Recently, a human monoclonal anti–lipid A IgM has been used successfully to reduce motality in septic patients [110]. Although incompletely understood, this approach may work by opsonizing bacteria for phagocytosis and thereby removing the infectious agent and free and complexed LPS, as well as the LPS source. Finally, several putative LPS receptor antagonists, i.e., *R. sphaeroides* lipid A, lipid IVA, lipid X, and lipid X analogues, have recently been isolated and/or synthesized that are effective at blocking LPS-induced TNF expression in vitro [51–54]. Furthermore, *R. sphaeroides* lipid A and lipid X are effective at blocking LPS-induced TNF and mortality in mice [51,111]. These studies encourage the further development of more potent antagonists with pharmacokinetic properties favorable for clinical use.

Currently, even with the use of anti–lipid A or the IL-1 receptor antagonists, mortality rates remain at or above 30%, and perseverance is needed to learn how and when to use the therapeutic interventions in hand. There is cause for optimism, however, that as new information is available regarding LPS receptor(s), signaling mechanisms, and the cytokine cascade, it will be possible to identify new and more effective approaches to treating gram-negative sepsis.

ACKNOWLEDGMENTS

This work was supported by USUHS Protocol #R07338 (SV) and NIH #AI-18797 (SV) and AI-08451 (CM). The opinions or assertions contained within are the private views of the authors and should not be construed as official or necessarily reflecting the views of the Uniformed Services University of the Health Sciences or the Department of Defense.

REFERENCES

1. Centers for Disease Control. Increase in national hospital discharge survey rates for septicemia—United States, 1979–1987. Morbidity and Mortality Weekly Report 1990; 39:31–4.
2. Rokke O, Revhaug A, Osterud B, Giercksky KE. Increased plasma levels of endotoxin and corresponding changes in circulatory performance in a porcine sepsis model: the effect of antibiotic administration. In: Levin J, Büller HR, ten Cate JW, van Deventer SJH, Sturk A, eds. Bacterial endotoxins: pathophysiological effects, clinical significance, and pharmacological control. New York: Alan R. Liss, 1988:247–62.
3. Vogel SN. The *Lps* gene: insights into the genetic and molecular basis of LPS responsiveness and macrophage differentiation. In: Beutler B, ed. Tumor necrosis factors: the molecules and their emerging role in medicine. New York: Raven Press, 1992:485–523.
4. Manthey CL, Vogel SN. The role of cytokines in host responses to endotoxin. Revs Med Microbiol 1992; 3:72–9.

5. Vogel SN, Hogan MM. Role of cytokines in endotoxin-mediated host responses. In: Oppenheim JJ, Shevach ER, eds. Immunophysiology. New York: Oxford University Press, 1990:238–58.
6. Rietschel ET, Brade L, Holst O, Kulshin VA, Lindner B, Moran AP, Schade UF, Zähringer U, Brade H. Molecular structure of bacterial endotoxin in relation to bioactivity. In: Nowotny A, Spitzer JJ, Ziegler EJ, eds. Endotoxin research series, volume 1. cellular and molecular aspects of endotoxin reactions. New York: Excerpta Medica, 1990:15–32.
7. Michalek SM, Moore RN, McGhee JR, Rosenstreich DL, Mergenhagen SE. The primary role of lymphoreticular cells in the mediation of host responses to bacterial endotoxins. J Infect Dis 1980; 141:55–63.
8. Freudenberg MA, Keppler D, Galanos C. Requirement for lipopolysaccharide-responsive macrophages in galactosamine-induced sensitivity to endotoxin. Infect Immun 1986; 51:891–5.
9. Beutler B, Milsark W, Cerami AC. Passive immunization against cachectin/tumor necrosis factor protects mice from lethal effects of endotoxin. Science 1985; 229:869–71.
10. Ohlsson K, Bjork P, Bergenfeldt M, Hageman R, Thompson RC. Interleukin 1 receptor antagonist reduces mortality in endotoxin shock. Nature 1985; 348:550–2.
11. Alexander HR, Doherty GM, Buresh CM, Venzon DJ, Norton JA. A recombinant human receptor antagonist to interleukin 1 improves survival after lethal endotoxemia to mice. J Exp Med 1991; 173:1029–32.
12. Lemaire G, Tenu J-P, Petit J-F, Lederer E. Effects of microbially derived products on mononuclear phagocytes. In: Hadden JW, Szentivanyi A, eds. The reticuloendothelial system: a comprehensive treatise vol. 8. Pharmacology. New York: Plenum Press, 1985:181–245.
13. Seydel U, Brandenburg K. Conformations of endotoxin and their relationship to biological activity. In: Nowotny A, Spitzer JJ, Ziegler EJ, eds. Endotoxin research series, volume 1. Cellular and molecular aspects of endotoxin reactions. New York: Excerpta Medica, 1990:61–71.
14. Morrison DC. Diversity of mammalian macromolecules which bind to bacterial lipopolysaccharides. In: Nowotny A, Spitzer JJ, Ziegler EJ, eds. Endotoxin research series, volume 1. Cellular and molecular aspects of endotoxin reactions. New York: Excerpta Medica, 1990:183–9.
15. Schumann RR, Leong SR, Flaggs GW, Gray PW, Wright SD, Mathison JC, Tobias PS, Ulevitch RJ. Structure and function of lipopolysaccharide binding protein. Science 1990; 249:1429–31.
16. Wright SD, Ramos RA, Patel M, Miller DS. Septin: a factor in plasma that opsinizes LPS-bearing particles for recognition by CD14 on phagocytes. J Exp Med 1992; 176:719–28.
17. Munford RS, Hall CL, Lipton JM, Dietschy JM. Biological activity, lipoprotein-binding behavior, and in vivo disposition of extracted and native forms of *Salmonella typimmurium* lipopolysacchrides. J Clin Invest 1982; 70:877–88.
18. Mathison JC, Tobias P, Wolfson E, Ulevitch R. Regulatory mechanisms of host responsiveness to endotoxin (lipopolysaccharide). Pathobiology 1991; 59:185–8.
19. Heumann D, Galley P, Barras C, Zaech P, Ulevitch RJ, Tobias PS, Glauser M-P, Baumgartner JD. Control of LPS binding and LPS-induced TNF secretion in human peripheral blood monocytes. J Immunol 1992; 148:3505–12.
20. Tobias PS, Ulevitch RJ. Control of lipopolysaccharide–high density lipoprotein binding by acute phase protein(s). J Immunol 1983; 131:1913–6.
21. Tobias PS, Soldau K, Ulevitch RJ. Isolation of a lipopolysaccharide-binding acute phase reactant from rabbit serum. J Exp Med 1986; 164:777–93.
22. Wright SD, Tobias PS, Ulevitch RJ, Ramos RA. Lipopolysaccharide (LPS) binding protein opsonizes LPS-bearing particles for recognition by a novel receptor on macrophages. J Exp Med 1989; 170:1231–41.
23. Wright SD, Ramos RA, Tobias PS, Ulevitch RJ, Mathison JC. CD14, a receptor for complexes of lipopolysaccharide (LPS) and LPS binding protein. Science 1990; 249:1431–3.
24. Gabay JE, Scott RW, Campanelli D, Griffith J, Wilde CG, Marra MN, Seeger M, Nathan CF. Antibiotic proteins of human polymorphonclear leukocytes. Proc Natl Acad Sci USA 1989; 86:5610–.
25. Marra MN, Wilde CG, Collins MS, Snable JL, Thorton MB, Scott RW. The role of bactericidal/permeability-increasing protein as a natural inhibitor of bacterial endotoxin. J Immunol 1992; 148:532–7.
26. Tobias PS, Soldau K, Ulevitch RJ. Identification of a lipid A binding site in the acute phase reactant lipopolysaccharide binding protein. J Biol Chem 1989; 264:10867–71.

27. Ramadori G, zum Buschenfelde M, Tobias PS, Mathison JC, Ulevitch RJ. Biosynthesis of lipopolysaccharide-binding protein in rabbit hepatocytes. Pathobiology 1990; 58:89–94.
28. Tobias PS, Soldau K, Hatlen L, Schumann R, Einhorn G, Mathison J, Ulevitch R. Lipopolysaccharide binding protein. J Cell Biochem 1992; (suppl 16C):151.
29. Freudenberg MA, Meier-Dieter U, Staehelin T, Galanos C. Analysis of LPS released from *Salmonella abortus equi* in human serum. Microbial Pathogenesis 1991; 10:93–104.
30. Goyert SM, Ferrero EM, Seremetis SV, Winchester RJ, Silver J, Mattison AC. Biochemistry and expression of myelomonocytic antigens. J Immunol 1986; 137:3909–14.
31. Ferrero E, Hsieh C-L, Francke U, Goyert SM. CD14 is a member of the family of leucine-rich proteins and is encoded by a gene syntenic with multiple receptor genes. J Immunol 1990; 145:331–6.
32. Haziot A, Chen S, Ferrero E, Low MG, Silber R, Goyert SM. The monocyte differentiation antigen, CD14, is anchored to the cell membrane by a phosphatidylinositol linkage. J Immunol 1988; 141:547–52.
33. Couturier C, Haeffner-Cavaillon N, Caroff M, Kazatchkine MD. Binding sites for endotoxins (lipopolysaccharides) on human monocytes. J Immunol 1991; 147:1899–1904.
34. Heumann D, Philippe G, Barras C, Zaech P, Ulevitch RJ, Tobias PS, Glauser M-P, Baumgartner JD. Control of LPS binding and LPS-induced TNF secretion in human peripheral blood monocytes. J Exp Med 1992; 148:3565–12.
35. Hara Kuge S, Amano F, Nishijima M, Akamatsu Y. Isolation of a lipopolysaccharide (LPS)-resistant mutant, with defective LPS binding, of cultured macrophage-like cells. J Biol Chem 1990; 265:6606–10.
36. Wollenweber H-W, Morrison DC. Synthesis and biochemical characterization of a photoactivatable, iodinatable, cleavable bacterial lipopolysaccharide derivative. J Biol Chem 1985; 260:15068–74.
37. Tobias RJ, Tobias PS, Soldau K, Kline L, Lee J-D, Kato K, Martin TP, Ulevitch RJ. Crosslinking of lipopolysaccharide to CD14 on THP-1 cells mediated by lipopolysaccharide binding protein. J Immunol 1993 (in press).
38. Lee J-D, Kato K, Tobias PS, Kirkland TN, Ulevitch RJ. Transfection of CD14 into 70Z/3 cells dramatically enhances the sensitivity of complexes of lipopolysaccharide (LPS) and LPS binding protein. J Exp Med 1992; 175:1697–1705.
39. Frey EA, Miller DS, Jahr GT, Sundan A, Bazil V, Espevik T, Finlay BB, Wright SD. Soluble CD14 participates in the response of cells to lipopolysaccharide. J Exp Med 1992 176:1665–71.
40. Lei M-G, Morrison DC. Specific endotoxic lipopolysaccharide-binding proteins on murine splenocytes. I. Detection of lipopolysaccharide-binding sites on splenocytes and splenocyte subpopulations. J Immunol 1988; 141:996–1005.
41. Halling JL, Hamill DR, Lei M-G, Morrison DC. Identification and characterization of lipopolysaccharide-binding proteins on human peripheral blood cell populations. Infect Immun 1992; 60:845–52.
42. Kirkland TN, Virca GD, Klus-Reichel T, Multer FK, Kim SY, Ulevitch RJ, Tobias PS. Identification of lipopolysaccharide-binding proteins in 70Z/3 cells by photoaffinity cross-linking. J Biol Chem 1990; 265:9520–5.
43. Bright SW, Chen T-Y, Flebbe LM, Lei M-G, Morrison DC. Generation and characterization of hamster-mouse hybridomas secreting monoclonal antibodies with specificity for lipopolysaccharide receptor. J Immunol 1990; 145:1–7.
44. Chen TY, Bright SW, Pace JL, Russell SW, Morrison DC. Induction of macrophage-mediated tumor cytotoxicity by a hamster monoclonal antibody with specificity for lipopolysaccharide receptor. J Immunol 1990; 145:8–12.
45. Green SJ, Chen T-Y, Crawford RM, Nacy CA, Morrison DC, Meltzer MS. Cytotoxic activity and production of toxic nitrogen oxides by macrophages treated with IFMγ and mAb against the 73 kDa LPS receptor. J Immunol 1992; 149:2069–75.
46. Perera PY, Chen T-Y, Morrison DC, Vogel SN. Detection and analysis of the 80-kd lipopolysaccharide receptor in macrophage derived from Lps^n and Lps^d mice. J Leukocyte Biol 1992; 51:501–6.
47. Morrison DC, Silverstein R, Bright SW, Chen T-Y, Flebbe LM, Lei M-G. Monoclonal antibody to mouse lipopolysaccharide receptor protects mice against the lethal effects of endotoxin. J Infect Dis 1990; 162:1063–8.
48. Wright SD, Jong MTC. Adhesion-promoting receptors on human macrophages recognize Escherichia coli by binding to lipopolysaccharide. J Exp Med 1986; 164:1876–88.

49. Hampton RY, Golenbock DT, Penman M, Krieger M, Raetz CRH. Recognition and plasma clearance of endotoxin by scavenger receptors. Nature 1991; 352:342–4.
50. Parent JB. Membrane receptors on rat hepatocytes for the inner core region of bacterial lipopolysaccharides. J Biol Chem 1990; 265:3455–61.
51. Qureshi N, Takayama K, Kurtz R. Diphosphoryl lipid A obtained from the nontoxic lipopolysaccharide of *Rhodopseudomas sphaeroides* is an endotoxin antagonist in mice. Infect Immun 1991; 59:441–4.
52. Henricson BE, Perera P-Y, Qureshi N, Takayama K, Vogel SN. *Rhodopseudomonas sphaeroides* lipid A derivatives block in vitro induction of TNF and endotoxin tolerance by smooth LPS and monophosphoryl lipid A. Infect Immun 1992; 60:4285–90.
53. Kovach NL, Yee E, Munford RS, Raetz CRH, Harlan JM. Lipid IVA inhibits synthesis and release of tumor necrosis factor induced by lipopolysaccharide in human whole blood ex vivo. J Exp Med 1990; 172:77–84.
54. Stuetz PL, Aschauer H, Hildebrandt J, Lam C, Loibner H, Acher I, Scholz D, Schuetze E, Vyplel H. Chemical synthesis of endotoxin analogues and some structure activity relationships. In: Nowotny A, Spitzer JJ, Ziegler EJ, eds. Endotoxin research series, volume 1. Cellular and molecular aspects of endotoxin reactions. New York: Excerpta Medica, 1990:129–44.
55. Jakway JP, DeFranco AL. Pertussis toxin inhibition of B cell and macrophage responses to bacterial lipopolysaccharide. Science 1986; 234:743–6.
56. Prpic V, Weiel JE, Somers SD, DiGuiseppi J, Gonias SL, Pizzo SV, Hamilton TA, Herman B, Adams DO. Effects of bacterial lipopolysaccharide on the hydrolysis of phosphatidylinositol-4,5-bisphosphate in murine peritoneal macrophages. J Immunmol 1987; 139:526–33.
57. Letari O, Nicosia S, Chiavaroli C, Vacher P, Schlegel W. Activation by bacterial lipopolysaccharide causes changes in the cytosolic free calcium concentration in single peritoneal macrophages J Immunol 1991; 147:980–3.
58. Ohmori Y, Hamilton TA. Ca^{2+} and calmodulin selectively regulate lipopolysaccharide-inducible cytokine mRNA expression in murine peritoneal macrophages. J Immunol 1992; 148:538–45.
59. Aderem AA, Albert KA, Keum MM, Wang JKT, Greengard P, Cohn ZA. Stimulus-dependent myristoylation of a major substrate for protein kinase C. Nature 1988; 332:362–4.
60. Law RE, Stimmel JB, Damore MA, Carter C, Clarke S, Wall R. Lipopolysaccharide-induced NF-KB activation in mouse 70Z/3 pre-B lymphocytes is inhibited by mevinolin and 5'-methylthioadenosine: roles of protein isoprenylation and carboxyl methylation reactions. Mol Cell Biol 1992; 12:103–11.
61. Weiel JE, Hamilton TA, Adams DO. LPS induces altered phosphate labeling of proteins in murine peritoneal macrophages. J Immunol 1986; 136:3012–8.
62. Shinomiya H, Hirata H, Nakano M. Purification and characterization of the 65-kDa protein phosphorylated in murine macrophages by stimulation with bacterial lipopolysaccharide. J Immunol 1991; 146:3617–25.
63. Weinstein SL, Gold MR, DeFranco AL. Bacterial lipopolysaccharide stimulates protein tyrosine phosphorylation in macrophages. Proc Natl Acad Sci USA 1991; 88:4148–52.
64. Weinstein SL, Sanghera JS, DeFranco AL, Pelech SL. LPS induces tyrosine phosphorylation and activation of MAP kinases in macrophages. J Cell Biochem 1992; (suppl 16C):169.
65. Manthey CL, Brandes ME, Perera P-Y, Vogel SN. Taxol increases steady-state levels of LPS-inducible genes and protein-tyrosine phosphorylation in murine macrophages. J Immunol 1992; 149:2459–2465.
66. Blenis J. Growth-regulated signal transduction by the MAP kinases and RSKs. Cancer Cells 1991; 3:445–9.
67. Ding AH, Porteu F, Sanchez E, Nathan CF. Shared actions of endotoxin and taxol on TNF receptors and TNF release. Science 1990; 248:370–2.
68. Ettehadieh E, Sanghera JS, Pelech SL, Hess-Bienz D, Watts J, Shastri N, Aebersold R. Tyrosyl phosphorylation and activation of MAP kinases by $p56^{lck}$. Science 1992; 255:853–5.
69. Nel AE, Hanekom C, Hultin L. Protein kinase C plays a role in the induction of tyrosine phosphorylation of lymphoid microtubule-associated protein-2 kinase: evidence for a CD3-associated cascade that includes $pp56^{lck}$ and that is defective in HPB-ALL. J Immunol 1991; 147:1933–9.
70. Kovacs EJ, Radzioch D, Young HA, Varesio L. Differential inhibition of IL-1 and TNF-α mRNA expression by agents which block second messenger pathways in murine macrophages. J Immunol 1988; 141:3101–5.

71. Pabst MJ, Johnston RB. Increased production of superoxide anion by macrophages exposed in vitro to muramyl dipeptide or lipopolysaccharide. J Exp Med 1980; 151:101–14.
72. Han J, Brown T, Beutler B. Endotoxin-responsive sequences control cachectin/tumor necrosis factor biosynthesis at the translational level. J Exp Med 1990; 171:465–75.
73. Shaw G, Kamen R. A conserved AU sequence from the 3' untranslated region of GM-CSF mRNA mediates selective mRNA degradation. Cell 1986; 46:659–66.
74. Yu S-F, Koerner TJ, Adams DO. Gene regulation in macrophage activataion: differential regulation of genes encoding for tumor necrosis factor, interleukin-1, JE, and KC by interferon-γ and lipopolysaccharide. J Leukocyte Biol 1990; 48:412–9.
75. Collart MA, Baeuerle P, Vassalli P. Regulation of tumor necrosis factor alpha transcription in macrophages: Involvement of four κB-like motifs and of constitutive and inducible forms of NF-κB. Mol Cell Biol 1990; 10:1498–506.
76. Mackman N, Korbinian B, Edgington TS. Lipopolysaccharide-mediated transcriptional activation of the human tissue factor gene in THP-1 monocytic cells requires both activator protein 1 and nuclear κB binding sites. J Exp Med 1991; 174:1517–26.
77. Ghosh S, Baltimore D. Activation in vitro of NF-κB by phosphorylation of its inhibitor IκB. Nature 1990; 344:678–82.
78. Schreck R, Rieber P, Baeuerle PA. Reactive oxygen intermediates as apparently widely used messengers in the activation of the NF-κB transcription factor and HIV-1. EMBO J 1991; 10:2247–58.
79. Schönthal A. Nuclear protooncogene products: fine-tuned components of signal transduction pathways. Cellular Signalling 1990; 2:215–25.
80. Introna M, Hamilton TA, Kaufman RE, Adams DO, Bast RC Jr. Treatment of murine peritoneal macrophages with bacterial lipopolysaccharide alters expression of c-Fos and c-Myc oncogenes. J Immunol 1986; 137:2711–5.
81. Auwerx J, Sassone-Corsi P. IP-1: A dominant inhibitor of Fos/Jun whose activity is modulated by phosphorylation. Cell 1991; 64:983–93.
82. McCarthy JB, Wahl SM, Rees J, Olsen CE, Sandberg AL, Wahl LM. Mediation of macrophage collagenase production by 3'-5' cyclic adenosine monophosphate. J Immunol 1980; 124:2405–10.
83. Taffet SM, Singhel KJ, Overholtzer JF, Shurtleff SA. Regulation of tumor necrosis factor expression in a macrophage-like cell line by lipopolysaccharide and cyclic AMP. Cell Immunol 1989; 120:291–300.
84. Ohmiri Y, Strassman G, Hamilton TA. cAMP differentially regulates expression of mRNA encoding IL-1α and IL-1β in murine peritoneal macrophages. J Immunol 1990; 145:3333–9.
85. Wahl LM, Mergenhagen SE. Regulation of monocyte/macrophage collagenase. J Oral Pathol 1988; 17:452–5.
86. Rosenstreich DL, Weinblatt AC, O'Brien AD. Genetic control of resistance to infection in mice. CRC Crit Rev Immunol 1982; 3:263–330.
87. Brandes ME, Wahl SM. Inflammatory cytokines: an overview. In: Schook LB, Laskin DL, eds. Xenobiotics induced inflammation: roles of cytokines and growth factors. Orlando-Florida: Academic Press, 1992: (in press).
88. Wright SD, Ramos RA, Hermanowski-Vosatka A, Rockwell P, Detmers PA. Activation of the adhesive capacity of CR3 on neutrophils by endotoxin: dependence on lipopolysaccharide binding protein and CD14. J Exp Med 1991; 173:1281–6.
89. Nathan C, Nogueira N, Juangbhanich C, Ellis J, Cohn Z. Activation of macrophages in vivo and in vitro: correlation between hydrogen peroxide release and killing of *Trypanosoma cruzi*. J Exp Med 1979; 149:1056–68.
90. Wiltrout RH, Varesio L. Activation of macrophages for cytotoxic and suppressor-effector functions. In: Oppenheim JJ, Shevach ER, eds. Immunophysiology. New York: Oxford University Press, 1990:365–85.
91. Sasada M, Pabst MJ, Johnston RB Jr. Activation of mouse peritoneal macrophages by lipopolysaccharide alters the kinetic parameters of the superoxide-producing NADPH oxidase. J Biol Chem 1983; 258:9631–5.
92. Thelen M, Rosen A, Nairn AC, Aderem A. Regulation by phosphorylation of reversible association of a myristoylated protein kinase C substrate with the plasma membrane. Nature 1991; 351:320–2.
93. Xie Q-W, Cho HJ, Calaycay J, Mumford RA, Swiderek KM, Lee TD, Ding A, Troso T, Nathan C.

Cloning and characterization of inducible nitric oxide synthase from mouse macrophages. Science 1992; 256:225–8.
94. Ding AH, Nathan CF, Stuehr DJ. Release of reactive nitrogen intermediates and reactive oxygen intermediates from mouse peritoneal macrophages: comparison of activating cytokines and evidence for independent production. J Immunol 1988; 141:2407–12.
95. Wahl SM, McCartney-Francis N, Hunt DA, Smith PD, Wahl LM, Katona IM. Monocyte interleukin 2 receptor gene expression and interleukin 2 augmentation of microbicidal activity. J Immunol 1987; 139:1342–7.
96. Brandes ME, Wakefield LM, Wahl SM. Modulation of monocyte type I transforming growth factor-β receptors by inflammatory stimuli. J Biol Chem 1991; 266:19697–703.
97. Chen BD-M, Lin H-S, Hsu S. Lipopolysaccharide inhibits the binding of colony-stimulation factor (CSF-1) to murine peritoneal exudate macrophages. J Immunol 1983; 130:2256–60.
98. Katona IM, Ohura K, Allen JB, Wahl LM, Chenoweth DE, Wahl SM. Modulation of monocyte chemotactic function in inflammatory lesions: role of inflammatory mediators. J Immunol 1991; 146:708–14.
99. Virca GD, Kim SY, Glaser KB, Ulevitch RJ. Lipopolysaccharide induces hyporesponsiveness to its own action in RAW 264.7 cells. J Biol Chem 1989; 264:21951–6.
100. Bazil V, Strominger JL. Shedding as a mechanism of down-modulation of CD14 on stimulated human monocytes. J Immunol 1991; 147:1567–74.
101. Greisman SE. Induction of endotoxin tolerance. In: Nowotny A, ed. Beneficial effects of endotoxins. New York: Plenum Press, 1983:149–78.
102. Freudenberg MA, Kumazawa Y, Meding S, Langhorne J, Galanos C. Gamma interferon production in endotoxin-responder and -nonresponder mice during infection. Infect Immun 1991; 59:3484–91.
103. Vogel SN, Kaufman E, Tate M, Neta R. Recombinant interleukin 1α and recombinant tumor necrosis factor α synergize in vivo to induce early endotoxin tolerance and associated hematopoietic changes. Infect Immun 1988; 56:2650–7.
104. Ribi EE. Beneficial modification of the endotoxin molecule. J Biol Response Modifiers 1984; 3:1–9.
105. Chase JJ, Kubey W, Dulek MH, Jolmes CJ, Salit MG, Pearson FC, Ribi EE. Effect of monophosphoryl lipid A on host resistance to bacterial infection. Infect Immun 1986; 53:711–2.
106. Henricson BE, Benjamin WR, Vogel SN. Differential cytokine induction by doses of lipopolysaccharide and monophosphoryl lipid A that result in equivalent early endotoxin tolerance. Infect Immun 1990; 58:2429–37.
107. Lam C, Schütze E, Hildebrandt J, Aschauer H, Liehl E, Macher I, Stütz P. SDZ MRL 953, a novel immunostimulatory monosaccharidic lipid A analog with an improved therapeutic window in experimental sepsis. Antimicrob Agents Chemother 1991; 35:500–5.
108. Septic Shock—Is MPL an answer? Controlled challenge testing of MPL completed. Ribi ImmunoChem Research, Inc. 1991 Annual Report. February 28, 1992. Vol 11:1109–11.
109. Ziegler E, Fisher CJ, Sprung CL, Straube RC, Sadoff JC, Foulke GE, Wortel CH, Fink MP, Dellinger RP, Teng NNH, Allen IE, Berger HJ, Knatterud GL, LoBuglio AF, Smith CR, HA-1A Sepsis Study Group. Treatment of gram-negative bacteremia and septic shock with HA-1A human monoclonal antibody against endotoxin. N Engl J Med 1991; 324:429–36.
110. Proctor RA, Will JA, Burhop KE, Raetz CRH. Protection of mice against lethal endotoxemia by a lipid A precursor. Infect Immun 1986; 52:905–7.

4

Transmembrane and Intracellular Signaling Events in Lipopolysaccharide-Stimulated Macrophages

Thomas A. Hamilton, Yoshihiro Ohmori, Julie M. Tebo, Shosaku Narumi,* and Charles S. Tannenbaum

Cleveland Clinic Foundation, Cleveland, Ohio

I. INTRODUCTION

Bacterial endotoxin or lipopolysaccharide (LPS) is ubiquitously distributed in biological settings and is responsible for a substantial proportion of the tissue injury that occurs following invasion of multicellular organisms by gram-negative microbes. Response to LPS exposure occurs in essentially all tissues. Under circumstances where the exposure to LPS is restricted to a small area following a localized injury or infectious process, the response to endotoxin will promote a beneficial inflammatory reaction at the site, one that will include the recruitment of additional inflammatory cells, the modulation of their phagocytic and cytotoxic activities, and the promotion of wound healing and tissue repair. In some instances chronic exposure to site-restricted LPS may result in localized chronic inflammation, seen histologically as formation of granulomas. When LPS exposure within the organism is not tissue-restricted, the massive host-wide inflammatory response that ensues may produce devastating effects on host physiology, effects that are often life-threatening. Because of the diverse nature of responses that may be initiated or modulated by LPS, an understanding of the mechanisms involved in host response to LPS is of obvious importance in clinical medicine.

Laboratory research focused on LPS and its action has recently begun to identify the basic features of response to LPS in molecular and cellular terms [1–3]. The following discussion is intended to focus on intracellular events through which the extracellular signal is transduced into changes in cellular behavior. By way of introducing this topic, the nature of the cellular and molecular responses to LPS will be discussed, with brief consideration of the cell types that exhibit sensitivity, the surface structures that may be involved in the functional interaction of LPS with responding cell types, and the molecular nature of the changes in cell behavior.

A. Lipopolysaccharide-Responsive Cell Types

As mentioned above, essentially all tissues are capable of mounting a localized inflammatory response to microbial invasion, much of which is related to endotoxin or endotoxin-like stimulation. Many cell types respond to LPS, though the magnitude and diversity of such response vary greatly. Studies in vivo must be interpreted loosely, as the responses seen may be mediated indirectly through the LPS-stimulated production of multiple secondary mediators. Studies in vitro demonstrate responses to LPS by a variety of different cell types [3–6]. Probably the most thoroughly studied cells are those that participate directly in the inflammatory responses. Resting B

**Current affiliation:* Japan Tobacco Inc., Kanagawa, Japan

lymphocytes are well known to exhibit a polyclonal proliferative and immunoglobulin (Ig)-secretory response following exposure to LPS in vitro [6]. The concentration of LPS required for such stimulation is relatively high (≈1 μg/ml), and the physiologic significance of this is uncertain. Among the more LPS-sensitive cell types are vascular endothelial cells that line the luminal surface of the circulatory system, the resident members of the mononuclear phagocyte family that are found in every tissue, and a subset of the cell types that infiltrate sites of developing or continuing inflammation [6]. Responses are seen in these cell types at levels ranging from 10 pg/ml to 1 μg/ml, and this sensitivity may be even higher in vivo. Qualitatively, macrophages and endothelial cells are capable of producing many of the same inflammatory factors (cytokines, arachidonic acid metabolites, reactive nitrogen intermediates, etc.); the magnitude of expression, however, may vary dramatically between cell types [1–5, 7].

B. Lipopolysaccharide

Despite much attention, the concept of a single LPS-specific receptor has yet to be rigorously supported by experimental data, though substantial evidence indicates that there are multiple structures that bind LPS with high affinity [3,8]. Conventional approaches to the study of receptor-ligand interactions using LPS or its active derivatives have been troubled by the unwieldy chemical characteristics of these materials. Lipopolysaccharide is amphipathic and thus is able to interact with a wide variety of biological materials, making the demonstration of binding specificity technically difficult. The very wide range of concentrations (six orders of magnitude) over which LPS can elicit functional responses implies greater complexity than is generally available through the action of a single binding site [6,9].

In the last several years, new directions have emerged that begin to provide some picture of the possible complexity involved in cellular detection of and response to LPS. By preparing photoactivatable derivatives of LPS or lipid A, David Morisson and coworkers have identified an 80-kDa protein present on a wide variety of cell types that has a relatively high affinity for lipid A [10,11]. The presence of this structure alone does not determine LPS responsiveness, though the absence of the structure is limited to species or cell types that are not LPS-responsive [10,11]. Furthermore, polyclonal and monoclonal antibodies raised against this structure have been shown to produce a biologic response in murine macrophages that is indistinguishable from that seen with LPS or lipid A, suggesting that the 80-kDa structure can signal intracellular events [12]. There are multiple structures with which LPS can and will associate on the cell surface, and at the present time it is not clear that all or even any of the responses to LPS are physiologically mediated through the 80-kDa structure [9,13].

A second major contribution in the understanding of physiologic response to LPS, particularly in vivo, has come from the work of Richard Ulevitch and his colleagues. A serum protein that itself binds LPS with high affinity (LPS-binding protein [LBP]) was identified and was subsequently found to react specifically (in complex with LPS) with the CD14 molecule expressed on mononuclear phagocytes [3,14–16]. The physiologic importance of the CD14-LBP-LPS interaction is indicated by the ability of antibodies to either LBP or CD14 to block much of the response to LPS both in vitro and in vivo [14,16]. However, CD14 does not appear to be a signaling receptor itself, as CD14-negative cells are responsive to LPS (though at generally higher concentrations), and CD14-positive cells can also respond to LPS in the absence of LBP [14,16]. Thus it appears that CD14 serves to concentrate LPS at the surface of phagocytes, thereby increasing the sensitivity of these cells to LPS; initiation of intracellular signaling pathways may require additional structures.

C. Lipopolysaccharide-Stimulated Activities

Defining the endpoint of response(s) to LPS is requisite to examination of the pathways through which such change occurs. Most responses of mononuclear phagocytes can be considered to fall into two separate categories, which can be distinguished from one another on the basis of time-dependence and/or a requirement for new gene expression [8,17,18]. The first category, which may be termed an acute response, occurs within seconds or minutes after stimulation, and generally results from the triggering of preexistent enzymatic activities. Such response systems include the NADPH oxidase complex responsible for the generation of O_2^- and ultimately H_2O_2 as well as the enzyme systems that metabolize arachidonic acid. As compared to stimuli such as complement factor fragments, immune complexes, and large particles subject to phagocytic uptake (all of which are strong stimuli of acute responses in macrophages), LPS is a relatively poor trigger of acute response activities. While there is detectable release of arachidonic acid metabolites such as prostaglandin E (PGE) and leukotriene C by macrophages exposed to LPS, the time course of such response is quite slow as compared with treatments with the above-mentioned materials. Of some interest, though, is the fact that some products of acute responses may be part of the intracellular signaling pathways through which later changes are produced.

The second category of stimulus-induced activational change in mononucleaer phagocytes is termed an adaptive response and takes place over hours to days [17,18]. The use of the term "adaptive" implies that the cell, in the course of its response, is adapting to the environment and will, as a consequence, acquire a new phenotype. Lipopolysaccharide is perhaps the most potent stimulus of adaptive responses in macrophages. As mentioned above, there are LPS-induced effects that apparently do not require new gene expression and that resemble, but occur more slowly than, acute-type responses. The release of arachidonic acid and its metabolism is probably the most thoroughly studied example of this kind of response. For the most part, LPS-stimulated adaptive change depends on the induced expression of new genes and their related protein products that are requisite to peformance of complex host defense functions. Induced gene expression occurs in temporal waves that are, in some cases, related to one another in cause-and-effect fashion [2,8]. Immediate-early genes, which are expressed even in the presence of protein synthesis inhibitors, often represent regulatory proteins that may control subsequent responses. Included among these are transcription factor genes such as *c-fos*, *c-jun*, and *c-myc* [2,8,18]. Also included in this early wave of gene expression are a number of secreted cytokine products that may act in either paracrine or autocrine fashion to regulate subsequent activities of the cell. Later waves of gene expression represent activities that contribute to the functional phenotype of the cell and appear approximately 6 to 24 hours after stimulation. Depending on the nature of the product, some changes are long-lived, while others disappear almost as quickly as they appear. Indeed the transient nature of induced gene expression represents an additional characteristic of the macrophage activation process; it is plastic or even elastic in that the cells transiently acquire new features but ultimately return to their resting phenotype and regain the capability of responding to the same stimulus.

In consideration of the nature of adaptive cellular responses to LPS (or any other stimulus, for that matter), one additional concept may be introduced; specific biochemical and molecular changes within the stimulated cell may contribute either to the acquired capability for performance of a new function or to transition of the cell from one state of activity to another [2,8,18]. Examples of the first category of change will include the expression of enzymes of cytokines that participate in the execution of effector function. The second category wìll include events considered to be part of the intracellular signaling pathways that are necessary to transduce the extracellular stimulus into altered functional behavior or potential. Clearly some events may be important to both processes. In the following sections, the intracellular events that occur in response to LPS and that constitute stimulus-response coupling pathways will be outlined. We will consider two temporal stages in the adaptive response. The first stage ocurs within seconds to minutes of LPS exposure and includes immediate transmembrane signaling events that generate second messengers as well as the changes in cytoplasmic components that result from the action of second messengers. The second temporal category of the signaling cascade occurs over a period of hours to days and involves transcriptional, post-transcriptional, translational, and post-translational events, all of which may participate in governing the quantity and functional character of newly induced gene products.

II. TRANSMEMBRANE AND CYTOPLASMIC SIGNALING EVENTS

There are a number of well-studied early transmembrane signaling pathways. For the purposes of the following discussion, consideration of transmembrane signaling events will include phospholipid metabolism, ion transport activities, cyclic nucleotide generation, and nucleotide regulatory protein signal transducing systems. Second messengers generated by such early events are often coupled, either directly or indirectly, to protein kinase systems that presumably represent the primary target of early second messenger systems. The following discussion will consider first the early transmembrane events and then the relationship(s) between second messengers and protein phosphorylation. The relationships involved in LPS response are outlined schematically in Figure 1.

A. Phospholipid Metabolism

Since the discovery, roughly a decade ago, that diacylglycerol and inositol polyphosphates were principal players in a major transmembrane signaling pathway [19,20], substantial effort has been made toward understanding the role that such events play in signaling responses of many cell types to a diverse array of extracellular stimuli. The common features of these responses are receptor-coupled phospholipases that generate a complex array of metabolic breakdown products from several different complex lipid substrates including phosphatidylinositolbisphosphate (PIP_2), phosphatidylcholine, and sphingomyelin [21,22]. In many cases, these are coupled with specific ligand occupancy of cell surface receptors via the trimolecular nucleotide regulatory (G) protein complexes whose activities are regulated by GTP binding and hydrolysis (see below) [23]. The products of phospholipase activation include fatty acid chains released from the phospholipid (particularly arachidonic acid), inositol trisphosphate (IP_3) (which can release Ca^{2+} ions from intracellular stores within endosomes by virtue of specific receptor interaction), diacylglycerol (DAG) (the principal activating stimulus for many forms of protein kinase C [PKC]), lysophosphatidic acid, and phosphatidic acid. Many of these products are *interelated* by a complex network of enzymatic activities.

A number of studies have demonstrated phospholipid hydrolysis in LPS-treated macrophages [1,2,8,17,18,23–26]. Several laboratories have reported that LPS, either alone or with other agents, promotes the generation of DAG via either phospholipase C or phospholipase D [23,24]. Some DAG certainly derives from the hydrolysis of PIP_2, as several different forms of polyphosphoinositol have been demonstrated in LPS-treated cells [26]. However, the magnitude of the DAG generated cannot be accounted for by the hydrolysis of PIP_2 alone; evidence indicating LPS-dependent activation of phospholipase D has also been reported [24,25]. As mentioned above, many of these initial breakdown products of phospholipids are also substrates for other enzymes, and the secondary products that emerge many have additional activities beyond those attributed to the major products mentioned above.

B. Ion Movement

One of the principal consequences of PIP_2 hydrolysis has been the release of Ca^{2+} from both intracellular stores (through IP_3) and extracellular stores (through IP_4) [27]. There are a number of Ca^{2+} channels through which extracellular Ca^{2+} may enter the cell. At least some of these channels are receptor-linked and may function in either voltage-dependent or voltage-independent

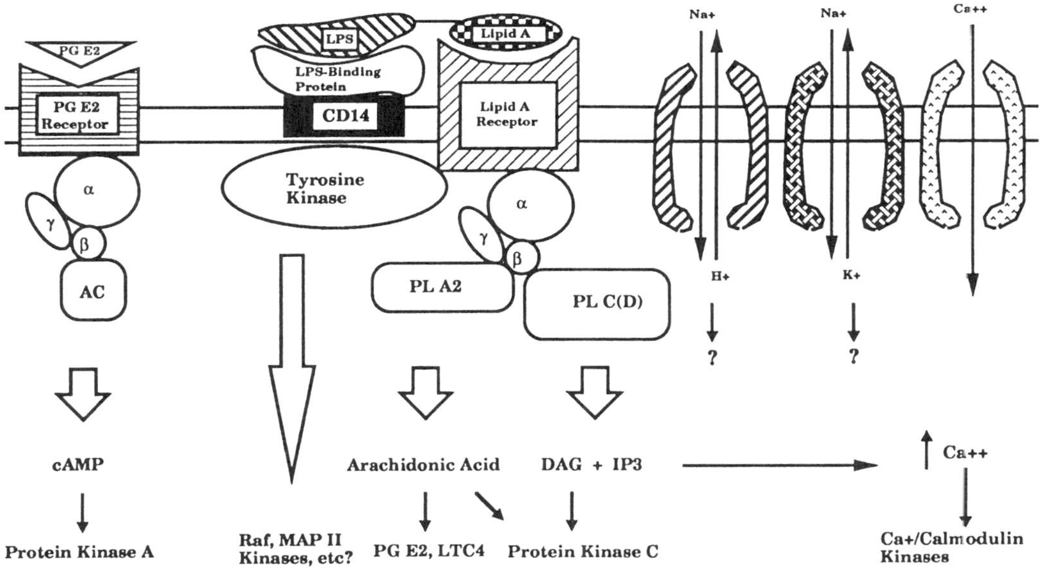

FIGURE 1 Schematic model of transmembrane signaling events in lipopolysacharide-treated macrophages. Lipopolysaccharide (LPS), in complex with the LPS-binding protein, can bind to CD14, perhaps positioning lipid A for interaction with a specific receptor. This may couple directly with phospholipases C, D, or A2, in some cases via nucleotide regulatory protein complexes, or with tyrosine kinase activities. These direct changes may influence serine-threonine–directed protein kinases, which may influence or be influenced by changes in mono- and divalent cation movements. Changes in intracellular ion concentrations can modulate other downstream effects of LPS. Prostaglandin (PG)E2, produced in response to LPS, may activate adenylate cyclase (AC) via a nucleotide regulatory protein coupled to cell surface receptors, and the elevation in intracellular cAMP can also modulate numerous intracellular events via protein kinase activation.

fashion. There is, however, only equivocal evidence regarding the ability of LPS to generate either transient or prolonged elevations in intracellular Ca^{2+} levels [20,28–30]. Furthermore, changing intracellular Ca^{2+} alone through the use of ionophores has not been reproducibly reported to be a sufficient signal to induce many of the adaptive changes that follow stimulation with LPS. Despite the lack of convincing evidence that Ca^{2+} levels are a primary signaling event in response to LPS, a number of studies have found that Ca^{2+} and calmodulin antagonists exhibit selective effects on LPS-induced gene expression [28,31,32]. Such findings lead to the conclusion that alterations in Ca^{2+} may represent a pathway through which the magnitude of an LPS response is controlled but are not directly responsible for initiating the response.

Alteration in Na^+/H^+ exchange has also been studied in a variety of cell types and has been closely linked with response to mitogen stimulation [33,34]. Stimulation of macrophages with interferon-γ (IFN-γ) has been shown to increase the activity of the Na^+/H^+ pump, and this change has been linked to IFN-γ–driven gene expression (class II MHC molecules) [35]. In cells stimulated with LPS, there is also evidence that the Na^+/H^+ pump is activated, though the change in pH_i occurs more slowly than with IFN-γ [1]. Changes similar to those seen with LPS can be produced by treatment of macrophages with phorbol ester [1]. Whether this slower pattern of intracellular pH change associated with response to LPS is linked to subsequent change in transcriptional control remains to be established.

In addition to Ca^{2+} and Na^+/H^+ exchange, modulation of Na^+/K^+ exchange may also be involved in control of LPS-induced gene expression [36]. Experimental manipulations including elimination of either extracellular Na^+ or K^+ or treatment of cells with ouabain, an inhibitor of Na^+/K^+-ATPase activity, have been shown to markedly enhance LPS-induced increases in tumor necrosis factor-α (TNF-α), interleukin-1 (IL-1), IP-10 and KC mRNA expression through transcriptional mechanisms [36]. Since many of the intracellular mechanisms that control monovalent and divalent cation concentrations and movement are closely linked with each other, many of these experimental manipulations may act through related pathways.

C. Cyclic Nucleotide Generation

Generation of cAMP is believed to be suppressive with respect to activation responses of mononuclear phagocytes [1,2,8]. Indeed, many changes in macrophages have been shown to be down-regulated or prevented in cells treated with agents that raise the intracellular levels of cAMP. These include the induced expression of class II MHC molecules, the acquisition of tumoricidal activity, and the expression of several (though not all) cytokines (e.g., TNF-α, MCP-1) [37,38]. These suppressive activities may be physiologically important and are often mediated by the action of PGE. In this regard LPS has recently been reported to antagonize the suppressive action of prostaglandins and the associated rise in cAMP [39]. Macrophages pretreated with LPS exhibit a markedly reduced elevation of cAMP following subsequent exposure to PGE. Interestingly, there are several reports demonstrating that elevations in intracellular cAMP can promote the expression and secretion of IL-1β [40,41]. What significance this in vitro observation has on the behavior of monocytes/macrophages at a site of infection is not clear.

D. Guanine Nucleotide Regulatory Proteins

Several studies have provided experimental evidence indicating that heterotrimeric guanine nucleotide regulatory or G proteins are involved in coupling between LPS binding to the cell surface and the subsequent generation of second messengers [25,42–45]. Most of these studies have relied on the use of bacterial toxins, such as cholera toxin and pertussis toxin, that act upon various G proteins to either disrupt or initiate signaling events. For example, several responses to LPS have been shown to be inhibited by pertussis toxin, and LPS has been shown to activate pertussis toxin–sensitive G proteins. Responses affected include B cell mitogenic activation, production of IL-1β, and the release and metabolism of arachidonic acid [42–45]. However, many LPS activities are not affected by pertussis toxin treatment, indicating that only some portion of the response to LPS utilizes G proteins and/or that some of the G proteins involved are insensitive to these particular bacterial toxins.

E. Protein Phosphorylation

A common feature of most intracellular signaling mechanisms is the activation of one or more protein kinase activities, generally mediated by the direct action of second messenger. Protein kinases show specificity for phosphorylation of proteins either on serine/threonine residues or on tyrosine residues; the majority of protein phosphorylation is serine-threonine directed, while tyrosine phosphorylation, though lower in magnitude, appears to be more closely linked with signal transduction [46,47]. At least some tyrosine-directed protein kinases represent the first step in transmembrane signaling, as they are either physically part of the receptor structure or associate closely with receptors and are activated upon receptor occupancy with specific ligands. Tyrosine-directed phosphorylation has been linked to the sequential activation of multiple biochemical and cellular activities including phospholipases and serine-threonine–directed protein kinases [46,47]. Much of the experimental support for the importance of protein phosphorylation as a general component of response to specific stimuli has been derived from studies using protein kinase inhibitors. Unfortunately, the specificity of such agents is often quite broad, and the results obtained may in some cases be misleading. More rigorous relationships have been established by identification of specific protein substrates and by demonstration that their function is modulated following the phosphorylation event.

Gross changes in protein phosphorylation and alteration of protein kinase activities in LPS-stimulated murine peritoneal macrophages have been observed in several laboratories. Lipopolysaccharide treatment induced the phosphorylation of several distinct proteins (33, 37, 65, 75, and 103 kDa) as detected by incorporation of radiolabeled phosphate into protein bands detected on one-dimensional SDS-PAGE (sodium dodecyl sulfate polyacrylamide gel electrophoresis) [48–51]. The induced pattern of protein phosphorylation resembled, but was not identical to, that induced by stimulation of macrophages with phorbol esters. These findings, along with several other lines of evidence, indicated a possible role for protein kinase C (PKC)-directed protein phosphorylation in mediating responses to LPS. This possibility is consistent with the documentation of increased DAG in LPS-treated cells. Furthermore, LPS may act to directly stimulate PKC [51]. While some components of response to LPS can be induced through pharmacologic activation of PKC with phorbol esters, much of the adaptive response of macrophages to LPS (i.e., altered gene expression) cannot be replicated with phorbol esters, suggesting that components of the signaling process beyond PKC-driven protein phosphorylation are required. Treatment of macrophages with platelet activating factor (PAF) results in the stimulation of PKC-mediated phosphorylation, which resembles that seen with LPS except that it occurs more rapidly and is more transient [53]. Platelet activating factor does not, however, mimic LPS as a stimulus of macrophage activation. Thus it may be that more prolonged phosphorylation is also necessary for the functional consequences of LPS treatment.

Lipopolysaccharide has been shown to stimulate phosphorylation of a specific protein of approximately 68 kDa identified as related to the 87-kDa PKC substrate common to a variety of cells [52,54]. This has recently been termed the MARCKS protein (myristoylated alanine rich C kinase substrate). It is specifically phosphorylated in LPS-treated macrophages and as a consequence is redistributed from the cytoplasm to the plasma membrane. This protein has been identified as a filamentous (F) actin crosslinking protein. The ability to form crosslinks with actin is regulated both by PKC-mediated phosphorylation and via interaction with calmodulin [54]. This demonstrates a direct relationship between the exposure of macrophages to LPS and an immediate change in cytoskeleton assembly that may be an important aspect of cellular mobility and/or shape.

Several reports have presented data supporting a role for tyrosine-directed protein phosphorylation in response to LPS stimulation [55,56]. Treatment of peritoneal macrophages or macrophage-like cell lines with LPS or lipid A results in stimulus-specific tyrosine-directed phosphorylation of a series of proteins that occurs rapidly (within 4 to 5 minutes) and declines within 30 to 60 minutes. Thus, tyrosine-directed protein phosphorylation should more appropriately be considered an early transmembrane signaling event. This pattern of phosphorylation was inhibited by cotreatment of the cells with herbimycin, a specific inhibitor of tyrosine kinases. Interstingly, the release of arachidonic acid metabolites from LPS-treated macrophages was also inhibited by herbimycin, suggesting that the activation of phospholipid hydrolysis (which produces arachidonic acid) may be linked to the activation of the tyrosine kinase. More recently, it has been demonstrated that the activation of tyrosine phosphorylation was related to interaction of LPS-LBP complexes with the macrophage cell surface CD14 molecule [56]. As with other responses linked to CD14, tyrosine-directed protein phosphorylation occurred at high doses of LPS in a CD14-independent fashion, suggesting that transmembrane signaling events are not initiated through the

CD14 molecule itself. Whether tyrosine phosphorylation links directly to adaptive changes in macrophage physiology and is mechanistically involved in the modulation of gene expression remains to be determined.

III. REGULATION OF TRANSCRIPTION IN LIPOPOLYSACCHARIDE-TREATED MACROPHAGES

Many of the effects of treating macrophages with LPS involve alterations in gene expression. Modulation of the amount of any particular gene product found within or secreted by macrophages may be controlled at numerous mechanistic levels including transcription of the gene, processing of the primary transcript, stability of cytoplasmic mRNA, translational activity of the mRNA, and post-translational processing of the polypeptide product [57–59]. Specific regulatory effects at any of these mechanistic levels will result from coupling between the intracellular signaling processes initiated in the cell following LPS treatment and the particular mechanistic components of the process in question. In many cases, molecular and cellular features of post-transcriptional control mechanisms are not well understood, and thus it is difficult to appreciate how they may interact with or be controlled by intracellular signaling events. In contrast, much progress has been made over the last decade in the understanding of transcriptional control in eukaryotic cells. The following section will deal specifically with recent experimental results on transcriptional control in LPS-treated macrophages.

The two predominant features of the control of eukaryotic transcription are *cis*-acting nucleotide sequences generally found in regions near transcription start sites and *trans*-acting nuclear factors that recognize and bind these regions in sequence-specific fashion. Broadly speaking, there are sequences and factors requisite to the general formation and function of RNA polymerase transcription complexes that act on many genes, and there are sequences and factors necessary to the expression of particular genes. Within this second category are factors requisite to stimulus-dependent expression of specific genes. The latter group contains sequence-factor combinations that are constitutively active and those that act only in response to specific stimulatory conditions. With regard to understanding stimulus-response coupling and its relationship to the control of gene transcription, it is this last group that has been of greatest interest, and the activation of such factors represents a control point for such processes. Indeed, specific changes in gene expression represent at least one end result of a stimulus-initiated intracellular signaling pathway, and the specific DNA-binding proteins are the final substrates and ultimate effector molecules in such pathways. Through analysis of specific nuclear factors and the processes whereby their transcriptional control activities are modulated, the connections between early transmembrane signaling changes, intermediate changes in cytoplasmic components, and ultimate changes in transcriptional activity in the nucleus can be established.

A. *Cis*-Acting Elements Regulating Lipopolysaccharide Sensitivity

One of the central concepts to emerge from many studies carried out in multiple cell types using a diverse array of stimuli is that specific regulatory sequences controlling transcriptional activity function in the context of the surrounding nucleotide sequences. Thus, particular elements may behave in functionally distinct ways depending on the nucleotide sequences immediately surrounding them and will be influenced by the presence of other regulatory elements that may or may not be located in relatively close proximity. To date, only a few LPS-responsive genes have been characterized with respect to the regulatory elements that control their LPS sensitivity. These include TNF-α, granulocyte colony-stimulating factor (G-CSF), Iα and Iϵ, and murine IP-10

[60–65; Ohmori and Hamilton, unpublished data]. The results from these studies have focused attention upon three major sequence elements: κB type elements (first defined as important for immunoglobulin κ chain expression in B cells [66,67]), CK-1 type elements [68,69], and finally IFN-responsive elements, specifically the interferon stimulus response element (ISRE).

An elegant series of studies examining the regulatory sequences governing the expression of the TNF-α gene in murine bone marrow–derived macrophages using a transient transfection protocol has defined four regulatory regions within an approximately 1000-base region 5' of the transcription start site of the TNF-α gene [60–62]. Both deletion and selective site-specific mutagenesis show that each of the four sites is contributory to the LPS response, though they are not quantitatively equivalent [62]. Two of the sites are clearly κB-like sites, while the other two sites are very similar to regulatory elements found in the Iα and G-CSF genes, where they have been shown to be important for LPS-initiated transcription [63–65]. The latter sites have been termed CK-1 sites (*c*yto*k*ine-1) [68,69]. Though the element resembles a κB motif, binding factors obtained from B cells are believed to recognize a sequence region distinct from the κB element that is necessary for positive transcriptional response to LPS [63,64]. Studies of the CK-1 element in cells responding to other stimuli have identified a factor now termed NF-GMa, which is distinct from NF-κB [68.69]. Studies with the TNF-α gene, however, indicate that the CK-1 elements are also able to bind NF-κB activities found in nuclear extracts from LPS-treated bone marrow macrophages [62]. These findngs indicate that the same elements may indeed function quite differently depending on the sequence context and the cell type. These results strongly support the importance of the κB sequence motif and of factors that can recognize this in determining the nature of transcriptional responses to LPS.

Recent work examining the expression of the murine IP-10 gene using transient transfection in a macrophage cell line (RAW 264.7) identified multiple regions in a 2000-base region located 5' to the transcription start site that control the relative magnitude of transcription in either negative or positive fashion (Ohmori et al., unpublished data). A 242-nucleotide fragment (–243 to –2) was able to mediate transcriptional activation of the CAT reporter gene in response to treatment of the cells with either LPS or IFN-γ. The reporter gene expression pattern resembled quite closely the behavior of the endogenous IP-10 gene in this cell line. This region contains two κB sites and a number of IFN-responsive elements (most importantly a full-length, highly conserved ISRE). Deletion analysis and site-directed mutation in both of the κB elements and in the ISRE illustrate that all three sites contribute toward the regulation of LPS-driven IP-10 gene transcription. Functionally, each of the sites could be distinguished from the others both qualitatively and quantitatively. These findings indicated that LPS might be able to drive transcription using regulatory elements that also control responses to IFNs. Furthermore, they confirm the picture that emerges from the studies of TNF-α expression indicating that multiple sites interact with one another to produce cooperative transcriptional control. The ability of independent sites to modulate the behavior of either related or unrelated sites present in the same gene provides for tremendous diversity in the patterns of gene expression in response to multiple forms of inflammatory stimulation.

B. Lipopolysaccharide-Stimulated Transacting Factors

The limited available evidence obtained in transfection studies with plasmid constructs containing sequences from LPS-responsive genes indicates that κB, CK-1, and ISRE motifs may be important for transriptional control in LPS-treated macrophages. The κB motif has been defined by examination of residues critical both for driving transcription and for interaction with transacting factor activities found in the nuclei of appropriately stimulated cells. Different κB sequences can be distinguished from one another, however, by the relative binding affinity for various forms of NF-κB and by differential transcriptional control function as measured by gene transfer experiments [70]. Most of the proteins that recognize the κB motif are members of a group of related genes referred to collectively as the rel family based on sequence homology to the *v-rel* oncogene. These include C-rel, B-rel, I-rel, p49, p50 (p105, KBF-1), and p65. These various products are able to interact with one another, forming a diverse array of heterodimeric complexes with quite distinct DNA-binding and transactivation functions. In resting cells, various forms of the rel family members exist in an inactive state, and following stimulation, sequence-specific DNA binding activity is acquired and nuclear localization domains are exposed [70,71]. In B cells and T cells it is believed that heterodimers of p65 with p50 are inhibited by interaction with a third subunit called IκB. Upon stimulation of the cell, IκB may be phosphorylated, thereby allowing activation of the NF-κB complex [70–73]. However, other mechanisms of control have also been postulated, and the multiple forms of κB binding activity suggest that regulation of these activities may be a highly diverse process mechanistically [70,71].

There is a limited but convincing body of evidence indicating that LPS can indeed stimulate the activation of nuclear factors that recognize and bind the κB sequences. Large DNA fragments as well as smaller oligonucleotides containing the exact sequences from the different sites in the TNF-α gene were shown to form complexes with LPS-sensitive sites in the TNF-α gene were shown to form complexes with LPS-sensitive nuclear factors [60–62]. That these activities are NF-κB–like is indicated by the ability to compete such complex formation with unlabled oligonucleotides containing defined κB sequence motifs [62,74]. Stimulation of NF-κB activity in macrophages seems, however, to be somewhat distinct from that seen in other cell lines, on the basis of several lines of evidence. For example, phorbol esters, which have been shown to be potent stimuli of NF-κB in several other cell types by virtue of their ability to stimulate PKC activity, do not activate NF-κB in primary macrophages or macrophage cell lines [75,76] (Narumi, Tebo, and Hamilton, unpublished data). The relative electrophoretic mobilities of protein-DNA complexes from macrophages are different from those described in B cells and T cells, and the polypeptide composition of the complexes also appears to be distinct [74]. While κB-binding complexes from 70Z/3 cells contain 65- and 75-kDa proteins, the macrophage-derived complexes have 2 proteins of approximately 39 and 49 kDa [74]. These preliminary observations raise many questions regarding the molecular nature of the nuclear activities in macrophages that bind the κB elements in the various LPS-responsive genes and the signal pathways through which the DNA binding and nuclear localization of these factors are controlled.

The finding that the ISRE motif in the IP-10 gene is a contributor to LPS-mediated transcription of the IP-10 gene suggests that LPS may act to induce ISRE-binding activity. Indeed, a promoter containing 3 repeats of the IP-10 ISRE sequence placed in front of a heterologous promoter is a strong driver of CAT expression in transient transfection assays in macrophage cell lines (Ohmori and Hamilton, unpublished data). Regulatory elements controlling response to IFN stimulation have been identified by examination of sequences found in a variety of IFN-inducible genes and by gene transfer experiments using specific promoter regions from IFN-inducible genes linked to reporter genes [77,78]. Collectively these studies have focused attention upon a 13- to 14-base motif called the interferon stimulus response element (ISRE). It has been shown that ISRE motifs mediate transcriptional responses to both type I and type II IFNs [77,78]. The ISRE sequences have been shown to bind a number of factors in various cell types, some of which are IFN-dependent while others are expressed constitutively [77]. These include IFN-stimulated gene factor (ISGF)-3, ISGF-2—also termed IFN response factor-1 (IRF-1)—IRF-2, IFN consensus binding protein (ICSBP), and IFN binding factor (IBF-1) as well as others (ISGF-2, IRF-2, and ICSBP are homologous members of a single gene family). The specific transcriptional function of most of these factors remains unknown, though accumulated evidence from a variety of experimental approaches indicates that ISGF-3 is the predominant transacting factor responsible for

IFN-α/β–driven transcription [77]. Peritoneal macrophages express LPS-dependent ISRE binding activity resembling that induced by IFN-α or IFN-β in the same cells [79]. That this activity may be similar to that of ISGF-3 is suggested by its relatively slow electrophoretic mobility and similar sensitivity to stimulation by IFN-γ, IFN-β or LPS. Lipopolysaccharide can induce expression of class I IFN in macrophages (predominantly IFN-β [80]), and it seems possible that the ability of LPS to stimulate transcription through ISRE elements may depend in part on the induced expression of endogenous IFN [81]. However, antibody to IFN-β does not prevent the response to LPS, though it does, as expected, block response to IFN-β, suggesting the possibility that the function of ISRE in mediating response to LPS is direct.

IV. CONCLUSIONS AND PERSPECTIVES

Most if not all of the LPS-induced changes in mononuclear phagocyte function are presumably initiated by virtue of interaction between LPS and cell surface structures. Indeed, there is a growing body of evidence indicating that LPS does bind to and act through specific receptor sites (see Figure 1 for a schematic summary of signaling events). While the character and functional significance of many potential LPS-binding sites have yet to be determined, responses of macrophages to LPS have been associated with alterations in most of the known second messenger systems. These include phospholipid hydrolysis, ion movements into or out of the cell, cyclic nucleotides (at least in negative fashion), and probably receptor-linked tyrosine protein phosphorylation. These early events are very likely followed by serine-threonine–directed protein phosphorylation, at least some of which is a consequence of the action of various second messenger products. Direct linkage between protein phosphorylation and functional changes in cell behavior has been established only in a very limited number of cases.

Many LPS-stimulated functions are dependent on induced gene expression. While this may be regulated at multiple mechanistic levels, the control of the transcriptional response to LPS is perhaps the best-characterized. Sequences in the regions 5' to transcription start sites of several LPS-responsive genes have been analyzed (Table 1). The studies to date implicate CK-1 motifs, κB motifs, and possibly the ISRE. Lipopolysaccharide-activated nuclear factors that may function as either positive or negative transcriptional regulators through these sequence elements have been identified in several circumstances. While specific details regarding the molecular changes that mediate the acquisition of DNA-binding activity and/or nuclear localization by these factors remain to be identified and characterized, protein phosphorylations are primary candidates.

Despite substantial effort and progress in elucidating generic pathways of stimulus-response coupling, many of which have been examined in one or more models of mononuclear phagocyte physiology, there remain major unanswered questions. One area that needs more rigorous defini-

Table 1 Lipopolysaccharide-Responsive Sequence Elements[a]

Element	Binding factors	References
kB: GGGRYTYYCC	NFkB (p49,p50,p65 C-Rel. B-Rel etc)	60–62,70,74
CK-1: GRGRTTYCCYN	NF-GMa	64,65,68,69
ISRE: GGAAAGTGAAACT (from IP-10 gene)	ISGF3, ISGF2 (IRF-1) IFR-2, ICSBP, IBF-1 etc	77,79

[a]Sequence elements given are the consensus sequences for either kB or CK-1, and the ISRE is derived from the IP-10 gene.
Y = pyrimidine; R = purine; N = any base.

tion is LPS-binding sites, with specific interest in those that may be responsible for intracellular signaling. Furthermore, substantial efforts also need to be focused on analysis of mechanistic linkages between receptor occupancy and cytoplasmic and nuclear consequences of LPS treatment. For example, examination of kinase activities and substrates within macrophages will help to provide additional tools. The presence of sarc homology (SH) domains within such substrates may provide additional leads. An area likely to be highly fruitful in the immediate future will be the molecular definition of LPS-responsive nuclear transcription factor activities. While transcription factor groups or families are emerging as involving highly complex interactions between multiple related polypeptides, understanding the link between cytoplasmic change and nuclear change will be critical to completing the pathway from cell surface to chromosome.

REFERENCES

1. Adams DO. LPS-initiated signal transduction pathways. In: Morrison DC, Ryan JL, eds. Macrophages. Bacterial endotoxic lipopolysaccharides. New York: CRC Press, 1992 pp. 285–310.
2. Adams DO, Hamilton TA. Molecular basis of macrophage activation: diversity and its origins. In: Lewis CE, McGee JO'D, eds. The natural immune system volume II: the macrophage. Oxford: Oxford University Press, 1992 pp. 75–114.
3. Raetz CRH, Ulevitch RJ, Wright SD, Sibley CH, Ding A, Nathan, CF. Gram negative endotoxin: an extraordinary lipid with profound effects on eukaryotic signal transduction. FASEB J 1991; 5:2652–60.
4. Tannenbaum CS, Major JA, Poptic EJ, DiCorleto PE, Hamilton TA. Lipopolysaccharide induces competence genes JE and KC in Balb/C 3T3 cells. J Cell Physiol 1990; 144:77–83.
5. Shen XI, Hamilton TA, DiCorleto PE. Lipopolysaccharide-induced expression of the competence genes KC in vascular endothelial cells is mediated through protein kinase C. J Cell Physiol 1989; 140:44–51.
6. Morrison DC, Ryan JL. Bacterial endotoxins and host immune responses. Adv Immunol 1979; 28:293–450.
7. Fajardo LF. The complexity of endothelial cells. Am J Clin Pathol 1989; 92:241–50.
8. Kirkland TN, Virca GD, Kuus-Reichel T, Multer FK, Kim SY, Ulevitch RJ, Tobias PS. Identification of lipopolysaccharide-binding proteins in 70/Z3 cells by photoaffinity crosslinking. J Biol Chem 1990; 265:9520–5.
9. Hamilton TA. Molecular mechanisms in the activation of mononuclear phagocytes. In: Rogers TJ, Gilman SC, eds. Immunopharmacology. Caldwell, New Jersey: Telford Press, 1988;213–52.
10. Lei M-G, Morrison DC. Specific endotoxic lipopolysaccharide-binding proteins on murine splenocytes. I. Detection of lipopolysaccharide-binding sites on splenocytes and splenocyte subpopulations. J. Immunol 1988; 141:996–1005.
11. Morrison DC. The case for specific lipopolysaccharide receptors expressed on mammalian cells. Microb Pathogen 1989; 7:389–98.
12. Chen TY, Bright SW, Pace JL, Russell, SW, Morrison DC. Induction of macrophage-mediated tumor cytotoxicity by a hamster monoclonal antibody with specificity for LPS receptor. J Immunol 1990; 145:8–12.
13. Golenbock DT, Hampton RY, Raetz CRH, Wright SD. Human phagocytes have multiple lipid A binding sites. Infect Immun 1990; 58:4069–75.
14. Schumann RR, Leong SR, Falggs GW, Gray PW, Wright SD, Mathison JC, Tobias PS, Ulevitch RJ. Structure and function of lipopolysaccharide binding protein. Science 1990; 249:1429–31.
15. Wright SD, Tobias PS, Ulevitch RJ, Ramos R. Lipopolysaccharide binding protein opsonizes LPS-bearing particles for recognition by a novel receptor on macrophages. J Exp Med 1989; 170:1231–41.
16. Wright SD, Ramos R, Tobias PS, Ulevitch RJ, Mathison JC. CD14: a receptor for complexes of lipopolysaccharide (LPS) and LPS binding protein. Science 1990; 249:1431–3.
17. Adams DO, Hamilton TA. The cell biology of macrophage activation. Annu Rev Immunol 1984; 2:283–319.
18. Hamilton TA, Adams DO. Molecular mechanisms of signal transduction in macrophages. Immunology Today 1987; 8:151–158.

19. Nishizuka Y. The molecular heterogeneity of protein kinase C and its implications for cellular regulation. Nature 1988; 334:661–5.
20. Berridge MJ, Irvine RF. Inositol trisphosphate, a novel second messenger in cellular signal transduction. Nature 1984; 312:315–21.
21. Dennis EA, Rhee SG, Billah MM, Hannun YA. Role of phospholipases in generating lipid second messengers in signal transduction. FASEB J 1991; 5:2068–77.
22. Hannun YA, Bell RB. Functions of sphingolipids and shingolipid breakdown products in cellular regulation. Science 1989; 243:500–7.
23. Sebaldt RJ, Prpic V, Hollenbach PW, Adams DO, Uhing RJ. IFNγ potentiates the accumulation of diacylglycerol in murine macrophages. J Immunol 1990; 145:645–89.
24. Grove RI, Allegretto NJ, Liener PA, Warr GA. Lipopolysaccharide (LPS) alters phosphatidylcholine metabolism in elicited peritoneal macrophages. J Leukocyte Biol 1990; 48:38–42.
25. Harris WE, Bursten SL. Lipid A stimulates phospholipase D activity in rat mesangial cells via a G-protein. Biochem J 1992; 281:675–82.
26. Prpic V, Weiel JE, Somers SD, DiGuiseppi J, Gonias SL, Pizzo SV, Hamilton TA, Herman B, Adams DO. Effects of bacterial lipopolysaccharide on the hydrolysis of phosphatidylinositol-4,5-bisphosphate in murine peritoneal macrophages. J Immunol 1987; 139:526–33.
27. Berridge, MJ. Inositol trisphosphate and diacylglycerol as second messengers. Biochem J 1984; 220:345–357.
28. Hamilton TA, Ohmori Y, Narumi S, Tannenbaum CS. Regulation of diversity in macrophage activation. In: Lopez-Berestein G, Klostergaard J, eds. Mononuclear phagocytes in cell biology. New York: CRC Press, 1992:47–70.
29. Neton RC. Lack of a central role for calcium in the induction and release of human interleukin 1. Biochem Biophys Res Commun 1987; 147:1027–33.
30. Klein JB, Schepers TM, Dean WL, Sopnnenfeld G, McLeish K. Role of intracellular calcium concentration and protein kinase C activation in IFNγ stimulation of U937 cells. J Immunol 1990; 144:4305–12.
31. Ohmori Y, Hamilton TA. Ca^{++} and calmodulin selectively regulate LPS-inducible cytokine mRNA expression in murine peritoneal macrophages. J Immunol 1992; 148:538–545.
32. Kovacs EJ, Brock B, Varesio L, Young HA. IL-2 induction of IL-1β mRNA expression in monocytes. Regulation by agents that block second messenger pathways. J Immunol 1989; 143:3532–7.
33. Gaidano G, Ghigo D, Schena M, Bergui L, Treves S, Turrini F, Caligaris C, Bosia M. Na^+/H^+ exchange activation mediates the lipopolysaccharide-induced proliferation of human B lymphocytes and is impaired in malignant B-chronic lymphocytic leukemia lymphocytes. J Immunol 1989; 142:913–8.
34. Cheung RK, Grinstein S, Gelfand EW. Mitogenic and nonmitogenic ligands trigger a calcium dependent acidification in human T lymphocytes. J Immunol 1988; 141:1648–51.
35. Prpic V, Yu S.-F, Figueiredo F, Hollenbach PW, Gawdi G, Herman B, Uhing RJ, Adams DO. Role of Na^+/H^+ exchange by interferon gamma in enhanced expression of JE and I-Abeta genes. Science 1989; 244:469–71.
36. Ohmori Y, Reynolds E, Hamilton TA. Modulation of Na^+/K^+ exchange potentiates LPS-induced gene expression in murine peritoneal macrophages. J Cell Physiol 1991; 148:96–105.
37. Kunkel SL, Spengler M, May MA, Spengler R, Larrick J, Remick D. Prostanglandin E2 regulates macrophages-derived tumor necrosis factor gene expression. J Immunol 1988; 263:5380–5389.
38. Tannenbaum CS, Hamilton TA. Lipopolysaccharide-induced gene expression in murine peritoneal macrophages is selectively suppressed by agents which elevate intracellular cAMP. J Immunol 1989; 142:1274–80.
39. Figueiredo F, Uhing, RJ, Okonogi K, Gettys TW, Johnson SW, Adams DO, and Prpic V. Activation of the cAMP cascade inhibits an early event involved in murine macrophage Ia expression. J Biol Chem 1990; 265:12317–23.
40. Ohmori Y, Strassmann G, Hamilton TA. cAMP differentially regulates expression of mRNA encoding IL-1α and IL-1β in murine peritoneal macrophages. J Immunol 1990; 145:3333–9.
41. Sung S-SJ, Walters JA. Increased cAMP levels enhance IL-1α and IL-β mRNA expression and protein production in human myelomonocytic cell lines and monocytes. J Clin Invest 1991; 88:1915–23.
42. Jakway JP, DeFranco AL. Pertussis toxin inhibition of B cell and macrophage responses to bacterial lipopolysaccharide. Science 1986; 234:743–6.

43. Daniel-Issakani S, Speigel AM, Strulovici B. Lipopolysaccharide response is linked to the GTP-binding protein Gi2, in the promonocytic cell line U937. J Biol Chem 1989; 264:20240–7.
44. Dziarski R. Correlation between ribosylation of pertussis toxin substrates and inhibition of peptidoglycan-, muramyl dipeptide-, and lipopolysaccharide-induced mitogenic stimulation in B lymphocytes. Eur J Immunol 1989; 19:125–30.
45. Wang J, Kester M, Dunn MJ. Involvement of a pertussis toxin–sensitive, G-protein coupled phospholipase A2 in lipopolysaccharide stimulated prostaglandin E2 synthesis in cultured rat mesangial cells. Biochim Biophys Acta 1988; 936:429–35.
46. Cantley LC, Auger KR, Carpenter C, Duckworth B, Graziani A, Kapeller R, Soltoff S. Oncogenes and signal transduction. Cell 1991; 164; 281–302.
47. Pawson T. Tyrosine kinases and their interactions with signalling protein. Curr Opin Genet Dev 1992; 2:4–12.
48. Weiel JE, Hamilton TA, Adams DO. LPS induces altered phosphate labeling of proteins in murine peritoneal macrophages. J. Immunol 1986; 136:3012–3020.
49. Shinomiya H, Hirata H, Nakano M. Purification and characterization of the 65 kDa protein phosphorylated in murine macrophages by stimulation with bacterial lipopolysaccharide. J Immunol 1991; 146:3617–25.
50. Rosen A, Nairn AC, Greengard, P, Cohn ZA, Aderem AA. Bacterial lipopolysaccharide regulates the phosphorylation of the 68K protein kinase C substrate in macrophages. J Biol Chem 1989; 264:9118–21.
51. Wrightman PD, Raetz CRH. The activation of protein kinase C by biologically active lipid moieties of lipolysaccharide. J Biol Chem 1984; 259:10048–56.
52. Bakouche O, Moreau JL, Lachman LB. Secretion of Il-1: role of protein kinase C. J Immunol 1992; 148:84–91.
53. Prpic V, Weiel JE, Somers SD, DiGiuseppi J, Gonias SL, Pizzo SV, Hamilton TA, Hermann B, and Adams DO. Biochemical and functional responses stimulated by platelet activating factor in murine peritoneal macrophages. J Cell Biol 1988; 107:363–72.
54. Hartwig JH, Thelen M, Rosen A, Janmey PA, Nairn AC, Aderem A. MARCKS is an actin filament crosslinking protein regulated by protein kinase C and calcium-calmodulin. Nature 1992; 356:618–22.
55. Weinstein SL, Gld MR, DeFranco AL. Bacterial lipopolysaccharide stimulates protein tyrosine phosphorylation in macrophages. Proc Natl Acad Sci USA 1991; 88:4148–52.
56. Weinstein SW, June CH, Defranco AL. LPS-stimulated protein tyrosine phosphorylation in human macrophages: involvement of CD14 (abstr). J Leukocyte Biol 1991; 2(suppl):41.
57. Han J, Brown T, Beutler B. Endotoxin-responsive sequences control cachectin/tumor necrosis factor biosynthesis at the translation level. J Exp Med 1990; 171:465–75.
58. Han J, Beutler B, Huez G. Complex regulation of tumor necrosis factor mRNA turnover in lipopolysaccharide-activated macrophages. Biochim Biophys Acta 1991; 1090:22–8.
59. Brand K, Fowler B, Edgington TS, Mackman N. Tissue factor mRNA in THP-1 monocytic cells is regulated at both transcriptional and posttranscriptional levels in response to lipopolysaccharide. Mol Cell Biol 1991; 11:4732–8.
60. Collart MA, Baeuerle P, Vassalli P. Regulation of tumor necrosis factor alpha transcription in macrophages: involvement of four κB-like motifs and of constitutive and inducible forms of NF-κB. Mol Cell Biol 1990; 10:1498–1506.
61. Shakhov AN, Collart MA, Vassalli P, Nedospasov SA, Jongeneel CV. κB-type enhancers are involved in lipopolysaccharide-mediated transcriptional activation of the tumor necrosis factor α gene in primary macrophages. J Exp Med 1990; 171:35–47.
62. Drouet C, Shakhov AN, Jongeneel CV. Enhancers and transcription factors controlling the inducibility of the tumor necrosis factorα promoter in primary macrophages. J Immunol 1991; 147;1694–1700.
63. Nishizawa M, Nagata S. Regulatory elements responsible for inducible expression of the granulocyte colony stimulating factor gene in macrophages. Mol Cell Biol 1990; 10:2002–11.
64. Gravallese EM, Boothby MR, Smas CM, Glimcher LH. A lipopolysaccharide-induced DNA-binding protein for a class II gene in B cells is distinct from NF-κB. Mol Cell Biol 1989; 9:3184–92.
65. Rothman P, Li SC, Gorham B, Glimcher L, Alt F, Boothby MR. Identification of a conserved lipopolysaccharide-plus-interleukin-4-responsive element located at the promoter of germ line ϵ Transcripts. Mol Cell Biol 1991; 11:5551–61.

66. Lenardo M, Pierce JW, Baltimore D. Protein-binding sites in Ig gene enhancers determine transcriptional activity and inducibility. Science 1987; 236:1573–6.
67. Lenardo MJ, Baltimore D. NF-κB: a pleiotropic mediator of inducible and tissue-specific gene control. Cell 1989; 58:227–39.
68. Shannon MF, Pell LM, Lenardo MJ, Kuczek ES, Occhiodoro FS, Dunn SM, Vadas MA. A novel tumor necrosis factor–responsive transcription factor which recognizes a regulatory element in hemopoietic growth factor genes. Mol Cell Biol 1990; 10:2950–60.
69. Kuczek ES, Shannon MF, Pell LM, Vadas MA. A granulocyte colony stimulating factor gene promoter element responsive to inflammatory mediators is functionally distinct from an identical sequence in the granulocyte-macrophage colony stimulating factor gene. J Immunol 1991; 146:2426–33.
70. Blank V, Kourilsky P, Israel A. NF-κB and related proteins: rel/dorsal homologies meet ankyrin-like repeats. TIBS 1992; 17:135–40.
71. Perkins ND, Schmid RM, Duckett CS, Leung K, Rice NR, Nabel GJ. Distinct combinations of NF-κB subunits determine the specificity of transcriptional activation. Proc Natl Acad Sci USA 1992; 89:1529–33.
72. Ghosh S, Baltimore D. Activation in vitro of NF-κB by phosphorylation of its inhibitor IkB. Nature 1990; 344:678–82.
73. Kerr LD, Inoue J-I, Davis MN, Link E, Baeuerle PA, Bose HR Jr, Verma IM. The rel-associated pp40 protein prevents DNA binding of rel and NFκB: relationship with IkBb and regulation by phosphorylation. Genes Dev 1991; 5:1464–76.
74. Narumi S, Tebo JM, Finke JH, Hamilton TA, IFNγ and IL-2 cooperatively activate NFκB in murine peritoneal macrophages. J Immunol 1992; 149:529–534.
75. Meichle A, Schutz S, Hensel G, Brunsing D, Kronke M. Protein kinase C–independent activation of nuclear factor kB by tumor necrosis factor. J Biol Chem 1990; 265:8339–43.
76. Vincenti MP, Burrell TA, Taffet SM. Regulation of NF-κB ativity in murine macrophages: effect of lipopolysaccharide and pharbol ester. J Cell Physiol 1992; 150:204–13.
77. Williams BRG. Transcriptional regulation of interferon stimulated genes. Eur J Biochem 1991; 200:1–11.
78. Watine DB, Israel A, Kourilsky P. Immunology Today 1990; 11:286–92.
79. Tebo JM, Hamilton TA. LPS induces DNA binding activity specific for the IFN-stimulated response element in murine peritoneal macrophages. J Immunol 1992; 149:2352–2357.
80. Gessani S, Belardelli F, Pecorelli A, Paddu P, Baglioni C. Bacterial lipopolysaccharide and gamma interferon induce transcription of beta interferon mRNA and interferon secretion in murine macrophages. J Virol 1989; 63:2785–9.
81. Riches DWH, Underwood GA. Expression of inteferon-β during the triggering phase of macrophage cytocidal activation. Evidence for an autocrine role in the regulation of this state. J Biol Chem 1992; 266:24785–92.

5

Receptors on Phagocytic Cells Involved in Microbial Recognition

David M. Mosser

Temple University School of Medicine, Philadelphia, Pennsylvania

The purpose of this chapter is to give a general overview of phagocytic receptors on leukocytes. I have limited the focus to those receptors that either have been associated with microbial recognition or have the potential to participate in this process. Many such recognition events will be discussed in detail in subsequent chapters in this book. In this chapter I will examine the structure, function, and cell biology of the various receptors and then give an overview of their role in host defense. Some liberties have been taken in grouping receptors according to a broad interpretation of the primary ligands they recognize.

The criteria for a receptor are that it binds ligand in a specific and saturable manner and then transmits a biological signal to the cell. This is in contrast to a "binding site," which binds ligand but elicits no biological signal. Recently many binding sites have been described that have the potential to serve as microbial recognition structures by cooperating with phagocytic receptors on leukocytes. A detailed description of all of these structures is beyond the scope of this chapter. One such class of binding sites, however, deserves mention. These are the cell surface proteoglycans, which have recently received a great deal of attention relating to the binding of a diverse array of microbes. For an excellent review of these structures and their biology, see Esko [1].

I. RECEPTORS FOR PARTICLES OPSONIZED WITH IMMUNOGLOBULIN G

There are three distinct classes of receptors for the Fc portion of immunoglobulin G (IgG) on leukocytes (Figure 1). These are collectively termed the Fcγ receptors (FcγRs) and individually designated FcγRI, FcγRII, and FcγRIII. These receptors provide a "crucial link between the phagocytic effector cells and the lymphocytes that secrete Ig" [2]. The Fcγ receptors are members of a large multigene family that mediate a series of diverse functions critical to cellular immune responses, including phagocytosis, antibody-dependent cellular cytotoxicity (ADCC), secretion of inflammatory mediators, generation of the respiratory burst, and clearance of immune complexes [2,3]. The functional diversity of this group of receptors is undoubtedly related to the structural diversity of the various members of this receptor family as well as to their wide cell and tissue distribution.

The FcγRI, designed CD64 and also referred to as the high-affinity Fcγ receptor, binds IgG with an apparent affinity of approximately 1×10^{-8} M [2]. The cDNAs for both the murine [4] and the human [5] receptors have been identified. The FcγRI are glycosylated monomeric proteins with three immunoglobulin-like extracellular domains [6] (Figure 1). Their apparent molecular mass is

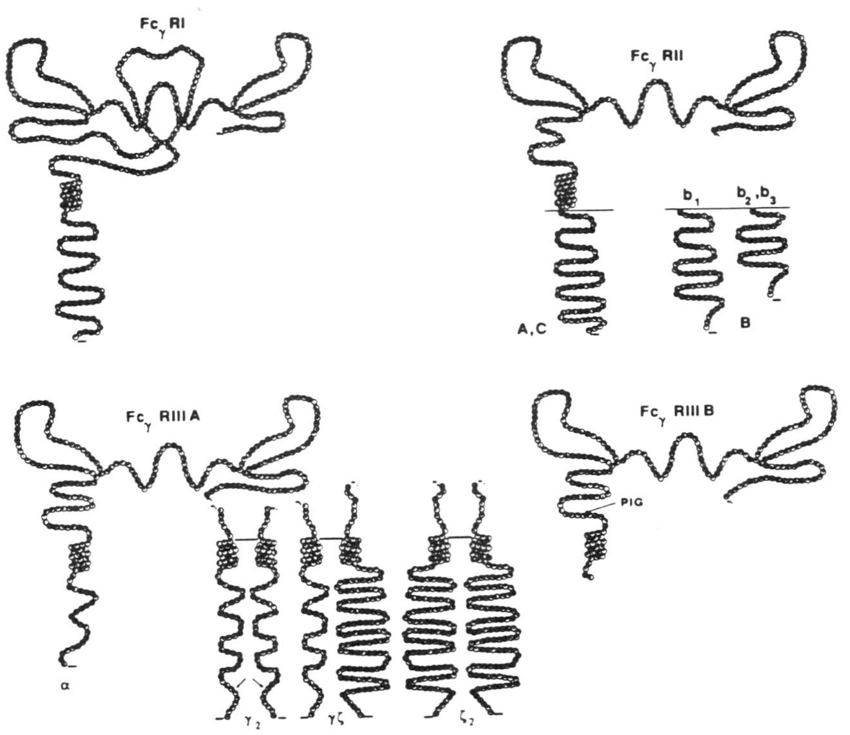

FIGURE 1 Schematic representation of the proteins encoded by the FcγR locus. (From Ref. 6.)

approximately 70 kDa, with a protein backbone of approximately 40 kDa. The FcγRI are limited in tissue distribution, being expressed primarily on cells of the monocyte/macrophage lineage. Resting cells express between 10,000 and 30,000 receptors per cell [2]. The FcγRI are also expressed on stimulated, but not resting, neutrophils. The most well-described stimulator of the FcγRI is interferon-γ (IFN-γ), which also up-regulates receptor expression on monocytes by as much as 20-fold [7].

The high affinity of the FcγRI for monomeric IgG raises questions about their biological function on cells circulating in the blood, where milligram concentrations of IgG in blood could compete with specific antibody for these receptors. The FcγRI is able to mediate ADCC [8]. An IFNγ-regulated monocyte ADCC against a hybridoma making monoclonal antibody (mAb) to the FcγRI has been demonstrated under conditions where ADCC against hybridomas making non-FcγR-specific mAb did not occur [8].

The FcγRII, designated CD32, are low-affinity receptors for IgG. These receptors bind monomeric IgG poorly, if at all, with an affinity some 10 times lower than that of the FcγRI. These receptors, however, are widely distributed, they bind immune complexes readily, and they efficiently mediate the endocytosis of bound ligand. The FcγRII are expressed on a variety of cells of myeloid and lymphoid lineage, including monocytes, neutrophils, lymphocytes, and platelets. Absent from this list are NK cells, which do not appear to express CD32. The cDNAs for both the murine [9,10] and the human [11] receptors have been identified. The FcγRII are glycosylated monomeric proteins with two Ig-like extracellular domains. Their apparent molecular mass ranges from 40 to 60 kDa; the variability is due to differences in glycosylation and alternative splicing, primarily in cytoplasmic exons. Murine macrophages express approximately 100,000 FcγRII, whereas platelets express approximately 1000 receptors/cell. Several mAbs to the murine and human FcγRII, many of which block ligand binding, have been described. For a complete list of mAbs to the FcγR, consult Ravetch [6].

Crosslinking the FcγRII with mAbs results in Ca^{2+} fluxes, superoxide generation, and neutrophil degranulation [12,13]. Most of these processes are inhibitable with pertussis toxin, indicating a role for G proteins in FcγRII signal transduction. The cytoplasmic portion of the FcγRII has been shown to be important for transducing phagocytic signals to cells [14]. Fibroblastoid cells transfected with the B2 form of the FcγRII localized ligated receptor in coated pits. The B1 form of the receptor, however, containing an insertion in the cytoplasmic domain, did not [14].

The FcγRIII, termed CD16, are also low-affinity receptors for IgG. They are expressed on neutrophils, NK cells, and tissue macrophages. They are not expressed on undifferentiated macrophage precursors and are expressed poorly, if at all, on freshly isolated monocytes. Monocyte-derived macrophages, however, begin to express FcγRIII after several days in culture [2]. The cDNAs for the FcγRIII have been identified [15,16]. The most striking feature of the human FcγRIII is the structural heterogeneity between the receptors expressed on NK cells and those expressed on neutrophils. The neutrophil receptor (FcγRIIIB) contains no transmembrane or cytoplasmic domains and is attached to the membrane by an inositol phosphate linkage [15]. As a consequence of this, the FcγRIIIB may not transmit the myriad of intracellular signals typically associated with IgG binding to leukocyte FcγRs. Antibody to the FcγRIII does not trigger calcium fluxes in neutrophils [17]. Furthermore, neutrophils fail to exhibit cytotoxicity to hybridomas secreting anti-CD16 mAb under conditions where NK cells efficiently lyse these cells [18]. The receptor on NK cells (FcγRIIIA) has a typical transmembrane segment and a short cytoplasmic domain. This receptor can associate with other integral membrane proteins, including the γ chain of the FcϵR or the ζ chain of the T cell receptor, and transduce cellular signals upon ligation [19]. In addition to mediating ADCC, the FcγRIII participate in the clearance of immune complexes. Infusion of mAb to FcγRIII causes a dramatic increase in the half-life of circulating immune complexes [20]. Binding of immune complexes in vitro by NK cells causes the secretion of cytokines, including IFN-γ and tumor necrosis factor (TNF) [21].

The cell biology of the FcγRs has been studied by several groups [reviewed in 2,6]. Ligation of the collective FcγRs has been shown to result in the secretion of a number of leukocyte-derived mediators involved in inflammation and host defense [12,13,22]. Preliminary study of the cell biology of the individual receptors suggests that while unique biological functions may be ascribed to each receptor, all three classes have redundant or overlapping functions relating to the cognition of complexed or monomeric IgG. The biology of the individual FcγRs is complicated by the fact that there are different forms of each of the three receptors. The functional diversity observed is a reflection of the structural heterogeneity, especially in the cytoplasmic portion of the FcγR. Thus, a single FcγR, expressed on two different cell types, may mediate different biological effects.

The contribution of the leukocyte FcγRs in host defense is best inferred from numerous in vivo observations documenting the contribution of antibody to host defense. For example, administration of exogenous gamma globulin has been shown to be protective in the pediatric hypogammaglobulinemic patient [23]. In addition, the collective experience with numerous vaccines has demonstrated that protection correlates with Ig titers, and vaccine failures are often associated with defective IgG responses [24]. In the case of *Streptococcus pneumoniae,* there is an inverse relationship between type-specific antibody titers and susceptibility to disease in the elderly [25]. These observations are supported by data from an experimental guinea pig model, in which type-specific antibody to pneumococcus promoted efficient bacterial clearance [26]. The protective effect of antibody in host defense to group B streptococcus has been established [27], and deficiencies in maternal antibody correlate with susceptibility to neonatal group B streptococcus infection [28]. Antibody to type b capsule protects humans and experimental animals from *Hemophilus influenzae* type b disease [24,29]. There is clearly an accessory role for complement in many of these processes, which will be discussed subsequently.

In vitro, antibody opsonization has been amply demonstrated. In general, particles opsonized with IgG bind efficiently to leukocytes, trigger a respiratory burst, and enter phagolysosomes, where they are degraded [30]. Early studies demonstrated that unopsonized toxoplasmas reside in unfused phagosomes, whereas their IgG-coated counterparts reside in phagolysosomes [31]. This approach was also used with mycobacteria [32], confirming that the FcRs mediate efficient particle internalization and typically target particles to phagolysosomal destruction in phagocytic cells. Joiner and colleagues have used Chinese hamster ovary cells transfected with intact FcγRs to demonstrate that these cells mediate the internalization of antibody-opsonized *Toxoplasma gondii*, providing that the transfected receptor contains an intact cytoplasmic domain [33].

In summary, the FcγRs are the prototypical "protective" receptors on leukocytes. They mediate the binding and internalization of IgG-opsonized particles. This internalization is usually associated with phagolysosomal fusion and the release of a variety of inflammatory mediators.

II. RECEPTORS FOR PARTICLES OPSONIZED WITH COMPLEMENT

Similar to the Fcγ receptors, there are multiple classes of receptors on leukocytes that recognize particles opsonized with complement. The primary component of complement recognized by the leukocyte complement receptors is the third component, termed C3. Macrophages and polymorphonuclear leukocytes have three distinct receptors, termed the CR1, CR3, and CR4, that can recognize some fragment of C3. Leukocytes have also been reported to recognize the collagen-like domains of the first complement component, C1 [34].

The CR1, designated CD35, is the receptor for C3b and C4b. It is expressed on human erythrocytes, macrophages, neutrophils, B cells, some T cells, dendritic cells, and glomerular epithelial cells [35]. Monocytes, B cells, and neutrophils express between 20,000 and 40,000 receptors per cell. In monocytes and neutrophils, the majority of receptors is in cytoplasmic pools and can be induced to be expressed on the surface by the addition of activating agents such as chemotactic peptides [36]. The human CR1 has been cloned [37]. It is a polymorphic, *N*-glycosylated, monomeric protein with a molecular mass ranging from 190 to 250 kDa, depending on the allotypic variant. The CR1 is composed of a series of 30 short consensus repeats (SCRs). Each SCR contains four invariant cysteins that form two disulfide bonds, conferring a "triple loop" structure. This structural configuration, consisting of cysteine-rich repeating units of 40 to 70 amino acids, is found in a number of C3/C4-binding proteins that compose what is now referred to as the regulator of complement activation (RCA) gene superfamily [38]. On the CR1, these loops are tandomly arranged into larger motifs called long homologous repeats (LHRs) [39] (Figure 2). The A allotype CR1 is composed of four LHRs termed A to D, each of which is composed of seven SCRs. The fact that the SCR loops are linked to each other by only a few amino acids presumably gives the molecule structural rigidity. In high-resolution electron micrographs, these molecules resemble a filamentous string of beads, a configuration that has been described as "an extended array of triple loops maintained by disulfide bonds" [36].

Deletion analysis and chimeric receptor constructs have been used to determine the C3b and C4b binding sites on CR1 [40,41]. The amino terminal LHR of the CR1 (LHR-A) binds C4b, while the next two LHRs, B and C, each possess C3b binding sites. The first two amino terminal SCRs in each of the LHRs are involved in ligand binding. These same regions have been implicated in the well-described Factor I cofactor activity of CR1 [36]. The CR1 binds monomeric C3b with relatively low affinity, but presumably the clustering of receptors on cells and the multiple ligand-binding sites per receptor confer functional activity on the molecule.

Three different murine proteins with homology to the human complement receptors have been identified [42,43]. A protein which is now designated Crry [42] was identified from a C3-sepharose eluate. It is a 60–65 kDa protein which cross reacts with polyclonal antibody to human CR1. The molecule was cloned using probes derived from human CR1 [42]. The biological role of this protein remains speculative. Transcripts for Crry are expressed on a wide variety of murine

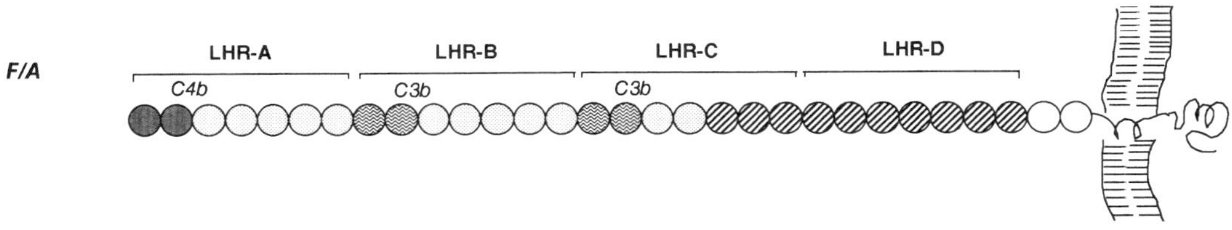

FIGURE 2 Schematic representation of the predicted structure of the F/A allotype CR1. (From Ref. 39.)

cells, many of which express no C3-binding activity, indicating that in situ the Crry protein may not function as a C3b receptor [43]. Cells transfected with the cDNA encoding this protein express intrinsic complement regulatory activity, suggesting that Crry may function in a manner analogous to human decay accelerating factor or membrane cofactor protein [43]. Two other murine proteins were eluted from C3 sepharose and have been designated MCR1 and MCR2 [43]. The MCR1 molecule is 190 kDa in size. It is composed of 21 SCRs and, like the human CR1, it expresses C3b/C4b binding activity and Factor I cofactor activity [43]. The MCR2 protein is 150 kDa in size. It is composed of 15 SCRs and, like the human CR2, it binds C3dg. The MCR2 protein has 70% homology to the human CR2 and the genomic structure is identical to the human CR2 [43]. Monoclonal antibodies to the mCR1 often cross-react with MCR2, suggesting a high degree of homology between these two proteins. It has subsequently been demonstrated that MCR1 and MCR2 arise from alternatively spliced transcripts of a single common gene [43]. In summary, there are clear structural distinctions between the murine and the human RCA family of proteins. However, both species utilize the tandem arrangement of SCRs to achieve both C3b/C4b binding and complement regulatory activity.

The cell biology of the CR1 is quite dynamic, and receptor physiology can vary depending on the environmental stimuli. On resting cells, the CR1 binds ligand, but fails to mediate internalization [44]. Exposure of these cells to T cell lymphokines, however, converts these receptors into ones that are mobile in the plane of the membrane and are capable of internalizing ligands. Likewise, fibronectin and other stimuli, including high doses of lipopolysaccharide (LPS) and treatment with phorbol esters, can convert the CR1 into an endocytic receptor [45]. The CR1, in contrast to the FcR's described above, does not trigger a respiratory burst upon particle ligation, nor does it lead to cellular prostaglandin synthesis [46]. Thus, the CR1 can be converted from a binding site on resting cells to a receptor capable of transmitting the cellular signals required for internalization. However, even under these conditions, ligation of this receptor does not typically lead to the cellular secretion of inflammatory mediators.

The role of the CR1 in microbial clearance has received a disproportionately small amount of attention, relative to its other functions. This inattention is due, in part, to the paucity of mAbs to the CR1 that block ligand binding, because of the multiple binding sites on CR1 for C3 and C4. The role for this receptor in microbial clearance, however, is inferred from studies citing the importance of complement to the in vivo clearance of encapsulated bacteria [26,47], which will be discussed below. It has been suggested that the main role of the CR1 in microbial recognition may be as cooperative receptor that facilitates both FcR- and CR3-mediated endocytosis [35].

The CR3, also termed Mac-1 or CD18/CD11b, is expressed on mononuclear phagocytes, neutrophils, and NK cells [48]. In contrast to the monomeric CR1, the CR3 is a heterodimeric structure composed of a M_r 95-kDa β chain (CD18) noncovalently associated with a M_r 165-kDa α chain (CD11b). The CR3 β chain is common to two other related receptors, LFA-1 (CD18/CD11a) and p150/95 (CD18/CD11c) [48] (Figure 3). These three receptors are commonly called the leukocyte integrins. The α chains of the leukocyte integrins are typical transmembrane proteins,

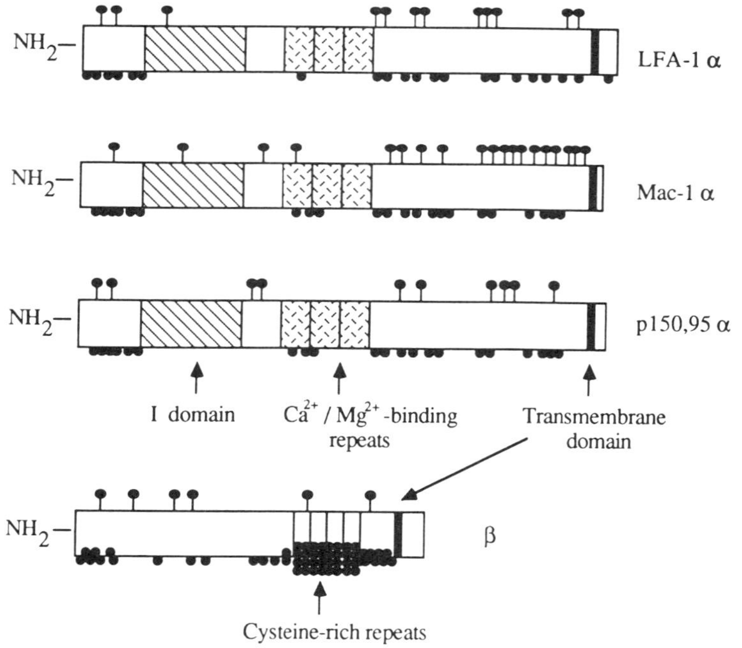

FIGURE 3 Schematic representation of the α subunits of the three leukocyte integrins and the common β subunit. (From Ref. 48.)

containing a relatively short (50– to 60–amino acid) carboxy-terminal cytoplasmic domain, and extracellular domains that include a divalent cation-binding domain and an "insertion region" that has homology to sequences in Von Willebrand factor [49] and other matrix proteins. The β chain contains an extracellular cysteine-rich domain.

Phosphorylation of the CD18 is thought to influence Mac-1 physiology. PMA-induced phosphorylation of the Mac-1 β-chain correlates with the ability of neutrophils to phagocytize iC3b-coated sheep red blood cells [50]. The cytoplasmic domain of CD18 contains a single potential site for tyrosine phosphorylation and multiple sites for serine or threonine phosphorylation [51].

Mac-1 preferentially recognizes the proteolytically inactivated fragment of fixed C3b, called iC3b [52]. In addition to iC3b, however, Mac-1 has also been reported to recognize multiple diverse ligands with no apparent homology. These ligands include coagulation factors, including Factor X [53] and fibrinogen [54]; cell-surface ligands, including ICAM-1 [55]; and various saccharides or polysaccharides [56], including LPS [57]. Several groups have proposed that Mac-1 has two distinct binding sites [56,57], one for iC3b and a second lectin-like site, which may bind bacterial LPS. The evidence in support of this is that several mAbs to Mac-1 that block iC3b binding do not affect the binding of zymosan [56] or LPS [57]. Conversely, other Mac-1 antibodies block LPS and zymosan binding while not blocking iC3b binding. The role of the tripeptide Arg-Gly-Asp (RGD) in ligand binding by the $\beta 2$ integrins remains equivocal. Inhibition of Mac-1 ligand binding with high concentrations of linear RGD-containing peptides has been demonstrated [58]. However, the elegant studies by Taniguchi-Sidle and Isenman have shown that RGD sequences in C3 are not required for the binding of that molecule to Mac-1 [59].

A second $\beta 2$ integrin, p150,95, has been implicated in the binding of the C3 fragments iC3b and C3dg and consequently termed the CR4 [60]. p150,95 has also been implicated in bacterial binding, and mAbs to p150,95 inhibit the binding of *Bordetella pertussis* to macrophages [61]. Other studies suggest that p150,95 plays a relatively minor role in complement fragment binding [62]. Thus, the biological role of p150,95 remains unclear and awaits the identification of other ligands for p150,95, especially cellular ligands that may be involved in leukocyte adherence or migration.

Phagocytic cells have also been shown to bind to a subunit of the first complement component, C1q [34]. Treatment of human monocytes with C1q caused an enhancement in IgG-mediated phagocytosis, and treatment of neutrophils results in an enhanced respiratory burst following phagocytosis. The phagocytosis of the infective stage of *Trypanosoma cruzi* by human mononuclear phagocytes was enhanced by C1q [63]. Both phagocytosis and killing of *Cryptococcus neoformans* was enhanced by adherence of monocytes to C1q-coated surfaces [64]. C1q has recently been implicated in the recognition of the human immunodeficiency virus [65].

The role of complement receptors in host defense is somewhat engimatic. There is ample and compelling in vivo evidence that complement is an important mediator of bacterial clearance by phagocytic cells and therefore is an important component of host defense. In the guinea pig model, complement was shown to be critical to the clearance of *S. pneumoniae* [26]. The early clearance of *H. influenzae* from the blood of mice was also shown to be complement-dependent [47]. In vitro, complement receptors alone or in cooperation with Fcγ receptors have been shown to participate in uptake of a number of bacterial species of macrophages. The synergy between these receptors has long been appreciated [66]. In the *H. influenzae* system, Noel et al. demonstrated that even the presence of opsonic antibody, optimal bacterial uptake depended on complement receptors [67]. The importance of the leukocyte integrins to host defense is best illustrated by a congenital leukocyte adhesion deficiency disease (LAD) [68] in which patients express no $\beta 2$ integrins because of a defect in the common CD18 gene. These patients suffer from a variety of frequently recurring, sometimes fatal, bacterial infections.

Complement opsonization has been amply demonstrated in vitro. Complement fixes to the LPS-containing gram-negative bacteria and opsonizes them for leukocyte phagocytosis. Encapsulated bacteria, however, may require antibody-directed complement fixation in order for complement to be opsonic. For example, *S. pneumoniae* fix complement by the alternative pathway [69] but require specific anticapsular antibody to promote phagocytosis [70]. Other bacteria, for example the group A streptococci, express a virulence factor, called M-protein [71], that inhibits complement fixation and prevents complement-mediated phagocytosis by neutrophils [72]. Thus, in vitro and in vivo studies support the importance of complement fixation and complement-mediated phagocytosis to host defense. Complement-mediated mechanisms are especially evident both in the early clearance of bacteria by nonimmune hosts and in the subsequent antibody-directed phagocytosis of the encapsulated bacteria.

In the restricted case of intracellular pathogens of macrophages, however, the complement receptors may be detrimental to host defense. A number of groups have demonstrated that one or more of the complement receptors can mediate the endocytosis of intracellular microbes of macrophages. We have demonstrated that leishmania opsonized with complement survive in macrophages better than their unopsonized counterparts (73). We have hypothesized that the complement receptors may represent a "permissive" receptor class that allows leishmania and other intracellular microorganisms to enter phagocytes without triggering the appropriate cellular microbicidal defense responses. *Listeria monocytogenes* may represent another example of this phenomenon [74]. *Legionella pneumoniae* and *Mycobacterium tuberculosis* are also intracellular pathogens of macrophages that depend, at least in part, on complement receptors for cellular entry [75]. Recently we have demonstrated that *Rhodococcus equi*, an intracellular pathogen of pulmonary macrophages, also relies on complement receptors to enter macrophages [76]. The CR1, and in some circumstances the CR3, can mediate endocytosis without triggering a respiratory burst.

Therefore, the complement receptors may represent a potential mechanism for entry into macrophages that is used by intracellular microorganisms to circumvent the microbicidal responses associated with endocytic events mediated by other receptors, such as the FcγRs.

III. RECEPTORS FOR LIPOPOLYSACCHARIDE

It is well established that exposure of macrophages to LPS causes a dramatic alteration in their physiology. Lipopolysaccharide induces the release of TNF, interleukin-1 (IL-1), IL-6, prostaglandins, and cytolytic proteases [77,78]. Macrophages become tumoricidal and microbicidal in response to high dose of LPS [77]. There are at least four different receptors expressed on macrophages with the potential to bind LPS. Not all of these receptors, however, transmit the macrophage-activating signals typically associated with endotoxemia.

Morrison and colleagues have isolated an 80-kDa protein from murine splenocytes that binds LPS in a specific and saturable manner [79]. This receptor not only binds LPS but also appears to convey the intracellular signals that result in the activated phenotype typical of LPS-exposed macrophages. Furthermore, mAbs to this receptor have been raised, and one of these, 5D3, mimics LPS activity and converts resident macrophages into tumoricidal cells. Thus, the 80-kDa protein appears to be a functional LPS receptor.

Krieger and colleagues [80] have described a "scavenger receptor" on macrophages and sinusoidal cells. The receptor exhibits ligand specificity for a variety of anionic lipids, including chemically modified low-density lipoprotein (LDL), and also certain polynucleotides, acidic phospholipids, and recently lipid IV_A. The receptor is an integral membrane glycoprotein with an apparent molecular mass of 220 kDa. It is composed of three subunits, each having an apparent molecular mass of M_r 77 kDa, 20% to 25% of which is due to *N*-linked glycosylation. Structurally, the receptor can be divided into five domains, including an *N*-terminal cytoplasmic region, an extracellular series of α-helical coiled coils, a collagen-like domain and, in the type I receptor, a cysteine-rich C-terminus [80] (Figure 4).

The observation that macrophages exhibited a "foam cell" appearance upon uptake of acetylated LDL suggested a role for this receptor in atherogenesis. The recent observation that this receptor can also mediate the binding of bacterial lipid A and its precursor suggests a possible role for this receptor in microbial recognition. In this regard, the presence of high receptor levels in the liver would suggest a potential role for this receptor in hepatic microbial clearance. The cell biology of this receptor is under current scrutiny, but it does not appear to be the receptor that transduces the LPS-activating signals to macrophages. Blockade of scavenger receptors with acetylated LDL does not inhibit cellular responses to LPS [81]. Conversely, this receptor may detoxify LPS and serve to decrease the host inflammatory response to circulating LPS [80].

Another set of receptors implicated in the binding of LPS are the CD18-containing leukocyte integrins, described above. Wright and Jong have observed that cultivating macrophages on LPS caused a removal of CD18 receptors from the apical surface of cells and, conversely, that cultivating macrophages on antibodies to CD18 inhibited *Escherichia coli* binding [57]. This group subsequently demonstrated that erythrocytes coated with lipid IV_A failed to bind to leukocytes from LAD patients genetically deficient in CD18 [81]. It is doubtful that the CD18-containing receptors mediate the activation effects of LPS. Monocytes from LAD patients respond to LPS, and treatment with mAb to CD18 does not prevent the LPS-stimulated secretory response of monocytes [81].

Lipopolysaccharide-binding protein (LPB) is a trace plasma glycoprotein that binds to the lipid A moiety of LPS [82]. The concentration of this acute-phase protein increases up to 100-fold shortly after the onset of bacteremia. Lipopolysaccharide-binding protein can opsonize LPS-bearing particles and mediate their attachment to macrophages [81]. Significantly, the macrophage

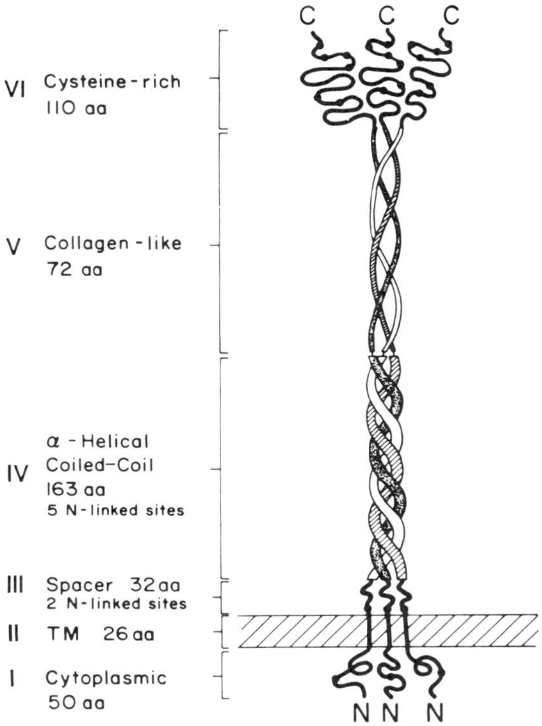

FIGURE 4 Schematic representation of the predicted trimeric structure of the type I bovine scavenger receptor. (From Ref. 80.)

activation response to LPS is enhanced approximately 100-fold in the presence of LPB [83]. The receptor for LPB appears to be CD14 [84]. The binding of LPS-LPB complexes does not occur to macrophages that have been incubated on immobilized mAb to CD14. Furthermore, treatment of macrophages with PI-PLC to partially remove CD14 decreases LPS-LPB binding. CD14 has been cloned [85]. It is a 55-kDa inositolphosphate-linked glycoprotein. It is composed of a series of leucine-rich residues, making it a member of a putative family of proteins with related structure, but as yet unresolved functions. The binding of LPS-LPB complexes to macrophages causes the release of TNF, a process that can be blocked by some mAb to CD14. Thus, the binding of LPB-LPS complexes to CD14 on macrophages may contribute to the inflammatory consequences typically associated with endotoxemia [83].

IV. RECEPTORS FOR SIMPLE AND COMPLEX SACCHARIDES

The rapid in vivo clearance of high mannose-containing glycoproteins such as β-glucuronidase led to the eventual identification of a macrophage receptor capable of binding terminal mannose [86]. The receptor binds mannose with an affinity of 5×10^{-9}M. It is a Ca^{2+}-dependent lectin that works optimally at pH 7 and minimally at pH 5 [86]. The receptor is expressed on tissue macrophages but not on circulating monocytes. The receptor is sometimes referred to as the mannose-fucose receptor because it is capable of recognizing fucose and to a lesser extent other hexoses.

The primary structure of the mannose receptor has recently been deduced from a cDNA clone isolated from placenta [87]. The mannose receptor is a 162-kDa glycoprotein containing a carboxyterminal cytoplasmic domain and a membrane-spanning region. Extracellularly, it has a membrane-proximal cysteine-rich domain followed by a domain that resembles the Type II repeats of fibronectin. Following this demain, there are eight tandem *N*-terminal domains that are related in sequence to the Ca^{2+}-dependent carbohydrate recognition domains (CRD) of other animal lectins [88] (Figure 5). Thus, the high-affinity binding exhibited by this receptor is probably the result of multiple clustered CRDs, each of which binds mannose with low affinity.

The cloned mannose receptor has been expressed in COS cells [88]. This receptor retains ligand binding and mediates efficient ligand internalization. Removal of the cytoplasmic domain of the receptor results in a receptor that retains ligand binding but is no longer able to internalize [88]. The mannose receptor has been reported to have two primary biological roles. The first is as a scavenger-type receptor capable of removing released neutrophil granule glycoproteins at the site of inflammation [86]. Its other primary role appears to be in microbial recognition and destruction. The majority of *N*-linked glycoproteins in higher mammals are processed either to remove mannoses or to cover them with terminal sugars. Lower organisms, however, have little processing capability and are therefore more likely to express glycoproteins with terminal mannose. Thus the mannose receptor is a candidate for a cell surface structure involved in mediating many of the previously described "nonspecific" macrophage recognition events that occur in the absence of antibody or complement. This receptor has been directly implicated in the recognition of yeasts [89] by binding to mannan in yeast cell walls. The protozoan *Leishmania donovani* binds to mannose receptors [90], and recently the binding of *Pneumocystis carinii* to human mononuclear cells has been shown to be mediated by the mannose receptor [91]. The mannose receptor has also been implicated in the binding of *Pseudomonas aeruginosa* [92]. Binding to the mannose receptor generally results in efficient particle internalization, even in transfected cells [88], and on phagocytic cells this internalization is usually accompanied by a brisk respiratory burst.

Other saccharide-specific receptors, which have not yet been cloned, have been identified on leukocytes. A β-glucan-inhibitable receptor has been identified on human monocytes [93] and elicited rodent peritoneal macrophages. This receptor has been reported to recognize carbohydrate polymers contained in yeast cell walls and a variety of particulate activators of the alternative complement pathway. Czop and colleagues [94] have identified two molecular weight species of 150 kDa and 170 kDa on the monocytic cell line U-937 with characteristics of the β-glucan receptor. Rodent stromal macrophages express a "sialoadhesin" [95] that is specific for sialic acid. This sialoadhesin mediates the binding (but not internalization) of sheep erythrocytes in hematopoietic organs. The receptor for advanced glycosylation end products [96] binds to proteins containing nonenzymatically modified glucose adducts, and this binding activity on tissue macrophages may contribute to microbial clearance.

A unique recognition mechanism involved in the interaction of some microbes with macrophages has been described. It is called "lectinophagocytosis" [97], and it involves the binding of bacterial lectins to glycoproteins expressed on the surface of leukocytes. The mannose-specific lectin from type 1 fimbriae is an example of such a lectin. Bacteria expressing this lectin activity

FIGURE 5 Schematic representation of the macrophage mannose receptor. (From Ref. 88.)

bind to mannose residues on macrophage glycoproteins. This mechanism has been implicated in the opsonin-independent binding of *E. coli* [97], *Salmonella typhi,* and *Klebsiella pneumoniae* [98]. The identity of the glycosylated receptors on monocytes/macrophages that mediate this effect is uncertain. The macrophage Fcγ receptors are mannosylated and have been shown to participate in the phagocytosis of *E. coli* [99]. Recently, the CD18-containing leukocyte adhesion molecules have been identified as the primary mediators of lectinophagocytosis of *E. coli* [100].

V. THE β1 AND β3 INTEGRIN RECEPTORS

In addition to the well-described β2 integrins, leukocytes express both β1 and β3 integrins. The β1 integrins, referred to as VLA antigens on lymphocytes, include receptors for fibronectin, collagen, and laminin [101]. Receptors for matrix proteins on leukocytes have been isolated and characterized [102]. The fibronectin receptor, a $\beta 1/\alpha 5$ heterodimer, recognizes the RGD-containing cell binding domain of fibronectin [101]. This receptor, in addition to binding to the extracellular matrix, binds microbes that have been opsonized with fibronectin or that have fibronectin-like proteins on their surface. Fibronectin recognition has been directly implicated in the recognition of a number of pathogens, including the protozoa *Trypanosoma cruzi* [103] and *L. major* [104]. The presence of RGD sequences in proteins expressed on the surface of *H. influenzae* [105] and *B. pertussis* [61] suggests the possibility of an integrin-RGD interaction in the initial microbial recognition event. The addition of fibronectin to macrophages also causes an increase in complement receptor-mediated phagocytosis [45].

Another recently described integrin on leukocytes is the so-called "leukocyte response integrin" [102]. This protein resembles the β3 cytoadhesions. It binds ligand via an oxidation-sensitive RGD or KGAGDV-inhibitable mechanism. Ligand binding to this receptor causes a general enhancement of neutrophil phagocytosis of unrelated targets, including IgG-coated erythrocytes. This receptor, which may be triggered during leukocyte chemotaxis, may be involved in the extracellular matrix-dependent stimulation of the phagocytosis of opsonized microbes by neutrophils.

VI. SUMMARY

There are two general concepts that we hope to have stressed concerning the recognition of microbes by phagocytic cells. The first is the concept of receptor redundancy and receptor cooperativity. Multiple receptors on leukocytes often participate in a given microbial recognition event. This concept can be illustrated by leishmania promastigotes that utilize both mannose receptors and Mac-1 to bind efficiently to macrophages. Likewise, macrophages use receptors for both IgG and complement to phagocytize encapsulated bacteria. This cooperativity between multiple receptors often changes the apparent affinity of the receptors for their ligand. Consequently the physiology of these receptors is altered. Fibronectin ligation, for example, results in the internalization of C3b-coated particles by the CR1. The second concept concerns the transduction of specific cellular signals following receptor ligation. Often, the receptor to which a microbe binds orchestrates many of the subsequent intracellular events during phagocytosis by transducing specific cellular signals. Some receptors, for example the mannose and Fcγ-receptors, are particularly well suited to direct particles to phagolysosomes and trigger a respiratory burst, whereas other receptors, for example the CR1, may not. Indeed, from this perspective, one can view the immune response as being designed to target microbes preferentially to those receptors on phagocytic cells capable of making the appropriate cellular responses. In the case of leishmania, phagocytosis mediated by the Fcγ receptors leads to parasite killing even by resident macrophages, while complement-mediated phagocytosis leads to parasite survival.

REFERENCES

1. Esko JD. Genetic analysis of proteoglycan structure, function and metabolism. Curr Opin Immunol 1991; 3:805–16.
2. Unkeless JC, Scigliano E, Freedman VH. Structure and function of human and murine receptors for IgG. Annu Rev Immunol 1988; 6:251–81.
3. Rosales CJ, Brown EJ. Signal transduction by neutrophil immunoglobulin G Fc receptors. J Biol Chem 1992; 267:5265–71.
4. Sears D, Osman N, Tate B, McKenzie IFC, Hogarth PM. Molecular cloning and expression of the mouse high affinity Fc receptor for IgG. J Immunol 1990; 144:371–8.
5. Allen JM, Seed B. Isolation and expression of functional high affinity Fc receptor complementary DNAs. Science 1989; 243:378–80.
6. Ravetch JV. Fc receptors. Annu Rev Immunol 1991; 9:457–92.
7. Schreiber RD. Identification of γ-interferon as a murine macrophage activating factor for tumor cytotoxicity. Cont Topics Immunobiol 1984; 13:171–98.
8. Graziano RF, Fanger MW. Human monocyte-mediated cytotoxicity: the use of Ig-bearing hybridomas as targets to detect trigger molecules on the monocyte cell surface. J Immunol 1987; 138:945–50.
9. Lewis VA, Koch T, Plutner H, Mellman, I. A complementary DNA clone for a macrophage lymphocyte Fc receptor. Nature 1986; 324:372–5.
10. Ravetch JV, Luster AD, Weinshank JK, Pavlovek A, Portnoy D, Hulmes J, Pan Y, Unkeless JC. Structural heterogeneity and functional domains of murine immunoglobulin G Fc receptors. Science 1986; 234:718–25.
11. Hibbs ML, Bonadonna L, Scott BM, McKenzie IF, Hogarth PM. Molecular cloning of the human immunoglobulin G Fc receptor. Proc Natl Acad Sci USA 1988; 85:2240–4.
12. Feister AJ, Browder B, Willis HE, Mohanakumar T, Ruddy S. Pertussis toxin inhibits human neutrophil responses mediated by the 42 kilodalton IgG receptor. J Immunol 1988; 141:228–33.
13. MacIntyre EA, Roberts PJ, Abdul-Gaffar R, O'Flynn K, Pilkington GR, Farace F, Morgan J, Linch DC. Mechanisms of human monocyte activation via the 40 kDa Fc receptor for IgG. J Immunol 1988; 141:4333–43.
14. Miettinen JM, Rose JK, Mellman I. Fc receptor isoforms exhibit distinct abilities for coated pit localization due to cytoplasmic domain heterogeneity. Cell 1989; 58:317–27.
15. Simmons D, Seed B. The Fcγ receptor of natural killer cells is a phospholipid linked membrane protein. Nature 1988; 333:568–70.
16. Ravetch JV, Perussia B. Alternative membrane forms of Fcγ RIII (CD16) on human NK cells and neutrophils: cell type specific expression of 2 genes which differ in single nucleotide substitutions. J Exp Med 1989; 170:481–97.
17. MacIntyre EA, Wallace DW, O'Flynn K, Abdul-Gaffar R, Tetteroo P, Morgan G, Linch DC. Binding of monoclonal antibody to CD16 causes calcium mobilization in large granular lymphocytes but inhibits NK killing. Immunology 1989; 66:459–64.
18. Lanier LL, Ruitenberg JJ, Phillips JH. Functional and biochemical analysis of CD16 antigen on natural killer cells and granulocytes. J Immunol 1988; 141:3478–85.
19. Ra C, Jouvin M, Blank V, Kinet JP. A macrophage Fcγ receptor and the mast cell receptor for immunoglobulin E share an identical subunit. Nature 1989; 341:752–4.
20. Clarkson SB, Kimberly RP, Valinsky JE, Witmer MD, Bussel JB, Nachman RL, Unkeless JC. Blockade of clearance of immune complexes by an anti-Fcγ receptor monoclonal antibody. J Exp Med 1986; 164:474–89.
21. Anegon I, Cuturi C, Trinchieri G, Perussia B. Interaction of Fc receptor (CD16) ligands induces transcription of interleukin 2 receptor (CD25) and lymphokine genes and expression of their products in human natural killer cells. J Exp Med 1988; 167:452–72.
22. Carsatella MA, Anegon I, Cuturi M, Griskey P, Trinchieri G, Perussia B. FcγR (CD16) interaction with ligand induces Ca^{2+} mobilization and phosphatide turnover in human natural killer cells. J Exp Med 1989; 169:549–68.
23. Baker CJ, Rench MA, Noya FJ, Garcia-Prats JA. Role of intravenous immunoglobulin in prevention of late-onset infection in low-birth-weight neonates. Rev Inf Dis 1990; 12:S463–8.
24. Siber GR, Santosham M, Reid GR, Thompson C, Almeido-Hill J, Morell A, deLange G, Ketcham JK,

Callahan EH. Impaired antibody response to Haemophilus influenzae type b polysaccharide and low IgG2 and IgG4 concentrations in Apache children. N Engl J Med 1990; 323:1387–92.
25. Fattal-German M, Taillandier J, Mathieu D, Bizzini B. Pneumococcal vaccination of elderly individuals. Vaccine 1991; 9:542–4.
26. Brown EJ, Hosea SW, Frank MM. The role of antibody and complement in the reticuloendothelial clearance of pneumococci from the bloodstream. Rev Inf Dis 1983; 5S:S797–805.
27. Baker CJ, Kasper DL, Tager IB, Paredes A, Alpert S, McCormack WM, Goroff D. Quantitative determination of antibody to capsular polysaccharide in infection with type III strains of group B Streptococcus. J Clin Invest 1977; 59:810–8.
28. Baker CJ, Kasper DL. Correlation of maternal antibody deficiency with susceptibility to neonatal group B streptococcal infection. N Engl J Med 1976; 294:753–6.
29. Daum RS, Watson B, Clark HF, Plotkin SA. New developments in vaccines. Adv Ped Inf Dis 1991; 6:1–57.
30. Mellman IS, Plutner H, Steinman RM, Unkeless JC, Cohn ZA. Internalization and degradation of macrophage Fc receptors during receptor-mediated phagocytosis. J Cell Biol 1983; 96:887–95.
31. Jones T, Hirsch J. The interaction between Toxoplasma gondii and mammalian cells II. The absence of lysosomal fusion with phagocytic vacuoles containing living parasites. J Exp Med 1972; 136:1173–86.
32. Armstrong J, D'Arcy-Hart P. Response of cultured macrophages to Mycobacterium tuberculosis with observations on fusion of lysosomes with phagosomes. J Exp Med 1971; 134:713–25.
33. Joiner KA, Fuhrman SA, Miettinen HM, Kasper LH, Mellman I. Toxoplasma gondii: fusion competence of parasitophorous vacuoles in Fc receptor-transfected fibroblasts. Science 1990; 249:641–6.
34. Tenner AJ, Cooper NR. Analysis of receptor-mediated C1q binding to human peripheral blood mononuclear cells. J Immunol 1980; 125:1658–64.
35. Fearon DT. Cellular receptors for the third component of complement. Immunol Today 1984; 5:105–10.
36. Fearon DT, Ahearn JM. Complement receptor type 1 (C3b/C4b receptor; CD35) and complement receptor type 2 (C3d/EBV receptor; CD21). Curr Topics Microbiol Immunol 1989; 153:83–98.
37. Wong WW, Klickstein LB, Smith JA, Weis JH, Fearon DT. Identification of a partial cDNA clone for the human receptor for complement fragments C3b/C4b. Proc Natl Acad Sci USA 1985; 82:7711–5.
38. Campbell RD, Law SK, Reed KBM, Sim RB. Structure, organization and regulation of the complement genes. Annu Rev Immunol 1988; 6:161–95.
39. Wong WW. Structural and functional correlation of the human complement receptor type 1. J Invest Dermatol 1990; 6:64S–67S.
40. Klickstein LB, Bartow TJ, Miletic V, Rabson LD, Smith JA, Fearon DT. Identification of distinct C3b and C4b recognition sites in the human C3b/C4b receptor (CR1, CD35) by deletion mutagenesis. J Exp Med 1988; 168:1699–1717.
41. Krych M, Hourcade D, Atkinson JP. Sites within the complement C3b/C4b receptor important for the specificity of ligand binding. Proc Natl Acad Sci USA 1991; 88:4353–7.
42. Paul MS, Aegerter M, O'Brien SE, Kurtz CB, Weis JH. The murine complement receptor gene family: Analysis of mCRY gene products and their homology to human CR1. J Immunol 1989; 142:582–9.
43. Holers VM, Kinoshita T, Molina H. The evolution of mouse and human complement C3 binding proteins: divergence of form but conservation of function. Immunol Today 1992; 13:231–6.
44. Griffin FM, Activation of macrophage C3 receptors by a unique lymphokine. Lymphokines 1987; 14:431–41.
45. Bohnsack JF, Takahashi T, Brown EJ. Interaction of culture-derived macrophages with the fibroblast binding domain of fibronectin is a necessary but insufficient signal for fibronectin enhancement of CR-1 mediated phagocytosis. J Immunol 136:3793–8.
46. Aderem AA, Wright SD, Silverstein SC, Cohn ZA. Ligated complement receptors do not activate the arachidonic acid cascade in resident peritoneal macrophages. J Exp Med 1985; 161:617–22.
47. Noel GJ, Katz S, Edelson PJ. Complement mediated early clearance of Haemophilus influenzae type b from blood is independent of lytic activity. J Inf Dis 1988; 157:85–90.
48. Kishimoto TK, Larson RS, Corbi AL, Dustin ML, Staunton DE, Springer TA. The leukocyte integrins: LFA-1, Mac-1, and P150,95. Adv Immunol 1989; 46:149–82.
49. Larson RS, Corbi AL, Berman L, Springer TS. Primary structure of the leukocyte function-associated molecule-1 alpha subunit: an integrin with an embedded domain defining a protein superfamily. J Cell Biol 1989; 1108:703–12.

50. Roubey RA, Ross GD, Merrill JT, Walton F, Reed W, Winchester RJ, Buyon JP. Staurosporine inhibits neutrophil phagocytosis but not iC3b binding by CR3 (CD11b/CD18). J Immunol 1991; 146:3557–62.
51. Kishimoto TK, O'Connor K, Lee A, Roberts TM, Springer TA. Cloning of the β subunit of the leukocyte adhesion protein: homology to the extracellular matrix receptors defines a novel supergene family. Cell 1987; 48:681–90.
52. Beller DI, Springer TA, Schreiber RD. Anti-Mac-1 selectively inhibits the mouse and human type three complement receptor. J Exp Med 1982; 156:1000–11.
53. Alterieri DC, Badger R, Mannucci PM, Edgington TS. Oligospecificity of the cellular adhesion receptor Mac-1 encompasses an inducible recognition specificity for fibrinogen. J Cell Biol 1988; 107:1893–1900.
54. Wright SD, Weitz JI, Huang AJ, Levin SM, Silverstein SC, Loike JD. Complement receptor type three (CD11b/CD18) of human polymorphonuclear leukocytes recognizes fibrinogen. Proc Natl Acad Sci USA 1988; 85:7734–8.
55. Diamond MS, Staunton DE, de Fougerolles AR, Stacker SA, Garcia-Aguilar J, Hibbs ML, Springer TA. ICAM-1 (CD54): a counter receptor for Mac-1 (CD11b/CD18). J Cell Biol 1990; 111:3129–39.
56. Ross GD, Cain JA, Lachmann PJ. Membrane complement receptor type three has lectin-like properties analogous to bovine conglutinin and functions as a receptor for zymosan and rabbit erythrocytes as well as a receptor for iC3b. J Immunol 1985; 134:3307–15.
57. Wright SD, Jong MTC. Adhesion promoting receptors on human macrophages recognize Escherichia coli by binding to lipopolysaccharide. J Exp Med 1986; 164:1876–88.
58. Wright SD, Reddy PA, Jong MTC, Erickson BW. C3bi receptor (complement receptor type 3) recognizes a region of complement protein C3 containing the sequence Arg-Gly-Asp. Proc Natl Acad Sci USA 1987; 84:1965–8.
59. Taniguchi-Sidle A, Isenman DE. Mutagenesis of the Arg-Gly-Asp triplet in human complement component C3 does not abolish binding of iC3b to the leukocyte integrin complement receptor type III (CR3, CD11b/CD18). J Biol Chem 1992; 267:635–43.
60. Myones BL, Dalzell JG, Hogg N, Ross GD. Neutrophil and monocyte cell surface p150,95 has iC3b receptor (CD4) activity resembling CR3. J Clin Invest 1988; 82:640–51.
61. Relman D, Tuomanen E, Falkow S, Golenbock DT, Saukkonen K, Wright SD. Recognition of a bacterial adhesion by an integrin: macrophage CR3 binds filamentous hemagglutinin of Bordatella pertussis. Cell 1990; 61:1375–82.
62. Mosser DM, Springer TA, Diamond MS. Leishmania require opsonic complement to bind to the human leukocyte integrin Mac-1 (CD11b/CD18). J Cell Biol 1991; 116:511–20.
63. Rimoldi MT, Tenner AJ, Bobak DA, Joiner KA. Complement component C1q enhances invasion of human mononuclear phagocytes and fibroblasts by Trypanosoma cruzi trypomastigotes. J Clin Invest 1989; 84:1982–9.
64. Bobak DA, Washburn RG, Frank MM. C1q enhances the phagocytosis of Cryptococcus neoformans blastospores by human monocytes. J Immunol 1988; 141:592–7.
65. Spear GT, Jiang HX, Sullivan BL, Gewurz H, Landay AL, Lint TF. Direct binding of complement component C1q to human immunodeficiency virus (HIV) and human T lymphotropic virus-I (HTLV-I) coinfected cells. Aids Res Hum Retrovirus 1991; 7:579–85.
66. Montovani B, Rabinovitch M, Nussenzweig V. Phagocytosis of immune complexes by macrophages. J Exp Med 1972; 135:780–92.
67. Noel GJ, Mosser DM, Edelson PJ. Role of complement in mouse macrophage binding of Haemophilus influenzae type b. J Clin Invest 1990; 85:208–18.
68. Anderson DC, Schmalsteig FC, Arnaout MA, Kohl S, Tosi MF, Dana N, Buffone GJ, Hughes BJ, Brinkley BR, Abramson JS, Springer TA, Boxer LA, Hollers JM, Smith CW. Abnormalities of polymorphonuclear leukocyte function associated with a heritable deficiency of high molecular weight surface glycoproteins (GP138): common relationship to diminished cell adherence. J Clin Invest 1984; 74:536–51.
69. Winkelstein JA, Tomasz A. Activation of the alternative complement pathway by pneumococcal cell wall teichoic acid. J Immunol 1978; 120:174–82.

70. Wilkinson BJ. Staphylococcal capsules and slime. In: Easmon CSF, Adlam C, eds. Staphylococci and staphylococcal infections. New York: Academic Press 1982:481–523.
71. Phillips GN, Flicker PF, Cohen C, Manjula BN, Fischetti VF. Streptococcal M protein: α-helical coiled coil structure and arrangement on the cell surface. Proc Natl Acad Sci USA 1981; 78:4689–93.
72. Weis JJ, Law SK, Levine RP, Cleary PP. Resistance to phagocytosis by group A streptococci: failure of deposited complement opsonins to interact with cellular receptors. J Immunol 1985; 134:500–5.
73. Mosser DM, Edelson PJ. The third component of complement is responsible for the intracellular survival of Leishmania major. Nature 1987; 327:329–31.
74. Drevets DA, Campbell PA. Roles of complement and CR3 in phagocytosis of Listeria monocytogenes by inflammatory mouse peritoneal macrophages. Infect Immun 1991; 59:2645–52.
75. Payne NR, Horowitz MA. Phagocytosis of Legionella pneumophila is mediated by human monocyte complement receptors. J Exp Med 1987; 166:1377.
76. Hondalus MK, Diamond MS, Mosser DM. The facultative intracellular bacteria Rhodococcus equi requires MAC-1 to adhere to mammalian cells (manuscript submitted).
77. Beutler B, Cermai A. Cachectin and tumor necrosis factor as two sides of the same biological coin. Nature 1986; 320:584–8.
78. Durum SK, Schmidt JA, Oppenheim JJ. Interleukin 1: An immunological perspective. Annu Rev Immunol 1985; 3:263–87.
79. Lei MG, Morrison DL. Specific endotoxic lipopolysaccharide-binding proteins on murine splenocytes. II. Membrane localization and binding characteristics. J Immunol 1988; 141:1006–11.
80. Kodama T, Freeman M, Rohrer L, Zabrecky J, Matsudaira P, Krieger M. Type I macrophage scavenger receptor contains αhelical and collagen-like coiled coils. Nature 1990; 343:531–5.
81. Wright SD. Multiple receptors for endotoxin. Curr Opin Immunol 1991;3:83–90.
82. Tobias PS, Soldau K, Ulevitch RJ. Isolation of a lipopolysaccharide-binding acute phase reactant from rabbit serum. J Exp Med 1986; 164:777–93.
83. Shumann RR, Leong SR, Flaggs GW, Gray PW, Wright SD, Mathison JC, Tobias PS, Ulevitch RJ. Structure and function of lipopolysaccharide binding protein. Science 1990; 249:1429–31.
84. Wright SD, Ramos RA, Tobias PS, Ulevitch RJ, Matheson JC. CD14, a receptor for complexes of lipopolysaccharide (LPS) and LPS binding protein. Science 1990;249:1431–3.
85. Ferrero E, Goyert SM. Nucleotide sequence of the gene encoding the monocyte differentiation antigen CD14. Nucleic Acids Res 1988; 16:4173–81.
86. Stahl PD. The macrophage mannose receptor: current status. Am J Resp Cell Mol Biol 1990; 2:317–8.
87. Taylor ME, Conary JT, Lennartz MR, Stahl PD, Drickamer K. Primary structure of the mannose receptor contains multiple motifs resembling carbohydrate recognition domains. J Biol Chem 1990; 265:12156–62.
88. Exekowitz RAB, Sastry K, Bailly P, Warner A. Molecular characterization of the mannose receptor: demonstration of multiple carbohydrate recognition-like domains and phagocytosis of yeasts in Cos-1 cells. J Exp Med 1990; 172:1785–94.
89. Sung SJ, Nelson RS, Silverstein SC. Yeast mannans inhibits binding and phagocytosis of zymosan by mouse peritoneal macrophages. J Cell Biol 1983; 96:160–6.
90. Blackwell JM, Ezekowitz RAB, Roberts MB, Channon JY, Sim RB, Gordon S. Macrophage complement and lectin like receptors bind leishmania in the absence of serum. J Exp Med 1985; 162:324–31.
91. Ezekowitz RAB, Williams DJ, Koziel H, Armstrong MYK, Warner A, Richards FF, Rose RM. Uptake of Pneumocystis carinii mediated by the macrophage mannose receptor. Nature 1991; 351:155–8.
92. Speert DP, Wright SD, Silverstein SC, Mah B. Functional characterization of macrophage receptors for in vitro phagocytosis of Pseudomonas aeruginosa. J Clin Invest 1988; 82:872–9.
93. Czop JK, Austin KF. A β-glucan inhibitable receptor on human monocytes: its identity with the phagocytic receptor for particulate activators of the alternative complement pathway. J Immunol 1985; 134:2588–93.
94. Czop JK, Kay J. Isolation and characterization of β-glucan receptors on human mononuclear phagocytes. J Exp Med 1991; 173:1511–20.
95. Crocker PR, Gordon S. Mouse macrophage hemagglutinin (sheep erythrocyte receptor) with specificity for sialylated glycoconjugates characterized by a monoclonal antibody. J Exp Med 1989; 169:1337–46.
96. Vlassara H, Brownlee M, Cerami A. High affinity receptor-mediated uptake and degradation of glucose

modified proteins: a potential mechanism for removing senescent molecules. Proc Natl Acad Sci USA 1981; 78:5190–5.
97. Ofek I, Sharon N. Lectinophagocytosis: a molecular mechanism of recognition between cell surface sugars and lectins. Infect Immun 1988; 56:539–47.
98. Athamna A, Ofek I, Keisari Y, Markowitz S, Dutton GGS, Sharon N. Lectinophagocytosis of encapsulated Klebsiella pneumoniae mediated by surface lectins of guinea pig alveolar macrophages and human monocyte-derived macrophages. Infect Immun 1991; 59:1673–82.
99. Salmon JE, Kapur S, Kimberly RP. Opsonin-independent ligation of the Fc gamma receptors. The 3G8-bearing receptors on neutrophils mediate the phagocytosis of conconavalin A–treated erythrocytes and nonopsonized Escherichia coli. J Exp Med 1987; 166:1798–1813.
100. Gbarah A, Gahmberg CG, Ofek I, Jacobi U, Sharon N. Identification of the leukocyte adhesion molecules CD11 and CD18 as receptors for type I fimbriated (mannose specific) Escherichia coli. Infect Immun 1991; 59:4524–30.
101. Hemler ME. VLA proteins in the integrin family: structures, functions, and their role on leukocytes. Annu Rev Immunol 1990; 8:365–400.
102. Gresham HD, Goodwin JL, Allen PM, Anderson DC, Brown EJ. A novel member of the integrin receptor family mediates Arg-Gly-Asp-stimulated neutrophil phagocytosis. J Cell Biol 1989; 108:1935–43.
103. Wirth JJ, Kierszenbaum F. Fibronectin enhances macrophage association with invasive forms of Trypanosoma cruzi. J Immunol 1984; 133:460–4.
104. Wyler DJ, Sypek JP, McDonald JA. In vitro parasite monocyte interactions in human leishmaniasis: possible role of fibronectin in parasite attachment. Infect Immun 1985; 49:305–11.
105. Barenkamp SJ, Leininger E. Cloning, expression, and DNA sequence analysis of genes encoding nontypable Haemophilus influenzae high-molecular-weight surface-exposed proteins related to filamentous hemagglutinin of Bordatella pertussis. Infect Immun 1992; 60:1302–13.

6

Antigen Processing and Presentation

Susan A. Safley and H. Kirk Ziegler

Emory University School of Medicine, Atlanta, Georgia

I. INTRODUCTION

A. Role of the Macrophage in the Immune Response

Monocytes and tissue macrophages are the major cellular components of the classical reticuloendothelial system, a ubiquitous and dynamic group of cells with vital roles in inflammation and host defense against foreign pathogens. Along with B cells and dendritic cells, phagocytic macrophages are responsible for internalizing and degrading proteins from the extracellular environment. Processing of antigens by macrophages is important because while some intact, unfolded proteins can associate with major histocompatibility complex (MHC) molecules for effective presentation, T cells usually recognize antigenic sites (epitopes) that are buried in the native protein structure. During antigen processing native proteins are denatured and cleaved into immunogenic peptides that bind to MHC molecules intracellularly as the biosynthetic pathways of class I and class II MHC intersect the protein-processing pathways. Peptide-MHC complexes are transported to the cell surface, where they are displayed for recognition by T cells bearing specific receptors. T cell–macrophage contact is augmented by binding of the CD4 molecule with class II MHC or CD8 with class I MHC and by other accessory molecules [1].

B. Role of T Cells in the Immune Response

T lymphocytes bearing alpha-beta T cell receptors can be divided into two classes: cytotoxic T cells (T_c) and helper T cells (T_H). Cytotoxic T cells recognize tumor cells or host cells that are infected with intracellular bacterial or viral pathogens. Cytotoxic T lymphocytes bind reversibly to class I MHC-peptide complexes on the surface of offending cells and directly lyse them. Mature T_c cells express cell surface CD8 coreceptors that interact with class I MHC molecules to facilitate formation of the TCR-MHC-peptide trimolecular complex. Cytotoxic T cells also secrete cytokines such as interferon-γ. One of the effects of interferon-γ on macrophages is the induction of macrophage-mediated cytolytic activity against other infected cells.

Helper T cells, on the other hand, recognize antigenic peptides in association with class II MHC molecules. While class I MHC molecules are found on almost all cell types, class II MHC molecules are expressed constitutively only by antigen-presenting cells such as B cells, dendritic cells, and macrophages. Helper T cells bear the CD4 coreceptor, which facilitates binding to class II MHC molecules. Helper T cells can be subdivided into two categories: T_{H1} and T_{H2}. T_{H1} lymphocytes secrete interleukin-2 (IL-2) and interferon-γ, two lymphokines that enhance cell-mediated immune responses. T_{H2} cells secrete IL-4, IL-5, and IL-6, which stimulate maturation of B cells into plasma cells for production of antibodies necessary for humoral immunity [2].

C. Different Antigen-Processing Pathways for Presentation of Peptides by Class I or Class II MHC Molecules

There are fundamental differences in macrophage catabolism of exogenous antigens, those introduced into the cell from the outside, and endogenous antigens, nascent proteins synthesized intracellularly. The processing pathway followed by these two kinds of antigens influences which epitopes are produced and whether peptides have the opportunity to bind to class I or to class II MHC molecules. The particular MHC-peptide complex formed will determine which type of T cell will be stimulated (CD4+ T_h cells or CD8+ T_c cells) and the type of immune response that ensues. The outcome is dependent to a large extent on the way in which the macrophage first encounters the antigenic protein: by the endocytic or the cytoplasmic route [3,4].

Macrophages are constantly surveying the extracellular environment by ingesting samples via pinocytosis or phagocytosis. Once inside the cell, exogenous antigens are partially degraded during passage through a series of intracellular compartments characterized by decreasing pH and processing enzymes unique to each compartment. The resultant antigenic fragments, oligopeptides of about 13 to 18 residues, are available for binding to newly synthesized class II MHC molecules or possibly to those that have been recycled from the cell surface.

A different intracellular trafficking pathway is followed by endogenous antigens such as viral or tumor proteins and those produced by intracellular bacteria. The currently accepted model is that endogenous antigens (cytosolic, membrane, or secreted antigens) are processed in the cytoplasm and that peptides are transported into the endoplasmic reticulum, where they bind to newly synthesized class I molecules. Complexes consisting of class I MHC and peptides of consistent length (8 to 9 residues) are exported to the cell surface for presentation to CD8+ T cells. Evidence suggests that some cytoplasmically processed peptides may occasionally be carried into the endocytic pathway for binding to class II MHC molecules [5].

D. Overview of This Chapter

The aim of this chapter is to review current understanding of antigen processing and presentation. The discussion will be divided into three parts: 1) class I MHC versus class II MHC molecules, including their assembly, peptide binding, intracellular trafficking, and proposed chaperones; 2) endocytic versus cytoplasmic processing of antigens, including recent progress in understanding the roles of proteasomes and peptide transporters; 3) T cell–macrophage interactions; and 4) the processing and presentation of antigens from the intracellular pathogen *Listeria monocytogenes* as an example of these different processing and presentation pathways.

II. CLASS I AND CLASS II MHC MOLECULE BIOSYNTHETIC PATHWAYS

A. Class I MHC Molecules

In recent years much insight has been gained into the nature of the interaction between antigenic peptides and MHC molecules and the importance of this interaction for transport of class I molecules to the surface of the cell, where presentation to CD8+ T cells occurs. Studies have shown that class I MHC molecules acquire antigenic peptides primarily in an intracellular compartment, rather than at the cell surface. Agents such as chloroquine that raise the pH of lysosomes or endosomes and neutralize acid proteases do not interfere with presentation by class I MHC molecules. (These agents do interfere with class II MHC presentation.) The fungal antibiotic brefeldin A, which inhibits exit of proteins from the endoplasmic reticulum (ER), blocks class I–mediated antigen presentation, as does treatment of antigen-presenting cells with inhibitors of protein synthesis such as cycloheximide. In addition, infection of cells with recombinant vaccinia virus expressing the E19 adenovirus glycoprotein, which retains class I MHC molecules in the ER, has been shown to prevent MHC class I presentation. This information suggests that newly synthesized class I molecules are involved in antigen binding. Extensive degradation of proteins within the ER itself is thought to supply some of the peptides, while others are delivered from the cytoplasm. Class I MHC molecules can effectively present antigens delivered to the cytosol by osmotic lysis of pinosomes or antigens that remain in the cytosol because of the lack of an ER targeting sequence [6]. (See Note in Proof.)

Experiments with the mutant mouse cell line RMA-S have provided evidence that peptides play an important role in class I MHC assembly, transport, and conformation. RMA-S cells synthesize both class I MHC heavy-chain and β_2-microglobulin, but the two molecules associate very weakly and move slowly to the cell surface. Surface expression is enhanced by reduced temperatures (27° C), at which "empty" class I MHC molecules may be more stable. The addition of exogenous peptide stabilizes the expression of MHC-peptide complexes at 37° C, enhances the association of newly synthesized MHC class I heavy-chain with β_2-microglobulin, and shifts class I chains from high mannose to the complex carbohydrate form, indicating movement from the ER through the Golgi complex [7–9]. These findings strongly suggest that stable assembly of class I MHC–β_2-microglobulin–peptide complexes and transport to the cell surface require the presence of processed peptide. As a result of this requirement, class I molecules may be retained in the ER until association with processed peptide has a chance to occur.

Studies of naturally processed peptides acid-eluted from class I molecules and purified by reversed-phase high-performance liquid chromatography (HPLC) have provided insight into the sizes and types of antigenic peptides that are bound. The data indicate that peptides are highly specific for binding to particular MHC alleles. For example, a peptide isolated from H-2K^d of cells infected with vesicular stomatitis virus does not bind to H-2K^b. In fact, sequencing of self-peptides eluted from the four MHC class I molecules H-2K^d, H-2K^b, H-2D^b, and HLA-A2 indicate a distinct allele-specific motif presented by each class I molecule [10,11]. The mechanism by which MHC-specific peptides are generated is not known but may involve MHC-dependent processing, for example trimming of an MHC-bound peptide by exopeptidases in the ER [12]. Protease activities and transport mechanisms independent of the MHC may also be involved, as discussed in the next section of this chapter.

Isolation of endogenous peptides bound to the human class I molecule HLA-B27 confirms earlier reports that class I–restricted epitopes are short peptides, octamers or nonamers. Self-peptides were identified by matching to proteins listed in sequence data bases, and all are derivatives of cytosolic or nuclear proteins, such as histone or ribosomal proteins and members of the 90K heat-shock protein family [13]. Probably any peptide resulting from normal cellular metabolism has the potential to be presented by class I molecules. X-ray crystallography of electron density in the binding site of HLA-B27 indicates that the peptides are in an extended conformation with side chains protruding in a roughly alternating pattern typical of beta strands [14]. Naturally processed peptides bound to HLA-A2.1 have also been sequenced and found to be identical in length to those associated with HLA-B27, possibly reflecting an identical processing mechanism, although the binding motifs are different [15].

In summary, the current view of class I antigen presentation is that most newly synthesized class I molecules acquire short, processed peptides (octamers or nonamers) intracellularly and are already "loaded" when they reach the cell surface. Evidence suggests that peptide binding occurs in a pre-Golgi compartment, most likely the ER, and that peptide association stabilizes the class I MHC–β_2-microglobulin complex for efficient and rapid transport through the secretory pathway to the cell surface. Complexes lacking peptide are thought to be in an inappropriate conformation and

TABLE 1 Characteristics of Processing and Presentation via Class I and Class II MHC

Characteristics	Class I MHC	Class II MHC
Inhibition by lysosomotropic agents	No	Yes
Inhibition by brefeldin A	Yes (sometimes no)	No (sometimes yes)
Antigen source	Endogenous (sometimes exogenous)	Exogenous (sometimes endogenous)
Site of processing	Cytoplasm, ER or vacuole	Endosome, lysosome
Processing enzymes	Proteosomes?	Cathepsin B and D and others
Site of peptide-MHC interaction	ER	Endosome
pH optimum of peptide-MHC binding	? $<$ pH 7	pH 5–6
Peptide size	Restricted, 8–9 amino acids	Variable, 10–18 amino acids
Peptide-MHC binding motifs	Class- and allele-specific	Class- and allele-specific
MHC de novo synthesis required	Yes	Yes, sometimes no
Peptide transporter required	Yes	Probably no
Coreceptor	CD8	CD4

therefore retained in the ER by a hypothetical chaperone [16–18]. According to this view, class I MHC molecules do not present exogenous antigens because they are already charged with endogenous peptides should their path intersect the endocytic route [19]. However, recent evidence indicates that the class I and class II intracellular trafficking pathways do not usually intersect [20]. The characteristics of processing and presentation via class I MHC are summarized in Table 1.

B. Class II MHC Molecules and Invariant Chain

Class II MHC molecules primarily present peptides derived from exogenous antigens that are taken into the macrophage from the extracellular fluid by endocytosis or phagocytosis. Class II antigen presentation to CD4+ T cells is inhibited by lysosomotropic agents such as chloroquine and ammonium chloride, indicating that antigen is degraded by the endocytic route. Studies on the intracellular trafficking and subcellular localization of class II MHC molecules indicate that nascent class II molecules are translocated into the ER before transport to the Golgi apparatus and then are delayed 1 to 3 hours in their movement from the *trans*-Golgi to the surface. During the delay, class II molecules become concentrated in an endocytic compartment distinct from the ER and *trans*-Golgi that may be the site where binding to antigenic peptides occurs [6].

The sequence of peptides naturally bound to class II MHC molecules has recently been determined. I-A^b and I-E^b were purified from the murine B cell lymphoma line LB27.4 (H-2^{bxd}), and peptides were eluted by treatment with acetic acid. Reversed-phase HPLC analysis revealed distinct sets of peptides associated with different class II molecules but no clear allele-specific "binding motif". All of the class II–associated peptides ranged in size from 13 to 18 residues, as opposed to peptides of 8 to 9 amino acids for class I MHC molecules, as mentioned above. Overlapping peptides corresponding to the same core determinant were identified and found to have precise *N*-terminal ends but varying lengths because of different C-terminal cleavage sites. Based on this information a reasonable model for peptide-class II MHC association would involve protection of the *N*-terminus in the peptide-binding groove, with the C-terminus dangling out of the groove for truncation by proteases [19,21]. (See Note in Proof.)

There is controversy as to whether presentation of exogenous antigens requires newly synthesized class II molecules or whether class II can be recycled from the cell surface. There is evidence for a rapidly recycling internal pool of class II that exceeds the number of nascent class II molecules by more than 60 times [22]. In some B cell lines peptide-MHC complexes appear to have a short half-life and rapid turnover, implying the existence of a mechanism to dissociate preexisting peptide-MHC class II complexes. In that case class II molecules might be recycled to present more than one immunogenic peptide in a lifetime [23]. However, other studies show that presentation of exogenous antigen by class II MHC is inhibited by brefeldin A. This finding implicates newly synthesized class II in antigen presentation [24].

Newly synthesized class II MHC molecules are polymorphic heterodimers consisting of a 34-kDa α chain and a 28-kDa β chain. Class II molecules in the ER are associated with a nonpolymorphic 31-kDa accessory molecule, known as the invariant chain (Ii). It has been hypothesized that class II MHC molecules do not compete with class I MHC molecules for acquisition of endogenous, cytoplasmically processed peptides in the ER because nascent class II chains quickly associate with Ii. Invariant chain is thought to block the peptide-binding site of class II MHC early in its biosynthetic path. This suppresses the presentation of peptides found in the ER and protects class II from associating with newly synthesized proteins that are still partially unfolded. Later, when Ii is proteolytically cleaved in an acidic compartment, such as an endocytic vesicle, the binding site becomes available to peptides [6,25,26].

Evidence that Ii is important for antigen processing and presentation has been provided by studies of L cell fibroblasts, which express only low levels of Ii, transfected with class II MHC molecules. Transfected cells could present peptide fragments but not native protein antigens [27]. Supertransfection with the Ii gene enabled the class II–bearing L cells to present intact antigens even more effectively than normal spleen cells or B lymphoma cells [28]. The biochemical basis for the effect of Ii on antigen presentation may involve its influence on the proper folding, conformation, and post-translational modification of class II molecules. In studies of cells transfected with class II MHC molecules with and without Ii, it was found that in the absence of Ii, class II dimers were misfolded and inefficiently transported from the ER to the Golgi. The class II molecules that were transported through the Golgi in the absence of Ii had an altered molecular mass associated with *N*-linked glycosylation [29,30].

Invariant chain may be responsible for directing class II MHC molecules to endocytic compartments because the amino-terminus of Ii contains a signal sequence that targets the molecule to post-Golgi vesicles. In the absence of class II molecules, Ii alone is sorted to post-Golgi compartments. But when the first 15 amino acids of the amino-terminus are deleted, Ii goes directly to the cell surface [31]. On the other hand, class II–positive L cell transfectants can process and present soluble protein antigens in the absence of Ii, suggesting that class II may contain the appropriate targeting signals. Possibly the endosomal localization signal in the cytoplasmic tail of Ii functions in retention, rather than in sorting [32]. Current evidence supports the hypothesis that class II MHC-Ii complexes are sorted to an endocytic compartment, where a retention signal on Ii delays further movement. When Ii is proteolytically cleaved by processing enzymes such as cathepsin B, the antigen-binding cleft of class II is made available to peptide fragments, and the class II–peptide complex travels to the cell surface [31]. The characteristics of processing and presentation via class II MHC are summarized in Table 1.

A nonendosomal Ii-independent pathway for class II MHC presentation has recently been described for some viral antigens. Studies have shown that cells infected with live influenza virus can be recognized by both class I– and class II–restricted T cells. The T cell response can be blocked by brefreldin-A but not by chloroquine, implying that nascent class II MHC molecules can acquire endogenously processed viral peptides early in the biosynthetic pathway, possibly in the ER. This could conceivably happen if Ii were displaced because of a large molar excess of antigenic peptides, as might happen in viral infection or with peptides having an extremely high affinity for class II [5].

III. ENDOGENOUS AND EXOGENOUS ANTIGEN-PROCESSING PATHWAYS

A. Endogenous Antigens: Processing, Proteasomes, and Peptide Transporters

Recently, much has been learned about how proteins may be degraded in the cytoplasm and subsequently transported by a signal sequence–independent route into the ER for binding to class I molecules (Figure 1). Work by a number of investigators has recently shown that the cytoplasmic proteases and the peptide transporters involved in class I MHC processing and presentation are encoded by a cluster of genes within the class II MHC coding region [33–41]. These genes along with those for class I and class II MHC molecules are up-regulated by interferon-γ, so all elements necessary for adequate antigen processing and presentation can be induced simultaneously by this cytokine.

Catabolism of endogenous antigens in the cytoplasm occurs outside of any membrane-bound compartment, such as the lysosome, and is dependent on ATP. Proteolysis is thought to be carried out by the 20S proteasome (multicatalytic proteinase complex), a 700-kDa nonlysosomal enzyme complex found in all eukaryotic cells. The 20S proteasome is composed of about 30 subunits stacked into a barrel-like complex of four ring-shaped layers surrounding a central pore. It has been suggested that proteolytic degradation occurs within the barrel, thereby preventing digestion of the surrounding cellular constituents. The 20S proteasome is similar in structure and DNA sequence to the low-molecular-mass polypeptide (LMP) of mice, encoded by the mouse MHC region between the class I K locus and class II MHC genes. Two genes (LMP-1 and LMP-2) were identified over a decade ago, and there has been speculation that they might have a role in antigen processing and presentation. Evidence now points to LMP as a special type of 20S proteasome [40–43].

Cytoplasmic degradation of proteins is believed to occur as follows. A protein with an abnormal conformation, for example, is first marked for proteolysis by covalent conjugation with ubiquitin, a step that requires ATP and at least three enzymes. The ubiquitin-tagged protein is then degraded by the 20S proteasome, which is complexed with two other proteins. (The larger complex is known as the 26S proteasome.) One of the proteins is thought to inhibit proteolysis by the 20S proteasome until arrival of an appropriately ubiquinated protein substrate, while the other can activate the peptidase activity of the 20S proteasome in the presence of ATP [43]. This system of proteins probably exists in macrophages for the cytoplasmic processing of endogenous antigens, although as yet direct experimental evidence for ubiquitin-conjugation or the 26S proteasome in antigen presentation is lacking. (See Note in Proof.)

In addition to cytoplasmic processing, endogenous antigens may be broken down within lysosomes or in the ER. Lysosomal degradation of cytosolic proteins is thought to occur by two chloroquine-sensitive pathways; one involves nonspecific uptake of cytoplasmic proteins into lysosomes by the process of autophagy, and the other involves selective translocation of peptides tagged with a sequence motif related to KFERQ by a 70-kDa heat shock protein (hsp) [6,44]. Evidence supporting a role for the 70-kDa hsp in antigen processing is its localization to endosomes containing internalized immunoglobulin and class II MHC molecules [45]. While peptides that are processed in the cytosol are presented by class I MHC to CD8+ T cells, those that are translocated into lysosomes for processing may travel through the endocytic pathway to the cell surface for presentation to CD4+ T cells in association with class II MHC, as has been demonstrated for some viral proteins.

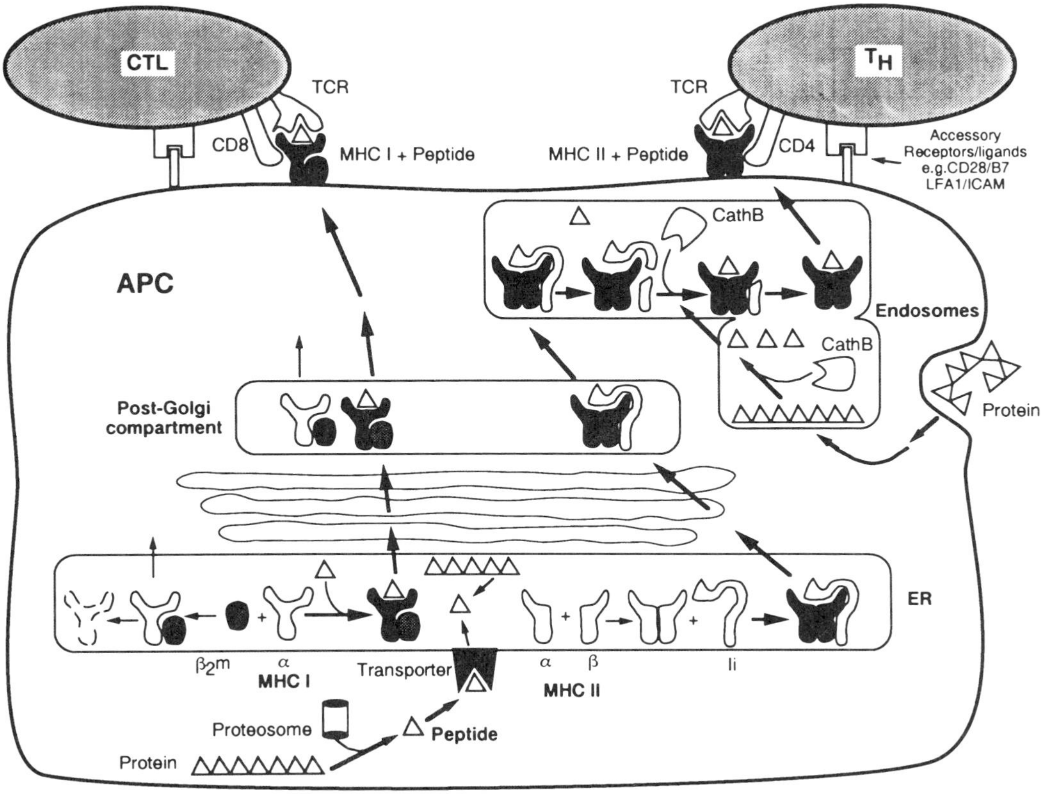

FIGURE 1 Endogenous and exogenous antigen processing pathways. As described in the text, endogenously synthesized proteins are believed to be processed by proteosomes in the cytoplasm and moved by peptide transporters into the endoplasmic reticulum, where binding to class I MHC molecules occurs. Exogenous antigens are taken up by the cell and processed by enzymes such as cathepsin B in membrane-bound compartments during transport through the endocytic pathway, where binding to class II MHC molecules occurs following cleavage of the invariant chain. CD8+ cytolytic T cells recognize processed peptides presented by class I MHC molecules, and CD4+ T helper cells recognize peptides presented by class II MHC molecules. Signals are transduced to T cells and antigen-presenting cells via accessory receptors/ligands such as CD28 on T cells and B7 on B cells.

Nascent proteins that are incorrectly folded or do not assemble properly may be degraded in the ER, where they bind to newly synthesized class I MHC molecules. It is currently believed, however, that most peptides associating with class I MHC molecules in the ER are derived from cytosolically processed antigens [6].

This leads to the question of how cytosolically processed peptides that lack signal sequences can be translocated across the ER membrane. It is possible that short peptides up to about 10 residues in length may not require signal sequences for active transport. Recently, however, genes encoding putative peptide transporters have been cloned: HAM1 and HAM2 in the mouse, *mtp*1 and *mtp*2 in the rat, and TAP1 (RING4 or PSF1) and TAP2 (RING11 or PSF2) in the human. These homologous genes are all located in the class II region of the MHC, and HAM1 and HAM2 are tightly linked to the murine proteasome subunit genes LMP1 and LMP2 implicated in cytoplasmic processing. DNA sequencing indicates that the protein products of these genes are closely related to a large group of prokaryotic and eukaryotic ATP-driven transport proteins, known as the ABC (ATP-binding cassette) superfamily [34–36,46–49].

Members of the ABC superfamily are specialized to pump a wide range of substrates, including peptides, across cell membranes. They have a common structure consisting of two ATP-binding domains and two hydrophobic transmembrane domains. The putative peptide transporter genes are closely linked in the MHC, and each encodes polypeptides with one ATP-binding domain and one transmembrane domain. Thus, HAM1 and HAM2 and their homologues may each make up half of a functional peptide pump [50]. Evidence supporting this idea has recently been supplied by several groups.

A number of studies have utilized the mutant murine cell line, RMA-S, which lacks cell surface expression of stable class I MHC and cannot present peptide epitopes derived from endogenous antigens. Since the normal phenotype could be restored by transfection with the rat *mtp*2 gene alone, the underlying defect in RMA-S cells was thought to lie in the mouse *mtp*2 homologue, HAM2 [47]. This idea was confirmed by a group of researchers who recently transfected RMA-S cells with the HAM-2 gene and rescued cell surface expression of class I MHC along with antigen presentation capability [48].

In similar studies a mutant human lymphoblastoid cell line (721.134) that has a transcriptionally inactive TAP1 gene was transfected with a normal TAP1 gene. This restored stable assembly of HLA-A2 class I MHC molecules and the ability to effectively present viral antigens. In addition, TAP1-TAP2 heterodimers were identified by immunoprecipitation, indicating that the two molecules normally associate with each other [46]. A different group used a mutant cell line, BM36.1, that has a phenotype similar to RMA-S. BM36.1 cells express normal TAP1 but have an aberrant form of TAP2. The mutant phenotype was reversed by transfection with the normal TAP2 gene, showing that TAP1 and TAP2 form a functional complex [51]. Recently used high-resolution immunoelectron microscopy has been used to demonstrate that TAP1 is located on membranes of the ER and is probably oriented with its ATP-binding domains in the cytosol [52]. This location of TAP1 is consistent with its proposed role of transporting peptides from the cytosol into the ER.

Although direct evidence is still lacking, it seems likely that the putative peptide transporters function in class I MHC antigen processing and presentation by moving antigenic peptides from the cytosol into the ER. Two pieces of evidence argue against this role for the ATP-dependent transporters, however. First, short peptides have been shown to enter microsomal vesicles in vitro in the absence of ATP. Experiments using an in vitro system for translating peptides across microsomal membranes suggest that ATP may be required for efficient assembly of class I MHC–peptide–β_2-microglobulin complexes, rather than for peptide transport. Second, peptides can be transported into microsomes made from human T2 cells, in which the TAP1 and TAP2 genes are both deleted [53].

While their peptide-transporting role awaits clarification, some far-reaching effects of these molecules on antigen presentation have been suggested by studies of the "*cim* phenomenon". The rat class I MHC molecule, RT1.A^a, undergoes drastic changes in its antigenicity when associated by genetic recombination with different alleles at the adjacent class II MHC region. The responsible locus in the class II MHC region was originally called *cim* (class I modifier), and it was eventually found to contain *mtp*1 and *mtp*2, the two rat members of the ABC transporter gene family. One explanation for the *cim* phenomenon is that after recombination with a different *mtp* allele, class I MHC molecules bind a uniquely different set of peptides transported into the ER. Class I MHC loaded with different peptides would stimulate different sets of T cells [54].

The *mtp*2 allele is in fact highly polymorphic, and each form of the molecule seems to be highly specific for certain peptides. Evidence for this specificity has been provided by reverse-phase HPLC analysis of radiolabeled peptides acid-eluted from the immunoprecipitated RT1.A^a

class I molecules of cells expressing different polymorphic forms of *mtp*2. The elution profiles showed very distinct sets of peaks indicating that remarkably different peptides were bound to the same class I molecule in the presence of different forms of *mtp*2. It is interesting to note that in rats class I MHC molecules seem to have coevolved with linked peptide transporters that will specifically pump the peptides that bind best to that particular class I allele [54].

No functional polymorphism of genes homologous to *mtp*2 has yet been observed in humans or mice. However, it is not difficult to see that heterozygous expression of transporters might be advantageous to an individual because a broader range of peptides would be made available for presentation. This could result in a better chance of developing immunity to pathogenic organisms. On the other hand, presentation of different self peptides by individuals with identical class I alleles, but different peptide transporter alleles, could change the development of the T cell repertoire and have profound effects on individual susceptibility to autoimmune disease.

B. Exogenous Antigens: The Endocytic Processing Pathway and Acid Hydrolases

Exogenous protein antigens are taken up by antigen-presenting cells, and after about 30 to 60 minutes discrete antigenic fragments are displayed on the surface of the cell in association with class II MHC molecules for recognition by specific CD4+ T cells. This section of the chapter will focus on events that occur along the endocytic pathway from antigen uptake to presentation of class II MHC-peptide complexes on the cell surface (Figure 1).

Extracellular proteins can enter the endocytic processing pathway in several ways, including phagocytosis, pinocytosis, or receptor-mediated endocytosis. Immunoglobulin on the surface of B cells binds monovalent antigens, and the antigen-antibody complexes are internalized in clathrin-coated pits. Macrophages bear surface Fc receptors that bind to multivalent antigens. When Fc receptors are crosslinked by antigen, they are quickly internalized and targeted to the endocytic processing pathway. Since macrophages are professional phagocytes, they can also ingest extracellular proteins by fluid-phase endocytosis without the need for any receptor.

The intracellular pathway followed by endocytosed antigens and the proteases utilized for their catabolism have been the subjects of intense investigation. Various kinds of antigen-presenting cells have been shown to handle antigens differently on the basis of a number of factors, including the types of intracellular compartments involved and the proteases they contain. In macrophages internalized protein antigens travel through a series of compartments beginning with the "early" and "late" endosomes and ending with lysosomes. It is unclear at present whether these compartments are truly separate organelles or if vesicular transport of antigen from one compartment to the other is required. In addition, it is still not known in which compartment antigenic peptides are finally produced and binding to class II MHC molecules occurs [6].

Endocytosed antigen can be found in peripheral or "early" endosomes just beneath the plasma membrane within 2 to 10 minutes of uptake, after the phagosome has fused with endosomal membranes. The interior of the endosome is acidic (pH 6.0 to 6.5) because of the action of a hydrogen ion (H^+) ATPase in the endosomal membrane that pumps H^+ from the cytosol into the compartment. Early endosomes contain the acid-dependent proteases cathepsin B and cathepsin D, which have been implicated in the production of antigenic peptides and the dissociation of invariant chain from class II MHC molecules [55]. Evidence indicates that cathepsin B is a major protease involved in antigen processing [56]. In addition, recent studies have shown that class II MHC–peptide association is greatly enhanced at H^+ concentrations in the range of pH 5.0 to 6.5. This observation is consistent with the idea that acidification of endocytic vesicles regulates the formation of peptide–class II MHC complexes [57].

Since experimental evidence indicates that early endosomes contain class II MHC molecules, antigen, proteases, and proper pH, all components necessary for antigen processing may be present there [58], but it is not certain that this is the site of peptide association with MHC molecules. Early endosomes are thought to be the main sorting compartments in the endocytic pathway. Vesicles containing recycling receptors, various ligands, and class II MHC molecules pinch off from the early endosomes and travel to the cell membrane, endosomes, or lysosomes [6].

By 10 to 20 minutes after endocytosis, antigen has progressed to the perinuclear region of the cell close to the Golgi apparatus within internal compartments that are slightly more acidic (pH 5.5) than those in the periphery. Unlike the early endosomes, these so-called "late" endosomes or prelysosomal compartments do not recycle vesicles to the cell surface. Instead, they receive vesicles carrying lysosomal enzymes from the *trans*-Golgi and sort them to the lysosomes. Although the pH of late endosomes is not optimal for lysosomal enzymes, there may be limited hydrolytic activity by lysosomal enzymes that further degrades antigen, in addition to proteolysis by the endosomal enzymes cathepsin B and cathepsin D [6].

The final endocytic compartment is the lysosome, which is extremely acidic (pH 4.5 to 5.0). Here proteins are reduced to single amino acids and dipeptides that are transported across the lysosomal membrane into the cytoplasm to be used in synthesizing new proteins. While there is speculation about the role of the lysosome in antigen processing, there is evidence for the presence of class II MHC molecules in a lysosome-like compartment that is only mildly acidic [59]. This compartment contains the lysosomal enzyme β-hexosaminidase as well as typical lysosome-associated membrane proteins, and it lacks markers that define late endosomes. It may be that class II MHC molecules encounter antigenic peptides along the endocytic pathway in an internal compartment with characteristics of a late endosome/early lysosome. There would certainly be sufficient proteolytic activity here to remove invariant chain and expose binding sites on class II MHC for antigenic peptides. If this is the case, then there must be a mechanism for transporting class II MHC–peptide complexes to the cell surface from this compartment. Recycling to the plasma membrane from lysosomes has in fact been observed for some lysosomal membrane proteins [6].

IV. REGULATION OF T CELL–MACROPHAGE INTERACTIONS

While the specific ligand for the T cell receptor (TCR) is the MHC-peptide complex, it is now clear that other molecular interactions regulate T cell and macrophage function during antigen presentation. Accessory molecules may reinforce adhesion between the T cell and the macrophage, while others may be involved in signal transduction. TCR engagement may be aided by CD4 or CD8 acting as coreceptors. T cell activation via the TCR can result in changes in the avidity of adhesion molecules expressed by the T cell. Also, the MHC molecules of antigen-presenting cells may transduce signals to the APC that increase the efficiency of antigen presentation.

T cell receptors and CD4 or CD8 molecules are believed to diffuse independently in the plasma membrane until they are brought into close contact by binding to the same peptide-MHC complex on the surface of the antigen presenting cell (APC). CD8 is a dimer, and the amino-terminal domain of each chain is thought to participate in binding to the membrane-proximal region of class I MHC molecules much as the heavy and light chains of antibody bind antigen. CD4 is a monomer consisting of four immunoglobulin-like domains that bind to the β chain of class II MHC molecules in an extended conformation [60]. In addition to their contribution to adhesion between T cell and APC, an important function for CD4 and CD8 is in signal transduction involving association of their cytoplasmic domains with a lymphocyte-specific protein tyrosine kinase ($p56^{lck}$) [61].

The interaction of lymphocyte function–associated antigen (LFA-1) on T cells and intercellular adhesion molecule-1 (ICAM-1) on APCs represents an example in which both receptor and ligand can be significantly regulated. The expression of ICAM-1 is induced on a wide variety of cell types by inflammatory mediators such as lipopolysaccharide (LPS) of gram-negative bacteria, interferon-γ, IL-1, and tumor necrosis factor (TNF). On resting T cells, LFA-1 has low avidity for its ligand ICAM-1 (and CD2 on the T cell has low avidity for LFA-3 on the APC). The affinity of an individual TCR for peptide-MHC is also believed to be low. When T cells encounter APCs, conjugate formation is probably initiated by interactions between accessory molecules. TCRs can then "survey" peptide-MHC complexes on the APC to find the appropriate antigen. More peptide-MHC complexes and accessory molecules can diffuse through the fluid membrane into the interface between T cell and APC, strengthening adhesion. If appropriate TCR-peptide-MHC matches reach a threshold density, a sufficient signal could be delivered for stimulation of the T cell [62]. This T cell activation results in an increase in the avidity of accessory molecules for their ligands, causing additional increases in the overall avidity of the T cell–APC interaction. This high avidity state is transitory, so after activation a T cell can be released from binding to the APC.

Studies of T cell–B cell interaction suggest that a signal may simultaneously be transduced into the APC via the cytoplasmic domains of class II MHC molecules. This signal increases the efficiency of antigen presentation so that less antigen is required for T cell stimulation [63]. Similarly, it has recently been shown that another accessory molecule on the surface of T cells, CD28, mediates adhesion to activated B cells by binding to the B7 antigen on B cells. A CD28-mediated signaling pathway is activated that results in increased T cell IL-2 production and proliferation. At the same time, the interaction between CD28 and B7 may result in B cell activation or differentiation [64].

V. PROCESSING AND PRESENTATION OF BACTERIAL ANTIGENS

Considering the roles of different processing and presentation mechanisms in immunity to infectious disease, the issue of how bacterial antigens can be presented by both class I and class II MHC proteins is well illustrated in the case of *Listeria monocytogenes*.

Listeria is a gram-positive foodborne intracellular pathogen capable of causing severe infections in immunocompromised individuals and pregnant women, although apparently healthy individuals may also become infected. After ingestion of contaminated food, virulent strains of *Listeria* are able to penetrate epithelial cells lining the intestinal tract. Following an initial period of infection and multiplication, the bacteria enter the bloodstream and travel to the liver and spleen, where they are ingested by macrophages. Strains of *Listeria* that secrete an exotoxin known as listeriolysin O (LLO) survive and grow inside macrophages, but strains that do not produce LLO are quickly cleared. Listeriolysin O has membrane-disrupting capability that enables the bacteria to escape from acidic host cell phagosomes, where they would otherwise be destroyed by degradative enzymes. *Listeria* multiply rapidly in the cytoplasm and begin to nucleate host cell actin, forming a "tail" of short actin filaments and actin-binding proteins. Actin polymerization allows the bacteria to move toward the surface of the macrophage, where they are projected in pseudopod-like structures. Cell-to-cell spread is accomplished when adjacent macrophages phagocytize the pseudopods. Pathogenic strains of *Listeria* are able to escape from the resulting double-walled vacuole into the cytoplasm of neighboring cells for further growth and replication [65,66].

Immunity to *Listeria* is exclusively cell-mediated. The currently accepted model is that *Listeria*-specific CD4+ T cells secrete cytokines such as interferon-γ that recruit and activate macrophages for increased microbial killing, while cytolytic CD8+ T cells eliminate infected cells. Besides functioning as a major virulence factor that promotes the intracellular growth of *Listeria* [67–69], LLO is one of the most important antigens in the immune response to *Listeria*. Evidence supporting the role of LLO in anti-*Listeria* immunity has been provided by several studies. Only immunization with viable, LLO-producing strains of *Listeria* or with purified LLO can induce protective immunity, and both CD4+ T cells and CD8+ T cells from *Listeria*-primed mice recognize epitopes of LLO [70,71]. We have identified a peptide fragment of listeriolysin-O, LLO 215-234, that contains the dominant epitope for class II MHC–restricted CD4+ T cells.

Competitive inhibition assays demonstrate that LLO 215-234 binds purified murine class II MHC molecules IE^k, IA^d, and IE^d. In addition, the majority of CD4+ T cell hybridomas made from peritoneal exudate cells of *Listeria*-immunized mice that recognize purified LLO are also specific for LLO 215-234. Using a nested set of synthetic peptides of LLO 215-234, we have identified the dominant minimal T cell epitope as LLO 215-226. It is interesting that a cut site for the major endocytic processing enzyme, cathepsin B, is located on the carboxy-side of residue 226 [72].

To identify the dominant epitope recognized by CD8+ T cells, naturally processed peptides were acid-eluted from spleen cells of mice heavily infected with *Listeria*. A CD8+ cytolytic T cell clone was found that killed target cells pulsed with one of these peptides in a class I MHC–restricted fashion (H-$2K^d$). Since the same clone killed target cells transfected with the LLO gene, the H-$2K^d$ binding motif was used to predict which nonamer peptides of LLO might contain the dominant epitope. One of the synthetic peptides tested, LLO 91-99, stimulated the T cell clone very strongly and was identified as the dominant class I MHC–restricted epitope for CD8+ T cells [71].

Thus, both CD4+ and CD8+ T cells can recognize discrete regions of LLO. Epitopes operative with class I and class II MHC molecules appear to be largely nonoverlapping. In light of the current understanding of antigen processing and presentation and the cell biology of *Listeria*-macrophage interactions, how can we account for the processing and presentation of LLO via both class I and class II MHC proteins?

Macrophages infected with LLO-expressing bacteria can process and present LLO epitopes to CD4+ T cells. However, at higher bacterial numbers and with increased time of infection an inhibition of processing and presentation via class II MHC proteins is observed [73]. It is likely that this inhibition corresponds to the time at which bacteria are causing disruption of the endosome membrane and entering the cytoplasm. As such, the inhibition of class II MHC–mediated presentation may result from the escape of bacteria from the endocytic route, where class II presentation predominates. It has been shown that LLO-expressing *Listeria* are actually required for presentation of antigens via class I MHC to CD8+ T cells [74]. A reasonable interpretation of these results is that bacteria that cannot make LLO remain trapped within endosomes, where their antigens are processed for presentation to CD4+ T cells in association with class II MHC molecules, while bacteria that make LLO enter the cytoplasm, where they use the processing pathway for presentation of antigens in association with class I MHC molecules. The different LLO epitopes operative with class I and II MHC may be due to either different MHC-peptide binding requirements or the possibility that cytosolic processing of LLO by proteasomes produces antigenic peptides that are different from those generated in the endocytic pathway by cathepsin B.

VI. CONCLUSION

Much has been learned about the mechanisms of antigen processing and presentation that should prove extremely useful for understanding infection and immunity to microbial pathogens. A better understanding of immunity, antigen processing, and MHC function should continue to permit more useful methods for predicting dominant antigenic epitopes based on primary structure. For the rational design of effective vaccines, careful attention will need to be paid to efficient delivery of the epitope to class I and/or class II MHC presentation routes depending on the desire to stimulate CD8+ and CD4+ T lymphocytes.

Additionally, studies on the mechanisms by which successful pathogens regulate and subvert immune defense reactions will continue to be an extremely fruitful area of research. This work not only will provide insights into the cell biology of the immune system and mechanisms of microbial pathogenesis, but also may suggest unique ways to intervene therapeutically in the treatment of immune disorders. Because antigen presentation may be an initial and determining event in immune responses, pathogenic microbes may have evolved virulence factors that focus on processing and presentation pathways.

For these reasons, continued research focused on the antigen-presenting macrophage should prove useful. With more information, it may be possible to understand how antigenic dominance is determined and how the TCR repertoire is shaped by the antigenic experience of microbial exposure. Considerations of both the infectious disease process and immune regulation should continue to lead to new insights that will make possible the development of effective vaccines and therapeutic treatments.

NOTE IN PROOF

An alternative class I processing pathway that is not blocked by brefeldin A or cyclohexamide has recently been proposed by Pfeifer et al (Nature 361:359–61). This group has shown that antigens from bacteria taken up by phagocytosis can be effectively presented in association with class I MHC molecules to CD8+ T cells, even though the bacteria may never gain access to the cytosol.

Rudensky et al (Nature 359:429–31) recently analyzed the sequences of peptides eluted from I-A^s, I-A^b, and I-E^b and suggested that there may be distinctive motifs for each class II MHC allele. In addition, Chicz et al (Nature 358:764–8) recently showed that twenty different peptides acid-eluted from class II MHC molecules (HLA-DR1) were derived from only five core T cell epitopes and were truncated at both ends, rather than at the carboxy terminus only as previously found by Rudensky et al (Nature 353:622–7). This suggests that both ends of a peptide could extend from the class II MHC peptide binding groove.

New evidence from two groups (Arnold et al, Nature 360:171–3; Momburg et al, Nature 360:174–7) challenges the assumption that proteasomes have essential roles in antigen processing. Instead, their experimental results indicate that proteasomes are not absolutely required for stable expression of MHC class I molecules or for the processing and presentation of antigenic peptides.

REFERENCES

1. Berzofsky JA, Berkower IJ. Immunogenicity and antigen structure. In: Paul WE, ed. Fundamental immunology. 2nd ed. New York: Raven Press, 1989:188–92.
2. Mosmann TR, Coffman RL. TH1 and TH2 cells: different patterns of lymphokine secretion lead to different functional properties. Annu Rev Immunol 1989; 7:145–73.
3. Germain RH. The ins and outs of antigen processing and presentation. Nature 1986; 322:687–9.
4. Bevan MJ. Class discrimination in the world of immunology. Nature 1987; 325:192–4.
5. Nuchtern JG, Biddison WE, Klausner RD. Class II MHC molecules can use the endogenous pathway of antigen presentation. Nature 1990; 343:74–6.
6. Brodsky FM, Guagliardi LE. The cell biology of antigen processing and presentation. Annu Rev Immunol 1991; 9:707–44.
7. Ljunggren H-G, Stam NJ, Ohlen C, Neefjes JJ, Hoglund P, Heemels M-T, Bastin J, Schumacher TNM, Townsend A, Karre K, Ploegh HL. Empty MHC class I molecules come out in the cold. Nature 1990; 346:476–80.
8. Schumacher TNM, Heemels M, Neefjes JJ, Kast WM, Melief CHM, Ploegh HL. Direct peptide binding to empty MHC class I molecules on intact cells and *in vitro*. Cell 1990; 62:563–7.
9. Ortiz-Navarrette V, Hammerling GJ. Surface appearance and instability of empty H-2 class I molecules under physiological conditions. Proc Natl Acad Sci USA 1991; 88:3594–7.
10. Rotzschke O, Falk K, Deres K, Schild H, Norda M, Metzger J, Jung G, Rammensee HG. Isolation and analysis of naturally processed viral peptides as recognized by cytotoxic T cells. Nature 1990; 348: 252–4.
11. Falk K, Rotzschke O, Stevanovic S, Jung G, Rammensee HG. Allele-specific motifs revealed by sequencing of self-peptides eluted from MHC molecules. Nature 1991; 351:290–6.

12. Falk K, Rotzschke O, Rammensee HG. Cellular peptide composition governed by major histocompatibility complex class I molecules. Nature 1990; 348:248–51.
13. Jardetzky TS, Lane WS, Robinson RA, Madden DR, Wiley DC. Identification of self peptides bound to purified HLA-B27. Nature 1991; 353:326–9.
14. Madden DR, Gorga JC, Strominger JL, Wiley DC. The structure of HLA-B27 reveals nonamer self-peptides bound in an extended conformation. Nature 1991; 353:321–5.
15. Hunt DF, Henderson RA, Shabanowitz J, Sakaguchi K, Michel H, Sevilir N, Cox AL, Appella E, Englehard V. Characterization of peptides bound to the class I MHC molecule HLA.A2.1 by mass spectrometry. Science 1992; 255:1261–3.
16. Townsend A, Ohlen C, Bastin J, Ljunggren HG, Foster L, Karre K. Association of class I major histocompatibility heavy and light chains induced by viral peptides. Nature 1989; 340:443–8.
17. Cox JH, Yewdell JW, Eisenlohr LC, Johnson PR, Bennink JR. Antigen presentation requires transport of MHC class I molecules from the endoplasmic reticulum. Science 1990; 247:715–8.
18. Otten GR, Bikorr E, Ribaudo RK, Kozlowski S, Margulies DH, Germain RN. Peptide and beta$_2$-microglobulin regulation of cell surface MHC class I conformation and expression. J Immunol 1992; 148:3723–32.
19. Germain RH. The second class story. Nature 1991; 353:605–7.
20. Neefjes JJ, Stollorz V, Peters PJ, Geuze HJ, Ploegh HL. The biosynthetic pathway of MHC class II but not class I molecules intersects the endocytic route. Cell 1990; 61:171–83.
21. Rudensky AY, Preston-Hurlburt P, Hong SC, Barlow A, Janeway CA. Sequence analysis of peptides bound to MHC class II molecules. Nature 1991; 353:622–7.
22. Reid P, Watts C. Cycling of cell-surface MHC glycoproteins through primaquine-sensitive intracellular compartments. Nature 1990; 364:655–7.
23. Harding CV. Pathways of antigen processing. Curr Opinion Immunol 1991; 3:3–9.
24. Adorini L, Ullrich SJ, Appella E, Fuchs S. Inhibition by brefeldin A of presentation of exogenous protein antigens to MHC class II–restricted T cells. Nature 1990; 346:63–6.
25. Roche PA, Cresswell P. Invariant chain association with HLA-DR molecules inhibits immunogenic peptide binding. Nature 1990; 345:615–18.
26. Teyton L, O'Sullivan D, Dickson PW, Lotteau V, Sette A, Fink P, Peterson PA. Invariant chain distinguishes between the exogenous and endogenous antigen presentation pathways. Nature 1990; 348:39–44.
27. Shastri N, Malissen B, Hood L. Ia-transfected L cell fibroblasts present a lysozyme peptide but not the native protein to lysozyme-specific T cells. Proc Natl Acad Sci USA 1985; 82:5885–9.
28. Stockinger B, Pessara U, Lin RH, Habicht J, Grez M, Koch N. A role of Ia-associated invariant chains in antigen processing and presentation. Cell 1989; 56:683–9.
29. Schaiff WT, Hruska KA, Bono C, Shuman S, Schwartz BD. Invariant chain influences posttranslational processing of HLA-DR molecules. J Immunol 1991; 147:603–8.
30. Anderson MS, Miller J. Invariant chain can function as a chaperone protein for class II major histocompatibility complex molecules. Proc Natl Acad Sci USA 1991; 89:2282–6.
31. Bakke O, Dobberstein B. MHC class II-associated invariant chain contains a sorting signal for endosomal compartments. Cell 1990; 63:707–16.
32. Peterson M, Miller J. Antigen presentation enhanced by the alternatively spliced invariant chain gene product p41. Nature 1992; 357:596–8.
33. Cerundolo V, Alexander J, Anderson K, Lamb C, Cresswell P, McMichael A, Gotch F, Townsend A. Presentation of viral antigen controlled by a gene in the major histocompatibility complex. Nature 1990; 345:449–52.
34. Deverson EV, Gow IR, Coadwell WJ, Monaco JJ, Butcher GW, Howard JC. MHC class II region encoding proteins related to the multidrug resistance family of transmembrane transporters. Nature 1990; 348:738–41.
35. Trowsdale J, Hanson I, Mockridge I, Beck S, Townsend A, Kelly A. Sequences encoded in the class II region of the MHC related to the 'ABC' superfamily of transporters. Nature 1990; 348:741–3.
36. Spies T, Bresnahan M, Bahram S, Arnold D, Blanck G, Mellins E, Pious D, De Mars R. A gene in the human major histocompatibility complex class II region controlling the class I antigen presentation pathway. Nature 1990; 348:744–7.

37. Brown MG, Driscoll J, Monaco JJ. Structural and serological similarity of MHC-linked LMP and proteasome (multicatalytic proteinase) complexes. Nature 1991; 353:355–60.
38. Glynne R, Powis SH, Beck S, Kelly A, Kerr L-A, Trowsdale J. A proteasome related gene between the two ABC transporter loci in the class II region of the human MHC. Nature 1991; 353:357–60.
39. Ortiz-Navarrete V, Seelig A, Gernold M, Frentzel S, Kloetzel PM, Hammerling GJ. Subunit of the "20S" proteasome (multicatalytic proteinase) encoded by the major histocompatibility complex. Nature 1991; 353:662–4.
40. Martinez CK, Monaco JJ. Homology of proteasome subunits to a major histocompatibility complex-linked LMP gene. Nature 1991; 353:664–7.
41. Kelly A, Powis SH, Glynne R, Radley E, Beck S, Trowsdale J. Second proteasome related gene in the human MHC class II region. Nature 1991; 353:667–8.
42. Robertson M. Proteasomes in the pathway. Nature 1991; 353:300–1.
43. Goldberg AL, Rock KL. Proteolysis, proteasomes and antigen presentation. Nature 1992; 357:375–9.
44. Dice JF. Peptide sequences that target cytosolic proteins for lysosomal proteolysis. Trends Biochem Sci 1990; 15:305–9.
45. Van Buskirk AM, Nagel CC, Guagliardi LE, Brodsky FM, Pierce SK. Cellular and sub-cellular distribution of PBP72/72, a peptide binding protein which plays a role in antigen processing. J Immunol 1991; 146:500–6.
46. Spies T, DeMars R. Restored expression of major histocompatibility class I molecules by gene transfer of a putative peptide transporter. Nature 1991; 351:323–4.
47. Powis SJ, Townsend ARM, Deverson EV, Bastin J, Butcher GW, Howard JC. Restoration of antigen presentation to the mutant cell line RMA-S by an MHC-linked transporter. Nature 1991; 354:528–31.
48. Attaya M, Jameson S, Martinez CK, Hermel E, Aldrich C, Forman J, Lindahl KF, Bevan MJ, Monaco JJ. *Ham-2* corrects the class I antigen-processing defect in RMA-S cells. Nature 1992; 355:647–9.
49. Parham P. Transporters of delight. Nature 1990; 348:674–5.
50. Parham P. Half of a peptide pump. Nature 1991; 351:271–2.
51. Kelly A, Powis SH, Kerr LA, Mockridge I, Elliott T, Bastin J, Uchanska-Ziegler B, Ziegler A, Trowsdale J, Townsend A. Assembly and function of the two ABC transporter proteins encoded in the human major histocompatibility complex. Nature 1992; 355:641–4.
52. Kleijmeer MJ, Kelly A, Geuze HJ, Slot JW, Townsend A, Trowsdale J. Location of MHC-encoded transporters in the endoplasmic reticulum and *cis*-Golgi. Nature 1992; 357:342–4.
53. Levy F, Gabathuler R, Larsson R, Kvist S. ATP is required for in vitro assembly of MHC class I antigens but not for transfer of peptides across the ER membrane. Cell 1991; 67:265–74.
54. Powis SJ, Deverson EV, Coadwell WJ, Ciruela A, Huskisson NS, Smith H, Butcher GW, Howard JC. Effect of polymorphism of an MHC-linked transporter on the peptides assembled in a class I molecule. Nature 1992; 357:211–5.
55. Diment S. Different roles for thiol and aspartyl proteases in antigen presentation of ovalbumin. J Immunol 1990; 145:417–22.
56. Takahashi H, Cease KB, Berzofsky JA. Identification of proteases that process distinct epitopes on the same protein. J Immunol 1989; 142:222–9.
57. Jensen P. Enhanced binding of peptide antigen to purified class II major histocompatibility glycoproteins at acidic pH. J Exp Med 1991; 174:1111–20.
58. Guagliardi LE, Koppelman B, Blum JS, Marks MS, Cresswell P, Brodsky FM. Co-localization of molecules involved in antigen processing and presentation in an early endocytic compartment. Nature 1990; 343:133–9.
59. Peters PJ, Neefjes JJ, Oorschot V, Ploegh HL, Geuze HJ. Segregation of MHC class II molecules from MHC class I molecules in the Golgi complex for transport to lysosomal compartments. Nature 1991; 349:669–76.
60. Parham P. The box and the rod. Nature 1992; 357:538–9.
61. Springer TA. Adhesion receptors of the immune system. Nature 1990; 346:425–34.
62. Matsui K, Boniface JJ, Reay PA, Schild H, De St Groth BF, Davis MM. Low affinity interaction of peptide-MHC complexes with T cell receptors. Science 1991; 254:1788–90.
63. St. Pierre Y, Watts TH. Characterization of the signaling function of MHC class II molecules during antigen presentation by B cells. J Immunol 1991; 147:2875–82.

64. Linsley PW, Brady W, Grosmaire L, Afuffo A, Damle NK, Ledbetter JA. Binding of the B cell activation antigen B7 to CD28 costimulates T cell proliferation and interleukin 2 mRNA accumulation. J Exp Med 1991; 173:721–30.
65. Cossart P, Mengaud J. *Listeria monocytogenes*, a model system for the molecular study of intracellular parasitism. Mol Biol Med 1989; 6:463–74.
66. Portnoy DA, Chakraborty T, Goebel T, Cossart P. Molecular determinants of *Listeria monocytogenes* pathogenesis. Infect Immun 1992; 60:1263–7.
67. Cossart P, Vicente MF, Mengaud J, Baquero F, Perez-Diaz JC, Berche P. Listeriolysin O is essential for virulence of *Listeria monocytogenes*: direct evidence obtained by gene complementation. Infect Immun 1989; 57:3629–36.
68. Gaillard J-L, Berche P, Sansonetti P. Transposon mutagenesis as a tool to study the role of hemolysin in the virulence of *Listeria monocytogenes*. Infect Immun 1986; 52:50–5.
69. Portnoy DA, Jacks PS, Hinrichs DJ. Role of hemolysin for the intracellular growth of *Listeria monocytogenes*. J Exp Med 1988; 167:1459–71.
70. Safley SA, Cluff CW, Marshall NE, Ziegler HK. Role of listeriolysin O in the T lymphocyte response to infection with *Listeria monocytogenes*: identification of T cell epitopes of LLO. J Immunol 1991; 146:3604–16.
71. Pamer EG, Harty JT, Bevan MJ. Precise prediction of a dominant class I MHC-restricted epitope of *Listeria monocytogenes*. Nature 1991; 353:852–5.
72. Safley S, Ziegler HK. Purified listeriolysin O is antigenic—the minimal T cell epitope is a 12 amino acid peptide with strong MHC class II binding activity. FASEB J 1992; 6:A1436.
73. Cluff WC, Garcia M, Ziegler HK. Intracellular hemolysin-producing *Listeria monocytogenes* strains inhibit macrophage-mediated antigen processing. Infect Immun 1990; 58:3601–12.
74. Brunt LM, Portnoy DA, Unanue ER. Presentation of *Listeria monocytogenes* to CD8+ T cells requires secretion of hemolysin and intracellular bacterial growth. J Immunol 1990; 145:3540–6.

7

Macrophages as Producers of Chemotactic Proinflammatory Cytokines

Bent Windelborg Nielsen,* Naofumi Mukaida, and Kouji Matsushima

Kanazawa University, Kanazawa, Japan

Tadashi Kasahara

Jichi Medical School, Tochigi, Japan

I. INTRODUCTION

Inflammation is a vital consequence of tissue injuries caused by various inciting stimuli such as physical trauma, bacterial and viral infections, excessive metabolites (such as ureic crystals), heavy metal inhalation, autoimmune diseases, malignant neoplasia, and ischemia-reperfusion syndrome. Acute inflammation is characterized by intravascular activation of platelets and movement of plasma and neutrophils from blood into extravascular tissues. Initiation of the inflammatory reaction occurs in the microvasculature, mainly at the level of postcapillary venules. The postcapillary venule is not only the primary site at which the vasoactive mediators, such as histamine, serotonin, bradykinin, anaphylatoxins, leukotrienes, prostaglandins, and platelet-activating factor induce endothelial contraction and gap formation leading to reversible plasma leakage, but also the site at which specific types of leukocytes marginate to endothelial cells in response to chemoattractants (complement factor C5a, leukotriene B4, formyl peptides, and chemotactic cytokines) and egress from the bloodstream. Following acute tissue injury, lymphocytes, monocytes, and plasma cells predominate as infiltrating cells. Chronic inflammation is mediated by both immunologic and nonimmunologic mechanisms, and macrophages are pivotal participants in regulating these reactions through producing various inflammatory mediators, such as active oxygen radicals, lysosomal enzymes, prostacyclins, procoagulants, proteases, and cytokines, as well as directly functioning as phagocytic cells to scavenge foreign particles, denatured substances, and dead cells, and as antigen-presenting cells to accelerate T lymphocyte–mediated immune reactions.

Cytokines are hormone-like active polypeptides that are produced by various types of somatic cells. As shown in Figure 1, the macrophage is a key player in the cytokine network, regulating host defensive immunological and inflammatory reactions by producing various cytokines including interleukin-1 (IL-1), tumor necrosis factor-α (TNF-α), IL-6, IL-8, monocyte chemotactic and activating factor (MCAF), interferons, and colony-stimulating factors. Macrophages also produce natural antagonists to cytokines and cytokine synthesis inhibitors such as transforming growth factor-β, IL-1 receptor antagonist, and soluble cytokine receptors. In this chapter, we focus on contemporary concepts of the mechanism of leukocyte infiltration, and on the characteristics and pathophysiological roles of the novel proinflammatory chemotactic cytokines, IL-8, and MCAF, which are produced by activated macrophages.

**Current affiliation:* Aarhus Kommunehospital, Aarhus, Denmark

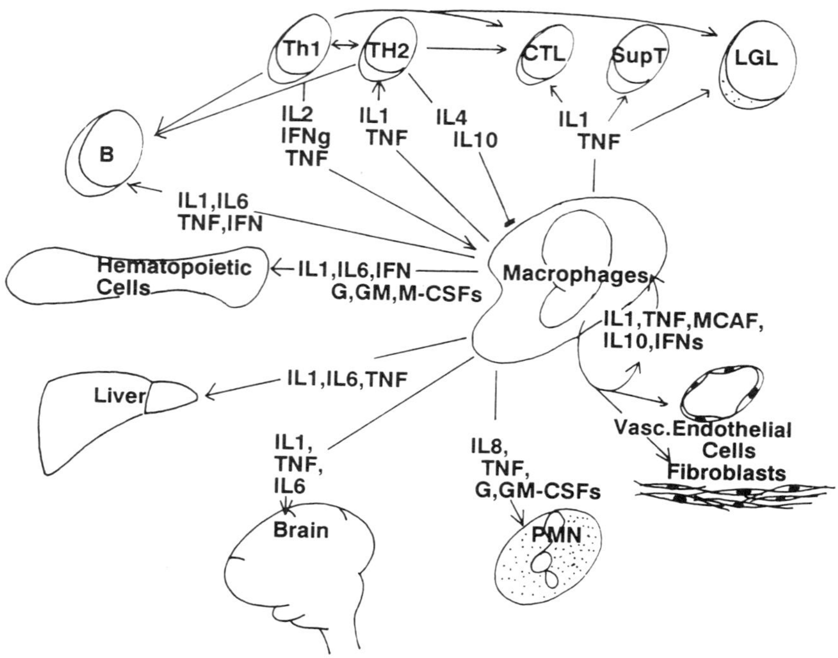

FIGURE 1 The macrophage in the cytokine network.

II. REGULATION OF LEUKOCYTE INFILTRATION

Leukocyte infiltration into inflamed tissue is a highly regulated process, consisting of at least two stages: leukocyte adhesion to endothelium and leukocyte migration into perivascular tissues. The principal components of these processes are shown in Figure 2, and the principal molecules involved are shown in Table 1. This scenario is based primarily on studies of neutrophil-endothelial interactions, but the principles may be extended to lymphocytes and monocytes as well [1]. However, different subtypes of adhesion molecules on different cell types confer cellular specificity to the adhesion and emigration in varying clinical states [2].

A. Leukocyte Margination, "Rolling," and Adhesion to Activated Endothelium

Activation of resident tissue macrophages causes production of proinflammatory cytokines, such as TNF and IL-1. These cytokines, along with substances secreted from other cell types, e.g., histamine from mast cells, activate endothelial cells lining the regional microvasculature [1].

Only seconds after stimulation by histamine or thrombin, the endothelial cells translocate P-selectin to their luminal surface. P-selectin belongs to a family of transmembrane glycoproteins, termed *selectins* because of their *N*-terminal lectin domain capable of selective binding to specific carbohydrates [3]. The binding of P-selectin to its ligand, an as yet uncharacterized sialylated glycoprotein constitutively expressed on leukocytes, provides only transient tethering of the leukocyte to the endothelium. Accordingly, direct observation in vivo has shown leukocytes "rolling" along activated endothelium.

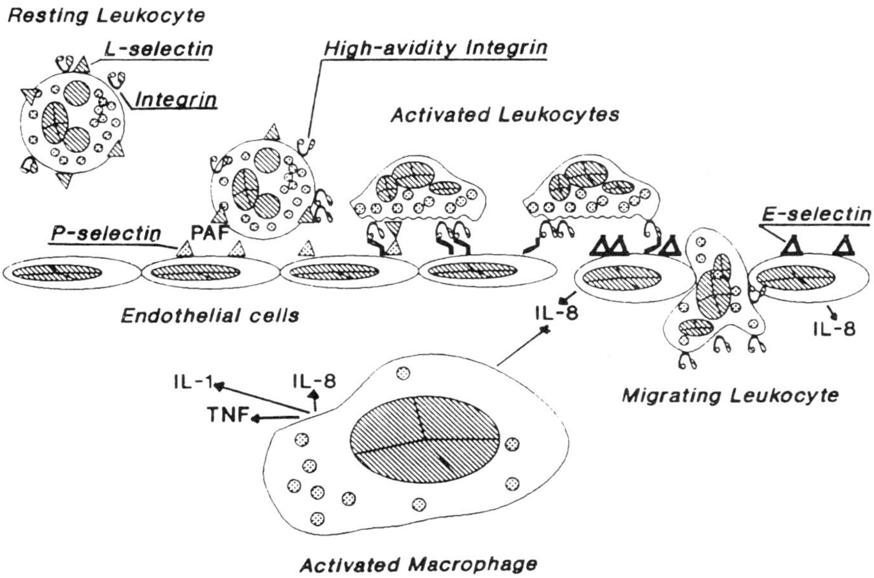

FIGURE 2 The key events in leukocyte–endothelial cell interaction and leukocyte infiltration of perivascular tissue. Different stages of the time course are represented by the leukocytes when viewed from left to right.

However, on a time scale of minutes after activation, the endothelial cells begin to secrete platelet activating factor (PAF). Apart from activating platelets, this phospholipid mediator also binds to the marginated, rolling leukocytes, causing rapid increase in avidity of the β_2-integrin (CD11a–CD18) on the leukocyte surface. The integrins are a family of about 20 proteins, all of which are heterodimers [4]. Most mediate adhesiveness to extracellular matrix components, but in the case of CD11–CD18, the substrate is ICAM-1 or -2, members of the immunoglobulin supergene family. Conformational change within CD11–CD18, possibly coupled to phosphorylation of its cytoplasmic domain, is thought to be responsible for the increase in avidity after leukocyte activation. Crosslinking of CD11–CD18 has been shown to induce Ca^{2+} mobilization in neutrophils, thus serving both adhesive and signaling purposes [5].

After stimulation with lipopolysaccharide (LPS), TNF, or IL-1, the endothelial cells begin to present E-selectin on the luminal surface after 1 hour, with expression maximal at 4 to 8 hours and gradually declining over 24 hours. Although this time course coincides with that observed for leukocyte adhesion to epithelium, the relative importance of E-selectin in mediating leukocyte adhesiveness seems to depend on the strength of the proinflammatory stimulus: At high concentrations of IL-1, the integrin-mediated adhesion plays the key role in this process. The adhesiveness mediated by E-selectin does not require activation of the leukocyte, and E-selectin has been shown to interact with L-selectin, which is found on villous extensions of the leukocyte surface but undergoes active and rapid shedding in conjunction with up-regulation of the leukocyte integrins.

Besides presenting adhesion molecules for leukocytes, the endothelial cells produce IL-1 and thromboplastin (activating thrombin), thereby activating themselves in an autocrine or paracrine fashion. Activated platelets, as well as thrombogenic basement membrane components laid bare by

TABLE 1 Important Adhesion Molecules in Leukocyte–Endothelial Cell Interactions

	Other names	Counter receptor(s)	Chemical characteristics
Endothelial Cells:			
Selectins:			
P-selectin	CD62 GMP-140 PADGEM	Sialylated glycoprotein	C-type lectin–like, EGF-like, and immunoglobulin-like domains, transmembrane proteins
E-selectin	ELAM-1	Sialyl-Lewis^x-antigen L-selectin?	
Ig-superfamily:			
ICAM-1	CD54	CD11a–CD18	Immunoglobulin-like repeat regions
ICAM-2	—	CD11a–CD18 CD11b–CD18	
Leukocytes:			
Integrins:			
CD11a–CD18	LFA-1	ICAM-1 and -2	Heterodimeric transmembrane proteins, Ca^{2+}-coupled signaling. Cell activation increases avidity
CD11b–CD18	Mac-1, Mo-1	ICAM-1	
CD11c–CD18	p150-95	ICAM-1	
Selectin:			
L-selectin	LAM-1, LECAM-1, LEC.CAM-1, Mel-14-, Leu-8-, TQ1-, and DREG.56-antigen	Glycoprotein, E-selectin?	Rapidly shed after cell activation

endothelial contraction, further augment the generation of an "adhesive cascade," to which even the vascular smooth muscle contributes by production of TNF [6].

B. Transendothelial Migration and Tissue Invasion of Leukocytes in Response to Chemotactic Stimuli

Beginning 4 to 24 hours after stimulation of the endothelium by cytokines (TNF or IL-1) or LPS, the endothelial cells secrete IL-8 and the closely related gro/MGSA, which accumulates primarily on the abluminal side of the endothelium. The gradient of IL-8 thus formed provides a strong chemotactic stimulus for the activated leukocytes adhering to activated endothelium, causing the leukocytes to exit the bloodstream by directed ameboid movement through gaps formed between the endothelial cells. The production of IL-8 by the endothelial cells seems to be the key regulatory element for the migration of leukocytes into the extravascular space [7].

C. Chemotactic Factors and Chemotactic Receptors

Once leukocytes are outside the bloodstream, a variety of chemotactic stimuli can direct their movements. Apart from the intercrine (IL-8) family of chemotactic cytokines, these substances differ widely in their chemical composition [8,9].

N-formyl-oligopeptides, the most potent of which is *N*-formyl-Met-Leu-Phe, have been shown to be produced in bacterial cultures. Specific receptors are found on neutrophils and monocytes [10].

Endogenous chemotactic factors include anaphylatoxin C5a, a 74-amino-acid cationic glycoprotein (m.w. 11 kDa) produced by activation of complement that causes vasodilation and increased vascular permeability in addition to being a potent chemoattractant for neutrophils. Two proinflammatory lipid mediators, Leukotriene B_4, produced by 5-lipoxygenase from arachidonic acid, and PAF, 1-*O*-alkyl-2-acetyl-sn-glyceryl-3-phosphorylcholine, share the proinflammatory and chemotactic activities of C5a and are produced by a number of cell types, including leukocytes and endothelial cells. Other proteins, such as extracellular matrix components (e.g., laminin), have shown neutrophil chemotactic properties for neutrophils.

Despite the structural diversity of these chemotactic factors, the chemoattractant receptors have provided several unifying features. So far, the IL-8, C5a, PAF, and fMLP receptors on human leukocytes have been characterized [10–14]. All of these belong to the superfamily of G protein–coupled receptors containing seven transmembrane domains. The amino acid homologies are in the range 18% to 33%, and the cytosolic C-terminals present glycosylation and phosphorylation sites.

The signal transduction mechanisms of the chemoattractant receptors include a rapid, G protein–dependent Ca^{2+} influx. Further, activation of phospholipase C causes polyphosphoinositide turnover, generating diacylglycerol and inositol-3-phosphate (IP_3). Diacylglycerol activates protein kinase C, which appears to be critically involved in chemotactic responses. Inositol-3-phosphate acts as a trigger for the release of Ca^{2+} from intracellular stores, but this mechanism contributes only a minor part of the rise in intracellular Ca^{2+} concentration observed after triggering of chemotactic receptors. Phospholipase A_2–mediated turnover of membrane lipids, releasing arachidonic acid, also increases after stimulation of PAF receptors [14,15].

The signal transduction mechanisms briefly outlined above are involved in a variety of cellular responses other than chemotactic movement, and triggering of chemoattractant receptors induces a plethora of responses in the leukocyte. These responses include increased capacity for phagocytosis, "respiratory burst," and secretion of granule contents. Combinations of chemotactic factors act synergistically with each other, and cellular responsiveness can be modulated by interaction with extracellular matrix components.

D. Molecular Basis for Cell Movement in Response to Chemoattractants

The central molecule involved in cell motility is actin, constituting as much as 50% of total cellular protein. In the resting cell, polymerized F-actin forms a continuous network beneath the cell surface, serving to maintain cell shape as well as anchoring a number of proteins in contact with the extracellular milieu, including the integrins. However, free monomeric G-actin is present at an estimated concentration of 50 μM, and a closely regulated actin polymerization and depolymerization forms the basis of changes in cell shape, i.e., the extension of pseudopods from the cell leading edge during ameboid movement (Figure 3). The G-actin is complexed with profilin, making it unavailable for polymerization, and a second actin-binding protein [16].

Although still a matter of dispute, an attractive hypothesis for the signaling events causing the directed movement toward a chemotactic gradient will be presented here. Binding of a chemotactic factor to its G protein–coupled receptor generates IP_3 and diacylglycerol (DAG), as well as causing an increased intracellular Ca^{2+} concentration. Together, these changes cause a localized dissociation of profilin, making G-actin available for polymerization (actin nucleation), and activate another actin-binding protein, gelsolin, which causes breakdown of the F-actin network into monomers, leading to a local remodeling of the cytoskeleton, and possibly increases osmotic pressure, causing formation of a membrane "bleb" [17,18]. The sum of these actions is the formation of a pseudopod at the site of chemoattractant binding. In the presence of a chemotactic gradient, pseudopod extensions will form preferentially toward the gradient.

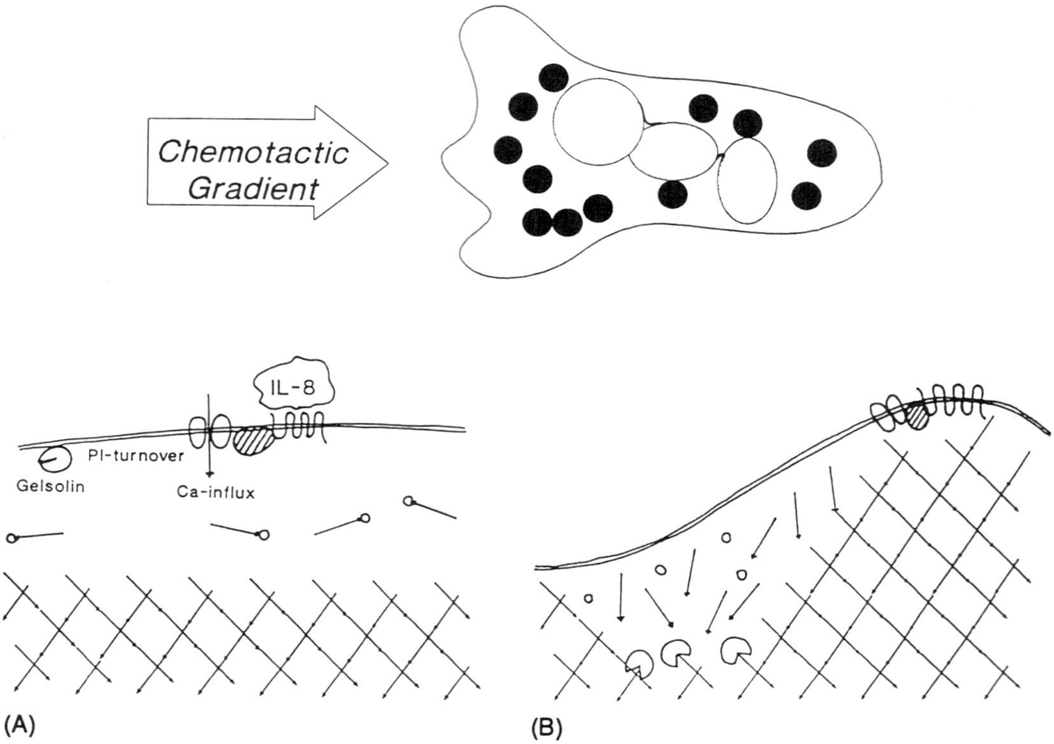

FIGURE 3 Schematic presentation of a contemporary concept of the molecular basis of directed ameboid leukocyte movement. Extension of pseudopods in the direction of movement occurs after binding of a chemotactic factor (IL-8) to its G protein–coupled receptor (A), causing calcium influx and phosphoinositide (PI) turnover. Profilin (○) is released from monomeric *G*-actin (←), and the breakdown of cytoskeletal F-actin into monomers by activated gelsolin makes more G-actin available for polymerization, leading to a local remodeling of the cytoskeleton and pseudopod extension (B).

Direct evidence has shown that such newly polymerized actin filaments remain fixed in relation to the extracellular space, as the cell body moves "around" it [19]. The force of movement seems to be provided by the Ca^{2+}-dependent interaction of actin with myosin I, a tailless myosin molecule, at the end of the cell opposite to the direction of movement.

III. NOVEL LEUKOCYTE CHEMOTACTIC FACTORS, IL-8 AND MCAF

A. Biochemical Characteristics of IL-8 and MCAF

As described above, several lipid or peptide factors, including leukotriene B_4, PAF, C5a, and bacteria-derived formyl peptides, have been shown to possess chemotactic activity for any kind of leukocytes. The types of infiltrating leukocytes, however, depend on the kind, degree, and timing of tissue injuries, suggesting the presence of cell type–specific leukocyte chemotactic factors [20]. Recently, IL-8 and MCAF were purified and molecularly cloned as neutrophil and monocyte chemotactic factors, respectively.

TABLE 2 Relationships of Intercrine Family Members

Species	Human	Murine	Major functions
Intercrine α subfamily			
Chromosome	4(q12-21)	?	
Structure	C-X-C	C-X-C	
Cytokines	IL-8	?	Neutrophil, T lymphocyte, basophil chemotaxis
	groα	MIP-2/KC	Neutrophil chemotaxis
	groβ	groβ	Neutrophil chemotaxis
	groγ	groγ	Neutrophil chemotaxis
	PF4	?	Neutrophil chemotaxis
	βTG	?	?
	IP-10	CRG-2	T lymphoyte chemotaxis
Intercrine β subfamily			
Chromosome	17(q11-12)	11	
Structure	C-C	C-C	
Cytokines	MCAF	JE	Monocyte chemotaxis
	LD78	MIP-1α	Stem cell growth inhibition
	ACT2	MIP-1β	?
	RANTES	?	Monocyte and T cell chemotaxis
	I-309	TCA-3	Neutrophil chemotaxis (?)

Monocytes massively produce both IL-8 and MCAF in response to LPS. The availability of antibodies and cDNA enabled us to identify the cellular sources and their stimulants. It is of interest that a wide variety of cell types can produce both IL-8 and MCAF in the presence of IL-1 and TNF, two well-known monokines [8]. Interleukin-1 and TNF induce leukocyte infiltration when given in vivo, whereas both lack leukocyte chemotactic activity in vitro. These results indicate that leukocyte infiltration induced by IL-1 or TNF in vivo is at least partly ascribable to induction of IL-8 and MCAF production. Thus, the effects of IL-1 and TNF on the inflammatory processes may be mediated by these leukocyte chemotactic factors produced locally.

It has become evident that IL-1 and TNF induce IL-8 and MCAF production through the activation of transcription in various types of cells, including human endothelial cells, a fibrosarcoma cell line, and an astrocytoma cell line [8]. Functional studies on the enhancer region of the IL-8 gene revealed that AP-1-, NF-κB- and C/EBP-like factor binding sites are involved in the activation of IL-8 gene transcription by IL-1 and TNF [21]. It is, however, unclear whether a similar mechanism works in MCAF gene transcription by IL-1 or TNF.

Interleukin-8 and MCAF have heparin-binding capacity, and their mature forms consist of 72 and 76 amino acids, respectively. Moreover, cysteine residues are located at well-conserved positions, suggesting that these two factors possess a similar overall structure [22]. Over the past several years, numerous cytokines with the above-mentioned characteristics have been identified and classified according to whether the first cysteine pair is separated by a single amino acid or in an adjacent position [8]. Among these cytokines, several have been shown to have cell type–specific leukocyte chemotactic activity (Table 2).

B. Biological Functions and Clinical Relevance of IL-8

Interleukin-8 activates a number of neutrophil functions in vitro, as shown in Table 3. In the presence of cytochalasin B, IL-8 induces release of lysosomal enzymes, including myeloperoxidase, α-mannosidase, and β-glucuronidase. Although a ten-times-higher concentration of IL-8 is needed to induce enzyme release than to induce neutrophil chemotaxis, no high-dose suppression is

TABLE 3 Biological Functions of Interleukin-8 (IL-8) Monocyte Chemotactic and Activating Factor (MCAF)

Cell target	Effects
IL-8	
In vitro	
Neutrophils	Chemotaxis
	Shape change
	Induction of degranulation
	Induction of respiratory burst
	Induction of lysosomal enzymes
	Increase of intracellular Ca^{2+}
	Induction of adherence to unstimulated endothelial cells
	Inhibition of adherence of cytokine-activated endothelial cells
	Induction of transendothelial migration
	Induction of binding of C3bi and lipopolysaccharide
	Induction of expression of complement receptor type 1
	Induction of expression of adhesion molecules (CD11a, b, c, and 18)
	Enhancement of growth inhibition of *Candida albicans*
	Activation of 5-lipoxygenase
	Induction of release of leukotriene B_4 and 5-HETE
T lymphocytes	Chemotaxis
Basophils	Chemotaxis
	Induction of release of histamine and leukotriene C_4
Monocytes	Increase of adhesion to endothelial cells
In vivo	
	Induction of neutrophil infiltration
	Induction of lymphocyte infiltration
	Increase of vascular permeability in the presence of prostaglandin E
	Neutrophilia
	Destruction of synovial membrane
MCAF	
In vitro	
Monocytes	Chemotaxis
	Induction of superoxide anion release
	Induction of release of lysosomal enzymes
	Augmentation of cytostatic activity
Basophils	Induction of degranulation
	Induction of release of histamine and leukotiene C_4
In vivo	
	Induction of macrophage infiltration

observed, in contrast to in vitro chemotaxis [8]. Although IL-8 induces only marginal superoxide/hydrogen peroxide in normal human neutrophils, IL-8 primes human neutrophils to generate more superoxide anions in response to granulocyte-macrophage colony-stimulating factor (GM-CSF), TNF, and fMLP [23].

Interleukin-8 shows complicated effects on neutrophil adhesion to endothelium. It increases shedding of L-selectin and expression of β_2-integrin (Mac-1) on neutrophils, thereby increasing the adhesion of neutrophils to unstimulated human umbilical vein endothelial cells [7]. On the other hand, Gimbrone and coworkers purified a leukocyte adherence inhibitory factor (LAI) that inhibits the adherence of neutrophils to umbilical endothelial cells preactivated with IL-1, TNF, or LPS

[24]. Surprisingly, the determination of the amino acid sequence of LAI revealed that this factor is identical to the 77-amino-acid intermediate form of IL-8. Thus, it is important to determine how IL-8 affects the interaction between neutrophils and endothelial cells, in particular, in vivo.

Human T cells seem to be ten times more sensitive to IL-8 than neutrophils in chemotactic response [25]. Both CD4+ and CD8+ subpopulations migrate in response to IL-8. It remains to be examined whether IL-8 has additional activities on lymphocytes besides chemotaxis. Interleukin-8 also induces the release of histamine and leukotriene C_4 by basophils pretreated with IL-3 as well as the chemotaxis of untreated basophils.

Injection of IL-8 into the skin induces a leukocyte infiltrate consisting predominantly of neutrophils at higher doses and of lymphocytes at lower doses [25]. This leukocyte infiltration is observed within 1 hour and reaches a maximal level at 3 hours. Another aspect of inflammation, increase in vascular permeability, is also induced in a neutrophil-dependent manner by the combination of IL-8 and prostaglandin E.

The IL-8–related molecule gro utilizes the same receptor as IL-8, as revealed by competitive binding assays [8]. Thus gro's also induce release of intracellular enzymes of neutrophils in vitro and leukocyte infiltration composed predominantly of neutrophils when injected subcutaneously. Gro's are produced by various kinds of cells in response to LPS, IL-1, and TNF [8], suggesting that IL-8 and gro's are produced simultaneously and act additively or synergistically to induce leukocyte infiltration in vivo.

Infections, particularly bacterial ones, are usually accompanied by neutrophilia and/or neutrophil infiltration. We observed that the gram-positive streptococcus-derived biological response modifier OK-432 induced a marked elevation of the IL-8 level followed by massive neutrophil infiltration when injected into the pleural cavity (unpublished data). Moreover, the urinary IL-8 level increased significantly without concomitant increase of serum level in all the urinary tract infections that we examined, whether caused by gram-negative or gram-positive bacteria or fungi (unpublished data). These results suggest that bacteria and fungi, by themselves, can induce the production of IL-8, which is involved in the neutrophil recruitment seen in the infections.

In addition to infections, neutrophilia and/or neutrophil infiltration is observed in a variety of types of disease, including rheumatoid arthritis (RA), gout, psoriasis, adult respiratory distress syndrome, and inflammatory bowel disease. Synovial cells of RA have the capacity to produce IL-8, and the IL-8 level is increased in joint fluids from RA patients [26]. Injection of IL-8 into rabbit knee joint cavities caused the destruction of synovial membranes accompanied by neutrophil extravasation, resembling the pathological features of RA joints [26]. These results suggest the involvement of IL-8 in the pathogenesis of RA.

Recently, it was demonstrated that neutrophil infiltration is involved in the reperfusion injury seen after hepatic ischemia [27]. Furthermore, Kunkel and his coworkers observed that human peripheral blood monocytes produced IL-8 in response to oxidant stress analogous to ischemia-reperfusion [28]. These results suggest that locally produced IL-8 is involved in the ischemia-reperfusion syndromes, including hepatic ischemia-reperfusion injury, myocardial infarction, and multiple organ failure. Thus, the development of specific IL-8 inhibitors, such as peptide antagonists, soluble receptors, and neutralizing antibodies against IL-8, might lead to a novel class of agents that could control the above-mentioned conditions.

On the other hand, since IL-8 augments cytostatic/killing activities against *Candida albicans* and *Mycobacterium* [29], IL-8 itself may be useful for the treatment of infections accompanied by granulocytopenia and neutrophil dysfunction in which antibiotics have little therapeutic efficacy.

C. Biological Functions and Clinical Relevance of MCAF

Reflecting the in vitro chemotactic activity for monocytes, the injection of MCAF into rat ear induces selective monocyte infiltration into the injection site [30]. The infiltration begins at 3 hours and reaches a maximal level at 18 hours without inducing neutrophil or lymphocyte infiltration.

Monocyte chemotactic and activating factor has a variety of actions on monocyte/macrophage functions besides induction of chemotaxis. It induces monocytes to generate and release superoxide anions and to release lysosomal enzymes [30]. Furthermore, MCAF activates monocytes to be cytostatic for several human tumor cell lines, including the A375 melanoma cell line, HTB2 rhabdomyosarcoma cell line, and MCF7 mammary carcinoma cell line, although the precise molecular mechanism of action remains to be examined [31]. Rollins and his coworkers reported that transfection of MCAF/JE gene into CHO cells resulted in the loss of their tumorigenicity in nude mice and that the coinjection of these transformed cells with nontransformed malignant cells suppressed tumor formation [32]. These results suggest that MCAF induces monocyte tumoricidal activity also in vivo.

In contrast to IL-8, MCAF induces histamine release from basophils in the absence of other stimuli [33]. This release is rapid and reaches maximum level within 1 minute, suggesting that MCAF interacts directly with basophils. Protracted histamine release and/or mast cell degranulation is seen in a variety of immunological diseases, including the late phase of allergic reactions, chronic urticaria, atopic dermatitis, asthma, and RA. The specific contribution of MCAF to histamine release in these diseases is a subject to be investigated.

Snyderman and his coworkers reported that a monocyte chemotactic factor (or factors) is produced by stimulating human peripheral blood mononuclear cells (PBMC) with purified protein derivative (PPD), and the production of this factor correlated well with the degree of a delayed hypersensitivity caused by PPD [34]. It remains to be examined whether MCAF is identical to the monocyte chemotactic factor(s) involved in the delayed-type hypersensitivity.

The first step of atherosclerosis is presumed to be the recruitment into the vessel wall of monocytes/macrophages that will transform into foam cells. Monocyte chemotactic and activating factor is produced by both endothelial cells and smooth muscle cells in the presence of minimally oxidized low-density lipoprotein, whose concentration increases in hypercholesterolemia [35]. Moreover, in atherosclerotic lesions in humans and hypercholesterolemia-prone rabbits, MCAF protein is detected immunohistochemically [36]. Thus, it is an attractive assumption that locally produced MCAF is involved in the recruitment and transformation of monocytes/macrophages in the lesion.

At present, there are no data concerning the level of MCAF in human body fluids, because of the limited availability of antibodies against MCAF. Thus, the precise role of MCAF in various diseases remains an open question.

REFERENCES

1. Zimmerman GA, Prescott SM, McIntyre TM. Endothelial cell interactions with granulocytes: tethering and signaling molecules. Immunol Today 1992; 13:93–100.
2. Butcher EC. Leukocyte–endothelial cell recognition: three (or more) steps to specificity and diversity. Cell 1991; 67:1033–6.
3. Bevilaqua M, Butcher E, Furie B, Furie B, Gallatin M, Gimbrone M, Harlan J, Kishimoto K, Lasky L, McEver R, Paulson J, Rosen S, Seed B, Siegelman M, Springer T, Stoolman L, Tedder T, Varki A, Wagner D, Weissman I, Zimmerman G. Selectins: a family of adhesion receptors. Cell 1991; 67:233.
4. Hynes RO. Integrins: versatility, modulation, and signaling in cell adhesion. Cell 1992; 69:11–25.
5. Ng-Sikorsky J, Andersson R, Patarroyo M, Andersson T. Calcium signaling capacity of the CD11b/CD18 integrin on human neutrophils. Exp Cell Res 1991; 195:504–8.
6. Pardi R, Inverardi L, Bender JR. Regulatory mechanisms in leukocyte adhesion: flexible receptors for sophisticated travelers. Immunol Today 1992; 13:224–30.
7. Huber AR, Kunkel SL, Todd RF III, Weiss SJ. Regulation of transendothelial neutrophil migration by endogenous interleukin-8. Science 1991; 254:99–102.
8. Oppenheim JJ, Zachariae COC, Mukaida N, Matsushima K: Properties of the novel proinflammatory supergene "intercrine" cytokine family. Annu Rev Immunol 1991; 9:617–48.

9. Harvath L. Neutrophil chemotactic factors. In: Goldberg ID, ed. Cell motility factors. Basel: Birkhäuser Verlag, 1991:35–52.
10. Boulay F, Tardif M, Brouchon L, Vignais P. Synthesis and use of a novel *N*-formyl peptide derivative to isolate a human *N*-formyl peptide receptor cDNA. Biochem Biophys Res Commun 1990; 168:1103–9.
11. Holmes WE, Lee J, Kuang W-J, Rice GC, Wood WI: Structure and functional expression of a human interleukin-8 receptor. Science 1991; 253:1278–80.
12. Murphy PM, Tiffany HL: Cloning of complementary DNA encoding a functional human interleukin-8 receptor. Science 1991; 253:1280–3.
13. Gerard NP, Gerard C. The chemotactic receptor for human C5a anaphylatoxin. Nature 1991; 349:614–17.
14. Shukla SD. Platelet-activating factor and receptor and signal transduction mechanisms. FASEB J 1992; 6:2296–301.
15. Sozzani S, Luini W, Molino M, Jílek P, Bottazzi B, Cerletti C, Matsushima K, Mantovani A. The signal transduction pathway involved in the migration induced by a monocyte chemotactic cytokine. J Immunol 1991; 147:2215–21.
16. Cunningham CC. Actin structural proteins in cell motility. Cancer and Metastasis Reviews 1992; 11:69–77.
17. Condeelis J, Jones J, Segall JE: Chemotaxis of metastatic tumor cells: Clues to mechanisms from the *Dictyostelium* paradigm. Cancer Metastasis Rev 1992; 11:55–68.
18. Lester BR, McCarthy JB. Tumor cell adhesion of the extracellular matrix and signal transduction mechanisms implicated in tumor cell motility, invasion and metastasis. Cancer Metastasis Rev 1992; 11:31–44.
19. Theriot JA, Mitchison TJ. Actin microfilament dynamics in locomoting cells. Nature 1992; 352:126–131.
20. Ryan GB. The origin and sequence of the cells found in the acute inflammatory response. J Exp Med Sci 1967; 145:149–59.
21. Mukaida N, Mahe Y, Matsushima K. Cooperative interaction of nuclear factor-κB- and *cis*-regulatory enhancer binding protein–like factor binding elements in activating the interleukin-8 gene by pro-inflammatory cytokines. J Biol Chem 1990; 266:21128–21133.
22. Clore GM, Gronenborn AM. NMR and X-ray analysis of the three-dimensional structure of interleukin-8. In: Baggiolini M, Sorg C, eds. Interleukin-8 (NAP-1) and related chemotactic cytokines, cytokines vol. 4. Basel: Karger, 1992:18–40.
23. Yuo A, Kitagawa S, Kasahara T, Matsushima K, Saito M, Takaku F. Stimulation and priming of human neutrophils by interleukin-8: cooperation with tumor necrosis factor and colony-stimulating factors. Blood 1991; 78:2708–14.
24. Gimbrone MA, Obin MS, Brock AF, Luis EA, Hass PE, Herbert CA, Yip YK, Leung DW, Lowe DG, Kohr WJ, Darbonne WC, Bechtol KB, Baker JB: Endothelial interleukin 8: a novel inhibitor of leukocyte–endothelial cell interactions. Science 1990; 246:1601–3.
25. Larsen CG, Anderson AO, Appella E, Oppenheim JJ, Matsushima K. Neutrophil activating protein (NAP-1) is also chemotactic for T lymphocytes. Science 1989; 243:1464–6.
26. Endo H, Akahoshi T, Takagishi K, Kashiwazaki S, Matsushima K. Elevation of interleukin-8 (IL-8) levels in joint fluids of patients with rheumatoid arthritis and the induction by IL-8 of leukocyte infiltration and synovitis in rabbit joints. Lymphokine Cytokine Res 1991; 10:245–52.
27. Hartnut J, Anwar F, Wayne S: Neutrophils contribute to ischemia/reperfusion injury in rat liver *in vivo*. FASEB J 1990; 4:3355–9.
28. Metinko A, Kunkel S, Standiford T, Strieter R. Monocyte expression of interleukin-8 in response to oxidant stress (abstr). FASEB J 1991; Abstracts part 1: A704.
29. Djeu JY, Matsushima K, Oppenheim JJ, Shiotsuhki K, Blanchard DK. Functional activation of human neutrophils by recombinant monocyte-derived neutrophil chemotactic factor/IL-8. J Immunol 1990; 144:2205–10.
30. Zachariae COC, Anderson AO, Thompson HL, Appella E, Mantovani A, Oppenheim JJ, Matsushima K. Properties of monocyte chemotactic and activating factor (MCAF) purified from a human fibrosarcoma cell line. J Exp Med 1990; 171:2177–82.
31. Matsushima K, Larsen CG, Dubois GC, Oppenheim JJ. Purification and characterization of a novel

monocyte chemotactic and activating factor produced by a human myelomonocytic cell line. J Exp Med 1988; 169:1485–90.
32. Rollins BJ, Sunday ME. Suppression of tumor formation in vivo by expression of the *JE* gene in malignant cells. Mol Cell Biol 1991; 11:3125–31.
33. Bischoff SC, Krieger M, Brunner T, Dahinden CA. Monocyte chemotactic protein 1 is a potent activator of human basophils. J Exp Med 1992; 175:1271–5.
34. Altman LC, Snyderman RS, Oppenheim JJ, Mergenhagen SE. A human mononuclear leukocyte chemotactic factor: characterization, specificity, and kinetics of production by homologous leukocytes. J Immunol 1973; 110:801–9.
35. Cushing SD, Berliner JA, Valente AJ, Territo MC, Navab M, Perhami F, Gerrity R, Schwartz CJ, Fogelman AM. Minimally modified low density lipoprotein induces monocyte chemotactic protein 1 in human endothelial cells and smooth muscle cells. Proc Natl Acad Sci USA 1990; 87:5134–8.
36. Ylae-Herttuala S, Lipton B, Rosenfeld MA, Saerkioja T, Yoshimura T, Leonard EJ, Witztum JL, Steinberg D. Expression of monocyte chemoattractant protein 1 in macrophage-rich areas of human and rabbit atherosclerotic lesions. Proc Natl Acad Sci USA 1991; 88:5252–6.

8

Macrophages and Acute-Phase Proteins

Richard F. Mortensen

The Ohio State University, Columbus, Ohio

I. INTRODUCTION

Homeostatic mechanisms that operate in all vertebrates under normal circumstances are altered substantially immediately after tissue injury or infection. Together these systemic alterations are termed the acute-phase response. The systemic inflammatory response consists of a series of coordinated metabolic and endocrine events that result in seemingly diverse physiological changes such as fever and leukocytosis, as well as the dramatic increase in the synthesis and secretion by the liver of several plasma proteins that have been termed acute-phase proteins, or more commonly acute-phase reactants (APRs) [1–3]. The characteristic increase in the level of APRs is coordinated with inflammatory cellular events such as neutrophil and monocyte chemotaxis and infiltration during the first 24 to 48 hours following tissue infection or damage. The function of the acute systemic reaction appears to be to generate heightened nonspecific protection against microbial invasion during the preimmune or early stage of infection. Consistent with this role are the numerous "immune-like" activities that have been ascribed to some of the APRs. Not surprisingly, several of the liver-derived proteins interact with leukocytes, especially macrophages (Mϕs). It is these APRs and their interaction with Mϕs that will be described in this brief review, with an emphasis on their role as mediators of nonspecific immunity.

A. Classification of Acute Phase-Reactants

The APRs are groups on the basis of either negative or positive changes in blood levels, as well as their relative increase in concentration by 24 hours after the onset of inflammation [reviewed in 1,4,5]. The list in Table 1 includes the more common APRs found in humans and most other mammals. It should be noted that the most dramatic of the APRs are C-reactive protein (CRP) and serum amyloid A (SAA) protein, which can increase by several thousand times from endogenous levels of only nanograms per milliliter. Many of the APRs that increase by approximately two to five times are protease inhibitors with the obvious role of modifying the untoward tissue-damaging effects of the inflammatory response. Several other plasma APRs with well-defined functions, such as fibronectin, fibrinogen, and the complement components, increase in concentration by only 50% (Table 1). Almost all of the APRs are glycoproteins except CRP and the SAA protein. It should be noted that there is considerable variability between species in the acute-phase behavior of these proteins; e.g., serum amyloid P-component (SAP) displays a 50-fold increase in the mouse and only a 0.5-fold increase in man, but the homologous protein CRP is only a minor APR in the mouse [5]. Nonetheless, the phylogenetic conservation of the APRs and their responsiveness to inflammatory stimuli among all vertebrates implies a fundamental role for these proteins in restoring homeostasis and augmenting nonspecific resistance mechanisms. Since the cornerstone

TABLE 1 Classification of Acute-Phase Proteins and Their Proposed Biological Function

Group	X-fold increase	Protein	Biological function
I	0.25–1.5	Complement: C3, C4	Opsonin, anaphylatoxin
		Ceruloplasmin	Scavenger of O_2-radicals
		Fibronectin	Cell attachment, opsonin
		Serum amyloid P-component (SAP)	Unknown
II	2 to 5	α_1-Proteinase inhibitor	Protease inhibitor
		α_1-Antichymotrypsin	Protease inhibitor
		α_2-Macroglobulin	Binding, protease inhibitor
		Fibrinogen	Clot formation
		Haptoglobin	Binds hemoglobin
		C4b-binding protein	C-inhibitor
		Mannose-binding protein	Opsonin, C activation
III	100 to 1000	α-1-Acid glycoprotein	Transport
		Serum amyloid A protein	?
		C-reactive protein (CRP)	Opsonin, macrophage activation, C-activation
Negative	<50% Decrease	Albumin	Transport
		Transferrin	Transport

of nonspecific resistance to microbial pathogens is a systemic inflammatory response that includes a greatly amplified synthesis of hepatic plasma proteins and the subsequent effector mechanisms of recruited leukocytes, there is considerable interest in the interaction of acute-phase proteins with neutrophils (PMNs) and Mϕs. In this chapter I briefly describe the properties of those APRs that function as either opsonins, leukocyte activators, or both. These proteins are CRP and mannose-binding protein (MBP).

B. Induction and Synthesis of Acute-Phase Proteins

Although inflammatory reactions are nonspecific with respect to their induction and effector phase, they are nevertheless part of the inducible host response. The magnitude of acute-phase protein production is usually proportional to the extent of tissue damage or tissue involvement in acute microbial infections, and therefore the levels of APRs are extensively used clinically as gauges of inflammation or underlying infection [5,6]. The prototype APR in humans is CRP, since it increases from levels of approximately 0.1 μg/ml in healthy individuals to several hundred μg/ml [7].

The rapid de novo synthesis of APRs by liver hepatocytes accounts for virtually all of the increase in blood levels of APRs; local production of APRs by infiltrating Mϕs may have a significant impact at a site of inflammation but contributes little to the blood levels [8–11]. The best example of local production of APRs by Mϕs is the synthesis of C3 and C4 [6]. The elevated plasma levels of APRs are the direct result of an increase in the rate of synthesis, since APR levels are not significantly influenced by either catabolism or a change in the half-life of the protein during the inflammatory state [8,9]. Organ perfusion studies and the use of primary hepatocyte cultures clearly established the phenomenon of progressive recruitment, by blood-borne mediators, of hepatocytes nearest the portal circulation into APR synthesis of CRP [5,10,11]. The earliest studies of the mechanism of APR synthesis at the cellular level implicated the fever-inducing leukocyte endogenous mediators or pyrogens [12], which upon more thorough analysis were found to consist primarily of interleukin-1 (IL-1) and IL-6, but not tumor necrosis factor-α (TNF-α)

[reviewed in 2,13]. Although these hepatocyte-stimulating factors were originally identified in the supernatants of activated monocytes/Mϕs, the use of recombinant cytokines clarified the identity of the most active cytokines [14–17]. The most active cytokine for induction at the transcriptional level of several, but not all, APRs is IL-6 [18–20]. Since the existence of a cytokine cascade is widely accepted [21–23], it seems likely that the in vivo sources of IL-6 may also depend on the response of cells (fibroblasts, endothelial cells) at inflamed sites to IL-1 [21,22]. Nevertheless, hepatocytes respond only to an exogenous source of a cytokine, as opposed to an autocrine source [2]. Along with the stimulation of hepatocytes by these cytokines, both IL-1 and IL-6 act on the central nervous system, resulting in fever and production of adrenocorticotropic hormone (ACTH) [5,22]. The eventual release of cortisol may be an effective negative feedback loop, since cortisol is a potent inhibitor of IL-1 and IL-6 synthesis [23]. Many other systemic effects of these two inflammatory cytokines augment host defense, including an increase in the production of granulocyte-macrophage colony-stimulating factor (GM-CSF), granulocyte colony-stimulating factor (G-CSF), T cells and B cells, as well as the localized action of increasing vascular permeability and the margination and chemotaxis of PMNs and monocytes [23,24]. The role in the hepatocyte APR response of additional cytokines or augmenting factors such as the glucorticoids is still under active investigation; however, steroids augment transcription of APR genes by cytokines in vitro [14]. Albumin synthesis appears to be sacrificed by the hepatocytes as part of the reorchestration of the pattern of protein synthesis by the hepatocytes [2].

The molecular genetic mechanism of human CRP gene expression is the best-studied of the APR genes [reviewed in 13,25,26]. A general scheme for the induction of synthesis of CRP and subsequent biological activities is shown in Figure 1. The emphasis in this schema is on the multiple sources of the inflammatory cytokines along with at least one potent inhibitor of CRP synthesis at the post-transcriptional level, transforming growth factor-β (TGF-β), a mediator also released into the blood by platelets during inflammation [27] (Figure 1). The short-lived intermediate nitric oxide (NO) inhibits hepatocyte APR synthesis also at the post-transcriptional level [28].

Interleukin-6–responsive elements in the 5'-flanking region of the CRP gene have been identified [18,19,26], along with a *trans*-acting, IL-6–inducible nuclear protein that binds to the IL-6–responsive consensus element [29]. Several studies also implicate IL-1 in human CRP gene expression already initiated by IL-6 [18,20,26]. Its mode of action depends on enhancer elements that are 3' to the IL-6–responsive element in the CRP [19,26]. Interestingly, TNF-α inhibits human hepatocyte production of both CRP and SAA protein [30]. The nature of the second signals after the interaction of IL-6 with its receptor and the IL-6 receptor–associated signal-transduction protein [21] remain unexplored. Since the half-life of CRP is only 18 hours and it becomes selectively deposited at sites of inflammation [5,10], a substantial commitment by the hepatocyte to synthesis of this protein suggests an essential function or functions. Transcription of the other APR genes also occurs in response to IL-6 [14,32]. Macrophages not only serve as the major source of the inflammatory cytokines, especially IL-6, but also respond to some of the APR during the effector phase of the response.

II. C-REACTIVE PROTEIN

Information on the direct involvement of acute-phase proteins in host resistance in experimental models of infectious disease is limited [33]. A few studies even show that certain APRs, such as fibronectin and α_2-macroglobulin, bind to the surface of protozoan parasites and prevent their recognition and destruction by Mϕs (3). However, most of the evidence gathered from experiments with CRP both in vitro and in vivo strongly supports the concept that this particular APR is involved in resistance to microbial pathogens. This proposed function for CRP is based not only on its well-established ability to recognize microbial pathogens and damaged host cells, but also on

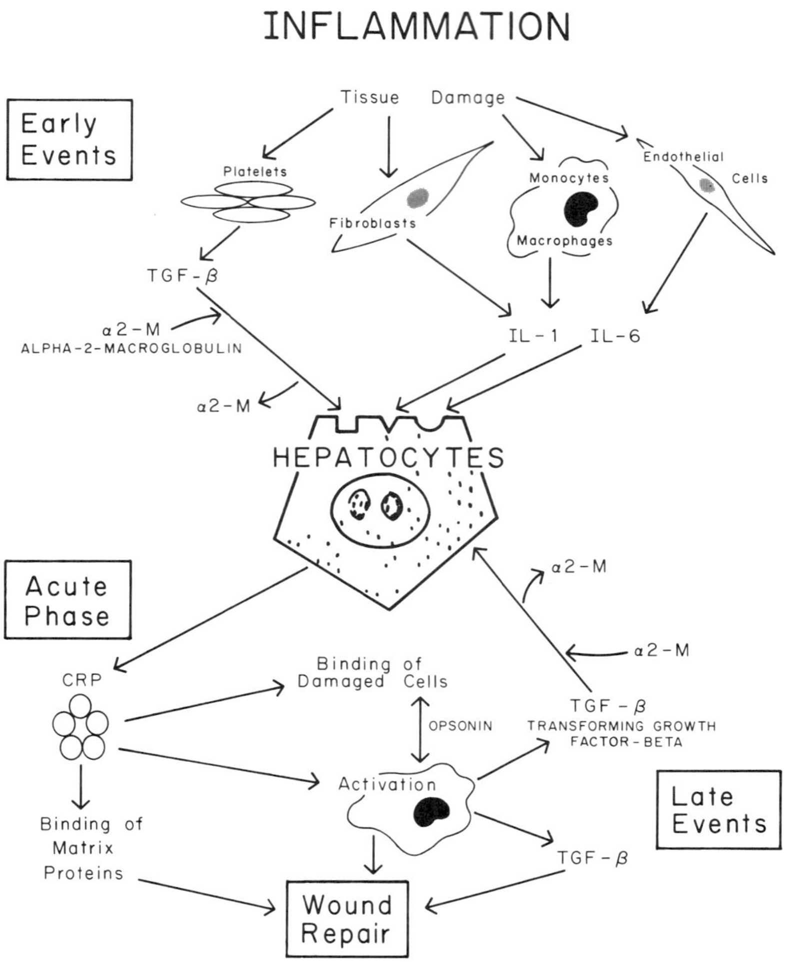

FIGURE 1 A general scheme for the induction of acute-phase protein synthesis by hepatocytes and subsequent activities mediated by C-reactive protein (CRP). The early events that occur within 12 hours of the onset of inflammation include the rapid production of cytokines (interleukin-1 [IL-1] and IL-6) that trigger the increased synthesis of CRP, and the release of mediators (transforming growth factor-β [TGF-β]) that inhibit synthesis of CRP. Activity of TGF-β is controlled by its binding to α_2-macroglobulin (α2-M). Acute-phase events mediated by CRP include binding to matrix proteins, opsonization, C-activation, and macrophage activation; these events eventually contribute to wound repair and enhanced host resistance.

its ability to eliminate these substances by subsequently interacting with either serum complement (C) or cellular effector systems [1,5,6,13]. Thus, CRP serves both recognition and effector functions in a manner analogous to specific immunoglobulin G (IgG) antibody. The properties of CRP and the mechanisms whereby it serves as a nonspecific effector molecule with Mϕs are described below.

A. Molecular Properties of C-Reactive Protein

C-reactive protein was initially described in 1930 as a C-precipitin in the serum of febrile patients because it could precipitate the C-polysaccharide of pneumococcal cell walls. For a historical description of CRP research, see Gottschlich [34]. C-reactive protein was rarely investigated after 1940, because specific antibody was associated with host resistance. C-reactive protein is a pentameric protein composed of 5 identical 23,500-dalton globular subunits noncovalently linked and arranged in cyclic symmetry in a single plane [7]. This unusual planar pentagonal arrangement of subunits is the basis for classifying CRP as the prototype of the pentraxin group of proteins [7,10]. The only other member of the pentraxin family is serum amyloid P-component (SAP), which circulates as two pentamers bound face-to-face [7]. Although SAP is structurally similar to CRP, it has distinct binding reactivities and biological activities [7]. Human CRP subunits are composed of 206 amino acids in a linear chain with a single intrachain disulfide bond [25]. C-reactive protein is neither glycosylated nor phosphorylated, whereas SAP is approximately 10% carbohydrate by weight [35]. The human CRP gene is encoded on chromosome 1 as a single copy, and its structure consists of two exons with a single long intron that interrupts the codon for the third residue in the mature protein [25]. The human SAP gene has a similar organization and also maps to the same region on chromosome 1 [25,36] and encodes a 204-amino-acid subunit that is approximately 70% homologous to the CRP sequence [25,35]. The hydrophobic residues of both proteins are predominantly conserved, suggesting that the pentraxin structure itself was conserved throughout vertebrate evolution [25]. Of interest is that this chromosome 1 region (q21) is syntenic with mouse chromosome 1 genes for Fc receptors and several of the cytokine-receptors [36]. Structurally homologous "CRP-like" proteins are widely distributed among the invertebrates as constitutive lectin-like molecules in the hemolymph. An example of the latter is limulin from the horseshoe crab, *Limulus polyphemus* [37]. Low resolution x-ray diffraction analysis of CRP crystals has confirmed the pentameric structure, but the three-dimensional structure has not yet been reported [38].

B. Binding Reactivities of C-Reactive Protein

The precipitation reaction between the heat-labile, non-Ig CRP and C-polysaccharide that served as the initial description of CRP was distinct from antibody-mediated reactions because it was dependent on Ca^{2+} ions [34]. Studies of this reaction revealed that the binding specificity was directed toward the phosphorylcholine (PC) moiety [39]. The PC-binding sites on CRP are present on each subunit on only one of the two planes of the pentraxin [40]. The highly conserved amino acids between residues 47 and 63 make up the binding site for PC and share the same idiotope as the PC-specific T-15 idiotype of the mouse Ig repertoire [41]. Other phosphate monoesters also bind to CRP through the PC-binding site with lower affinity [42]. The range of substrates recognized by CRP via its PC-binding site includes fibronectin [43], laminin [44], phospholipids [45], chromatin [46,47], histones [48], and the U1 snRNP [48]. The PC-hapten itself is widely distributed on bacterial cell walls and is potentially available on disrupted eukaryotic cell membranes. The Ca^{2+} requirement for all of the binding reaction depends on the induction of an allosteric change in the CRP pentraxin upon binding of two Ca^{2+} ions per subunit [49]. The Ca^{2+}-dependent binding to PC serves as the basis for the affinity chromatography purification procedure for CRP [50] and functionally defines the CRP-like molecules present in vertebrates and invertebrates [7]. The Ca^{2+}-dependent binding to a wide variety of physiological substrates is also characteristic of the more "primitive" lectin-mediated recognition events that account for nonspecific resistance among the invertebrates and plants. The significance of the binding of CRP to components of damaged tissues and cells, as well as microbial pathogens, is that these substances could be removed from sites of tissue damage by subsequent activation of the complement (C) cascade and/or phagocytosis.

C. Antibody-Like Activities of C-Reactive Protein

C-reactive protein displays two distinct activities that are often associated with the host defense properties of IgG antibody: activation of the classical complement cascade and opsonization.

1. *Complement Activation*

The realization almost 20 years ago that ligand-complexed CRP was able to bind to and then activate C1 as efficiently as specific IgG antibody was the incentive for all subsequent studies of the molecular properties and biology of CRP [51,52]. Although the requirement for CRP to react with any one of its many substrates via its PC-binding site before C-activation is analogous to triggering C by antigen-antibody complexes [51,52], the molecular interaction between CRP and C1q occurs with the collagen-like stalks rather than the globular heads of C1q [53]. There is some evidence that CRP-ligand complexes may actually inhibit the alternative C pathway by inhibiting the formation of a C5 convertase and thus assembly of the membrane-attack complex [54]; nonetheless, the inflammatory cleaved C components, the anaphylatoxins C3a and C4a and the opsonic fragments C3b, C3b, and iC3b, would all be generated. Therefore, unlike antibody, the C-activating function of CRP appears to be focused on the generation of those mediators that clearly contribute directly to localized host resistance.

Another IgG antibody–like property of the CRP in CRP-ligand complexes is that the alternative C pathway–dependent solubilization of the complexes is dependent on covalent bond formation between CRP and C3b [55]. Similar findings of solubilization of CRP-chromatin complexes also occur [56]. The CRP-driven C activation is not dependent on the presence of antibody molecules, nor are they incorporated into CRP complexes [51,52]. C-reactive protein also binds to a variety of cationic substances, including the contents of neutrophil granules, through a site distinct from the PC-binding site, a reaction that also triggers classical C cascade activation [52]. The studies of CRP-dependent C activation led directly to studies of its opsonic potential because of the obvious sharing of the biological properties with the Fc fragment of IgG antibody.

2. *Opsonization and Phagocytosis*

The very first indication that CRP might serve as an opsonin came from experiments by G. Löfström, who demonstrated nearly 50 years ago that human acute-phase serum containing CRP provided temporary protection of mice against pneumococcal infection and mediated capsular swelling of the bacteria [57]. Later studies using partially purified CRP indicated that CRP nonspecifically increased the phagocytic rate of leukocytes [58,59]. More definitive experiments with purified human CRP clearly demonstrated that the CRP molecule itself when coated onto erythrocytes was required for ingestion by both mouse $M\phi s$ and human blood monocytes [60]. These in vitro findings were extended to in vivo protection experiments with mice challenged with lethal inocula of pneumococci, experiments that clearly demonstrated the protective effects of CRP [61,62]. These studies implicated both the opsonic and the C-activating properties of CRP in the clearance of the bacteria [63]. C-reactive protein–opsonized particles are also ingested by human PMNs; however, there may be a requirement for an additional signal or signals derived from activated monocytes [64]. A comparative study by Briles et al. of the protective ability of CRP vs. PC-specific antibody in mice [65] shows that capsular antibody is more protective; however, the half-life of IgG antibody is more than 20 times longer than that of CRP, and this may explain the relative protection afforded by each [9–11]. C-reactive protein in the absence of any ligand or aggregating reagent is also able to trigger its own uptake by Mϕs [66]. Obviously, the nature of the protection afforded by CRP at sites of a localized or systemic infection will depend on the type of leukocyte present and the efficiency of the CRP-initiation of phagocytosis and classical C pathway activation. Passive protection experiments with mice bearing an inducible transgenic gene for human CRP are likely to provide new information on the role of CRP in host protection.

3. Precipitation and Agglutination

The first antibody-like property of CRP described was its potent precipitating activity with the C-polysaccharide or the teichoic acid of the capsule and cell walls of streptococci [7,34]. The agglutinating activity of CRP was later quantitated with CPS or PC-coated erythrocytes and found to be more potent than that of IgG antibody to CPS, probably because CRP is pentavalent [50]. C-reactive protein has the ability to bind and agglutinate a variety of bacteria, especially the gram-positive pathogens [67], and like IgG-complexed bacteria, the CRP-coated bacteria are readily removed by cells of the monocyte/Mϕ lineage [68]. Although such a mechanism lacks specificity, CRP substantially augments host resistance during acute bacterial infections.

The presence of CRP or CRP complexes as discrete bodies in synovial membranes of patients with autoimmune diseases [69] led to the discovery by Robey et al. [46] that CRP binds chromatin with even higher affinity than it binds PC, and that this occurs via the PC-binding site of CRP in a Ca^{2+}-dependent reaction. Each CRP subunit binds the equivalent of 160 bp of dsDNA [46]. Binding of CRP to chromatin is also directed toward the histones H1, H2A, and H2B as well as the DNA component [48]. The histones are thought to supply the basic residues, and the monophosphate group is contributed by the DNA sugar backbone and therefore mimics the choline and PO_4 groups of PC [46,48]. Since CRP-chromatin complexes should be readily phagocytized, a potentially unique physiological role for CRP is to serve as a scavenger for chromatin fragments, preventing autosensitization and effecting eventual clearance of the DNA, since Mϕs rapidly digest DNA.

4. Platelet Activation

Several groups have proposed that CRP has a protective role in the initial stages of an inflammatory reaction by binding to and inhibiting the actions of the phospholipid platelet-activating factor (PAF) [70]. Platelet-activating factor contains the PC moiety and therefore is one of the few CRP-binding substrates available in the extracellular environment. C-reactive protein as a single pentraxin inhibits the PAF-triggered aggregation and release reactions of platelets [71] as well as the PAF-induced degranulation of PMNs [72]. A related finding is that CRP also binds to platelets in the rat and mediates the cytolysis of schistosomes [73]. The interaction between CRP and sporozoite components of *Plasmodium falciparum* and *P. yoelii* interferes with their ability to penetrate hepatocytes, a protective mechanism that is dependent on both platelets and Mϕs [74]. The precise mechanism for CRP-mediated nonspecific resistance to protozoan parasites remains unexplored.

D. Macrophage Activation by C-Reactive Protein

The opsonic activity of CRP and its associated ability to passively protect mice against pneumococci led to studies of direct activation of monocytes/Mϕs by CRP. The studies of Mϕ and PMN functions affected by CRP, such as chemotaxis, generation of reactive O_2-intermediates, and extracellular killing, have not yielded a consensus concerning the mechanisms of activation. This lack of consensus may be the result of the very different effects of CRP on PMNs and on Mϕs, although CRP is opsonic with both cell types.

1. Tumoricidal Activation

The initial findings that implicated CRP as a Mϕ activator were obtained from experiments with solid tumors in mice by Deodhar and his colleagues, who initially documented tumor shrinkage and inhibition of metastasis as a result of therapy with liposomes incorporating CRP [75,76]. These studies were extended to show that the native CRP pentamer itself was able to generate human monocyte tumoricidal activity and formation of reactive O_2-intermediates in vitro [77]. Mouse resident and elicited Mϕs, and certain mouse Mϕ cell lines, became tumoricidal upon exposure to CRP under in vitro conditions that precluded involvement of any T cell lymphokine or lipopolysaccharide (LPS) [78]. C-reactive protein is capable of triggering monokine synthesis, as was shown by one group documenting enhanced IL-1 production [79]; however, production of other cytokines has not been examined. The homologous pentraxin SAP was shown to augment IL-1 production by mouse Mϕs, thereby providing a potential route for amplifying the synthesis of APRs [80]. C-reactive protein conceivably might also participate in human monocyte-mediated destruction of autologous tumor cells based on its presence at elevated levels in cancer patients and occasionally in solid tumors [81]. C-reactive protein has no direct cytolytic effect on nucleated cells even in the presence of C [82]. It is therefore a nonimmune activator of Mϕs.

2. Antimicrobial Activation

The interaction between the cellular and humoral components of the acute inflammatory response often leads to an increase in the antimicrobial activities of both PMNs and Mϕs. Studies of the interaction between CRP and monocytes have usually shown an enhanced capacity to generate and release antimicrobial substances. Purified mouse SAP, and APR, was shown to generate a modest listericidal response with elicited Mϕ in vitro [83]. Later studies revealed that Mϕ receptors for SAP recognize the terminal mannose residues [84]. Both intact CRP and CRP-derived peptides function as a chemotactic stimulus for monocytes, especially at low concentrations of only 1 to 10 μg/ml [66,85,86]. By contrast, CRP either did not alter [86,87] or even inhibited [88] the chemotactic response of PMNs. C-reactive protein in the form of complexes or in an aggregated form enhanced the respiratory burst of monocytes generating the superoxide anion O_2^- [85]. The effect of CRP or CRP-complexes on PMN superoxide generation was also the opposite of that observed with monocytes, i.e., it inhibited O_2^- production triggered by an agonist [89]. The very different response of monocytes and PMNs to a CRP stimulus may occur at the level of the second signal events triggered upon binding.

C-reactive protein is selectively deposited at sites of inflammation, where it can be digested by neutrophils into biologically active peptides [86,90]. The PMNs use neutral or acidic proteases [87,90] to generate peptides that greatly increase monocyte activities such as chemotaxis, IL-1 production, and formation of reactive O_2-intermediates [86]. Whether only CRP-derived peptides or intact CRP, or both, have the capacity to stimulate antimicrobial responses of monocytes is not clear. Nonetheless, the notion that CRP or its peptides can focus the monocyte response to sites of infection is important for heightened host defense, especially at sites where antimicrobial activities may give rise to elements leading to antigen presentation and sensitization of T cells.

The various Mϕ functions triggered by either CRP or its peptides are listed in Table 2. A comparison with the immune mediators that activate Mϕ, also listed in Table 2, illustrates the range of functions enhanced or triggered by CRP.

E. Monocyte/Macrophage Receptors for C-Reactive Protein

The characterization of specific receptors for CRP on leukocytes is of considerable interest for two reasons. The first is that there is some controversy about the participation of IgG Fc receptors (FcRs) in CRP binding and uptake. The second is that the molecular signaling aspects of the CRP receptor (CRP-R) must be understood to modify the effects of CRP. The initial investigations of the monocyte/Mϕ CRP-R suggested that CRP mediated phagocytosis through IgG FcRs, because

TABLE 2 Macrophage Functions Mediated by C-Reactive Protein (CRP) Versus Immune Mediators

Function	Immune mediators	CRP and/or CRP-Peptides
Phagocytosis	Fc of IgG, C3b/C3bi/C4b	Intact CRP—opsonin
Chemotaxis	C5a, Bacterial peptides	CRP-peptides
Microbicidal	IFN-γ, Lymphokines, H_2O_2, O_2^-, NO/NO_2^-	CRP, CRP-peptides, CRP-aggregates
Tumoricidal	IFN-γ, IL-4, IL-2, ADCC, (IgG) NO/NO_2^-	CRP CRP-peptides
Antigen presentation	MHC-II increase	MHC-II increase
Secretion of	Cytokines: IL-1, IL-6, TNF-α, IFN-α IFN-β, TGF-β_1 C proteins Proteases, Protease inhibitors PGs, LTs	IL-1, IL-6 (?) TNF-α (?)

IgG = immunoglobulin G; IFN - interferon; IL = interleukin; ADCC = antibody-dependent cell-mediated cytotoxicity; MHC = major histocompatibility complex; TNF = tumor necrosis factor; TGF = transforming growth factor; PG = prostaglandin; LT = leukotriene; C = complement.

IgG-immune complexes blocked CRP-dependent ingestion [91]. Later binding studies with peripheral blood mononuclear cells and ^{125}I-CRP revealed a correlation between the small fraction of CRP-binding lymphocytes and the presence of IgG FcRs [92]. Receptors distinct from FcRs for CRP on mouse Mϕs and mouse Mϕ cell lines were found to have a K_D of 10 nM with a density of 10^5 sites/cell [93]. The CRP-Rs isolated by ligand-affinity chromatography from solubilized membrane proteins were of a different size than FcRs, failed to bind to IgG, and did not react with mAb to FcRs [93]. Subsequent studies with human blood monocytes by three different groups provided convincing evidence that the CRP-R was indeed distinct from all three classes of the human IgG FcR [94–96]. The CRP-R on human PMNs displays a similar affinity for CRP and is also distinct from the FcRII and FcRIII on these cells [64]. C-reactive protein binds in a saturable, specific manner to monocytes and PMNs; however, the binding is reversible and appears to be Ca^{2+}-dependent.

Some limited information on the biochemical properties of the monocyte CRP-R has been gathered by using the conventional approach of ligand-affinity chromatography. A 40-kDa membrane glycoprotein has been isolated from the human promonocytic cell line U-937 by two groups [94,97]. An additional 60-kDa membrane glycoprotein has also been claimed to be part of the CRP-R [97]. Neither of these two proteins appears to share the size or epitopes of the FcRI or FcRII of human monocytes. Crowell et al. [97] have recently shown that the higher-molecular-weight CRP-binding protein from U-937 cell membranes may be the high-affinity IgG FcRI present on Mϕs. This finding may explain the earlier observations showing partial inhibition of CRP binding to monocytes by IgG or IgG complexes, but a lack of inhibition of IgG-binding to FcR by CRP in the converse experiment [93,94]. The resolution of the relationship between the CRP-R and IgG FcR awaits cloning of the gene(s) for the CRP-R and determining the extent of sequence similarity between these two opsonic receptors. It is possible that the CRP-R arose earlier in evolution as a phagocytic-receptor for the CRP-like molecules present constitutively in the hemolymph of invertebrates.

The second signal mechanism triggered by CRP-R engagement is largely unknown; however, a few observations suggest that the nature of the second signaling events differs significantly between PMNs and monocytes. The monocytic cell line U-937 internalizes and degrades receptor-bound CRP in the endosomal compartment followed by recycling of the CRP-R to the membrane [66]. C-reactive protein itself at inflammatory levels of more than 100 μg/ml appears to inhibit intracellular Ca^{2+}-mobilization and subsequent superoxide production triggered by inflammatory agonists—e.g., fMLP—in alveolar Mϕs [88] and also human PMNs [98]. By contrast, elicited Mϕs, human monocytes, and human monocytic cell lines have all been shown to produce reactive O_2^- intermediates in response to similar concentrations of CRP [66,94,95,99]. Like other Mϕ activators such as interferon-γ (IFN-γ), CRP was capable of driving the differentiation of the promonocytic U-937 cell line to become capable of tumoricidal activity [66]. Since superoxide generation by PMNs and Mϕs in response to agents such as fMLP requires Ca^{2+} mobilization and protein kinase C (PKC) activation [64], it seems likely that CRP may exert a triggering or modifying effect on such responses at the level of the second signal events. A consistent finding is that CRP at levels greater than 10μg/ml down-regulates the PMN respiratory burst in response to inflammatory stimuli at the level of second signal events, since it inhibits phosphorylation of several proteins associated with PMN activation [100]. At least one study suggests that CRP at concentrations of more than 100 μg/ml prevents PMA-induced PKC translocation and its activation on the membrane of neutrophils [100]. Yet CRP at less than 1 μg/ml elicits neutrophil chemotaxis and phagocytosis [86]. The outcome of the interaction of CRP with leukocytes may be influenced by the tissue source of the responding Mϕs and PMNs and whether the CRP is present as the intact pentamer or biologically active peptides, and their relative concentration at inflamed sites. Perhaps the CRP found in inflammatory lesions with infiltrating PMNs protects host tissues by down-regulation, but when present with Mϕs at the same sites elicits a host-protective response.

III. MANNOSE-BINDING PROTEIN

Blood clearance and cell uptake experiments in the 1970s documented the existence of a Mϕ receptor for glycoproteins bearing high mannose side chains [101]. Many subsequent studies of the Mϕ mannose receptor (MR) characterized the receptor as a lectin-like binding protein selectively expressed on the surface of tissue Mϕs, but not monocytes [102]. The Mϕ MR plays a major role in the removal of yeasts and protozoan parasites by direct engulfment, with subsequent production of many Mϕ secretory products, including proteases and monokines [103]. Since blood monocytes do not express the MR, it seemed unlikely that these receptors were essential for inflammatory events. However, experiments since 1988 suggest that a circulating mannose-binding protein (MBP) synthesized by hepatocytes during the acute phase is the functional equivalent of the MR on tissue Mϕs, and this MBP was predicted by Stahl to participate in inflammatory events [102,104]. A brief description of the molecular properties of MBP and its host-protective role as an opsonin follows.

A. Molecular Properties

The MBP was purified first from rabbit serum [105] and subsequently from the serum and liver of both humans and rats [106,107]. The human MBP is an acute-phase serum glycoprotein of 300 to 300 kDa composed of multimers of identical 32-kDa monomers [108–110]. The MBP exists in mammalian sera as at least a trimer [111]. The monomer itself has three domains: the NH_2-terminal domain with multiple cysteine residues involved in interchain covalent bonds; a collagen α-helical domain consisting of up to 19 repeats of a Gly-x-Y; and a globular C-terminal domain that recognizes mannose oligosaccharides [111,112]. Both the MR and the MBP are members of a large family of animal lectins that possess invariant residues in their carbohydrate recognition domains and display Ca^{2+}-dependent ligand-binding activities at neutral pH [112]. The human MBP gene is very similar to the human pulmonary surfactant A gene, since the four exons in the gene appear to have evolved from a recombination between an ancestral collagen gene and a lectin-carbohydrate recognition gene [113]. The structural gene for the MBP receptor on Mϕs has not yet been elucidated. The pentraxin proteins CRP and SAP, described earlier, also are Ca^{2+}-dependent lectin-like sugar-binding proteins, but they appear to be unrelated to MBP on the basis of sequence and structural properties.

Cloning of the two genes for the mouse MBPs (MBP-A and MBP-C) led to the discovery that MBP-C is an APR in this species [114]. Both of the genes in humans (MBP-A and MBP-C) contain IL-1– and IL-6–responsive elements as well as glucocorticoid-responsive elements in the 5'-regulatory region and are considered APRs [113]. The molecular genetic mechanism of the functional activity of the cytokine-responsive elements has yet to be determined for the cloned MBP genes. The kinetics of induction of gene expression for the acute-phase MBP genes is very similar to that of the human CRP and mouse SAP genes [114].

B. Host Defense Activities

The MBP shares not only similar binding characteristics and specificity with the MR, but also epitopes [reviewed in 102,112]. Therefore, it is not surprising that many of the host defense activities exerted by Mϕs through the MR have also been ascribed to the MBP [115]. The most important host defense activity of MBP is its opsonic capability that enhances the clearance of yeasts and Gram-negative bacteria, and perhaps other microorganisms [116]. In addition, MBP binds to the high mannose glycans of the envelope of HIV, preventing infection of T cell lines in vitro [117]. Mannose-binding protein probably mediates the natural resistance of mice to influenza virus [118]. However, the most noteworthy host protection provided by MBP may be derived from its ability to activate both the classical [109] and the alternative [119] C pathways when complexed to its ligand. This activity would ensure coating of microbial pathogens with additional opsonins as well as the generation of cleaved C components with phlogistic activities. Low serum levels of the MBP opsonin in humans are associated with recurrent infections, illustrating the direct involvement of MBP in host protection [120].

The activities displayed by MBP are very similar, but not identical, to those of CRP in terms of the interaction with both the C system and phagocytic cells, especially Mϕs. Despite this functional similarity, the molecular properties of MBP and CRP are sufficiently different to suggest that the genes for each were conserved during vertebrate evolution, since among the vertebrate species most use the proteins as part of the inducible repertoire of APRs.

REFERENCES

1. Kushner I. The acute phase response: an overview. Methods Enzymol 1988; 163:373–83.
2. Fey G, Gauldie J. The acute phase response of the liver in inflammation. In: Popper H, Schaffner F, eds. Progress in liver diseases. Philadelphia: W.B. Saunders, 1990:89–116.
3. Stadnyk A, Gauldie J. The acute phase protein response during parasitic infection. Immunoparasitol Today 1991; 7:A7–A12.
4. Gauldie J, Lamontagne L, Stadnyk A. Acute phase response in infectious disease. Surv Synth Path Res 1985; 4:126–51.
5. Ballou S, Kushner I. C-Reactive protein and the acute phase response. Adv Intern Med 1992; 37:313–36.
6. Schultz DR, Arnold PI. Properties of four acute phase proteins: C-reactive protein, serum amyloid A protein, γ_1-acid glycoprotein, and fibrinogen. Semin. Arth-Rheum. 20:129–147.
7. Pepys M, Baltz M. C-Reactive protein and related pentraxins: serum amyloid P-component, serum amyloid A protein. Adv Immunol 1983; 34:141–210.
8. Chelladurai M, Macintyre S, Kushner I. In vivo studies of serum c-reactive protein turnover in rabbits. J Clin Invest 1983; 71:604–10.
9. Baltz M, Rowe I, Pepys MB. In vivo turnover studies of C-reactive protein. Clin Exp Immunol 1985; 59:243–51.

10. MacIntyre S. C-Reactive protein. Methods Enzymol 1988; 163:383–99.
11. Kushner I, Ribich W, Blair J. Control of the acute-phase response. J Lab Clin Med 1980; 96:1037–42.
12. Merriman CA, Pulliam LA, Kampshscmidt RF. Effect of leukocyte endogenous mediator on C-reactive protein in rabbits. Proc Soc Exp Biol Med 1975; 149:782–4.
13. Kilpatrick J, Volanakis J. Molecular genetics, structure, and function of C-reactive protein. Immunol Res 1991; 10:43–53.
14. Baumann H, Richards C, Gauldie J. Interaction among hepatocyte-stimulating factors, interleukin-1 and glucocorticoids for regulation of acute phase plasma proteins in human hepatoma (Hep G2) cells. J Immunol 1987; 139:4122–8.
15. Darlington G, Wilson D, Lachman L. Monocyte-conditioned medium, interleukin-1, and tumor necrosis factor stimulate the acute phase response in human hepatoma cells *in vitro*. J Cell Biol 1986; 103:787–93.
16. Goldman N, Liu T. Biosynthesis of human C-reactive protein in cultured hepatoma cells is induced by a monocyte factor(s) other than IL 1. J Biol Chem 1987; 262:2363–8.
17. Castell J, Gomez-Lechon M, David M, Hirano T, Kishimoto T, Heinrich P. Recombinant human IL-6 regulates the synthesis of acute phase proteins in human hepatocytes. FEBS Lett 1988; 232:347–50.
18. Ganter U, Arcone R, Toniatti C, Morrone G, Ciliberto G. Dual control of C-reactive protein gene expression by IL-1 and IL-6. EMBO J 1989; 8:3773–9.
19. Li S, Liu T, Goldman N. Cis-acting elements responsible for IL-6 inducible C-reactive protein gene expression. J Biol Chem 1990; 265:4136–42.
20. Steel D, Whitehead A. Heterogeneous modulation of acute-phase-reactant mRNA levels by interleukin-1β and interleukin-6 in the human hepatoma cell line PLC/PRF/5. Biochem J 1991; 277:477–82.
21. Akira S, Hirano T, Taga T, Kishimoto T. Biology of multifunctional cytokines: IL-6 and related molecules (IL-1 and TNF). FASEB J 1990; 4:2860–7.
22. Le J, Vilcek J. Interleukin 6: a multifunctional cytokine regulating immune reactions and the acute phase response. Lab Invest 1989; 61:588–602.
23. Balkwill FR, Burke F. The cytokine network. Immunol Today 1989; 10:299–303.
24. Wong GC, Clark SC. Multiple actions of IL-6 within a cytokine network. Immunol Today 1988; 9:137.
25. Whitehead A. Organization, structure and expression of pentraxin genes. In: Acute phase proteins and the acute phase response. Proceedings of the XIII Argenteuil symposium. Berlin: Springer-Verlag, Ed. by MB Pepys 1989:151–72.
26. Toniatti C, Arcone R, Majello B, Ganter U, Arpaia G, Ciliberto G. Regulation of the human C-reactive protein gene, a major marker of inflammation and cancer. Mol Biol Med 1990; 7:199–212.
27. Taylor AW, Ku N-O, Mortensen RF. Regulation of cytokine-induced human C-reactive protein production by transforming growth factor β1. J Immunol 1990; 145:2507–13.
28. Curran RD, Ferrari FK, Kispert PH, Stadley J, Stuehr DJ, Simmons RL, Billiar TR. Nitric oxide and nitric-oxide generating compounds inhibit hepatocyte protein synthesis. FASEB J 1991; 5:2085–92.
29. Poli V, Cortese R. Interleukin 6 induces a liver-specific nuclear protein that binds to the promoter of acute-phase genes. Proc Natl Acad Sci USA 1989; 86:8202–6.
30. Yap SH, Moshage AJ, Hazenburg PBC, Aarden LA. Tumor necrosis factor inhibits IL-1 and/or IL-6 stimulated synthesis of C-reactive protein and SAA in primary cultures of human hepatocytes. Biochim Biophys Acta 1991; 1091:405–8.
32. Baumann H, Prowse KR, Marinkovic S, Won KA, Jahreis GP. Stimulation of hepatic acute phase response by cytokines and glucocorticoids. Ann NY Acad Sci 1989; 557:280–96.
33. Kolbachofen V. A review of the biological properties of C-reactive protein. Immunobiology 1991; 183:133–45.
34. Gottschlich EC. C-reactive protein. A historical overview. Ann NY Acad Sci 1989; 557:9–18.
35. Skinner M, Cohen AS. Serum amyloid P-component. Methods Enzymol 1988; 163:523–36.
36. Yunis I, Whitehead AS. The mouse C-reactive protein gene maps to distal chromosome 1, and like its human counterpart, is closely linked to the SAP gene. Immunogenetics 1990; 32:361–365.
37. Robey F, Liu T. Limulin: a C-reactive protein from Limulus polyphemus. J Biol Chem 1981; 256:969–75.
38. Myles D, Rule S, Delucas, Babu Y, Xu Y, Volanakis J, Bugg C, Bailey S, Greenhough T. Rotation function studies of human C-reactive protein. J Mol Biol 1990; 216:491–6.

39. Volanakis J, Kaplan M. Specificity of C-reactive protein for choline phosphate residues of pneumococcal C-polysaccharide. Proc Soc Exp Biol 1971; 36:612–4.
40. Roux KH, Kilpatrick JM, Volanakis JE, Kearney JE. Localization of the phosphocholine-binding sites on C-reactive protein by immunoelectron microscopy. J Immunol 1983; 131:2411–5.
41. Swanson SJ, Lin BF, Mullenix MC, Mortensen RF. A synthetic peptide corresponding to the PC-binding region of human CRP possesses the T-15 myeloma PC-idiotype. J Immunol 1991; 146:1596–1601.
42. Gotschlich EC, Edelman GM. Binding properties and specificity of C-reactive protein. Proc Natl Acad Sci USA 1967; 57:706–710.
43. Tseng J, Mortensen RF. Binding of human C-reactive protein (CRP) to plasma fibronectin occurs via the phosphorylcholine-binding site. Mol Immunol 1988; 25:679–86.
44. Swanson SJ, McPeek MM, Mortensen RF. Characteristics of the binding of human C-reactive protein (CRP) to laminin. J Cell Biochem 1989; 40:121–32.
45. Pepys MB, Rowe IF, Baltz ML. C-reactive protein: binding to lipids and lipoproteins. Int Rev Exp Pathol 1981; 27:83–111.
46. Robey F, Jones K, Tanaka T, Liu T. Binding of C-reactive protein to chromatin and nucleosome core particles. J Biol Chem 1984; 259:7311–6.
47. DuClos TW, Zlock LT, Rubin RL. Analysis of the binding of C-reactive protein to histones and chromatin. J Immunol 1988; 141:4266–70.
48. DuClos T, Zlock L, Marnell L. Definition of a C-reactive protein binding determinant on histones. J Biol Chem 1991; 266:2167–71.
49. Kilpatrick JM, Kearney JF, Volanakis JE. Demonstration of calcium-induced conformational change(s) in C-reactive protein by using monoclonal antibodies. Mol Immunol 1982; 19:1159–65.
50. Osmand A, Mortensen R, Siegel J, Gewurz H. Interactions of C-reactive protein with the complement system III. Complement-dependent passive hemolysis initiated by CRP. J Exp Med 1975; 142:1065–77.
51. Kaplan M, Volanakis J. Interaction of C-reactive protein complexes with the complement system I. Consumption of human complement associated with the reaction of CRP with pneumococcal CPS and with the choline phosphatides lecithin and sphingomyelin. J Immunol 1974; 112:2135–47.
52. Siegel J, Rent R, Gewurz H. Interactions of C-reactive protein with the complement system I. Protamine-induced consumption of complement in acute phase sera. J Exp Med 1974; 140:631–47.
53. Jiang H, Siegel J, Gewurz H. Binding and complement activation by C-reactive protein via the collagen-like region of C1q and inhibition of these reactions by monoclonal antibodies to C-reactive protein and C1q. J Immunol 1991; 146:2324–30.
54. Mold C, Gewurz H. Inhibitory effect of C-reactive protein on alternative C pathway activation by liposomes and *S. pneumoniae*. J Immunol 1981; 127:2089–92.
55. Volanakis J. Complement-induced solubilization of C-reactive protein C-polysaccharide precipitates: evidence for covalent binding of complement proteins to CRP and to CPS. J Immunol 1982; 128:2745–50.
56. Robey FA, Jones KD, Steinberg AD. C-reactive protein mediates the solubilization of nuclear DNA by complement *in vitro*. J Exp Med 1985; 161:1344–56.
57. Löfström G. Comparison between the reactions of acute phase serum with pneumococcus C-polysaccharide and with pneumococcus type XXVII. Br J Exp Pathol 1944; 25:21–6.
58. Hokama Y, Coleman MK, Riley RF. In vitro effects of C-reactive protein on phagocytosis. J Bacteriol 1962; 83:1017–24.
59. Kindmark C. Stimulating effect of C-reactive protein on phagocytosis of various species of pathogenic bactera. Clin Exp Immunol 1971; 8:941–8.
60. Mortensen R, Osmand A, Lint T, Gewurz H. Interactions of C-reactive protein with lymphocytes and monocytes: complement dependent adherence and phagocytosis. J Immunol 1976; 117:774–81.
61. Mold C, Nakayama S, Holzer T, Gewurz H, DuClos T. C-reactive protein is protective against Streptococcus pneumoniae infection in mice. J Exp Med 1981; 154:1703–8.
62. Yother J, Volanakis J, Briles D. Human C-reactive protein is protective against fatal Streptococcus pneumoniae infection in mice. J Immunol 1982; 128:2374–7.
63. Nakayama S, Mold C, Gewurz H, DuClos T. Opsonic properties of C-reactive protein *in vivo*. J Immunol 1982; 128:2435–8.

64. Kilpatrick J, Gresham H, Griffin F, Volanakis J. Peripheral blood mononuclear leukocytes release a mediator(s) that induces phagocytosis of C-reactive protein-coated cells by polymorphonuclear leukocytes. J Leukocyte Biol 1987; 41:150–5.
65. Briles D, Forman C, Horowitz J, Volanakis J, Benjamin W, McDaniew L, Eldridge J, Brooks J. Antipneumonococcal effects of C-reactive protein and monoclonal antibodies to pneumococcal cell wall and capsular antigens. Infect Immun 1989; 57:1457–64.
66. Tebo J, Mortensen R. Internalization and degradation of receptor bound C-reactive protein by U-937 cells: induction of H_2O_2 production and tumoricidal activity. Biochim Biophys Acta 1991; 1095:210–6.
67. Mold C, Rodgers CP, Kaplan RL, Gewurz H. Binding of human C-reactive protein to bacteria. Infect Immun 1982; 38:392–5.
68. Nakayama S, Gewurz H, DuClos T, Mold C. The role of the spleen in the protective effect of C-reactive protein in Streptococcus pneumoniae infection. Immunology 1983; 34:319–28.
69. Gitlin JD, Gitlin JI, Gitlin D. Localization of C-reactive protein in synovium of patients with rheumatoid arthritis. Arthritis Rheum 1977; 20:1491–9.
70. Vigo C. Effect of C-reactive protein on platelet-activating factor–induced aggregation and membrane stabilization. J Biol Chem 1985; 260:3418–22.
71. Fiedel B. Platelet response to aggregated C-reactive protein: fibrinogen-dependent and independent signals. Immunology 1988; 63:169–72.
72. Kilpatrick JM, Virella G. Inhibition of platelet activating factor by rabbit C-reactive protein. Clin Immunol Immunopathol 1985; 37:267–81.
73. Bout D, Joseph M, Pontet M, Vorng H, Deslee D, Capron A. Rat resistance to schistosomiasis: platelet-mediated cytotoxicity induced by C-reactive protein. Science 1986; 231:153–6.
74. Pied S, Nussler A, Pontet M, Miltgen F, Matile H, Lambert PH, Mazier D. C-reactive protein protects against the preerythrocytic stages of malaria. Infect Immun 1989; 57:278–82.
75. Deodhar S, James K, Chiang T, Edinger M, Barna B. Inhibition of lung metastases in mice bearing a malignant fibrosarcoma by treatment with liposomes containing human C-reactive protein. Cancer Res 1982; 42:5084–8.
76. Barna D, Deodhar S, Gautam S, Yen-lieberman B, Roberts D. Macrophage activation and generation of tumoricidal activity by liposome-associated human C-reactive protein. Cancer Res 1984; 44:305–10.
77. Barna B, James K, Deodhar S. Activation of human monocyte tumoricidal activity by C-reactive protein. Cancer Res 1987; 47:3959–63.
78. Zahedi K, Mortensen R. Macrophage tumoricidal activity induced by human C-reactive protein. Cancer Res 1986; 46:5077–83.
79. Hokama Y, Honda S, Hanakahi L, Terada K. C-reactive protein and platelet activating factor complexes induce production and release of interleukin-1 by human monocytes. J Clin Lab Analysis 1988; 2:155–60.
80. Sarlo KT, Mortensen RF. Enhanced IL-1 production mediated by mouse SAP. Cell Immunol 1985; 93:398–406.
81. Cooper EA, Stone J. Acute phase reactant proteins in cancer. Adv Cancer Res 1979; 30:1–44.
82. Berman S, Gewurz H, Mold C. Binding of C-reactive protein to nucleated cells leads to complement activation without cytolysis. J Immunol 1986; 136:1354–9.
83. Singh PP, Gervais F, Skamene E, Mortensen RF. Serum amyloid P-component-induced enhancement of macrophage listericidal activity. Infect Immun 1986; 52:688–96.
84. Siripont J, Tebo JM, Mortensen RF. Receptor mediated binding of mouse serum amyloid P-component to macrophages. Cell Immunol 1988; 117:239–51.
85. Zeller J, Landay A, Lint T, Gewurz H. Enhancement of human peripheral blood monocyte respiratory burst activity by aggregated C-reactive protein. J Leukocyte Biol 1986; 40:769–83.
86. Robey F, Ohura K, Futaki S, Fujii N, Yajima H, Goldman N, Jones K, Wahl S. Proteolysis of human C-reactive protein produces peptides with potent immunoldulating activity. J Biol Chem 1987; 262:7053–7.
87. Shephard EG, Anderson R, Rosen O, Myer MS, Fridkin M, Strachan AF, DeBeer FC. Peptides generated from C-reactive protein (CRP) by a neutrophil membrane protease. Amino acid sequence and effects of peptides on neutrophil oxidative metabolism and chemotaxis. J Immunol 1990; 145:1469–76.
88. Kew R, Hyers T, Webster R. Human C-reactive protein inhibits neutrophil chemotaxis *in vitro*: possible implications for the adult respiratory distress syndrome. J Lab Clin Med 1990; 115:339–45.

89. Buchta R, Fridkin M, Pontet M, Contessi E, Scaggiante B, Romeo D. Modulation of human neutrophil function by C-reactive protein. Eur J Biochem 1987; 163:141–6.
90. Shephard EG, Beer SM, Anderson R, Strachan AF, Nel AE, DeBeer F. Generation of biologically active C-reactive protein peptides by a neutral protease on the membrane of phorbol myristate acetate–stimulated neutrophils. J Immunol 1989; 143:2974–81.
91. Mortensen R, Duskiewicz J. Mediation of CRP-dependent phagocytosis through mouse macrophage Fc-receptors. J Immunol 1977; 119:1611–6.
92. James K, Hansen B, Gewurz H. Binding of C-reactive protein to human lymphocytes. Interaction with a subset of cells bearing Fc receptor. J Immunol 1981; 127:2545–50.
93. Zahedi K, Tebo J, Siripont J, Klimo G, Mortensen R. Binding of human C-reactive protein to mouse macrophages is mediated by distinct receptors. J Immunol 1989; 142:2384–92.
94. Tebo J, Mortensen R. Characterization and isolation of a C-reactive protein receptor from the human monocytic cell line U-937. J Immunol 1990; 144:231–8.
95. Zeller J, Kubak B, Gewurz H. Binding sites for C-reactive protein on human monocytes are distinct from IgG Fc receptors. Immunology 1989; 67:51–5.
96. Ballou S, Cleveland R. Binding of human C-reactive protein to monocytes: analysis by flow cytometry. Clin Exp Immunol 1991; 84:329–35.
97. Crowell R, DuClos T, Montoya G, Heaphy E, Mold C. 1991; C-reactive protein receptors on the human monocytic cell line U-937. J Immunol 1991; 147:3445–51.
98. Müller H, Fehr J. Binding of C-reactive protein to human polymorphonuclear leukocytes: evidence for association of binding sites with Fc receptors. J Immunol 1986; 136:2202–7.
99. Miyagawa N, Okamoto Y, Nakano H. Effect of C-reactive protein on peritoneal macrophages. II. Human CRP activates peritoneal macrophages of guinea pigs to release superoxide anion in vitro. Microbiol Immunol 1988; 32:721–31.
100. Buchta R, Gennaro R, Pontet M, Fridkin M, Romeo D. C-reactive protein decreases protein phosphorylation in stimulated human neutrophils. FEBS Lett 1988; 257:173–7.
101. Stahl RD, Schlessinger P. Receptor-mediated pinocytosis of mannose/*N*-acetylglucosamine-terminated glycoproteins and lysosomal enzymes by macrophages. Trends Biochem Sci 1980; 5:194–200.
102. Ezekowitz RAB, Stahl PD. The structure and function of vertebrate mannose lectin-like proteins. J Cell Sci 1988; 9(suppl):121–125.
103. Ezekowitz RAB, Sastry K, Bailly P, Warner A. Molecular characterization of the human macrophage mannose receptor: demonstration of multiple carbohydrate recognition-like domains and phagocytosis of yeasts in Cos-1 cells. J Exp Med 1990; 172:1785–97.
104. Stahl PD, Wileman TE, Diment S, Shepherd VL. Mannose-specific oligosaccharide recognition by mononuclear phagocytes. Biol Cell 1984; 51:215–8.
105. Mizuno Y, Kozutsumi Y, Kawasaki T, Yamashina I. Isolation and characterization of a mannose-binding protein from rat liver. J Biol Chem 1981; 256:4247–53.
106. Summerfield JA, Taylor ME. Mannose-binding proteins of human serum: identification of mannose-specific immunoglobulins and a calcium dependent lectin. Biochim Biophys Acta 1986; 883:197–208.
107. Townsend R, Stahl P. Isolation and characterization of a mannose/*N*-acetylglucosamine/fucose binding protein from rat liver. Biochem J 1981; 194:209–19.
108. Wild J, Robinson D, Winchester B. Isolation of mannose-binding proteins from human liver. Biochem J 1983; 210:167–75.
109. Ideda K, Sannoh T, Kawasaki N, Kawasaki T, Yamashina I. Serum lectin with known structure activates complement through the classical pathway. J Biol Chem 1987; 262:7451–7.
110. Ezekowitz RAB, Day L, Herman G. A human mannose-binding protein is an acute phase reactant that shares sequence homology with other vertebrate lectins. J Exp Med 1988; 167:1034–48.
111. Drickamer K, Dordal MS, Reynolds L. Mannose-binding proteins isolated from rat liver contain carbohydrate-recognition domains linked to collagenous tail. J Biol Chem 1986; 261:6878–86.
112. Drickamer K. Two distinct classes of carbohydrate recognition domains in animal lectins. J Biol Chem 1988; 263:9557–66.
113. Sastry K, Herman G, Day L, Deignan E, Burns G, Morton C, Ezekowitz RAB. The human mannose-binding protein gene. Exon structure reveals its evolutionary relationship to a human pulmonary surfactant gene and its localization to chromosome 10. J Exp Med 1989; 170:1175–90.

114. Sastry K, Zahedi K, Lelias J-M, Whitehead AS, Ezekowitz RAB. Molecular characterization of the mouse mannose-binding proteins. J Immunol 1991; 147:692–7.
115. Taylor ME, Brickell PM, Craig RK, Summerfield JA. The structure and evolutionary origin of the gene encoding a human mannose-binding protein. Biochem J 1989; 262:763–9.
116. Kuhlman M, Joiner K, Ezekowitz RAB. The human mannose-binding protein functions as an opsonin. J Exp Med 1989; 169:1733–44.
117. Ezekowitz RAB, Kuhlman M, Groopman J, Byrn R. A human serum mannose-binding protein inhibits *in vitro* infection by the human immunodeficiency virus. J Exp Med 1989; 169:185–95.
118. Anders EM, Hartley CA, Jackson DC. Bovine and mouse serum B inhibitors of influenza A viruses are mannose-binding lectins. Proc Natl Acad Sci USA 1990; 87:4485–89.
119. Schweinle J, Ezekowitz RAB, Tenner A, Joiner K. Human mannose-binding protein activates the alternative complement pathway and enhances serum bactericidal activity on a monnose-rich isolate of Salmonella. J Clin Invest 1989; 84:1821–9.
120. Super M, Thiel S, Lu J, Turner MW. Association of low levels of mannan-binding protein with a common defect of opsonization. Lancet 1989; 2:1236–8.

9

Production of Active Oxygen Species by Phagocytic Leukocytes

John M. Robinson

The Ohio State University, Columbus, Ohio

John A. Badwey

Boston Biomedical Research Institute and Harvard Medical School, Boston, Massachusetts

I. INTRODUCTION AND BACKGROUND

Certain types of leukocytes (e.g., activated macrophages, neutrophils) produce massive amounts of superoxide (O_2^-) and hydrogen peroxide (H_2O_2) during phagocytosis or upon stimulation with a variety of agents. This response results in a large increase in the consumption of oxygen and has therefore been termed the "respiratory burst" [1]. The enzyme system that is responsible for these phenomena is an NADPH oxidase that catalyzes the following reaction:

(Rx 1) $NADPH + 2O_2 \rightarrow 2O_2^- + NADP^+ + 2H^+$ [e.g., 2]

Hydrogen peroxide is produced by the dismutation of O_2^-. Both O_2^- and H_2O_2 are important components of the oxygen-dependent antimicrobial and cytocidal arsenal of these cells [1,2]. The cytotoxic effects of these compounds are due, in part, to their ability to form the more reactive hydroxyl radical ($OH^\bullet$) and singlet oxygen ($^1O_2$) (for a review of the chemistry and assay of these species, see Badwey et al. [3]).

Hydroxyl radicals and perhaps $^1O_2$ can be formed in vitro from O_2^- and H_2O_2 by the transition metal–catalyzed Haber-Weiss reaction (Rx2) (e.g., 2,4):

(Rx 2) $O_2^- + H_2O_2 \rightarrow OH^\bullet + OH^- + {}^1O_2$

Whether this reaction occurs in vivo is uncertain because of questions about the nature and availability of the metal catalyst (e.g., iron) [5]. However, phagocytic leukocytes may produce $OH^\bullet$ by reaction of O_2^- with either hypochlorous acid (Rx 3) [5] or nitric oxide

(Rx 3) $HOCl + O_2^- \rightarrow OH^\bullet + OH^- + Cl^-$

(Rxs 4–6) [6]. In the latter case, $OH^\bullet$ is generated during the decomposition of peroxynitrous acid (Rx 6), which is formed by the reaction of NO with O_2^- followed by protonation [6]:

(Rx 4) $NO + O_2^- \rightarrow ONOO^-$ (peroxynitrite anion)

(Rx 5) $ONOO^- + H^+ \rightarrow ONOOH$ (peroxynitrous acid)

(Rx 6) $ONOOH \rightarrow OH^\bullet + NO_2$

Neither of these routes of synthesis for $OH^\bullet$ requires a transition metal. Hypochlorous acid (HOCl) and NO are formed in phagocytic cells by the action of myeloperoxidase and nitric oxide synthetase, respectively [7,8]. Hypochlorous acid can also react with H_2O_2 to generate $^1O_2$ [9].

Apart from these "short-lived" products of oxygen reduction and excitation, phagocytic leukocytes also produce a number of highly reactive stable oxidants (e.g., monochloramine) that participate in the antimicrobial mechanisms of these cells [10]. Formation of these stable oxidants is dependent on H_2O_2 formed during the respiratory burst [10].

In this chapter, we will review the structure of the NADPH oxidase complex that produces O_2^-, the stimulatory pathways that activate this system, and the levels of O_2^- and H_2O_2 produced by various types of phagocytic cells. We will further discuss how the levels of O_2^-/H_2O_2 produced by macrophages depend on their state of activation and/or exposure to cytokines. Other chapters in this volume will concentrate on the reactive nitrogen species produced by phagocytic leukocytes and on the roles of these compounds in the cytotoxic properties of these cells.

II. THE NADPH-OXIDASE SYSTEM

The enzyme that catalyzes the production of O_2^- by Rx 1 is a complex that contains several different proteins. This oxidase system is dormant and dissociated in unstimulated cells, with the components residing in the membrane (cytochrome b_{558}), cytoplasm (p47, Rac-2), and cytoskeleton (p67) [11–15]. The cytoskeleton is operationally defined here as those proteins that are insoluble in 1.0% (v/v) Triton X-100 [13]. During stimulation of the cells, there is a translocation of the cytoplasmic and cytoskeletal subunits to the plasmalemma, where the functional oxidase is assembled [11–14]. The plasmalemma invaginates to produce the phagosome. This situation is outlined in Figure 1. Superoxide formed by the oxidase is transported vectorially into the phagosome.

The various components of the oxidase were identified over a period of some 33 years (1959–1992). Their elucidation was largely a result of two events: 1) uncovering the molecular basis of chronic granulomatous disease (CGD) and 2) describing and characterizing a cell-free system from phagocytic cells in which large quantities of O_2^- could be generated by various manipulations. Chronic granulomatous disease is an inherited syndrome characterized by the inability of neutrophils, monocytes, and macrophages to produce O_2^- upon stimulation [2]. The phenotype of CGD can result from the absence of, or mutations in, cytochrome b_{558}, p67, or p47 [2]. The fact that alterations in any one of these proteins affect O_2^- in all of the cell types listed indicates that the oxidase system is essentially similar in all of these cases.

The original cell-free system for O_2^- production required both the particulate and soluble fractions of unstimulated cells, Mg^{2+}, GTP, and an anionic lipid (e.g., arachidonate, phosphatidate) [15,16]. A recent study has reconstituted O_2^- production utilizing purified cytochrome b_{558}, p47, p67, and a Rac-related GTP-binding protein [17]. While the cell-free system has been extremely useful in identifying components of the oxidase, the situation in vivo may be more complicated (see below). In particular, the role of the anionic lipid in this system remains a mystery. An unanswered question here is whether this lipid obviates the need for additional proteins that may be required in the intact cell. Evidence for the involvement of several other proteins has been reported [18–20]. Some of these proteins are shown in Figure 1.

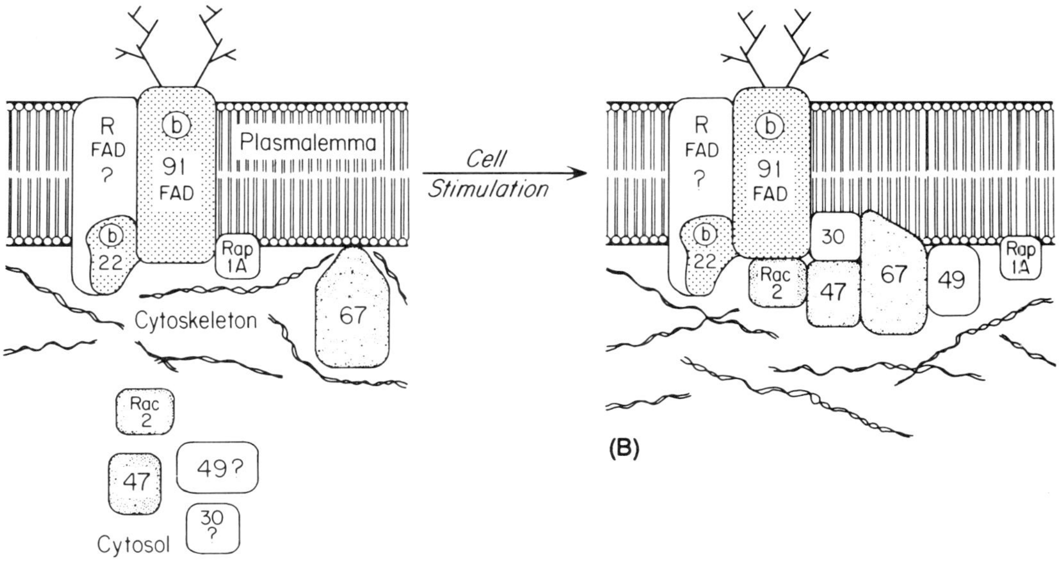

FIGURE 1 Assembly/activation of the NADPH oxidase system. In unstimulated cells (A) the oxidase is disassembled with components residing in the plasmalemma, cytoplasm, and cytoskeleton. Upon stimulation (B), there is a translocation of the various subunits to the plasmalemma, where the active oxidase is assembled on the *b* cytochrome (i.e., the 91- and 22-kDa membrane-bound components). Components whose involvement has not been firmly established are labeled with question marks.

A. The Membrane-Bound Oxidase Component Cytochrome b_{558}

A long-standing question in this field is how electrons extracted from NADPH in the cytoplasm are transferred to the opposite side of the plasmalemma or the phagocytic vacuole. Recent insights into the structure of cytochrome b_{558} provide some answers to this problem. Cytochrome b_{558} is a heterodimer that is located in the plasmalemma and specific granules [21–23]. The subunits have molecular masses of about 91 and 23 kDa [22]. Both proteins have been cloned and sequenced [24,25]. Predicted structural features from these sequences (e.g., hydrophilicity index) [26,27] and biochemical studies [27,28] indicate that both subunits have transmembrane helical spans. For example, the 91-kDa protein exhibits 5 hydrophobic regions (residues 6–25, 46–76, 100–119, 150–172, and 168–190) that may span the bilayer [26]. Recent data indicate that cytochrome *b* is a bi-heme or tri-heme molecule and that these hemes probably lie within the membrane [29,30]. Histidine-100, -110, -117, -207, -208 and -220 of the large subunit and -94 of the small subunit are spaced similar to heme-coordinating ligands in other cytochromes [26]. Moreover, the large subunit contains a pair of histidines (residues 208 and 209) and three cysteines (residues 58, 63, and 64) that constitute a potential Cu binding site [27].

The carboxy terminus of the large subunit resides on the cytoplasmic side of the membrane [26,28]. The sequence motif Gly-Ala-Gly-Ile-Gly-Val-Thr-Pro-Phe (residues 407–415) is present in this area and is highly analogous to a sequence present in a variety of NADPH-binding proteins (Gly-x-Gly-x-Gly-x-x-Pro-Phe) [17,27]. Thus, this region of the cytochrome may contain an NADPH-binding site [17,27]. Consistent with this hypothesis, cytochrome b_{558} was found to bind a small amount of flavin-adenine dinucleotide (FAD) (about 20%) when reconstituted into lipid vesicles [17,27]. Mutation of proline-114 to histidine is known to result in CGD [31]. Thus, all of the prosthetic groups and binding sites may be present in cytochrome b_{558} to extract electrons from NADPH and to pass these electrons to oxygen on the opposite side of the membrane. Cytochrome b_{558} is the first flavocytochrome to be described in eukaryotic cells [17,27].

Strong evidence indicates that the C-terminus of the large subunit has a second function in the generation of O_2^-: it provides the site on which the oxidase is assembled. Neutrophils deficient in cytochrome b do not exhibit translocation of p47 or p67 to the plasmalemma upon stimulation [11,14]. Moreover, synthetic peptides with amino acid sequences that correspond to certain regions of the C-terminus inhibited O_2^- production in the cell-free system and in permeabilized neutrophils stimulated with phorbol 12-myristate 13-acetate (PMA) or the chemotactic peptide fMet-Leu-Phe (fMLP) [28,32]. The most efficient inhibitor had the sequence Arg-Gly-Val-His-Phe-Ile-Phe, which corresponds to residues 559 through 565 [32].

B. The Soluble Oxidase Components p47, p67 and Rac-2

Studies in the early 1980s reported that neutrophils stimulated with PMA exhibited an intense phosphorylation of a 47-kDa protein (p47). The kinetics of this phosphorylation reaction were consistent with a role for this protein in the respiratory burst [33,34]. It was subsequently reported that neutrophils from four patients with the cytochrome *b*–positive, autosomal recessive form of CGD failed to phosphorylate this 47-kDa protein upon stimulation [35]. A major breakthrough in this area was achieved when a polyclonal antibody was raised to a partially purified cytosolic fraction from human neutrophils and found to react with two soluble proteins with molecular masses of 47 and 67 kDa [36]. This antibody did not detect p47 in neutrophils from patients with CGD who lacked the 47-kDa phosphoprotein but did recognize the 67-kDa species in these patients. In contrast, the antibody recognized the 47-kDa protein, but not the 67-kDa entity, in a rare group of cytochome *b*–positive patients with the autosomal recessive form of CGD who exhibited *normal* phosphorylation of p47 [36]. These data established p47 and p67 as components of the oxidase (see below). Moreover, this polyclonal antibody provided the probe that enabled both of these proteins to be cloned and sequenced [37–39].

Little is known about the function or functions of p67 or p47 in the generation of O_2^-. Neither of these proteins contain FAD or any other redox components that would indicate a role in electron transport [2]. The major structural features of p67 are a highly acidic C-terminus and two "SH3" or "A" domains. These domains are located in the C-terminus and midsection of p67 [39]. In contrast, p47 has an extremely basic C-terminus but also contains two SH3 domains that are located in tandem in the middle of the molecule [37,38]. The SH3 regions may govern the binding of p67 and p47 to the cytoskeleton [39].

A particularly noteworthy feature of p47 is the large number of potential phosphorylation sites in the C-terminus. For example, serine residues 303, 304, 315, 320, 328, and 359 are located in the vicinity or flanked by basic groups [37]. This arrangement constitutes consensus sequences that are highly favorable for phosphorylation by protein kinase C (PKC) and several other protein kinases [40]. In addition, p47 contains three serine residues (residues 215, 345, and 348) with the consensus sequence -P-X_n-S/T-P- ($n = 1$ or 2) that have the potential to be phosphorylated by the microtubule-associated protein-2 kinase and other "proline-directed kinases" [41]. We will return to the phosphorylation of p47 in Section III.

Studies in the mid-1980s reported that O_2^- production by the cell-free system was stimulated by GTP and inhibited by GDP [42]. This immediately indicated a role for a GTP-binding protein in this system. Subsequent studies identified a 20-kDa protein that mediated these effects. This protein is most likely Rac-2 [43,44; but see 45], which is a member of the Ras superfamily. The ability of Rac-2 to enhance O_2^- production was the first demonstration of a direct effect of a member of this family on a cellular response [43].

Rac-2 shuttles between an active GTP-bound form and an inactive GDP-bound species. Shuttling between these forms can be mediated by a number of "exchange proteins" such as smgGDS and rhoGDI [44]. A provocative report has demonstrated that the binding of [^{35}S]GTP to Rac-2 *and O_2^- production* in the cell-free system are stimulated and inhibited by smgGDS and rhoGDI, respectively [44]. These investigators suggested that rhoGDI was responsible for the inhibition of the oxidase in unstimulated cells and that activation was mediated by alterations in smgGDS and rhoGDI during cell stimulation. Rac-2 is isolated from guinea pig macrophages in a complex with rhoGDI [45]. It should be noted here that there are data suggesting that a small GTP-binding protein (e.g., Rac-2?) is *not* necessary for O_2^- production by permeabilized neutrophils stimulated with PMA [46]. Thus, it is possible that Rac-2 is not a mandatory component of the oxidase but functions in the stimulatory pathway by which the anionic lipid activates O_2^- production in the cell-free system.

What can be the function of p67, p47, and Rac-2? While there are no data on this subject, a modest amount of speculation may be appropriate here. As noted above, cytochrome *b* appears to contain all of the redox components necessary to produce O_2^-. However, the cells do not generate O_2^- unless stimulated, and this involves translocation of p67 and p47 to the membrane. Perhaps association of these proteins with the cytochrome induces a conformational change that allows the electrons to be extracted from NADPH by FAD and/or the transfer of these elections to O_2 by the hemes. The recent availability of the purified proteins in reasonable quantities should facilitate investigations in this area.

III. SIGNAL TRANSDUCTION PATHWAYS THAT TRIGGER ACTIVATION OF THE OXIDASE

A. Overview

Despite a decade of research in this area, the steps involved in the signal transduction pathways for O_2^- production are not entirely known for any stimulus. Part of the problem reflects a general lack of knowledge of how the cytoskeleton and G proteins are regulated in phagocytic and other cells. Investigations into the nature of these pathways is further complicated by the existence of multiple routes of assembly/activation of the oxidase. Moreover, the stimulatory pathway utilized by a particular agent can vary depending on the presence of cytokines [47] and/or if the cells are attached to a substratum [48]. However, illuminating some of these mechanisms has been facilitated by the existence of a large number of soluble and quasisoluble agonists that can trigger O_2^- production. These stimuli include tumor-promoters that activate PKC (e.g., phorbol esters, mezerein) [49,50], chemotactic peptides (e.g., fMLP) [51], and various agents that increase the liquid crystalline fraction of lipid in the plasmalemma (e.g., *cis*-unsaturated fatty acids, retinoids) [52,53]. In some cases, these agents preferentially utilize a single stimulatory pathway and are thus useful in sorting out and dissecting the various mechanisms.

Three stimulatory pathways of O_2^- release can be identified with antagonists of protein kinases. As intimated above, another important criterion for differentiating these pathways may be a requirement for Rac-2 [cf. 46]. However, more data are needed to substantiate this point. The most widely employed antagonists of protein kinases in these studies are 1-(5-isoquinolinylsulfonyl)-2-methylpiperazine (H-7) [54] and staurosporine [55]. Both of these drugs inhibit a wide variety of protein kinases [54–56]. A topical problem in these investigations is the lack of a suitable cell-free system to investigate the phosphorylation and/or Ca^{2+}-dependent events in the stimulation of the oxidase. For example, the cell-free system described above is insensitive to H-7, staurosporine, and antagonists of Ca^{2+}-binding proteins (e.g., trifluoperazine) [57,58]. In contrast, these antagonists block O_2^- release in vivo with a wide array of stimuli (e.g., PMA, fMLP, opsonized zymosan) [52,53,59,60]. A more detailed description of the components of the stimulatory pathways may enable us to reconstitute these events in vitro. In the next section, we will briefly discuss our present knowledge concerning the various stimulatory pathways in neutrophils for O_2^- release.

B. Stimulatory Pathways with Certain Agonists: Some Examples

The most topical and intensely investigated pathway is that utilized by PMA and other tumor-promoters that activate PKC [reviewed in 61]. Neutrophils stimulated with optimal amounts of PMA [49], or stimulated synergistically with suboptimal amounts of PMA and a Ca^{2+} ionophore [62], release quantities of O_2^- comparable to those observed with physiological agonists. In the case of synergy, Ca^{2+} is thought to increase the activity of PKC by enhancing the sensitivity of this kinase for the tumor promoter [50,63]. As noted earlier, several potential phosphorylation sites for PKC are present in the C-terminus of p47. Neutrophils stimulated with PMA exhibit several isoforms of p47 [64] that differ only in the stoichiometry of phosphorylation [65]. However, the actual sites phosphorylated and the relationship between the stoichiometry of phosphorylation and O_2^- production are not known. Phosphorylation of the C-terminus of p47 will alter the large positive charge in this area, which may enable this molecule to interact with the *b* cytochrome. Treatment of PMA-stimulated neutrophils with staurosporine or H-7 blocks phosphorylation of p47 [50,60,63,66], translocation of this protein to the cytoskeleton and plasmalemma [13], and O_2^- release [13,50,60,63,66]. It should be noted here that phosphoproteins other than p47 and protein kinases other than PKC may also be involved in this response. The cytoskeleton appears to be responsible for the assembly of the oxidase [13,67]. Furthermore, several cytoskeletal proteins are substrates for PKC [68]. Moreover, cells treated with PMA exhibit enhanced activities of several protein kinases (e.g., MAP-2 kinases) subsequent to the stimulation of PKC [69]. The latter kinases may be active against the "proline-directed" sites on p47 noted above.

Superoxide release from neutrophils stimulated with fMLP is also sensitive to H-7 and staurosporine [60,66]. However, a good deal of evidence indicates that the signal-transduction pathway utilized by fMLP requires a protein kinase that is different from that activated by PMA. For example, the production of O_2^- with these two agonists is different with regard to temperature-dependence for H-7 [70] and sensitivity toward staurosporine [60]. Moreover, Wortmannin blocks O_2^- release from neutrophils stimulated with fMLP but not PMA [71]. Wortmannin inhibits myosin light-chain kinase [72] and perhaps other protein kinases. The nature of this putative protein kinase and its substrates is not known. Candidates include a tyrosine protein kinase [73], a Ca^{2+}-independent myosin light-chain kinase [72], and an isozyme of PKC that is different from that activated when PMA is the stimulus.

A second major difference between the stimulatory pathways utilized by fMLP and PMA is that O_2^- generated by fMLP is highly dependent on the action of phospholipase D [e.g., 74,75]. Aliphatic alcohols (e.g., ethanol, butanol) inhibit the production of phosphatidic acid by phospholipase D and block O_2^- production by neutrophils stimulated with fMLP but not with PMA [75]. The reason for this requirement for phosphatidic acid is not known. However, it is interesting to note with regard to the previous discussion that a phosphatidate-activated protein kinase has been described in a variety of cells [76].

Finally, O_2^- release from neutrophils stimulated with retinoids, *cis*-unsaturated fatty acids, and high concentrations of the synthetic diglyceride sn-1,2-dioctanoyl-glycerol (DiC8) is largely insensitive to staurosporine and H-7 [46,66,77]. Could the signal-transduction pathway utilized by these agonists be analogous to that which operates in the cell-free system? While retinoids and DiC8 are not anionic lipids, it is likely that neutrophils treated with these compounds will generate phosphatidate [66,76]. Phosphatidate or DiC8 can interact synergistically with sodium dodecyl sulfate (SDS) (and presumably arachidonate) to enhance O_2^- production in the cell-free system [57,74] and to promote translocation of p47 to the membrane [78]. The initial site of action of the anionic lipid appears to be the cytosol [78]. Binding of arachidonate or phosphatidate to p47 would reduce the overall positive charge on this protein. This may obviate the need for phosphorylation of p47 along with the requirement for a protein kinase.

While some common themes may be slowly emerging in understanding these signal-transduction pathways, major problems remain. In particular, it is important to note here that neither the "soluble" stimuli nor the cell-free system provide a complete paradigm for all of the physiologically relevant responses (e.g., O_2^- production during phagocytosis). For example, the signal-transduction pathway for O_2^- release from neutrophils ingesting immunoglobulin G (IgG)-coated erythrocytes in the presence of certain cytokines exhibits a number of similarities to the pathway utilized by PMA [47]. In contrast, neutrophils ingesting yeast particles opsonized with C3b/bi or IgG can exhibit a respiratory burst that does not require diglyceride, arachidonate, or phosphatidate [79] and is insensitive to staurosporine [80]! The mechanism of this response is clearly different from that of any of the models discussed above and provides a challenge for future work in this field.

IV. THE RESPIRATORY BURST IN MACROPHAGES AND MACROPHAGE-LIKE CELLS

A major function of phagocytic leukocytes, including monocytes/macrophages, is the stimulus-dependent consumption of molecular oxygen and the release of toxic oxygen products (see above). This respiratory burst has received a great deal of attention because of its importance in the oxidative microbicidal and cytocidal properties of these cells. We will discuss the production of O_2^-/H_2O_2 by various types of macrophages and factors that modulate the amounts produced.

The respiratory burst in monocytes has been compared to that in neutrophils. It has been reported that there are differences in the rates of phagocytosis, oxidative metabolism, and microbiocidal activity between human peripheral blood monocytes and neutrophils. Oxygen consumption and formate oxidation during phagocytic stimulation are equivalent or higher in monocytes as compared with neutrophils [81,82]. Monocytes, however, appear to convert a smaller part of the consumed oxygen to O_2^- and H_2O_2 than do neutrophils; this difference is related to a large increase in mitochondrial respiration by phagocytosing monocytes but not neutrophils [83]. Release of O_2^- and chemiluminescence was lower in monocytes than in neutrophils during phagocytosis or contact with immobilized IgG [84].

The quantity of toxic oxygen products released by macrophages depends on the species from which the cells are derived and the nature of the stimulus employed. Consumption of oxygen and release of oxygen-derived metabolites by macrophages also depends on the degree of cellular differentiation, the site of isolation, and the state of activation of the cells. Specific experimental conditions may also affect the quantities of these reactive oxygen species that are measured.

Much of the experimental work related to the analysis of the respiratory burst in macrophages has been carried out with murine cells. Different categories of murine peritoneal macrophages have been defined (i.e., resident [normal]; elicited [inflammatory]; cytokine-activated [activated]) [for earlier reviews see 85–87]. These categories of macrophages display a number of phenotypic differences including the quantity of reactive oxygen species released upon stimulation. Resident peritoneal macrophages release little O_2^- or H_2O_2 in response to stimulation when compared with elicited or activated macrophages [88,89] (Table 1). Moreover, the chemical agent that induces macrophages to migrate into the peritoneal cavity can affect the ability of these macrophages to produce O_2^- [90] (Table 1).

The levels of O_2^- released by macrophages from other anatomic sites have been recorded. Alveolar macrophages are relatively easy to obtain [91] and have been widely studied for their ability to mount a respiratory burst [92]. Alveolar macrophages typically release lower quantities of O_2 metabolites than do blood monocytes [93,94]. Analyses of the properties of the respiratory burst in macrophages from other tissues such as the liver (Kupffer cells) and central nervous system

TABLE 1 Superoxide Release by Murine Peritoneal Macrophages upon Stimulation with PMA

Macrophages	Treatment	O_2^- Release (nmol O_2^-/min/10^7 cells) Mixed cell preparation	"Pure" macrophages (calc.)[a]
Normal	—	1.1 ± 1.9 (3)	1.1 ± 2.0 (3) [NS]
Activated	*L. monocytogenes* (5–7 days)	14.0 ± 3.2 (7)	17.7 ± 3.5 (7) [$p = 0.05$]
	BCG-PPD	13.6 ± 2.7 (4)	14.1 ± 6.6 (4) [NS]
Elicited	Sodium caseinate	27.2 ± 4.7 (5)	33.4 ± 3.6 (5) [$p < 0.05$]
	Proteose-peptone	29.1 ± 7.7 (3)	35.0 ± 8.7 (3) [NS]
	Thioglycolate	3.9 ± 1.4 (5)	4.1 ± 2.4 (5) [NS]

Superoxide release was monitored by the cytochrome c reduction method and the data are expressed as the mean ± standard deviation. The numbers of cell preparations employed (experiments) are given in parentheses. Each experiment was performed in triplicate. In no case was release of O_2^- detected in unstimulated cells. The statistical significance of differences between the value for O_2^- release by the mixed cell preparations and those calculated for pure macrophages are shown in square brackets. NS = not significant.

[a]Calculated for macrophages by subtracting the rate of O_2^- release contributed by neutrophils (PMNs) and attributing all of the remaining O_2^- release exclusively to macrophages, since the lymphocytes were incapable of releasing O_2^-. For these calculations, PMNs were assumed to produce O_2^- at a rate of 32 nmol/min/10^7 cells, except in the case of thiogylycolate-elicited macrophage preparations, where a value of 21 nmol O_2^-/min/10^7 cells was employed. Modified from reference 90.

(microglia) have been hampered by the difficulty of obtaining homogeneous cell preparation in sufficient quantities. However, it has been shown that Kupffer cells and microglia release O_2^- upon appropriate stimulation [95–98].

While monocytes and macrophages produce less O_2^-/H_2O_2 than do neutrophils on a per-cell basis, the mechanism by which these cells produce O_2^- appears to be identical to that in neutrophils. This conclusion is drawn from studies of CGD patients (see above) and from the identification of components of the NADPH oxidase system in macrophages (i.e., cytochrome b_{558}) [99] and mRNAs for these proteins in human monocytes (i.e., p22, p47, p91) [100].

As noted above, the majority of investigations related to macrophages have been conducted with murine peritoneal macrophages or human peripheral blood monocytes. Studies with these cells are hampered by the paucity of cells available and the functional heterogeneity of freshly isolated monocytes and macrophages [101,102]. Continuous cell lines that display monocytic or macrophagic properties represent an alternative source of material for studies of the respiratory burst. A number of lines have been established with cells derived from patients with leukemias or lymphomas [103–107]. Work with these cell lines may, however, be problematic, since they can be functionally and phenotypically unstable [108]. Stable clones, as well as mutants, of monocytic cell lines have been developed to compensate for this problem [108–110]. Moreover, clones that produce O_2^- and clones that fail to do so have been isolated [108–110]. Monocytic cell lines have been useful for correlation of the induction of the respiratory burst and expression of components of the NADPH oxidase system [111–113]. Cell lines that vary in the properties of their metabolic burst should continue to be beneficial in the analysis of signal-transduction mechanisms leading to the release of active oxygen species.

V. MODULATION OF THE RESPIRATORY BURST IN MACROPHAGES

There is a wide variety of compounds that can modulate the respiratory burst in macrophages, both positively and negatively. Physiologically the most important of these are certain cytokines and other polypeptides, hormones, and biologically active lipids; we will discuss these compounds in this chapter. In addition there are a number of structurally dissimilar agents and environmental insults that affect the stimulus-dependent release of active oxygen species from macrophages [114–117]. An analysis of these agents and conditions is beyond the scope of this chapter.

A. The Influence of Cytokines on the Respiratory Burst

Proliferation and differentiation of hemopoietic cells are controlled by extracellular signals. A variety of specific polypeptides (collectively referred to as cytokines) are crucial in this regard; these include interleukin-1 (IL-1); interleukin-2 (IL-2); interleukin-3 (IL-3); interleukin-4 (IL-4); interleukin-5 (IL-5); interleukin-6 (IL-6); interleukin-7 (IL-7); interferon-α (IFN-α); interferon-β (IFN-β); interferon-γ (IFN-γ); tumor necrosis factors (TNF-α and- β) transforming growth factors (TGF-α and -β); granulocyte-macrophage colony stimulating factor (GM-CSF); granulocyte-colony stimulating factor (G-CSF); and macrophage-colony stimulating factor (M-CSF). An account of the multiple roles of these cytokines has been given in several recent reviews [118–127].

In addition to affecting proliferation and differentiation of leukocytes, a number of these cytokines are capable of modulating functional activities, including the respiratory burst, in mature phagocytes. As discussed earlier, the ability of macrophages to produce O_2^-/H_2O_2 depends on the state of "activation" or "priming." Macrophages that are activated in vivo or primed in vitro produce more active oxygen species than do resident or unprimed cells when challenged with appropriate stimuli.

Early studies had established that macrophage activation could be induced by sensitized T cells or their secretory products [128]. The next step in understanding this process was the identification of chemically defined cytokines capable of macrophage activation. Interferon-γ was the first cytokine identified that enhanced the respiratory burst in macrophages [129–130]. Furthermore, incubation of crude lymphokine preparations from T cell cultures with a neutralizing antibody to IFN-γ eliminated the ability of these lymphokines to enhance H_2O_2 release [129–130].

Colony-stimulating factors can also activate mature phagocytes. The cytokines GM-CSF and M-CSF enhance the secretion of H_2O_2 by macrophages incubated with either phorbol esters or zymosan [131–133]. Recent work has shown that mice transgenic for murine GM-CSF expressed from a retroviral promoter have high levels of GM-CSF [134]. These animals display lesions that are associated with accumulation of macrophages at certain anatomic sites. Such transgenic mice have subsequently been used to study the role increased in vivo exposure to GM-CSF has on macrophage function. Enhanced production of O_2^- by peritoneal and pleural macrophages challenged with zymosan or phorbol esters was noted (there was a tenfold increase in phorbol ester–induced O_2^- release in macrophages from transgenic animals compared with normal litter mates) [135]. While the underlying mechanism for up-regulation of this cellular function in macrophages from these transgenic animals is not completely understood, this work illustrates the potential for this experimental approach.

Tumor necrosis factor (TNF-α) is another cytokine capable of enhancing the levels of reactive oxygen products released by macrophages [136]. Moreover TNF-α acts synergistically with IFN-γ to increase macrophage responsiveness [137,138]. In other studies TNF-α synergized with immune complexes to maximally activate alveolar macrophages [139]. The priming effect of IL-1 on monocytes and macrophages has also been noted [140,141].

The molecular mechanism by which cytokines effect the ability of macrophages to undergo an enhanced respiratory burst is not well understood. However, recent experiments with human monocyte–derived macrophages suggests that induction of the p91 gene and protein product by IFN-γ is required for the enhancement of macrophage oxidative metabolism [100]. What other molecular alterations may be required for the cytokine-induced enhancement of the oxidative burst in phagocytes remains to be determined.

Most studies related to cytokine effects on macrophage function have examined the stimulatory properties of these polypeptides. There is increasing evidence that some cytokines play an opposite role and suppress the oxidative burst in macrophages. Interleukin-4 inhibits H_2O_2 and O_2^- release from monocytes cultured with IFN-γ [142,143]. Inhibition of TNF-α–induced enhancement of the respiratory burst by IL-4 has also been demonstrated [144]. However, it has been noted that IL-4 does not have a similar inhibitory effect on murine macrophages activated in vivo [124,145]. Interleukin-10 has recently been added to the group of cytokines capable of inhibiting macrophage [146] and monocyte [147] function. While the major effect of IL-10 in this regard appears to be inhibition of cytokine synthesis, it also suppresses H_2O_2 release [148]. A factor designated macrophage-deactivating factor (MDF), found in tumor cell conditioned medium, has been shown to inhibit the macrophage respiratory burst [149]. This factor has been purified and is a polypeptide of approximately 13 kDa that is distinct from IL-4 [150]. Other cytokines capable of suppressing the macrophage respiratory burst are TGF-β1 and TGF-β2 [151]. The calcitonin gene–related peptide also inhibits H_2O_2 release from macrophages [152]. The molecular mechanism by which MDF and TGF-β exert their inhibitory effect is not completely understood; however, it has been reported that MDF alters the affinity of the NADPH oxidase for the substrate NADPH [153]. These inhibitory cytokines are likely to be important in diminishing the oxidative burst so as to protect normal tissue from untoward damage from toxic oxygen products during chronic infections or inflammation.

B. The Effects of Certain Polypeptides on the Respiratory Burst in Macrophages

In addition to the cytokines previously mentioned there are a number of other factors, including specific polypeptides, that are capable of modulating the respiratory burst in phagocytes. Three of these compounds are discussed in this section. Growth hormone (somatotropin) has been shown to prime porcine alveolar and rat peritoneal macrophages for augmented O_2^- release in response to zymosan particles; growth hormone was as effective as IFN-γ in this regard [154]. More recently, growth hormone and prolactin have been shown to prime human neutrophils for enhanced O_2^- release [155,156]. Evidence has been presented that this priming effect in human neutrophils is related to the binding of growth hormone to the prolactin receptor [156]. Suppression of prolactin secretion in mice has been shown to prevent T cell–dependent induction of macrophage tumoricidal activity after intraperitoneal injection of *Listeria monocytogenes* or *Mycobacterium bovis* [157]. While these findings are intriguing, the underlying mechanism(s) by which these hormones modulate macrophage function remains obscure.

Endothelin, a polypeptide derived from endothelial cells [158], has marked effects on vasoconstriction [159] and has been reported to increase O_2^- release in human alveolar macrophages [160]. Receptors for endothelin on murine peritoneal macrophages have been demonstrated; occupancy of this receptor induces mobilization of intracellular Ca^{2+} [161]. Endothelin also induces release of arachidonic acid from guinea pig alveolar macrophages [162]. The potential priming effect of endothelin on the macrophage respiratory burst remains to be elucidated fully.

C. The Effects of Some Other Biologically Relevant Molecules on the Respiratory Burst in Macrophages

Macrophages are known to be activated by a number of natural products, including certain biologically active lipids and lipid-containing molecules. We will briefly discuss some of these in the context of modulation of the macrophage burst.

The hormone 1α,25-dihydroxyvitamin D_3 [1,25-$(OH)_2D_3$] induces maturational changes in monocytes and monocytic cell lines [163]. Moreover, 1,25-$(OH)_2D_3$ enhances the release of H_2O_2 from monocytes [164] and O_2^- from U937 cells [165] when challenged with a stimulus. More recently it has been suggested that the maturational effects of 1,25-$(OH)_2D_3$ on HL-60 cells are independent of its capacity to prime these cells for the respiratory burst; the Ca^{2+} ionophoretic effect of 1,25-$(OH)_2D_3$ has been implicated as playing a major role in priming cells for this response [166]. On the other hand, 1,25-$(OH)_2$ D_3, which induces HL-60 cells to differentiate into monocyte-like cells, caused a time-dependent increase in NADPH oxidase components (i.e., p22, p47, p67, p91) [113].

While the effects of lipopolysaccharide (LPS) are covered in more detail in other portions of this volume, a few comments about its effect on the macrophage respiratory burst are in order. Peritoneal macrophages that are elicited by injection of LPS release greater amounts of O_2^- than do resident cells following stimulation with phorbol esters. This may, in part, be explained by alterations in kinetic parameters of the NADPH oxidase; the oxidase from LPS-elicited cells displays a higher V_{max} and a lower K_m for NADPH than that from resident cells [167]. Recent experiments have demonstrated that in vitro treatment of monocyte-derived macrophages with LPS increased the levels of transcripts of NADPH oxidase components (i.e., p22, p91) [100,168] and that this occurred, at least in part, at the transcriptional level [168]. Induction of an LPS response (i.e., IL-1 production) in the monocytic cell line U937 has been linked to a pertussis toxin–sensitive G protein [169]. In addition, LPS stimulates GTPase activity in membranes of the macrophage-like cell line RAW 264.7 [170]. These findings may be of relevance to the respiratory burst, since a role for a GTP-binding protein is indicated (see above). Lipopolysaccharide induced the expression of *hck* and *lyn* tyrosine kinases in murine bone marrow–derived macrophages [171]. The expression of these tyrosine kinases may be involved in the regulation of the respiratory burst [171]. Last, an LPS-resistant mutant isolated from the macrophage cell line J774.1 does not exhibit elevated O_2^-/H_2O_2 generation after LPS treatment [172]. The interaction of LPS with macrophages and the complex series of molecular events that ensue are now being unraveled. The results mentioned here, along with anticipated future discoveries, should provide increased insight into the regulation of the respiratory burst in these cells.

Another lipid that exerts dramatic physiological effects on macrophages is platelet-activating factor (PAF) [173]. Space limitations preclude a full discussion of the pathophysiological effects of PAF on phagocytic cells. The reader is therefore referred to some excellent reviews [e.g., 174–176].

D. The Effect of Surface Adherence on the Respiratory Burst

Macrophages are monocyte-derived cells that form resident populations within various tissues and display functional differences that depend on a variety of factors including their interactions with other tissue cells and the surface or substrate to which they bind. The role of surface adherence on monocyte maturation has been observed. For example, peroxidase reactivity appears in the rough endoplasmic reticulum of blood monocytes soon after surface adherence [177]. Maturation into macrophages occurs in tissues where these cells are in intimate contact with a variety of surfaces and extracellular matrix proteins. Mature macrophages can also be modulated by adherence to surfaces. For example, adherence of thioglycolate-elicited murine macrophages to plastic induces the accumulation of c-*sis* mRNA, while adherence to fibronectin-coated surfaces is a stronger c-*sis* mRNA inducer [178]. Moreover, adherence to a fibronectin-coated surface induces GM-CSF mRNA in these macrophages, while soluble fibronectin does not evoke this response [178].

Adherence and extracellular matrix proteins also affect the respiratory burst in phagocytic leukocytes and can modulate other processes that are associated with the burst (i.e., phagocytosis). The role of fibronectin and laminin in the control of phagocytosis has been reviewed [179]. Surface adherence also plays a role in the quantities of O_2 metabolites released by macrophages. Spontaneously adherent human peripheral blood monocytes released almost twice as much O_2^- as nonadherent monocytes [180]. The nature of the substratum also appears to be important, since rat inflammatory peritoneal macrophages attached to endothelial cell–derived extracellular matrix release twice as much O_2^- as these cells do when attached to denatured collagen/fibronectin [181].

Likewise, attachment of neutrophils to various surfaces greatly enhanced the respiratory burst in response to certain stimuli [48]. The mechanism by which adherence induces alterations in phagocytic leukocytes, including increased release of active oxygen species, is not well understood. It is, however, of considerable importance in pathophysiological situations where these cells are in contact with various surfaces at sites of inflammation.

ACKNOWLEDGMENTS

Research from our laboratories was supported by NIH grants AI-24321 (to J.M.R.) and AI-28342 (to J.A.B.) and grant 2065 from the Council for Tobacco Research (to J.M.R.).

REFERENCES

1. Badwey JA, Karnovsky ML. Active oxygen species and the functions of phagocytic leukocytes. Annu Rev Biochem 1980; 49:695–726.
2. Smith RM, Curnutte JT. Molecular basis of chronic granulomatous disease. Blood 1991; 77:673–86.
3. Badwey JA, Robinson JM, Karnovsky MJ, Karnovsky ML. Reduction and excitation of oxygen by phagocytic leukocytes: biochemical and cytochemical techniques. In: Weir DM, Herzenberg LA, Blackwell C, Herzenberg LA, eds. Handbook of experimental immunology, vol. II. 4th ed. Edinburgh: Blackwell Scientific Publications, 1986:50.1–50.16.
4. Beauchamp C, Fridovich I. A mechanism for the production of ethylene from methional. The generation of the hydroxyl radical by xanthine oxidase. J Biol Chem 1970; 245:4641–6.
5. Ramos CL, Pou S, Britigan BE, Cohen MS, Rosen GM. Spin trapping evidence for myeloperoxidase-dependent hydroxyl radical formation by human neutrophils and monocytes. J Biol Chem 1992; 267:8307–12.
6. Beckman JS, Beckman TW, Chen J, Marshall PA, Freeman BS. Apparent hydroxyl radical production by peroxynitrite: implications for endothelial injury from nitric oxide and superoxide. Proc Natl Acad Sci USA 1990; 87:1620–4.
7. Harrison JE, Schultz J. Studies on the chlorinating activity of myeloperoxidase. J Biol Chem 1976; 251:1371–4.
8. Xie Q, Cho HJ, Calaycay J, Mumford RA, Swiderek KM, Lee TD, Ding A, Trosco T, Nathan C. Cloning and characterization of inducible nitric oxide synthetase from mouse macrophages. Science 1992; 256:225–8.
9. Steinbeck ML, Khan AU, Karnovsky MJ. Intracellular singlet oxygen generation by phagocytosing neutrophils in response to particles coated with a chemical trap. J Biol Chem 1992; 267:13425–33.
10. Grisham MB, Jefferson MM, Melton DF, Thomas EL. Chlorination of endogenous amines by isolated neutrophils. Ammonia dependent bactericidal, cytotoxic and cytolytic activities of the chloramines. J Biol Chem 1984; 259:10404–13.
11. Heyworth PG, Shrimpton CF, Segal AW. Localization of the 47 kDa phosphoprotein involved in the respiratory burst NADPH oxidase of phagocytic cells. Biochem J 1989; 260:243–8.
12. Clark RA, Volpp BD, Leidal KG, Nauseef WM. Two cytosolic components of the human neutrophil respiratory burst oxidase translocate to the plasma membrane during cell activation. J Clin Invest 1990; 85:714–21.
13. Nauseef WM, Volpp BD, McCormick S, Leidal KG, Clark RA. Assembly of the neutrophil respiratory burst oxidase. Protein kinase C promotes cytoskeletal and membrane association of cytosolic oxidase components. J Biol Chem 1991; 266:5911–7.
14. Heyworth PG, Curnutte JT, Nauseef WM, Volpp BD, Pearson DW, Rosen H, Clark RA: Neutrophil NADPH oxidase assembly. Translocation of p47-phox and p67-phox requires interaction between p47-phox and cytochrome b_{558}. J Clin Invest 1991; 87:352–6.
15. Bromberg Y, Pick E. Unsaturated fatty acids stimulate NADPH-dependent superoxide production by cell-free system derived from macrophages. Cell Immunol 1984; 88:213–21.
16. Heyneman RA, Vercauteren RE. Activation of a NADPH-oxidase from horse polymorphonuclear leukocytes in a cell-free system. J Leukocyte Biol 1984; 36:751–9.

17. Rotrosen D, Yeung CL, Leto TL, Malech HL, Kwong CH. Cytochrome b_{558}: the flavin-binding component of the phagocyte NADPH oxidase. Science 1992; 256:1459–62.
18. Someya A, Yomogida S, Nagaoka I, Iwabuchi K, Yamashita T. Purification of the 28.5 kDa cytosolic protein involved in the activation of the NADPH-oxidase from guinea pig neutrophils. FEBS Lett 1992; 302:69–72.
19. Yea CM, Cross AR, Jones OTG. Purification and some properties of the 45 kDa diphenylene iodonium-binding flavoprotein of the neutrophil NADPH-oxidase. Biochem J 1990; 265:95–100.
20. Umei T, Babior BM, Curnutte JT, Smith RM. Identification of the NADPH-binding subunit of the respiratory burst oxidase. J Biol Chem 1991; 266:6019–20.
21. Segal AW, Jones OTG. Novel cytochrome b system in phagocytic vacuoles from human granulocytes. Nature 1978; 276:515–7.
22. Parkos CA, Allen RA, Cochrane CG, Jesaitis AJ. Purified cytochrome *b* from human granulocyte plasma membrane is comprised of two polypeptides with relative molecular weights of 91,000 and 22,000. J Clin Invest 1987; 80:732–42.
23. Borregaard N, Heiple JM, Simmons ER, Clark RA. Subcellular localization of the *b*-cytochrome component of the human neutrophil microbicidal oxidase: translocation during activation. J Cell Biol 1983; 97:51–62.
24. Royer-Pokora B, Kunkel LM, Monaco AP, Goff SC, Newburger PE, Baehner RL, Cole FS, Curnutte JT, Orkin SH. Cloning the gene for an inherited human disorder—chronic granulomatous disease—on the basis of its chromosomal location. Nature 1986; 322:32–8.
25. Parkos CA, Dinauer MC, Walker LE, Allen RA, Jesaitis AJ, Orkin SH. Primary structure and unique expression of the 22-kilodalton light chain of human neutrophil cytochrome *b*. Proc Natl Acad Sci USA 1988; 85:3319–23.
26. Imajoh-Ohm S, Tokita K, Ochiai H, Nakamura M, Kanegasaki S. Topology of cytochrome b_{558} in neutrophil membrane analyzed by anti-peptide antibodies and proteolysis. J Biol Chem 1992; 267: 180–4.
27. Segal AW, West I, Wientjes F, Nugent JHA, Chavan AJ, Haley B, Garcia RC, Rosen H, Scrace G. Cytochrome *b*-245 is a flavocytochrome containing FAD and the NADPH binding site of the microbicidal oxidase of phagocytes. Biochem J 1992; 284:781–8.
28. Rotrosen D, Kleinberg ME, Nunoi H, Leto T, Gallin JI, Malech HL. Evidence of a functional domain of phagocyte oxidase cytochrome b_{558}. J Biol Chem 1990; 265:8745–50.
29. Hurst JK, Loehr TM, Curnutte JT, Rosen H. Resonance Raman and electron paramagnetic resonance structural investigations of neutrophil cytochrome b-$_{558}$. J Biol Chem 1991; 266:1627–34.
30. Quinn MT, Mullen ML, Jesaitis AJ. Human neutrophil cytochrome *b* contains multiple hemes. Evidence for heme associated with both subunits. J Biol Chem 1992; 267:7303–9.
31. Dinauer MC, Curnutte JT, Rosen H, Orkin SH. A missense mutation in the neutrophil cytochrome *b* heavy chain in cytochrome-positive X-linked chronic granulomatous disease. J Clin Invest 1989; 84:2012–6.
32. Kleinberg ME, Mital D, Rotrosen D, Malech HL. Characterization of a phagocyte cytochrome b_{558} 91-kilodalton subunit functional domain: identification of peptide sequence and amino acids essential for activity. Biochemistry 1992; 31:2686–90.
33. Schneider C, Zanetti M, Romeo D. Surface reactive stimuli selectively increase protein phosphorylation in human neutrophils. FEBS Lett 1981; 127:4–8.
34. White JR, Huang C-K, Hill JM, Naccache PH, Becker EL, Sha'afi RI. Effect of phorbol 12-myristate 13-acetate and its analogue 4-α-phorbol 12,13-didecanoate on protein phosphorylation and lysosomal enzyme release in rabbit neutrophils. J Biol Chem 1984; 259:8605–11.
35. Segal AW, Heyworth PG, Cockcroft S, Barrowman MM. Stimulated neutrophils from patients with autosomal recessive chronic granulomatous disease fail to phosphorylate a Mr-44,000 protein. Nature 1985; 316:547–9.
36. Volpp BD, Nauseef WM, Clark RA. Two cytosolic neutrophil oxidase components absent in autosomal chronic granulomatous disease. Science 1988; 242:1295–7.
37. Volpp BD, Nauseef WM, Donelson JE, Moser DR, Clark RA. Cloning of the cDNA and functional expression of the 47-kilodalton cytosolic component of human neutrophil respiratory burst oxidase. Proc Natl Acad Sci USA 1989; 86:7195–9. Correction Proc Natl Acad Sci USA 1989; 86:9563.

38. Lomax KJ, Leto TL, Nunoi H, Gallin JI, Malech HL. Recombinant 47-kilodalton cytosol factor restores NADPH oxidase in chronic granulomatous disease. Science 1989; 245:409–12. Correction Science 1989; 246:987.
39. Leto TL, Lomax KL, Volpp BD, Nunoi H, Sechler JMG, Nauseef WM, Clark RA, Gallin JI, Malech HL. Cloning of a 67-kD neutrophil oxidase factor with similarity to a noncatalytic region of $p60^{c\text{-}src}$. Science 1990; 248:727–30.
40. Kennelly PJ, Krebs EG. Consensus sequences as substrate specificity determinants for protein kinases and protein phosphatases. J Biol Chem 1991; 266:15555–8.
41. Gonzalez FA, Raden DL, Davis RJ. Identification of substrate recognition determinants for human ERK1 and ERK2 protein kinases. J Biol Chem 1991; 266:22159–63.
42. Seifert R, Rosenthal W, Schultz G. Guanine nucleotides stimulate NADPH oxidase in membranes of human neutrophils. FEBS Lett 1986; 205:161–5.
43. Knaus UG, Heyworth PG, Evans T, Curnutte JT, Bokoch GM. Regulation of phagocyte oxygen radical production by the GTP-binding protein Rac 2. Science 1991; 254:1512–5.
44. Mizuno T, Kaibuchi K, Ando S, Musha T, Hiraoka K, Takaishi K, Asada M, Nunoi H, Matsuda I, Takai Y. Regulation of the superoxide generating NADPH oxidase by a small GTP-binding protein and its stimulatory and inhibitory GDP/GTP exchange proteins. J Biol Chem 1992; 267:10215–8.
45. Abo A, Pick E, Hall A, Totty N, Teahan CG, Segal AW. Activation of the NADPH oxidase involves the small GTP-binding protein p21-rac1. Nature 1991; 353:668–70.
46. Lu DJ, Grinstein S. ATP and guanine nucleotide dependence of neutrophil activation. Evidence for the involvement of two distinct GTP-binding proteins. J Biol Chem 1990; 265:13721–9.
47. Gresham HD, Zheleznyak A, Mormol JS, Brown EJ. Studies on the molecular mechanisms of human neutrophil Fc receptor–mediated phagocytosis. J Biol Chem 1990; 265:7819–26.
48. Nathan C, Srimal S, Faber C, Sanchez E, Kabbash L, Asch A, Gailit J, Wright SD. Cytokine-induced respiratory burst of human neutrophils: Dependence on extracellular matrix proteins and CD11/CD18 integrins. J Cell Biol 1989; 109:1341–9.
49. Robinson JM, Badwey JA, Karnovsky ML, Karnovsky MJ. Release of superoxide and change in morphology by neutrophils in response to phorbol esters: antagonism by inhibitors of calcium binding proteins. J Cell Biol 1985; 101:1052–8.
50. Badwey JA, Robinson JM, Horn W, Soberman RJ, Karnovsky MJ, Karnovsky ML. Synergistic stimulation of neutrophils. Possible involvement of 5-hydroxy-6,8,11,14-eicosatetraenoate in superoxide release. J Biol Chem 1988; 263:2779–86.
51. Williams LT, Snyderman R, Pike MC, Lefkowitz RJ. Specific receptor sites for chemotactic peptides on human polymorphonuclear leukocytes. Proc Natl Acad Sci USA 1977; 74:1204–8.
52. Badwey JA, Curnutte JT, Robinson JM, Berde CB, Karnovsky MJ, Karnovsky ML. Effects of free fatty acids on the release of superoxide and on change of shape by human neutrophils. Reversibility by albumin. J Biol Chem 1984; 259:7870–7.
53. Robinson JM, Badwey JA, Karnovsky ML, Karnovsky MJ. Cell surface dynamics of neutrophils stimulated with phorbol esters or retinoids. J Cell Biol 1987; 105:417–26.
54. Hidaka H, Inagaki M, Kawamoto S, Sasaki Y. Isoquinolinesulfonamides, novel and potent inhibitors of cyclic nucleotide dependent protein kinases and protein kinase C. Biochemistry 1984; 23:5036–41.
55. Tamaoki T, Nomoto H, Takahashi I, Kato Y, Morimoto M, Tomita F. Staurosporine, a potent inhibitor of phospholipid/Ca^{2+} dependent protein kinase. Biochem Biophys Res Commun 1986; 135:397–402.
56. Badwey JA, Erickson RE, Curnutte JT. Staurosporine inhibits the soluble and membrane-bound protein tyrosine kinases of human neutrophils. Biochem Biophys Res Commun 1991; 178:423–9.
57. Burnham DN, Uhlinger DJ, Lambeth DJ. Diradlyglycerol synergizes with an anionic amphiphile to activate superoxide generation and phosphorylation of p47-phox in a cell-free system from human neutrophils. J Biol Chem 1990; 265:175550–9.
58. Cox JA, Jeng AY, Blumberg PM, Tauber AI. Comparison of subcellular activation of the human neutrophil NADPH-oxidase by arachidonic acid, SDS, and phorbol myristate acetate. J Immunol 1987; 138:1884–8.
59. Curnutte JT, Badwey JA, Robinson JM, Karnovsky MJ, Karnovsky ML. Studies on the mechanism of superoxide release from human neutrophils stimulated with arachidonate. J Biol Chem 1984; 259: 11851–7.

60. Robinson JM, Heyworth PG, Badwey JA. Utility of staurosporine in uncovering differences in the signal transduction pathways for superoxide production in neutrophils. Biochim Biophys Acta 1990; 1055:55–62.
61. Heyworth PG, Badwey JA. Protein phosphorylation associated with the stimulation of neutrophils. Modulation of superoxide production by protein kinase C and calcium. J Bioenerg Biomemb 1990; 22:1–26.
62. Robinson JM, Badwey JA, Karnovsky ML, Karnovsky MJ. Superoxide release by neutrophils: synergistic effects of a phorbol ester and a calcium ionophore. Biochem Biophys Res Commun 1984; 122:734–9.
63. Heyworth PG, Karnovsky ML, Badwey JA. Protein phosphorylation associated with the synergistic stimulation of neutrophils. J Biol Chem 1989; 264:14935–9.
64. Okamura N, Curnutte JT, Roberts RL, Babior BM. Relationship of protein phosphorylation to the activation of the respiratory burst in human neutrophils. Defects in the phosphorylation of a group of closely related 48-kDa proteins in two forms of chronic granulomatous disease. J Biol Chem 1986; 263:6777–82.
65. Rotrosen D, Leto TL. Phosphorylation of neutrophil 47-kDa cytosolic oxidase factor. Translocation to membrane is associated with distinct phosphorylation events. J Biol Chem 1990; 265:19910–5.
66. Badwey JA, Horn W, Heyworth PG, Robinson JM, Karnovsky ML. Paradoxical effects of retinal in neutrophil stimulation. J Biol Chem 1989; 264:14947–53.
67. Quinn MTC, Parkos CA, Jesaitis AJ. The lateral organization of components of the membrane skeleton and superoxide generation in the plasma membrane of stimulated human neutrophils. Biochim Biophys Acta 1989; 987:83–94.
68. Hornbeck P, Paul WE. Anti-immunoglobulin and phorbol ester induce phosphorylation of proteins associated with the plasma membrane and cytoskeleton in murine B lymphocytes. J Biol Chem 1986; 261:14817–24.
69. Thomas SM, DeMarco M, D'Arcangelo G, Halegoua A, Brugge JS. Ras is essential for nerve growth factor and phorbol ester–induced tyrosine phosphorylation of MAP kinases. Cell 1992; 68:1031–41.
70. Nath J, Powledge A, Wright DG. Studies of signal transduction in the respiratory-burst associated stimulation of fmet-leu-phe–induced tubulin tyrosinolation and phorbol 12-myristate 13-acetate-induced posttranslational incorporation of tyrosine into multiple proteins in activated neutrophils and HL-60 cells. J Biol Chem 1989; 264:848–55.
71. Dewald B, Thelen M, Baggiolini M. Two transduction sequences are necessary for neutrophil activation by receptor agonists. J Biol Chem 1988; 263:16179–84.
72. Nakanishi S, Kakita S, Takahashi I, Kawahara K, Tsukuda E, Sano T, Yamada K, Yoshida M, Kase H, Matsuda Y, Hashimoto Y, Nonomura Y. Wortmannin, a microbial product inhibitor of myosin light chain kinase. J Biol Chem 1992; 267:2157–63.
73. Gomez-Cambronero J, Huang C-K, Bonak VA, Wang E, Casnellie JE, Shiraishi T, Sha'afi RI: Tyrosine phosphorylation in human neutrophils. Biochem Biophys Res Commun 1989; 162:1478–85.
74. Agwu DE, McPhail LC, Sozzani S, Bass DA, McCall CE. Phosphatidic acid as a second messenger in human polymorphonuclear leukocytes. Effects on activation of NADPH oxidase. J Clin Invest 1991; 88:531–9.
75. Bonser RW, Thompson NT, Randall RW, Garland LG. Phospholipase D activation is functionally linked to superoxide generation in the human neutrophil. Biochem J 1989; 264:617–20.
76. Bocckino SB, Wilson PB, Exton JH. Phosphatidate-dependent protein phosphorylation. Proc Natl Acad Sci USA 1991; 88:6210–3.
77. Badwey JA, Robinson JM, Heyworth PG, Curnutte JT. 1,2-Dioctanoyl-sn-glycerol can stimulate neutrophils by different mechanisms. Evidence for a pathway that does not involve phosphorylation of the 47-kDa protein. J Biol Chem 1989; 264:20676–82.
78. Tyagi SR, Neckelmann N, Uhlinger DJ, Burnham DN, Lambeth JD. Cell-free translocation of recombinant p47-phox, a component of the neutrophil NADPH oxidase: effects of guanosine 5'-O-(3-thiotriphosphate), diacylglycerol, and an anionic amphiphile. J Biol Chem 1992; 31:2765–74.
79. Della Bianca V, Grzeskowiak M, Rossi F. Studies on the molecular regulation of phagocytosis and activation of the NADPH-oxidase in neutrophils. IgG- and C3b-mediated ingestion and associated respiratory burst independent of phospholipid and Ca^{2+} transients. J Immunol 1990; 144:1411–7.

80. Della Bianca V, Grzeskowiak M, Lissandrini D, Rossi F. Source and role of diacylglycerol formed during phagocytosis of opsonized yeast particles and associated respiratory burst in human neutrophils. Biochim Biophys Res Commun 1991; 177:948–55.
81. Baehner RL, Johnston RB, Jr. Monocyte function in children with neutropenia and chronic infections. Blood 1972; 40:31–41.
82. Sagone AL, King GW, Metz EM. A comparison of the metabolic response to phagocytosis in human granulocytes and monocytes. J Clin Invest 1976; 57:1352–8.
83. Reiss M, Roos D. Differences in oxygen metabolism of phagocytosing monocytes and neutrophils. J Clin Invest 1978; 61:480–7.
84. Johnston RB Jr, Lehmeyer JE, Guthrie LA. Generation of superoxide and chemiluminescence by human monocytes during phagocytosis and on surface bound immunoglobulin G. J Exp Med 1976; 143:1551–6.
85. North RJ. The concept of the activated macrophage. J Immunol 1978; 121:806–9.
86. Karnovsky ML, Lazdins JK. Biochemical criteria for activated macrophages. J Immunol 1978; 121:809–13.
87. Cohn ZA. The activation of mononuclear phagocytosis: fact, fancy, and future. J Immunol 1978; 121:813–6.
88. Nathan CF, Root RK. Hydrogen peroxide release from mouse peritoneal macrophages. Dependence on sequential activation and triggering. J Exp Med 1977; 146:1648–62.
89. Johnston RB, Jr, Godzik CA, Cohn ZA. Increased superoxide anion production by immunologically activated and chemically elicited macrophages. J Exp Med 1978; 148:115–27.
90. Badwey JA, Robinson JM, Lazdins JK, Briggs RT, Karnovsky MJ, Karnovsky ML. Comparative biochemical and cytochemical studies on superoxide and peroxide in mouse macrophages. J Cell Physiol 1983; 115:208–16.
91. Myrik QN, Leake ES, Fariss B. Studies on pulmonary alveolar macrophages from the normal rabbit: a technique to procure then in a high state of purity. J Immunol 1961; 86:128–36.
92. Fantone JC, Ward PA. Mechanism of lung parenchymal injury. Am Rev Respir Dis 1984; 130:484–90.
93. Greening AP, Lowrie DB. Extracellular release of hydrogen peroxide by human alveolar macrophages: the relationship to cigarette smoking and lower respiratory tract infections. Clin Sci 1983; 65:661–4.
94. Fels AOS, Cohn ZA. The alveolar macrophage. J Appl Physiol 1986; 60:353–69.
95. Colton AC, Gilbert DL. Production of superoxide by a CNS macrophage, the microglia. FEBS Lett 1987; 223:284–8.
96. Banati RB, Rothe G, Valet G, Kreutzberg GW. Respiratory burst in brain macrophages: a flow cytometric study on cultured rat microglia. Neuropathol Appl Neurobiol 1991; 17:223–30.
97. Latocha G, Dieter P, Schulze-Specking A, Decker K. Fc receptors mediate prostaglandin and superoxide synthesis in cultured rat Kupffer cells. Biol Chem Hoppe Seyler 1989; 370:1055–61.
98. Bautista AP, Schuler A, Spolarics Z, Spitzer JJ. *In vivo* latex phagocytosis primes the Kupffer cells and hepatic neutrophils to generate superoxide anion. J Leukocyte Biol 1992; 51:39–45.
99. Knoller S, Shpungin S, Pick E. The membrane-associated component of the amphiphile-activated, cytosol-dependent superoxide-forming NADPH oxidase of macrophages is identical to cytochrome b559. J Biol Chem 1991; 266:2795–804.
100. Cassatella MA, Bazzoni F, Flynn RM, Dusi S, Trinchieri G, Rossi F. Molecular basis of interferon-γ and lipopolysaccharide enhancement of phagocyte respiratory burst capability. Studies on the gene expression of several NADPH oxidase components. J Biol Chem 1990; 265:20241–6.
101. Lee KC. On the origin and mode of action of functionally distinct macrophage subpopulations. Mol Cell Biochem 1980; 30:39–55.
102. Bursuker I, Goldman R. On the origin of macrophage heterogeneity: A hypothesis. J Reticuloendothel Soc 1983; 33:207–20.
103. Harris P, Ralph P. Human leukemic models of myelomonocytic development: A review of HL-60 and U937 cell lines. J Leukocyte Biol 1985; 37:407–22.
104. Collins SJ. The HL-60 promyelocytic leukemia cell line: proliferation, differentiation, and cellular oncogene expression. Blood 1987; 70:1233–44.
105. Sundström C, Nilsson K. Establishment and characterization of a human histiocytic lymphoma cell line (U-937). Int J Cancer 1976; 17:565–77.

106. Txuchiya S, Yamabe M, Yamaguchi Y, Kobayashi Y, Konno T, Tada K. Establishment and characterization of a human acute monocytic cell line (THP-1). Int J Cancer 1980; 26:171–6.
107. Tucker KA, Lily MB, Heck L, Rado TA. Characterization of a new human diploid myeloid leukemia cell line (PLB-985) with granulocytic and monocytic differentiating capacity. Blood 1987; 70:372–8.
108. Garotta G, Thelen M, Delia D, Kamber M, Baggiolini M. GM-1, a clone of the monoblastic phagocyte U937 that expresses a large respiratory burst capacity upon activation with interferon-γ. J Leukocyte Biol 1991; 49:294–301.
109. Damiani G, Kiyotaki C, Soeller W, Sasada M, Peisach J, Bloom BR. Macrophage variations in oxygen metabolism. J Exp Med 1980; 152:808–22.
110. Kiyotaki C, Peisach J, Bloom BR. Oxygen metabolism in cloned macrophage cell lines; glucose dependence of superoxide production, metabolic and spectral analysis. J Immunol 1984; 132:857–66.
111. Newburger PE, Ezekowitz RAB, Whitney C, Wright J, Orkin SH. Induction of phagocytic cytochrome *b* heavy chain gene expression by interferon gamma. Proc Natl Acad Sci USA 1988; 85:5215–9.
112. Goldberg M, Belkowski LS, Bloom BR. Regulation of macrophage function by interferon-γ. Somatic cell genetic approaches in murine macrophage cell lines to mechanisms of growth inhibition, the oxidative burst, and expression of the chronic granulomatous disease gene. J Clin Invest 1990; 85:563–9.
113. Levy R, Rotrosen D, Nagauker O, Leto TL, Malech HL. Induction of the respiratory burst in HL-60 cells. Correlation of function and protein expression. J Immunol 1990; 145:2595–601.
114. Pabst MJ, Gross JM, Borzna JP, Goren MB. Inhibition of macrophage priming by sulfatide from *Mycobacterium tuberculosis*. J Immunol 1988; 140:634–40.
115. Calhoun WJ, Stevens CA, Lambert SB. Modulation of superoxide production of alveolar macrophage and peripheral blood mononuclear cells by β-agonists and theophylline. J Lab Clin Med 1991; 117:514–22.
116. Becker S, Madden MC, Newman SL, Devlin RB, Koren HS. Modulation of human alveolar macrophage properties by ozone exposure *in vitro*. Toxicol Appl Pharmacol 1991; 110:403–15.
117. Reddy MV, Gangadharam PRJ. Heat shock treatment of macrophages causes increased release of superoxide anion. Infect Immun 1992; 60:2386–90.
118. Arai K-I, Lee F, Miyajima A, Miyatake S, Arai N, Yokota T. Cytokines: coordinators of immune and inflammatory responses. Annu Rev Biochem 1990; 59:783–836.
119. Sherr CJ. Colony-stimulating factor-1 receptor. Blood 1990; 75:1–12.
120. West MA. Role of cytokines in leukocyte activation: phagocytic cells. Curr Top Memb Transport 1990; 35:537–70.
121. Metcalf D. Control of granulocytes, and macrophages: molecular, cellular, and clinical aspects. Science 1991; 254:529–33.
122. Gregory SH, Magee DM, Wing EJ. The role of colony-stimulating factors in host defense. Proc Soc Exp Biol Med 1991; 197:349–60.
123. Gordon MY. Hemopoietic growth factors and receptors: bound and free. Cancer Cells 1991; 3:127–33.
124. Nathan C. Mechanisms and modulation of macrophage activation. Behring Inst Mitt 1991; 88:200–7.
125. Oppenheim JJ, Zachariae COC, Mukaida N, Matsushima K. Properties of the novel proinflammatory supergene "intercrine" cytokine family. Annu Rev Immunol 1991; 9:617–48.
126. Deuel TF, Kawahara RS, Mustoe TA, Pierce GF. Growth factors and wound healing: platelet-derived growth factor as a model cytokine. Annu Rev Med 1991; 42:567–84.
127. Vassalli P. The pathophysiology of tumor necrosis factors. Annu Rev Immunol 1992; 10:411–52.
128. North RJ. Introduction to macrophage activation. Lymphokines 1981; 3:1–10.
129. Nathan CF, Murray HW, Wiebe ME, Rubin BY. Identification of interferon-γ as the lymphokine that activates human macrophage oxidative metabolism and antimicrobial activity. J Exp Med 1983; 158:670–89.
130. Murray HW, Rubin BY, Rothermol CD. Killing intracellular *Leishmania donovani* by lymphokine-stimulated human mononuclear phagocytes. Evidence that interferon-γ is the stimulating lymphokine. J Clin Invest 1983; 72:1506–10.
131. Reed SG, Nathan CF, Pihl DL, Rodricks P, Shanebeck K, Conlon PJ, Grabstein KH. Recombinant granulocyte/macrophage colony stimulating factor activates macrophages to inhibit *Trypanosoma cruzi* and release hydrogen peroxide: comparison with interferon gamma. J Exp Med 1987; 166:1734–46.

132. Coleman DL, Chodakewitz JA, Bartiss AH, Mellors JW. Granulocyte-macrophage colony-stimulating factor enhances selective effector functions of tissue-derived macrophages. Blood 1988; 72:573–8.
133. Wing EJ, Ampel NM, Waheed A, Shadduck RS. Macrophage colony-stimulating factor (M-CSF) enhances the capacity of murine macrophages to secrete oxygen reduction products. J Immunol 1985; 135:2052–6.
134. Lang RA, Metcalf D, Cuthbertson RA, Lyons I, Stanley E, Kelso A, Kannourakis G, Williamson DJ, Klintworth GK, Gonda TJ, Dunn AR. Transgenic mice expressing a hemopoietic growth factor gene (GM-CSF) develop accumulations of macrophages, blindness, and a fatal syndrome of tissue damage. Cell 1987; 51:675–86.
135. Elliot MJ, Strasser A, Metcalf D. Selective up-regulation of macrophage function in granulocyte-macrophage colony-stimulating factor transgenic mice. J Immunol 1991; 147:2957–63.
136. Hoffman M, Weinberg JB. Tumor necrosis factor-α induces increased hydrogen peroxide production and Fc receptor expression, but not increased Ia antigen expression by peritoneal macrophages. J Leukocyte Biol 1987; 42:704–7.
137. Ding AH, Nathan CF, Stuehr DJ. Release of reactive nitrogen intermediates and reactive oxygen intermediates from mouse peritoneal macrophages. Comparison of activating cytokines and evidence for independent production. J Immunol 1988; 141:2407–12.
138. Bogdan C, Moll H, Solbach W, Rölinghoff M. Tumor necrosis factor-α in combination with interferon-γ, but not with interleukin-4 activates murine macrophages for elimination of *Leishmania major* amastigotes. Eur J Immunol 1990; 20:1131–5.
139. Warren JS, Kunkel SL, Cunningham TW, Johnson KJ, Ward PA. Macrophage-derived cytokines amplify immune complex–triggered O_2^- responses by rat alveolar macrophages. Am J Pathol 1988; 130:489–95.
140. Kharazmi A, Nielson H, Bendtzen K. Recombinant interleukin 1α and β prime human monocyte superoxide production but have no effect on chemotaxis and oxidative burst response of neutrophils. Immunobiology 1988; 177:32–9.
141. Phillips WA, Hamilton JA. Phorbol ester–stimulated superoxide production by murine bone marrow–derived macrophages requires preexposure to cytokines. J Immunol 1989; 142:2445–9.
142. Lehn M, Weiser WY, Engelhorn S, Gillis S, Remold HG. IL-4 inhibits H_2O_2 production and anti-leishmanial capacity of human cultured monocytes mediated by IFN-γ. J Immunol 1989; 143:3020–4.
143. Abramson SL, Gallin JI. IL-4 inhibits superoxide production by human mononuclear phagocytes. J Immunol 1990; 144:625–30.
144. Ho JL, He SH, Rios MJC, Wick EA. Interleukin-4 inhibits human macrophage activation by tumor necrosis factor, granulocyte-monocyte colony-stimulating factor, and interleukin-3 for antileishmanial activity and oxidative burst capacity. J Infect Dis 1992; 165:344–51.
145. Phillips WA, Croatto M, Hamilton JA. The effect of interleukin-4 on the macrophage respiratory burst is species dependent. Biochem Biophys Res Comm 1992; 182:727–32.
146. Fiorentino DF, Zlotnik A, Mosmann TR, Howard M, O'Garra A. IL-10 inhibits cytokine production by activated macrophages. J Immunol 1991; 147:3815–22.
147. De Wall Malefyt R, Abrams J, Bennett B, Figdor CG, de Vries JE. Interleukin 10 (IL-10) inhibits cytokine synthesis by human monocytes: an autoregulatory role of IL-10 produced by monocytes. J Exp Med 1991; 174:1209–20.
148. Bogdan C, Vodovotz Y, Nathan C. Macrophage deactivation by interleukin 10. J Exp Med 1991; 174:1549–55.
149. Szuro-Sudol A, Nathan CF. Suppression of macrophage oxidative metabolism by products of malignant and nonmalignant cells. J Exp Med 1983; 156:945–61.
150. Scrimal S, Nathan C. Purification of macrophage deactivating factor. J Exp Med 1990; 171:1347–61.
151. Tsunawaki S, Sporn M, Ding A, Nathan C. Macrophage deactivation by transforming growth factor-β. Nature 1988; 334:260–2.
152. Nong Y, Titus RG, Ribeiro JM, Remold HG. Peptides encoded by the calcitonin gene inhibit macrophage function. J Immunol 1989; 143:45–9.
153. Tsunawaki S, Nathan CF. Macrophage deactivation: altered kinetic properties of the superoxide-producing enzyme after exposure to tumor cell–conditioned medium. J Exp Med 1986; 164:1319–31.

154. Edwards CK, Ghiasuddin SM, Schepper JM, Yunger LM, Kelley KW. A newly defined property of somatotropin: priming of macrophages for production of superoxide anion. Science 1988; 239:769–71.
155. Wiedermann CJ, Niedermuhlbichler M, Geissler D, Beimpold H, Braunsteiner H. Priming of normal human neutrophils by recombinant human growth hormone. Br J Haematol 1991; 78:19–22.
156. Fu YK, Arkins S, Fuh G, Cunningham BC, Wells JA, Fong S, Cronin MJ, Dantzer R, Kelley KW. Growth hormone augments superoxide anion secretion by human neutrophils by binding to the prolactin receptor. J Clin Invest 1992; 89:451–7.
157. Bernton EW, Meltzer MS, Holaday JW. Suppression of macrophage activation and T-lymphocyte function in hypoprolactinemic mice. Science 1988; 239:401–4.
158. Yanagisawa M, Kurihara H, Kimura S, Tomobe Y, Kobayashi M, Mitsui Y, Yazaki Y, Goto K, Masaki T. A novel potent vasoconstrictor peptide produced by vascular endothelial cells. Nature 1988; 332:411–5.
159. Kasuya Y, Ishikawa T, Yanagisawa M, Kimura S, Goto K, Masaki T. Mechanism of contraction to endothelin in isolated porcine coronary artery. Am J Physiol 1989; 257:H1828–35.
160. Haller H, Schaberg T, Lindschau C, Lode H, Distler A. Endothelin increases $[Ca^{2+}]_i$, protein phosphorylation, and O_2^- production in human alveolar macrophages. Am J Physiol 1991; 261:L478–84.
161. Kishino J, Hanasaki K, Kato T, Arita H. Endothelin-induced intracellular Ca^{2+} mobilization through its specific receptors in murine peritoneal macrophages. FEBS Lett 1991; 280:103–6.
162. Millul V, Lagente V, Gillardeaux O, Boichot E, Dugas B, Mencia-Huerta J-M, Béréziat G, Braquet P. Activation of guinea pig alveolar macrophages by endothelin-1. J Cardiovasc Pharmacol 1991; 17(suppl 7):S233–5.
163. Abe E, Miyaura C, Sakagami H, Takeda M, Konno K, Yamazaki T, Yoshiki S, Suda T. Differentiation of mouse myeloid leukemia cells induced by 1α,25-dihydroxyvitamin D_3. Proc Natl Acad Sci USA 1981; 78:4990–4.
164. Cohen MS, Mesler DE, Snipes RG, Gray TK. 1,25-dihydroxyvitamin D_3 activates secretion of hydrogen peroxide by human monocytes. J Immunol 1986; 136:1049–53.
165. Roux-Lombard P, Cruchaud A, Dayer J-M. Effect of interferon-γ and 1α,25-dihydroxyvitamin D_3 on superoxide anion, prostaglandins E_2, and mononuclear cell factor production by U937 cells. Cell Immunol 1986; 97:286–96.
166. Tanaka H, Hruska KA, Seino Y, Malone JD, Nishii Y, Teitelbaum SL. Disassociation of the macrophage maturational effects of vitamin D from respiratory burst priming. J Biol Chem 1991; 266:10888–92.
167. Sasada M, Pabst MJ, Johnston RB Jr. Activation of mouse peritoneal macrophages by lipopolysaccharide alters the kinetic parameters of the superoxide-producing NADPH oxidase. J Biol Chem 1983; 258:9631–5.
168. Newburger PE, Dai Q, Whitney C. *In vitro* regulation of human phagocyte cytochrome *b* heavy and light chain gene expression by bacterial lipopolysaccharide and recombinant human cytokines. J Biol Chem 1991; 266:16171–7.
169. Daniel-Issakani S, Spiegel AM, Strulovici B. Lipopolysaccharide response is linked to the GTP-binding protein, Gi_2, in the promonocytic cell line U937. J Biol Chem 1989; 264:20240–7.
170. Tanke T, van de Loo J-W, Rhim H, Leventhal PS, Proctor RA, Bertics PJ. Bacterial lipopolysaccharide-stimulated GTPase activity in RAW 264.7 macrophage membranes. Biochem J 1991; 277:379–85.
171. Boulet I, Ralph S, Stanley E, Lock P, Dunn AR, Green SP, Phillips WA. Lipopolysaccharide- and interferon-gamma-induced expression of *hck* and *lyn* tyrosine kinases in murine bone marrow–derived macrophages. Oncogene 1992; 7:703–10.
172. Amano F, Akamatsu Y. A lipopolysaccharide (LPS)-resistant mutant isolated from a macrophage-like cell line, J774.1, exhibits an altered activated-macrophage phenotype in response to LPS. Infect Immunol 1991; 59:2166–74.
173. Prpic V, Uhing RJ, Weiel JE, Jakoi L, Gawdi G, Herman B, Adams DO. Biochemical and functional responses stimulated by platelet-activating factor in murine peritoneal macrophages. J Cell Biol 1988; 107:363–72.
174. Braquet P, Touqui L, Shen TY, Vargaftig BB. Perspectives in platelet-activating factor research. Pharmacol Rev 1987; 39:97–145.

175. Braquet P, Rola-Pleszczynski M. Platelet-activating factor and cellular immune responses. Immunol Today 1987; 8:345–52.
176. Snyder F. Platelet-activating factor and related acetylated lipids as potent biologically active cellular mediators. Am J Physiol 1990; 259:C697–708.
177. Bodel PT, Nichols BA, Bainton DF. Appearance of peroxidase reactivity within the rough endoplasmic reticulum of blood monocytes after surface adherence. J Exp Med 1977; 145:264–74.
178. Thorens B, Mermod J-J, Vassalli P. Phagocytosis and inflammatory stimuli induce GM-CSF mRNA in macrophages through post-transcriptional regulation. Cell 1987; 48:671–9.
179. Brown EJ. The role of extracellular matrix proteins in the control of phagocytosis. J Leukocyte Biol 1986; 39:579–91.
180. Owen CA, Campbell EJ, Stockley RA. Monocyte adherence to fibronectin: role of CD11/CD18 integrins and relationship to other monocyte functions. J Leukocyte Biol 1992; 51:400–8.
181. Gudewicz PW, Frewin MB. Surface contact modulation of inflammatory macrophage dependent cytotoxicity and prostanoid release. J Cell Physiol 1991; 149:195–201.

10

Antiparasitic Effects of Nitric Oxide in an In Vitro Murine Model of *Chlamydia trachomatis* Infection and an In Vivo Murine Model of *Leishmania major* Infection

Marion L. Woods II, Jeanmarie Mayer, and Thomas G. Evans

The University of Utah School of Medicine, Salt Lake City, Utah

John B. Hibbs, Jr.

Veterans Affairs Medical Center and The University of Utah School of Medicine, Salt Lake City, Utah

I. ROLE OF NITRIC OXIDE

Until 1987, the only inducible biochemical mechanism that could explain activated macrophage cytotoxicity was synthesis of reactive oxygen intermediates (ROIs) by NADPH oxidase. However, it was clear that other unrecognized cytotoxic mechanisms of major importance must exist. For example, antioxidant molecules did not inhibit expression of activated macrophage cytotoxicity in many experimental systems [1], and mammalian macrophages with a genetic deficiency of the NADPH oxidase–mediated respiratory burst (a deficiency that is the basis of chronic granulomatous disease in humans) were still capable of expressing cytotoxicity for tumor cells as well as for intracellular protozoa such as *Leishmania* and *Toxoplasma* [2,3]. In 1987, it was reported that expression of cytokine-inducible cytotoxicity by activated macrophages was L-arginine–dependent [4,5]. L-Arginine was the substrate for an enzyme, inducible nitric oxide synthase, that synthesized inorganic nitrogen oxides and L-citrulline [5–7] (Figure 1). Nitric oxide (NO), a paramagnetic gas with a short half-life, was found to be the inorganic nitrogen oxide that mediated L-arginine–dependent cytotoxicity [7–9]. It is now clear that NO has significant cytotoxic effects on a variety of facultative and obligate intracellular pathogens including *Mycobacterium tuberculosis* [10,11], *M. avium* [12], plasmodia [13,14], *Leishmania* [15–19], *Cryptococcus neoformans* [20], *Toxoplasma gondii* [21], *M. leprae* [22], and now *Chlamydia trachomatis* [100], in addition to schistosomula of the helminth *Schistosoma mansoni* [23].

These very diverse intracellular pathogens are inhibited or killed by a cytotoxic mechanism that is mediated by NO derived from the enzymatic oxidative deimination of L-arginine by the cytokine-inducible enzyme nitric oxide synthase [7,8]. This enzyme is induced by cytokines in murine macrophages as well as in a number of other nonmacrophage cell types [24–28]. Since

[1]

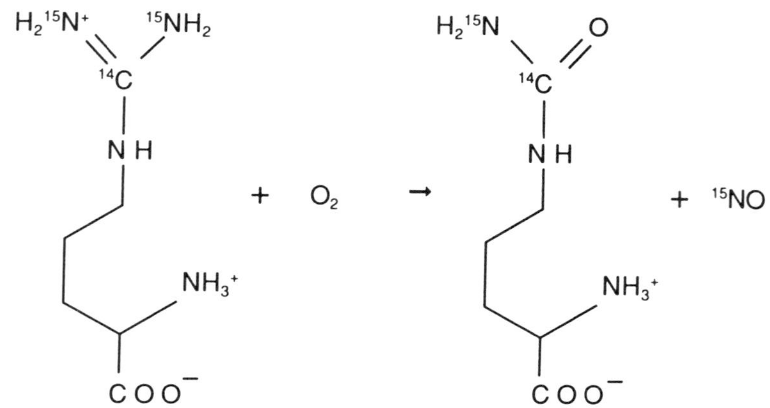

[2] $2\,^{15}NO + O_2 \rightarrow 2\,^{15}NO_2$

[3] $2\,^{15}NO_2 + H_2O \rightarrow {}^{15}NO_2^- + {}^{15}NO_3^- + 2H^+$

[4] $2\,^{15}NO$ + enzyme [4Fe-4S] →

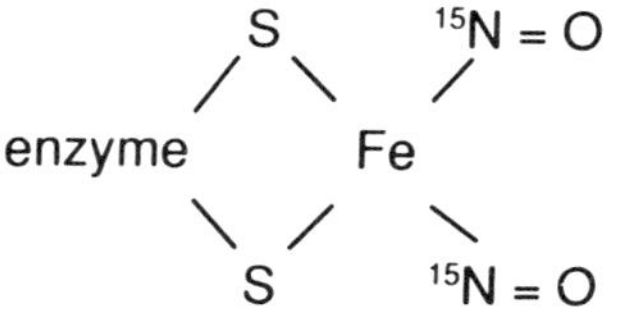

[5] $4HbO_2 + 4NO_2^- + 4H^+ \rightarrow 4Hb^+ + 4NO_3^- + O_2 + 2H_2O$

FIGURE 1 Precursor and products of the biological synthesis of inorganic nitrogen oxides and L-citrulline from L-arginine. ^{15}N-containing products derived from L-[^{15}N]arginine were identified by GC/MS [6–8,29] electron paramagnetic resonance spectroscopy [41], except for nitrogen dioxide (NO_2), which was detected by another method [7]. The direct synthesis of L-citrulline from L-arginine has been identified with several techniques [5–7,24]. The experiments utilizing L-[guanidino-^{14}C]arginine [7,24] are illustrated in the figure. Nitric formed by reaction [1] can undergo oxidative degradation in aqueous solution (reactions [2] and [3]) or react with non-heme iron associated with sulfur atoms to form nitrosyl-iron-sulfur complexes (reaction [4]). Although not shown, certain other forms of intracellular iron also complex with NO (Hibbs et al., unpublished data). When NO_2^- enters the vascular system it reacts rapidly with oxyhemoglobin [99]. This results in the stoichiometric formation of methemoglobin and NO_3^- from oxyhemoglobin and NO_2^- reaction [5]. Adapted from reference 31.

many intracellular pathogens invade and proliferate in nonmacrophage cells (e.g., fibroblasts, endothelial cells, hepatocytes, etc.) the cytokine-inducible NO synthase potentially represents an efficient biochemical defense of the intracellular environment. The identification of a specific nontoxic competitive inhibitor of the inducible and constitutive enzymes, *N*-guanidino monomethyl L-arginine (MLA), has made study of the nitric oxide pathways readily achievable in living systems [4,5,29,30]. Furthermore, there is growing evidence that cytokine-inducible NO synthesis is operative in humans as well [12,31,32].

II. EFFECTOR FUNCTION

The discovery of the L-arginine:NO pathway provided a biochemical explanation for an unusual pattern of metabolic perturbation described previously in tumor target cells of cytotoxic activated macrophages. Prominent among the metabolic changes induced in target cells by cytotoxic activated macrophages were inhibition of DNA replication [33] and inhibition of mitochondrial respiration [34]. Subsequent studies demonstrated that cytotoxic activated macrophages caused inhibition of complex I and complex II of the mitochondrial electron transport system as well as the citric acid cycle enzyme aconitase [35–37] and that the then-unknown cytotoxic activated macrophage cytotoxic mechanism targeted intracellular iron [38–41]. These enzymes all have [4Fe-4S] prosthetic groups that are essential for their catalytic function. The studies showed that enzymatic inhibition of aconitase involved the [4Fe-4S] prosthetic group rather than the apoenzyme [36,37]. It was observed in these experiments that tumor cells cocultivated with cytotoxic activated macrophages released a significant fraction of their intracellular iron in parallel with development of inhibition of DNA synthesis and mitochondrial respiration [38–40]. Discovery of the cytokine induced L-arginine: NO pathway provided an effector mechanism capable of causing the biochemical changes that had been observed [4,5,7,9]. Indeed, electron paramagnetic resonance (EPR) spectroscopic identification of $[Fe(NO)2(SR)2]^-$ complex formation links inhibition of enzymes with [4Fe-4S] prosthetic groups and NO synthesis by cytotoxic activated macrophages [41–43].

Early work studying resistance to both tumor cells and intracellular pathogens suggested that control of tumor and parasite growth could be effected nonspecifically, that is, with the target cells being immunologically naive to the effector immune cells. Macrophages activated by exposure to *Mycobacterium bovis*, strain Calmette-Guérin (BCG), had increased ability to kill *Listeria monocytogenes* [44,45]. Likewise, cytotoxic activated murine macrophages were cytotoxic for both syngeneic and allogeneic tumor cells, but in a nonspecific fashion [46–48]. The cytotoxic activity required two signals for induction [1,49–53], and close contact between the effector activated macrophages and the target tumor cells was necessary for expression of cytotoxicity [40,47,48].

It is now known that L-arginine–dependent synthesis of NO is required for an important component of the nonspecific cytotoxicity/cytostasis that macrophages exert following activation with BCG and other agents that induce a strong cell-mediated immune response [4,5,7,9,21]. The nature of several of the physiological stimuli that induce the high-output NO synthase are now known. Interferon-γ (IFN-γ) is utilized as a priming signal, and bacterial components such as lipopolysaccharide (LPS) or muramyl dipeptide induce macrophages to synthesize tumor necrosis factor-α (TNF-α), which is the physiological second signal triggering final expression of high-output NO synthase activity in this cell type [40,54,55]. Interferon-γ–primed nonmacrophage cells respond to both interleukin 1 (Il-1) and TNF as a second signal [24,27,28,32,40], suggesting that many cell types in a variety of tissues participate in the cell-mediated immune response to intracellular pathogens. This is important because it implies that under usual conditions, induction of the high-output NO synthase would be contained and localized to local areas of cytokine stimulation.

A well-documented metabolic perturbation that NO induces in target cells is mediated by formation of nitrosylated complexes with non-heme iron molecules, i.e., iron-sulfur complexes that are necessary for enzyme activity (heme-NO complexes also occur). For example, ribonucleotide reductase, which requires non-heme iron for its activity [56], is a target of NO. The inactivation of this enzyme by cytokine-induced NO results in inhibition of DNA synthesis in the target cell [57,58]. Cytotoxic activated macrophage activity can be inhibited by macrophage endocytosis of erythrocytes or hemoglobin, or by treatment with ferrous and ferric iron [59], presumably by providing competitive targets for NO produced by nitric oxide synthase.

Three enzymes that are essential for production of energy for the cell have been identified as targets of nitric oxide. Each enzyme has an iron-sulfur complex that is required for catalytic activity. One of these enzymes is aconitase, which catalyzes citrate conversion to isocitrate [36]. Aconitase activity is also inhibited by exposure to authentic NO [7]. Enzymatic activity of aconitase can be restored if either ferrous iron or ferrous iron with a reducing agent is added to target cells whose aconitase activity has been inhibited by NO synthesis by activated macrophages (36).

Two mitochondrial electron-transport chain enzymes containing iron-sulfur prosthetic groups, complex I (NADH:ubiquinone oxidoreductase) and complex II (succinate:ubiquinone oxidoreductase), are also targets for cytokine-induced NO. This results in inhibition of oxidative phosphorylation [4,5,7,9,38–40]. Because the glycolytic pathway does not depend for activity on enzymes that contain iron, glycolysis becomes the major pathway for energy production in target cells that are exposed to NO. Nitric oxide–injured target cells of the nonlytic phenotype, such as L1210 cells, when deprived of glucose have no alternative means of energy production and ultimately die [4,5,34]. Lysis of NO-injured cells with the nonlytic phenotype can be prevented by providing adequate glycolytic substrate for the glycolytic pathway [34].

All experiments to date suggest that intracellular iron is the primary target of cytokine-induced NO (5–7,9,24,31,35–43,57–59). In addition to inhibition of certain enzymes with iron-dependent catalytic activity, cytokine-induced NO causes a dramatic transfer of iron from the intracellular environment to the extracellular environment (24,31,38,39). Therefore, NO-mediated restriction of iron availability within the interior of infected mammalian host cells could be a major effector mechanism of cell-mediated immunity. Nitric oxide-induced intracellular iron deficiency could limit proliferation of prokaryotic and eukaryotic intracellular pathogens.

III. OVERVIEW OF NITRIC OXIDE EFFECTS ON INTRACELLULAR PATHOGENS

T. gondii growth is effectively inhibited by activated macrophages that have been obtained from mice infected with *Corynebacterium parvum* or from macrophages activated in vitro with LPS and interferon-γ [3]. Inducible NO synthesis can be correlated with inhibition of growth of *Toxoplasma* [21]. Conversely, MLA, an effective inhibitor of the cytokine-inducible NO synthase, progressively inhibited the anti-*Toxoplasma* activity of activated macrophages, with complete inhibition reached at MLA concentrations of 100 μM. Addition of supplemental L-arginine to macrophages incubated with 100 μM MLA partially restored anti-*Toxoplasma* effects when L-arginine levels reached 1.6 mM and completely restored anti-*Toxoplasma* effect at 4.8 mM L-arginine [21]. Other mechanisms, such as enhanced lysosome-phagosome fusion [60,61], are also probably important in defense of the intracellular environment against *Toxoplasma*. In addition, reactive oxygen intermediates may contribute to the documented activity of NO to produce anti-*Toxoplasma* activity [62,63].

Cryptococcus neoformans is significantly inhibited by L-arginine–dependent NO production in activated murine macrophages, with fungistasis being directly correlated with level of NO production [20]. Both MLA and arginase can block the fungistasis activity of these activated macrophages, through competitive inhibition of the enzyme NO synthase or depletion of the L-arginine substrate of NO synthase, respectively.

Schistosomula of *S. mansoni* derive energy from oxidative phosphorylation during the 24-hour interval following transformation from the free-living cercarial phase. During this interval schistosomula are very susceptible to NO generated by activated murine macrophages [23]. N-guanidino monomethyl L-arginine effectively inhibits NO production by activated macrophages, which results in loss of schistosomicidal activity of activated macrophages. It appears that iron depletion of schistosomula is related to the larvicidal effects of NO. This is supported by the observation that addition of ferrous sulfate to reactions containing larva incubated with activated macrophages rescued the larva from the cidal action of activated macrophages. The authors suggested that iron-containing enzymes from mitochondria of schistosomula are the targets of NO [23].

Although *Mycobacterium leprae* is defined as an obligate intracellular pathogen because it cannot be propagated outside living cells in the laboratory, its viability/metabolic rate can be readily assessed by its ability to oxidize ^{14}C-palmitic acid to $^{14}CO_2$ [22]. Using this means of assessing *M. leprae* viability, it is apparent that murine macrophages that have been activated either in vitro with IFN-γ and LPS or in vivo following infection with *C. parvum* exert potent antimycobacterial activity. The degree of this antimycobacterial activity is correlated with L-arginine–derived NO synthesis. This effect can be inhibited with 100 μM MLA, which in turn can be reversed by addition of excess L-arginine (4.8 mM). Addition of arginase, which converts L-arginine to ornithine and urea, to macrophages containing *M. leprae* effectively inhibits the antimycobacterial activity of these activated macrophages [22]. The target (host cell, bacterium, or both) of the metabolic perturbations induced by NO resulting in the observed antimycobacterial effects are not yet known.

A recent report of an in vitro murine model to study the development of the exoerythrocytic stage of *Plasmodium berghei* in hepatocytes demonstrated that L-arginine is necessary for IFN-γ–induced inhibition of liver schizont development [14]. The anti-*Plasmodium* effect was directly proportional to the dose of interferon used. The anti-*Plasmodium* effect of IFN-γ was inhibited by MLA in a dose-responsive fashion. Arginase also blocked anti-*Plasmodium* effect when this enzyme was used at high concentrations. The authors determined that the anti-*Plasmodium* effect was not secondary to elaboration of TNF by hepatocytes. This finding was supported by the observation that neutralizing amounts of antibody to TNF-α had no effect on anti-*Plasmodium* activity seen in this model. Although NO metabolites were not detected in culture supernatants in which only 1% of cells were infected, the authors concluded that the low number of cells infected may have produced NO levels below the detection limits of the Greiss reaction.

IV. HISTORIC AND CONTEMPORARY USE OF NITRIC OXIDE–GENERATING SALTS IN MEAT PRESERVATION

Nitric oxide has had significant effects on civilization for some 3000 years, although its identity was unknown to its early users [64]. Nitrite salt treatment of meats allowed storage of this perishable food source in the prerefrigeration era. Indeed, nitrite-containing salts are still important in the preservation of meats such as pork, sausage, and bacon. Nitrite at a concentration of 200 ppm was inhibitory for a wide variety of bacteria when the pH was 6.0. Shank et al. concluded that the increased inhibitory effect of nitrite toward *Clostridium botulinum* at low pH was due to undissociated nitrous acid [65] (NO is generated from undissociated nitrous acid at low pH). *C. botulinum* was treated with an NO generation system (ascorbate and nitrite) in an attempt to determine the antibacterial mechanism of nitrite added to meat-curing solutions [66]. A strong EPR signal at $g = 2.039$ characteristic of $[(RS)_2Fe(NO)_2]^-$ complexes was detected. This is the same $g = 2.039$ signal that develops after induction of activity of the high-output NO synthase in activated macrophages and their target cells [40–43]. Iron-nitrosyl complex formation was associated with inhibition of the phosphoroclastic system, which contains the iron-sulfur redox proteins ferredoxin and pyruvate ferredoxin oxidoreductase and hydrogenase [67–69]. Nitrite also inhibited active transport, O_2 uptake, and oxidative phosphorylation by *Pseudomonas aeruginosa* [70]. It is of interest that the lactobacillus, which does not contain enzymes with iron-dependent catalytic activity, was not inhibited by NO-generating curing solutions [65].

V. CHLAMYDIA/LEISHMANIA

Examples of how NO effects viability and disease caused by two major pathogens for humans, *Chlamydia trachomatis* and *Leishmania major*, will be presented below. These studies suggest that the new awareness of cytokine-induced NO synthesis from L-arginine promises to have a significant contribution to our understanding of host/parasite interactions and efforts to control and eradicate disease caused by these and other intracellular parasites.

C. trachomatis is the etiologic agent of trachoma, which affects 500,000,000 persons worldwide and remains the leading cause of preventable blindness in the world [71]. *C. trachomatis* is also the leading cause of sexually transmitted disease in the United States, with as many as 4,000,000 new cases annually [72]. The observation that *C. trachomatis* is very sensitive to the effects of IFN-γ [73–87] led us to investigate the effect that inducible NO might have on this intracellular parasitic bacterium, which requires host cell ATP for growth and replication. At least part of the observed IFN-γ antichlamydial effect in human cells is dependent on enzymatic degradation of tryptophan by indoleamine 2,3-dioxygenase [74,84,85]. This cytotoxic mechanism acting in human cells can be reversed by addition of extracellular tryptophan. However, antichlamydial IFN-γ effects are not mediated by tryptophan depletion in murine-derived cell lines [87].

Chlamydia are normally parasites of mucosal epithelial tissues such as the conjunctiva, respiratory tract, and genital tracts of humans. Recent work by Daynes et al. [83] has demonstrated that murine T lymphocytes resident in lymph nodes draining mucosal surfaces are phenotypically relatively low-level producers of IFN-γ and IL-2, compared with T lymphocytes draining skin surfaces. Although chlamydial predilection for infecting mucosal surfaces is likely to be determined by ligand/receptor specificity between chlamydia and mucosal cell types, this specificity may be selectively favored by low-level concentrations of lyphokines such as IFN-γ found in local circulation of mucosal surfaces.

A. Chlamydia-Induced Cytokines

A number of reports have detailed that chlamydia can induce cytokines and monokines both in vitro and in vivo. Interleukin-1 can be induced in human monocytes in vitro by exposure to the lymphogranuloma venereum (LGV) strains of *Chlamydia trachomatis* (L2 biovar) [88]. The LGV strains are among the few chlamydia strains that produce systemic illnesses, with fever, myalgia, and other constitutional symptoms. It will be important to determine if non-LGV *C. trachomatis* strains induce similar types of responses in monocyte-derived macrophages. Both major outer membrane protein (MOMP) and LPS from this LGV strain, as well as intact ultraviolet-killed elementary bodies of this strain, were sufficient to induce Il-1 in human monocytes.

Tumor necrosis factor-α, a monocyte/macrophage monokine, has been detected as a product of mouse spleen cell preparations that are enriched with macrophages, when these cells are exposed to the mouse pneumonitis strain of *C. trachomatis* that has been previously killed with ultraviolet irradiation [89].

Interferon-γ has been detected in endocervical secretions of women infected with *C. trachomatis* [90]. This study suggested that chlamydial infection may result in local (endocervical) IFN-γ production, although in nearly half of the women (49.9%) infected with chlamydia no IFN-γ was detected, a finding that the authors believed may have been related to assays and specimen collection. Patients with culture-proven chlamydial infection had local IFN-γ levels five times greater than those of women without chlamydial infection, although the plasma levels of IFN-γ from infected and uninfected patients were virtually the same.

Cytokines detected during *Chlamydia* infection could function in concert to stimulate cells of murine origin (macrophages [IFN-γ and TNF-α] and fibroblasts [IFN-γ, Il-1, and TNF-α]) as well as cells of human origin (hepatocytes [IFN-γ, Il-1, endotoxin, and TNF-α]) to generate high levels of NO [24,32,54,55]. It is plausible that cytokine-induced NO synthesis is an important defense mechanism for control and eradication of chlamydial infection in humans.

B. Nitric Oxide Inhibits Chlamydia in a McCoy Cell Model

Because *Chlamydia* infection is primarily an infection of epithelial cells rather than of macrophages, it is important to determine if cytokine inducible NO production with antichlamydial effects could be reproduced in nonmacrophage cell lines. We determined that McCoy cells (ATCC CRL 1696), a murine cell line of fibroblast origin commonly used for propagation of *Chlamydia*, were very efficient producers of NO when primed with IFN-γ and then exposed to agents acting as second signals—including Il-1, TNF-α, and LPS, singly or in combination—or exposed singly to *C. trachomatis* type H (Figure 2) [100]. This finding was not surprising, since Amber et al. determined that EMT6 cells, a murine adenocarcinoma cell line, and normal murine fibroblasts were capable of producing NO in response to IFN-γ and a variety of second signaling agents [24]. This was an important observation because it suggested that cytokine-induced NO synthesis, a

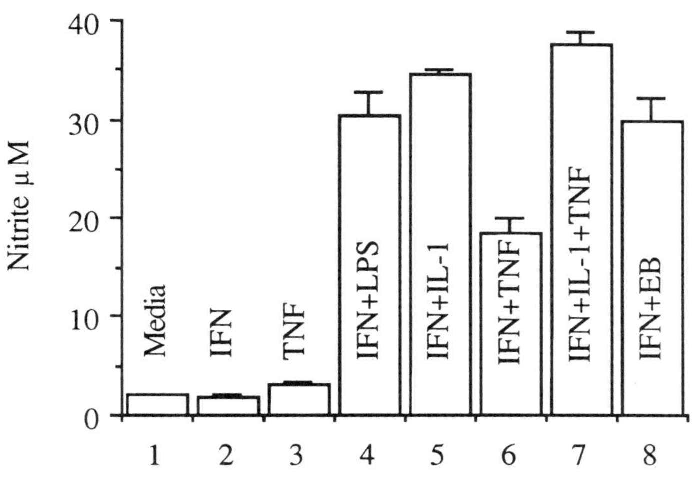

FIGURE 2 Effects of cytokines and lipopolysaccharide (LPS) on nitrite production in McCoy cells. Confluent McCoy cell monolayers (5×10^5 cells/well) received the following treatment: 1) media control; 2) interferonγ (IFN-γ) 20 U/ml alone; 3) tumor necrosis factor-α (TNF-γ) 50 U/ml alone; 4) IFN-γ 20 U/ml plus LPS 10 ng/ml; or 5) IFN-γ 20 U/ml plus IL-1α 10 U/ml, or 6) IFN-γ 20 U/ml plus TNF-α 25 U/ml, or 7) IFN-γ 20 U/ml plus IL-1α 10 U/ml and TNF-α 25 U/ml, or 8) IFN-γ 20 U/ml plus elementary bodies (EB) of *Chlamydia trachomatis* using 5×10^5 EBs (multiplicity of infection [MOI] of one EB/one McCoy cell). Values expressed are the mean total concentration of nitrite $\pm$ 2 SEM (95% confidence intervals) present in supernatants at 72 hours after cytokine/LPS/EB treatment from three determinations of a representative experiment. The differences in experimental groups were termed significant when values fell outside the error bars ($p < 0.05$).

potential biochemical defense against intracellular pathogens, was not restricted to macrophages. This finding allowed us to test the hypothesis that IFN-γ induced its antichlamydial effect in murine cells through the effector molecule NO.

Interferon-γ priming of McCoy cells has a very dramatic effect on chlamydial viability [73–85]. Chlamydial viability was assessed in our studies by determining surviving chlamydia present 72 hours post-infection (viability was defined as the ability of chlamydia-infected cell lysates to infect fresh McCoy cell monolayers in the presence of 1 μg/ml cycloheximide). Interferon-primed cells (50 U/ml) had 4 $\log_{10}$ fewer viable chlamydia compared with control cells infected with identical inocula in the absence of interferon (Figure 3). This pronounced anti-chlamydial effect was associated with an interferon dose-dependent rise in NO production by the infected cells measured 72 hours post-infection (Figure 4A). When MLA was added to the media coincident with infection, there was MLA dose–dependent reduction in NO synthesis. At the 0.8-mM MLA dose, the antichlamydial effect of IFN-γ (over a range of 1–50 U/ml) was completely inhibited. L-arginine concentration in the media for these experiments was 0.6 mM. Uninfected control cells had virtually identical low-level NO production at all levels of IFN-γ and MLA used (Figure 4B).

Whether *Chlamydia* organisms induce NO production via autocrine/paracrine production of cytokines is not yet known. For example, murine macrophages infected with *Leishmania* produce TNF-α, and this monokine is the second signal that induces NO synthesis [18]. Although *Chlamydia* can induce TNF production in murine macrophages, anti-TNF antibodies do not affect the ability of *Chlamydia* to act as second signals in our McCoy cell model (Woods and Mayer, unpublished data), suggesting that other or additional effector molecules synthesized by the McCoy cell in response to chlamydial infection may be involved.

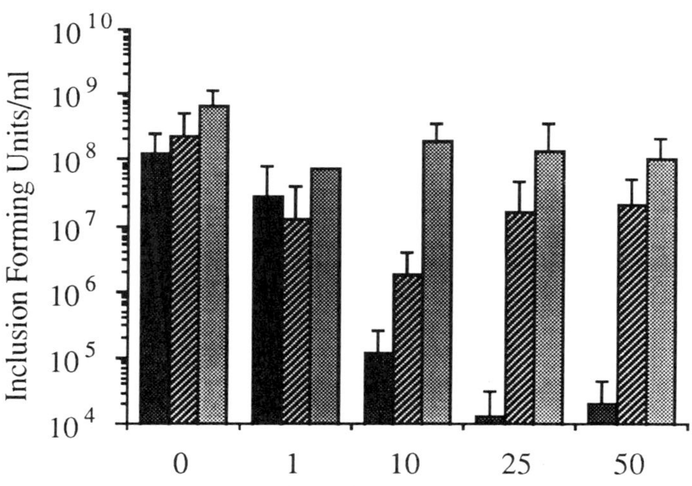

FIGURE 3 Effects of interferon-γ (IFN-γ) priming and *N*-guanidino monomethyl L-arginine (MLA) on chlamydial viability 72 hours after infection with a multiplicity of infection of 1. McCoy cell monolayers were primed with IFN-γ (0–50 U/ml) 24 hours before infection with *Chlamydia trachomatis* type H. At the time of infection McCoy cell monolayers received either no MLA as a control (black stippled bars), 0.2 mM MLA (hatched bars), or 0.8 mM MLA (grey bars). L-arginine concentration in the media was 0.6 mM. Values expressed are the mean inclusion-forming units ± 2 SEM (95% confidence intervals) for three experiments done in duplicate except for one experiment at IFN-γ 1 U/ml. The differences in experimental groups were termed significant when values fell outside the error bars ($p < 0.05$).

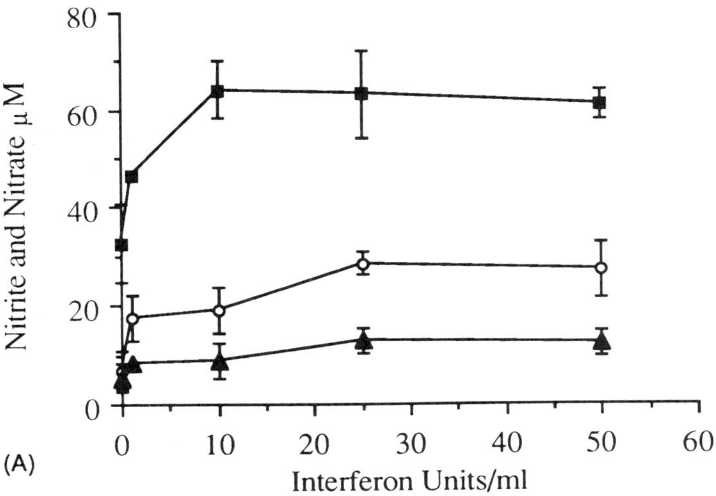

(A)

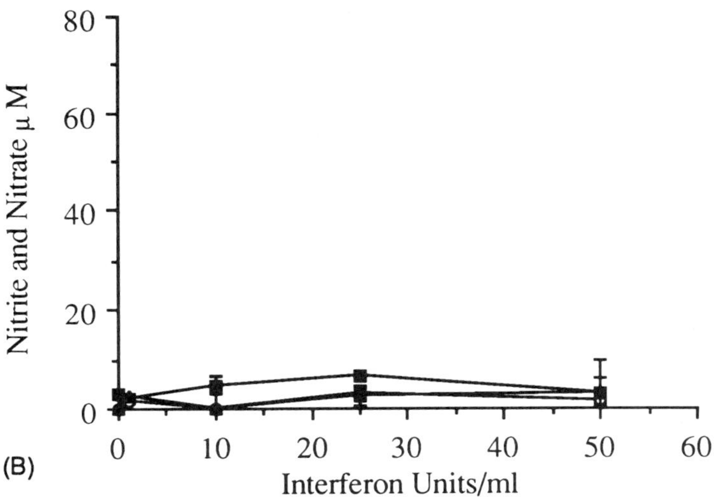

(B)

FIGURE 4 Effects of *N*-guanidino monomethyl L-arginine (MLA) and interferon-γ (IFN-γ) priming on production of nitric oxide metabolites in infected McCoy cell monolayers with *Chlamydia trachomatis* type H at a multiplicity of infection of 1 (panel A) and uninfected McCoy cells (panel B). Total concentrations of nitrite plus nitrate, expressed as μmoles/L present in supernatants at 72 hours after infection, from cultures treated with no MLA as a control (solid squares), 0.2 mM MLA (open circles), and 0.8 mM MLA (solid triangles) are shown at IFN-γ concentrations ranging from 0 to 50 U/ml. Values expressed are the mean total concentrations of nitrite plus nitrate as μmoles/L $\pm$ 2 SEM (95% confidence intervals) for three experiments done in duplicate except for one experiment at IFN-γ 1 U/ml. The differences in experimental groups were termed significant when values fell outside the error bars ($p < 0.05$).

Although we determined that *C. trachomatis* was a very effective second signaling agent for production of NO following IFN-γ priming, we have not determined the identity of the chlamydial molecule or molecules responsible for stimulating cells to produce NO. The chlamydial factor appears to be preformed, since neither doxycycline nor chloramphenicol affects the ability of *Chlamydia* to act as a second signaling agent. This chlamydial factor is partially heat-sensitive, because heat treatment at 60° C and 100° C reduces NO production by 25% to 40%. Whether this heat-sensitivity is related to diminished endocytosis of chlamydial elementary bodies (EBs) that have been heat-treated is not yet clear, since previous investigators suggest that heat treatment reduces EB uptake by 50% [91]. Cytochalasin B has no effect on second-signaling activity of chlamydia; however, previous investigators have shown that cytochalasin B reduces *C. trachomatis* uptake by only 28%, and this reduction may be insufficient to show an effect in our model [92]. Because *Chlamydia* organisms possess LPS in their outer membranes, it is logical to assume that the chlamydial molecule responsible for stimulating NO production is LPS, but polymixin B has no effect on inhibition of *Chlamydia*-induced NO production in our model, whereas polymixin B effectively blocks NO production induced by *Escherichia coli* LPS controls. However, polymixin B may not have been in high enough concentration to inhibit chlamydial LPS present in intact EBs. Our current work is directed at determining the identity of cytokines that may be involved in providing the second signal for NO synthase induction and trying to identify the chlamydial molecule or molecules that are responsible for this NO synthesis.

The mechanism through which NO synthesis affects chlamydial viability is not yet understood. In studies of macrophage antitumor effects induced by IFN-γ plus TNF or by IFN-γ plus bacterial products such as LPS, nitric oxide production is associated with tumor cell cytostasis activity through its ability to nitrosylate non-heme iron of cellular enzymes [41–43]. Additionally, NO has a known effect on mitochondrial respiration that may limit the ability of host cells to produce ATP [34–37,40]. If host cell ATP production is significantly decreased by NO synthesis in our model, chlamydial growth, which is dependent on ATP parasitism, would probably be impaired. Alternatively, NO could have direct bactericidal or bacteriostatic effects on *Chlamydia*, as has been demonstrated with *Leishmania* [15–19], *Toxoplasma* [21], mycobacteria [10–12,22], and other intracellular pathogens. Our unpublished work indicates that NO, which can be efficiently generated from nitrite (100 mM) plus ascorbic acid (100 mM), reduces EB infectivity after 30-minute exposure by 4 $\log_{10}$ compared with nitrite (100 mM) alone and by 2 $\log_{10}$ compared with ascorbic acid (100 mM). Finally, NO could affect both the host cell and *Chlamydia* directly to produce the antichlamydial effect that we have observed.

C. In Vivo Model of Nitric Oxide Effects on *Leishmania major*

Human infection with *Leishmania* is major cause of morbidity and mortality in many areas of the world. It affects millions of people and causes disfiguring cutaneous lesions and life-threatening visceral complications. Standard therapy for the diseases caused by *Leishmania* is not without side effects, and efforts to enhance natural immune responses with immune-modulating agents such as IFN-γ may have a role in future therapy strategies. Knowledge provided by this and other in vivo murine models may ultimately contribute to therapy for human disease.

A more extensive discussion of the effects of cytokine-induced NO synthesis on *Leishmania* infection appears elsewhere in this book, but we will briefly discuss recent work from our laboratories dealing with an in vivo murine model of *Leishmania* infection (Evans et al., unpublished data). In vivo NO production has been shown to have a marked effect on experimental *L. major* footpad infection in naturally susceptible BALB/c and naturally resistant C_3H/HeN mice. BALB/c mice infected with *Leishmania* are low-level producers of both IFN-γ and NO compared with C_3H/HeN mice, and this may explain the increased susceptibility of BALB/c mice to *Leishmania* infection (Figure 5). Treatment with MLA (50 mM in the drinking water) made

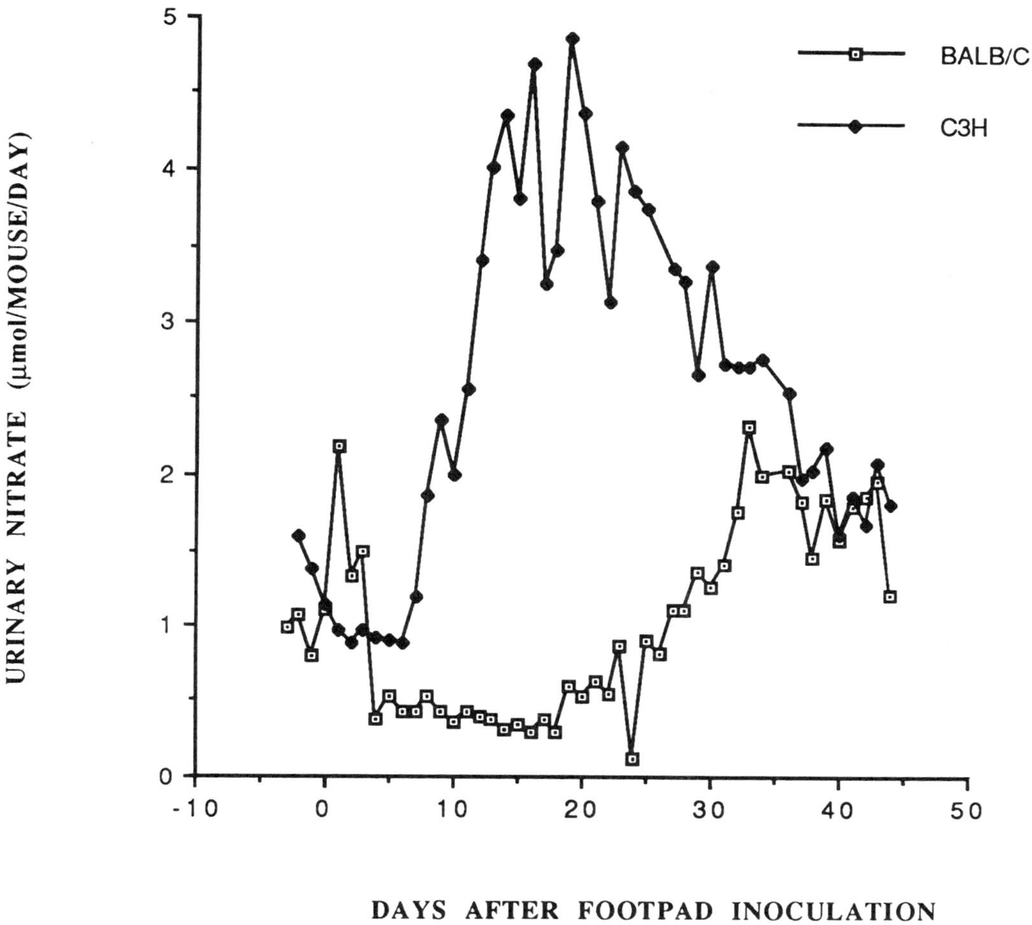

FIGURE 5 Comparison of urinary nitrate production in BALB/c mice versus C3H mice after inoculation with stationary phase promastigotes of *Leishmania major*. The BALB/c mice have little initial nitrate production, inherent susceptibility with poor control of infection, and late and large lesion development. Symbols for the BALB/c and C3H/HeN groups of mice are given at the top of the figure.

BALB/c mice more susceptible than usual to footpad challenge with *Leishmania*. Animals treated with MLA had a 10 times greater regional lymph node parasite load than infected controls not receiving MLA. The footpad lesions of MLA-treated animals were 25% larger at 7 weeks as well. An observation as yet unexplained is that MLA-treated animals with *Leishmania* infection developed cachexia, weight loss, and hair loss, unlike controls with either *Leishmania* infection alone or MLA treatment alone. The results show that susceptible Balb/c mice become even more susceptible to *Leishmania* infection when treated with MLA and that inhibition by MLA treatment of low-level NO synthesis that occurs in response to *Leishmania* infection in Balb/c mice results in cachexia.

Naturally resistant C_3H/HeN mice synthesized significantly more NO in response to *Leishmania* infection than susceptible Balb/c mice (Figure 5). Lesions occurring in C_3H/HeN mice were small and resolved after a few weeks following injection of 10^7 stationary phase *L. major* promastigotes. However, MLA treatment, which effectively blocked NO production in these mice, was associated with progressive growth of the *Leishmania* amastigotes in footpads of C_3H/HeN mice (Figure 6). Regional lymph nodes from these MLA-treated animals were enlarged, abscessed, and easily ruptured and contained innumerable degenerated amastigote forms. However, early dissemination to spleens and livers of these C_3H/HeN mice that were subjected to footpad infection was not found in MLA-treated mice following 24 days of MLA treatment. These studies show that NO is an essential effector molecule whose presence is required for controlling proliferation of *L. major* in the naturally resistant C_3H/HeN strain. Like the susceptible Balb/c strain, the resistant C_3H/HeN strain developed cachexia when infected with *Leishmania*, and endogenous NO synthesis was simultaneously inhibited by MLA treatment.

D. Relevance of Animal Studies Demonstrating Actions of Cytokine-Induced Nitric Oxide to Human Physiology and Disease

Early studies with germ-free rats indicated that urinary nitrate was endogenously synthesized and not produced by bacterial flora [93]. There is now experimental evidence that both the constitutive

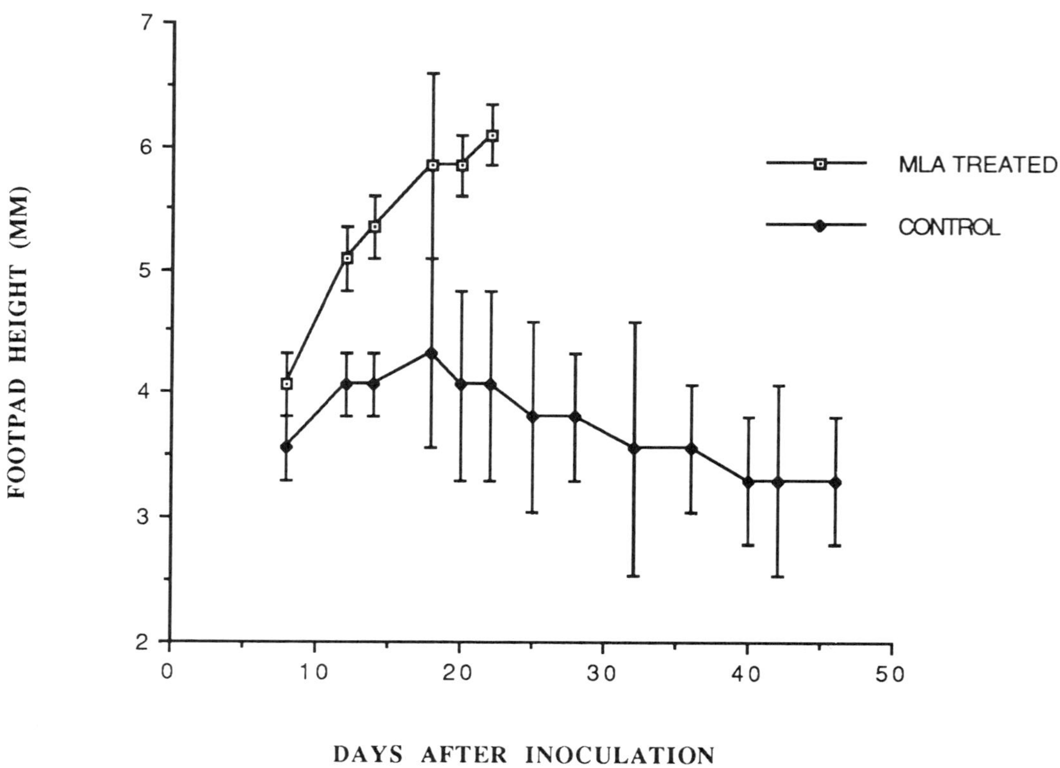

FIGURE 6 Administration of *N*-guanidino monomethyl L-arginine (MLA) to C_3H/HeN mice with *Leishmania major* footpad infection results in progressive lesion growth. Treatment with MLA was started 24 hours after *L. major* infection and discontinued on day 21 because of massive lesion size and autoamputation of infected footpads in the experiments illustrated. Mice were inoculated with 10^7 stationary phase promastigotes of *L. major* on day 0 and then were administered either sterile water (N = 6) or water containing MLA 50 mM (N = 5). The mice received MLA through day 21 and were then switched to sterile water. Because of autoamputation, footpad height could not be measured after day 22 in the MLA treated group.

and the cytokine-inducible NO synthases exist in humans. Green and co-workers showed that healthy humans excreted endogenously synthesized nitrate constitutively during metabolic studies while on low-nitrate diets [94]. Serendipitously, one of the subjects developed a febrile illness with diarrhea and a ninefold increase in urinary nitrate excretion [95]. Two in vitro models examining NO synthesis using human cell lines have demonstrated NO synthesis. It has been difficult, for reasons that are not clear at present, to detect activity of the cytokine-inducible NO synthase in human cells in vitro [96]. One report indicates that human monocyte–derived macrophages produce NO through the cytokine-inducible pathway in response to in vitro *M. avium* infection [12]. The second report details inducible NO synthesis in human hepatocytes in response to cytokine stimulation using a combination of IFN-γ, TNF, IL-1, and *E. coli* (011:B_4) LPS [32].

Evidence for constitutive NO synthesis from a terminal guanidino nitrogen atom of L-arginine was provided by a report that two normal human volunteers excreted [^{15}N]nitrate in urine following oral administration of L-[guanidino-$^{15}N_2$]arginine [97]. Other studies indicated that MLA administered via the brachial artery in normal human volunteers reduced blood flow to 50% of baseline as well as reducing responsiveness to acetylcholine but not to glycerol trinitrate [98].

Interleukin-2 is known to stimulate cellular production of IL-1, IFN-γ, and TNF, the three cytokines known to induce high-output NO synthesis in a variety of murine cell lines. Hibbs and coworkers studied the effect of IL-2 administration to patients undergoing therapy for malignant melanoma (n = 7) and renal cell carcinoma (n = 5) to determine if high-level inducible NO synthesis occurred in this patient population following administration of IL-2 [31]. By maintaining the patients on low-nitrite/nitrate oral diets and nitrite/nitrate-free total parenteral nutrition formulas, they demonstrated that basal NO production as measured by nitrate/nitrite excreted in urine was 521 $\pm$ 71 μmoles during 24 hours for renal cell carcinoma patients and 600 $\pm$ 80 μmoles per 24 hours for metastatic melanoma patients. Six to seven days after IL-2 therapy, 24-hour urinary excretion of nitrate/nitrite rose to 3376 $\pm$ 470 μmoles and 5401 $\pm$ 1436 μmoles, respectively. In this report the authors demonstrated that endogenously synthesized urinary nitrate excreted in urine was derived from a terminal guanidino nitrogen atom of L-arginine by using tracer studies with L-[guanidino-$^{15}N_2$]arginine. Basal excretion of [^{15}N]nitrate in urine over a 24-hour period was increased approximately sevenfold after IL-2 therapy and exactly paralleled the pattern seen in measuring total urinary excretion of nitrate/nitrite, described above.

In summary, the experimental evidence shows that induction of high-output NO synthesis from L-arginine by cytokines or by inhibition of cytokine-induced NO synthesis with MLA has dramatic effects on cell-mediated resistance to intracellular pathogens. The in vitro *C. trachomatis* model and the in vivo *L. major* model described in this review provide further support for the important role of cytokine-induced NO in defense of the intracellular environment. Knowledge of the L-arginine:NO pathway should provide a basis for the development of novel strategies for the management of infections caused by intracellular pathogens. The recent demonstration of NO synthesis in human subjects receiving IL-2 for treatment of melanoma and renal cell carcinoma suggests that cytokine-induced synthesis of NO from L-arginine is also an important biochemical component of the cell-mediated immune response in humans. This is evidence that information obtained in animal models will have relevance for eventual better understanding of the role of the cell-mediated immune response in humans.

REFERENCES

1. Weinberg JB, Chapman HA Jr, Hibbs JB Jr. Characterization of the effects of endotoxin on macrophage tumor cell killing. J Immunol 1978; 121:72–80.
2. Pearson RD, Wheeler DA, Harrison LH, Kay HD. The immunobiology of leishmaniasis. Rev Infect Dis 1983; 5:907–27.
3. Sibley LD, Krahenbuhl JL, Weidner E. Lymphokine activation of J774.G8 cells and mouse peritoneal macrophages challenged with *Toxoplasma gondii*. Infect Immun 1985; 49:760–4.

4. Hibbs JB Jr, Vavrin Z, Taintor RR. L-arginine is required for expression of the activated macrophage effector mechanism causing selective metabolic inhibition in target cells. J Immunol 1987; 138:550–65.
5. Hibbs JB Jr, Taintor RR, Vavrin Z. Macrophage cytotoxicity: role for L-arginine deiminase activity and imino nitrogen oxidation to nitrite. Science 1987; 235:473–6.
6. Iyengar R, Stuehr DJ, Marletta MA. Macrophage synthesis of nitrite, nitrate, and N-nitrosamines: Precursors and role of the respiratory burst. Proc Natl Acad Sci USA 1987; 84:6369–73.
7. Hibbs JB Jr, Taintor RR, Vavrin Z, Rachlin EM. Nitric oxide: a cytotoxic activated macrophage effector molecule. Biochem Biophys Res Commun 1988; 157:87–94. (Erratum Biochem Biophys Res Commun 1989; 158:624.)
8. Marletta MA, Yoon PS, Iyengar R, Leaf CD, Wishnock JS. Macrophage oxidation of L-arginine to nitrite and nitrate: nitric oxide is an intermediate. Biochemistry 1988; 27:8706–11.
9. Stuehr DJ, Nathan CF. Nitric oxide. A macrophage product responsible for cytostasis and respiratory inhibition in tumor target cells. J Exp Med 1989; 169:1543–55.
10. Denis M. Interferon-gamma-treated murine macrophages inhibit growth of tubercle bacilli via the generation of reactive nitrogen intermediates. Cell Immunol 1977; 132:150–7.
11. Chan J, Xing Y, Magliozzo RS, Bloom BR. Killing of virulent *Mycobacterium tuberculosis* by reactive nitrogen intermediates produced by activated murine macrophages. J Exp Med 1992; 175:1111–22.
12. Denis M. Tumor necrosis factor and granulocyte macrophage-colony stimulating factor stimulate human macrophages to restrict growth of virulent *Mycobacterium avium* and kill avirulent *M. Avium*: killing effector mechanism depends on the generation of reactive nitrogen intermediates. Leukocyte Biol 1991; 49:380–7.
13. Nussler A, Drapier J-C, Renia L, Pied S, Miltgen F, Gentilini M, Maxier D. L-arginine-dependent destruction of intrahepatic malaria parasites in response to tumor necrosis factor and/or interleukin 6 stimulation. Eur J Immunol 1991; 21:227–30.
14. Mellouk S, Green SJ, Nacy CA, Hoffman SL. IFN-γ inhibits development of *Plasmodium berghei* exoerythrocytic stages in hepatocytes by an L-arginine–dependent effector mechanism. J Immunol 1991; 146:3971–6.
15. Liew FY, Millott S, Parkinson C, Palmer RMJ, Moncada S. Macrophage killing of *Leishmania* parasite *in vivo* is mediated by nitric oxide from L-arginine. J Immunol 1990; 144:4794–7.
16. Mauel J, Ransijn A, Buchmuller-Rouiller Y. Killing of *Leishmania* parasites in activated murine macrophages is based on an L-arginine–dependent process that produces nitrogen derivatives. J Leukocyte Biol 1991; 49:73–82.
17. Green SJ, Meltzer MS, Hibbs JB Jr, Nacy CA. Activated macrophage destroy intracellular *Leishmania major* amastigotes by an L-arginine dependent killing mechanism. J Immunol 1990; 144:278–83.
18. Green SJ, Crawford RM, Hockmeyer JT, Meltzer MS, Nacy CA. *Leishmania major* amastigotes initiate the L-arginine–dependent killing mechanism in IFNγ stimulated macrophages by induction of tumor necrosis factor. J Immunol 1990; 145:4290–7.
19. Liew FY, Li Y, Severn A, Millott S, Schmidt J, Salter M, Moncada S. A possible novel pathway of regulation by murine T helper type-2 (T_h2) cells of a T_h1 cell activity via the modulation of the induction of nitric oxide synthase on macrophages. Eur J Immunol 1991; 21:2489–94.
20. Granger D, Hibbs JB Jr, Perfect JR, Durack DT. Metabolic fate of L-arginine in relation to microbiostatic capability of murine macrophages. J Clin Invest 1990; 85:264–73.
21. Adams LB, Hibbs JB Jr, Taintor RR, Krahenbuhl JL. Microbiostatic effect of murine-activated macrophages for *Toxoplasma gondii*. J Immunol 1990; 144:2725–9.
22. Adams LA, Franzblau AG, Vavrin Z, Hibbs JB Jr, Krahenbuhl JL. L-arginine–dependent macrophage effector functions inhibit metabolic activity of *Mycobacterium leprae*. J Immunol 1991; 147:1642–6.
23. James SL, Glaven J. Macrophage cytotoxicity against schistosomula of *Schistosoma mansoni* involved arginine-dependent production of reactive nitrogen intermediates. J Immunol 1989; 143:4208–12.
24. Amber IJ, Hibbs JB Jr, Taintor RR, Vavrin Z. Cytokines induce an L-arginine–dependent effector system in nonmacrophage cells. J Leukocyte Biol 1988; 44:58–65.
25. Curran RD, Billiar TR, Stuehr DJ, Simmons RL. Hepatocytes produce nitrogen oxides from L-arginine in response to inflammatory products of Kupffer cells. J Exp Med 1989; 170:1769–74.
26. Lepoivre M, Boudbid H, Petit JF. Antiproliferative activity of γ-interferon combined with lipopolysaccharide on murine adenocarcinoma: dependence on an L-arginine metabolism with production of nitrite and citrulline. Cancer Res 1989; 49:1970–6.

27. Kilbourn RG, Belloni P. Endothelial cell production of nitrogen oxides in response to interferon-γ in combination with tumor necrosis factor, interleukin-1, or endotoxin. J Natl Cancer Inst 1990; 82:772–6.
28. Werner-Felmayer G, Werner ER, Fuchs D, Hausen A, Reibnegger G, Wachter H. Tetrahydrobiopterin-dependent formation of nitrite and nitrate in murine fibroblasts. J Exp Med 1990; 172:1599–1607.
29. Palmer RMJ, Ashton DS, Moncada S. Vascular endothelial cells synthesize nitric oxide from L-arginine. Nature 1988; 153:1251–6.
30. Granger DL, Hibbs JB Jr, Broadnax LM. Urinary nitrate excretion in relation to murine macrophage activation. J Immunol 1991; 146:1294–1302.
31. Hibbs JB Jr, Westenfelder C, Taintor R, Vavrin Z, Kablitz C, Baronowski RL, Ward JH, Menlove RL, McMurray MP, Kushner JP, Samlowski WE. Evidence for cytokine-inducible nitric oxide synthesis from L-arginine in patients receiving interleukin-2 therapy. J Clin Invest 1992; 89:867–77.
32. Nussler AK, Di Silvio M, Billiar TR, Hoffman RA, Geller DA, Selby R, Madariaga J, Simmons RL. Stimulation of the nitric oxide synthase pathway in human hepatocytes by cytokines and endotoxin. J Exp Med 1992; 176:261–4.
33. Keller R. Cytostatic elimination of syngeneic rat tumor cells *in vitro* by nonspecifically activated macrophages. J Exp Med 1973; 138:625–44.
34. Granger DL, Taintor RR, Cook JL, Hibbs JB Jr. Injury of neoplastic cells by murine macrophages leads to inhibition of mitochondrial respiration. J Clin Invest 1980; 65:357–70.
35. Granger DL, Lehninger AL. Sites of inhibition of mitochondrial electron transport in macrophage-injured neoplastic cells. J Cell Biol 1982; 95:527–35.
36. Drapier J-C, Hibbs JB Jr. Murine cytotoxic activated macrophage inhibit aconitase in tumor cells. Inhibition involves the iron-sulfur prosthetic group and is reversible. J Clin Invest 1986; 78:790–7.
37. Drapier J-C, Hibbs JB Jr. Differentiation of murine macrophages to express nonspecific cytotoxicity for tumor cells results in L-arginine–dependent inhibition of mitochondrial iron-sulfur enzymes in the macrophage effector cells. J Immunol 1988; 140:2829–38.
38. Hibbs JB Jr, Taintor RR, Vavrin Z. Iron depletion: possible cause of tumor cell cytotoxicity induced by activated macrophage. Biochem Biophys Res Commun 1984; 123:716–23.
39. Wharton M, Granger DL, Durack DT. Mitochondrial iron loss from leukemia cells injured by macrophages. A possible mechanism for electron transport chain defects. J Immunol 1988; 141:1311–7.
40. Hibbs JB Jr, Taintor RR, Vavrin Z, Granger DL, Drapier J-C, Amber IJ, Lancaster JR Jr. Synthesis of nitric oxide from a terminal guanidino nitrogen atom of L-arginine: a molecular mechanism regulating cellular proliferation that targets intracellular iron. In: Moncada S, Higgs EA, eds. Nitric oxide from L-arginine: a bioregulatory system. New York: Elsevier, 1990:189–223.
41. Lancaster JR Jr, Hibbs JB Jr. EPR demonstration of iron-nitrosyl complex formation by cytotoxic activated macrophage. Proc Natl Acad Sci USA 1990; 87:1223–7.
42. Pellat C, Henry Y, Drapier JC. IFNγ activated macrophage: detection by electron paramagnetic resonance of complexes between L-arginine–derived nitric oxide and non-heme iron proteins. Biochem Biophys Res Commun 1990; 166:119–25.
43. Drapier J, Pellat C, Henry Y. Generation of EPR-detectable nitrosyl-iron complexes in tumor target cells cocultured with activated macrophage. J Biol Chem 1991; 266:10162–7.
44. Mackaness GB. The influence of immunologically committed lymphoid cells on macrophage activity *in vivo*. J Exp Med 1969; 129:973–92.
45. Mackaness GB. Resistance to intracellular infection. J Infect Dis 1971; 123:439–45.
46. Hibbs JB Jr, Lambert LH Jr, Remington JS. Resistance to murine tumors conferred by chronic infection with intracellular protozoa, Toxoplasma gondii and Bensnoitia jellisoni. J Infect Dis 1971; 124:587–92.
47. Hibbs JB Jr, Lambert LH Jr, Remington JS. Possible role of macrophage mediated nonspecific cytotoxicity in tumor resistance. Nature New Biol 1972; 235:48–50.
48. Hibbs JB Jr, Lambert LH Jr, Remington JS. Control of carcinogenesis: a possible role for the activated macrophage. Science 1972; 177:998–1000.
49. Hibbs JB Jr, Taintor RR, Chapman HA Jr, Weinberg JB. Macrophage tumor killing: influence of the local environment. Science 1977; 197:279–82.
50. Chapman HA Jr, Hibbs JB Jr. Modulation of macrophage tumoricidal capability by components of normal serum: a central role for lipid. Science 1977; 197:282–4.
51. Russel SW, Doe WF, McIntosh AT. Functional characterization of a stable, noncytolytic stage of macrophage activation in tumors. J Exp Med 1977; 146:1511–20.

52. Ruco LP, Meltzer MS. Macrophage activation for tumor cytotoxicity: development of macrophage cytotoxic activity requires completion of a sequence of short-lived intermediary reactions. J Immunol 1978; 121:2035–42.
53. Pace JL, Russel SW, Torres BA, Johnson HM, Gray PW. Recombinant mouse gamma interferon induces the priming step in macrophage activation for tumor cell killing. J Immunol 1983; 130:2011–3.
54. Drapier J-C, Wietzerbin J, Hibbs JB Jr. Interferon-γ and tumor necrosis factor induce the L-arginine–dependent cytotoxic effector mechanism in murine macrophages. Eur J Immunol 1988; 18:1587–92.
55. Ding AH, Nathan CF, Stuehr DJ. Release of reactive nitrogen intermediates and reactive oxygen intermediates from mouse peritoneal macrophages. J Immunol 1988; 141:2407–14.
56. Reichard P, Ehrenberg A. Ribonucleotide reductase—a radical enzyme. Science 1983; 221:514–9.
57. Lepoivre M, Chenais B, Yapo A, Lemaire B, Thelander L, Tenu J-P. Alterations of ribonucleiotide reductase activity following induction of the nitrite-generating pathway in adenocarcinoma cells. J Biol Chem 1990; 265:14143–9.
58. Kwon NS, Stuehr DJ, Nathan CF. Inhibition of tumor cell ribonucleotide reductase by macrophage-derived nitric oxide. J Exp Med 1991; 174:761–7.
59. Weinberg JB, Hibbs JB Jr. Endocytosis of red blood cells of haemoglobin by activated macrophages inhibits their tumoricidal effect. Nature 1977; 269:245–7.
60. Jones TC, Lem L, Hirsch JC. Assessment of *in vitro* immunity against *Toxoplasma gondii*. J Exp Med 1975; 141:466–82.
61. Sibley LD, Krahenbuhl JL. Phagosome acidification blocked by intracellular *Toxoplasma gondii*. Nature 1985; 315:416–20.
62. Wilson CB, Tsai V, Remington JS. Failure to trigger the oxidative metabolic burst by normal macrophages. J Exp Med 1980; 151:328–46.
63. Murray HW, Cohn ZA. Macrophage oxygen-dependent antimicrobial activity. I. Susceptibility of *Toxoplasma gondii* to oxygen intermediates. J Exp Med 1979; 150:938–49.
64. Pierson MD, Smoot LA. Nitrite, nitrite alternatives, and the control of *Clostridium botulinum* in cured meats. CRC Critical Reviews in Food Science and Nutrition 1981; 17:141–87.
65. Shank JL, Silliker JH, Harper RH. The effect of nitric oxide on bacteria. Appl Microbiol 1962; 10:185–9.
66. Reddy D, Lancaster JR Jr, Cornforth DP. Nitrite inhibition of *Clostridium botulinum*: electron spin resonance detection of iron–nitric oxide complexes. Science 1983; 221:769–70.
67. Woods LFJ, Wood JM, Gibbs PA. The involvement of nitric oxide in the inhibition of the phosphoroclastic system in *Clostridium sporogenes* by sodium nitrite. J Gen Microbiol 1981; 125:399–406.
68. Woods LFJ, Wood JM. A note on the effect of nitrite inhibition on the metabolism of *Clostridium botulinum*. J Appl Bacteriol 1982; 52:109–10.
69. Carpenter CE, Reddy DSA, Corthforth DP. Inactivation of clostridial ferredoxin and pyruvate-ferredoxin oxidoreductase by sodium nitrite. Appl Environment Microbiol 1987; 53:549–52.
70. Yarbrough JM, Rake JB, Eagon RG. Bacterial inhibitory effects of nitrite: inhibition of active transport, but not of group translocation, and of intracellular enzymes. Appl Environment Microbiol 1980; 39:831–4.
71. Jones BR. The prevention of blindness from trachoma. Trans Ophthalmol Soc UK 1975; 95:16.
72. Centers for Disease Control. *Chlamydia trachomatis* infections: policy guidelines for prevention and control. MMWR 1985; 34(suppl 3):53–74.
73. Byrne GI, Grubbs B, Marshall TJ, Schachter J, Williams DM. Gamma interferon–mediated cytotoxicity related to murine *Chlamydia trachomatis* infection. Infect Immun 1988; 56:2023–7.
74. Byrne GI, Lehmann LK, Landry GJ. Induction of tryptophan catabolism is the mechanism for gamma-interferon–mediated inhibition of intracellular *Chlamydia psittaci* replication in T24 cells. Infect Immun 1986; 53:347–51.
75. Byrne GI, Schobert CS, Williams DM, Krueger DA. Characterization of gamma interferon–mediated cytotoxicity to *Chlamydia*-infected fibroblasts. Infect Immun 1989; 57:870–4.
76. De La Maza LM, Plunkett MJ, Carlson EJ, Peterson EM, Czarneicki CW. Ultrastructural analysis of the anti-chlamydial activity of recombinant murine interferon-γ. Exp Mol Pathol 1987; 47:13–25.
77. De La Maza LM, Peterson EM, Goebel JM, Fennie CW, Czarniecki CW. Interferon-induced inhibition of *Chlamydia trachomatis:* dissociation from antiviral and antiproliferative effects. Infect Immun 1985; 47:719–22.

78. Huebner RE, Byrne GI. *In vivo*–activated mononuclear phagocytes and protective immunity to *Chlamydiae* in mice. Infect Immun 1988; 56:1492–9.
79. Shemer-Avni Y, Sarov I. Inhibition of growth of *Chlamydia trachomatis* by human gamma interferon. Infect Immun 1985; 48:592–6.
80. Williams DM, Byrne GI, Grubbs B, Marshal TJ, Schachter J. Role *in vivo* for gamma interferon in control of pneumonia caused by *Chlamydia trachomatis* in mice. Infect Immun 1988; 56:3004–6.
81. Zhong GM, Peterson EM, Czarniecki CW, De La Maza LM. Recombinant murine gamma interferon inhibits *Chlamydia trachomatis* serovar L1 *in vivo*. Infect Immun 1988; 56:283–6.
82. Zhong G, Peterson EM, Czarniecki CW, Schreiber RD, De La Maza LM. Role of endogenous gamma interferon in host defense against *Chlamydia trachomatis* infections. Infect Immun 1989; 57:152–7.
83. Daynes RA, Araneo BA, Dowell TA, Huang K, Dudley D. Regulation of murine lymphokine production *in vivo*. IV. The lymphoid tissue microenvironment exerts regulatory influences over T-helper function. J Exp Med 1990; 171:979–96.
84. Byrne GI, Lehmann LK, Landry GJ. Induction of tryptophan catabolism is the mechanism for gamma-interferon–mediated inhibition of intracellular *Chlamydia psittaci* replication in T24 cells. Infect Immun 1986; 53:347–51.
85. Murray HW, Szuro-sudol A, Wellner D, Oca MJ, Granger AM, Libby DM, Rothermel CD, Rubin BY. Role of tryptophan degradation in respiratory burst–independent antimicrobial activity of gamma interferon–stimulated human macrophages. Infect Immun 1989; 57:845–9.
86. Shemer-Avni Y, Wallach D, Sarov I. Reversion of the antichlamydial effect of tumor necrosis factor by tryptophan and antibodies to beta interferon. Infect Immun 1989; 57:3484–90.
87. De La Maza LM, Peterson EM, Czariecki CW. The anti-chlamydial and anti-proliferative activities of recombinant murine interferon-γ are not dependent on tryptophan concentrations. J Immunol 1985; 135:4198–200.
88. Rothermel CD, Schachter J, Lavrich P, Lipsitz EC, Francus T. *Chlamydia trachomatis*–induced production of interleukin-1 by human monocytes. Infect Immun 1989; 57:2705–11.
89. Williams DM, Bonewald LF, Roodman GD, Byrne GI, Magee DM, Schachter J. Tumor necrosis factor alpha is a cytotoxin induced by murine *Chlamydia trachomatis* infection. Infect Immun 1989; 57:1351–5.
90. Arno JN, Ricker VA, Batteiger BE, Katz BP, Caine VA, Jones RB. Interferon-γ in endocervical secretions of women infected with Chlamydia trachomatis. J Infect Dis 1990; 162:1385–9.
91. Byrne G. Requirements for ingestion of Chlamydia psittaci by mouse fibroblasts (L cells). Infect Immun 1976; 14:645–51.
92. Ward ME, Murray A. Control mechanisms governing the infectivity of Chlamydia trachomatis for HeLa cells: mechanisms of endocytosis. J Gen Microbiol 1984; 130:1765–80.
93. Green LC, Tannenbaum SR, Goldman P. Nitrate synthesis in the germfree and conventional rat. Science 1981; 212:56–8.
94. Green LC, de Luzuriaga KR, Wagner DA, Rand W, Istfan N, Young VR, Tannenbaum SR. Nitrate biosynthesis in man. Proc Natl Acad Sci USA 1981; 78:7764–8.
95. Wagner DA, Tannenbaum SR. Enhancement of nitrate biosynthesis by *E. coli* LPS. In: Magee PN, ed. Nitrosamines and human cancer. Banbury Rep. 12:437–43.
96. Cameron ML, Granger DL, Weinberg JB, Kozumbo WJ, Koren HS. Human alveolar and peritoneal macrophages mediate fungistasis independently of L-arginine oxidation to nitrite or nitrate. Am Rev Respir Dis 1990; 142:1313–9.
97. Leaf CD, Sishnok JS, Tannenbaum SR. L-arginine is a precursor for nitrate biosynthesis in humans. Biochem Biophys Res Commun 1989; 163:1032–7.
98. Vallance P, Collier J, Moncada S. Effects of endothelium-derived nitric oxide on peripheral arteriolar tone in man. Lancet 1989; 2:997–1000.
99. Kosaka H, Imaizumi K, Imai K, Tyuma I. Stoichiometry of the reaction of oxyhemoglobin with nitrite. Biochim Biophys Acta 1979; 581:184–8.
100. Mayer J, Woods ML, Vavrin Z, Hibbs JB, Jr. Gamma interferon-induced nitric oxide production reduces *Chlamydia trachomatis* infectivity in McCoy cells. Infect Immun 1993; 61:491–7.

11

Antimicrobial Mechanisms: Antimicrobial Polypeptides of Mononuclear Phagocytes

Pieter S. Hiemstra and Ralph van Furth

University Hospital, Leiden, The Netherlands

I. INTRODUCTION

The effector cells of the mononuclear phagocyte cell line are monocytes, which in inflamed tissues are called exudate macrophages, and resident macrophages. These cells can phagocytose nonopsonized and opsonized microorganisms. Macrophages are able to restrict the growth of or to kill ingested microorganisms. This antimicrobial activity is mediated by processes that can be grouped into oxidative and nonoxidative mechanisms. Oxidative mechanisms, which are mediated by the production of reactive oxygen intermediates (ROIs) and, in rodents, also by reactive nitrogen intermediates, are discussed in the previous two chapters of this volume. Nonoxidative mechanisms include phagosomal acidification, nutrient deprivation, and the action of antimicrobial polypeptides (Figure 1). These antimicrobial polypeptides can act alone, or in concert with the other antimicrobial mechanisms, to create an environment that is unfavorable for the multiplication of ingested microorganisms, or lethal to them.

II. ANTIMICROBIAL POLYPEPTIDES IN MACROPHAGES

Antimicrobial polypeptides are present in granulocytes (neutrophils and eosinophils), monocytes, macrophages, specialized epithelial cells in exocrine glands, and mucous secretions of humans and other mammals. In addition, antimicrobial polypeptides other than lysozyme are present in the skin of frogs and hemolymph of insects and the horseshoe crab [1]. Much of our knowledge of antimicrobial polypeptides comes from the study of neutrophils. The azurophilic granules of human neutrophils contain a variety of antimicrobial polypeptides: bactericidal/permeability-increasing protein (BPI), cathepsin G, elastase, proteinase 3, azurocidin, and defensins [reviewed in 2–4]. The specific granules of neutrophils contain lactoferrin, whereas lysozyme is present in both azurophilic and specific granules. Following phagocytosis of microorganisms by the neutrophil, these polypeptides are transferred from the cytoplasmic granules to the phagolysosome, where they can inhibit the growth of the ingested microorgainisms or kill them. Several differences exist between neutrophils and macrophages: macrophages contain fewer cytoplasmic granules, and they continue to actively synthesize granule proteins such as lysozyme and defensins [5–7]. Under certain circumstances macrophages may contain antimicrobial polypeptides from neutrophils, because they have ingested senescent or dead neutrophils [8–10]. Although this ingestion of phagocytes by other phagocytes was noted almost a century ago by Metchnikoff [11], it is unknown whether the neutrophil-derived polypeptides contribute to the antimicrobial activity of macrophages at sites of inflammation and infection.

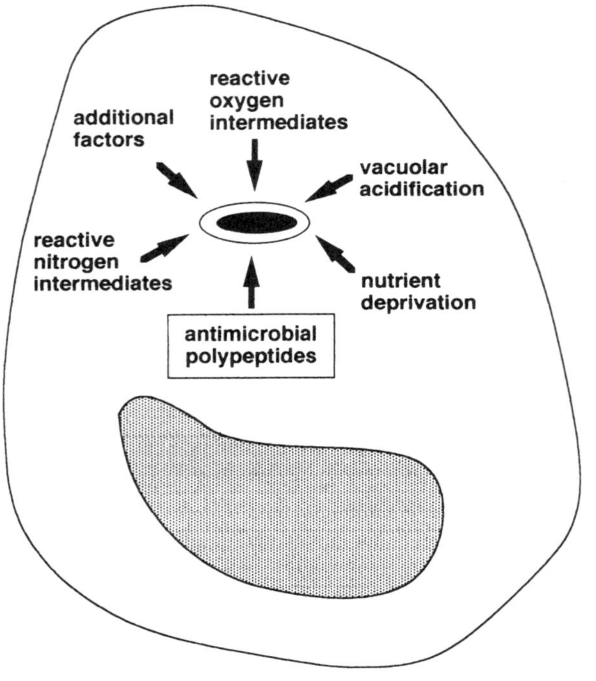

FIGURE 1 Intracellular antimicrobial mechanisms in macrophages.

TABLE 1 Antimicrobial Polypeptides in Monocytes and Macrophages

Component	M_r (kDa)	Expression	Reference
Lysozyme	14	Monocytes and macrophages from various species	5
Defensins (MCP-1, MCP-2)	3–4	Rabbit alveolar macrophages	25–27
Cathepsin G, elastase, proteinase 3, azurocidin	25–29	Expression lost upon differention into macrophages	34,36–38
Histone-like proteins	16,31, and 36	Mouse macrophages	43,44

A. Lysozyme

Lysozyme is a small (M_r 14,500) cationic muramidase that hydrolyzes glycosidic bonds in the peptidoglycan layer of the bacterial cell wall. In addition to this muramidase activity, the cationic nature of lysozyme may also contribute to its antimicrobial activity [12]. Lysozyme is present in granules of neutrophils, macrophages, and specialized epithelial cells of exocrine glands (e.g., Paneth cells). Macrophages secrete large amounts of lysozyme, representing up to 2.5% of the total amount of protein synthesized in vitro [5].

Although macrophages synthesize and secrete lysozyme constitutively, recent studies have shown that in vivo the content of mRNA for lysozyme is higher in exudate and/or activated macrophages than in resident macrophages [13,14]. In contrast to macrophages, in neutrophils and Paneth cells lysozyme is stored in cytoplasmic granules.

B. Defensins

Defensins are small (M_r < 4000) cationic antimicrobial polypeptides with a characteristic cysteine motif organized into three intramolecular disulfide bridges. They display antimicrobial activity against gram-positive and gram-negative bacteria, fungi, and certain enveloped viruses. In addition to this broad-spectrum antimicrobial activity, other functions have been described, including cytotoxic activity for eukaryotic cells [15,16], chemotactic activity [17], opsonic activity [18], and inhibition of the ACTH-mediated release of glucocorticosteroids by adrenal cells [19,20]. Defensins are present in human neutrophils and in neutrophils from rats, rabbits, and guinea pigs. Murine neutrophils do not contain defensins [21]. Several defensin-related peptides, called cryptdins, are present in murine intestinal Paneth cells [22–24]. These Paneth cells are specialized epithelial cells present at the bottom of the crypts of the small intestine, and they are thought to play a role in the defense against microorganisms in the small intestine. The only mononuclear phagocytes in which the presence of defensins has been demonstrated are rabbit alveolar macrophages [25]. These cells contain macrophage cationic peptide (MCP)-1 and MCP-2, which constitute up to 2% of the cellular protein [26–31]. Macrophage cationic peptide-1 and MCP-2 are identical to the rabbit neutrophil defensin (NP)-1 and NP-2. In addition to NP-1 and NP-2, rabbit neutrophils also contain the defensins NP-3a, NP-3b, NP-4, and NP-5, which are not present in rabbit alveolar macrophages [30,31]. Defensins are absent from rabbit monocytes, peritoneal macrophages, and alveolar macrophages from newborn rabbits [6,32]. In rabbit alveolar macrophages, the defensin content increases to adult levels during the first 3 weeks of life [6]. Intravenous injection of rabbits with complete Freund's adjuvant results in a threefold increase in defensin concentration in adult rabbit alveolar macrophages [33].

C. Antimicrobial Proteinases and Noncatalytic Analogues

At an early stage of the differentiation of mononuclear phagocytes, their cytoplasmic granules contain, in addition to myeloperoxidase, the antimicrobial polypeptides cathepsin G, elastase, proteinase 3, and azurocidin. These antimicrobial proteinases and noncatalytic analogues [3] are also known as serprocidins [34]. Upon differentiation of promonocytes into monocytes and subsequently into macrophages, the capacity to synthesize these polypeptides, which are also present in neutrophils, is lost [34–38].

D. Other Antimicrobial Polypeptides

It has previously been demonstrated that human and mouse mononuclear phagocytes contain antimicrobial polypeptides that are distinct from lysozyme and defensins, but these compounds have only been partially characterized [39–42]. Recently we have isolated and characterized three murine microbicidal proteins (M_r 16, 31, and 36 kDa) from the granule fraction of interferon-γ (IFN-γ)–activated cells from the mouse macrophage cell line RAW264.7 [43,44]. These proteins are microbicidal in vitro for various microorganisms, e.g., *Salmonella typhimurium*, *Escherichia coli*, *Staphylococcus aureus*, *Listeria monocytogenes*, and *Cryptococcus neoformans*. Evidence was obtained that these polypeptides are also present in nonstimulated RAW 264.7 cells and in resident, exudate, and activated peritoneal macrophages of mice. Characterization of the proteins by amino acid composition and sequence analysis, and by analysis of their reactivity with monoclonal antibodies, demonstrated that the 16-kDa protein was identical to murine histone H2B, whereas the 31- and 36-kDa proteins were homologous to H1 histones. In 1958 the late J. R. Hirsch already demonstrated that histones have bactericidal activity [45]. In addition, two small polypeptides (M_r 7 and 9 kDa) with antimicrobial activity against *S. typhimurium* and *L. monocytogenes* were purfied from the cytosol fraction of the IFN-γ–activated RAW264.7 cells. N-terminal amino acid sequence analysis of these polypeptides indicated that these were identical to ribosomal subunit proteins (P. S. Hiemstra, M. Th. van den Barselaar, M. Roest, and R. van Furth, unpublished data). Interestingly, indirect evidence was obtained that these polypeptides were present only in IFN-γ–activated RAW264.7 macrophages and activated peritoneal macrophages from CBA/J mice that were infected with *L. monocytogenes,* and not in nonactivated RAW264.7 macrophages or resident peritoneal macrophages. Whether the histone-related granule polypeptides are transferred to the phagolysosome and there kill ingested microorganisms, and whether the cytosol proteins are involved in killing microorganisms that have escaped from the phagolysosome (e.g., *L. monocytogenes*), is unknown as yet.

III. CLINICAL RELEVANCE

Congenital defects in phagocytes could offer insight into the relative contribution of their oxidative and nonoxidative microbicidal mechanisms. Considering the nonoxidative mechanisms of human monocytes and macrophages, it must be realized that at present it is still unknown whether these cells contain antimicrobial polypeptides, besides lysozyme and the components previously mentioned, that are lost upon differentiation of monocytes into macrophages. In this connection it is of interest that cow monocytes and macrophages do not contain lysozyme, yet their microbicidal functions are still potent. In humans a lysozyme deficiency has not been reported as yet. Since macrophages secrete lysozyme, an increase of lysozyme level in the circulation or urine can serve as a diagnostic marker in diseases with an increase of monocytes and macrophages, such as monocytic leukemia, and granulomatous diseases such as sarcoidosis and Crohn's disease. The most frequent congenital disorder of phagocytes is chronic granulomatous disease (CGD). The phagocytes of these patients, which cannot form ROIs, cannot kill catalase-positive microorganisms, but they kill catalase-negative bacteria rather well. The contribution of nonoxidative mechanisms, in particular of antimicrobial polypeptides, in the function of CGD-cells is not known, however. Information about defects in the nonoxidative killing mechanism of monocytes is also lacking in two other congenital disorders of phagocytes, i.e., neutrophil specific-granule deficiency and Chédiak-Higashi syndrome, in which the neutrophils contain almost no elastase and cathepsin G [46]. The contribution of antimicrobial polypeptides in conjunction with ROIs in neutrophils and monocytes is unknown as well. Whether a (relative) deficiency of such polypeptides plays a role in the increased frequency of infections in certain patients is also not known. Persons with myeloperoxidase deficiency, a relatively common disorder, do not suffer from more frequent infections. Myeloperoxidase is also absent in rabbit monocytes, and these cells have normal microbicidal functions. It is conceivable that activation of macrophages, besides their increase of oxidative killing mechanisms, results in an increase of nonoxidative microbicidal activity of these cells, but quantitative data are still lacking.

IV. CONCLUDING REMARKS

Various polypeptides, including lysozyme, are involved in the nonoxidative microbicidal mechanisms of monocytes and macrophages. However, these antimicrobial polypeptides are not abundantly present in these cells as compared with neutrophils. Monocytes and macrophages do not contain lactoferrin and BPI, whereas defensins are present only in rabbit alveolar macrophages. In mouse macrophages several antimicrobial polypeptides are found, but it is not yet known whether they are biologically active in the phagolysosome. Information about their presence in human monocytes and macrophages is also lacking, and thus their clinical relevance cannot yet be established.

REFERENCES

1. Boman HG. Antibacterial peptides: key components needed in immunity. Cell 1991; 65:205–7.
2. Lehrer RI, Ganz T. Antimicrobial polypeptides of human neutrophils. Blood 1990; 76:2169–81.
3. Elsbach P, Weiss J. Oxygen-independent antimicrobial systems of phagocytes. In: Gallin JI, Goldstein IM, Snyderman R, eds. Inflammation: basic principles and clinical correlates. New York: Raven Press. 1992:603–36.
4. Spitznagel JK. Antibiotic proteins of human neutrophils. J Clin Invest 1990; 86:1381–6.
5. Gordon S, Todd J, Cohn ZA. In vitro synthesis and secretion of lysozyme by mononuclear phagocytes. J Exp Med 1974; 139:1228–48.
6. Ganz T, Sherman MP, Selsted ME, Lehrer RI. Newborn rabbit alveolar macrophages are deficient in two microbicidal cationic peptides, MCP-1 and MCP-2. Am Rev Respir Dis 1985; 132:901–4.
7. Skudlarek MD, Swank RT. Turnover of two lysosomal enzymes in macrophages. J Biol Chem 1981; 256:10137–44.
8. Newman SL, Henson JE, Henson PM. Phagocytosis of senescent neutrophils by human monocyte-derived macrophages and rabbit inflammatory macrophages. J Exp Med 1982; 156:430–42.
9. Savill JS, Wyllie AH, Henson JE, Walport MJ, Hesnon PM, Haslett C. Macrophage phagocytosis of aging neutrophils in inflammation. Programmed cell death in the neutrophil leads to its recognition by macrophages. J Clin Invest 1989; 83:865–75.
10. Grigg J, Savill JS, Sarraf C, Haslett C, Silverman M. Neutrophil apoptosis and clearance from neonatal lungs. Lancet 1991; 338:720–2.
11. Metchnikoff E. Lecture VII. In: Lectures on the comparative pathology of inflammation. Delivered at the Pasteur Institute in 1891. Translated by Starling FA and Starling EH. New York: Dover Publications, Inc., 1968:94–110.
12. Laible NJ, Germaine GR. Bactericidal activity of human lysozyme, muramidase-inactive lysozyme, and cationic polypeptides against *Streptococcus sanguis* and *Streptococcus faecalis*: inhibition by chitin oligosaccharides. Infect Immun 1985; 48:720–8.
13. Chung LP, Keshav S, Gordon S. Lysozyme gene expression in macrophages and specialized epithelial cells. In: Mononuclear phagocytes. Biology of monocytes and macrophages. Van Furth R, ed. Dordrecht, Boston, London: Kluwer Academic Publishers, 1992:397–403.
14. Keshav S, Chung P, Milon G, Gordon S. Lysozyme is an inducible marker of macrophage activation in murine tissues as demonstrated by in situ hybridization. J Exp Med 1991; 174:1049–58.
15. Lichtenstein A, Ganz T, Selsted ME, Lehrer RI. In vitro tumor cell cytolysis mediated by peptide defensins of human and rabbit granulocytes. Blood 1986; 68:1407–10.
16. Okrent DG, Lichtenstein AK, Ganz T. Direct cytotoxicity of polymorphonuclear leukocyte granule proteins to human lung-derived cells and endothelial cells. Am Rev Respir Dis 1990; 141:179–85.
17. Territo MC, Ganz T, Selsted ME, Lehrer RI. Monocyte-chemotactic activity of defensins from human neutrophils. J Clin Invest 1989; 84:2017–20.
18. Fleischmann J, Selsted ME, Lehrer RI. Opsonic activity of MCP-1 and MCP-2, cationic peptides from rabbit alveolar macrophages. Diagn Microbiol Infect Dis 1985; 3:233–42.
19. Sing A, Bateman A, Zhu QZ, Shimasaki S, Esch F, Solomon S. Structure of a novel human granulocyte peptide with anti-ACTH activity. Biochem Biophys Res Commun 1988; 155:524–9.
20. Zhu QZ, Hu J, Mulay S, Esch F, Shimasaki S, Solomon S. Isolation and structure of corticostatin peptides from rabbit fetal and adult lung. Proc Natl Acad Sci USA 1988; 85:592–6.
21. Eisenhauer PB, Lehrer RI. Mouse neutrophils lack defensins. Infect Immun 1992; 60:3446–7.
22. Eisenhauer PB, Harwig SSL, Lehrer RI. Cryptdins: antimicrobial defensins of the murine small intestine. Infect Immun 1992; 60:3556–65.
23. Ouelette AJ, Greco RM, James M, Frederick D, Naftilan J, Fallon JT. Developmental regulation of cryptdin, a corticostatin/defensin precursor mRNA in mouse small intestinal crypt epithelium. J Cell Biol 1989; 108:1687–95.
24. Selstedt ME, Miller SI, Henschen AH, Ouelette AJ. Enteric defensins: antibiotic peptide components of intestinal host defense. J Cell Biol 1992; 118:929–36.
25. Ganz T, Selsted ME, Lehrer RI. Defensins in macrophages. In: Van Furth R, ed. Mononuclear phagocytes. Biology of monocytes and macrophages. Dordrecht, Boston, London: Kluwer Academic Publishers, 1992:268–72.

26. Patterson Delafield J, Martinez RJ, Lehrer RI. Microbicidal cationic proteins in rabbit alveolar macrophages: a potential host defense mechanism. Infect Immun 1980; 30:180–92.
27. Patterson Delafield J, Szklarek D, Martinez RJ, Lehrer RI. Microbicidal cationic proteins of rabbit alveolar macrophages: amino acid composition and functional attributes. Infect Immun 1981; 31:723–31.
28. Lehrer RI, Selsted ME, Szklarek D, Fleischmann J. Antibacterial activity of microbicidal cationic proteins 1 and 2, natural peptide antibiotics of rabbit lung macrophages. Infect Immun 1983; 42:10–4.
29. Selsted ME, Szklarek D, Lehrer RI. Purification and antibacterial activity of antimicrobial peptides of rabbit granulocytes. Infect Immun 1984; 45:150–4.
30. Selsted ME, Brown DM, DeLange RJ, Lehrer RI. Primary structures of MCP-1 and MCP-2, natural peptide antibiotics of rabbit lung macrophages. J Biol Chem 1983; 258:14485–9.
31. Selsted ME, Brown DM, DeLange RJ, Harwig SS, Lehrer RI. Primary structures of six antimicrobial peptides of rabbit peritoneal neutrophils. J Biol Chem 1985; 260:4579–84.
32. Ganz T, Rayner JR, Valore EV, Tumolo A, Talmadge K, Fuller F. The structure of the rabbit macrophage defensin genes and their organ-specific expression. J Immunol 1989; 143:1358–65.
33. Lehrer RI, Szklarek D, Selsted ME, Fleischmann J. Increased content of microbicidal cationic peptides in rabbit alveolar macrophages elicited by complete Freund adjuvant. Infect Immun 1981; 33:775–8.
34. Nathan CF, Gabay J. Antimicrobial mechanisms of macrophages. In: Mononuclear phagocytes. Biology of monocytes and macrophages. Van Furth R, ed. Dordrecht, Boston, London: Kluwer Academic Publishers, 1992:259–67.
35. Van Furth R, Raeburn JA, van Zwet ThL. Characteristics of human mononuclear phagocytes. Blood 1979; 54:485–500.
36. Csernok E, Ludemann J, Gross WL, Bainton DF. Ultrastructural localisation of proteinase 3, the target antigen of anti-cytoplasmic antibodies circulating in Wegener's granulomatosis. Am J Pathol 1990; 137:1113–20.
37. Senior RM, Campbell EJ, Landis JA, Cox RF, Kuhn C, Koren HS. Elastase of U-937 monocytelike cells. Comparisons with elastases derived from human monocytes and neutrophils and murine macrophagelike cells. J Clin Invest 1982; 69:384–93.
38. Senior RM, Campbell EJ. Cathepsin G in human mononuclear phagocytes: comparisons between monocytes and U937 monocyte-like cells. J Immun 1984; 132:2547–51.
39. Flesch IEA, Schwamberger G, Kaufmann SHF. Fungicidal activity of IFN-γ–activated macrophages. Extracellular killing of *Cryptococcus neoformans*. J Immunol 1989; 142:3219–24.
40. Lehrer RI. The fungicidal mechanisms of human monocytes. I. Evidence for myeloperoxidase-linked and myeloperoxidase-independent candidacidal mechanisms. J Clin Invest 1975; 55:338–46.
41. Watanabe K, Kagaya K, Yamada Y, Fukazawa Y. Mechanism for candidacidal activity in macrophages activated by recombinant gamma interferon. Infect Immun 1991; 59:521–8.
42. Rosenman M, Harwig SSL, Eisenhauer PB, Ganz T, Lehrer RI. Antibacterial components of human alveolar macrophages. Clin Res 1991; 39:73A (abstract).
43. Hiemstra PS, Eisenhauer PB, Harwig SSL, van Furth R, Lehrer RI. Antimicrobial polypeptides in mouse macrophages. In: Mononuclear phagocytes. Biology of monocytes and macrophages. Van Furth R, ed. Dordrecht, Boston, London: Kluwer Academic Publishers, 1992:273–8.
44. Hiemstra PS, Eisenhauer PB, Harwig SSL, van den Barselaar MTh, van Furth R, Lehrer RI. Antimicrobial proteins of murine macrophages. 1993; submitted for publication.
45. Hirsch JG. Bactericidal action of histone. J Exp Med 1958; 108:925–44.
46. Ganz T, Metcalf JA, Gallin JI, Boxer LA, Lehrer RI. Microbicidal/cytotoxic proteins of neutrophils are deficient in two disorders: Chédiak-Higashi syndrome and "specific" granule deficiency. J Clin Invest 1988; 82:552–6.

12

Suppressor Macrophages

Toby K. Eisenstein

Temple University School of Medicine, Philadelphia, Pennsylvania

In infection, macrophages are needed for processing of microbial antigens to stimulate the afferent arm of the immune system. For many infectious agents, particularly those described in this volume, macrophages serve as host cells for pathogen replication. Eradication of the pathogen can occur by activation of macrophages to enhanced microbicidal capacity. Thus macrophages, in particular activated macrophages, are also of major importance in the efferent arm of the immune system. Paradoxically, however, conditions that lead to macrophage activation frequently also induce immunosuppression. Macrophage-mediated immunosuppression has been observed concomitant with infection by a number of microbes in humans and in experimental animals, including fungi, *Mycobacterium tuberculosis*, *M. leprae*, *M. lepraemurium*, atypical mycobacteria, Bacillus Calmette-Guérin (BCG), *Toxoplasma*, *Trypanosoma* species, *Schistosoma*, *Salmonella*, *Histoplasma*, *Leishmania*, *Listeria*, *Chlamydia*, *Rickettsia*, *Plasmodium*, and *Treponema*. In addition, microbial extracts such as peptidoglycan and lipopolysaccharide have been shown to induce suppressor macrophages. Some killed organisms such as *Corynebacterium parvum* have a similar capacity. Normal alveolar, splenic, and peritoneal macrophages have been shown to be immunosuppressive, as have neonatal spleen cells. Macrophages associated with tumors down-regulate various parameters of immune function, as do those obtained from patients with sarcoidosis. This review will examine the conditions that generate suppressor macrophages, the nature of the suppressor factors, and the relationship between expression of suppression and activation by macrophages, with particular emphasis on suppressor macrophages generated by microbial infection.

I. MICROBE-INDUCED MACROPHAGE-MEDIATED IMMUNOSUPPRESSION

A. Live Organisms

1. Mycobacterium tuberculosis

Kleinhertz and Ellner [1] showed that adherent peripheral blood mononuclear cells from patients with pulmonary tuberculosis inhibited purified protein derivative (PPD)-induced blastogenesis in vitro. The suppression was antigen-specific, as streptokinase-streptodornase reactivity was not affected. The suppression was not indomethacin-inhibitable, suggesting suppression is not mediated by prostaglandins. The suppressor cells were shown to be positive for C3bi receptor (OKM1-positive) and not able to rosette with sheep red blood cells (non–T cell). Suppression was abrogated by exposure to 1500 rads. These studies extended earlier observations [2] that attributed the suppression in some patients to adherent macrophages. Of note is the fact that suppressor factors were not found in the supernatants of adherent cells from those patients. Parra et al. [3] described a protein extract of *Mycobacterium tuberculosis* that could induce adherent cells obtained from normal human peripheral blood to produce a supernatant factor that could inhibit lectin stimulation of T cells. The mouse macrophage cell line J774A.1 was also shown to produce a suppressor factor in the supernatant. The factor was only characterized as being heat-labile. Wadee's group found that heat-killed *M. tuberculosis* or *M. leprae* induced a heat-stable, nondialyzable suppressor factor from human peripheral blood–adherent mononuclear cells [4]. A suppressor factor was also found in splenic macrophages from *M. tuberculosis*–infected guinea pigs [5].

2. Atypical Mycobacteria

Mason et al. [6] showed that patients with atypical mycobacterial infection had peripheral blood mononuclear cells with suppressed responses to PPD, *Candida* antigen, and phytohemagglutinin (PHA) compared with normal controls. Indomethacin partially alleviated the suppression. The cellular source of the prostaglandin was not identified. Studies on *M. avium* and *M. simiae* in mice by Orme and Collins [7] showed that infection with these organisms inhibited the mitogenic response of spleen cells to PHA and blastogenic responses to allogenic cells in a mixed lymphocyte reaction. Suppression was alleviated by passage of cells over nylon wool or a Sephadex G-10 column but not by anti-Thy1.2 plus complement. The in vitro suppression correlated with suppressed delayed-type hypersensitivity (DTH) responses to sheep red blood cells (SRBC) in vivo. This in vivo lack of reactivity was attributed to more vigorous degradation of antigen in the infected animals, as suppression could be overcome by increasing the dose of antigen. Further, in vivo studies showed that infected mice that were at the peak period of suppression as judged by in vitro data could nonetheless reject an allogeneic tumor inoculated into the footpad. Edwards et al. [8] reported that *M. intracellulare* infection of mice partially suppressed mitogenic responses to ConA, but profoundly suppressed responses to *M. intracellulare*–derived mitogens. The suppression was indomethacin-inhibitable.

3. Mycobacterium leprae *and* Mycobacterium lepraemurium

Bullock et al. [9] found that mice inoculated intravenously with *M. lepraemurium* showed depressed in vitro splenic responses to SRBC in Mishell-Dutton cultures using the direct plaque-forming cell assay. Two populations of cells mediated suppression. From the fifth to the tenth week post-infection, nylon wood-adherent, glass-adherent, phagocytic, nonspecific esterase-positive cells were shown to mediate the suppression. Suppression at this time was not affected by treatment with Thy1.2 plus complement or by enriched T cell populations in the nylon wool–nonadherent cell fraction. Direct cytotoxicity by residual *M. lepraemurium* present in the Mishell-Dutton cultures was ruled out as a cause of the suppression. Further, suppression was not due to a cytotoxic effect on lymphocytes, as cell viability was not decreased. Splenic cells or macrophages from infected animals could suppress the plaque-forming cell (PFC) responses of normal cells in coculture. Suppression did not appear to be due to antigen catabolism. Later in infection, at 10 to 11 weeks, a second population of suppressor cells was evident, and these were T lymphocytes.

Mehra et al. [10] investigated the capacity of peripheral blood mononuclear cells of leprosy patients to respond to ConA in the presence or absence of lepromin. Patients with lepromatous or borderline leprosy had uniformly suppressed responses, whereas tuberculoid patients were generally not suppressed or were minimally suppressed. Cells from normal persons were not suppressed. Adherent cells from infected but not normal donors suppressed the ConA response of nonadherent cells from normal donors. As with *M. lepraemurium*, a second suppressor population consisting of T cells was also identified. Studies from Nath's laboratory have also shown that leprosy patients had *M. leprae*–mediated suppressed responses to ConA [11], and that suppression was mediated by monocyte-rich adherent cells from lepromatous leprosy patients. A soluble, indomethacin-resistant, heat-stable factor (56° C, 30 min) with a molecular weight above 25,000 kDa was implicated [12].

4. Toxoplasma gondii

A mouse model of toxoplasmosis was used to investigate the mechanism of immunosuppression observed in infected patients. Strickland and colleagues [13] showed that 3 weeks after *T. gondii* infection, mitogenic responses to T cell mitogens, PHA and ConA, and to a B cell mitogen, lipopolysaccharide (LPS), were suppressed, as were antibody responses to SRBC [14]. Suzuki et al. [15] reported that 7 days following intraperitoneal infection with *T. gondii*, mice showed profound suppression of their in vivo response to SRBC as measured by hemagglutinating titer. In vitro, the PFC response to SRBC, a T-dependent antigen, and to DNP-Ficoll, a T-independent antigen, were suppressed, but less strongly. Plastic-adherent cells were the suppressive cells, as they could suppress normal spleen cell responses.

5. Schistosoma mansoni

Fitch's laboratory [16] described suppression of alloantigen-specific cytotoxic T lymphocytes by an adherent cell population from schistosome-infected mice. Depletion of cells that phagocytized carbonyl iron with a magnet relieved suppression. T cell depletion with antisera to T markers was without effect. Studies from Colley's laboratory (17) investigated similar phenomena in peripheral blood mononuclear cells from humans infected with *Schistosoma mansoni* and found also that adherent, phagocytic cells that ingested carbonyl iron mediated suppression to schistisoma worm or cercarial antigens. Their removal potentiated blastogenic responses. Interestingly, as in leprosy patients, the suppression was specific for schistosomal antigens, as responses to *Candida albicans* extract and PHA were not affected. Ottensen [18] obtained similar results and also demonstrated that purified adherent cells prepared from infected patients could suppress the patient's nonadherent cells. Stadecker et al. reported that macrophages from a schistosomal egg granuloma inhibited stimulation of pigeon cytochrome C peptides by a Th cell clone [19].

6. Trypanosoma

African trypanosomiasis (sleeping sickness) causes profound immunosuppression [20]. Experimental infection in mice caused by *Trypanosoma brucei* [21] or by *Trypanosoma rhodesiense* [22] has been shown to induce suppressor macrophages [21–23] and T cells [21] that inhibit normal macrophages from initiating an immune response to SRBC or to mitogens. Further investigation by Wellhausen and Mansfield [24] into the nature of the cellular interactions mediating suppression induced by *T. brucei* showed that if suppressor cells and normal cells were separated by a dialysis membrane in a dual-chamber Marbrook vessel, no suppression occurred. A soluble factor was not found in culture supernatants, leading the investigators to conclude that cell-cell contact was required for the inhibition. The factor was not a prostaglandin, as indomethacin did not block suppression. Treatment of suppressor macrophages with silica, which is toxic to them, abrogated suppression. However, suppression was resistant to mitomycin C treatment and exposure to 1500 rads. Suppressor macrophages were not cytotoxic for the target cells. In an earlier study Mansfield and Wallace [25] showed that *Trypanosoma* infection also suppressed cell-mediated immunity. In experiments of Grosskinsky and Askonas [26], lethally irradiated *T. brucei* were used to infect peptone-elicited peritoneal macrophages by intraperitoneal injection. The infected peritoneal macrophage population, depleted of T cells by Thy1.2 and complement, was shown to be able to transfer suppression to naive hosts, as measured by the immune response to SRBC injected in vivo 4 days later. Kuhn's laboratory has used an experimental mouse model of *T. cruzi* infection. This group has shown that macrophages from infected mouse spleens mediate suppression, as their removal and replacement with normal macrophages restores direct PFC responses to SRBC and T-independent antigens [27]. The macrophage-mediated suppression could be overcome by increasing the number of antigen-specific T helper cells, and the augmented responses could be again suppressed by addition of more suppressor macrophages [28]. Harel-Bellan et al. found that *T. cruzi*–infected mice had impaired responses to ConA and impaired production of interleukin-2 (IL-2) [29]. Supernatants would not transfer suppression, and addition of exogenous interleukin-2 (IL-2) did not reverse suppression. Depletion of Thy1$^+$ cells enhanced suppression. It was concluded that suppression was mediated by an adherent cell that was not a T cell. In this system, T helper cells were not depleted, simply less functional. Flynn and Sileghem [30] have recently reported that cattle infected with *T. congolense* have depressed mitogen responses in peripheral blood leukocytes, spleen, and lymph nodes. Depletion of monocytes and macrophages from lymph node cells by cell sorting restored responses. Although the monoclonal antibody used for sorting also recognizes polymorphonuclear leukocytes, the suppressor cell was interpreted as being in the macrophage lineage. The suppression was not indomethacin-inhibitable, nor was it abrogated with catalase, ruling out prostaglandins and peroxide as the suppressor factors elaborated by the macrophages.

7. Listeria monocytogenes, Histoplasma capsulatum, Chlamydia psittaci, Rickettsia tsutsugamushi, Leishmania tropica, Treponema pallidum, *and* Plasmodium chabaudi

Unanue's laboratory reported in 1974 that mouse adherent peritoneal exudate cells obtained 7 days after giving *Listeria monocytogenes* intraperitoneally inhibited replication of EL-4, a leukemia cell line [31], Kongshaven et al. [32] reported in 1977 that spleen cells taken from mice injected intravenously with a virulent strain of *L. monocytogenes* were suppressed in their capacity to mount an in vitro response to SRBC and that *Listeria*-"primed" cells could suppress normal cells in coculture. The effect was shown not to be mediated by T cells, as irradiated, thymectomized animals that were reconstituted with bone marrow depleted of T cells could still generate suppressor cells. The suppressor cell was postulated to be an activated macrophage, and it was noted that the peak of suppression, at the 6th to 8th day post-infection, corresponds with the peak of antibacterial activity. In this study, the suppressor effect did not cross a 0.45-μm membrane.

Bullock's laboratory used a murine model of histoplasmosis to show that infection of mice induced an adherent suppressor cell that inhibited PFC responses. The supernatant contained no suppressor factor, and suppression was not indomethacin-inhibitable [33,34]. These observations are to be distinguished from later work showing involvement of a T suppressor cell network that is regulated in part by macrophages [35].

Jerrells [36] found that *Rickettsia tsutsugamushi* also induces suppressor macrophages that appear at the time of maximal macrophage activation. Responses to T and B cell mitogens were markedly suppressed, as were PFC responses to SRBC. Responses to the T-independent antigen TNP-Ficoll were only partially inhibited. Suppression correlated with the appearance of inflammatory, nonspecific esterase-positive cells in the spleen. Adherent spleen cells were tumoricidal and inhibited rikettsial multiplication. Sephadex G-10 passage of cells and indomethacin reversed mitogenic suppression, indicating that prostaglandin-secreting macrophages mediated this phenomenon.

Scott and Farrell [37] reported that spleens of mice infected with *Leishmania tropica* 6 to 15 weeks previously were suppressed to mitogenic stimulation with ConA, PHA, and LPS. Passage over a G-10 column removed the suppression, and indomethacin reversed it, leading to the conclusion that prostaglandin-producing macrophages were mediators of the immunosuppression. Petersen et al. [38] investigated the immune responses of peripheral blood monocytes of patients with cutaneous leishmaniasis and found them unresponsive to leishmanial antigens but not to other antigens. Suppression was relieved by removal of adherent cells by passage over nylon wool.

Studies by Byrne et al. [39] were carried out using a sublethal dose of *Chlamydia psittaci* in mice. Immunosuppression of spleen cells to mitogenic stimulation occurred, and capacity to make an in vivo PFC response to SRBC was suppressed. It was noted that suppression, rather than signaling disease progression, occurred concomitant with protection against a lethal challenge.

Treponema pallidum was found to result in rabbit splenic macrophages that could suppress ConA-induced IL-2 synthesis [40,41]. Prostaglandins mediate this suppression [40,41].

Suppressor macrophages were also found in spleens of mice infected with *Plasmodium chabaudi* [42]. Human peripheral blood macrophages from patients with disseminated fungal infections suppressed responses to mitogens and antigens [43].

8. Salmonella

Studies on the organisms reviewed up to this point examined immune responses in infected humans or used virulent organisms in animal models of infection. The suppression was usually interpreted to be a detrimental effect of the microbe on the immune system of the host [44]. However, Kongshaven et al [32], Jerrells [36], and Byrne et al. [39] observed that suppression was occurring concomitant with antibacterial activity. Orme and Collins [7] had also observed that suppression induced by atypical mycobacteria existed concomitant with capacity to resist a tumor implant. Studies from Eisenstein's laboratory showed that an attenuated vaccine strain of *Salmonella typhimurium* with a lesion in the *aroA*$^-$ pathway induced high levels of protection against challenge with virulent *Salmonella* and *Listeria* but [45,46] nevertheless induced suppression of spleen cells to B and T cell mitogens [45,47]. The suppression was not inhibited by indomethacin and was partially alleviated by depletion of macrophages by plastic adherence followed by passage over a Sephadex G-10 column. Depletion of B or T cells had no effect. Coculture of immune cells with normal cells was suppressive [47]. Fibronectin-adherent spleen cells from immune mice, which were thought to be macrophages, transferred protection against virulent challenge [46]. Suppressed mice also had tumoricidal and leishmaniacidal peritoneal macrophages [48]. Mitogenic suppression was evident by 3 days after immunization, maximal at 7 days and waning by 21 days [47]. Immune responses to SRBC were shown to be suppressed in vivo and in vitro, again with maximal onset at 7 days after inoculation of the attenuated strain. Responses to SRBC were suppressed as long as 30 days later [49]. Suppression of both in vivo and in vitro responses to SRBC was dependent on the dose of *Salmonella* [50]. Humoral, secondary immune responses to tetanus toxoid were also suppressed [50]. Nauciel's group [51] also observed that a temperature-sensitive strain of *S. typhimurium*, C5TS, induced suppression to mitogens and impairment of IL-2 production. Removal of adherent cells alleviated the majority of the suppression, and indomethacin was only marginally effective in reversing the nonresponsive state [51]. Further, *Salmonella*-infected mice had depressed DTH responses to SRBC [52].

Eisenstein's laboratory explored further the mechanisms of immunosuppression of the PFC response to SRBC in Mishell-Dutton cultures in vitro. Depletion of mature macrophages by one round of plastic adherence and passage of the nonadherent cells over Sephadex G-10 alleviated suppression approximately 60% [53]. Depletion of T cells with anti-Thy1.2 had no effect. Thus, more than half of the suppression was mediated by the adherent macrophage population in murine spleens. Splenic macrophage precursors were also found to be suppressive [53]. Suppression of the PFC response was not H-2–restricted, was not reversed by indomethacin or catalase, and was accompanied by a marked depression of IL-2 levels in spleen cell cultures. Interleukin-4 but not IL-2, reversed suppression in vitro [54].

9. Bacillus Calmette-Guérin

Bacillus Calmette-Guérin is a strain of mycobaceria used for vaccination against *M. tuberculosis*. It has also been used as an immunomodulator to augment tumor immunity. There is a considerable literature showing that BCG can induce immunosuppression as well as immunostimulation.

Klimpel and Henney [55] showed that mice given BCG intraperitoneally developed suppression to mitogens and to induction of cytotoxic T cells by alloantigenic stimulation. Suppression was evident from 5 to 40 days after immunization. Suppression was not H-2–restricted and was mediated by an adherent cell that phagocytized carbonyl iron and was Thy-1 negative. Nylon wool passage removed the cell. The cell differed from adherent cells present in normal spleens, which were not suppressive. Studies by Allen and Moore [56] support the conclusion that BCG-induced suppressor cells are macrophages. Turcotte [57] found suppressor macrophages and also suppressor T cells in spleens of mice 14 to 40 days after intravenous administration of BCG. Kato et al. [58] showed that BCG given subcutaneously resulted in peritoneal exudate cells that suppressed the macrophage migration of peritoneal exudate cells from mice immunized with BCG cell walls in oil. Removal of B or T cells did not abolish suppression. Suppressor cells were adherent. Suppression was not H-2–restricted and was indomethacin-sensitive, implicating prostaglandins [59].

BCG given intravenously has been found to inhibit the development of insulitis and overt diabetes in nonobese diabetic mice [60]. The suppressor cell was identified as a macrophage by abolition of suppression following depletion of plastic-adherent cells and by the regaining of suppression by addition of Mac-1^+ cells from BCG-immune spleens obtained by cell sorting and flow cytometry. Macrophage-enriched, but not T-enriched, BCG-immune spleen cell fractions could transfer the suppressive capacity and prevent spontaneous onset of diabetes [61].

B. Nonviable Organisms or Microbial Extracts

It has been recognized for some time that many microbial products, including exotoxins, endotoxins, enzymes, cell wall fractions, and whole killed organisms, are immunomodulatory. In many cases microbes and their products or subcellular fractions can be immunostimulatory or immunosuppressive, depending on the dose and time of administration with respect to the immune parameter examined. An excellent review of this area was published by Schwab [62]. Selected aspects of this literature will be discussed. The focus will be studies providing clear evidence that macrophages mediate the suppressive phenomena.

1. Corynebacterium parvum

Perhaps the most widely recognized microbial preparation that can induce macrophage-dependent suppression is killed *C. parvum*. In a series of studies Scott showed that *C. parvum* suppresses mitogen responses to PHA, delayed-type hypersensitivity, and graft-versus-host responses [63]. Using the response to PHA as the indicator system, macrophages were shown to be the suppressor cell. Their removal by a magnet after ingestion of carbonyl iron restored responses [63]. Wing and Remington [64] showed that *C. parvum*–activated allogeneic or syngeneic peritoneal macrophages suppressed ConA responses of normal spleen cells. Kirchner et al. [65] found that macrophages activated by *C. parvum* inhibited secondary in vitro cytotoxic responses to Moloney sarcoma virus–induced tumors. Macrophage depletion alleviated suppression, and the suppressive effect was resistant to 2500 rads. The dual activity of *C. parvum* as either an enhancer or suppressor of tumor immunity, depending on the time of administration, was clearly delineated in this study. A potential mechanism for *C. parvum*–induced suppression was suggested by the report of Grimm et al. [66] showing that this immunomodulator could induce synthesis of prostaglandin E (PGE) by macrophages. Metzger et al. [67] found that *C. parvum* or thioglycolate-elicited peritoneal macrophages inhibited ConA proliferation, and that indomethacin and catalase each gave partial reversal or suppression. The effects of the two were additive.

2. Lipopolysaccharide

It was recognized for some time that lipopolysaccharides extracted from gram-negative bacteria could be immunosuppressive as well as immunostimulatory [68,69]. A mechanism for the immunosuppression was first suggested by studies of Nelson [70]. Injection of *S. typhimurium* LPS

24 hours before harvest of peritoneal macrophages induced macrophages whose supernatants inhibited PHA-induced DNA synthesis in normal spleen cells. Ellner and Spagnuolo [71] established that LPS added in culture to human peripheral blood mononuclear cells inhibited responses to PHA and to streptokinase-streptodornase. Adherent cells mediated the suppression, and they were monocytes. The cells stimulated by LPS produced prostaglandin E_2, and the suppression was indomethacin inhibitable.

3. Peptidoglycan

Dziarski [72] found that staphylococcal peptidoglycan, which is like LPS in that it has adjuvant properties, can also be suppressive. Animals given peptidoglycan 24 hours previously had depressed PFC responses to SRBC. Spleen cells from treated animals transferred suppression. Macrophages mediated suppression and could be made suppressive by in vitro exposure to peptidoglycan.

4. Pseudomonas aeruginosa

Petit and collaborators observed that administration of live or killed *P. aeruginosa* or of supernatants of heat-killed cultures 24 hours before infection with *L. monocytogenes* sensitized to infection [73]. The *Pseudomonas* preparations also inhibited delayed hypersensitivity to SRBC. Adherent cells mediated suppression, and they were active even when heat-killed. The suppressor factor was concluded to be prostaglandin, as aspirin or indomethacin blocked the effect [74]. Prostaglandin E_1 given in vivo stimulated the suppression.

II. NONMICROBE-INDUCED MACROPHAGE-MEDIATED IMMUNOSUPPRESSION

A. Suppression by Normal or Elicited Macrophages

As early as 1966, Parkhouse and Dutton [75] observed that addition of normal rabbit alveolar or peritoneal macrophages to spleen cell cultures inhibited immune responses to antigens to which they were primed. Waldman and Gottlieb observed that oil-elicited rat peritoneal macrophages could suppress spleen cell DNA synthesis [76]. Keller reported [77] that peptone-elicited peritoneal exudate cells could suppress mitogen responses. Good's laboratory observed that depletion of adherent cells from human subjects with abnormally low mixed lymphocyte culture responses augmented their reactivity [78]. Goodwin et al. [79] found that glass wool–adherent cells suppressed PHA and ConA responses. Mattingly and Kemp showed that indomethacin could block the suppressive activity of normal rat splenic adherent cells in a secondary plaque-forming cell response in vitro [80]. Webb and Nowowiejski [81] demonstrated a cell circuit in which normal splenic macrophages could activate T cells to be suppressive. The suppressor factor passed through a membrane and was inhibited by indomethacin. Normal rat spleens were recognized to contain cells that inhibit mixed lymphocyte reaction (MLR) cultures and cytotoxic T lymphocyte (CTL) generation. The suppressor cells were adherent macrophages [82,83]. Rat oil-elicited peritoneal exudate cells displayed similar activities [82]. The suppressor cell for MLR and CTL assays in mouse splenocytes cultured for 3 to 5 days in vitro was shown to be a macrophage [84], as was the rat spleen cell that suppresses CTL to Gross virus–induced lymphoma [85]. Heat-killed thioglycolate-elicited macrophages were identified as suppressor cells for in vitro responses to SRBC [86]. Rat thioglycolate-elicited peritoneal exudate cells were also shown to be inhibitory to cytotoxic reactions generated in vitro [87]. More recently, Herscowitz's laboratory has shown that the normal mouse alveolar macrophages are as suppressive as BCG-indu[illegible]hes, and that suppression was not due to prostaglandins or oxygen metabolites [88].

B. Suppression by Tumor-Associated Macrophages

Spleen cells of mice bearing tumors induced by injection with Moloney sarcoma virus (MSV) suppressed mitogen and PFC responses, and suppressor cells were identified as macrophages [89,90]. Spleens of mice bearing methylcholanthrene-induced tumors had macrophages that suppressed in vitro cytotoxicity [91] and mitogen responses [92]. Spleens of mice with SV40-induced tumors had macrophages that suppressed secondary cytotoxic reactivity [93]. Similarly, mice bearing plasmocytomas had suppressor macrophages that inhibited the PFC response to SRBC [94]. Patients with Hodgkin's disease were shown to have decreased responses of peripheral blood mononuclear cells to PHA because of prostaglandin-secreting, glass-adherent cells [95].

C. Suppression of Immune Responses by Macrophages in Experimental Uremia and Sarcoidosis

In an experimental model in rats of uremia induced by nephrectomy, splenic and alveolar macrophages were found to be more suppressive than normal rat macrophages for PHA-stimulated blastogenesis. In contrast to the suppressive effect of the macrophage from the normal rat spleen or lung, the suppressive activities of splenic and alveolar macrophages from uremic rats were not inhibited by indomethacin [96,97].

Katz and Fauci showed that peripheral blood lymphocytes of patients with sarcoidosis inhibited pokeweed mitogen–driven polyclonal B cell activation [98].

D. Suppressor Macrophages in Liver-Grafted Rats

When liver grafts from DA rats are given to PVG rats, they are not rejected. Investigation of the mechanism of "tolerance" has shown that there is a population of suppressor macrophages in the spleen 5 to 28 days after surgery. Later, allospecific T suppressor cells appear [99]. The suppression by macrophages is sensitive to indomethacin [100].

E. Suppressor Macrophages Induced by Pyran Copolymer

Pyran copolymer, an immunomodulator, was shown by Baird and Kaplan to induce macrophages that inhibited blastogenesis to LPS or PHA [101,102]. Macrophage supernatants had little activity. Macrophages could be suppressive when cultured with lymphocytes for as little as 24 hours.

F. Neonatal "Natural" Suppressor Cells

Newborn mice have depressed immune responses. At least one factor in their failure to respond to antigen is the presence in the neonatal spleen of a large number of macrophage precursors that can proliferate in vitro and differentiate into macrophages. Adherent cells from neonatal spleens were shown to suppress adult spleen cell responses to SRBC in vitro. The suppression was abrogated by silica treatment and by indomethacin [103]. Snyder et al. [104] pursued the observation that neonates fail to respond to injection of live *L. monocytogenes* with production of Ia-positive macrophages in the peritoneal cavity. Suppressor cells found in neonatal spleen, characterized as precursors in the monocyte-macrophage lineage, were found to mediate the inability to express Ia antigen. They could block Ia expression when adoptively transferred into adult mice along with Ia-inducing adult T cells sensitized to *Listeria*, plus *Listeria* antigen. Indomethacin blocked suppression. The suppressor cells were only poorly adherent to glass in this study [104] and in one by Piguet et al. [103], reflecting the immaturity of the macrophage. In the studies by both groups, the suppressor cells differentiated into glass-adherent macrophages in culture. Oseroff et al. also described neonatal suppressor cells that could suppress a mixed leukocyte reaction [105]. They were characterized as "null" cells, since they were Thy-1⁻, IgG⁻, Mac-1⁻, F4/80⁻, nonspecific esterase–, and poorly adherent to glass. This description is compatible with that of a macrophage precursor, but these cells did not differentiate in culture [105]. Jadus and Peck [106] and Jadus and Parkman [107] also described two different types of neonatal suppressor cells. Freshly isolated neonatal suppressors secreted protein suppressor factors found in their own culture supernatants [106], whereas neonatal suppressors expanded in supernatants of Wehi-3 cells did not secrete supernatant suppressor factors. Suppressor cells were Ly-5^+ and J11d^+ [107]. Whether these cells are macrophages or their precursors is not clear.

Granulocyte-macrophage colony-stimulating factor has also been shown to enhance the suppresive capacity of neonatal spleen cells in PFC assays using SRBC. The suppressor cells were G-10- and glass-adherent and radiation-sensitive. Granulocyte-macrophage colony-stimulating factor increased their PGE_2 production and also increased the number of macrophages in culture [108].

G. Natural Suppressor Cells in Spleen and Bone Marrow

Natural suppressor cells have been identified in human [109], rat [110], and mouse [111] bone marrow. The human cell suppresses the primary in vitro response of normal tonsillar cells to SRBC and had a phenotype compatible with precursor natural killer cells [109]. In rats [110], the cell could suppress graft-versus-host disease and had characteristics of a macrophage precursor. The mouse natural suppressor cell inhibited T-dependent and T-independent responses to antigens and was described as an immature cell in the myeloid series. It was phagocytic for iron carbonyl and weakly adherent. Snyder et al. [104] also identified suppressor cells in adult murine bone marrow that could block Ia induction by T cells. Maier et al. characterized suppressor cells from irradiated mice undergoing a graft-versus-host reaction across minor histocompatibility lines [112]. The cells were non-T, non-B, non–natural killer, and non–plastic-adherent. They were thought to be similar to natural suppressor cells.

III. TRAFFICKING OF SUPPRESSOR MACROPHAGES

An interesting question concerns the origin of suppressor macrophages that are induced by microbes, microbial products, tumors, or grafts. There are a few studies addressing this issue. Bennett et al. observed that normal bone marrow cells inhibited the development of allogeneic cytotoxic cells by normal mouse spleen cells against P815-Y tumor cells [113]. Administration of intravenous BCG 2 days before harvest of bone marrow cells potentiated their suppressive activity. Both the bone marrow and the splenic suppressor cells were nylon wool–adherent, and the adherent fraction alone was suppressive. Splenic suppressor cells did not develop for 7 days, and it was hypothesized that bone marrow might be their source. A relationship was noted between the capacity of BCG to stimulate bone marrow and spleen granulocyte-macrophage colony-forming units [114]. Kato et al. [114] examined the origin of suppressor macrophages found in the peritoneal cavities of mice that were previously sensitized to BCG cell walls and that gave a positive reaction with use of the in vitro macrophage migration inhibition assay. Suppression was induced when mice were sensitized by intravenous injection, 19 days before harvest of the peritoneal cells, with bone marrow cells from donor mice immunized with live BCG 12 days before bone marrow harvest. Peritoneal cells from recipients of the BCG-stimulated bone marrow cells suppressed migration inhibition of peritoneal cells obtained from BCG-sensitized mice. The capacity of bone marrow cells from BCG-inoculated mice to transfer suppression was abrogated by prior treatment of those animals with methotrexate. Normal bone marrow cells did not transfer suppression. The suppressor cell in BCG-treated bone marrow was shown to be in the monocyte/

macrophage lineage, as it had Fc receptors, phagocytized carbonyl iron, and was adherent to Sephadex G-10. It was postulated that these BCG-stimulated cells migrated to the peritoneal cavity after intravenous injection of BCG. Further, the investigators hypothesized that the suppressor cells were the "natural suppressor" cells previously described as residing in bone marrow.

Volkman and colleagues investigated the origin of splenic suppressor cells induced by injection of *C. parvum* into mice 7 days previously. The contribution of bone marrow cells to the suppressor pool was investigated by pretreating mice with strontium 89 to ablate radiosensitive cells [115–117]. Suppressive activity was assessed by inhibition of ConA mitogenic stimulation. In these very careful studies, suppressor cells in the spleen were shown to be prostaglandin-secreting and to have Fc-γ2b receptors. Treatment with ^{89}Sr markedly reduced the suppressor activity and number of cells with the aforementioned characteristics. Resident peritoneal macrophages were not affected by ^{89}Sr treatment. This observation does not contradict the work of Kato et al., as they were examining inflammatory macrophages in the peritoneal cavity. Volkman et al. did not find a relationship between macrophage colony-forming cells (M-CFCs) and suppression. In fact, ^{89}Sr treatment increased the number of M-CFCs but decreased suppression. Bennett and Marsh also found a correlation between GM-CFCs elicited by BCG and suppression [118].

Shibata et al. [119] found that mice bearing a subcutaneous methylcholanthrene tumor had adherent, phagocytic suppressor cells in the spleen and also a second population of cells that formed macrophage colonies (M-CFCs) upon culture with L cell–conditioned medium. The mature macrophages blocked development of M-CFC by an indomethacin-inhibitable pathway. The cells that grew out in the macrophage colonies were themselves suppressive for ConA-stimulated spleen cells, and suppression was indomethacin-sensitive.

Al-Ramadi et al. found that mice given an attenuated strain of *Salmonella* have two populations of suppressor cells in the spleen [53]. One was the mature, adherent, phagocytic macrophage, and the other was nonadherent, nonphagocytic, Thy1.2$^-$, and IgG$^-$, differentiated into typical macrophages in culture, and was thought to be a macrophage precursor. The suppression in this system was not inhibited by indomethacin [120]. The macrophage precursors were enriched by negative selection in which adherent cells were removed by plastic adherence followed by G-10 filtration. The resulting population was depleted of mature macrophages and had lost 60% of its suppressive capacity. However, when cells were panned on IgG-coated plates to remove B cells, suppression was regained (>80%). The interpretation of this experiment was that a second population of suppressor cells was concentrated in the postpanning fraction or that the panning step removed a factor that blocked suppression. The postpanning cells were further purified with anti-J11D to remove polymorphonuclear leukocytes. The resulting cells were highly enriched for the macrophage precursors (75%) and were highly suppressive in Mishell-Dutton cultures [53]. The origin of these cells, whether from bone marrow or in situ from spleen, has not yet been determined.

Cerrone and Kuhn [121], studying suppression in *T. cruzi* in mice, also found two populations of suppressor macrophages in spleens by limiting dilution analysis and suggested that they were part of a complex immunoregulatory pathway in which T cells blocked suppression at intermediate concentrations of suppressor cells. Their results would be compatible with a model in which one of the macrophage suppressor cell populations is mature, and the other is a precursor.

IV. NATURE OF MACROPHAGE SUPPRESSOR FACTORS

In the description of many of the experiments cited in preceding sections, suppressor factors were observed to mediate the suppressive activity of macrophages. The nature of these factors will be considered in a more systematic fashion in this section. Reviews of some of the earlier work have appeared [122–124].

A. Thymidine

In many of the experimental systems where macrophage-mediated suppression was demonstrated, the indicator system was the inhibition of uptake of tritiated thymidine by mitogen- or antigen-stimulated lymphocytes. In 1975 Opitz and colleagues proposed that the heat-stable, dialyzable factor was cold thymidine released by dying macrophages [125,126]. It inhibited thymidine uptake, but not DNA synthesis. Since publication of this observation, investigators have sought to rule out this artifactual explanation for suppression. In many suppressor cell systems it was shown not only that ^{3}H-thymidine uptake was inhibited, but that blastogenesis [6] and DNA replication [47] were also curtailed. In addition, macrophages were shown to suppress cytotoxicity of T lymphocytes, PFC responses, and other immune functions not requiring uptake of thymidine. Further, blockage of suppression with specific agents such as indomethacin or catalase pointed to factors other than thymidine in mediating suppression. Most investigators have ruled out thymidine as the cause of suppression.

B. Arginase

Arginine was depleted in medium of mouse spleen cells in mixed leukocyte cultures that were suppressed by addition of peritoneal macrophages [127]. Macrophages activated by culture or thioglycolate had elevated arginase [127], although the accuracy of the arginase technique has been questioned [124].

C. Prostaglandins

In the preceding portions of this review a number of studies were cited in which the suppression by macrophages could be abrogated by indomethacin, thus implicating prostaglandins as the suppressor factor elaborated by these cells (Table 1). Prostaglandin secretion appears to be one mechanism by which macrophages can be suppressive. Further, a regulatory circuit has been described in which adherent human peripheral blood monocytes secrete both colony-stimulating factor (CSF) and prostaglandin. The latter inhibits the activity of CSF on normal human bone marrow cells, and indomethacin augments colony-forming units (CFUs) [128]. Resident murine peritoneal cells were shown to act in a similar fashion to augment or inhibit bone marrow CFUs. Augmentation was via secretion of macrophage CSF and suppression via prostaglandin elaboration. Macrophage CSF augmented PGE production, completing the feedback loop [129].

D. Oxygen-Derived Products

Metzger et al. showed that catalase could reverse suppression of ConA-activated T cells by *C. parvum*–induced mouse peritoneal macrophages [67], implying that peroxide produced by them is suppressive. Boraschi et al. confirmed that catalase reduced mouse peritoneal macrophage suppression of mouse spleen cells to ConA [130].

Aune and Pierce described a soluble immune response suppressor factor (SIRS) produced by murine suppressor T cells of 45,000–50,000 m.w. This factor was activated by peroxide produced by macrophages. As peroxide potentiates suppression, the reducing agent 2-mercaptoethanol potentiated suppression [131].

E. Macromolecular Suppressors

Waldmann's laboratory [132] reported identification of a suppressor factor in supernatants of ConA-activated human peripheral blood mononuclear cells that inhibits PHA and streptokinase-streptodornase-induced T cell proliferation and is produced by adherent cells. The factor has a molecular weight of 30,000 to 45,000 and is blocked by *N*-acetyl-D-glucosamine. Krakauer [133]

Table 1 Characteristics of Suppressor Factors Produced by Macrophages

Prostaglandin	ref.	Not prostaglandin (indomethacin-insenstive)[a]	ref.	Other suppressor factors	ref.	Nitric oxide	ref.
Mycobacteria intracellulare	6,8	*Mycobacterium tuberculosis*	1	*M. tuberculosis*: protein	3–5	*C. parvum*	137
Rickettsia	36	*Mycobacterium leprae*	12	*M. leprae*: protein	12	Normal rat spleen macrophages	137
Leishmania tropica	37	*Trypanosoma rhodesiense*	24	*C. parvum*: peroxide	67	*Salmonella*	143
Treponema pallidum	40	*Trypanosoma congolense*	30	Neonatal spleen macrophages: protein	106	Bacillus Calmette-Guérin	141
Bacillus Calmette-Guérin	59	*Histoplasma capsulatum*	33.34	Human peripheral blood–adherent cells: protein	132		
C. parvum	66,67,115	*Salmonella typhimurium*	120	THP-1 cell line: protein	133		
Lipopolysaccharide	71	Normal mouse alveolar macrophages	88	U937 cell line: protein	134–136		
Pseudomonas aeruginosa	73,74	Spleen cells from uremic rats	96.97				
Tumor-bearing spleen macrophages	92,93,119	*C. parvum*	61				
Peripheral blood-adherent cells in Hodgkin's disease	95						
Splenic macrophages in liver grafted rats	99,100						
Newborn mouse spleen macrophages	103,104,108						

[a]Suppressor factor not identified further.

purified a 60,000–70,000 m.w. factor from the human leukemia cell line THP-1 that suppresses ConA and PHA mitogenic responses of human peripheral blood mononuclear cells, IL-1–driven proliferation of mouse thymocytes, and proliferation of an IL-2–dependent cell line. It is heat-labile and sensitive to proteases. Wilkins et al. [134] and Ellner's laboratory [135,136] have both purified suppressor factors from U937 cells, a histiocytic human lymphoma–derived cell line. The Wilkins factor has a molecular weight of 40,000 and inhibits mitogen responses. The Ellner factor has biologic activity similar to that described by Krakauer for THP-1 cells, but its molecular weight is 85,000. It inhibited IL-2 production by PHA- and PPD-stimulated human peripheral blood monocytes, and IL-2 could not restore responsiveness. Interleukin-2 receptor expression was also depressed by the factor.

F. Nitric Oxide

Recently (1991), Mills [137] published evidence that macrophage suppression of ConA or other mitogen responses by rat spleen cells is mediated by nitric oxide (NO). *N*-monomethyl-L-arginine (NMLA), an inhibitor of the NO pathway, inhibited suppression. NMLA also reversed the splenic suppression induced by intravenously administered *C. parvum* in mice. Nitric oxide is produced by macrophages by metabolism of arginine to yield citrulline and NO when they are induced to make NO synthase [138,139]. Albina et al. [140] also reported that mouse splenic macrophages inhibited mitogen responses to ConA. ConA supernatants induced macrophage NO production to provide a feedback loop. Indomethacin and catalase only partially inhibited ConA suppression, but NMLA gave complete reversal. In contrast, LPS-stimulated cultures did not make NO and were inhibitable by indomethacin. This study shows that there are two pathways to suppressor macrophages and two mediators, PGE and NO. As interferon-γ (IFN-γ) is the most potent inducer of NO, it is postulated that ConA supernatants activate NO production in macrophages via IFN-γ [140]. It is interesting that heat-killed BCG and LPS have been shown to induce NO in thioglycolate-induced mouse peritoneal macrophages [141,142]. Even nonstimulated thioglycolate macrophages made measurable NO [142].

Eisenstein's laboratory has shown that spleens of mice infected with attenuated *Salmonella* make NO, and that NMLA reverses macrophage-mediated suppression of the primary PFC response to SRBC [143]. Further, IL-4 and anti-IFN-γ reversed suppression and inhibited NO secretion [54,143]. A model has been proposed [144] in which IFN-γ activates macrophages to produce NO, and IL-4 down-regulates the macrophage so that it produces less NO. Interleukin-4 has been shown by others to down-regulate other macrophage functions, including peroxide production [145]. As Th1 cells produce IFN-γ and Th2 cells make IL-4, it has been proposed that NO levels may be regulated in accordance with whichever T cell subset is functioning [146]. If Th2 cells are upregulated, then NO will be suppressed, but Th1 activation will result in NO production. This differential activation of NO, by augmenting one of the T cell subsets, could explain many of the phenomena observed in intracellular infections. For example, NO levels might govern whether mouse strains are progressors or regressors during *Leishmania* infection [147]. Mouse macrophage–generated NO has been shown to be cytostatic or cytotoxic for *Leishmania* [148], *Toxoplasma* [149], *Schistosoma* [150], *Trypanosoma* [151], and *Mycobacterium* [152–154], but not *Salmonella* [155]. Thus, infection that results in microbicidal activity of macrophages could also result in macrophage-mediated immunosuppresion.

V. OVERVIEW, MODELS, AND CONCLUSIONS

It is clear that a variety of stimuli induce suppressor macrophages, including live microbes, microbial products, tumors, and autoimmune diseases. These stimuli also induce inflammatory responses characterized by monocyte/macrophage accumulation. It is paradoxical that highly activated macrophages are also suppressive. Both mature macrophages and macrophage precursors can participate in suppressive reactions, although it is not clear if the macrophage precursors need to differentiate during the assay to be suppressive.

Two main factors have been identified as mediating suppression: prostaglandin and nitric oxide. Protein suppressor factors have also been documented, but not cloned or named. Peroxide has also been implicated in mediating immunosuppression. Discrepancies between studies regarding presence of suppressor factors in macrophage supernatants can probably be explained according to whether the suppression was being mediated by prostaglandin or NO. Prostaglandin would be expected to be present in culture supernatants, but NO would not. Nitric oxide is extremely labile and lasts less than a minute in culture before it is converted to nitrite or other nitrogen intermediates [156]. Thus, it would not be retrievable in culture supernatants. The lability of this molecule also probably explains why suppression would not cross a membrane in some systems, but would in others. Prostaglandin would be expected to cross membranes used in Marbrook chambers. Nitric oxide would be predicted to cross on the basis of size, but depending on the geometry of the chambers and the NO concentration, it might or might not have been able to reach the target cell before being inactivated in the culture medium.

In some studies, removal of the suppressor cells reversed suppression, whereas in others suppression was irreversible. An explanation for these differences can be postulated on the basis of the suppressor factor involved. Prostaglandin-induced suppression may be reversible. However, NO poisons the target cell [156,158], so that removal of the supernatant would not be expected to reverse target cell immunocompetence. Nitric oxide is known to bind to iron-sulphur–containing enzymes in mitochondria [157], which metabolically incapacitates the target cell but does not necessarily kill it. Nitric oxide has also been reported to deaminate DNA [159] and inhibit ribonucleotide reductase [160]. Target cells exposed to NO would be nonfunctional but would still stain with trypan blue, which would explain why in many of the systems demonstrating suppressor macrophages, the target lymphocytes were viable as judged by failure to take up trypan blue, but were nonetheless inhibited.

The precise relationship between PGE_2 and NO production is to be clarified. However, several investigators have observed that suppression can be expressed independent of tumoricidal activity [161–163]. It can be hypothesized that in these cases prostaglandin is mediating the suppression, since NO has both antitumor and immunosuppressive capacity. Interferon-γ is reported to reduce PGE_2 secretion by macrophages [41,162], but it is also a powerful signal for induction of NO synthase [139–141]. Thus, the important question of whether suppressor macrophages and tumoricidal/microbicidal macrophages are the same populations is not yet answered. It may be that the same cells can mediate both phenomena if NO is produced. Activators that only trigger PGE_2 production may appear to induce suppression without activation. Further investigation of this question is needed.

The rationale for the host to produce factors from macrophages that down-regulate immune responses is open to speculation. Several investigators have suggested that suppressor factors are part of natural feedback loops by which the macrophage activation induced by cytokines is damped. Another explanation is suggested by the observation that suppression is not induced by microbes or tumors until some time after their injection or transfer, usually at least 3 days and more commonly 7 days to several weeks (for slow-growing organisms and tumors). Frequently, suppression is accompanied by splenomegaly and appearance of macrophage precursors in the spleen. Thus, suppression results from the inflammatory response to the foreign invader, microbe or tumor. Bennett and Marsh showed that induction of suppression by BCG is strain-dependent [118]. Only a strain that replicates in vivo induced suppression. A correlation between virulence and immunosuppression was observed also in trypanosomiasis [164]. Eisenstein's laboratory [50] showed that induction of suppression by attenuated *Salmonella* was dose-dependent, and there was a threshold of about 4×10^3 cells.

A new way of viewing NO-mediated suppression is that it results from an attempt by the host to rid itself of the microbe or tumor. The suppression is due to bystander injury of lymphocytes in the vicinity of the NO-secreting macrophages. Nitric oxide production by macrophages can be viewed as an "alarm" response of the host to a dangerous level of microbes or tumor cells. In an attempt to oust the invaders, the host cells also suffer damage. The toxicity to surrounding lymphocytes is absorbed as part of the cost of the battle between macrophages and the microbes. This scenario would also explain why suppression occurs to exogenous antigens later in the infection, when inflammatory response is as its peak. The observation made by some investigators, that suppression is stronger to mitogenesis induced by the microbial antigen derived from the infecting organism than to other microbial antigens, could be due to the release of IFN-γ by T cells when the homologous antigen is used, and subsequent up-regulation of NO production.

What might be the consequences to the host of switching-on the NO pathway? It might make reduced immune responses to the invading organism itself [165–167]. If other pathogens invade close behind the first one, the immune response to them might be impaired. Thus, NO production, like delayed-type hypersensitivity, is a double-edged sword. Overproduction of NO leaves the host vulnerable through immunosuppression. Yet its production may be necessary to restrain the growth of pathogens. These hypotheses imply that macrophage-mediated immunosuppression due to NO is not a state that is beneficial to the microbe. Instead it is a manifestation of a powerful host defense mechanism that is attempting to control microbial growth.

ACKNOWLEDGMENTS

This work was supported by grant AI15613 from the National Institutes of Health. Thanks to Joseph Meissler, Jr., and Duan Huang for helping in gathering the references and to Greg Harvey for editorial assistance and typing of the manuscript. Special thanks to Basel Al-Ramadi for our stimulating conversations about nitric oxide.

REFERENCES

1. Kleinhenz ME, Ellner JJ. Immunoregulatory adherent cells in human tuberculosis: radiation-sensitive antigen-specific suppression by monocytes. J Infect Dis 1985; 152:171–6.
2. Ellner JJ. Suppressor adherent cells in human tuberculosis. J Immunol 1978; 121:2573–9.
3. Parra C, Montaño LF, Huesca M, Rayón I, Willms K, Goodsaid F. Inhibition of mitogenesis induced by phytohemagglutinin and *Lens culinaris* lectin in adherent-cell supernatants treated with protein extract of *Mycobacterium tuberculosis*. Infect Immun 1986; 52:309–13.
4. Wadee AA, Sher R, Rabson AR. Production of a suppressor factor by human adherent cells treated with mycobacteria. J Immunol 1980; 125:1380–6.
5. Wadee AA, Rabson AR. Production of a suppressor factor by adherent cells from *Mycobacterium tuberculosis*–infected guinea-pigs. Clin Exp Immunol 1981; 45:427–32.
6. Mason UG III, Greenberg LE, Yen SS, Kirkpatrick CH. Indomethacin-responsive mononuclear cell dysfunction in "atypical" mycobacteriosis. Cell Immunol 1982; 71:54–65.
7. Orme IM, Collins FM. Immune response to atypical mycobacteria: immunocompetence of heavily infected mice measured in vivo fails to substantiate immunosuppression data obtained in vitro. Infect Immun 1984; 43:32–7.
8. Edwards CK III, Hedegaard HB, Zlotnik A, Gangadharam PR, Johnston RB Jr, Pabst MJ. Chronic infection due to *Mycobacterium intracellulare* in mice: association with macrophage release of prostaglandin E_2 and reversal by injection of indomethacin, muramyl dipeptide, or interferon-γ. J Immunol 1986; 136:1820–7.
9. Bullock WE, Carlson EM, Gershon RK. The evolution of immunosuppressive cell populations in experimental mycobacterial infection. J Immunol 1978; 120:1709–16.
10. Mehra V, Mason LH, Fields JP, Bloom BR. Lepromin-induced suppressor cells in patients with leprosy. J Immunol 1979; 123:1813–17.

11. Nath I, Singh R. The suppressive effect of *M. leprae* on the in vitro proliferative responses of lymphocytes from patients with leprosy. Clin Exp Immunol 1980; 41:406–14.
12. Sathish M, Bhutani LK, Sharma AK, Nath I. Monocyte-derived soluble suppressor factor(s) in patients with lepromatous leprosy. Infect Immun 1983; 42:890–9.
13. Strickland GT, Ahmed A, Sell KW. Blastogenic response of *Toxoplasma*-infected mouse spleen cells to T- and B-cell mitogens. Clin Exp Immunol 1975; 22:167–76.
14. Strickland GT, Sayles PC. Depressed antibody responses to a thymus-dependent antigen in toxoplasmosis. Infect Immun 1977; 15:184–90.
15. Suzuki Y, Watanabe N, Kobayashi A. Nonspecific suppression of primary antibody responses and presence of plastic-adherent suppressor cells in *Toxoplasma gondii*–infected mice. Infect Immun 1981; 34:30–5.
16. Coulis PA, Lewert RM, Fitch FW. Splenic suppressor cells and cell-mediated cytotoxicity in murine schistosomiasis. J Immunol 1978; 120:1074–6.
17. Todd CW, Goodgame RW, Colley DG. Immune responses during human Schistosomiasis mansoni. V. Suppression of schistosome antigen–specific lymphocyte blastogenesis by adherent/phagocytic cells. J Immunol 1979; 122:1440–6.
18. Ottesen EA. Modulation of the host response in human schistosomiasis. I. Adherent suppressor cells that inhibit lymphocyte proliferative responses to parasite antigens. J Immunol 1979; 123:1639–44.
19. Stadecker MJ, Kamisato JK, Chikunguwo SM. Induction of T helper cell unresponsiveness to antigen by macrophages from schistosomal egg granulomas. J Immunol 1990; 145:2697–700.
20. Goodwin LG, Green DR, Guy NW, Voller A. Immunosuppression during trypanosomiasis. Br J Exp Pathol 1972; 53:40–3.
21. Jayawardena AN, Waksman BH. Suppressor cells in experimental trypanosomiasis. Nature 1977; 265:539–41.
22. Wellhausen SR, Mansfield JM. Lymphocyte function in experimental African trypanosomiasis. II. Spenic suppressor cell activity. J Immunol 1979; 122:818–24.
23. Corsini AC, Clayton C, Askonas BA, Ogilvie BM. Suppressor cells and loss of B-cell potential in mice infected with *Trypanosoma brucei*. Clin Exp Immunol 1977; 29:122–31.
24. Wellhausen SR, Mansfield JM. Characteristics of the splenic suppressor cell–target cell interaction in experimental African trypanosomiasis. Cell Immunol 1980; 54:414–24.
25. Mansfield JM, Wallace JH. Suppression of cell-mediated immunity by African trypanosomiasis. Infect Immun 1974; 10:335–9.
26. Grosskinsky CM, Askonas BA. Macrophages as primary target cells and mediators of immune dysfunction in African trypanosomiasis. Infect Immun 1981; 33:149–55.
27. Cunningham DS, Kuhn RE. *Trypanosoma cruzi* induced suppression of the primary immune response in murine cell culture to T cell dependent and independent antigens. J Parasitol 1980; 66:16–27.
28. Ritter DM, Kuhn RE. Antigen-specific T-helper cells abrogate suppression in *Trypanosoma cruzi*–infected mice. Infect Immun 1990; 58:3248–56.
29. Harel-Bellan A, Joskowicz M, Fradelizi D, Eisen H. Modification of T-cell proliferation and interleukin 2 production in mice infected with *Trypanosoma cruzi*. Proc Natl Acad Sci USA 1983; 80:3466–9.
30. Flynn JN, Sileghem M. The role of the macrophage in induction of immunosuppression in *Trypanosoma congolense*–infected cattle. Immunology 1991; 74:310–6.
31. Calderon J, Williams RT, Unanue ER. An inhibitor of cell proliferation released by cultures of macrophages. Proc Natl Acad Sci USA 1974; 71:4273–7.
32. Kongshaven PAL, Ho A, Sebaldt RJ. Suppression of *in vitro* antibody responses by *Listeria* primed spleen cells. Cell Immunol 1977; 28:284–97.
33. Nickerson DA, Havens RA, Bullock WE. Immunoregulation in disseminated histoplasmosis: characterization of splenic suppressor cell populations. Cell Immunol 1981; 60:287–97.
34. Deepe GS Jr, Kravitz GR, Bullock WE. Pharmacological modulation of suppressor cell activity in mice with disseminated histoplasmosis. Infect Immun 1983; 41:114–20.
35. Deepe GS Jr, Watson SR, Bullock WE. Cellular origins and target cells of immunoregulatory factors in mice with disseminated histoplasmosis. J Immunol 1984; 132:2064–71.
36. Jerrells TR. Immunosuppression associated with the development of chronic infections with *Rickettsia tsutsugamushi*: adherent suppressor cell activity and macrophage activation. Infect Immun 1985; 50:175–82.

37. Scott PA, Farrell JP. Immunodepression in BALB/c mice infected with *Leishmania tropica*. In: Eisenstein TK, Actor A, Friedman H, eds. Host defenses to intracellular pathogens. New York: Plenum Press, 1983:151–5.
38. Petersen EA, Neva FA, Oster CN, Diaz HB. Specific inhibition of lymphocyte-proliferation responses by adherent suppressor cells in diffuse cutaneous leishmaniasis. N Engl J Med 1982; 306:387–92.
39. Byrne GI, Guagliardi LE, Huebner RE, Paulnock DM. Immunomodulation and *Chlamydia*: immunosuppression and the protective immune response to *C. psittaci* in mice. In: Eisenstein TK, Bullock WE, Hanna N, eds. Host defenses and immunomodulation to intracellular pathogens. New York: Plenum Press, 1988:343–52.
40. Fitzgerald TJ, Tomau MA, Trachte GJ, Rice T. Prostaglandins in experimental syphilis: treponemes stimulate adherent spleen cells to secrete prostaglandin E_2, and indomethacin upregulates immune functions. Infect Immun 1991; 59:143–9.
41. Tomai MA, Fitzgerald TJ. Splenic macrophage function in early syphilitic infection is complex. Stimulation versus down-regulation. J Immunol 1991; 146:3171–6.
42. Corrêa M, Narayanan PR, Miller HC. Suppressive activity of splenic adherent cells from *Plasmodium chabaudi*–infected mice. J Immunol 1980; 125:749–54.
43. Stobo JD. Immunosuppression in man: suppression by macrophages can be mediated by interactions with regulatory T cells. J Immunol 1977; 119:918–24.
44. Playfair JHL. Workshop report: suppressor cells in infectious disease. Parasitol Immunol 1982; 4:229–304.
45. Eisenstein TK, Killar LM, Stocker BAD, Sultzer BM. Cellular immunity induced by avirulent *Salmonella* in LPS-defective C3H/HeJ mice. J Immunol 1984; 133:958–61.
46. Killar LM, Eisenstein TK. Immunity to *Salmonella typhimurium* infection in C3H/HeJ and C3H/HeNCrlBR mice: studies with an aromatic-dependent live *S. typhimurium* strain as a vaccine. Infect Immun 1985; 47:605–12.
47. Lee J-C, Gibson CW, Eisenstein TK. Macrophage-mediated mitogenic suppression induced in mice of the C3H lineage by a vaccine strain of *Salmonella typhimurium*. Cell Immunol 1985; 91:75–91.
48. Schafer R, Nacy CA, Eisenstein TK. Induction of activated macrophages in C3H/HeJ mice by avirulent *Salmonella*. J Immunol 1988; 140:1638–44.
49. Eisenstein TK, Dalal N, Killar L, Lee J-C, Schafer R. Paradoxes of immunity and immunosuppression in Salmonella infection. In: Eisenstein TK, Bullock WE, Hanna N, eds. Host defenses and immunomodulation to intracellular pathogens. New York: Plenum Press, 1988:353–66.
50. Al-Ramadi BK, Greene JM, Meissler JJ Jr, Eisenstein TK. Immunosuppression induced by attenuated *Salmonella*: effect of LPS responsiveness on development of suppression. Microbial Pathogenesis 1992; 12:267–78.
51. Deschenes M, Guenounou M, Ronco E, Vacheron F, Nauciel C. Impairment of lymphocyte proliferative responses and interleukin-2 production in susceptible (C57BL/6) mice infected with *Salmonella typhimurium*. Immunology 1986; 58:225–30.
52. Nauciel C, Ronco E, Guenet J-L. Genetic control of *Salmonella typhimurium*–induced depression of delayed-type hypersensitivity to sheep erythrocytes in mice. Infect Immun 1988; 56:310–3.
53. Al-Ramadi BK, Brodkin MA, Mosser DM, Eisenstein TK. Immunosuppression induced by attenuated *Salmonella*. Evidence for mediation by macrophage precursors. J Immunol 1991; 146:2737–46.
54. Al-Ramadi BK, Meissler JJ Jr, Huang D, Eisenstein TK. Immunosuppression induced by nitric oxide and its inhibition by IL-4. Eur J Immunol, 1992; 2249–54.
55. Klimpel GR, Henney CS. BCG-induced suppressor cells. I. Demonstration of a macrophage-like suppressor cell that inhibits cytotoxic T cell generation *in vitro*. J Immunol 1978; 120:563–9.
56. Allen EM, Moore VL. Suppression of phytohemagglutinin and lipopolysaccharide responses in mouse spleen cells by Bacillus Calmette-Guerin. J Reticuloendothel Soc 1979; 26:349–56.
57. Turcotte R. Evidence for two distinct populations of suppressor cells in the spleens of *Mycobacterium bovis* BCG–sensitized mice. Infect Immun 1981; 34:315–22.
58. Kato K, Yamamoto K-I, Kakinuma M, Ishihara C. Suppression of BCG cell wall induced delayed-type hypersensitivity by BCG pre-treatment. I. Induction of adherent suppressor cells by live BCG injection and their characterization. Immunology 1981; 42:259–66.
59. Kato K, Yamamoto K-I. Involvement of prostaglandin E_1 in delayed-type hypersensitivity suppression induced with live *Mycobacterium bovis* BCG. Infect Immun 1982; 36:426–9.

60. Yagi H, Matsumoto M, Suzuki S, Misaki R, Suzuki R, Makino S, Harada M. Possible mechanism of the preventive effect of BCG against diabetes mellitus in NOD mouse. I. Generation of suppressor macrophages in spleen cells of BCG-vaccinated mice. Cell Immunol 1991; 138:130–41.
61. Yagi H, Matsumoto M, Kishimoto Y, Makino S, Harada M. Possible mechanism of the preventive effect of BCG against diabetes mellitus in NOD mouse. II. Suppression of pathogenesis by macrophage transfer from BCG-vaccinated mice. Cell Immunol 1991, 138:142–9.
62. Schwab JH. Suppression of the immune response by microorganisms. Bacteriological Rev 1975; 39:121–43.
63. Scott MT. Biological effects of the adjuvant. *Corynebacterium parvum*. II. Evidence for macrophage–T-cell interaction. Cell Immunol 1972: 5:469–79.
64. Wing EJ, Remington JS. Studies on the regulation of lymphocyte reactivity by normal and activated macrophages. Cell Immunol 1977; 30:108–21.
65. Kirchner H, Chused TM, Herberman RB, Holden HT, Lavrin, DH. Evidence of suppressor cell activity in spleens of mice bearing primary tumors induced by Maloney sarcoma virus. J Exp Med 1974; 139:1473–87.
66. Grimm W, Seitz M, Kirchner H, Gemsa D. Prostaglandin synthesis in spleen cell cultures of mice injected with *Corynebacterium parvum*. Cell Immunol 1978; 40:419–26.
67. Metzger A, Hoffeld JT, Oppenheim JJ. Macrophage-mediated suppression. I. Evidence for participation of both hydrogen peroxide and prostaglandins in suppression of murine lymphocyte proliferation. J Immunol 1980; 124:983–8.
68. Johnson AG, Jacobs A, Abrams G, Merritt K. Comparative changes in the mouse spleen during immuno-stimulation or immunosuppression. In: Cottier H, Odortchenko ND, Schindler R, Congden CC, eds. Germinal centers in immune response. New York: Springer-Verlag, 1967:214–20.
69. Franzl RE, McMaster PD. The primary immune response in mice. I. The enhancement and suppression of hemolysin production by a bacterial endotoxin. J Exp Med 1968; 127:1087–107.
70. Nelson DS. Production by stimulated macrophages of factors depressing lymphocyte transformation. Nature 1973; 246:306–7.
71. Ellner JJ, Spagnuolo PJ. Suppression of antigen and mitogen induced human T lymphocyte DNA synthesis by bacterial lipopolysaccharide: mediation by monocyte activation and production of prostaglandins. J Immunol 1979; 123:2689–94.
72. Dziarski R. Splenic macrophages: mediators of immunosuppressive activity of staphylococcal peptidoglycan. J Reticuloendothel Soc 1979; 26:239–47.
73. Petit JC, Richard G, Albert B, Daguet GL. Depression by *Pseudomonas aeruginosa* of two T-cell–mediated responses, anti-*Listeria* immunity and delayed-type hypersensitivity to sheep erythrocytes. Infect Immun 1982; 35:900–8.
74. Petit JC, Burghoffer B, Daguet GL. Suppression of cellular immunity to *Listeria monocytogenes* by activated macrophages: mediation by prostaglandins. Infect Immun 1985; 49:383–8.
75. Parkhouse RME, Dutton RW. Inhibition of spleen cell DNA synthesis by autologous macrophages. J Immunol 1966; 97:663–9.
76. Waldman SR, Gottlieb AA. Macrophage regulation of DNA synthesis in lymphoid cells: effects of a soluble factor from macrophages. Cell Immunol 1973; 9:142–56.
77. Keller R. Major changes in lymphocyte proliferation evoked by activated macrophages. Cell Immunol 1975; 17:542–51.
78. Berlinger NT, Lopez C, Good RA. Facilitation or attenuation of mixed leukocyte culture responsiveness by adherent cells. Nature 1976; 260:145–6.
79. Goodwin JS, Bankhurst AD, Messner RP. Suppression of human T-cell mitogenesis by prostaglandin, Existence of a prostaglandin-producing suppressor cell. J Exp Med 1977; 146:1719–34.
80. Mattingly JA, Kemp JD. Suppression of secondary plaque-forming cell responses by rat splenic adherent cells: evidence for dependence on prostaglandin production. Cell Immunol 1979; 48:195–200.
81. Webb DR, Nowowiejski I. Control of suppressor cell activation via endogenous prostaglandin synthesis: the role of T cells and macrophages. Cell Immunol 1981; 63:321–8.
82. Oehler JR, Herberman RB, Campbell DA Jr, Djeu JY. Inhibition of rat mixed lymphocyte cultures by suppressor macrophages. Cell Immunol 1977; 29:238–50.
83. Weiss A, Fitch FW. Macrophages suppress CTL generation in rat mixed leukocyte cultures. J Immunol 1977; 119:510–6.

84. Rollwagen FM, Stutman O. Culture-generated suppressor cells: evidence for an adherent cell component. Cell Immunol 1981; 64:371–80.
85. Veit BC. Immunoregulatory activity of culture-induced suppressor macrophages. Cell Immunol 1982; 72:14–27.
86. Ptak W, Gershon RK. Immunosuppression effected by macrophage surfaces. J Immunol 1975; 115:1346–50.
87. Lonai P, Feldman M. Studies on the effect of macrophages in an *in vitro* graft reaction system. Immunology 1971; 21:861–7.
88. Mbawuike IN, Herscowitz HB. Role of activation in alveolar macrophage–mediated suppression of the plaque-forming cell response. Infect Immun 1988; 56:577–81.
89. Kirchner H, Chused TM, Herberman RB, Holden HT, Lavrin DH. Evidence of suppressor cell activity in spleens of mice bearing primary tumors induced by Maloney sarcoma virus. J Exp Med 1974; 139:1473–87.
90. Kirchner H, Muchmore AV, Chused TM, Holden HT, Herberman RB. Inhibition of proliferation of lymphoma cells and T lymphocytes by suppressor cells from spleens of tumor-bearing mice. J Immunol 1975; 114:206–10.
91. Eggers AE, Wunderlich JR. Suppressor cells in tumor-bearing mice capable of nonspecific blocking of in vitro immunization against transplant antigens. J Immunol 1975; 114:1554–6.
92. Pelus LM, Bockman RS. Increased prostaglandin synthesis by macrophages from tumor-bearing mice. J Immunol 1979; 123:2118–25.
93. Glaser M. Indomethacin-sensitive suppressor cells regulate the cell-mediated cytotoxic response to SV 40–induced tumor-associated antigens in mice. Eur J Immunol 1990; 10:489–95.
94. Kolb J-P, Arrian S, Zolla-Pazner S. Suppression of the humoral immune response by plasmacytomas: mediation by adherent mononuclear cells. J Immunol 1977; 118:702–9.
95. Goodwin JS, Messner RP, Bankhurst AD, Peake GT, Saiki JH, Williams RC Jr. Prostaglandin-producing suppressor cells in Hodgkin's disease. N Engl J Med 1977; 297:963–8.
96. Alevy YG, Slavin RG. Immune response in experimentally induced uremia. II. Suppression of PHA response in uremia is mediated by an adherent, Ia-negative and indomethacin-insensitive suppressor cells. J Immunol 1981; 126:2007–10.
97. Alevy YG, Hutcheson P, Mueller KR, Slavin RG. Suppressor alveolar macrophages in experimentally induced uremia. J Reticuloendothel Soc 1983; 33:11–20.
98. Katz P, Fauci AS. Inhibition of polyclonal B-cell activation by suppressor monocytes in patients with sarcoidosis. Clin Exp Immunol 1978; 32:554–62.
99. Yoshimura S, Gotoh S, Kamada N. Immunological tolerance induced by liver grafting in the rat: splenic macrophages and T cells mediate distinct phases of immunosuppressive activity. Clin Exp Immunol 1991; 85:121–7.
100. Goto S, Kamada N, Kim YI, Kobayashi M. Induction of immunosuppressive activity by splenic macrophages in liver-grafted rats. Transpl Proc 1991; 23:2017–9.
101. Baird LG, Kaplan AM. Macrophage regulation of mitogen-induced blastogenesis. I. Demonstration of inhibitory cells in the spleens and peritoneal exudates of mice. Cell Immunol 1977; 28:22–35.
102. Baird LG, Kaplan AM. Macrophage regulation of mitogen-induced blastogenesis. II. Mechanism of inhibition. Cell Immunol 1977; 28:36–50.
103. Piguet P-F, Irle C, Vassali P. Immunosuppressor cells from newborn mouse spleen are macrophages differentiating *in vitro* from monoblastic precursors. Eur J Immunol 1981; 11:56–61.
104. Snyder DS, Lu CY, Unanue ER. Control of macrophage Ia expression in neonatal mice—role of a splenic suppressor cell. J Immunol 1982; 128:1458–65.
105. Oseroff A, Okada S, Strober S. Natural suppressor (NS) cells found in the spleen of neonatal mice and adult mice given total lymphoid irradiation (TLI) express the null surface phenotype. J Immunol 1984; 132:101–10.
106. Jadus MR, Peck AB. Naturally occurring, spleen-associated suppressor activity of the newborn mouse. Biochemical and functional identification of three monokines secreted by newborn suppressor-inducer monocytes. Scand J Immunol 1986; 23:35–44.
107. Jadus MR, Parkman R. The selective growth of murine newborn-derived suppressor cells and their probable mode of action. J Immunol 1986; 136:783–92.

108. Morrissey PJ, Ireland R. GM-CSF augments the immunosuppressive capacity of neonatal spleen cells *in vitro*. Cell Immunol 1991; 136:462–71.
109. Mortari F, Bains MA, Singhal SK. Immunoregulatory activity of human bone marrow. Identification of suppressor cells possessing OKM1, SSEA-1, and HNK-1 antigens. J Immunol 1986; 137:1133–7.
110. Noga SJ, Wagner JE, Horwitz LR, Donnenberg AD, Santos GW, Hess AD. Characterization of the natural suppressor cell population in adult rat bone marrow. J. Leukocyte Biol 1988; 43:279–87.
111. Corvese JS, Levy EM, Bennett M, Cooperband SR. Inhibition of an *in vitro* antibody response by a suppressor cell in normal bone marrow. Cell Immunol 1980; 49:293–306.
112. Maier T, Holda JH, Claman HN. Graft-vs-host reactions (GVHR) across minor murine histocompatibility barriers. II. Development of natural suppressor cell activity. J Immunol 1985; 135:1644–51.
113. Bennett JA, Rao VS, Mitchell M. Systemic Bacillus Calmette-Guérin (BCG) activates natural suppressor cells. Proc Natl Acad Sci USA 1978; 75:5142–4.
114. Kato K, Yamamoto K-I, Kimura T. Migration of natural suppressor cells from bone marrow to peritoneal cavity by live BCG. J Immunol 1985; 135:3661–8.
115. Shibata Y, Volkman A. The effect of bone marrow depletion on prostaglandin E–producing suppressor macrophages in mouse spleen. J Immunol 1985; 135:3897–904.
116. Shibata Y, Volkman A. The effect of hemopoietic microenvironment on splenic suppressor macrophages in congenitally anemic mice of genotype $S1/S1^{dl}$. J Immunol 1985; 135:3905–10.
117. Shibata Y, Dempsey WL, Morahan PS, Volkman A. Selectively eliminated blood monocytes and splenic suppressor macrophages in mice depleted of bone marrow by strontium 89. J Leukocyte Biol 1985; 38:659–69.
118. Bennett JA, Marsh JC. Relationship of *Bacillus Calmette-Guérin*–induced suppressor cells to hematopoietic precursor cells. Cancer Res 1980; 40:80–5.
119. Shibata Y, Shibuya E, Ishida N. Relationships between suppressor macrophages and macrophage precursors in the spleens from tumor-bearing mice. Cell Immunol 1984; 85:45–57.
120. Al-Ramadi BK, Chen Y-W, Meissler JJ Jr, Eisenstein TK. Immunosuppression induced by attenuated *Salmonella*. Reversal by IL-4. J Immunol 1991; 147:1954–61.
121. Cerrone MC, Kuhn RE. Macrophage regulation of immune responses of spleen cells from mice infected with *Trypanosoma cruzi*. Cell Immunol 1991; 138:423–36.
122. Hume DA, Weidemann MJ. Immunological functions of activated lymphocytes: interactions between leukocyte subpopulations during mitogenesis. In: Research Monographs in Immunology. Vol. 2. Amsterdam: Elsevier/North-Holland Biomedical Press, 1980:21–76.
123. Allison AC. Mechanisms by which activated macrophages inhibit lymphocyte responses. Immunological Rev 1978; 40:3–21.
124. Unanue ER. The regulatory role of macrophages in antigenic stimulation. Part two: symbiotic relationship between lymphocytes and macrophages. In: Kunkel HG, Dixon FJ, eds. Advances in Immunology. Vol. 31. New York: Academic Press, 1981:1–136.
125. Opitz HG, Niethammer D, Lemke H, Flad HD, Huget R. Inhibition of ^{3}H-thymidine incorporation of lymphocytes by a soluble factor from macrophages. Cell Immunol 1975; 16:379–88.
126. Opitz H-G, Niethammer D, Jackson RC, Lemke H, Huget R, Flad H-D. Biochemical characterization of a factor released by macrophages. Cell Immunol 1975; 18:70–5.
127. Kung JT, Brooks SB, Jakway JP, Leonard LL, Talmage DW. Suppression of in vitro cytotoxic response by macrophages due to induced arginase. J Exp Med 1977; 146:665–72.
128. Kurland JI, Bockman RS, Broxmeyer HE, Moore MAS. Limitation of excessive myelopoiesis by the intrinsic modulation of macrophage-derived prostaglandin E. Science 1978; 199:552–5.
129. Pelus LM, Broxmeyer HE, Kurland JI, Moore MAS. Regulation of macrophage and granulocyte proliferation. J Exp Med 1979; 150:277–92.
130. Boraschi D, Ghezzi P, Pasqualetto E, Salmona M, Nencioni L, Soldateschi D, Villa L, Tagliabue A. Interferon decreases production of hydrogen peroxide by macrophages: correlation with reduction of suppressive capacity and of antimicrobial activity. Immunology 1983; 50:359–68.
131. Aune TM, Pierce CW. Preparation of soluble immune response suppressor and macrophage-derived suppressor factor. J Immunolog Methods 1982; 53:1–14.
132. Greene WC, Fleisher TA, Waldmann TA. Soluble suppressor supernatants elaborated by concanavalin A–activated human mononuclear cells. I. Characterization of a soluble suppressor of T cell proliferation. J Immunol 1981; 126:1185–91.

133. Krakauer T. A macrophage-derived factor that inhibits the production and action of interleukin-2. J Leukocyte Biol 1985; 38:429–39.
134. Wilkins JA, Sigurdson L, Jordon Y, Rutherford WJ, Warrington RJ. Immunoregulatory factors from a human macrophage-like cell line. II. A human T-cell lymphokine-induced suppressor factor for lymphocyte proliferation. Cell Immunol 1983; 77:329–37.
135. Fujiwara H, Ellner JJ. Spontaneous production of a suppressor factor by the human macrophage-like cell line U937. I. Suppression of interleukin 1, interleukin 2, and mitogen-induced blastogenesis in mouse thymocytes. J Immunol 1986; 136:181–5.
136. Fujiwara H, Toossi Z, Ohnishi K, Edmonds K, Ellner JJ. Spontaneous production of a suppressor factor by a human macrophage-like cell line U937. II. Suppression of antigen- and mitogen-induced blastogenesis, IL 2 production and IL 2 receptor expression in T lymphocytes. J Immunol 1987; 138:197–203.
137. Mills CD. Molecular basis of "suppressor" macrophages. Arginine metabolism via the nitric oxide synthetase pathway. J Immunol 1991; 146:2719–23.
138. Hibbs JB Jr, Taintor RR, Vavrin Z. Macrophage cytotoxicity: role for L-arginine deiminase and imino nitrogen oxidation to nitrite. Science 1987; 235:473–5.
139. Nathan CF, Hibbs JB Jr. Role of nitric oxide synthesis in macrophage antimicrobial activity. Curr Opinion Immunol 1991; 3:65–70.
140. Albina JE, Abate JA, Henry WL Jr. Nitric oxide production is required for murine resident peritoneal macrophages to suppress mitogen-stimulated T cell proliferation. J Immunol 1991; 147:144–8.
141. Stuehr DJ, Marletta MA. Induction of nitrite/nitrate synthesis in murine macrophages by BCG infection, lymphokines, or interferon-γ. J Immunol 1987; 139:518–25.
142. Stuehr DJ, Marletta MA. Mammalian nitrate biosynthesis: mouse macrophages produce nitrite and nitrate in response to *Escherichia coli* lipopolysaccharide. Proc Natl Acad Sci USA 1985; 82:7738–42.
143. Al-Ramadi BK, Meissler JJ Jr, Huang D, Eisenstein TK. Immunosuppression induced by nitric oxide and its inhibition by IL-4. Eur J Immunol 1992; 2249–2254.
144. Eisenstein TK, Al-Ramadi BK, Huang D, Schafer R, Killar L, Lee J-C, Meissler JJ Jr. Immunity and immunosuppression induced by attenuated *Salmonella:* the role of nitric oxide. In: Cabello F, Hormaeche C, Mastroenni P, and Bonina L, eds. Proceedings of advanced NATO workshop on *Salmonella*. New York: Plenum Press, 1993.
145. Lehn M, Weiser WY, Engelhorn S, Gillis S, Remold HG. IL-4 inhibits H_2O_2 production and antileishmanial capacity of human cultured monocytes mediated by IFN-γ. J Immunol 1989; 143:3020–4.
146. Liew FY, Li Y, Severn A, Millott S, Schmidt J. Salter M, Moncada S. A possible novel pathway of regulation by murine T helper type-2 (T_h2) cells of a T_h1 cell activity via the modulation of the induction of nitric oxide synthase on macrophages. Eur J Immunol 1991; 21:2489–94.
147. Liew FY, Li Y, Moss D, Parkinson C, Rogers MV, Moncada S. Resistance to *Leishmania major* infection correlates with the induction of nitric oxide synthase in murine macrophages. Eur J Immunol 1991; 21:3009–14.
148. Green SJ, Meltzer MS, Hibbs JB Jr, Nacy CA. Activated macrophages destroy intracellular *Leishmania major* amastigotes by an L-arginine–dependent killing mechanism. J Immunol 1990; 144:278–83.
149. Adams LB, Hibbs JB Jr, Taintor RR, Krahenbuhl JL. Microbiostatic effect of murine-activated macrophages for *Toxoplasma gondii*. J Immunol 1990; 144:2725–9.
150. James SL, Glaven J. Macrophage cytotoxicity against schistosomula of *Schistosoma mansoni* involves arginine-dependent production of reactive nitrogen intermediates. J Immunol 1989; 143:4208–12.
151. Vincendeau P, Daulouède S. Macrophage cytostatic effect on *Trypanosoma musculi* involves an L-arginine–dependent mechanism. J Immunol 1991; 146:4338–43.
152. Denis M. Interferon-gamma-treated murine macrophages inhibit growth of tubercle bacilli via the generation of reactive nitrogen intermediates. Cell Immunol 1991; 132:150–7.
153. Flesch IEA, Kaufmann SHE. Mechanisms involved in mycobacterial growth inhibition by gamma interferon-activated bone marrow macrophages: role of reactive nitrogen intermediates. Infect Immun 1991; 59:3213–8.
154. Chan J, Xing Y, Magliozzo RS, Bloom BR. Killing of virulent *Mycobacterium tuberculosis* by reactive nitrogen intermediates produced by activated murine macrophages. J Exp Med 1992; 175:1111–22.
155. Saito S, Onozuka K, Shinomiya H, Nakano M. Sensitivity of bacteria to $NaNO_2$ and to L-arginine–dependent system in murine macrophages. Microbial Immunol 1991; 35:325–9.

156. Moncada S, Palmer RMJ, Higgs EA. Nitric EA. Nitric oxide: physiology, pathophysiology, and pharmacology. Pharmacolog Rev 1991; 43:109–42.
157. Hibbs JB Jr, Vavrin Z, Taintor RR. L-Arginine is required for expression of the activated macrophage effector mechanism causing selective metabolic inhibition in target cells. J Immunol 1987; 138:550–65.
158. Stuehr DJ, Nathan CF. Nitric oxide. A marophage product responsible for cytostasis and respiratory inhibition in tumor target cells. J Exp Med 1989; 169:1543–55.
159. Wink DA, Kasprzak KS, Maragos CM, Elespuru RK, Misra M, Dunams TM, Cebula TA, Koch WH, Andrews AW, Allen JS, Keefer LK. DNA deaminating ability and genotoxicity of nitric oxide and its progenitors. Science 1991; 254:1001–3.
160. Kwon NS, Stuehr DJ, Nathan CF. Inhibition of tumor cell ribonucleotide reductase by macrophage-derived nitric oxide. J Exp Med 1991; 174:761–7.
161. Boraschi D, Pasqualetto E, Ghezzi P, Salmona M, Bartalini M, Barbarulli G, Censini S, Soldateschi D, Tagliabue A. Dissociation between macrophage tumoricidal capacity and suppressive activity: analysis with macrophage-defective mouse strains. J Immunol 1983; 131:1707–13.
162. Boraschi E, Censini S, Tagliabue A. Interferon-γ reduces macrophage-suppressive activity by inhibiting prostaglandin E_2 release and inducing interleukin 1 production. J Immunol 1984; 133:764–8.
163. Taramelli D, Holden HT, Varesio L. *In vitro* induction of tumoricidal and suppressor macrophages by lymphokines: possible feedback regulation. J Immunol 1981; 126:2123–8.
164. Sacks DL, Selkirk M, Ogilvie BM, Askonas BA. Intrinsic immunosuppressive activity of different trypanosome strains varies with parasite virulence. Nature 1980; 283:476–8.
165. Tarleton RL, Kuhn RE. Measurement of parasite-specific immune responses *in vitro*: evidence for suppression of the antibody response to *Trypanosoma cruzi*. Eur J Immunol 1985; 15:845–50.
166. Suzuki Y, Joh K, Kobayashi A. Macrophage-mediated suppression of immune responses in *Toxoplasma*-infected mice. Cell Immunol 1987; 110:218–25.
167. Dempsey WL, Mansfield JM. Lymphocyte function in experimental African trypanosomiasis. VI. Parasite-specific immunosuppression. J Immunol 1983; 130:2896–8.

13

Macrophage–Natural Killer Cell Interactions in Resistance to Mycobacteria

Julie Y. Djeu, M. Beatriz Michelini-Norris, and D. Kay Blanchard

University of South Florida College of Medicine, and H. Lee Moffitt Cancer Center and Research Institute, Tampa, Florida

I. INTRODUCTION

The monocyte/macrophage forms a niche for a number of bacterial pathogens that, after initial ingestion by the phagocyte, can survive and multiply in the intracellular environment [1]. The infected macrophage can respond to the intracellular pathogens, but efficient killing of virulent strains requires macrophage activation by cytokines. T cells participate in resistance to intracellular bacteria by release of cytokines that are capable of activating the infected macrophage. Such cytokine production requires specific antigen recognition by T cells. Another key function of T cells is antigen-specific lysis of infected macrophages, either to eliminate the source of bacteria or to release bacteria from the host cell, which is insufficently equipped for intracellular killing, and to provide access to other phagocytes—e.g., neutrophils—for more efficient microbicidal action. However, nonspecific natural killer (NK) cells can also substitute for T cells in providing both cytokines and lytic function against infected macrophages. The chapter will focus on an important pathogen in AIDS patients, *Mycobacterium avium-intracellulare* complex (MAC), to demonstrate the interaction of NK cells and macrophages for elimination of intracellular bacteria. This interaction may constitute the first line of defense against MAC, producing rapid clearance from the host system without the need for antigen presentation and T cell activation. Data from our laboratory and others will be presented to support the hypothesis that resistance to opportunistic microbes such as MAC, which usually cause little morbidity and mortality in immunocompetent hosts, may rely heavily on the NK/macrophage axis, and that this axis may have as much weight as specific T cell immunity in the repertoire of normal host defense mechanisms.

II. ASSOCIATION OF VIRULENCE OF *MYCOBACTERIUM AVIUM-INTRACELLULARE* COMPLEX WITH FAILURE OF THE INFECTED HUMAN MONOCYTE TO RELEASE INTERLEUKIN-1 AND INTERLEUKIN-6

Nontuberculous MAC occurs in two morphological forms during in vitro culture. The transparent colonial variant has long been associated with in vivo virulence in animals and has recently been shown to be the primary form isolated from AIDS patients [2–5]. On the other hand, the opaque morphology is seen in laboratory strains and cultivated clinical isolates that are avirulent. The transparent isolates are associated with greater ability to multiply in normal human macrophages in vitro than the opaque variety. For careful evaluation of factors associated with virulence and morphological change in MAC, we have developed a pair of isogenic transparent and opaque colonial variants from the same clinical MAC isolated from a patient, in order to avoid strain variation and difference in strain-related virulence of MAC recovered from different sources.

Direct comparison of the infectivity and survival in human monocytes of transparent and opaque variants from the same MAC isolate has given us significant insight into the pathogenicity of MAC [6]. The transparent variant was found to be highly efficient in establishing infection and rapid proliferation within normal human monocytes in vitro, increasing almost tenfold by day 4 of infection. The opaque variant was capable of infecting monocytes but was unable to divide and was maintained at the initial infective dose throughout the 4 days of culture of the infected monocytes. Further analysis indicated that monocytes infected with the virulent, transparent variant had deficient secretion of interleukin (IL)-1α, IL-1β, and IL-6 as well as deficient expression of membrane-associated IL-1α when compared with monocytes from the same donor infected with the opaque morphotype. Reduced monokine induction by the transparent variant was not due to a delayed response or a suboptimal stimulator: response ratio, because significantly lower levels of IL-1α, IL-1β, and IL-6 were recovered with the transparent variant than with the opaque variant at all times and all ratios tested. This decreased induction of the protein levels of monokines was confirmed at the mRNA level by Northern blot analysis with specific cDNA probes for IL-1α, IL-1β, and IL-6. These monokines are key factors released by antigen-exposed monocytes/ macrophages and are involved in the proper initiation of an immune response against the antigen. Specific T cell responses to an antigen require not only triggering via the T cell receptor for antigen but also binding to accessory molecules such as IL-1 and IL-6 [7–9]. Thus failure of the monocyte to produce these factors when infected with virulent MAC could partly be responsible for their escape from the host immune system.

III. LYSIS OF INFECTED MONOCYTES BY INTERLEUKIN-2–ACTIVATED KILLER CELLS DERIVED FROM NATURAL KILLER CELLS

Monocytes harboring intracellular bacteria can come under the control of lymphocytes, in particular natural killer cells. NK cells were originally discovered to nonspecifically and spontaneously lyse tumor cells, and an extensive literature has accumulated to indicate that NK cells may play a key role in immunosurveillance against tumors and prevent tumor development and metastasis in normal hosts [10,11]. Subsequent studies have shown, however, that NK cells are capable of lysing virus-infected target cells but, unlike specific T cells, do so in a major histocompatibility complex (MHC)–unrestricted manner [12,13]. Thus NK cells, particularly when activated by interferon or IL-2, can lyse a variety of tumor cells and target cells infected with different viruses. How NK cells and IL-2–activated NK cells, which are also termed lymphokine-activated killer (LAK) cells [14], recognize such a wide range of target cells is not understood, and the receptor on NK cells for recognition of such diverse antigens has not been identified.

This body of work led us to examine whether monocytes infected with intracellular bacteria could also be recognized by NK/LAK cells. Our first attempt to explore this possibility was made with human monocytes infected with *Legionella pneumophila* [15]. Human NK cells, purified from peripheral blood by Ficoll/Hypaque gradient centrifugation followed by discontinuous Percoll gradient centrifugation, were found to have lytic activity against autologous monocytes infected with live *L. pneumophila*, and this activity could be greatly enhanced by incubation of NK cells with IL-2. Monocytes infected in vitro with *L. pneumophila* were washed to remove uningested bacteria and cultured 2 days at 37° C to allow replication of intracellular bacteria. At the same time, lymphocytes from the same donor were incubated for 2 days at 37° C with medium or IL-2. The infected monocytes were then labeled with ^{51}Cr and mixed with IL-2–activated autologous LAK cells at various effector/target ratios for 4 hr at 37° C. Assessment of radioactive Cr released into the supernatant of the effector/target cultures indicated the level of lysis of the radiolabeled infected target cells. By this 4-hour ^{51}Cr release assay, it was possible to show that NK cells could lyse infected monocytes and that incubation of NK cells with IL-2 up-regulated their lytic property. These effector cells lacked T cell–specific markers such as CD4 and CD8 and monocyte-specific antigen such as CD14. However, they did display CD16 and CD2, which are also seen on tumoricidal NK/LAK cells.

We next examined whether monocytes infected with other intracellular bacteria, i.e., MAC, could also be recognized by NK/LAK cells [16]. As expected, human NK/LAK cells could lyse MAC-infected monocytes. Of additional importance was the observation that fewer MAC could be recovered from infected monocytes after exposure to LAK cells in comparison with infected monocytes untreated with LAK cells. Thus LAK cells may not only lyse infected monocytes but also prevent intracellular bacterial growth. These observations were confirmed by Katz et al., who also recovered less MAC from LAK-treated MAC-infected monocytes [17]. In their hands, the NK killing of MAC-infected monocytes was reduced by addition of K562 tumor cells to the reaction mixture, suggesting that these targets shared common ligands recognized by NK cells. The mechanism of lysis of tumor cells and infected monocytes also appeared to be similar, in that killing of both targets required accessory signaling molecules, defined to be CD11a/CD18, on NK cells [16]. Antibodies specific for this heterodimer blocked effector lysis not only of tumor cells but also of MAC-infected monocytes.

Other in vitro studies have added to the importance of NK cells in control of intracellular bacteria, which include *Mycobacterium leprae*, *Shigella flexneri*, and typhus group rickettsiae [18–20]. Lysis of monocytes infected with these microbes has been seen with NK cells. One subtle difference that emerges from this group of studies is in the work by Steinhoff et al. with *M. leprae*, where IL-2–activated LAK cells were found capable of lysing only monocytes and Schwann cells that have ingested nonviable mycobacteria [18]. Target cells infected with viable *M. Leprae* were resistant to LAK lysis. Thus *M. leprae* may present a more complex problem, possibly with the ability to induce a resistant state in infected monocytes by induction of inhibitory factors. Candidates for such inhibition are interferon-α and prostaglandin E_2, which are monocyte products known to interfere with NK/LAK lysis [21,22]. The investigators did provide some clues to the mechanism of resistance and suggest that resistance is due to induction of heat shock proteins [18]. This conclusion was reached because heat shock, known to protect tumor cells from lysis [23], protected monocytes treated with dead *M. leprae* from LAK lysis.

IV. RELEASE OF MONOCYTE-ACTIVATING CYTOKINES FROM NK/LAK CELLS

Several recent reports have indicated that NK cells may have other functions that could also influence the survival of intracellular bacteria in monocytes. It has been found that IL-2–activated human NK cells can induce microbistatic and microbicidal action in MAC-infected monocytes [24]. Exposure of monocytes to IL-2–activated NK cells for 24 hours before infection with MAC caused a dramatic decrease of MAC survival, as compared with MAC-infected monocytes that had not been pretreated with activated NK cells. Similar protective effects were observed when monocytes were infected with MAC and incubated for 24 hours to establish the infection before activated NK cells were added to the culture. The supernatants of IL-2–activated NK cells could substitute for the cells themselves and induce microbistatic activity in MAC-infected monocytes. Further examination indicated that antibodies against tumor necrosis factor (TNF) could partially neutralize the protective effect of the activated NK supernatants [24]. Thus TNF may be an important factor induced in IL-2–activated NK cells that could activate monocytes for intracellular killing of MAC.

This work is similar to ours in that less MAC was recovered from infected human monocytes after exposure to LAK cells [16]. We have further defined, however, that other cytokines may also be induced from NK/LAK cells by the mycobacterium itself. Percoll gradient–purified NK cells, when incubated with MAC, were found to produce granulocyte-macrophage colony-stimulating factor (GM-CSF) [25]. This cytokine was detectable in the supernatants of MAC-stimulated NK cells within 2 days of culture and steadily increased, with peak levels of production seen even at day 7. It is important to emphasize that highly purified T cells, in the absence of monocytes, were unable to produce GM-CSF in response to *M. avium-intracellulare*, while NK cells required no accessory cell. However, the NK cells that were responsible for GM-CSF production appeared to be a subset to NK cells in being DR+, CD2+, and CD16+. In light of these findings, recombinant human GM-CSF was examined for its ability to modulate MAC survival in monocytes. When monocytes were infected with a virulent, transparent strain of MAC in the presence of GM-CSF, these monocytes were able to kill the intracellular microbe in a dose-dependent manner. As little as 1 U/ml of GM-CSF already had a positive effect on monocytes, with initial microbistatic activity seen at the low doses and microbicidal activity gaining ground by 100 U/ml. Thus both microbistatic and microbicidal function could be achieved with GM-CSF, depending on the dose. Similar observations have also been reported by others using recombinant GM-CSF to activate human monocytes against MAC [26]. Thus NK cells may play a key role in normal resistance to MAC by producing monocyte-activating factors that could protect monocytes and cause intracellular killing of the pathogen.

It should be noted that we have also observed GM-CSF induction from monocytes after stimulation with an avirulent laboratory strain of MAC [25]. It is possible that monocytes can produce autocrine activating factors that could activate internal microbicidal action, particularly against an avirulent pathogen. We have earlier shown that only the avirulent MAC variant and not the virulent variant could induce optimal levels of IL-1 and IL-6 from human monocytes [6]. Whether the virulent MAC variant also fails to induce GM-CSF from infected monocytes remains to be investigated. If this is the case, it might be one of the causes for preferential survival of the virulent variant in monocytes. Other cytokines that human DR+ NK cells released, when stimulated by MAC, were TNF and IL-6 [27,28]. Some data already points to the involvement of TNF in NK-mediated intracellular killing of MAC in monocytes [24]. Further experiments will indicate what role IL-6 plays in MAC resistance and whether monocytes can be activated by IL-6 for intracellular MAC killing.

Of further significance is the finding that GM-CSF can sensitize monocytes to LAK lysis [29]. Normal monocytes treated with GM-CSF became increasingly susceptible to LAK lysis, with peak susceptibility maintained up to 7 days in culture. Thus GM-CSF has dual actions, first to activate intracellular bacterial killing in monocytes and second to differentiate the infected monocytes to become LAK-sensitive. Such LAK/monocyte interactions via cytokines may operate to control opportunistic intracellular pathogens in the normal host.

V. ANIMAL MODELS FOR NATURAL KILLER CELL–MACROPHAGE INTERACTIONS FOR RESISTANCE TO MYCOBACTERIA

Murine infections with *M. lepraemurium* and MAC have provided strong in vivo evidence of the involvement of NK cells in resistance to these intracellular pathogens [30,31]. In the study with *M. lepraemurium*, it was shown that murine NK cells needed to be activated by IL-2 and could then function as in human NK cells, first in limiting intracellular growth of the mycobacteria in murine macrophages in vitro when low infectious doses were used, and second in directly lysing macrophages infected with high doses of mycobacteria [30]. It was of interest that macrophages infected with low levels of *M. lepraemurium* were not lysed by IL-2–activated NK cells; yet they could be induced to control the intracellular pathogen once they had been exposed to NK cells. These results suggest that macrophages infected with low infectious doses may not express sufficient surface ligands that are recognized by activated NK cells, but that high infectious doses may induce changes in macrophages, including up-regulation of NK-specific surface structures. However, IL-2–activated NK cells apparently could produce factors that caused intracellular killing in macrophages, as seen in the human system. The factors, however, were not related to TNF or interferon-γ, because the presence of antibodies against these two cytokines did not interfere with NK activation of intracellular killing in infected macrophages. On the basis of our human studies, the putative macrophage activating factor from IL-2–stimulated NK cells could be related to GM-CSF. However, other cytokines capable of activating macrophages may also be involved and may yet need to be identified.

In vivo, infusion of mice twice a week with anti-NK 1.1, which specifically reacts with murine NK cells [32], was found to cause reduced survival of mice infected with *M. lepraemurium* [30]. Administration of an irrelevant antibody had no effect on survival of infected mice. Evidence for in vivo relevance of intact NK cells in resistance to intracellular bacteria was also provided in mice infected with MAC [31]. In this study, mice were repeatedly treated with either anti-NK 1.1 [32] or anti-asialo-GM1 [33], either of which can specifically eliminate NK cells in vivo. Such treatment was first confirmed to have depleted NK cells by testing for the lack of NK function against tumor cells from the spleens of mice treated with anti-NK 1.1 or anti-asialo-GM1. In such NK-depleted mice, MAC appeared to survive better than in normal mice, as measured by colony counts recovered from isolated spleens after culture in Middlebrook 7H11 agar. The enhanced proliferation of MAC in spleens of NK-depleted mice was comparable to that seen in MAC-infected beige mice, which are genetically deficient in NK function against tumor cells. An interesting observation was the inability of in vivo administration of poly I:C, which is known to enhance tumoricidal function and interferon production from NK cells [34], to increase survival of the mycobacteria in the spleens of MAC-infected mice. This might be due to the earlier reported negative effect of interferon on MAC survival in human macrophages [35–37]. It has been a paradox that interferon-γ, which is capable of activating numerous macrophage functions including bactericidal and tumoricidal activity [36,40,41], has now been shown by several investigators to enhance intracellular survival of mycobacteria in macrophages in vitro. Why this should occur is as yet unknown.

VI. CONCLUSIONS

Macrophage activation by cytokines has generally been accepted as the mechanism of acquired resistance to intracellular bacteria. There is now increasing evidence that lysis of infected macrophages is also an important arm in the host defense against these pathogens. From the results gathered from various sources on mycobacteria, it is evident that NK cells are key players in normal resistance to bacterial pathogens that parasitize macrophages and proliferate within them. Natural killer cells, without activation, apparently have little to moderate activity against infected monocytes, but upon activation with IL-2 they can acquire high levels of cytotoxicity against macrophages infected with mycobacteria and other intracellular bacteria. In addition, NK cells can be activated by IL-2 to produce monocyte-activating factors that can induce intracellular killing. In addition to IL-2, NK cells can respond to mycobacteria themselves by production of monocyte-activating factors, which include GM-CSF, TNF, and IL-6. The dual role of NK cells in control of intracellular bacteria 1) by cytokine release to activate intracellular killing in infected monocytes and 2) by lysis of infected monocytes probably to release intracellular pathogens for easy targeting to more efficient phagocytes may represent an alternative host defense system, in case infected monocytes fail to produce autocrine activating factors or cytokines that could initiate specific T and B cell immunity. The increased incidence of mycobacterial infections in AIDS patients may thus be related to a deficiency of NK cells as well as of CD4+ helper cells.

ACKNOWLEDGMENTS

The work conducted in our laboratory was supported by USPHS Grant AI26560. We thank Ann Taylor, Susan McMillen and Carolyn Pearson for their technical support.

REFERENCES

1. Kaufmann SHE. Immunity against intracellular bacteria: biological effector functions and antigen specificity of T lymphocytes. Curr Top Microbiol Immunol 1988; 138:141–76.
2. Schaefer WB, Davis CL, Cohn ML. Pathogenicity of transparent, opaque and rough variants of *Mycobacterium avium* in chickens and mice. Am Rev Resp Dis 1970; 102:499–506.
3. Moehring JM, Solotorovsky MR. Relationship of colonial morphology to virulence for chicken of *Mycobacterium avium* and the nonphotochromogens. Am Rev Resp Dis 1965; 92:704–13.
4. Meylan PR, Richmann DD, Kornbluth RS. Characterization and growth of *Mycobacterium avium* complex strains isolated from the blood of patients with acquired immunodeficiency syndrome. Infect Immun 1990; 58:2564–8.
5. Crowle AJ, Tsang AY, Vatter AE, May MH. Comparison of 15 laboratory and patient-derived strains of *Mycobacterium avium* for ability to infect and multiply in cultured human macrophages. J Clin Microbiol 1986; 24:812–21.
6. Michelini-Norris MB, Blanchard DK, Pearson CA, Djeu JY. Differential release of interleukin (IL)-1α, IL-1β, and IL-6 from normal human monocytes stimulated with a virulent and an avirulent isogenic variant of *Mycobacterium avium intracellulare* complex. J Infect Dis 1992; 165:702–9.
7. Houssiau FA, Coulie PG, Olive D, van Snick J. Synergistic activation of human T cells by IL-1 and IL-6. Eur J Immunol 1988; 18:653–6.
8. Durum SK, Schmidt SA, Oppenheim JJ. Interleukin-1: an immunological perspective. Am Rev Immunol 1985; 3:263–87.
9. Akira S, Hirano T, Taga T, Kishimoto T. Biological of multifunctional cytokines: IL-6 and related molecules (IL-1 and TNF). FASEB J 1990; 4:2860–76.
10. Whiteside T, Herberman RB. The role of natural killer cells in human disease. Clin Immunol Immunopathol 1989; 53:1–23.
11. Trinchieri G. Biology of natural killer cells. Adv Immunol 1989; 47:187–376.
12. Welsh RM, Hallenbeck, LA. Effect of virus infections on target susceptibility of natural killer cell–mediated lysis. J Immunol 1980; 124:2491–8.
13. Ching C, Lopez C. Natural killing of herpes simplex virus type 1–infected target cells: normal human responses and influence of anti-viral antibody. Infect Immun 1980; 26:49–58.
14. Grimm EA, Mazumder A, Chang A, Rosenberg SA. Lymphokine-activated killer cell phenomenon. Lysis of natural killer-resistant fresh solid tumors cells by interleukin 2–activated autologous human peripheral blood lymphocytes. J Exp Med 1982; 155:1823–30.
15. Blanchard DK, Stewart WE, Klein TW, Friedman H, Djeu JY. Cytolytic activity of human peripheral blood leukocytes against *Legionella pneumophila*–infected monocytes: characterization of the effector cell and augmentation by interleukin-2. J Immunol 1987; 139:551–6.
16. Blanchard DK, Michelini-Norris MB, Friedman H, Djeu JY. Lysis of mycobacteria-infected monocytes by IL-2–activated killer cells: role of LFA-1. Cell Immunol 1989; 119:402–4011.
17. Katz P, Yeager H, Whalen G, Evans M, Swartz RP, Roecklein J. Natural killer cell–mediated lysis of *Mycobacterium-avium* complex–infected monocytes. J Clin Immunol 1990; 10:71–7.
18. Steinhoff U, Wand-Wurttenberger A, Bremerich A, Kaufmann SHE. *Mycobacterium leprae* renders Schwann cells and mononuclear phagocytes susceptible or resistant to killer cells. Infect Immun 1991; 59:684–8.
19. Klimpel GR, Nisel DW, Klimpel KD. Natural cytotoxic effector cell activity against *Shigella flexneri*–infected HeLa cells. J Immunol 1986; 136:1081–6.
20. Carl M, Dasch GA. Characterization of human cytotoxic lymphocytes directed against cells infected with typhus group rickettsiae: evidence for lymphokine activation of effectors. J Immunol 1986; 136:2654–61.

21. Trinchieri G, Granato D, Perussia B. Interferon-induced resistance of fibroblasts to cytolysis mediated by natural killer cells: specificity and mechanism. J Immunol 1981; 126:335–42.
22. Brunda MJ, Herberman RB, Holden HT. Inhibition of murine natural killer cell activity by prostaglandins. J Immunol 1980; 124:222682–8.
23. Jaattela M. Effects of heat shock on cytolysis mediated by NK cells, LAK cells, activated monocytes and TNFs α and β. Scand J Immunol 1990; 31:175–82.
24. Bermudez LEM, Young LS. Natural killer cell–dependent mycobacteriostatic and mycobactericidal activity in human macrophages. J Immunol 1991; 146:265–70.
25. Blanchard DK, Michelini-Norris MB, Pearson CA, McMillen S, Djeu JY. Production of granulocyte-macrophage colony stimulating factor (GM-CSF) by monocytes and large granular lymphocytes stimulated by *Mycobacterium avium-intracellulare*: activation of bactericidal activity by GM-CSF. Infect Immun 1991; 59:2396–2401.
26. Bermudez LEM, Young LS. Recombinant granulocyte-macrophage colony-stimulating factor activates human macrophages to inhibit growth or kill *Mycobacterium avium* complex. J Leukocyte Biol 1990; 48:67–73.
27. Michelini-Norris MB, Blanchard DK, Friedman H, Djeu JY. TNF induction by Mycobacterium avium-intracellulare: involvement of HLA-DR+ large granular lymphocytes and monocytes. J Leukocyte Biol 1991; 50:529–36.
28. Blanchard DK, Michelini-Norris MB, Pearson CA, Freitag CS, Djeu JY. *Mycobacterium avium-intracellulare* induces interleukin-6 from human monocytes and larger granular lymphocytes. Blood 1991; 77:2218–24.
29. Blanchard DK, Serbousek D, Djeu JY. Induction of monocyte susceptibility to lymphokine-activated killer cell lysis by granulocyte-macrophage colony stimulating factor. Cancer Res 1989; 49:5037–42.
30. Denis M. Activated murine natural killer cells control growth of *Mycobacterium lepraemurium* in mouse macrophages: in vitro and in vivo evidence. *Int J Immunopharm* 1991; 13:881–7.
31. Hashan KV, Gangadharam PRJ. In vivo depletion of natural killer cell activity leads to enhance multiplication of *Mycobacterium avium* complex in mice. Infect Immun 1991; 59:2818–21.
32. Seaman WE, Leisenger M, Erickson E, Koo GC. Depletion of natural killer cells in mice monoclonal antibody to NK1.1. J Immunol 1987; 138:4539–44.
33. Stein-Streilen J, Guffee J. In vivo treatment of mice and hamsters with antibodies to asialo-GM1 increases morbidity and mortality to pulmonary influenza infection. J Immunol 1986; 136:1435–41.
34. Djeu JY, Heinbaugh JA, Holden HT, Herman RB. Augmentation of mouse natural killer cell activity by interferon and interferon inducers. J Immunol 1979; 122:175–81.
35. Blanchard DK, Michelini-Norris MB, Djeu JY. Interferon decreases the growth inhibition of *Mycobacterium avium-intracellulare* complex by fresh human monocytes but not by culture-derived macrophages. J Infect Dis 1991; 164:152–7.
36. Douvas GS, Looker DL, Vatter AE, Crowle AJ. Gamma interferon activates human macrophages to become tumoricidal and leishmanicidal but enhances replication of macrophage-associated mycobacteria. Infect Immun 1985; 50:1–8.
37. Squires KE, Murphy WF, Madoff LC, Murray HW. Interferon gamma and *Mycobacterium avium-intracellulare* infection. J Infect Dis 1989; 159:599–605.
38. Bermudez LEM, Young LS. Tumor necrosis factor, alone or in combination with IL-2, but not IFNγ, is associated with macrophage killing of *Mycobacterium avium* complex. J Immunol 1989; 140:3006–12.
39. Mor N, Goren MB, Corwle AJ. Enhancement of growth of *Mycobacterium lepraemurium* in macrophages by gamma interferon. Infect Immun 1989; 57:2586–7.
40. Bhardwaj N, Nash TW, Horowitz MA. Interferon-gamma activated human monocytes to inhibit the intracellular multiplication of *Legionella pneumophila*. J Immunol 1986; 137:2662–9.
41. Murray HW, Rubin BY, Rothermal CD. Killing of intracellular *Leishmania donovani* by lymphokine-stimulated human mononuclear phagocytes. Evidence that interferon-γ is the activating lymphokine. J Clin Invest 1983; 72:1506–10.

14

Macrophage Resistance Genes: *Bcg/Ity/Lsh*

Bruce S. Zwilling and Mary E. Hilburger

The Ohio State University, Columbus, Ohio

I. INTRODUCTION

Genetic resistance to infection by microorganisms is one of several factors that play an important role in the ability of a host to resist microbial growth. The macrophage plays a critical role in the defense against many microbial pathogens and has been implicated in the genetic resistance of mice to several unrelated groups of intracellular pathogens. The ability of various inbred strains of mice to limit the growth of *Mycobacterium* species [1–5], *Salmonella typhimurium* [6,7], and *Leishmania donovani* [8–10] is controlled by a gene, or a set of closely linked genes, referred to as *Bcg/Ity/Lsh*.

II. RESISTANCE IN VIVO

Early studies in the mouse defined resistance as the ability to survive a challenge dose of virulent organisms ($LD_{50} > 10^5$) or to control the growth of these microorganisms in the liver [9–11] or spleen [1,3,12,13]. Susceptible animals were those that could not control the growth of the parasites or with an LD_{50} of less than 100 organisms. A clear division of inbred strains of mice into groups of resistant or susceptible animals resulted [1,4,6,8,9]. Resistance to each of the different microorganisms followed a similar pattern. Six of the mouse strains compared by Plant and Glynn [6] and Bradley and Kirkley [9] were correspondingly resistant to *S. typhimurium* and to *L. donovani*. Skamene et al. [4] determined that the distribution of Bcg^r and Bcg^s alleles among 14 inbred and 38 recombinant inbred strains exactly matched that for the resistance and susceptible alleles of *Lsh* and *Ity*. The possibility that the growth of these diverse microorganisms was controlled by the same gene gained further support when the locus controlling resistance to *L. donovani* [11], *S. typhimurium* [14], and later *Mycobacterium bovis* (strain BCG) [4,15] was mapped to chromosome 1 at a position between *Idh* 1 and *Pep* 3.

III. THE ROLE OF THE RETICULOENDOTHELIAL SYSTEM IN *BCG/ITY/LSH*-MEDIATED RESISTANCE

The three groups of microorganisms whose growth is regulated by *Bcg/Ity/Lsh* are intracellular parasites and grow within mononuclear phagocytes. The role of the macrophage in mediating resistance was therefore examined. Silica treatment selectively inactivated macrophages from resistance mice, resulting in a 100-fold decrease of the LD_{50} for *S. typhimurium* when the silica was given 1 day before infection [16]. Similarly, Gros et al. [17] found that silica abrogated resistance to the in vivo growth of BCG. Their results differed because multiple silica injections were required to abrogate resistance to BCG. These differences probably reflect differences in the growth rate of the microorganisms. Silica treatment did not alter the growth of BCG in susceptible mice. Further experiments by Crocker et al. [18] and by Gros et al. [17] showed that susceptible mice could become phenotypically resistant following transfer to bone marrow cells from resistant donors into irradiated susceptible recipients. The participation of T cells in mediating resistance was eliminated by showing that the growth rate of *S. typhimurium* and BCG was identical in congenitally athymic nude mice and in their euthymic nu/+ littermate controls [17,19]. Similarly, treatment of *Lsh* resistant mice with cyclophosphamide did not alter the resistance phenotype [20]. However, these experiments did not rule out the participation of natural killer (NK) cells or the role of interferon-γ (IFN-γ) or other cytokines in mediating resistance in vivo. Thus, experiments suggesting that NK cells do not participate in mediating resistance because BCG grew equally well in irradiated and nonirradiated recipients do not take into account the radioresistance of these cells [21].

IV. IN VIVO ANALYSIS OF *BCG/ITY/LSH*

The mechanism by which *Bcg/Ity/Lsh* controls the growth of three unrelated microorganisms remains unknown. An analysis of the fate of the microorganisms immediately following infection indicates that macrophages from resistant and susceptible mice take up equivalent numbers. Similarly, the majority of the microorganisms taken up by the macrophages do not survive. However, the organisms that survive initial infection apparently trigger the resistance mechanisms. The ability of the surviving microorganisms to trigger resistance appears to be related to the generation time of the microorganisms. Thus, only 24 hours are required to detect differences in the growth of *S. typhimurium* in the spleen and liver of infected mice; 2 to 3 days must elapse before the resistance is detectable following injection of *L. donovani*, and as long as 1 to 3 weeks are required to detect differences in the growth of BCG. Crocker et al. [22] have suggested that the cause of this difference in the delay of expression of *Bcg/Ity/Lsh* may be that the induction of the gene can occur only during a specific period in the growth cycle of *Leishmania*.

This possibility is reinforced by the observation that the *Ity*^{r} phenotype requires the growth of *S. typhimurium* [23]. Using an aroA mutation of *S. typhimurium*, which does not replicate in vivo, Benjamin et al. [23] found that bacterial recovery from infected mice was independent of the *Ity* phenotype. In fact, these authors provided very convincing evidence that the predominant effect of *Ity* may be to inhibit the growth of *Salmonella*, rather than superior ability of the macrophages to kill the bacteria. To accomplish this, *Salmonella* containing a temperature-sensitive plasmid, which did not replicate at 37°C, was used to infect mice. Thus, replication of the bacteria would result in a decrease in the frequency of plasmid-containing bacteria. When the bacteria were recovered from resistance mice, there was no decrease in the frequency of plasmid-containing *Salmonella*, indicating that the bacteria failed to replicate. In contrast, there was approximately 18 times more growth of *Salmonella* in the susceptible animals. Briles et al. [24] further found that prior activation of the macrophages by injection of *Corynebacterium parvum* did not result in increased killing of *Salmonella*. Similarly, the injection of recombinant IFN-γ (rIFN-γ) did not affect the replication rate of *S. typhimurium* during the first 2 days after infection [25].

Despite the observation that injection of the macrophage-activating cytokine did not affect the replication of *Salmonella*, Morrissey and Charrier found that the injection of granulocyta-macrophage colony-stimulating factor (GM-CSF) increased the survival of resistant mice but not of susceptible mice following infection with *S. typhimurium* [26]. Similar observations have been made following the injection of interleukin-1 (IL-1) before infection with *Salmonella* (P.J. Morrissey, personal communication). No significant differences in the number or composition of cells in the peritoneal cavity of control and GM-CSF–treated mice was noted. A recent report by Peterson et al. [27] has noted that a sharp increase in myeloid precursor cells occurs following the injection of *S. typhimurium* in susceptible mice but that this increase is transient, with no detectable progenitor cells present 7 days after injection of the bacteria. In contrast, myeloid progenitor cells increased at a much slower rate in resistant mice, and the increase was sustained over a much longer period. The relationship of these observations to resistance is not clear, especially since the expression of *Ity* appears to require mature macrophages (see below).

V. MACROPHAGE-MEDIATED GROWTH INHIBITION

The results of studies following the transfer of resistance with bone marrow cells from resistant to susceptible mice [5,18], together with observations that treatment of resistant animals with silica resulted in a susceptible phenotype [16,17], indicated that *Bcg/Ity/Lsh* was expressed at the level of the macrophage. Studies by Lissner et al. [28], by Stach et al. [29], and by Crocker et al. [30] reported that macrophages from resistant mice controlled the growth of *S. typhimurium*, BCG, or *L. donovani*, respectively, while macrophages from susceptible mice were permissive for the growth of the microorganisms. Resident peritoneal, splenic, and liver macrophages from resistant mice controlled the growth of *S. typhimurium* better than the macrophages from the susceptible mice [28,31,32]. The growth of BCG, *M. smegmatis*, and *M. avium* was shown to be controlled by resident and peptone-elicited inflammatory peritoneal macrophages or by splenic macrophages from resistant mice but not by macrophages from susceptible mice [29,33,34]. Results from our laboratory have been somewhat different. We have never observed a differential effect of peptone-elicited peritoneal macrophages from congenic BCG-resistant and -susceptible mice in the in vitro growth of BCG or of *M. avium* using the tritiated uracil incorporation assay. However, splenic adherent cells as well as lung macrophages differentially affect the growth of the mycobacteria. Thus, spleen and lung macrophages from resistant mice restrict the growth of the mycobacteria, while macrophages from susceptible mice do not (Figure 1). Thioglycolate-elicted macrophages from resistant or susceptible mice do not support the growth of mycobacteria.

Differences in the effect of the macrophage populations of the growth of *L. donovani* have also been observed. Thus, Crocker et al. [22,30] and Davies et al. [35] found that liver and lung macrophages as well as bone marrow-derived macrophages from resistant mice controlled the growth of the microorganism, while macrophages from susceptible mice were permissive. These findings were later confirmed by Roach et al. [36], who also found that peritoneal macrophages from resistant or susceptible mice were not different in their capacity to control the growth of *Leishmania*. Interestingly, splenic adherent cells and replicating bone marrow-derived macrophages also did not control the growth of the parasite.

The differences in the effect of the different macrophage populations may be related to the ability of the cells from the different anatomical locations to support the growth of the microorganisms. Thus, thioglycolate-elicited peritoneal macrophages do not support mycobacterial growth, and the growth of *L. donovani* appears to be limited to some resident macrophage populations. Neither resident peritoneal macrophages nor bone marrow macrophages maintained in CSF-1 support parasite replication. The observation that splenic adherent cells from resistant mice do not control the growth of *Leishmania* is in marked contrast to their effect on the growth of BCG and *Salmonella*. This difference has been attributed to the possibility that the splenic macrophage population is heterogenous and includes cells that will support the growth of *Leishmania* as well as those that will not support leishmanial growth [22]. Indeed, the growth of *L. major* appears not to be affected by *Lsh* primarily because it requires less mature macrophage populations for replication [35].

During the course of studies that evaluated the differential effect of macrophages from resistant mice on the growth of *Salmonella*, Lissner et al. reported that these macrophages also controlled the growth of *Escherichia coli, Corynebacterium diptheriae*, and *Staphylococcus aureus*, while macrophages from susceptible mice were more permissive for the growth of these organisms [31]. These observations have served as the basis, in part, for the interpretation that

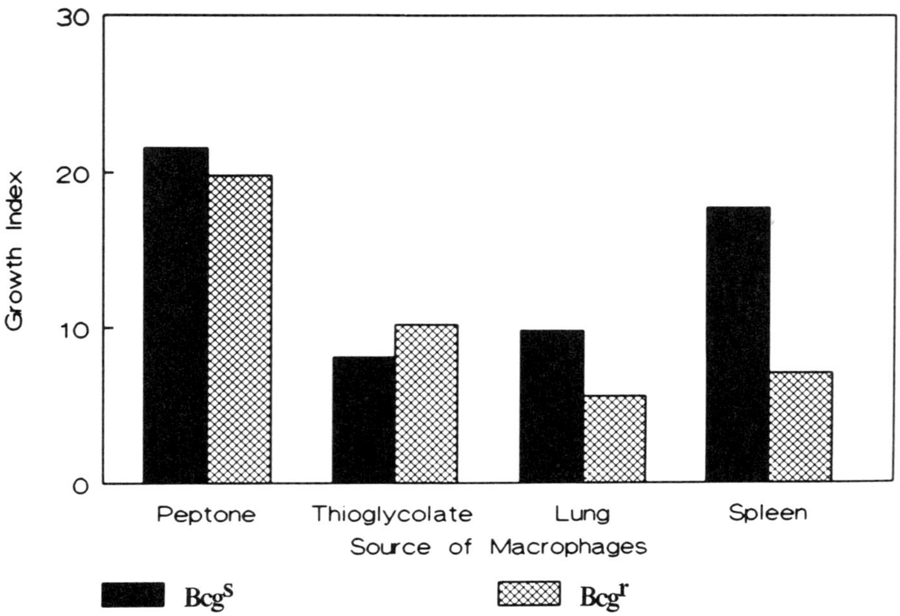

FIGURE 1 Mycobacterial growth in macrophages from BCG-resistant and -susceptible mice. Peptone-elicited, thioglycolate-elicited, lung, or splenic macrophage monolayers containing 1×10^5 cells were infected with 4×10^5 *M. avium*. The cultures were incubated overnight to allow for phagocytosis and then washed to remove any unphagocytized mycobacteria. In order to determine the viability of organisms taken up by the macrophages cultures, replicate cultures were lysed immediately with medium (Iscoves Modified Dulbeccos Medium) containing 5×10^{-5} M 2-mercaptoethanol and 0.1% saponin and pulsed with ^{3}H-uracil (Amersham, Sp. Act. 1 uCi/well) for 12 hours and harvested onto glass fiber filters using a MASH unit or were incubated with fresh medium for 5 days at 37°C before the addition of radiolabel. Other infected cultures were incubated for 5 days before lysis of the macrophages. These cultures were then pulsed with ^{3}H-uracil. The amount of radioactivity incorporated by the viable bacteria was determined by scintillation counting. The data are expressed as the growth index, which reflects the radioactivity incorporated by the mycobacteria released from macrophages after the 5 days of culture divided by that incorporated by the mycobacteria initially after phagocytosis.

macrophages from resistant mice may be more activated than are macrophages from susceptible mice. However, similar studies by Van Dissel et al. [37–39] do not lend support to this interpretation. These investigators showed that the growth of *E. coli, Listeria monocytogenes*, and *S. aureus* in macrophages from resistant mice and susceptible mice was not different, although the growth of *Salmonella typhimurium* was differentially affected.

VI. DIFFERENCES IN MAJOR HISTOCOMPATIBILITY COMPLEX CLASS II EXPRESSION

The first indication that macrophages from resistant and susceptible mice may differ functionally was a report by Johnson and Zwilling that major histocompatibility complex (MHC) class II was stably expressed by macrophages from BCG-resistant mice but was expressed transiently by macrophages from BCG-susceptible mice [40]. The difference in I-A expression was linked to *Bcg* by showing that I-A could be induced to be differentially expressed by macrophages from BALB/c.Bcg^r and BALB/c.Bcg^s congenic mice that are resistant or susceptible to mycobacterial growth [41]. An extensive characterization of the macrophages from the two congenic strains of mice indicated that the percentage of peritoneal macrophages that expressed I-A, 7 days after the intraperitoneal injection of BCG, did not differ, nor did the level of expression. Similarly, the percentage of cells that expressed I-A following in vitro treatment with rIFN-γ did not differ. Similar observations have been reported by Kaye et al. [42], who showed that splenic adherent cells from resistant and susceptible mice did not differ with regard to the percentage of cells expressing I-A. In contrast, Denis et al. found that a higher percentage of splenic adherent cells from resistant mice expressed I-A than of splenic adherent cells from susceptible mice [43]. The differences in these observations can be attributed to differences in the method used to prepare macrophages. Thus, Kaye et al. [42] used splenic adherent cells that were tested within 24 hours, while the study by Denis et al. [43] compared Ia expression after 48 hours of culture, a time when Ia expression has already begun to decline. Infection with Leishmania either in vivo or in vitro appeared to result in an enhanced expression of I-A by macrophages from resistant mice [42]. The enhanced expression of I-A cannot be attributed to different responses to IFN-γ, since macrophages from resistant and susceptible mice respond identically to stimulation by rIFN-γ; no differences the number and affinity of the receptors for rIFN-γ have been reported [44,45].

VII. ACCESSORY CELL ACTIVITY

The differences in the expression of I-A that have been reported by the different laboratories have correlated with reports of differences in antigen presentation. We have not found that elicited peritoneal macrophages differ in their ability to present native ovalbumin or the ovalbumin peptide 323-339 [46] to an I-A^d restricted T cell hybridoma (Figure 2). Similarly, resident peritoneal cells [47] as well as splenic adherent cells [42] from resistant and susceptible mice have been shown to present antigen equally well. In contrast, these workers [47] have also found that peritoneal macrophages elicited with bovine serum albumin from BCG-resistant mice, previously infected with BCG, presented antigen better then macrophages from susceptible mice. Splenic adherent cells from resistant mice infected with BCG [47] or *Leishmania* [42] were also found to present antigens better then splenic adherent cells from susceptible mice. The results of these experiments have been interpreted as reinforcing observations that the macrophage populations also differ in I-A expression. However, these must be interpreted with caution because only a very small percentage of macrophages within a population need to express MHC class II in order to effectively present antigen. Thus, a twofold difference in I-A expression may not be sufficient to detect differences in the cell populations that could account for the reported differences in accessory cell activity. This raises the possibility that differences in antigen processing or cytokine production may account for the differences in accessory cell activity reported by some laboratories. Perhaps a better indication of the differences in antigen-presenting capacity of macrophages from resistant or susceptible mice is to compare their ability to stimulate histoincompatible responding lymphocytes. Spleen cells from C57BL/6 mice responded equally to both Balb/c and congenic BALB/c.Bcg^r stimulator cells [43], indicating that differences other than the expression of I-A could account for the differences in accessory cell activity.

Interleukin-1 (IL-1) production by macrophages has been shown to be an important cofactor in the response of T cells to antigen, and differences in IL-1 production could account for the observed differences in accessory cell activity by macrophages from resistant and susceptible mice. However, Hoffenbach et al. did not find any correlation of *Bcg/Ity/Lsh* phenotype with IL-1 production following stimulation of *M. lepraemurium*–infected or –uninfected splenic adherent cells with lipopolysaccharide (LPS) [48]. Similarly, Bahr et al. found that stimulation of resident or peritoneal exudate macrophages from BALB/c (susceptible) mice with muramyl dipeptide resulted in a greater amount of membrane and supernatant IL-1 activity of macrophage cultures from BALB/c mice than by macrophages from DBA/2 (resistant) mice [49]. These findings were

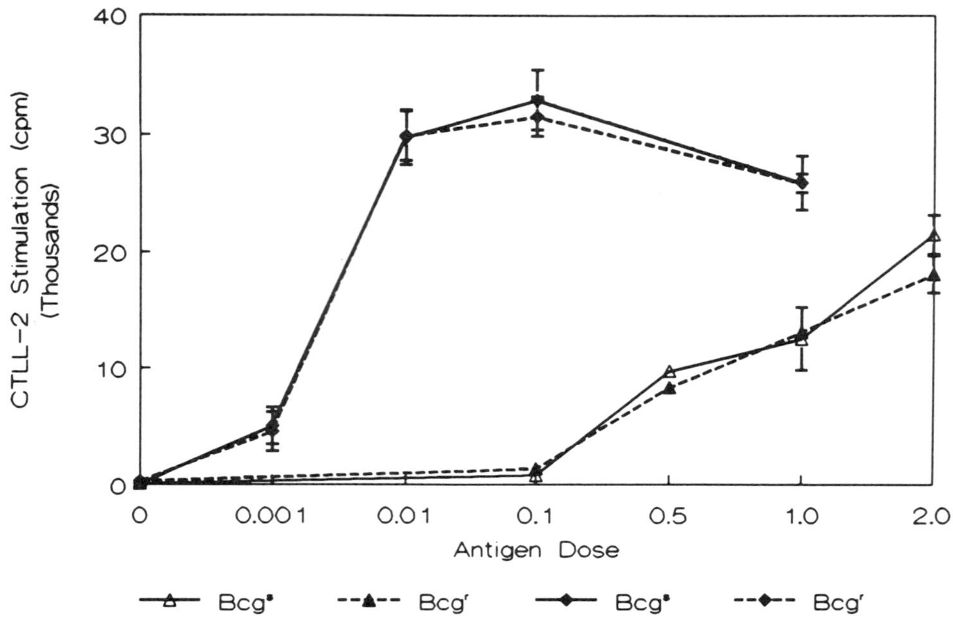

FIGURE 2 Antigen presentation by macrophages from BCG-resistant and -susceptible mice. Thioglycolate-elicited peritoneal macrophages (1×10^5/well) were pulsed with native ovalbumin or with I-A^d–restricted ovalbumin peptide 323-339 [46]. The ability of the cell to present antigen to the I-A^d–restricted T cell hybridoma D0.11-10/S4.4 was evaluated by assessing the ability of the supernatant fluids to stimulate the replication of the IL-2–dependent CTLL-2. The data are expressed as the counts per minute of ^{3}H-thymidine incorporated by the CTLL-2 cell line. Cultures stimulated with native OVA are identified by triangles, those stimulated with OVA 323-339 by diamonds.

supported by those of Denis et al., who showed that culture supernatants of resistant and susceptible mice contained equal quantities of IL-1 [43].

VIII. HYDROGEN PEROXIDE AND REACTIVE OXYGEN PRODUCTION

Considerable effort has been made to determine whether macrophages from resistant and susceptible mice differ with regard to the antimicrobial biochemical pathways that can account for the differences in their ability to differentially limit the growth of *Mycobacterium* species, *S. typhimurium*, *L. donovani*. The antimicrobial pathways that have received the most attention have included the production of hydrogen peroxide and reactive oxygen intermediates and, more recently, the production of reactive nitrogen intermediates. Splenic adherent cells from resistant or susceptible mice produced equal quanities of H_2O_2 following stimulation with phorbol myristate acetate (PMA) or with lymphokine [50]. However, following ingestion of mycobacteria, the macrophages from the resistant mice produced greater quantities of H_2O_2 than did the macrophages from the susceptible mice. Splenic macrophages from resistant mice fed BCG in vitro also produced more O_2^- than did the macrophages from the susceptible mice. These observations are in agreement with those of Blackwell et al., who showed that macrophages from resistant mice produced more O_2^- following activation with PMA or IFN-γ than did macrophages from susceptible mice [51]. Others, however, have not been able to correlate differences in the production of either H_2O_2 or O_2^- with the resistant or susceptible phenotype [25,52,53]. It should be pointed out, however, that resistance to the growth of *Mycobacterium* or *Leishmania* does not appear to be mediated by H_2O_2

or O_2^-, since neither the addition of catalase nor that of superoxide dismutase affects the outcome of the interaction of the macrophages with the microorganism. Additionally, those studies that have not correlated increased H_2O_2 or O_2^- production with the resistant phenotype have been done not with macrophages from congenic mice but with cells from inbred strains of mice that are resistant or susceptible to the in vivo growth of the three groups of organisms [25,52,53]. One study did compare the ability of the BALB/c congenics infected with *M. avium* to produce H_2O_2 following stimulation with PMA and found that they did not differ [54].

IX. TUMOR NECROSIS FACTOR AND REACTIVE NITROGEN INTERMEDIATES

Recent evidence has implicated an L-arginine–dependent nitrogen oxide (NO) synthase pathway in the destruction of a variety of intracellular pathogens, including *Mycobacterium* and *Leishmania* [36,45,55–57]. The pathway appears to be dependent on IFN-γ and on tumor necrosis factor-α (TNF-α) and can be triggered by priming macrophages with IFN-γ followed by LPS or by the infecting microorganisms. Thus, it is possible that the differences in the ability of macrophages from resistant or susceptible mice to control the growth of the microorganisms may be the result of differences in TNF-α production or in the production of reactive nitrogen intermediates.

However, there does not appear to be a good correlation between the resistant phenotype and an increased production of TNF-α or NO. Several reports have appeared that show no apparent differences between the level of TNF-α released by macrophages from resistant or susceptible mice following stimulation of *M. leparemurium*–infected macrophages with LPS [58], or stimulation with lipoarabinomannan [59] or *Leishmania* [60,45]. In contrast, Blackwell et al. [61] reported that *Lsh*^{r} macrophages produce more TNF-α than do *Lsh*^{s} macrophages and that infected macrophages from resistant mice produced more TNF-α when stimulated with IFN-γ and LPS than did macrophages from susceptible mice.

The results of experiments that compared the ability of the macrophages from resistant and susceptible mice to produce reactive nitrogen intermediates have been inconclusive. Thus, Appelberg and Sarmento found that macrophages from BALB/c.Bcg^r and BALB/c.Bcg^s mice produced similar amounts of NO_2 [54]. In contrast, Liew et al. reported that macrophages from resistant mice, despite producing levels of TNF-α comparable to those produced by macrophages from susceptible mice, produced greater levels of reactive nitrogen intermediates following activation with rIFN-γ and *Leishmania* than did macrophages from susceptible mice [45]. When NO production by bone marrow and peritoneal macrophages was compared, Roach et al. [36] found that bone marrow macrophages from resistant mice produced more NO than did comparable cells from susceptible mice. In contrast, peritoneal macrophages from susceptible mice produced greater quantities of NO than did peritoneal macrophages from resistant mice. Studies in our laboratory indicate that peritoneal macrophages from BALB/c.Bcg^r and BALB/c.Bcg^s mice produce equal quantities of reactive nitrogen intermediates when primed with rIFN-γ and stimulated with either LPS or *M. avium* (Figure 3).

X. HUMAN-MOUSE HOMOLOGIES

The gene that controls resistance to the growth of *Mycobacterium* species, *S. typhimurium*, and *L. donovani* maps to chromosome 1 in mice [4,11,14,15]. A syntenic group of genes also maps to chromosome 2q in humans [62]. Recent epidemiologic evidence indicates that groups of people differ in their susceptibility to infection with *Mycobacterium tuberculosis* [63]. Stead and colleagues [63] found that among comparably housed residents of nursing homes, blacks were twice as likely as whites to become tuberculin-positive. Similar relationships have been found in three outbreaks of tuberculosis in two prisons [63]. According to Stead, the differences in the sus-

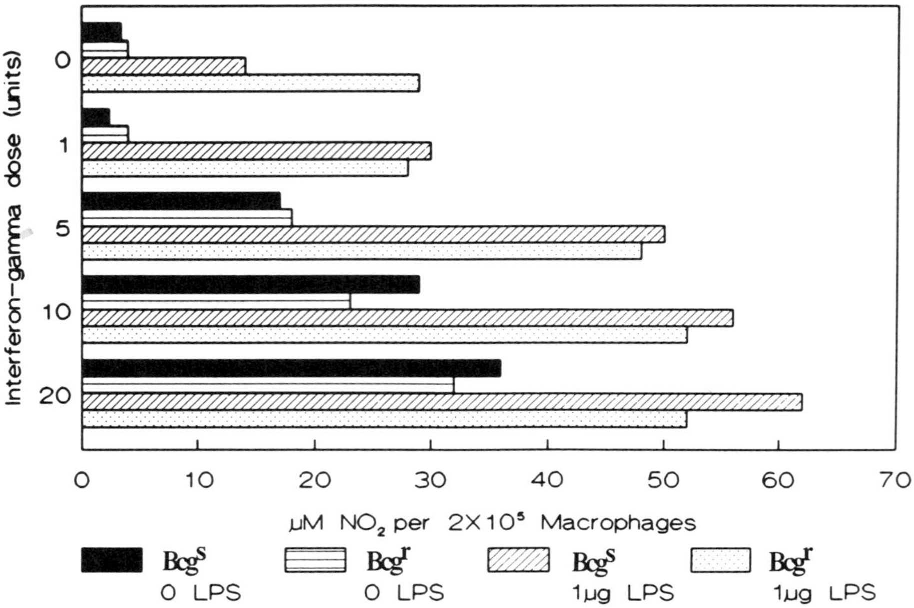

FIGURE 3 Reactive nitric oxide production by macrophages from BCG-resistant and -susceptible mice. Cultures containing 2×10^5 peptone-elicited macrophages were stimulated with recombinant interferon-γ and lipopolysaccharide (LPS) as indicated for 72 hours. The amount of NO_2^- present in the supernatant was determined by the Griess reaction [57]. The production of NO_2^- was suppressed by the addition of N^G monomethyl L-arginine to the cultures.

ceptibility of people probably arose because of the differences in the historic exposure of the populations to *M. tuberculosis* [64]. Thus, individuals living in densely populated areas were exposed to tuberculosis, and so developed resistance, while those living in rural communities were not. The disease was virtually unknown in sub-Saharan Africa until the middle of the 19th century [64].

Recent evidence also indicates that macrophages from susceptible and resistant populations also differ in their permissiveness to mycobacterial growth. Crowle and Elkins have shown that macrophages from white donors are less permissive for the growth of *M. tuberculosis* than are macrophages from black donors [65]. Similarly, work in our laboratory has shown that the expression of HLA-DR by monocytes from a majority of white donors and a majority of black donors is different [66]. The differences mimic those that we have defined using macrophages from resistant and susceptible mice [67]. Thus, in mice, the persistence of I-A expression by macrophages from resistant mice is not dependent on the synthesis of new protein and is therefore not affected by the addition of cycloheximide (CHX). In contrast, I-A expression by macrophages from susceptible mice is transient, and the continued expression of the MHC class II glycoproteins is dependent on its continued synthesis. Thus, the addition of CHX results in a loss of surface I-A. The expression of HLA-DR by human monocytes is constitutive. In vitro culture of the cells results in a large spontaneous increase in surface HLA-DR [68,69]. The addition of CHX to cultures of monocytes from black donors prevents the increase in HLA-DR expression. In contrast, the addition of CHX to cultures of monocytes from a majority of white donors does not affect the increase in HLA-DR expression. This differential sensitivity of MHC class II expression is analogous to the differences that we have defined in the mouse that distinguish macrophages from resistant and susceptible mice. In humans, the inhibition of the increase in HLA-DR expression by

CHX occurs in 70% of the historically susceptible population, while HLA-DR expression is not affected by CHX in 70% of the resistant population. These observations are consistent with the finding that resistance is about twice as common in whites as in blacks [63] and closely approximates the 62% penetrance of a major recessive locus for susceptibility to leprosy [70]. Interestingly, recent studies by Newman et al. [71] have found that the survival of monocytes following infection by *M. avium* is dependent on their ability to produce TNF-α and IL-6. Monocytes from approximately 30% of the blood donors did not survive *M. avium* infection and were deficient in their ability to produce the cytokines. Monocytes from the 70% of the donors that survived *M. avium* infection produced increased amounts of IL-6 and TNF-α.

XI. CONCLUSIONS

We have reviewed work concerning resistance to three unrelated microorganisms that is controlled by a gene alternatively referred to as *Bcg/Ity/Lsh*. Resistance to the growth of these microorganisms can be reliably demonstrated in vivo. We believe that the majority of evidence indicates that resistance is mediated by the macrophage, and this resistance appears to be expressed primarily by mature tissue macrophages. Thus, some of the inconsistencies we have noted are due to the use of different macrophage populations. Clearly more work needs to be done to understand how macrophages can control the growth of the microorganisms. Resistance appears to be an inducible phenomenon that requires IFN-γ and TNF-α. No work has been done in vivo to implicate any of the cytokines as important mediators of resistance. Finally, resistance appears to have some specificity in that it appears to be induced following exposure of the macrophage, in vivo or in vitro, to the microorganism or components of the microorganisms.

How can one gene control resistance to the growth of three groups of unrelated microorganisms? Clearly, they must have something in common, yet differ from other intracellular parasites. Mycobacteria and *Leishmania donovani* are susceptible to reactive nitrogen–mediated growth inhibition [36,45,55–57]. Other microorganisms are also susceptible to reactive nitrogen intermediates [72–77], yet their growth is not controlled by *Bcg/Ity/Lsh*. The uptake of several intracellular parasites is mediated via C3bi receptor [78–80], yet the growth of only some is controlled by *Bcg/Ity/Lsh*. The intracellular compartment of the microorganism, following its uptake, may ultimately determine its fate. While conventional wisdom holds that the group of microorganisms that are susceptible to *Bcg/Ity/Lsh*-mediated growth regulation are found in phagosomes or phagolysosomes, other intracellular compartments may also play an important role in controlling the growth of the microorganisms. The linkage of *Bcg/Ity/Lsh* with a family of cytoskeletal proteins [81], together with the recent demonstration that the intracellular compartments may be connected by a series of tubules [82,83], lends support to the possibility that internalized microorganisms are chaperoned to compartments that are unfavorable for growth (resistance) or favor replication (susceptibility).

REFERENCES

1. Forget A, Skamene E, Gros P, Miailhe AC, Turcotte R. Differences in response among inbred mouse strains to infection with small doses of *Mycobacterium bovis* BCG. Infect Immun 1981; 32:42–7.
2. Brown IN, Glynn AA, Plant J. Inbred mouse strain resistance to *Mycobacterium lepraemurium* follows the *Ity/Lsh* pattern. Immunology 1982; 47:149–56.
3. Gros P, Skamene E, Forget A. Genetic control of natural resistance to *Mycobacterium bovis* (BCG) in mice. J Immunol 1981; 127:2417–21.
4. Skamene E, Gros P, Forget A, Kongshavn PAL, St Charles C, Taylor BA. Genetic regulation of resistance to intracellular pathogens. Nature 1982; 297:506–9.
5. Orme IM, Stokes RW, Collins FM. Genetic control of natural resistance to nontuberculous mycobacterial infections in mice. Infect Immun 1986; 54:56–62.

6. Plant J, Glynn AA. Genetics of resistance to infection with *Salmonella typhimurium* in mice. J Infect Dis 1976; 133:72–8.
7. Swanson RN, O'Brien AD. Genetic control of the innate resistance of mice to *Salmonella typhimurium: Ity* gene is expressed *in vivo* by 24 hours after infection. J Immunol 1983; 131:3014–20.
8. Bradley DJ, Kirkley J. Variation in susceptibility of mouse strains to *Leishmania donovani* infection. Trans Roy Soc Trop Med Hyg 1972; 66:527–8.
9. Bradley DJ, Kirkley J. Regulation of *Leishmania* populations within the host. I. The variable course of *Leishmania donovani* infections in mice. Clin Exp Immunol 1977; 30:119–29.
10. Bradley DJ. Regulation of *Leishmania* populations within the host. II. Genetic control of acute susceptibility of mice to *Leishmania donovani* infection. Clin Exp Immunol 1977; 30:130–40.
11. Bradley DJ, Taylor BA, Blackwell J, Evans EP, Freeman J. Regulation of *Leishmania* populations within the host. III. Mapping the locus controlling susceptibility to visceral leishmaniasis in the mouse. Clin Exp Immunol 1979; 37:7–14.
12. O'Brien AD, Rosentreich DL, Taylor BA. Control of natural resistance to *Salmonella typhimurium* and *Leishmania donovani* in mice by closely linked but distinct genetic loci. Nature 1980; 287:440–2.
13. Taylor BA, O'Brien AD. Position on mouse chromosome 1 of a gene that controls resistance to *Salmonella typhimurium*. Infect Immun 1982; 36:1257–60.
14. Plant J, Glynn AA. Locating salmonella resitance gene on mouse chromosome 1. Clin Exp Immunol 1979; 37:1–6.
15. Schurr E, Skamene E, Nesbitt M, Hynes R, Gros P. Identification of a linkage group including the *Bcg* gene by restriction length polymorphism analysis. Curr Top Microbiol Immunol 1988; 137:310–5.
16. O'Brien AD, Scher I, Formal SB. Effect of silica on the innate resistance of inbred mice to *Salmonella typhimurium* infection. Infect Immun 1979; 25:513–20.
17. Gros P, Skamene E, Forget A. Cellular mechanisms of genetically controlled host resistance to *Mycobacterium bovis* (BCG). J Immunol 1983; 131:1966–72.
18. Crocker PR, Blackwell JM, Bradley DJ. Transfer of innate resistance and susceptibility to *Leishmania donovani* infection in mouse radiation bone marrow chimaeras. Immunology 1984; 52:417–22.
19. O'Brien AD, Metcalf ES. Control of early *Salmonella typhimurium* growth in innately *Salmonella*-resistant mice does not require functional T lymphocytes. J Immunol 1982; 129:1349–51.
20. Ulczak OM, Blackwell JM. Immunoregulation of genetically controlled acquired responses to *Leishmania donovani* infection in mice: the effects of parasite dose, cyclophosphamide and sublethal irradiation. Parasite Immunol 1983; 5:449–463.
21. Zwilling B. Comments on the paper by E. Buschman *et al.* Res Immunol 1989; 140:818.
22. Crocker PR, Davies EV, Blackwell JM. Variable expression of the murine natural resistance gene *Lsh* in different macrophage populations infected *in vitro* with *Leishmania donovani*. Parasite Immunol 1987; 9:705–19.
23. Benjamin WH, Hall P, Roberts SJ, Briles DE. The primary effect of the *Ity* locus is on the rate of growth of *Salmonella typhimurium* that are relatively protected from killing. J Immunol 1990; 144:3143–51.
24. Briles DE, Benjamin W, Posey B, Michalek SM, McGhee JR. Independence of macrophage activation and expression of the *Ity* (immunity to *typhimurium*) locus. Microbial Pathogenesis 1986; 1:33–41.
25. Van Dissel JT, Stikkelbroeck JJM, Michel BC, Van den Barselaar MTH, Leijh PCJ, Van Furth R. Inability of recombinant interferon-γ to activate the antibacterial activity of mouse peritoneal macrophages against *Listeria monocytogenes* and *Salmonella typhimurium*. J Immunol 1987; 139:1673–8.
26. Morrissey PJ, Charrier K. GM-CSF administration augments the survival of *Ity*-resistant A/J mice, but not *Ity*-susceptible C57/BL6 mice, to a lethal challenge with *Salmonella typhimurium*. J Immunol 1990; 144:557–61.
27. Peterson VM, Madonna GS, Vogel SN. Differential myelopoietic responsiveness of Balb/c (Ity^s) and C.D2 (Ity^r) mice to lipopolysaccharide administration and *Salmonella typhimurium* infection. Infect Immun 1992; 60:1375–84.
28. Lissner CR, Swanson RN, O'Brien AD. Genetic control of innate resistance of mice to *Salmonella typhimurium:* expression of the *Ity* gene in peritoneal and splenic macrophages isolated *in vitro*. J Immunol 1983; 131:3006–13.
29. Stach JL, Gros P, Forget A, Skamene E. Phenotypic expression of genetically controlled natural resistance to *Mycobacterium bovis* (BCG). J Immunol 1984; 132:888–92.

30. Crocker PR, Blackwell JM, Bradley DJ. Expression of the natural resistance gene *Lsh* in resident liver macrophages. Infect Immun 1984; 43:1033–40.
31. Lissner CR, Weinstein DL, O'Brien AD. Mouse chromosome 1 *Ity* locus regulates microbicidal activity of isolated peritoneal macrophages against a diverse group of intracellular and extracellular bacteria. J Immunol 1985; 135:544–7.
32. Harrington KA, Hormaeche CE. Expression of the innate resistance gene *Ity* in mouse Kupffer cells infected with *Salmonella typhimurium in vitro*. Microbial Pathogenesis 1986; 1:269–74.
33. Denis M, Forget A, Pelletier M, Gervais F, Skamene E. Killing of *Mycobacterium smegmatis* by macrophages from genetically susceptible and resistant mice. J Leukocyte Biol 1990; 47:25–30.
34. Stokes RW, Orme IM, Collins FM. Role of mononuclear phagocytes in expression of resistance and susceptibility to *Mycobacterium avium* infections in mice. Infect Immun 1986; 54:811–9.
35. Davies EV, Singleton AMT, Blackwell JM. Differences in *Lsh* gene control over systemic *Leishmania major* and *Leishmania donovani* or *Leishmania mexicana mexicana* infections are caused by differential targeting to infiltrating and resident liver macrophage populations. Infect Immun 1988; 56:1128–34.
36. Roach TIA, Kiderlen AF, Blackwell JM. Role of inorganic nitrogen oxides and tumor necrosis factor alpha in killing *Leishmania donovani* amastigotes in gamma interferon-lipopolysaccharide-activated macrophages from Lsh^s and Lsh^r congenic mouse strains. Infect Immun 1991; 59:3935–44.
37. Van Dissel JT, Leijh PCJ, Van Furth R. Differences in initial rate of intracellular killing of *Salmonella typhimurium* by resident peritoneal macrophages from various mouse strains. J Immunol 1985; 134:3404–10.
38. Van Dissel JT, Stikkelbroeck JJM, Michel BC, Leijh PCJ, Van Furth R. *Salmonella typhimurium*–specific difference in rate of intracellular killing by resident peritoneal macrophages from salmonella-resistant CBA and salmonella-susceptible C57/BL10 mice. J Immunol 1987; 138:4428–34.
39. Van Dissel JT, Stikkelbroeck JJM, Van den Barselaar MTH, Sluiter W, Leijh PCJ, Van Furth R. Divergent changes in antimicrobial activity after immunologic activation of mouse peritoneal macrophages. J Immunol 1987; 139:1665–72.
40. Johnson SC, Zwilling BS. Continuous expression of I-A antigen by peritoneal macrophages from mice resistant to *Mycobacterium bovis* (BCG). J Leukocyte Biol 1985; 38:635–45.
41. Zwilling BS, Vespa L, Massie M. Regulation of I-A expression by murine peritoneal macrophages: differences linked to the *Bcg* gene. J Immunol 1987; 138:1372–6.
42. Kaye PM, Patel NK, Blackwell JM. Acquisition of cell-mediated immunity to *Leishmania* II. *Lsh* gene regulation of accessory cell function. Immunology 1988; 65:17–22.
43. Denis M, Buschman E, Forget A, Pelletier M, Skamene E. Pleiotropic effects of the *Bcg* gene II. Genetic restriction of responses of mitogens and allogeneic targets. J Immunol 1988; 141:3988–93.
44. Kwasniewski MN, Zwilling BS. Induction of continuous I-A expression by murine peritoneal macrophages: lack of correlation with T cell responses of Bcg^r and Bcg^s mice. J Leukocyte Biol 1988; 44:192–7.
45. Liew FY, Li Y, Moss D, Parkinson C, Rogers MV, Moncada S. Resistance to *Leishmania major* infection correlates with the induction of nitric oxide synthase in murine macrophages. Eur J Immunol 1991; 21:3009–14.
46. Shimonkevitz R, Colon S, Kappler JW, Marrack P, Grey HM. Antigen recognition by H-2-restricted T cells II. A tryptic ovalbumin peptide that substitutes for processed antigen. J Immunol 1984; 133:2067–74.
47. Denis M, Forget A, Pelletier M, Skamene E. Pleitropic effects of the *Bcg* gene I. Antigen presentation in genetically susceptible and resistant congenic mouse strains. J Immunol 1988; 140:2395–400.
48. Hoffenbach A, Lagrange PH, Bach MA. Strain variation of lymphokine production and specific antibody secretion in mice infected with *Mycobacterium lepraemurium*. Cell Immunol 1985; 91:1–11.
49. Bahr GM, Chedid LA, Behbehani K. Induction, *in vivo* and *in vitro*, of macrophage membrane interleukin-1 by adjuvant-active synthetic muramyl peptides. Cell Immunol 1987; 107:443–54.
50. Denis M, Forget A, Pelletier M, Skamene E. Pleiotropic effects of the *Bcg* gene: III. Respiratory burst in *Bcg*-congenic macrophages. Clin Exp Immunol 1988; 73:370–5.
51. Blackwell JM, Toole S, King M, Dawda P, Roach TIA, Cooper A. Analysis of *Lsh* gene expression in congenic B10.L-Lsh^r mice. Curr Top Microbiol Immunol 1988; 137:301–9.
52. Stokes RW, Collins FM. Growth of *Mycobacterium avium* in activated macrophages harvested from

inbred mice with differing innate susceptibilites to mycobacterial infection. Infect Immun 1988; 56:2250–4.
53. Brett SJ, Butler R. Macrophage activity in resistant and susceptible mouse strains infected with *Mycobacterium lepraemurium*. Immunology 1988; 63:701–6.
54. Appelberg R, Sarmento AM. The role of macrophage activation and of *Bcg*-encoded macrophage function(s) in the control of *Mycobacterium avium* infection in mice. Clin Exp Immunol 1990; 80:324–31.
55. Denis M. Tumor necrosis factor and granulocyte macrophage-colony stimulating factor stimulate human macrophages to restrict growth of virulent *Mycobacterium avium* and to kill avirulent *M. avium*: killing effector mechanism depends on the generation of reactive nitrogen intermediates. J Leukocyte Biol 1991; 49:380–7.
56. Liew FY, Li Y, Millott S. Tumor necrosis factor-α synergized with IFN-γ in mediating killing of *Leishmania major* through the induction of nitric oxide. J Immunol 1990; 145:4306–10.
57. Green SJ, Crawford RM, Hockmeyer JT, Meltzer MS, Nacy CA. *Leishmania major* amastigotes initiate the L-arginine–dependent killing mechanism in IFN-γ stimulated macrophages by induction of tumor necrosis factor-α. J Immunol 1990; 145:4290–7.
58. Ha DKK, Gardner ID, Lawton JWM. Characterization of macrophage function in *Mycobacterium lepraemurium*–infected mice: sensitivity of mice to endotoxin and release of mediators and lysosomal enzymes after endotoxin treatment. Parasite Immunol 1983; 5:513–26.
59. Chatterjee D, Roberts AD, Lowell K, Brennan PJ, Orme IM. Structural basis of capacity of lipoarabinomannan to induce secretion of tumor necrosis factor. Infect Immun 1992; 60:1249–53.
60. Theodos CM, Povinelli L, Molina R, Sherry B, Titus RG. Role of tumor necrosis factor in macrophage leishmanicidal activity *in vitro* and resistance to cutaneous leishmaniasis *in vivo*. Infect Immun 1991; 59:2839–42.
61. Blackwell JM, Roach TIA, Kiderlen A, Kaye PM. Role of *Lsh* in regulating macrophage priming/activation. Res Immunol 1989; 140:798–805.
62. Schurr E, Skamene E, Forget A, Gros P. Linkage analysis of the *Bcg* gene on mouse chromosome 1. Identification of a tightly linked marker. J Immunol 1989; 142:4507–13.
63. Stead WW, Senner JW, Reddick WT, Lofgren JP. Racial differences in susceptibility to infection by *Mycobacterium tuberculosis*. N Engl J Med 1990; 322:422–7.
64. Stead WW. Genetics and resistance to tuberculosis. Could resistance be enhanced by genetic engineering? Ann Intern Med 1992; 116:937–41.
65. Crowle AJ, Elkins N. Relative permissiveness of macrophages from black and white people for virulent tubercle bacilli. Infect Immun 1990; 58:632–8.
66. McPeek M, Salkowitz J, Laufman H, Pearl D, Zwilling BS. The expression of HLA-DR by monocytes from black and from white donors; different requirements for protein synthesis. Clin Exp Immunol 1992; 87:163–8.
67. Zwilling BS, Massie M, Vespa L, Kwasniewski M, Nath J. The duration of class II MHC glycoprotein expression by mononuclear phagocytes is regulated by the Bcg gene. Prog Leukocyte Biol 1988; 7:201–10.
68. Gonwa TA, Stobo JD. Differential expression of Ia molecules by human monocytes. J Clin Invest 1984; 74:859–66.
69. Moore ME, Rossman MD, Schreiber AD, Douglas SD. DR framework antigens and 63D3 antigens on human peripheral blood monocytes and alveolar macrophages. Clin Immunol Immunopathol 1983; 29:119–28.
70. Abel L, Demenais F. Detection of major genes for susceptibility to leprosy and its subtypes in a Caribbean island: Desirade Island. Am J Hum Genet 1988; 42:256–66.
71. Newman GW, Gan HX, McCarthy PL, Remold HG. Survival of human macrophages infected with *Mycobacterium avium intracellulare* correlates with increased production of tumor necrosis factor-α and IL-6. J Immunol 1991; 147:3942–8.
72. Fortier AH, Polsinelli T, Green SJ, Nacy CA. Activation of macrophages for destruction of *Francisella tularensis*: identification of cytokines, effector cells, and effector molecules. Infect Immun 1992; 60:817–25.
73. Anthony LSD, Morrissey PJ, Nano FE. Growth inhibition of *Francisella tularensis* live vaccine strain by

IFN-γ activated macrophages is mediated by reactive nitrogen intermediates derived from l-arginine metabolism. J Immunol 1992; 148:1829–34.

74. Keller R, Gehri R, Keist R, Huf E, Kayser FH. The interaction of macrophages and bacteria: a comparative study of the induction of tumoricidal activity and reactive nitrogen intermediates. Cell Immunol 1991; 134:249–56.
75. Nathan CF, Hibbs JB. Role of nitric oxide synthesis in macrophage antimicrobial activity. Curr Opin Immunol 1991; 3:65–70.
76. Langermans JAM, Van der Hulst MEB, Nibbering PH, Hiemstra PS, Fransen L, Van Furth R. IFN-γ induced l-arginine–dependent toxoplasmastatic activity in murine peritoneal macrophages is mediated by endogenous tumor necrosis factor-α. J Immunol 1992; 148:568–74.
77. Munoz-Fernandez MA, Fernandez M, Fresno M. Synergism between tumor necrosis factor-α and interferon-γ on macrophage activation for the killing of intracellular *Trypanosoma cruzi* through a nitric oxide–dependent mechanism. Eur J Immunol 1992; 22:301–7.
78. Mosser DM, Springer TA, Diamond MS. Leishmania promastigotes required opsonic complement to bind to human leukocyte integrin Mac-1 (CD11b/CD18). J Cell Biol 1992; 116:511–20.
79. Drevets DA, Campbell PA. Roles of complement and complement receptor type 3 in phagocytosis of *Listeria monocytogenes* by inflammatory mouse peritoneal macrophages. Infect Immun 1991; 59:2645–52.
80. Schlesinger LS, Horwitz MA. Phenolic glycolipid-1 of *Mycobacterium leprae* binds complement components C3 in serum and mediates phagocytosis by human monocytes. J Exp Med 1991; 174:1031–8.
81. Malo D, Schurr E, Epstein DJ, Vekemans M, Skamene E, Gros P. The host resistance locus *Bcg* is tightly linked to a group of cytoskeleton-associated protein genes that include villin and desmin. Genomics 1991; 10:356–64.
82. Kobayashi T, Robinson JM. A novel intracellular compartment with unusual secretory properties in human neutrophils. J Cell Biol 1991; 113:743–56.
83. Luo Z, Robinson JM. Co-localization of an endocytic marker and acid phosphatase in a tubular/reticular compartment in macrophages. J Histochem Cytochem 1992; 40:93–103.

II

MACROPHAGE–BACTERIAL INTERACTIONS

15

Macrophages and *Mycobacterium tuberculosis*: The Key to Pathogenesis

Graham A. W. Rook

University College London Medical School, London, England

I. INTRODUCTION

The breakdown of primary health care programs because of political and economic collapse, immunosuppression secondary to human immunodeficiency virus (HIV) infection, and the increasing incidence of strains of *Mycobacterium tuberculosis* that are resistant to known treatments have combined in the last few years to give rise to a new epidemic of tuberculosis. This threatens not only the developing world, but even the wealthy countries of the West. Unfortunately, our attempts to combat tuberculosis are hampered by the fact that we still do not understand the mechanism of immunity to it in humans. However, it seems clear that interactions between *M. tuberculosis* and macrophages play a key role at every stage, including the induction of the response, immunity, and the necrotic immunopathology.

II. INTERACTIONS BETWEEN MYCOBACTERIA AND MACROPHAGES DURING THE INITIATION OF INFECTION; ADJUVANTICITY AND THE SELECTION OF EFFECTOR MECHANISMS

When confronted with an organism, the immune response must make two types of "decisions." First it must select the epitopes against which the response will be directed, and second it must choose the type of response mechanism that it will direct against the selected epitopes. The target of most laboratories working on the immunology of the mycobacterioses has been the first of these "decisions." However, the evidence that the second one is at least as important has been growing rapidly over the last few years, thanks largely to the pioneering work of Mosmann and his colleagues, who have shown that CD4+ helper T cells are not all the same, and that they differ in lymphokine output [1]. Similar functional subdivisions are seen among CD8+ cells responding to mycobacterial antigens [2]. The importance of these observations to the mycobacterioses has been confirmed by studies of T cells in leprosy lesions, where dominance of T_{H2} cells seems to correlate with lepromatous disease [3]. It is tempting to speculate that some similar imbalance of T_{H1} and T_{H2} responses could occur in tuberculosis and be associated with the tissue-damaging pathology. For instance, it is usually assumed that *M. tuberculosis* is an adjuvant that drives the T_{H1} (interferon-γ–producing) response, yet T_{H1} cells alone fail to mediate tuberculin-type delayed hypersensitivity reactions in mice [1], and many tuberculosis patients have raised levels of immunoglobulin E (IgE) [4]; some even have an eosinophilia. These facts suggest T_{H2} activity. In practice it may turn out to be more complex, with yet more T_H types secreting distinct blends of cytokines. How is the choice of mechanism made as the response is initiated? The answer appears to lie in the adjuvant properties of microbial components, for which mycobacteria have been noted since Freund described his adjuvant. Thus, when it encounters an organism the immune system undertakes a complex "taxonomic" exercise [5], using phylogenetically ancient recognition systems that can alert the host to the presence of a wide variety of common bacterial products. For some of these, such as mannans, glucans, and formyl peptides, there are well-defined receptors (mannan-binding protein, mannose-fucose receptors, CR3, fMet-Leu-Phe receptors) through which the macrophages and other cells can become activated. For other components such as muramyl peptides, polyanions, lipoarabinomannan, trehalose dimycolate, etc., receptors are difficult to define, but exposure of macrophages to these substances activates the cells and induces release of cytokines. (The ability of *M. tuberculosis* to trigger release of cytokines is discussed in greater detail below.) Interestingly, many of these cytokines—interferon-γ (IFN-γ), tumor necrosis factor-α (TNF-α), interleukin-1 (IL-1), IL-2, IL-3, IL-6—can be shown to have adjuvant properties, presumably due to regulatory effects on antigen-presenting cell function and T cells. Consequently, the properties of antigen-processing cells are likely to be modified and the selection of effector pathways altered.

III. UPTAKE OF MYCOBACTERIA BY MACROPHAGES AND OTHER CELLS

There is some evidence that uptake of *M. tuberculosis* by macrophages can involve complement receptors (CR3) [6]. However, this work fails to address one of the neglected paradoxes of the biology of tuberculosis, discovered by the late Charles Shepard in the 1950s [7]. In vitro, *M. tuberculosis* readily enters fibroblasts, HeLa cells, human amnion cells, and a variety of other "nonprofessional" phagocytic cells that do not express CR3. Why then is *M. tuberculosis* not seen inside these cell types in vivo? This is particularly odd because *M. leprae* is regularly found inside a large range of "nonprofessional" cell types in infected tissues [8]. Possible explanations for this point are discussed in Section VI. D. The entry of mycobacteria into such nonprofessional phagocytes in vitro may involve the two types of fibronectin-binding molecule produced by the bacteria [9]. These can be associated with the surface of the organism. Fibronectin facilitates uptake of the Bacillus Calmette-Guérin (BCG) by bladder epithelium, apparently by binding to the cell wall–associated fibronectin-binding proteins [10], which may then cross-link the bacteria onto integrins expressed on the cell membrane (T. Ratliff, personal communication). It is possible that the organism can use this pathway to enter other cells, and Shepard noted that a serum factor was involved [7]. It is not known whether a similar pathway can enable the organisms to enter macrophages without triggering killing mechanisms.

IV. CONTROL OF PROLIFERATION OF INTRACELLULAR *M. TUBERCULOSIS*

A. The Technical Problems Associated with Attempts to Study Antimycobacterial Effects of Macrophages

It is widely assumed that immunity to *M. tuberculosis* is mediated by lymphokine-activated macrophages, though as outlined below, this pathway is readily demonstrated with murine cells, but not with human ones. Moreover, the literature is conflicting. This is due to the technical difficulties of assessing growth of mycobacteria within macrophages during the prolonged periods of culture required. *M. tuberculosis* tends to kill macrophages that contain more than about five bacilli, so these detach, and are lost if the wells are rinsed out before the end of the experiment. Moreover, the released organisms grow in the medium at very variable rates, determined above all by the availability of iron. Such growth is rapid in fetal calf serum, but very slow in human serum until the organisms are numerous. The organisms also tend to clump, so colony counts can be misleading unless sonication or detergents are used. Finally, reports of mediators causing 50% to 90% reductions in bacterial (or isotope incorporation) counts compared with controls, over culture periods of 4 days or more, may not represent protective mechanisms. Assuming a normal replication time of about 24 hours, a deficit of 50% after 4 days in culture represents a reduction of one generation in four, while a deficit of 87.5% still means only that the organisms are doubling in 4 days rather than in 1. These problems are usually ignored, and much literature on this topic is suspect.

B. Murine Macrophages and *M. tuberculosis*

1. *Control of Proliferation of* M. tuberculosis *by Murine Macrophages*

In murine peritoneal macrophages total stasis of *M. tuberculosis*, possibly accompanied by some killing, can be induced by crude lymphokine-rich T cell supernatants [11], recombinant IFN-γ (rIFNγ) [12,13], or class II MHC–restricted T cell lines responding to mycobacterial antigen [11]. The effect is abrogated by steroids [14] and does not seem to involve oxygen reduction products [15]. More recent work has suggested that killing of *M. tuberculosis* can be induced by sequentially exposing murine macrophages to IFN-γ and TNF-α. This correlates with formation of nitric oxide (NO) [16,17] by a pathway described in detail in Chapter 10. It is probable that the NO is directly toxic to the bacteria, though it is also conceivable that the NO may act as a secondary signal within the cell, leading to the activation of some other unknown killing mechanism. In murine bone marrow–derived macrophages minor degrees of inhibition of *M. tuberculosis* can be evoked by other cytokines [18], though this may be attributable to maturation rather than activation. Evidence that IL-4, macrophage-deactivating factor (MDF), and the transforming growth factor-β (TGF-β) family can oppose the tendency of IFN-γ to activate the NO pathway suggests that the precise balance of cytokines within the lesion, and the balance of T_{H1} to T_{H2} activity, may play a critical role [19,20].

2. *The* Bcg *Gene and Anti-Mycobacterial Activity*

The susceptibility of mice to infection with several intracellular pathogens, including *Leishmania*, *Salmonella*, and *Mycobacterium*, is partially determined by a single autosomal dominant gene (designated *Lsh*, *Ity*, or *Bcg*, respectively) located on the centromeric portion of chromosome I, within a linkage group that shows structural homology with a conserved region of human chromosome 2 [21,22]. It is linked to genes for cytoskeleton-associated proteins such as villin and desmin, but the gene itself has not been identified, and its precise role remains unknown. Recent data suggest that it influences both positive and negative regulatory pathways at a very early stage in macrophage activation. There is not yet any convincing evidence that the human homologue of the gene is an important determinant of susceptibility to tuberculosis in humans.

3. *Macrophages, Tumor Necrosis Factor-α, and Granuloma Formation*

A further role of TNF-α in the response of mice to mycobacteria and other pathgens has been revealed by the use of neutralizing antibody in vivo. Mice were first rendered tolerant to rabbit IgG. They then received the IgG fraction from normal rabbit serum, or from a polyclonal antibody to TNF-α, and were infected with BCG. The BCG proliferated more in the recipients of anti-TNF-α, and formation of granulomata was inhibited in these animals [23]. The increased bacterial proliferation could be explained by inhibition of the triggering of the NO pathway, so it was possible that the failure of granuloma formation was secondary to a failure to kill bacilli. However, recent studies in another system have confirmed that TNF-α is indeed directly involved in granuloma formation [24]. It is not clear whether the same is true in mice infected with virulent *M. tuberculosis*. The only published report used mice that were not rendered tolerant to rabbit IgG, and the resulting protocol was unacceptable, since such mice develop severe serum sickness. Nevertheless, it seems that in mice the balance between the protective and immunopathological roles of TNF-α is heavily in favor of the former. As explained below, the situation may be different in humans.

C. Human Macrophages and *M. tuberculosis*

The situation in humans is less well understood. As in the mouse, a major macrophage-activating lymphokine in human tuberculosis is IFN-γ [25], but incubation of human macrophages or monocytes with this lymphokine does not cause inhibition of *M. tuberculosis* and often causes increased proliferation of the organism [12,26]. In humans an additional pathway further amplifies the activation induced by IFN-γ. Thus when human (but not murine) macrophages are exposed to IFN-γ they develop a 1-hydroxylase that converts the circulating inactive form of vitamin D_3 (25-hydroxy-cholecalciferol) into the active 1-hydroxylated derivative (1,25-dihydroxy-cholecalciferol, or calcitriol) [27–29]. We are familiar with this enzyme in the kidney, and with the role of calcitriol in the regulation of calcium balance. However, there is less awareness of this quite different function of calcitriol in the regulation of macrophage activation during human cell-mediated immune responses. Recent data have confirmed that this pathway is a striking feature of human tuberculosis lesions [30–32]. The fact that it is of fundamental importance to both immunity and immunopathology in tuberculosis is illustrated by the many reports that treatment of tuberculosis with vitamin D in the 1940s caused the lesions to undergo necrosis. When superficial skin lesions were treated in this way, the result was cure. Pulmonary lesions liquified, with disastrous consequences [33–35].

If human monocytes or monocyte-derived macrophages are incubated with a combination of IFN-γ and calcitriol at a physiological concentration of 10^{-7} to 10^{-9}M, a modest degree of inhibition of proliferation of *M. tuberculosis* is achieved, but the inhibition seen is still too limited to be convincing as a protective pathway [28,29]. However, there is one unconfirmed report that strong inhibition of *M. tuberculosis* is seen if the macrophages are cultured with IFN-γ, calcitriol, *and* TNF-α [36]. This result is not obtained in other laboratories.

A major difference between human and murine macrophages seems to be the inability of the former to make NO. Because the enzyme 6-pyruvoyl tetrahydrobiopterin synthase is present at very low levels in human macrophages [37], they lack tetrahydrobiopterin, which is an essential cofactor for the NO pathway. Clearly it will be important to discover whether the combination of IFN-γ, calcitriol, and TNF-α enables NO formation in human macrophages infected with *M. tuberculosis*. Unfortunately, the presence of a nitrate reductase in *M. tuberculosis* means that nitrite formed in such cultures may be derived from nitrate rather than from NO, so the very simple assay for nitrite using the Griess reagent cannot be employed as an indication of NO formation in this system.

D. Avoidance of Macrophage Killing Mechanisms

There are reports that *M. tuberculosis* can inhibit the fusion of lysosomes and phagosomes [38,39]. The effect may be due to release of polyanions or ammonia and correlates with changes in the "saltatory" movements of lysosomes within the cell [39]. However, the importance of this mechanism is unclear. It will depend on the relevance of lysosomal contents to the killing pathways, which remains unknown in human cells. Some strains of *M. tuberculosis* release catalase, and the phosphatidylinositol mannoside lipoarabinomannan (LAM) is able to scavenge free radicals [40]. These features may contribute to the apparent irrelevance of reactive oxygen intermediates in the killing of mycobacteria by macrophages [15,17]. Lipoarabinomannan is also able to inhibit transcriptional activation of genes regulated by IFN-γ [40] or by lipopolysaccharide (LPS) or phytohemagglutinin [41]. Its ability to inhibit protein kinase C activity is likely to be relevant to these effects [40].

E. Cytotoxic Cells Heat Shock Proteins and γ/δ Cells

It remains possible that protection from *M. tuberculosis* is not only a matter of appropriate macrophage activation, but also involves killing of infected cells, perhaps by cytotoxic T cells. The kinetics of the emergence of these cells after infection of C57B1/6 mice with the Erdman strain of *M. tuberculosis* has been documented in detail [42]. The early protective response in this model was associated with CD4+ (and a minor population of CD8+) cells that secrete IFN-γ in response to a broad range of mycobacterial antigens. This is compatible with the view that in the mouse, protection is mediated by the IFN-γ/TNF-α/NO pathway discussed above. The IFN-γ-secreting cells peaked at about 10 days and declined by 30 days. Cytotoxic CD4+ cells began to emerge at this time. They lysed antigen-presenting cells pulsed with bacterial culture filtrate, and it is suggested that these cytotoxic cells detect the presence of persistent dormant intracellular organisms that, if not eliminated, would lead to future relapse.

Peripheral blood mononuclear cells from human tuberculosis patients or normal donors also generate cytotoxic CD4+ T cells in response to mycobacterial antigens [43,44]. If macrophages infected with *M. tuberculosis* are used as the targets, the lytic event may be accompanied by unidentified microbicidal effects [45], though this observation requires confirmation. One of the mycobacterial antigens that are recognized by such cytotoxic responses is the 65-kDa heat shock protein [46]. Some T cell clones responsive to this protein can also recognize the very similar human homologue (60-kDa mitochondrial heat shock protein) and, in the mouse at least, can apparently lyse stressed autologous cells [47,48]. It is therefore possible that when infected cells are beginning to succumb to infection, the stressed state of the cell is detected, and the cell is killed so that the bacilli can be taken up by a new macrophage. It is also possible that this mechanism can lead to immunopathology. This hypothesis has given rise to much discussion in relation to mechanisms of autoimmunity, which are outside the scope of this review [49].

A relatively large percentage of those human peripheral blood T cells that bear γ/δ receptors will proliferate in response to *M. tuberculosis.* [50] Similary, the lymphocytes that proliferate in murine lymph nodes draining sites primed with mycobacterial antigen are enriched for γ/δ T cells [51]. There is some evidence that this is due to a small protease-resistant component that may be a superantigen [52]. That is to say, it may cross-link class II MHC (though without conventional MHC restriction) and Vγ9-positive γ/δ T cells. These cells can be cytotoxic, but it is not clear whether they play a role in immunity to *M. tuberculosis.*

F. The Possible Role of Neutrophils

Since neutrophils form a large proportion of the cells in early infiltrates in experimental tuberculosis lesions, it is important to know whether they have any ability to kill *M. tuberculosis*. There are authors who claim that they do [53], or that after incubation in IFN-γ they can kill *M. fortuitum* [54]. Others deny that neutrophils can kill *M. tuberculosis*, with or without preactivation [55]. We were quite unable to demonstrate killing of *M. tuberculosis* by neutrophils (unpublished data) using ^{3}H-uracil uptake in an assay that readily detected the small antimycobacterial effects of calcitriol-activated monocytes [28]. Similarly, there is evidence that the sulfolipid-1 from *M. tuberculosis* can activate neutrophils to release superoxide and, at lower concentrations, prime them for subsequent enhanced responses to other stimuli [56], but this may contribute to pathology rather than protection.

V. SYSTEMIC CONSEQUENCES OF MACROPHAGE ACTIVATION IN TUBERCULOSIS

The activation of calcitriol formation by macrophages is such a major pathway in human tuberculosis [30–32] that calcitriol can spill over from the granulomata into the circulation and cause subtle disturbances of vitamin D_3 metabolism, or even frank hypercalcemia. Such changes are detectable in all diseases characterized by chronic granulomatous T cell–mediated inflammation in humans [35].

The activation of macrophages by IFN-γ and other lymphokines, further amplified by the autocrine effect of calcitriol, leads to a high degree of priming for subsequent release of cytokines such as TNF-α [28]. In our hands, a sensitive bioassay rarely finds detectable levels of TNF-α in the serum of tuberculosis patients, but this is not surprising in view of the short half-life of this cytokine (6 to 7 minutes). (Since cytokines are endogenous pyrogens, it would be interesting, but difficult, to look at samples taken shortly before night sweats.) However, there is good indirect evidence that both local [25] and systemic [57–59] TNF-α release is occurring. Macrophages derived from alveolar lavage samples from tuberculosis patients release TNF-α apparently spontaneously in vitro [60], though it is possible that they become triggered during the lavage process. However, they release more TNF-α spontaneously than alveolar macrophages from normal donors can be induced to release in the presence of an optimal concentration of endotoxin from *Escherichia coli* [60]. Similarly, peripheral blood monocytes from tuberculosis patients release the cytokine in vitro [57–59]. Moreover, patients with tuberculosis (and other granulomatous diseases) have high circulating levels of inhibitors of the toxicity of TNF-α in a bioassay [61]. These appear to be the extracellular domains of TNF-α receptors, which are known to be released in response to release of TNF-α.

The chronic release of TNF may be attributable to the ability of several components of *M. tuberculosis* to trigger cytokine release. The most potent is LAM [62,63]. Other substances such as peptidoglycan fragments [64], trehalose dimycolate [65], sulfolipid-1 [66], and possibly unidentified protein constituents may also have this property [67].

It is likely that consequent "leakage" of TNF-α into the circulation plays a role in the weight loss and fever that are often seen in tuberculosis patients, and TNF-α and other cytokines such as IL-1 and IL-6 drive the acute-phase response, and are probably involved in the striking increase in the level of agalactosyl IgG seen in the serum of these patients. The probable immunopathological role of this glycoform of IgG is beyond the scope of this review, but it has been discussed elsewhere [68].

The cytokines, acting via the hypothalamus and pituitary, also explain the modest rise in plasma cortisol characteristic of tuberculosis [69–71]. It may be that the recently described ability of TNF-α to oppose the effect of adrenocorticotropic hormone (ACTH) on cortisol output from adrenal cells [72] can account for the poor response to ACTH found in severe tuberculosis and the risk of sudden death in such patients if treated with powerfully bactericidal drug combinations [71,73,74]. These rapidly release cytokine-triggering bacterial components and consequently flood the system with cytokines, leading to a form of "tuberculin shock" analogous in some ways to the Jarisch-Herxheimer reaction induced by penicillin treatment in patients with syphilis.

VI. MACROPHAGES, CYTOKINES, AND THE KOCH PHENOMENON

A. The Role of the Koch Phenomenon

It was Robert Koch in the 1890s [75] who first described the necrotic local response to intradermal challenge of tuberculous guinea-pigs with cultures or soluble antigens of *M. tuberculosis*. Koch also noted that subcutaneous injections of larger quantities of culture filtrate (old tuberculin) caused necrosis to appear in tuberculous lesions at other sites, and he attempted to use this as a form of immunotherapy for tuberculosis in humans. There are explicit descriptions of necrosis induced in this way in long-standing lesions of skin tuberculosis (lupus vulgaris)[76]. There is some confusion in the literature as to whether the local necrosis at the site of challenge or the necrosis elicited in the distant tuberculous lesions is the eponymous "Koch phenomenon." There are good reasons, outlined below, for supposing that the mechanism is similar in the two sites.

We suggested elsewhere that these phenomena are related to the contemporary observation that the tuberculin skin test is frequently necrotic in past or present cases of human tuberculosis [77]. Tuberculin itself is nontoxic, and it does not evoke necrotic skin test responses in past recipients of BCG or in tuberculoid leprosy patients, even when there is vigorous induration. Therefore, necrosis is clearly not an inevitable component of a positive cell-mediated response to mycobacterial antigen. After many years of controversy about the true role of the necrosis, Wilson and his colleagues pointed out that the necrotic reaction can protect guinea pigs from reinfection in the superficial layers of the skin, because the challenged tissue undergoes necrosis and is sloughed off [78]. On the other hand, the necrosis leads to increased susceptibility to infection induced by injection of small numbers of viable bacilli into deep tissues [78]. According to this reasoning, the Koch phenomenon is detrimental to the host because the organisms proliferate readily in the necrotic tissue. However, it can also be argued that a late consequence of the Koch phenomenon is the "walling off" and eventual calcification of the lesion, in which case, even if it is not the "first-line" mechanism of protection, which logically ought to be the killing of bacilli, it could play a role as a second line defense. At present we cannot resolve these dilemmas because of several large gaps in our knowledge. First, as already outlined above, we have not identified this hypothetical "first-line" mechanism of immunity to tuberculosis in humans. Second, we do not know the relationship between the necrosis characteristic of the Koch phenomenon, which takes 24 to 48 hours to occur, and the much slower "walling off" or caseation seen in human disease. However, although we are uncertain as to the pathogenetic role of the rapid necrosis described by Koch, we can now provide an explanation of its mechanism, in which the activated macrophage plays the crucial role.

B. Parallels Between Koch's Immunotherapy and the Shwartzman Reaction

The striking aspect of Koch's attempts at immunotherapy is that a subcutaneous injection of soluble mycobacterial products evoked necrosis in distant tuberculous lesions. This suggests a parallel with the Shwartzman reaction, in which intravenous LPS-rich bacterial suspensions evoked hemorrhagic necrosis in skin sites "prepared" 24 hours previously by an injection of similar bacterial material. The current interpretation of the Shwartzman reaction suggests that certain types of inflammation render the tissues, particularly the endothelial cells and capillaries, sensitive to a necrotizing effect of subsequently introduced cytokines. The cytokines can be released systemically by, for instance, intravenous LPS (as in Shwartzman's original studies) [79] or injected directly into the site [80]. Thus, Koch was injecting old tuberculin, rich in cytokine triggers such as LAM and peptidoglycan fragments into tuberculosis patients with macrophages primed for TNF-α release, as already outlined in Section V. It is not surprising that this caused necrosis in the tuberculous lesions. We know in rabbits and guinea pigs, mycobacterial lesions are "prepared sites" in the sense used by Shwartzman, because several early authors showed that intravenous injections of gram-negative bacteria into tuberculous animals caused necrosis in the infected tissues [81,82]. This was similar to that seen in Shwartzman's original studies. Interestingly, in guinea pigs an injection of muramyl dipeptide, which is a moderately effective trigger of cytokine release, will cause necrosis in a distant site previously "prepared" with complete Freund's adjuvant [83]. Similarly, the classical paper by Carswell and Old showed that when intraveous LPS is administered to mice with granulomata caused by a previous injection of BCG, there is massive production of the activity subsequently identified as TNF-α [84], and the mycobacterial granulomata in these animals undergo necrosis [85]. Intravenous injection of mycobacterial sonicates will trigger similar systemic TNF release (Rook and Taverne, unpublished data). These results imply that as in the classical Shwartzman reaction, TNF-α may play a role in the tissue damage in the mycobacterial lesions.

C. Tumor Necrosis Factor-α Causes Necrosis in Sites Primed by Mycobacterial Components

In order to test the hypothesis that TNF-α plays a role in the necrosis seen in mycobacterial lesions, we established a model allowing us to examine the effect of direct introduction of TNF-α into sites of mycobacterium-induced inflammation. In C57B1/GrFa mice, 1 μg of recombinant TNF-α will cause hemorrhagic necrosis when injected directly into sites "prepared" 24 hours previously with soluble mycobacterial antigen [86]. This necrosis occurs even in unimmunized mice and is quantitatively comparable to that seen in sites optimally primed with LPS [80]. However, the same protocol in preimmunized animals results in dramatic swelling and necrosis, with as much as 20 mg of hemoglobin detectable in the challenged footpad. This greatly enhanced TNF-α sensitivity is dependent on CD4+ T cells and is inhibited if these cells are removed with appropriate depleting antibodies [86]. The hemorrhagic necrosis is decreased by inhibitors of platelet-activating factor (PAF) or by depletion of complement. It is also strongly inhibited by procedures that deplete circulating macrophages or block their passage from the circulation into the lesion [87].

Further studies have revealed, however, that not all delayed hypersensitivity responses show the same exquisite sensitivity to TNF-α [88]. The factors that determine this property are under investigation and may involve the balance of T_{H1} to T_{H2} cells.

D. *M. tuberculosis* Distorts the Function of Tumor Necrosis Factor-α

One factor that influences the sensitivity to TNF-α of cells and tissues exposed to mycobacterial components is a product of the organisms themselves. This product enhances both the toxic and the regulatory effects of the cytokine [89]. It was discovered during investigation of the paradoxical ability of *M. tuberculosis* to enter fibroblasts, HeLa cells, human amnion cells [7], and a variety of other "nonprofessional" phagocytic cells and proliferate intracellularly without killing them (outlined in Section III). We observed that fibroblasts infected with *M. tuberculosis* become exquisitely sensitive to the toxic effects of TNF-α. Indeed, a cell line that is inherently resistant to TNF-α can be killed when infected with this organism. It subsequently emerged that invasion of the cells by *M. tuberculosis* is not a prerequisite for this effect. Similar increases in sensitivity to TNF-α are seen when cells are incubated in crude supernatants of *M. tuberculosis*. The levels of this TNF-α–enhancing activity (TEA) are very low in the three BCG strains tested. Normal human cell types such as endothelial cells and fibroblasts are also affected by TEA (Filley et al., unpublished data). For instance, TNF-α usually stimulates the growth of fibroblasts, but in the presence of TEA the TNF-α inhibits or even kills them. Also, TEA increases the ability of TNF-α to enhance expression of ICAM-1 by endothelial cells or secretion of IL-6 by fibroblasts.

It is possible, but not proven, that the tendency of *M. tuberculosis* to kill macrophages that engulf it is attributable to TEA, since it may render the macrophages sensitive to the toxic effect of the cytokine that they are themselves secreting. Similarly, it is possible that *M. tuberculosis* is not seen inside nonprofessional phagocytes in vivo, in spite of its rapid entry into such cells in vitro [7], because such cells are rapidly killed by the TNF-α acting in synergy with other cytokines such as IFN-γ that are also present in the lesions [25].

VII. *M. TUBERCULOSIS*, CYTOKINES, AND ACTIVATION OF HUMAN IMMUNODEFICIENCY VIRUS

Although peripheral to the main argument of this review, another consequence of cytokine release in tuberculosis has become of enormous importance. Reactivation of tuberculosis is an early event in HIV-infected persons. This is often followed by rapid progression of the HIV infection to AIDS. Tumor necrosis factor-α, and other cytokines such as IFN-γ that are potent inducers of TNF-α, can serve as autocrine growth factors for HIV replication. Thus in vitro studies show that neutralizing antibodies to TNF-α can inhibit HIV replication in CD4+ T cells and monocytes [90,91]. Therefore, the effects of *M. tuberculosis* on cytokine release and function discussed in previous sections provide several potential reasons why established tuberculosis infection could accelerate virus production. For instance, we have shown that LAM will increase production of HIV from normal human peripheral blood lymphocytes (Vyakarnam, Moreno, and Rook, unpublished data). It may do this by enhancing cytokine release (see section V). We have not yet tested the effect of TEA (see Section VI.D.). Similarly, calcitriol at physiological levels (see Section IV.C.) can increase replication of monocyte- and lymphocyte-tropic HIV strains in peripheral blood mononuclear cells by up to 10,000 times [92], so the striking release of this steroid both in the lesions and systemically in tuberculosis may promote progression to AIDS.

This is disagreement as to whether coinfection of macrophages with mycobacteria and HIV leads to increased replication of either organism [93,94], and further studies are needed.

VIII. CONCLUSIONS

It is remarkable that despite so much sophisticated immunology we still do not know the mechanism of immunity to *M. tuberculosis* in man. We do not even have formal proof that it is mediated by killing pathways within activated macrophages. Not do we fully understand the regulatory defects that lead to tissue-destructive rather than sterilizing T cell–mediated responses. It has been suggested in this chapter that at least some of the answers lie in disturbances of the role of macrophage-derived cytokines, so that in humans the toxic functions of TNF-α dominate over the protective ones. These protective roles are readily demonstrated in the mouse, but humans seem surprisingly different, with a lack of NO production by IFN-γ–activated macrophages and striking formation of calcitriol by these cells. In the opinion of the author, our understanding of the pathogenesis of tuberculosis requires more attention to the physiology of the disease, particularly to the macrophage and the cytokine/pituitary/adrenal axis. Progress is unlikely to emerge from the fashionable identification of epitopes recognized by T cells in a disease where the T cell response is clearly intact, even if dysregulated, and MHC associations so weak as to be meaningless.

REFERENCES

1. Mosmann TE, Moore KW. The role of IL-10 in cross-regulation of TH1 and TH2 responses. In: Ash C, Gallagher RB, eds. Immunoparasitology Today Vol. Elsevier Trends Journals. 1991:A49–A53.
2. Bloom BR, Salgame P, Diamond B. Revisiting and revising suppressor T cells. Immunol Today 1992; 13:131–6.
3. Yamamura M, Uyemura K, Deans RJ, Weinberg K, Rea TH, Bloom BR, Modlin RL. Defining protective responses to pathogens: cytokine profiles in leprosy lesions. Science 1991; 254:277–9.
4. Yong AJ, Grange JM, Tee RD, Swanson Beck J, Bothamley GH, Kemeny DM, Kardjito T. Total and anti-mycobacterial IgE levels in serum from patients with tuberculosis and leprosy. Tubercle 1989; 70:273–9.
5. Rook GAW. Mobilizing the appropriate T cell subset: the immune response as taxonomist? Tubercle 1991; 72:253–4.
6. Schlesinger LS, Bellinger-Kawahara CG, Payne NR, Horwitz MA. Phagocytosis of Mycobacterium tuberculosis is mediated by human monocyte complement receptors and complement component C3. J Immunol 1990; 133:2771–80.
7. Shepard CC. A comparison of the growth of selected mycobacteria in HeLa, monkey kidney, and human amnion cells in tissue culture. J Exp Med 1958; 107:237–43.
8. Esiri MM, Weddell AGM, Rees RJW. Infection of murine striated muscle with *Mycobacterium leprae*: a study by light and electron microscopy. J Pathol 1972; 106:73–80.
9. Abou-Zeid C, Ratliff TL, Wiker HG, Harboe M, Bennedsen J, Rook GAW. Characterization of fibronectin-binding antigens released by Mycobacterium tuberculosis and Mycobacterium bovis BCG. Infect Immun 1988; 56:3046–51.

10. Kavoussi LR, Brown EJ, Ritchey JK, Ratliff TL. Fibronectin-mediated Calmette-Guérin bacillus attachment to murine bladder mucosa. Requirement for the expression of an antitumor response. J Clin Invest 1990; 85:62–7.
11. Rook GAW, Champion BR, Steele J, Varey AM, Stanford JL. I-A restricted activation by T cell lines of anti-tuberculosis activity in murine macrophages. Clin Exp Immunol 1985; 59:414–20.
12. Rook GAW, Steele J, Ainsworth M, Champion BR. Activation of macrophages to inhibit proliferation of Mycobacterium tuberculosis: comparison of the effects of recombinant gamma-interferon on human monocytes and murine peritoneal macrophages. Imumunology 1986; 59:333–8.
13. Flesch IE, Kaufmann SH. Activation of tuberculostatic macrophage functions by gamma interferon, interleukin-4, and tumor necrosis factor. Infect Immun 1990; 58:2675–7.
14. Rook GAW, Steele J, Ainsworth M, Leveton C. A direct effect of glucocorticoid hormones on the ability of human and murine macrophages to control the growth of *Mycobacterium tuberculosis*. Eur J Respir Dis 1987; 71:286–91.
15. Flesch IE, Kaufmann SH. Attempts to characterize the mechanisms involved in mycobacterial growth inhibition by gamma- interferon-activated bone marrow macrophages. Infect Immun 1988; 56:1464–9.
16. Denis M. Interferon-gamma-treated murine macrophages inhibit growth of tubercle bacilli via the generation of reactive nitrogen intermediates. Cell Immunol 1991; 132:150–7.
17. Chan J, Xing Y, Magliozzo RS, Bloom BR. Killing of virulent *Mycobacterium tuberculosis* by reactive nitrogen intermediates produced by activated murine macrophages. J Exp Med 1992; 175:1111–22.
18. Kaufmann SH, Flesch IE. Antimycobacterial functions in bone-marrow–derived macrophages. Res Microbiol 1990; 141:244–8.
19. Liew FY, Li Y, Severn A, Millott S, Schmidt J, Salter M, Moncada S. A possible novel pathway of regulation by murine T helper type-2 (Th2) cells of a Th1 cell activity via the modulation of the induction of nitric oxide synthase on macrophages. Eur J Immunol 1991; 21:2489–94.
20. Ding A, Nathan CF, Graycar J, Derynck R, Stuehr DJ, Srimal S. Macrophage deactivating factor and transforming growth factors-$\beta 1$, -$\beta 2$ and -$\beta 3$ inhibit induction of macrophage nitrogen oxide synthesis by IFN-gamma. J Immunol 1990; 145:940–4.
21. Blackwell JM, Roach TI, Atkinson SE, Ajioka SE, Barton CH, Shaw MA. Genetic regulation of macrophage priming/activation: the Lsh gene story. Immunol Lett 1991; 30:241–8.
22. Schurr E, Malo D, Radzioch D, Buschman E, Morgan K, Gros P, Skamene E. Genetic control of innate resistance to mycobacterial infections. Immunol Today 1991; 12:A42–A45.
23. Kindler V, Sappino AP, Grau GE, Piguet PF, Vassalli P. The inducing role of tumor necrosis factor in the development of bactericidal granulomas during BCG infection. Cell 1989; 56:731–40.
24. Amiri P, Locksley RM, Parslow TG, Sadick M, Rector E, Ritter D, McKerron JM. Tumour necrosis factor alpha restores granulomas and induces parasite egg-laying in schistosome-infected SCID mice. Nature 1992; 356:604–7.
25. Barnes PF, Fong SJ, Brennan PJ, Twomey PE, Mazumder A, Modlin RL. Local production of tumor necrosis factor and IFN-gamma in tuberculosis pleuritis. J Immunol 1990; 145:149–54.
26. Douvas GS, Berger EM, Repine JE, Crowle AJ. Natural mycobacteriostatic activity in human monocyte–derived adherent cells. Am Rev Respir Dis 1986; 134:44–8.
27. Koeffler HP, Riechel H, Bishop JE, Norman AW. Gamma interferon stimulates production of 1,25 dihydroxy vitamin D3 by normal human macrophages. Biochem Biophys Res Commun 1985; 127:596–603.
28. Rook GA, Taverne J, Leveton C, Steele J. The role of gamma-interferon, vitamin D3 metabolites and tumour necrosis factor in the pathogenesis of tuberculosis. Immunology 1987; 62:229–34.
29. Crowle AJ, Ross EJ, May MH. Inhibition by 1,25(OH)2-vitamin D3 of the multiplication of virulent tubercle bacilli in cultured human macrophages. Infect Immun 1987; 55:2945–50.
30. Adams JS, Modulin RL. Diz MM, Barnes PF. Potentiation of the macrophage 25-hydroxyvitamin D-1-hydroxylation reaction by human tuberculous pleural effusion fluid. J Clin Endocrinol Metab 1989; 69:457–60.
31. Barnes PF, Modlin RL, Bikle DD, Adams JS. Transpleural gradient of 1,25-dihydroxyvitamin D in tuberculous pleuritis. J Clin Invest 1989; 83:1527–32.
32. Cadranel J, Hance AJ, Milleron B, Paillard F, Akoun GM, Garabedian M. Vitamin D metabolism in tuberculosis. Production of 1,25(OH)2D3 by cells recovered by bronchoalveolar lavage and the role of this metabolite in calcium homeostasis. Am Rev Respir Dis 1988; 138:984–9.

33. Macrae DE. Calciferol treatment of lupus vulgaris. Br J Dermatol 1947; 59:333–8.
34. Brincourt J. Le calciférol a-t-il une action liquéfiante sur le caséum? Poumon Coleur 1967; 23:841–51.
35. Rook GAW. The role of vitamin D in tuberculosis. Am Rev Respir Dis 1988; 138:768–70.
36. Denis M. Killing of Mycobacterium tuberculosis within human monocytes: activation by cytokines and calcitriol. Clin Exp Immunol 1991; 84:200–6.
37. Werner ER, Werner-Felmayer G, Fuchs D, Hausen A, Reibnegger G, Yim JJ, Pfleiderer W, Wachter H. Tetrahydrobiopterin biosynthetic activities in human macrophages, fibroblasts. THP-1, and T 24 cells. GTP-cyclohydrolase I is stimulated by interferon-gamma, and 6-pyruvoyl tetrahydropterin synthase and sepiapterin reductase are constitutively present. J Biol Chem 1990; 265:3189–92.
38. Gordon AH, Hart PD, Young MR. Ammonia inhibits phagosome-lysosome fusion in macrophages. Nature 1980; 286:79–80.
39. Hart PD, Young MR, Gordon, AH, Sullivan KH. Inhibition of phagosome-lysosome fusion in macrophages by certain mycobacteria can be explained by inhibition of lysosomal movements observed after phagocytosis. J Exp Med 1987; 166:933–46.
40. Chan J, Fan XD, Hunter SW, Brennan PJ, Bloom BR. Lipoarabinomannan, a possible virulence factor involved in persistence of Mycobacterium tuberculosis within macrophages. Infect Immun 1991; 59:1755–61.
41. Chujor CS, Kuhn B, Schwerer B, Bernheimer H, Levis WR, Bevec D. Specific inhibition of mRNA accumulation for lymphokines in human T cell line Jurkat by mycobacterial lipoarabinomannan antigen. Clin Exp Immunol 1992; 87:398–403.
42. Orme IM, Miller ES, Roberts AD, Furney SK, Griffin JP, Dobos KM, Chi D, Rivoire B, Brennan PJ. T lymphocytes mediating protection and cellular cytolysis during the course of *Mycobacterium tuberculosis* infection. Evidence for different kinetics and recognition of a wide spectrum of protein antigens. J Immunol 1992; 148:189–96.
43. Mustafa AS, Godal T. BCG induced CD4+ cytotoxic T cells from BCG vaccinated healthy subjects: relation between cytotoxicity and suppression in vitro. Clin Exp Immunol 1987; 69:255–62.
44. Lorgat F, Keraan MM, Lukey PT, Ress SR. Evidence for in vivo generation of cytotoxic T cells. PPD-stimulated lymphocytes from tuberculous pleural effusions demonstrate enhanced cytotoxicity with accelerated kinetics of induction. Am Rev Respir Dis 1992; 145:418–23.
45. Kumararatne DS, Pithie AS, Drysdale P, Gaston JSH, Kiessling R, Iles PB, Ellis CJ, Innes J, Wise R. Specific lysis of mycobacterial antigen–bearing macrophages by class II MHC–restricted polyclonal T cell lines in healthy donors, or patients with tuberculosis. Clin Exp Immunol 1990; 80:314–23.
46. Ottenhoff THM, Ab BK, van-Embden JDA, Thole JER, Kiessling R. The recombinant 65kDa heat shock protein of *Mycobacterium bovis*, bacillus Calmette Guérin/*Mycobacterium tuberculosis* is a target molecule for CD4+ cytotoxic T lymphocytes that lyse human monocytes. J Exp Med 1988; 168:1947–52.
47. Koga T, Wand-Wurttenberger A, Debruyn J, Munk ME, Schoel B, Kaufmann SH. T cells against a bacterial heat shock protein recognize stressed macrophages. Science 1989; 245:1112–5.
48. Steinhoff U, Schoel B, Kaufmann SH. Lysis of interferon gamma–activated Schwann cells by cross-reactive CD8+ alpha/beta T cells with specificity for the mycobacterial 65kDa heat shock protein. Int Immunol 1990; 2:279–84.
49. Winfield J, Jarjour W. Do stress proteins play a role in arthritis and autoimmunity? Immunol Rev 1991; 121:193–220.
50. Kabelitz D, Bender A, Schondelmaier S, Schoel B, Kaufmann SH. A large fraction of human peripheral blood gamma/delta + T cells is activated by Mycobacterium tuberculosis but not by its 65-kD heat shock protein. J Exp Med 1990; 171:667–9.
51. Janis EM, Kaufmann SH, Schwartz RH, Pardoll DM. Activation of gamma delta T cells in the primary immune response to *Mycobacterium tuberculosis*. Science 1989; 244:713–6.
52. Pfeffer K, Schoel B, Plesnila N, Lipford GB, Kromer S, Deusch K, Wagner H. A lectin-binding, protease-resistant mycobacterial ligand specifically activates V gamma 9+ human gamma delta T cells. J Immunol 1992; 148:575–83.
53. Jones GS, Amirault HJ, Andersen BR. Killing of Mycobacterium tuberculosis by neutrophils: a nonoxidative process (see comments). J Infect Dis 1990; 162:700–4.
54. Geertsma MF, Nibbering PH, Pos O, Van-Furth R. Interferon-gamma–activated human granulocytes kill ingested Mycobacterium fortuitum more efficiently than normal granulocytes. Eur J Immunol 1990; 20:869–73.

55. Denis M. Human neutrophils activated with cytokines or not, do not kill virulent Mycobacterium tuberculosis (letter; comment). J Infect Dis 1991; 163:919–20.
56. Zhang L, English D, Andersen BR. Activation of human neutrophils by Mycobacterium tuberculosis–derived sulfolipid-1. J Immunol 1991; 146:2730–6.
57. Ogawa T, Uchida H. Kusumoto Y, Mori Y, Yamamura Y, Hamada S. Increase in tumor necrosis factor alpha– and interleukin-6–secreting cells in peripheral blood mononuclear cells from subjects infected with Mycobacterium tuberculosis. Infect Immun 1991; 59:3021–5.
58. Takashima T, Ueta C, Tsuyuguchi I, Kishimoto S. Production of tumor necrosis factor alpha by monocytes from patients with pulmonary tuberculosis. Infect Immun 1990; 58:3286–92.
59. Cadranel J, Philippe C, Perez J, Milleron B, Akoun G, Ardaillou R, Baud L. In vitro production of tumour necrosis factor and prostaglandin E2 by peripheral blood mononuclear cells from tuberculosis patients. Clin Exp Immunol 1990; 81:319–24.
60. Rook GAW, Foley NM, Meager A. What mediates the immunopathological component of the immune response to Mycobacterium tuberculosis? Can it be switched off? Bull Int Union Tuberc Lung Dis 1990; 65:23–6.
61. Foley N, Lambert C, McNicol M, Johnson N, Rook GA. An inhibitor of the toxicity of tumour necrosis factor in the serum of patients with sarcoidosis, tuberculosis and Crohn's disease. Clin Exp Immunol 1990; 80:395–9.
62. Moreno C, Taverne J, Mehlert A, Bate CA, Brealey RJ, Meager A, Rook GAW, Playfair JH. Lipoarabinomannan from Mycobacterium tuberculosis induces the production of tumour necrosis factor from human and murine macrophages. Clin Exp Immunol 1989; 76:240–5.
63. Chatterjee D, Roberts AD, Lowell K, Brennan PJ, Orme IM. Structural basis of capacity of lipoarabinomannan to induce secretion of tumor necrosis factor. Infect Immun 1992; 60:1249–53.
64. Noso Y, Parant M, Parant F, Chedid L. Production of tumour necrosis factor in nude mice by muramyl peptides associated with bacterial vaccines. Cancer Res 1988; 48:5766–9.
65. Silva CL, Tincani I, Brandao-Filho SL, Faccioli LH. Mouse cachexia induced by trehalose dimycolate from Nocardia asteroides. J Gen Microbiol 1988; 134:1629–33.
66. Brozna JP, Horan M, Rademacher JM, Pabst KM, Pabst MJ. Monocyte responses to sulfatide from Mycobacterium tuberculosis: inhibition of priming for enhanced release of superoxide, associated with increased secretion of interleukin-1 and tumor necrosis factor alpha, and altered protein phosphorylation. Infect Immun 1991; 59:2542–8.
67. Valone SE, Rich EA, Wallis RS, Ellner JJ. Expression of tumor necrosis factor in vitro by human mononuclear phagocytes stimulated with whole Mycobacterium bovis BCG and mycobacterial antigens. Infect Immun 1988; 56:3313–5.
68. Rook GAW, Stanford JL. Slow bacterial infections or autoimmunity. Immunol Today 1992; 13:160–4.
69. Barnes DJ, Naraqi S, Temu P, Turtle JR. Adrenal function in patients with active tuberculosis. Thorax 1989; 44:422–4.
70. Ellis ME, Tayoub F. Adrenal function in tuberculosis. Br J Dis Chest 1986; 80:7–12.
71. Sarma GR, Chandra I, Ramachandran G, Krishnamurthy PV, Kamaraswami V, Prabhakar R. Adrenocortical function in patients with pulmonary tuberculosis. Tubercle 1990; 71:277–82.
72. Jaattela M, Ilvesmaki V, Voutilainen R, Stenman UH, Saksela E. Tumour necrosis factor as a potent inhibitor of adrenocorticotropin-induced cortisol production and steroidogenic P450 enzyme gene expression in cultured foetal adrenal cells. Endocrinology 1991; 128:623–9.
73. Scott GM, Murphy PG, Gemidjioglu ME. Predicting deterioration of treated tuberculosis by corticosteroid reserve and C-reactive protein. J Infect 1990; 21:61–9.
74. Onwubalili JK, Scott GM, Smith H. Acute respiratory distress related to chemotherapy of advanced pulmonary tuberculosis: a study of two cases and review of the literature. Q J Med 1986; 59:599–610.
75. Koch R. Fortsetzung uber ein Heilmittel gegen Tuberculose. Deutsch Med Wochenschr 1891; 17:101–2.
76. Anderson M. On Koch's treatment. Lancet 1891; i:651–2.
77. Rook GAW, Taverne J, Moreno C. The role of tumour necrosis factor in the pathogenesis of tuberculosis and leprosy, and the relationship between tissue-damaging pathology and the production of agalactosyl IgG. In: Bonavida B, Granger G, eds. Tumour necrosis factor: structure, mechanism of action, role in disease and therapy. Bazel: Karger, 1990: 168–76.
78. Wilson G, Schwabacher H, Maier I. The effect of the desensitisation of tuberculous guinea-pigs. J Path Bact 1940; 50:89–98.

79. Shwartzman G. Phenomenon of local tissue reactivity and its immunological, pathological and clinical significance. New York: Paul B. Hoeber, 1937.
80. Rothstein J, Schreiber H. Synergy between tumor necrosis factor and bacterial products causes hemorrhagic necrosis and lethal shock in normal mice. Proc Natl Acad Sci USA 1988; 85:607–11.
81. Debonera G, Tzortakis N, Falchetti E. Inflammation et phénomène de Shwartzman. CR Seanc Soc Biol 1932; 109:24–26.
82. Bordet P. Contribution à l'étude de l'allergie. CR Seanc Soc Biol 1931; 107:622–3.
83. Nagao S, Tanaka A. Necrotic inflammatory reaction induced by muramyl dipeptide in guinea-pigs sensitised by tubercle bacilli. J Exp Med 1985; 162:401–12.
84. Carswell EA, Old LJ, Green S, Fiore N, Williamson B. An endotoxin-induced serum factor that causes necrosis of tumors. Proc Natl Acad Sci USA 1975; 72:3666–70.
85. Shands JW, Senterfitt VC. Endotoxin-induced hepatic damage in BCG-infected mice. Am J Pathol 1972; 67:23–40.
86. Al Attiyah R, Moreno C, Rook GAW. TNFα-mediated tissue damage in mouse foot-pads primed with mycobacterial preparations. Res Immunol 1992; 143:601–10.
87. Al Attiyah R, Rosen H, Rook GAW. A model for the investigation of factors influencing haemorrhagic necrosis mediated by tumour necrosis factor in tissue sites primed with mycobacterial antigen preparations. Clin Exp Immunol 1992; 88:537–42.
88. Rook GAW, Al Attiyah R. Cytokines and the Koch phenomenon. Tubercle 1991; 72:13–20.
89. Filley EA, Rook GAW. Effect of mycobacteria on sensitivity to the cytotoxic effects of tumor necrosis factor. Infect Immun 1991; 59:2567–72.
90. Poli G, Kinter A, Justment J, Kehrl JH, Bressler A, Stanley S, Fauci A. Tumor necrosis factor alpha functions in an autocrine manner in the induction of human immunodeficiency virus expression. Proc Natl Acad Sci USA 1990; 87:782–5.
91. Vyakarnam A, McKeating J, Meager A, Beverley PCL. Tumor necrosis factors (a,b) induced by HIV in peripheral blood mononuclear cells potentiate virus replication. AIDS 1990; 4:21–7.
92. Skolnik PE, Jahn B, Wang MZ, Rota TR, Hirsch MS, Krane SM. Enhancement of human immunodeficiency virus 1 replication in monocytes by 1,25-dihydroxycholecalciferol. Proc Natl Acad Sci USA 1991; 88:6632–6.
93. Meylan PR, Munis JR, Richman DD, Kornbluth RS. Concurrent human immunodeficiency virus and mycobacterial infection of macrophages in vitro does not reveal any reciprocal effect. J Infect Dis 1992; 165:80–6.
94. Shattock R, Griedland JS, Griffin GE. Phagocytosis of *M. tuberculosis* stimulates transcription of HIV types 1 and 2 in macrophages. Proc 7th Int Conf AIDS 1991; Abstr. MA 1222.

16

Monocyte–*Mycobacterium avium* Complex Interactions: Studies of Potential Virulence Factors for Humans

John L. Johnson and Jerrold J. Ellner

Case Western Reserve University and University Hospitals of Cleveland, Cleveland, Ohio

Hiroe Shiratsuchi

Case Western Reserve University, Cleveland, Ohio

I. EPIDEMIOLOGIC ASPECTS OF *MYCOBACTERIUM AVIUM-INTRACELLULARE* COMPLEX DISEASE

Disseminated infection with *Mycobacterium avium-intracellulare* complex is the most common late opportunistic infection in patients with AIDS [1–3] and is the initial AIDS-defining condition in 3% to 4% of AIDS cases [4]. As most typable isolates (97%) from AIDS patients are *M. avium* strains [5], the designation *M. avium* will be used henceforth. Clinical syndromes related to disseminated *M. avium* infection in patients with AIDS include fever, weight loss, fatigue, severe anemia, hepatosplenomegaly and abdominal lymphadenopathy, and chronic diarrhea.

Evidence of disseminated *M. avium* infection is present in 47% to 56% of patients with AIDS undergoing autopsy [6,7]. This extremely high prevalence of disease in patients with AIDS is in striking contrast to the rarity of clinically significant infection with *M. avium* in immunocompetent hosts. Before the onset of the global human immunodeficiency virus (HIV) pandemic disseminated *M. avium* infection was extremely rare, with fewer than 100 cases reported before 1985. A slowly progressive necrotizing pulmonary infection similar to reactivation pulmonary tuberculosis occurred in elderly patients with severe underlying emphysema [8], but disseminated infection was uncommon. Recently, pulmonary disease without preexisting chronic obstructive lung disease has been described [9].

The prevalence of disseminated *M. avium* disease is greatly increased in patients with AIDS. There is also some evidence that the prevalence of disseminated *M. avium* infection is increasing in other categories of patients who are not HIV-infected. Patients with AIDS in developed nations now live longer because of effective antiretroviral therapy and prophylaxis of other opportunistic infections, including *Pneumocystis carinii* pneumonia. The increased prevalence may also be related to improved diagnosis due to growing physician awareness of the importance of the organism as a significant pathogen, lower clinical thresholds for culturing HIV-infected patients for mycobacterial disease, and improved laboratory isolation and identification techniques such as lysis centrifugation blood cultures.

M. avium infection occurs late in the course of HIV infection and is associated with severe levels of immunosuppression [10]. The major risk factor for disseminated *M. avium* infection in patients with AIDS is a severe depression in the peripheral CD4 lymphocyte count. The median CD4 count at the time of diagnosis in a recent study was 13/μl, and disseminated *M. avium* infection has been reported only rarely in patients with CD4 counts above 100/μl [11]. The aggregate lifetime risk of disseminated *M. avium* disease in patients with HIV infection is high; the risk of developing disseminated *M. avium* infection is linear after the diagnosis of AIDS, and *M. avium* bacteremia developed in 39% of patients who survived 1 year after peripheral blood CD4 counts fell below 10/μl [11]. As better therapy evolves to treat HIV infection and to treat and prevent other AIDS-related opportunistic infections such as *Pneumocystis carinii* pneumonia and cryptococcal meningitis, disseminated *M. avium* infection may become almost universal in the terminal stages of AIDS. Disseminated *M. avium* infection contributes significantly to morbidity and mortality in AIDS [12–14]. Although *M. avium* has significant in vitro resistance to many antimicrobial agents, treatment with newer four- and five-drug chemotherapy regimens for disseminated *M. avium* infection has led to improvement in symptoms, clearing of *M. avium* bacteremia, and improved survival in some cases [10,15,16]. Clarithromycin and azithromycin also appear to be extremely active chemotherapeutic agents. Regimens designed to prevent disseminated *M. avium* infection are under investigation.

II. MICROBIOLOGY OF *M. AVIUM* COMPLEX

M. avium is a ubiquitous environmental organism that has been recovered from foodstuffs, water, and soil sources [8,17–21]. *M. avium* have been traditionally grouped by seroagglutination. The serogroup differences are due to variations in the haptenic oligosaccharide of the surface glycopeptidolipid (GPL) layer of the mycobacterium. Most isolates from patients with AIDS are from a small number of serogroups—1,4, and 8 [22]. This phenomenon is poorly understood, given the widespread presence of the organism in nature. Marked geographic differences exist in he occurrence of disseminated *M. avium* infection in AIDS, with very high prevalence rates reported in the United States [13], whereas no *M. avium* isolates were recovered in cultures of blood from patients with advanced AIDS in Uganda, a sub-Saharan African nation with an extremely high prevalence of AIDS [23]. Further studies in Uganda identified no *M. avium* isolates from respiratory secretions of AIDS patients in Uganda, despite high levels of the organism in the environment [24]. The basis for this geographic variation is poorly understood, but it has been postulated as being due to earlier deaths (at a point where CD4 counts are not as severely depressed) in patients with AIDS in the developing world or to differences in colonization with *M. avium* related to increased exposure to waterborne *M. avium* by aerosolization in bath showers [25]. Alternatively, environmental *M. avium* isolates may be less pathogenic for AIDS patients. In addition to serogroup restriction, *M. avium* isolates from patients with AIDS differ from non-AIDS isolates in plasmid content, antimicrobial resistance, and virulence for the beige mouse infection model [26–28].

III. DETERMINANTS OF VIRULENCE

M. avium grows on solid media with two distinct colonial morphologies: a smooth, flat, and transparent colony type with serrated edges (SmT) and a smooth, domed, and more opaque colonial morphotype (SmD). Fresh clinical isolates from AIDS patients with disseminated *M. avium* disease are of the SmT morphotype and are highly virulent for intracellular growth in cultured human monocyte–derived macrophages [29]. With serial passage in culture, the SmT morphotype is transformed into the SmD morphotype, which is less virulent in chickens, mice, and humans [30–32]. AIDS-associated *M. avium* isolates show a similar transition to SmD morphology after serial passage in vitro [29].

We also find that the factor most consistently related to *M. avium* virulence for intracellular growth in human monocytes is the colonial morphotype (SmT versus SmD) (Figure 1) [32]. Mechanisms underlying the transition from SmT to SmD colonial morphology are not well understood but may involve quantitative or qualitative alterations in the mycobacterial surface

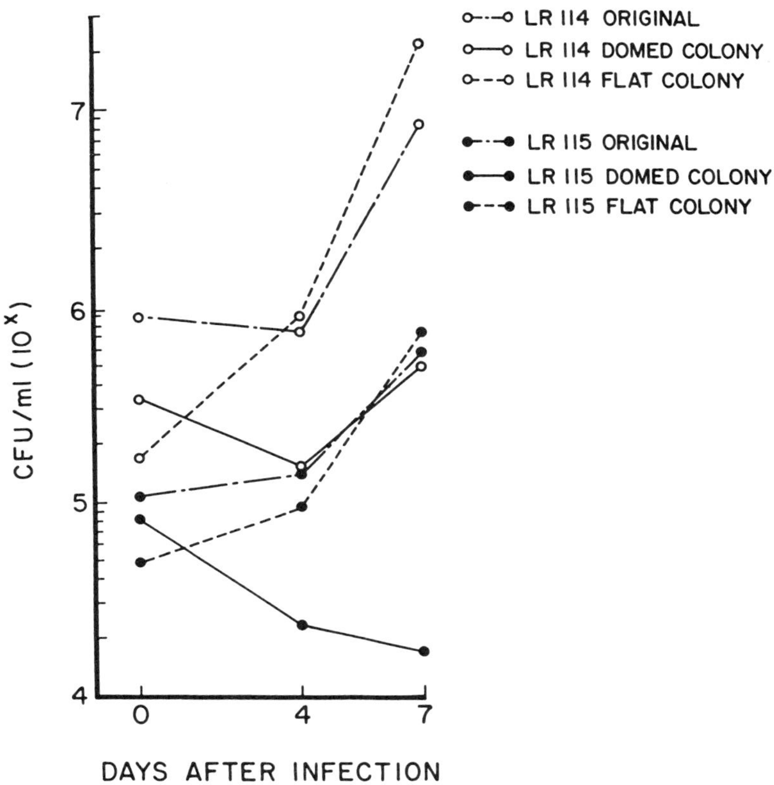

FIGURE 1 Growth of *M. avium* strains with different colonial morphotypes in human monocytes from a single healthy donor. The number of viable bacilli is plotted as mean colony-forming units (CFU) of *M. avium* per milliliter of monocyte lysate at each time point. The descriptors "original" refer to the parent *M. avium* strain and to isogenic SmD ("domed") and SmT ("flat", transparent) colonial morphotypes. (From Ref. 32, copyright 1990, The University of Chicago.)

component GPL. All *M. avium* possess a common singly glycosylated nonspecific GPL moiety. Serotypic differences between strains are due to individual variations in GPL via multiglycosylation at the threonine substituent of a conserved lipopeptide core [33,34]. Purified GPL may itself have important immunomodulating properties. Glycopeptidolipid from *M. avium* serogroup 4 inhibits mitogenesis of murine splenic lymphocytes [35]. Glycopeptidolipid and related substances may also be important in the intracellular survival of the pathogen. In chronically infected mice, mycobacteria within phagosomes are surrounded by a multilamellar electron-transparent zone (ETZ) [36]. This multilamellar coating is located between the outer surface of the mycobacterial cell wall and the phagosome membrane and structurally resembles GPL. It is possible that the ETZ represents a barrier inside the phagosome that inhibits intracellular killing of the organism.

IV. REGULATION OF MONOCYTE UPTAKE OF *M. AVIUM*

Complement receptor binding has been shown to be important in the initial attachment of *M. leprae* and *M. tuberculosis* to phagocytic cells [37,38]. Preincubation with heated or cobra venom factor–depleted serum [39] or anti-CR3 antibody [40] decreased *M. avium* attachment or phagocytosis. Selective pretreatment with EGTA and magnesium chloride resulted in levels of *M. avium* phagocytosis comparable to incubation with normal serum, whereas preincubation with EDTA or heat-inactivated serum decreased *M. avium* uptake, suggesting that the alternative complement pathway is essential for *M. avium* phagocytosis [39]. Our group also has evidence of alternative complement pathway activation with deposition of C'3 on *M. avium* and has identified a synergistic role for CR1 and CR3 complement receptors in *M. avium* uptake by freshly isolated peripheral blood monocytes (Figure 2). Preincubation with anti-CR1 or anti-CR3 antibody alone decreased *M. avium* uptake by 20% to 40% and 10% to 20%, respectively, whereas incubation with both antibodies blocked phagocytosis by 50% to 80%. Bermudez et al. have also demonstrated that CR3, fibronectin, and mannosyl-fucosyl receptors are relevant for *M. avium* uptake [40]. *M. avium* binds to extracellular matrix proteins of eukaryotic cells, including fibronectin, laminin, and collagen I, through a β_1 integrin protein on the mycobacterial surface [41]; these bacterial surface constitutents may play an important role in the initial adherence of the pathogen to basement membrane proteins in host tissues.

Certain cytokines such as interferon-γ (IFN-γ) have been shown to decrease *M. avium* uptake and phagocytosis in a dose-dependent manner [42]. This effect may be due to down-regulation of surface C'3 receptor expression by monocytes after stimulation with IFN-γ. Interferon-γ has been shown to decrease C3bR expression [43] and to decrease binding of C3b and C3bi [44]. These interactions may be important in limiting the uptake of a pathogen that is poorly killed in the intracellular environment.

V. FACTORS INFLUENCING *M. AVIUM* GROWTH IN ANIMAL AND HUMAN MONOCYTES

Intracellular pathogens such as mycobacteria must evade several host defense systems after phagocytosis and then replicate intracellularly before dissemination and production of systemic

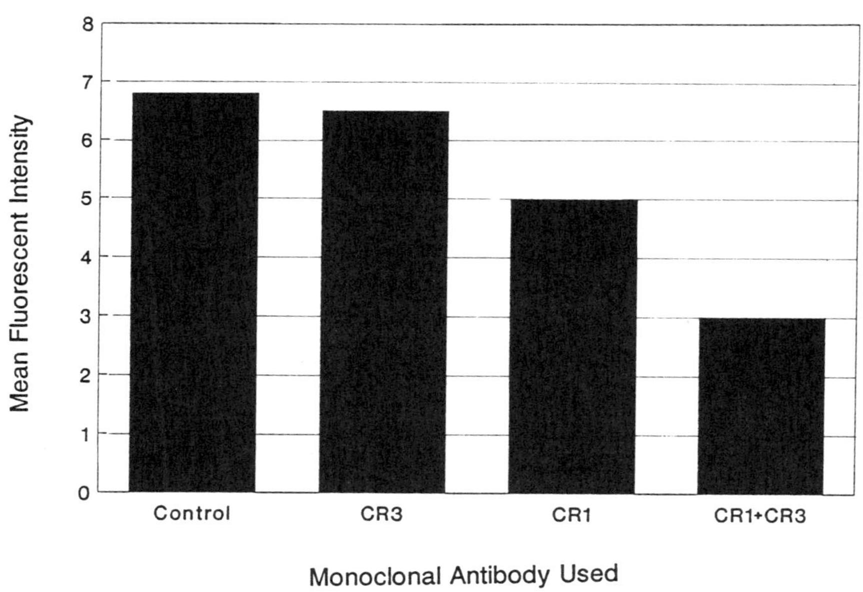

FIGURE 2 Effect of anti–complement receptor antibodies on phagocytosis of *M. avium* by human monocytes. Representative data using monocytes from a healthy subject are shown. Monocytes were incubated in the presence or absence of monoclonal anti-CR1 (3D9) or anti-CR3 (Mo1) antibody alone, or a combination of both antibodies, for 30 minutes and were then infected with *M. avium* previously stained with Lucifer yellow and opsonized with unheated serum. Uptake of *M. avium* by monocytes was analyzed by flow cytometry. Data are expressed as mean fluorescent intensity of monocytes under each condition.

disease. Growth of *M. avium* in cultured human monocytes results in a characteristic biphasic growth curve, with an early period during the first 4 days of culture after infection of bacteriostasis or killing followed by subsequent rapid intracellular replication. The biological basis for this growth pattern remains obscure; however, several observations have yielded new insights into the regulation of intracellular mycobacterial growth. After phagocytosis, *M. avium* reside in nonacidic vesicles within monocytes [45]. These vesicles contain *M. avium* surrounded by an ETZ that structurally resembles GPL [36]. Inhibition of phagosomal-lysosomal fusion occurs after phagocytosis of virulent strains of *M. avium* by murine bone marrow–derived macrophages and correlates with capacity for rapid intracellular replication of the pathogen [46].

Macrophages activated by cytokines express potent antimicrobial and tumoricidal activities. Body fluids in infected tissues also contain increased amounts of cytokine mediators that have important biological activities leading to either enhancement or suppression of intracellular and extracellular *M. avium* growth. Data concerning the effects of cytokines on mycobacterial killing have often been discordant. This probably reflects several factors—strain-to-strain and donor-to-donor differences in the virulence of *M. avium* [32,47], species-to-species differences in innate resistance to *M. avium*, and differences in culture conditions, especially with regard to the duration of cell culture before infection with *M. avium* [48,49]. *M. avium*–infected macrophages may become more permissive to intracellular mycobacterial replication after longer periods in culture [50]. Cytokines also differ in functional activities, targets, and interaction with other cytokines. Certain cytokines may act synergistically with other cytokines to activate monocytes [51], may stimulate monocytes to produce other cytokines [52–56], or may modulate surface cytokine receptor expression on target cells [57,58].

Interferon-γ is the prototypic agent with macrophage activating factor (MAF) activity and enhances intracellular killing of *Toxoplasma gondii* [59–61], *Leishmania* species [62,63], and other intracellular pathogens [64–66]. The effects of IFN-γ on mycobacterial killing are divergent. The addition of IFN-γ to murine macrophages enhances intracellular *M. tuberculosis* killing [67,68]; however, IFN-γ has been shown by different investigators under different conditions to enhance or suppress growth of this mycobacterium in human monocytes [69]. Bermudez and Young have shown IFN-γ to be ineffective in augmenting *M. avium* killing in human monocytes in vitro [70]. We have demonstrated modest degrees of MAF activity by IFN-γ for certain strains of *M. avium* in studies of fresh peripheral blood monocytes from healthy and HIV-infected persons [32,71].

We and other investigators have shown that selected cytokines alter the capacity of monocytes/macrophages for intracellular *M. avium* killing. We examined the effects of a panel of recombinant, highly purified cytokines for potential macrophage activating effects against *M. avium*. The effects of coculture with indomethacin were also examined, on the basis of earlier reports that *M. avium*–infected murine macrophages produced prostaglandin E_2 (PGE_2), which inhibited macrophage function [72]. Monocytes were cultured continuously in the presence of interleukin-1α (IL-1α), IL-1β, IL-2, IL-3, IL-4, IL-6, IFN-γ, granulocyte-macrophage colony-stimulating factor (GM-CSF), or macrophage colony-stimulating factor (M-CSF). Culture in the presence of IFN-γ decreased intracellular *M. avium* growth, with further enhancement of growth inhibition when indomethacin was added to IFN-γ (Figure 3) [73]. Pretreatment with transforming growth factor-β (TGF-β) for 2 days before monocyte infection increased the MAF activity of IFN-γ (data not shown). Coculture with migration inhibitory factor (MIF) strongly decreased intracellular mycobacterial growth (data not shown). No other cytokine or combination of cytokines consistently suppressed intracellular *M. avium* growth; indeed, IL-3, IL-6, and M-CSF increased the growth of one and IL-1α of both of the *M. avium* strains used in these experiments (Table 1, Figure 4) [73]. Interestingly, IL-1 and IL-6 enhanced the growth of *M. avium* directly in media alone, an observation confirmed by Denis and Gregg [74].

Denis has shown that exogenous TNF-α decreases intracellular *M. avium* growth, whereas IL-6 enhances such growth in human monocytes [74]. Bermudez and Young have reported that TNF alone or the combination of TNF and IL-2 decreased *M. avium* growth inside human

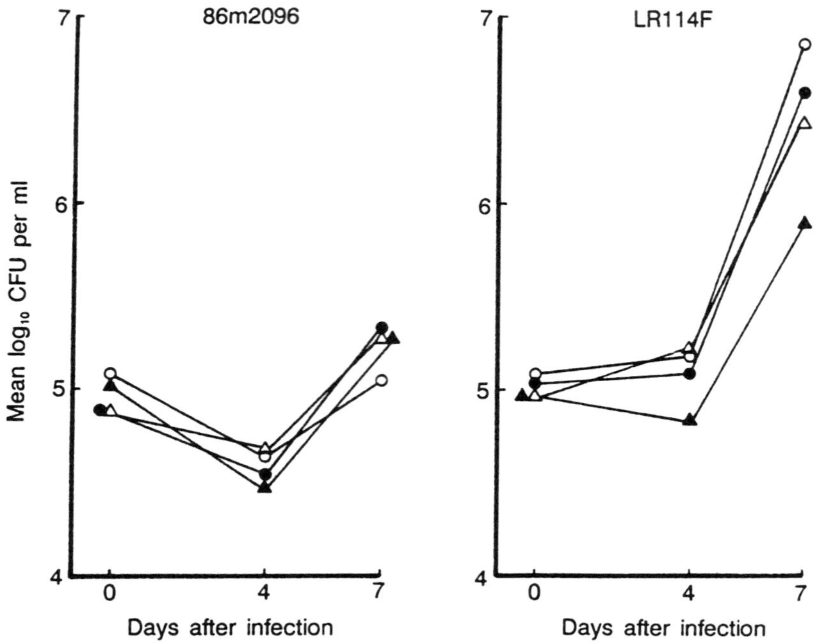

FIGURE 3 Effect of interferon-γ (IFN-γ) alone and in combination with indomethacin (IM) on the intracellular growth of *M. avium* in cultured monocytes. Two-day precultured monocytes were infected with two strains of *M. avium*. After infection, monocytes were cultured continuously in the presence of IFN-γ 300 U/ml final concentration (●——●), IM 1 μg/ml (△——△), IFN-γ 300 U/ml + IM 1 μg/ml (▲——▲), or without added IFN or IM (○——○). Data are shown as the mean CFU/ml of *M. avium* in monocyte lysates from three donors at each time point. The combination of IFN-γ and IM decreased intracellular *M. avium* growth at 7 days in culture 10-fold ($p < 0.01$, paired *t*-test compared with control culture). This growth-inhibitory effect exceeded the effect of either IFN-γ or IM alone. Growth inhibition in the presence of IFN + IM was apparent only in experiments using the more virulent LR114F (SmT) strain for infection. (From Ref. 73, copyright 1991, *The Journal of Immunology*.)

macrophages [70]. Interleukin-2 alone was incapable of enhancing *M. avium* killing in murine splenic macrophages; however, IL-2 did increase *M. avium* killing in the presence of autologous sensitized T cells or soluble *M. avium* antigens [75]. Granulocyte-macrophage colony-stimulating factor alone [76] or in combination with TNF-α [76–78] also augmented *M. avium* killing in human monocyte–derived macrophages. The interpretation of reported differences in the effects of cytokines on intracellular *M. avium* growth is complicated by donor-to-donor, strain-to-strain, and technical differences in the assay.

Mycobacteria and mycobacterial cell surface components also directly induce cytokine production by monocytes. Stimulation of human monocytes with purified lipoarabinomannan (LAM) from *M. tuberculosis* induces release of TNF [79]. Peripheral blood mononuclear cells from healthy purified protein derivative (PPD)-positive donors and patients with active pulmonary tuberculosis produce higher levels of IL-6 and TNF-α than normal monocytes from PPD-negative subjects [80]. *M. avium* infection also induces production of GM-CSF [76], TNF-α, and IL-6 [47] from human monocytes. Interleukin-6 and TNF-α production by *M. avium*–stimulated monocytes differs among healthy donors; high levels of expression of these cytokines are positively correlated with the ability of monocytes to limit intracellular *M. avium* growth [47]. Surface constituents of *M. avium* also may be important in protecting the bacilli from digestion in lysosomes.

As discussed earlier, different strains and different colonial morphotypes of the same strain of

TABLE 1 Effect of Colony Stimulating Factors on the Growth of *M. avium* Strain LR114F (SmT) in Human Monocytes

	Mean *M. avium* CFU/ml (× 10^5) in monocyte lysate		
	Day 0	Day 4	Day 7
(–)	2.15 ± 0.54[a]	3.06 ± 1.14	33.2 ± 11.8
IL-3 100 U/ml	2.56 ± 0.60	3.07 ± 1.10	79.5 ± 30.4[b]
(–)	1.09 ± 0.24	1.25 ± 0.32	41.7 ± 29.9
M-CSF (1:3000)	1.27 ± 0.42	1.66 ± 0.67	64.8 ± 32.8[c]
(–)	1.19 ± 0.36	1.36 ± 0.44	16.4 ± 7.51
GM-CSF 5000 U/ml	1.40 ± 0.46	1.70 ± 0.32	36.4 ± 28.0

[a]Data shown is the mean ± SE *M. avium* CFU/ml in monocyte lysates harvested 0, 4, and 7 days after infection. Data are from studies using monocytes from 4 to 6 healthy donors.
[b]$p < 0.01$ compared with control (–) culture.
[c]$p < 0.02$ compared with control (–) culture.
IL-3 = interleukin-3; M-CSF = macrophage colony-stimulating factor; GM-CSF = granulocyte-macrophage colony-stimulating factor.
Source: Ref. 73, copyright 1991, *The Journal of Immunology*.

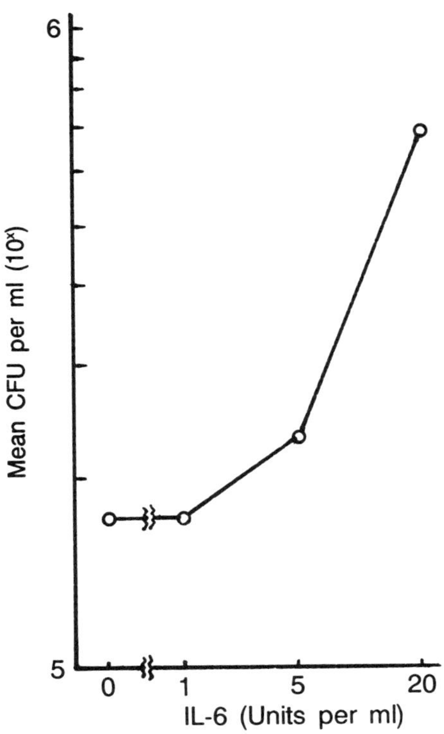

FIGURE 4 Effect of interleukin-6 (IL-6) on intracellular growth of *M. avium* in monocytes from healthy subjects. Two-day precultured monocytes were infected with *M. avium* and then were cocultured continuously in the presence of human recombinant IL-6 (doses ranging from 0 to 20 U/ml final concentration). Data are expressed as the mean CFU/ml of *M. avium* in cell lysates harvested 7 days after infection. (From Ref. 73, copyright 1991, *The Journal of Immunology*.)

M. avium show striking differences in their ability to grow in human monocytes. Michelini-Norris et al. recently demonstrated that the more virulent SmT morphotype stimulated monocytes to express less extracellular IL-1 and IL-6 and less membrane-associated IL-1 than the avirulent SmD morphotype of the same *M. avium* strain [81]. Using a different *M. avium* strain, we studied the induction of immunoreactive and bioactive cytokine expression by fresh human monocytes stimulated with SmT and SmD morphotypes of *M. avium*. Stimulation with the SmD morphotype resulted in greater expression of TNF and IL-1 activity than stimulation with the SmT morphotype; IL-6 production was comparable after stimulation with either colonial type (Table 2). Levels of specific cytokine mRNA induced by stimulation with SmT and SmD morphotypes of *M. avium* were similar, suggesting that the differences in cytokine expression are related to post-translational events or cytokine processing (data not shown). The significance of this association of cytokine induction with colonial morphotype remains speculative. It is possible that the rapid induction of cytokine production after infection with SmD activates the infected phagocyte or adjacent cells in an autocrine or paracrine manner to augment monocyte mycobacteriostatic/mycobactericidal activity.

Differences in surface glycolipid components of SmD and SmT morphotypes may ultimately account for differential triggering of cytokine expression. SmT colonial morphotypes of *M. avium* grow faster in human monocytes than SmD variants; as noted, SmT and SmD morphotypes differ in mycobacterial surface polysaccharides of the cell wall. Whereas these surface differences in *M. avium* have yet to be studied in detail, dissimilarities in the expression of LAM have been found to be relevant in differential cytokine expression after stimulation with *M. tuberculosis*. Lipoarabinomannan from the avirulent H37Ra stain of *M. tuberculosis* was 100 times more potent in inducing TNF secretion by murine macrophages than LAM from the virulent Erdman *M. tuberculosis* strain [82].

VI. INTERACTIONS OF *M. AVIUM* WITH MONOCYTES FROM HIV-INFECTED PERSONS

Disseminated *M. avium* infection in patients with AIDS is characterized by a continuous high-grade bacteremia and enormous tissue burdens of the organism. At the time of initial diagnosis, an average of 500 CFU/μl are present in the bloodstream. Involved viscera are filled with phagocytes engorged with acid-fast bacilli; tissue burdens may be as high as 10^{10} CFU/gm tissue [83]. Studies of this extraordinarily exuberant intracellular mycobacterial growth have centered on the adequacy of phagocytic effector cell functions against mycobacteria, the appropriateness of activation of the macrophage by other cells and cytokines, and innate characteristics of the microbe itself. Levels of HIV infection of circulating monocytes are believed to be low [84,85]; however, macrophage-associated HIV antigens are readily demonstrable in lymph nodes and brain and lung tissue [86].

TABLE 2 Cytokine Production by Human Monocytes Cocultured with *M. avium*[a]

	n	IL-1α (pg/ml)	IL-1β (ng/ml)	IL-6 (ng/ml)	TNF-α (ng/ml)
(−)	4	0	0	0.8 ± 0.5	0
LPS 10 μg/ml	4	543 ± 224	16.6 ± 2.0	70.8 ± 11.3	10.4 ± 1.4
SmT 10^7/ml	4	34 ± 24[b]	2.1 ± 0.7[b]	47.4 ± 13.7	1.8 ± 0.8[c]
SmD 10^7/ml	4	150 ± 61[b]	4.7 ± 1.3[b]	47.1 ± 9.0	3.3 ± 1.0[c]

[a]Monocytes obtained from healthy donors were cultured at 10^6/ml in 24-well tissue culture plates with or without the SmT or SmD variant of *M. avium*, or LPS, for 24 hours. Culture supernatants were harvested and cytokine concentrations were measured by immunoassay (ELISA).
[b]$p < 0.05$ for differences between SmD and SmT by paired *t*-test.
[c]$p < 0.01$ for differences between SmD and SmT by paired *t*-test.
IL = interleukin; TNF-α = tumor necrosis factor-α; LPS = lipopolysaccharide.

Data have been discordant regarding effector functions of monocytes from HIV-infected persons. Whereas many monocyte effector functions such as intracellular killing or inhibition of *T. gondii* and *Chlamydia psittaci* [60], killing of fungi [87], and expression of antibody-dependent cellular cytotoxicity against WEHI 164 sarcoma cells [88] are intact, other functions including monokine secretion [89,90] and Fc and C3 receptor–mediated phagocytosis [91] are abnormal. Phagocytosis of latex beads [88] and aspergilli and cryptococci [87] is normal in monocytes from patients with AIDS; however, monocyte chemotaxis to several chemoattractants, including lymphocyte-derived chemotactic factor and *N*-formyl-methionyl-leucyl-phenylalanine, is deficient [88]. The general capacity of monocytes for oxidative burst generation assessed by nitroblue tetrazolium reduction is depressed in patients with early asymptomatic HIV infection and worsens in later stages of the disease [92]. In vitro monocyte survival in patients with AIDS is decreased and directly correlated with progressive immunosuppression in AIDS [93].

In studies utilizing 2-day precultured monocytes from patients with AIDS, phagocytosis of *M. avium* by monocytes did not differ among healthy subjects, patients with AIDS, and patients with AIDS and prior known disseminated *M. avium* infection [71,94]. Intracellular *M. avium* growth inhibition at 7 days in culture was comparable between patients with AIDS and healthy donors for all 4 strains (AIDS-associated and non-AIDS-associated isolates) tested (Figure 5). Decreased *M.*

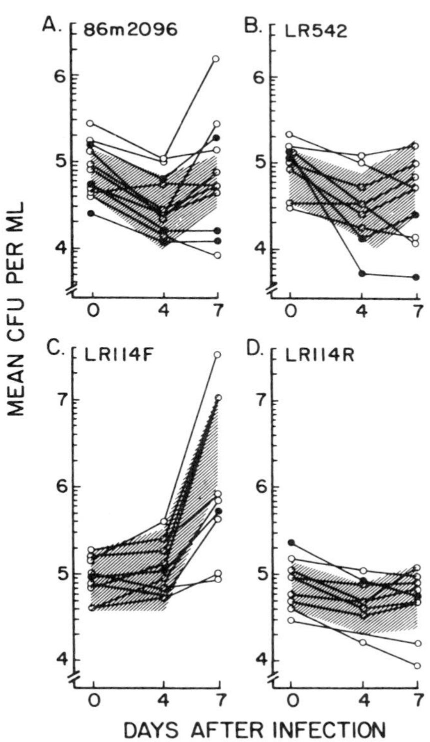

FIGURE 5 Intracellular growth of 4 strains of *M. avium* in monocytes from AIDS patients (n = 11) and healthy donors (n = 13). *M. avium* growth curves in monocyte lysates from AIDS patients without known prior *M. avium* infection (○——○) and AIDS patients with prior known disseminated *M. avium* infection (●——●) are shown for each patient. Shaded areas represent the mean ± 1 SD of *M. avium* CFU/ml in monocyte lysates of all healthy donors. *M. avium* strains 86m2096 and LR542 are AIDS-associated isolates; LR114F and LR114R are isogenic SmT and SmD colonial morphotypes, respectively, from a parent non-AIDS associated strain. (From Ref. 71, copyright 1991, *Infection and Immunity*.)

avium uptake and a modest augmentation of intracellular growth inhibition occurred after stimulation in vitro with IFN-γ, similar to effects previously noted in healthy subjects [71].

Schnittman et al. also have reported similar capacity of monocytes from patients with AIDS to limit *M. avium* growth in vitro [94]. In contrast, using 7-day precultured monocyte-derived macrophages, Crowle et al. found monocytes from patients with AIDS to be highly permissive for *M. avium* growth [95]. *M. avium*, but not *M. intracellulare*, strains replicate more rapidly in monocytes from patients with AIDS than in those from healthy subjects [96]. Selectively increased susceptibility to *M. avium* infection may help to explain the much higher incidence of *M. avium* disease in HIV-infected patients than disease due to closely related *M. intracellulare* strains [96].

The source of serum used for monocyte culture also markedly affects intracellular *M. avium* replication. Serum from patients with AIDS appears to lack an inhibitory factor contained in normal serum that limits intracellular *M. avium* growth [95]. Increased intracellular *M. avium* growth occurred when monocytes from healthy subjects were cultured using sera from AIDS patients. Incubation of *M. avium* in tissue culture medium with AIDS patient sera in the absence of monocytes did not alter *M. avium* growth patterns [96], suggesting that the sera do not contain a factor that directly promotes *M. avium* growth. The nature of this serum inhibitory activity is at present unknown.

Coinfection with HIV could conceivably influence intracellular *M. avium* replication at several levels. Concomitant infection of monocytes with mycobacteria and HIV could lead directly to exuberant *M. avium* growth, or HIV could have indirect effects on mycobacterial parasitism through its effects on other cells or by induction of cytokine expression and the creation of a local tissue milieu conducive to the growth of *M. avium*. Viral products might also directly promote *M. avium* growth. Meylan et al. tested the first of these hypotheses by infecting macrophages sequentially with a macrophage-tropic HIV strain and virulent *M. tuberculosis* H37Rv strain or *M. avium*. Phagocytosis and intracellular mycobacterial growth were comparable between healthy and HIV-infected cells, and HIV replication as assessed by release of p24 antigen into culture supernatants was not enhanced by mycobacterial infection [97]. Dual HIV and *M. avium* infection of phagocytes did not alter *M. avium* growth kinetics in this short-term model; however, increased *M. avium* growth in HIV-infected monocytes has been shown to occur after 14 days in culture and was associated with decreased IL-1α and IL-1β release [50].

Infection with HIV is associated with the spontaneous activation of other immune effector cells. Alveolar macrophages from patients with AIDS or AIDS-related complex spontaneously release TNF-α [98]. Peripheral blood monouclear cells from HIV-infected individuals spontaneously produce IL-6 [89]. Interleukin-6 induces activated B cells to mature into immunoglobulin-secreting cells and may underlie the polyclonal hypergammaglobulinermia commonly seen in AIDS. As noted, IL-6 is also a growth factor for *M. avium*.

Viral proteins are abundant in many tissues in AIDS, and local effects of these proteins might adversely affect macrophage function directly or by altering the local balance of MAF and mycobacterial growth–enhancing mediators. Data regarding the effects of HIV and HIV envelope protein gp120 on monocytes are conflicting. Earlier reports demonstrated that live or killed HIV and recombinant gp120 induced IL-1β, IL-6, and TNF production by monocytes [99–101]. Other studies using lipopolysaccharide-free reagents have not reproduced these effects [102]. The source of gp120 may be relevant, as stimulation of monocytes with native gp120, but not recombinant gp120, increased cytokine production; this effect could be blocked by soluble CD4 [103].

We studied the effects of recombinant gp120 on phagocytosis and intracellular growth inhibition of *M. avium* using monocytes from healthy donors. Stimulation with gp120 decreased *M. avium* uptake and increased intracellular *M. avium* growth (Table 3) [104]. Also, gp120 induced release of cytokines by *M. avium*–infected monocytes and blocked multinucleated giant cell formation in response to IL-3 and IFN-γ (data not shown). Interference with multinucleated

TABLE 3 Effect of Recombinant HIV-1 gp120 Protein on Phagocytosis and Intracellular Growth Inhibition of *M. avium* strain LR114F[a]

gp120 (μg/ml)	% Infected MN	AFB/100 MN	Growth index[b]
0	43 ± 3.7	165 ± 19	29 ± 8.8
0.2	35 ± 4.1	131 ± 19	52 ± 7.8
p value	<0.01	<0.001	<0.05

[a]Monocytes (MN) were precultured with recombinant gp120 (0–0.2 μg/ml) (Repligen, Boston, Massachusetts) for 2 days and then infected with *M. avium* LR114F (SmT morphotype). Phagocytosis of *M. avium* was assessed by Kinyoun staining and light-microscopic enumeration of acid-fast bacilli (AFB) in fixed monocyte preparations. Intracellular *M. avium* growth was measured by CFU assay of monocyte lysates 0 and 7 days after infection. Data are expressed as mean ± SE.
[b]Growth index was defined as *M. avium* CFU/ml in day 7 monocyte lysate divided by *M. avium* CFU/ml in day 0 monocyte lysate.
Source: Ref. 104, copyright 1992, Editions Scientifiques Elsevier.

giant cell formation is relevant, as this capacity has been proposed as a marker of MAF activity paralleling granuloma formation, a critical step in effective local control of mycobacterial infection.

Up to this point we have described how HIV infection may alter host responses to *M. avium* infection. It is also conceivable that *M. avium* infection may alter the course of HIV infection. Lymphocytes and other cells capable of harboring HIV are exposed to cytokines in the circulation and at the tissue level. Infection of quiescent T cells by HIV results in incomplete HIV replication; cellular activation is required for complete transcription of retroviral RNA [105]. Interleukin-1, IL-6, and TNF alone and the combination of IL-6 and TNF enhance HIV replication in vitro in cells infected with HIV [106–108]. As described earlier, *M. avium* induces IL-1, IL-6, and TNF-α production by monocytes (Table 2) [47,81,109]; cytokine induction due to *M. avium* infection in human tissues could, therefore, theoretically accelerate the course of HIV infection.

VII. CURRENT HYPOTHESES AND FUTURE DIRECTIONS

Although much has been learned about host and phagocyte interactions with *M. avium*, many issues remain unresolved. It is not yet clear why the prevalence of disseminated *M. avium* disease is so much higher in HIV-infected patients than in other groups of immunosuppressed persons and what factors lead to the enormous tissue burden of *M. avium* present in many patients during the advanced stages of AIDS. Factors regulating the intracellular growth of *M. avium* are incompletely understood. Intracellular *M. avium* growth is initially contained; however, this brief phase is followed by rapid, uncontrolled intracellular replication.

The work of our group and others has provided new insights into the pathophysiology of *M. avium* infection. Strain, species, and colonial morphotype are important determinants of the potential for intracellular growth of *M. avium* complex isolates. *M. avium* isolates with smooth, flat, and transparent colonial morphology are phagocytosed less well but undergo more rapid intracellular replication than isolates with smooth, domed, and opaque colonies. Phagocytosis of *M. avium* occurs via complement deposition on the surface of the organism and is mediated by CR1 and CR3 receptors on the monocyte. Additional adhesion molecules including an integrin-like protein have also been found on the surface of the mycobacterium and may be important for *M. avium* uptake by certain tissues.

Certain cytokines (e.g., IFN-γ, GM-CSF, and TGF-β) also modulate phagocytosis of the organism and have complex, bidirectional effects on intracellular *M. avium* growth. In addition, IL-1α and IL-6 promote extracellular growth of *M. avium*. Cytokines have important local autocrine and paracrine effects that regulate inflammation and granuloma formation. Colonial morphotypes differ in their capacity to induce cytokine production, with the less virulent SmD morphotype inducing higher IL-1α, IL-1β, and TNF-α activity than the SmT morphotype. Variations in mycobacterial surface components including GPL and LAM may underlie these differences and await further study. Given the evidence for bidirectional effects of cytokines on *M. avium* growth, the overall balance between macrophage-activating factor activity and *M. avium* growth–enhancing activity may ultimately be a major determinant of virulence. The differential induction of cytokines by *M. avium* colonial morphotypes, possibly as modulated by HIV products, may determine the growth characteristics of *M. avium* in the mononuclear phagocyte and in the tissues of the HIV-infected patient.

Direct and indirect interactions between *M. avium*, HIV, and viral proteins such as gp120 must be studied further in order to elucidate the pathophysiologic mechanisms underlying uncontrolled intracellular *M. avium* growth and disseminated infection. Whether or not cytokine production induced by *M. avium* or other mycobacteria increases HIV replication in otherwise latently infected cells and accelerates the natural history of HIV infection is another important question.

Future studies in these areas should provide new insights into the pathogenesis of disseminated *M. avium* in patients with AIDS and the mechanisms whereby mycobacterial surface components, viral products, and induction of host cytokines contribute to successful parasitism. This knowledge may lead to the inception of new therapeutic approaches to this important opportunistic pathogen.

REFERENCES

1. Greene JB, Sidhu GS, Lewin S, Levine JF, Masur H, Simberkoff MS, Nicholas P, Good RC, Zolla-Pazner SB, Pollock AA, Tapper ML, Holzman RS. *Mycobacterium avium-intracellulare*: a cause of disseminated life-threatening infection in homosexuals and drug abusers. Ann Intern Med 1982; 97:539–46.
2. Hawkins CC, Gold JWM, Whimbey E, Kiehn TE, Brannon P, Cammarata R, Brown AE, Armstrong D. *Mycobacterium avium* complex infections in patients with the acquired immunodeficiency syndrome. Ann Intern Med 1986; 105:184–8.
3. Macher AM, Kovacs JA, Gill V, Roberts GD, Ames J, Park CH, Straus S, Lane, HC, Parrillo JE, Fauci AS, Masur H. Bacteremia due to *Mycobacterium avium-intracellulare* in the acquired immunodeficiency syndrome. Ann Intern Med 1983; 99:782–5.
4. Chaisson RE, Hopewell PC. Mycobacteria and AIDS mortality. Am Rev Respir Dis 1989; 139:1–3.
5. Yakrus MA, Good RC. Geographic distribution, frequency, and specimen source of *Mycobacterium avium* complex serotypes isolated from patients with acquired immunodeficiency syndrome. J Clin Microbiol 1990; 28:926–9.
6. Fauci AS. Acquired immunodeficiency syndrome: epidemiologic, clinical, immunologic, and therapeutic considerations. Ann Intern Med 1984; 100:92–106.
7. Wallace JM, Hannah JB. *Mycobacterium avium* complex infection in patients with the acquired immunodeficiency syndrome. A clinicopathologic study. Chest 1988; 93:926–32.
8. Wolinsky E. Nontuberculous mycobacteria and associated diseases. Am Rev Respir Dis 1979; 119:107–59.
9. Prince DS, Peterson DD, Steiner RM, Gottlieb JE, Scott R, Israel HL, Figueroa WG, Fish JE. Infection with *Mycobacterium avium* complex in patients without predisposing conditions. N Engl J Med 1989; 321:863–8.
10. Horsburgh CR Jr. *Mycobacterium avium* complex infection in the acquired immunodeficiency syndrome. N Engl J Med 1991; 324:1332–8.
11. Nightingale SD, Byrd LT, Southern PM, Jockusch JD, Cal SX, Wynne BA. Incidence of *Mycobacteri-*

um avium-intracellulare complex bacteremia in human immunodeficiency virus–positive patients. J Infect Dis 1992; 165:1082–5.
12. Jacobson MA, Hopewell PC, Yajko DM, Hadley WK, Lazarus E, Mohanty PK, Modin GW, Feigal DW, Cusick PS, Sande MA. Natural history of disseminated *Mycobacterium avium* complex infection in AIDS. J Infect Dis 1991; 164:994–8.
13. Horsburgh CR Jr, Selik RM. The epidemiology of disseminated nontuberculous mycobacterial infection in the acquired immunodeficiency syndrome (AIDS). Am Rev Respir Dis 1989; 139:4–7.
14. Horsburgh CR Jr, Havlik JA, Ellis DA, Kennedy E, Fann SA, Dubois RE, Thompson SE. Survival of patients with acquired immune deficiency syndrome and disseminated *Mycobacterium avium* complex infection with and without antimycobacterial chemotherapy. Am Rev Respir Dis 1991; 144: 557–9.
15. Hoy J, Mijch A, Sandland M, Grayson L, Lucas R, Dwyer B. Quadruple-drug therapy for *Mycobacterium avium-intracellulare* bacteremia in AIDS patients. J Infect Dis 1990; 161:801–5.
16. Chiu J, Nussbaum J, Bozzette S, Tilles JG, Young LS, Leedom J, Heseltine PNR, McCutchan JA, and The California Collaborative Treatment Group. Treatment of disseminated *Mycobacterium avium* complex infection in AIDS with amikacin, ethambutol, rifampin, and ciprofloxacin. Ann Intern Med 1990; 113:358–61.
17. Horsburgh CR Jr, Mason UG III, Fahri DC, Iseman MD. Disseminated infection with *Mycobacterium avium-intracellulare*. Medicine (Baltimore) 1985; 64:36–48.
18. Stine TM, Harris AA, Levin S, Rivera N, Kaplan RL. A pseudoepidemic due to atypical mycobacteria in a hospital water supply. JAMA 1987; 258:809–11.
19. Du Moulin GC, Stottmeier KD, Pelletier PA, Tsang AY, Hedley-Whyte J. Concentration of *Mycobacterium avium* by hospital hot water systems. JAMA 1988; 260:1599–601.
20. Wendt SL, George KL, Parker BC, Gruft H, Falkinham JO III. Epidemiology of infection by nontuberculous mycobacteria. III. Isolation of poentially pathogenic mycobacteria from aerosols. Am Rev Respir Dis 1980; 122:259–63.
21. Graham L Jr, Warren NG, Tsang AY, Dalton HP. *Mycobacterium avium* complex pseudobacteriuria from a hospital water supply. J Clin Microbiol 1988; 26:1034–6.
22. Kiehn TE, Edwards FF, Brannon P, Tsang AY, Maio M, Gold JWM, Whimbey E, Wong B, McClatchy JK, Armstrong D. Infections caused by *Mycobacterium avium* complex in immunocompromised patients: diagnosis by blood culture and fecal examination, antimicrobial susceptibility tests, and morphological and seroagglutination characteristics. J Clin Microbiol 1985; 21:168–73.
23. Okello DO, Sewankambo N, Goodgame R, Aisu TO, Kwezi M, Morrissey A, Ellner JJ. Absence of bacteremia with *Mycobacterium avium-intracellulare* in Ugandan patients with AIDS. J Infect Dis 1990; 162:208–10.
24. Morrissey AB, Aisu TO, Falkinham JO III, Eriki PP, Ellner JJ, Daniel TM. Absence of *Mycobacterium avium* complex disease in patients with AIDS in Uganda. J AIDS 1992; 5:477–8.
25. Iseman M. *Mycobacterium avium* complex and the normal host. N Engl J Med 1989; 321:896–8.
26. Crawford JT, Bates JH. Analysis of plasmids in *Mycobacterium avium-intracellulare* isolates from persons with acquired immunodeficiency syndrome. Am Rev Resp Dis 1986; 134:659–61.
27. Gangadharam PRJ, Perumal VK, Crawford JT, Bates JH. Association of plasmids and virulence of *Mycobacterium avium* complex. Am Rev Respir Dis 1988; 137:212–4.
28. Horsburgh CR Jr, Cohn DL, Roberts RB, Masur H, Miller RA, Tsang AY, Iseman MD. *Mycobacterium avium–M. intracellulare* isolates from patients with or without acquired immunodeficiency syndrome. Antimicrob Agents Chemother 1986; 30:955–7.
29. Meylan PR, Richman DD, Kornbluth RS. Characterization and growth in human macrophages of *Mycobacterium avium* complex strains isolated from the blood of patients with acquired immunodeficiency syndrome. Infect Immun 1990; 58:2564–8.
30. Dunbar FP, Pejovic I, Cacciatore R, Peric-Golia L, Runyon EH. *Mycobacterium intracellulare* maintenance of pathogenicity in relationship to lyophilization and colony form. Scand J Respir Dis 1968; 49:153–62.
31. Schaefer WB, Davis CL, Cohn ML. Pathogenicity of transparent, opaque, and rough variants of *Mycobacterium avium* in chickens and mice. Am Rev Respir Dis 1970; 102:499–506.
32. Shiratsuchi H, Johnson JL, Toba H, Ellner JJ. Strain- and donor-related differences in the interaction of

Mycobacterium avium with human monocytes and its modulation by interferon-γ. J Infect Dis 1990; 162:932–8.
33. Brennan PJ, Goren MB. Structural studies on the type-specific antigens and lipids of the *Mycobacterium avium–Mycobacterium intracellulare–Mycobacterium scrofulaceum* serocomplex. J Biol Chem 1979; 254:4205–11.
34. Camphausen RT, Jones RL, Brennan PJ. Structure and relevance of the oligosaccharide hapten of *Mycobacterium avium* serotype 2. J Bacteriol 1986; 168:660–7.
35. Brownback PE, Barrow WW. Modified lymphocyte response to mitogens after intraperitoneal injection of glycopeptidolipid antigens from *Mycobacterium avium* complex. Infect Immun 1988; 56:1044–50.
36. Rulong S, Aguas AP, daSilva PP, Silva MT. Intramacrophagic *Mycobacterium avium* bacilli are coated by a multiple lamellar structure: freeze fracture analysis of infected mouse liver. Infect Immun 1991; 59:3895–902.
37. Schlesinger LS, Horwitz MA. Phagocytosis of leprosy bacilli is mediated by complement receptors CR1 and CR3 on human monocytes and complement component C3 in serum. J Clin Invest 1990; 85:1304–14.
38. Schlesinger LS, Bellinger-Kawahara CG, Payne NR, Horwitz MA. Phagocytosis of *Mycobacterium tuberculosis* is mediated by human monocyte complement receptors and complement component C3. J Immunol 1990; 144:2771–80.
39. Swartz RP, Naai D, Vogel CW, Yeager H Jr. Differences in uptake of mycobacteria by human monocytes: a role for complement. Infect Immun 1988; 56:2223–7.
40. Bermudez LE, Young LS, Enkel H. Interaction of *Mycobacterium avium* complex with human macrophages: roles of membrane receptors and serum proteins. Infect Immun 1991; 59:1697–702.
41. Rao SP, Gehlsen KR, Catanzaro A. Identification of a β_1 integrin on *Mycobacterium avium–Mycobacterium intracellulare*. Infect Immun 1992; 60:3652–7.
42. Toba H, Crawford JT, Ellner JJ. Pathogenicity of *Mycobacterium avium* for human monocytes: absence of macrophage-activating factor activity of gamma interferon. Infect Immun 1989; 57:239–44.
43. Esparza I, Fox RI, Schreiber RD. Interferon-γ–dependent modulation of C3b receptors (CR1) on human peripheral blood monocytes. J Immunol 1986; 136:1360–5.
44. Wright SD, Detmers PA, Jong MTC, Meyer BC. Interferon-γ depresses binding of ligand by C3b and C3bi receptors on cultured human monocytes, an effect reversed by fibronectin. J Exp Med 1986; 163:1245–59.
45. Crowle AJ, Dahl R, Ross E, May MH. Evidence that vesicles containing living virulent *Mycobacterium tuberculosis* or *Mycobacterium avium* in cultured human macrophages are not acidic. Infect Immun 1991; 59:1823–31.
46. Frehel C, deChastellier C, Lang T, Rastogi N. Evidence for inhibition of fusion of lysosomal and prelysosomal compartments with phagosomes in macrophages infected with pathogenic *Mycobacterium avium*. Infect Immun 1986; 52:252–62.
47. Newman GW, Gan HX, McCarthy PL, Remold HG. Survival of human macrophages infected with *Mycobacterium avium intracellulare* correlates with increased production of tumor necrosis factor-α and IL-6. J Immunol 1991; 147:3942–8.
48. Douvas GS, Berger EM, Repine JE, Crowle AJ. Natural mycobacteriostatic activity in human monocyte-derived adherent cells. Am Rev Respir Dis 1986; 134:44–8.
49. Blanchard DK, Michelini-Norris MB, Djeu JY. Interferon decreases the growth inhibition of *Mycobacterium avium-intracellulare* complex by fresh human monocytes but not by culture-derived macrophages. J Infect Dis 1991; 164:152–7.
50. Källenius G, Koivula T, Rydgård KJ, Hoffner SE, Valentin A, Åsjo B, Ljungh C, Sharma U, Svenson SB. Human immunodeficiency virus type 1 enhances intracellular growth of *Mycobacterium avium* in human macrophages. Infect Immun 1992; 60:2453–8.
51. Belosevic M, Davis CE, Meltzer MS, Nacy CA. Regulation of activated macrophage antimicrobial activities: identification of lymphokines that cooperate with IFN-γ for induction of resistance to infection. J Immunol 1988; 141:890–6.
52. Numerof RP, Aronson FR, Mier JW. IL-2 stimulates the production of IL-1α and IL-1β by human peripheral blood mononuclear cells. J Immunol 1988; 141:4250–7.
53. Zucali JR, Dinarello CA, Oblon DJ, Gross MA, Anderson L, Weiner RS. Interleukin 1 stimulates

fibroblasts to produce granulocyte-macrophage colony-stimulating activity and prostaglandin E_2. J Clin Invest 1986; 77:1857–63.
54. Oster W, Lindemann A, Horn S, Mertelsmann R, Herrmann F. Tumor necrosis factor (TNF)-alpha but not TNF-beta induces secretion of colony stimulating factor for macrophages (CSF-1) by human monocytes. Blood 1987; 70:1700–3.
55. Sisson SD, Dinarello CA. Production of interleukin-1α, interleukin-1β, and tumor necrosis factor by human mononuclear cells stimulated with granulocyte-macrophage colony stimulating factor. Blood 1988; 72:1368–74.
56. Lindemann A, Riedel D, Oster W, Meuer SC, Blohm D, Mertelsmann RH, Herrmann F. Granulocyte/ macrophage colony-stimulating factor induces interleukin 1 production by human polymorphonuclear neutrophils. J Immunol 1988; 140:837–9.
57. Hancock WW, Pleau ME, Kobzik L. Recombinant granulocyte-macrophage colony-stimulating factor down-regulates expression of IL-2 receptor on human mononuclear phagocytes by induction of prostaglandin E. J Immunol 1988; 140:3021–5.
58. Ruggiero V, Tavernier J, Fiers W, Baglioni C. Induction of the synthesis of tumor necrosis factor receptors by interferon-γ. J Immunol 1986; 136:2445–50.
59. Murray HW, Gellene RA, Libby DM, Rothermel CD, Rubin BY. Activation of tissue macrophages from AIDS patients: in vitro response of AIDS alveolar macrophages to lymphokines and interferon-γ. J Immunol 1985; 135:2374–7.
60. Murray HW, Rubin BY, Masur H, Roberts RB. Impaired production of lymphokines and immune (gamma) interferon in the acquired immunodeficiency syndrome. N Engl J Med 1984; 310:883–9.
61. Murray HW, Scavuzzo D, Jacobs JL, Kaplan MH, Libby DM, Schindler J, Roberts RB. *In vitro* and *in vivo* activation of human mononuclear phagocytes by interferon-γ. Studies with normal and AIDS monocytes. J Immunol 1987; 138:2457–62.
62. Douvas GS, Looker DL, Vatter AE, Crowle AJ. Gamma interferon activates human macrophages to become tumoricidal and leishmanicidal but enhances replication of macrophage-associated mycobacteria. Infect Immun 1985; 50:1–8.
63. Passwell JH, Shor R, Shoham J. The enhancing effect of interferon-β and -γ on the killing of *Leishmania tropica major* in human mononuclear phagocytes *in vitro*. J Immunol 1986; 136:3062–6.
64. Bhardwaj N, Nash TW, Horwitz MA. Interferon-γ-activated human monocytes inhibit the intracellular multiplication of *Legionella pneumophila*. J Immunol 1986; 137:2662–9.
65. Rothermel CD, Rubin BY, Murray HW. γ-Interferon is the factor in lymphokine that activates human macrophages to inhibit intracellular *Chlamydia psittaci* replication. J Immunol 1983; 131:2542–4.
66. Esparza I, Mannel D, Ruppel A, Falk W, Krammer PH. Interferon γ and lymphotoxin or tumor necrosis factor act synergistically to induce macrophage killing of tumor cells and schistosomula of *Schistosoma mansoni*. J Exp Med 1987; 166:589–94.
67. Flesch I, Kaufmann SHE. Mycobacterial growth inhibition by interferon-γ–activated bone marrow macrophages and differential susceptibility among strains of *Mycobacterium tuberculosis*. J Immunol 1987; 138:4408–13.
68. Flesch IEA, Kaufmann SHE. Attempts to characterize the mechanisms involved in mycobacterial growth inhibition by gamma-interferon–activated bone marrow macrophages. Infect Immun 1988; 56:1464–9.
69. Rook AW, Steele J, Ainsworth M, Champion BR. Activation of macrophages to inhibit proliferation of *Mycobacterium tuberculosis*: comparison of the effects of recombinant gamma-interferon on human monocytes and murine peritoneal macrophages. Immunology 1986; 59:333–8.
70. Bermudez LEM, Young LS. Tumor necrosis factor, alone or in combination with IL-2, but not IFN-γ, is associated with macrophage killing of *Mycobacterium avium* complex. J Immunol 1988; 140:3006–13.
71. Johnson JL, Shiratsuchi H, Toba H, Ellner JJ. Preservation of monocyte effector functions against *Mycobacterium avium–M. intracellulare* in patients with AIDS. Infect Immun 1991; 59:3639–45.
72. Edwards CK III, Hedegaard HB, Zlotnick A, Gangadharam PR, Johnston RB Jr, Pabst MJ. Chronic infection due to *Mycobacterium intracellulare* in mice: association with macrophage release of prostaglandin E_2 and reversal by injection of indomethacin, muramyl dipeptide, or interferon-γ. J Immunol 1986; 136:1820–7.

73. Shiratsuchi H, Johnson JL, Ellner JJ. Bidirectional effects of cytokines on the growth of *Mycobacterium avium* within human monocytes. J Immunol 1991; 146:3165–70.
74. Denis M, Gregg EO. Recombinant tumour necrosis factor-alpha decreases whereas recombinant interleukin-6 increases growth of a virulent strain of *Mycobacterium avium* in human macrophages. Immunol 1990; 71:139–41.
75. Hubbard RD, Collins FM. Immunomodulation of mouse macrophage killing of *Mycobacterium avium in vitro*. Infect Immun 1991; 59:570–4.
76. Blanchard DK, Michelini-Norris MB, Pearson CA, McMillen S, Djeu JY. Production of granulocyte-macrophage colony-stimulating factor (GM-CSF) by monocytes and large granular lymphocytes stimulated with *Mycobacterium avium–M. intracellulare*: activation of bactericidal activity by GM-CSF. Infect Immun 1991; 59:2396–402.
77. Bermudez LEM, Young LS. Recombinant granulocyte-macrophage colony-stimulating factor activates human macrophages to inhibit growth or kill *Mycobacterium avium* complex. J Leukocyte Biol 1990; 48:67–73.
78. Denis M. Tumor necrosis factor and granulocyte macrophage-colony stimulating factor stimulate human macrophages to restrict growth of virulent *Mycobacterium avium* and to kill avirulent *M. avium*: killing effector mechanism depends on the generation of reactive nitrogen intermediates. J Leukocyte Biol 1991; 49:380–7.
79. Moreno C, Taverne J, Mehlert A, Bate CAW, Brealey RJ, Meager A, Rook GAW, Playfair JHL. Lipoarabinomannan from *Mycobacterium tuberculosis* induces the production of tumor necrosis factor from human and murine macrophages. Clin Exp Immunol 1989; 76:240–5.
80. Ogawa T, Uchida H, Kusumoto Y, Mori Y, Yamamura Y, Hamada S. Increase in tumor necrosis factor alpha- and interleukin-6–secreting cells in peripheral blood mononuclear cells from subjects infected with *Mycobacterium tuberculosis*. Infect Immun 1991; 59:3021–5.
81. Michelini-Norris MB, Blanchard DK, Pearson CA, Djeu JY. Differential release of interleukin (IL)-1α, IL-1β, and IL-6 from normal human monocytes stimulated with a virulent and an avirulent isogenic variant of *Mycobacterium avium-intracellulare* complex. J Infect Dis 1992; 165:702–9.
82. Chatterjee D, Roberts AD, Lowell K. Brennan PJ, Orme IM. Structural basis of capacity of lipoarabinomannan to induce secretion of tumor necrosis factor. Infect Immun 1992; 60:1249–53.
83. Wong B, Edwards FF, Kiehn TE, Whimbey E, Donnelly H, Bernard EM, Gold JWM, Armstrong D. Continuous high-grade *Mycobacterium avium-intracellulare* bacteremia in patients with the acquired immune deficiency syndrome. Am J Med 1985; 78:35–40.
84. Schnittman SM, Psallidopoulos MC, Lane HC, Thompson L, Baseler M, Massari F, Fox CH, Salzman NP, Fauci AS. The reservoir for HIV-1 in human peripheral blood is a T cell that maintains expression of CD4. Science 1989; 245:305–8.
85. Massari FE, Poli G, Schnittman SM, Psallidopoulos MC, Davey V, Fauci AS. In vivo T lymphocyte origin of macrophage-tropic strains of HIV. J Immunol 1990; 144:4628–32.
86. Gendelman HE, Orenstein JM, Baca LM, Weiser B, Burger H, Kalter DC, Meltzer MS. The macrophage in the persistence and pathogenesis of HIV infection. AIDS 1989; 3:475–95.
87. Washburn RG, Tuazon CU, Bennett JE. Phagocytic and fungicidal activity of monocytes from patients with acquired immunodeficiency syndrome. J Infect Dis 1985; 151:565.
88. Poli G, Bottazzi, Acero R, Bersani L, Rossi V, Introna M, Lazzarin A, Galli M, Mantovani A. Monocyte function in intravenous drug abusers with lymphadenopathy syndrome and in patients with acquired immunodeficiency syndrome: selective impairment of chemotaxis. Clin Exp Immunol 1985; 62:136–42.
89. Breen EC, Rezai AR, Nakajima K, Beall GN, Mitsuyasu RT, Hirano T, Kishimoto T , Martínez-Maza O. Infection with HIV is associated with elevated IL-6 levels and production. J Immunol 1990; 144:480–4.
90. Wright SC, Jewett A, Mitsuyasu R, Bonavida B. Spontaneous cytotoxicity and tumor necrosis factor production by peripheral blood monocytes from AIDS patients. J Immunol 1988; 141:99–104.
91. Bender BS, Davidson BL, Kline R, Brown C, Quinn TC. Role of the mononuclear phagocyte system in the immunopathogenesis of human immunodeficiency virus infection and the acquired immunodeficiency syndrome. Rev Infect Dis 1988; 10:1142–54.
92. Müller F, Rollag H, Froland SS. Reduced oxidative burst responses in monocytes and monocyte-derived macrophages from HIV-infected subjects. Clin Exp Immunol 1990; 82:10–15.

93. Müller F, Rollag H, Gaudernack G, Froland SS. Impaired *in vitro* survival of monocytes from patients with HIV infection. Clin Exp Immunol 1990; 81:25–30.
94. Schnittman S, Lane HC, Witebsky FG, Gosey LL, Hoggan MD, Fauci AS. Host defense against *Mycobacterium-avium* complex. J Clin Immunol 1988; 8:234–43.
95. Crowle AJ, Cohn DL, Poche P. Defects in sera from acquired immunodeficiency syndrome (AIDS) patients and from non-AIDS patients with *Mycobacterium avium* infection which decrease macrophage resistance to *M. avium*. Infect Immun 1989; 57:1445–51.
96. Crowle AJ, Ross ER, Cohn DL, Gilden J, May MH. Comparison of the abilities of *Mycobacterium avium* and *Mycobacterium intracellulare* to infect and multiply in cultured human macrophages from normal and human immunodeficiency virus–infected subjects. Infect Immun 1992; 60:3697–703.
97. Meylan PRA, Munis JR, Richman DD, Kornbluth RS. Concurrent human immunodeficiency virus and mycobacterial infection of macrophages in vitro does not reveal any reciprocal effect. J Infect Dis 1992; 165:80–6.
98. Agostini C, Zambello R, Trentin L, Garbisa S, DiCelle PF, Bulian P, Onisto M, Poletti V, Spiga L, Raise E, Foa R, Semenzato G. Alveolar macrophages from patients with AIDS and AIDS-related complex constitutively synthesize and release tumor necrosis factor alpha. Am Rev Respir Dis 1991; 144:195–201.
99. Merrill JE, Koyanagi Y, Chen ISY. Interleukin-1 and tumor necrosis factor α can be induced from mononuclear phagocytes by human immunodeficiency virus type 1 binding to the CD4 receptor. J Virol 1989; 63:4404–8.
100. Nakajima K, Martínez-Maza O, Hirano T, Breen EC, Nishamian PG, Salazar-Gonzalez JF, Fahey JL, Kishimoto T. Induction of IL-6 (B cell stimulatory factor-2/IFN-β2) production by HIV. J Immunol 1989; 142:531–6.
101. Wahl LM, Corcoran ML, Pyle SW, Arthur LO, Harel-Bellan A, Farrar WL. Human immunodeficiency virus glycoprotein (gp120) induction of monocyte arachidonic acid metabolites and interleukin-1. Proc Natl Acad Sci USA 1989; 86:621–5.
102. Molina JM, Scadden DT, Amirault C, Woon A, Vannier E, Dinarello CA, Groopman JE. Human immunodeficiency virus does not induce interleukin-1, interleukin-6, or tumor necrosis factor in mononuclear cells. J Virol 1990; 64:2901–6.
103. Clouse KA, Cosentino CM, Weih KA, Pyle SW, Robbins PB, Hochstein HD, Naturajan V, Farran WL. The HIV-1 gp120 envelope protein has the intrinsic capacity to stimulate monokine secretion. J Immunol 1991; 147:2892–901.
104. Wallis RS, Ellner JJ, Shiratsuchi H. Macrophages, mycobacteria and HIV: the role of cytokines in determining mycobacterial virulence and regulating viral replication. Res Microbiol 1992; 143:398–405.
105. Zack JA, Arrigo SV, Weitsman SR, Go AS, Haislip A, Chen ISY. HIV-1 entry into quiescent primary lymphocytes: molecular analysis reveals a labile, latent viral source. Cell 1990; 61:213–22.
106. Poli G, Bressler P, Kinter A, Ouh E, Timmer C, Rabson A, Justement AS, Stanley S, Fauci AS. Interleukin 6 induces human immunodeficiency virus expression in infected monocytic cells alone and in synergy with tumor necrosis factor α by transcriptional and posttranscriptional mechanisms. J Exp Med 1990; 172:151–8.
107. Osborn LO, Kunkel S, Nabel GJ. Tumor necrosis factor-α and interleukin 1 stimulate the human immunodeficiency virus enhancer by activation of the nuclear factor κB. Proc Natl Acad Sci USA 1989; 89:2336–40.
108. Clouse KA, Powell D, Washington I, Poli G, Strebel K, Farrar W, Barstad P, Kovacs J, Fauci AS, Folks TM. Monokine regulation of human immunodeficiency virus-1 expression in a chronically infected human T cell clone. J Immunol 1989; 142:431–8.
109. Blanchard DK, Michelini-Norris MB, Pearson CA, Freitag CS, Djeu JY. *Mycobacterium avium-intracellulare* induces interleukin-6 from human monocytes and large granular lymphocytes. Blood 1991; 77:2218–24.

17

The Role of the Macrophage in Resistance to the Leprosy Bacillus

James L. Krahenbuhl and Linda B. Adams

G. W. Long Hansen's Disease Center, Laboratory Research Branch at Louisiana State University, United States Public Health Service, Baton Rouge, Louisiana

I. INTRODUCTION

Leprosy is predominantly a disease of skin, mucous membranes, and peripheral nerves. Active infection with *Mycobacterium leprae* is characterized by great variability in clinical course of infection, ranging from paucibacillary disease, in which few organisms are present, to multibacillary disease, in which enormous numbers of bacilli are present in the lesions [1]. Nerve damage is attributable either to overwhelming bacterial proliferation or to the localized host response to the presence of relatively few bacilli in areas of the dermis adjacent to peripheral nerves. The variability in clinical leprosy is mirrored by an equally broad spectrum of histopathological features that stem from variations in host humoral and cell-mediated immunity (CMI) response to the leprosy bacillus.

In tuberculoid leprosy (TT) disease is localized; few, if any, *M. leprae* are seen in well-organized skin granulomas consisting of epithelioid cells, Langhans' giant cells, and a discrete organization of CD4+ and CD8+ T cells [2]. Nerve damage in TT may be due to CMI response to *M. leprae* in sites adjacent to nerves. Individuals with TT manifest a strong delayed-type hypersensitivity to antigens but produce relatively low levels of antibody.

Opposite TT in the clinical spectrum is lepromatous leprosy (LL), in which widespread proliferation of bacilli results in skin lesions diffusely distributed in the skin or localized to nodules. Enormous numbers of *M. leprae* (up to 10^{10} per gram of tissue) are present in these lepromatous granulomas, and few T cells are found in the lesions. Unlike TT, there is a potent antibody response in LL and, strikingly, a clearly identified progressive anergy in T cell response that is limited to specific antigens of the leprosy bacillus. Several hypotheses are under investigation to explain specific CMI anergy in LL, including a genetic predisposition, clonal deletion of *M. leprae*–specific T cells, and a role for specific suppressor T cells [1].

Thus LL presents to the immunologist the opportunity to investigate defective T cell interactions in a chronic nonfatal human immunodeficiency disease. However, the principal host cells for *M. leprae*, an obligate intracellular pathogen, are mononuclear phagocytes (MNPs) or macrophages (Mϕs). The failure of the Mϕ to cope effectively with *M. leprae* is a conspicuous characteristic of LL. Macrophages gorged with bacilli characterize LL granulomas, and even after prolonged chemotherapy, the bacilli are only slowly cleared. In contrast, in paucibacillary leprosy or in disease-free individuals exposed to *M. leprae*, the ability of the Mϕ to kill or inhibit bacilli is probably protective, with nerve damage an unfortunate complication. Thus, although the mechanisms underlying defective CMI in LL are not clearly understood, the ability of the Mϕ to cope with *M. leprae* is also an issue central to understanding the mechanism or mechanisms of host resistance to leprosy.

II. THE MACROPHAGE IN GENERAL

A general discussion of the origin and development of Mϕs, their phenotypic, functional, and secretory heterogeneity, and their afferent and efferent roles in CMI is beyond the scope of this chapter and is covered earlier in the present volume.

The MNP system represents a cellular lineage in which cells that originate in the bone marrow enter the circulation and eventually transform into tissue Mϕs [3]. A few Mϕs (<5%) divide locally in the tissues. Descendants of bone marrow–derived monoblasts and promonocytes include circulating and marginating blood monocytes, free Mϕs in the peritoneal and pleural cavities, the Mϕs in splenic sinusoids and lymph nodes, connective tissue histiocytes, Kupffer cells, alveolar Mϕs, and the microglial cells of the brain.

The maturation of tissue Mϕs is influenced by a number of stimuli and receptor-ligand interactions, giving them a constitutive and inducible diversity in functional activity in the steady state as well as in an inflammatory response. They are unique in their endocytic and phagocytic capacity for soluble substances and particulate matter and produce a vast array of secretory products to regulate the immune, hematopoietic, and inflammatory responses [4]. In regulating the immune response to infection with intracellular pathogenic microorganisms (such as *M. leprae*), the Mϕ not only plays a key role in initiating and intensifying CMI but is the principal cellular defensive arm that is stimulated and called on to cope with the pathogen.

In their afferent immunoregulatory role, Mϕs function as antigen-presenting cells (APCs) in their interaction with T cells. All mammalian cells can endocytose and process soluble (protein) antigens and some also express surface class II major histocompatibility complex (MHC) molecules. Many cells other than Mϕs (i.e., dendritic cells, Langerhans cells, B cells, endothelial cells) express MHC class II antigens and can function as APCs. However, Mϕs are uniquely efficient APCs because they are not only capable of endocytosis of soluble antigens but can phagocytose particulate antigens, including pathogenic microorganisms, and degrade, process, and present antigens relevant for stimulation of CMI in an MHC class II–restricted context. Moreover, by secretion of T cell–amplifying monokines such as interleukin I (IL-1) and tumor necrosis factor-α (TNF-α), Mϕs can markedly upregulate CMI, especially in a local site.

Primarily through the work of Mackaness and his colleagues in the 1960s [5] and numerous workers since, we recognize the important role of the activated Mϕ in host resistance to obligate and facultative intracellular pathogens. Activation modifies numerous biochemical, phenotypic, and functional properties of the Mϕ, but chief among these changes is the induction of an enhanced microbicidal capacity. Whereas normal Mϕs support the growth of intracellular microorganisms, activated Mϕs kill or inhibit not only the pathogen that incited the CMI response, but a broad spectrum of phylogenetically unrelated intracellular microorganisms as well [6]. Thus, Mϕs activated by lymphokines released from sensitized T cells [7] responding specifically to antigen play a major role in resistance to a wide variety of obligate and facultative intracellular pathogens. The key Mϕ-activating lymphokine appears to be interferon-γ (IFN-γ) [8].

III. THE MACROPHAGE IN LEPROSY

A. The Leprosy Spectrum from the Perspective of the Macrophage

Across the leprosy spectrum the local lesions are represented by collections of lymphocytes, Mϕs, and other cell types organized within distinct three-dimensional architectural structures (granulomas). The level of host resistance hinges on complex interactions between lymphocytes of the helper and suppressor phenotype, *M. leprae* antigens and constituents, and maturity of the Mϕs in the lesions, as well as modulating cytokines and other factors elaborated by the cells composing the lesion. Nevertheless, ultimately it is the relationship between the Mϕ and the leprosy bacillus that mirrors the clinical spectrum of leprosy. From the perspective of the Mϕ, this spectrum ranges from successful management of infection to the complete incapacity of the Mϕ to cope with the growth of the leprosy bacillus.

1. Ability of Macrophages to Cope in Leprosy

In the vast majority of cases, it is safe to presume that CMI events follow infection with the leprosy bacillus and that the Mϕ plays a critical, albeit undetected, role in events that quietly culminate in the elimination of the organism without production of clinically detectable inflammation or disease. On another plane, the events leading to the formation of an epithelioid cell granuloma and the subsequent limitation and elimination of the local bacilli in TT leprosy or a positive lepromin reaction are also likely to be the consequences of such a scenario occurring in a localized focus of inflammation. Of the numerous morphological and metabolic criteria that characterize activated Mϕs [9], two enhanced metabolic pathways stand out: the enhanced production of oxidative metabolites [10] and the increased production of L-arginine–derived reactive nitrogen intermediates [11,12]. These pathways may underlie the enhanced microbicidal capacity of the activated Mϕ for *M. leprae* and will be discussed in more detail below.

2. Inability of Macrophages to Cope in Leprosy

Regardless of the lack of a clear understanding of the underlying mechanisms of T cell anergy and defective CMI in LL, the failure of the Mϕ to cope with *M. leprae* is a conspicuous characteristic of the lepromatous form of the disease and is an issue central to understanding the mechanisms of host resistance to the leprosy bacillus. Accumulating data suggest that *M. leprae* and some of its constituents may be able to subvert Mϕ function and, in addition to T cell anergy, play a role in the down-regulation of CMI associated with LL (see below).

B. Macrophage Acquisition of *M. leprae*

M. leprae has an apparent predilection for the nasal mucous membranes and the skin, where the destruction of peripheral sensory and motor nerves results in most serious forms of clinical disease in leprosy. The Mϕ is clearly the predominant host cell for *M. leprae*, although it is not yet clear how the bacilli reach these sites. Although aerosol invasion of the nasal mucosa [13] and skin penetration [14,15] are possibilities, the actual route of infection is not known. The presence of bacilli in the skin and mucous membranes may be the result more of suitable local growth conditions than of the proximity of these sites to the route of infection. Infected blood monocytes [16] or bone marrow promonocytes [17] would be an ideal transport vehicle for hematogenous spread and deposition of bacilli in distant skin sites adjacent to peripheral nerve endings. Thus, we have little knowledge about the first encounter between the Mϕ and *M. leprae*, but in vitro studies allow exploration of the unique relationship between the leprosy bacillus and its host cell, the Mϕ.

Studies of Mϕ function largely employ murine peritoneal Mϕs and human peripheral blood monocyte–derived macrophages (MdMs), because they are readily obtainable from easily accessible compartments, and Mϕ cell lines, which can be conveniently maintained in continuous culture. Comparatively little is known about the functions and properties of Mϕs in other anatomical compartments, and even less is known about the function of Mϕs isolated from granulomas such as those that characterize TT or LL.

C. Are Macrophages Defective in Leprosy?

The question of whether there are defects in the innate microbicidal capacity of human Mϕs has been raised in a number studies employing blood MNPs. Contradictory evidence supports a lysosomal enzyme deficiency in Mϕs from LL patients compared with TT patients [18], with some evidence that Mϕs from LL patients are unable to digest killed *M. leprae* [19], whereas cells from lepromin-positive normal individuals and TT patients could destroy leprosy bacilli. Not all workers have found these deficiencies [20]. More recently, Marolia et al [21] reported that MdMs of leprosy patients from across the clinical spectrum were unable to "kill" (fluorescein diacetate:ethidium bromide staining [FDA:EB]) *M. leprae*. Only MdMs from normal subjects were able to kill *M. leprae*, i.e., leprosy patients' MdMs were "defective."

However, other studies [20] found no correlation between lepromin activity and the ability of an individual's blood MNPs to kill or destroy *M. leprae*. Drutz et al. [22] employed several intracellular pathogens, including *Listeria monocytogenes* and *Candida albicans*, and found no baseline defect in antimicrobial capacity of blood MdMs from LL patients. Similarly, Horwitz et al. [23] found no intrinsic defect in the ability of monocytes from LL and TT patients to inhibit *Legionella pneumophila*. Sharp and Banerjee [24] studied the oxidative metabolism of peripheral blood MNPs of leprosy patients and found no differences in the production of hydrogen peroxide and superoxide anion between TT and LL patients and normal controls, suggesting that in leprosy these Mϕs were competent. However, Nathan et al. [25] reported that MNPs from LL patients *were* deficient in oxidative metabolism and, presumably, microbicidal function as well. They equated their findings with the defective MNP function of patients with chronic granulomatous disease, a life-threatening syndrome in which patients are unduly susceptible to infection with opportunistic intracellular pathogens.

It is important to stress that LL patients are not immunocompromised hosts, i.e., cancer-prone or susceptible to infection with the opportunistic intracellular pathogens that plague patients with immunodeficiency diseases. The immunologic anergy of LL is specific for antigens of *M. leprae*. The inability of Mϕs in these patients to cope with *M. leprae* is linked more with the failure of the T cells to respond to *M. leprae* antigens by secretion of Mϕ-activating lymphokines (IFN-γ) than to an innate failure of the Mϕ to kill *M. leprae*. Blood MNPs from patients with LL or TT could be activated by lymphokines to strongly inhibit the multiplication of *L. pneumophila*, indicating that, if they received the appropriate signals, MNPs from leprosy patients could function normally [23].

The sometimes conflicting findings obtained with blood MNPs could be due to the wide variation in methods employed, and the relevance of studying Mϕs solely from readily available compartments is questionable. Macrophages in the leprosy lesions themselves may be under paracrine or autocrine influences that regulate their capacity to cope with the leprosy bacillus.

D. Leprosy Granuloma Macrophages

Leprosy granuloma Mϕs represent the influx of recruited bone marrow–derived monocytes or a relatively small population of proliferating resident Mϕs. Freeing intact and functional Mϕs from leprosy lesions is difficult. Granuloma Mϕs possess delicate cytoplasmic extensions that intertwine with similar extensions of other cells. These Mϕ plasma membrane extensions are easily ruptured by mechanical isolation procedures, resulting in destruction of the cells.

Because the *M. leprae*–infected athymic (nu/nu) mouse footpad offers a model system for a lepromatous granuloma, we have employed methods developed to disperse and isolate cells from solid tumors to isolate the lepromatous Mϕs from these footpad granulomas [26,27]. Digestion of the footpad granulomas with DNase and collagenase yielded millions of viable lepromatous Mϕs, gorged with leprosy bacilli. These studies compared the phenotypic characteristics of these lepromatous Mϕs with those of (enzyme-treated) Mϕs from another anatomical compartment in the same mice (the peritoneal cavity), and included a study of the baseline and IFN-γ–inducible efferent and afferent effector functions of Mϕs from these two compartments. That these granuloma Mϕs were incapable of coping with *M. leprae* was clearly evident. However, except for being gorged with bacilli these Mϕs were phenotypically indistinguishable from peritoneal Mϕs from the same nu/nu mouse or from immunologically normal heterozygote (nu/+) mice. Like normal Mϕs, footpad Mϕs were tightly adherent to plastic, avidly phagocytized latex beads, *C. albicans* and *Toxoplasma gondii*, stained positive for nonspecific esterase (negative for peroxidase), possessed normal numbers of Fc receptors, and stained heavily for MAC-1 monoclonal antibody, specific for the Mϕ C3Bi receptor. *T. gondii* multiplied normally in both peritoneal and footpad Mϕs, but whereas the former Mϕs could be activated by IFN-γ to kill *T. gondii* and tumor target cells, footpad Mϕ could not be activated by IFN-γ. Similarly, superoxide levels were not enhanced by stimulation with phorbol myristate acetate (PMA), and IFN-γ did not up-regulate class II MHC (Ia) antigen expression in footpad Mϕs as was the case in peritoneal Mϕs. These findings clearly point to defective Mϕ function localized to the experimental lepromatous granuloma coinciding with normal Mϕ function in a different anatomical compartment in the same individual.

Human leprosy granuloma cells have been examined phenotypically. Munro et al. [28] examined sections of human TT granulomas using monoclonal antibodies to distinguish between epithelioid cells and macrophages. Epithelioid cells develop from Mϕs, and their organization into a granuloma characterizes the successful CMI seen in the tuberculoid end of the spectrum [29]. Although epithelioid cells appear to be activated, as shown by abundant cytoplasmic organelles and increased enzymatic and secretory activity, they are poorly phagocytic and have fewer immunoglobulin and complement receptors than Mϕs [30].

Narayanan and his colleagues have studied human LL granuloma Mϕs. They digested human LL granulomas and obtained low numbers of large, adherent *M. leprae*–infected cells that were both esterase- and peroxidase-positive [31]. Although these cells did not possess C3b receptors, Narayanan and colleagues concluded that they were Mϕs. Unlike the murine Mϕs, in which MHC class II expression is IFN-γ–inducible, human MNPs constituitively express Ia. Ridley and Ridley [32] described Ia-like expression on Mϕs more at the TT end of the spectrum than at the LL end. With more sensitive techniques, Narayanan's group observed MHC class II expression on Mϕs from both TT and LL granulomas [33].

IV. INFECTION OF THE MACROPHAGE WITH THE LEPROSY BACILLUS

A large body of evidence allows us to partially describe the encounter and subsequent intracellular relationship that develops between the Mϕ and the leprosy bacillus.

A. Phagocytosis/Entry

Abnormal phagocytosis of *M. leprae*, characterized by excessive uptake of bacilli, was demonstrated in MdMs from LL patients in comparison with cells from TT patients or normal subjects [34]. The role of the abundant anti–*M. leprae* antibodies present in LL sera was not considered. Evidence exists that *M. leprae* invade a subpopulation of Mϕs that are prone to active entry by the bacillus [35], suggesting these Mϕs may be efficient in phagocytosis but less efficient in other Mϕ functions such as antigen presentation.

Horwitz et al. [36,37] described the importance of heat-labile human serum components to ensure maximum adherence and phagocytosis of *M. leprae*. The phagocytosis of *M. leprae* was mediated by complement receptors CR1 and CR3 on the surface of monocytes and CR1, CR3, and CR4 on the surface of the more mature MdMs. Phagocytosis was facilitated by the interacting of these receptors with complement component C3 fixed to the surface phenolic glycolipid (PGL-1) antigen on *M. leprae* [38]. The activation of C3 in nonimmune serum by PGL-1 occurred by the classical complement pathway and was triggered by low amounts of natural or cross-reacting IgG or IgM antibody in the serum.

B. The Nonprofessional Phagocyte

The Mϕ is not the only host cell for *M. leprae*. In spite of its special affinity for cells of the MNP system, the leprosy bacillus invades and multiplies in a variety of poorly phagocytic or nonphagocytic host cells. With light and electron microscopy, Job et al. [39] studied *M. leprae* infection in the armadillo and found bacilli in a variety of MNPs and parenchymal cells, including peripheral blood monocytes, Mϕs from the spleen and lymph node, Kupffer cells, hepatocytes, striated and cardiac muscle cells, endothelial cells, and even cortical and medullary cells from the adrenal glands. Cultures of mouse myocardial cells readily became infected with *M. leprae* [40]. The apparent predilection of *M. leprae* for the Schwann cell probably serves as the basis for the peripheral nerve destruction that is the chief pathological change in leprosy at both poles of the clinical spectrum.

C. Phagosome/Lysosome Fusion

A successful intracellular pathogen must cope with the host Mϕ's innate or enhanced capacity to kill or impair the organism. Once a pathogen is ingested by the Mϕ and is encased in a phagosome, fusion of the phagosome with nearby lysosomes (P-L fusion) brings into a number of potent antimicrobial factors into direct contact with the microbe. D'Arcy Hart [41] describes three potential responses to the interaction between phagosome and lysosome in normal (cultured mouse) Mϕs: 1) P-L fusion, in which nonpathogens and attenuated pathogens are killed, or virulent pathogens survive and multiply in phagolysosomes (*L. monocytogenes, Leishmania mexicana, Salmonella typhimurium*); 2) nonfusion, in which the pathogens multiply in unfused phagosomes (*T. gondii*, virulent *M. tuberculosis*); and 3) escape of the organism from the phagosome into the cytoplasm (*Trypanosoma cruzi* and possibly *M. leprae*). The correlation between P-L fusion and microbial death is only circumstantial. The existence of quick-acting nonlysosomal oxidative and nonoxidative antimicrobial mechanisms raises a circular argument. Does P-L fusion kill the organism, or is P-L fusion the culmination of other antimicrobial events in normal Mϕs. Lysosomal contents are increased in activated Mϕs, P-L fusion seems to be enhanced, and pathogens are more readily killed, but oxidative and nonoxidative antimicrobial factors are also enhanced and the circular argument is not mitigated.

M. leprae can be found in Mϕs in membrane-bound compartments containing lysosomal enzymes such as acid phosphatase [42]. We have recently studied by electron microscopy the intracellular fate of *M. leprae* in cultured normal and activated mouse Mϕs [43]. Pure suspensions of bacilli were employed to avoid intracellular responses to contaminating mouse tissue rather than the leprosy bacillus. Killed or freshly harvested (nu/nu mouse footpad) live organisms were used, enriched for purity (free of host cell components) and viability (determined by ATP activity) by density gradient centrifugation. Secondary lysosomes were labeled with an electron-dense marker, Thoria-Sol. The results suggest that phagosomes containing live bacilli resist fusion with lysosomes in normal Mϕs but that fusion is the rule if killed bacilli are employed. Thus *M. leprae* seem similar to *M. tuberculosis, M. bovis, M. microti,* and *M. avium*, which previous ultrastructural studies have shown to resist lysosome fusion [44,45]. Phagosome/lysosome fusion was enhanced after opsonization of live bacilli with leprosy patients' sera, a finding also observed when *M. leprae* surface antigens were coated with a variety of antimycobacterial antibody preparations [46]. Finally, in IFN-γ-activated Mϕs, phagosomes containing viable *M. leprae* underwent extensive fusion with secondary lysosomes.

Rastogi and his colleagues [47] employed immunogold labeling techniques to study the distribution of *M. leprae* antigens in defined bacterial structures in ultrathin sections of infected Mϕs. They reported that bacterial death and/or lysis was not a prerequisite for the processing of antigens and their subsequent transportation to the surface membrane of the Mϕ. Their evidence also suggested that surface antigens were processed early but that cell wall antigens remained unreleased even after 7 days inside the Mϕ.

The possibility that *M. leprae* may possess the ability to evade Mϕ antimicrobial function by escaping the phagosome to exist freely in the cytoplasm was raised by observations in footpad biopsy specimens [48,49]. However, in vitro work shows that time and dose of infection clearly affect the observed intracellular fate of organisms, and the inability to control these variables in biopsy specimens makes accurate interpretation of the events in Mϕs in tissue sections difficult. In a nonprofessional phagocyte such as the infected cardiac muscle cell [40] the absence of a phagosomal membrane around leprosy bacilli was also observed. Mistry et al. [34] reported the uptake of leprosy bacilli by cytocholasin B–treated Mϕs, suggesting internalization of *M. leprae* by Mϕs occurs by mechanisms distinct from typical phagocytosis. Whether the bacilli may thus end up outside the Mϕ phagosome in the cytoplasm was not examined. Chandi and Job [50] observed cultures of *M. leprae*–infected mouse Mϕs for 14 weeks and failed to see bacilli free in the cytoplasm. In our own in vitro studies we observed over 300 infected cells for up to 6 days. All organisms remained bounded by a cytoplasmic membrane; none were seen free in cytoplasm.

D. The Course of Infection

1. *Multiplication of* M. leprae *in the Macrophage*

Numerous attempts have been made to demonstrate that *M. leprae* can multiply in tissue cultures of Mϕs. Sharp and Banerjee [51] infected Mϕs from animal hosts highly susceptible to in vivo growth of *M. leprae* (athymic mice and rats and armadillos), observed these cultures for over 200 days, and reported no growth of leprosy bacilli. Several characteristics of *M. leprae* present obstacles to this approach. First, the generation time of *M. leprae* during logarithmic growth (mouse footpad) is approximately 2 weeks [52]. Second, even after the organism has been killed, the prominent glycolipid-containing capsule of *M. leprae* is thought to afford protection from intracellular degradation by forming an undigestible hydrophobic interface between the bacillus and the host cell [53,54]. Capsule and cell wall constituents may also function more actively in the intracellular survival of leprosy bacilli, as discussed below (see "Virulence Mechanisms"). Third, even under the most ideal conditions for in vivo propagation (armadillo, nu/nu mouse footpad), only a small proportion ($<1\%$) of the harvested bacilli are thought to be viable. Conclusive demonstration of a tenfold increase in total numbers of bacilli would require months of culture. Observation of even a modest increase in the number of *M. leprae* per Mϕ must be cautiously interpreted because of the possibility that dead Mϕs only offer their released bacilli for subsequent phagocytosis by remaining Mϕs. Convincing evidence that *M. leprae* multiplies in the Mϕ in vitro remains to be demonstrated.

2. *Killing of the Leprosy Bacillus in the Macrophage*

In 1974 we [55] showed that growth of *M. leprae* in the mouse footpad was markedly restricted in mice with potent populations of activated Mϕs; however, we could only imply a role for activated Mϕs. Local and systemic treatment of infected footpads with the Mϕ-activating immunopotentiator *Corynebacterium parvum* [56] showed that only the local footpad injections affected *M. leprae* growth and viability (subsequent footpad titration). Again a role for the activated Mϕ was suspected but not proven. In vitro evidence was needed.

The fundamental obstacle in directly addressing the role of the Mϕ in resistance to leprosy is the inability to routinely quantitate the viability of *M. leprae*. These organisms can obviously survive and multiply in Mϕs in vivo, but confirming that they even remain viable in vitro in Mϕs is difficult. A highly subjective evaluation [52] of acid-fast staining patterns of the bacilli (morphological index) has been employed, and viability of *M. leprae* can also be determined by titration in the mouse footpad, as mentioned above [56], an expensive method requiring hundreds of mice and a year or more to complete [57].

We recently evaluated five alternative viability assays to examine the in vitro effects of activated mouse Mϕs on the leprosy bacillus [58]. Only freshly harvested, highly viable nu/nu mouse–derived bacilli were employed to infect normal peritoneal Mϕs, Mϕs harvested from mice chronically infected with *C. parvum* or *T. gondii*, or Mϕs activated by pretreatment in vitro with IFN-γ and lipopolysaccharide (LPS).

Viability staining with FDA:EB suggested killing by activated Mϕs. Others have concluded that this technique was unreliable as a measure of viability in a suspension of leprosy bacilli [59,60]. Our findings were marked, but interpretation of the slides was subjective and we had reservations about the high baseline level of "viables."

Intracellular ATP content of *M. leprae* is a more objective measure of viability [61], and we showed that *M. leprae* released from activated Mϕs 6 to 10 days after infection possessed markedly lower levels of bacterial ATP than organisms from control Mϕs.

Synthesis of phenolic glycolipid-1 (PGL-1), a biochemically unique and immunologically specific *M. leprae* constituent, was quantitated in infected Mϕs by measuring incorporation of radiolabeled palmitic acid [62]. The data clearly revealed that this anabolic pathway was almost completely curtailed in bacilli residing in activated Mϕs for as little as 24 hours.

Oxidation of palmitic acid by *M. leprae* released from Mϕs is a rapid and sensitive catabolic measure of metabolism and, most probably, viability, and this assay allows study of multiple experimental variables. Using both the BACTEC and Buddemeyer [61] adaptations of this technique to quantitate CO_2 production, we observed a profound loss in viability of *M. leprae* released from activated Mϕs (Figure 1).

These studies were designed to explore the ability of the activated Mϕ to cope with relatively few bacilli (i.e., <10/Mϕ), unlike the studies we described previously in which Mϕs were so

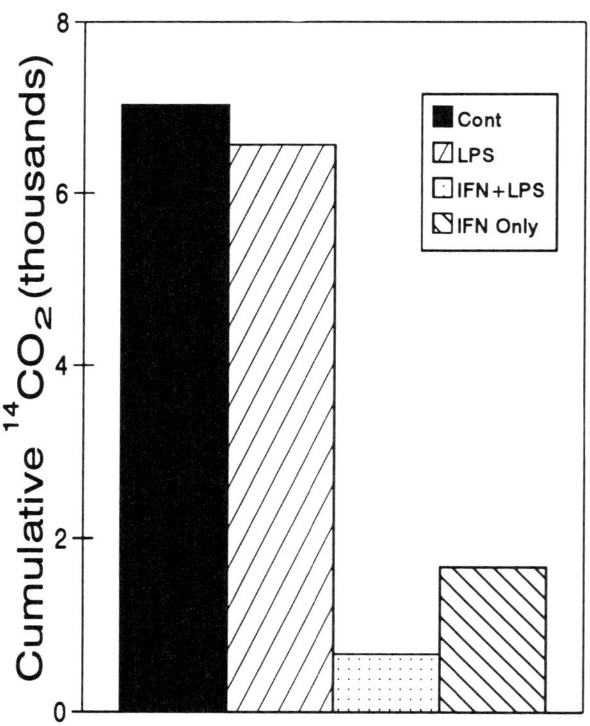

FIGURE 1 Mouse macrophages vs. *M. leprae*. Monolayers of mouse peritoneal macrophages were pretreated with interferon-γ (IFN-γ) with or without lipopolysaccharide (LPS) for 24 hours before challenge with viable *M. leprae*. After 24 hours *M. leprae* released from macrophages were placed in Buddemeyer vials, and $^{14}CO_2$ production was measured at 7 days (see Ref. 58 for more detail).

heavily burdened with *M. leprae* that they become refractory to activation with IFN-γ [26,27,63,64]. With these five in vitro methods of measuring *M. leprae* metabolism (and a sixth, phagosome-lysosome fusion [43]), we concluded that, in contrast to normal Mϕs, activated murine Mϕs have a marked deleterious effect on the leprosy bacillus.

Previous reports have demonstrated a role for the enhanced microbicidal activity of the activated mouse Mϕ against a variety of mycobacterial pathogens, including *M. tuberculosis* [65], *M. bovis* [35,66], *M. microti* [67], and *M. avium* [68]. Unlike the murine system, IFN-γ does not appear to activate human MdMs to kill or inhibit mycobacteria. In well-controlled studies, human IFN-γ efficiently activated MdMs to kill *Leishmania*, but in the case of *M. tuberculosis* it decreased phagocytosis of the bacilli and appeared to actually enhance intracellular replication [69]. Similar findings with human Mϕs were reported by Rook et al [65], also using *M. tuberculosis*, and by Toba et al. [70] using *M. avium*. Cytokines other than IFN-γ have been employed with some success. Bermudez and Young [71] employed TNF-α alone or in combination with IL-2 to enhance *M. avium* killing by human Mϕs, and Denis [72] reported that TNF-α and granulocyte-Mϕ colony-stimulating factor (GM-CSF) stimulated human Mϕs to inhibit virulent *M. avium* and kill avirulent *M. avium*. In recent reports by Denis and Gregg [73] and Shiratsuchi et al. [74], the cyclooxygenase inhibitor indomethacin was combined with IFN-γ to demonstrate activation of human Mϕs to inhibit *M. avium*. In our own on-going studies with *M. leprae*, we have shown that, unlike the murine system, human MdMs treated with combinations of IFN-γ, TNF-α, LPS, and indomethacin do not have a deleterious effect on the leprosy bacillus (Figure 2). Other cytokine combinations are being evaluated.

E. Immunoregulatory Macrophage Products

As a secretory cell the Mϕ is also the source of such potent immunoregulatory compounds as arachidonic acid, cyclooxygenase products, and cytokines such as IL-1, TNF-α, IL-6, IFN-α, and IFN-β. By attracting additional inflammatory cells and by modulating the function of these new, as well as existing, Mϕs, Mϕ secretory products play important roles in the initiation and maintenance of the local leprosy lesion.

Interleukin-1 and TNF-α may be especially important in host defense and the pathogenesis of mycobacterial disease. Both cytokines have multiple biologic activities too numerous to discuss here; in many cases they have overlapping pleiotropic effects on other cell types. Both are produced and secreted rapidly by MNPs in response to various stimuli, especially LPS. In the activated Mϕ, production of these cytokines is markedly up-regulated.

Both TNF-α and IL-1β seem elevated in the sera of LL patients undergoing erythema nodosum leprosum (ENL) reactions [75], although the data conflict in sera in uncomplicated multibacillary and paucibacillary disease [76,77]. Barnes et al. [78] measured TNF-α production, induced by *M. leprae* or its constituents, and reported TNF-α levels were higher from cells of TT patients than from those of patients with LL. Cells from LL patients with ENL released the highest levels of TNF-α. An important role for TNF-α in the pathogenesis of ENL was recently pinpointed by demonstrating that thalidomide treatment of patients with ENL inhibited TNF-α production by human blood monocytes [79].

An important role for TNF-α in resistance to mycobacterial disease was shown in Bacillus Calmette-Guérin (BCG)-infected mice by Kindler et al. [80], who employed a TNF-α–neutralizing antibody to demonstrate the critical role that TNF-α secretion plays in the development and maintenance of an epithelioid cell granuloma. They proposed that within the microenvironment TNF-α is released from Mϕs and acts in a paracrine or autocrine manner to regulate its own synthesis and modulate granuloma development by acting on endothelial cells to trigger mononuclear cell accumulation and by influencing the differentiation of Mϕs into epithelioid cells. A poor

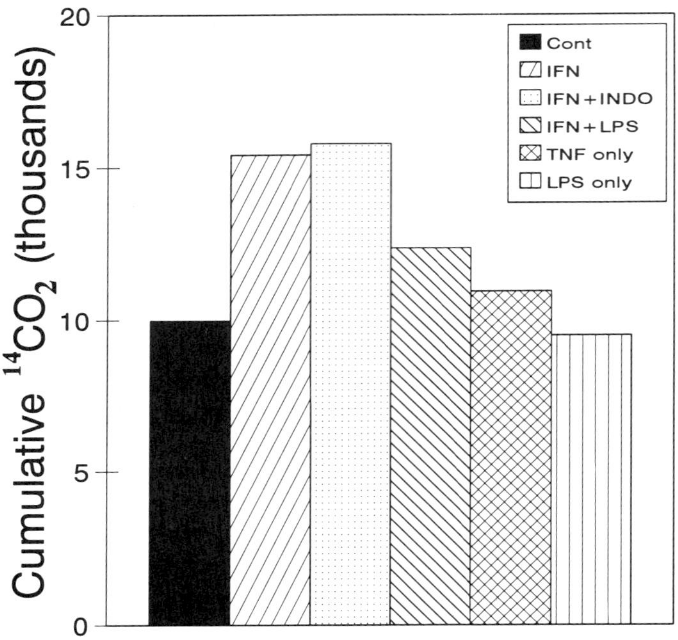

FIGURE 2 Human macrophages vs. *M. leprae*. Human peripheral blood monocyte pretreated with human interferon-γ (IFN-γ) (500 U/ml) or tumor necrosis factor-α (TNFα) (1000 U/ml) with or without lipopolysaccharide (LPS) (5 ng/ml) or indomethacin (1 μg/ml) for 24 hours before challenge with viable *M. leprae* for 2 hours. MdM were washed and the cytokines replaced for 24 hours when *M. leprae* were released and placed in Buddemeyer vials (see legend Fig. 1). Results shown are representative of seven separate experiments.

counterpart to the well-defined BCG liver granuloma in mice is the growth of *M. leprae* in the mouse footpad model. It remains to be seen whether anti–TNF-α neutralizing antibody could similarly influence growth of leprosy bacilli, but the protracted nature of the infection in the latter model may present an obstacle to such an experiment.

However, in vitro, the ability of *M. leprae* and certain mycobacterial constituents to stimulate TNF-α production by mouse Mϕs has been examined [81,82]. Purified *M. leprae* and mycobacterial lipoarabinomannan (LAM) were potent stimulators of TNF-α. Production of TNF-α is dependent on the dose of bacilli or LAM and is distinct from TNF-α stimulation by LPS, since the TNF-α-stimulatory properties of LPS were abrogated by treatment with polymyxin B, while those of *M. leprae* and LAM were not. Interferon-γ–activated Mϕs produced markedly elevated levels of TNF-α. In vivo, a continuous low level of TNF-α was elaborated by cultures of lepromatous Mϕs isolated from the *M. leprae*–infected footpads of nu/nu mice.

As an important regulatory factor in the expression of enhanced antimicrobial activity displayed by IFN-γ–primed Mϕs, TNF-α plays a varying role, depending in part on the pathogen. Under low-endotoxin conditions, IFN-γ alone failed to activate Mϕs to kill *T. gondii* [83]. As little as 0.1ng LPS/ml served as the second signal for IFN-γ and triggered Mϕs to the activated state. In the absence of LPS, TNF-α will also trigger IFN-γ–primed Mϕs to kill *T. gondii*. In our *M. leprae* studies, there was no need for a second signal for IFN-γ–primed mouse Mϕs. *M. leprae* itself, or certain of its constituents, triggered IFN-γ–primed Mϕs to kill or inhibit LPS (Figure 1). The induction of TNF-α by *M. leprae* may provide the second signal to IFN-γ in the absence of LPS [81].

Rook et al. [84] have recently discussed some of the circumstances in which TNF-α either affords protection to the host or is detrimental to the host defenses against mycobacterial disease. The apparently contradictory in vivo findings presented by the ENL reaction, the BCG lesion, and the *M. leprae*–infected nu/nu mouse footpad cells suggest that local production of TNF-α in a multibacillary lesion modulates cell functions in an entirely different manner than it does in a paucibacillary epithelioid cell granuloma. These findings underscore the difficulty of assigning simple cause-effect relationships between cytokine production and host response and add to the already pleotropic list of functions of this potent cytokine.

F. Microbicidal Mechanisms

The mechanisms that can be brought to bear by Mϕs against intracellular pathogens are broadly divided into oxygen-dependent and oxygen-independent antimicrobial pathways.

1. Reactive Oxygen Intermediates

Depending on their maturity, location, and state of activation, appropriate stimulation of Mϕs (treatment with PMA or phagocytosis of microorganisms) is associated with a respiratory burst that leads to the production of reactive oxygen intermediates (ROIs) formed by the partial reduction of oxygen [85]. The principal ROIs are superoxide anion (SO), hydrogen peroxide (H_2O_2), and hydroxyl radical (OH·). The molecular excitation of oxygen yields a fourth ROI, singlet oxygen. The baseline capacity of blood monocytes to produce ROIs is greater than that of more mature tissue Mϕs, due in part to the presence of cytoplasmic peroxidase granules in the former. However, the IFN-γ–activated Mϕ has an enhanced ability to produce ROIs [8], which underlies, in part, the enhanced nonspecific microbicidal activity of the activated Mϕ.

Although ROIs (H_2O_2, SO) produced in the respiratory burst have been implicated as a major antimicrobial effector mechanism [86,87], there are a number of reports [88–90] of oxygen-independent antimicrobial effects of activated Mϕs on a wide variety of intracellular bacterial (including mycobacterial [24,65]), fungal, and protozoal pathogens, suggesting that an alternative, nonoxidative mechanism may operate as well. In *M. leprae*, the innate ability to neutralize the toxic effects of ROIs may be an important virulence factor (see below).

2. Reactive Nitrogen Intermediates

Recent reports by Hibbs's group [11,12] described the importance of a single amino acid, L-arginine, as a necessary substrate for activated Mϕ–mediated cytotoxic activity for tumor cells and microbiostatic activity against a variety of intracellular pathogens including fungi [91], bacteria [92], and protozoa [93,94]. These Mϕ-mediated functions depended on a novel biochemical pathway synthesizing nitrogen oxides from the terminal guanidino nitrogen atom of L-arginine. L-citrulline was another by-product of this pathway. Nitric oxide gas causes the same pattern of metabolic inhibition in tumor cells as is caused by activated Mϕs [95] and has a direct antifungal [91] and antimycobacterial [96] function as well. These data suggest that nitric oxide, a stable free radical with a relatively short half-life, is the actual effector molecule of this pathway.

We have already explored the role of the L-arginine–dependent pathway (AdP) [97] in the detrimental effects of activated murine Mϕs on *M. leprae*. Five clear lines of evidence support the role of this pathway in the ability of activated Mϕs to inhibit the metabolism of *M. leprae*. First, in tissue culture medium depleted of L-arginine, activated Mϕs were unable to inhibit *M. leprae*. Second, the guanidine-methylated derivative of L-arginine, monomethyl arginine (N^gMMA), served as a competitive inhibitor in the pathway synthesizing RNI and L-citrulline from L-arginine and blocked the ability of the activated Mϕ to inhibit *M. leprae*. Third, supplementation of the medium with excess L-arginine reversed the block induced by N^gMMA. Fourth, arginase, which competes with nitric oxide synthase for L-arginine by converting L-arginine to L-ornithine and urea [11], blocked the Mϕ anti–*M. leprae* effect. Finally, analysis of culture supernatant medium for nitrite (NO_2^-), a stable end product of nonenzymatic degradation of nitric oxide in aqueous solution, correlated directly with the ability of the activated Mϕ to inhibit the metabolism of *M. leprae*.

There are also indirect lines of evidence hinting at the importance of the AdP in resistance to leprosy in humans. Katoch's group described a direct correlation between serum arginase levels and degree of impairment of CMI to *M. leprae* across the clinical spectrum. Local concentrations of arginase could have a substantial inhibitory effect on the synthesis of nitric oxide from L-arginine. The source of arginase appeared to be the circulating lymphocytes [98]. Of course, LL granulomas are characterized by a paucity of lymphocytes, so whether the circulating levels of arginase are sufficient to contribute to the inability of lepromatous Mϕs to cope with *M. leprae* remains unknown. The link between the AdP and antileprosy Mϕ effects in humans remains unclosed because of the difficulty of demonstrating an antimicrobial function for reactive nitrogen intermediates (RNIs) in human MNPs [98] using cytokine-induction procedures successful in mice [91,93,97,99]. Tumor necrosis factor-α and GM-CSF are reported to activate human MdMs to inhibit *M. avium* [72], and the high-output production of RNIs was implicated. More recently, Hibbs et al. [100] studied cancer patients receiving IL-2 immunotherapy and reported that a cytokine-inducible, high-output L-arginine/nitric oxide pathway exists in humans.

V. REGULATORY MACROPHAGES IN LEPROSY

A. Down-Regulation of Cell-Mediated Immunity

It is beyond the scope of this chapter to discuss the potential role of suppressor T cells in the immunologically specific anergy that characterizes the lepromatous form of leprosy. However, there is evidence that Mϕs themselves may act as "suppressors" and play a role in down-regulating afferent or efferent CMI. Interestingly, L-arginine–dependent production of RNIs may also be one mechanism whereby Mϕs suppress T cell proliferation [101].

T cell suppressor factors produced by Mϕs heavily loaded with mycobacteria have been described in mouse [102] and human [103] Mϕs. In human leprosy Mϕs have also been shown to suppress T cell responses [104,105] and may be defective in presenting *M. leprae* antigens to T cells [106]. In LL, Mϕs may also be defective in the production of IL-1, the cytokine that can amplify the production of IL-2 by T cells [107,108]. The uptake of live *M. leprae* may decrease Mϕ protein metabolism [109] and reduce expression of surface Fc receptors [34]. However, none of these studies addressed the effects of *M. leprae* infection on the principal afferent and efferent functions of the Mϕ itself, antigen presentation and enhanced microbicidal capacity.

To further explore our findings of defective Mϕ function in cells from an LL granuloma, we recently explored the IFN-γ–mediated activation of resident mouse peritoneal Mϕs infected with various numbers of viable and nonviable *M. leprae*. Unresponsiveness to IFN-γ was evident by the failure of IFN-γ (and LPS) to induce enhanced efferent (oxidative burst, killing of *T. gondii*, and cytotoxicity for tumor target cells) and afferent (enhanced expression of surface Ia antigen) effector function in *M. leprae*–infected Mϕs (26,27,63). A high intracellular burden of *M. leprae* was required ($>50:1$), and only viable bacilli induced the defect. The defective responses to IFN-γ were not observed in recently infected Mϕs (48 hours or less) but were consistently observed when Mϕs harbored leprosy bacilli for 3 to 5 days before stimulation with IFN-γ. The development of defective activation was closely correlated with elevated prostaglandin E_2 (PGE_2) production by Mϕs infected with live *M. leprae*, which peaked at 48 to 72 hours after infection at a level of approximately 10^{-8} M. Others have shown that elevated levels of PGE_2 can dramatically down-regulate a number of T cell and Mϕ functions, including T cell proliferation, MHC class II expression, and Mϕ cytotoxicity for tumor cells.

Ridel et al. [107] showed that *M. leprae* stimulated human monocytes to release PGE_2 in vitro, and skin-biopsied tissues from the sites of LL or borderline lepromatous leprosy lesions released PGE_2 in culture [27]. As discussed above, IFN-γ may activate human Mϕs to inhibit or kill mycobacteria if indomethacin is included in the system [73,74]. In our mouse studies, inclusion of indomethacin in the culture medium blocked PGE_2 production and restored the Mϕ response to IFN-γ. Interestingly, although the *M. leprae*–gorged Mϕs isolated from the infected nu/nu mouse footpad produce high levels of PGE_2 [27], indomethacin treatment in vitro failed to restore IFN-γ activation of these lepromatous Mϕs, suggesting that nonprostanoid mechanisms of inhibition are also involved in *M. leprae*–induced refractory response to IFN-γ.

B. Virulence Mechanisms

A variety of evidence supports the role of certain mycobacterial glycolipids in subverting CMI and innate and inducible Mϕ effector function. Lipoarabinomannan, discussed briefly above as a potent TNF-α inducer, is the highly immunogenic major constituent of the outer cell wall of *Mycobacterium* sp., including *M. leprae* [110]. Soluble LAM inhibited antigen responsiveness of human peripheral blood lymphocytes and antigen-induced proliferation of CD4+ T cell clones [111,112]. Purified LAM from *M. leprae* or *M. tuberculosis* not only blocks IFN-γ–induced enhancement of microbicidal capacity in murine and human Mϕs but down-regulates IFN-γ–induced MHC class II presentation in murine Mϕs as well [64,113]. Pretreatment with relatively high doses of LAM is necessary. As dead *M. leprae*, in contrast to viable bacilli, do not induce unresponsiveness [63], this LAM effect may be the consequence of the intracellular accumulation of a threshold dose of LAM. The potentially important role of contaminating gram-negative bacterial LPS (or endotoxin) in these LAM-induced effects has been ruled out. These effects are dependent on the lipid portion of the molecule and do not appear to be the result of PGE_2 production, as was the case with viable bacilli. Altered IFN-γ receptor function (binding, internalization) was not responsible, suggesting that LAM blocks IFN-γ signal transduction [114].

Recent studies by Chatterjee et al. [115] defined molecular differences in LAM isolated from virulent and avirulent strains of *M. tuberculosis* that may underlie a major biological difference between LAM from these two strains. Lipoarabinomannan from the avirulent $H37_{Ra}$ strain induced TNF-α by Mϕs, but LAM from the virulent Erdman strain failed to do so, suggesting the ability to circumvent TNF-α production may be a virulence factor encouraging survival of certain mycobacteria in Mϕs. We have confirmed and extended these findings (Figure 3) [116]. Production of TNF-α is only one of a myriad of Mϕ functions. The difference between these two LAM preparations is much less striking when their effects on other Mϕ functions are compared. Both LAMs served as a second signal, distinct from LPS, for IFN-γ activation of Mϕs and enhancement of NO_2^- production, and pretreatment with both LAMs blocked the subsequent ability of IFN-γ to activate Mϕs. Of interest, LAM from *M. leprae* was a poor inducer of TNF-α in comparison with LAM $H37_{Ra}$ (Figure 3) [116], although, like Erdman LAM, pretreatment of Mϕ and *M. leprae* LAM induced a refractory response to IFN-γ [64]. It is noteworthy that the LAM employed by Moreno et al. [112] to inhibit T cell proliferation *was* prepared from a virulent strain, $H37_{Rv}$.

Other mycobacterial constituents have now been purified [53] and are being investigated for their ability to modulate macrophage function or CMI. Preliminary studies with lipomannan (LM) and phosphatidylinositol mannoside (PIM) show that these mycobacterial components as well as LAM can serve as the triggering signal to IFN-γ–induced activation of Mϕs, in place of bacterial endotoxin (Adams and Krahenbuhl, unpublished data). As in the case of LAM, pretreatment of Mϕs with LM and PIM also down-regulates IFN-γ–induced activation.

In addition to these *M. leprae*– and mycobacterial component–induced defective responses, the leprosy bacillus may also possess a variety of virulence mechanisms that permit it to reside and flourish within Mϕs. These virulence factors represent the evolution of defense mechanisms that

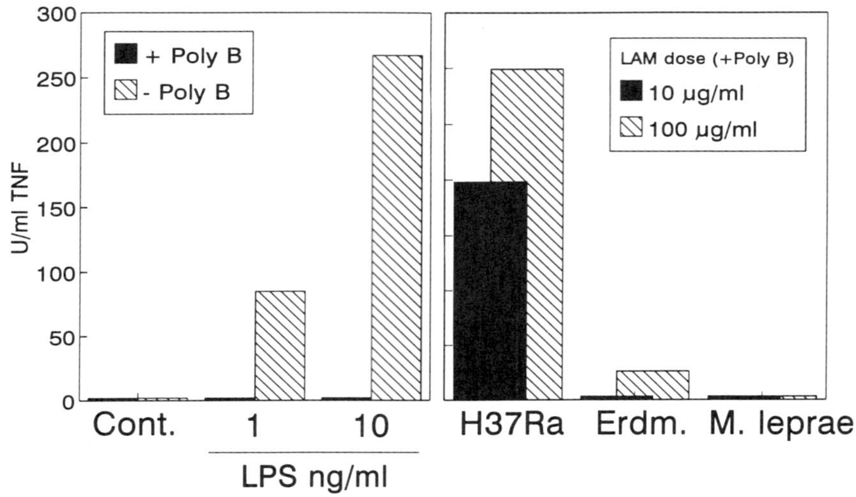

FIGURE 3 TNF induction by lipoarabinomannan (LAM) from different sources. Monolayers of mouse macrophages were treated with lipopolysaccharide (LPS) or LAM with or without 10 μg/ml polymyxin B. After 4 hours supernatant medium was harvested and tested for tumor necrosis factor-α (TNFα) using the L-cell assay, as described in Ref. 83.

circumvent the efforts of the antimicrobial function(s) of the Mϕ. *M. leprae* have been shown to be only a weak stimulus of the oxidative burst by Mϕs [117,118], possibly because of a specific modulation of superoxide generation by a leprosy bacillus component, PGL-1 [119,120]. In addition, *M. leprae* possesses a superoxide dismutase [119], and PGL-1 itself may be an efficient scavenger of superoxide [119]. Thus, *M. leprae* seem to be well equipped to cope with ROIs generated by the Mϕ.

Phenolic glycolipid-1 is not the only virulence factor possessed by the leprosy bacillus. As discussed above, *M. leprae* LAM, produced in copious amounts in infected Mϕs, may contribute directly to intracellular survival by down-regulating IFN-γ–induced afferent (Ia expression) and efferent (killing) Mϕ functions [64,113]. But Chan et al. [96] have recently shown that LAM is also an efficient scavenger of ROIs and blocks protein kinase C activity (an important regulatory pathway in the transduction of Mϕ-activating signals).

VI. TURNOVER OF BONE MARROW–DERIVED MACROPHAGES IN LESIONS

A. Influx of New Macrophages

The total population of Mϕs in a given tissue, under steady-state conditions, is determined by the influx of new bone marrow–derived monocytes and the local division of a small number of Mϕs. If this influx is even slightly greater than the loss of existing cells, there is a steady accumulation of Mϕs, as is seen in the lepromatous granuloma that develops during the increase in nu/nu footpad size that accompanies *M. leprae* growth from 10^9 to more than 10^{10} late in the course of experimental infection.

A single intravenous pulse of ^{3}H-thymidine will label the DNA of dividing promonocytes in the bone marrow [3]. These labeled cells mature and traffic into the blood and eventually into the tissues, where they can be identified as newly arrived by autoradiography. A small percentage of blood monocytes (2% to 5%) divide a single time once they have arrived in the tissues and are identified by a "flash pulse" of ^{3}H-thymidine given 1 hour before harvest.

We have explored the steady-state influx of Mϕs into the footpads of untreated *M. leprae*–infected nu/nu mice and compared these findings with an experimental condition designed to induce death of bacilli (dietary rifampin) or enhance host Mϕ-mediated defense mechanisms (multiple subcutaneous injections of IFN-γ [121]). Peak numbers of labeled monocytes appeared in the blood 48 to 72 hours after a single ^{3}H-thymidine pulse, 72 hours for peritoneal Mϕs. These data generally agree with findings in normal immunocompetent mice [3]. To study Mϕ turnover in the lepromatous granuloma, a ^{3}H-thymidine pulse interval of 5 to 7 days was employed [122]. Cyto-centrifuge preparations of enzyme-dispersed Mϕs were prepared, and only Mϕs burdened with *M. leprae* were scored for labeled nuclei. In untreated *M. leprae*–infected nu/nu mice, a large proportion (15% to 20%) of the footpad granuloma Mϕs burdened with bacilli had labeled nuclei and thus could be identified as newly arrived. Subcutaneous treatment with IFN-γ appeared to enhance significantly the influx of Mϕs into the footpad granuloma; 25% to 35% of the Mϕs infected with *M. leprae* had labeled nuclei. A 1-hour flash pulse of ^{3}H-thymidine labeled fewer than 1% of the footpad Mϕs. These data suggest a remarkably high rate of influx of new bone marrow–derived Mϕs.

B. Mechanisms

Although lymphocytes do recirculate from the tissues, evidence for the recirculation of tissue Mϕs via peripheral blood or the lymphatic system is lacking [3]. Macrophages probably die in the tissues and are replaced by new Mϕs, thus raising a fundamental question: in LL how do the new Mϕs acquire a burden of leprosy bacilli? At least four possibilities are supported by experimental evidence in mycobacterial infections. Chiplunkar et al. [123] demonstrated lysis of *M. leprae*–infected mouse Mϕs by CD8+ cytotoxic T cells, and Hancock et al. [124] have shown that human CD4+ cytotoxic T cells are capable of lysing human MdMs infected with *M. leprae* by MHC class II mechanisms. Blanchard et al. [125] showed that mycobacteria-infected human MdMs serve as targets for IL-2–driven natural killer (NK) cells, an observation confirmed in our own laboratory with *M. leprae*-infected mouse Mϕs (Gu et al., unpublished data). Finally, our studies show that normal Mϕs do acquire *M. leprae* from infected target Mϕs. More important, activated Mϕs clearly can acquire and kill the *M. leprae* harbored by normal infected Mϕ target cells (Adams et al., unpublished data). Current studies are aimed at determining whether the newly arrived Mϕs engulf the old infected Mϕs or whether the new Mϕs scavenge extracellular bacilli released from the old Mϕs by other effector cells.

VII. LOCAL REGULATION OF THE LEPROSY LESION—THE KEY TO RESISTANCE

The manifestations of clinical leprosy are essentially limited to localized permissive sites in the mucous membranes, skin, and peripheral nerves. As suggested by much of the experimental evidence discussed in this chapter, resistance to *M. leprae*, including the role of T cell recognition or anergy to specific antigens and the subsequent progression of disease to a position in the leprosy spectrum, is also largely exercised and controlled at the local level.

The experimental "immunotherapy" of leprosy underscores the need to up-regulate CMI at the site of infection. In humans, local injection of *M. leprae* mixed with BCG [125,126], purified protein derivative [127], recombinant human IFN-γ [26,128,129], or IL-2 [130], or local sensitization with dinitrochlorobenzene [131], resulted in marked local changes in LL lesions characterized by induration from the accumulation of fresh lymphocytes and monocytes and changes that suggested a shift in histopathological classification toward the tuberculoid end of the spectrum.

Multiple local (intradermal) administration of IFN-γ and IL-2 into LL lesions has been especially well studied [132] and has shown an influx into the dermis of CD4+ and CD8+ lymphocytes and monocytes, with evidence of a local reduction in leprosy bacilli. This approach also enhanced local CMI, as shown by increased expression of MHC class II antigens on keratinocytes overlying the treated site. In earlier studies, systemic administration of allogeneic lymphocytes [133] or transfer factor [134] also led to local changes interpreted as an upgrading of lesion classification. Systemic immunotherapy with IFN-γ in LL patients led to marked local changes in dermal LL lesions and the widespread accumulation of T lymphocytes and monocytes [135].

Clearance of leprosy bacilli from the LL lesion as a consequence of local or systemic immunotherapy probably depends on the influx of new competent Mϕs rather than the local activation of resident lepromatous Mϕs. While the circulating blood monocytes of leprosy patients are functionally normal in their innate microbicidal capacity and response to IFN-γ, upon their arrival in a granulomatous lesion they confront local conditions that modulate their function. For example, Mϕs infected with *M. leprae* or treated with mycobacterial constitutents such as LAM are a principal source of TNF-α, a cytokine with pleotropic effects on a variety of other cell types and a major modulator of the local inflammatory response. The role of TNF-α production in host resistance in paucibacillary leprosy is open to speculation. Whether the compelling evidence that demonstrated the important role of local TNF-α elaboration in the formation and maintenance of the epithelioid granuloma in mouse BCG infections [80] has a counterpart in mouse models of leprosy or human paucibacillary disease is unknown and worthy of investigation. Local TNF-α elaboration may also be a potent mediator of pathogenesis in the reactional episode associated with multibacillary disease [79]. In LL the continuous local elaboration of low levels of TNF-α may have a profound local effect on maintenance of the lesion by virtue of the potent effects that TNF-α has on blood vessel endothelial cells, perhaps underlying the large influx of new Mϕs into the lesion.

Preactivated Mϕs appear able to cope with newly acquired leprosy bacilli, but *M. leprae* possess a number of apparent virulence factors that can neutralize or at least frustrate the antimicrobial mechanisms of the Mϕ. However, a high intracellular burden of *M. leprae* may block cytokine-induced activation of the infected Mϕs and down-regulate other Mϕs in the lesion via secretion of arachidonic acid metabolites such as PGE_2. Cohn's group [136] have emphasized Mϕ regulation of arachidonic acid metabolism in localized anatomical compartments. Clearly the influence of local conditions on Mϕ function within the tissue lesion is an extremely relevant but poorly characterized feature of leprosy.

REFERENCES

1. Hastings RC, Gillis TP, Krahenbuhl JL, Franzblau SG. Leprosy. Clin Micro Rev 1988; 1:330–48.
2. Modlin RL, Melancon-Kaplan J, Young SM, Pirmez C, Kino H, Convit J, Rea TH, Bloom BR. Learning from lesions: patterns of tissue inflammation in leprosy. Proc Natl Acad Sci USA 1988; 85:1213–7.
3. Van Furth R. Phagocytic cells: development and distribution of mononuclear phagocytes in normal steady state and inflammation. In: Gallin JI, Goldstein IM, Snyderman R, eds. Inflammation: basic principles and clinical correlates. New York: Raven Press, 1988:281–95.
4. Nathan CF. Macrophage secretory products. J Clin Invest 1987; 79:319–26.
5. Mackaness GB. Resistance to intracellular infection. J Infect Dis 1971; 123:439–45.
6. Krahenbuhl JL, Remington JS, McLeod R. Cytotoxic and microbicidal properties of macrophages. In: van Furth R, ed. Mononuclear phagocytes—functional aspects: proceedings of the Third Conference on Mononuclear Phagocytes. Amsterdam: Martinus Nijhoff, 1980:1631–53.
7. Krahenbuhl JL, Rosenberg L, Remington J. The effects of anti-thymocyte serum on in vitro activation of macrophages to kill *Listeria monocytogenes*. J Immunol 1973; 111:992–5.
8. Nathan CF, Murray HW, Wiebe ME, Rubin BY. Identification of interferon gamma as the lymphokine that activates human macrophage oxidative metabolism and antimicrobial activity. J Exp Med 1983; 158:670–4.

9. Karnovsky M, Lazdins J. Biochemical criteria for activated macrophages. J Immunol 1978; 121:809–13.
10. Nathan CF, Root RK. Hydrogen peroxide release from mouse peritoneal macrophages: dependence on sequential activation and triggering. J Exp Med 1977; 146:1648–62.
11. Hibbs JB Jr, Taintor RR, Vavrin Z. Macrophage cytotoxicity: role for L-arginine deiminase and imino nitrogen oxidation to nitrate. Science 1987; 235:473–6.
12. Hibbs JB Jr, Vavrin Z, Taintor RR. L-arginine is required for expression of the activated macrophage effector mechanism causing selective metabolic inhibition in target cells. J Immunol 1987; 138:550–65.
13. Chehl S, Job CK, Hastings RC. Transmission of leprosy in nude mice. Am J Trop Med Hyg 1985; 34:1161–6.
14. Job CK, Chehl S, Hastings RC. New findings on the mode of entry of *Mycobacterium leprae* in nude mice. Int J Lepr 1990; 58:726–9.
15. Job CK, Harris EB, Allen JL, Hastings, RC. Thorns in armadillo ears and noses and their role in the transmission of leprosy. Arch Pathol Lab Med 1986; 110:1025–8.
16. Drutz DJ, O'Neill SM, Levy L. Viability of blood-borne *Mycobacterium leprae*. J Infect Dis 1974; 130:288–92.
17. Sen R, Sehgal PK, Singh U, Yadav MS, Chaudhary SD, Sika R. Bacillaemia and bone marrow involvement in leprosy. Ind J Lepr 1989; 61:445–52.
18. Avila JL, Convit J. Studies on cellular immunity in leprosy. I. Lysosomal enzymes. Int J Lepr 1970; 38:359–64.
19. Beiguelman B. Leprosy and genetics. A review of past research with remarks concerning future investigations. Bull WHO 1967; 37:461–76.
20. Samuel PR, Godal T, Myrvang B, Song YK. Behavior of *Mycobacterium leprae* in human macrophages in vitro. Infect Immun 1973; 8:446–9.
21. Marolia J, Robinson P, Mahadevan PR. A complex component modulating immune-deficient cells in leprosy patients leading to loss of viability of *Mycobacterium leprae*—a possible vaccine. Clin Exp Immunol 1990; 79:7–14.
22. Drutz DJ, Cline MJ, Levy L. Leukocyte antimicrobial function in patients with leprosy. J Clin Invest 1974; 53:380–6.
23. Horwitz MA, Levis WR, Cohn ZA. Defective production of monocyte-activating cytokines in lepromatous leprosy. J Exp Lepr 1984; 159:666–78.
24. Sharp AK, Banerjee DK. Hydrogen peroxide and superoxide production by peripheral blood monocytes in leprosy. Clin Exp Immunol 1985; 60:203–6.
25. Nathan CF, Kaplan G, Levis WR, Nusrat A, Witmer MD, Sherwin SA, Job CK, Horowitz CR, Steinman RM, Cohn ZA. Local and systemic effects of intradermal recombinant interferon-gamma in patients with lepromatous leprosy. N Engl J Med 1986; 315:6–15.
26. Sibley LD, Krahenbuhl JL. *Mycobacterium leprae*-burdened macrophages are refractory to activation by gamma-interferon. Infect Immun 1987; 55:446–50.
27. Sibley LD, Krahenbuhl JL. Defective activation of granuloma macrophages from *Mycobacterium leprae*-infected nude mice. J Leukocyte Biol 1988; 43:60–6.
28. Munro CS, Campbell DA, Collings LA, Poulter LW. Monoclonal antibodies distinguish macrophages and epithelioid cells in sarcoidosis and leprosy. Clin Exp Immunol 1987; 68:282–7.
29. Ridley MJ. The mononuclear cell series in leprosy: an ultrastructural report. Lepr Rev 1981; 52:35–50.
30. Turk JL, Narayanan RB. The origin, morphology, and function of epithelioid cells. Immunobiol 1982; 161:274–8.
31. Narayanan R, Girdhar B, Sengupta U, Desikan K. In vitro studies on dermal granulomas of human leprosy—cellular characteristics. Int J Lepr 1985; 53:39–44.
32. Ridley MJ, Ridley DS. Unique expression of HLA-DR (Ia-like) antigens in the lesions of polar tuberculoid leprosy. Lepr Rev 1982; 53:249–52.
33. Kumar V, Narayanan R, Girdhar B. Comparison of the characteristics of infiltrates in skin and nerve granulomas of leprosy. Acta Leprol 1989; 7:19–24.
34. Mistry NF, Birdi TJ, Antia NH. *M. leprae* phagocytosis and its association with membrane changes in macrophages from leprosy patients. Parasite Immunol 1986; 8:129–38.
35. Stach JL, Delgado G, Strobel M, Millan J, LaGrange PH. Preliminary evidence of natural resistance to *Mycobacterium bovis* (BCG) in lepromatous leprosy. Int J Lepr 1984; 52:140–6.

36. Schlesinger LS, Horwitz MA. Phagocytosis of *Mycobacterium leprae* by human monocyte derived macrophages is mediated by complement receptors CR1 (CD35), CR3 (CD11b/CD18), and CR4 (CD11C/CD187) and interferon gamma activation inhibits complement receptor function and phagocytosis of this bacterium. J Immunol 1991; 147:1983–4.
37. Schlesinger LS, Horwitz MA. Phenolic glycolipid-1 of *Mycobacterium leprae* binds complement component C3 in serum and mediates phagocytosis by human monotyes. J Exp Med 1991; 174:1031–8.
38. Hunter SW, Brennan PJ. A novel phenolic glycolipid from *Mycobacterium leprae* possibly involved in immunogenicity and pathogenicity. J Bacteriol 1981; 147:728–35.
39. Job CK, McCormick GT, Hastings RC. Intracellular parasitism of parenchymal cells by *M. leprae*. Int J Lepr 1989; 57:659–70.
40. Job CK, McCormick GT, Moses RL, Hastings RC. Phagocytosis of *Mycobacterium leprae* by cardiac muscle cells—a preliminary report. Ind J Lepr 1990; 62:55–9.
41. D'Arcy Hart P. Lysosome fusion responses of macrophages to infection: behavior and significance. In Karnovsky M, Bolis L, eds. Phagocytosis—past and future. New York: Academic Press, 1982:437–47.
42. Kaplan G, Cohn ZA. The immunobiology of leprosy. Exp Pathol 1986; 28:45–78.
43. Sibley LD, Franzblau SG, Krahenbuhl JL. Intracellular fate of Mycobacterium leprae in normal and activated mouse macrophages. Infect Immun 1987; 55:680–5.
44. Frehl C, Chastellier C, Lang T, Rastogi N. Evidence for inhibition of fusion of lysosomal and prelysosomal compartments in macrophages infected with pathogenic Mycobacterium avium. Infect Immun 1986; 51:252–62.
45. D'Arcy Hart P, Armstrong JA, Brown CA, Draper P. Ultrastructural study of the behavior of macrophages toward parasitic mycobacteria. Infect Immun 1972; 5:803–7.
46. Rastogi N, Frehl C. Evidence that coating of *Mycobacterium leprae* surface antigens reduces its ability to hinder host microbicidal functions. Int J Med Microbiol 1990; 272:337–46.
47. Rastogi N, Cadou S, Hellio R. Differential handling of bacterial antigens in macrophages infected with *Mycobacterium leprae* as studied by immunogold labeling of ultrathin sections. Int J Lepr 1991; 59:278–91.
48. Evans MJ, Levy L. Ultrastructural changes in cells of the mouse foot pad infected with *Mycobacterium leprae*. Infect Immun 1972; 5:238–47.
49. Levy L, Ng H, Evans M, Krahenbuhl JL. Susceptibility of thymectomized and irradiated mice to challenge with several organisms and the effect of dapsone on infection with *Mycobacterium leprae*. Infect Immun 1975; 11:1122–32.
50. Chandi SM, Job CK. The early cellular response to *M. leprae*—an ultrastructural study. Lepr Ind 1978; 50:345–51.
51. Sharp AK, Banerjee DK. Attempt at cultivation of Mycobacterium leprae in macrophages from susceptible animal hosts. Int J Lepr 1984; 52:189–97.
52. Shepard CC, McRae DH. *Mycobacterium leprae* in mice: minimal infectious dose, relationship between staining quality and infectivity and effects of cortisone. J Bacteriol 1965; 89:365–70.
53. Brennan PJ. The phthiocerol-containing surface lipids of *M. leprae*—a perspective of past and present work. Int J Lepr 1983; 51:387–96.
54. Draper P, Rees RFW. Electron transparent zone of mycobacteria may be a defense mechanism. Nature 1970; 228:860–1.
55. Krahenbuhl JL, Levy L, Remington JS. Resistance to *Mycobacterium leprae* in mice infected with *Toxoplasma gondii* and *Besnoitia jellisoni*. Infect Immun 1974; 10:1068–71.
56. Krahenbuhl JL, Humphres RC, Henika PC. Effects of *Propionobacterium acnes* treatment on the course of *Mycobacterium leprae* infection in mice. Infect Immun 1982; 37:183–8.
57. Colston MJ, Hilson GRF, Banerjee DK. The "proportional bactericidal test", a method for assessing bactericidal activity of drugs against *Mycobacterium leprae* in mice. Lepr Rev 1978; 49:7–15.
58. Ramasesh N, Adams LB, Franzblau SG, Krahenbuhl JL. Effects of activated macrophages on *Mycobacterium leprae*. Infect Immun 1991; 59:2864–9.
59. Katoch VM, Katoch K, Ramanathan U, Sharma V, Shivannavar C, Datta A, Bharadwaj V. Effect of chemotherapy on viability of *Mycobacterium leprae* as determined by ATP content, morphological index, and FDA:EB fluorescent staining. Int J Lepr 1989; 57:615–21.
60. Odinsen T, Nilson T, Humber D. Viability of *Mycobacterium leprae*: a comparison of morphological

index and fluorescent staining techniques in slit-skin smears and *M. leprae* suspensions. Int J Lepr 1986; 54:403–8.
61. Franzblau SG. Oxidation of palmitic acid by *Mycobacterium leprae* in an axenic medium. J Clin Microbiol 1988; 26:18–21.
62. Ramasesh N, Hastings RC, Krahenbuhl J. Metabolism of *Mycobacterium leprae* in macrophages. Infect Immun 1987; 55:1203–6.
63. Sibley LD, Krahenbuhl JL. Induction of unresponsiveness to gamma interferon in macrophages infected with *Mycobacterium leprae*. Infect Immun 1988; 56:1912–9.
64. Sibley LD, Hunter SW, Brennan PJ, Krahenbuhl JL. Mycobacterial lipoarabinomannan inhibits gamma interferon–mediated activation of macrophages. Infect Immun 1988; 56:1232–6.
65. Rook GAW, Steel J, Ainsworth M, Champion BR. Activation of Mϕ to inhibit proliferation of *Mycobacterium tuberculosis*: comparison of the effects of recombinant gamma-interferon on human monocytes and murine peritoneal macrophages. Immunology 1986; 19:333–8.
66. Flesch I, Kaufman S. Mycobacterial growth inhibition by interferon gamma-activated bone-marrow Mϕ and differential susceptibility among strains of *Mycobacterium tuberculosis*. J Immunol 1987; 138:4400–13.
67. Walker L, Lowrie D. Killing of *Mycobacterium microti* by immunologically activated macrophages. Nature 1981; 293:69–70.
68. Alfes C, Steele J, Stanford J, Rook G. The effect of lymphokines on the ability of macrophages to protect mycobacteria from a bactericidal antibiotic. Tubercle 1985; 1:1–7.
69. Douvas GS, Looker DL, Vatter AE, Crowle AF. Gamma interferon activites human Mϕ to become tumoricidal and leishmanicidal but enhances replication of Mϕ associated mycobacteria. Infect Immun 1986; 50:1–8.
70. Toba H, Crawford J, Ellner J. Pathogenicity of *Mycobacterium avium* for human monocytes: absence of macrophage-activating factor activity of human gamma interferon. Infect Immun 1989; 57:239–44.
71. Bermudez LE, Young LS. Tumor necrosis factor alone or in combination with IL-2, but not IFN-g, is associated with macrophages killing of *Mycobacterium avium* complex. J Immunol 1988; 140:3006–13.
72. Denis M. Tumor necrosis factor and granulocyte-macrophage colony stimulating factor stimulate human macrophages to inhibit the growth of virulent *Mycobacterium avium* and to kill avirulent *M. avium*: killing effector mechanism depends on the generation of reactive nitrogen intermediates. J Leukocyte Biol 1991; 49:380–7.
73. Denis M, Gregg EO. Modulation of *Mycobacterium avium* in murine macrophages. Reversal of unresponsiveness to interferon gamma with indomethacin by interleukin-4. J Leukocyte Biol 1991; 49:65–72.
74. Shiratsuchi H, Johnson JL, Ellner JJ. Bidirectional effects of cytokines on the growth of *Mycobacterium avium* within human macrophages. J Immunol 1991; 146:3165–70.
75. Sarno EN, Grau GE, Vieira LM, Nery JA. Serum levels of tumor necrosis factor-alpha and interleukin-1 beta during leprosy reactional states. Clin Exp Immunol 1991; 84:103–8.
76. Silva CL, Foss NT. Tumor necrosis factor in leprosy patients. J Infect Dis 1989; 159:787–92.
77. Pisa P, Gennene A, Soder O, Ottenhoff T, Hansson M, Kiessling R. Inverse relationship between antigen specific cellular immunity and serum tumor necrosis factor in leprosy and leishmaniasis. In: Seventh International Congress of Immunology. Stuttgart: Gustav Fisher, 1989:655.
78. Barnes PF, Chatterjee D, Brennan PJ, Rea TH, Modlin RL. Tumor necrosis factor production in patients with leprosy. Infect Immun 1992; 60:1441–6.
79. Sampaio EP, Sarno EN, Galilly R, Cohn ZA, Kaplan G. Thalidomide selectively inhibits tumor necrosis factor alpha production by stimulated human monocytes. J Exp Med 1991; 173:699–703.
80. Kindler V, Sappino AP, Grau GE, Piquet P-F, Vassalli P. The inducing role of tumor necrosis factor in the development of bacterial granulomas during BCG infection. Cell 1989; 56:731–40.
81. Fukutomi Y, Adams L, Krahenbuhl J. Macrophages infected with *Mycobacterium leprae* or treated with mycobacterial constituents release factors that modulate afferent and efferent macrophage effector function. Int J Lepr 1991; 59:711–712.
82. Moreno C, Taverne J, Mehlert A, Bate CAW, Brealey RJ, Meager A, Rook GAW, Playfair JHL. Lipoarabinomannan from *M. tuberculosis* induces the production of tumor necrosis factor from human and murine macrophages. Clin Exp Immunol 1989; 76:240–5.

83. Sibley LD, Adams LB, Fukutomi Y, Krahenbuhl JL. Tumor necrosis factor-α triggers antitoxoplasmal activity of IFNγ primed macrophages. J Immunol 1991; 147:2340–5.
84. Rook GA, Al Attiyah R, Filley E. New insights into the immunopathology of tuberculosis. J Pathobiol 1991; 59:148–52.
85. Klebanoff SJ. Oxygen-dependent cytotoxic mechanisms of phagocytes. Adv Host Def Mech 1982; 1:111–62.
86. Klebanoff SJ, Shepard CC. Toxic effect of the peroxidase–hydrogen peroxide–halide antimicrobial system on *Mycobacterium leprae*. Infect Immun 1984; 44:534–6.
87. Sharp AK, Colston MJ, Banerjee DK. Susceptibility of *Mycobacterium leprae* to the bactericidal activity of mouse peritoneal macrophages and to hydrogen peroxide. J Med Microbiol 1985; 19:77–84.
88. Murray HW, Rubin BY, Carriero SM, Harris AM, Jaffee EA. Human mononuclear phagocyte antiprotozoal mechanisms: oxygen-dependent vs oxygen-independent activity against intracellular *Toxoplasma gondii*. J. Immunology. 1985; 134:1982–1988.
89. Sibley LD, Krahenbuhl JL, Weidner E. Lymphokine activation of J774.G8 cells and mouse peritoneal macrophages challenged with *Toxoplasma gondii*. Infect Immun 1985; 49:760–4.
90. Chatterall JR, Sharma SD, Remington JS. Oxygen-independent killing by alveolar macrophages. J Exp Med 1986; 163:1113–31.
91. Granger DL, Hibbs JB Jr, Perfect JR, Durack, DT. Specific amino acid (L-arginine) requirements for the microbiostatic activity of murine macrophages. J Clin Invest 1988; 81:1129–36.
92. Denis M. Interferon gamma–treated murine macrophages inhibit growth of tubercle bacilli via the generation of reactive nitrogen intermediates. Cell Immunol 1991; 132:150–157.
93. Adams LB, Hibbs JB Jr, Taintor RR, Krahenbuhl JL. Microbiostatic effect of murine-activated macrophages for *Toxoplasma gondii*. Role for synthesis of inorganic nitrogen oxides from L-arginine. J Immunol 1990; 144:2725–9.
94. Green SJ, Meltzer MS, Hibbs JB Jr, Nacy CA. Activated macrophages destroy intracellular *Leishmania major* amastigotes by an L-arginine–dependent killing mechanism. J Immunol 1990; 144:278–84.
95. Hibbs JB, Taintor RR, Vavrin Z, Rachlin E. Nitric oxide: a cytotoxic activated macrophage effector molecule. Biochem Biophys Res Commun 1988; 157:87–94.
96. Chan J, Xing Y, Magliozzo RS, Bloom B. Killing of virulent *Mycobacterium tuberculosis* by reactive nitrogen intermediates produced by activated murine macrophages. J Exp Med 1992; 175:1111–2.
97. Adams LB, Franzblau SG, Vavrin Z, Hibbs JB Jr, Krahenbuhl JL. L-arginine–dependent macrophage effector functions inhibit metabolic activity of *Mycobacterium leprae*. J Immunol 1991; 147:1642–6.
98. Suri Babu CSS, Kannan KB, Bharadwaj VP, Katoch VM. Lymphocyte arginase activity in leprosy—a preliminary report. Ind J Med Res 1990; 91:193–6.
99. Cameron ML, Granger DL, Weinberg JB, Kozumbo WJ, Koren HS. Human alveolar and peritoneal macrophages mediate fungistasis independently of L-arginine oxidation of nitrite to nitrate. Am Rev Respir Dis 1990; 142:1313–9.
100. Hibbs JB, Westenelder C, Taintor R, Vavrin Z, Kablitz C, Baranowski RL, Ward JH, Menlove RL, McMurry MP, Kushner JP, Samlowski WE. Evidence for cytokine-inducible nitric oxide synthesis from L-arginine in patients receiving interleukin-2 therapy. J Clin Invest 1992; 867–77.
101. Mills CD. Molecular basis of "suppressor" macrophages. Arginine metabolism via the nitric oxide synthetase pathway. J Immunol 1991; 146:2719–23.
102. Watson SR, Sljivic VJ, Brown IN. Defect of macrophage function in the antibody response to sheep erythrocytes in systemic Mycobacterium lepraemurium infection. Nature 1975; 256:206–8.
103. Wadee AA, Sher R, Rabson AR. Production of a suppressor factor by human adherent cells treated with mycobacteria. J Immunol 1980; 125:1380–6.
104. Nath I, van Rood RR, Mehra NK, Vaidya MC. Natural suppressor cells in human leprosy: the role of HLA-D-identical peripheral lymphocytes and macrophages in the in vitro modulation of lymphoproliferative responses. Clin Exp Immunol 1980; 42:203–10.
105. Salgame PR, Mahadevan PR, Antia NH. Mechanism of immunosuppression in leprosy: presence of suppressor factor(s) from macrophages of lepromatous patients. Infect Immun 1983; 40:1119–26.
106. Hirschberg H. The role of macrophages in the lymphoproliferative response to *M. leprae* in vitro. Clin Exp Immunol 1978; 34:46–51.

107. Ridel PR, Jamet P, Robon Y, Bach M-A. Interleukin-1 released by blood monocyte–derived Mϕ from patients with leprosy. Infect Immun 1986; 52:303–8.
108. Watson S, Bullock W, Nelson K, Schauf V, Gelber R, Jacobson R. Interleukin-1 production by peripheral blood mononuclear cells from leprosy patients. Infect Immun 1984; 45:787–9.
109. Birdi TJ, Salgame PR, Antia NH. The role of macrophages in leprosy as studied by protein synthesis of macrophages from resistant and susceptible hosts—a mouse and human study. Lepr Ind 1979; 51:23–42.
110. Gaylord H, Brennan PJ, Young DR, Buchanan TM. Most *Mycobacterium leprae* carbohydrate-reactive monoclonal antibodies are directed to lipoarabinomannan. Infect Immun 1987; 55:2860–3.
111. Kaplan G, Ghandi RR, Weinstein DE, Levis WR, Patarroyo ME, Brennan PJ, Cohn ZA. *Mycobacterium leprae* antigen-induced suppression of T cell proliferation in vitro. J Immunol 1987; 138:3028–34.
112. Moreno C, Mehlert A, Lamb J. The inhibitory effects of mycobacterial lipoarabinomannan and polysaccharides upon polyclonal and monoclonal human T cell proliferation. Clin Exp Immunol 1988; 74:206–10.
113. Sibley LD, Adams LB, Krahenbuhl JL. Inhibition of interferon-gamma–mediated activation in mouse macrophages treated with lipoarabinomannan. Clin Exp Immunol 1990; 80:141–8.
114. Chan J, Fan X, Hunter SW, Brennan PJ, Bloom BR. Lipoarabinomannan, a possible virulence factor involved in persistence of *Mycobacterium tuberculosis* within macrophages. Infect Immun 1991; 59:1755–61.
115. Chaterjee D, Roberts AD, Lowell K, Brennan PJ, Orme IA. Structural basis of capacity of lipoarabinomannan to induce secretion of tumor necrosis factor. Infect Immun 1992; 60:1249–53.
116. Adams LB, Fukutomi Y, Krahenbuhl JL. Regulation of murine macrophage effector functions by lipoarabinomannan from mycobacterial strains of differing virulence. Submitted for publication, 1993.
117. Holzer H, Nelson KE, Schauf V, Crispen R, Anderson B. Mycobacterium leprrae fails to stimulate phagocytic cell superoxide anion generation. Infect Immun 1986; 51:514–20.
118. Launois PB, Maillere A, Dieye JL, Sarthou JL, Bach MA. Human phagocyte oxidative burst activation by BCG, *M. leprae*, and atypical mycobacteria: defective activation by *M. leprae* is not reversed by interferon-gamma. Cell Immunol 1989; 124:168–174.
119. Neill MA, Klebanoff SJ. The effect of phenolic glycolipid-1 from *Mycobacterium leprae* on the antimicrobial activity of human macrophages. J Exp Med 1988; 167:30–42.
120. Vachula M, Holzer TJ, Andersen BR. Suppression of monocyte oxidative response by phenolic glycolipid I of *Mycobacterium leprae*. J Immunol 1989; 142:1696–1701.
121. Wheeler PR, Gregory D. Superoxide dismutase, peroxidatic activity and catalase in *Mycobacterium leprae* purified from armadillo liver. J Gen Microbiol 1980; 121:457–464.
122. Krahenbuhl JL, Sibley LD, Chae GT. Gamma interferon in experimental leprosy. Diag Microbiol Infect Dis 1990; 13:405–9.
123. Chiplunkar S, DeLibero G, Kaufman S. Mycobacterium leprae-specific Lyt-2_+ T lymphocytes with cytotoxic activity. Infect Immun 19868; 54:793–7.
124. Hancock GE, Cohn ZA, Kaplan G. The generation of antigen-specific, major histocompatibility complex–restricted cytotoxic T lymphocytes of the CD4+ phenotype. Enhancement by the cutaneous administration of interleukin-2. J Exp Med 1989; 169:909–19.
125. Blanchard D, Michelini-Norris N, Friedman H. Lysis of mycobacteria-infected monocytes for IL-2 activated killer cells: role of LFA-1. Cell Immunol 1989; 119:402–11.
126. Convit J, Aranzazu N, Zuniga M, Ulrich M, Pinardi ME, Castellazi Z, Alvarado J. Immunotherapy and immunoprophylaxis of leprosy. Lepr Rev 1983; 54:47–60.
127. Kaplan G, Sheftel G, Job CK, Mathur NK, Nath I, Cohn Z. Efficacy of a cell-mediated reaction to the purified protein derivative of tuberculin in the disposal of *Mycobacterium leprae* from human skin. Proc Natl Acad Sci USA 1988; 85:5210–5214.
128. Samuel NM, Grange JM, Samuel S, Lucas S, Owilli OM, Adalla S, Leigh IM, Navarette C. A study of the effects of intradermal administration of recombinant gamma interferon in lepromatous leprosy patients. Lepr Rev 1987; 58:389–400.
129. Kaplan G, Mathur N, Job C, Nath I, Cohn Z. Effect of multiple interferon gamma injections on the disposal of *Mycobacterium leprae*. Proc Natl Acad Sci USA 1989; 86:8073–7.
130. Kaplan G, Kiessling R, Teklemariam S, Hancock G, Sheftel G, Job CK, Converse P, Ottenhoff TH,

Becx-Bleumink M, Dietz M, Cohn ZA. The reconstitution of cell-mediated immunity in the cutaneous lesions of lepromatous leprosy by recombinant interleukin-2. J Exp Med 1989; 169:893–907.
131. Volc-Platzer B, Kremsner P, Stemberger H, Widermann G. Restoration of defective cytokine activity within lepromatous leprosy lesions. Int J Med Microbiol 1990; 272:458–66.
132. Kaplan G, Cohn ZA. Leprosy and cell-mediated immunity. Curr Opin Immunol 1991; 3:91–6.
133. Lim SK, Fusaro R, Good RA. Leprosy. VI. The treatment of leprosy patients with intravenous infusions of leukocytes from normal persons. Clin Immunol Immunopathol 1972; 1:22.
134. Hastings RC, Job CK. Reversal reactions in lepromatous leprosy follow transfer factor therapy. Am J Trop Med Hyg 1978; 27:995–1004.
135. Nathan C, Squires K, Griffo W, Levis, W, Varghese M, Job CK, Nusrat AR, Sherwin S, Rappoport S, Sanchez E, Burkhardt RA, Kaplan G. Widespread intradermal accumulation of mononuclear leokocytes in lepromatous leprosy patients treated systemically with recombinant interferon gamma. J Exp Med 1990; 172:1509–12.
136. Fels A, Pawlowski N, Abraham E, Cohn Z. Compartmentalized regulation of macrophage arachidonic acid metabolism. J Exp Med 1986; 163:752–7.

18

The Intracellular Nature of Salmonella Infection During the Early Stages of Mouse Typhoid

Nancy E. Dunlap, William H. Benjamin, Jr., and David E. Briles

University of Alabama at Birmingham, Birmingham, Alabama

I. INTRODUCTION

Salmonellae can cause typhoid-like diseases (enteric fever) in many different mammalian species. In most cases, however, the host range in which a particular salmonella can cause invasive typhoid-like disease is quite narrow. This stands in marked contrast to gastroenteritis, which can be induced in most species by a much larger nunber of different types of salmonella. In humans, typhoid fever is primarily caused by *Salmonella typhi* and *Salmonella paratyphi*. *Salmonella typhimurium* and *Salmonella enteritidis* are major etiological agents of mouse typhoid but generally are not causes of invasive disease in immunocompetent humans. In nature, salmonellae causing typhoid infections are generally acquired as a contaminant with ingested food or water. The organisms quickly enter the lymph and blood, and most of those that survive killing during blood clearance are found in the liver and spleen. Disease caused by these enteric bacteria is almost exclusively limited to these two reticuloendothelial organs [1–3]. Salmonellae that cause enteric fever have been described as facultative intracellular parasites that reside within macrophages. However, there has recently been more than a little controversy over the major location of *S. typhimurium* multiplication in vivo [4–8].

A mouse model of enteric fever involves the injection of salmonella directly into the blood. Intravenous (IV) injection of salmonella allows the infecting salmonellae to be introduced into the blood as a single bolus, thus allowing a kinetics study of the fate of injected salmonellae, without ambiguity about the duration of time during which they might reach the bloodstream following an oral or intraperitoneal injection. However, since mucosal immunity may be important in modifying the host's defense against salmonellae, and since the mucosal route might signal activation of salmonella virulence genes, it has been important to confirm results obtained with the IV route, whenever possible, by experiments using the oral route of infection.

The course of murine salmonellosis following IV infection can be divided into different phases, each with its own relationship to host defenses. The first phase of host defense would be the killing of salmonellae during blood clearance. During this phase of infection, the salmonellae that survive become localized to the liver and spleen [4,9,10]. Once the organisms are in the reticuloendothelial system (RES), the second phase of the infection begins. Host killing of salmonellae is almost nonexistent, from 2 to 24 hours after infection [4]. The ability of salmonellae within the spleen and liver to grow, in the relative absence of killing, suggests that the surviving salmonellae have reached a "safe site" that is probably instrumental in their ability to cause disease.

If a mouse survives this growth phase of salmonellosis, it becomes resistant to subsequent challenge and gradually eliminates the growing salmonellae of the first inoculum. In the later phase of murine salmonellosis, both the humoral and the cellular limbs of the immune system appear to play the important roles [11–15]. However, the specific functions of the different components of the immune system in the early, middle, and late phases of infection have yet to be fully worked out.

II. INFLAMMATORY RESPONSE TO SALMONELLA INFECTION

A. There Appears to Be a Critical Number of Salmonellae Within the Reticuloendothelial System of the Mouse that Is Necessary to Elicit an Inflammatory Response

An interesting facet of salmonella infections is that as few as 1 to 10 bacteria injected IV into a susceptible mouse proceed to kill the animal, whereas when 10^6 bacteria are injected IV, over 90% of the inoculum is killed within 4 hours. On the basis of these results it seemed likely that mice kill a higher percentage of a large inoculum than of a small one. In order to examine the dose response kinetics of salmonella infections in susceptible mice, varying inoculums of virulent salmonella (range: log 0 to 5) were injected IV, and mice were sacrificed 4 days later. Spleens were removed and the numbers of salmonellae present within the spleen were calculated.

As the inoculum was raised from 10^1 to 10^2 colony-forming units (CFUs), there was a proportional increase in CFUs at 4 days (10^4 and 10^5, respectively). Inoculums of 10^2 and 10^3 CFUs both yielded about 10^5 CFUs in the spleen and liver at 4 days after inoculation. Inocula above 10^3 CFUs again showed proportional increases in CFUs at 4 days. Our interpretation of these data is that at some critical load of salmonellae the mouse is able to make an inflammatory response that blunts the course of the infection. One way this might occur would be if higher percentages of the high-dose inoculum than of the low-dose inoculum were killed during the initial blood clearance. Another possibility might be that as the numbers of salmonellae in the liver and spleen increase beyond a few thousand, the growth rate of the salmonellae in the spleen decreases. Either explanation would have the effect of allowing the numbers of CFUs in mice infected with low numbers of CFUs to catch up with those in mice infected with larger numbers of CFUs.

The hypothesis that a critical number of salmonellae are required to initiate the inflammatory response is supported by our experiments examining the effect of infection on the density of polymorphonuclear cells (PMNs) in the spleen. Mice were inoculated with 10^2 CFUs of virulent salmonellae and sacrificed on days 3 and 4 after infection. Three days after infection there was no change from baseline in the numbers or densities of PMNs or monocytes within the infected spleens. There were approximately 10^3 CFUs of salmonellae within the spleen at this point. Four days after infection, when the numbers of bacteria had reached approximately 10^5 CFUs, the density of PMNs decreased (as determined by fractionation on Percoll density gradients). No influx of less dense PMNs into the spleen, however, was apparent at this time. We thus hypothesize that between 10^4 and 10^5 CFUs the numbers of salmonellae reach a level that can provide an inflammatory signal to the PMN.

B. There Is a Rapid Salmonella-Induced Inflammatory Response in the Spleen, As Evidenced by Changes in Splenic Polymorphonuclear Cell Populations

By 2 hours after injection of 10^6 salmonellae the average density of PMNs had decreased. By 6 hours, PMNs had increased in total number, and a peak of dense PMNs was again seen, suggesting that there was an influx of new PMNs into the spleen. At 24 hours, the total number of PMNs remained elevated. In contrast to PMNs, monocytes/macrophages did not appear to increase in number during the first 24 hours after infection, although there was a tendency for the monocytes/macrophages present to be of lower density than those in uninfected spleens [8].

C. Prior Infection with Mouse Hepatitis Virus Resulted in a Temporal Increase in Resistance to Salmonella Infection

This increase in resistance was apparent 6 days following nonfatal mouse hepatitis virus (MHV) infection in *Ity*^{s}, and was the result of an MHV-mediated decrease in the salmonella growth rate. The increase in resistance was temporary and was not observed at 3, 10, or 14 days following MHV infection. The mechanisms of the increased resistance is not known, but it is probably a response to inflammatory signals arising from the MHV infection [3,16].

III. LOCATION OF *SALMONELLA TYPHIMURIUM* IN THE LIVER AND SPLEEN OF MICE

It was long assumed that the bulk of salmonella growth in the spleen and liver of the mouse is inside of macrophages. Much of the data for this supposition was indirect, however. A major piece of evidence was that while immunity to both live and killed salmonellae elicited antibody to salmonellae, immunity to live salmonellae elicited the most effective protection and was the only form of immunity that elicited cell-mediated immunity [17]. The association between the elicitation of protection and cell-mediated immunity was reminiscent of a similar association observed with *Listeria monocytogenes* and *Mycobacterium* infection in the mouse, where it seemed clear that the infection was in macrophages and that protection was by transfer of immune T cells [18,19]. The conclusion that infecting salmonellae were intracellular also fit well with the observations that they can grow in macrophages in vitro [11,20,21] and that passive antibody could be shown to be protective only under special circumstances [15,22–25].

As pointed out by Hsu et al., there was little or no direct evidence supporting the intracellular location of salmonella in vivo [26,27]. In order to be able to readily visualize salmonellae in the tissue, Hsu et al. examined mice late in infection, when many of the animals were probably already in terminal sepsis. These workers reported that at these late stages of infection most of the salmonellae in the liver and spleen were either extracellular or in nonprofessional phagocytes [26]. Because of the study design used, the results of these workers may not be applicable to the question of the cellular versus extracellular location of salmonellae during the early stages of enteric fever.

Likewise, the location of salmonella growth within the cells is not firmly established. On the basis of the in vitro studies with cultured macrophages it has been shown that salmonella growth occurs in phagocytic vacuoles that fail to fuse with lysosomes [28]. In other studies in which infected cells were removed from acutely ill mice it has been reported that salmonellae reach the cytoplasm of ingesting phagocytes [29]. The in vivo situation, after infection with moderate numbers of salmonellae, is still unknown.

A. A Safe Site for *S. typhimurium* in Mice

We have been able to study separately the growth and killing of salmonella for the first few days after IV infection by using *S. typhimurium* carrying a temperature-sensitive plasmid that shows minimal replication in vivo. During the first few hours after infection almost all of the injected salmonellae are cleared from the blood. Most of them are killed, but about 5% to 20% survive in the liver and spleen. After the first 2 hours the salmonella in the liver and spleen were able to grow, with little subsequent killing [4]. Thus, the salmonellae appear to reach some type of "safe site" where they are protected from killing.

1. *Survival of Salmonellae Within the Liver and Spleen Is Not Due to Acquisition by Salmonellae of the Ability to Resist Killing*

The safe site might be a privileged location that makes the salmonellae relatively inaccessible to killing. To test the hypothesis that the salmonellae adapt to the host by the acquisition of enhanced antibacterial mechanisms, mice were infected with salmonellae recovered from other mice after 24 hours of infection. We observed the same degree of killing (approximately 90%) of the salmonella grown in vivo and of the media-grown salmonella.

2. *Survival of Salmonellae Within the Liver and Spleen Is not Due to the Host's Becoming Unable to Kill Salmonellae*

Alternatively, it could be the result of a suppression of nonspecific immunity in the mouse or the activation of salmonella virulence genes shortly after infection that protect the surviving salmonellae from subsequent killing. To examine the possibility that host defenses are overwhelmed or down-regulated shortly after infection, mice were infected with wild-type salmonellae and then reinfected 24 hours later with the same bacteria carrying a kanamycin-resistance cassette. Control mice, not previously infected, were also infected with the kanamycin-resistant salmonella. Although slightly faster clearance from the blood was observed in the previously infected animals, the numbers of kanamycin-resistant salmonellae surviving in the spleen and liver was essentially the same in the infected and the uninfected mice.

3. *Evasion of Host Defense Mechanisms*

Our results described above suggest that the "safe site" for salmonellae is a privileged intracellular location, rather than a change in the resistance of the mouse or the virulence of salmonellae. Evidence that salmonellae resist both oxygen-dependent and oxygen-independent killing by phagocytic cells is compatible with the hypothesis that some of the salmonellae simply evade the normal killing mechanisms. Salmonellae have several enzyme activities such as superoxide dismutase and catalase that may help protect them against the oxidants [30]. Recently, an outer membrane protein has been identified that protects salmonellae from oxidative killing by PMNs [31]. Other microbial structures have been identified that protect salmonellae from nonoxidative killing [32–34]. Electron-microscopic pictures of splenic phagocytes in mice with septic salmonella infections show many bacteria within the cytoplasm, thus providing another mechanism of escaping the destructive mechanisms of the lysosome [29].

B. Virtually All Viable Salmonellae Are Intracellular

Because of the lack of direct evidence for an intracellular or extracellular location for salmonellae during the early stages of enteric fever, we sought to confirm previous studies suggesting that salmonella growth probably occurs within cells. A first piece of evidence indicating that salmonellae probably were growing intracellularly was the demonstration that mutations in iron acquisition pathways, used to acquire iron in the presence of serum, altered salmonella growth in serum in vitro but failed to effect the pathogenesis of typhoid infections caused by *S. typhimurium* in mice [35,36]. These results strongly suggested that salmonellae growing in mice were not exposed to serum and were consistent with the hypothesis that they were growing intracellularly.

To study the cellular versus extracellular location of in vivo salmonella by a totally different method, we treated the infected mice with gentamicin [37]. Gentamicin quickly kills salmonellae outside cells but is inefficient in killing intracellular salmonellae. Gentamicin was injected either before or after IV injection of salmonellae. In all cases, exposure to gentamicin was only 30 minutes. When given before infection, gentamicin reduced the numbers of salmonella in the blood 100-fold but reduced the numbers of salmonellae in the liver and spleen only 10-fold. However, when gentamicin was injected 2 hours after salmonella infection, we observed no reduction in the numbers of salmonellae in the liver and spleen. These results suggest that the salmonellae in the liver and spleen were intracellular. When the spleens of infected mice were teased and the cells exposed to gentamicin in vitro, approximately half of the salmonellae remained resistant to gentamicin, suggesting that they remained intracellular. Sonication of these cells rendered the salmonellae sensitive to gentamicin [37]. Thus, the results of both the iron metabolism studies and the gentamicin treatment studies strongly supported an intracellular location of virtually all salmonellae during the early stages of enteric fever in mice.

C. Splenic Polymorphonuclear Cells Contain the Majority of Intracellular Salmonellae 24 Hours after Infection

In an attempt to identify the cellular location of salmonellae in the spleen 24 hours after infection, we fractionated splenic cells by Ficoll-Hypaque separation and enriched for salmonella-containing cells in the Ficoll band and pellet by fluorescence-activated cell sorting. The majority of salmonella-containing cells were found in the Mac-1+/J-11d+ cell fractions of the band and pellet. By direct visualization of the enriched cell populations it was observed that virtually all of the salmonellae were in PMNs. Polymorphonuclear cells containing salmonellae in the Ficoll band generally contained a single bacterium, whereas those from the pellet generally contained several salmonellae [37].

IV. *Ity* LOCUS OF THE MOUSE

During the early phase of murine salmonellosis, the *Ity*^{r} (*immunity to typhimurium*, *resistant*) and *Ity*^{s} (*immunity to typhimurium*, *sensitive*) alleles of the *Ity* locus on chromosome 1 of the mouse play a major role in the resistance to systemic *S. typhimurium* infection [4,9,38]. This locus (or closely linked loci) also controls resistance to infections with certain strains of *Mycobacterium* and *Leishmania* [39–41], two organisms known to grow within host macrophage phagolysosomes [42,43]. The *Ity* gene appears to be responsible for much of the difference in salmonella resistance among inbred strains [44–46]. About half of inbred mice are *Ity*^{r} and half *Ity*^{s} [47]. Other, less well defined genes that are commonly variant in mice also affect murine resistance to *S. typhimurium* [48,49]. Three other genes, identified by their mutant alleles *nu*, *Lps*^{d}, and *xid*, also have measurable effects on salmonella virulence [50–53].

The effects of the *Ity* locus on the host's ability to control the salmonella infection can be detected as early as 24 hours after injection and can alter the LD_{50} of salmonella from 10^1 to 10^3 or more [9,54]. When *Ity*^{s} mice are injected IV with 10 to 100 CFUs of virulent *S. typhimurium*, the number of salmonellae in the liver and spleen increases exponentially to levels of 10^8 to 10^9 CFUs before death at 5 to 8 days after infection [46,55]. In *Ity*^{r} mice, the numbers of salmonellae in the liver and spleen also increase exponentially, but at a slower rate than in *Ity*^{s} mice. The number of CFUs in *Ity*^{r} mice is seldom higher than 10^4 or 10^5 (after injection with 10^1 to 10^3 CFUs) by the time the numbers of salmonellae are held in check between days 7 and 14 [56,57]. Subsequently, the numbers gradually decline over a period of a couple of months. The only other murine locus with a comparable effect on salmonella LD_{50} values is the *nu* locus, which affects induced immunity but does not play a significant role until as late as 2 weeks after infection [51].

Several observations have led to the idea that the effector cell of *Ity* expression and the cell in which growing salmonellae reside is the resident macrophage [38,45,46,57]. Indirect findings that support this theory include the fact that *S. typhimurium* is a facultative intracellular pathogen that can survive within phagocytes [20], and the observation that the numbers of viable salmonellae increase more rapidly in the spleens and livers of *Ity*^{s} mice than in the same organs of *Ity*^{r} animals. More direct in vivo data that implicate a central role for macrophages in *Ity*-controlled events include the fact that cells involved in conferring early resistance are bone marrow–derived [58] and silica-sensitive [59]. *Ity*^{r} nude mice (CD-1 *nu*/*nu*) are able to restrict early bacterial growth as efficiently as do euthymic mice (CD-1 *nu*/+), indicating that mature T cells are not required for the expression of the *Ity* phenotype [51].

Swanson and O'Brien [9] compared the blood clearance and RES uptake during the first 24 hours of *S. typhimurium* infection in *Ity*^{r} and *Ity*^{s} mice. Their data indicated that the blood clearance and early RES uptake was similarly completed in less than 2 hours in *Ity*^{s} and *Ity*^{r} mice. Furthermore, there was a similar decrease in the numbers of splenic bacteria between 2 and 5 hours after infection in *Ity*^{r} and *Ity*^{s} strains. At 12 hours, there were net increases in splenic bacterial content in both strains (as compared with inoculating dose). By 24 hours, the net increase was significantly higher in the *Ity*^{s} strain. Their data indicated that while the *Ity* phenotype influences the pathogenesis of murine typhoid within 24 hours after IV innoculation, it does not influence the initial blood clearance, RES uptake, or RES killing within the first few hours after infection. Other studies by these investigators demonstrated that in vitro cultured macrophages from *Ity*^{r} mice were able to kill intracellular salmonellae more efficiently than macrophages from *Ity*^{s} mice [11]. These authors thus postulated that differences in macrophage killing, rather than salmonella growth rate, were responsible for the disparity in the net growth that they had observed in vivo in *Ity*^{r} and *Ity*^{s} mice [9,11].

Data by other investigators indicate that the primary in vivo effect of *Ity* is on salmonella growth. The first studies by Hormaeche in 1980 comparing inbred mouse strains, suggested that the *Ity* locus controls the rate of salmonella growth rather than the rate of killing [60]. Since those studies used relatively avirulent salmonellae and failed to show linkage of the difference to the *Ity* locus, the results were not considered conclusive [11].

Our own recent studies have also compared the killing and growth of salmonellae in mouse strains congenic for the *Ity*^{r} locus [4]. We used an *S. typhimurium aro A* mutant, which does not grow in vivo, and demonstrated that growth of the infecting salmonellae is necessary for manifestation of the *Ity* phenotype. To study the effects of the *Ity* locus on salmonella growth and killing, we used fully virulent salmonellae carrying the temperature-sensitive plasmid pHSG422. This plasmid does not replicate at body temperature and thus is diluted out during salmonella growth in vivo. It is therefore possible, from the total number of salmonellae and the number of plasmid-containing salmonellae recovered from the mice, to separately estimate the amount of salmonella killing and the amount of salmonella growth. Using such plasmid-containing salmonellae, we found that there is approximately three times more killing of salmonellae in *Ity*^{r} mice than in *Ity*^{s} mice during the 4 hours after infection. However, by 48 hours after infection, there is about 18 times more growth of salmonellae in *Ity*^{s} mice than in *Ity*^{r} mice. These results show that the major in vivo effect of the *Ity* locus is on the regulation of salmonella growth [4].

Studies by Skamene and others have reported that resident *Ity*^{r} macrophages are in a higher state of activation than *Ity*^{s} macrophages [61,62]. It would seem likely that this could play a role in the greater salmonella killing and lower rate of salmonella growth in the *Ity*^{r} versus *Ity*^{s} mice. These data may not be relevant to the situation with salmonella, however, since our data indicate that at least during the first 48 hours of *Ity* effect on salmonella pathogenesis the amount of killing of salmonella by *Ity*^{s} and *Ity*^{r} mice is very similar, and that all of the surviving (growing) salmonellae are in PMNs [37]. Furthermore recent studies by Langermans et al. [63] demonstrated that macrophages from *Ity*^{r} mice that become activated as a result of a 12-day sublethal infection with *S. typhimurium* do not display enhanced bactericidal activity in vitro as compared with *Ity*^{r} macrophages from noninfected mice. Also, they reported no significant in vivo effect of prior infection on the numbers of salmonellae recovered from the liver and spleen for up to 48 hours after challenge of the previously infected mice with a second strain of salmonella. They concluded that activated macrophages are probably not very important in host defense during infections with *S. typhimurium*. Also, there is no direct in vivo evidence to indicate the degree to which macrophages play a major role in salmonella killing or harboring growing salmonella.

The above studies are all consistent with a hypothesis that the major effect of *Ity*^{s} on virulence is the regulation of the intracellular growth rate of salmonellae. Initially this *Ity* effect is apparently mediated largely in PMNs. Whether the *Ity*-mediated effect on growth rate is also mediated in macrophages or other cell types later in infection is not known.

V. THE *mviA* LOCUS OF SALMONELLA REGULATES THE ABILITY OF AT LEAST CERTAIN SALMONELLAE TO EXPLOIT THE SALMONELLA SUSCEPTIBILITY OF MICE WITH THE $Ity^{s/s}$ GENOTYPE

The virulent *S. typhimurium* strain WB600 carries the *mivA* allele of the gene mouse virulence A. We have shown that the virulent phenotype of WB600 is dependent on a nonfunctional *mviA* gene. As compared to the functional allele, $mviA^+$, *mviA* increases virulence in Ity^s mice but not in Ity^r mice. A specific BglII site, *mviA4185*, located at about 35 min on the salmonella chromosome (between *osmZ* and *galU*), was within *mviA*. It was clear that virulence resulted from an a lack of $mviA^+$ function, because chromosomal insertion of an antibiotic cassette in the *mviA4185* site of $mviA^+$, as well as the homologus *mviA4093* site of *mviA* DNA, resulted in virulence. When *mviA* and $mviA^+$ were both expressed in the same strain, with one carried in the chromosome and the other on a plasmid, avirulence was always dominant. We have recently shown that the $mviA^+$ gene product is a protein of about 37 kDa. Surprisingly, we also observed that independent wild mouse-virulent isolates of *S. typhimurium* all carried the avirulent $mviA^+$ allele. We assume that in wild strains the expression of this gene is shut off by a regulatory mechanism, and that this mechanism must be defective in the laboratory strain WB600. Studies are currently under way to identify this suspected regulatory mechanism as well as the mechanism of action of the *mviA* gene product. Hopefully, studies to elucidate the mechanism of action of *mviA* will provide insight into the mechanism of action of *Ity*.

VI. SUMMARY

In mouse enteric fever (typhoid) infection with *S. typhimurium*, the bacteria appear to grow intracellularly, and at least during the early phase of infection they are in splenic PMNs rather than macrophages. Inflammation caused by salmonella infection and other infections, such as MHV, results in inflammatory responses that enhance resistance to salmonella infection. At least in the case of MHV, this effect is most pronounced on the rate of salmonella growth. Since the effects of the *Ity* locus on salmonella growth rate are readily seen during the first few days of infection, when salmonella are primarily within PMNs, the *Ity* locus is able to mediate its effect on salmonella pathogenesis in PMNs in vivo. Whether or not macrophages play a predominant role in salmonella pathogenesis later in infection is not yet known.

REFERENCES

1. Carter PB, Collins FM. Growth of typhoid and paratyphoid bacilli in intravenously infected mice. Infect Immun 1974; 10:816–22.
2. Carter PB, Collins FM. The route of enteric infection in normal mice. J Exp Med 1974; 139:1189–1203.
3. Fallon MT, Benjamin WH, Schoeb TR, Briles DE. Mouse hepatitis virus strain UAB infection enhances resistance to *Salmonella typhimurium* by inducing suppression of bacterial growth. Infect Immun 1991; 59:852–6.
4. Benjamin WH Jr, Hall P, Roberts SI, Briles DE. The primary effect of the *Ity* locus is on the growth rate of *Salmonella typhimurium* that are relatively protected from killing. J Immunol 1990; 144:3143–51.
5. Guo Y-N, Hsu HS, Mumaw VR, Nakoneczna I. Electromicroscopy studies on the bacterial action of inflammatory leukocytes in murine salmonellosis. J Med Microbiol 1986; 21:151–9.
6. Carrol MEW, Jackett PS, Aber VR, Lowrie DB. Phagolysosome formation, cyclic adenosine 3':5'-

monophosphate and the fate of *Salmonella typhimurium* within mouse peritoneal macrophages. J Gen Microbiol 1979; 110:421–9.
7. Schurr E, Skamene E, Forget A, Gros P. Linkage analysis of the *Bcg* gene on mouse chromosome 1. Identification of a tightly linked marker. J Immunol 1989; 142:4507–13.
8. Dunlap NE, Benjamin WH Jr, McCall RD, Tilden AB, Briles DE. A "safe site" for *Salmonella typhimurium* is within splenic cells during the early phase of infection in mice. Microbial Pathogenesis 1991; 10:297–310.
9. Swanson RN, O'Brien AD. Genetic control of the innate resistance of mice to *Salmonella typhimurium*: *Ity* gene is expressed in vivo by 24 hours after infection. J Immunol 1983; 131:3014–20.
10. Benjamin WH Jr, Posey BS, Briles DE. Effects of *in vitro* growth phase on the pathogenesis of *Salmonella typhimurium* in mice. J Gen Microbiol 1986; 132:1283–95.
11. Lissner CR, Swanson RN, O'Brien AD. Genetic control of the innate resistance of mice to *Salmonella typhimurium*: expression of the *Ity* gene in peritoneal and splenic macrophages isolated in vitro. J Immunol 1983; 131:3006–13.
12. Briles DE, Lehmeyer J, Forman C. Phagocytosis and killing of *Salmonella typhimurium* by peritoneal exudate cells. Infect Immun 1981; 33:380–8.
13. Mitsuhashi S, Sato I, Tanka T. Experimental salmonellosis: Intracellular growth of *Salmonella enteritidis* ingested in mononuclear phagocytes of mice and cellular basis of immunity. J Bacteriol 1961; 81:863–68.
14. Chandler R, Sainis KB, Lewis NF. Role of thymus-derived lymphocytes in acquired immunity in salmonellosis in mice. Microbiol Immunol 1986; 30:1299–1306.
15. Colwell DE, Michalek SM, Briles DE, Jirillo E, McGhee JR. Monoclonal antibodies to salmonella lipopolysaccharide: anti-O-polysaccharide antibodies protect C3H mice against challenge with virulent *Salmonella typhimurium*. J Immunol 1984; 133:950–7.
16. Fallon MT, Schoeb TR, Benjamin WH Jr, Lindsey JR, Briles DE. Modulation of resistance to *Salmonella typhimurium* infection in mice by mouse hepatitis virus (MHV). Microbial Pathogenesis 1989; 6:81–91.
17. Collins FM, Mackaness GB. Delayed hypersensitivity and Arthus reactivity in relation to host resistance in salmonella-infected mice. J Immunol 1968; 101:830–45.
18. Lane FC, Unanue ER. Requirement of thymus (T) lymphocytes for resistance to listerisis. J Exp Med 1972; 135:1104–12.
19. Orm IM, Collins FM. Protection against *M. tuberculosis* infection by adoptive immunotherapy. J Exp Med 1983; 158:74–83.
20. Lowrie DB, Abner VR, Carrol MEW. Division and death rates of *Salmonella typhimurium* inside macrophages: use of penicillin as a probe. J Gen Microbiol 1979; 110:409–19.
21. Fields PI, Swanson RV, Haidaris CG, Heffron F. Mutants of *Salmonella typhimurium* that cannot survive within the macrophage are avirulent. Proc Natl Acad Sci USA 1986; 83:5189–93.
22. Angerman C, Eisenstein T. Comparative efficacy and toxicity of a ribosomal vaccine, acetone-killed cells, lipopolysaccharide, and a live cell vaccine prepared from *Salmonella typhimurium*. Infect Immun 1978; 19:575–82.
23. Muotiala A, Hovi M, Måkelå PH. Protective immunity in mouse salmonellosis: comparison of smooth and rough live and killed vaccines. Microbial Pathogenesis 1989; 6:51–60.
24. Rowley D, Turner JK. Number of molecules of antibody required to promote phagocytosis of one bacterium. Nature 1966; 1:496–8.
25. Saxen H. Mechanism of the protective action of anti-Salmonella IgM in experimental mouse salmonellosis. J Gen Microbiol 1984; 130:2277–83.
26. Nakoneczna I, Hsu HS. The comparative histopathology of primary and secondary lesions in murine salmonellosis. Br J Exp Pathol 1980; 61:76–84.
27. Lin FR, Wang XM, Hsu HS, Mumaw VR, Nakoneczna I. Electron microscopic studies on the location of bacterial proliferation in the liver in murine salmonelosis. Br J Exp Pathol 1987; 68:539–50.
28. Buckmeier NA, Heffron F. Inhibition of macrophage phagosome-lysosome fusion by *Salmonella typhimurium*. Infect Immunol 1991; 59:2232–8.
29. Lin FR, Hsu HS, Mumaw VR, Moncure CW. Confirmation of destruction of salmonellae within murine peritoneal exudate cells by immunocytochemical technique. Immunology 1989; 67:394–400.
30. Fridovich I. The biology of oxygen radicals. Science 1975; 201:875–80.
31. Stinavage PS, Martin LE, Spitznagel JK. A 59 kilodalton outer membrane protein of *Salmonella*

typhimurium protects against oxidative intraleukocytic killing due to human neutrophils. Mol Microbiol 1990; 4:283–93.
32. Fields PI, Groisman EA, Heffron F. A Salmonella locus that controls resistance to microbicidal proteins from phagocytic cells. Science 1989; 243:1059–62.
33. Kim K, Kim IH, Rhee SG, Stadtman ER. The isolation and purification of a specific "protector" protein which inhibits enzyme activation by a thiol/Fe(III)O_2 mixed-function oxidation system. J Biol Chem 1988; 263:4704–11.
34. Miller SI, Rulkkinen WS, Selsted ME, Mekalanos JJ. A characterization of defensin resistance phenotypes associated with mutations in the phoP virulence regulon of *Salmonella typhimurium*. Infect Immun 1990; 58:3706–10.
35. Benjamin WH Jr, Turnbough CL Jr, Posey BS, Briles DE. The ability of *Salmonella typhimurium* to produce the siderophore enterobactin is not a virulence factor in mouse typhoid. Infect Immun 1985; 50:392–7.
36. Stocker BAD, Måkelä PH. Genetic determination of bacterial virulence, with a special reference to *Salmonella*. Curr Top Microbiol Immunol 1986; 124:149–72.
37. Dunlap NE, Benjamin WHJ, Berry AK, Eldridge JH, Briles DE. A "safe-site" for *Salmonella typhimurium* is within splenic polymorphonuclear cells. Microbial Pathogenesis 1992; 13:181–90.
38. Plant J, Glynn AA. Locating salmonella resistance gene on mouse chromosome 1. Clin Exp Immunol 1979; 37:1–6.
39. Skamene E, Gros P, Forget A, Kongshavn PAL, Charles CS, Taylor BA. Genetic regulation of resistance to intracellular pathogens. Nature 1982; 297:506–9.
40. Skamene E. Genetic regulation of host resistance to bacterial infection. Rev Infect Dis 1983; 5:S823–S832.
41. Plant JE, Blackwell JM, O'Brien AD, Bradley DJ, Glynn AA. Are the *Lsh* and *Ity* disease resistance genes at one locus on mouse chromosome 1? Nature 1982; 297:510–1.
42. Bradley DJ, Taylor BA, Blackwell J, Evans EP, Freeman J. Regulation of Leishmania populations within the host. III. Mapping of the locus controlling susceptibility to visceral leishmaniasis in the mouse. Clin Exp Immunol 1979; 37:7–14.
43. Hart PD, Armstrong JA, Brown CA, Draper P. Ultrastructural study of the behavior of macrophages toward parasitic mycobacteria. Infect Immun 1972; 5:803–7.
44. Plant J, Glynn AA. Natural resistance to salmonella infection, delayed hypersensitivity and Ir genes in different strains of mice. Nature 1974; 248:345–7.
45. Plant J, Glynn AA. Genetics of resistance to infection with *Salmonella typhimurium*. J Infect Dis 1976; 133:72–8.
46. Robson HG, Vas SI. Resistance of inbred mice to *Salmonella typhimurium*. J Infect Dis 1972; 126:378–86.
47. O'Brien AD, Rosenstreich DL. Genetic control of the susceptibility of C3HeB/FeJ mice to *Salmonella typhimurium* is regulated by a locus distinct from known Salmonella response genes. J Immunol 1983; 131:2613–5.
48. Briles DE, Benjamin WH Jr, Hustler WJ, Posey B. Genetic approaches to the study of disease resistance: with special emphasis on the use of recombinant inbred mice. Curr Top Microbiol Immunol 1986; 124:21–35.
49. Benjamin WH Jr, Turnbough CL Jr, Posey BS, Briles DE. *Salmonella typhimurium* virulence genes necessary to exploit the $Ity^{s/s}$ genotype of the mouse. Infect Immun 1986; 51:872–8.
50. O'Brien AD, Scher I, Metcalf EE. Genetically conferred defect in antisalmonella antibody formation renders CBA/N mice innately susceptible to *Salmonella typhimurium* infection. J Immunol 1981; 126:1368–72.
51. O'Brien AD, Metcalf ES. Control of early *Salmonella typhimurium* growth in innately salmonella-resistant mice does not require functional T-lymphocytes. J Immunol 1982; 129:1349–51.
52. Hormaeche CE, Maskell DJ, Harrington K, Joysey H, Brock J. Mechanisms of natural resistance to mouse typhoid. Bull Eur Physiopath Resp 1983; 19:137–42.
53. O'Brien AD, Weinstein DA, Soliman MY, Rosenstreich DL. Additional evidence that the *Lps* gene locus regulates natural resistance to *S. typhimurium* in mice. J Immunol 1985; 134:2w820–3.
54. Briles DE, Benjamin WH Jr, Williams CA, Davie JM. A genetic locus responsible for salmonella

susceptibility in BSVS mice is not responsible for the limited T-dependent immune responsiveness of BSVS mice. J Immunol 1981; 127:906–11.
55. Hormaeche CE, Brock J, Pettifor RA. Natural resistance to mouse typhoid: possible role of the macrophage. New York: Academic Press, 1980:121–32.
56. O'Brien AD, Taylor BA, Rosenstreich DL. Genetic control of natural resistance to *Salmonella typhimurium* in mice during the late phase of infection. J Immunol 1984; 133:3313–8.
57. Hormaeche CE. Natural resistance to *Salmonella typhimurium* in different inbred mouse strains. Immunology 1979; 37:311–8.
58. Hormaeche CE. The natural resistance of radiation chimeras to *S. typhimurium* C5. Immunol 1979; 37:329–32.
59. O'Brien AD, Scher I, Formal SB. Effects of silica on the innate resistance of inbred mice to *Salmonella typhimurium* infection. Infect Immun 1979; 25:513–20.
60. Hormaeche CE. The in vivo division and death rates of *S. typhimurium* in the spleens of naturally resistant and susceptible mice measured by the superinfecting phage technique of Meynell. Immunology 1980; 41:973–9.
61. Buschman E, Taniyama T, Nakamura R, Skamene E. Functional expression of the *Bcg* gene in macrophages. Res Immunol 1989; 140:767–828.
62. Zwilling BS. MHC class II glycoprotein expression and resistance to mycobacterial growth. Res Immunol 1989; 140:806–9.
63. Langermans JAM, van der Hulst MEB, Nibbering PH, van Furth R. Activation of mouse peritoneal macrophages during infection with *Salmonella typhimurium* does not result in enhanced intracellular killing. J Immunol 1990; 144:4340–6.

19

Macrophage–*Listeria* Interactions

Priscilla A. Campbell

National Jewish Center for Immunology and Respiratory Medicine, and University of Colorado Health Sciences Center, Denver, Colorado

I. INTRODUCTION

Facultative intracellular bacteria live and grow in both professional phagocytes and nonphagocytic target cells. In nonphagocytic cells, they must only devise a method of entry; entry alone is sufficient to evade host defense mechanisms, which usually are ineffective against organisms hiding within nonbactericidal cells. In phagocytic cells, however, the story is different. These cells, particularly immunologically stimulated monocytes and macrophages, are important effector cells against bacteria that can live intracellularly—thus strategies have been devised by both host and invader to potentiate their own survival.

One such organism is the facultative intracellular bacterium *Listeria monocytogenes*. This bacterium is a gram-positive rod that grows quite well extracellularly as well as intracellularly. It is ubiquitous, found in over 60 animal species, and is easily cultured from soils [1]. Because it grows at temperatures down to 4° C, *Listeria* can contaminate foodstuffs such as coleslaw, soft cheeses, and pasteurized milk: each of these has been implicated in recent outbreaks of listeriosis [reviewed in 2,3]. It rarely causes disease in normal, healthy individuals but is a significant pathogen in immunocompromised individuals, and in pregnant women as well [1–3].

L. monocytogenes is perhaps unique among facultative intracellular bacteria, because quite a bit is known about its cell biology. In this chapter, how this organism interacts with macrophages, both those that can kill *Listeria* and those that cannot, will be discussed in some detail. Receptors used by macrophages to phagocytose *Listeria*, as well as mechanisms by which *Listeria* enters nonphagocytic cells, will be described. Current knowledge regarding how macrophages kill *Listeria*, and cytokines that regulate this process, will also be reviewed. Finally, the heterogeneity of macrophages with regard to their ability to kill this facultative bacterium will also be discussed.

II. CELL BIOLOGY OF *LISTERIA* INFECTION

A. Invasion of Cells by *Listeria*

In the mammalian host, *Listeria* can enter and live in nonphagocytic as well as phagocytic cells. In natural infection, *Listeria* is consumed orally and enters the host by invading the epithelial cells of the gut. It seems to take up residence in hepatocytes, endothelial cells, and perhaps spleen cells as well, and then grows intracellularly. A recent study indicates that to invade a nonphagocytic cultured epithelial cell, a bacterial gene called *inlA* encodes a 95-kDa protein, internalin [4]. This protein becomes expressed on the bacterial surface and allows binding of the organism to cells, followed by internalization [4]. In this manner, colonization of a host with *L. monocytogenes* may begin.

Another protein secreted by *Listeria* may also contribute to its ability to invade nonphagocytic cells. The protein is a 60-kDa molecule encoded by a gene designated *iap* [5,6]. Fairly large quantities of this protein are produced by all virulent strains of *L. monocytogenes* [7]. Evidence for a role of p60 as a virulence factor comes from experiments in which mutants with impaired synthesis of p60 were shown to have lost the ability of invade a fibroblast cell line [5]. Whether this invasion protein also contributes to invasion of phagocytes is not yet known.

As the host's immune system becomes aware of the invader, it mobilizes the effector cells that can kill it, the professional phagocytes, neutrophils and macrophages. Both these cells have been shown to express high levels of listericidal activity in mice and in humans [8,9]. While not much is known yet about the cell biology of *Listeria*:neutrophil interactions, exciting experiments from several groups have defined many of the intracellular events that occur as *Listeria* invades macrophages and nonphagocytic cells. Tilney and Portnoy have used the electron microscope to monitor the progress of *Listeria* through cells [10]. Using a mouse macrophage cell line, J774, that cannot kill *Listeria* (P. A. Campbell, unpublished data), they and others have identified the following sequence of events as *Listeria* invade these and other cells (Figure 1).

Infection of phagocytic and nonphagocytic cells begins with binding of bacteria to the cell surface. Bacteria are then phagocytosed and initially enclosed within a membrane-bound phagosome. The specific cellular events that mediate phagocytosis may be different in phagocytic and nonphagocytic cells. In nonprofessional phagocytes, entry is mediated by internalin and is pathogen-directed [4]. In phagocytric cells, however, the host is an active participant in this initial step and, as discussed in Section II.A., specific phagocyte receptors may play an important role in regulating whether the bacterium lives or dies within the cell.

Within 30 minutes after the onset of phagocytosis, the virulent organism has secreted a hemolytic enzyme that begins to dissolve the phagosome membrane [10]. This enzyme, described in more detail below, is required for escape of the bacterium from the phagosome into the cytoplasm, and for virulence [11,12]. Once in the cytoplasm *Listeria* can divide and mobilize around themselves short, randomly oriented microfilaments of F-actin [10,13]. These filaments appear random at first but quickly become polarized and move to one end of the bacterium, where they form a long comet-like tail. Evidence suggests that these actin filaments are nucleated on a fine fibrillar material secreted by *L. monocytogenes* [14,15], which is encoded by the Listeria *actA* gene [15]. Actin-coated *Listeria* become motile within the cells, probably because of actin polymerization [16], and move to the cell surface. Then the bacteria extend from the cell in long pseudopod-like structures in which the *Listeria* are at the head and the actin filaments form a cytoskeletal tail that probably is responsible for formation of the pseudopod. This structure may be as long as the macrophage itself. The *Listeria*-containing pseudopod then contacts an adjacent macrophage, which soon phagocytoses the *Listeria* and the membrane surrounding it [10,13,17]. Now the newly infected macrophage contains *Listeria* surrounded by a double membrane composed of the cell membrane of the first macrophage and that of the new host cell. The organism then dissolves both membranes and has successfully parasitized the second cell. In this manner, *Listeria* are thought to travel from one permissive, nonlistericidal cell to another, without extracellular exposure to phagocytes that can kill them, or to other elements of a successful host defense. This method of cell-to-cell spread also seems to occur in nonprofessional phagocytes, such as the enterocyte-like human colonic carcinoma cell line Caco-2 [11,13].

The model described above clearly implicates the *Listeria*-derived hemolysin as an important virulence factor. It was first identified as a factor required for growth of *Listeria* in the livers and spleens of infected mice by transposon mutagenesis experiments [18,19]: *L. monocytogenes* containing the gene encoding the hemolysin were virulent and grew in recipient mice, while

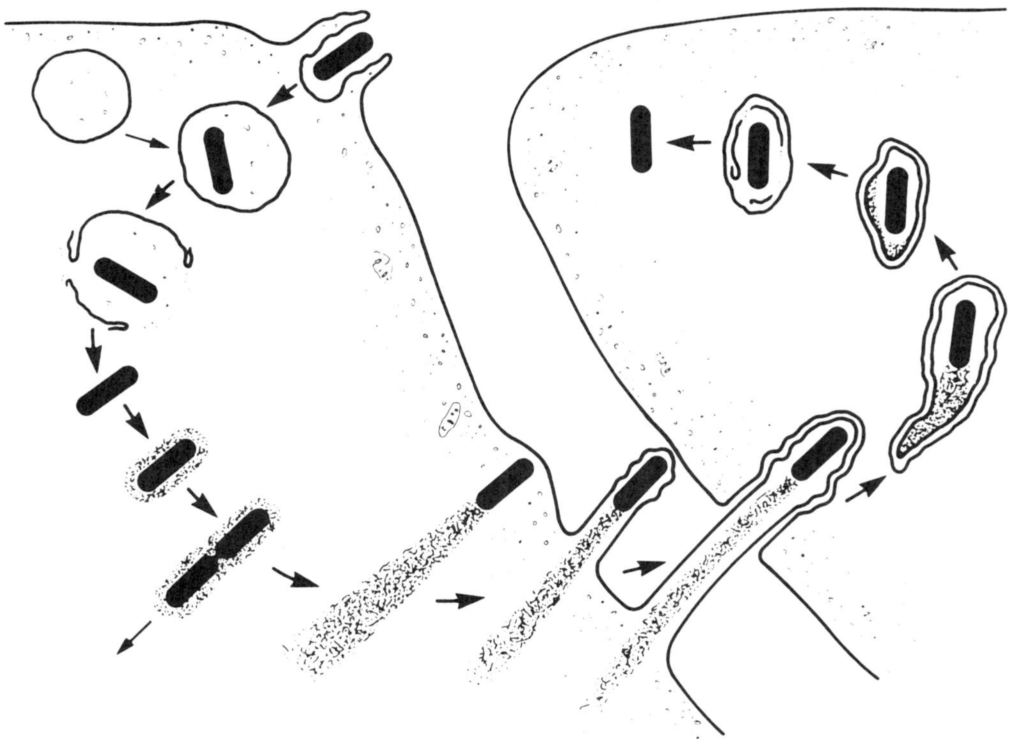

FIGURE 1 Cell biology of *Listeria* within cells that cannot kill it. The organism is phagocytosed into a phagosome. *Listeria* secretes listeriolysin O, dissolving the phagosome membrane, and escapes into the cytoplasm. F-actin is then mobilized around the bacterium, which may divide in the cytoplasm. The organism then polymerizes the actin, forming a comet-like structure with *Listeria* at one pole, trailing polymerized actin. From this arises a pseudopod containing *Listeria* at the tip. This pseudopod extends toward an adjacent cell, is recognized, and is internalized. The bacterium dissolves the double membrane, and now has successfully parasitized a new cell without being exposed to the extracellular environment. (From Ref. 10. Copyright of the Rockefeller University Press.)

hemolysin-negative mutants did not. The 60-kDa hemolysin has been characterized and named listeriolysin O [20], and the gene encoding it, *hlyA*, has been cloned and sequenced [21,22]. Listeriolysin O, which cross-reacts with streptolysin O, is a β-hemolysin, one of the extracellular proteins that belong to the group of thiol-activated, cholesterol-inhibited cytolysins [23]. It is produced in small amounts by *L. monocytogenes* growing under normal conditions, but heat shock can induce increased synthesis [24]. Perhaps surprisingly, the degree of *Listeria* virulence is not correlated with the levels of hemolysin expressed [24,25]. A very recent report suggests that listeriolysin O is the major antigen against which protective immunity is generated in mice [26], and the specific peptide responsible for immunity has been identified [27].

Listeriolysin O plays an important role in growth of *Listeria* in phagocytic macrophages as well as in cells that are not normally phagocytic. In Caco-2, and in the mouse fibroblast line 3T6, *Listeria* seems to induce its own phagocytosis [11,28]. This event is not dependent on secretion of the hemolysin, since hemolysin-negative mutants and hemolysin-positive bacteria enter cells equally well [28]. Once inside the cell, listeriolysin O–secreting bacteria escape into the cytoplasm, while listeriolysin O–negative mutants remain in the phagocytic vesicle [11,12]. Similarly, in the J774 mouse macrophage cell line, in bone marrow–derived macrophages, and in mouse peritoneal macrophages, intracellular growth of *Listeria* requires hemolysin production [28,29]. However, listeriolysin O–negative mutants can replicate in human epithelial cells or fibroblasts [29]. Other experiments have shown that the common soil bacterium *Bacillus subtilis*, which normally is incapable of intracellular growth, can escape from the phagosome and grow in the cytoplasm when transfected with the gene encoding listeriolysin O [30]. These experiments document the importance of this hemolytic enzyme in the invasion of cells by *Listeria*.

Listeriolysin O is not the only product of *L. monocytogenes* that may be involved in the pathogenesis of this organism. Recently, a phosphatidylinositol-specific phospholipase C (PI-PLC) secreted by *Listeria* has been implicated as a virulence factor [31]. Using a mutant with a transposon insertion in the structural gene *plcA*, which encodes PI-PLC, Portnoy and colleagues showed that this mutant was less virulent in mice: the LD_{50} was roughly 1000-fold less than that of the wild type. Similarly, the PI-PLC–negative mutant did not grow well in resident peritoneal macrophages or the J774 macrophage cell line [31,32]. These experiments suggest that PI-PLC as well as listeriolysin O is an essential determinant of virulence. While the stage in the intracellular life cycle at which the PI-PLC–negative mutant is defective is not yet known, a good guess is that it, too, is defective in its ability to leave the phagosome and enter the cytoplasm.

Other components in the cytoplasmic life cycle of *Listeria* in macrophages also have been described. A nonvirulent mutant has been generated that can escape from the phagosome into the cytoplasm and induce actin polymerization, but it cannot rearrange the actin filaments to generate cytoplasmic motility [33]. Thus this mutant remains trapped in the cytoplasm within an actin cloud and cannot travel from cell to cell. This demonstrates that escape into the cytoplasm by itself is not sufficient to ensure that the bacterium can move from one cell to another.

It is clear from the experiments described above that *L. monocytogenes* has devised unique mechanisms to evade host defenses, and that it is an active participant in its own intracellular survival. It secretes extracellular products that allow it to invade cells that cannot kill it, and in which it can hide from microbicidal effector cells. Even after it has been internalized by a potentially lethal macrophage, *Listeria* secretes an enzyme that allows it to escape into the cytoplasm, a site that once again seems to protect it from host defense mechanisms. Finally, it has devised methods that allow it to move from one cell to another without being exposed to either cells or body fluids that might mediate host defense to this organism.

B. Effect of *Listeria* Infection on Host Macrophage Functions

Infection with virulent, hemolysin-secreting *Listeria* may affect various functions of the host cell differently from infection with hemolysin-negative variants. For instance, mice infected with even very small doses of hemolysin-secreting *Listeria* developed excellent T cell–mediated immunity, measured as delayed-type hypersensitivity reactions, resistance to infection, and induction of immunity [34]. In contrast, mice infected with hemolysin-negative variants of the same organism developed none of the above, regardless of the dose of *Listeria* administered. A possible explanation for this is provided by experiments testing the ability of bone marrow–derived macrophages that had been fed heat-killed *Listeria*, live hemolysin-negative *Listeria*, or live hemolysin-positive *Listeria*, to stimulate CD4+ and CD8+ T cells [35]. CD8+ T cells recognized only bone marrow macrophages pulsed with live hemolysin-positive *Listeria*, while CD4+ T cells responded to macrophages pulsed with any of these three antigens. Thus, hemolysin-positive organisms seem required to stimulate CD8+ T cells, a subset of cells that has been shown to play an important role in vivo in resistance to *Listeria* infection [36,37].

Other experiments suggest not only that *Listeria* can avoid normal macrophage defense mechanisms against itself, as described above, but also that it is able to suppress certain other macrophage functions. Intracellular production of listeriolysin O by hemolysin-producing *Listeria* inhibits the ability of macrophages to process and present antigens in vitro [38]. The ability of T cells to secrete interleukin-2 (IL-2) in response to antigen was inhibited more with respect to antigens that required intracellular processing and presentation than with respect to those antigens, such as ovalbumin peptides, that did not require a processing event.

III. MACROPHAGE MICROBICIDAL MECHANISMS

The discussion above describes the intracellular life cycle of *Listeria* when the organism successfully parasitizes nonphagocytic or phagocytic cells. But an entirely different consequence of infection of macrophages by *Listeria* can occur: the phagocyte can contain and kill the bacterium. This happens not only with nonhemolytic *Listeria*, but also when macrophages successfully thwart infection by virulent, hemolysin-secreting organisms. In this section, mechanisms by which macrophages may kill *Listeria* and differences between listericidal and nonlistericidal macrophages will be discussed.

A. The Receptor Used by *Listeria* to Enter a Cell May Influence Its Intracellular Fate

Early evidence suggested that *Listeria* is phagocytosed in a complement-dependent manner. For instance, *Listeria* can activate the alternative complement pathway, and heat-labile serum factors, most likely complement components, have been implicated as opsonins for this bacterium [39–42]. Recently, it has been demonstrated that macrophages that can kill *Listeria*, that is, proteose peptone–elicited macrophages, phagocytose *Listeria* primarily through complement receptor type 3 (CR3) rather than through other receptors with the potential to mediate phagocytosis [43]. In this study, adherence versus internalization of bacteria was examined using fluorescence microscopy, and the data showed that CR3 mediated both binding and internalization of serum-opsonized *Listeria* [43,44].

There are other ways in which *Listeria* could enter cells. For instance, in the absence of serum and the complement components it contains, peptone-elicited macrophages still phagocytose *Listeria*, but much less well [43]. An interesting corollary is that macrophages that cannot kill *Listeria* appear to phagocytose the bacterium primarily through receptors other than CR3. That is, thioglycolate-elicited mouse peritoneal exudate macrophages, which cannot kill *Listeria* [45,46], do not use CR3 as the major receptor for the phagocytosis of *Listeria*, even in the presence of serum [47]. What receptor they do use is not yet known. It is rather interesting that about 30% of the phagocytosis of *Listeria* by these cells in the absence of serum can be inhibited by laminarin, an inhibitor of the β-glucan receptor (Drevets, unpublished data). Peptone-elicited macrophages, on the other hand, appear not to use the β-glucan receptor for serum-independent phagocytosis. Thus there seem to be inherent differences among different cell populations, even among different macrophage populations, in the receptor used to internalize *Listeria*.

Taken together, these experiments indicate that macrophages that can kill the facultative intracellular bacterium *Listeria monocytogenes* internalize this organism primarily through CR3, while nonlistericidal macrophages do not use CR3 as the primary receptor for phagocytosis of *Listeria*. This suggests that the receptor a bacterium uses to enter a macrophage may influence the intracellular fate of the organism: internalization through one receptor may lead to bacterial killing, while internalization through a different receptor may favor bacterial growth. In addition, we are currently completing experiments that provide direct evidence that CR3 mediates not only phagocytosis of *Listeria*, but also intracellular killing of this organism by macrophages (Drevets and Campbell, unpublished data).

B. Mechanisms by Which Macrophages Kill *L. monocytogenes*

As with many other bacterial pathogens, for years it was thought that the most important listericidal products produced by macrophages were the toxic oxygen intermediates, particularly superoxide anion, hydrogen peroxide, and the hydroxyl radical. Yet there were few experiments that directly implicated these metabolites in the killing of *Listeria*. More recently, several studies have provided evidence that reactive oxygen intermediates may indeed not be important effector molecules used by macrophages to kill *L. monocytogenes*. For instance, minimal levels of superoxide and hydrogen peroxide were generated when antigen-elicited macrophages from *Listeria*-immune mice were stimulated with opsonized virulent *Listeria*, even though these macrophages appeared to kill *Listeria* [48]. In addition, when these macrophages were incubated with scavengers of superoxide, hydroxyl radical, or singlet oxygen, virtually no inhibition of killing was seen. Only the H_2O_2 scavenger catalase caused any inhibition of listericidal activity, and this inhibition may not have been physiologically significant [49]. Moreover, in other studies, virulence of *Listeria* did not correlate with levels of superoxide dismutase, an enzyme that can neutralize the bactericidal effects of superoxide [50]. Finally, patients with chronic granulomatous disease of childhood, who show significant deficiency in oxygen-dependent microbicidal mechanisms, have not been reported to have increased incidences of infection with *Listeria* or other facultative intracellular bacteria. These data suggest that the primary way macrophages kill *Listeria* may not be by toxic oxygen intermediates.

Other macrophage-derived molecules with potent antilisterial activity have been described. For instance, defensins, small microbicidal cationic peptides found in human alveolar macrophages, may have potent antilisterial activity. Microbicidal cationic protein 1, one such defensin isolated from rabbit alveolar macrophages, was reported to be highly active against *Listeria* [50].

More recently, evidence has been accumulating that reactive nitrogen intermediates are important in many macrophage-mediated effector functions. Mouse macrophages use the enzyme nitric oxide synthase to metabolize arginine to citrulline and the short-lived intermediate nitric oxide [reviewed in 51,52,53]. Then, in the presence of O_2 and H_2O, nitric oxide is metabolized to nitrates and nitrites. Biologically active products of arginine metabolism that are likely to be bactericidal include NO, NO_2, and HNO_2 [53,54]. Various effector functions of macrophages, including killing of tumor targets and of bacteria such as *Mycobacterium* sp. and *Franciscella tularensis*, have been shown to be arginine-dependent, and thus dependent on generation of toxic products of nitrogen metabolism [52–57].

L. monocytogenes, too, might be killed by reactive nitrogen intermediates. One report [58] implicated arginine-dependent mechanisms when interferon-γ (IFN-γ) and bacterial lipopolysaccharide (LPS) were used to stimulate bone marrow–derived macrophages to kill *Listeria*. We are conducting experiments that also indicate that there may be a role for products of arginine metabolism in killing of *Listeria*, but these experiments suggest that there may be some important differences in the roles of arginine metabolites in killing of *Listeria* and other facultative intracellular bacteria. The reason for the differences may be that *Mycobacteria* and *F. tularensis* are facultative intracellular bacteria that do not escape the phagosome, but live there in macrophages unable to kill them, while *Listeria*, as described above, escapes into the cytoplasm if it successfully establishes residence in a host macrophage. Experiments in progress should clarify the role of nitric oxide and its products in macrophage listericidal activity.

Restriction of essential bacterial growth factors is another mechanism by which cells could be bactericidal or bacteriostatic. *L. monocytogenes* is a bacterium that requires iron for growth [59]. Thus hosts with iron overload may be at higher risk for developing infection caused by *Listeria*. Patients with chronic iron overload have an increased incidence of listeriosis [60], and mice homozygous for a genetic defect that causes β-thalassemia have increased susceptibility to *Listeria* [61]. The reason excess iron favors the bacterium may be in part that its growth requires iron, and in part that excess iron alters host metabolism.

Iron is transported into macrophages primarily via the transferrin receptor, which binds and internalizes transferrin containing two molecules of ferric iron [62]. Once inside the cell, the endosome containing the transferrin receptor–iron-saturated transferrin complex is acidified, releasing the iron, which is now available to the cell and any bacteria it contains [63]. We recently reported that cell surface transferrin receptor expression and intracellular iron levels correlated with the ability of a macrophage to kill *Listeria* [48]. That is, resident mouse peritoneal macrophages, which cannot kill *Listeria*, expressed low levels of transferrin receptor and had low intracellular iron levels. Proteose peptone– or fetal calf serum–elicited macrophages, which had intermediate levels of intracellular iron and intermediate expression of the transferrin receptor, could kill *Listeria*. On the other end of the spectrum, thioglycolate-elicited peritoneal exudate cells, with high levels of transferrin receptor expression and high levels of intracellular iron, also could not kill *Listeria*. Thus, intracellular iron levels appeared to influence macrophage listericidal activity.

Perhaps most compelling, the high–transferrin receptor, high–intracellular iron, thioglycolate-elicited macrophage, which could not kill *Listeria*, became listericidal if transferrin receptor expression and thus intracellular iron levels were lowered. Similarly, the listericidal peptone-elicited macrophage, expressing middle levels of transferrin receptor and of iron, lost listericidal activity if its intracellular iron stores were lowered. As illustrated in Figure 2, this implicates intracellular iron levels directly in macrophage listericidal mechanisms. Perhaps macrophages must have enough available iron to support listericidal mechanisms, but in the presence of too much iron, the balance is tipped in favor of bacterial growth. It is also possible that iron is compartmentalized differently in different cell populations, resulting in alterations in availability to the bacterium and/or host defense mechanisms. Thioglycolate-elicited macrophages, for example, have been reported to put more intracellular iron into a relatively inert intracellular pool than into ferritin, whereas *Corynebacterium parvum*–elicited macrophages do the opposite [64]. Perhaps in nonlistericidal thioglycolate-elicited macrophages, iron is unavailable for host defense mechanisms

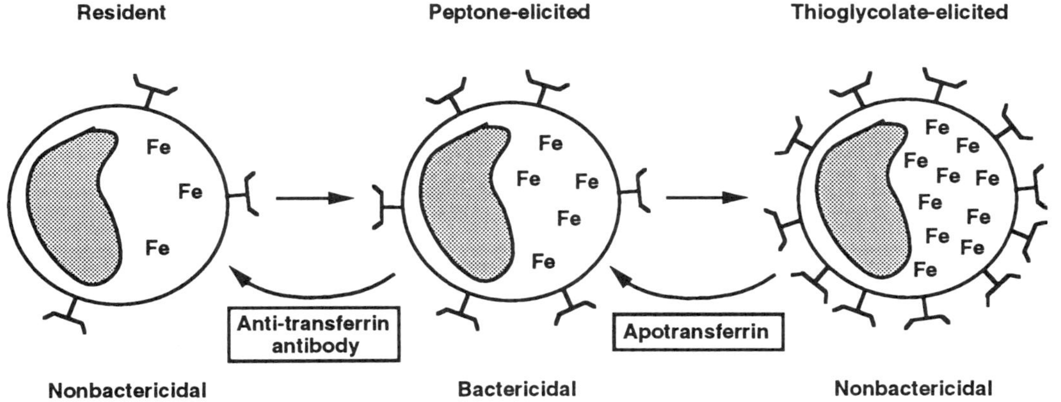

FIGURE 2 Transferrin receptor expression and intracellular iron levels predict listericidal activity of macrophages. Resident peritoneal macrophages, with low transferrin receptor expression and low intracellular iron, are not listericidal; neither are thioglycolate-elicited peritoneal exudate macrophages, which have high transferrin receptor expression and high levels of intracellular iron. Only cells with intermediate levels of transferrin receptor expression and intracellular iron, peptone-elicited cells, are bactericidal. Importantly, cells can be converted from one phenotype to another by altering iron content: that is, thioglycolate-elicited cells acquire bactericidal activity if their iron levels are lowered by overnight incubation in iron-free apotransferrin, while peptone-elicited cells loose bactericidal activity if they are incubated for 1 hour with anti-transferrin antibody. These experiments implicate intracellular iron levels in macrophage listericidal activity.

that require iron-dependent enzymes, and this might contribute to successful growth of bacteria. Whether macrophage listericidal mechanisms require iron-dependent enzymes, for instance, is not yet known.

Experiments discussed in this section indicate that the biochemical basis of macrophage listericidal activity has not yet been characterized completely. Products of nitric oxide metabolism might be important mediators of macrophage listericidal activity, as they seem to be for certain other microbes. Moreover, intracellular iron levels are apparently important regulatory elements of macrophage listericidal mechanisms, although the biochemical events regulated by iron levels are not yet identified.

IV. CYTOKINES INVOLVED IN *LISTERIA*–MACROPHAGE INTERACTIONS

Not all macrophages can kill *L. monocytogenes*, but it is not yet entirely clear what distinguishes those that can from those that cannot; nor are the mediators and signals that drive a nonlistericidal cell to express listericidal activity yet completely characterized. A number of studies have identified certain T cell– and monocyte–derived cytokines as being important for resistance in the whole animal, while other studies have identified cytokines that can influence the listericidal activity of macrophages in vitro.

Some of the most convincing data come from in vivo studies in which antibodies to certain mediators decreased the ability of an animal to resist *Listeria* infection. In this manner, an importrant role for two mediators in particular, IFN-γ and tumor necrosis factor-α (TNF-α), has been identified. Interferon-γ is a T cell and natural killer (NK) cell product that acts primarily on macrophages, while TNF-α is an important secretory product of macrophages that could have autocrine function. The effects of these cytokines, alone and in combination, on resistance to listeriosis will be discussed below.

A. Interferon-γ

Buchmeier and Schreiber, in 1985, presented compelling data demonstrating that injection of an antibody against IFN-γ decreased in vivo resistance to *Listeria* in mice [65]. In these experiments, intraperitoneal injection of a hamster monoclonal antibody to mouse IFN-γ 1 day after infection with *Listeria* increased the number of bacteria in the spleens and peritoneal cavities 6 days later approximately 1000-fold. This large increase in bacterial number induced by anti–IFN-γ antibody implicated IFN-γ as an important mediator of resistance to *Listeria*. Nakane and colleagues [66] confirmed these data, but more important, their studies emphasized that IFN-γ acts early in infection, since the antibody had to be given on days 0 or 1 of infection to decrease resistance. This notion was supported by the studies of Dunn and North [67], who confirmed that early neutralization of IFN-γ with antibodies exacerbated in vivo infection. Using a spot-ELISA technique to measure IFN-γ production, peak levels were found to occur at 24 to 48 hours after infection. Rather intriguingly, these investigators also reported that NK cells, rather than CD4+ or CD8+ T cells, were the primary producers of this early IFN-γ. Another study confirmed that IFN-γ is produced early but not late in infection, with peak levels reached at day 2 [68].

The experiments described above demonstrate that IFN-γ is required for mice to mount successful resistance to *Listeria* in vivo, but they provide little evidence regarding its mechanism of action. Because IFN-γ has well-described effects on macrophages, such as its ability to induce expression of class II major histocompatability complex (MHC) antigens [69] and its ability to induce these cells to kill tumor targets and certain intracellular protozoan parasites [70–72], several investigators have studied the ability of this cytokine to stimulate macrophages to express listericidal activity in vitro.

In a few studies, investigators have reported that nonlistericidal populations of macrophages can be stimulated to kill *Listeria*. Using a colorimetric assay to quantify bacteria, Peck showed that human peripheral blood monocyte–derived macrophages cultured for 3 days with recombinant IFN-γ (rIFN-γ) were able to kill *Listeria*, while those cultured without rIFN-γ were not [73]. Experiments using resident mouse macrophages incubated with rIFN-γ suggested that IFN-γ enhanced macrophage killing of *Listeria* by limiting the organism's access to the cytoplasm [74]. In contrast, we as well as others found that stimulation of mouse peritoneal exudate macrophages with rIFN-γ induced them to express tumoricidal but not listericidal activity [75,76]. Furthermore, we reported that peritoneal exudate macrophages that expressed listericidal activity lost the ability to kill *Listeria* but gained the ability to kill tumor cells following overnight incubation in rIFN-γ [75]. The reason for this may be that IFN-γ decreases transferrin receptor expression [77] and thus intracellular iron to levels too low to allow macrophages to be listericidal [46]. Another possibility is that IFN-γ could decrease CR3-mediated phagocytosis of *Listeria*, since IFN-γ has been shown to deactivate CR3, thus decreasing its ability to mediate phagocytosis [78]. Taken together, the experiments described above suggest that stimulation of mouse macrophages by rIFN-γ in ways sufficient to induce them to express tumoricidal activity does not induce them to express listericidal activity, and furthermore, that existing listericidal activity may be inhibited by rIFN-γ.

If IFN-γ increases resistance to *Listeria* in the whole animal by stimulating macrophages to high levels of listericidal activity, then it is reasonable to expect this to be demonstrable in vitro. The paragraph above demonstrates that very few studies have shown enhancement of macrophage listericidal activity in vitro by rIFN-γ, and other studies indicate that this lymphokine may either fail to induce, or inhibit, existing listericidal activity. Clearly the mechanism or mechanisms of action of IFN-γ in resistance to listeriosis are not yet well understood.

B. Tumor Necrosis Factor-α

Tumor necrosis factor-α, sometimes called cachectin, is a macrophage product that can kill some tumor target cells and is thought to cause wasting diseases in animals and humans with tumors or infections [79]. Recent evidence obtained with monoclonal antibodies to this monokine suggests that it also plays an important role in resistance to infection by *L. monocytogenes*. When the immunoglobulin G (IgG) portion of a rabbit anti–TNF-α antiserum was injected into mice that were then challenged with 0.1 LD_{50} *Listeria*, all the mice died within 7 days [66,80–82]. Antibody had to be given early in infection, suggesting that TNF-α exerted its effects early [82]. Measurement of TNF-α levels in mice infected with *L. monocytogenes* supported this concept, since TNF-α peaked early in infection [81,82].

While it is not known how TNF-α increases resistance to *Listeria*, one group suggested that it might aid in the recruitment of listericidal monocytes into infectious foci [82]. This could be a direct effect, since TNF-α has been reported to be chemotactic for monocytes [83]; however, while TNF-α is chemotactic for human monocytes, there are no reports that it is chemotactic for mouse monocytes, and we find that it is not (Hiester and Campbell, unpublished data). Therefore, in the mouse system, TNF-α might function indirectly by inducing secretion of other factors chemotactic for monocytes, for instance IL-1 [84], or have other as yet undefined effects. A report that mice treated with anti–TNF-α have normal levels of IFN-α and IFN-γ suggests that TNF-α does not enhance resistance by increasing levels of these cytokines; by implication this suggests it may influence macrophage listericidal activity directly [81]. In contrast, in vivo experiments indicate that the hepatocyte, not the monocyte, allows uncontrolled growth of *Listeria* in mice treated with anti–TNF-α [82], suggesting TNF-α's effects may not be directly on the macrophage. Thus the mechanisms by which TNF-α contributes to resistance against *Listeria* also remain to be determined.

C. Tumor Necrosis Factor-α and Interferon-γ

A number of reports suggest that TNF-α and IFN-γ, which independently have been shown to be important in resistance to *Listeria* in vivo, as described above, may have synergistic effects as well. Injection of anti–IFN-γ antibody into mice injected 1 day earlier with *Listeria* decreased production of TNF-α, but interestingly, not of IFN-γ itself [66]. These antibodies did not cause a synergistic decrease in resistance during a primary response to *L. monocytogenes*, but did during a secondary response [66], suggesting that TNF-α and IFN-γ might act differently during primary and secondary infection.

The in vivo experiments described above using specific antibodies have demonstrated important roles for IFN-γ and TNF-α in resistance to infection by *L. monocytogenes*, but have shed little light on the mechanisms by which these cytokines, one T cell– or NK cell–derived and the other macrophage-derived, exert their effects. Clarification of mechanisms of action may require in vitro experiments using carefully defined effector cell populations. We are conducting experiments suggesting that TNF-α and IFN-γ act directly on the macrophage and in concert can induce these cells to kill *Listeria* (Campbell, Canono, and Leenen, unpublished data). These experiments use an immature, nonlistericidal macrophage precursor hybrid generated by Pieter Leenen and collaborators [85]. Following incubation first with TNF-α, then with IFN-γ, these cells acquire high levels of listericidal activity. Experiments are in progress to define the changes these macrophages undergo as they become able to kill the bacterium, and which of these changes are induced by TNF-α and which by IFN-γ. These experiments should help define the mechanisms by which TNF-α and IFN-γ induce nonbactericidal cells to express bactericidal activity.

D. Other Cytokines

Compelling data also implicate other cytokines known to be secreted by macrophages and/or to affect macrophage function in resistance against infection by *L. monocytogenes*. For instance, mice injected for 5 consecutive days with recombinant human macrophage colony-stimulating factor (M-CSF) were not killed by injection of an otherwise lethal dose of *Listeria* and showed increased numbers of peripheral blood mononuclear cells [86]. Furthermore, interleukin-6 (IL-6), made by T cells, macrophages, and other cells, is also secreted during *Listeria* infection but is not regulated by TNF-α [87]. Interestingly, in this study, the more severe the infection was, the higher were the IL-6 levels in the spleen and blood. A potential role for the cytokine IFN-β is unclear, since it enhanced resistance of mice to infection with *L. monocytogenes* in one study [88] but not in another [81].

Another macrophage-derived cytokine often implicated as a component of resistance to infection by *Listeria* is interleukin-1 (IL-1). This important mediator of inflammation could act at many different levels during the host response against *Listeria*. An important role for IL-1 was suggested by experiments showing that mice injected simultaneously with rIL-1α and *Listeria* were more resistant to *Listeria* than mice receiving bacteria alone [89]. The effect of IL-1 on specific cell types has also been studied. One report suggested that an important effect of IL-1α, which enhanced resistance, was directly on T cells, promoting their differentiation and their ability to mediate delayed-type hypersensitivity, but not their anti-*Listeria* effector functions [90]. In another study, the ability of an intraperitoneal injection of *Listeria* to induce peritoneal macrophages to express class II MHC antigens was inhibited by a cocktail of antibodies against IL-1α, IL-1β, and the type I IL-1 receptor [91]. This suggests that IL-1 might act indirectly on macrophages, perhaps by inhibiting their ability to present antigens and initiate T cell–dependent resistance to *Listeria*.

As the review of the literature above indicates, there is compelling evidence that a number of cytokines contribute to host resistance to listeriosis. Harder to clarify is which ones directly affect macrophage function, particularly the ability of these cells to kill *Listeria*, and which ones indirectly affect the host immune and/or inflammatory responses. In vitro studies, in which direct effects on macrophages can be separated from indirect effects on other cells, clearly show important roles for both IFN-γ and TNF-α. What remains to be clarified is what each of these mediators does and how they cause a single macrophage to convert from a nonlistericidal to a listericidal cell. At the moment, many possibilities exist, ranging from altering function and/or expression of relevant surface receptors, such as the transferrin receptor and CR3, to altering intracellular metabolism, for instance by preventing listeriolysin O from mediating the bacterium's escape from the phagosome into the cytoplasm. Clarification of which mechanisms occur awaits further experimentation.

V. NOT ALL MACROPHAGES ARE EQUAL . . .

It is becoming increasingly clear that macrophages are an extremely heterogeneous cell population at both the functional and the phenotypic level [reviewed in 92]. In this regard, some macrophages are able to kill bacteria, while others are not. Perhaps surprisingly, listericidal macrophages often seem to be those recruited most recently to a site of infection, that is inflammatory cells, rather than highly differentiated and stimulated macrophages [9,93,94]. Whether cells express particular functional properties may reflect their elicitation with different inflammatory agents [46] or stimulation with cytokines that alter their function [75]. Similarly, macrophages induced to express tumoricidal activity by defined ligands do not necessarily express microbicidal activity, including listericidal activity [75,76,95,96]. Whether heterogeneity reflects different macrophage maturational stages, stimulation of different macrophage precursor subsets to express different effector functions, or another mechanism is not yet known. What seems clear is that effector properties of macrophages are highly regulated and are probably dictated by local cytokines and mediators, and even by phagocytosed products, such as bacteria.

Certain elements of macrophage functional heterogeneity are becoming clarified. As described above, thioglycolate-elicited macrophages are not able to kill *Listeria*, while those elicited with a similar inflammatory agent, proteose peptone, are [45,46]. Recent experiments from our laboratory show that there are differences in the way in which these cells deal with intracellular *Listeria*. Peptone-elicited cells, which internalize *Listeria* primarily through CR3, allow few organisms to escape from the phagosome into the cytoplasm, while thioglycolate-elicited cells, which phagocytose Listeria primarily through receptors other than CR3, permit the bacterium to become cytoplasmic and to become coated with F-actin [46]. This indicates that escape of the organism into the cytoplasm is *de facto* evidence that it has successfully parasitized a macrophage which cannot contain and kill it.

It will also be interesting to determine whether there are differences either in innate bactericidal activity of macrophages from genetically susceptible and genetically resistant mice or in the ability to induce these cells to express bactericidal activity. It has been known for some time that mice, differing at a single locus, can be identified as either genetically susceptible or genetically resistant to *Listeria* [97,98]. Although mice can be identified as genetically susceptible or resistant, there is not yet evidence that this is reflected in the ability of the individual macrophage to kill *Listeria*. Studies so far indicate that macrophages from genetically susceptible and resistant mice kill *Listeria* equally well [99,100]. What they differ in is their ability to mobilize macrophages to a site of infection [97,99,100]. Whether there are other important differences in function of these cells remains to be determined.

VI. SUMMARY AND CONCLUSIONS

It is clear that the interaction between macrophages and the facultative intracellular bacterium *Listeria monocytogenes* is complex and dictated by the needs of both organisms. On the one hand, the invading pathogen has devised strategies to locate itself intracellularly in a site where host defense mechanisms are minimal; thus it escapes the phagosome and enters the cytoplasm, from which it travels to the next cell without exposing itself to the extracellular environment. On the other hand, the infected cell, potentially a highly efficient killer of microbes, is influenced by cytokines and other mediators, autocrine and exocrine, that enhance its ability to thwart the invader. Thus a macrophage, under the influence of at least IFN-γ and TNF-α, can progress from being a nonlistericidal cell to one that can kill the intracellular organism. It probably does this by preventing escape of *Listeria* from the phagosome into the cytoplasm. The bacterium, retained in the phagosome, is killed by unknown mechanisms, which appear to be regulated by iron and may involve reactive nitrogen intermediates.

ACKNOWLEDGMENTS

The author thanks Drs. Doug Drevets and Pieter Leenen for their thoughtful and critical review of the manuscript.

REFERENCES

1. Gray ML, Killinger AH. *Listeria monocytogenes* and listeric infections. Bacteriol Rev 1966; 30:309–82.
2. Farber JM, Peterkin PI. *Listeria monocytogenes*, a food-borne pathogen. Microbiol Rev 1991; 55:476–511.
3. Schuchal A, Deaver LA, Wenger JD, Plikaytis BD, Mascola L, Pinner RW, Reingold AL, Broome CV, and the Listeria Study Group. Role of foods in sporadic listeriosis. I. Case-control study of dietary risk factors. JAMA 1992; 267:2041–5.
4. Gaillard J-L, Berche P, Frehel C, Gouin E, Cossart P. Entry of L. monocytogenes into cells is mediated by internalin, a repeat protein reminiscent of surface antigens from gram-positive cocci. Cell 1991; 65:1127–41.
5. Kuhn M, Goebel W. Identification of an extracellular protein of *Listeria monocytogenes* possibly involved in intracellular uptake by mammalian cells. Infect Immun 1989; 57:55–61.
6. Kohler S, Leimeister-Wachter M, Chakraborty T, Lottspeich F, Goebel W. The gene coding for protein p60 of *Listeria monocytogenes* and its use as a specific probe for *Listeria monocytogenes*. Infect Immun 1990; 58:1943–50.
7. Goebel W, Kathariou S, Kuhn M, Sokolovic Z, Kreft J, Köhler S, Funke D, Chakraborty T, Leimeister-Wächter M. Hemolysin from Listeria—biochemistry, genetics and function in pathogenesis. Infection 1988; 16(suppl 2):S149–56.
8. Czuprynski CJ, Campbell PA, Henson PM. Killing of Listeria monocytogenes by human neutrophils and monocytes, but not by monocyte-derived macrophages. J Reticuloendothel Soc 1983; 34:29–43.
9. Czuprynski CJ, Henson PM, Campbell PA. Killing of Listeria monocytogenes by inflammatory neutrophils and mononuclear phagocytes from immune and nonimmune mice. J Leukocyte Biol 1984; 35:193–208.
10. Tilney LG, Portnoy DA. Actin filaments and the growth, movement, and spread of the intracellular bacterial parasite, *Listeria monocytogenes*. J Cell Biol 1989; 109:1597–1608.
11. Gaillard J-L, Berche P, Mounier J, Richard S, Sansonetti P. In vitro model of penetration and intracellular growth of *Listeria monocytogenes* in the human enterocyte-like cell line Caco-2. Infect Immun 1987; 55:2822–9.
12. Cossart P, Vicente MF, Mengaud J, Baquero F, Perez-Diaz JC, Berche P. Listeriolysin O is essential for virulence of *Listeria monocytogenes*: direct evidence obtained by gene complementation. Infect Immun 1989; 57:3629–36.
13. Mounier J, Ryter A, Coquis-Rondon M, Sansonetti PJ. Intracellular and cell-to-cell spread of *Listeria monocytogenes* involves interaction with F-actin in the enterocytelike cell line Caco-2. Infect Immun 1990; 58:1048–58.
14. Tilney LG, Connelly PS, Portnoy DA. Actin filament nucleation by the bacterial pathogen, *Listeria monocytogenes*. J Cell Biol 1990; 111:2979–88.
15. Kocks C, Gouin E, Tabouret M, Berche P, Ohayon H, Cossart P. L. monocytogenes–induced actin assembly requires the *actA* gene product, a surface protein. Cell 1992; 68:521–31.
16. Theriot JA, Mitchison TJ, Tilney LG, Portnoy DA. The rate of actin-based motility of intracellular *Listeria monocytogenes* equals the rate of actin polymerization. Nature 1992; 357:257–60.

17. Dabiri GA, Sanger JM, Portnoy DA, Southwick FS. *Listeria monocytogenes* moves rapidly through the host-cell cytoplasm by inducing directional actin assembly. Proc Natl Acad Sci USA 1990; 87:6068–72.
18. Gaillard JL, Berche P, Sansonetti P. Transposon mutagenesis as a tool to study the role of hemolysin in the virulence of *Listeria monocytogenes*. Infect Immun 1986; 52:50–5.
19. Kathariou S, Metz P, Hof H, Goebel W. Tn916-induced mutations in the hemolysin determinant affecting virulence of *Listeria monocytogenes*. J Bacteriol 1987; 169:1291–7.
20. Geoffroy C, Gaillard J-L, Alouf JE, Berche P. Purification, characterization, and toxicity of the sulfhydryl-activated hemolysin listeriolysin O from *Listeria monocytogenes*. Infect Immun 1987; 55:1641–6.
21. Vicente MF, Baquero F, Perez-Diaz J-C. Cloning and expression of the *Listeria monocytogenes* haemolysin in *E. coli*. FEMS Microbiol Lett 1985; 30:77–9.
22. Mengaud J, Vicente M-F, Chenevert J, Pereira JM, Geoffroy C, Gicquel-Sanzey B, Baquero F, Perez-Diaz J-C, Cossart P. Expression in *Escherichia coli* and sequence analysis of the listeriolysin O determinant of *Listeria monocytogenes*. Infect Immun 1988; 56:766–72.
23. Chakraborty T, Kathariou S, Hacker J, Hof H, Huhle B, Wagner W, Kuhn M, Goebel W. Molecular analysis of bacterial cytolysins. Rev Infect Dis 1987; 9(suppl 5):S456–66.
24. Sokolovic Z, Fuchs A, Goebel W. Synthesis of species-specific stress proteins by virulent strains of *Listeria monocytogenes*. Infect Immun 1990; 58:3582–7.
25. Kathariou S, Rocourt J, Hof H, Goebel W. Levels of *Listeria monocytogenes* hemolysin are not directly proportional to virulence in experimental infections of mice. Infect Immun 1988; 56:534–6.
26. Bouwer HGA, Nelson CS, Gibbins BL, Portnoy DA, Hinrichs DJ. Listeriolysin O is a target of the immune response to *Listeria monocytogenes*. J Exp Med 1992; 175:1467–71.
27. Henry JT, Bevan MJ. CD8+ T cells specific for a single nonamer epitope of *Listeria monocytogenes* are protective in vivo. J Exp Med 1992; 175:1531–8.
28. Kuhn M, Kathariou S, Goebel W. Hemolysin supports survival but not entry of the intracellular bacterium *Listeria monocytogenes*. Infect Immun 1988; 56:79–82.
29. Portnoy DA, Jacks PS, Hinrichs DJ. Role of hemolysin for the intracellular growth of *Listeria monocytogenes*. J Exp Med 1988; 167:1459–71.
30. Bielecki J, Youngman P, Connelly P, Portnoy DA. *Bacillus subtilis* expressing a haemolysin gene from *Listeria monocytogenes* can grow in mammalian cells. Nature 1990; 345:175–6.
31. Camilli A, Goldfine H, Portnoy DA. *Listeria monocytogenes* mutants lacking phosphatidylinositol-specific phospholipase C are avirulent. J Exp Med 1991; 173:751–4.
32. Sun AN, Camilli A, Portnoy DA. Isolation of *Listeria monocytogenes* small-plaque mutants defective for intracellular growth and cell-to-cell spread. Infect Immun 1990; 58:3770–8.
33. Kuhn M, Prévost M-C, Mounier J, Sansonetti PJ. A nonvirulent mutant of *Listeria monocytogenes* does not move intracellularly but still induces polymerization of actin. Infect Immun 1990; 58:3477–86.
34. Berche P, Gaillard J-L, Sansonetti PJ. Intracellular growth of *Listeria monocytogenes* as a prerequisite for in vivo induction of T cell–mediated immunity. J Immunol 1987; 138:2266–71.
35. Brunt LM, Portnoy DA, Unanue ER. Presentation of Listeria monocytogenes to CD8+ T cells requires secretion of hemolysin and intracellular bacterial growth. J Immunol 1990; 145:3540–6.
36. Bishop DK, Hinrichs DJ. Adoptive transfer of immunity to *Listeria monocytogenes*. The influence of in vitro stimulation on lymphocyte subset requirements. J Immunol 1987; 139:2005–9.
37. Mielke MEA, Ehlers S, Hahn H. T-cell subsets in delayed-type hypersensitivity, protection, and granuloma formation in primary and secondary *Listeria* infection in mice: superior role of Lyt-2+ cells in acquired immunity. Infect Immun 1988; 56:1920–5.
38. Cluff CW, Garcia M, Ziegler HK. Intracellular hemolysin-producing *Listeria monocytogenes* strains inhibit macrophage-mediated antigen processing. Infect Immun 1990; 58:3601–12.
39. Baker LA, Campbell PA, Hollister JR. Chemotaxigenesis and complement fixation by *Listeria monocytogenes* cell wall fractions. J Immunol 1977; 119:1723–6.
40. Harrington-Fowler L, Henson PM, Wilder MA. Fate of *Listeria monocytogenes* in resident and activated macrophages. Infect Immun 1981; 33:11–6.
41. Van Kessel KPM, Antonissen, ACJM, van Dijk H, Rademaker PM, Willers JMN. Interactions of killed *Listeria monocytogenes* with the mouse complement system. Infect Immun 1981; 34:16–9.

42. Weinberg DS, Unanue ER. Antigen-presenting function of alveolar macrophages: uptake and presentation of *Listeria monocytogenes*. J Immunol 1981; 126:794–9.
43. Drevets DA, Campbell PA. Roles of complement and complement receptor type 3 in phagocytosis of *Listeria monocytogenes* by inflammatory mouse peritoneal macrophages. Infect Immun 1991; 59:2645–52.
44. Drevets DA, Campbell PA. Macrophage phagocytosis: use of fluorescence microscopy to distinguish between extracellular and intracellular bacteria. J Immunol Met 1991; 142:31–8.
45. Baker LA, Campbell PA. Thioglycollate medium decreases resistance to bacterial infection in mice. Infect Immun 1980; 27:455–60.
46. Alford CE, King TE Jr, Campbell PA. Role of transferrin, transferrin receptors, and iron in macrophage listericidal activity. J Exp Med 1991; 174:459–66.
47. Drevets DA, Canono BP, Campbell PA. Listericidal and nonlistericidal mouse macrophages differ in complement receptor type 3-mediated phagocytosis of *L. monocytogenes* and in preventing escape of the bacteria into the cytoplasm. J Leukocyte Biol 1992; 52:70–9.
48. Godfrey RW, Wilder MS. Relationships between oxidative metabolism, macrophage activation, and antilisterial activity. J Leukocyte Biol 1984; 36:533–43.
49. Welch DF, Sword CP, Brehm S, Dusanic D. Relationship between superoxide dismutase and pathogenic mechanisms of *Listeria monocytogenes*. Infect Immun 1979; 23:863–72.
50. Lehrer RI, Selsted ME, Szklarek D, Feischman J. Antibacterial activity of microbicidal cationic proteins 1 and 2, natural peptide antibiotics of rabbit lung macrophages. Infect Immun 1983; 42:10–4.
51. Green SJ, Nacy CA, Meltzer MS. Cytokine-induced synthesis of nitrogen oxides in macrophages: a protective host response to *Leishmania* and other intracellular pathogens. J Leukocyte Biol 1991; 50:93–103.
52. Hibbs JB Jr, Taintor RR, Vavrin Z. Macrophage cytotoxicity: role for L-arginine deiminase and imino nitrogen oxidation to nitrite. Science 1987; 235:473–6.
53. Steuhr DJ, Nathan CF. Nitric oxide. A macrophage product responsible for cytostasis and respiratory inhibition in tumor targets. J Exp Med 1989; 169:1543–55.
54. Chan J, Zing Y, Magliozzo RS, Bloom BR. Killing virulent *Mycobacterium tuberculosis* by reactive nitrogen intermediates produced by activated murine macrophages. J Exp Med 1992; 175:1111–22.
55. Adams LB, Franzblau SG, Vavrin Z, Hibbs JB Jr, Krahenbuhl JL. L-arginine–dependent macrophage effector functions inhibit metabolic activity of *Mycobacterium leprae*. J Immunol 1991; 147:1642–6.
56. Flesch IEA, Kaufmann SHE. Mechanisms involved in mycobacterial growth inhibition by gamma interferon–activated bone marrow macrophages: role of reactive nitrogen intermediates. Infect Immun 1991; 59:3213–8.
57. Fortier AH, Polsinelli T, Green SJ, Nacy CA. Activation of macrophages for destruction of *Francisella tularensis*: identification of cytokines, effector cells, and effector molecules. Infect Immun 1992; 60:817–25.
58. Rutherford MS, Schook LB. Differential immunocompetence of macrophages derived using macrophage or granulocyte-macrophage colony-stimulating factor. J Leukocyte Biol 1992; 51:69–76.
59. Sword CP. Mechanisms of pathogenesis in *Listeria monocytogenes* infection. I. Influence of iron. J Bacteriol 1966; 92:536–42.
60. Mossey RT, Sondheimer J. Listeriosis in patients with long-term hemodialysis and transfusional iron overload. Am J Med 1985; 79:397–400.
61. Ampel NM, Van Wyck DB, Aguirre ML, Willis DG, Popp RA. Resistance to infection in murine β-thalassemia. Infect Immun 1989; 57:1011–7.
62. Van Renswoude J, Bridges KR, Harford JB, Klausner RD. Receptor-mediated endocytosis of transferrin and the uptake of Fe in K562 cells: identification of a nonlysosomal acidic compartment. Proc Natl Acad Sci USA. 1982; 79:6186–90.
63. Jacobs A. Low molecular weight intracellular iron transport compounds. Blood 1977; 50:433–9.
64. Alvarez-Hernandez X, Felstein MV, Brock JH. The relationship between iron release, ferritin synthesis and intracellular iron distribution in mouse peritoneal macrophages. Evidence for a reduced level of metabolically available iron in elicited macrophages. Biochim Biophys Acta 1986; 886:214–22.
65. Buchmeier NA, Schreiber RD. Requirement of endogenous interferon-γ production for resolution of *Listeria monocytogenes* infection. Proc Natl Acad Sci USA 1985; 82:7404–8.

66. Nakane A, Minagawa T, Kohanawa M, Chen Y, Sato H, Moriyama M, Tsuruoka N. Interactions between endogenous gamma interferon and tumor necrosis factor in host resistance against primary and secondary *Listeria monocytogenes* infections. Infect Immun 1989; 57:3331–7.
67. Dunn PL, North RJ. Early gamma interferon production by natural killer cells is important in defense against murine listeriosis. Infect Immun 1991; 59:2892–900.
68. Nakane A, Numata A, Asano M, Kohanawa M, Chen Y, Minagawa T. Evidence that endogenous gamma interferon is produced early in *Listeria monocytogenes* infection. Infect Immun 1990; 58:2386–8.
69. Steeg PS, Moore RN, Johnson HM, Oppenheim JJ. Regulation of murine macrophage Ia antigen expression by a lymphokine with immune interferon activity. J Exp Med 1982; 156:1780–93.
70. Adams DO, Hamilton TA. The cell biology of macrophage activation. Ann Rev Immunol 1984; 2:283–318.
71. Murray HW, Spitalny GL, Nathan CF. Activation of mouse peritoneal macrophages in vitro and in vivo by interferon-γ. J Immunol 1985; 134:1619–22.
72. Turco J, Winkler HH. Effect of mouse lymphokines and cloned mouse interferon-γ on the interaction of *Rickettsia prowazekii* with mouse macrophage-like RAW264.7 cells. Infect Immun 1984; 45:303–8.
73. Peck R. A one-plate assay for macrophage bactericidal activity. J Immunol Met 1985; 82:131–40.
74. Portnoy DA, Schreiber RD, Connelly P, Tilney LG. γ interferon limits access of *Listeria monocytogenes* to the macrophage cytoplasm. J Exp Med 1989; 170:2141–6.
75. Campbell PA, Canono BP, Cook JL. Mouse macrophages stimulated by recombinant gamma interferon to kill tumor cells are not bactericidal for the facultative intracellular bacterium *Listeria monocytogenes*. Infect Immun 1988; 56:1371–5.
76. Van Dissel JT, Stikkelbroeck JJM, Michel BC, van den Barselaar MTh, Leijh PCJ, van Furth R. Inability of recombinant interferon-γ to activate the antibacterial activity of mouse peritoneal macrophages against *Listeria monocytogenes* and *Salmonella typhimurium*. J Immunol 1987; 139:1673–8.
77. Byrd TF, Horwitz MA. Interferon gamma–activated human monocytes downregulate transferrin receptors and inhibit the intracellular multiplication of *Legionella pneumophila* by limiting the availability of iron. J Clin Invest 1989; 83:1457–65.
78. Wright SD, Detmers PA, Jong MTC, Meyer BC. Interferon-γ depresses binding of ligand by C3b and C3bi receptors on cultured human monocytes, an effect reversed by fibronectin. J Exp Med 1986; 163:1245–59.
79. Beutler B, Cerami A. The biology of cachectin/TNF—a primary mediator of the host response. Ann Rev Immunol 1989; 7:625–55.
80. Havell EA. Production of tumor necrosis factor during murine listeriosis. J Immunol 1987; 139:4225–31.
81. Nakane A, Minagawa T, Kato K. Endogenous tumor necrosis factor (cachectin) is essential to host resistance against *Listeria monocytogenes* infection. Infect Immun 1988; 56:2563–9.
82. Havell EA. Evidence that tumor necrosis factor has an important role in antibacterial resistance. J Immunol 1989; 143:2894–9.
83. Ming WJ, Bersani L, Mantovani A. Tumor necrosis factor is chemotactic for monocytes and polymorphonuclear leukocytes. J Immunol 1987; 138:1469–74.
84. Dinarello CA, Cannon JG, Wolff SM, Bernheim HA, Beutler B, Cerami A, Figari IS, Palladino Jr MA, O'Connor JV. Tumor necrosis factor (cachectin) is an endogenous pyrogen and induces production of interleukin 1. J Exp Med 1986; 163:1433–50.
85. Leenen PJM, Slieker WAT, Melis M, van Ewijk W. Murine macrophage precursor characterization. I. Production, phenotype and differentiation of macrophage precursor hybrids. Eur J Immunol 1990; 20:15–25.
86. Kayashima S, Tsuru S, Shinomiya N, Katsura Y, Motoyoshi K, Rokutanda M, Nagata N. Effects of macrophage colony-stimulating factor on reduction of viable bacteria and survival of mice during *Listeria monocytogenes* infection. Characteristics of monocyte subpopulations. Infect Immun 1991; 59:4677–80.
87. Havell EA, Sehgal PB. Tumor necrosis factor–independent IL-6 production during murine listeriosis. J Immunol 1991; 146:756–61.
88. Fujiki T, Tanaka A. Antibacterial activity of recombinant murine beta interferon. Infect Immun 1988; 56:548–51.

89. Czuprynski CJ, Brown JF. Recombinant murine interleukin-1α enhancement of nonspecific antibacterial resistance. Infect Immun 1987; 55:2061–5.
90. Igarashi K-I, Mitsuyama M, Muramori K, Tsukada H, Nomoto K. Interleukin-1–induced promotion of T-cell differentiation in mice immunized with killed *Listeria monocytogenes*. Infect Immun 1990; 58:3973–9.
91. Rogers HW, Sheehan KCF, Brunt LM, Dower SK, Unanue ER, Schreiber RD. Interleukin 1 participates in the development of anti-*Listeria* responses in normal and SCID mice. Proc Natl Acad Sci USA 1992; 89:1011–5.
92. Leenen PJM, Campbell PA. Heterogeneity of mononuclear phagocytes—an interpretive review. In: Horton MA, ed. Blood cell biochemistry, vol. 5. Macrophages and related cells. New York: Plenum, 1993; in press.
93. Gervais F, Stevenson M, Skamene E. Genetic control of resistance to *Listeria monocytogenes*: regulation of leukocyte inflammatory responses by the *Hc* locus. J Immunol 1984; 132:2078–83.
94. Stevenson MM, Kongshavn PAL, Skamene E. Genetic linkage of resistance to *Listeria monocytogenes* with macrophage inflammatory responses. J Immunol 1981; 127:402–7.
95. Wing EJ, Gardner ID, Ryning FW, Remington JS. Dissociation of effector functions in populations of activated macrophages. Nature 1977; 268:642–4.
96. Haidaris CG, Bonventre PF. Elimination of *Leishmania donovani* amastigotes by activated macrophages. Infect Immun 1981; 33:918–26.
97. Cheers C, McKenzie IFC. Resistance and susceptibility of mice to bacterial infection: genetics of listeriosis. Infect Immun 1978; 19:755–62.
98. Skamene EP, Kongshavn PAL, Sachs DH. Resistance to *Listeria monocytogenes* in mice is genetically controlled by genes which are not linked to the H-2 complex. J Infect Dis 1979; 139:228–31.
99. Czuprynski CJ, Canono BP, Henson PM, Campbell PA. Genetically determined resistance to listeriosis is associated with increased accumulation of inflammatory neutrophils and macrophages which have enhanced listericidal activity. Immunol 1985; 55:511–8.
100. Gervais F, Morris-Hooke A, Tran TA, Skamene E. Analysis of macrophage bactericidal function in genetically resistant and susceptible mice by using the temperature-sensitive mutant of *Listeria monocytogenes*. Infect Immun 1986; 54:315–21.

20

Legionella and Macrophages

Yoshimasa Yamamoto, Thomas W. Klein, and Herman Friedman

University of South Florida College of Medicine, Tampa, Florida

I. INTRODUCTION

Legionella pneumophila was isolated as a causative agent of legionellosis, first recognized in 1976 at the American Legion convention in Philadelphia when 221 participants contracted pneumonia and 34 persons died [1,2]. Subsequently, many *Legionella* species were isolated not only from humans but also from other sources. There are now 32 known species with 51 serogroups in the Legionellaceae family [3]. However, most species have been isolated from environmental sources, and their pathogenicity for humans is unknown [4]. Most cases of legionellosis in man are caused by *L. pneumophila*, especially serogroups 1, 4, and 6 [5]. The usual clinical presentation of legionnaires' disease is pneumonia, occurring in more than 95% of cases [6]. Once the bacteria enter the respiratory tract and cause pneumonia, most are located within inflammatory phagocytic cells such as polymorphonuclear cells and macrophages [7,8]. Electron-microscopic studies have shown that these bacteria in phagocytic cells are intact and appear viable [9,10]. Thus, phagocytic cells, especially macrophages, are thought by most investigators to be the preferential site for growth of these bacteria in humans. That cellular immunity is important in resistance to *Legionella* infection seems likely, since it has been pointed out by many investigators that impairment of cellular immunity, especially that involving phagocytic cell functions, results in high risk for infection with these bacteria. For example, legionellosis often occurs in immunocompromised hosts such as patients with hairy-cell leukemia, a malignancy associated with monocyte deficiency and dysfunction [11–14]. Such clinical observations suggest the importance of phagocytic cells in resistance to *Legionella* infection, as is the case with many other intracellular pathogens.

L. pneumophila is a gram-negative pleiomorphic bacillus, and for growth on artificial medium it requires several ingredients, such as L-cysteine, keto acids, and ferric iron, and a narrow pH range (pH 6.9–7.0)[15,16]. Therefore, *Legionella* cannot grow on regular bacteriologic medium such as nutrient agar. In general, these bacteria are cultured on buffered charcoal yeast extract (BCYE) agar medium. A remarkable feature of this organism's growth in vitro is that the bacteria cannot multiply in either regular tissue culture medium or conditioned medium without cells present [17–19]. However, when *Legionella* are cultured in macrophages or various cell lines such as U937, large numbers of the bacteria are recovered in lysates of the cells 24 or 48 hours after culture initiation. Other experimental evidence also indicates that intracellular localization is essential for growth of these bacteria in culture [20]. Therefore, it is widely acknowledged that the number of *Legionella* recovered from in vitro cultures reflects growth in cells only. Thus, in vitro studies concerning the interaction between the bacteria and phagocytic cells seem appropriate, since extracellular growth of *Legionella* in tissue culture medium does not occur. In contrast, study of other bacteria, such as *Salmonella* or *Listeria*, requires use of a sublethal concentration of antibiotics to prevent extracellular growth, since these bacteria can easily replicate in tissue culture medium and can cause damage to the macrophages directly or by their products [21,22]. Furthermore, frequent medium changes or the use of temperature-sensitive mutants is usually required for the study of such bacteria in vitro [22,23]. Therefore, *Legionella* seems to be a useful bacterium for analysis of the interactions between bacteria and macrophages in vitro, as compared with other intracellular bacteria.

As mentioned above, on the basis of pathologic studies of legionellosis in humans the majority of these bacteria have been seen mainly within inflammatory phagocytic cells in infected patients. Furthermore, by electron-microscopic studies of experimental legionellosis [9,10], it similarly seems clear that *Legionella* are phagocytized by neutrophils at the site of an infection, such as the lung, and most phagocytized bacteria appear to degenerate quickly. In contrast, macrophages contain numerous *Legionella*, most of which appear intact and many of which are replicating. These findings concerning *Legionella* infection indicate that these bacteria replicate well in macrophages and are in fact an intracellular parasite. In this chapter, we will discuss the interaction of *Legionella*, mainly *L. pneumophila*, and macrophages.

II. *LEGIONELLA*– MACROPHAGE INTERACTIONS

A schematic summary of what is known about the interaction between *Legionella* and macrophages is given in Figure 1.

A. Phagocytosis

This interaction can be briefly summarized or characterized as phagocytosis of *Legionella* by macrophages in two stages, namely, 1) binding or attachment and 2) entering of the bacteria into the cell.

1. *Binding/Attachment*

The phagocytosis of *Legionella* is preceded by the binding or attachment of the bacteria to macrophages. In general, it is well accepted that serum factors such as complement and antibodies are important for phagocytosis and killing of the bacteria following phagocytosis. However, it is still not completely known what correlations exist between opsinization and killing.

a. *Opsonization.* Studies on opsonization of *Legionella* have shown that human monocytes bind approximately three times more *Legionella* in the presence of both antibody and complement than in the presence of complement alone, suggesting phagocytosis of this organism is enhanced by opsonization, similar to phagocytosis of other bacteria [24]. Furthermore, in the absence of antibody, the presence of fresh human serum enhanced the adherence of *Legionella* to macrophages approximately seven times more than the presence of heat-inactivated serum [25]. However, the process of opsonization of *Legionella* appears to be mediated by the classical rather than the alternative pathway of complement activation [26]. This view is supported by the findings that *Legionella* is highly resistant to the opsonic activity of normal pooled human serum [27] and fixes complement in nonimmune serum approximately 10 times less than *E. coli* [28]. However, studies concerning the C3 receptor molecules on the surface of *Legionella* indicate that the major outer membrane protein (MOMP) of these bacteria fixes C3 exclusively via the alternative pathway of complement activation [28]. Furthermore, when purified MOMP is incorporated into liposomes, the MOMP liposomes avidly fix C3 and are efficiently phagocytized by monocytes. From these observations, it seems clear that *Legionella* has C3 acceptor molecules that activate the alternative pathway of the complement system, but that the potency for activation is weak compared with other bacteria such as *E. coli*. *Legionella* MOMP has been isolated from several *L. pneumophila*

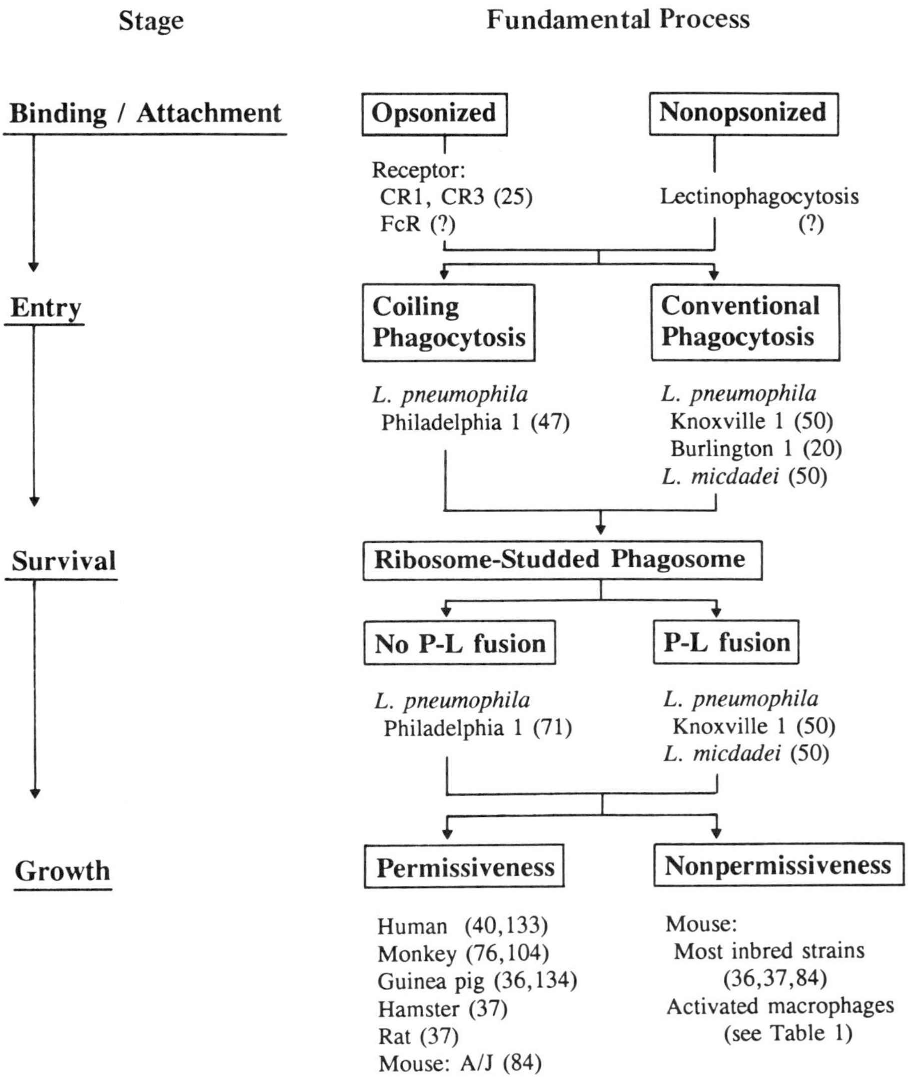

FIGURE 1 Process of *Legionella* phagocytosis in macrophages. P-L = phagosome-lysosome. Number in parenthesis shows reference.

serogroups and characterized [29–31]. From such studies, it has been shown that *Legionella* MOMP has a molecular mass between 24 and 29 kDa and is a cation-selective porin capable of being inserted into membranes and opening channels through which ions can pass. However, the role of MOMP in pathogenesis is less clear, since this component is present in not only virulent but also avirulent *Legionella* strains [29]. Alternatively, recent research concerning complement activation by *L. pneumophila* has indicated the importance of *Legionella* lipopolysaccharide (LPS) as a complement activator through the classical pathway rather than through activation by MOMP [32]. Therefore, MOMP is possibly not the only complement acceptor molecule on the surface of these bacteria.

In contrast to studies with *L. pneumophila*, it is clear that *L. micdadei*, a pathogenic *Legionella* that can also cause legionellosis in humans, fixes complement by the alternative pathway in the absence of antibody, and that normal serum is sufficient for opsonization of these bacteria by phagocytes [33,34]. On the other hand, when heat-inactivated fetal calf serum (FCS) is used in place of fresh serum, *L. pneumophila* is still taken up and multiplied rapidly in macrophages [35–37]. The comparative study of effects of heat-inactivated FCS versus fresh serum on the multiplication of *Legionella* in macrophages shows there is no significant difference in the number of viable bacteria in macrophage lysates up to 48 hours after infection [38]. In this regard, using a human macrophage-like cell line (U937), it was found that *L. pneumophila* is also taken up rapidly and multiplies intracellularly in the absence of fresh serum and antibody [39].

The effect of opsonization on intracellular killing by macrophages has also been studied using monocytes and alveolar macrophages. The presence of antibody and complement did not enhance intracellular killing, i.e., monocytes were capable of killing only a limited portion (0.25 log) of the number of *Legionella* inoculated [24]. In the case of alveolar macrophages, *L. pneumophila* were phagocytized efficiently in the presence of antibody and complement, but only 10% of the bacteria were killed even in the presence of both antibody and complement [40]. This minimum enhancement of intracellular killing after opsonization of the bacteria was observed not only with macrophages but also with neutrophils [26].

From these observations it can be stated in summary that *L. pneumophila* is opsonized by both the alternative and the classical pathway of complement activation. However, it is clear that the ability to activate complement through the alternative pathway is lower for *Legionella* than for other gram-negative bacteria. Therefore, phagocytosis of *Legionella* opsonized without antibody is relatively weak, and relatively few bacteria are taken up by the cells in the absence of antibody. Furthermore, after phagocytosis, opsonization with antibody and complement still has only minimal effects on the intracellular killing of *Legionella*, markedly in contrast with other bacteria. Such findings suggest that antibody is not an essential factor for intracellular killing of *L. pneumophila* by macrophages.

b. *Receptors*. The possible role of receptors in the phagocytosis of *L. pneumophila* has been examined by Payne and Horwitz [25]. Their studies with monoclonal antibodies (mAbs) against CR1 and CR3, which recognize complement components C3b and C3bi, respectively, indicate that adherence of *L. pneumophila* to monocytes in the presence of nonimmune fresh serum is inhibited with these mAbs. Furthermore, these mAbs also inhibit intracellular multiplication of *Legionella*. Such studies indicate that specific receptors that mediate phagocytosis of intracellular pathogens may regulate the intracellular fate of bacteria following phagocytosis. However, CR1 and CR2 mediate phagocytosis not only of *Legionella*, but also of other opsonized bacteria, including an extracellular pathogen such as *Streptococcus* [41]. Therefore, several questions remain regarding receptor-mediated entry of *Legionella* into phagocytic cells. Does a complement receptor–mediated entry of the bacteria into phagocytic cells affect their intracellular fate? For example, does phagocytosis due to CR3 lead to survival of the pathogen, whereas entry by a different means leads to intracellular killing?

Recent studies concerning complement receptor–mediated phagocytosis have shown that phagocytosis of either C3- or immunoglobulin G (IgG)-sensitized bacteria can result in the presence of azurophil granular components in the phagosomes of human neutrophils. In contrast, secondary granular constituents were incorporated in phagosomes following the phagocytosis of only IgG-opsonized bacteria [42]. In addition, CR3-mediated phagocytosis by macrophages of erythrocytes opsonized with IgM and complement led to rapid intracellular degradation. Such degradation could be blocked by anti-CR3 monoclonal antibodies and contrasted with the fate of unopsonized erythrocytes, which were much less efficiently lysed following phagocytosis [43]. These diverse effects on CR3-mediated phagocytosis indicate that the fate of *Legionella* after entry into the cell may be the result of a complex series of interactions between the pathogen and the macrophage. Other factors, especially bacterial-derived factors, might participate in these interactions and lead to survival and multiplication of *Legionella*. Therefore, the uptake and multiplication of these bacteria in the absence of heat-labile complement and antibody may be important.

Phagocytosis in the absence of serum suggests a direct cell-bacterium interaction, similar to a lectin-ligand interaction, in which phagocyte carbohydrate-binding proteins mediate binding and phagocytosis [44]. A lectin-binding site on CR3 has been identified and may play a role in the serum-independent phagocytosis of bacteria such as *Listeria monocytogenes* [45]. Therefore, the possibility of CR3-mediated phagocytosis of *Legionella* in the absence of complement is still likely. However, possible receptors involved in phagocytosis of *Legionella* under these conditions have not yet been studied.

2. Entry Coiling Phagocytosis

Legionella do not actively penetrate macrophages and do not passively enter a cell through hydrophobic interactions between the membranes of the bacterium and the macrophage [20,46]. Details of phagocytosis of *L. pneumophila* by phagocytic cells in vitro have been examined by Horwitz [47]. He reported that *Legionella* are phagocytized by coiling phagocytosis, characterized by a long phagocyte pseudopod coiled around the bacterium as the organism is internalized. This relatively unique phagocytosis is, however, also observed when *Chlamydia psittaci* or *Leishmania donovani* are phagocytized [48,49]. The coiling phagocytosis is observed not only with monocytes, but also with alveolar macrophages and polymorphonuclear leukocytes. Furthermore, treatment of the bacteria with anti-*Legionella* antibody abolishes coiling phagocytosis. Such findings suggest the possibility that surface components of the bacteria mediate the coiling phagocytosis and are involved in interaction with receptors of phagocytic cells. Also, structural studies with *L. micdadei* showed more detail concerning the differences in phagocytosis between *Legionella* strains and other species [50]. For example, *L. pneumophila*, Philadelphia 1 strain, is taken up by monocytes by a coiling phagocytosis, and no phagosome-lysosome fusions occurred after such phagocytosis. In contrast, *L. pneumophila*, Knoxville 1 strain, and *L. micdadei* were both phagocytized in the classical manner, which does not involve coiling phagocytosis. Furthermore, *L. pneumophila*, Burlington 1, is also phagocytized predominantly by the classical process [20]. Such findings indicate that the manner of *Legionella* phagocytosis by macrophages differs according to the strain or species interacting with macrophages. Nevertheless, some of the species of these pathogens, such as *L. pneumophila*, Philadelphia 1, *C. psittaci*, and *L. donovani*, are apparently taken up by such relatively unique coiling phagocytosis. However, whether the mode of entry of these oganisms into phagocytes contributes to their intracellular survival is not known.

3. Bacterial Factors

It is widely believed that factors associated with a parasite affect the manner in which it is phagocytized. However, the actual role of bacterial factors in phagocytosis is still unclear. *L. pneumophila* produces several types of exoproduct and also expresses on its surface biologically active antigens that appear to modulate phagocytosis. For instance, *L. pneumophila* possesses a fairly typical smooth LPS that has endotoxin activity as well as toxicity for macrophages [51,52]. Various cytotoxins [53,54], proteases [55,56], hemolysins [57], phospholipases [58], and leukocyte toxins [54] have all been reported in *Legionella* cultures, and these products are thought to affect the function of phagocytic cells such as modulation of oxidative burst and chemotaxis of phagocytes [54,59,60]. However, the most distinctive factor in *L. pneumophila* that may affect phagocytosis is macrophage infectivity potentiator (Mip) [61]. Macrophage infectivity potentiator is an *L. pneumophila* species-specific 24-kDa surface antigen for which the gene *mip* has been cloned. When the *mip* gene is deleted by site-specific mutagenesis, the mutant is significantly impaired in its ability to infect transformed human U937 cells and human alveolar macrophages [62]. An 80-fold-greater inoculum of the mutant strain than of the parent strain is required to infect U937 cells. The mutant strain evinces full infectivity following introduction of the *mip* gene. However, the mutant strain is thought to remain capable of intracellular multiplication, since the growth rates of the mutant and parent strains in U937 cells are similar. These results indicate that the *mip*$^-$ mutant strain is less capable of initiating infection but remains capable of intracellular multiplication. The mechanism by which the Mip protein regulates the infectivity of *Legionella* is unknown. Strains representing all 14 serotypes of *L. pneumophila* contain the *mip* gene and express a 24-kDa Mip protein [63]. However, coiling phagocytosis and inhibition of phagosome-lysosome fusion are not evident for all of the *L. pneumophila* strains interacting with macrophages. From these findings, it seems likely that the Mip protein is not directly correlated with the relatively unique coiling phagocytosis of *L. pneumophila* and inhibition of phagosome-lysosome fusion.

4. *Molecular Events in Phagocytosis*

It is widely accepted that many intracellular bacteria, including *L. pneumophila* as well as other pathogenic microorganisms, have biologically active surface components that affect phagocyte function. On the other hand, it is also acknowledged that phagocytic cells can recognize invading microorganisms and respond by evincing antimicrobial activity or secreting soluble mediators such as cytokines that prevent the interaction of the infecting microorganisms with host cells. Nevertheless, the recognition mechanism by which phagocytic cells interact with invading microbes, especially at the molecular level, is poorly understood. In this regard, recent studies concerning protein phosphorylation in *Legionella*-infected macrophages have provided a new approach to understanding the mechanisms whereby bacteria and macrophages interact at the molecular level [64]. Phosphorylation and dephosphorylation of proteins have been widely recognized as major processes in regulating cellular functions of macrophages in response to extracellular signals such as that provided by bacterial lipopolysaccharides or growth factors from other host cells responding to the microorganism [65–67]. In this regard, it was reported that macrophages may respond to bacterial infection by protein phosphorylation as one of the steps in molecular recognition of the microbe by the macrophage. In fact, when macrophages were infected with a virulent strain of *L. pneumophila*, a 76-kDa macrophage protein was found to be phosphorylated (Figure 2). Interestingly, however, infection of macrophages with avirulent *L. pneumophila*, as well as *Pseudomonas aeruginosa* or *Salmonella typhimurium*, did not result in phosphorylation of this 76-kDa protein. Phosphorylation of this protein was also not observed when macrophages were stimulated with either *E. coli* LPS or phorbol myristate acetate. These studies indicated that the 76-kDa protein phosphorylation was unique in that macrophages infected only with a virulent strain of *L. pneumophila* showed this effect. Furthermore, phosphorylation of the 76-kDa protein did occur in macrophages infected with virulent *Legionella* treated either with erythromycin to inhibit growth of the bacteria [68] or with cytochalasin D to inhibit uptake of *Legionella* but not attachment of bacteria to macrophages [20]. Such results support the view that phosphorylation of this protein occurs during attachment or binding of *Legionella* to macrophages. Furthermore, these results suggest that macrophages are capable of distinguishing between virulent and avirulent bacteria at the stage of binding or attachment. However, the role of this specific protein in recognition, cellular uptake, and growth of *Legionella* in macrophages is not yet known.

B. Intracellular Events—Survival and Growth

1. *Phagosomes*

Following phagocytosis of *L. pneumophila*, but not *L. micdadei* [69], the bacteria were rapidly located in membrane-bound vacuoles studded with ribosomes (Figure 3) [10,70–72]. Such relatively unique phagosomes were present not only in cultured monocytes and macrophages, but also in cultures of established cell lines such as MRC5, Hep 2, and Vero cells [73]. The ribosome-studded phagosomes were also present in pulmonary macrophages from *Legionella*-infected guinea pigs [10]. The role of ribosome-studded phagosomes in the fate of intracellular *Legionella* is unknown. However, such phagosomes do not appear to interact with any primary or

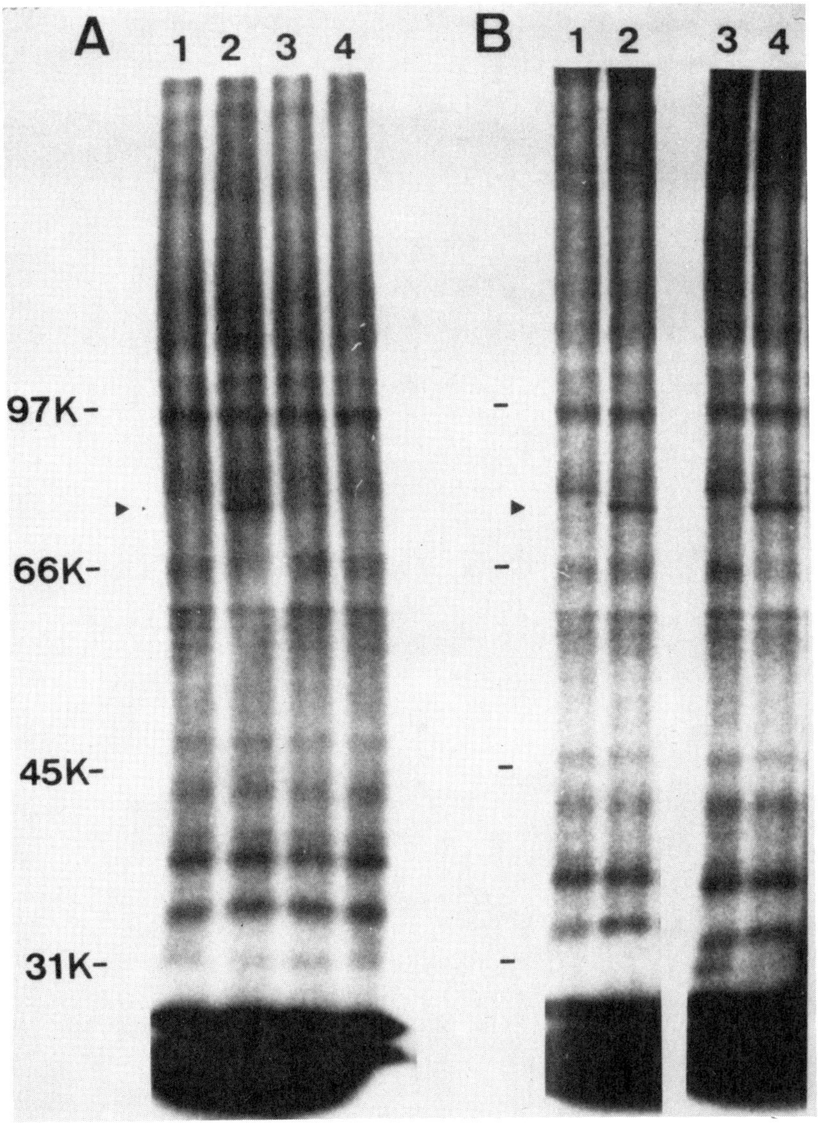

FIGURE 2 *L. pneumophila*–induced phosphorylation of A/J mouse macrophage proteins. SDS-polyacrylamide (8%) gel electrophoresis of ^{32}P-labeled cellular phosphoproteins prepared from macrophage monolayers 2 hours after infection with virulent *L. pneumophila*. (A) Effect of different infectivity ratios on protein phosphorylation. Lane 1, uninfected; lanes 2 through 4, 10^6 macrophages infected with 10^8, 10^7, or 10^6 bacteria, respectively. (B) Effect of serum on protein phosphorylation. Lanes 1 and 2, uninfected and infected with 10^7 bacteria in the absence of 10% fetal calf serum, respectively; Lanes 3 and 4, uninfected and infected with 10^7 bacteria in the presence of 10% fetal calf serum, respectively. Molecular weight markers were phosphorylase B (97,400), serum albumin (66,200), ovalbumin (45,000), and carbonic anhydrase (31,000). Arrow heads indicate the 76-kDa phosphoprotein. (From Ref. 64.)

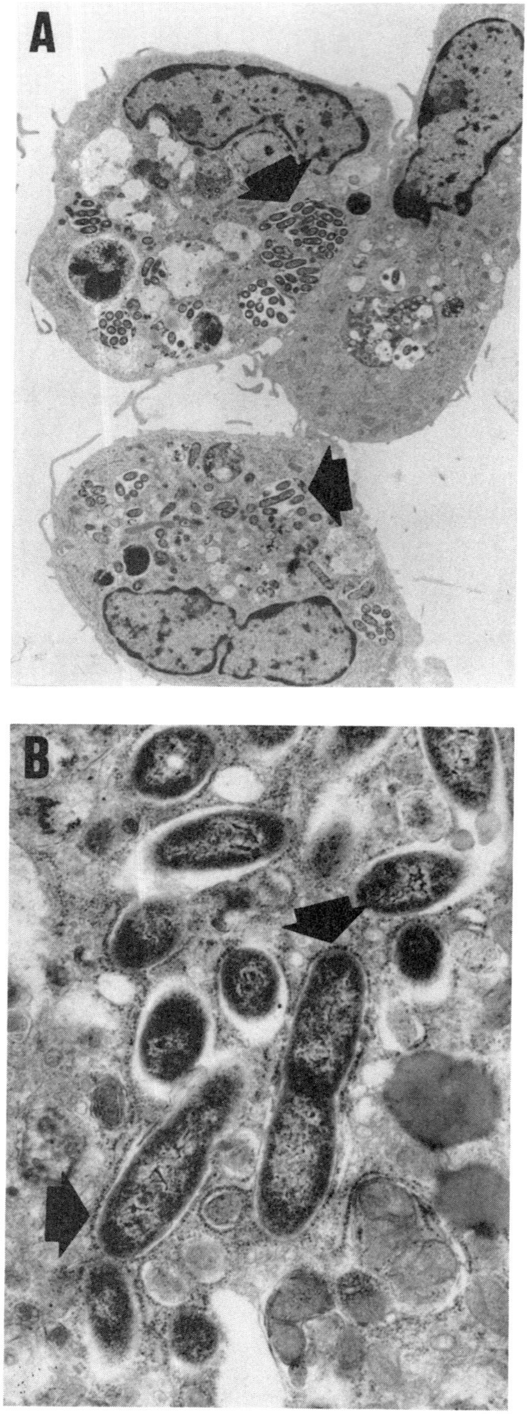

FIGURE 3 Electron micrographs of guinea pig macrophages infected with *L. pneumophila*. (A) 24 hours after infection, showing clusters of *Legionella* (arrows) in phagosomes. (B) 48 hours after infection, showing enclosed in phagosomes studded with ribosomes (arrows). (From Ref. 72.)

secondary lysosomes and, furthermore, no phagosomal acidification is evident [74]. Erythromycin, an inhibitor of bacterial protein synthesis that can enter macrophages, had no effect on phagosome-lysosome (P-L) fusion at concentrations that completely inhibited intracellular multiplication of *Legionella*. This suggests that ongoing bacterial protein synthesis is not required for inhibition of P-L fusion [71]. The coating of viable *L. pneumophila* with antibody promotes P-L fusion of only a portion of the bacteria containing phagosomes [71]. However, the mechanism by which *Legionella* inhibits P-L fusion is unknown.

Recent studies in this laboratory concerning *Legionella*-infected mouse macrophages showed that differences occur in the number of ribosome-studded phagosomes between A/J and BDF_1 mouse macrophages, which are permissive and nonpermissive for intracellular *Legionella* growth, respectively [74,75]. That is, when phagosomes of *Legionella*-infected permissive (A/J) and nonpermissive (BDF_1) mouse macrophages were analyzed by electron microscopy, the permissive macrophages evinced *Legionella* in phagosomes surrounded with ribosomes. In contrast, nonpermissive macrophages showed few phagosomes that contained *Legionella*, and only a few of these phagosomes were surrounded by ribosomes. Phagosome-lysosome fusions were not seen in macrophages from either strain of mice. However, when *S. typhimurium* temperature-sensitive mutants were used as a reference bacterium, macrophages from both strains of mice showed evidence of P-L fusion of phagosomes that contained *Salmonella*. These results suggest that macrophages from both strains of mice have the ability to undergo P-L fusion after phagocytizing bacteria.

Inhibition of P-L fusion has not been seen with all *L. pneumophila* strains used for infection of macrophages. For example, P-L fusion was observed when bacteria of the Knoxville strain of *L. pneumophila* were examined [50]. Such results suggest that inhibition of P-L fusion may not be essential for intracellular growth of *L. pneumophila*. The appearance of ribosome-studded phagosomes seems to be a more critical event in the processing of *L. pneumophila* in macrophages and may be more important for multiplication of the bacteria than inhibition of P-L fusion. However, there is no direct evidence concerning this possibility.

2. *Growth*

Once *Legionella* is taken up and enters macrophages and then escapes from the antimicrobial activity of the cells, such as P-L fusion, the bacteria can grow exponentially within the phagosomes. Electron microscopy studies with cultures of monkey alveolar macrophages infected with *Legionella* for 24 hours showed that many of the macrophages contained vacuoles filled with bacteria [76]. The doubling time for *L. pneumophila* growth in human monocyte cultures was found to be approximately 2 hours at the mid-log phase [77]. The intracellular replication of *L. pneumophila* was found not only in macrophage cultures, but also in nonphagocytic cells, such as HeLa cells [78], MRC5 and MRC9 cells [79], Vero cells [72], McCoy cells [78], chicken embryo fibroblasts [80], and mouse L929 cells [81], as well as U939 [39] and HL-60 cells [82], which originate in human histiocytic and peripheral blood cells, respectively. Conversely, this vigorous multiplication of *Legionella* did not occur in macrophages from the peritoneal cavity of mice. Recent studies show that BDF_1 mouse macrophages restrict the intracellular multiplication of *Legionella*, even when cultured with high numbers of virulent bacteria [36]. In contrast, guinea pig macrophages cultured with similar virulent *Legionella* rapidly replicate the bacteria, as well as exhibiting marked phagocytosis as shown by electron microscopy [73]. Elicited macrophages from other animals such as hamsters and rats were also found to support *Legionella* growth, similar to growth of the bacteria in macrophages from the guinea pigs [37]. Therefore, it appears that only mouse macrophages are not permissive for growth of *Legionella* as compared with macrophages from different animal species examined. Such differences concerning permissiveness for *Legionella* growth between macrophages from different animal species lead to important questions concerning the interactions between *Legionella* and macrophages.

3. Genetic Regulation

It is widely known that mice are relatively resistant to *Legionella* infection as compared with susceptible guinea pigs [83]. However, recent study showed that thioglycolate-induced peritoneal macrophages from A/J mice are permissive for growth of *Legionella* in vitro and, furthermore, that A/J mice are relatively permissive for *Legionella* infection as compared with other inbred mouse strains, such as BALB/c, C57BL/6, DBA/2, C3H/HeN, or BDF_1 mice [84]. These findings suggest that permissiveness of macrophages for *Legionella* growth may be regulated by genetic mechanisms similar to the susceptibility of different mouse strains to intracellular pathogens such as *Salmonella*, *Mycobacterium*, or *Listeria*, which is regulated by certain genes, namely *Ity*, *Bcg*, and *Lr* [85–87]. However, the genes for susceptibility or resistance to these bacteria were found to regulate susceptibility of intact animals to infection by mechanisms that involved the mobilization of mononuclear phagocytes to the site of infection [88], rather than at the cellular level.

The study of genetic regulation of macrophage susceptibility to *Legionella* infection has been investigated using hybrid mice of two strains, namely, A/J and BALB/c, whose macrophages are permissive and nonpermissive for growth of *Legionella*, respectively (Figure 4) [89]. For example, macrophages from all A/J mice support the growth of *Legionella*, whereas macrophages from BALB/c mice are nonpermissive. Furthermore, the macrophages from approximately 50% of offspring of ACF_1 mice, hybrid of A/J and BALB/c, backcrossed to A/J were found to be permissive for *Legionella* growth, whereas macrophages from all offspring of ACF_1 mice backcrossed to BALB/c were found to be nonpermissive. Macrophages from approximately 25% of ACF_2 mice, offspring of brother and sister ACF_1 mice, were found permissive for growth of *L.*

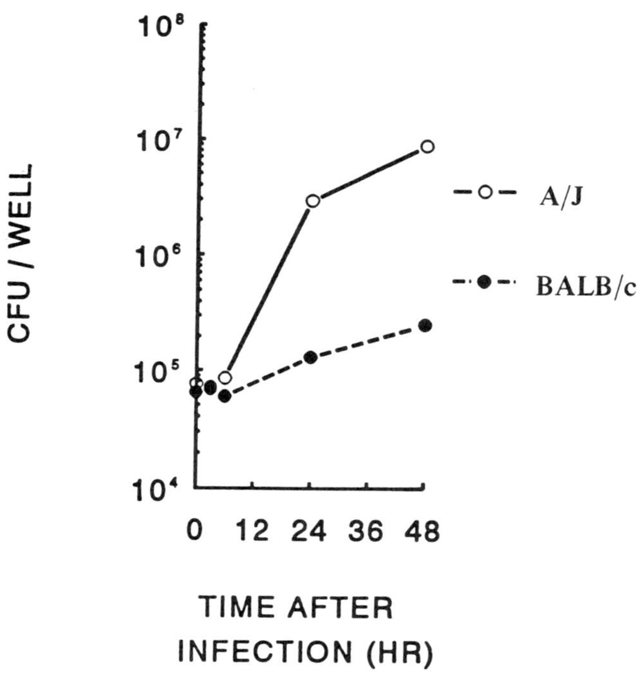

FIGURE 4 Growth of *L. pneumophila* in A/J versus BALB/c mice macrophages. Macrophage monolayers were infected with *L. pneumophila* (infectivity ratio 10 bacteria per macrophage) for 30 minutes at 37° C, washed to remove nonphagocytized bacteria, and then incubated a further 48 hours in 10% heat-inactivated fetal calf serum–RPMI 1640 medium. Number of viable bacteria (CFU) in macrophage lysates was determined by plate count method. Each point represents the mean ± SD of triplicate macrophage cultures.

pneumophila. These results indicate that the permissive nature of the macrophages is controlled by genetic mechanisms with a single gene or clustered genes involving a recessive phenotype, but differs from the resistance genes such as *Ity* [90]. Yoshida et al. [91] also reported similar results that showed the genetic regulation of permissiveness of macrophages to the intracellular growth of *Legionella*. From such data it can be concluded that the intracellular growth of *L. pneumophila* in macrophages, even when a virulent strain is used, it is regulated by a cellular factor or factors controlled by a single gene or clustered genes.

C. Macrophage Activation

The influence of macrophage activation on the intracellular multiplication of *L. pneumophila*, as well as other intracellular bacteria, has been studied by many investigators using human monocytes, alveolar macrophages, and mouse or guinea pig peritoneal macrophages (Table 1) [92–97]. Macrophages pretreated with culture supernatants from lymphocytes stimulated with mitogens such as concanavalin A or *Legionella* antigens and then infected with *L. pneumophila* have markedly inhibited bacterial growth as compared with nontreated controls. However, anti-*Legionella* activity of macrophages induced by lymphocyte culture supernatants is diminished by monoclonal anti–interferon-γ (IFN-γ) antibody [97]. Furthermore, recombinant IFN-γ is a powerful activating factor for anti-*Legionella* activity in macrophages [93–96]. These observations indicate that IFN-γ is one of the major macrophage activating factors (MAFs) produced by lymphocytes for induction of anti-*Legionella* activity. On the other hand, tumor necrosis factor (TNF) has also been shown to promote anti-*Legionella* activity of neutrophils [98]. Tumor necrosis factor is known as an immunomodulating factor for macrophages in general in terms of inducing functional activity. Therefore, it could be expected that TNF should influence the intracellular growth of *Legionella* in macrophages. However, there have been no reports concerning the effect of TNF on *Legionella*-infected macrophages.

Bacterial LPS is known to be a strong macrophage activator, since LPS modulates many macrophage functions including induction of interleukin-1 (IL-1), TNF, and colony-stimulating factor (CSF). Recent studies in this laboratory showed that *E. coli* LPS can induce anti-*Legionella* activity in mouse macrophages [99], and such activity could be inhibited by dibutyryl cAMP or agents that induce intracellular cAMP activity [100]. Therefore, it seems likely that the mechanism of induction of anti-*Legionella* activity by LPS may be mediated by cAMP cascade.

The mechanisms whereby IFN-γ–activated macrophages inhibit *L. pneumophila* multiplication intracellularly are not fully understood. In general, it seems plausible that the enhanced antimicrobial activity induced by IFN-γ may occur in parallel with increased production of toxic oxygen metabolites [101–103]. *L. pneumophila* is known to trigger a respiratory burst in alveolar

TABLE 1 Macrophage-Activating Factors for Induction of Anti-*Legionella* Activity

Agent	Macrophage	Reference
Activated or stimulated lymphocyte culture supernatants	Human monocyte/macrophage	92
	Guinea pig macrophage	36,97
	Mouse macrophage	97
Interferon-γ	Human monocyte/macrophage	93
	Human alveolar macrophage	94,95
	Mouse macrophage	96
Lipopolysaccharide	Mouse macrophage	99
Tumor necrosis factor	Mouse polymorphonuclear cell	98
	Macrophage (?)	

macrophages [104] and to be sensitive to toxic oxygen metabolites in cell-free systems [105,106]. However, *L. pneumophila* has also been shown to inhibit oxygen radical production by neutrophils and monocytes [54,107,108]. In addition, recent studies have shown that anti-*Legionella* activity in IFN-γ–activated macrophages was not diminished by reactive oxygen scavengers [96]. Nonoxidative processes have also been implicated in controlling intracellular multiplication of pathogens in IFN-γ–treated macrophages. For instance, tryptophan deprivation has been studied as an antimicrobial mechanism of IFN-γ–activated cells [109,110]. However, up to the present time, there has been no direct evidence that tryptophan deprivation affects anti-*Legionella* activity of macrophages.

Byrd and Horwitz [111] recently proposed a novel mechanism of anti-*Legionella* activity of human monocytes induced by IFN-γ in which down-regulation of transferrin receptors reduced the intracellular availability of iron. This conclusion was reached after demonstration that deferoxamine, an iron chelator, inhibited *L. pneumophila* replication within unstimulated monocytes and the capacity of IFN-γ–activated monocytes to inhibit *Legionella* correlated with regulation of transferrin receptor expression and was reversed by iron-saturated transferrin. Similarly, Skerrett and Martin [112] reported expression of transferrin receptors on alveolar macrophages elicited from rats by viable *L. pneumophila*. Exudate macrophages from resolving legionellosis showed increased resistance to the intracellular growth of *L. pneumophila*, increased Ia expression, diminished superoxide production, and down-regulation of transferrin receptors. However, alveolar exudate macrophages elicited by *E. coli* also inhibited growth of *L. pneumophila* and exhibited a modest increase in Ia expression without a change in transferrin receptors. From such reports, it appears that in activated macrophages that become nonpermissive for growth of *Legionella*, transferrin receptors are not always down-regulated. Furthermore, recent studies concerning transferrin receptor expression in IFN-γ–activated monocytes and macrophages showed contrasting results. That is, IFN-γ treatment down-regulated transferrin receptors on murine macrophages [113], but IFN-γ promoted human monocyte/macrophage transferrin receptors [114]. If these receptors are considered a key factor for anti-*Legionella* activity, it can be postulated that the mechanism of anti-*Legionella* activity induced by different factors such as IFN-γ or *E. coli* stimulation might not be the same. Limiting the availability of intracellular iron following down-regulation of transferrin receptor expression may be one of the restriction mechanisms for *Legionella* multiplication intracellularly, but not broadly applicable to antimicrobial activity in general by which activated macrophages exert control over other intracellular pathogens. A recent study by Murray et al. [115] support this view, since they found that limiting the availability of intracellular iron is not required for effective antimicrobial activity against *Toxoplasma gondii*, *Chlamydia psittaci*, and *Leishmania donovani*, all of which have high iron requirements.

D. Cytokine Production

Macrophages are one of the major types of cells that secrete cytokines in response to various stimuli involving bacteria. The cytokines secreted by macrophages, especially IL-1, are thought to have a critical role in the immune network of host responses to infection. Various reports have been published concerning the production of cytokines by macrophages stimulated or infected with bacteria during infection. In this regard, *L. pneumophila* has been examined in detail concerning its potential ability to induce IL-1 in macrophages. For example, formalin-killed *L. pneumophila* elicits IL-1–like activity in both murine peritoneal macrophages and human peripheral blood monocytes as measured by the thymocyte comitogenic assay system [116]. The molecular size of this factor is 15.5 kDa, as measured by gel filtration, similar to the molecular size reported for murine IL-1 induced by other stimuli. Production of IL-1 by macrophages involves both the membrane-associated and the secreted IL-1 variety in both *Legionella*-infected macrophage cultures and cultures stimulated with killed *Legionella* [117]. The ability of *L. pneumophila* to induce cytokine production by macrophages may be related to the marked adjuvant properties of *Legionella* as compared with *Mycobacterium tuberculosis* [118]. However, the mechanism of IL-1 induction by infection with *Legionella* is poorly understood. Recent studies using LPS, a component of bacteria first reported to induce IL-1 in macrophage cultures [119], suggest that protein kinase C– and calmodulin kinase–dependent pathways are central for induction of IL-1 [120]. However, up to the present there have been no reports indicating whether the mechanisms of induction of IL-1 in macrophages infected with gram-negative bacteria are similar to those that occur with LPS, although this antigen of bacteria is a major cell surface component. Recent studies in this laboratory concerning the mechanisms of IL-1 induction by *L. pneumophila* infection in macrophages indicate that production of this cytokine may be related to 1) increased antigen load, 2) qualitative differences in surface ligands involved in uptake of the bacteria, 3) cellular alterations in the concentration of bacterial metabolites, or 4) the biochemical effects of endosomes packed with dividing bacteria [121,122]. These conclusions were derived from experimental studies using the assay for biological activity and mRNA expression of IL-1. 1) Infection of macrophages with virulent *L. pneumophila* induced IL-1 but infection with avirulent *Legionella* did not. 2) Treatment with erythromycin, which inhibited multiplication of *Legionella* in macrophages, resulted in reduced production of IL-1 biologic activity in the culture supernatants but failed to diminish IL-1 production completely. 3) Cytochalasin D treatment, which inhibits the uptake of *Legionella* by macrophages, did not affect the binding of *Legionella* or diminish expression of IL-1 mRNA in macrophages. Furthermore, IL-1 β mRNA expression occurred even 2 hours after infection. These results strongly support the view that IL-1 induction mechanisms in *Legionella*-infected macrophages involve unique steps, including binding of *Legionella* with certain surface antigen(s), that may be different from those that occur in LPS stimulation of macrophages and directly triggered IL-1 synthesis.

L. pneumophila antigens also induce IFN-γ and TNF production by lymphocytes including natural killer cells, both in vivo and in vitro [98]. These lymphokines are thought to be important for regulation of intracellular growth of bacteria not only in macrophages but also in neutrophils. In nonimmune human leukocyte cultures, *Legionella* antigens induce IFN-γ and TNF, and these cytokines are able to stimulate the bactericidal activity of isolated PMNs against *L. pneumophila* in vitro. Furthermore, *Legionella*-infected monocytes have been shown to produce TNF in a dose-dependent response to the infecting bacteria, and addition of recombinant IFN-γ to infected monocytes resulted in augmented production of TNF in a synergistic manner [98]. The production of CSF by *Legionella*-infected macrophages has also been investigated [123]. *L. pneumophila* infection of murine macrophage cultures induced production of low levels of CSF as measured by an assay using a CSF-dependent promyelocyte cell (FDCP-1). However, treatment of these cultures with IFN-α or IFN-β, but not IFN-γ, enhanced production of CSF remarkably. Since IFN-α and IFN-β, unlike IFN-γ, do not enhance induction of anti-*Legionella* activity by macrophages, the regulating activity of these interferons in the response of macrophages to bacterial infection, in terms of cytokine production, seems important in acquired immunity to infection with these bacteria.

III. IMMUNE CELLS TO INFECTED MACROPHAGES

Besides the activation of macrophages with MAF to restrict intracellular growth of *Legionella*, the lysis of infected macrophages is also believed to be an effective way of eliminating intracellular bacteria and/or exposing the bacteria to other bactericidal cell such as neutrophils. Recently there have been several reports demonstrating the ability of lymphokine-activated killer (LAK) cells to mediate lysis of cells infected with *Rickettsia* [124,125] or *Listeria* [126,127]. Similarly, cloned T cell lines that lyse *Listeria*-infected peritoneal exudate macrophages were cultured from mice infected with *L. monocytogenes* [126]. Natural killer (NK) cells seem also to play a role in the lysis of certain bacteria-infected cells [128]. In this regard, *L. pneumophila*–infected macrophages seemed to be lysed in vitro. Actually, recent studies in this laboratory have shown that *L. pneumophila*–infected human monocytes become a target for NK cell activity [129]. Monocytes from human peripheral blood leukocytes were infected in vitro with *L. pneumophila*, cultured for 2 days to allow intracellular replication of the bacteria, and then labeled with ^{51}Cr and used as a target in a ^{51}Cr release assay. It was found that autologous nonadherent leukocytes effectively lysed the infected monocytes, and this activity was enhanced when the infected cells were precultured with IL-2. The IL-2–activated killer cells were also cytolytic against uninfected cultured monocytes, but cytotoxicity was much higher against *Legionella*-infected target cells. The phenotype of the effector cells activated by IL-2 was determined to be OKM1$^+$, OKT11$^+$, partially Leu-11$^+$ and negative for Leu-M1, OK T4, OKT8 and Leu-7, indicating that the effector cells were possibly in the NK subsets. Similar observations were reported for the IL-2–activated killer cells [130]. Murine lymphocytes activated by IL-2 have been shown to be cytotoxic also against syngeneic elicited peritoneal macrophages infected with *L. pneumophila*. These data, therefore, demonstrate that the effector cells are NK cells and that LAK cells are important in controlling *L. pneumophila*, besides direct macrophage activation by cytokines.

IV. ROLE OF MACROPHAGES IN RESISTANCE

The importance of macrophages in host resistance to infection with intracellular pathogens is well recognized. However, the definitive role of macrophages in resistance is still not clear. As indicated above, macrophages might be the harbor for dispersal and growth of *Legionella* in a host, since pathogenic and electron-microscopic studies showed the presence of numerous *Legionella* in macrophages in infected individuals. Macrophages obtained from immunized guinea pigs or guinea pigs with resolving infection are resistant to a second challenge with *Legionella*, exhibiting a strong restriction to intracellular growth of these bacteria [131,132]. Such results indicate that macrophage activation related to the immune status of the host has an important role in initial resistance to *Legionella* growth in the susceptible individuals. Recent studies of the role of macrophages and lymphocytes in resistance to *Legionella* infection demonstrated more clearly their role in resistance, using in a murine model system [97]. Guinea pigs are well known to be highly susceptible to *Legionella* infection, and their macrophages are permissive for the intracellular growth of these bacteria. BDF_1 mice are resistant to *Legionella* infection, and their macrophages are not permissive. In contrast, macrophages from the A/J mouse strains are highly permissive for the growth of *Legionella*, similar to the permissiveness of macrophages from guinea pigs. However, the susceptibility of A/J mice is only moderately greater than that of BDF_1 mice. These differences in susceptibility to *Legionella* infection between animal species and strains could be caused by permissiveness of macrophages and responsiveness of lymphocytes to *Legionella*. That is, infected *Legionella* may easily grow in guinea pig macrophages. Guinea pig lymphocytes do not produce MAF in response to *Legionella* and so do not activate macrophages to restrict the growth of bacteria, and, finally, *Legionella* growth is not restricted in the site of infection of nonimmune guinea pigs. In contrast, *Legionella* can grow in macrophages of A/J mice, whose lymphocytes produce MAF in response to *Legionella* stimulation. Therefore, the total growth of *Legionella* in macrophages is readily regulated in these mice. *Legionella* are not capable of growing in macrophages of genetically resistant BDF_1 mice, since the macrophages from these animals are nonpermissive for growth of these bacteria and, furthermore, lymphocytes from BDF_1 mice produce MAF in response to stimulation with *Legionella*. Thus, the final outcome is the resistance of the BDF_1 mice to *Legionella* infection. The results of such studies with these strains of animals infected with *Legionella* show the central role of macrophages and the regulatory role of lymphocyte on activation of macrophages in resistance to infection with *Legionella*.

V. CONCLUSIONS

It seems clear that *L. pneumophila* has many characteristics that distinguish it from other intracellular bacteria in terms of interaction with macrophages. For example, inoculated *Legionella* are capable of multiplying between 100 and 1000 times in infected macrophages 48 to 72 hours after infection in vitro. This vigorous multiplication in macrophages is remarkable as compared with other bacteria. Furthermore, when macrophages are activated by cytokines, such as INF-γ, restriction of *Legionella* multiplication is marked. For example, 100 times fewer bacteria are recovered from activated macrophages than from nontreated infected macrophages. The distinctive differences in *Legionella* growth between activated and nonactivated macrophages also are remarkable in comparison with growth of other bacteria that infect macrophages, and the difference is of value in analyzing interaction between bacteria and macrophages. The interactions between *Legionella* and macrophages may be different from those of macrophages with other intracellular bacteria, such as *Salmonella*, *Listeria*, or *Mycobacterium*, with respect to mechanisms of phagocytosis and bacterial survival such as coiling phagocytosis, ribosome-studded phagosomes, and genetic regulations dependent on resistant genes. Therefore, it appears likely that *L. pneumophila* will be useful in analyzing the interaction of bacteria and macrophages.

REFERENCES

1. Fraser DW, Tsai TR, Ovenstein W, Pardin WE, Beecham HJ, Sharrar RG, Harris CC, Brachman PS, the Field Investigation Team. Legionnaires' disease: description of an epidemic of pneumonia. N Engl J Med 1977; 297:1189–97.
2. McDade JE, Shepard CC, Fraser DW, Tsai TR, Redus MA, Dowdle WR, the Laboratory Investigation Team. Legionnaire's disease: isolation of a bacterium and demonstration of its role in other respiratory disease. N Engl J Med 1977; 297:1197–1203.
3. Fang GD, Yu VL, Vickers RM. Disease due to the *Legionellaceae* (other than *Legionella pneumophila*): historical, microbiological, clinical and epidemiological review. Medicine 1989; 68:116–39.
4. Dowling JH, Asha AK, Glew RH. Virulence factors of the family *Legionellaceae*. Microbiol Rev 1992; 56:32–60.
5. Reingold AL, Thomason BM, Brake BJ, Thacker L, Wilkinson HW, Kuritsky JN. *Legionella* pneumonia in the United States: the distribution of serogroups and species causing human illness. J Infect Dis 1984; 149:819.
6. Nguyen MH, Stout JE, Yu VL. Legionellosis. Infect Dis Clin N Am 1991; 5:561–84.
7. Glavin FL, Winn WC Jr, Craighead JE. Ultrastracture of lung in Legionnaires' disease. Observations of three biopsies done during the Vermont epidemic. Ann Intern Med 1979; 90:555–9.
8. Winn WC Jr, Myerowitz RL. The pathology of the Legionella pneumonias. Hum Pathol 1981; 12:401–422.
9. Chandler FW, McDade JE, Hicklin MD, Blackmon JA, Thomason BM, Ewing EP Jr. Pathologic findings in guinea pigs inoculated with the legionnaires' disease bacterium. Ann Intern Med 1979; 90:671–5.
10. Katz SM, Hashemi S. Electron microscopic examination of the inflammatory response to *Legionella pneumophila* in guinea pigs. Lab Invest 1982; 46:24–32.
11. Cordonnier C, Farcet JP, Desforges L. Legionnaires' disease and hairy-cell leukemia. Arch Intern Med 1984; 144:2373–5.
12. Fang GD, Stout JE, Yu VL, Goetz A, Rihs JD, Vickers RM. Community-acquired pneumonia caused by *Legionella dumoffii* in a patient with hairy cell leukemia. Infection 1990; 18:383–5.
13. Lang R, Miller I, Manon J, Kazak R, Boldur I. *Legionella longbeachae* in a splenectomized hairy-cell leukemia patient. Infection 1990; 18:31–2.
14. Voirot P, Melet M, Aymand J, Gerard J, Lion C, Gayet S, Buisine J, Canton P, Dureux J. Legionnaires' disease in hairy cell leukemia. Two new cases. Ann Med Interne (Paris) 1987; 138:287–8.
15. Feeley JC, Gorman GW, Weaver RE, Mackel DC, Smith HW. Primary isolation media for Legionnaires' disease bacterium. J Clin Microbiol 1978; 8:320–5.

16. Isenberg HD. Microbiology of Legionnaires' disease bacterium. Ann Intern Med 1979; 90:502–5.
17. Daisy JA, Benson CE, McKitrick J, Friedman HM. Intracellular replication of *Legionella pneumophila*. J Infect Dis 1981; 143:460–4.
18. Horowitz MA. Symbiotic interactions between *Legionella pneumophila* and human leukocytes. Int Rev Cytol 1983; (suppl 14):307–28.
19. Oldham LJ, Rodgers FG. Adhesion, penetration and intracellular replication of *Legionella pneumophila*: an *in vitro* model of pathogenesis. J Gen Microbiol 1985; 131:697–706.
20. Elliott JA, Winn WC Jr. Treatment of alveolar macrophages with cytochalasin D inhibits uptake and subsequent growth of *Legionella pneumophila*. Infect Immun 1986; 51:31–6.
21. Gahring LC, Heffron F, Finlay BB, Falkow S. Invasion and replication of *Salmonella typhimurium* in animal cells. Infect Immun 1990; 58:443–8.
22. Rhodes MW, Hsu HS. Effect of kanamycin on the fate of *Salmonella enteritidis* within cultured macrophages of guinea pigs. J Reticuloendothel Soc 1974; 15:1–12.
23. Gervais F, Morris-Hooke A, Trn TA, Skamene E. Analysis of macrophage bactericidal function in genetically resistant and susceptible mice by using the temperature sensitive mutant of *Listeria monocytogenes*. Infect Immun 1986; 54:315–21.
24. Horwitz MA, Silverstein SC. Interaction of the Legionnaires' disease bacterium (*Legionella pneumophila*) with human phagocytes. II. Antibody promotes binding of *L. pneumophila* to monocytes but does not inhibit intracellular multiplication. J Exp Med 1981; 153:398–406.
25. Payne NR, Horwitz MA. Phagocytosis of *Legionella pneumophila* is mediated by human monocyte complement receptors. J Exp Med 1987; 166:1377–89.
26. Horwitz MA, Silverstein SC. Interaction of the Legionnaires' disease bacterium (*Legionella pneumophila*) with human phagocytes. I. *L. pneumophila* resists killing by polymorphonuclear leukocytes, antibody, and complement. J Exp Med 1981; 153:386–97.
27. Verbrugh HA, Lee DA, Elliot GR, Keane WF, Hoidal JR, Peterson PK. Opsonization of *Legionella pneumophila* in human serum: key roles for specific antibodies and the classical complement pathway. Immunology 1985; 54:643–53.
28. Bellinger-Kawahara C, Horwitz MA. Complement component C3 fixes selectively to the major outer membrane protein (MOMP) of *Legionella pneumophila* and mediates phagocytosis of liposome-MOMP complexes by human monocytes. J Exp Med 1990; 172:121–10.
29. Butler CA, Street ED, Hatch TP, Hoffman PS. Disulfide-bonded outer membrane proteins in the genus *Legionella*. Infect Immun 1985; 48:14–8.
30. Gabay JE, Blake M, Niles WD, Horwitz MA. Purification of *Legionella pneumophila* major outer membrane protein and demonstration that it is a porin. J Bacteriol 1985; 162:85–91.
31. Hindahl MS, Iglewski BH. Outer membrane proteins from *Legionella pneumophila* serogroups and other *Legionella* species. Infect Immun 1986; 51:94–101.
32. Mintz CS, Schultz DR, Arnold PI, Johnson W. *Legionella pneumophila* lipopolysaccharide activates the classical complement pathway. Infect Immun 1992; 60:2769–76.
33. Steffensen DO, Weinbaum DL, Dowling JN. Opsonic requirements for phagocytosis of *Legionella micdadei* by polymorphonuclear leukocytes. Infect Immun 1985; 49:695–699.
34. Weinbaum DL, Bailey J, Benner RR, Pasculle AW, Dowling JN. The contribution of human neutrophils and serum to host defense against *Legionella micdadei*. J Infect Dis 12983; 148:510–7.
35. Elliott JA, Winn WC Jr. Treatment of alveolar macrophages with cytochalasin D inhibits uptake and subsequent growth of *Legionella pneumophila*. Infect Immun 1986; 51:31–6.
36. Yamamoto Y, Klein TW, Newton CA, Widen R, Friedman H. Differential growth of *Legionella pneumophila* in guinea pig versus mouse macrophage cultures. Infect Immun 1987; 55:1369–74.
37. Yoshida S, Mizuguchi Y. Multiplication of *Legionella pneumophila* Philadelphia-1 in cultured peritoneal macrophages and its correlation to susceptibility of animals. Can J Microbiol 1986; 32:438–42.
38. Yamamoto Y, Klein TW, Friedman H. Unpublished data.
39. Pearlman E, Jiwa AJ, Engleberg NC, Eisenstein BI. Growth of *Legionella pneumophila* in a human macrophage-like (U937) cell line. Microbial Pathogenesis 1988; 5:87–95.
40. Nash TW, Libby DM, Horwitz MA. Interaction between the Legionnaires' disease bacterium (*Legionella pneumophila*) and human alveolar macrophages. Influence of antibody, lymphokines, and hydrocortisone. J Clin Invest 1984; 74:771–82.

41. Noel GJ, Katz SL, Edelson PJ. The role of C3 in mediating binding and ingestion of group B *Streptococcus* serotype III by murine macrophages. Pediatric Res 1991; 30:118–23.
42. Joiner KA, Ganz T, Albert J, Rotrosen D. The opsonizing ligand on *Salmonella typhimurium* influences incorporation of specific, but not azurophil, granule constituents into neutrophil phagosomes. J Cell Biol 1989; 109:2771–82.
43. Rothlein R, Springer TA. Complement receptor type three–dependent degradation of opsonized erythrocytes by mouse macrophages. J Immunol 1985; 135:2668–72.
44. Ofek I, Sharon N. Lectinophagocytosis: a molecular mechanism of recognition between cell surface sugars and lectins in the phagocytosis of bacteria. Infect Immun 1988; 56:539–47.
45. Ross GD, Cain JA, Lachmann PJ. Membrane complement receptor type three (CR3) has lectin-like properties analogous to bovine conglutinin and functions as a receptor for zymosan and rabbit erythrocytes as well as a receptor for iC3b. J Immunol 1985; 134:3307–15.
46. Van Oss CJ. Phagocytosis as a surface phenomenon. Annu Rev Microbiol 1978; 32:19–39.
47. Horwitz MA. Phagocytosis of the Legionnaires' disease bacterium (*Legionella pneumophila*) occurs by a novel mechanism: engulfment within a pseudopod coil. Cell 1984; 36:27–33.
48. Wyrick PB, Brownridge EA. Growth of *Chlamydia psittaci* in macrophages. Infect Immun 1978; 19:1054–60.
49. Chang K-P. *Leishmania donovani*: promastigote-macrophage surface interactions in vitro. Exp Parasitol 1979; 48:175–89.
50. Rechnitzer C, Blom J. Engulfment of the Philadelphia strain of *Legionella pneumophila* within pseudopod coils in human phagocytes. Comparison with other *Legionella* strains and species. APMIS 1989; 97:105–14.
51. Wong KH, Feeley JC. Antigens and toxic components of *Legionella* in pathogenesis and immunity. Zentralbl Bakteriol Microbiol Hyg 1983; A255:132–7.
52. Friedman M, Klein TW, Friedman H. *Legionella pneumophila*–induced suppression of macrophage spreading *in vitro*. Infect Immun 1983; 42:421–3.
53. Friedman RL, Iglewski BH, Miller RD. Identification of a cytotoxin produced by *Legionella pneumophila*. Infect Immun 1980; 29:271–4.
54. Friedman RL, Lockner JE, Bigley RH, Iglewski BH. The effects of *Legionella pneumophila* toxin on oxidative processes and bacterial killing of human polymorphonuclear leukocytes. J Infect Dis 1982; 146:328–34.
55. Dreyfus LA, Iglewski BH. Purification and characterization of an extracellular protease of *Legionella pneumophila*. Infect Immun 1986; 51:736–43.
56. Conlan JW, Baskerville A, Ashworth LA. Separation of *Legionella pneumophila* proteases and purification of a protease which produces lesions like those of Legionnaires' disease in guinea pig lung J Gen Microbiol 1986; 132:1565–74.
57. Bornstein N, Nowicki M, Fleurette J. Hemolytic activity in the genus Legionella. Ann Inst Pasteur Microbiol 1988; 139:325–9.
58. Baine WB. A phospholipase C from the Dallas 1E strain of *Legionella pneumophila* serogroup 5: purification and characterization of conditions for optimal activity with an artificial substrate. J Gen Microbiol 1988; 134:489–98.
59. Rechnitzer C, Kharazmi A, Nielson H. Effects of *Legionella pneumophila* sonicate on human neutrophil granulocyte and monocyte chemotaxis. Eur J Clin Invest 1986; 16:368–75.
60. Rechnitzer C, Kharazmi A, Nielsen H, Baek L. Modulation of human neutrophil and monocyte oxidative burst by *Legionella pneumophila* sonic extract. Eur J Clin Microbiol 1987; 6:646–52.
61. Cianciotto NP, Eisenstein BI, Mody CH, Toews GB, Engleberg NC. A *Legionella pneumophila* gene encoding a species-specific surface protein potentiates initiation of intracellular infection. Infect Immun 1989; 57:1255–62.
62. Cianciotto NP, Eisenstein BI, Mody CH, Engleberg NC. A mutation in the *mip* gene results in an attenuation of *Legionella pneumophila* virulence. J Infect Dis 1990; 162:121–6.
63. Cianciotto NP, Bangsborg JM, Eisenstein BI, Engleberg NC. Identification of *mip*-like genes in the genus *Legionella*. Infect Immun 1990; 58:2912–8.
64. Yamamoto Y, Klein TW, Shinomiya H, Nakano M, Friedman H. Infection of macrophages with *Legionella pneumophila* induces phosphorylation of a 76kDa protein. Infect Immun 1992; 60:3452–3455.

65. Cohen P. The role of protein phosphorylation in neural and hormonal control of cellular activity. Nature 1982; 296:613–20.
66. Hunter T. A thousand and one protein kinases. Cell 1987; 50:823–9.
67. Krebs EG, Beavo JA. Phosphorylation-dephosphorylation of enzymes. Annu Rev Biochem 1979; 48:923–59.
68. Vilde JL, Dournon E, Rajagopalan P. Inhibition of *Legionella pneumophila* multiplication within human macrophages by antimicrobial agents. Antimicrob Agents Chemother 1986; 30:743–8.
69. Winbaum DL, Benner RR, Dowling JN, Alpern A, Pasculle AW, Donowitz GR. Interaction of *Legionella micdadei* with human monocytes. Infect Immun 1984; 46:68–73.
70. Rodgers FG. The role of structure and invasiveness on the pathogenicity of Legionella. Zentralbl Bakteriol Microbiol Hyg 1983; A255:138–44.
71. Horwitz MA. Formation of a novel phagosome by the Legionnaires' disease bacterium (*Legionella pneumophila*) in human monocytes. J Exp Med 1983; 158:1319–31.
72. Yamamoto Y, Klein TW, Brown K, Friedman H. Differential morphologic and metabolic alterations in permissive vs. nonpermissive murine macrophages infected with *Legionella pneumophila*. Infect Immun 1992; 60:3231–37.
73. Oldham LJ, Rodgers FG. Adhesion, penetration and intracellular replication of *Legionella pneumophila*: an *in vitro* model of pathogenesis. J Gen Microbiol 1985; 131:697–706.
74. Horwitz MA, Maxfield FR. *Legionella pneumophila* inhibits acidification of its human monocytes. J Cell Biol 1984; 99:1936–43.
75. Yamamoto Y, Klein TW, Brown K. Friedman H. Unpublished data.
76. Kishimoto R, Kastello MD, White JD, Shirey FG, McGann VG, Larson EW, Hedlund KW. *In vitro* interaction between normal cynomolgus monkey alveolar macrophages and legionnaires disease bacteria. Infect Immun 1979; 25:761–3.
77. Horwitz MA. The immunology of *Legionella pneumophila*. In: Moulder JW, ed. Intracellular parasitism. Boca Raton, Florida: CRC Press, 1989:141–56.
78. Daisy JA, Benson CE, McKitrick J, Friedman HM. Intracellular replication of *Legionella pneumophila*. J Infect Dis 1981; 143:460–4.
79. Wang MC, Peacock WL, Mckinney RM, Wong K-W. *Legionella pneumophila*: avirulent to virulent conversion through passage in cultured human embryonic lung fibroblasts. Curr Microbiol 1981; 5:31–4.
80. Ormsbee RA, Peacock MG, Bickel WD, Fiest P. A comparison of some biological characteristics of isolates of the Legionnaires' disease bacterium. Ann Clin Lab Sci 1981; 11:53–62.
81. Fernandez RC, Lee SHS, Haldane D, Sumarah R, Rozee KR. Plaque assay for virulent *Legionella pneumophila*. J Clin Microbiol 1989; 27:1961–4.
82. Marra A, Horwitz MA, Shuman HA. The HL-60 model for the interaction of human macrophages with the Legionnaires' disease bacterium. J Immunol 1990; 144:2738–44.
83. Katz SM, Matus JP. Animal models of legionellosis. In: Katz SM, ed. Legionellosis. Vol. II. Boca Raton, Florida: CRC Press, 1985:119–32.
84. Yamamoto Y, Klein TW, Newton CA, Widen R, Friedman H. Growth of *Legionella pneumophila* in thioglycolate-elicited peritoneal macrophages from A/J mice. Infect Immun 1988; 56:370–5.
85. Plant J, Glynn AA. Locating Salmonella resistance gene on chromosome 1. Clin Exp Immunol 1979; 37:1–6.
86. Bradley DJ, Taylor BA, Blackwell J, Evans EP, Freeman J. Regulation of Leishmania populations within the host. III. Mapping of the locus controlling susceptibility to visceral Leishmaniasis in mice. Clin Exp Immunol 1979; 37:7–14.
87. Cheers C, McKenzie IFC. Resistance and susceptibility of mice to bacterial infection: genetics of listeriosis. Infect Immun 1978; 19:755–62.
88. Stevenson MM, Konshaun PAL, Skamene E. Genetic linkage of resistance to *Listeria monocytogenes* with macrophage inflammatory responses. J Immunol 1981; 127:402–7.
89. Yamamoto Y, Klein TW, Friedman H. *Legionella pneumophila* growth in macrophages from susceptible mice is genetically controlled. Proc Soc Exp Biol Med 1991; 196:405–9.
90. Yamamoto Y, Klein TW, Friedman H. Genetic control of macrophage susceptibility to infection by *Legionella pneumophila*. FEMS Microbiol Immunol 1992; 89:137–46.
91. Yoshida S, Goto Y, Mizuguchi Y, Nomoto K, Skamene E. Genetic control of natural resistance in

mouse macrophages regulating intracellular *Legionella pneumophila* replication *in vitro*. Infect Immun 1991; 59:428–32.
92. Horwitz MA, Silverstein SC. Activated human monocytes inhibit the intracellular multiplication of the Legionnaires' disease bacterium. J Exp Med 1981; 154:1618–35.
93. Bhardwaj N, Nash TW, Horwitz MA. Interferon-gamma–activated human monocytes inhibit the intracellular multiplication of *Legionella pneumophila*. J Immunol 1986; 137:2662–9.
94. Jensen WA, Rose RM, Wasserman AS, Kalb TH, Anton K, Remold HG. *In vitro* activation of the antibacterial activity of human pulmonary macrophages by recombinant gamma interferon. J Infect Dis 1987; 155:574–7.
95. Nash TW, Libby DM, Horwitz MA. IFN-gamma activated human alveolar macrophages inhibit the intracellular multiplication of *Legionella pneumophila*. J Immunol 1988; 140:3978–81.
96. Klein TW, Yamamoto Y, Brown K, Friedman H. Interferon-gamma induced resistance to *Legionella pneumophila* in susceptible A/J mouse macrophages. J Leukocyte Biol 1991; 49:98–103.
97. Yamamoto Y, Klein TW, Newton CA, Friedman H. Differing macrophage and lymphocyte roles in resistance to *Legionella pneumophila* infection. J Immunol 1992; 148:584–9.
98. Blanchard DK, Friedman H, Klein TW, Djeu JY. Induction of interferon-gamma and tumor necrosis factor by *Legionella pneumophila*: augmentation of human neutrophil bactericidal activity. J Leukocyte Biol 1989; 45:538–45.
99. Egawa K, Klein TW, Yamamoto Y, Newton CA, Friedman H. Enhanced growth restriction of *Legionella pneumophila* in endotoxin treated macrophages. Proc Soc Exp Biol Med 1992; 200:338–42.
100. Egawa K, Klein TW, Yamamoto Y, Newton CA, Friedman H. Cyclic AMP inhibition of lipopolysaccharide-induced restriction of *Legionella pneumophila* growth in macrophage cultures. Infect Immun 1992; 60:1936–40.
101. Nathan CF, Murray HW, Weibe ME, Rubin BY. Identification of interferon-gamma as the lymphokine that activates human macrophage oxidative metabolism and antimicrobial activity. J Exp Med 1983; 158:670–89.
102. Rothermel CD, Rubin BY, Murray HW. Gamma-interferon is the factor in lymphokine that activates human macrophages to inhibit intracellular *Chlamydia psittaci* replication. J Immunol 1983; 131:2542–4.
103. Murray HW, Rubbin BY, Rothermel CD. Killing of intracellular *Leishmania donovani* by lymphokine-stimulated human mononuclear phagocytes. Evidence that interferon-gamma is the activating lymphokine. J Clin Invest 1983; 72:1506–10.
104. Jacobs RF, Locksley RM, Wilson CB, Haas JE, Klebanoff SJ. Interaction of primate alveolar macrophages and *Legionella pneumophila*. J Clin Invest 1984; 73:1515–23.
105. Locksley RM, Jacobs RF, Wilson CB, Weaver WM, Klebanoff SJ. Susceptibility of *Legionella pneumophila* to oxygen-dependent microbicidal systems. J Immunol 1982; 129:2192–7.
106. Lochner JE, Friedman RL, Bigley RH, Iglewski BH. Effect of oxygen-dependent antimicrobial systems on *Legionella pneumophila*. Infect Immun 1983; 39:487–9.
107. Lochner JE, Bigley RH, Iglewski BH. Defective triggering of polymorphonuclear leukocyte oxidative metabolism by *Legionella pneumophila* toxin. J Infect Dis 1985; 151:42–6
108. Rechnitzer C, Kharazami A, Nielsen H, Baek L. Modulation of human neutrophil and monocyte oxidative burst by *Legionella pneumophila* sonic extract. Eur J Clin Invest 1987; 6:646–52.
109. Murray HW, Szuro-Sudol A, Wellner D, Oca MJ, Grager AM, Libby DM, Rothermel CD, Rubin BY. Role of tryptophan degradation in respiratory burst–independent antimicrobial activity of gamma-interferon–stimulated human macrophages. Infect Immun 1989; 57:845–9.
110. Byrne GI, Lehmann LK, Landry GJ. Induction of tryptophan catabolism is the mechanism for gamma-interferon mediated inhibition of intracellular *Chlamydia psittaci* replication in T24 cells. Infect Immun 1986; 53:347–51.
111. Byrd TF, Horwitz MA. Interferon gamma–activated human monocytes downregulate transferrin receptors and inhibit the intracellular multiplication of *Legionella pneumophila* by limiting the availability of iron. J Clin Invest 1989; 83:1457–65.
112. Skerrett SJ, Martin TR. Alveolar macrophage activation in experimental legionellosis. J Immunol 1991; 147:337–45.
113. Hamilton TA, Gray PW, Adams DO. Expression of the transferrin receptor on murine peritoneal macrophages is modulated by in vitro treatment with interferon gamma. Cell Immunol 1984; 89:478–88.

114. Taetle R, Honeysett JM. Gamma-interferon modulates human monocyte/macrophage transferrin receptor expression. Blood 1988; 71:1590–5.
115. Murray HW, Granger AM, Teitelbaum RF. Gamma interferon–activated human macrophages and *Toxoplasma gondii*, *Chlamydia psittaci*, and *Leishmania donovanii*: antimicrobial role of limiting intracellular iron. Infect Immun 1991; 59:4684–6.
116. Klein TW, Newton CA, Blanchard K, Widen R, Friedman H. Induction of interleukin 1 by *Legionella pneumophila* antigens in mouse macrophage and human mononuclear leukocyte cultures. Zentralbl Bakteriol Hyg Microbiol 1987; A265:462–71.
117. Widen R, Klein TW, Newton CA, Friedman H. Induction of interleukin 1 by *Legionella pneumophila* in murine peritoneal macrophage cultures. Proc Soc Exp Biol Med 1989; 191:304–8.
118. McMaster PRB, Tsang VCW, Wong KH, Feeley JC, Gann DS. Comparative adjuvant activities of *Legionella pneumophila* and *Mycobacterium tuberculosis*. Int Arch Allergy Appl Immunol 1984; 73:357–62.
119. Gery I, Gershon RK, Waksman BH. Potentiation of the T-lymphocyte response to mitogens. J Exp Med 1972; 136:128–42.
120. Kovacs EJ, Radzioch D, Young HA, Varesio L. Differential inhibition of IL-1 and TNF-alpha mRNA expression by agents which block second messenger pathways in murine macrophages. J Immunol 1988; 141:3101–5.
121. Yamamoto Y, Wilson S, Klein TW, Friedman H. Induction of interleukin 1 in macrophages is associated with the intracellular growth of *Legionella pneumophila*. International Union of Microbiological Societies (IUMS) Congress, 1990; Osaka, Japan.
122. Klein TW, Yamamoto Y, Wilson S, Newton C, Friedman H. *Legionella pneumophila* infection and cytokine production, In: Friedman H, Klein TW, Yamaguchi Y, eds. Microbiol infections: role of biological response modifiers. New York: Plenum, 1992: 97–104.
123. Yamamoto Y, Klein TW, Friedman H. Mechanisms of immune responses affected by bacterial endotoxin: cytokine production in macrophages infected with *Legionella pneumophila*. In: Tsuchiya M, et al. ed. Frontiers of mucosal immunology. Vol. 2. Cambridge: Elsevier Sci. Publishers, 1991:363–75.
124. Carl M, Dasch GA. Characterization of human cytotoxic lymphocytes directed against cells infected with typhus group *Rickettsiae*: evidence for lymphokine activation of effectors. J Immunol 1986; 136:2654–61.
125. Rollwagen FM, Dasch GA, Jerrells TR. Mechanisms of immunity to Rickettsiae infection: characterization of cytotoxic effector cell. J Immunol 1986; 136:1418–21.
126. DeLibero G, Kaufmann SHE. Antigen-specific Lyt-2+ cytolytic T lymphocytes from mice infected with the intracellular bacterium *Listeria monocytogenes*. J Immunol 1986; 137:2688–94.
127. Chen-Woan M, McGregor DD, Noonan SK. Isolation and characterization of protective T cells induced by *Listeria monocytogenes*. Infect Immun 1986; 52:401–7.
128. Klimpel GR, Niesel DW, Klimpel KD. Natural cytotoxic effector cell activity against *Shigella flexneri*–infected HeLa cells. J Immunol 1986; 136:1081–6.
129. Blanchard DK, Stewart II WE, Klein TW, Friedman H, Djeu JY. Cytolytic activity of human peripheral blood leukocytes against *Legionella pneumophila*–infected monocytes: characterization of the effector cell and augmentation by interleukin 2. J Immunol 1987; 139:551–6.
130. Resnick M, Roguel N, Bercovier H, Enk C, Frankenburg S, Kedar E. Lysis of murine macrophages infected with intracellular pathogens by interleukin 2–activated killer (LAK) cells *in vitro*. Cell Immunol 1988; 113:214–219.
131. Yamamoto Y, Klein TW, Friedman H. *Legionella pneumophila* intracellular growth in normal vs. immune guinea pig macrophage cultures. Curr Microbiol 1988; 16:333–6.
132. Yoshida S, Mizuguchi Y, Nikaido Y, Mitsuyama M, Nomoto K. Fate of *Legionella pneumophila* Philadelphia-1 strain in resident, elicited, activated, and immune peritoneal macrophages of guinea pigs. Infect Immun 1987; 55:2477–82.
133. Horwitz MA, Silverstein SC. The Legionnaires' disease bacterium (*Legionella pneumophila*) multiplies intracellularly in human monocytes. J Clin Invest 1980; 66:441–50.

21

Life and Death of an Intracellular Pathogen: *Francisella tularensis* and the Macrophage

Anne H. Fortier, Shawn J. Green, Tammy Polsinelli, Tonya R. Jones, Robert M. Crawford, David A. Leiby, Karen L. Elkins, Monte S. Meltzer, and Carol A. Nacy

Walter Reed Army Institute of Research, Washington, D.C.

I. INTRODUCTION

Intracellular microorganisms, as a class of infectious agents, are among the most difficult pathogens for mammalian hosts to eradicate: not only do they remain sequestered inside cells most of their lives, hidden from view of the immune system, but they dramatically amplify their numbers in this habitat. Thus, when cells finally release their infectious particles, the host is exposed to an overwhelming antigenic challenge. Many of the clinical symptoms of diseases caused by these agents are not due to toxicity by the pathogen itself, but to the overreaction of the body, particularly the immune system, as it attempts to restore homeostasis. The resulting pathology, therefore, is largely of host origin.

A facultative intracellular pathogen is a microorganism that can grow in vitro in the absence of cells but can also grow inside cells. A limited number of bacteria are classified as facultatively or obligately intracellular: *Rickettsia* and *Chlamydia* are two genera recognized as obligate intracellular pathogens; *Salmonella, Shigella, Brucella, Yersinia,* and *Listeria* are facultative intracellular pathogens that spend some, if not all, of their existence within cells of an infected host. In this chapter, we review the data that identifies *Francisella* as a facultative intracellular organism in vitro but an obligate intracellular pathogen of macrophages in vivo.

A. The Bacterium

Francisella tularensis is a tiny pleomorphic gram-negative bacterium that can infect mammals through either the dermal or the respiratory route. It is the etiologic agent of tularemia, or trapper's fever, and the usual method of infection is through creation of an aerosol or accidental laceration during the skinning of infected game. Depending on the route of infection, failure to treat tularemia results in up to 60% mortality in humans [1]. Approximately 130 cases of naturally transmitted tularemia were reported in the United States in 1991 [2].

Francisella is a fastidious organism and requires a complex medium for isolation and growth in vitro [3]. Both cysteine and a source of iron are nutritional requirements for extracellular replication of this pathogen [3]. Although *Francisella* can be cultured on synthetic medium without cells, there are no data to support its extracellular growth in vivo in experimental animals or humans [4]. Polymerase chain reaction (PCR) analysis of blood collected over time from *F. tularensis*–infected mice suggests that there is little to no bacteremia during the course of lethal disease with this pathogen: organisms are not found in serum, but only within circulating white blood cells [5]. Further, electron microscopy studies of peritoneal cells from mice infected intraperitoneally showed *Francisella* only within macrophages (Figure 1). In vivo, then, *Francisella* is an intracellular pathogen, and its principal host cell for intracellular replication is the macrophage [6,7]. The factors for successful growth of *Francisella* within macrophages are reviewed in a later section.

B. An Experimental Animal Model of Tularemia

The live vaccine strain (LVS) of *F. tularensis* is pathogenic for several different strains of mice and produces infection in these experimental animals similar phenotypically (pathology, symptoms, course of infection) to that seen in humans infected with the wild-type organisms [6,8]. Disease in

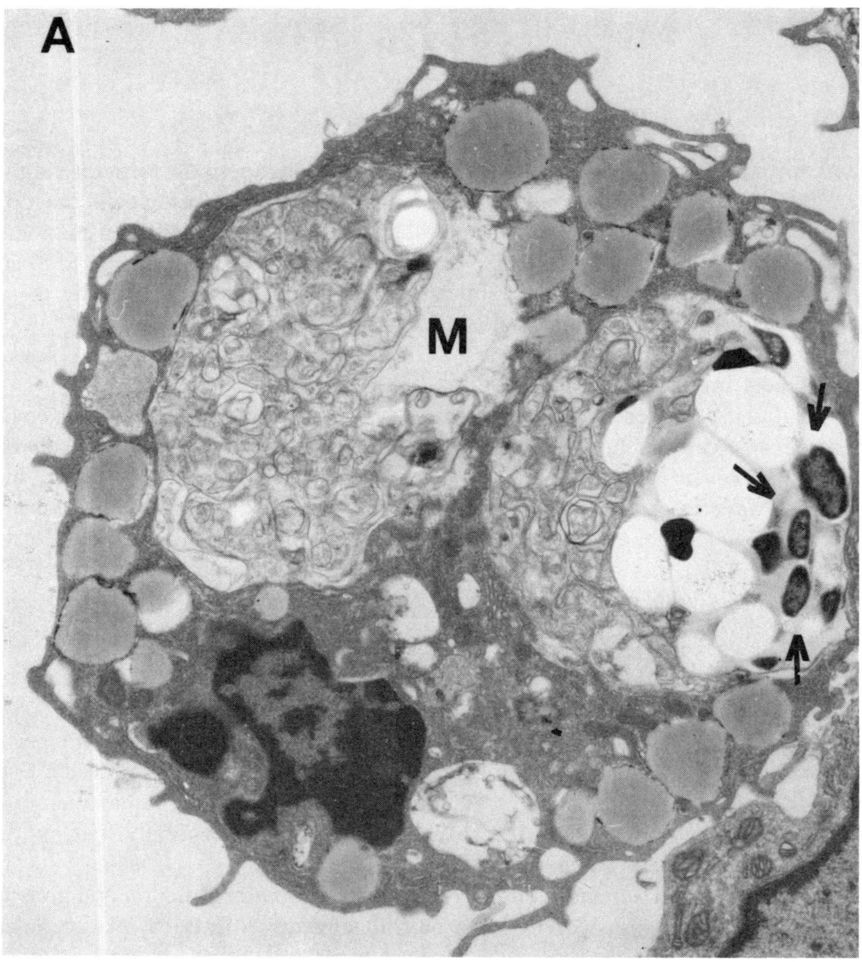

FIGURE 1 Electron micrographs of murine macrophages infected in vivo with *Francisella tularensis* live vacine strain. Mice were inoculated intraperitoneally with 10^2 CFU, and peritoneal cells were harvested 3 days after infection. Cells were processed for electron microscopy by standard procedures. A. Peritoneal macrophage (M) infected with *F. tularensis* (arrows). B. Close up of *F. tularensis* (Ft) in membrane-bound vesicles (arrows).

both humans and mice depends on route of inoculation and ranges from ulceroglandular infection (entry through the skin), to respiratory tularemia (inhalation), to a typhoidal disease. Regardless of the initial site of infection, pneumonia is a common sequela [6].

Bacteria are isolated from all reticuloendothelial organs of experimental animals within 3 days of *Francisella* introduction by any route. Routes of infection (intraperitoneal, intranasal, intravenous) that lead to fatal disease in mice result in overwhelming pneumonitis by day 5 and death of the animal shortly thereafter. Routes of infection that are not lethal (introduction into any skin site, [9]) also show dissemination of organisms to the lung; the time of appearance in this tissue is delayed in comparison with all other routes of inoculation, however, and the bacteria do not replicate [8]. Complete clearance of *Francisella* from all tissues of mice recovering from intradermal infection occurs within 3 weeks [8].

The LD_{50} in mice for the different sites of infection with *F. tularensis* reflects this fortuitous dichotomy of route-dependent natural resistance: the LD_{50} is fewer than 10 organisms for inoculation of LVS intraperitoneally and fewer than 100 organisms for intravenous and intranasal inoculation, but for intradermal inoculation it is more than 500,000 organisms. Furthermore, mice infected intradermally with a sublethal dose of LVS develop immunity to reinfection at any other

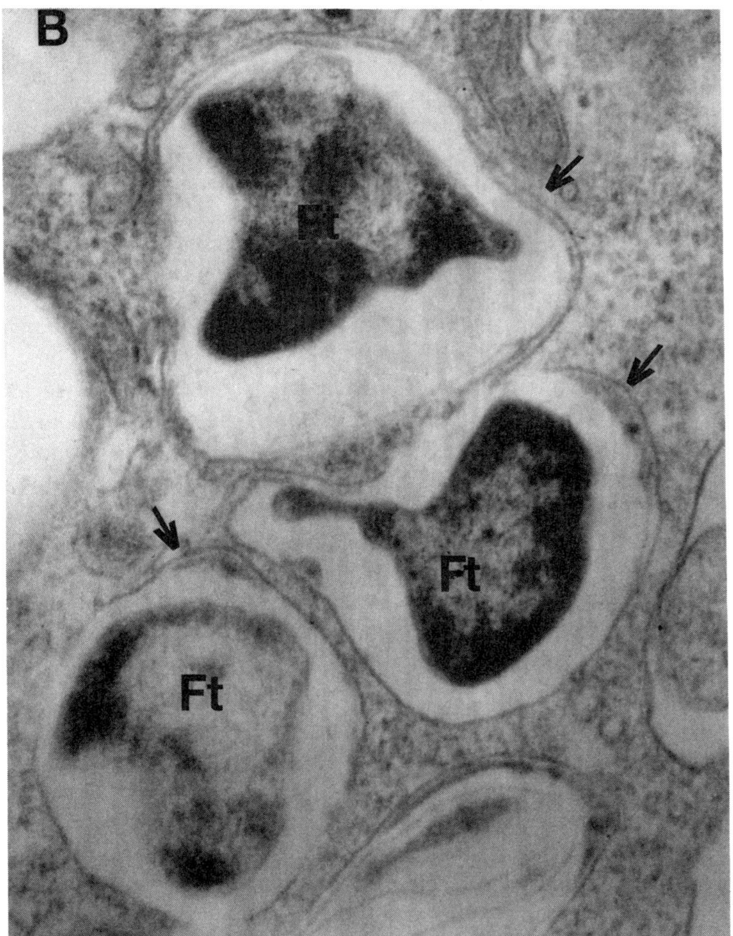

FIGURE 1 *(continued)*

site. For example, the LD_{50} changes from less than 10 organisms intraperitoneally in C3H/HeN mice to more than 10,000 organisms intraperitoneally following recovery from inoculation of LVS in the skin [8]. Both cells and serum passively transfer this immunity to naive recipients [8].

By simply altering the route of entry, it is possible to analyze both protective immune mechanisms and ineffective responses induced in a single animal strain with a single bacterium. Mice susceptible to the lethal effects of *F. tularensis* LVS infections clearly possess the machinery to generate an appropriate and protective immune response, and we exploited the difference in route susceptibility to assess the cells infected with *Francisella,* the immunoregulatory cytokines that influence in vivo progressive disease, and the cellular and humoral elements responsible for protection against this pathogen. That macrophages play a role in both the pathology and the resolution of *Francisella* infections is now clear [6–8].

II. MACROPHAGES AND *FRANCISELLA* IN VITRO

In any infectious process, macrophages have multiple functions: they act as a natural barrier in early interactions to prevent establishment of the pathogen in tissue using their natural antimicrobial armamentarium, they prevent the spread of organisms throughout tissue during inflammation by secretion of a variety of substances that enhance lytic activity of humoral factors or that enhance granuloma formation, and they initiate development of specific immunity while acting as immune effector cells once immunity is established. In this section, we summarize what is known about the interaction between differentiated murine tissue and inflammatory macrophages and *F. tularensis* LVS. We explore the capacity of these cells to support replication of the pathogen and to be activated for suppression of bacterial growth. We compare this with the interaction of *Francisella* and human monocytes and show distinct similarities between murine and human macrophage experiences with this pathogen.

A. Infection and Replication in Murine Macrophages

Since it is essential for intracellular pathogens to enter cells, they develop unique and effective mechanisms to effect entry. Macrophages ingest most particulates and many bacteria by endocytosis or phagocytosis: both processes mobilize microfilaments to bring things into the cell. Cytochalasin B, a microfilament inhibitor [10], blocks uptake of intracellular bacteria *(Salmonella, Shigella, Listeria)* that are dependent on host cell phagocytosis for entry [11,12]. In contrast, macrophages treated with cytochalasin B take up *Francisella* as well as macrophages cultured in the absence of the inhibitor, although phagocytosis of latex beads is decreased by over 70%. Thus, entry of *Francisella* into macrophages, unlike that of other intracellular pathogens of a similar phenotype, does not require a phagocytic event dependent on microfilament mobilization. *Chlamydia* also enter cells by a mechanism insensitive to cytochalasin B [13,14]: elementary bodies (EBs), the infectious particle for *Chlamydia,* are apparently too small to be susceptible to cytochalasin B inhibition [10,14]. Several of the *Rickettsia* use an active penetration mechanism for entry into cells, and phagocytosis is not involved [15]. We have no information at this time on the similarity of *Francisella* entry to that of *Chlamydia* EBs or *Rickettsia,* but LVS has clearly evolved a mechanism of entry into macrophages that circumvents the need for a classic phagocytic event.

Intracellular pathogens also develop interesting and diverse strategies for survival in macrophages. Some fail to trigger a respiratory burst as they enter the cell [16]; some neutralize the potent reactive oxygen intermediates released in the respiratory burst during phagocytosis [16]. Some evade destruction in the phagolysosome, the digestive organelle of the macrophage, by replicating in the cytoplasm [15,17]; some actively prevent lysosome fusion [18,19]; and some replicate in the phagolysosome, using the low pH of this organelle to their enzymatic advantage [20,21]. The most successful of the intracellular pathogens acquire multiple evasive strategies to ensure survival. Like many obligate intracellular microorganisms, LVS does not initiate a respiratory burst during its ingestion by macrophages. However it enters these cells, LVS clearly replicates within a vacuole in macrophages infected in vitro [22] and in vivo (Figure 1). Many fewer LVS-containing than *E. coli*–containing vacuoles fuse with prelabeled lysosomes (30% and 70%, respectively) [22], although LVS does require an acidic environment for intracellular growth [23]. Using lysosomotropic agents (NH_4Cl, chloroquine, or ouabain) to raise the intracellular pH, we found that LVS growth is inhibited in cells. Each of these reagents blocks acidification by a different mechanism [24]: regardless of the mechanism, however, LVS cannot replicate. The requirement for low pH may reflect LVS's strict iron requirement for growth [25]. Transferrin receptors at the cell surface transport transferrin-bound iron into the cell and release iron when endocytic vesicles containing internalized transferrin receptors become acidified. Low pH is the catalyst for this event [26,27]. That the acidiphilic nature of LVS is related to its iron requirements is supported indirectly by the fact that deferoxamine (soluble iron chelator) added to LVS-infected macrophage cultures suppresses intracellular bacterial growth [23]. Infective *Francisella,* then, localize in an acidified vesicle, but not necessarily a phagolysosome, within macrophages. Whether they can, like *Mycobacterium* [19] and *Toxoplasma* [18], actively prevent lysosomal fusion is not yet clear.

All macrophage populations are equivalently infected with LVS in vitro following exposure to similar inocula (Table 1), and all are capable of sustaining intracellular replication of this pathogen. Unlike *Listeria* and *Salmonella,* however, organisms whose numbers are depleted substantially in the initial 30 minutes of interaction with macrophages [28,11], LVS is resistant to this natural antimicrobial arsenal. *Francisella* has developed broadly effective protective strategies against the innate antimicrobial activities of macrophages, both oxidative and nonoxidative. *Francisella* have, among other protective defenses, catalase, which eliminates H_2O_2 produced by macrophages as an effector molecule.

Francisella is, however, susceptible to oxidative killing mechanisms of polymorphonuclear leukocytes (PMNs), particularly hypochlorous acid [29]. During the initial inflammatory response to *Francisella,* or any other pathogen, the first cells to arrive at the site of infection are PMNs. Extracellular bacteria exposed to antibody will be phagocytosed and killed by these cells, but *Francisella* that enter tissue macrophages are likely to survive. Perhaps the capacity to enter macrophages derived from the need of this pathogen to escape the lethal consequences of interactions with PMNs. The PMN infiltrate is followed within hours by inflammatory macrophages newly immigrated into tissue from circulation; these cells rapidly replace PMNs as the predominant cell type at the site of infection.

Inflammatory macrophages, compared with differentiated tissue macrophages, have high levels of certain enzymes, increased oxidative capacity, and are excellent cytotoxic cells for extracellular pathogens or tumor targets [30,31]. They also efficiently kill ingested *Listeria* without immune stimulation [28]. However, LVS infects and replicates in inflammatory macrophages (Table 1). In fact, the generation time of LVS is shorter in inflammatory macrophages (2 to 4 hours) than in any differentiated tissue macrophages (6 to 8 hours), perhaps a reflection of the higher metabolic activity of inflammatory cells. The implication of these in vitro studies, yet to be confirmed by in vivo experiments, is that inflammation, normally beneficial to the host, supplies the host cell in which *Francisella* replicates with greatest efficiency.

B. Death and Destruction in Murine Macrophages

If the host's natural barrier defenses and first-line inflammatory responses are unable to control infection, the host then unleashes its immune arsenal. Memory and helper T cells react to antigenic challenge with the production of cytokines that act on macrophages to change their physiology and function. The macrophage acquires new skills as a potent, nonspecific cytotoxic effector cell.

Macrophages exposed in vitro to cytokines in antigen-stimulated spleen cell culture fluids

TABLE 1 Interaction of *Francisella tularensis* LVS and Macrophages

Source of macrophages	Cells treated 72 h after infection with	Log_{10} CFU LVS/ml 2 hr	Log_{10} CFU LVS/ml 72 hr	Doubling Time (hr)	Nitrites (μM) at 72 hr
Murine					
Resident	Medium	3	8	6	0
	IFN-γ[a]	3	4		8
	IFN-γ + N[G]MMLA[b]	3	8		1
Inflammatory	Medium	2	7	3	0
	IFN-γ[a]	3	3		8
	IFN-γ + N[G]MMLA[b]	3	8		1
Human					
PBLC	Medium	3	6	6	0
	IFN-γ[c]	3	4		0
	IFN-γ + N[G]MMLA[b]	3	4		0
Monocytes	Medium	3	7	6	0
	IFN-γ[c]	3	4		0
	IFN-γ + N[G]MMLA[b]	3	4		0

[a]Macrophages were harvested by peritoneal lavage from untreated mice (resident cells), mice inoculated with 1.0 ml 2% colloidal starch 5 days before lavage, or mice inoculated with 1.0 ml 3% proteose peptone 3 days before lavage. Cell suspensions were adjusted to 10^6 macrophages/ml in Dalbecco's minimal essential medium with 5% fetal bovine serum and cultured as cell pellets in polypropylene tubes. Cells were infected with LVS at a multiplicity of infection of 1 for 2 hours at 37°C. Macrophages were stimulated with 50 U/ml recombinant murine IFN-γ. All cells were sampled by lysing with 0.01% sodium dodecyl sulfate, serially diluting the lysate, and plating on Meuller Hinton Agar plates to assess bacterial CFU.
[b]500 μM of the L-arginine analogue N^G-monomethyl-L-arginine, added to IFN-γ stimulated cultures at time of infection.
[c]Peripheral blood lympocytes or elutriated human monocytes were cultured in RPMI with 10% FBS at 10^5 cells/well in microtiter plates and infected with 10^3 LVS. Macrophages were stimulated with 50 U/ml recombinant human IFN-γ. All cells were sampled by lysing with 0.01% SDS, serially diluting the lysate, and plating on Meuller Hinton agar plates to assess bacterial CFU.
LVS = live vaccine strain; CFU = colony-forming units; IFN=γ = interferon-γ; PBLC = peripheral blood lymphocytes.

develop the capacity to kill intracellular *Francisella,* and the predominant cytokine responsible for this antimicrobial activity is interferon-γ (IFN-γ) [32]. All inflammatory or differentiated macrophages acquire this capacity, but the sensitivity of the cells to activation factors differ. Thus, as little as 5 U/ml IFN-γ induces 50% maximal intracellular killing activity in inflammatory macrophages, whereas 20 U/ml IFN-γ induces an equivalent amount of intracellular killing in differentiated tissue cells (Figure 2) [32]. Although inflammatory macrophages support faster replication of *Francisella,* these cells are exquisitely sensitive to IFN-γ for elimination of the intracellular pathogen.

Cytokine-activated macrophages have many potential antimicrobial effector activities not present in unstimulated macrophages, including the down-regulation of transferrin receptors for decreasing intracellular iron [33], degradation of tryptophan [34,35], and production of such effector molecules as reactive oxygen and nitrogen intermediates [36,37]. Recent studies with inhibitors of nitric oxide (NO) synthesis suggest that this molecule is involved in murine macrophage destruction of a number of intracellular and extracellular microbial targets [38–40]; this short-lived reactive nitrogen intermediate is rapidly oxidized to NO_2^- [41], which can be measured by a simple colorimetric test [42]. Thus, NO_2^- in culture fluids is a quantitative index of

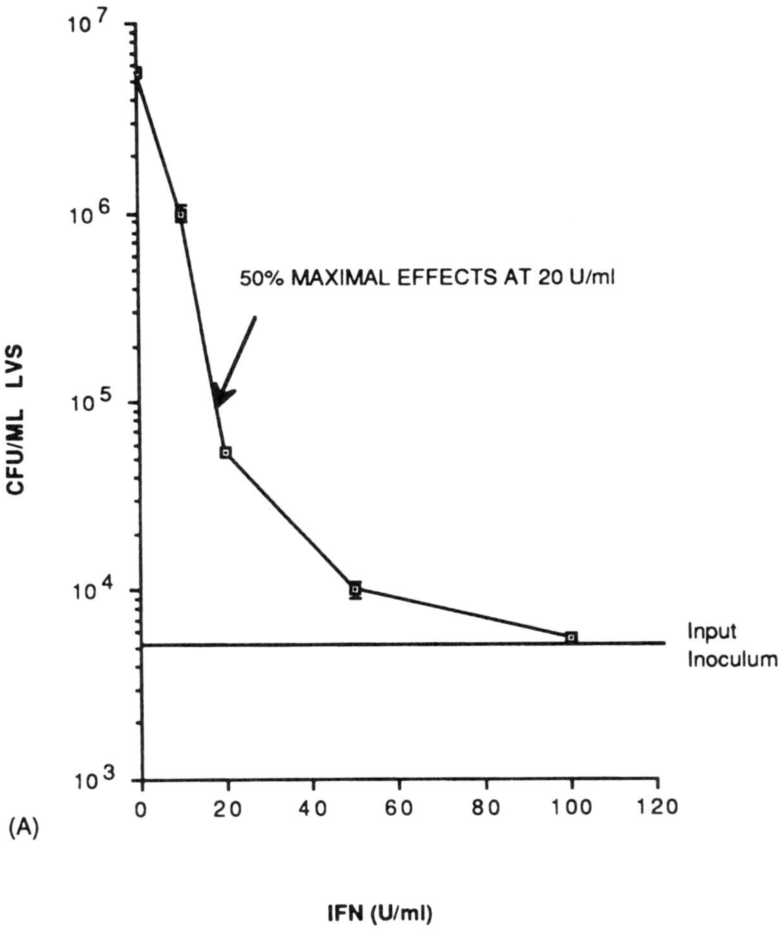

FIGURE 2 Growth of *Francisella tularensis* live vaccine strain (LVS) in macrophages exposed to interferon-γ (IFN). A. Resident peritoneal macrophages infected with LVS and cultured for 72 hours. B. Proteose peptone–elicited macrophages infected with LVS and cultured for 72 hours. Data are reported in both panels as mean CFU/ml ± SEM.

macrophage activation and killing capacity. Production of NO_2^- correlates with anti-LVS activity in murine macrophages activated in vitro with IFN-γ (Table 1) [32,43] or in vivo following infection of mice with *Mycobacterium bovis,* strain BCG [7]. A competitive inhibitor of L-arginine–derived NO, N^GMMLA, blocks IFN-γ–induced anti-LVS activity and production of No_2^- in all murine macrophage populations (Table 1). Mice treated with N^GMMLA in vivo fail to produce urinary nitrates, and macrophages recovered from these animals are no longer able to kill LVS in vitro. The conclusion of these in vitro and in vivo studies is that NO is the effector molecule used by activated macrophages to eliminate intracellular *Francisella*.

L-arginine–derived NO from activated macrophages exerts its effects by reacting with fee iron, forming iron-nitrosyl complexes, and inactivating enzymes requiring iron for activity in both the target cell and the macrophage itself [44,45]. Paralysis of iron-dependent enzymes involved in such essential processes as mitochondrial respiration, DNA synthesis, and Krebs' citric acid cycle enables macrophage-derived NO to affect a number of eukaryotic and prokaryotic targets [38–40]. *Francisella*'s absolute nutritional requirement for iron suggests that the iron-nitrosyl complexes

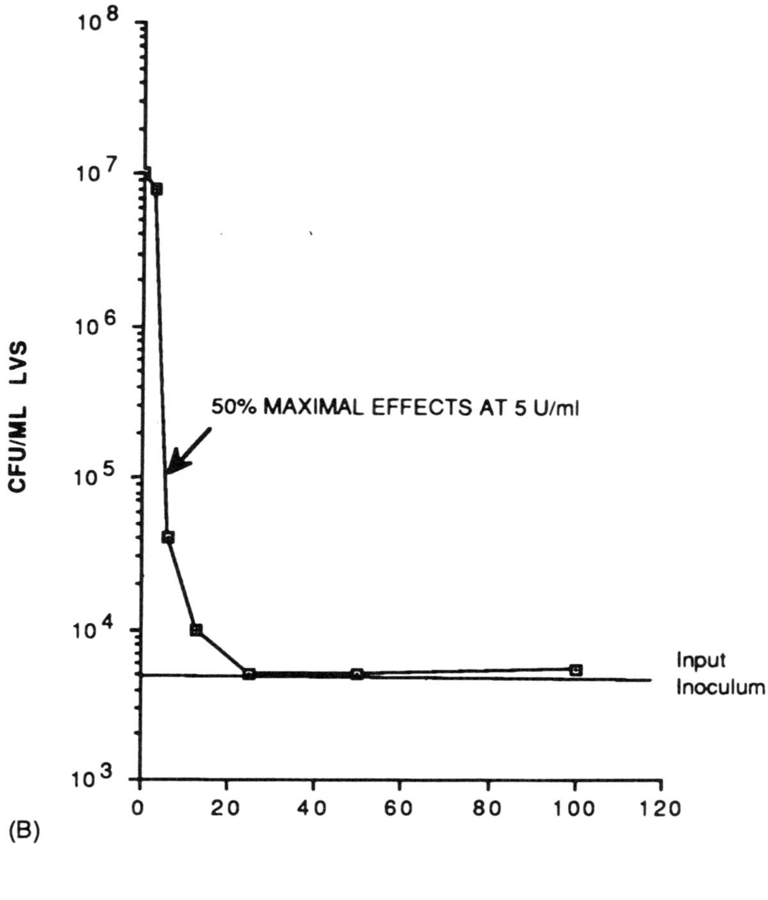

(B)

FIGURE 2 *(continued)*

prevent the bacterium from acquiring this essential nutrient. Thus, IFN-γ induction of NO synthase in activated macrophages results in production of NO, complexing of free iron, and cytostasis of the bacterium. With continual deprivation of iron, *Francisella* eventually dies [32]. Supporting this concept is the fact that cytostatic and cytocidal activity of IFN-γ-activated macrophages for *Francisella* can be completely reversed by supplying an unlimited source of exogenous iron.

In summary, murine macrophages are infected by and support the replication of *F. tularensis* LVS in vitro and can be activated by IFN-γ to eliminate this intracellular pathogen by limiting iron, which is required for replication.

C. Human Monocytes and *Francisella*

Cultures of human peripheral blood mononuclear cells (PBMCs) support the growth of *F. tularensis* LVS with a doubling time of about 6 hours (Table 1). Separation of PBMCs into lymphocyte- and monocyte-rich fractions shows that bacteria grow only in unfractionated PBMCs and in monocytes. Flow-cytometric examination of fixed and saponin-permeabilized cells with a mouse mAb against *F. tularensis* and fluorescein isothiocyanate (FITC)-labeled goat anti–mouse immunoglobulin M (IgM) demonstrates that up to 50% of monocytes are infected by 72 hours.

Virtually all bacteria visualized by this technique are inside monocytes. As in murine macrophages, then, *Francisella* is obligately intracellular in monocyte cultures.

Replication of *F. tularensis* in human monocytes is markedly inhibited by treatment of these cells with human IFN-γ, and the ID_{50} for this activating factor is 10 IU/ml (Table 1). Antibacterial activity occurs both in cells treated with IFN-γ at the time of and continuously after infection and in cells treated 24 hours before infection and washed free of the cytokine. Monocytes treated with human cytokines other than IFN-γ (IFN-α, interleukin-1β [IL-1β], IL-2, IL4, IL-6, macrophage colony-stimulating factor, granulocyte-macrophage colony-stimulating factor, tumor necrosis factor-α [TNF-α], and transforming growth factor-β) show no antibacterial activity at $\leq$ 500 U/ml. Monocytes treated with LPS or PMA also fail to inhibit the replication of *F. tularensis*.

Interferon-γ–induced intracellular killing occurs in freshly isolated human monocytes, but not in cells cultured for 5 to 7 days and then treated with IFN-γ. The mechanism for IFN-γ–induced antibacterial activity is independent of reactive oxygen and nitrogen metabolites and of depletion of tryptophan in the culture medium. Addition of deferoxamine to cultures of human monocytes and *F. tularensis,* however, completely inhibits the intracellular replication of this bacterium. Interferon-γ–induced antimicrobial activity is also completely abrogated by addition of iron-saturated transferrin.

In summary, *Francisella* infects and replicates in human monocytes, and these cells can be activated by IFN-γ to kill the intracellular pathogen. Although IFN-γ acts to limit intracellular iron available to *Francisella* in monocytes, as it does in murine macrophages, the mechanism for iron sequestration is different: it is not NO. Thus the same cytokine (IFN-γ) acts on both murine and human macrophages to induce antimicrobial activities by the same mechanism (depletion of intracellular iron), but through different intracellular biochemical pathways.

III. MACROPHAGES AND *FRANCISELLA* IN VIVO

Intraperitoneal inoculation of mice with 1 to 100 CFU of LVS results in progressive disease and death within 7 days. Peritoneal cells harvested from these infected animals harbor up to 10^9 *Francisella* [7], and virtually all of the bacteria are within macrophages (Figure 1). The distribution of bacteria throughout the reticuloendothelial system suggests hematogenous spread, and PCR determination of *Francisella* in blood from infected animals confirms that bacteria are not free in circulation, but are found within cells [5]. That IFN-γ can markedly disrupt the intracellular survival of *Francisella* in cells of the monocytic lineage in vitro (see above) suggests that a deficiency of this cytokine is the underlying basis for suseptibility to disease. The rapid decline of mice inoculated intraperitoneally makes in vivo analysis of this hypothesis technically problematic, but others, using a different route of inoculation, confirm the possibility [8,46,47]. In our studies, we exploited the relative resistance of mice infected intradermally to explore the role of cells, antibodies, and cytokines in both primary infections and challenge infections of mice exposed to immune cells or serum.

A. Humoral Immune Factors

Mice infected intradermally with *Francisella* produce antibody that can passively transfer protection against LVS challenge infection to naive recipients [8]. The antibody is composed of two major classes, IgM and IgG2a [48]; both subclasses independently confer immunity when inoculated into naive mice at the time of infection. Although antibody can opsonize *Francisella* for uptake and killing by PMNs [29], immune serum does not facilitate increased ingestion of *Francisella* by macrophages or affect subsequent intracellular replication of the bacterium. Immune serum also does not render the microorganism susceptible to complement-mediated lysis. Clearance of the pathogen from infected organs is accelerated, however, in mice treated at the time of infection with this serum. How this is accomplished is not yet clear. The only discernible difference between mice infected with *Francisella* in the presence or absence of antibody is a prolonged inflammatory PMN response in the peritoneum. Of interest, however, is the fate of mice treated with immune serum and at the same time exposed to anti–IFN-γ antibodies: these animals die of fulminant disease in 5 to 7 days [49]. The mechanism underlying enhanced clearance by antibodies, as yet unidentified, has a strict requirement for IFN-γ.

B. Cellular Immune Factors

Sublethal *Francisella* infection in vivo in experimental animals results in the expansion of immune T cells of both the CD4+ and CD8+ lineages, each of which contribute to resolution of disease and transfer of protection [50]. Elimination of one or the other subtype is insufficient to abrogate protection, but elimination of both renders the animal susceptible to lethal disease [50]. In addition to T lymphocytes, two cytokines, IFN-γ and TNF-α, present shortly after the infection with *Francisella* determine the ultimate fate of the animal. It should be noted that both CD4+ and CD8+ cells can produce IFN-γ [51]. Administration of antibodies directed against either of these immunoregulatory factors completely blocks the development of immunity to reinfection in mice inoculated intradermally with *Francisella,* and the mice die at the same time, and with the same pathology, as mice inoculated intraperitoneally [46]. These cytokines are most important in the first 2 days of infection [46], and the actual events initiated during this time that are dependent on the presence of the cytokines (influences on antigen presentation by macrophages, cytotoxic activity, etc.) are not yet clear. In passive-transfer studies, immunity to challenge can be transferred to naive recipients with splenic or lymph node T cells from immune mice [8,46]. Again, anticytokine antibodies directed against either IFN-γ or TNF-α completely abrogate passive protection [46]. This protective immunity is clearly derived from the extrinsic interactions between T cells and cells present at the site of infection, and not from direct cytotoxic effects of the T cells themselves.

Interferon-γ and TNF-α influence a number of cellular functions in PMNs and macrophages, but one in which they clearly cooperate is the induction of synthesis of NO, the effector molecule used by activated macrophages to limit intracellular replication of *Francisella* [32,43]. We explored the nonspecific effects of activated macrophages and NO in vivo using BCG [7]. Mice inoculated with BCG and then infected intraperitoneally with *Francisella* 8 days later survive this normally lethal infection. Survival correlates with the secretion of urinary nitric oxides, and macrophages recovered from BCG-treated animals produce NO and actively kill intracellular *Francisella* in vitro [7]. Treatment of mice with anti–IFN-γ or anti–TNF-α antibodies at the time of BCG inoculation (day 0) or at the time of *Francisella* infection (day 8) totally abrogates BCG-induced protection. Further, macrophages recovered from BCG-infected mice treated with the anticytokine antibodies are not activated to kill *Francisella* in vitro. Thus, the nonspecific protection afforded by BCG, like antigen-specific protection, is dependent on IFN-γ and TNF-α. Whether the protective mechanisms induced by specific and nonspecific immunologic events are similar (i.e., whether both induce activated macrophages that are responsible for elimination of the pathogen) is currently being analyzed.

IV. CONCLUSIONS

To identify therapies for infectious agents, particularly intracellular pathogens, one must have a firm grasp of the biologic idiosyncrasies of the effector cell–target interaction. Through its efforts to maintain a homeostatic balance and its ubiquitous presence in tissues, the macrophage unwittingly becomes the target of many intracellular pathogens. Pathogens have developed diverse strategies for intracellular survival in these cells, and it is clear that host-target interactions are refined and diversified by pathogen-specific characteristics. That *Francisella* replicates in all types of macrophages suggests that it too has developed a number of evasive strategies for survival in these inhospitable cells. Elucidation of these strategies will identify critical areas to explore in our experimental models of tularemia.

All evidence to date suggests that the macrophage is a pivotal cell in the pathology and resolution of infections with *Francisella*. Any attempt to modify the course of disease with this pathogen will have to take into consideration the nature of *Francisella* interaction with this cell.

REFERENCES

1. Tarnvik A. Nature of protective immunity to *Francisella tularensis*. Rev Infect Dis 1989; 11:440–51.
2. Statistics of infectious diseases of interest to dermatologists. In: Annual Review of Dermatology. Vol. 10. Academic Press, 1991:13. (Compiled by CDC, Atlanta, GA.)
3. Chamberlain RE. Evaluation of live tularemia vaccine prepared in a chemically defined medium. Appl Microbiol 1965; 13:232–5.
4. White JD, Rooney JR, Prickett PA, Derrenbacher EB, Beard, CW, Griffith WR. Pathogenesis of experimental respiratory tularemia in monkeys. J Infect Dis 1964; 114:277–83.
5. Long GW, Oprandy JJ, Narayanana RB, Fortier AH, Nacy CA. Detection of *Francisella tularensis* in blood by polymerase chain reaction. J Clin Micro 1993; 31:152–4.
6. McCrumb FR Jr. Aerosol infection of man with *Pasteurella tularensis*. Bacterial Rev 1961; 25:262–7.
7. Green SJ, Nacy CA, Schreiber RD, Granger DL, Crawford RM, Meltzer MS, Fortier AH. Administration of monoclonal antibodies to IFN-γ and TNF-α inhibits nitric oxide synthase mRNA, production of reactive nitrogen intermediates and protection against *Francisella tularensis* infections in BCG-treated mice. Infect. Immun. 1993; 61:689–98.
8. Fortier AH, Slayter MV, Ziemba R, Meltzer MS, Nacy CA. Live vaccine strain of *Francisella tularensis*: infection and immunity in mice. Infect Immun 1991; 59:2922–8.
9. Elkins KL, Winegar RK, Nacy CA, Fortier AH. Introduction of *Francisella tularensis* at skin sites induces resistance to infection and generation of protective immunity. Microbial Pathogenesis 1992; 13:417–21.
10. Davis MW, Allison AC. Effect of cytochalasin B on endocytosis and exocytosis. In: Tannenbaum SW, ed. Cytochalasins: biochemical and cell-biological aspects. Amsterdam: Elsevier/North-Holland Biomedical Press, 1978:143–60.
11. Finlay BB, Falkow S. Comparison of the invasion strategies used by *Salmonella cholerae suis, Shigella flexneri,* and *Yersinia enterocolitica* to enter cultured animal cells: endosome acidification is not required for bacterial invasion or intracellular replication. Biochimie 1988; 70:1089–99.
12. Kuhn M, Kathariou S, Goebbel W. Hemolysin supports survival but not entry of the intracellular bacterium *Listeria monocytogenes*. Infect Immun 1988; 56:79–82.
13. Gregory WW, Byrne GL, Gardner M, Moulder JW. Cytochalasin B does not inhibit ingestion of Chlamydia psittaci by mous fibroblasts (L cells) and mouse peritoneal macrophages. Infect Immun 1979; 25:463–9.
14. Ward ME, Murray A. Control mechanisms governing the infectivity of *Chlamydia trachomatis* for HeLa cells: mechanisms of endocytosis. J Gen Microbiol 1984; 130:1765–80.
15. Walker TS. Rickettsial interactions with human endothelial cells in vitro: adherence and entry. Infect Immun 1984; 44:205–10.
16. Murray HW. Pretreatment with phorbol myristate acetate inhibits macrophage activity against intracellular protozoa. J Reticuloendothel Soc 1982; 31:479–87.
17. Tilney LG, Portnoy DA. Actin filaments and the growth, movement and spread of the intracellular bacterial parasite, *Listeria monocytogenes*. J Cell Biol 1989; 109:1597–1608.
18. Jones TC, Hirsch JG. The interaction between *Toxoplasma gondii* and mammalian cells. II. The absence of lysosomal fusion with phagocytic vacuoles containing living parasites. J Exp Med 1972; 136:1173–94.
19. Armstrong JA, Hart PD. Response of cultured macrophages to *Mycobacterium tuberculosis,* with observations on fusion of lysosomes with phagosomes. J Exp Med 1971; 134:713–40.

20. Chang KP, Dwyer DM. Multiplication of a human parasite *(Leishmania donovani)* in phagolysosomes of hamster macrophages *in vitro*. Science 1976; 193:678–80.
21. Channon JY, Blackwell JM. A study of the sensitivity of *Leishmania donovani* promastigotes and amastigotes to hydrogen peroxide. II. Possible mechanisms involved in protective H_2O_2 scavenging. Parasitology 1985; 91:207–13.
22. Anthony LSD, Burke RD, Nano FE. Growth of *Francisella* spp. in rodent macrophages. Infect Immun 1991; 59:3291–6.
23. Fortier AH, Nacy CA, Meltzer MS. Growth of *Francisella tularensis* Live Vaccine Strain (LVS) in macrophages requires a low pH compartment. J Leukocyte Biol 1992; (s3):28.
24. Mehlman I, Fuchs R, Helenius A. Acidification of the endocytic and exocytic pathways. Annu Rev Biochem 1986; 55:663–700.
25. Traub A, Mager J, Grossowicz N. Studies on the nutrition of *Pasteurella tularensis*. J Bacteriol 1955; 70:60–9.
26. Brezkorovainy A, Zschocke RH. Structure and function of transferrins. Arzneimitttelforsch 1974; 24:476–85.
27. Dautry-Varsat A, Lodish HF. How receptors bring proteins and particles into cells. Scientific American 1984; 250:52–8.
28. Czuprynski C, Henson PM, Campbell PA. Killing of *Listeria monocytogenes* by inflammatory neutrophils and mononuclear phagocytes from immune and nonimmune mice. J Leukocyte Biol 1984; 35:193–208.
29. Lofgren S, Tarnvik A, Thore M, Carlsson J. A wild and an attenuated strain of *Francisella tularensis* differ in susceptibility to hypochlorous acid: a possible explanation of their different handling by polymorphonuclear leukocytes. Infect Immun 1984; 43:730–4.
30. Cohn ZA. The maturation and activation of mononuclear phagocytes: fact, fancy, and future. J Immunol 1978; 121:813–4.
31. Nacy CA, Oster CN, James SL, Meltzer MS. Activation of macrophages to kill *Rickettsiae* and *Leishmania*: dissociation of intracellular microbicidal activities and extracellular destruction of neoplastic and helminth targets. Cont Topics Immunobiol 1984; 13:147–70.
32. Fortier AH, Polsinelli T, Green SJ, Nacy CA. Activation of macrophages for destruction of *Francisella tularensis*: identification of cytokines, effector cells, and effector molecules. Infect Immun 1992; 60:817–25.
33. Byrd T, Horwitz M. Interferon gamma–activated human monocytes down-regulate transferrin receptors and inhibit the intracellular multiplication of *Legionella pneumophila* by limiting the availability of iron. J Clin Invest 1989; 83:1457–65.
34. Carlin JM, Borden EC, Byrne GI. Interferon-induced indoleamine 2,3-dioxygenase activity inhibits *Chlamydia psittaci* replication in human macrophages. J Int Res 1989; 9:329–37.
35. Murray H, Szuro-Sudol A, Wellner D, Oca J, Granger A, Libby D, Rothermel C, Rubin B. Role of tryptophan degradation in respiratory burst–independent antimicrobial activity of gamma interferon–stimulated human macrophages. Infect Immun 1989; 57:845–9.
36. Sasada M, Johnston Jr RB. Macrophage microbicidal activity: correlation between phagocytosis-associated oxidative metabolism and the killing of candida by macrophages. J Exp Med 1980; 152:85–91.
37. Green SJ, Meltzer MS, Hibbs JB, Nacy CA. Activated macrophages destroy intracellular *Leishmania major* amastigotes by an L-arginine–dependent killing mechanism. J Immunol 1990; 144:278–83.
38. James SL, Glaven J. Macrophage cytotoxicity against schistosomula of *Schistosoma mansoni* involves arginine-dependent production of reactive nitrogen intermediates. J Immunol 1989; 143:4208–14.
39. Green SJ, Nacy CA, Meltzer MS. Cytokine-induced synthesis of nitrogen oxides in macrophages: a protective host response to Leishmania and other intracellular pathogens. J Leukocyte Biol 1991; 50:93–103.
40. Denis M. Interferon-gamma–treated murine macrophages inhibit growth of tubercle bacilli via the generation of reactive nitrogen intermediates. Cell Immunol 1991; 132:150–7.
41. Granger DL, Hibbs JB, Prefect JR, Durack DT. Metabolic fate of L-arginine in relation to microbistatic capability of murine macrophages. J Clin Invest 1990; 85:264–8.

42. Green LC, Wagner DA, Glogowski J, Skipper PL, Wishnok JS, Tannenbaum SR. Analysis of nitrate, nitrite, and ^{15}nitrate in biological fluids. Anal Biochem 1982; 126:131–4.
43. Anthony LSD, Morrissey PJ, Nano FE. Growth inhibition of *Francisella tularensis* live vaccine strain by IFN-γ-activated macrophages is mediated by reactive nitrogen intermediates derived from L-arginine metabolism. J Immunol 1992; 148:1829–34.
44. Lancaster JR Jr, Hibbs JB Jr. EPR demonstration of iron-nitrosyl complex formation by cytotoxic activated macrophages. Proc Natl Acad Sci USA 1990; 87:1223–8.
45. Hibbs JB Jr, Vavrin Z, Taintor RR. L-arginine is required for expression of the activated macrophage effector mechanism causing selective metabolic inhibition in target cells. J Immunol 1987; 138:550–9.
46. Leiby DA, Fortier AH, Crawford RM, Schreiber RD, Nacy CA. In vivo modulation of the murine immune response to *Francisella tularensis* (LVS) by administration of anti-cytokine antibodies. Infect Immun 1992; 60:84–9.
47. Anthony LSD, Ghadirian E, Nestel FP, Kongshavn PAL. The requirement for gamma interferon in resistance of mice to experimental tularemia. Microbial Pathogenesis 1989; 7:421–8.
48. Rhinehart TR, Fortier AH, Elkins KL. Characterization of the protective antibody response to *Francisella tularensis* strain LVS in mice. Abstracts from the 92nd General Meeting of the American Society for Microbiology 1992.
49. Jones TR, Elkins KL, Nacy CA, Fortier AH. Protection against lethal *Francisella tularensis* (LVS) infection by passive transfer of LVS-immune serum requires endogenous IFN-γ production by the recipients. J Leukocyte Biol 1992; (s3):34.
50. Fortier AH, Nelson BJ, Elkins KL. Cellular immunity to *Francisella tularensis* in mice. J Leukocyte Biol 1991; (s2):66. (Abstract.)
51. Fortier AH, Nacy CA, Sitkovsky MV. Similar molecular requirements for antigen receptor–triggered secretion of interferon and granule enzymes by cytolytic T lymphocytes. Cell Immunol 1989; 124:64–76.

22

Macrophages and *Brucella*

Cynthia L. Baldwin

The Ohio State University, Columbus, Ohio

Alexander J. Winter

Cornell University, Ithaca, New York

I. INTRODUCTION

A. The Organism

The genus *Brucella* is composed of facultative intracellular bacteria that produce chronic infections in humans and animal species [1–4]. Brucellae are gram-negative non-spore-forming coccobacillary rods with a cell wall that is characteristic of gram-negative bacteria but possessing certain distinctive features in the noncovalent bonding forces that bridge the lipopolysaccharide (LPS), outer membrane proteins, and peptidoglycan [5]. The major outer membrane proteins of *B. abortus* are homologous to those of the Enterobacteriaceae [6], whereas the lipid A differs in some essential respects from that of *Escherichia coli* [7], and the O-polysaccharide has been characterized as a homopolymer of 4,6-dideoxy-4-formamido-α-D-mannopyranosyl residues [7,8]. The genus *Brucella* includes species, e.g., *B. abortus*, *B. melitensis*, and *B. suis*, in which smooth strains are virulent, while rough strains, which are deficient in or lack the O-polysaccharide side chain of the LPS, are either attenuated or avirulent. In other species of *Brucella*, e.g., *B. ovis* and *B. canis*, virulent strains are naturally occurring rough forms that lack the O-polysaccharide.

Since brucellae infect and multiply in professional [9–13], as well as in nonprofessional [14], phagocytes and localize in macrophages [15,16], survival in macrophages may be the basis for establishment of chronic infections. The liver and spleen are effective at clearing brucellae from the blood [15–17], and brucellae are found localized in organs that are rich in reticuloendothelial system elements, including the spleen, liver, bone marrow, and lymph nodes [2,18]. It has been shown in vitro that murine macrophages kill intracellular brucellae during the first 12 to 24 hours after infection, but thereafter the surviving bacteria multiply [12,13]. Similar events may occur in vivo, resulting in establishment of chronic infections even in the face of macrophage activation. Early studies by Mackaness [19] showed that while infection with brucellae clearly resulted in activation of macrophages, demonstrated by the ability of brucella-infected mice to clear superinfections with *Listeria*, the infection nevertheless persisted.

B. Epidemiology and Pathology

Brucellosis in humans was originally described in 1894 by Bruce on the island of Malta as Mediterranean (undulant) fever. Today it occurs throughout the world. Human disease is caused, in order of virulence, by *B. melitensis*, *B. abortus*, *B. suis*, or *B. canis* [20,21]. Brucellosis in humans is a debilitating disease with diverse pathological manifestations including fever, chills, sweats, weakness, malaise, body ache, and anorexia [3]. Chronic brucellosis may also lead to endocarditis [22,23], arthritis [24], meningitis [25–27], osteoarticular complications [28], pancreatitis [29], pulmonary empyema [30], spondylitis [31], and neurological involvement [32]. Untreated brucellosis in humans caused by *B. melitensis* may be fatal.

Brucella species also infect a large number of animal species, including bison, buffalo, camels, cattle, dogs, elk, goats, horses, pigs, and sheep [2]. Because of the propensity of the organism to infect agricultural species of animals, zoonosis is an important component of human brucellosis [21]. Increases in the incidence of human brucellosis in the United States and Kuwait have been documented as correlating directly with increases in brucellosis in domestic animals [1,2,21]. Zoonotic infections occur from contaminated dairy products and during the parturition or slaughter of infected animals [20,21,33,34]. In addition, infection of pregnant cows may result in abortion, whereupon large numbers of bacteria are excreted, creating a tremendous source of infection [2]. Currently, human infections are caused primarily by *B. melitensis* and have reached epidemic proportions in Middle Eastern countries. Sheep, goats, and camels are the main sources of human infection with *B. melitensis* [20]. Bovine brucellosis due to infection with *B. abortus* has previously occurred worldwide [20], and before compulsory pasteurization of cows' milk it presented a significant human health problem in the United States. *B. abortus* no longer represents a threat to human health in most parts of the world, although it does result in economic losses to the cattle industry. *B. suis* presents a threat to human health in Indonesia and possibly other Asian countries, where infection occurs in swine. The precise incidence of human disease due to brucellae is unreported but is estimated to range from less than 0.1 case/100,000 people in the United States to over 100 cases/100,000 people in the Middle East [35].

C. Control of Infection

There is no widely accepted vaccine against brucellosis for use in humans, although populations deemed to be at great risk have been vaccinated with live attenuated *B. abortus* strain 19 [36]. Strain 19 is an attenuated strain of *B. abortus* used for vaccinating cattle throughout the world. Strain 19 is not pathogenic in nonpregnant cattle, but it can cause human disease [37,38]. Its efficacy for inducing protective immunity in people is unestablished. Chemotherapy for people with brucellosis relies on large doses of antibiotics, such as combinations of tetracycline and streptomycin [3,39,40].

Because of the zoonotic aspects of brucellosis and because there is no widely accepted vaccine for humans [41], control of brucellosis in agricultural animals is important for its control in people. Currently, control in animals relies on vaccination with *B. abortus* strain 19 or *B. melitensis* strain Rev-1 [21,42] and serological diagnosis of infected animals [43,44]. Strain 19 vaccine is used worldwide and has a 65% to 75% efficacy rate in cattle, measured in terms of preventing abortion. The efficacy of Rev-1 has been demonstrated by trials in China [42] and more recently in Kuwait [21]. An important shortcoming of vaccination, in addition to its limited effectiveness, lies in compromising identification of animals infected with virulent field strains. Vaccination stimulates production of antibodies that cross-react with field strains. The development of serological tests that distinguish field strain–infected cattle from vaccinated cattle has been a major focus of research in recent years, and significant progress has been made in this area [43,44]. Infection of wild animals [45–48] may also contribute to the persistence of the disease, although wildlife are more often infected from domestic animals than the reverse [2]. In developing and Middle Eastern countries, control of zoonotic infections is hampered by traditions of food consumption, such as nonpasteurization of milk and cheese, and by lack of veterinary services [21].

D. Acquired Immunity to *Brucella abortus*

The great majority of animals that abort following *B. abortus* infection do so only once [2]. This reflects the acquisition of an immune response that controls the proliferation and spread of the brucellae, since most animals remain infected in the mammary gland, the supramammary lymph nodes, and various other lymphoid tissues [2]. Using inbred strains of mice, acquired resistance to *B. abortus* has been shown to involve both antibodies and T cells. Cheers and Ho [49] have shown that antibodies to *B. abortus* facilitate the clearance of the organism during the first hours after infection. Antibodies, however, did not influence the rate of growth of uncleared bacteria in vivo, nor did they confer resistance to susceptible mice [49]. Winter and his colleagues have demonstrated in mice that protective antibodies are specific for the O-polysaccharide of the LPS [50,51]; that antibodies are significantly more protective against the vaccine strain 19 than against virulent strain 2308 [52]; and that antibodies alone can prevent the establishment of infection with a virulent strain if the challenge dose was sufficiently low [53]. The specificity of protective antibodies for the O-polysaccharide has been amply confirmed [54,55].

Some disagreement exists on the respective roles of CD4 and CD8 T cells in conferring protective immunity. It was concluded from passive transfer and in vivo depletion experiments in CBA mice that immunity was provided only by CD8 T cells [56], and the same conclusion was reached by Mielke [57] through depletion experiments in BALB/c mice. However, Araya et al. [58] demonstrated with purified populations of T cells in BALB/c mice that CD4 and CD8 T cell populations transferred equivalent levels of protection independently of each other, and that protection provided by CD4 T cells could not be ascribed to a helper function for B cells.

E. Innate Resistance and Susceptibility to Brucellosis

In inbred mice, resistance to the attenuated *B. abortus* strain 19 occurs in the *Bcg/Ity/Lsh*-susceptible mouse strains C57B1/10 and BALB/c, while susceptibility to brucellosis occurs in the *Bcg/Ity/Lsh*-resistant CBA mice (Figure 1) [59]. In mice, susceptibility refers to establishment of a protracted infection and resistance refers to the ability to clear the infection. Cheers and colleagues showed that resistance is a multigenic trait manifested by bone marrow cells [59]. CBA mice could

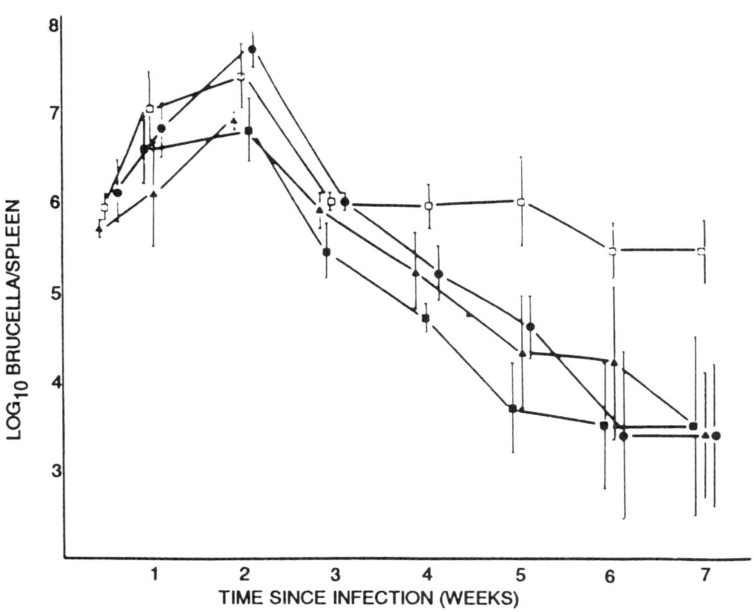

FIGURE 1 Growth of *Brucella abortus* strain 19 in the spleens of CBA/H (open squares), BALB/c (closed circles), C57B1/10 (closed triangles), and B10Br (closed squares) mice. The mice were injected intravenously with 5×10^5 *B. abortus* organisms. Each point represents the geometric mean for five mice; vertical bars—standard deviations. (From Ref. 59.)

be rendered resistant by transfer of bone marrow cells from the innately resistant B10.Br mice [59]. B10.Br mice are matched at the major histocompatibility complex with CBA mice (H-2^k) but have a C57B1/10 background. Hence, resistance is mediated by cells of the lymphohemopoietic lineage, but it does not correlate with expression of major histocompatibility complex antigens.

Strains of mice can also be categorized as resistant or susceptible in their response to challenge with virulent *B. abortus* strain 2308. BALB/c mice are susceptible to infection with strain 2308, while C57B1/10 mice are relatively resistant (Figure 2) [60].

1. Temporal Expression of Resistance

Expression of resistance to *B. abortus* strain 19, the attenuated vaccine strain, in mice did not correlate in an obvious fashion with events that control bacterial replication during the first 7 days after infection [59], since numbers of *B. abortus* in the spleens and livers of susceptible and resistant mice were similar at that time (Figure 1). Furthermore, at the peak of infection 2 weeks after inoculation of strain 19, the number of bacteria in the spleens of the susceptible CBA mice was similar to that in resistant BALB/c mice [59]. BALB/c mice, however, subsequently cleared the infection, while the number of brucellae in the spleens and livers of susceptible mice remained at a plateau level for several weeks. As a result, by 7 weeks after infection the CBA mice had 100 times more bacteria in their spleens than did the resistant mice [59]. The bacteria that persisted in chronically infected mice were not more resistant to in vivo clearance. When they were used to infect naive mice they were cleared as rapidly as bacteria derived from the parent stock [61].

When mice are infected with the virulent *B. abortus* strain 2308, resistance correlates with

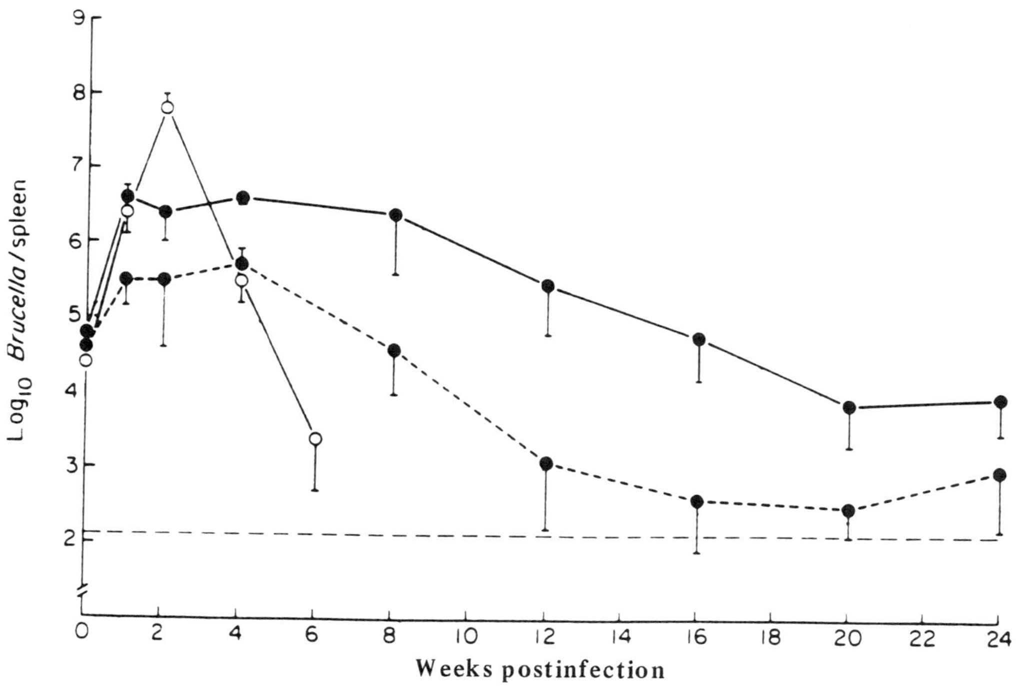

FIGURE 2 Course of infection by *B. abortus* 2308 (closed circles) and 19 (open circles) in spleens of BALB/c (solid lines) and C57B1/10 (broken line) mice. Bars indicate standard deviations. Dashed line at $\log_{10}$ 2.12 represents the limit of detection of organisms in the spleen. (From Ref. 60.)

lower numbers of bacteria in the spleens of resistant mouse strains from the onset of infection and again 4 weeks after infection (Figure 2) [60]. During the first week after infection, resistant C57B1/10 mice control the replication of brucellae more effectively than the susceptible BALB/c mice. Between 1 and 4 weeks the number of organisms in the spleens of both strains of mice does not change, but after 4 weeks the C57B1/10 mice clear the organisms more rapidly than the BALB/c mice.

2. *Immune Suppression and Suppressor Macrophages*

At 2 to 3 weeks after infection of mice with strain 19, high concentrations of suppressor macrophages are found in the spleen [62,63]. There is strong evidence that these macrophages originate from precursors in the bone marrow and spleen that proliferate in response to colony-stimulating factors secreted during the course of *B. abortus* infection [64,65]. Suppressor activity can be demonstrated by the ability of the macrophages to inhibit mitogen-stimulated [62,63] or antigen-stimulated [63] proliferative responses in vitro. The highest concentration of suppressor macrophages occurs in the spleen in direct association with peak numbers of brucellae [58,62]; drastic reductions in the number of detectable T cells (particularly CD4 cells) [58,63]; a profound depletion of periarteriolar lymphoid sheaths [66]; and an absence of demonstrable protective T cells [58]. Although these suppressor cells did not cause generalized immunosuppression [67] or even interfere with enhanced killing of *Listeria monocytogenes* within the spleen [58,67], they may lack brucellicidal capabilities and may delay the generation of a protective cell-mediated immune response by contributing to the depletion of splenic T cells [58,63,66]. A number of products secreted by activated macrophages also suppress T cell proliferation, including reactive oxygen intermediates [68], reactive nitrogen intermediates [69], prostaglandin E-2 [70], and transforming growth factor-β [71].

II. SURVIVAL OF BRUCELLAE IN MACROPHAGES

A. Relationship to Virulence in Vivo

Peritoneal macrophages from guinea pigs were used in early studies to demonstrate a correlation between the virulence of strains of *B. abortus* and their ability to survive in macrophages. Braun et al. [72] found that rough strains of *B. abortus* were ingested by guinea pig macrophages more rapidly than smooth strains but grew more slowly once inside the cells. However, Freeman and Vana [9] were unable to find differences in the growth rates of rough and smooth strains of *B. abortus* in guinea pig macrophages. Since that time, a number of studies have shown that the virulence of *B. abortus* in vivo and its ability to survive in macrophages in vitro correlated inversely with the degree of "roughness" of the strain (i.e., the length of the O-polysaccharide side chain). In BALB/c mice, the smooth *B. abortus* strain 2308 establishes a chronic infection, whereas the semi-rough mutant strain 19 is cleared much more rapidly [60]. In accordance with this, Jones and Winter have shown that resident peritoneal macrophages from BALB/c mice eliminated intracellular strain 19 much more effectively than strain 2308 [12]. Furthermore, contrary to earlier results [72], they found that following opsonization the virulent smooth strain 2308 was more readily phagocytosed by macrophages than was strain 19 [12].

An early study employing a mixed population of bovine phagocytes (peripheral blood monocytes and neutrophils) conducted by Smith and Fitzgeorge [73] also demonstrated that smooth strains of *B. abortus* were more resistant to killing by phagocytes than rough strains. Subsequent investigators substantiated these results with both purified populations of bovine macrophages [11,74] and neutrophils [75]. Bacteria were opsonized with antibody in the studies employing bovine macrophages. The same correlation occurred with human neutrophils infected with smooth and rough strains of *B. abortus* [75,76].

In addition, Young et al. [76] found a correlation between killing by neutrophils and the degree of "roughness" among strains of *B. melitensis*. Thus, human neutrophils were not able to kill a virulent strain of *B. melitensis* isolated from the blood of a patient, but they did kill the attenuated strain *B. melitensis* Rev-1. Moreover, they found a correlation between species of *Brucella* with regard to virulence and killing by human neutrophils. Neutrophils were able to kill a smooth strain of *B. abortus* isolated from the blood of a patient, but not an isolate of *B. melitensis* [72], which is more virulent in humans than *B. abortus*.

B. Mechanism of Survival in Phagocytes

1. *Inhibition of Host Cell Function and Virulence Factors*

A number of studies have been conducted to elucidate the mechanism or mechanisms by which *Brucella* resists killing by professional phagocytes and to isolate the bacterial components responsible. While some of these studies have employed neutrophils, the results are applicable for understanding survival of brucellae in macrophages, since antimicrobial mechanisms of neutrophils and macrophages are overlapping, e.g., phagolysosomal fusion, oxidative burst, and the myeloperoxidase-hypohalide system. Using neutrophils, it has been shown that virulent strains of *B. abortus* inhibit both the myeloperoxidase–hydrogen peroxide–halide system (by inhibiting degranulation [75,77]) and the oxidative burst [78]. Canning et al. showed that the cyclic nucleotides adenine and guanosine-monophosphate of *B. abortus* strain 2308 were responsible for inhibition of neutrophil degranulation and the myeloperoxidase–hydrogen peroxide–halide system, but not for inhibition of the oxidative burst [79,80]. Studies conducted with macrophages also have shown that *B. abortus* strain 2308 (but not attenuated strain 19) inhibited phagolysosomal fusion [81]. Strain 2308 was also directly toxic to macrophages [81], suggesting that survival strategies may include destruction of the host cell. Whereas Hinsdill and Berman [82] isolated components of smooth *B. abortus* strain 2308 that were directly toxic to guinea pig macrophages, Braun et al. [72] found that rough strains of *B. abortus* were more toxic to macrophages than smooth strains. This does not rule out macrophage toxicity as a general survival mechanism, but it indicates that this property is unlikely to contribute to the differences observed between rough and smooth strains of *B. abortus*. It is noteworthy that Frenchick et al. [81] found that the water-soluble component of strain 2308 that was responsible for inhibiting phagolysosomal fusion in murine peritoneal macrophages was absent from attenuated strain 19.

Smooth LPS has been proposed as a virulence factor for *B. abortus*, on the basis of the association of virulence and intracellular survival with the presence of the O-polysaccharide side chain (see above). In addition, Riley and Robertson showed that smooth strains were more resistant than rough strains to killing by extracts of bovine neutrophil granules [83], and Soto et al. [84] recently demonstrated that smooth LPS from *B. abortus* or *B. melitensis* enhanced the intracellular survival of the rough mutant *B. abortus* strain 1119 in bovine neutrophils. Lipopolysaccharide is unlikely to exert this effect by inhibiting the oxidative burst or phagolysosmal fusion, since Kreutzer et al. [78] showed that the component of *B. abortus* responsible for inhibiting the oxidative burst was not contained in the cell wall, and Frenchick et al. [81] found that LPS did not inhibit phagolysosomal fusion. Rasool and colleagues [85] argued that smooth LPS was unlikely to be a virulence factor, given the fact that both smooth and rough strains of *B. abortus* have equivalent amounts of lipid A, the biological mediator of many LPS effects, and that in some species of *Brucella* (*B. ovis*, *B. canis*) fully virulent strains are rough and so lack O-polysaccharide side chains. Rasool et al. [85] found that the lipid A of *B. abortus* is a poor inducer of superoxide and lysozyme by neutrophils, compared with lipid A of *Salmonella*, and proposed that this property might be advantageous for the intracellular survival of the genus *Brucella*.

Recent evaluations of an LPS-deficient transposon mutant of strain 2308 indicated that LPS was not the only factor correlated with survival in macrophages, as demonstrated by the ability of a rough mutant of the virulent strain 2308 to survive in bovine macrophages [86]. Two-dimensional gel electrophoretic analyses of the LPS-deficient *B. abortus* mutants performed by Sowa et al. [87] demonstrated that the mutants differed from the parent strains by 28 proteins, some of which may be involved in LPS biosynthesis. Fourteen proteins were identified as potential virulence components by paired comparisons between virulent and avirulent strains [87]. Proteins of *Brucella* that may have a role in enhancing survival within phagocytes include superoxide dismutase (SOD), the genes for which have been cloned [88], and catalase [89]. Superoxide dismutase and catalase are presumably involved in protecting brucellae from superoxide and hydrogen peroxide killing, respectively. Although Fitzgeorge et al. [90] demonstrated a correlation of virulence of *B. abortus* strains with their resistance to hydrogen peroxide killing, they could not demonstrate such a correlation for *B. melitensis* or *B. suis*. Perhaps amounts of SOD and catalase differ among strains of *B. abortus* but not among strains of other *Brucella* species. A phosphomonoesterase of *B. abortus*, which has intense acid phosphatase activity, has also been evaluated for its role in blocking production of superoxide anion [91]. Although an acid phosphatase of *Legionella* had been shown to have this function [92], the *B. abortus* acid phosphatase did not. Heat-shock proteins, involved in bacterial resistance to host killing, have also been cloned from *Brucella* [93–95]. In *Brucella* synthesis of heat-shock proteins is not initiated in response to oxidative stress, but these proteins do have an extended expression period relative to those of other bacteria, suggesting an important physiological role [95].

2. *Intracellular Localization*

B. abortus has been shown to localize in the phagosomes and rough endoplasmic reticulum of trophoblasts during in vivo infections as well as after infection of trophoblasts in vitro [96,97]. Using Vero cells [14,98], evidence was obtained that brucellae are transported to the lumen of the rough endoplasmic reticulum from phagosomes. Localization of *B. suis* and *B. melitensis* also occurred within the rough endoplasmic reticulum [98] and was necessary for large-scale intracellular replication of brucellae [14]. Other intracellular pathogens have been described in ribosome-studded vesicles, although the endosomal compartment has not been designated as rough endoplasmic reticulum [99]. In macrophages and neutrophils, brucellae have been observed only in phagosomes or phagolysosomes [74,100,101]. It remains to be established whether translocation into the rough endoplasmic reticulum may under some circumstances occur within these cell types. In macrophage phagosomes, brucellae become surrounded by fibronectin secreted from nonlysosomal vesicles [103]. This is a property peculiar to *Brucella*, but its relevance to virulence or to host defense is unknown.

C. Relationship of Innate Host Resistance to Macrophage Infection

Using guinea pig macrophages, Stinebring and Kessel [104] found that monocytes from innately resistant animals restricted the multiplication of intracellular *B. abortus*. Cattle have also been identified that are innately resistant to brucellosis, as determined by their failure to abort when infected during pregnancy [11]. Eighty percent of resistant cattle had macrophages that inhibited intracellular replication of *B. abortus*, whereas the percentage of cattle with "nonpermissive" macrophages in a susceptible population was much lower [11,102]. It has not been established whether resistance and susceptibility among inbred mouse strains correlate with the ability of their macrophages to regulate intracellular growth of brucellae in macrophages.

D. Influence of Opsonization

It has been documented that efficient phagocytosis of *B. abortus* by macrophages requires opsonization by antibody [12,53,74,76]. Young et al. [76] found that *B. abortus* also had to be opsonized to be phagocytosed by human neutrophils, although Riley and Robertson reported that antibodies are unnecessary [83]. Since antibody is produced in response to infection with *B. abortus* and is involved in clearance [49,52,53,60], in vitro evaluations made with antibody-opsonized *B. abortus* are likely to reflect events that occur in vivo after formation of antibodies. Opsonization of brucellae may also alter their ability to survive intracellularly. Early studies by Ralston and Elberg showed that immune serum from rabbits contained "macrophage-stimulating factor," which mediated more efficient ingestion of *B. melitensis* Rev 1 by macrophages from both normal and immune animals [105,106]. Moreover, they found that treatment of both naive and immune macrophages with the immune serum increased their ability to control replication of intracellular brucellae. Treatment of naive macrophages with the antiserum rendered them as effective in controlling intracellular growth of brucellae as were macrophages from immune rabbits that had not been treated with antiserum [106]. This effect was presumably due to cytophilic antibody. Harmon et al. [74] also found that both the virulent *B. abortus* strain 2308 and the rough mutant 45/20 stimulated an oxidative burst in bovine mammary gland macrophages when the bacteria were opsonized with antibody, and that opsonized strain 2308 could no longer inhibit phagolysosomal fusion. Opsonization did not, however, alter the relationship between virulence and the ability to survive in either macrophages or neutrophils [12,74,76], nor did it alter the relative ability of macrophages from resistant and susceptible animals to control intracellular brucellae [11].

III. ROLE OF CYTOKINES IN MEDIATING ANTI-*BRUCELLA* ACTIVITIES

A. Activation of Macrophages in Vivo

Macrophages that are activated during *Brucella* infections have enhanced anti-*Brucella* activity. It was noted in early studies conducted in guinea pigs infected with either *B. abortus* [107] or *B. suis* [108], in cattle infected with *B. abortus* [109], and in rabbits infected with *B. melitensis* [105,106,110] that macrophages from infected animals were better at killing and controlling replication of intracellular brucellae than macrophages from uninfected animals. Cheers and Pagram [61] found that macrophages in chronically infected CBA mice had an increased antimicrobial activity compared with macrophages from uninfected controls, as shown by enhanced killing of *Listeria monocytogenes* in vivo. This activity began to wane by 21 days and returned to baseline by 42 days after infection [61]. Using *Brucella*-resistant C57B1/6 mice, Birmingham and Jeska [111] showed that macrophages from *B. abortus* strain 19–infected mice exhibited enhanced spreading, increased immunoglobulin G (IgG) Fc-receptor expression, and enhanced activities in phagocytosis of sheep red blood cells by C3 receptors, in chemotactic responsiveness, and in random locomotion. Monocytes taken from patients undergoing chemotherapy for *B. melitensis* infections and simultaneously being given killed brucellae intravenously were also activated as assessed by phagocytosis, bactericidal assays, and morphological characteristics [112]. Finally, Gay et al. [113] showed that production of superoxide and turnover of NADPH oxidase was higher in macrophages from *Brucella*-immunized mice, and that a higher percentage of macrophages from immunized mice phagocytosed brucella in comparison with macrophages from naive mice (95% versus 80%, respectively).

B. Cytokines That Activate Macrophages

Recent work has focused on identifying the cytokines that activate macrophages for anti-*Brucella* activity [12,13,114]. It has been shown that interferon-γ (IFN-γ) [12,13,114], and to a much lesser extent interleukin-2 (IL-2) [13], stimulate macrophages for brucellistatic activity. Other cytokines including granulocyte-macrophage colony-stimulating factor (GM-CSF), tumor necrosis factor

(TNF), IL-4, and IL-1, shown elsewhere to activate macrophages for antimicrobial activities, induced no demonstrable effects against *B. abortus* when used either alone or as cofactors with other cytokines [13]. Macrophages treated with either IFN-γ or IL-2 were able to inhibit growth of strains of *B. abortus* that differed substantially in virulence, including attenuated strain 19, virulent strain 2308, and strain RB51, a rough avirulent mutant of strain 2308 [12,13,114]. Interferon-γ was effective in inducing brucellistatic activity in resident and peptone- or thioglycolate-induced peritoneal macrophages, and in splenic macrophages [13]. Moreover, the ability of *B. abortus* to survive in IFN-γ–stimulated macrophages from BALB/c mice was directly correlated with the virulence of the strain, as it was in non-cytokine-activated macrophages [12,114].

C. Cytokines and Macrophage Recruitment in Vivo

Administration of recombinant IL-1 before infection has been shown to play a role in vivo in preventing chronic infections with *B. abortus*, as shown by fewer colony-forming units in the spleen by 14 days after infection [115]. This effect was obviated if administration was delayed for even 24 hours after infection, nor did IL-1 alter the course of an established infection [115]. It was proposed that the action of IL-1 in vivo was to induce production of colony-stimulating activity, thereby increasing neutrophils and myeloperoxidase-positive macrophages in the spleen [115]. A recent study by Doyle et al. [65] showed that if recombinant colony-stimulating factor (CSF) was given to mice before infection with *B. abortus*, the number of bacteria was not reduced at 7 days after infection. Unfortunately, the mice were not evaluated at 14 days after infection, the time when differences were apparent following administration of IL-1 [115]. Treatment of mice with recombinant CSF during the first 14 days of infection, however, resulted in an influx of macrophages into the spleen and liver, and fewer bacteria were detected in the spleen but not in the liver [65]. Together these results demonstrate a role for both IL-1 and CSF in protection but also indicate that susceptibility did not necessarily or exclusively result from a quantitative deficit in recruitment of macrophages, since the increase of macrophages in the liver was of no consequence. Doyle and co-workers [65] further suggested that macrophages recruited just before infection were not brucellicidal. This is in agreement with results of Spector et al. [116], who found that radiolabeled *B. abortus* were degraded during the first 48 hours after infection but not over the next 8 days, when monocytes would be recruited from the bone marrow.

Experiments with sublethal irradiation further support the view that initial clearance of *B. abortus* is dependent on radioresistant fixed tissue macrophages rather than on newly recruited, radiosensitive bone marrow–derived macrophages. It was demonstrated by Cheers and Waller [117] that, in contrast to murine listeriosis, in which sublethal irradiation resulted in fatal infection with *L. monocytogenes*, *B. abortus* strain 19 was cleared more effectively from irradiated CBA mice during the first week after infection. Elzer et al. [118] confirmed and extended this finding in BALB/c mice infected with virulent strain 2308. Peak resistance was manifested in the spleen at 3 weeks after infection, when numbers of brucellae were 1000 times lower than in controls [118]. Thereafter, numbers of organisms rose, so that by 5 weeks only small differences remained between spleen counts of irradiated and control groups [118]. The kinetics of infection in the livers of irradiated mice were similar, but delayed by about 1 week [118]. These reductions in bacterial numbers were not caused by aberrant patterns of localization in organs of irradiated mice and were associated with reduced numbers of circulating monocytes and of nonspecific esterase staining cells in the spleen, and with an increase in activated macrophages in the spleen (based on enhanced killing of *Listeria*) [118]. In other experiments it was demonstrated that the kinetics of infection with strain 2308 are not altered during the first 11 days in mice in which treatment with azathioprine before infection had produced a profound depletion of circulating monocytes and neutrophils [118]. To explain these findings, it was hypothesized [118] that in irradiated mice *B. abortus* infection is effectively controlled by radioresistant [119] resident macrophages, known to be activated by ionizing radiation [120,121]. Control of the infection was proposed to be highly efficient in the absence of newly recruited bone marrow monoctyes, which, in the absence of immune T cells, would fail to become activated to a brucellicidal state and would serve to harbor the organism [118]. It was suggested [118] that these macrophages might still demonstrate enhanced killing of *Listeria*, and might also represent the suppressor macrophage population characterized earlier [62] (see above).

IV. MECHANISMS BY WHICH MACROPHAGES CONTROL INTRACELLULAR BRUCELLAE

A. Non-Cytokine-Activated Macrophages

Jiang and Baldwin [122; unpubished data, 1992] recently evaluated the mechanism by which non-cytokine-activated macrophages kill brucellae during the first 24 hours after infection. Superoxide and hydrogen peroxide were involved in the brucellicidal effects, since addition of either catalase or superoxide dismutase inhibited killing. Addition of the hydroxyl radical scavengers thiourea, mannitol, and dimethylsulfoxide; N^G-monomethyl-L-arginine to inhibit nitric oxide formation; or tryptophan to overcome indoleamine 2,3-dioxygenase catabolism of tryptophan, had no effect on intracellular killing. This is in agreement with results of Gay et al. [113], which showed that phagocytosis of *B. suis* by mouse macrophages induced superoxide production. Curiously, however, they found that NADPH oxidase was rarely detected in phagosomes after phagocytosis of *Brucella*, and they suggested that perhaps reactive oxygen intermediates played a greater role in extracellular killing of brucellae than in intracellular killing [113]. In support of the role of oxygen intermediates, Gay et al. found that when the NADPH oxidase was present in the phagosome membrane it was tightly coalesced with the phagocytosed brucellae [113]. Moreover, macrophages from *Brucella*-resistant cattle also demonstrated a higher oxidative burst than those from susceptible cattle [102].

B. Mechanisms of Cytokine-Induced Anti-*Brucella* Activities

Addition of IFN-γ to macrophage cultures increased the level of initial killing of intracellular brucellae above that which occurred in non-cytokine-activated macrophages until approximately 24 hours after infection [12,13,114]. Thereafter, IFNγ-activated macrophages inhibited the replication of surviving bacteria and/or increased brucellicidal activity [12,13,114]. Interferon-γ–induced antibrucella activities could be prevented by addition of superoxide dismutase and catalase to the macrophage cultures [122]. Other inhibitors described in the preceding section as well as inhibitors of hypochlorous acid had no effect on cytokine-induced stasis, with the exception of hydroxyl radical scavengers, which in some instances inhibited antibrucella activities. How superoxide and hydrogen peroxide mediate brucellistatic or brucellicidal activities is unclear. It is possible that the appearance of stasis results from equivalent levels of bacterial death and replication. Killing by IFN-γ–activated macrophages was also inhibited by addition to the cultures of transforming growth factor-β (TGF) or antibodies that neutralized TNF [122]. Although endogenous TNF is known to be a mediator for production of nitrogen intermediates by macrophages [123], no role for reactive nitrogen intermediates in controlling intracellular replication of *B. abortus* was evident. This, therefore, suggests that its role in this instance is to mediate increased production of reactive oxygen intermediates. Since [Jiang and Baldwin, unpublished data, 1992] TGF is unable to inhibit the brucellicidal effects of non-cytokine-activated macrophages, initial killing does not appear to involve activation of macrophages by pathways utilized for cytokine-induced antibrucella activities.

C. Role of Iron in Mediating Brucellicidal Activity

Inhibition of growth of *Legionella pneumophila* in human monocytes activated with IFN-γ has been shown to result from iron starvation [124]. Inhibition of growth of *L. pneumophila* can be reversed by supplementing macrophage cultures with iron-transferrin. However, in similar experiments with *B. abortus*, Jiang and Baldwin found that iron-loaded macrophages activated with IFN-γ had a strikingly increased capacity to kill intracellular brucellae and frequently eliminated all intracellular brucellae [125]. This observation was not without precedent. It has been reported that *Borrelia burgdorferi* [126] and *Staphylococcus aureus* [127] are more susceptible to hydrogen peroxide killing following incubation of the bacteria in iron-supplemented medium. Iron-loaded *S. aureus* are also marginally more susceptible to macrophage killing [127].

Increased bactericidal activity in the presence of iron was postulated to result from the ability of iron to catalyze the Haber-Weiss reaction [127]. In the Haber-Weiss reaction hydroxyl radicals are generated from hydrogen peroxide in the presence of Fe^{2+}, which is oxidized to Fe^{3+} [128]. Superoxide and Fe^{3+} then react to generate more hydroxyl radicals and Fe^{2+}. As much as 50% of the iron/IFN-γ–mediated brucellicidal activity was blocked by addition of the hydroxyl radical scavengers mannitol and thiourea to the macrophage cultures, supporting this hypothesis [125].

V. STIMULATION OF ANTIGEN-SPECIFIC T CELLS

It is known that injection of mice with killed whole cells of *B. abortus* strain 1119, used in diagnostic tests, causes a polyclonal induction of T_{H1} cells [129,130]. Elzer et al. [131] have now demonstrated that infection of BALB/c mice with *B. abortus* strains 19 or 2308 causes a large and sustained polyclonal increase in IgG2a, as well as a bias toward the formation of IgG2a antibodies against outer membrane proteins. Thus, infection with *B. abortus* represents a potent stimulus for the formation of T_{H1} cells. Other data indicated that strain 19, but not strain 2308, may have induced a transient T_{H0} response, or simultaneous T_{H1} and T_{H2} responses, during the first 6 weeks of infection [131]. The importance of the latter property in accounting for differences in virulence between the two strains remains speculative. Araya et al. [58] showed that BALB/c mice infected with strain 19 had developed populations of protective CD4+ as well as CD8+ T cells by 4 weeks after infection. The CD4 population was presumably composed of T_{H1} cells, since they did not protect by providing help to B cells [58]. Araya and Winter [52] demonstrated by passive transfer assays that mice infected with virulent strain 2308 had also generated protective T cells by 8 weeks after infection, but the phenotypes of these cells were not established. Although protective T cell populations are thus expanded following infection of mice with *B. abortus* strains to which they are susceptible as well as resistant, such responses may be subverted in vivo during chronic infections, by mechanisms such as the suppressor macrophages described above. Jiang and Baldwin have shown that murine macrophages that are supporting exponential replication of *B. abortus* can stimulate an antigen-specific T cell proliferative response when the macrophages have been either gamma-irradiated or glutaraldehyde-fixed (X. Jiang, D. Fernanadez, and C. L. Baldwin, unpublished data, 1992). This indicates that *Brucella* antigens and costimulatory molecules are expressed on the surface of macrophages supporting growth of *B. abortus* but does not preclude production of suppressive molecules by *Brucella*-infected macrophages.

Under certain circumstances, infected macrophages may fail to express appropriate costimulatory molecules. Splitter and colleagues demonstrated that bovine monocytes that had phagocytosed acetone-killed *B. abortus* could present antigens to peripheral blood lymphocytes from *B. abortus*–immunized cattle, as measured by induction of lymphocyte proliferation [132]. When similarly pulsed monocytes were treated with IFN-γ, however, a decrease in membrane expression of IL-1 occurred, and the monocytes were unable to stimulate proliferation of lymphocytes [133]. Reduction in IL-1 was proposed as being responsible for the lack of stimulatory ability. The relevance of this phenomenon to the pathogenesis of brucellosis is unclear at this time.

VI. SUMMARY

The studies reviewed here indicate that during the course of infection with *B. abortus* protective immune responses occur [52,53,56–58,60] and that macrophages are activated for antimicrobial activity [61,111,113] in both susceptible and resistant animals [105–107,109,110]. Yet in susceptible animals chronic infections with *B. abortus* become established [60,61]. There is circumstantial evidence from experiments in vitro that the outcome of infection is related to macrophage/ *Brucella* interactions. This includes the following: 1) the virulence of strains of *B. abortus* is proportional to their ability to inhibit host phagocyte functions [75,77,78,81] and to survive in murine [12], bovine [11,74], and guinea pig [72] macrophages; and 2) the ability of guinea pig and bovine macrophages to inhibit intracellular growth of brucellae is directly proportional to the innate resistance of the host from which the macrophages were derived [11,102,104].

In the murine model, the relative resistance of C57B1/10 mice to infection with virulent *B. abortus* strain 2308 correlated with more efficient clearance during the first week following infection, compared with the susceptible BALB/c strain [60]. It is not known whether this was due to innate differences in the recruitment or bactericidal activities of macrophage populations in the two mouse strains, as demonstrated with *L. monocytogenes* [134–137], or to other nonimmune factors such as natural killer cells secreting IFN-γ and lysing infected macrophages [138] following direct interaction with the brucellae, as shown with other intracellular bacteria [139]. It is also of interest that clearance of strain 2308 following the plateau phase was slower in BALB/c than in C57B1/10 mice [60]. Since at this time protective T cells are known to be present in the spleens of BALB/c mice [52], this difference may have been due either to more efficient clearance of brucellae by C57B1/10 macrophages following activation with T cell cytokines, or to the inability of immune T cells to interact with *Brucella*-infected macrophages in the susceptible BALB/c strain. CD8 T cells have been shown to have an important role in clearance of brucellae following the peak of infection [56–58] and may act by lysing chronically infected macrophages. Although protective T cells have been demonstrated in susceptible strains of mice [52,57], chronically infected macrophages in these animals may fail to act as targets or may down-regulate T cell functions. As a result, infected macrophages could persist in the presence of *Brucella*-specific T cells. Alternatively, infected macrophages in susceptible animals may exist in immunologically privileged sites inaccessible to immune T cells, although this seems unlikely at least in the case of mice, since the majority of bacteria are found in the spleen in both resistant and susceptible animals.

The role of antibodies in modulating macrophage/*Brucella* interactions may be illustrated by the differential clearance of *B. abortus* strains 19 and 2308 in BALB/c mice. When these mice were infected with strain 19 or strain 2308, numbers of strain 19 in the spleen at 1 week after infection equaled, and at 2 weeks consistently exceeded, numbers of strain 2308 [60]. Nevertheless, strain 19 was thereafter cleared rapidly, while strain 2308 produced chronic infection [60]. Winter and co-workers proposed that during the first 2 weeks large populations of nonbrucellicidal macrophages accumulated in the spleens and livers, in response to either strain, resulting in a period of uncontrolled bacterial proliferation [66]. They hypothesized that with the emergence of IgG opsonins, strain 19 would be rapidly eliminated within activated macrophages, whereas some strain 2308 organisms would survive within these cells despite opsonization [66]. Recent studies support this view, including the findings that strain 19 opsonized by a hyperimmune antiserum was killed by macrophages in vitro much more effectively than similarly opsonized strain 2308 [12], that O-polysaccharide–specific antibodies of the IgG3 isotype were first detected at 3 weeks after infection [131], and that in passive transfer assays high levels of protection against strain 19 were provided by antisera taken from donors infected 3 weeks, but not 2 weeks, earlier [131].

REFERENCES

1. Centers for Disease Control. Brucellosis in the United States, 1965–1974. J Infect Dis 1977; 136:312–6.
2. Nicoletti P. The epidemiology of bovine brucellosis. Adv Vet Science Comp Med 1980; 24:69–98.
3. Young EJ. Human brucellosis. Rev Infect Dis 1983; 5:821–42.
4. Joint FAO/WHO Expert Committee on Brucellosis. WHO Tech Rep 1986; 740:1–132.
5. Moriyon I, Gamazo C, Diaz R. Properties of the outer membrane of *Brucella*. Ann Inst Pasteur Microbiol 1987; 138:89–91.
6. Winter AJ. Outer membrane proteins of *Brucella*. Ann Inst Pasteur Microbiol 1987; 138:87–9.
7. Moreno E, Borowiak D, Mayer H. *Brucella* lipopolysaccharides and polysaccharides. Ann Inst Pasteur Microbiol 1987; 138:102–5.
8. Bundle DR, Cherwonogrodsky JW, Caroff M, Perry MB. The lipopolysaccharides of *Brucella abortus* and *B. melitensis*. Ann Inst Pasteur Microbiol 1987; 138:92–8.
9. Freeman BA, Vana LR. Host-parasite relationships in brucellosis. I. Infection of normal guinea pig macrophages in tissue culture. J Infect Dis 1957; 102:258–67.
10. Ralston DJ, Elberg SS. Serum-mediated immune cellular responses to *Brucella melitensis* I. Role of a macrophage-stimulating factor in promoting ingestion of *Brucella* by streptomycin-protected cells. J Bacteriol 1968; 96:24–38.
11. Price RE, Templeton JW, Smith R, Adams LG. Ability of mononuclear phagocytes from cattle naturally resistant or susceptible to brucellosis to control *in vitro* intracellular survival of *Brucella abortus*. Infect Immun 1990; 58:879–86.
12. Jones SM, Winter AJ. Survival of virulent and attenuated strains of *Brucella abortus* in normal and gamma interferon–activated murine peritoneal macrophages. Infect Immun 1992; 60:3011–4.
13. Jiang X, Baldwin CL. Effects of cytokines on the intracellular growth of *Brucella abortus*. Infect Immun 1993; 61:124–34
14. Detilleux PG, Deyoe BL, Cheville NF. Penetration and intracellular growth of *Brucella abortus* in nonphagocytic cells *in vitro*. Infect Immun 1990; 58:2320–8.
15. Braude AI. Studies in the pathology and pathogenesis of experimental brucellosis. II. The formation of the hepatic granuloma and its evolution. J Infect Dis 1951; 89:87–94.
16. Guerra H, Deter RL, Williams RP. Infection at the subcellular level. I. Localization of intravenously injected brucellae in the vacuolar apparatus of cells of guinea pig liver. Infect Immun 1972; 5:513–23.
17. Guerra H, Deter RL, Williams RP. Infection at the subcellular level. II. Distribution and fate of intravenously injected brucellae within phagocytic cells of guinea pigs. Infect Immun 1973; 8:694–9.
18. Spink WW. The nature of brucellosis. Minneapolis: University of Minnesota Press, 1956:145–90.
19. Mackaness MB. The immunological basis of acquired cellular resistance. J Exp Med 1964; 120:105–20.
20. Young EJ. Brucellosis: A model zoonosis in developing countries. AMPIS Suppl 1988; 3:17–29.
21. Nicoletti P. The control of brucellosis—a veterinary responsibility. Saudi Med J 1992; 13:10–3.
22. Ben-Chetrit E, Nashif M, Levo Y. Infective endocarditis caused by uncommon bacteria. Scand J Infect Dis 1983; 15:179–83.
23. Valliattu J, Shuhaiber H, Kiwan Y, Araj G, Chugh T. Brucella endocarditis. Report of one case and review of the literature. J Cardiovasc Surg (Torino) 1989; 30:782–5.
24. Rheumatism Review Committee. Arthritis associated with known infectious agents. Arthritis Rheum 1970; 13:459–61.
25. Mugerwa RD, D'Arbela PG. Brucella meningitis—a case report and review of the literature. E Afr Med J 1976; 53:266–9.
26. Challoner KR, Riley KB, Larsen RA. Brucella meningitis. Am J Emerg Med 1990; 8:40–2.
27. Peacock JE Jr. Persistent neutrophilic meningitis. Infect Dis 1990; 4:747–67.
28. Mousa AR, Muhtaseb SA, Almudallal DS, Khodeir SM, Marafie AA. Osteoarticular complications of brucellosis: a study of 169 cases. Rev Infect Dis 1987; 9:531–43.

29. Awadhi NZ, Ashkenani F, Khalaf ES. Acute pancreatitis associated with brucellosis. Am J Gastroenterol 1989; 84:1570–4.
30. Garcia-Rodriguez JA, Garcia-Sanchez JE, Munoz-Bellido JL, Ortiz de la Tabla V, Bellido-Barbero J. Review of pulmonary brucellosis: a case report on brucellar pulmonary empyema. Diagn Microbiol Infect Dis 1988; 11:53–60.
31. Ariza J, Gudiol F, Valverde J, Pallares R, Fernandez-Viladrich P, Rufi G, Espadaler L, Fernandez-Nogues F. Brucellar spondylitis: a detailed analysis based on current findings. Rev Infect Dis 1985; 7:656–64.
32. Murrell TG, Matthews BJ. Multiple sclerosis—one manifestation of neurobrucellosis? Med Hypotheses 1990; 33:43–8.
33. Davies G, Casey A. The survival of *Brucella abortus* in milk and milk products. Br Vet J 1973; 129:345–53.
34. Buchanan TM, Hendricks SL, Patton CM, Feldman RA. Brucellosis in the United States, 1960–1972; an abattoir-associated disease. Part III. Epidemiology and evidence for acquired immunity. Medicine (Baltimore) 1974; 53:427–39.
35. Young EJ. Worldwide brucellosis—1991. A conference report. VA Medical Center, Houston, Texas, April 1991.
36. Vershilova PA. The use of live vaccine for vaccination of human beings against brucellosis in the USSR. Bull WHO 1961; 24:85–9.
37. Andrew H, Love J, Leahy PN, Miller A. Contracting brucellosis from S19 vaccine (letter). Vet Rec 1973; 93:380.
38. Nicoletti P, Ring J, Boysen B, Buczek J. Illness in a veterinary student following accidental inoculation of *Brucella abortus* strain 19. J Am Coll Health 1986; 34:236–7.
39. Francke EL, Neu HC. Chloramphenicol and tetracyclines. Med Clin North Am 1987; 71:1155–68.
40. Hall WH. Modern chemotherapy for brucellosis in humans. Rev Infect Dis 1990; 12:1060–99.
41. Dubray G. Protective antigens in brucellosis. Ann Inst Pasteur Microbiol 1987; 138:84–7.
42. Kolar J. Control of *Brucella melitensis* brucellosis in developing countries. Ann Inst Pasteur Microbiol 1987; 138:122–6.
43. Cherwonogrodzky JW, Nielsen KH. Brucella abortus 1119-3 O-chain polysaccharide to differentiate sera from *B. abortus* S-19-vaccinated and field-strain-infected cattle by agar gel immunodiffusion. J Clin Microbiol 1988; 26:1120–3.
44. Nielsen K, Cherwonogrodzky JW, Duncan JR, Bundle DR. Enzyme-linked immunosorbent assay for differentiation of the antibody response of cattle naturally infected with *Brucella abortus* or vaccinated with strain 19. Am J Vet Res 1989; 50:5–9.
45. Boeer WJ, Crawford RP, Hidalgo RJ, Robinson RM. Small mammals and white-tailed deer as possible reservoir hosts of *Brucella abortus* in Texas. J Wildl Dis 1980; 16:19–24.
46. Moore CG, Schurrenberger PR. A review of naturally occurring *Brucella abortus* infections in wild mammals. J Am Vet Med Assoc 1981; 179:1105–12.
47. McCorquodale SM, DiGiacomo RF. The role of wild North American ungulates in the epidemiology of bovine brucellosis: a review. J Wildl Dis 1985; 21:351–7.
48. Schurrenberger PR, Brown RR, Hill EP, Scanlan CM, Altiere JA, Wykoff JT. *Brucella abortus* in wildlife on selected cattle farms in Alabama. J Wildl Dis 1985; 21:132–6.
49. Cheers C, Ho M. Resistance and susceptibility of mice to bacterial infection. IV. Functional specificity in natural resistance to facultative intracellular bacteria. J Reticuloendothel Soc 1983; 34:299–309.
50. Montaraz JA, Winter AJ, Hunter DM, Sowa BA, Wu AM, Adams LG. Protection against *Brucella abortus* in mice with O-polysaccharide specific monoclonal antibodies. Infect Immun 1986; 51:961–3.
51. Winter AJ, Rowe GE, Duncan JR, Eis MJ, Widom J, Ganem B, Morein B. Effectiveness of natural and synthetic complexes of porin and O-polysaccharide as vaccines against *Brucella abortus* in mice. Infect Immun 1988; 56:2808–17.
52. Araya LN, Winter AJ. Comparative protection of mice against virulent and attenuated strains of *Brucella abortus* by passive transfer of immune T cells or serum. Infect Immun 1990; 58:254–6.
53. Winter AJ, Duncan JR, Santisteban CG, Douglas JT, Adams LG. Capacity of passively administered antibody to prevent establishment of *Brucella abortus* infection in mice. Infect Immun 1989; 57:3438–44.

54. Limet J, Plommet A-M, Dubray G, Plommet M. Immunity conferred upon mice by anti-LPS monoclonal antibodies in murine brucellosis. Ann Inst Pasteur Immunol 1987; 138:417–24.
55. Limet JN, Bosseray N, Garin-Bastuji B, Dubray G, Plommet M. Humoral immunity in mice mediated by monoclonal antibodies against the A and M antigens of *Brucella*. J Med Microbiol 1989; 30:37–43.
56. Pavlov H, Hogarth M, McKenzie IFC, Cheers C. *In vivo* and *in vitro* effects of monoclonal antibody to Ly antigens on immunity to infection. Cell Immunol 1982; 71:127–38.
57. Mielke MEA. T cell subsets in granulomatous inflammation and immunity to *L. monocytogenes* and *B. abortus*. Behring Inst Mitt 1991; 88:99–111.
58. Araya LN, Elzer PH, Rowe GE, Enright FM, Winter AJ. Temporal development of protective cell-mediated and humoral immunity in BALB/c mice infected with *B. abortus*. J Immunol 1989; 143:3330–7.
59. Ho M, Cheers C. Resistance and susceptibility of mice to bacterial infection. IV. Genetic and cellular basis of resistance to chronic infection with *Brucella abortus*. J Infect Dis 1982; 146:381–7.
60. Montarez JA, Winter AJ. Comparison of living and nonliving vaccines for *Brucella abortus* in BALB/c mice. Infect Immun 53:245–51.
61. Cheers C, Pagram F. Macrophage activation during experimental murine brucellosis: a basis for chronic infection. Infect Immun 1979; 23:197–205.
62. Riglar C, Cheers C. Macrophage activation during experimental murine brucellosis. Cell Immunol 1980; 49:154–67.
63. Lauderdale T-L, Jones SM, Winter AJ. Response of murine T cells to antigens of *Brucella abortus* at sequential periods after infection. Immunol Infect Dis 1990; 1:59–66.
64. Cheers C, Young AM. Colony stimulating activity and colony forming cells in murine brucellosis: relationship to immunopathology. Microbial Pathogenesis 1987; 3:185–94.
65. Doyle AG, Halliday WJ, Barnett CJ, Dunn TL, Hume DA. Effect of recombinant human macrophage colony-stimulating factor 1 on immunopathology of experimental brucellosis in mice. Infect Immun 1992; 60:1465–72.
66. Enright FM, Araya LN, Rowe GE, Winter AJ. Comparative histopathology in BALB/c mice infected with virulent and attenuated strains of *Brucella abortus*. Vet Immunol Immunopathol 1990; 26:171–82.
67. Cheers C, Pavlov H, Riglar C, Madraso E. Macrophage activation during experimental murine brucellosis. III. Do macrophages exert feedback control during brucellosis? Cell Immunol 1980; 49:168–77.
68. Lipsky PE. Imunosuppression by D-penicillamine in vitro: inhibition of human T lymphocyte proliferation by copper- or ceruloplasmin-dependent generation of hydrogen peroxide and protection by monocytes. J Clin Invest 1984; 73:53–65.
69. Ala-Ramadi BK, Meissler JJ, Huang D, Eisenstein TK. Immunosuppression induced by nitric oxide and its inhibition by IL-4. Eur J Immunol 1992: 22:2249–58.
70. Goodwin JS, Webb DR. Prostaglandin suppression of mitogen-stimulated lymphocytes in vitro. J Clin Invest 1978; 62:753–60.
71. Kehrl JH, Wakefield LM, Roberts AB, Jakowlew S, Alvarez-mon M, Derynck R, Sporn MB, Fauci AS. Production of transforming growth factor-beta by human T lymphocytes and its potential role in the regulation of T cell growth. J Exp Med 1986; 163:1037–1050.
72. Braun W, Pomales-Lebron A, Stinebring WR. Interactions between mononuclear phagocytes and *Brucella abortus* strains of different virulence. Proc Soc Exp Biol Med 1958; 97:393–7.
73. Smith H, Fitzgeorge RB. The chemical basis of the virulence of *Brucella abortus*. V. The basis of intracellular survival and growth in bovine phagocytes. Br J Exp Pathol 1964; 45:174–85.
74. Harmon BG, Adams LG, Frey M. Survival of rough and smooth strains of *Brucella abortus* in bovine mammary gland macrophages. Am J Vet Res 1988; 49:1092–7.
75. Riley LK, Robertson DC. Ingestion and intracellular survival of *B. abortus* in human and bovine polymorphonuclear leukocytes. Infect Immun 1984; 46:224–30.
76. Young EJ, Borchert M, Kretzer FL, Musher DM. Phagocytosis and killing of *Brucella* by human polymorphonuclear leukocytes. J Infect Dis 1985; 151:682–90.
77. Bertram TA, Canning PC, Roth JA. Preferential inhibition of primary granule release from bovine neutrophils by a *Brucella abortus* extract. Infect Immun 1986; 52:285–92.
78. Kreutzer DL, Dreyfus LA, Robertson DC. Interaction of polymorphonuclear leukocytes with smooth and rough strains of *Brucella abortus*. Infect Immun 1979; 23:737–42.

79. Canning PC, Roth JA, Tabatabai LB, Deyoe BL. Isolation of components of *Brucella abortus* responsible for inhibition of function in bovine neutrophils. J Infect Dis 1985; 152:913–21.
80. Canning PC, Roth JA, Deyoe BL. Release of 5'-guanosine monophosphate and adenine by *B. abortus* and their role in the intracellular survival of the bacteria. J Infect Dis. 1986; 154:464–70.
81. Frenchick PJ, Markham RJF, Cochrane AH. Inhibition of phagosome-lysosome fusion in macrophages by soluble extracts of virulent *Brucella abortus*. Am J Vet Res 1985; 46:332–5.
82. Hinsdill RD, Berman DT. Antigens of *Brucella abortus*. II. Toxicity for macrophages in culture. J Infect Dis 1968; 118:307–16.
83. Riley LK, Robertson DC. Brucellacidal activity of human and bovine polymorphonuclear leukocyte granule extracts against smooth and rough strains of *B. abortus*. Infect Immun 1984; 46:231–6.
84. Soto L, Rojas X, Alonso O. Estudie *in vitro* del efecto de fracciones de pared celular de *Brucella* sobre la actividad de leucocoitos polimorfonucleares bovinos. Arch Med Vet 1991; 23:27–33.
85. Rasool O, Freer E, Moreno E, Jarstrand C. Effect of *Brucella abortus* lipopolysaccharide on oxidative metabolism and lysozyme release by human neutrophils. Infect Immun 1992; 60:1699–702.
86. Price RE, Templeton JW, Adams LG. Survival of smooth, rough and transposon mutant strains of *Brucella abortus* in bovine mammary macrophages. Vet Immunol Immunopathol 1990; 26:353–65.
87. Sowa B, Kelly KA, Ficht TA, Adams LG. Virulence associated proteins of *Brucella abortus* identified by paired two-dimensional gel electrophoretic comparisons of virulent, vaccine and LPS deficient strains. Appl Theor Electrophoresis 1992; 2:1–8.
88. Beck BL, Tabatabai LB, Mayfield JE. A protein isolated from *Brucella abortus* is a Cu-Zn superoxide dismutase. Biochemistry 1990; 29:372–6.
89. Sha Z, Stabel TJ, Mayfield JE. Purification and characterization of catalase from *B. abortus*. Reported at the 44th Annual Brucellosis Research Conference, Chicago, Illinois, November 1991.
90. Fitzgeorge RB, Keppie J, Smith H. The relation between resistance to hydrogen peroxide and virulence in brucellae. J Pathol 1965; 89:745–7.
91. Saha AK, Mukhopadhyay NK, Dowling JN, Ficht TA, Adams LG, Glew RH. Characterization of a phosphomonoesterase from *Brucella abortus*. Infect Immun 1990; 58:1153–8.
92. Saha AK, Dowling JN, LaMarco KL, Das S, Remaley AT, Olomu N, Pope MT, Glew RH. Properties of acid phosphatase from *Legionella micdadei* which blocks superoxide anion production by human neutrophils. Arch Biochem Biophys 1985; 243:150–60.
93. Gor D, Mayfield JE. Cloning and nucleotide sequence of the *Brucella abortus* groE operon. Biochem Biophys 1992; 1130:120–2.
94. Roop II RM, Price LM, Dunn BE, Boyle SM, Sriranganathan N, Schurig GG. Molecular cloning and nucleotide sequence analysis of the gene encoding the immunoreactive *Brucella abortus* Hsp60 protein. Microbial Pathogenesis 1992; 12:47–62.
95. Lin J, Adams LG, Ficht TA. Characterization of the heat shock response in *Brucella abortus* and isolation of the genes encoding the GroE heatr-shock proteins. Infect Immun 1992; 60:2425–31.
96. Anderson TD, Cheville NF. Ultrastructural morphometric analysis of *Brucella abortus*–infected trophoblasts in experimental placentitis. Am J Pathol 1986; 124:226–37.
97. Anderson TD, Cheville NF, Meador VP. Pathogenesis of placentitis in the goat inoculated with *Brucella abortus*. II. Ultrastructural studies. Vet Pathol 1986; 23:227–39.
98. Detilleux PG, Deyoe BL, Cheville NF. Entry and intracellular localization of *Brucella* spp. in vero cells: fluorescence and electron microscopy. Vet Pathol 1990; 27:317–28.
99. Horwitz MA. Formation of a novel phagosome of the Legionnaire's disease bacterium (*Legionella pneumophila*) in human monocytes. J Exp Med 1983; 158:1319–31.
100. Detilleux PG, Cheville NF, Deyoe BL. Pathogenesis of *Brucella abortus* in chicken embryos. Vet Pathol 1988; 25:138–46.
101. Meador VP, Deyoe BL. Intracellular localization of *Brucella abortus* in bovine placenta. Vet Pathol 1989; 26:513–5.
102. Harmon BG, Adams LG, Templeton JW, Smith R III. Macrophage function in mammary glands of *Brucella abortus*–infected cows and cows that resisted infection after inoculation of *Brucella abortus*. Am J Vet Res 1989; 50:459–65.
103. Gay B, Mauss H, Sanchez-Teff S. Identification of fibronectins in peritoneal macrophages during the phagocytosis of *Brucella*. An immunocytochemical study by electron microscopy. Virchows Arch B—Cell Pathol 1986; 52:169–76.

104. Stinebring WR, Kessel R. Continuous growth of *Brucella abortus* in mononuclear phagocytes of rats and guinea pigs. Proc Soc Exp Biol Med 1959; 101:412–5.
105. Ralston DJ, Elberg SS. Serum-mediated immune cellular responses to *Brucella melitensis* REV 1. III. Infection of macrophages fixed-to-glass. Br J Exp Pathol 1968; 49:586–96.
106. Ralston DJ, Elberg SS. Serum-mediated immune cellular responses to *Brucella melitensis* REV 1. II. Restriction of *Brucella* by immune sera and macrophages. J Reticuloendothel Soc 1969; 6:109–39.
107. Pomales-Lebron A, Stinebring WR. Intracellular multiplication of *Brucella abortus* in normal and immune mononuclear phagocytes. Proc Soc Exp Biol Med 1957; 94:78–83.
108. McGhee JR, Freeman BA. Osmotically sensitive *Brucella* in infected normal and immune macrophages. Infect Immun 1970; 1:146–50.
109. Fitzgeorge RB, Solotorovsky M, Smith H. The behaviour of *Brucella abortus* within macrophages separated from the blood of normal and immune cattle by adherence to glass. Br J Exp Path 1967; 48:522–8.
110. Ralston DJ, Elberg SS. Sensitization and recall of anti-*Brucella* immunity in guinea-pig macrophages by attenuated and virulent *Brucella*. Infect Immun 1971; 3:200–8.
111. Birmingham JR, Jeska EL. Characterization of macrophage functions in mice infected with *Brucella abortus*. Infect Immun 1981; 32:1079–83.
112. Magliulo E, Carosi G, Scevola DF, Filice G, Saladino I. Macrophage activation in human brucellosis. Infection 1973; 1:144–50.
113. Gay B, Sanchez-Teff S, Caravano R. Ultrastructural localization of NADPH-oxidase activity in murine peritoneal macrophages during phagocytosis of *Brucella*. Virchows Arch B 1984; 45:147–55.
114. Bagchi T. Ph.D. thesis. Virginia Polytechnic Institute and State University, Blacksburg, Virginia 1990.
115. Zhan YF, Stanley ER, Cheers C. Prophylaxis or treatment of experimental brucellosis with interleukin-1. Infect Immun 1991; 59:1790–4.
116. Spector WG, Reichhold N, Ryan GB. Degradation of granuloma-inducing microorganisms by macrophages. J Pathol 1970; 101:339–54.
117. Cheers C, Waller R. Activated macrophages in congenitally athymic "nude" mice and in lethally irradiated mice. J Immunol 1975; 115:844–7.
118. Elzer PH, Rowe GE, Enright FM, Winter AJ. Effects of gamma radiation and azathioprine on *Brucella abortus* infection in BALB/c mice. Am J Vet Res 1991; 52:838–44.
119. Brecher G, Endicott KM, Gump H, Brawner HP. Effects of x-ray on lymphoid and hemopoietic tissues of albino mice. Blood 1948; 1259–74.
120. Schultz RM, Pavlidis NA, Chirigos MA, Weiss JF. Effects of whole body x-irradiation and cyclophosphamide treatment on induction of macrophage tumoricidal function in mice. Cell Immunol 1978; 38:302–9.
121. Swartz RP, Saluk PH. Functional modifications of macrophage activity after sublethal whole-body irradiation. Proc Soc Exp Biol Med 1981; 167:20–4.
122. Jiang X, Leonard B, Benson R, Baldwin CL. Inhibition of growth of *Brucella abortus* by noncytokine and interferon-γ–activated macrophages is mediated by reactive oxygen intermediates. Cell Immunol 1993; (in press).
123. Green SJ, Crawford RM, Hockmeyer JT, Meltzer MS, Nacy CA. *Leishmania major* amastigotes initiate the L-arginine–dependent killing mechanism in IFN-γ–stimulated macrophages by induction of tumor necrosis factor-alpha. J Immunol 1990; 145:4290–7.
124. Byrd TF, Horwitz MA. Interferon gamma–activated human monocytes downregulate transferrin receptors and inhibit the intracellular multiplication of *Legionella pneumophila* by limiting the availability of iron. J Clin Invest 1989; 83:1457–62.
125. Jiang X, Baldwin CL. Iron augments macrophage-mediated killing of *Brucella abortus* alone and in conjunction with interferon-γ. Cell Immunol 1993; (in press).
126. Sambri V, Cevenini R, LaPlaca M. Susceptibility of iron-loaded *Borrelia burgdorferi* to killing by hydrogren peroxide and human polymorphonuclear leucocytes. FEMS Microbiol Lett 1991; 81:67–71.
127. Hoepelmann IM, Bezemer WA. Vanderbroucke-grauls CMJE, Marx JJM, Verhoef J. Bacterial iron enhances oxygen radical–mediated killing of *Staphylococcus aureus* by phagocytes. Infect Immun 1990; 58:26–31.
128. Babior BM. Oxidants from phagocytes: space agents of defense and destruction. Blood 1984; 64:959–66.

129. Finkelman FD, Katona KM, Mosmann TR, Coffman RL. IFN-gamma regulates the isotypes of Ig secreted during *in vivo* humoral immune responses. J Immunol 1988; 140:1022–7.
130. Street NE, Schumacher JH, Fong AT, Bass H, Fiorentino DF, Leverah JA, Mosmann TR. Heterogeneity of mouse helper T cells. Evidence from bulk cultures and limiting dilution cloning for precursors of Th1 and Th2 cells. J Immunol 1990; 144:1629–39.
131. Elzer PH, Jacobson RH, Jones SM, Nielsen KH, Douglas JT, Schurig GG, Winter AJ. Isotype distribution and functional analyses of O-polysaccharide-specific antibodies from BALB/c mice infected with *Brucella abortus*. Submitted for publication.
132. Splitter GA, Everlith KM. Collaboration of bovine T lymphocytes and macrophages in T-lymphocyte response to *Brucella abortus*. Infect Immun 1986; 51:776–83.
133. Splitter GA, Everlith KM. *Brucella abortus* regulates bovine macrophage–T-cell interaction by major histocompatibility complex class II and interleukin-1 expression. Infect Immun 1989; 57:1151–7.
134. Czuprynski CJ, Canono BP, Henson PM, Campbell PA. Gentically determined resistance to listeriosis is associated with increased accumulation of inflammatory neutrophils and macrophages which have enhanced listericidal activity. Immunology 1985; 55:511–8.
135. Young AM, Cheers C. Colony-forming cells and colony-stimulating activity during listeriosis in genetically resistant or susceptible mice. Cell Immunol 1986; 97:227–37.
136. Wood PR, Spanidis V, Frangos K, Cheers C. The *in vitro* bactericidal activity of peritoneal and spleen cells from *Listeria* resistant and susceptible mouse strains. Cell Immunol 1986; 99:160–9.
137. Cheers C, Stanley ER. Macrophage production during murine listeriosis: colony-stimulating factor 1 (CSF-1) and CSF-1-binding cells in genetically resistant and susceptible mice. Infect Immun 1988; 56:2972–8.
138. Katz P. Natural killer cell mediated lysis of *Mycobacterium avium* complex infected monocytes. J Clin Immunol 1990; 10:71–77.
139. Guo Y, Niesel DW, Ziegler HK, Klimpel GR. *Listeria monocytogenes* activation of human peripheral blood lymphocytes: induction of non–major histocompatibility complex–restricted cytotoxic activity and cytokine production. Infect Immun 1992; 60:1813–9.

23

Interactions Between Macrophages and Chlamydiae

David La Verda and Gerald I. Byrne

University of Wisconsin—Madison, Madison, Wisconsin

I. INTRODUCTION

Mononuclear phagocytes are believed to be important in diverse ways during chlamydial infections. It appears that the clearance of an infection depends on the activation of macrophages to express activities that are directed toward the eradication of chlamydiae in infected host cells as well as within the macrophage itself. However, these activities do not immediately lead to killing. Viable *Chlamydia* may persist within macrophages and other host cells in spite of the immune response, and chlamydial antigen may not be fully eradicated. Macrophages may even be the target cell for replication in a number of chlamydial diseases. Thus macrophages may contribute to the control of an acute chlamydial infection, but also may be involved with maintaining the chronicity and immunopathology characteristic of virtually all chlamydial diseases.

A. Biology of the Chlamydiae

The chlamydiae are a group of obligate intracellular bacteria that are perhaps the most common human pathogens [1]. They appear to be incapable of synthesizing a number of essential compounds such as ATP, and thus are bound to a life inside a eukaryotic host cell [2]. All chlamydiae undergo a characteristic developmental cycle involving the alternation between two morphologically and functionally distinct forms (Figure 1). The compact elementary body (EB) is rigid and environmentally resistant. Its primary function is to transfer from an infected cell to an uninfected cell. Although EBs are metabolically inert, they have the capacity to attach to and trigger uptake by susceptible host cells. After the EB enters the host cell, it remains enclosed in an endosome throughout the entire developmental cycle. Phagosome-lysosome function is inhibited. Inside the host cell, the EB reorganizes into the larger, metabolically active reticulate body (RB). Unlike the EBs, the RBs are noninfectious and unable to survive outside the host cell. The RBs divide by binary fission. After a series of divisions, the RBs reorganize to once again form EBs. The EBs are released from the host cell and go on to repeat the cycle in other cells.

Morphologically, chlamydiae resemble gram-negative bacteria, with distinct outer and inner membranes [2]. However, chlamydiae do not possess a peptidoglycan layer in their cell envelope, and *N*-acetyl muramic acid appears to be absent completely. Still, chlamydiae do possess penicillin-binding proteins and are affected by both penicillin and cycloserine. Perhaps a tetrapeptide-linked structure exists that performs a function essential for the reorganization of RBs to infectious EBs. Such a structure need not be associated with the chlamydial cell envelope.

The outer membrane is composed of lipopolysaccharide (LPS), a major outer membrane

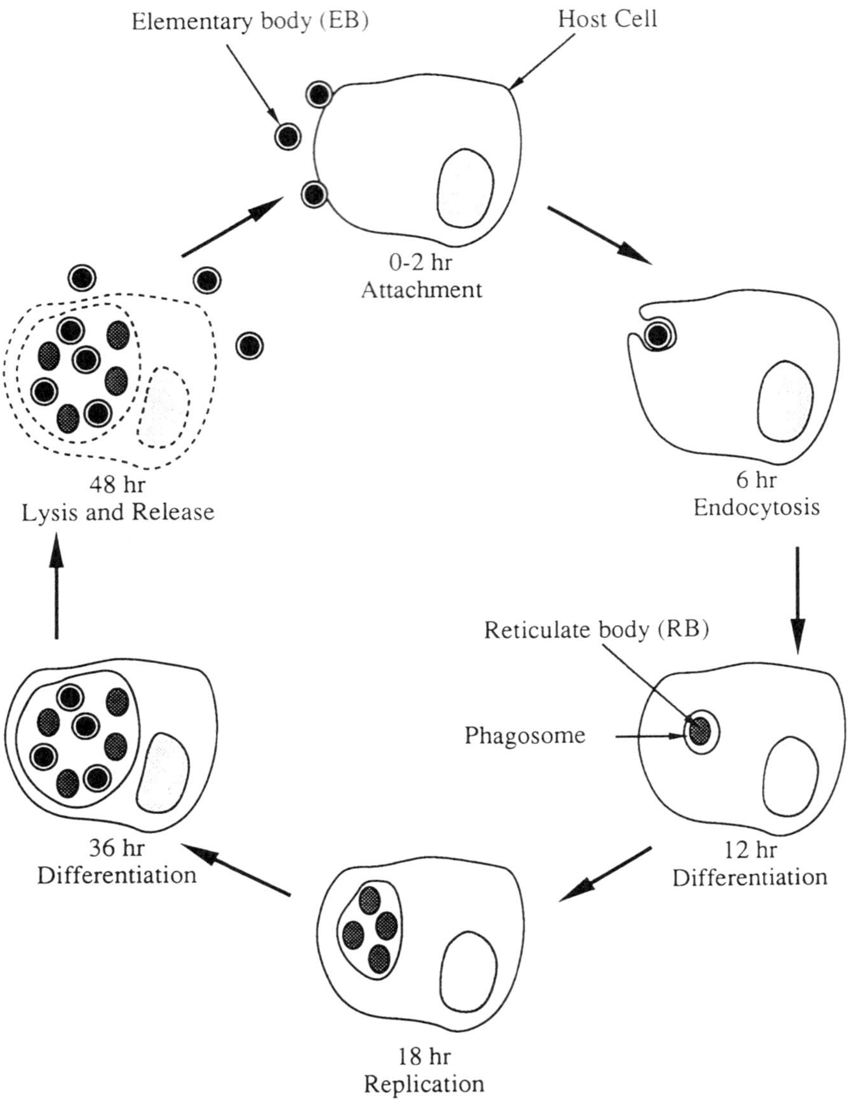

FIGURE 1 Developmental cycle of *Chlamydia psittaci* 6BC.

protein (MOMP), and a number of other lipid and protein components. The major outer membrane protein makes up about 70% of the total outer envelope protein and contains a number of serovar-specific epitopes. The structural rigidity exhibited by EBs appears to be related to cross-linking of MOMP with two other cystine-rich outer membrane proteins (12-kDa and 60-kDa) via disulfide bonds [2]. On differentiation from EB to RB, these bonds become reduced, thereby imparting the flexible, fragile characteristics to the RB. Though RBs synthesize MOMP throughout the cell cycle, the other cystine-rich outer membrane proteins are not expressed until the RB starts to reorganize into an EB. Then the disulfide cross-linking reforms.

B. Chlamydial Disease and Epidemiology

There are four characterized species within the genus *Chlamydia*: *C. trachomatis*, *C. pneumoniae*, *C. psittaci*, and *C. pecorum*. *C. trachomatis* is further subdivided into three distinct biovars: the oculogenital biovar, the lymphogranuloma venereum (LGV) biovar, and the mouse pneumonitis biovar [2]. *C. pecorum* is a recently differentiated species [3] causing pneumonia, polyarthritis, encephalomyelitis, and diarrhea in ruminants. It is not yet known to infect humans. Table 1 compares the *Chlamydia* species and biovars.

Twelve serovars of *C. trachomatis* constitute the oculogenital biovar. Serovars A, B, Ba, and C cause trachoma, a severe chronic keratoconjunctivitis that is the leading cause of preventible blindness in the world [1]. Serovars D through K are the most common causes of sexually transmitted disease in the world, including urethritis, cervicitis, and pelvic inflammatory disease (PID). In addition, these serovars cause neonatal pneumonia and inclusion conjunctivitis of both adults and infants. Furthermore, *C. trachomatis* has been implicated in the etiology of Reiter's syndrome, a complicated urethritis that also includes uveitis and arthritis. Members of the oculogenital biovar principally infect columnar epithelial cells. Infections are usually limited to the conjunctiva or genital tract, where these cells predominate.

There are three serovars of the LGV biovar: L1, L2, and L3. Lymphogranuloma venereum is a more invasive sexually transmitted disease characterized by involvement of the inguinal or pelvic lymphatics. Biovar LGV organisms have a broader host cell range than oculogenital biovar organisms and may infect both lymphoid and endothelial cells [1]. Although the initial lesion may be asymptomatic, the infection progresses with considerable necrosis and the formation of stellate abscesses. Infection by biovar LGV organisms stimulates granuloma formation in which mononuclear phagocytes undergo transformation into epithelioid cells. Giant cells also may be present. Humans are the principal host for both the oculogenital and the LGV biovars of *C. trachomatis*, while the mouse pneumonitis biovar is not known to infect humans [1].

C. pneumoniae, with no known animal reservoir, is an important cause of acute respiratory disease in humans [4]. It has only recently been established as a leading cause of atypical pneumonia, largely because of difficulty in cultivating this organism in cell culture and the absence of a specific serological test in the past [5]. Clinical manifestations of *C. pneumoniae* disease range from asymptomatic infection to a life-threatening pneumonia. Cases of bronchitis, pharyngitis, and sinusitis may frequently occur either alone or in association with one another. One interesting morphological feature is that the EBs of *C. pneumoniae* have a distinctive pear shape, in contrast to the round EBs of the other chlamydial species.

C. psittaci is a heterologous species that is ubiquitous among birds and nonhuman mammals [1]. It appears that different strains of *C. psittaci* can cause infections of varying degrees of severity. Most often humans are infected with strains of avian origin [6]. Human disease caused by avian *C. psittaci*, psittacosis, often manifests itself as a respiratory infection ranging from an atypical pneumonia to a severe pneumonitis [5]. Fever, chills, and severe headache usually occur. Alternately, *C. psittaci* may cause a febrile disease characterized by high fever and severe headache in the absence of respiratory involvement. Visualization of the growth of these chlamydiae in macrophages and a history of avian contact are classical diagnostic criteria distinguishing psittacosis from nonchlamydial pneumonias. *C. psittaci* is transmitted most commonly by aerosols from dried bird droppings or by handling of contaminated meat in the poultry industry. Human-to-human transmission is uncommon. The organism usually enters via inhalation, whereby it enters the lymphatic system and invades the liver and spleen. The lungs are then seeded from these sites, possibly involving transportation by macrophages. While *C. psittaci* is also an important cause of disease for both feral and domesticated mammals, mammalian strains seldom infect humans. However, mammalian strains of *C. psittaci* have been implicated as the cause of spontaneous abortion in a number of pregnant women who were exposed to *Chlamydia*-infected sheep [7,8].

II. CHLAMYDIAE AND MACROPHAGES AS HOST CELLS

A. Description of Biovars as They Relate to Host Cell Specificity

Unfortunately, there is no unifying theme concerning the growth of chlamydiae in mononuclear phagocytes. Some biovars seem to do rather well in at least certain types of macrophages, whereas

Table 1 Comparison of the *Chlamydia* Species and Biovars

	C. trachomatis oculogenital biovar	*C. trachomatis* LGV biovar	*C. pneumoniae*	*C. psittaci* avian strains	*C. psittaci* mammalian strains	*C. pecorum*
Example strain	Serovar B/TW5	Serovar L2/434	TWAR	6BC	GPIC	Bo/E58
Inclusion morphology	Oval, vacuolar	Oval, vacuolar	Oval, dense	Irregular, dense	Irregular, dense	Oval, dense
Morphology of elementary bodies	Round	Round	Pear-shaped	Round	Round	Round
Glycogen-containing inclusions	Yes	Yes	No	No	No	No
Susceptible to sulfonamides	Yes	Yes	No	No	No	No
Possess plasmid	Yes	Yes	No	Yes	Yes	Yes
Number of serovars	12	3	1[a]	Unknown	Unknown	3
Natural hosts	Humans	Humans	Humans	Birds	Nonhuman mammals	Ruminant mammals
Human diseases caused	Trachoma, conjunctivitis, genital diseases, infant pneumonia	Lymphogranuloma venereum	Atypical pneumonia, bronchitis	Psittacosis, pneumonia, fever	Pneumonia, fever, abortion	Unknown

[a]Recent evidence suggests there may be others [93].
Source: Refs. 2,3,4,91,92.

other biovars appear not to grow in any monocyte or macrophage population. In addition, there are some unresolved controversies related to chlamydial growth. Studies of interactions between various chlamydial strains and mononuclear phagocytes have indicated a differential capacity for intracellular growth among the species and biovars of chlamydiae (Table 2). In addition, experimental infections using macrophages at different stages of development, or macrophages from different tissue sites, have yielded different and sometimes conflicting results.

1. Chlamydia trachomatis

In vitro experiments have demonstrated that neither the oculogenital biovar nor the LGV biovar of *C. trachomatis* produces a productive infection in freshly isolated monocytes [9–11]. However, there is a marked difference between these two biovars in their capacity to grow in either human monocyte-derived macrophages or mouse peritoneal macrophages [12,13]. Kuo [13] originally noted that oculogenital serovar B grew to a much more limited extent than LGV serovar L2 in mouse peritoneal macrophages. Yong et al. [9] further investigated the growth of oculogenital serovars B, C, E, and F and LGV serovars L1 and L2 in human mononuclear phagocytes. Both biovars were completely inhibited by fresh monocytes. However, after the monocytes had matured into macrophages (monocyte-derived macrophages [MdMs]), only the LGV organisms survived and replicated intracellularly. Oculogenital biovar organisms, on the other hand, were unable to replicate in the macrophages kept in culture for as long as 21 days. The capacity of biovar LGV organisms to grow in macrophages correlates with their increased capacity to disseminate in comparison with the oculogenital biovar.

In addition to the difference between growth in monocytes and growth in MdMs, there appears to be a difference between growth observed in different types of macrophages. Whereas MdMs supported biovar LGV growth in vitro [9,12], Nakajo et al. [14] showed that explanted human alveolar macrophages from healthy adults readily killed both LGV and oculogenital biovar organisms. This chlamydicidal activity of alveolar macrophages may be a reason why healthy adults generally do not develop *C. trachomatis* pneumonia [14]. *C. trachomatis* is, however, a major cause of pneumonia in the newborn [15]. While information on the antimicrobial activity of infant alveolar macrophages is limited, it is believed that their antimicrobial capacity is low at birth and increases with age [14]. This difference may not be due only to the increased phagocytosis of lung surfactant observed in infant alveolar macrophages, since in vitro engorgement of healthy adult alveolar macrophages with lung surfactant did not affect chlamydial killing [14,16]. Interestingly, biovar LGV organisms could sustain a productive infection in unactivated alveolar macrophages from patients with AIDS [17]. This corresponds with the observation that most cases of adult *C. trachomatis* pneumonia occur in immunocompromised hosts [14,16].

2. Chlamydia pneumoniae

Little information is currently available concerning the interaction of *C. pneumoniae* with mononuclear phagocytes. For chlamydial infections caused by the mouse pneumonitis biovar of *C.*

Table 2 Capacity of the Various Species and Biovars of *Chlamydia* to Grow Within Mononuclear Phagocytes

	Fresh monocytes	Unactivated macrophages	Activated macrophages
C. trachomatis oculogenital biovar	No	No	No
C. trachomatis LGV biovar	Minimal	Yes	No
C. pneumoniae	Unknown	Unknown	Unknown
C. psittaci 6BC avian strains	Minimal	Yes	No
C. psittaci GPIC mammalian strains	No	No	No

trachomatis, which also preferentially infects the respiratory tract, inclusions have been found in both bronchial epithelial cells and alveolar macrophages [18]. It would be interesting to know if *C. pneumoniae* infection follows a similar pattern. Studies are currently being conducted in several laboratories to define the interaction of *C. pneumoniae* with alveolar macrophages and the contribution of this interaction to both pathogenesis and protection in humans.

3. Chlamydia psittaci

Unlike *C. trachomatis*, *C. psittaci* is capable of limited growth in monocytes, though at yields considerably less than in macrophages or control "nonprofessional phagocyte" cell lines [10,19,20]. In addition *C. psittaci* readily grows in unactivated human MdMs [21,22] at a level comparable to that of *C. trachomatis* biovar LGV [10]. In contrast to results seen for *C. trachomatis* biovar LGV [14], *C. psittaci* is capable of growth in healthy unactivated explanted alveolar macrophages [23].

B. Uptake of Chlamydiae by Macrophages

Internalization of chlamydiae by mononuclear phagocytes can lead to either a productive infection or processing of particulate antigens. It is likely that distinct uptake mechanisms are required for these very different events. Initiation of a productive infection is a pathogen-directed process that involves adhesins recognizing host cell plasma membrane ligands in such a way that internalization occurs without damaging the pathogen. Uptake for the purposes of antigen processing and presentation involves sequestering the internalized particle in a phagolysosome such that protein degradation, binding of peptides to class II major histocompatability (MHC) complexes, and expression of these bound protein fragments on the surface of the cell occur. This last event is host cell–driven and requires an entirely different repertoire of signaling than is thought to occur in initialization of a productive infection.

All chlamydiae bind to and are taken up by mononuclear phagocytes by a process that is not inhibited by cytochalasin B [14,24]. Once internalized, the intracellular fate of these organisms depends on two major factors: the multiplicity of infection (MOI), i.e., the *Chlamydia*-to-phagocyte ratio; and the capacity of the EB to inhibit phagosome-lysosome fusion [21,25]. Table 3 summarizes the in vitro interaction of the various chlamydial species and biovars with host cells under a variety of experimental conditions.

At a low MOI, internalized EBs of *C. psittaci* survive, differentiate, and multiply in macrophages, and phagosome-lyosome fusion does not occur [26]. At high MOI, both *C. psittaci* and *C. trachomatis* EBs cause direct damage to the macrophage resulting in cell death [25,27]. This immediate cytotoxicity requires ingestion of the EBs [27]. The cytotoxic effect appears to be a consequence of the EB–host cell interaction, since a high MOI of polystyrene latex spheres does not induce immediate cytotoxicity [25]. It is believed that each EB disrupts the host cell membrane to some degree on entry. At high MOI, a large number of disruptions would be occurring simultaneously, possibly leading to host cell lysis. Interestingly, noninfectious RBs, which do not inhibit phagosome-lysosome fusion and are not taken up by "nonprofessional phagocytes," are phagocytosed by macrophages at high MOI and are destroyed with no apparent cytotoxicity to the macrophage [28].

Like all successful intracellular pathogens, chlamydial strains that successfully replicate within mononuclear phagocytes must have a way of subverting the defenses of the host cell to survive [29]. Chlamydiae have the capacity to inhibit fusion of the EB-containing endosome with macrophage lysosomes. This property is destroyed by heating the EBs before infection, but it remains intact after ultraviolet irradiation [25], implying a heat-labile surface component as a key element in mediating this process. However, there are few differences in membrane protein composition between the endosomes of infectious and heat-killed EBs [30]. The inhibition of phagosome-lysosome fusion is localized to the chlamydial inclusion [31]. Furthermore, isolated

Table 3 *Chlamydia*–Host Cell Interaction in Cell Culture

	C. trachomatis oculogenital biovar	*C. trachomatis* LGV biovar	*C. pneumoniae*	*C. psittaci* avian strains	*C. psittaci* mammalian strains	*C. pecorum*
Treatment of host cells with cytochalasin B	No effect	No effect	?	No effect	No effect	?
Inactivation of chlamydiae by ultraviolet radiation	No effect	No effect	?	No effect	No effect	?
Inactivation of chlamydiae by heat	Inhibition	Inhibition	?	Inhibition	Inhibition	?
Centrifugation of chlamydiae onto host cells	Pronounced enhancement of infection	Minimal enhancement of infection	Pronounced enhancement of infection	Minimal enhancement of infection	Pronounced enhancement of infection	?
Pretreatment of host cells with DEAE dextran	Pronounced enhancement of infection	Minimal enhancement of infection	Pronounced enhancement of infection	Minimal enhancement of infection	Pronounced enhancement of infection	Pronounced enhancement of infection
Treatment of host cells with trypsin	No effect	Inhibition	?	Inhibition	?	?
Treatment of elementary bodies with trypsin	Inhibition	No effect	?	Inhibition	?	?

Source: Refs, 2,3,4,90,91.

EB envelopes also inhibit phagosome-lysosome fusion [26]. This may be a mechanism by which antigens may persist in the host, and it may contribute to the immunologically mediated pathology associated with chlamydial disease. Partially degraded EB walls internalized by phagocytes could persist with minimal degradation over long periods, inducing the production of cytokines, such as tumor necrosis factor-α (TNF-α) and interleukin-1 (IL-1), that can keep the local immune system stimulated and lead to nonspecific tissue damage and subsequent scarring characteristic of chlamydial diseases.

Ingestion of infectious EBs by macrophages does not induce lysosomal enzyme synthesis. However, heat-killed EBs, which do not inhibit phagosome-lysosome fusion, do induce three key lysosomal enzymes: acid phosphatase, β-glucuronidase, and cathepsin D. These data indicate that the phagosome-lysosome fusion event itself may signal the up-regulation of lysosomal enzymes [32].

Opsonization of chlamydial EBs increases the extent of uptake into phagocytes [21,22,33]. Phagosome-lysosome fusion is no longer inhibited [21], and the opsonized chlamydiae are readily destroyed [34]. The binding of opsonized chlamydiae to human macrophages is rapid, saturable, and temperature-independent [34]. In addition, serum from chlamydial antibody–positive and –negative donors opsonized equally well [22,34]. This latter result suggests that *C. psittaci* [22] and *C. trachomatis* [35] nonspecifically bind antibody.

Recently, Lundemose et al. [36] demonstrated that the LGV biovar of *C. trachomatis* contains a membrane protein similar in sequence to the *Legionella pneumophila mip* gene product. In *L. pneumophila,* the Mip (macrophage infectivity potentiator) protein allows an 80-fold difference in infectivity for U937 human macrophage-like cells and explanted human alveolar macrophages over that of a Mip– mutant [37]. Furthermore, isogenic Mip– mutants have attenuated virulence in the guinea pig model of legionnaires' disease [38]. Mip does not seem to affect the intracellular growth rate, as those Mip– mutants that caused infection had a rate comparable to that of the parent. Thus Mip is believed to influence the capacity of the microorganism to initiate infection, possibly involving the induction of phagocytosis or the inhibition of phagosome-lysosome fusion [37]. Whether the chlamydial Mip-like protein performs a comparable function is currently under investigation.

C. The Activation of Macrophages to Inhibit Chlamydial Growth

The endocytosis of chlamydiae by the activated macrophage does not necessarily lead to immediate killing. Not only is phagosome-lysosome fusion inhibited, but many of the macrophage's oxidative killing mechanisms do not seem to play a role in its antichlamydial activity (Table 4). Instead, nonoxidative, or oxygen-independent, mechanisms appear to be responsible for the inhibition of *Chlamydia* in mononuclear phagocytes and polymorphonuclear leukocytes (PMNs), as well as "nonprofessional phagocytes" such as epithelial cells and fibroblasts.

At least for the first few days after internalization by an activated macrophage, chlamydiae appear to remain viable though not infectious [39]. The antichlamydial mechanisms of the macrophage initially exert a microbistatic effect that can be reversed. During this time, chlamydiae may persist within the phagocyte, continuously inducing the release of inflammatory monokines that may lead to the overstimulation of the local immune response and the subsequent immunological damage and scarring characteristic of chlamydial disease. In addition, the chlamydiae may even utilize the macrophage for transportation to nearby tissues. The antichlamydial effect becomes irreversible only after an extended period within the activated macrophage [39].

Oxygen-dependent mechanisms of microbial killing have been detected during chlamydial infection. Infection with a *Chlamydia*-to-phagocyte ratio greater than 10:1 [33] or 60:1 [40] induces a luminol-dependent chemiluminescent response in PMNs and monocytes [40], and *C. trachomatis* is degraded within peroxidase-positive phagolysosomes in both PMNs [41] and

Table 4 Antimicrobial Mechanisms of Mononuclear Phagocytes

Mechanism	Oxygen requirement	Effective against chlamydiae
Myeloperoxidase system	Dependent	Inconclusive
O_2^- generation	Dependent	No
H_2O_2 Production	Dependent	No
·OH Synthesis	Dependent	No
Limitation of essential nutrients:		
Tryptophan (human)	Independent	Yes
Arginine (mouse)	Independent	Inconclusive
Iron	Independent	No
Nitric oxide (mouse)	Independent	Inconclusive
Phagosome-lysosome fusion	Independent	No
susceptible to degradation		Yes

monocytes [9]. In addition, chlamydiae do appear to be susceptible in vitro killing by a cell-free myeloperoxidase (MPO) system [42]. However, PMNs from patients deficient in oxidative killing mechanisms still killed chlamydiae as well as controls [42], and at a low MOI, *C. psittaci* did not induce a chemiluminescent response in activated mouse peritoneal macrophages even though inhibition was apparent [43]. Byrne and Faubion [44] showed that inhibition of the respiratory burst by glucose starvation, pretreatment with phorbol myristate acetate (PMA), or addition of exogenous catalase did not reverse the lymphokine-induced inhibition of *C. psittaci*. Using oxidatively deficient macrophages, Murray et al. [45] further determined that chlamydiae were inhibited by an oxygen-independent mechanism mediated by lymphokine. These studies were significant because before these findings it was thought that toxic oxygen products were the principal antimicrobicidal mechanism of activated macrophages.

Additionally, chlamydiae do not induce superoxide anion (O_2^-) generation in PMNs [33] or alveolar macrophages [14,16]. This lack of induction seems to be passive, since the macrophages retain the capacity to produce O_2^- on stimulation with PMA [43]. Since O_2^- generation is one of the primary oxidative microbial mechanisms of mononuclear phagocytes [46], this lack of induction may contribute to the capacity of certain strains of chlamydiae to utilize the macrophage as a host cell.

Subsequently, Rothermel et al. [47] and Byrne and Krueger [48] attempted to identify the component of lymphokine responsible for inducing inhibition of chlamydiae in macrophages. Pretreatment of macrophages with lymphokine plus anti–interferon-γ (anti–IFN-γ) antibodies prevented activation to inhibit chlamydial growth [47]. Furthermore, purified human IFN-γ [47] or recombinant human IFN-γ [49] alone could induce macrophages to inhibit intracellular chlamydiae.

The enzyme indoleamine 2,3-dioxygenase (indoleamine:oxygen 2,3-oxidoreductase [decyclizing]; IDO) was found to be induced in human peripheral blood mononuclear phagocytes on stimulation with IFN-γ [50,51]. This enzyme catalyzes the initial pyrrole ring oxidative cleavage step in the degradation of L-tryptophan to *N*-formylkynurenine [51] (Figure 2). The result of the cleavage is the depletion of intracellular pools of tryptophan, which cannot be synthesized by chlamydiae [2]. Interferon-γ induces IDO in MdMs, alveolar macrophages, and oxidatively deficient macrophages [53], as well as a variety of epithelial cells and fibroblasts [52] (Table 5). Interferon-γ–induced monocytes have IDO activity, but IFN-γ–induced MdMs possess a sevenfold greater activity per cell [50]. In addition, IFN-α and IFN-β can induce IDO activity in macrophages, but not epithelial cells [50,51,53]. Nevertheless, the level of activity of IFN-γ–induced

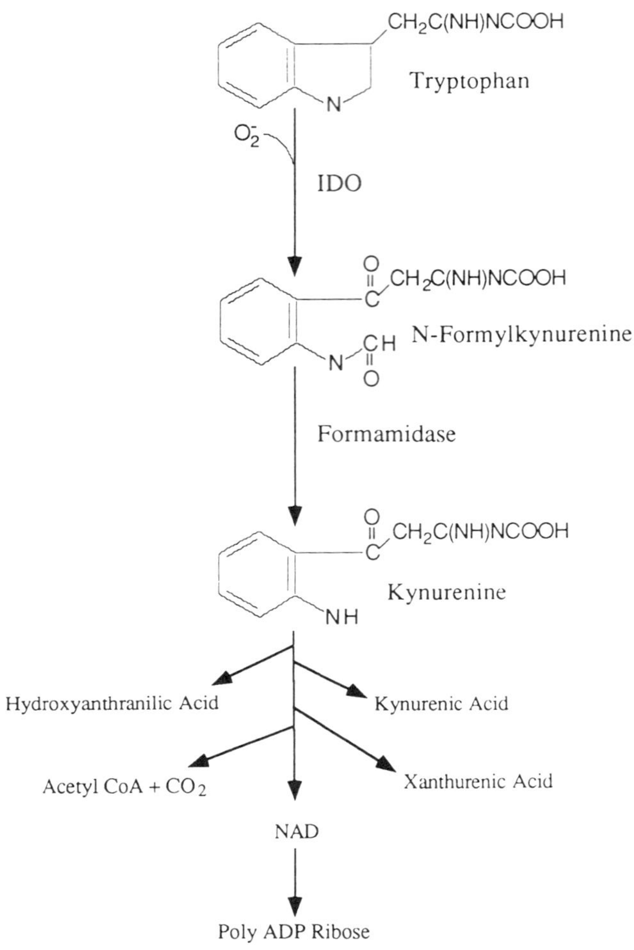

FIGURE 2 Pathway of tryptophan catabolism.

Table 5 Summary of Treatments Known to Induce IDO Activity

Treatment	IDO Induction	Cell types	Other relevant inductions
IFN-γ	++++	Many	—
IFN-β	+	Macrophages	—
IFN-β + LPS	++++	Macrophages	—
PMA	+	U937	Protein kinase C
PMA + LPS	++++	U937	Protein kinase C
LPS (<1 ng/ml)	–	Macrophages	Interferon
LPS (>10 ng/ml)	++++	Macrophages	Interferon

IDO = indoleamine 2,3-dioxygenase; IFN = interferon; LPS = lipopolysaccharide; PMA = phorbol myristate acetate.

IDO is greater than that of IFN-β alone [54]. The induction of activity by IFN-β, however, could reach levels comparable to those of IFN-γ when used in conjunction with concentrations of LPS that were insufficient to induce IDO activity alone [52,54]. Apparently IFN-γ and IFN-β induce IDO in macrophages via different pathways, with that of IFN-β being protein kinase C (PKC)-dependent and that of IFN-γ being PKC-independent [55].

Studies by Byrne with *C. psittaci* [56] and Shemer and Sarov with *C. trachomatis* [49] showed that the addition of exogenous tryptophan to the culture medium was sufficient to reverse the IFN-γ–induced inhibition in human cell lines. Carlin et al. [50,54] demonstrated that IDO is induced and restricts chlamydial growth in mononuclear phagocytes by tryptophan starvation. The degree to which tryptophan can reverse the inhibition is dependent on the concentration of IFN-γ [49]. Thus, there appears to be a direct relationship between the amount of IFN-γ and the amount of IDO activity induced [52].

Byrne et al. [39] noted that the potential to reverse inhibition of *C. psittaci* in IFN-γ–activated macrophages by exogenous tryptophan was limited to the first 3 days after infection. Thereafter, infectious chlamydiae could not be recovered after addition of tryptophan. These data indicate that the microbistatic mechanism of inhibition is followed by a microbicidal activity that is irreversible by modulation of essential nutrient levels. Similarly, Shemer et al. [57] noted an analaogous situation with *C. trachomatis* infection. When *C. trachomatis*–infected cells were exposed to high doses of IFN-γ, inhibition could not be reversed by the addition of tryptophan to the culture medium.

The actual mechanism involved in the change from inhibition to killing is not clear. One possibility is that prolonged exposure or exposure to large doses of interferon irreversibly blocks a step in chlamydial metabolism. Perhaps the mechanism by which phagosome-lysosome fusion is inhibited may be overcome, resulting in the destruction of the chlamydiae. Alternatively, there may be an accumulation of a product of tryptophan catabolism that is toxic for the chlamydiae. This product could accumulate inside the host cell over time or accumulate rapidly, depending on the level of IFN-γ–stimulated IDO activity. While the end product of tryptophan catabolism is kynurenine in most cells, macrophages convert kynurenine into hydroxyanthranilic acid, which is toxic for both prokaryotic and eukaryotic cells [52]. The role of hydroxyanthranilic acid production in the chlamydicidal activity of activated macrophages is yet to be determined.

A distinct situation exists in infection of murine cells. De la Maza et al. [58] initially noted that tryptophan did not reverse the antiproliferative effect of murine IFN-γ–induced murine fibroblasts. Further investigation has shown that IDO is not induced in murine cells by IFN. Yet purified recombinant murine IFN-γ activated murine macrophages to inhibit the growth of chlamydiae both in vitro and in vivo [59]. Thus, a different IFN-γ–induced oxygen-independent antichlamydial mechanism must be at work.

Activation of murine macrophages by IFN-γ or TNF-α leads to the induction of nitric oxide (NO) synthase, which catalyzes the production of NO from L-arginine [60]. Nitric oxide synthase induced in murine macrophages has been shown to inhibit a variety of intracellular pathogens, including *Toxoplasma gondii* [61], *Leishmania major* [60], *Plasmodium falciparum* [62], *Trypanosoma musculi* [63], *Cryptococcus neoformans* [64], and *Mycobacterium bovis* [65]. The inhibitory effect of NO can be prevented by addition of arginine analogues such as N^G-monomethyl-L-arginine [61]. It is possible that in addition to control of intracellular growth by toxic NO radicals, arginine depletion by NO synthase may starve the intracellular parasites for an essential amino acid. Further investigation is under way to determine the contribution of NO synthase to chlamydial inhibition.

Another oxygen-independent mechanism by which macrophages may inhibit intracellular pathogens is limitation of intracellular iron. Interferon–activated human macrophages down-regulate transferrin receptor expression and are able to inhibit the intracellular multiplication of *L. pneumophila* by limiting the availability of iron inside the cell [66]. This inhibition was totally reversible by addition of excess iron-saturated transferrin, and deferoxamine, an iron chelator, was able to inhibit *L. pneumophila* replication in unstimulated macrophages [66]. Murray et al. [67], however, found that *C. psittaci* growth was not inhibited by deferoxamine, nor was excess iron-saturated transferrin able to reverse chlamydial inhibition in IFN-γ–activated human MdMs. Therefore, it is unlikely that chlamydiae are inhibited in macrophages by this mechanism.

III. EFFECTOR ACTIVITIES OF THE MACROPHAGE IN CHLAMYDIAL INFECTION

A. Activated Macrophages and Monokines in Chlamydial Inhibition

In addition to their role as host cells for *C. trachomatis* biovar LGV, and many strains of *C. psittaci,* macrophages are important effector cells involved in the host defense against chlamydial infection. Williams et al. [68,69] have developed a mouse pneumonitis model to study the cell-mediated immune response to *C. trachomatis*. They showed that alveolar macrophages were stimulated in vivo to control the infection by a T cell–dependent mechanism. Subsequently, IFN-γ, as had been shown to be important in vitro, was shown to be this T cell–derived factor in the animal model [70].

Further investigation by Huebner and Byrne [71] using a murine model of *C. psittaci* infection revealed that peritoneal macrophages collected from *C. psittaci* 6BC–immunized mice dramatically restricted the intracellular growth of chlamydiae. While in vitro IFN-γ–treated macrophages showed a transient microbistatic activity that was lost after removal of the IFN-γ, macrophages from the immune-boosted mice showed a consistent inhibition of chlamydial growth. Furthermore, addition of exogenous tryptophan or treatment with anti–IFN-γ antisera could not abrogate the antichlamydial effect of the immunized macrophages. These results suggested that a factor other than or in addition to IFN-γ was required for macrophages to become fully activated to kill intracellular chlamydiae [71]. This factor may be produced by the macrophage itself.

A number of cytokines have been reported to be produced by mononuclear phagocytes during a chlamydial infection. Coculture of human monocytes with untreated, heat-killed, or UV-treated *C. trachomatis* biovar LGV triggers the release of IL-1 by a mechanism independent of internalization [72]. Monocyte-derived macrophages have been reported to produce TNF-α and prostaglandin E_2 (PGE_2) in response to *C. trachomatis* LGV antigen, but no IL-1 was detected [12]. While this discrepancy is still under investigation, the level of maturity of the mononuclear phagocyte may be responsible for this result [12]. Purified chlamydial LPS can be an effective inducer of IL-1 in monocytes, whereas purified MOMP is a poor inducer [72]. Chlamydial LPS alone could activate peritoneal macrophage to produce PGE_2, but the LPS was generally not active and did not induce classical endotoxin characteristics such as lethal toxicity, pyrogenicity, or local Shwartzman reaction [73].

Shemer-Avni et al. [74] demonstrated an antichlamydial effect of TNF-α in vitro using HEp-2 cells, a human epidermoid carcinoma cell line. Treatment of *C. trachomatis* biovar LGV–infected HEp-2 cells with recombinant TNF-α resulted in marked decrease in a chlamydial yield. This inhibition was accompanied by an increase in tryptophan catabolism [75], reminiscent of the inhibition induced by IFN-γ. Indeed, addition of excess tryptophan to the cell culture medium could reverse its inhibitory effect. Furthermore, a subinhibitory dose of IFN-γ could act synergistically with TNF-α to enhance antichlamydial activity [75]. The addition of anti–IFN-β antibodies could reduce, but not fully eliminate, the inhibition induced by TNF-α. This suggested that the process of TNF-α–induced inhibition is mediated, at least in part, through an autocrine function of IFN-β [75].

Further studies showed that addition of exogenous tryptophan not only interfered with inhibition by TNF-α but also blocked induced PGE_2 formation [76]. However, the addition of antibodies to IFN-β, which reduced TNF-α inhibition, did not affect PGE_2 production [76,77]. Prostaglandin E_2 itself, known to be a modulator of the immune system [78,79], has been shown to have an antichlamydial effect linked to endogenous production of cAMP by the infected host cell [80]. Exogenous cAMP inhibits *C. trachomatis* biovar LGV protein-synthesis at the transcriptional level [81]. Therefore, PGE_2 production may contribute to the observed TNF-α–mediated inhibition.

Using the mouse pneumonitis model of *C. trachomatis* infection, Williams et al. [82–84] confirmed that TNF-α is produced in vivo in response to chlamydial challenge. Exogenous neutralizing antibody to TNF-α significantly accelerated mortality of infected mice, indicating a protective role of TNF-α [84]. In addition, mice treated with recombinant murine TNF-α or LPS were protected against an otherwise lethal challenge. Since anti–TNF-α antibody could also block the protection conferred by LPS, TNF-α was implicated in its protection [84]. Recently, experiments utilizing the mouse pneumonitis model of *C. trachomatis* infection have also confirmed the induction of IL-1 and IL-6, which may be produced by mononuclear phagocytes [85]. Furthermore, both live and heat-killed *C. trachomatis* were found to induce the T cell–dependent production of colony-stimulating factors (CSFs) [86].

Manor and Sarov [87] used an in vitro model to investigate the interaction between mononuclear phagocytes and infected host cells. Both monocytes and MdMs were assayed for their effect on *C. trachomatis* biovar LGV–infected HEp-2 cells. Treatment of the infected cells with fresh monocytes produced a mean five-fold reduction in yield of infectious *Chlamydia*, while treatment with MdMs reduced chlamydial yield by a mean of 20-fold [87]. In order to characterize this inhibition, further experiments were conducted to evaluate the effect of direct contact between the MdMs and the infected HEp-2 cells. Incubation of MdMs in direct contact with infected HEp-2 cells resulted in a considerable decrease in *C. trachomatis* DNA production. When the MdMs and infected HEp-2 cells were separated by a polycarbonate membrane, a significant decrease was also seen, but this decrease was less dramatic than that of the cells in direct contact [87]. In addition, the contribution of TNF-α was assessed by the ability of anti-TNF-α antisera to prevent inhibition. Incubation with anti–TNF-α antisera resulted in a statistically significant reduction in the inhibitory effect of the MdMs both when they were in direct contact with infected cells and when they were separated from them by a membrane. However, it did not fully abrogate the inhibitory effect. No interferon could be detected in the culture media. These results suggest that macrophage-produced factors other than IFN and TNF-α are involved in their extrinsic inhibitory activity [87].

Recently, Shemer-Avni et al. [77] demonstrated that recombinant IL-1α can act directly to inhibit the development of *C. trachomatis* biovar LGV in HEp-2 cells by a mechanism that appears to be similar to that of TNF-α. Inhibition of chlamydial growth by IL-1α, like that of TNF-α but not like that of IFN-γ, is accompanied by increased PGE_2 production [77]. The antichlamydial effects of IL-1α, like those of TNF-α, are reversed by addition of antibody to IFN-β, suggesting an IFN-β–mediated process of inhibition. Still, PGE_2 levels remain high even though TNF-α– and IL-1–mediated inhibition is reversed by anti–IFN-β [77].

Interferon-β, like IFN-γ, can induce IDO in macrophages, but requires LPS to attain a level of activity similar to that of IFN-γ [50,54]. Also, LPS alone in high doses could induce IDO activity [50,54]. The contribution of the added LPS may be to induce TNF-α in the macrophage, which then acts synergistically with the IFN to inhibit the intracellular chlamydia. Therefore, it is possible that IDO induction plays a role in the antichlamydial effect of TNF-α and IL-1α. Indeed, high concentrations of tryptophan can reverse the antichlamydial effect of TNF-α and IL-1α [75–77]. Interestingly, though, inhibition by IL-1α was not accompanied by demonstrable production of tryptophan degradation products [77]. In this case, the protective effect of tryptophan may have been to inhibit PGE_2 synthesis, as excess tryptophan also could prevent the inhibition of chlamydiae in HEp-2 cells treated with exogenous PGE_2 [77]. It is apparent that a complex web of multiple pathways is induced by these cytokines (Figure 3).

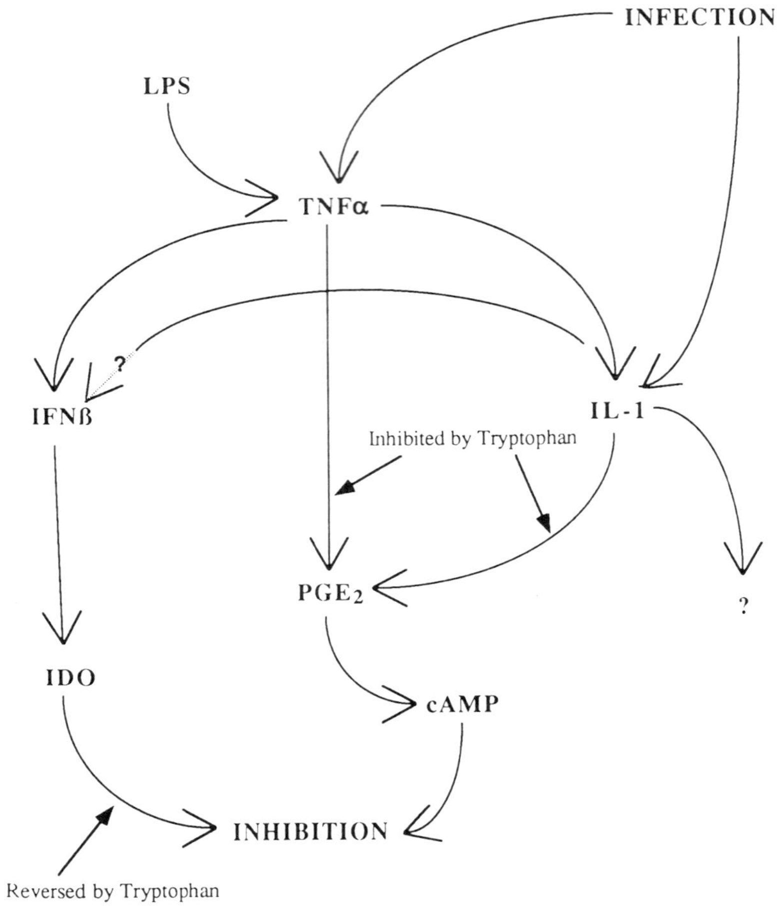

FIGURE 3 The complex web of monokine interactions in chlamydial inhibition. IDO = indoleamine 2,3-dioxygenase; IFNβ = interferon-β; IL-1 = interleukin-1; LPS = lipopolysaccharide; PGE_2 = progstaglandin E_2; TNFα = tumor necrosis factor-α.

B. Processing and Presentation of Chlamydial Antigen by Macrophages

While the aspects of antigen presentation in chlamydial disease have not been worked out, it would not be surprising to find that macrophages make an important contribution. Peritoneal macrophages from mice chronically infected with *C. psittaci* had increased expression of class I MHC antigens (H-2 K, D) and class II MHC antigens (I-A and I-E) [88]. Zhong and de la Maza [59] showed that peritoneal macrophages activated in vitro with recombinant murine IFN-γ to restrict the growth of *C. trachomatis* biovar LGV had a parallel expression of Ia^d on the macrophages. Moreover, mice receiving these Ia^d-bearing activated macrophages showed a significant decrease in the yield of inclusion-forming units from their spleens and peritoneal fluids [59]. Stagg et al. [89] showed that dendritic cells could stimulate both synovial fluid–derived and peripheral blood–derived T cells that were selected with *C. trachomatis* biovar LGV, but not T cells selected with an irrelevant antigen. Further investigation is necessary to characterize chlamydial antigen processing and presentation by macrophages.

IV. PROSPECTUS

As more is discovered about the chlamydial disease process, the contribution of macrophages to both pathogenesis and immunity becomes more apparent. In the mouse, clearance of a primary chlamydial infection corresponds to the appearance of an influx of activated macrophages, and macrophages are the principal responding phagocytic cell in subsequent infections. Cytokines produced by monocytes and macrophages during an infection may serve to stimulate host resistance. Alternatively, there are a number of ways macrophages may contribute to damage during a chlamydial infection. During chronic infection, sustained release of cytokines, especially IL-1 and TNF-α, can lead to nonspecific immunological damage and scarring. Macrophages also may serve to transport chlamydiae from an initial site of infection to deeper tissues. This is apparent in lymphogranuloma venerum, where otherwise nonmotile *C. trachomatis* biovar LGV organisms, which replicate in unactivated macrophages, reach the regional lymph nodes. In *C. psittaci* infection, inhaled chlamydiae are transported to the spleen and liver via lymphatics, probably by macrophages. Thus macrophages may be exploited by the chlamydiae for dissemination in certain types of chlamydial disease.

Another important factor may be the macrophage's contribution to the persistence of a chlamydial infection. Often chlamydial infections persist over long periods, and cultures may be negative between recurrences. Since the antichlamydial effect of macrophages is initially only microbistatic, there is a window where chlamydiae may persist in the face of the immune response. Furthermore, in cases where cytokine stimulation may be weak, it is possible that chlamydiae may remain within macrophages in this persistent state indefinitely.

REFERENCES

1. Schachter J. Chlamydial infections. West J Med 1990; 153:523–34.
2. Moulder JW. Interaction of Chlamydiae and host cells in vitro. Microbiol Rev 1991; 55:143–90.
3. Fukushi H, Hirai K. Proposal of *Chlamydia pecorum* sp. nov. for *Chlamydia* strains derived from ruminants. Int J Syst Bacteriol 1992; 42:306–8.
4. Thom DH, Grayston JT. Infections with *Chlamydia pneumoniae* strain TWAR. Clin Chest Med 1991; 12:245–56.
5. Grayston JT, Thom DH. The chlamydial pneumonias. Curr Clin Top Infect Dis 1991; 11:1–18.
6. Crosse BA. Psittacosis: a clinical review. J Infect 1990; 21:251–9.
7. Wong SY, Gray ES, Buxton D, Finlayson J, Johnson FWA. Acute placentitis and spontaneous abortion caused by *Chlamydia psittaci* of sheep origin: a histological and ultrastructural study. J Clin Pathol 1985; 38:700–11.
8. Caul EO, Roome APCH, McGivern D. Spontaneous abortion caused by *Chlamydia psittaci* of ovine origin. Com Dis Rep 1986; 45:3–4.
9. Yong EC, Chi EY, Kuo C-C. Differential antimicrobial activity of human mononuclear phagocytes against the human biovars of *Chlamydia trachomatis*. J Immunol 1987; 139:1297–1302.
10. Manor E, Sarov I. Fate of *Chlamydia trachomatis* in human monocytes and monocyte-derived macrophages. Infect Immun 1986; 54:90–5.
11. Bard J, Levitt D. *Chlamydia trachomatis* (L2 serovar) binds to distinct subpopulations of human peripheral blood leukocytes. Clin Immunol Immunopathol 1986; 38:150–60.
12. Manor E, Schmitz E, Sarov I. TNF-α, IL-1, and PGE_2 production by human monocyte–derived macrophages infected with *C. trachomatis*. In: Bowie WR, Caldwell HD, Jones RP, Mardh P-A, Ridgeway GL, Schachter J, Stamm WE, Ward ME, eds. Chlamydial infections. Cambridge, England: Cambridge University Press, 1990:209–12.
13. Kuo C-C. Cultures of *Chlamydia trachomatis* in mouse peritoneal macrophages: factors affecting growth. Infect Immun 1978; 20:439–45.
14. Nakajo MN, Roblin PM, Hammerschlag MR, Smith P, Nowakowski M. Chlamydicidal activity of human alveolar macrophages. Infect Immun 1990; 58:3640–4.
15. Hammerschlag MR. Clamydial infections. J Pediatr 1989; 114:727–34.

16. Nakajo MM, Roblin PM, Hammerschlag MR. Interaction of *Chlamydia trachomatis* with human alveolar macrophages. In: Bowie WR, Caldwell HD, Jones RP, Mardh P-A, Ridgeway GL, Schachter J, Stamm WE, Ward ME, eds. Chlamydial infections. Cambridge, England: Cambridge University Press, 1990: 28–31.
17. Nakajo M, Roblin P, Hammerschlag M, Steiner P, Nowakowski M. Interaction of alveolar macrophages with *Chlamydia trachomatis* (abstr). Pediatr Res 1988;23:377A.
18. Gogolak FM. The histopathology of murine pneumonitis infection and the growth of the virus in the mouse lung. J Infect Dis 1953; 92:254–72.
19. Rothermel CD, Rubin BY, Jaffe EA, Murray HW. Oxygen-independent inhibition of intracellular *Chlamydia psittaci* growth by human monocytes and interferon-γ–activated macrophages. J Immunol 1986; 137:689–92.
20. Rothermel CD. Chlamydia and human mononuclear phagocytes: relationship between parasite growth and host cell differentiation. In: Oriel D, Ridgeway G, Schachter J, Taylor-Robinson D, Ward M, eds. Chlamydial infections. Cambridge, England: Cambridge University Press, 1986; 437–40.
21. Wyrick PB, Brownridge EA. Growth of *Chlamydia psittaci* in macrophages. Infect Immun 1978; 19:1054–60.
22. Rothermel CD. Binding of *Chlamydia psittaci* to human monocyte-derived macrophages. In: Bowie WR, Caldwell HD, Jones RP, Mardh P-A, Ridgeway GL, Schachter J, Stamm WE, Ward ME, eds. Chlamydial infections. Cambridge, England: Cambridge University Press, 1990: 24–27.
23. Murray HW, Gellene RA, Libby DM, Rothmel CD, Rubin BY. Activation of tissue macrophages from AIDS patients: in vitro response of AIDS alveolar macrophages to lymphokines and interferon-γ. J Immunol 1985; 135:2374–7.
24. Gregory WW, Byrne GI, Gardner M, Moulder JW. Cytochalasin B does not inhibit ingestion of *Chlamydia psittaci* by mouse fibroblasts (L-cells) and mouse peritoneal macrophages. Infect Immun 1979; 25:463–6.
25. Wyrick PB, Brownridge EA, Ivins BE. Interaction of *Chlamydia psittaci* with mouse peritoneal macrophages. Infect Immun 1978; 19:1061–7.
26. Eissenberg LG, Wyrick PB, Davis CH, Rump JW. *Chlamydia psittaci* elementary body envelopes: ingestion and inhibition of phagolysosome fusion. Infect Immun 1983; 40:741–51.
27. Kuo C-C. Immediate cytotoxicity of *Chlamydia trachomatis* for mouse peritoneal macrophages. Infect Immun 1978; 20:613–18.
28. Brownridge E, Wyrick PB. Interaction of *Chlamydia psittaci* reticulate bodies with mouse peritoneal macrophages. Infect Immun 1979; 24:697–700.
29. Moulder JW. Comparative biology of intracellular parasitism. Micobiol Rev 1985; 49:298–337.
30. Zeichner SL. Isolation and characterization of macrophage phagosomes containing infectious and heat-inactivated *Chlamydia psittaci:* two phagosomes with different intracellular behaviors. Infect Immun 1983; 40:956–66.
31. Eissenberg LG, Wyrick PB. Inhibition of phagolysosome fusion is localized to *Chlamydia psittaci*–laden vacuoles. Infect Immun 1981; 32:889–96.
32. Mørland B, Byrne GI, Jones TC. the effect of intracellular *Chlamydia psittaci* on Lysosomal enzyme activities in mouse peritoneal macrophages. Acta Pathol Microbiol Immunol Scand Sect C 1987; 95:291–3.
33. Hammerschlag M, Suntharalingam K, Fikrig S. The effect of *Chlamydia trachomatis* on luminol-dependent chemiluminescence of human polymorphonuclear leukocytes: requirements for opsonization. J Infect Dis 1985; 151:1045–51.
34. Fitzpatrick DR, Wie J, Webb D, Boufiglioi R, Gardner ID, Mathews JD, Bielefeldt-Ohmann H. Preferential binding of *Chlamydia trachomatis* to subsets of human lymphocytes and induction of interleukin-6 and interferon-gamma. Immunol Cell Biol 1991; 69:337–48.
35. Newhall WJ, Batteiger B, Jones RB. Analysis of the human serological response to proteins of *Chlamydia trachomatis*. Infect Immun 1982; 38:1181–9.
36. Lundemose AG, Birkelund S, Fey SJ, Mose Larsen P, Christiansen G. *Chlamydia trachomatis* contains a protein similar to the *Legionella pneumophila mip* gene product. Mol Microbiol 1991; 5:109–15.
37. Cianciotto NP, Eisenstein BI, Mody CH, Toews GB, Engleberg NC. A *Legionella pneumophila* gene encoding a species-specific surface protein potentiates initiation of intracellular infection. Infect Immun 1989; 57:1255–62.

38. Cianciotto NP, Eisenstein BI, Mody CH, Engleberg NC. A mutation in the mip gene results in an attenuation of *Legionella pneumophila* virulence. J Infect Dis 1990; 162:121–6.
39. Byrne GI, Carlin JM, Merkert TP, Arter DL. Long-term effects of gamma interferon on *Chlamydia*-infected host cells: microbicidal activity follows microbistasis. Infect Immun 1989; 57:1318–20.
40. Robinson P, Wakefield D, Graham DM, Easter JF, Penny R. The chemiluminescence response of normal human leukocytes to *Chlamydia trachomatis*. Diag Immunol 1985; 3:119–25.
41. Yong EC, Chi EY, Chen W-J, Kuo C-C. Degradation of *Chlamydia trachomatis* in human polymorphonuclear leukocytes: an ultrastructural study of peroxidase-positive phagolysosomes. Infect Immun 1986; 53:427–31.
42. Yong EC, Klebanoff SJ, Kuo CC. Toxic effect of human polymorphonuclear leukocytes on *Chlamydia trachomatis*. Infect Immun 1982; 37:422–6.
43. Byrne GI, Faubion CL. Inhibition of *Chlamydia psittaci* in oxidatively active thioglycolate-elicited macrophages: distinction between lymphokine-mediated oxygen-dependent and oxygen-independent macrophage activation. Infect Immun 1983; 40:464–71.
44. Byrne GI, Faubion CL. Lymphokine-mediated microbistatic mechanisms restrict *Chlamydia psittaci* growth in macrophages. J Immunol 1982; 128:469–74.
45. Murray HW, Byrne GI, Rothermel CD, Cartelli DM. Lymphokine enhances oxygen-independent activity against intracellular pathogens. J Exp Med 1983; 158:234–9.
46. Murray HW. Survival of intracellular pathogens within human mononuclear phagocytes. Semin Hematol 1988; 25:101–11.
47. Rothermel CD, Rubin BY, Murray HW. γ-Interferon is the factor in lymphokine that activates human macrophages to inhibit intracellular *Chlamydia psittaci* replication. J Immunol 1983; 121:2542–4.
48. Byrne GI, Krueger DA. Lymphokine-mediated inhibition of *Chlamydia* replication in mouse fibroblasts is neutralized by anti–gamma interferon immunoglobulin. Infect Immun 1983; 142:1152–8.
49. Shemer Y, Sarov I. Inhibition of growth of *Chlamydia trachomatis* by human gamma interferon. Infect Immun 1985; 48:592–6.
50. Carlin JM, Borden EC, Sondel PM, Byrne GI. Interferon-induced indoleamine 2,3-dioxygenase activity in human mononuclear phagocytes. J Leukocyte Biol 1989; 45:29–34.
51. Carlin JM, Borden EC, Sondel PM, Byrne GI. Biologic response modifier–induced indoleamine 2,3-dioxygenase activity in human peripheral blood mononuclear cell cultures. J Immunol 19487; 139:2414–8.
52. Taylor MW, Feng G. Relationship between interferon-γ, indoleamine 2,3-dioxygenase, and tryptophan catabolism. FASEB J 1991; 5:2516–22.
53. Murray HW, Szuro-Sudol A, Wellner D, Oca MJ, Granger AM, Libby DM, Rothermel CD, Rubin BY. Role of tryptophan degradation in respiratory burst–independent antimicrobial activity of gamma interferon–stimulated human macrophages. Infect Immun 1989; 57:845–9.
54. Carlin JM, Borden EC, Byrne GI. Interferon-induced indoleamine 2,3-dioxygenase activity inhibits Chlamydia psittaci replication in human macrophages. J Interferon Res 1989; 9:329–37.
55. Radzioch D, Varesio L. Protein kinase C inhibitors block the activation of macrophages by IFN-β but not IFN-γ. J Immunol 1988; 140:1259–63.
56. Byrne GI. Mechanism of gamma interferon–mediated inhibition of *Chlamydia psittaci* growth in human uroepithelial cells. In: Oriel D, Ridgeway G, Schachter J, Taylor-Robinson D, Ward M, eds. Chlamydial infections. Cambridge, England: Cambridge University Press, 1986: 445–8.
57. Shemer Y, Kol R, Sarov I. Tryptophan reversal of recombinant human gamma-interferon inhibition of *Chlamydia trachomatis* growth. Curr Microbiol 1987; 16:9–13.
58. De la Maza LM, Peterson EM, Fennie CW, Czarniecki CW. The anti-chlamydial and anti-proliferative activities of recombinant murine interferon-γ are not dependent on tryptophan concentrations. J Immunol 1985; 135:4198–200.
59. Zhong G, de la Maza LM. Activation of mouse peritoneal macrophages in vitro or in vivo by recombinant murine gamma interferon inhibits the growth of *Chlamydia trachomatis* serovar L1. Infect Immun 1988; 56:3322–5.
60. Liew FY. Role of cytokines in killing of intracellular pathogens. Immunol Lett 1991; 30:193–8.
61. Adams LB, Hibbs JB, Taintor RR, Krahenbuhl JL. Microbistatic effect of murine activated macrophages for *Toxoplasma gondii*: role for synthesis of inorganic nitrogen oxides from L-arginine. J Immunol 1990; 144:2725–9.

62. Rockett KA, Awburn MM, Cowden WB, Clark IA. Killing of *Plasmodium falciparum* in vitro by nitric oxide derivatives. Infect Immun 1991; 59:3280–3.
63. Vincendeau P, Daulouède S. Macrophage cytostatic effect on *Trypanosoma musculi* involves an L-arginine–dependent mechanism. J Immunol 1991; 146:4338–43.
64. Alspaugh JA, Granger DL. Inhibition of *Cryptococcus neoformans* replication by nitrogen oxides supports the role of these molecules as effectors of macrophage-mediated cytostasis. Infect Immun 1991; 59:2291–6.
65. Flesch IEA, Kaufmann SHE. Mechanisms involved in mycobacterial growth inhibition by gamma interferon–activated bone marrow macrophages: role of reactive nitrogen intermediates. Infect Immun 1991; 59:3213–8.
66. Byrd TF, Horwitz MA. Interferon-gamma–activated human monocytes downregulate transferrin receptors and inhibit intacellular multiplication of *Legionella pneumophila* by limiting the availability of iron. J Clin Invest 1989; 83:1457–65.
67. Murray HW, Granger AM, Teitelbaum RF. Gamma interferon–activated human macrophages and *Toxoplasma gondii, Chlamydia psittaci*, and *Leishmania donovani*: antimicrobial role of limiting intracellular iron. Infect Immun 1991; 59:4684–6.
68. Williams DM, Schachter J, Coalson JJ, Grubbs B. Cellular immunity to the mouse pneumonitis agent. J Infect Dis 1984; 149:630–9.
69. Williams DM, Schachter J. Role of cell-mediated immunity in Chlamydial infection: implications for ocular immunity. Rev Infect Dis 1985; 7:754–9.
70. Byrne GI, Grubbs B, Dickey TJ, Schachter J, Williams DM. Interferon in the recovery from pneumonia due to *Chlamydia trachomatis* in the mouse. J Infect Dis 1987; 156:993–6.
71. Huebner RE, Byrne GI. In vivo-activated mononuclear phagocytes and protective immunity to chlamydiae in mice. Infect Immun 1988; 56:1492–9.
72. Rothermel CD, Schachter J, Lavrich P, Lipsitz EC, Francus T. *Chlamydia trachomatis*–induced production of interleukin-1 by human monocytes. Infect Immun 1989; 57:2705–11.
73. Brade L, Schramek S, Schade U, Brade H. Chemical, biological, and immunochemical properties of the *Chlamydia psittaci* lipopolysaccharide. Infect Immun 1986; 54:568–74.
74. Shemer-Avni Y, Wallach D, Sarov I. Inhibition of *Chlamydia trachomatis* growth by recombinant tumor necrosis factor. Infect Immun 1988; 56:2503–6.
75. Shemer-Avni Y, Wallach D, Sarov I. Reversion of the antichlamydial effect of tumor necrosis factor by tryptophan and antibodies to beta interferon. Infect Immun 1989; 57:3484–90.
76. Holtmann H, Shemer-Avni Y, Wessel K, Sarov I, Wallach D. Inhibition of growth of *Chlamydia trachomatis* by tumor necrosis factor is accompanied by increased prostaglandin synthesis. Infect Immun 1990; 58:3168–72.
77. Shemer-Avni Y, Holtmann H, Wallach D, Sarov I. Cytokine-cell interactions resulting in *Chlamydia trachomatis* inhibition–mechanisms involved. In: Bowie WR, Caldwell HD, Jones RP, Mardh P-A, Ridgeway GL, Schachter J, Stamm WE, Ward ME, eds. Chlamydial infections. Cambridge, England: Cambridge University Press, 1990:173–6.
78. Phipps RP, Stein SH, Roper RL. A new view of prostaglandin E regulation of the immune response. Immunol Today 1991; 12:349–52.
79. Chouaib S, Bertoglio JH. Prostaglandins E as modulators of the immune response. Lymphokine Res 1988; 7:237–45.
80. Ward ME, Salari H. Control mechanisms governing the infectivity of *Chlamydia trachomatis* for HeLa cells: modulation by cyclic nucleotides, prostaglandins and calcium. J Gen Microbiol 1982; 128:639–50.
81. Kaul R, Tao S, Wenman WM. Cyclic AMP inhibits protein synthesis in *Chlamydia trachomatis* at a transcriptional level. Biochim Biophys Acta 1990; 1053:106–12.
82. Williams DM, Bonewald LF, Roodman GD, Byrne GI, Magee DM, Schachter J. Tumor necrosis factor is a cytotoxin induced by murine *Chlamydia trachomatis* infection. Infect Immun 1989; 57:1351–5.
83. Williams DM, Magee DM, Bonewald LF, Smith JG, Bleicker CA, Schachter J. Tumor necrosis factor alpha induction by MoPn. In: Bowie WR, Caldwell HD, Jones RP, Mardh P-A, Ridgeway GL, Schachter J, Stamm WE, Ward ME, eds. Chlamydial infections. Cambridge, England: Cambridge University Press, 1990:217–20.
84. Williams DM, Magee DM, Bonewald LF, Smith JG, Bleicker CA, Byrne GI, Schachter J. A role in vivo

for tumor necrosis factor alpha in host defense against *Chlamydia trachomatis*. Infect Immun 1990; 58:1572–6.
85. Magee DM, Smith JG, Bleicker CA, Carter CJ, Bonewald LF, Schachter J, Williams DM. *Chlamydia trachomatis* pneumonia induces in vivo production of interleukin-1 and -6. Infect Immun 1992; 60:1217–20.
86. Magee DM, Williams DM, Wing EJ, Bleicker CA, Schachter J. Production of colony-stimulating factors during pneumonia caused by *Chlamydia trachomatis*. Infect Immun 1991; 59:2370–5.
87. Manor E, Sarov I. Inhibition of Chlamydia trachomatis replication in HEp-2 cells by human monocyte-derived macrophages. Infect Immun 1988; 56:3280–4.
88. Guagliardi LE, Paulnock DM, Albrecht RM. Expression of major histocompatability complex antigens on macrophages: correlative study using flow cytometry, radioimmunoassay, and colloidal gold immunolabeling. Scanning Microscopy 1987; 1:705–11.
89. Stagg AJ, Harding B, Hughees R, Keat A, Ward ME, Knight SC. Cell mediated immune response to chlamydial antigens in sexually acquired reactive arthritis. In: Bowie WR, Caldwell HD, Jones RP, Mardh P-A, Ridgeway GL, Schachter J, Stamm WE, Ward ME, eds. Chlamydial infections. Cambridge, England: Cambridge University Press, 1990:189–92.
90. Moulder JW. The relation of basic biology to pathogenic potential in the genus *Chlamydia*. Infection 1982; 10:S10–S18.
91. Kuo C-C, Grayston JT. Factors affecting viability and growth in HeLa 229 cells of *Chlamydia* sp. strain TWAR. J Clin Microbiol 1988; 26:812–5.
92. Schachter J. Pathogenesis of chlamydial infections. Pathol Immunopathol Res 1989; 8:206–20.
93. Black CM, Johnson JE, Farshy CE, Brown TM, Berdal BP. Antigenic variation among strains of Chlamydia pneumoniae. J Gen Microbiol 1991; 29:1312–6.

24

Rickettsiae and Macrophages

Herbert H. Winkler and Jenifer Turco

University of South Alabama College of Medicine, Mobile, Alabama

I. INTRODUCTION

Rickettsiae are obligate intracellular bacteria and the etiological agents of epidemic, endemic, and scrub typhus and various spotted fevers. These organisms morphologically and chemically resemble gram-negative bacteria, except for *Rickettsia tsutsugamushi*, the agent of scrub typhus [reviewed in 1]. Rickettsiae enter their host through the bite of an arthropod vector, a vector in which the rickettsiae grow intracytoplasmically as either parasites or symbiotes. Subsequently, the rickettsiae enter the cytoplasm of various cells within their host and grow free in this cytoplasm unbounded by any vacuole. Rickettsial pathology is the result of the parasites' growth in and destruction of their host cells. The rickettsiae have a tropism for vascular endothelial cells.

Our focus and fundamental question in this review is, What is the outcome of the interaction of a macrophage and an obligate intracytoplasmic bacterial parasite, a rickettsia? The response is the standard answer to an apparently simple question: It depends on both the rickettsia and the macrophage. Therefore, the review is divided into separate sections to deal with rickettsial interactions with 1) native macrophages, 2) antibody and macrophages, and 3) cytokines and macrophages. It should be noted that, although there are a large number of possible interactions, the possible outcomes of these interactions are limited: the rickettsia will either live or die, as will the macrophage. To understand the mechanisms involved in these enigmatic associations of professional phagocytes and rickettsiae is not only a daunting challenge, but also a guaranteed source of excitement for the scientist on this quest.

II. NATIVE MACROPHAGES

The etiologic agent of epidemic typhus, *Rickettsia prowazekii*, whether untreated or pretreated with normal serum, proliferates within the cytoplasm of human monocyte–derived macrophages [2–4], guinea pig and cotton rat peritoneal macrophages [5,6], and at least four mouse, and one human, macrophage-like cell lines [7]. The other member of the typhus group of rickettsiae, *Rickettsia typhi* (whether untreated, pretreated with normal serum, or pretreated with normal serum plus complement), also proliferates within human monocyte-derived macrophages [2,8]. Similarly, growth of normal serum–treated *Rickettsia rickettsii*, the etiologic agent of Rocky Mountain spotted fever, is supported by guinea pig peritoneal macrophages [9]. Another member of the spotted fever group of rickettsiae, *Rickettsia conorii*, is able to grow in cultures of the peritoneal macrophages of DBA/2 mice and guinea pigs [10] and is observed in the splenic and lymph node macrophages of patients with severe infections [11]. Yokota et al. demonstrated that normal human peripheral macrophages support the in vitro growth of *R. tsutsugamushi*, the etiologic agent of scrub typhus [12]. Growth of *R. tsutsugamushi* (Gilliam) in both BALB/c and C3H/He mouse resident peritoneal macrophage cultures has been demonstrated [13–15]. It is noteworthy that in vivo the BALB/c mouse strain is resistant to infection, while the C3H/He strain is susceptible. However, in studies with the agent of rickettsial pox, *Rickettsia akari*, a definite positive correlation is seen between the in vivo sensitivity of two mouse strains and the growth of the agent in the splenic macrophages of these strains following in vitro infection [16].

The lack of a suitable animal model that reproduces the features of human epidemic typhus has limited the study of host defenses against *R. prowazekii*. Animal models have been more extensively used in studying host defenses against other rickettsial species. For example, whole-animal studies with *R. tsutsugamushi* show that when susceptible mice are infected the macrophages are either full of rickettsiae or have succumbed to this rickettsial burden; however, such infected macrophages are not found in resistant mice [14,17,18]. These whole-animal observations contrast with those dealing with *R. tsutsugamushi* in cultured native macrophages discussed in the paragraph above. This points out the difficulty, if not the impossiblity, of defining a "native" macrophage in the complex whole animal.

The naive host is confronted with an organism that can grow not only in the vascular endothelium but also in macrophages, professional phagocytes specially adapted for the phagocytosis and destruction of bacteria. Since human rickettsial infections, even without antibiotic treatment, are fatal less than 20% of the time, the host must be able to mount an effective defense against this intracellular bacterium. The significance of growth or the lack of growth in macrophages to the outcome of a rickettsial infection may be inferred from the fact that an avirulent strain of *R. prowazekii* exists that, to the best of our knowledge, differs from a virulent strain most significantly in that the avirulent strain is killed by native macrophages or macrophage-like cells [2,7]. That this ability of macrophages to distinguish the virulent from the avirulent rickettsiae can occur at a stage after the entry of the rickettsiae into the macrophage is suggested by the observation that the avirulent E strain of *R. prowazekii* can be killed by undefined mechanisms within the cytoplasm of heterokaryons formed from the fusion of uninfected macrophage-like cells and infected L929 cells [19].

III. ANTIBODY AND MACROPHAGES

A. In Vitro Studies

The destruction of immune serum–treated rickettsiae by macrophages has been documented in in vitro studies with several *Rickettsia* species. This is in contrast with the interaction of rickettsiae and native macrophages, in which the rickettsiae grow and the macrophages die. For example, if epidemic typhus rickettsiae are pretreated with immune serum, they are phagocytosed and killed by human monocyte-derived macrophages [2–4] and mouse macrophage-like cells [7]. Human monocyte-derived macrophages also can destroy *R. typhi* organisms that are pretreated with immune serum [8]. Pretreatment of viable *R. typhi* with immune serum plus complement enhances uptake of the rickettsiae and leads to killing of many of the macrophages. This complement-associated cytotoxic effect presages the discovery of cytokine-associated cytotoxicity in macrophages (see below). Kekcheeva et al. [5] found that macrophages obtained from guinea pigs after infection with virulent *R. prowazekii* restrict the growth of immune serum–treated *R. prowazekii* in comparison with normal serum–treated *R. prowazekii*. Similarly, *R. rickettsii* organisms that are pretreated with immune serum are destroyed by guinea pig peritoneal macrophages [9]. In a study of the scrub typhus agent, destruction of immune serum–treated *R. tsutsugamushi* and growth of normal serum–treated *R. tsutsugamushi* in mouse and human peritoneal macrophages was demonstrated [12,13].

On the basis of ultrastructural observations, immune serum–treated epidemic typhus rickettsiae are apparently destroyed within the phagolysosomes of human [20,21] and guinea pig [22,23] macrophages and mouse macrophage-like cell lines [24], since abnormal rickettsiae are found within these vacuoles. The reason that immune serum–coated rickettsiae, but not naked or normal serum–coated rickettsiae, are destroyed in macrophages and macrophage-like cells is unknown. The first hypothesis would be that the route of entry is different. Antibody-treated rickettsiae would be taken up by macrophage Fc receptor–mediated phagocytosis (so that the rickettsiae would be in a phagolysosome), rather than by the normal mechanism, i.e., rickettsia-induced phagocytosis and concerted lysis of the phagosome [25] (in which case the rickettsiae would be free in the cytoplasm). The alternative hypothesis would be that the antibody neutralizes an activity of the rickettsiae that is necessary to prevent intracellular destruction of the rickettsiae by the macrophage. Although the first hypothesis may seem the more attractive, our attraction to it probably has been conditioned in us by the use of the word "opsonized"; in fact, the available evidence contradicts this hypothesis. Macrophages bearing cytophilic antirickettsial antibody phagocytize unopsonized rickettsiae to a greater extent than macrophages without cytophilic antibody; however, these rickettsiae, which apparently enter the macrophages in large part through an Fc receptor–mediated process, are not killed [3,4]. These data indicate that Fc receptor–mediated phagocytosis is not sufficient for the destruction of the rickettsiae in the macrophage. Whether Fc receptor–mediated phagocytosis is required for destruction of virulent *R. prowazekii* in mouse macrophage-like cells was examined by infecting macrophage-like cells with rickettsiae pretreated with $F(ab')_2$ fragments prepared from immunoglobulin G (IgG) that was purified from hyperimmune rabbit serum. Whereas the untreated rickettsiae grow well in the macrophage-like cells and the untreated, IgG-treated, and $F(ab')_2$-treated rickettsiae all grow at similar rates in L929 fibroblastic cells, both the IgG-treated and the $F(ab')_2$-treated rickettsiae are destroyed in the macrophage-like cells [24]. These data indicate that the Fc portion of the antibody molecule (and therefore Fc receptor–mediated phagocytosis) is not necessary for destruction of virulent *R. prowazekii* in mouse macrophage-like cells.

The mechanisms involved in the phagocytosis and destruction of $F(ab')_2$-coated rickettsiae (or, for that matter, IgG-coated rickettsiae) in mouse macrophage-like cells are unknown. These $F(ab')_2$-coated rickettsiae, like IgG-coated rickettsiae, are found in vacuoles [24] and thus apparently are killed within Fc-independent phagosomes or phagolysosomes in the macrophage-like cells. These vacuoles may be those induced by normal rickettsial uptake from which the rickettsiae fail to escape. The mode of killing within these vesicles does not involve a respiratory burst. This was shown experimentally not only by demonstrating that *R. prowazekii* does not induce a respiratory burst, but also by using a variant macrophage-like cell line that is unable to give a respiratory burst, a cell line in which the IgG-coated rickettsiae are still killed [24,26]. Alternatively, the coated rickettsiae may not be killed within these phagosomes. Instead, they may escape from these phagosomes to the cytoplasm and be killed there. However, since these rickettsiae are clearly found within vesicles, one would have to extend the argument to include an autophagic-like engulfing of the dead rickettsiae. There is a precedent for the killing of *R. prowazekii* in the cytoplasm of macrophage-like cells [19].

B. In Vivo Studies

The definitive experiments to demonstrate the in vivo relevance of the in vitro killing of opsonized rickettsiae by nonactivated macrophages and to establish if, in fact, antibody and nonactivated macrophages are acting together to kill rickettsiae in vivo have not been done. Reports concerning the effects of passive transfer of antibody have been mixed, and even when protection was observed a role for the macrophage could not be assessed. Early studies indicated that passive transfer of immune serum is protective against *R. prowazekii* [27–30]. More recently, Crist et al.

[31] found that passive transfer of immune serum to mice 6 hours after challenge with *R. typhi* does not control the infection in the spleen. In contrast, immune serum administered to guinea pigs before intradermal challenge with *R. typhi* suppresses the level of infection in the kidney, but not the formation of lesions at the inoculation sites [32]. Shirai et al. [33] found that although mice immunized with the Gilliam strain of *R. tsutsugamushi* are resistant to challenge with the lethal Karp strain, passive transfer of serum obtained from the Gilliam-immunized mice does not protect recipient mice from subsequent challenge with the Karp strain. In contrast, passive transfer of serum that contains antibodies against the Karp strain of *R. tsutsugamushi* protects recipient mice from subsequent challenge with the Karp strain of *R. tsutsugamushi*. Kokorin et al. [34] reported that administration of immune serum does not protect mice from lethal infection with *R. conorii*.

Jerrells suggested that the in vivo role of nonactivated macrophages and opsonized rickettsiae could be ascertained if antibody were passively transferred to athymic mice (mice that contain essentially normal macrophages) and the degree of protection of the mice determined after they were challenged with rickettsiae. He states, "In preliminary experiments [with *R. tsutsugamushi*], antibody produced in thymus-bearing mice given to athymic mice offered no protection" [35]. However, Kobayashi et al. [36] observed that the administration of immune serum not only protects thymus-bearing mice against lethal challenge with *R. tsutsugamushi*, but also prolongs the survival time of lethally challenged athymic mice.

The capacity of immune serum to neutralize the ability of viable rickettsiae to cause early (toxic) death or lethal infection in animals has been known for many years [37–39]. If macrophages are involved in these neutralization phenomena at all, then they probably are killing the immune serum–coated rickettsiae. Monoclonal antibodies to rickettsial epitopes have, to various extents, the ability to neutralize the lethal effects of rickettsiae in animals [21,40–53]. A well-designed in vitro study of the interaction of macrophages with rickettsiae that have been coated with these, and other, monoclonal antibodies could provide an important tool to help elucidate those rickettsial components necessary for survival in macrophages.

IV. CYTOKINES, OTHER MEDIATORS, AND MACROPHAGES

A. In Vitro Studies of Lymphokines

Lymphokines can dramatically alter the outcome of rickettsia-macrophage interactions in vitro. Crude lymphokine preparations (supernatant fluids collected from spleen cells cultured with concanavalin A or supernatant fluids collected from immune spleen cells cultured with antigen) activate mouse peritoneal macrophages in vitro to express antimicrobial activity against *R. tsutsugamushi* Gilliam [13,14,54,55], *R. conorii* Casablanca [56], and *R. akari* Kaplan [57]. Yokota et al. [12] found that culture supernatant fluids from T cell–enriched fractions of mononuclear cells obtained from humans convalescing after *R. tsutsugamushi* disease suppress growth of *R. tsutsugamushi* in normal human macrophages. The growth of *R. prowazekii* Breinl in human monocyte-derived macrophages is inhibited when the macrophages are treated with supernatant fluids collected from cultures of phytohemagglutinin- or antigen-stimulated human leukocytes [58]. In addition, mouse macrophage-like RAW264.7 cells do not support growth of *R. prowazekii* Breinl after treatment of the cells with crude mouse lymphokines [59].

Nacy and Meltzer [54] found that lymphokines induce two distinct antirickettsial effects in mouse macrophages: 1) treatment of macrophages with lymphokines before infection suppresses the ability of the macrophages to become initially infected with rickettsiae, and 2) intracellular rickettsiae are killed within lymphokine-treated macrophages. These two antirickettsial effects have been observed in macrophages harvested from appropriate mouse strains, treated with lymphokines in vitro, and then infected with either *R. tsutsugamushi* Gilliam or *R. akari* Kaplan [54,57]. Macrophages from rickettsia-challenged or BCG-challenged mice are activated in vivo and do not require lymphokine treatment in vitro to display the antirickettsial effects. Both effects have been documented in macrophages harvested from appropriate mouse strains 7 or 8 days after inoculation of the mice with *R. tsutsugamushi* Gilliam or *R. akari* Kaplan [57,60], and in macrophages harvested from C3H/HeN mice 10 days after intraperitoneal inoculation with viable *Mycobacterium bovis* strain BCG [60]. Treatment of macrophage-like RAW264.7 cells with high concentrations of lymphokines also inhibits the ability of these cells to be initially infected with *R. prowazekii* [59]. Macrophages harvested from mice and guinea pigs convalescent after *R. conorii* (strain M-1) infection are activated in that they restrict rickettsial growth and more actively digest intracellular rickettsiae when compared with macrophages from control animals [10]. Macrophages from guinea pigs infected 1 month earlier with *R. prowazekii* Breinl suppress the growth of *R. prowazekii* Breinl compared with macrophages from control guinea pigs, and the former exhibit a much greater capacity for killing rickettsiae pretreated with immune serum than the latter [5]. Although macrophages collected from guinea pigs 2 months after infection with *R. prowazekii* Breinl support of *R. prowazekii* Breinl in vitro, these macrophages contain larger phagolysosomes than macrophages from control guinea pigs, and the phagolysosomes contain more rickettsiae exhibiting morphological signs of destruction [23].

A third effect of lymphokines on rickettsia-macrophage interactions has been described [59,61]. Treatment of mouse macrophage-like RAW264.7 cells with low concentrations of lymphokines and infection of the cells with viable *R. prowazekii* results in killing of a substantial percentage of the macrophage-like cells within 4 to 6 hours after infection [59]. Neither the lymphokine treatment alone nor the rickettsial infection alone kills the RAW264.7 cells during this period. Kodama et al. [61] found that BALB/c mouse macrophages infected with *R. tsutsugamushi* Gilliam and treated with lymphokines die after 4 days. In this study, as in the experiments with *R. prowazekii* and macrophage-like cells, both the rickettsial infection and treatment with lymphokines were required in order for the macrophages to be killed.

B. Macrophage Activation: In Vivo Studies

After intradermal inoculation of *R. typhi* into guinea pigs, monocytes and macrophages are recruited to the site of inoculation. Numbers of rickettsiae in the skin increase initially and then decline as the inflammatory cells condense into granulomatous lesions that contain lymphocytes [62]. In immunized guinea pigs inoculated intradermally with *R. typhi*, the inflammatory response is greater, formation of granulomatous lesions occurs sooner, and the numbers of rickettsiae in the skin decrease [62]. Crist et al. [31] noted that the numbers of *R. typhi* organisms in the spleens of infected BALB/c mice are lower at the time when the liver macrophages are activated as evidenced by their enhanced microbicidal activity against *Listeria monocytogenes*.

Several studies have provided information about factors important in host resistance to rickettsial infection by examining the susceptibility of various mouse strains. C3H/HeJ mice are susceptible to infection with *R. akari* Kaplan, whereas several other C3H sublines (including C3H/HeN) are resistant to this rickettsia [63–65]. C3H/HeJ mice are also more susceptible to infection with *R. conorii* Malish than other mouse strains [66]. C3H/HeJ mice have defects that are controlled by the *Lps* gene: these mice are unresponsive to bacterial lipopolysaccharides, and they are defective in the ability to develop macrophage tumoricidal activity [67]. Another mouse strain that is susceptible to infection with *R. akari* is A/J [65], which has a defect in development of macrophage tumoricidal and microbicidal activities that is not controlled by the *Lps* gene [68–71]. Resistance of mice to *R. akari* does not correlate with ability of macrophages to develop tumoricidal activity [65,72]. NZW/N mice, which develop normal macrophage tumoricidal activity, are very susceptible to infection with *R. akari* [65], whereas P/J mice, which are defective in development of macrophage tumoricidal activity [73], are resistant to *R. akari* infection [65]. Furthermore, genetic studies of A and B10.A mice indicate that the abilities of macrophages to develop tumoricidal activity and to develop resistance to challenge with *R. akari* are controlled by distinct genes [72]. The ability of lymphokines to activate mouse macrophages in vitro to express anti–*R. akari* activity correlates well with the resistance of particular mouse strains to challenge with *R. akari* in vivo [57]. In contrast, this correlation failed with *R. tsutsugamushi* Gilliam, since Nacy and Groves [14] observed no difference in the ability of lymphokines to induce anti–*R. tsutsugamushi* activity in macrophages from resistant BALB/c mice and susceptible C3H/He mice.

Resistance of mice to intraperitoneal inoculation with *R. tsutsugamushi* Gilliam (but not *R. tsutsugamushi* Karp) is influenced by a single autosomal, dominant gene, the *Ric* gene, which has been mapped to chromosome 5 [74,75]. Mice carrying the dominant Ric^R allele (e.g., BALB/c and C3H/RV) are resistant to intraperitoneal challenge with *R. tsutsugamushi* Gilliam, whereas mice that are homozygous for the Ric^S allele (e.g., C3H/He, C3H/HeDub, C3H/HeN, and C3H/HeJ) are sensitive [74–76].

The Ric^S mouse strains, C3H/HeN and C3H/HeJ, are susceptible to intraperitoneal inoculation with *R. tsutsugamushi* Gilliam [75]. However, only C3H/HeN mice are protected from intraperitoneal challenge with *R. tsutsugamushi* Gilliam by prior intraperitoneal inoculation with viable *Mycobacterium bovis* BCG [60]. Macrophages harvested from BCG-treated C3H/HeN mice display rickettsicidal activity in vitro, whereas macrophages harvested from BCG-treated C3H/HeJ mice exhibit minimal rickettsicidal activity. Nacy and Meltzer [60] suggested that the presence of activated macrophages during a prolonged period is required for protection of C3H/HeN mice. They found that the acute immune response elicited by intraperitoneal inoculation of PPD into C3H/HeN mice previously given BCG by the intradermal route did not protect the mice from intraperitoneal challenge with *R. tsutsugamushi,* even though macrophages harvested at the time of rickettsial challenge exhibited rickettsicidal and tumoricidal activities [60].

In studies with resistant BALB/c mice and susceptible C3H/He mice, Nacy and Groves [14] demonstrated that macrophages from *R. tsutsugamushi* Gilliam–infected BALB/c mice were chronically activated beginning on the fourth day after infection; in contrast, macrophages from C3H/He mice were not activated on the sixth and seventh days after infection. Only 1% to 2% of the peritoneal macrophages in the BALB/c mice were infected with rickettsiae; in contrast, the percentage of peritoneal macrophages infected with rickettsiae increased to 40% in C3H/He mice as the infection progressed. Although intraperitoneal inoculation of *R. tsutsugamushi* elicits an inflammatory response in both mouse strains, the appearance of peroxidase-positive macrophages (which are more responsive to lymphokine-induced activation [77]) in the peritoneal exudates occurs later in C3H/He mice [14].

Kokorin et al. [17] examined the peritoneal exudates and abdominal wall scrapings of mice resistant to intraperitoneal inoculation with *R. tsutsugamushi* Gilliam and found numerous monocytes and macrophages and little or no evidence of rickettsial replication or damage. In contrast, in sensitive random-bred and highly susceptible Swiss mice, damage to the peritoneal exudate cells is a prominent feature. Mesothelial cells containing numerous rickettsiae are found in the random-bred mice; however, cells in the Swiss mice contain few rickettsiae [17]. These data indicate that a vigorous macrophage response and the absence of damage to the peritoneal exudate cells are associated with resistance to *R. tsutsugamushi* infection in mice.

Other researchers have examined the inflammatory responses of genetically similar C3H/HeDub (Ric^S) and C3H/RV (Ric^R) mice inoculated intraperitoneally with *R. tsutsugamushi* Gilliam [76,78]. The percentage of peritoneal cells infected with rickettsiae remains very low in the resistant mice; in contrast, it increases progressively to about 75% in the sensitive mice, in which both macrophages and lymphocytes become infected. The inflammatory response of the resistant strain consists mostly of mononuclear cells, whereas the response of the sensitive strain (which is greater in magnitude than that of the resistant strain) also includes an early influx of polymorphonuclear leukocytes [76]. More important, the inflammatory response of the sensitive strain is deficient in Ia antigen–bearing macrophages in comparison with that of the resistant strain [78], and the influx of Ia antigen–positive macrophages observed in resistant mice is reduced by treatment of the mice with monoclonal antibody against interferon γ (IFN-γ) [79]. Interferon-γ enhances the expression of Ia antigens on macrophages [80], and these Ia antigens are important in the development of immune responses because their expression on the surfaces of macrophages and other antigen-presenting cells is required for presentation of foreign antigens to T helper lymphocytes. Resistance of C3H/RV mice to intraperitoneal challenge with *R. tsutsugamushi* Gilliam is eliminated by sublethal irradiation, and this finding is consistent with the hypothesis that macrophages derived from bone marrow precursors play a role in protection [81].

From investigations of *R. tsutsugamushi* and various mouse strains one can conclude that lethal infections resulting from intraperitoneal inoculation of Ric^R mice with *R. tsutsugamushi* Karp and Ric^S mice with *R. tsutsugamushi* Gilliam have commonalities. In both pairings there is a similar rickettsial burden in the peritoneal macrophages and a similar deficiency in Ia antigen–positive macrophages in the peritoneal exudates [18,76,78,82].

Experimental evidence suggests that the cytokine eta-1 plays a major role in protection of mice from intraperitoneal inoculation with *R. tsutsugamushi* Gilliam [83]. This protein, which is secreted by activated T lymphocytes, is encoded by a gene (*Eta-1*) that maps to the *Ric* locus on mouse chromosome 5. Mouse strains that express the $Eta\text{-}1^b$ allele are susceptible to intraperitoneal challenge with *R. tsutsugamushi* Gilliam; whereas mouse strains that express the $Eta\text{-}1^a$ allele (with one exception) are resistant [83]. Macrophages express receptors for eta-1, and inoculation of mice with eta-1 is associated with production of a cellular infiltrate consisting mostly of macrophages [84]. Peritoneal cells of CBA/CaJ (Ric^R,$Eta\text{-}1^a$) and CBA/J (Ric^S, $Eta\text{-}1^b$) mice differ strikingly in production of *Eta-1* mRNA and rickettsial burden after intraperitoneal inoculation of these mice with *R. tsutsugamushi* Gilliam. Within 24 hours after infection, the amount of *Eta-1* mRNA is increased in peritoneal cells of CBA/CaJ mice, and it remains high through the seventh day after infection. The rickettsial burden is very low (0–1 rickettsiae per cell) in the peritoneal cells of these mice on the fifth day after infection. In contrast, in CBA/J mice, the amount of *Eta-1* mRNA is not increased until the fifth day after infection, and the peritoneal cells at this time contain large numbers of rickettsiae [83]. An increase in the amount of IFN-γ–specific mRNA in the peritoneal cells is detected in both mouse strains 5 to 7 days after infection, and the magnitude of the increase is similar in the two strains [83].

C. Role of Interferon-γ

Many of the effects of lymphokine preparations on macrophage-rickettsiae interactions can be reproduced by recombinant IFN-γ. For example, the effects of IFN-γ treatment are similar to the effects of lymphokine treatment in experiments with *R. prowazekii* and macrophage-like RAW264.7 cells [59]. Treatment of the macrophage-like cells with low concentrations of IFN-γ and infection with *R. prowazekii* lead to killing of some of the RAW264.7 cells (within 4 hours after infection) and inhibition of rickettsial growth in the surviving cells. Furthermore, high concentrations of recombinant IFN-γ suppress the ability of the RAW264.7 cells to be initially infected with *R. prowazekii* [59]. Like lymphokine preparations, recombinant IFN-γ inhibits the growth of *R. conorii* in mouse macrophages; and treatment of the lymphokines with monoclonal anti–murine IFN-γ antibodies conjugated to Sepharose beads removes the ability of the lymphokines to restrict the growth of *R. conorii* in mouse macrophages [56]. Thus IFN-γ is certainly a most important factor for induction of macrophage antirickettsial activity.

However, Nacy et al. [55] provided evidence that factors other than IFN-γ induce mouse macrophages to kill *R. tsutsugamushi*. They showed that fractionation of lymphokines that induce intracellular killing of *R. tsutsugamushi* in mouse macrophages by Sephadex G-200 column chromatography yields three distinct peaks of activity with molecular weights of less than 10,000, about 40,000, and about 120,000 [55]. In contrast, the lymphokine activity that suppresses the ability of macrophages to be initially infected with *R. tsutsugamushi* and the lymphokine activity that induces activation of macrophages for tumor cytotoxicity both elute only in the 45,000-dalton region [55], which also has IFN-γ activity. Treatment with crude lymphokine preparations, but not IFN-γ alone, suppresses the initial *R. prowazekii* infection in fibroblastic L929 cells [85,86]. In contrast, IFN-γ at high concentrations suppresses the ability of mouse macrophage-like RAW264.7 cells to become initially infected with *R. prowazekii* [59]. However, the suppression in L929 cells is eliminated by treatment of the lymphokines with a monoclonal antibody against IFN-γ [85,86]. One possible explanation for these data is that both IFN-γ and another cytokine are required for suppression of the initial *R. prowazekii* infection in L929 cells.

The mechanisms responsible for the killing of rickettsiae in IFN-γ–treated macrophages and the killing of macrophage-like cells by IFN-γ treatment and rickettsial infection have not been defined. Two types of evidence indicate that the macrophage respiratory burst is not required for the killing of macrophage-like cells by IFN-γ and *R. prowazekii* [26]. First, *R. prowazekii* does not induce a measurable respiratory burst in macrophage-like cells; and second, a macrophage-like cell line that lacks the ability to undergo a respiratory burst is still killed by the combination of IFN-γ treatment and *R. prowazekii* infection. Although dead *R. prowazekii* are unable to kill the IFN-γ–treated RAW264.7 cells [59], antibiotics that inhibit rickettsial growth do not prevent killing of the cells [26]. Treatment of the rickettsiae with rabbit anti–*R. prowazekii* IgG (a treatment that results in destruction of rickettsiae within untreated RAW264.7 cells [7,24]) suppresses killing of IFN-γ–treated macrophage-like RAW264.7 cells by the rickettsiae [26]. These anti–*R. prowazekii* antibodies did not inhibit association of the rickettsiae with RAW264.7 cells. Thus a reasonable hypothesis is that the antibodies neutralize a particular rickettsial antigen or function that is required for killing of the IFN-γ–treated macrophage-like cells [26].

The crucial role of IFN-γ in host defense against *R. conorii* (Malish 7 strain) infection in vivo was clearly demonstrated by Li et al. [87]. When *R. conorii*–resistant Swiss Webster mice are treated with a monoclonal antibody against IFN-γ, the mice become susceptible to *R. conorii* [87]. This study as well as other in vivo studies concerning the role of IFN-γ in rickettsial infections have been reviewed [88]. Although the mechanisms by which IFN-γ is protective against rickettsial infections in vivo have not been defined, IFN-γ-induced activation of macrophages for destruction of rickettsiae and IFN-γ–induced stimulation of Ia antigen expression on macrophages are both likely to be important.

D. Production of Cytokines and Other Mediators

Macrophages and other cells produce cytokines and other mediators in response to various agents, and some studies have examined production of such mediators in response to rickettsiae. Prostaglandin E_2 (PGE_2) is found in ascites obtained from BALB/c mice inoculated intraperitoneally with *R. tsutsugamushi* Karp, and adherent peritoneal cells obtained from Karp-infected BALB/c mice produce approximately twice as much PGE_2 as those from Gilliam-infected BALB/c mice [82]. Interferon-α/β is produced in vivo and in vitro in response to infection with rickettsiae [89–94]. Culture supernatant fluids collected from adherent peritoneal cells from *R. tsutsugamushi*–infected mice also contain acid-stable antiviral activity, which is presumably IFN-α/β [82]. Induction of Ia antigen expression on macrophages by crude lymphokines is suppressed by culture supernatant fluids collected from adherent cells obtained from Karp-infected BALB/c mice [82]. In addition, macrophages from Karp-infected BALB/c mice do not express Ia antigens in response to lymphokine treatment in vitro. Jerrells [88] has suggested a role for prostaglandins in suppressing the expression of Ia antigens on these macrophages because Ia antigen expression is partially restored by indomethacin, an inhibitor of prostaglandin synthesis. There is also evidence that prostaglandin production by macrophages plays a role in the immunosuppressive effects observed during the chronic infection that results from subcutaneous inoculation of BALB/c mice with *R. tsutsugamushi* Karp [95]. Manor and Sarov [96] demonstrated that human monocyte-derived macrophages produce tumor necrosis factor-α (TNF-α) and PGE_2 when they are infected with *R. conorii* Casablanca or Israeli spotted fever rickettsiae (strain G-212). Addition of dexamethasone suppresses the production of both TNF-α and PGE_2 and increases the yield of rickettsiae from the macrophages [96]. In a study with *R. prowazekii* Madrid E, Walker et al. [97] found that mouse peritoneal macrophages and mouse alveolar macrophages secrete increased amounts of leukotriene B_4 during incubation with rickettsiae, and that alveolar macrophages also produce increased amounts of PGE_2 during incubation with rickettsiae.

E. Studies with Other Host Cells

A complete discussion of studies in host cells other than macrophages is beyond the scope of this review; accordingly, the discussion herein is limited to those findings most pertinent to macrophages. In addition to its ability to activate macrophages to restrict the growth of rickettsiae, IFN-γ limits the growth of rickettsiae in some, but not all, cell lines [58,86,94,98–102]. The combination of treatment with IFN-γ and infection with *R. prowazekii* can also damage fibroblastic cells [58,86]. Similarly, IFN-γ treatment and infection with *R. tsutsugamushi* Gilliam (which is sensitive to IFN-γ in BALB/3T3 cells) are toxic for BALB/3T3 cells; however, IFN-γ treatment and infection with *R. tsutsugamushi* Karp (which is resistant to IFN-γ in BALB/3T3 cells) are not toxic for BALB/3T3 cells [100,101].

Interferon-α/β has the ability to restrict growth of *R. akari* [103] and *R. prowazekii* Madrid E in L929 cells [92,93]; however, the inhibition of *R. prowazekii* by IFN-α/β is less pronounced than the inhibition observed with IFN-γ. Interferon-α/β either does not inhibit growth, or slightly and variably inhibits growth, of *R. tsutsugamushi* and *R. rickettsii* [101]. The combination of treatment with IFN-α/β and infection with *R. prowazekii* damages mouse L929 cells [92,93]. In addition, IFN-α/β also can prepare macrophage-like RAW264.7 cells for rapid destruction by viable *R. prowazekii* [94]; however, treatment of the RAW264.7 cells with very high concentrations of IFN-α/β is required.

R. prowazekii Madrid E organisms resistant to IFN have been isolated [93,94,104]. Of the six resistant isolates characterized, five are resistant to both IFN-α/β and IFN-γ and one is resistant to IFN-γ but not IFN-α/β. All six IFN-resistant *R. prowazekii* isolates (like the virulent Breinl strain of *R. prowazekii*) grow well in macrophage-like RAW264.7 cells; whereas the parent strain (the avirulent Madrid E strain) from which the IFN-resistant strains were isolated grows poorly in macrophage-like RAW264.7 cells [7,94]. Resistance to murine IFN-γ does not correlate with ability of *R. prowazekii* strains to grow in macrophage-like RAW264.7 cells, since the Breinl strain of *R. prowazekii* is quite sensitive to IFN-γ [94]. Resistance to murine IFN-α/β also does not correlate with ability to grow in macrophage-like RAW264.7 cells, because one of the IFN-γ-resistant strains is not significantly resistant to IFN-α/β compared with the Madrid E strain [94]. Perhaps resistance to IFN-α/β correlates with virulence, since the virulent Breinl strain of *R. prowazekii* is resistant to IFN-α/β compared with the avirulent Madrid E strain [94], and resistance to IFN-α/β among *R. tsutsugamushi* strains correlates with the virulence of these rickettsiae in mice [101]. Additional studies are needed to define the role of IFN-α/β in host defense against rickettsial infections.

Addition of human monocyte-derived macrophages to *R. conorii* Casablanca–infected HEp-2 cells suppresses the rickettsial yield from the cultures, and TNF-α and interleukin-1 (IL-1) are detected in media collected from the cultures [105]. In contrast, when similar experiments are conducted with human monocyte-derived macrophages and *R. conorii*–infected BGM cells, the rickettsial yield is only slightly decreased, and the amounts of TNF-α and IL-1 detected in the culture media are markedly lower [105]. Although IFN-γ alone does not markedly inhibit growth of *R. conorii* Casablanca in HEp-2 cells, the ability of TNF-α to inhibit growth of *R. conorii* in HEp-2 cells and the ability of IFN-γ to enhance the inhibition have been documented by Manor and Sarov [106]. The inhibition of *R. conorii* growth in HEp-2 cells treated with TNF-α and

IFN-γ is partially reversed by supplementation of the culture medium with additional tryptophan [106]. Tryptophan supplementation, however, is ineffective in reversing the effect of IFN-γ on *R. prowazekii* in human and murine cell lines [99].

REFERENCES

1. Winkler HH. *Rickettsia* species as organisms. Annu Rev Microbiol 1990; 44:131–53.
2. Gambrill MR, Wisseman CL Jr. Mechanisms of immunity in typhus infections. II. Multiplication of typhus rickettsiae in human macrophage cell cultures in the nonimmune system: influence of virulence of rickettsial strains and of chloramphenicol. Infect Immun 1973; 8:519–27.
3. Beaman L, Wisseman CL Jr. Mechanisms of immunity in typhus infections. VI. Differential opsonizing and neutralizing action of human typhus rickettsia–specific cytophilic antibodies in cultures of human macrophages. Infect Immun 1976; 14:1071–6.
4. Beaman L, Wisseman CL Jr. Mechanisms of immunity in typhus infections. V. Demonstration of *Rickettsia mooseri*–specific antibodies in convalescent mouse and human serum cytophilic for mouse peritoneal macrophages. Infect Immun 1976; 14:1065–70.
5. Kekcheeva NG, Vovk OA, Abrosimova GE, Chereshkova EA. Activity of macrophages from guinea pigs at various periods after infection with *Rickettsia prowazeki*. Acta Virol 1981; 25:150–4.
6. Kekcheeva NG, Vovk OA, Chereshkova EA, Abrosimova GE. Multiplication of *Rickettsia prowazekii* in cotton rat macrophage cultures. Acta Virol 1983; 27:268–72.
7. Turco J, Winkler HH. Differentiation between virulent and avirulent strains of *Rickettsia prowazekii* by macrophage-like cell lines. Infect Immun 1982; 35:783–91.
8. Gambrill MR, Wisseman CL Jr. Mechanisms of immunity in typhus infections. III. Influence of human immune serum and complement on the fate of *Rickettsia mooseri* within the human macrophages. Infect Immun 1973; 8:631–40.
9. Kenyon RH, Ascher MS, Kishimoto RA, Pedersen CE Jr. *In vitro* guinea pig leukocyte reactions to *Rickettsia rickettsii*. Infect Immun 1977; 18:840–6.
10. Kokorin IN, Kabanova EA, Shirokova EM. Role of macrophages in infection with *Rickettsia conorii*. Acta Virol 1980; 24:137–43.
11. Walker DH, Gear JH. Correlation of the distribution of *Rickettsia conorii*, microscopic lesions, and clinical features in South African tick bite fever. Am J Trop Med Hyg 1985; 34:361–71.
12. Yokota T, Tachibana N, Shishime E, Okayama A, Tsuda K. A. study on immunity in tsutsugamushi disease. Kansenshogaku Zasshi (Japan) 1989; 63:1149–59.
13. Nacy CA, Osterman JV. Host defenses in experimental scrub typhus: role of normal and activated macrophages. Infect Immun 1979; 26:744–50.
14. Nacy CA, Groves MG. Macrophages in resistance to rickettsial infections: early host defense mechanisms in experimental scrub typhus. Infect Immun 1981; 31:1239–50.
15. Nacy CA, Meltzer MS, Leonard EJ, Stevenson MM, Skamene E. Activation of macrophages for killing of rickettsiae: analysis of macrophage effector function after rickettsial inoculation of inbred mouse strains. In: Eisenstein TK, Actor P, Friedman H, eds. Host defenses to intracellular pathogens. New York and London: Plenum Press, 1983:335–53.
16. Kokorin IN, Kabanova EA, Chyong-din-Kyet, Miskarova ED, Abrosimova GE. Differences in the susceptibility of mouse lines to the rickettsial pox agent. Acta Virol 1978; 22:497–501.
17. Kokorin IN, Chyong-din-Kyet, Kekcheeva NG, Miskarova ED. Cytological investigation of *Rickettsia tsutsugamushi* infection of mice with different allotypic susceptibility to the agent. Acta Virol 1976; 20:147–51.
18. Catanzaro PJ, Shirai A, Hildebrandt PK, Osterman JV. Host defenses in experimental scrub typhus: histopathological correlates. Infect Immun 1976; 13:861–75.
19. Winkler HH, Daugherty RM. Cytoplasmic distinction of avirulent and virulent *Rickettsia prowazekii*: fusion of infected fibroblasts with macrophage-like cells. Infect Immun 1986; 40:1245–7.
20. Meyer WA III. Morphological and physiological studies of the interaction of *Rickettsia prowazekii* with human monocyte-derived macrophages (dissertation). University of Maryland, 1982.
21. Wisseman CL Jr. Selected observations on rickettsiae and their host cells. Acta Virol 1986; 30:81–95.
22. Popov VL, Prozorovsky SV, Vovk OA, Kekcheeva NK, Smirnova NS, Barkhatova OI. Electron

microscopic analysis of in vitro interaction of *Rickettsia prowazekii* with guinea pig macrophages. I. Macrophages from nonimmune animals. Acta Virol 1987; 31:53–8.
23. Popov VL, Prozorovsky SV, Vovk OA, Kekcheeva NK, Smirnova NS, Barkhatova OI. Electron microscopic analysis of in vitro interaction of *Rickettsia prowazekii* with guinea pig macrophages. II. Macrophages from immune animals. Acta Virol 1987; 31:59–64.
24. Keysary A, McCaul TF, Winkler HH. Roles of the Fc receptor and respiratory burst in killing of *Rickettsia prowazekii* by macrophagelike cell lines. Infect Immun 1989; 57:2390–6.
25. Winkler HH. Early events in the interaction of the obligate intracytoplasmic parasite, *Rickettsia prowazekii*, with eucaryotic cells: entry and lysis. Ann Inst Pasteur Microbiol 1986; 137:333–6.
26. Turco J, Keysary A, Winkler HH. Interferon-γ and rickettsia-induced killing of macrophage-like cells is inhibited by anti-rickettsial antibodies and does not require the respiratory burst. J Interferon Res 1989; 9:615–29.
27. Wyckoff RWG, Bohnel E. Therapeutic effect in guinea pigs of hyperimmune epidemic typhus antiserum. Proc Soc Exp Biol Med 1942; 49:712–5.
28. Wolman M. Treatment of typhus with anti-typhus horse serum. Lancet 1944; 210–2.
29. Stevens RS. Louse-borne typhus fever trial of serum treatment. Lancet 1945; 106–9.
30. Yeomans A, Snyder JC, Gilliam AG. The effects of concentrated hyperimmune rabbit serum in louse borne typhus. JAMA 1945; 129:19–24.
31. Crist AE Jr, Wisseman CL Jr, Murphy JR. Characteristics of *Rickettsia mooseri* infection of normal and immune mice. Infect Immun 1984; 43:38–42.
32. Murphy JR, Wisseman CL Jr., Fiset P. Mechanisms of immunity in typhus infection: analysis of immunity to *Rickettsia mooseri* infection of guinea pigs. Infect Immun 1980; 27:730–8.
33. Shirai A, Catanzaro PJ, Phillips SM, Osterman JV. Host defenses in experimental scrub typhus: role of cellular immunity in heterologous protection. Infect Immun 1976; 14:39–46.
34. Kokorin IN, Kabanova EA, Shirokova EM, Abrosimova GE, Rygkina NN, Pushkareva VI. Role of T lymphocytes in *Rickettsia conorii* infection. Acta Virol 1982; 26:91–7.
35. Jerrells TR. Mechanisms of immunity to *Rickettsia* species and *Coxiella burnetii*. In: Walker DH, ed. Biology of rickettsial diseases. Vol. 2. Boca Raton, Florida: CRC Press, 1988:79–100.
36. Kobayashi Y, Kawamura S, Oyama T. Immunological studies of experimental tsutsugamushi disease in congenitally athymic (nude) mice. Am J Trop Med Hyg 1985; 34:568–77.
37. Bell EJ, Pickens EG. A toxic substance associated with the rickettsias of the spotted fever group. J Immunol 1953; 70:461–72.
38. Snyder JC, Anderson CR. The susceptibility of the eastern cotton rat, *Sigmodon hispidus hispidus*, to European typhus. Science 1942; 95:23.
39. Henderson RG, Topping NH. Studies of typhus fever. Epidemic typhus: neutralization of the toxic substance. Natl Inst Health Bull 1945; 183:41–56.
40. Black CM, Tzianabos T, Roumillat LF, Redus MA, McDade JE, Reimer CB. Detection and characterization of mouse monoclonal antibodies to epidemic typhus rickettsiae. J Clin Microbiol 1983; 18:561–8.
41. Lange JV, Walker DH. Production and characterization of monoclonal antibodies to *Rickettsia rickettsii*. Infect Immun 1984; 46:289–94.
42. Anacker RL, List RH, Mann RE, Hayes SF, Thomas LA. Characterization of monoclonal antibodies protecting mice against *Rickettsia rickettsii*. J Infect Dis 1985; 151:1052–60.
43. Anacker RL, List RH, Mann RE. Wiedbrauk DL. Antigenic heterogeneity in high- and low-virulence strains of *Rickettsia rickettsii* revealed by monoclonal antibodies. Infect Immun 1986; 51:653–60.
44. Anacker RL, McDonald GA, List RH, Mann RE. Neutralizing activity of monoclonal antibodies to heat-sensitive and heat-resistant epitopes of *Rickettsia rickettsii* surface proteins. Infect Immun 1987; 55:825–7.
45. Feng HM, Walker DH, Wang JG. Analysis of T-cell–dependent and –independent antigens of *Rickettsia conorii* with monoclonal antibodies. Infect Immun 1987; 55:7–15.
46. Hechemy KE, Fox J, Samsonoff WA, Anacker R, Silverman D, Eisemann C, Green IS. Production of antibody to and cellular localization of erythrocyte-sensitizing substance from *Rickettsia rickettsii*. J Clin Microbiol 1989; 27:377–84.
47. Li H, Lenz B, Walker DH. Protective monoclonal antibodies recognize heat-labile epitopes on surface proteins of spotted fever group rickettsiae. Infect Immun 1988; 56:2587–93.

48. Murata M, Yoshida Y, Osono M, Ohashi N, Oyanagi M, Urakami H, Tamura A, Nogami S, Tanaka H, Kawamura A Jr. Production and characterization of monoclonal strain-specific antibodies against prototype strains of *Rickettsia tsutsugamushi*. Microbiol Immunol 1986; 30:599–610.
49. McDonald GA, Anacker RL, Mann RE, Milch LJ. Protection of guinea pigs from experimental Rocky Mountain spotted fever with a cloned antigen of *Rickettsia rickettsii*. J Infect Dis 1988; 158:228–31.
50. Vishwanath S, McDonald GA, Watkins NG. A recombinant *Rickettsia conorii* vaccine protects guinea pigs from experimental boutonneuse fever and Rocky Mountain fever. Infect Immun 1990; 58:646–53.
51. Anderson BE, McDonald GA, Jones DC, Regnery RL. A protective protein antigen of *Rickettsia rickettsii* has tandemly repeated, near-identical sequences. Infect Immun 1990; 58:2760–9.
52. Carl M, Dobson ME, Ching W-M, Dasch GA. Characterization of the gene encoding the protective paracrystalline-surface-layer protein of *Rickettsia prowazekii*: presence of a truncated identical homolog in *Rickettsia typhi*. Proc Natl Acad Sci USA 1990; 87:8237–41.
53. Ching W-M, Carl M, Dasch GA. Mapping of monoclonal antibody binding sites on CNBr fragments of the S-layer protein antigens of *Rickettsia typhi* and *Rickettsia prowazekii*. Mol Immunol 1992; 29:95–105.
54. Nacy CA, Meltzer MS. Macrophages in resistance to rickettsial infection: macrophage activation in vitro for killing of *Rickettsia tsutsugamushi*. J Immunol 1979; 123:2544–9.
55. Nacy CA, Leonard EJ, Meltzer MS. Macrophages in resistance to rickettsial infections: characterization of lymphokines that induce rickettsiacidal activity in macrophages. J Immunol 1981; 126:204–7.
56. Jerrells TR, Turco J, Winkler HH, Spitalny GL. Neutralization of lymphokine-mediated antirickettsial activity of fibroblasts and macrophages with monoclonal antibody specific for murine interferon gamma. Infect Immun 1986; 51:355–9.
57. Nacy CA, Meltzer MS. Macrophages in resistance to rickettsial infection: strains of mice susceptible to the lethal effects of *Rickettsia akari* show defective macrophage rickettsicidal activity in vitro. Infect Immun 1982; 36:1096–101.
58. Wisseman CL Jr, Waddell A. Interferonlike factors from antigen- and mitogen-stimulated human leukocytes with antirickettsial and cytolytic actions on *Rickettsia prowazekii*–infected human endothelial cells, fibroblasts, and macrophages. J Exp Med 1983; 157:1780–93.
59. Turco J, Winkler HH. Effect of mouse lymphokines and cloned mouse interferon-gamma on the interaction of *Rickettsia prowazekii* with mouse macrophage-like RAW264.7 cells. Infect Immun 1984; 45:303–8.
60. Nacy CA, Meltzer MS. Macrophages in resistance to rickettsial infections: protection against lethal *Rickettsia tsutsugamushi* infections by treatment of mice with macrophage-activating agents. J Leukocyte Biol 1984; 35:385–96.
61. Kodama K, Yasukawa M, Kobayashi Y. Effect of rickettsial antigen-specific T cell line on the interaction of *Rickettsia tsutsugamushi* with macrophages. Microbiol Immunol 1988; 32:435–9.
62. Murphy JR, Wisseman CL Jr., Fiset P. Mechanisms of immunity in typhus infection: some characteristics of intradermal *Rickettsia mooseri* infection in normal and immune guinea pigs. Infect Immun 1978; 22:810–20.
63. Anderson GW Jr, Osterman JV. Host defenses in experimental rickettsialpox: Resistance of C3H mouse sublines. Acta Virol 1980; 24:294–6.
64. Anderson GW Jr, Osterman JV. Host defenses in experimental rickettsialpox: Genetics of natural resistance to infection. Infect Immun 1980; 28:132–6.
65. Meltzer MS, Nacy CA. Macrophages in resistance to rickettsial infection: susceptibility to lethal effects of *Rickettsia akari* infection in mouse strains with defective macrophage function. Cell Immunol 1980; 54:487–90.
66. Eisemann CS, Nypaver MJ, Osterman JV. Susceptibility of inbred mice to rickettsiae of the spotted fever group. Infect Immun 1984; 43:143–8.
67. Ruco LP, Meltzer MS, Rosenstreich DL. Macrophage activation for tumor cytotoxicity: control of macrophage tumoricidal capacity by the LPS gene. J Immunol 1978; 121:543–8.
68. Boraschi D, Meltzer MS. Defective tumoricidal capacity of macrophages from A/J mice. II. Comparison of the macrophage cytotoxic defect of A/J mice with that of lipid A-unresponsive C3H/HeJ mice. J Immunol 1979; 122:1592–7.
69. Boraschi D, Meltzer MS. Defective tumoricidal capacity of macrophages from A/J mice. I. Characterization of the macrophage cytotoxic defect after *in vivo* and *in vitro* activation stimuli. J Immunol 1979; 122:1587–91.

70. Boraschi D, Meltzer MS. Defective tumoricidal capacity of macrophages from A/J mice. III. Genetic analysis of the macrophage defect. J Immunol 1980; 124:1050–3.
71. Nacy CA, Oster CN, James SL, Meltzer MS. Activation of macrophages to kill rickettsiae and Leishmania; dissociation of intracellular microbicidal activities and extracellular destruction of neoplastic and helminth targets. In: Adams DO, Hanna MG Jr, eds. Contemporary topics in immunobiology. Vol. 13. New York: Plenum Press, 1984:147–70.
72. Meltzer MS, Nacy CA, Stevenson MM, Skamene E. Macrophages in resistance to rickettsial infections: genetic analysis of susceptibility to lethal effects of *Rickettsia akari* infection and development of activated, cytotoxic macrophages in A and B10.A mice. J Immunol 1982; 129:1719–23.
73. Boraschi D. Meltzer MS. Macrophage activation for tumor cytotoxicity: genetic variation in macrophage tumoricidal capacity among mouse strains. Cell Immunol 1979; 45:188–94.
74. Groves MG, Rosenstreich DL, Taylor BA, Osterman JV. Host defenses in experimental scrub typhus: mapping the gene that controls natural resistance in mice. J Immunol 1980; 125:1395–9.
75. Groves MG, Osterman JV. Host defenses in experimental scrub typhus: genetics of natural resistance to infection. Infect Immun 1978; 19:583–8.
76. Jerrells TR, Osterman JV. Host defenses in experimental scrub typhus: inflammatory response of congenic C3H mice differing at the *Ric* gene. Infect Immun 1981; 31:1014–22.
77. Ruco LP, Meltzer MS. Macrophage activation for tumor cytotoxicity: increased lymphokine responsiveness of peritoneal macrophages during acute inflammation. J Immunol 1978; 120:1054–62.
78. Jerrells TR. Association of an inflammatory I region–associated antigen-positive macrophage influx and genetic resistance of inbred mice to *Rickettsia tsutsugamushi*. Infect Immun 1983; 42:549–57.
79. Jerrells TR, Li H, Walker DH. *In vivo* and *in vitro* role of gamma interferon in immune clearance of *Rickettsia* species. Adv Exp Med Biol 1988; 239:193–200.
80. Steeg PS, Moore RN, Johnson HM, Oppenheim JJ. Regulation of murine macrophage Ia antigen expression by a lymphokine with immune interferon activity. J Exp Med 1982; 156:1780–93.
81. Jerrells TR, Osterman JV. Role of macrophages in innate and acquired host resistance to experimental scrub typhus infection of inbred mice. Infect Immun 1982; 37:1066–73.
82. Jerrells TR, Hickman CJ. Down regulation of Ia antigen expression on inflammatory macrophages by factors produced during acute infections with *Rickettsia tsutsugamushi*. In: Goldstein G, Bach J-F, Wigzell H, eds. Immune regulation by characterized polypeptides. New York: Alan R. Liss, 1987:659–68.
83. Patarca R, Freeman GJ, Singh RP, Wei FY, Durfee T, Blattner F, Regnier DC, Kozak CA, Mock BA, Morse HC III, Jerrells TR, Cantor H. Structural and functional studies of the early T lymphocyte activation 1 (*Eta-1*) gene. Definition of a novel T cell–dependent response associated with genetic resistance to bacterial infection. J Exp Med 1989; 170:145–61.
84. Singh RP, Patarca R, Schwartz J, Singh P, Cantor H. Definition of a specific interaction between the early T lymphocyte activation 1 (Eta-1) protein and murine macrophages in vitro and its effect upon macrophages in vivo. J Exp Med 1990; 171:1931–42.
85. Turco J, Winkler HH. Lymphokines, interferon-γ and *Rickettsia prowazekii*. In: Kazar J, ed. Rickettsiae and rickettsial diseases, proceedings of the third international symposium. Bratislava, Czechoslovakia: Slovak Academy of Sciences Publishing House, 1985:211–8.
86. Turco J, Winkler HH. Interactions between *Rickettsia prowazekii* and cultured host cells: Alterations induced by gamma interferon. In: Byrne GI, Turco J, eds. Interferon and nonviral pathogens. New York: Marcel Dekker, 1988:95–115.
87. Li H, Jerrells TR, Spitalny GL, Walker DH. Gamma interferon as a crucial host defense against *Rickettsia conorii* in vivo. Infect Immun 1987; 55:1252–5.
88. Jerrells TR. Gamma interferon as an effector molecule in antirickettsial immunity. In: Byrne GI, Turco J, eds. Interferon and nonviral pathogens. New York: Marcel Dekker, 1988:117–29.
89. Kazar J. Interferon-like inhibitor in mouse sera induced by rickettsiae. Acta Virol 1966; 10:277.
90. Hopps HE, Kohno S, Kohno M, Smadel JE. Production of interferon in tissue cultures infected with *Rickettsia tsutsugamushi*. Bact Proc 1964:115–6.
91. Kohno S, Kohase M, Sakata H, Shimizu Y, Hikita M, Shishido A. Production of interferon in primary chick embryonic cells infected with *Rickettsia mooseri*. J Immunol 1970; 105:1553–8.
92. Turco J, Winkler HH. Interferon-α/β and *Rickettsia prowazekii:* induction and sensitivity. Ann NY Acad Sci 1990; 590:168–86.

93. Turco J, Winkler HH. Selection of alpha/beta interferon- and gamma interferon-resistant ricketsiae by passage of *Rickettsia prowazekii* in L929 cells. Infect Immun 1990; 58:3279–85.
94. Turco J, Winkler HH. Comparison of properties of virulent, avirulent, and interferon-resistant *Rickettsia prowazekii* strains. Infect Immun 1991; 59:1647–55.
95. Jerrells TR. Immunosuppression associated with the development of chronic infections with *Rickettsia tsutsugamushi:* adherent suppresor cell activity and macrophage activation. Infect Immun 1985; 50:175–82.
96. Manor E, Sarov I. Tumor necrosis factor alpha and prostaglandin E_2 production by human monocyte-derived macrophages infected with spotted fever group rickettsiae. Ann NY Acad Sci 1990; 590:157–67.
97. Walker TS, Dersch MW, White WE. Effects of typhus rickettsiae on peritoneal and alveolar macrophages: rickettsiae stimulate leukotriene and prostaglandin secretion. J Infect Dis 1991; 163:568–73.
98. Turco J, Winkler HH. Cloned mouse interferon-gamma inhibits the growth of *Rickettsia prowazekii* in cultured mouse fibroblasts. J Exp Med 1983; 158:2159–64.
99. Turco J, Winkler HH. Gamma interferon–induced inhibition of the growth of *Rickettsia prowazekii* in fibroblasts cannot be explained by the degradation of tryptophan or other amino acids. Infect Immun 1986; 53:38–46.
100. Hanson B. Susceptibility of *Rickettsia tsutsugamushi* Gilliam to gamma interferon in cultured mouse cells. Infect Immun 1991; 59:4125–33.
101. Hanson B. Comparative susceptibility to mouse interferons of *Rickettsia tsutsugamushi* strains with different virulence in mice and of *Rickettsia rickettsii*. Infect Immun 1991; 59:4134–41.
102. Turco J, Winkler HH. D-98 Growth of *Rickettsia prowazekii* in cell lines which differ in their responses to interferon (abstr). In: Abstracts of the 92nd General Meeting of the American Society for Microbiology. Washington, DC: American Society for Microbiology, 1992:112.
103. Kazar J, Krautwurst PA, Gordon FB. Effect of interferon and interferon inducers on infections with a nonviral intracellular microorganism, *Rickettsia akari*. Infect Immun 1971; 3:819–24.
104. Turco J, Winkler HH. Isolation of *Rickettsia prowazekii* with reduced sensitivity to gamma interferon. Infect Immun 1989; 57:1765–72.
105. Manor E. The effect of monocyte-derived macrophages on the growth of *Rickettsia conorii* in permissive cells. In: Kazar J, Raoult D, eds. Rickettsiae and rickettsial diseases, proceedings of the fourth international symposium. Bratislava, Czechoslovakia: Slovak Academy of Sciences Publishing House, 1991:312–7.
106. Manor E, Sarov I. Inhibition of *Rickettsia conorii* growth by recombinant tumor necrosis factor alpha: enhancement of inhibition by gamma interferon. Infect Immun 1990; 58:1886–90.

III

MACROPHAGE–PARASITE INTERACTIONS

25

Alveolar Macrophage Interaction with *Pneumocystis carinii*

Henry Koziel,[a]* Deirdre O'Riordan,[a†] Angeline Warner,[b] Richard M. Rose[a‡]

[a]*New England Deaconess Hospital and Harvard Medical School;* [b]*Harvard School of Public Health, Boston, Massachusetts*

I. INTRODUCTION

Pneumocystis carinii (PC) is a unique opportunistic pathogen that has a particular predilection for producing pneumonia in persons with concomitant human immunodeficiency virus type I (HIV-1) infection. In contrast to many other AIDS-related pathogens, infection with PC is characteristically not disseminated, but rather remains localized to the alveolar surface of the lungs. At this anatomical locus in the respiratory tract, the alveolar macrophage is the primary defender against infectious agents. For this reason, understanding the interplay of macrophages with *Pneumocystis* is an important step in characterizing the pathophysiology of infection, and may lead to the development of new strategies for the prevention and treatment of disease based on macrophage-directed interventions.

The goals of this chapter are

1. To review recent advances in the understanding of the biology of PC with respect to its interaction with host defense cells.
2. To discuss the immunopathogenesis of PC pneumonia, with particular emphasis on disease occurring in patients with AIDS.
3. To describe laboratory techniques that have been developed to assist in the study of the macrophage-pneumocystis interaction.
4. To critically review recent studies in this area of research.
5. To suggest future approaches to understanding the macrophage response to *Pneumocystis*.

II. BIOLOGY OF *PNEUMOCYSTIS CARINII*

Advances in understanding the pathobiology of PC infection have been limited by the lack of a satisfactory in vitro cultivation system capable of consistently supporting the growth of human PC. However, organisms derived from several rodent species have been maintained in short-term culture by overlaying them on a variety of feeder layer cell lines [1–4]. In these in vitro culture systems, attachment to the feeder layer appears to be required for growth. However, unlike most opportunistic organisms that have an obligate intracellular phase in their life cycle, PC does not appear to grow intracellularly in in vitro systems.

---

*Parker B. Francis Foundation Pulmonary Research Fellow
Current affiliations:
†Mayo Clinic, Rochester, Minnesota
‡Cytel Corporation, San Diego, California

Although it has been a subject of intense study, the precise metabolic requirements for *Pneumocystis* growth are yet to be elucidated. Investigation in this area has been hampered by the general finding that, in vitro, *Pneumocystis* organisms are relatively inert metabolically and lack the energy-generating organelles typically found in other eukaryotic cells.

The requirement for cell culture systems to maintain *Pneumocystis* in vitro also forms a serious barrier to understanding the molecular biology of the organism, which is estimated to have a relatively small complement of genetic material (approximately 10^6 kilobase pairs) relative to mammalian cells. Thus, in a preparation of 10^3 culture-derived *Pneumocystis*, the presence of a single contaminating mammalian cell with approximately 10^9 kilobase pairs of DNA means that genomic material prepared from this sample will be composed of 50% non-*Pneumocystis* DNA. This limitation has seriously impaired the understanding of the pathogenesis of PC infection at the molecular level.

Despite these considerations, recent insights into the biology of PC have been valuable in establishing the taxonomic classification of this organism, its life cycle and immunobiology (Table 1).

Table 1 Characteristics of *Pneumocystis carinii*

Morphology	Trophozoite (1–4 μm) pleomorphic, most abundant form in vivo. Cyst (4–8 μm) spherical, thick wall; contains 8 sporozoites. Precyst, intermediate size and shape
Antigens	Major surface antigen is 95–120 kDa MW glycoprotein, high glycosyl-mannosyl and *N*-acetylglycosamine content. 40–55 kDa protein common in BALF in *P. carinii* pneumonia. Species cross-reactivity and species specificity
Metabolism	Extracellular parasite Optimal growth conditions unknown Limited uptake of RNA/DNA precursors No ergosterol present in cell membrane β-1,3-glucan synthase cell wall enzyme
Ultrastructure	Trophozoite pellicle 20–30 nm, electron-dense Cyst pellicle 100–160 nm; electron-lucent middle layer Few mitochondria Abundant endoplasmic reticulum Abundant ribosomes
Molecular analysis	Approx 10^6 kb pairs of DNA Strain differences based on DNA analysis DHFR/thymidylate synthetase on separate genes[a] 18S rRNA DNA sequence homology with fungi[a]

[a]Similar to fungi

A. Classification

The taxonomy of PC remains controversial. Previously, most investigators have favored a protozoan classification based on certain morphologic characteristics that show the organism to be a eukaryote with cyst and trophozoite phases. Also, the organism's known susceptibility to antiprotozoal drugs lends support to this viewpoint.

Recent work, however, indicates that PC may be best classified as a fungus. The most compelling evidence comes from studies of two of the organism's major enzymes, dihydrofolate reductase and thymidylate synthetase [5,6]. In protozoa these enzymes are encoded by a single gene. In PC these enzymes are genetically and physically distinct, and in this respect PC is comparable to fungi and mammalian cells. Another important piece of evidence favoring a fungal classification comes from the analysis of PC-derived DNA sequences coding for its small ribosomal RNA. These sequences exhibit far greater homology with certain fungi than they do with RNA of protozoan origin [7,8].

There seems no doubt that PC is related to fungi. However, unlike fungi, the cell membrane of PC does not contain ergosterol. As a result, commonly used antifungal drugs such as amphotericin B and ketoconazole, which act by altering the functional integrity of membrane ergosterol-containing lipids, have no effect against PC pneumonia.

Another implication of these taxonomical considerations relates to understanding the interaction of PC with host defense cells such as the alveolar macrophage. One distinguishing characteristic of fungi is the presence of a distinct cell wall that typically contains chitins and complex glycoproteins composed of glucans and mannans. These surface structures are likely to play a critical role in the process of pathogen recognition by macrophages, as discussed in subsequent sections.

B. Life Cycle and Ultrastructural Characteristics

Most of the information on the life cycle of PC has come from microscopic analysis of the organism in lungs or in culture. Light microscopy as well as phase and electron microscopy have been employed. Three principal developmental stages have been identified. The trophozoite, the most abundant form, is small (1–4 μm) and pleomorphic (Figure 1a). The prominent cyst (5–8 μm) has a thick wall and contains up to eight daughter forms or intracystic bodies (sporozoites) (Figure 1b). The precyst appears to be an intermediate stage. In the current view of this complex life cycle of PC the trophozoites multiply asexually, probably by binary fusion [1]. A sexual stage also appears to occur, which leads to development of the mature cyst [9]. The cyst liberates intracystic bodies, which then develop into trophic forms (Figure 2). The conditions that promote encystation and excystation are not fully understood,

The ultrastructure of PC has been studied by transmission electron microscopy, freeze-fracture technique, and scanning electron microscopy [10–12]. The external shape of the trophozoite is spherical to oval in young organisms just released from the cyst, and it becomes large and pleomorphic with further development. The trophozoite possesses a wall or pellicle 20 to 30 nm across that is composed of an outer electron-dense layer and an inner membrane. The precyst is defined as an intermediate form between the trophozoite and the mature cyst, on the basis of its ultrastructural properties and development. As the precyst develops and progresses from the early to the late stage, the thin doubled-layered pellicle of the trophozoite becomes thickened by the appearance of a new electron-lucent middle layer. The pellicle is very thick (100–160 nm) and composed of an electron-dense outer layer, an electron-lucent middle layer, and an innermost plasma membrane. The mature cyst contains 8 intracystic bodies, each with a single nucleus and mitochondrion and abundant endoplasmic reticulum and ribosomes. It is interesting to note that the ultrastructural characteristics of PC from rat lung are strikingly similar to those of organisms recovered from human lungs. Yet despite the morphologic similarity, recent research suggests that PC derived from rodents and from humans are antigenically distinct (cf. next section).

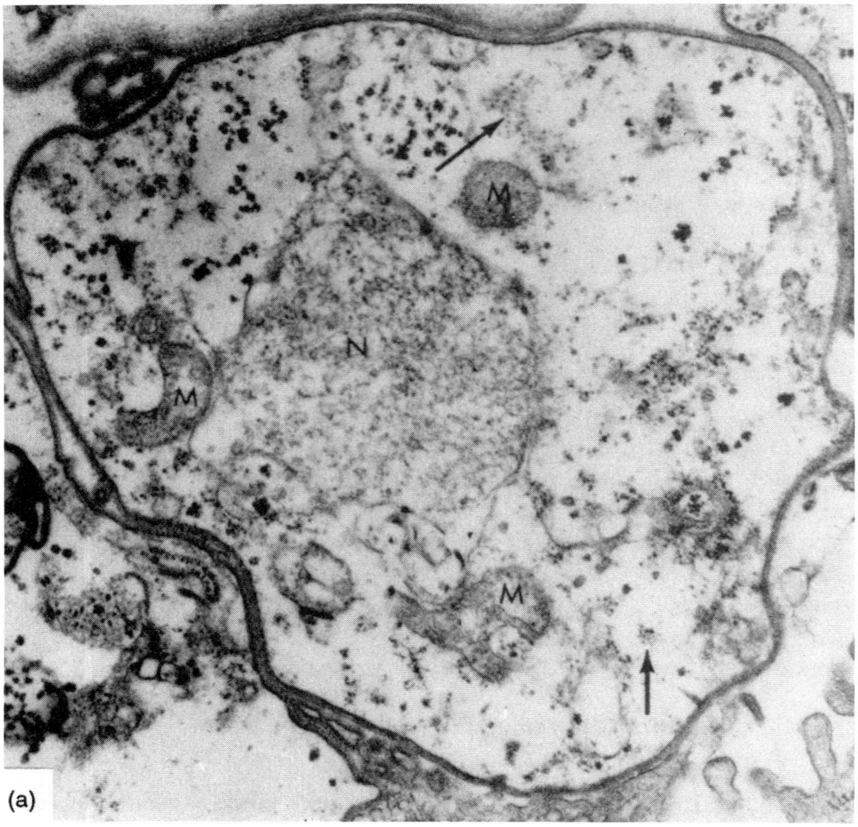

FIGURE 1a Electron photomicrograph of *Pneumocystis carinii* trophozoite in association with other organisms. The pleomorphic form is approximately 3 μm in diameter. The cytoplasm contains few mitochondria (M) and scattered small, electron-opaque granular particles that probably represent free ribosomes (arrows). The nucleoplasm (N) is contained within a fragmentary nuclear membrane. (Magnification × 41,000). (Reprinted with permission from Barton and Campbell, *Am J Pathol* 1969; 54:209.)

It is unclear whether effective host defense against PC requires macrophages to interact with all the various life forms of PC. Although the trophozoite stage is the one most closely associated with lung injury during experimental rodent infection, it is likely that other forms (e.g., the cyst) are important in establishing persistent low-level or latent infection. Therefore, the ability of macrophages to recognize and eradicate several stages of the PC life cycle is probably necessary for the host to control this organism.

C. Immunobiology

The surface of PC is highly antigenic. A number of investigators have analyzed PC antigens by Western immunoblotting with polyclonal and monoclonal antibodies [13,14]. Most attention to date has focused on the major surface glycoprotein, which is present on PC derived from humans as well as rodents. This "immunodominant" glycoprotein, which ranges in size from 95 to 120 kDa depending on the host species from which the organism is isolated, is a ligand for the binding of PC

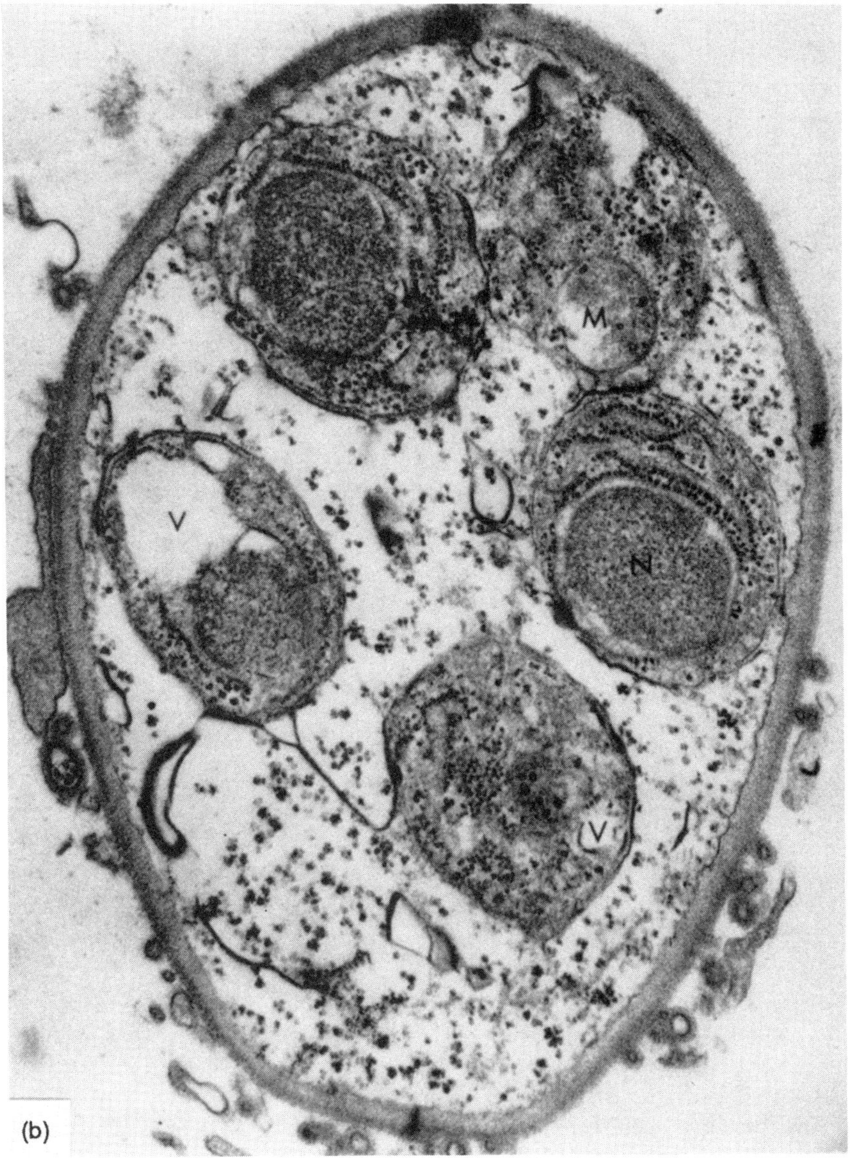

FIGURE 1b Electron photomicrograph of *Pneumocystis carinii* cyst. This oval-shaped thick-walled form contains several sporozoite forms, each confined by a thin membrane. The sporozoites demonstrate mitochondria (M), vacuoles (V), and a nucleus (N). (Magnification × 40,000.) (Reprinted with permission from Barton and Campbell, *Am J Pathol* 1969; 54:209.)

to host cells [15]. Purification of this major surface glycoprotein has been hindered by the need to use infected lung tissue and by the remarkable adherent properties of PC. Studies with lectin probes have demonstrated that the carbohydrate component of the antigen is rich in glucosyl-mannosyl and *N*-acetylglucosamine residues [16]. Attempts are currently under way to sequence this mannosylated surface glycoprotein of PC and to isolate and clone the gene encoding it. The considerable data accrued so far suggest that the mannosylated surface glycoprotein has a con-

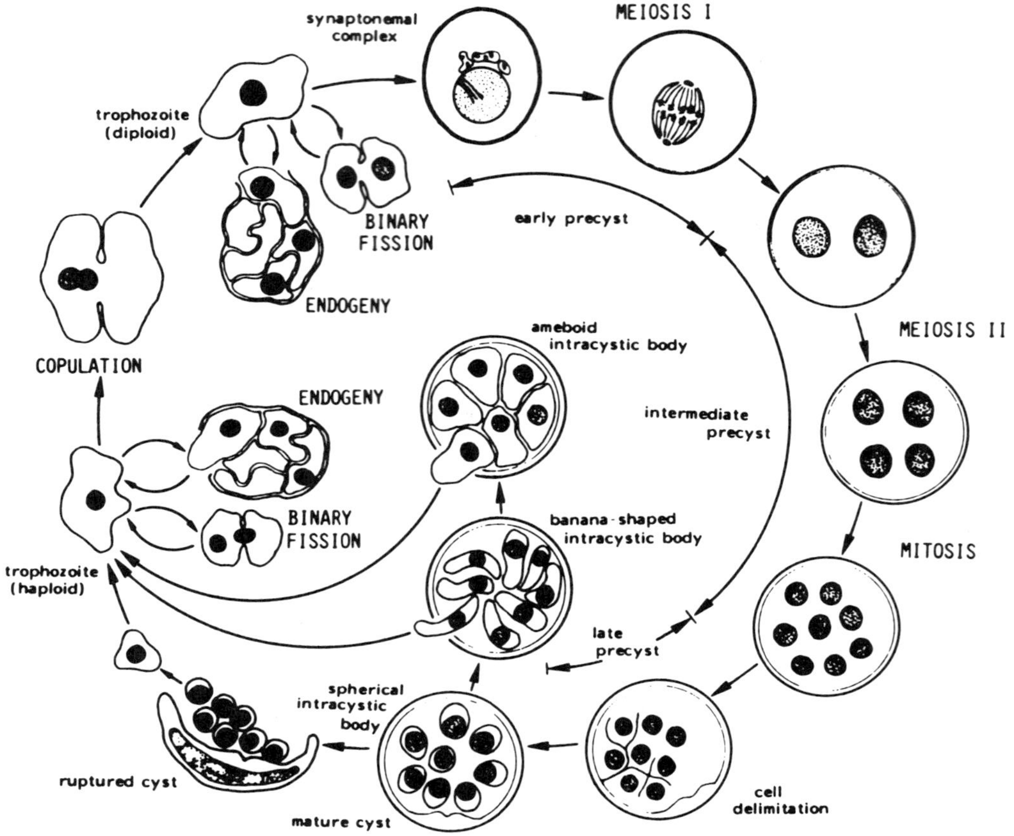

FIGURE 2 Proposed life cycle for *Pneumocystis carinii*. In addition to the cyst-forming sexual cycle, the trophozoite is also capable of asexual propagation by binary fission and by endogeny by either the haploid or diploid forms. (From Ref. 10.)

served epitope but also exhibits host species–specific antigenic variation [17–19]. It has recently been proposed that this glycoprotein be referred to as PC surface glycoprotein A, since it is the first surface molecule of PC to be isolated [20].

Other major antigens that have been recognized are bands of 45 to 50 kDa in rat PC and of about 40 kDa in human PC. These antigens are also glycoproteins, but their relationship to glycoprotein A and their location on the organism are as yet unknown. The 40-kDa band is the antigen most commonly detected in the lungs and bronchoalveolar fluid of patients with *Pneumocystis* pneumonia [14].

The question of antigenic variation and possible strain differences has also been studied. Pulsed field gel electrophoresis, a method used to analyze large DNA molecules, has been used to compare the karyotype of various PC isolates [21–23]. This technique demonstrates differences in the chromosomal band pattern of PC isolates from two different rat colonies as well as among the few human isolates studied to date.

These insights suggest that antigenic surface structures such as glycoprotein A are likely to be critically important in the recognition of PC by phagocytic cells, as discussed in greater detail in subsequent sections.

III. CLINICAL INSIGHTS INTO *PNEUMOCYSTIS* PNEUMONIA: MODES OF TRANSMISSION

Although the environmental reservoir for PC is unknown, a variety of experimental and clinical data support a respiratory mode of transmission [24]. In view of seroepidemiological studies indicating that subclinical infection is common in young children throughout the general population [25–27], it is likely that *Pneumocystis* disease results from reactivation of a preexisting latent focus of infection. This hypothesis is consistent with the fact that immunosuppression of outbred rodents virtually always induces disease. In humans, the occurrence of *Pneumocystis* pneumonia is directly correlated with the severity of immunosuppression [28]. In the clinical setting of HIV-1 infection, between 60% and 85% of persons seropositive for HIV-1 will develop *Pneumocystis* pneumonia unless chemoprophylaxis is instituted [29]. The occasional occurrence of isolated "outbreaks" of *Pneumocystis* pneumonia in closed populations suggests that an additional mechanism of disease is primary infection or reinfection from an exogenous focus.

IV. IMMUNOPATHOGENESIS OF *PNEUMOCYSTIS* PNEUMONIA

A variety of defects in host defense may predispose to *Pneumocystis* infection. These include protein calorie malnutrition [30, 31], organ transplantation, administration of immunosuppressive drugs (particularly corticosteroids), and congenital as well as acquired immunodeficiency states affecting cellular and/or humoral immunity. In experimental animal models, CD4+ T lymphocytes appear to be a critically important component of the normal host response to this pathogen [32–36]. Consistent with this research is the observation that in persons infected with HIV-1, the risk of developing *Pneumocystis* pneumonia is inversely related to the number of circulating CD4 lymphocytes [29]. The precise role of CD4 cells in defense against *Pneumocystis*, however, is unclear. Immune T cells are capable of proliferative responses to pneumocystis antigen [37, 38]. PC-reactive CD4 lymphocytes could contribute to the host's response against this pathogen via secretion of macrophage-activating cytokines as well as by the production of signals that foster expansion of the antibody-forming pool of B cells and cytotoxic CD8 lymphocytes. Although the precise role of the CD4 lymphocyte is not yet clear, it is noteworthy that one CD4-derived factor, interferon-γ, has been shown to reduce the intensity of *Pneumocystis* infection in a murine model of infection [39].

Specific antibody also plays a role in the host defense against *Pneumocystis*. This contention is supported by the observation that children with congenital B cell defects are predisposed to *Pneumocystis* pneumonia. Also, the administration of a monoclonal antibody directed against major surface epitope of pneumocystis has been shown to be beneficial in an experimental model of *Pneumocystis* pneumonia [17].

Further research is necessary to elucidate the precise mechanisms by which host defenses limit the spread of *Pneumocystis* on the alveolar surface of the lungs (Table 2). Such studies will have to account for the unique biology of the alveolar surface and its impact on the host interaction with this parasite.

Ultrastructural analysis indicates that a principal event in the development of *Pneumocystis* pneumonia is the interplay of the trophozoite form of *Pneumocystis* with the type 1 alveolar epithelial cell. Physical attachment of *Pneumocystis* to this cell type can be mediated, in part, by fibronectin and/or *Pneumocystis* glycoprotein A antigen [40,41]. This attachment process may represent a unique form of extracellular parasitism in which *Pneumocystis* derives essential growth factors from the host. However, the physical proximity of the *Pneumocystis* trophozoite form to the respiratory epithelium results in dysplastic changes in the epithelial cell layer and eventual disruption of the integrity of the epithelial border. As a result, material from the underlying interstitium and vascular space can gain access to the alveolus and disrupt normal lung function.

Table 2 Factors Involved in Host Resistance to *Pneumocystis carinii* infection

	Protective	Helpful	Not helpful
Cellular	CD4 lymphocyte	AM[a] B-lymphocyte	PMN NK cells CD8 lymphocyte
Noncellular	(?) IFN-γ (?) M-CSF (?) IL-1	Antibody Fibronectin[b] (?) C3 (?) TNF-α (?) IL-4	SP-A Fibronectin[c] (?) IFN-γ[d]

[a]Alveolar macrophage binding mediated through the mannose receptor.
[b]Mediates binding to alveolar macrophage.
[c]Mediates binding to alveolar epithelial cell.
AM = alveolar macrophage, PMN = polymorphonuclear cell; NK = natural killer; IFN-γ = interferon-γ; SP-A = surfactant protein A; M-CSF = macrophage colony-stimulating factor; IL-1 = interleukin-1; TNF-α = tumor necrosis factor-α; IL-4 = interleukin-4.
[d]Down regulation of aleveolar macrophage mannose receptor.

This process accounts for the typical histopathologic features of *Pneumocystis* pneumonia, in which the alveolar spaces are filled with a foamy exudate containing numerous cysts and trophozoite forms of the organism. Interestingly, there are few inflammatory cells at the site of infection, even though the interstitial spaces exhibit a mononuclear cell inflammatory response.

The pathophysiology of PC infection is complex. On the basis of current concepts of this process, presented above, it is possible to generate the following hypothesis about the macrophage-pneumocystis interaction: Conditions predisposing to infection may be mediated, at least in part, by impairing an effective alveolar macrophage response. Such conditions may affect macrophage host defense function directly, or may act indirectly by altering the microenvironment in which the macrophage-PC interaction occurs.

This hypothesis emphasizes the notion that a comprehensive understanding of the role of macrophages in *Pneumocystis* infection must account for the impact of local factors in the alveolus (e.g., surfactant), as well as immune factors (e.g., cytokines and antibody), on macrophage recognition and destruction of this pathogen.

V. LABORATORY METHODS FOR INVESTIGATION OF THE MACROPHAGE–*PNEUMOCYSTIS* INTERACTION

Efforts to clarify the interplay of macrophages with this pathogen have relied almost exclusively on in vitro systems. The majority of these methods utilize cultured rodent rather than human alveolar macrophages and PC organisms isolated from the lungs of corticosteroid-treated rats maintained on normocaloric low-protein diets. Following inoculation onto monolayers, organisms are allowed to interact with macrophages under experimental conditions. Upon removal of nonadherent organisms, a variety of methods have been used to visually examine the macrophage–PC interaction, including inverse-phase, light, and electron microscopy [42,43]. Nonvisual methods employing labeled organisms (e.g., with ^{51}Cr) have also been used to characterize macrophage handling of this organism [15].

Each of these methods attempts to determine the ability of macrophages to bind and internalize PC. However, estimates of binding versus internalization of PC by macrophage monolayers using light or inverse-phase microscopy techniques are limited in that these methods are subject to observer bias, and in most circumstances the distinction of bound versus internalized organisms cannot be established with certainty. As a result, these methods permit only a semiquantitative assessment of the macrophage–PC interaction. Although electron microscopy is capable of distinguishing internalized organisms from those that are bound to the surface (Figure 3), this method involves considerable sample preparation time, requires the availability of specialized instrumentation and trained personnel, and is not, therefore, appropriate for quantitative analysis. Efforts to quantitate this interaction have been successful utilizing ^{51}Cr-labeled PC [44]. However, this technique has the disadvantage of requiring radioactive labeling and assumes that under experimental conditions, the radiolabel remains tightly associated with the PC organisms.

Recently, our group has developed a method for investigating this interaction that is not dependent on the use of radioactive materials or electron microscopy [45]. Furthermore, this method circumvents the uncertainty of distinguishing bound from internalized organisms with techniques requiring direct observation, thus permitting quantitative characterization of the macrophage–PC interaction.

To facilitate the identification of PC, organisms are labeled with fluorescein isothiocyanate (FITC), which binds tightly to the major surface glycoprotein A of both trophozoite and cysts forms of the organism (as determined by SDS-PAGE analysis). Following macrophage exposure to FITC-PC, the total number of cell-associated organisms (representing bound and internalized organisms) is estimated by determination of cell-associated fluorescence utilizing a fluorescence

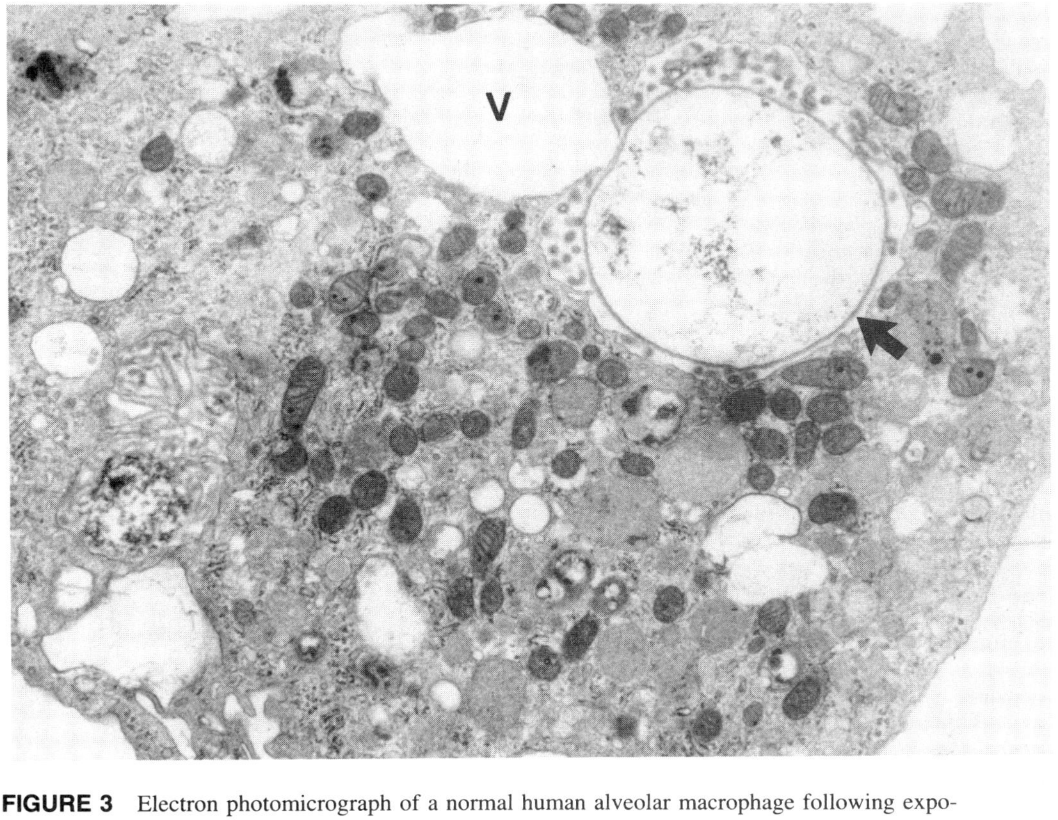

FIGURE 3 Electron photomicrograph of a normal human alveolar macrophage following exposure to *Pneumocystis carinii* in vitro. An internalized *P. carinii* trophozoite form is shown (arrow) adjacent to a large vacuole (V). (Magnification × 46,000.)

measurement system (e.g., a plate fluorimeter). An estimate of internalized organisms can then be determined by quenching extracellular fluorescence via the addition of trypan blue dye [46].

VI. MACROPHAGE RECOGNITION OF *PNEUMOCYSTIS CARINII*

Several investigators have demonstrated that normal rodent alveolar macrophages are capable of binding and internalizing PC [42,43]. However, these observations are often conflicting. Von Behren and Pesanti demonstrated that cultured alveolar macrophages more readily ingest PC than freshly harvested cells [42], while other investigators found no difference [43]. Some investigators have demonstrated binding of organisms to macrophage monolayers without observing significant internalization during short-term culture [44].

Utilizing the fluorescence detection method described above, we have directed our investigation to the interaction of PC with human alveolar macrophages (Figure 4). Normal human alveolar macrophages are fully capable of binding and internalizing PC in vitro. In the absence of potential opsonins contained in serum, the ability of macrophages to bind and ingest PC appears to be mediated by the macrophage mannose receptor [47]. This is not surprising in view of the mannose-rich composition of PC glycoprotein A. Since blood monocytes and some macrophage cell lines do not express functional mannose receptors [48], the use of these cells is of uncertain utility in the study of the PC interaction with host defense cells.

Several lines of evidence implicate the mannose receptor as a major pathway for macrophage

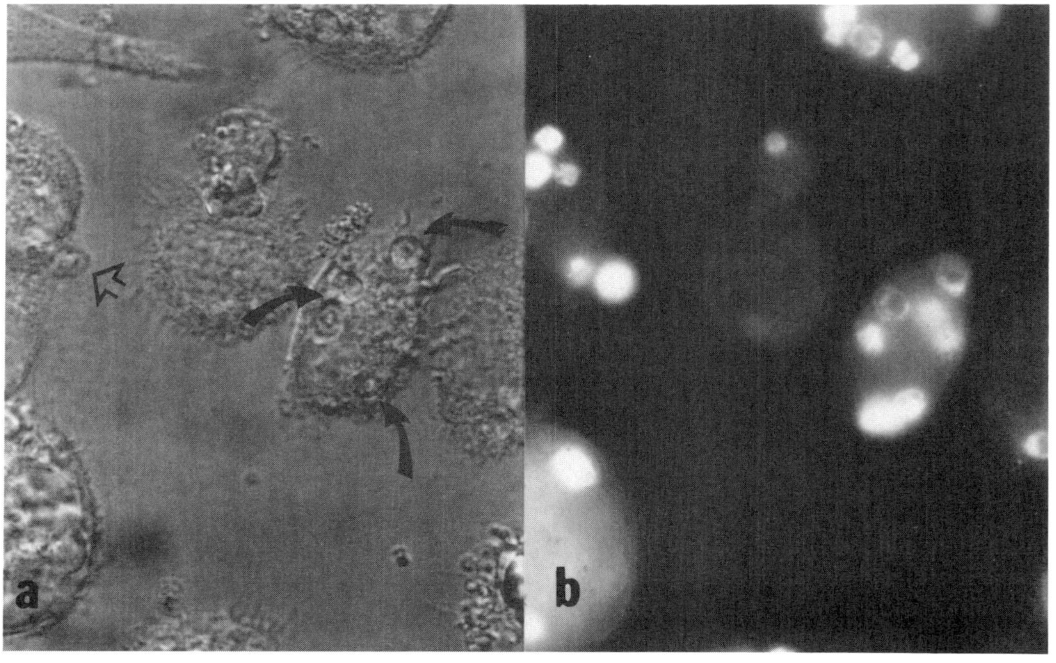

FIGURE 4 Photomicrograph of adherent normal alveolar macrophages viewed under high-contrast inverse-phase light microscopy (panel a) following in vitro exposure to FITC-labeled *Pneumocystis carinii*. Several thick-walled cyst forms are identified bound (open arrow) and internalized (curved arrows) by the macrophages. The same microscopic field viewed under epifluorescence microscopy (panel b) readily confirms these forms as *P. carinii*. Furthermore, with epifluorescence, numerous additional organisms are readily identified by the intense fluorescent signal.

recognition of PC. The introduction of competitive inhibitors (e.g., mannosyl–bovine serum albumin or yeast mannan) or modulation of the mannose receptor expression (e.g., by redistributing surface mannose receptor via adherence of cells to mannan-coated glass cover slips) significantly reduces the ability of normal alveolar macrophages to bind and internalize this organism. Transfection of the nonphagocytic COS cell line with the cDNA for the human mannose receptor renders this cell fully capable of binding and internalizing PC [47] (Figure 5).

A. Opsonins and Adhesive Molecules

The presence of the Fc receptor and complement receptor complex on the macrophage membrane provides this cell with several pathways for the efficient recognition and ingestion of opsonized particles. Levels of potential opsonins such as immunoglobulin G (IgG), IgA, and IgM directed against PC have been measured in the bronchoalveolar lavage fluid of animals with active infection [49]. However, the role of these opsonins in macrophage processing of PC is not clear. Masur and Jones demonstrated that immune serum was necessary for the internalization of PC cysts by rodent alveolar macrophages [43], whereas von Behren and Pesanti observed, as has our group, that efficient phagocytosis of PC trophozoites can occur in the absence of serum [42]. In addition, a role for complement components in macrophage recognition of PC is suggested by the observation that heat inactivation of serum reduces the avidity with which cells bind and internalize PC (H. Koziel, unpublished data).

Certain extracellular adhesive molecules have been implicated in the binding of PC to host cells. Fibronectin can mediate attachment of PC to alveolar macrophages as well as alveolar epithelial cells. Interestingly, fibronectin does not appear to subserve the role of a classic opsonin in that it does not facilitate macrophage ingestion of PC [50]. Although other extracellular adhesins such as laminin and vitrinectin have been demonstrated to affect binding of the organism to epithelial cells [51], their role in the interaction with alveolar macrophages has not yet been elucidated.

B. Intracellular Fate of *Pneumocystis carinii*

The intracellular fate of PC is not well characterized. Investigators have described rapid degradation of PC occurring shortly after internalization as determined by direct visualization utilizing inverse-phase microscopy [42]. We have also observed efficient destruction of PC occurring within several hours of exposure to normal macrophage monolayers (unpublished data). Electron microscopic analysis of macrophage monolayers infected with PC demonstrates numerous internalized organisms within several hours of inoculation (Figure 3). However, only few recognizable organisms are detected by 12 hours, and no morphologically intact organisms are apparent by 24 hours. This underscores previous observations that there does not appear to be a persistently intracellular phase in the life cycle of this organism. Therefore, unlike other opportunistic pathogens of macrophages, PC is an obligate *extracellular* rather than *intracellular* parasite.

The mechanism or mechanisms by which PC is killed are not completely understood. Several investigators have demonstrated that macrophage monolayers release hydrogen peroxide and other oxygen radicals following exposure to PC [47,52] (Figure 6). Pesanti proposed that oxygen radicals are directly toxic to the organism, as PC possesses limited antioxidant defenses [53]. The possible role of other pathways of macrophage killing, such as the arginine-dependent nitrous oxide pathway, have not yet been systematically examined [54].

Evaluation of the macrophage's ability to eradicate PC has been limited by the absence of a reliable assay for distinguishing viable from nonviable organisms. Metabolic markers of nucleotide synthesis demonstrate only a low level of incorporation of radiolabeled precursors [43], presumably as a reflection of the organism's paucity of energy-generating organelles and/or limited metabolic requirements. Recently a β-1,3-glucanase, an enzyme involved in PC cell wall synthe-

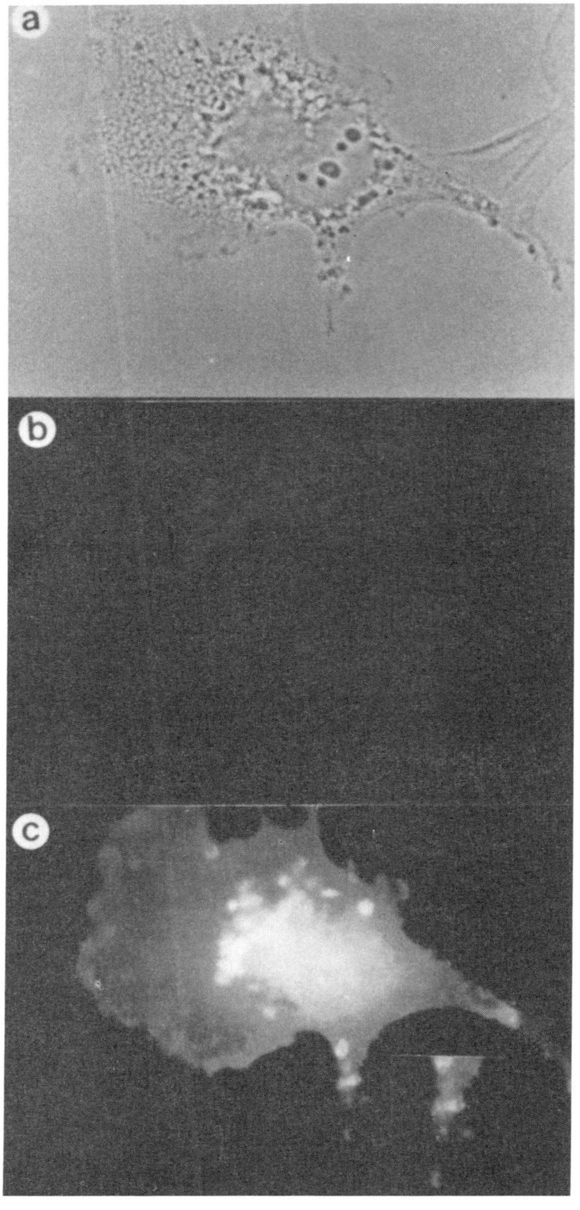

FIGURE 5 Transfection of the nonphagocytic COS-1 cell with the mannose receptor renders the cell capable of binding and internalizing *Pneumocystic carinii*. All three panels represent the same microscopic field. Phase-contrast photomicrograph (panel a) of an adherent COS-1 cell, transfected with full-length cDNA for the mannose receptor, following exposure to FITC-labeled *P. carinii*. Epifluorescence microscopy (panel b) reveals several fluorescent organisms associated with the transfected COS-1 cell. Further, counterstaining the COS-1 cell with primary anti–mannose receptor antibody and a secondary rhodamine-coupled immunoglobulin and viewing under epifluorescence (panel c) confirms surface expression of mannose receptor as evidenced by the homogeneous label. Only those COS-1 cells that express mannose receptor are capable of binding and phagocytosing organisms. (From Ref. 47.)

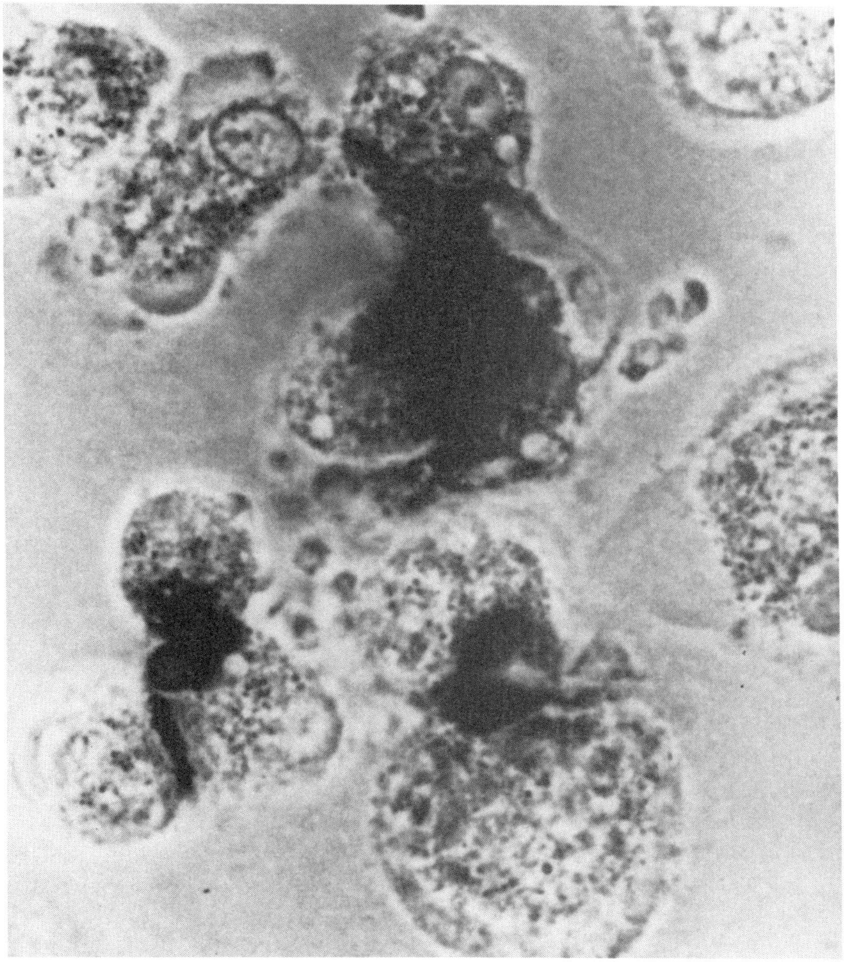

FIGURE 6 Exposure of alveolar macrophages to *Pneumocystis carinii* in vitro triggers release of superoxide by the macrophages. In the presence of nitroblue tetrazolium, release of superoxide is detected by the conversion of this histochemical dye to a dark-colored precipitant immediately adjacent to the cell. (From Ref. 47.)

sis, has been identified [55]. Activity of this enzyme correlates directly with the increase in PC organisms grown in short-term culture [56] and could therefore be useful as biological marker of viability.

Other methods to assess viability have utilized techniques involving vital dye fluorescence staining [57]. Exclusion of DNA-binding fluorescent dyes can assist in identifying organisms with intact cellular membranes in contrast to those which are nonviable or have disrupted cell membrane integrity. However, this determination may be difficult using direct visualization by epifluorescence microscopy because of limited visual sensitivity and variability of fluorescence staining intensity between the various life forms of PC.

Some of the inherent difficulty in viability determination may be overcome by utilizing the sensitivity of flow cytometry to identify the population of organisms that incorporate the dye [58]. However, this method does not distinguish between "nonviable" and "injured" organisms, nor does it distinguish "viable" organisms from those that are devoid of intracellular organelles ("ghosts").

To circumvent these problems, we have modified a well-described double-labeling viability assay [59,60] that uses fluorescein diacetate (FDA; a substrate that is converted to a fluorophore by cellular nonspecific esterase) and propridium iodide (PI; a DNA-binding protein excluded by an intact cell membrane), and adapted this technique to flow cytometry [56]. Using organisms propogated in short-term culture (which tends to allow the clusters of organisms to disperse and become singular trophozoite and cyst forms), we have described four populations of organisms (Figure 7): 1) those that convert FDA but exclude PI (viable); 2) those that are unable to convert FDA but incorporate PI (nonviable); 3) those that both convert FDA and incorporate PI ("injured"); and 4) those that neither convert FDA nor incorporate PI ("ghosts" or fragments of organisms). Furthermore, these observations correlate well with the activity of β-1,3-glucan synthase, a biological marker for viability.

The availability of this type of viability assay should be valuable in rapidly assessing the impact of numerous factors on PC survival. For instance, this system could be applied to screening of antimicrobials, evaluating the impact of various culture conditions on PC growth, and examining the effect of host defense factors (e.g., cytokines) on PC survival.

VII. EFFECT OF LOCAL ALVEOLAR FACTORS

Pneumocystis carinii infection occurs within the confines of alveolar lining material, which includes surfactant components, immunglobulins, complement, serum proteins, and secretory products of immune cells. Alterations in the composition of this material may contribute to the pathobiology of PC infection by affecting the functional integrity of alveolar macrophages.

This theme is exemplified by recent studies characterizing the impact of surfactant protein A (SP-A) on macrophage handling of PC. Surfactant protein A, a glycoprotein component of pulmonary surfactant, is elevated for unknown reasons in bronchoalveolar lavage fluid obtained from patients with AIDS-related PC pneumonia [61]. Interestingly, SP-A retards binding and internalization of PC by alveolar macrophages in vitro [62]. Surfactant protein A exerts this effect by binding to the surface of the organism rather than by interacting with the macrophage directly. It is capable of binding to the mannose-rich glycoprotein A of PC [63]. This association with glycoprotein A may be explained in part by the fact that SP-A demonstrates considerable homology with the mannose-binding protein [64], a soluble analogue of the mannose receptor that binds to mannose-rich ligands. In view of the importance of the macrophage mannose receptor in the recognition of PC, binding of SP-A to the surface of the organism may alter the ability of the macrophage to normally recognize this organism as foreign, This mechanism for evading macrophage host defenses, which takes advantage of a lung-specific factor present only in pulmonary surfactant, may also help to account for the unexplained tropism of PC for the alveolar surface of the lungs.

VIII. ROLE OF CYTOKINES

The role of cytokines in the interaction of PC with alveolar macrophages is an area of active investigation that has not yet produced a definitive picture of how these immune factors act. Since normal alveolar macrophages are fully capable of destroying PC in the absence of immune factors, it remains to be seen in what ways, if any, cytokines influence this interaction. Classically, cytokines augment macrophage mechanisms for restricting the growth of opportunistic pathogens. These mechanisms include augmenting oxidative and nonoxidative modes of intracellular killing and enhancing modes of extracellular antimicrobial activity.

In the case of PC, cytokines may also influence certain other aspects of macrophage host

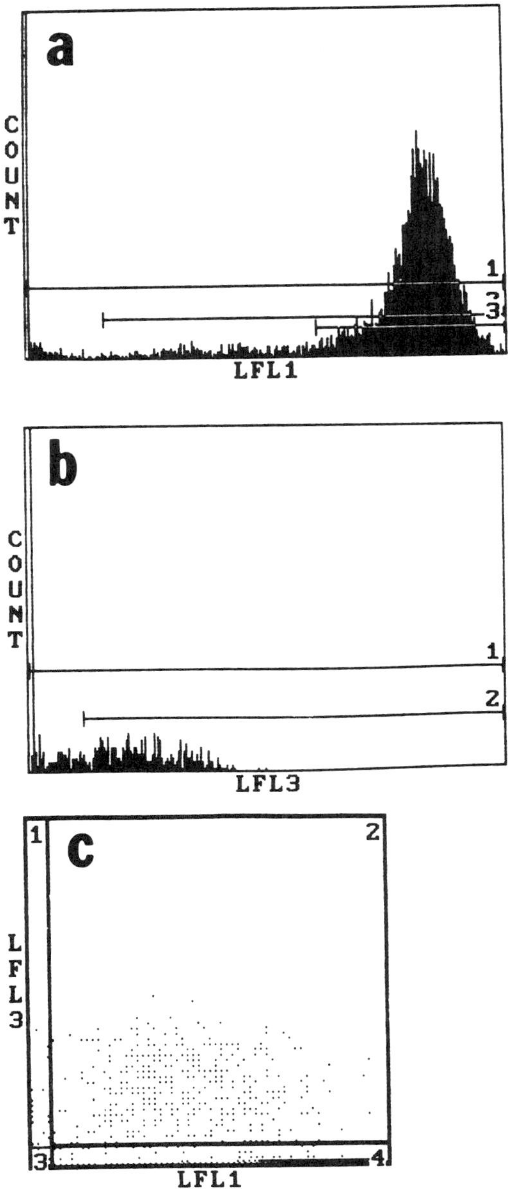

FIGURE 7 Flow cytometry analysis of *Pneumocystis carinii* following double-labeling with fluorescein diacetate (FDA) and propridium iodide (PI). The representative histograms identify organisms that convert FDA to a yellow-green fluorochrome, labeled as LFL1 (panel a), and organisms that incorporate the red fluorochrome PI, labeled as LFL3 (panel b). Combining the two histograms (LFL1 vs. LFL3 in panel c) identifies four populations of organisms that convert FDA to fluorescein (quadrant 1); organisms that incorporate PI (quadrant 4); organisms capable of *both* incorporating PI and converting FDA (quadrant 2); and organisms that *neither* incorporate or convert fluorescent dyes. (Refer to text for explanation.)

defense function. For instance, cytokines such as interferon-γ, which are known to decrease macrophage mannose receptor expression, or macrophage colony-stimulating factor and interleukin-4 (IL-4) [65], which enhance receptor expression, are capable of influencing the process of PC recognition by cells. Additionally, cytokines such as granulocyte-macrophage colony-stimulating factor, which stimulate directed movement of phagocytes through a chemotactic gradient, may be important in the attraction of cells to the site of PC infection in vivo.

At the present time, there is no evidence that cytokines directly alter binding and internalization of PC by normal cells. However, aerosolized interferon-γ can reduce the histologic intensity of *Pneumocystis* pneumonia in an experimental rodent model [39]. Similarly, in the SCID mouse model partial immune reconstitution with IL-1 has recently been shown to be beneficial [66]. Whether this beneficial effect is mediated via cytokine-induced macrophage activation or by other mechanisms (e.g., enhancement of cytotoxic T or natural killer cell activity) is unclear.

Cytokines may also exert a direct effect on PC. As recently suggested by Pesanti, tumor necrosis factor-α may be directly toxic to PC [67]. Confirmation of this observation will be necessary to determine whether macrophages may also be able to influence the survival of PC by secretion of monokines.

IX. ROLE OF HUMAN IMMUNODEFICIENCY VIRUS INFECTION OF ALVEOLAR MACROPHAGES

In view of the exceptional impact of HIV as a risk factor for *P. carinii* infection, we have attempted to further define the mechanisms responsible for this unusual host susceptibility. Ex vivo study of macrophages obtained from AIDS patients reveals a variety of functional deficiencies including diminished chemotactic response [68], decreased expression of major histocompatibility complex class II molecules [69], decreased IL-1 secretion [70], altered Fc-mediated ingestion [71], and defective bactericidal capacity [68]. However, most of these observations were made using peripheral blood monocyte-derived macrophages.

Using alveolar macrophages obtained from persons with active AIDS-related lung disorders, we observed an impaired ability to adequately bind and internalize PC in comparison with normal cells [72]. This interaction appears to be pathogen-specific, since the AIDS-derived cells were fully capable of internalizing a phylogenetically related organism, *Candida albicans*.

In view of the previously described role for the mannose receptor in the macrophage-PC interaction, it was of interest to examine the possibility that this defect in macrophage processing is related to altered expression of a functional mannose receptor. Preliminary studies in a small cohort of patient samples lend support to this hypothesis [73]. This idea is also supported by studies in which in vitro infection of normal alveolar macrophages with a monocytotropic strain of HIV-1 results in down-regulation of the mannose receptor, and it correlates with a tendency toward reduction of binding and internalization of PC.

In the setting of active HIV replication, high levels of HIV components in the alveolar microenvironment may potentially affect the interaction of PC with macrophages. For instance, the HIV-associated gp120 envelope glycoprotein, which is rich in mannose residues, is capable of binding to the alveolar macrophage mannose receptor in a competitive manner [73]. Thus, high levels of gp120 in the alveolar lining material could impede the macrophage recognition of PC through the mannose receptor pathway.

X. FUTURE STUDIES ON THE *PNEUMOCYSTIS CARINII*–MACROPHAGE INTERACTION

On the basis of the information presented in this chapter, several aspects of the macrophage–PC interaction have been identified and described. Normal alveolar macrophages are capable of binding, internalizing, and destroying this pathogen. Importantly, this can occur in the absence of serum or T lymphocyte factors, suggesting that additional deficiencies involving macrophage function may be necessary for the development of clinical pneumonia in the appropriate host. In this regard, the importance of the macrophage mannose receptor in the recognition and internalization of PC is underscored. Alterations in the expression or function of this receptor may be permissive for the development of pneumonia in the immunocompromised host. The absence of cells that are necessary for amplification of the host immune response and concurrent alterations of local host factors at the site of infection in situ may be additive. Taken together, these factors may be responsible for determining whether reactivation of latent PC infection occurs, or in determining the course of PC infection. Additionally, certain lung-specific alterations in local host factors, such as SP-A, may in part explain the tropism of this organism for the alveolus.

Although significant knowledge has been gained over the past decade, it is apparent that numerous questions remain concerning the factors involved in the interaction of this organism with pulmonary host defenses.

Suggested areas where future research should be valuable include

1. Further characterization of the role of other macrophage receptors that may be involved in this interaction. This would include Fc and CR3 receptors, in addition to adhesion molecule receptors.
2. Further definition of the role of various cytokines that may be protective in the normal host.
3. Examination of factors influencing the macrophage chemotatic response to PC infection.
4. Extending the characterization of surfactant component alterations to determine role of other surfactant glycoprotein and lipid components on phagocytic properties of normal macrophages.
5. Further investigations to delineate the effect of HIV infection of alveolar macrophages in the observed deficiency for phagocytosis. This would include identification of the alteration of receptor function, with emphasis on the mannose receptor.

Many of these questions are the focus of active investigation. Further identification of those factors that are responsible for normal host defense, and that may be altered in the immunocompromised patient, will provide insight into the pathogenesis of the infection and will hopefully lead to development of novel therapeutic approaches.

REFERENCES

1. Cushion MT. In vitro studies of *Pneumocystis carinii* (review) J Protozool 1989; 36:45–52.
2. Cushion MT, Walzer PD. Growth and serial passage of *Pneumocystis carinii* in the A549 cell line. Infect Immun 1984; 44:245–51.
3. Latorre CR, Sulzer AJ, Norman LG. Survival of *Pneumocystis carinii* in cell line cultures. Appl Environ Microbiol 1977; 33:1204–06.
4. Bartlett MS, Verbanic PA, Smith JW. Cultivation of *Pneumocystis carinii* with WI-38 cells. J Clin Microbiol 1979; 10:796–99.
5. Edman JC, Edman U, Cao M, Lungren B, Kovacs JA, Santi DV. Isolation and expression of the *Pneumocystis carinii* dihydrofolate reductase gene. Proc Natl Acad Sci USA 1989; 86:8625–9.
6. Edman U, Edman JC, Lundgren B, Santi DV. Isolation and expression of *Pneumocystis carinii* thymidylate synthase gene. Proc Natl Acad Sci USA 1989; 86:6503–7.
7. Edman JC, Kovacs JA, Masur H, Santi D, Elwood HJ, Sogin ML. Ribosomal RNA sequence shows *Pneumocystis carinii* to be a member of the fungi. Nature 1988; 334:519–22.
8. Stringer SL, Stringer JR, Blase MA, Walzer PD, Cushion MT, *Pneumocystis carinii*: sequence from ribosomal RNA implies a close relationship with fungi. Exp Parasitol 1989; 68:450–61.
9. Matsumoto Y, Yoshida Y. Sporogony in *Pneumocystis carinii*: synaptonemal complexes and meiotic nuclear divisions observed in precysts. J Protozool 1984; 31:420–28.

10. Yoshida Y. Ultrastructural studies of *Pneumocystis carinii*. J Protozool 1989; 36:53–60.
11. Barton EG, Campbell WG Jr. Further observation on the ultrastructure of Pneumocystis. Arch Pathol 1967; 83:527–34.
12. Vavra J, Kucera K. *Pneumocystis carinii Delanoe*, its ultrastructure and ultrastructural affinities. J Protozool 1970; 17:463–83.
13. Graves DC, McNabb SJ, Whorley MA, Downs TD, Iveh MH. Analysis of rat Pneumocystis carinii antigens recognized by human and rat antibodies using Western immunoblotting. Infect Immun 1986; 54:96–103.
14. Walzer PD, Linke MJ. A comparison of the antigenic characteristics of rat and human *Pneumocystis carinii* by immunoblotting. J Immunol 1987; 138:2257–65.
15. Pottratz ST, Paulsrud J, Bartlett M, Queener S, Smith J, Martin WJ II. *Pneumocystis carinii* membrane glycoprotein GP120 mediates attachment of Pneumocystis to host cells. Am Rev Respir Dis 1990; 141:A275. (Abtract.)
16. Cushion MT, DeStefano JA, Walzer PD. *Pneumocystis carinii*: surface reactive carbohydrates detected by lecitin probes. Exp Parisitol 1988; 67:137–47.
17. Gigliotti F, Hughes WT. Passive immunoprophylaxis with specific monoclonal antibody confers partial protection against *Pneumocystis carinii* pneumonitis in animal models. J Clin Invest 1988; 81:1666–8.
18. Tenabe K, Takasaki S, Watanabe JI, Kobata A, Egawa K, Nakamura Y. Glycoproteins composed of major surface immunodeterminants of *Pneumocystis carinii*. Infect Immun 1989; 57:1363–68.
19. Walzer PD, Stanforth D, Linke MJ, Cushion MT. *Pneumocystis carinii*: Immunoblotting and immunofluorescent analysis of serum antibodies during experimental rat infection and recovery. Exp Parasitol 1987; 63:319–28.
20. Gigliotti F. Antigenic variation of a major surface glycoprotein of *Pneumocystis carinii*. J Protozool 1991; 38:4S–5S, Supplement.
21. Yogonthan T, Lin H, Buck GA. An electrophoretic karyotype and assignment of ribosomal genes to resolved chromosomes of *Pneumocystis carinii*. Mol Microbiol 1989; 3:1473–80.
22. Lundgren B, Cotton R, Lunggren JD, Edman JC, Kovacs JA. Identification of *Pneumocystis carinii* chromosomes and mapping of five genes. Infect Immun 1990; 58:1705–10.
23. Hong ST, Steele P, Stringer S, Cushion MT, Walzer PD, Stringer JR. Molecular karyotyping of rat and human *Pneumocystis carinii*. J Clin Microbiol 1990; 28:1785–95.
24. Hughes WT. Natural mode of acquisition for de novo infection with *Pneumocystis carinii*. J Infect Dis 1982; 145:842–48.
25. Meuwissen J, Tauber I, Leeuwenberg A, Beckers P, Sieben M. Parasitological and serologic observations of infection with pneumocystis in humans. J Infect Dis 1977; 136:43–49.
26. Pifer LL, Hughes WT, Murphy MJ. Cultivation of *Pneumocystis carinii* in vitro. Pediatr Res 1977; 11:305–16.
27. Peglow SL, Smulian GA, Linke MJ, Pogue CL, Nurre S, Crisler J, Phair J, Gold JWM, Armstrong D, Walzer PD. Serologic responses to specific *Pneumocystis carinii* antigens in health and disease. J Infect Dis 1990; 161:296–306.
28. Browne MJ, Hubbard SM, Longo DL, Fisher R, Wesley R, Ihde DC, Young RC, Pizzo PA. Excess prevalence of *Pneumocystis carinii* pneumonia in patients treated for lymphoma with combination chemotherapy. Ann Intern Med 1986; 104:338–44.
29. Phair J, Munoz A, Detels R, Kaslow R, Rinaldo C, Saah A, and the Multicenter AIDS Cohort Study Group. The risk of *Pneumocystis carinii* pheumonia among men infected with human immunodeficiency virus type I. N Engl J Med 1990; 322:161–5.
30. Hughes WT, Price RA, Sisko F, Havron WS, Kafatos AG, Schonland M, Smythe PM. Protein-calorie malnutrition: a host determinant for *Pneumocystis carinii* infection. Am J Dis Child 1974; 128:44–52.
31. Gleason WA, Roodman ST. Reversible T-cell depression in malnourished infants with Pneumocystis pneumonia. J Pediatr 1977; 90:1022–28.
32. Walzer PD, LaBine M, Redington BJ, Cushion MT. Lymphocyte changes during chronic administration and withdrawal of corticosteroids: relevance to *Pneumocystis carinii* pneumonia. J Immunol 1984; 133:2502–8.
33. Walzer PD, Kim CK, Linke MJ, Pogue CL, Huerkamp MJ, Chrisp CE, Lerro AV, Wixson SK, Hall E, Shultz LD. Outbreaks of *Pneumocystis carinii* pneumonia in colonies of immunodeficient mice. Infect Immun 1989; 57:62–70.

34. Roth JB, Marshall JD, Allen RD, Carlson GA, Sidman CL. Spontaneous *Pneumocystis carinii* pneumonia in immunodeficient scid mice. Am J Pathol 1990; 136:1173–86.
35. Shilleto J, Suzura VV, Blumenfeld W, Beck JM, Steger HJ, Ermak TH. A new model of *Pneumocystis carinii* infection in mice selectively depleted of helper T-lymphocytes. J Clin Invest 1990; 85:1686–93.
36. Harmson AG, Stankiewicz M. Requirement for CD4+ cells in resistance to *Pneumocystis carinii* pneumonia in mice. J Exp Med 1990; 172:937–45.
37. Herrod HC, Valenski WR, Woods DR, Pifer LL. The in vitro response of human lymphocytes to *Pneumocystis carinii* antigen. J Immunol 1981; 126:59–61.
38. Hagler DN, Deepe GS, Pogue CL, Walzer PD. Blastogenic responses to *Pneumocystis carinii* among patients with human immunodeficiency virus (HIV) infection. Clin Exp Immunol 1988; 74:7–13.
39. Beck JM, Liggitt HD, Brunette EN, Fuchs HJ, Shellito JE, Debs RJ. Reduction in intensity of *Pneumocystis carinii* pneumonia in mice by aerosol administration of gamma interferon. Infect Immun 1991; 59:11:3859–62.
40. Pottratz ST, Martin WJ. Role of fibronectin in *Pneumocystis carinii* attachment to cultured lung cells J Clin Invest 1990; 85:351–56.
41. Limpert AH, Martin WJ II. *Pneumocystis carinii*: inhibition of lung cell growth mediated by parasite attachment. J Clin Invest 1990; 85:391–6.
42. Von Behren LA, Pesanti EL. Uptake and degradation *Pneumocystis carinii* by macrophages *in vitro*. Rev Respir Dis 1978; 118:1051–1059.
43. Masur H, Jones TC. The interaction *in vitro* of *Pneumocystis carinii* with macrophages and L-cells. J Exp Med 1978; 147:157–70.
44. Pottratz ST, Martin WJ. Mechanism of *Pneumocystis carinii* attachment to cultured rat alveolar macrophages. J Clin Invest, 1990; 86:1678–83.
45. Koziel H, Williams DJ, Armstrong MYK, Richards FF, Fishman JA, Ezekowitz RAB, Warner A, Fuglestad J, Rose RM. New rapid method for the study of *Pneumocystis carinii* interaction with alveolar macrophages. J Protozool 1991; 38:6:173S–174S.
46. Hed J. The use of fluorescence quenching in flow cytofluorometry to measure the attachment and ingestion phases in phagocytosis in peripheral blood without prior cell separation. J Immunol Methods 1987; 101:119–25.
47. Ezekowitz RAB, Williams DJ, Koziel H, Armstrong MYK, Warner A, Richards FF, Rose RM. Uptake of *Pneumocystis carinii* mediated by the macrophage mannose receptor. Nature 1991; 351:155–8.
48. Stahl PD. The macrophage mannose receptor: current status. Am J Respir Cell Mol Biol 1990; 2:317–8.
49. Walzer PD, Rutledge ME. Humoral immunity in experimental *Pneumocystis carinii* infection: I. Serum and bronchial lavage fluid antibody responses in rats. J Lab Clin Med, 1981; 97:820–33.
50. Pottratz ST, Paulsrud J, Smith JS, Martin WJ. *Pneumocystis carinii* attachment to cultured lung cells by Pneumocystis gp120, a fibronectin binding protein. J Clin Invest 1991; 88:403–7.
51. Fishman JA, Samia JA, Fuglestad J, Rose RM. The effects of extracellular matrix (ECM) proteins on the attachment of *Pneumocystis carinii* to lung cell lines in vitro. J Protozool 1991; 38:34S–37S, Supplement.
52. Hidalgo HA, Helmke RJ, German VF, Mangos JA. *Pneumocystis carinii* induces an oxidative burst in alveolar macrophages. Infect Immun 1992; 60:1:1–7.
53. Pesanti EL. *Pneumocystis carinii*: oxygen uptake, antioxidant enzymes, and susceptibility to oxygen-mediated damage. Infect Immun 1984; 44:7–11.
54. Sherman MP, Hidalgo HA, Aeberhard EE, Wong VZ, Gilliam MB, Helmke RJ. *Pneumocystis carinii* and interferon gamma induce rat alveolar macrophages to produce nitric oxide. Am Rev Respir Dis 1992; 145:A246. (Abstract.)
55. Williams DJ, Radding JA, Dell A, Khoo K-Y, Rogers ME, Richards FF, Armstrong MYK. Glucan synthesis in *Pneumocystis carinii*. J Protozool, 1991; 38:427–37.
56. Armstrong MRK, Koziel H, Rose RM, Arena CA, Richards FF. Indicators of *Pneumocystis carinii* viability in short-term cell culture. J Protozool 1991; 38:6:88S–90S, Supplement.
57. Kaneshiro ES, Wu Y-P, Cushion MT. Assays for testing *Pneumocystis carinii* viability. J Protozool 1991; 38:85S–86S, Supplement.
58. Lapinsky SE, Glencross D, Car NG, Kallenback JM, Zwi S. Quantification and assessment of viability of *Pneumocystis carinii* organisms by flow cytometry. J Clin Microbiol 1991; 29:911–15.
59. Schupp DG, Erlandsen SL. A new method to determine *Giardia* cyst viability: correlation of fluorescein

diacetate and propridium iodide staining with animal infectivity. Appl Environment Microbiol 1986; 53:4:704–7.

60. Jones KH, Senft JA, An improved method to determine cell viability by simultaneous staining with fluorescein diacetate–propridium iodide, J Histochem Cytochem 1985; 33:77–9.
61. Phelps DS, Rose RM. Increased recovery of surfactant protein A in AIDS-related pneumonia. Am Rev Respir Dis 1991; 143:1072–5.
62. Koziel H, O'Riordan D, Phelps D, Fishman JA, Armstrong MYK, Richards FF, Rose RM. Surfactant protein-A inhibits binding and internalization of *Pneumocystis carinii* by alveolar macrophages. Am Rev Respir Dis 1992; 145:A247. (Abstract.)
63. Zimmerman PE, Voelker DR, McCormack FX, Paulsrud JR, Martin WJ. 120-kD Surface glycoprotein of *Pneumocystis carinii* is a ligand for surfactant protein A. J Clin Invest 1992; 89:143–9.
64. Sastry K, Herman GA, Day L, Deignan E, Bruns G, Morton CC, Ezekowitz RAB. The human mannose binding protein gene exon structure reveals its evolutionary relationship to human pulmonary surfactant gene and localization to chromosome 10. J Exp Med 1989; 170:1175–89.
65. Stein M, Keshav S, Harris N, Gordon S, Interleukin 4 potently enhances murine macrophage mannose receptor activity: A marker of alternative immunologic macrophage activation. J Exp Med, 1992; 176:287–92.
66. Harmsen AG, Chen W. Resolution of *Pneumocystis carinii* pneumonia in CD4+ lymphocyte-depleted mice given aerosols of heat-treated *Escherichia coli*. J Exp Med, 1992; 176:881–6.
67. Pesanti EL. Interaction of cytokines and alveolar cells with *Pneumocystis carinii in vitro*. J Infect Dis 1991; 163:611–6.
68. Smith PD, Ohura K, Masur H, Lane HC, Fauci AS, Wahl SM. Monocyte function in the acquired immunodeficiency syndrome: defective chemotaxis. J Clin Invest 1984; 74:2121–8.
69. Belsito DV, Sanchez MR, Baer RL, Valentine F, Thorbecke GJ. Reduced langerhans cell Ia antigen and ATPase activity in patients with the acquired immunodeficiency syndrome. N Engl J Med 1984; 310:1279–82.
70. Roy S, Fitz-Gibbon L, Poulin L, Wainberg MA. Infection of human monocytes/macrophages by HIV-1: effect on secretion of IL-1 activity. Immunology 1988; 64:233–9.
71. Bender BS, Frank MM, Lawley TJ, Smith WJ, Brickman CM, Quinn TC. Defective reticuloendothelial system Fc-receptor function in patients with acquired immunodeficiency syndrome. J Infect Dis 1985; 152:409–12.
72. Koziel H, Williams DJ, Fugelstad J, Armstrong MYK, Richards FF, Warner A, Rose RM. Impact of HIV infection on human alveolar macrophage interaction with *Pneumocystis carinii*. Am Rev Resp Dis 1991; 143:4:A207. (Abstract.)
73. Koziel H, Kruskal BA, Ezekowitz RAB, Rose RM. Human immunodeficiency virus impairs alveolar macrophage mannose receptor function against *Pneumocystis carinii*. Chest 1993; 103:3:111S–2S, Supplement.

26

Interactions Between *Leishmania major* and Macrophages

Richard G. Titus, Cynthia M. Theodos, Anuraj H. Shankar, and Laurie R. Hall

Harvard School of Public Health, Boston, Massachusetts

I. INTRODUCTION

Members of the genus *Leishmania* are protozoan parasites that exist as flagellated promastigotes in the vector for the parasite, the phlebotomine sand fly. Human beings become infected with the parasite when bitten by an infected sand fly. Within humans, the parasite is an obligate intracellular parasite of macrophages (Mϕs), in which it transforms into its nonflagellated amastigote form. Amastigotes are found in phagolysosomes in the Mϕ, where they not only survive but replicate until parasite-specific T cell–mediated immunity intervenes and activates the Mϕ to destroy the parasite [1–3].

It is estimated that the incidence of leishmaniasis in the developing world is 1,000,000 cases/year [4]. Leishmaniasis is a complex of diseases caused by different species of *Leishmania*. The clinical manifestations of each disease are determined by the species of *Leishmania* that infects the host and by the response of the host to the parasite [1]. *Leishmania* are classified according to the nature of the disease they produce and the geographic region in which the disease occurs. Cutaneous leishmaniasis (CL) can be caused by *L. major (Lm)* in the Old World and *L. mexicana* or *L. braziliensis* in the New World; mucocutaneous leishmaniasis (MCL) can be caused by *L. braziliensis* in the New World; and visceral leishmaniasis (VL) can be caused by *L. donovani* in the Old World (Table 1) [1].

As mentioned above, *Leishmania* is able to survive and replicate within the phagolysosome of Mϕs. Since one of the primary functions of Mϕs is to destroy invading microorganisms, the fact that *Leishmania* lives in Mϕs is one of the mysteries of the biology of the parasite. In addition, although amastigote forms of each species of *Leishmania* are morphologically similar, the disease that each induces in human beings presents with widely different clinical manifestations. This could in part be due to 1) the type of Mϕ invaded by a given species of *Leishmania*, 2) the location of the Mϕ, and 3) the manner in which the parasite alters the functions of that Mϕ. Indeed, in animal models of leishmaniasis, genes that control resistance/susceptibility to different species of *Leishmania* [reviewed in 5] often, but not always [6], control functions that are related to the activities of Mϕs.

This review will focus on the interaction between *Leishmania* and Mϕs in a well-studied experimental system for leishmaniasis, namely, experimental murine CL induced with *Lm*. Reference will be made to data obtained with other species of *Leishmania* either when data are lacking for *Lm* or when there is controversy in the literature. The purpose of the review will be to

TABLE 1 Clinical Forms of Leishmaniasis and the Organisms Responsible

Human disease	Organism
Visceral leishmaniasis	*L. donovani*
Cutaneous leishmaniasis (Old World)	*L. major*
	L. aethiopica
Cutaneous leishmaniasis (New World)	*L. mexicana*
	L. amazonensis
	L. panamensis
	L. guyanensis
Mucocutaneous leishmaniasis	*L. braziliensis*

examine 1) the mechanisms by which *Lm* enters the Mϕ and survives, 2) the effect that infection has on the functions of the Mϕ, and 3) the influence that parasite-specific T cells (in particular, the cytokine products of those cells that either activate Mϕs to destroy *Lm* or interfere with Mϕ activation) and components of sand fly saliva [7] have on the interaction of *Lm* with the Mϕ.

II. THE ENTRY OF AND SURVIVAL OF *LEISHMANIA MAJOR* IN MACROPHAGES

A. Entry into the Vertebrate Host

The mouthparts of sand flies are too short to cannulate vessels in the skin, and their feeding stylets are wider than the superficial capillary loops that are within their reach. Therefore, sand flies appear to feed from pools of blood that are created by lacerating the capillary loops. Thus, when the fly injects *Lm*, it injects the parasite into the tissue spaces where it probes for a blood meal and potentially into the pool of blood from which it has fed. The parasite has to contend with all of these components of its new environment: tissue fluids containing serum proteins including complement (C), tissue phagocytic cells, and blood cells (including monocytes and neutrophils) that are present because of the sand fly's feeding.

L. major promastigotes are known to be exquisitely susceptible to lysis by C [8–10]. However, it is now known that highly infective metacylic forms of *Lm* develop in the gut of the sand fly [11], and it is this form of the parasite that is thought to be injected by the sand fly. Work by Sacks and colleagues has shown that metacyclic forms of *Leishmania* can be generated in vitro by allowing cultures of the parasite to enter a stationary phase of growth. These metacyclic parasites are more resistant to C lysis than logarithmically growing ("log phase") parasites. It should be stressed, however, that they are by no means completely resistant to C lysis ($\geq$25% fresh human serum kills 100% of *Lm* metacyclics) and that the degree of resistance varies with the species of *Leishmania* examined [10,12]. Recent work by Sacks and co-workers has revealed that the increased resistance of metacyclic forms of *Lm* is due to shedding of C5b-9 from the surface of the metacyclic parasite [13]. The mechanism underlying this shedding phenomenon is probably related to the fact that the surface lipophosphoglycan (LPG) molecule of *Lm* undergoes dramatic changes as the parasite transforms into its metacyclic form. Specifically, the LPG molecule elongates and changes its composition [14], which is thought to shield the parasite from lysis by C.

B. Attachment to Macrophages

The mechanism by which *Leishmania* attach to and are engulfed by Mϕs is being intensively studied by several laboratories. It is often difficult to make generalizations regarding this work because of differences in the techniques employed or the parasite studied by each laboratory. For example, some laboratories have studied interactions between *Leishmania* and Mϕs in the presence of serum, whereas others don't use serum. Some laboratories use metacyclic forms of *Leishmania*, while others don't. Different laboratories study different species of *Leishmania* and examine the interaction of their parasite with Mϕs obtained by differing techniques and from varying sources. The latter may be of the utmost importance; for example, since CL is a disease of the skin, is it possible that phagocytic cells of the skin (e.g., Langerhans cells) are in fact the most relevant cells to study. Thus, rather than extensively reviewing the subject of parasite attachment to Mϕs, which has recently been done by others [3,15,16], this topic will only be summarized here.

Several molecules on the surface of Mϕs have been shown to be involved in mediating the attachment and uptake of *Leishmania* (Table 2) [17–29]. What is apparent in Table 2 is that C receptors are thought to play a major role in the uptake of *Leishmania* by Mϕs. In fact, it is well known that *Leishmania* fix C and that opsonizing the parasite with C enhances its survival in Mϕs [30].

1. Lipophosphoglycan, Complement Receptor Type 1, and Leishmania major

Da Silva and colleagues [17] argued convincingly that interactions between *Lm* and Mϕs should be analyzed in the presence of fresh serum with metacyclic parasites, since it is likely that when the sand fly infects the mammalian host, the metacyclic form of *Lm* is delivered to the skin and that subsequent infection of Mϕs occurs in the presence of serum. Using this approach, the authors found that metacyclic *Lm* requires C opsonization for entry into human Mϕs, and that attachment to the Mϕ occurs via CR1. This same group had previously shown that the major acceptor molecule for C on *Lm* was LPG [12], and LPG has been shown to be necessary for the attachment of promastigotes to Mϕ [31]. Thus, with metacyclic *Lm*, C3 (primarily in the form of C3b, but also C3) is deposited on LPG molecules of the parasite. The C cascade is unable to proceed further to lyse *Lm*, since, as discussed above, the metacyclic LPG shields the parasite from attack by C. Therefore, with its coat of C3b, the metacyclic *Lm* interacts with CRl and enters the Mϕ. Finally, da Silva et al. [17] feel that entry of *Lm* via CR1 is in fact advantageous for the parasite, since this does not trigger an oxygen burst by the Mϕ [32].

2. Promastigote Surface Protease (gp63), Complement Receptor Type 3 and Leishmania mexicana

There are several fundamental differences between *Lm* (found in the Old World) and *L. mexicana* (found in the New World). First, the LPG molecule of *L. mexicana* is less complex than that of *Lm* [33] and does not undergo extensive alterations when *L. mexicana* enters a stationary phase of growth [3]. This observation may in part be responsible for the fact that, again unlike *Lm*,

TABLE 2 Macrophage Surface Molecules Involved in Attachment and Uptake of *Leishmania*

Macrophage molecule	Species of *Leishmania* that interact with molecule	References
Complement receptor type 1 (CR1)	*L. major*	17
Complement receptor type 3 (CR3)	*L. major*	18–25
	L. donovani	
	L. mexicana	
Mannose/fucose receptor (MFR)	*L. donovani*	21,26
Fibronectin receptor	*L. amazonensis*	27,28
	L. major	
	L. chagasi	
Advanced glycosylated end product (AGE) receptor	*L. major*	29

L. mexicana metacyclic parasites do not require C opsonization for attachment to M*φ*s [3]. However, it should be noted that no one has examined the interaction of *L. mexicana* metacyclic parasites with M*φ*s in the presence of serum (i.e., the approach of da Silva [17] with *Lm* described above), and therefore the exact role of C in the attachment of *L. mexicana* to M*φ*s is unknown. Another difference between *L. mexicana* and *Lm* is that while *L. mexicana* metacyclic parasites activate C via the alternative pathway, *Lm* metacyclics do not [12,19]. It is perhaps because of these differences between *L. mexicana* and *Lm* that *L. mexicana* has been found to utilize different surface molecules to interact with different receptors on the M*φ*.

Promastigote surface protease (PSP or gp63) [34,35], is a metalloproteinase and the major surface glycoprotein of *Leishmania*. Work by Russell and Wright [18] showed that PSP of *L. mexicana* interacts with CR3 on M*φ*s. In addition, PSP is the major C3 acceptor site on *L. mexicana*; the C3 is deposited primarily as C3b [19]. Recent work by Talamas-Rohana et al. [20] showed that the LPG molecule of *L. mexicana* also binds to CR3. It should be mentioned that, analogous to interacting with CR1, entering a M*φ* via the CR3 receptor also does not trigger an oxygen burst [32]. Thus, it is theoretically advantageous for *L. mexicana* to enter M*φ*s via the CR3 receptor.

C. Engulfment by Macrophages and Survival in the Parasitophorous Vacuole

After entering the M*φ*, *Leishmania* replicate in organelles of phagolysosomal origin, termed parasitophorous vacuoles (PVs) [reviewed in 36]. It is surprising that *Leishmania* survives in PVs; other intracellular parasites of M*φ*s have chosen to avoid the harsh environment of the PV. For example, *Toxoplasma gondii* lives in specialized intracellular compartments that neither fuse with lysosomes nor undergo acidification, and *Trypanosoma cruzi* escapes to the cytoplasm of the M*φ* [36]. It is still largely a mystery how *Leishmania* survive in PVs, especially given that the PVs in which the parasite lives appear to contain all of the elements necessary for destruction of *Leishmania*. Parasitophorous vacuoles containing *Leishmania* maintain an acidic pH [37] and fuse with lysosomes [38,39]. In addition, the PVs contain the lysosomal membrane proteins LAMP1 and LAMP2 [40], cathepsins B, L, H, and D [41], and the proteases dipeptidyl peptidase I and dipeptidyl peptidase II [41]. Indeed, lysosomal proteases in infected M*φ*s appear to be active, since infected M*φ*s degrade hen egg ovalbumin and bovine serum albumin as effectively as uninfected M*φ*s [41]. However, Kutish and Janovy [42] doubly infected M*φ*s with *L. donovani* and *Leptomonas costoris*. While *L. donovani* replicated in the M*φ*s and *Leptomonas* did not, *Leptomonas* was destroyed at a significantly slower rate in M*φ*s coinfected with *L. donovani* than in M*φ*s infected with *Leptomonas* alone. These results suggest that *Leishmania* partially inhibits lysosomal enzyme activity, but that it is still inherently more resistant to the actions of these enzymes than *Leptomonas*. In contrast, in work by Scott et al. [43], M*φ*s were activated with M*φ*-activating factors and infected with either *Lm* or *L. amazonensis* (Maria strain). The M*φ*s are able to kill *Lm* but not the Maria parasite. The M*φ*s were then activated and coinfected with the Maria parasite and *Toxoplasma gondii*. These doubly infected M*φ*s exhibited normal tumoricidal activity and were able to kill the *Toxoplasma* but not the Maria parasite. These results suggest that in activated M*φ*s, *Leishmania amazonensis* (Maria strain) does not inhibit the killing mechanisms of M*φ*s, but simply resists them.

Therefore, the mechanism by which *Leishmania* survives in the PV is not clear; however, some factors that may play a role are listed in Table 3. Different species of *Leishmania* may have evolved to survive in certain M*φ* types at certain temperatures. It has long been known that *Leishmania* that cause CL survive better at skin temperature (i.e., 34° C) than at a higher temperature (37° C), whereas *Leishmania* that cause VL survive at both temperatures [43–45]. These observations probably explain why CL-inducing *Leishmania* remain in the skin while VL-inducing parasites are able to invade the viscera. Temperature may also determine the sites on

TABLE 3 Possible Mechanisms by Which *Leishmania* Survives in the Macrophage

Evasion strategy	References
Tropism for appropriate host macrophage at appropriate temperature	43–52
Lipophosphoglycan prevents parasite degradation	54–56
Impairs oxyen burst	54,57
Scavenges toxic oxygen radicals	58
Inhibits lysosomal enzyme β-galactosidase	59
Promastigote surface protease (gp63) attacks lysosomal enzymes	60,61
gp63 shows optimal activity under acidic conditions	62
Proton pumps maintain intracellular pH and aid in transport of sugars and amino acids	68,69
Amastigotes show enhanced metabolism at acidic pH	70
Production of acid phosphatases	71–78
Prevention of, or resistance to, oxygen burst	79,80
Production of superoxide dismutase	81

a host where *Lm* can establish infection. Kirkpatrick et al. [46] showed that *Lm*-induced lesions develop only on certain areas of the shaved skin of mice. One explanation for this phenomenon was that those areas where lesions developed were areas with a lower temperature. Interestingly, BALB/c mice, which are susceptible to *Lm*, had a significantly lower skin temperature than mice that are resistant to the parasite. The exact reason or reasons for enhanced survival of *Lm* at lower temperatures are unknown, but it has been shown that the ability of Mϕs to destroy *Leishmania* decreases as the temperature decreases [47].

It is also possible that various forms of *Leishmania* not only prefer certain temperatures, but also prefer certain Mϕ types as host cells. With *L. donovani*, Mϕs isolated from the spleen, liver, and lung support the growth of the parasite, whereas Mϕs from the peritoneum do not [48]. In addition, Mϕs from the same site but from animals with different susceptibilities to infection can differ in their ability to control growth of *Leishmania* [49,50]. Moreover, Gorczynski and MacRae [51] showed that adherent cells isolated from the skin of BALB/c mice supported growth of intracellular *Lm* in vitro much more effectively than adherent cells isolated from mice that are resistant to infection with *Lm*. This phenomenon was most apparent when subpopulations of skin adherent cells were used, implying that even at the same site in a given mouse, there are differences in the susceptibility of phagocytic cells to infection with *Lm*. In this context, it should be mentioned that Will et al. [52] have recently examined the interaction of *Lm* with murine Langerhans cells (LCs), an important member of the phagocytic/antigen presenting cell (APC) system of the skin. These studies demonstrated that LCs could present antigen (either a lysate of *Lm* promastigotes or live organisms) to *Lm*-specific T cells. In fact, they were more efficient in their APC function than irradiated spleen cells. Furthermore, the LC-elicited cytokine production (interleukin-2 [IL-2] and IL-4) by the responding T cells. Finally, the authors have preliminary evidence that LC can be infected by *Lm* both in vitro and in vivo. This is in contrast to a study by Locksley et al. [53] using human LCs in which it was found that the LCs did not engulf *Lm*.

As is apparent in Table 3, the two major surface molecules of *Leishmania*, LPG and PSP, have several activities that may help to promote survival of the parasite in Mϕs [54–62]. For example, LPG may inhibit the oxygen burst of Mϕs, and PSP may degrade lysosomal enzymes. However, for these molecules to protect the parasite in Mϕs, they must be expressed by amastigotes, which are the form of *Leishmania* found in the Mϕs. The degree of expression of LPG and PSP by amastigotes, and the differences between the structures of LPG and PSP on promastigote and amastigote forms of *Leishmania,* are not clear. Amastigotes of *Lm* have been shown to express an LPG molecule, but it is biochemically and antigenically distinct from the LPG of promastigotes

[63,64]. Moreover, while metacyclic *Lm* has a thick coat of elongated LPG molecules, this coat is lacking on amastigotes [65]. Promastigote surface protease has been detected on *Lm* amastigotes. For example, in a study by Frommel et al. [66] using an anti-PSP antibody and immunofluorescence, *Lm* amastigotes were shown to express PSP. However, these studies could not address the exact amount and enzymatic activity of the PSP. In recent work by Schneider et al. [67], PSP expression by *Lm* amastigotes was assessed by Western and Northern blot, immunoprecipitation of metabolically labeled or surface-labeled amastigotes, and quantification of PSP's proteolytic activity using substrate-containing polyacrylamide gels. The authors found that PSP is scarce in amastigotes at both the mRNA and the protein levels.

A surface proton-translocating ATPase has been identified in promastigotes of *L. donovani* [68]. This enzyme may be critical for survival of *Leishmania* in the Mϕs since it may regulate the intracellular pH homeostasis of the parasite in the acidic environment of the PV. In addition, the proton gradient created by the ATPase may contribute to the active transport of sugars and amino acids by *Leishmania* [69].

Gottlieb and Dwyer [71] showed that *L. donovani* expresses an acid phosphatase on its surface and secretes a second acid phosphatase. Subsequent work revealed that acid phosphatase is produced by a number of species of *Leishmania* and is produced by both the promastigote and the amastigote stages of the parasite [72]. However, with the exception of *L. panamensis*, metacyclic and log phase *Leishmania* contain approximately equal amounts of acid phosphatase activity [73]. Work by Glew and colleagues showed that the membrane acid phosphatase of *L. donovani* is a potent inhibitor of the neutrophil superoxide burst [74–76], which prompted the authors to suggest that acid phosphatase might promote survival of *Leishmania* in the host. However, others have shown that neutrophils rapidly phagocytize and kill either promastigotes or amastigotes of *L. donovani* via the production of oxygen metabolites [77,78].

Early work by Murray [80,82,83] showed that when promastigotes of *L. donovani* and *Lm* were phagocytized by nonactivated Mϕs, an oxygen burst was elicited that killed the promastigotes. In contrast, when these promastigotes were engulfed by the J774 line, which does not mount an oxygen burst, the parasites were not killed. On the other hand, using amastigotes of *L. donovani*, Murray found that these parasites elicited less of an oxygen burst and were not killed unless the Mϕs were activated with lymphokine, which resulted in more of an oxygen burst. Others obtained similar results [84–86]. In addition, Murray found that *L. donovani* amastigotes have three times more catalase and 14 times more glutathione peroxidase than *L. donovani* promastigotes. Indeed, it has been reported [79] that Mϕs infected with *Leishmania* are less capable of generating an oxygen burst; however, this is not always the case [80]. Finally, Mauel and co-workers reported that when murine Mϕs were activated with lymphokine plus lipopolysaccharide (LPS), the cells destroyed intracellular *Leishmania* and that parasite killing always correlated with oxygen burst activity in the cells. Moreover, genetically susceptible BALB/c mouse Mϕs generated less of an oxygen burst than resistant CBA Mϕs [87,88].

Taken as a whole, the findings discussed above offer compelling evidence that the oxygen burst of Mϕs is involved in destruction of *Leishmania*. However, early work by Scott et al. [89] showed that *Leishmania* could be killed in Mϕs that were unable to generate an oxygen burst. In addition, Mauel and colleagues showed that the intensity of the oxygen burst was the same in Mϕ activated with either LPS or LPS plus lymphokine, yet only the Mϕs treated with LPS plus lymphokine were able to kill *Leishmania* [87,88]. These results suggested that a second mechanism for parasite killing existed. Recently, several laboratories have shown that Mϕs can kill *Leishmania* via the production of nitric oxide (NO) derived from L-arginine [90–92, and reviewed extensively in 93–95]. In addition, when Mϕs from mice that are resistant to infection with *Lm* were tested for their ability to produce NO and NO synthase in response to infection with the parasite in vitro, these Mϕs produced significantly more NO and NO synthase than Mϕs from mice that are susceptible to infection with *Lm* [96]. In fact, it has been proposed that the oxygen burst may inhibit NO production, since H_2O_2, a product of the oxygen burst, can interfere with a cofactor necessary for the synthesis of NO [97]. However, it is important to note that NO has not been shown to be consistently produced by human Mϕs in response to infection with microbes; thus, the importance of this killing pathway in human Mϕs is unclear [98–101]. Therefore, it is possible that *both* the oxygen burst and NO are important in killing of *Leishmania* by Mϕs. Liew and Cox [94] have proposed that the oxygen burst may be capable of killing *Leishmania* when nonactivated Mϕs are infected with promastigotes of the parasite, but that once *Leishmania* becomes established in the Mϕs as an amastigote, the parasite can be killed only by activating the Mϕ to produce NO.

III. THE EFFECTS OF *LEISHMANIA MAJOR* ON MACROPHAGE FUNCTIONS

A. Arachidonic Acid Metabolism in *Leishmania*-Infected Macrophages

Early work by Reiner and Malemud showed that arachidonic acid metabolism in *Leishmania*-infected Mϕs was altered. For example, splenic Mϕs isolated from BALB/c mice infected with *L. donovani* produced two to five times more prostaglandin E_2 (PGE_2) when stimulated with phytohemagglutinin (PHA) in vitro compared with Mϕs from normal mice [102,103]. Work by Farrell and Kirkpatrick [104] also showed that spleen cells from *Lm*-infected genetically susceptible BALB/c mice produced elevated levels of PGE_2. In addition, treating *Lm*-infected BALB/c mice with indomethacin inhibited the development of cutaneous lesions of the parasite, suggesting that PGE_2 plays a detrimental role in CL. The authors proposed that PGE_2 might exacerbate disease by inhibiting I-A expression on Mϕs or lymphokine production by parasite-specific T cells.

Work by Mauel and colleagues has revealed that PGE_2 can also play a protective role in leishmaniasis. As discussed above (see Section II.C), NO production by Mϕs correlates with destruction of *Leishmania*. Using this indicator system, it was found that PGE_2 enhanced NO production by CBA Mϕs that had been infected in vitro with *L. enriettii* and activated with either interferon-γ (IFN-γ) or tumor necrosis factor (TNF). Moreover, adding indomethacin to the cultures inhibited NO production [105]. Recent work by this same group indicates that PGE_2 also enhances intracellular killing of *Leishmania* by Mϕs [106]. Using CBA or C57Bl/6 Mϕs infected with *L. enriettii* and activated with IFN-γ plus TNF, it was shown that when PGE_2 is present in the cultures at a concentration of 10^{-7} to 10^{-9} M, there is a three- to fivefold increase in killing of the parasite by the Mϕs, and that the Mϕs produced larger amounts of NO. Furthermore, adding arachidonic acid to the cultures enhanced parasite killing, while indomethacin inhibited killing [107]. Similar to these observations, work in this laboratory has shown that when CBA mice are infected with *Lm* and treated with indomethacin in a manner identical to that of Farrell and Kirkpatrick [104], the course of lesion development is enhanced rather than inhibited, as was the case with BALB/c mice in the work of Farrell and Kirkpatrick (R. G. Titus, unpublished data). Taken together, these observations suggest that PGE_2 plays a detrimental role in BALB/c mice (which are susceptible to infection with *Lm*) but a beneficial role in strains such as CBA (which are resistant to infection). Indeed, in all of the work by Mauel and colleagues described above, Mϕs from either CBA or C57Bl/6 mice were used, and both of these mice are resistant to infection with *Lm*.

B. Major Histocompatibility Gene Expression and Interleukin-1 Production by *Leishmania*-Infected Macrophages and the Role of Interleukin-1 in Cutaneous Leishmaniasis

Work by Reiner and colleagues showed that when Mϕs were infected in vitro with *L. donovani*, there was a down-regulation of expression of both class I and class II major histocompatibility complex (MHC) expression on the cells, and the cells produced less IL-1 in response to stimuli such as *Listeria monocytogenes* [108,109]. Further work revealed that suppression of class II expression was due to inhibition of class II gene transcription, class II mRNA accumulation, and class II protein expression on the cells [110,111]. In addition, suppression of the Mϕ IL-1 response could be overcome by pretreating the Mϕs with IFN-γ [112]. The ability of *L. donovani* to inhibit both class II expression and IL-1 production by Mϕs was thought to be potentially beneficial to the parasite, since the Mϕs would thus be less capable of activating parasite-specific T cell immunity.

It should be mentioned that the effect of *Leishmania* on IL-1 production by Mϕs depends on the species of *Leishmania* used and the source of the Mϕs. In the work of Reiner discussed above, *L. donovani* caused suppression of IL-1 production in both mouse and human Mϕs. Similar results were obtained using *Lm* and human Mϕs [113] and *L. donovani* or *Lm* LPG and human Mϕs [114]. However, Mϕs from genetically susceptible BALB/c mice were shown to produce very large amounts of IL-1 when infected with *Lm* in vitro [115]. In addition, in contrast to the work discussed above in which IFN-γ enhanced IL-1 production by *L. donovani*–infected Mϕs [112], IFN-γ was found to suppress IL-1 production by *Lm*-infected Mϕs [115]. These conflicting results may be partially explained by work of Wagner et al. [116]. These authors found that splenic Mϕs from *Lm*-infected susceptible BALB/c mice produced more IL-1 than Mϕs from resistant C57Bl/6 mice when the cells were restimulated with antigens of *Lm* in vitro. In contrast, antigens of *L. donovani* not only failed to elicit IL-1 production by Mϕs but blocked the ability of *Lm* antigen to do so. Thus, with respect to IL-1 production, *Lm* and *L. donovani* appear to have very different effects on the Mϕ.

Because of the observations described above, and because IL-1 is thought to be a costimulator of T cell activation, we have treated both genetically susceptible BALB/c mice and resistant C57Bl/6 mice with an anti–IL-1 receptor monoclonal antibody (anti–IL-1R mAb) to determine whether this would modify the course of infection with *Lm* in the mice [117]. The anti–IL-1R mAb prevents binding of IL-1 to both the T cell and B cell/Mϕ forms of the IL-1R and does not itself cause activation of any of these cell types. Compared to control mice challenged with *Lm* and treated with normal immunoglobulin, lesion development was significantly less severe in both BALB/c and C57Bl/6 mice treated with the anti–IL-1R mAb. However, the number of *Lm* present in the lesions of mice treated with anti–IL-1R mAb was no different from the number in control lesions. Thus, in mice infected with *Lm*, IL-1 appears to play a role only in the development of lesion pathology.

C. Tumor Necrosis Factor Production by *Leishmania*-Infected Macrophages and the Role of Tumor Necrosis Factor in Cutaneous Leishmaniasis

Work from this laboratory has shown that TNF plays a protective role in infection with *Lm*. The injection of mice with TNF proved beneficial (i.e., lesions were significantly smaller and contained 10 to 20 times fewer parasites), while treatment with a neutralizing anti-TNF antibody exacerbated the disease [118]. This protective effect of TNF was confirmed by Liew and colleagues [119]. Moreover, when the lymph node cells draining the lesions on *Lm*-infected genetically resistant C57Bl/6 and susceptible BALB/c mice were restimulated with the parasite in vitro, the C57Bl/6 cells produced TNF while the BALB/c cells did not [118]. However, when susceptible BALB/c mice were vaccinated against CL by pretreating with an avirulent form of *Lm*, treating the mice with anti-TNF was able to overcome partially the protective effect of the vaccination protocol [120]. This implies that if susceptible BALB/c mice are vaccinated, the animals can be induced to produce TNF, a hypothesis that was confirmed by establishing parasite-specific CD4$^+$ T cell lines from the lymph node cells draining the lesions on the vaccinated mice and showing that the cells produced TNF [121]. Some authors, however, have not been able to demonstrate a correlation between TNF production by *Lm*-infected mice and resistance to CL. Moll et al. [122] showed that when spleen cells from *Lm*-infected resistant and susceptible mice were restimulated with *Lm* in vitro, there was no difference in TNF production. The reason for this discrepancy may be that spleen cells rather than lymph node cells were used. The results of Moll et al. [122] were recently confirmed (I. Muller, personal communication) with respect both to the levels of secretion of biologically active TNF and to TNF mRNA expression.

Since these original observations, the role of TNF in CL has been intensively investigated, and these studies have shown that TNF plays an instrumental role in the activation of Mϕs to kill *Leishmania*. Tumor necrosis factor (both murine and human) activates Mϕs to kill *Lm* [120], and this is associated with greater NO production by the treated Mϕs [123]. Furthermore, when Mϕs were activated with IFN-γ, both destruction of the parasite and NO production could be prevented by including an anti-TNF antibody in the cultures, implying that TNF acts in an autocrine fashion in the activation of Mϕs [124]. Others have obtained similar results and have found that TNF can synergize with IFN-γ for Mϕ activation and that TNF accounts for the antiparasitic effect seen when Mϕs are activated by IFN-γ plus IL-4 [125–127]. Moreover, when Mϕs are infected with *Lm* in vitro and stimulated with LPS, the cells produce 10 to 30 times more TNF than uninfected control Mϕs, suggesting that *Lm* infection primes the Mϕs for TNF production [120]. Finally, TNF can reverse the disease-exacerbating effect that is observed when attempting to immunize against CL by the subcutaneous route [128].

While the preceding discussion implies that activation of Mϕs for the destruction of *Leishmania* might be simple—that is, IFN-γ elicits TNF production by the infected Mϕ, which leads to NO production and the destruction of the parasite—recent work by Betz-Corradin et al. [105,129] suggests that activation may be far more complex. These authors showed that phagocytosis of particles, including *Leishmania*, by Mϕs treated with IFN-γ led to increased NO production by the Mϕs and increased killing of *Leishmania*. This killing could be blocked by an anti-TNF antibody; however, the same anti-TNF antibody did not block killing by Mϕs activated with IFN-γ and LPS. In addition, exogenously added TNF did not enhance NO production in Mϕs activated with IFN-γ plus LPS, but phagocytosis of *Leishmania* (a particle) by the Mϕs did enhance NO production. This suggested that other products of Mϕs might be involved in activation of the cells. Interestingly, our laboratory has been investigating Mϕ inflammatory proteins 1 and 2, two major secreted products of Mϕs stimulated with LPS [130], for their ability to activate Mϕs to destroy *Lm*; indeed, we find that they too are potent inducers of *Lm* destruction by Mϕs.

D. Macrophage Activation-Suppressing Factor

Work by Engelhorn et al. [131] has shown that when Mϕs are infected with *L. donovani*, they secrete a small peptide, termed Mϕ activation-suppressing factor (ASF), that blocks the ability of IFN-γ to activate the cells. Activation-suppressing factor appears to be a unique factor whose structure is, at present, unknown. Activation-suppressing factor may be responsible for a finding made by many laboratories, namely, that although Mϕs can kill *Leishmania* when *preactivated* with cytokines such as IFNγ, it is often difficult or even impossible to activate Mϕs to kill *Leishmania after* the cell is infected.

E. Transforming Growth Factor-β

Transforming growth factor-β (TGF-β), produced by a variety of cells including Mϕs, has been found to inhibit *Lm* killing by Mϕs in vitro when the Mϕs were pretreated with TGF-β before the addition of activating cytokines. However, TGF-β had no effect when it was added to the Mϕs after the cells were activated with cytokine. Transforming growth factor-β also inhibited NO production by Mϕs [132].

F. Signal Transduction in *Leishmania*-Infected Macrophages

Work by Descoteaux et al. [133] showed that *L. donovani* LPG had diverse effects on Mϕ activation markers. Exposure of Mϕs to LPG alone induced *c-fos* and TNF gene expression. However, pretreating Mϕs with LPG blocked induction of *c-fos* gene expression in response to certain activators of *c-fos* (LPS, OAG, A23187, TNF), but not others (dibutyryl cAMP). In addition, while pretreating with LPG augmented TNF gene expression in Mϕs, it down-regulated TNF receptor expression on the cells. The authors suggested that these observations might be explained by the ability of LPG (or of the proteins associated with LPG) to interfere with the activation of protein kinase C (PKC). Inhibition of PKC by LPG had, in fact, been shown in a cell-free system [57]. In related work, Olivier et al. [134] recently showed that there was a defect in PKC activation in *L. donovani*–infected Mϕs, and that this was possibly due to covalent modification of the enzyme. However, the authors did not believe that these effects were due to the LPG of the parasite. Finally, in a previous study by Olivier et al. [135] it was shown that in *L. donovani*–infected Mϕs there was also a defect in the regulation of the levels of intracellular Ca^{2+} and of calcium-dependent signaling in the cells. In addition, agonist-induced transients of calcium were attenuated, and this appeared to be due to decreased generation of the endogenous calcium-mobilizing agent inositol 1,4,5-triphosphate.

G. Antigen Presentation by *Leishmania*-Infected Macrophages

Little work has been done to analyze antigen presentation by Mϕs infected with *Leishmania*. However, Kaye and co-workers have begun to look at this phenomenon and have reported that presentation of *L. donovani* antigens occurs via a brefeldin A–sensitive pathway [136]. This observation implies that *L. donovani* antigens associate with newly synthesized MHC class II molecules; however, the authors believed that this association did not occur in the phagolysosomes containing the parasite, but rather, that parasite antigens were transported to early endosomal compartments for class II binding. Unfortunately, in these experiments if was difficult to exclude the possibility that during phagocytosis of the parasite by Mϕs, some parasite antigens, especially those exposed on the parasite's surface, were in fact the antigens transported to the early endosomes. To exclude this possibility more formally, this same group has recently employed *Leishmania* parasites transfected either with chicken ovalbumin (OVA) or β-galactosidase (β-gal). These antigens were shown to be retained as intracytoplasmic molecules in the parasites and to be neither secreted nor expressed at the surface or in the flagellar pocket. Thus, OVA and β-gal are acting as "reporter antigens" that are "bottled up" in the parasite. When Mϕs were infected with these genetically engineered parasites, it was found that the Mϕs presented OVA and β-gal to antigen-specific T cells as effectively as uninfected Mϕs that had been cultured with soluble OVA or β-gal. These studies offered direct proof that antigens derived from *Leishmania*-containing phagolysosomes are accessible to the class II antigen processing pathway of Mϕs [137, 138]. These transfected *Leishmania* are currently also being examined in this laboratory to determine whether they will induce a response to OVA or β-gal in mice infected with the parasite. Preliminary experiments indicate that this is the case, and thus we shall be able to use this "reporter antigen" system to determine the nature of the response to OVA and β-gal. Specifically, since susceptible BALB/c mice can develop a T_{H2} T cell response to infection with *Lm* and resistant mice can develop a T_{H1} response (discussed in detail in section IV.A below), we will determine whether the T cell response to OVA and β-gal is identical to the response to *Lm*. If so, this would suggest that it is not a particular antigen that elicits a T_{H2} or T_{H1} response, but rather the manner in which the Mϕ presents the antigen that determines the type of responding T cell.

IV. THE INFLUENCE OF T CELLS AND SAND FLY SALIVA ON THE INTERACTION BETWEEN *LEISHMANIA MAJOR* AND MACROPHAGES

A. T Cells

It is now generally accepted that *Lm*-specific T cells play a central role in determining whether the host will develop a healing or progressing disease with the parasite. Of the T cell lineages, CD8+ cells play a protective role in CL, perhaps having their greatest effect in resistance to reinfection with *Lm* [139–145]. However, the cell that has the greatest influence on the course of CL is the CD4+ T cell. Upon adoptive transfer, CD4+ parasite-specific T cells have been shown to have either a beneficial [146–149] or a detrimental [146,149–151] effect on the outcome of infection with *Lm* in susceptible BALB/c mice. In addition, depletion of more than 95% of CD4+ T cells in normally resistant CBA mice renders them highly susceptible to infection, again suggesting that CD4+ T cells are critical for resistance in CL [139,152].

Subsets exist in the CD4+ T cell compartment of mice [153]. One subset, the T_{H1}, produces IL-2 and IFN-γ, while the other subset, T_{H2}, produces IL-4 and IL-5. In CL, resistant C57Bl/6 mice infected with *Lm* produce IL-2 and IFN-γ mRNA, whereas susceptible BALB/c mice produce IL-4 and some IL-2 but no IFN-γ mRNA [154]. In addition, in the work of Scott [146,147], Muller [148], and Holaday [149], adoptive transfer of *Lm*-specific T_{H1} lines and clones protected mice from infection with *Lm*, while adoptive transfer of *Lm*-specific T_{H2} lines exacerbated disease. Taken together, these results have led to the current dogma that *Lm*-specific T_{H1} cells and the cytokines they produce play a protective role in CL, whereas T_{H2} T cells and the cytokines they produce exacerbate disease. Interferon-γ is thought to be a principal mediator of resistance to *Leishmania*, because of its ability to activate Mϕs to destroy the parasite [155–158], and treating *Lm*-infected resistant mice with a neutralizing anti–IFN-γ antibody exacerbates the course of infection [159]. In addition, IL-4 can block the ability of IFN-γ to activate Mϕs to destroy *Leishmania* [160,161], and treating *Lm*-infected susceptible mice with a neutralizing anti-IL-4 antibody allows lesions on the animals to heal [162,163].

However, there is as much evidence that argues that immunity to *Leishmania* is much more complex than a simple T_{H1}/T_{H2}, IFN-γ/IL-4 dichotomy (Table 4). *Lm*-specific T cell lines and clones have been described that produce large amounts of IFN-γ (i.e., these are T_{H1} T cells), yet the cells exacerbate the course of CL in either susceptible BALB/c or resistant CBA mice [150,151]. In addition, as discussed in the previous paragraph, IFN-γ mRNA could not be detected in susceptible BALB/c mice during the course of CL; however, *Lm*-specific T cells isolated from infected BALB/c mice at varying times during the course of infection have been shown to produce significant amounts of biologically active IFN-γ [143,144]. Work by Theodos and Titus [164] has shown that when CD4+ *Lm*-specific T cell clones are derived from resistant C3H mice, a T_{H1} clone protects, but it is a T_{H0}-like T cell [165] that exacerbates CL. Moll and Rollinghoff [166] showed that while protective CD4+ T cells in CL were T_{H1} cells, exacerbative T cells were not T_{H2} cells. In addition, adoptive transfer of a T_{H2}-type T cell has been shown to protect mice against CL [167]. As discussed above, CD8+ T cells also play a protective role in CL, especially in resistance to reinfection, and mast cells have been recently shown to play an exacerbative role in CL [168]. Boom et al. [121] showed that when BALB/c mice were vaccinated against CL, CD4+ *Lm*-specific T cells isolated from the animals no longer produced IL-4, but, surprisingly, did not produce IFN-γ. Instead the cells produced large amounts of TNF. As discussed in the previous paragraph, infusing anti–IFN-γ antibody is detrimental, while anti–IL-4 is protective for the course of CL [159,162,163]; however, infusing neutralizing antibody against IL-2 (which is a cytokine associated with resistance in CL) is surprisingly also protective [169]. Along the same lines, it has

TABLE 4 Effects of T_{H1}, T_{H2}, and Other Cell Types and of Interferon-γ, Interleukin-2, and Interleukin-4 in leishmaniasis

Question/answers	References
T_{H1} cells protect?	
Yes	146–149,164,166
No	150,151
T_{H2} cells exacerbate?	
Yes	146,149
No	166,167
T_{H0} cells exacerbate?	
Yes	164
CD8+ T cells protect?	
Yes	139–145
Mast cells exacerbate?	
Yes	168
IFN-γ protects?	
Yes	121,154,158,159
No	121,143,144,162
IL-4 exacerbates?	
Yes	121,154,160–163
No	172–176
IL-2 protects?	
Yes	121,154
No	169,170

been reported that IL-2 can enhance replication of *L. amazonensis* promastigotes in vitro and can exacerbate lesions induced with the parasite in mice [170]. In addition, although injecting anti–IFN-γ or anti–IL-4 antibody has dramatic effects on the course of CL, injecting the cytokines themselves (i.e., IFN-γ or IL-4) has little effect on disease [162,163,171,172], unless the cytokines are coadministered with adjuvant, in which case both are protective [171,173,174]. Finally, as mentioned above, IL-4 can block the ability of IFN-γ to activate Mϕs to destroy *Leishmania* [160,161]; however, depending on the protocol used and the timing of the addition of IL-4 to the Mϕs, IL-4 can also synergize with IFN-γ to enhance greatly the killing of *Lm* [175,176].

Thus, a complex picture is emerging with respect to the cytokines produced and the preferential outgrowth of protective or exacerbative T cells following *Lm* infection of mice. In fact, many cytokines are involved in activation of Mϕs to kill *Leishmania*. The ability of IFN-γ and TNF to activate Mϕs to kill *Lm* was discussed above. Work by Nacy and colleagues has also implicated at least IL-2, IL-4, and granulocyte-monocyte colony-stimulating factor (GM-CSF) in the killing process [177–179]. In the remainder of this section, work that has been performed on the role of other cytokines in CL is discussed.

1. Tumor Necrosis Factor

Although TNF was discussed above, there is an intriguing recent report from the group of Wyler and co-workers regarding membrane TNF. For several years these workers have described a phenomenon in which contact was necessary between the effector CD4+ T cell and Mϕ for effective killing of *Lm* by the Mϕ. The authors now believe that contact was necessary because activation is mediated by membrane TNF expressed on the surface of the effector T cell [180].

2. *Interleukin-3*

Interleukin-3 has been shown to be produced by mice susceptible to infection with *Lm* and to also exacerbate infection with *Lm* in mice [181,182]. The exact mechanism behind this effect is not known, but may be related to the ability of IL-3 to increase the number of immature monocytes present in the developing lesion. Such cells may be "safe targets" (i.e., permissive cells) in which the parasite can replicate, a hypothesis advanced by Mirkovich et al. [183]. Alternatively, the exacerbative effect of IL-3 may be due to its ability to block IFN-γ activation of Mϕs [161].

3. *Granulocyte-Monocyte Colony-Stimulating Factor*

Granulocyte-monocyte colony-stimulating factor has been shown in various settings to have either a beneficial [184] or detrimental [185] effect or no effect [186] on infection with *Lm* in mice. More work is required to elucidate its exact role in the disease. However, it is interesting that GM-CSF has been found to be a growth factor for *L. amazonensis* [187].

4. *Migration Inhibition Factor*

Human migration inhibition factor (MIF) has been cloned [188]. Functional studies have revealed that MIF is able to activate human monocytes to destroy *L. donovani* in vitro [189]. In addition, MIF is a potent adjuvant in mice and up-regulates expression of HLA-DR and IL-1β and elaboration of IL-1β by human monocyte–derived Mϕs [190]. Thus, if the molecule is produced in leishmanial lesions, it may dramatically influence the character of the immune response engendered against the parasite. In fact, we have evidence that supernatants from protective *Lm*-specific CD4+ T cells display MIF activity when the cells are activated with the parasite in vitro. The cells that produce MIF are cloned *Lm*-specific T cells that were isolated from genetically resistant C3H mice [164].

5. *Interleukin-10*

Interleukin-10 was originally described as a factor secreted by T_{H2} cells that inhibited IFN-γ secretion by T_{H1} cells. However, it is now known that IL-10 is produced by many cells, including B cells and Mϕs, and thus may have multiple effects on many cell types [191,192]. In addition, it inhibits the production of a number of inflammatory mediators, including IL-1, IL-6, IL-8, GM-CSF, TNF, and lymphotoxin [193] as well as IFN-γ. Because IL-10 was originally thought to be a T_{H2}-specific cytokine, Heinzel et al. [194] investigated its production by mice infected with *Lm* and found that whereas susceptible BALB/c mice produced IL-10, resistant mice did not. This result would appear to contradict previous results from the same laboratory, since elevated IL-10 production should be accompanied by decreased levels of IL-1, yet Heinzel et al. [154] also reported that infected BALB/c mice produce more IL-1. In addition, if IL-10 plays a detrimental role in CL, it is curious that treating BALB/c mice with a neutralizing anti–IL-10 antibody did not alter the course of CL in the animals [169]. Finally, it has been reported that IL-10 inhibits the induction of NO synthase by IFN-γ in murine Mϕs [195].

B. Sand Fly Saliva

We have found that the sand fly, the vector for *Leishmania*, is not simply a "flying syringe." An immunosuppressive substance in the saliva of the sand fly *Lutzomyia longipalpis* markedly enhances infection with *Lm* in mice [196]. Further work indicated that the exacerbative effect of saliva applies not only to *Lm*, but also to *L. amazonensis* [197] and *L. braziliensis braziliensis* [198], and that the saliva of another species of sand fly, *Phlebotomus papatasi* [197], also exacerbates CL. In addition, lesions induced at skin sites where *L. longipalpis* had probed for a blood meal exhibited an exacerbated course of CL, which demonstrated that the bite of a sand fly delivers the salivary immunosuppressive substance [197].

The mechanism of the phenomenon is currently under investigation. The immunosuppressive substance may be related to the mammalian neuropeptide calcitonin gene–related peptide (CGRP), since there is a CGRP-like substance in sand fly saliva [199] and CGRP itself is an immunosuppressive substance [200]. This "fly CGRP" has been molecularly cloned [201] and is currently being investigated. *L. longipalpis* saliva inhibits the presentation of leishmanial antigen by Mϕs in vitro [202], H_2O_2 production by the cells [Y.-H. Nong and H. G. Remold, unpublished data], and the activation of Mϕs to kill *Lm* (L. R. Hall, R. G. Titus, unpublished data). Finally, we know that sand fly saliva has an effect on the host, since it can be injected into mice as much as 4 days in advance of *Lm* and exacerbation of CL still occurs [202]. However, we have recently found that saliva also has a direct effect on the infectivity of *Lm*. When *Lm* are preincubated in saliva for 24 hours, the saliva is washed away, and the parasites are injected into mice, the saliva-exposed *Lm* are far more infective for mice than *Lm* incubated in medium alone (L. R. Hall, R. G. Titus, unpublished data).

When low numbers of *Lm* (10 to 100) are injected into mice, the parasite either survives poorly or does not survive at all. However, when these numbers of parasites are injected with sand fly saliva, survival of the parasites is greatly enhanced [196]. Since it is believed that 10 to 100 is approximately the number of parasites that the sand fly injects into the mammalian host [203], it is possible that sand fly saliva is essential for the parasite to establish an infection in human beings. These observations offer the exciting prospect that immunization of man against vector saliva may represent a means for controlling leishmaniasis. Specifically, in persons vaccinated against sand fly saliva, neutralization of the immunosuppressive effect of the saliva might occur and infection by *Leishmania* might be prevented [7]. Vector-borne parasites may be so well adapted to their vertebrate host that it will prove very difficult to control them by vaccination against the parasite itself. Thus, vector-based vaccines may not only be a novel approach to control the parasitic diseases, but may be the preferred method.

ACKNOWLEDGMENTS

The authors are indebted to the following colleagues for supplying preprints, submitted manuscripts, and unpublished data; for useful discussions; or for reading this manuscript: Linda Button, Robert Etges, Shawn Green, Keith Joiner, Paul Kaye, Eddy Liew, Jacques Mauel, Ingrid Muller, Carol Nacy, Carl Nathan, Neil Reiner, Martin Rollinghoff, David Russell, and David Sacks. Without their assistance, this review would have been much less complete and much more out-of-date by the time it reached the printed page. This work was supported by NIH grants R29 A124511 and RO1 A127511.

REFERENCES

1. Neva F, Sacks D. Leishmaniasis. In: Warren KS, Mahmoud AAS, eds. Tropical and geographic medicine. 2nd ed. New York: McGraw-Hill, 1990:296–308.
2. Titus RG, Theodos CM, Kimsey PB, Shankar A, Hall L, McGurn M, Povinelli L. Role of T cells in immunity to the intracellular pathogen, *Leishmania major*. Subcell Biochem 1992; 18:99–129.
3. Alexander J, Russell DG. The interaction of *Leishmania* species with macrophages. Adv Parasitol 1992; 31:175–253.
4. Walsh JA. Estimating the burden of illness in the tropics. In Warren KS, Mahmoud AAS, eds. Tropical and geographic medicine. 2nd ed. New York: McGraw-Hill, 1990: 296–308.
5. Blackwell JM. Protozoal infections. In: Wakelin DM, Blackwell JM, eds. Genetics of resistance to bacterial and parasitic infection. London: Taylor and Francis, 1988:103–52.
6. Fortier AH, Tong A, Nacy CA. Susceptibility of inbred mice to *Leishmania major* infection: genetic analysis of macrophage activation and innate resistance to disease in individual progeny of P/J (susceptible) and C3H/HeN (resistant) mice. Infect Immun 1990; 58:4149–52.

7. Titus RG, Ribeiro JMC. The role of vector saliva in transmission of arthropod-borne disease. Parasitol Today 1990; 6:157–60.
8. Mosser DM, Burke SK, Coutavas EE, Wedgwood JF, Edelson PJ. *Leishmania* species: mechanisms of complement activation by five strains of promastigotes. Exp Parasitol 1986; 62:394–404.
9. Mosser DM, Edelson PJ. Activation of the alternative complement pathway by *Leishmania* promastigotes: parasite lysis and attachment to macrophages. J Immunol 1984; 132:1501–5.
10. Franke ED, McGreevy PB, Katz SP, Sacks DL. Growth cycle–dependent generation of complement-resistant *Leishmania* promastigotes. J Immunol 1985; 134:2713–8.
11. Sacks DL, Perkins P. Identification of an infective stage of *Leishmania* promastigotes. Science 1984; 223:1417–9.
12. Puentes SM, Sacks DL, da Silva RP, Joiner KA. Complement binding by two developmental stages of *Leishmania major* promastigotes varying in expression of a surface lipophosphoglycan. J Exp Med 1988; 167:887–902.
13. Puentes SM, da Silva RP, Sacks DL, Hammer CH, Joiner KA. Serum resistance of metacyclic stage *Leishmania major* promastigotes is due to release of C5b-9. J Immunol 1990; 145:4311–6.
14. Sacks DL, Brodin TN, Turco SJ. Developmental modification of the lipophosphoglycan from *Leishmania major* promastigotes during metacyclogenesis. Mol Biochem Parasitol 1990; 42:225–34.
15. Sacks DL. Metacyclogenesis in *Leishmania* promastigotes. Exp Parasitol 1989; 69:100–3.
16. Russell DG, Talamas-Rohana P. *Leishmania* and the macrophage: a marriage of inconvenience. Immunol Today 1989; 10:328–33.
17. Da Silva RP, Hall BF, Joiner KA, Sacks DJ. CR1, the C3b receptor, mediates binding of infective *Leishmania major* metacyclic promastigotes to human macrophages. J Immunol 1989; 143:617–22.
18. Russell DG, Wright SD. Complement receptor type 3 (CR3) binds to an Arg-Gly-Asp-containing region of the major surface glycoprotein, gp63, of *Leishmania* promastigotes. J Exp Med 1988: 168:279–92.
19. Russell DG. The macrophage-attachment glycoprotein gp63 is the predominant C3-acceptor site on *Leishmania mexicana* promastigotes. Eur J Biochem 1987; 164:213–21.
20. Talamas-Rohana P, Wright SD, Lennartz MR, Russell DG. Lipophosphoglycan from *Leishmania mexicana* promastigotes binds to members of the CR3, p150,95 and LFA-1 family of leukocyte integrins. J Immunol 1990; 144:4817–24.
21. Blackwell JM, Ezekowitz RAB, Roberts MB, Channon JY, Sim RB, Gordon S. Macrophage complement and lectin-like receptors bind *Leishmania* in the absence of serum. J Exp Med 1985; 162:324–31.
22. Wozencraft AO, Sayers G, Blackwell JM. Macrophage type 3 complement receptors mediate serum-independent binding of *Leishmania donovani*. J Exp Med 1986; 164:1332–7.
23. Wozencraft AO, Blackwell JM. Increased infectivity of stationary-phase promastigotes of *Leishmania donovani*: correlation with enhanced C3 binding capacity and CR3-mediated attachment to host macrophages. Immunology 1987; 60:559–63.
24. Cooper A, Rosen H, Blackwell JM. Monoclonal antibodies that recognize distinct epitopes of the macrophage type three complement receptor differ in their ability to inhibit binding of *Leishmania* promastigotes harvested at different phases of their growth cycle. Immunology 1988; 65:511–4.
25. Mosser DM, Edelson PJ. The mouse macrophage receptor of C3bi (CR3) is a major mechanism in the phagocytosis of Leishmania promastigotes. J Immunol 1985; 135:2785–9.
26. Wilson ME, Pearson RD. Evidence that *Leishmania donovani* utilizes a mannose receptor on human mononuclear phagocytes to establish intracellular parasitism. J Immunol 1986; 136:4681–8.
27. Wyler DJ, Sypek JP, McDonald JA. In vitro parasite-monocyte interactions in human leishmaniasis: possible role of fibronectin in parasite attachment. Infect Immun 1985; 49:305–11.
28. Rizvi FS, Ouassi MA, Marty B, Santoro F, Capron A. The major surface protein of *Leishmania* promastigotes is a fibronectin-like molecule. Eur J Immunol 1988; 18:473–6.
29. Mosser DM, Vlassara H, Edelson PJ, Cerami A. *Leishmania* promastigotes are recognized by the macrophage receptor for advanced glycosylation end products. J Exp Med 1987; 165:140–5.
30. Mosser DM, Edelson PJ. The third component of complement (C3) is responsible for the intracellular survival of *Leishmania major*. Nature 1987; 327:329–31.
31. Handman E, Goding JW. The *Leishmania* receptor for macrophages is a lipid containing glycoconjugate. EMBO J 1985; 4:329–36.
32. Wright SD, Silverstein SC. Receptors for C3 and C3bi promote phagocytosis but not release of toxic oxygen from human phagocytes. J Exp Med 1983; 158:2016–23.

33. Ilg T, Etges, R, Overath P, McConville MJ, Thomas-Oates J, Thomas J, Homans SW, Ferguson MAJ. Structure of *Leishmania mexicana* lipopolysaccharide. J Biol Chem 1992; 267:6834–40.
34. Etges RD, Bouvier J, Hoffman R, Bordier C. Evidence that the major surface protein of three *Leishmania* species are structurally related. Mol Biochem Parasitol 1985; 14:141–9.
35. Bordier C. The promastigote surface protease of *Leishmania*. Parasitol Today 1986; 3:151–3.
36. Hall FB, Joiner KA. Strategies of obligate intracellular parasites for evading host defences. Immunol Today/Parasitol Today (Combined Issue) 1991; 12:A22–A27.
37. Antoine J-C, Prina E, Jouanne C, Bongrand P. Parasitophorous vacuoles of *Leishmania amazonensis*–infected macrophages maintain an acidic pH. Infect Immun 1990; 58:779–87.
38. Alexander J, Vickerman K. Fusion of host cell secondary lysosomes with the parasitophorous vacuoles of *Leishmania mexicana*–infected macrophages. J Protozool 1975; 22:502–8.
39. Chang K-P, Dwyer DM. Multiplication of a human parasite (*Leishmania donovani*) in phagolysosomes of hamster macrophages in vitro. Science 1976; 193:678–80.
40. Russell DG, Mendina-Acosta E, Golubev A. The interface between the *Leishmania*-infected macrophage and the host's immune system. Behring Institut Mitteilungen 1991; 88:68–79.
41. Prina E, Antoine J-C, Wiederanders B, Kirschke H. Localization and activity of various lysosomal proteases in *Leishmania amazonensis*–infected macrophages. Infect Immun 1990; 58:1730–7.
42. Kutish GF, Janovy J. Inhibition of *in vitro* macrophage digestion capacity by infection with *Leishmania donovani*. J Parasitol 1981; 67:457–62.
43. Scott P, Sacks D, Sher A. Resistance to macrophage-mediated killing as a factor influencing the pathogenesis of chronic cutaneous leishmaniasis. J Immunol 1983; 131:966–71.
44. Berman JD, Neva FA. Effect of temperature on multiplication of *Leishmania* amastigotes within human monocyte-derived macrophages in vitro. Am J Trop Med Hyg 1981; 30:318–21.
45. Sacks DL, Barral A, Neva FA. Thermosensitivity patterns of Old vs. New World cutaneous strains of *Leishmania* growing within mouse peritoneal macrophages in vitro. Am J Trop Med Hyg 1983; 32:300–4.
46. Kirkpatrick CE, Nolan TJ, Farell JP. Rate of *Leishmania*-induced skin-lesion development in rodents depends on the site of inoculation. Parasitology 1987; 94:451–65.
47. Scott P. Impaired macrophage leishmanicidal activity at cutaneous temperature. Parasite Immunol 1985; 7:277–88.
48. Crocker PR, Davies EV, Blackwell JM. Variable expression of the natural resistance gene *Lsh* in different macrophage populations infected *in vitro* with *Leishmania donovani*. Parasite Immunol 1987; 9:705–19.
49. Behin R, Mauel J, Sordat B. *Leishmania tropica*: pathogenicity and *in vitro* macrophage function in strains of inbred mice. Exp Parasitol 1979; 48:81–91.
50. Nacy CA, Fortier AH, Pappas MG, Henry RR. Susceptibility of inbred mice to *Leishmania tropica* infection. Correlation of susceptibility with in *vitro* defective macrophage microbicidal pathways. Cell Immunol 1983; 77:298–307.
51. Gorczynski RM, MacRae S. Analysis of subpopulations of glass-adherent mouse skin cells controlling resistance/susceptibility to infection with Leishmania tropica, and correlation with the development of independent proliferative signals to Lyt-1^+/Lyt-2^+ T lymphocytes. Cell Immunol 1982; 67:74–89.
52. Will A, Blank C, Rollinghoff M, Moll H. Murine epidermal Langerhans cells are potent stimulators of an antigen-specific T-cell response to *Leishmania major*, the cause of cutaneous leishmaniasis. Eur J Immunol 1992; 22:1341–47.
53. Locksley RM, Heinzel FP, Fankhauser JE, Nelson CS, Sadick MD. Cutaneous host defense in leishmaniasis: interaction of isolated dermal macrophages and epidermal Langerhans cells with the insect-stage promastigote. Infect Immun 1988; 56:336–42.
54. McNeely TB, Turco SJ. Requirement of lipophosphoglycan for intracellular survival of *Leishmania donovani* within human monocytes. J Immunol 1990; 144:2745–50.
55. Eilam Y, El-On J, Spira DT. *Leishmania major:* excreted factor, calcium ions and survival of amastigotes. Exp Parasitol 1985; 59:161–8.
56. Handman E, Schnur LF, Spithill TW, Mitchell GF. Passive transfer of Leishmania lipopolysaccharide confers parasite survival in macrophages. J Immunol 1986; 137:3608–13.
57. McNeely TB, Rosen G, Londner MV, Turco SJ. Inhibitory effects on protein kinase C by lipophosphoglycan and glycosyl-phosphatidylinositol antigens of *Leishmania*. Biochem J 1989; 259:601–4.

58. Chan J, Fujiwara T, Brennan P, McNeil M, Turco SJ, Sibille J, Snapper M, Aisen P, Bloom BR. Microbial glycolipids: possible virulence factors that scavenge oxygen radicals. Proc Natl Acad Sci USA 1989; 86:2453–7.
59. El-On J, Bradley DJ, Freeman JC. *Leishmania donovani*: action of excreted factor on hydrolytic enzyme activity of macrophages from mice with genetically different resistance to infection. Exp Parasitol 1980; 49:167–74.
60. Chaudhuri G, Chaudhuri M, Pan A, Change K-P. Surface acid protease (gp63) of *Leishmania mexicana*. A metalloenzyme capable of protecting liposome-encapsulated proteins from phagolysosomal degradation by macrophages. J Biol Chem 1989; 264:7483–89.
61. Kweider M, Lemesre J-L, Darcy F, Kusnierz J-P, Capron A, Santoro F. Infectivity of *Leishmania braziliensis* promastigotes is dependent on the increasing expression of a 65,000-dalton surface antigen. J Immunol 1987; 138:299–305.
62. Chaudhuri G, Chang K-P. Acid protease activity of a major surface membrane glycoprotein (gp63) from *Leishmania mexicana* promastigotes. Mol Biochem Parasitol 1988; 27:43–52.
63. Turco SJ, Sacks DL. Expression of a stage-specific lipophosphoglycan in *Leishmania major* amastigotes. Mol Biochem Parasitol 1991; 45:91–100.
64. Glaser TA, Moody SF, Handman E, Bacic A, Spithill TW. An antigenically distinct lipophosphoglycan on amastigotes of *Leishmania major*. Mol Biochem Parasitol 1991; 45:337–44.
65. Pimenta PFP, Saraiva EMB, Sacks DL. The comparative fine structure and surface glycoconjugate expression of three life stages of *Leishmania major*. Exp Parasitol 1991; 72:191–204.
66. Frommel TO, Button LL, Fujikura Y, McMaster WR. The major surface glycoprotein (gp63) is present in both life stages of *Leishmania*. Mol Biochem Parasitol 1990; 38:25–32.
67. Schneider P, Rosat J-P, Bouvier J, Louis J, Bordier C. *Leishmania major*: Differential regulation of the surface metalloprotease in amastigote and promastigote stages. Exp Parasitol 1992; 75:196–206.
68. Zilberstein D, Dwyer DM. Identification of a surface membrane proton-translocating ATPase in promastigotes of the parasitic protozoan *Leishmania donovani*. Biochem J 1988; 256:13–21.
69. Zilberstein D, Dwyer DM. Protonmotive force–driven active transport of D-glucose and L-proline in the protozoan parasite *Leishmania donovani*. Proc Natl Acad Sci USA 1985; 82:1716–20.
70. Mukkuda A, Meade JC, Glaser TA, Bonventre PF. Enhanced metabolism of *Leishmania donovani* amastigotes at acid pH: an adaptation for intracellular growth. Science 1985; 229:1099–1101.
71. Gottlieb M, Dwyer DM. *Leishmania donovani* surface membrane acid phosphatase activity of promastigotes. Exp Parasitol 1981; 52:117–28.
72. Bates PA, Hermes I, Dwyer DM. *Leishmania donovani*: immunochemical localization and secretory mechanism of soluble acid phosphatase. Exp Parasitol 1989; 68:335–46.
73. Mukhopadhyay NK, Saha AK, Lovelace JK, daSilva R, Sacks DL, Glew RH. Comparison of the protein kinase and acid phosphatase activities of five species of *Leishmania*. J Protozool 1988; 35:601–7
74. Remaley AT, Glew RH, Kuhns DB, Basford RE, Waggoner AS, Ernst LA, Pope M. *Leishmania donovani* surface membrane acid phosphatase blocks neutrophil oxidative metabolite production. Exp Parasitol 1985; 60:331–41.
75. Das S, Saha AK, Remaley AT, Glew RH, Dowling JN, Kajiyoshi M, Gottlieb M. Hydrolysis of phosphoproteins and inositol phosphates by cell surface phosphatase of *Leishmania donovani*. Mol Biochem Parasitol 1986; 20:143–53.
76. Glew RH, Saha AK, Das S, Remaley AT. Biochemistry of the *Leishmania* species. Microbiol Rev 1988; 52:412–32.
77. Chang K-P. Leishmanicidal mechanisms of human polymorphonuclear phagocytes. Am J Trop Med Hyg 1981; 30:322–33.
78. Pearson RD, Steigbigel RT. Phagocytosis and killing of the protozoan *Leishmania donovani* by human polymorphonuclear leukocytes. J Immunol 1981; 127:1438–43.
79. Buchmuller-Rouiller Y, Mauel J. Impairment of the oxidative metabolism of mouse peritoneal macrophages by intracellular *Leishmania* spp. Infect Immun 1987; 55:587–93.
80. Murray HW. Cell-mediated immune response in experimental visceral leishmaniasis. II. Oxygen-dependent killing of intracellular *Leishmania donovani* amastigotes. J Immunol 1982; 129:351–7.
81. Meshnick SR, Eaton JW. Leishmanial superoxide dismutase: a possible target for chemotherapy. Biochem Biophys Res Comm 1981; 102:970–6.

82. Murray HW. Susceptibility of *Leishmania* to oxygen intermediates and killing by normal macrophages. J Exp Med 1981; 153:1302–15.
83. Murray HW. Interaction of *Leishmania* with a macrophage cell line. Correlation between intracellular killing and the generation of oxygen intermediates. J Exp Med 1981; 153:1690–5.
84. Haidaris CG, Bonventre PF. A role for oxygen-dependent mechanisms in killing of *Leishmania donovani* tissue forms by activated macrophages. J Immunol 1982; 129:850–5.
85. Pearson RD, Harcus JL, Roberts D, Donowitz GR. Differential survival of *Leishmania donovani* amastigotes in human monocytes. J Immunol 1983; 131:1994–9.
86. Channon JY, Roberts MB, Blackwell JM. A study of the differential respiratory burst activity elicited by promastigotes and amastigotes of *Leishmania donovani* in murine resident peritoneal macrophages. Immunology 1984; 53:345–55.
87. Buchmuller-Rouiller Y, Mauel J. Correlation between enhanced oxidative metabolism and leishmanicidal activity in activated macrophages from healer and nonhealer mouse strains. J Immunol 1986; 136:3884–90.
88. Mauel J, Buchmuller-Rouiller Y. Effect of lipopolysaccharide on intracellular killing of *Leishmania enriettii* and correlation with macrophage oxidative metabolism. Eur J Immunol 1987; 17:203–8.
89. Scott P, James S, Sher A. The respiratory burst is not required for killing of intracellular and extracellular parasites by a lymphokine-activated macrophage cell line. Eur J Immunol 1985; 15:553–8.
90. Green SJ, Meltzer MS, Hibbs JB Jr, Nacy CA. Activated macrophages destroy intracellular *Leishmania major* amastigotes by an L-arginine–dependent killing mechanism. J Immunol 1990; 144:278–83.
91. Liew FY, Millott S, Parkinson C, Palmer RMJ, Moncada S. Macrophage killing of *Leishmania* parasite in vivo is mediated by nitric oxide from L-arginine. J Immunol 144:4794–7.
92. Mauel J, Ransijn A, Buchmuller-Rouiller Y. Killing of *Leishmania* parasites in activated murine macrophages is based on an L-arginine–dependent process that produces nitrogen derivatives. J Leukocyte Biol 1991; 49:73–82.
93. Green SJ, Nacy CA, Meltzer MS. Cytokine-induced synthesis of nitrogen oxides in macrophages: a protective host response to *Leishmania* and other intracellular pathogens. J Leukocyte Biol 1991; 50:93–103.
94. Liew FY, Cox FEG. Nonspecific defence mechanism: the role of nitric oxide. Immunol Today/Parasitol Today (Combined Issue) 1991; 12:A17–A21.
95. Mauel J, Betz-Corradin S, Buchmuller-Rouiller Y. Nitrogen and oxygen metabolites and the killing of *Leishmania* by activated murine macrophages. Res Immunol 1991; 142:577–80.
96. Liew FY, Li Y, Moss D, Parkinson C, Rogers MV, Moncada S. Resistance to *Leishmania major* infection correlates with the induction of nitric oxide synthase in murine macrophages. Eur J Immunol 1991; 21:3009–14.
97. Li Y, Severn A, Rogers MV, Palmer RMJ, Moncada S, Liew FY. Catalase inhibits nitric oxide synthesis and the killing of intracellular *Leishmania major* in murine macrophages. Eur J Immunol 1991; 22:441–6.
98. Cameron ML, Granger DL, Weinberg JB, Kozumbo WJ, Koren HJ. Human alveolar and peritoneal macrophages mediate fungistasis independently of L-arginine oxidation to nitrite and nitrate. Am Rev Respir Dis 1990; 142:1313–9.
99. James SL, Cook KW, Lazdins JK. Activation of human monocyte-derived macrophages to kill schistosomula of *Schistosoma mansoni* in vitro. J Immunol 1990; 145:2686–90.
100. Denis M. Tumor necrosis factor and granulocyte monocyte-colony stimulating factor stimulate human macrophages to restrict growth of virulent *Mycobacterium avium* and to kill avirulent *M. avium*: killing effector mechanism depends on the generation of reactive nitrogen intermediates. J Leukocyte Biol 1990; 49:380–7.
101. Murray HW, Teitelbaum RF. L-arginine-dependent reactive nitrogen intermediates and the antimicrobial effect of activated human mononuclear phagocytes. J Infect Dis 1992; 165:513–7.
102. Reiner NE, Malemud CJ. Arachidonic acid metabolism in murine leishmaniasis (*Donovani*): *Ex-vivo* evidence for increased cyclooxygenase and 5-lipoxygenase activity in spleen cells. Cell Immunol 1984; 88:501–10.
103. Reiner NE, Malemud CJ. Arachidonic acid metabolism by murine peritoneal macrophages infected with *Leishmania donovani*: *In vitro* evidence for parasite-induced alterations in cyclooxygenase and lipoxygenase pathways. J Immunol 1985; 134:556–63.

104. Farrell JP, Kirkpatrick CE. Experimental cutaneous leishmaniasis. II. A possible role for prostaglandins in exacerbation in disease in Leishmania major–infected BALB/c mice. J Immunol 1987; 138:902–7.
105. Betz-Corradin S, Buchmuller-Rouiller Y, Mauel J. Phagocytosis enhances murine macrophage activation by interferon-γ and tumor necrosis factor-α. Eur J Immunol 1991; 21:2553–8.
106. Mauel J. Macrophage activation in vitro by tumor necrosis factor-alpha. J Leukocyte Biol 1990; 1 (suppl):96.
107. Mauel J, Ransijn A, Betz-Corradin S, Buchmuller-Rouiller Y. Effect of PGE_2 and of agents that raise cAMP levels on macrophage activation induced by IFN-γ and TNFα. Submitted for publication.
108. Reiner NE, Ng W, McMaster WR. Parasite-accessory cell interactions in murine leishmaniasis. II. *Leishmania donovani* suppresses macrophage expression of class I and class II major histocompatibility complex gene products. J Immunol 1987; 138:1926–32.
109. Reiner NE. Parasite-accessory cell interactions in murine leishmaniasis. I. Evasion and stimulus-dependent suppression of the macrophage interleukin 1 response by *Leishmania donovani*. J Immunol 1987; 138:1919–25.
110. Reiner NE, Ng W, Ma T, McMaster WR. Kinetics of γ interferon binding and induction of major histocompatibility complex class II mRNA in *Leishmania*-infected macrophages. Proc Natl Acad Sci USA 1988; 85:4330–4.
111. Kwan WC, McMaster WR, Wong N, Reiner NE. Inhibition of expression of major histocompatibility complex class II molecules in macrophages infected with *Leishmania donovani* occurs at the level of gene transcription via a cyclic AMP-independent mechanism. Infect Immun 1992; 60:2115–20.
112. Reiner NE, Ng W, Wilson CB, McMaster WR, Burchett SK. Modulation of in vitro monocyte cytokine responses to *Leishmania donovani*. Interferon-γ prevents parasite-induced inhibition of interleukin 1 production and primes monocytes to respond to *Leishmania* by producing both tumor necrosis factor-α and interleukin 1. J Clin Invest 1990; 85:1914–24.
113. Crawford GD, Wyler DJ, Dinarello CA. Parasite-monocyte interactions in human leishmaniasis: production of interleukin-1 in vitro. J Infect Dis 1985; 152:315–22.
114. Frankenburg S, Leibovici V, Mansbach N, Turco SJ, Rosen G. Effect of glycolipids of *Leishmania* parasites on human monocyte activity. Inhibition by lipophosphoglycan. J Immunol 1990; 145:4284–9.
115. Cillari E, Dieli M, Maltese E, Milano S, Salerno A, Liew FY. Enhancement of macrophage IL-1 production by *Leishmania major* infection in vitro and its inhibition by IFN-γ. J Immunol 1989; 143:2001–5.
116. Wagner HM, Beuscher HU, Rollinghoff M, Solbach W. Interferon-γ inhibits the efficacy of interleukin 1 to generate a Th2-cell biased immune response induced by *Leishmania major*. Immunobiology 1991; 182:292–306.
117. Theodos CM, Shankar A, Glasebrook AL, Roeder WD, Titus RG. Treatment with an anti–interleukin-1 receptor antibody is therapeutic for the course of experimental murine cutaneous leishmaniasis. Submitted for publication.
118. Titus RG, Sherry B, Cerami A. Tumor necrosis factor plays a protective role in experimental murine cutaneous leishmaniasis. J Exp Med 1989; 170:2097–104.
119. Liew FY, Parkinson C, Millott S, Severn A, Carrier M. Tumor necrosis factor (TNFα) in leishmaniasis. I. TNFα mediates host protection against cutaneous leishmaniasis. Immunology 1990; 69:570–3.
120. Theodos CM, Povinelli L, Molina R, Sherry B, Titus RG. Role of tumor necrosis factor in macrophage leishmanicidal activity in vitro and resistance to cutaneous leishmaniasis in vivo. Infect Immun 1991; 59:2839–42.
121. Boom WH, Liebster L, Abbas AK, Titus RG. Patterns of lymphokine secretion in murine leishmaniasis: correlation with disease progression or resolution. Infect Immun 1990; 58:3863–70.
122. Moll H. Binoder K, Bogdan C, Solbach W, Rollinghoff M. Production of tumour necrosis factor during murine cutaneous leishmaniasis. Parasite Immunol 1990; 12:483–94.
123. Liew FY, Li Y, Millott S. Tumor necrosis factor (TNF-α) in leishmaniasis. II. TNF-α-induced macrophage leishmanicidal activity is mediated by nitric oxide from L-arginine. Immunology 1990; 71:556–9.
124. Green SJ, Crawford RM, Hockmeyer JT, Meltzer MS, Nacy CA. *Leishmania major* amastigotes initiate the L-arginine–dependent killing mechanism in IFN-γ–stimulated macrophages by induction of tumor necrosis factor-α. J Immunol 1990; 145:4290–7.

125. Liew FY, Li Y, Millott S. Tumor necrosis factor-α synergizes with IFN-γ in mediating killing of *Leishmania major* through the induction of nitric oxide. J Immunol 145:4306–10.
126. Bogdan C, Moll H, Solbach W, Rollinghoff M. Tumor necrosis factor-α in combination with interferon-γ, but not with interleukin 4, activates murine macrophages for elimination of *Leishmania major* amastigotes. Eur J Immunol 1990; 20:1131–5.
127. Stenger S, Solbach W, Rollinghoff M, Bogdan C. Cytokine interactions in experimental cutaneous leishmaniasis. II. Endogenous tumor necrosis factor-α production by macrophages is induced by the synergistic action of interferon (IFN)-γ and interleukin(IL) 4 and accounts for the antiparasitic effect mediated by IFN-γ and IL 4. Eur J Immunol 1991; 21:1669–75.
128. Liew FY, Li Y, Yang DM, Severn A, Cox FEG. TNF-α reverses the disease-exacerbating effect of subcutaneous immunization against murine cutaneous leishmaniasis. Immunology 1991; 74:304–9.
129. Betz-Corradin S, Mauel J. Phagocytosis of *Leishmania* enhances macrophage activation by IFN- γ and lipopolysaccharide. J Immunol 1991; 146:279–85.
130. Wolpe SD, Cerami A. Macrophage inflammatory proteins 1 and 2: members of a novel superfamily of cytokines. FASEB J 1989; 3:2565–73.
131. Engelhorn S, Bruckner A, Remold HG. A soluble factor produced by inoculation of human monocytes with *Leishmania donovani* promastigotes suppresses the IFN-γ–dependent monocyte activation. J Immunol 1990; 145:2662–8.
132. Nelson BJ, Ralph P, Green SJ, Nacy CA. Differential susceptibility of activated macrophages cytotoxic effector reactions to the suppressive effects of transforming growth factor-β1. J Immunol 1991; 146:1849–57.
133. Descoteaux A, Turco SJ, Sacks DL, Matlashewski G. *Leishmania donovani* lipophosphoglycan selectively inhibits signal transduction in macrophages. J Immunol 1991; 146:2747–53.
134. Olivier M, Brownsey RW, Reiner NE. Defective stimulus-response coupling in human monocytes infected with *Leishmania donovani* is associated with altered activation and translocation of protein kinase C. Proc Natl Acad Sci USA 1992; 89:7481–85.
135. Olivier M, Baimbridge KG, Reiner NE. Stimulus-response coupling in monocytes infected with *Leishmania*. Attenuation of calcium transients is related to defective agonist-induced accumulation of inositol phosphates. J Immunol 1992; 148:1188–96.
136. Lang T, Kaye PM. Presentation of *Leishmania donovani* promastigotes occurs via a brefeldin A–sensitive pathway. Eur J Immunol 1991; 21:2407–13.
137. Kaye P, Cuomo C, Coburn C, Beverley SM. Processing of antigen targeted to the phagolysosome. J Cell Biochem 1992; 16D(suppl):34.
138. Kaye PM, Coburn, C, Cuomo C, LeBowitz J, Beverley SM. Antigens targeted to the *Leishmania* phagolysosome are processed for CD4+ T cell recognition. Manuscript submitted for publication.
139. Titus RG, Milon G, Marchal G, Vassalli P, Cerottini J-C, Louis JA. Involvement of specific Lyt2+ T cells in the immunologic control of experimentally induced murine cutaneous leishmaniasis. Eur J Immunol 1987; 17:1429–33.
140. Farrell JP, Muller I, Louis JA. A role for Lyt-2+ T cells in resistance to cutaneous leishmaniasis in immunized mice. J Immunol 1989; 142:2052–6.
141. Hill JO, Awwad M, North RJ. Elimination of CD4+ suppressor T cells from susceptible BALB/c mice releases CD8+ T lymphocytes to mediate protective immunity against *Leishmania*. J Exp Med 169:1819–27.
142. Smith LE, Rodrigues M, Russell DG. The interaction between CD8+ cytotoxic T cells and *Leishmania*-infected macrophages. J Exp Med 1991; 174:499–505.
143. Muller I, Pedrazzini T, Kropf P, Louis J. Establishment of resistance to *Leishmania major* infection in susceptible BALB/c mice requires parasite-specific CD8+ T cells. Intl Immunol 1991; 3:587–97.
144. Muller I, Kropf P, Milon G, Louis JA. CD8+ T lymphocytes in experimental murine cutaneous leishmaniasis. In: Romagnani S, Mosman TR, Abbas AK, eds. New advances on cytokines. New York: Raven Press, 1992:265–70.
145. Muller I, Fruth U, Louis JA. Immunobiology of experimental leishmaniasis. Med Microbiol Immunol 1992; 181:1–12.
146. Scott P, Natovitz P, Coffman RL, Pearce E, Sher A. Immunoregulation of cutaneous leishmaniasis. T cell lines that transfer protective immunity or exacerbation belong to different T helper subsets and respond to distinct parasite antigens. J Exp Med 1988; 168:1675–84.

147. Scott P, Caspar P, Sher A. Protection against *Leishmania major* in BALB/c mice by adoptive transfer of a T cell clone recognizing a low molecular weight antigen released by promastigotes. J Immunol 1990; 144:1075–9.
148. Muller I, Louis JA. Immunity to experimental infection with Leishmania major: generation of protective L3T4+ T cell clones recognizing antigen(s) associated with live parasites. Eur J Immunol 1989; 19:865–71.
149. Holaday BJ, Sadick MD, Wang Z-E, Reiner SL, Heinzel FP, Parslow TG, Locksley RM. Reconstitution of *Leishmania* immunity in severe combined immunodeficiency mice using Th1- and Th2-like cell lines. J Immunol 1991; 147:1653–8.
150. Titus RG, Lima GC, Engers HD, Louis JA. Exacerbation of murine cutaneous leishmaniasis by adoptive transfer of parasite-specific helper T cell populations capable of mediating *Leishmania major*–specific delayed-type hypersensitivity. J Immunol 1984; 133:1594–1600.
151. Titus RG, Muller I, Kimsey P, Cerny A, Behin R, Zinkernagel RM, Louis JA. Exacerbation of experimental murine cutaneous leishmaniasis with CD4+ *Leishmania major*–specific T cell lines or clones which secrete interferon-gamma and mediate parasite-specific delayed-type hypersensitivity. Eur J Immunol 1991; 21:559–67.
152. Titus RG, Ceredig R, Cerottini J-C, Louis JA. Therapeutic effect of anti-L3T4 monoclonal antibody GK1.5 on cutaneous leishmaniasis in genetically susceptible BALB/c mice. J Immunol 1985; 135:2108–14.
153. Mosmann TR, Coffman RL. Th1 and Th2 cells: different patterns of lymphokine secretion lead to different functional properties. Annu Rev Immunol 1989; 7:145–73.
154. Heinzel FP, Sadick MD, Holaday BJ, Coffman RL, Locksley RM. Reciprocal expression of interferon γ or interleukin 4 during the resolution or progression of murine leishmaniasis. Evidence for expansion of distinct helper T cell subsets. J Exp Med 1989; 169:59–72.
155. Murray HW, Rubin BY, Rothermel CD. Killing of intracellular *Leishmania donovani* by lymphokine-stimulated human monocytes. Evidence that interferon γ is the activating lymphokine. J Clin Invest 1983; 72:1506–10.
156. Titus RG, Kelso A, Louis JA. Intracellular destruction of *Leishmania tropica* by macrophages activated with macrophage activating factor/interferon. Clin Exp Immunol 1984; 55:157–65.
157. Nacy CA, Fortier AH, Meltzer MS, Buchmeier NA, Gray PW. Macrophage activation to kill *Leishmania major:* activation of macrophages for intracellular destruction of amastigotes can be induced by both recombinant interferon-γ and non-interferon lymphokines. J Immunol 1985; 135:3505–11.
158. Sadick MD, Locksley RM, Tubbs C, Raff HV. Murine cutaneous leishmaniasis: Resistance correlates with the capacity to generate interferon-γ in response to Leishmania antigens in vitro. J Immunol 1986; 136:655–61.
159. Belosevic M, Finbloom DS, van der Meide PH, Slayter MV, Nacy CA. Administration of monoclonal anti-IFN-gamma antibodies in vivo abrogates natural resistance of C3H/HeN mice to infection with *Leishmania major*. J Immunol 1989; 143:266–74.
160. Lehn M, Weiser WY, Englehorn S, Gillis S, Remold HG. IL-4 inhibits H_2O_2 production and antileishmanial capacity of human cultured monocytes mediated by IFN-γ. J Immunol 1989; 143:3020–4.
161. Liew FY, Millott S, Li Y, Lelchuk R, Chan WL, Ziltener H. Macrophage activation by interferon-γ from host-protective T cells is inhibited by interleukin (IL) 3 and IL-4 produced by disease-promoting T cells in leishmaniasis. Eur J Immunol 1989; 19:1227–32.
162. Sadick MD, Heinzel FP, Holaday BJ, Pu RT, Dawkins RS, Locksley RM. Cure of murine leishmaniasis with anti–interleukin 4 monoclonal antibody. Evidence for a T cell-dependent, interferon γ-independent mechanism. J Exp Med 1990; 171:115–27.
163. Chatelain R, Varkila K, Coffman RL. IL-4 induces a Th2 response in *Leishmania major*–infected mice. J Immunol 1992; 148:1182–7.
164. Theodos CM, Titus RG. Analysis of *Leishmania major*–specific T cell clones derived from genetically resistant C3H mice. FASEB J 1991; 5:A1672.
165. Firestein GS, Roeder WD, Laxer JA, Townsend KS, Weaver CT, Hom JT, Linton J, Torbett BE, Glasebrook AL. A new murine CD4+ T cell subset with an unrestricted cytokine profile. J Immunol 1989; 143:518–25.
166. Moll H, Rollinghoff M. Resistance to murine cutaneous leishmaniasis is mediated by Th1 cells, but disease-promoting CD4+ cells are different from Th2 cells. Eur J Immunol 1990; 20:2067–74.

167. Sypek JP, Matzilevich MM, Wyler DJ. Th2 lymphocyte clone can activate macrophage antileishmanial defense by a lymphokine-independent mechanism *in vitro* and can augment parasite attrition *in vivo*. Cell Immunol 1991; 133:178–86.
168. Wershil BK, Galli SJ, Titus RG. Mast cells augment the lesions seen during experimental murine cutaneous leishmaniasis. Submitted for publication.
169. Coffman RL, Varkila K, Scott P, Chatelain R. Role of cytokines in the differentiation of CD4+ T-cell subsets in vivo. Immunol Rev 1991; 123:189–207.
170. Mazingue C, Cottrez-Detoeuf F, Louis J, Kweider M, Auriault C, Capron A. In vitro and in vivo effects of interleukin 2 on the protozoan parasite leishmania. Eur J Immunol 1989; 19:487–91.
171. Scott P. IFN-γ modulates the early development of Th1 and Th2 responses in a murine model of cutaneous leishmaniasis. J Immunol 1991; 147:3149–55.
172. Sadick MD, Street N, Mosmann TR, Locksley RM. Cytokine regulation of murine leishmaniasis: interleukin 4 is not sufficient to mediate progressive disease in resistant C57BL/6 mice. Infect Immun 1991; 59:4710–4.
173. Carter KC, Gallagher G, Baillie AJ, Alexander J. The induction of protective immunity to Leishmania major in the BALB/c mouse by interleukin 4 treatment. Eur J Immunol 1989; 19:779–82.
174. Lezama-Davila CM, Williams DM, Gallagher G, Alexander J. Cytokine control of Leishmania infection in the BALB/c mouse: enhancement and inhibition of parasite growth by local administration of IL-2 or IL-4 is species and time dependent. Parasite Immunol 1992; 14:37–48.
175. Scott P, Pearce E, Cheever AW, Coffman RL, Sher A. Role of cytokines and CD4+ T-cell subsets in the regulation of parasitic immunity and disease. Immunol Rev 1989; 112:161–82.
176. Bogdan C, Stenger S, Rollinghoff M, Solbach W. Cytokine interactions in experimental cutaneous leishmaniasis. Interleukin 4 synergizes with interferon-γ to activate murine macrophages for killing of *Leishmania major* amastigotes. Eur J Immunol 1991; 21:327–33.
177. Davis CE, Belosevic M, Meltzer MS, Nacy CA. Regulation of activated macrophage antimicrobial activities. Cooperation of lymphokines for induction of resistance to infection. J Immunol 1988; 141:627–35.
178. Belosevic M, Davis CE, Meltzer MS, Nacy CA. Regulation of activated macrophage antimicrobial activities. Identification of lymphokines that cooperate with IFN-γ for induction of resistance to infection. J Immunol 1988; 141:890–6.
179. Belosevic M, Finbloom DL, Meltzer MS, Nacy CA. IL-2: a cofactor for induction of activated macrophage resistance to infection. J Immunol 1990; 145:831–9.
180. Sypek JP, Wyler DJ. Antileishmanial defense in macrophages triggered by tumor necrosis factor expressed on CD4+ T lymphocyte plasma membrane. J Exp Med 1991; 174:755–9.
181. Feng ZY, Louis J, Kindler V, Pedrazzini T, Eliason JF, Behin R, Vassalli P. Aggravation of experimental cutaneous leishmaniasis in mice by administration of interleukin 3. Eur J Immunol 1988; 18:1245–51.
182. Lelchuk R, Graveley R, Liew FY. Susceptibility to murine cutaneous leishmaniasis correlates with the capacity to generate interleukin 3 in response to *Leishmania* antigen in vitro. Cell Immunol 1988; 111:66–74.
183. Mirkovich AM, Galelli A, Allison AC, Modabber FZ. Increased myelopoesis during L. major infection in mice: generation of "safe targets" a possible way to evade the effector immune mechanism. Clin Exp Immunol 1986; 64:1–7.
184. Handman E, Burgess AW. Stimulating by granulocyte-macrophage colony-stimulating factor of *Leishmania tropica* killing by macrophages. J Immunol 1979; 122:1134–7.
185. Greil J, Bodenhofer B, Rollinghoff M, Solbach W. Application of recombinant granulocyte-macrophage colony-stimulating factor has a detrimental effect in experimental murine leishmaniasis. Eur J Immunol 1988; 18:1527–33.
186. Corcoran LM, Metcalf D, Edwards SJ, Handman E. GM-CSF produced by recombinant vaccinia virus on in GM-CSF transgenic mice has no effect in vivo on murine cutaneous leishmaniasis. J Parasitol 1988; 74:763–7.
187. Charlab R, Blaineau C, Schechtman D, Barcinski MA. Granulocyte-macrophage colony-stimulating factor is a growth factor for promastigotes of *Leishmania mexicana amazonensis*. J Protozool 1990; 37:353–7.
188. Weiser WY, Temple PA, Witek-Giannotti J, Remold HG, Clark S, David JR. Molecular cloning of a

cDNA encoding a human macrophage migration inhibitory factor. Proc Natl Acad Sci USA 1989 86:7522–6.
189. Weiser WY, Pozzi L-AM, David JR. Human recombinant migration inhibition factor activates human macrophages to kill *Leishmania donovani*. J Immunol 1991; 147:2006–11.
190. Weiser WY, Pozzi L-AM, Titus RG, David JR. Recombinant human migration inhibitory factor has adjuvant activity. Proc Natl Acad Sci USA 1992; 89:8049–52.
191. O'Garra A, Stapleton G, Dhar V, Pearce M, Schumacher J, Rugo H, Barbis D, Stall A, Cupp J, Moore K, Vieira P, Mosmann T, Whitmore A, Arnold L, Haughton G, Howard M. Production of cytokines by mouse B cells: B lymphomas and normal B cells produce interleukin 10. Intl Immunol 1990; 2:821–32.
192. De Waal Malefyt R, Abrams J, Bennett B, Figdot CG, de Vries JE. Interleukin 10 (IL-10) inhibits cytokine synthesis by human monocytes: An autoregulatory role of IL-10 produced by monocytes. J Exp Med 1991; 174:1209–20.
193. Howard M, O'Garra A. Biological properties of interleukin 10. Immunol Today 1992; 13:198–200.
194. Heinzel FP, Sadick MD, Mutha SS, Locksley RM. Production of interferon γ, interleukin 2, interleukin 4 and interleukin 10 by CD4+ lymphocytes in vivo during healing and progressive murine leishmaniasis. Proc Natl Acad Sci USA 1991; 88:7011–5.
195. Cunha FQ, Moncada S, Liew FY. Interleukin-10 (IL-10) inhibits the induction of nitric oxide synthase by interferon-γ in murine macrophages. Biochem Biophys Res Comm 1992; 182:1155–9.
196. Titus RG, Ribeiro JMC. Salivary gland lysates from the sand fly *Lutzomyia longipalpis* enhance *Leishmania* infectivity. Science 1988; 239:1306–8.
197. Theodos CM, Ribeiro JMC, Titus RG. Analysis of enhancing effect of sand fly saliva on *Leishmania* infection in mice. Infect Immun 1991; 59:1592–8.
198. Samuelson J, Lerner E, Tesh R, Titus R. A mouse model of *Leishmania braziliensis braziliensis* infection produced by coinjection with sand fly saliva. J Exp Med 1991; 173:49–54.
199. Ribeiro JMC, Vachereau A, Modi GB, Tesh RB. A novel vasodilatory peptide from the salivary glands of the sand fly *Lutzomyia longipalpis*. Science 1989; 243:212–4.
200. Nong Y-H, Titus RG, Ribeiro JMC, Remold HG. Peptides encoded by the calcitonin gene inhibit macrophage function. J Immunol 1989; 143:45–9.
201. Lerner EA, Shoemaker CB. Maxadilan. Cloning and functional expression of the gene encoding this potent vasodilator peptide. J Biol Chem 1992; 267:1062–6.
202. Theodos CM, Titus RG. Salivary gland material from the sand fly *Lutzomyia longipalpis* has an inhibitory effect on macrophage function *in vitro*. Parasite Immunol. 1993; In press.
203. Warburg A, Schlein Y. The effect of post-bloodmeal nutrition of *Phlebotomus papatasi* on the transmission of *Leishmania major*. Am J Trop Med Hyg 1986; 36:926–30.

27

Immune Effector Role of Macrophages in Experimental Schistosomiasis mansoni

Stephanie L. James

National Institutes of Health, Bethesda, Maryland

Dov L. Boros

Wayne State University School of Medicine, Detroit, Michigan

I. INTRODUCTION

Schistosomiasis mansoni is a tropical helminthic disease affecting over 100 million people in 51 countries. The worms, also called blood flukes, live in pairs in the mesenteric veins of the infected host. The life cycle of the parasites alternates between snails of fresh water (intermediate host) and humans (definitive host). Humans acquire the infection through water exposure (irrigation, planting, fishing, washing, swimming) when the larval-stage cercariae that are released by the snails penetrate the intact skin. After penetration cercariae transform to schistosomula that are suited to exist in the high osmotic environment of the body. The presence of schistosomula may evoke a local transient dermatitis. Subsequently, schistosomula leave the skin and migrate via the blood and lymphatics into the lungs. Male and female larvae mature during the migratory period and then descend to the liver and mate. The adult worm pairs live for 5 to 10 years attached to the wall of the mesenteric veins. Each female produces about 300 eggs daily, half of which penetrate the intestinal wall, reach the lumen, and leave the body during defecation. Those eggs upon reaching fresh water release the embryo (miracidium) that penetrates the snails and initiates the asexual reproductive cycle. Eggs that are not evacuated are trapped within the intestinal wall or are carried by the portal system into the liver. Each egg induces by secreted soluble antigens a T lymphocyte–mediated granulomatous focus that may last 4 to 6 weeks. The incessant egg deposition, granulomatous inflammation, and fibrosis are responsible for the pathology of the disease [1].

The available evidence indicates that infected adults manifest a slowly developing acquired immunity against reinfecting larvae, without eliminating the indwelling worms (premunition). The protective host response is directed to the early larval stages (cercariae, schistosomula) of the infecting worms. A variety of resistance mechanisms have been described over the years. In humans and rats an antibody-dependent cell-mediated cytotoxicity (ADCC) that involves immunoglobulin E (IgE) or IgG2a antibodies and macrophages, eosinophils, platelets, etc., has been described [2]. In the mouse model, resistance is mediated by the activated macrophage that is signaled by sensitized T cells. Thus macrophages may play a major role in schistosomiasis, both in antihelminth resistance and in the isolation of eggs by the formation of the inflammatory granuloma. This chapter intends to describe this dual role of host macrophages against schistosome worms.

II. ACTIVATED MACROPHAGES AS IMMUNE EFFECTOR CELLS AGAINST SCHISTOSOMES

In schistosome infection, whereas the eggs are the principal stimuli of immunopathology, protective immunity is directed against the migrating larvae (schistosomula). In vitro, cytotoxicity against schistosome larvae is measured microscopically, by assessing loss of motility and disruption of larval tegument and/or internal organ structure with time in culture. The "skin-stage" larvae of the parasite, the form that is produced by cercarial transformation immediately following skin penetration, has been shown by this method to be susceptible to a variety of cytotoxic mechanisms involving antibody, complement, and effector cells including eosinophils, macrophages, and platelets [3]. The larvae seem to lose their susceptibility to most of these in vitro killing mechanisms as they mature [4]. Interestingly, however, in vivo experiments in some animal models of *Schistosoma mansoni* infection suggest that older larval forms migrating through the lungs are the primary target of immune attack [5,6]. In vivo, immunologically mediated resistance is best assessed directly by counting the number of adult parasites recovered from the bloodstream by perfusion of immunized and infected animals.

A. In Vitro Evidence: Macrophage Cytotoxicity Against Schistosome Larvae

Lymphokine-activated murine peritoneal macrophages kill the "skin-stage" larvae, as well as older (2- to 3-week) migrating forms, in an antibody-independent fashion [7,8]. Early experiments indicated that, as with extracellular killing of tumor cell targets, inflammatory macrophages are more susceptible than resident cells to lymphokine activation for larval killing. Interferon-γ (IFN-γ) appears to be the major lymphokine involved in macrophage activation. Thus, while supernatant fluids from antigen- or mitogen-stimulated spleen cells were able to activate macrophages for larval killing, this activity is abrogated by prior depletion of IFN-γ from the fluids. Moreover, recombinant IFN-γ alone is able to activate macrophages for in vitro larvicidal activity [7]. Addition of bacterial endotoxin, however, significantly enhances the stimulatory activity of IFN-γ, as is the case for cytotoxicity against other targets. Preliminary experiments suggest that other mediators in the supernatant fluids from antigen-stimulated spleen cells may serve as cofactors for macrophage activation (M. Williams, unpublished data), and the identity of these factors is currently under investigation. Neither interleukin-4 (IL-4) nor tumor necrosis factor-α (TNF-α) alone has been observed to activate macrophages for larval killing (I. Oswald et al., unpublished data).

Ultrastructural studies of schistosomula cocultured with activated macrophages showed that, unlike granulocyte-mediated killing, macrophage-mediated cytotoxicity is not directed against the parasite surface. Within the first hour of incubation, perturbation of the subtegumental muscle cells and mitochondria was noted. Progressive disintegration of parasite internal structures followed, such that by 48 hours of incubation the organisms were reduced to an intact tegumental "shell" surrounding severely vacuolated and disorganized internal tissues [9]. A similar pattern of macrophage-mediated destruction was also observed with older larval targets [8]. Studies of the effector mechanism of larvicidal activity showed that products of the macrophage respiratory burst are not required [10]. Killing was most effective when contact between effector cell and target was allowed, but long-term attachment of the cells to the larvae was not observed [9,11]. Subsequently, it was found that two macrophage products, TNF-α and nitric oxide, are toxic to schistosomula [12,13]. Current understanding of the effector mechanism is that IFN-γ treatment induces endogenous production of TNF-α by the macrophages. Although TNF-α is directly toxic to the parasite at high concentrations, its primary role under these conditions appears to be as a cofactor for macrophage activation. Thus, treatment of IFN-γ–stimulated macrophages with anti–TNF-α antibodies prevents larval killing, and this can be reversed by addition of exogenous TNF-α at levels that are not directly toxic to the parasite. Endogenous TNF-α production is required for induction of the arginine-dependent production of reactive nitrogen oxides (NO), particularly nitric oxide, that serve as the effector molecules of parasite killing [13a]. The principal targets of these molecules appear to be enzymes containing a catalytically active Fe-S group, such as those involved in mitochondrial electron transport, the citric acid cycle, and DNA synthesis, which become inactivated as a result of formation of nitrosyl-iron-sulphur complexes [14]. This conclusion is supported by the observation that larval killing is arginine-dependent and inhibited by addition of excess iron to the culture media [13]. The importance of endogenous TNF-α production in this process is indicated by the fact that addition of anti–TNF-α antibodies also inhibits NO production [13a]. The pattern of target damage previously observed in ultrastructural studies is consistent with the possibility that inhibition of larval metabolism is involved in macrophage-mediated parasite killing.

Recent studies have shown that IFN-γ activation of macrophages for larval killing is inhibitable by IL-10, IL-4, and transforming growth factor-β (TGF-β), cytokines that have demonstrated immunomodulatory activity in other systems as well [15,15a]. While these cytokines are individually active at relatively high levels, very low concentrations of any two of these mediators inhibit macrophage larvicidal activity as well as NO production in a synergistic fashion. Interleukin-10 inhibition appears to be manifested at the level of TNF-α production. Treatment of IFN-γ–stimulated cells with IL-10 suppresses their production of TNF-α, as measured at both the mRNA and the protein levels, and addition of exogenous TNF-α reverses the inhibitory effects of IL-10 [13a]. The mechanism of action of IL-4 and TGF-β in this system is not yet understood, although it does not appear to involve blockage of TNF-α production.

The observation that IFN-γ, a TH_1 lymphokine, activates a potent immune effector mechanism of parasite killing, while two TH_2 lymphokines (IL-10 and IL-4) suppress this mechanism, suggests that induction of TH_1 immune responses may be beneficial for resistance to schistosome infection. This hypothesis has been further investigated in a series of in vivo studies of protective immunity.

B. In Vivo Evidence: Murine Models of Protective Immunity

Two models of protective immunity against *S. mansoni* in the mouse provide evidence suggesting that lymphokine-activated macrophages may play a role in the immune mechanism of resistance. Infection with heavily irradiated larvae induces solid immunity against subsequent infection with nonattenuated parasites. In some strains of mice, the level of protection achieved by a single vaccination with radiation-attenuated larvae approaches 90%, quantitated as the percentage of challenge organisms that fail to mature in immunized as compared with control animals [16]. Following percutaneous infection, radiation-attenuated parasites live for only a few weeks, dying for the most part during their migration through the lungs [5]. A second model involves the induction of resistance by intradermal vaccination using nonliving schistosome antigens, derived either from larval or adult worm stages. In this model, using crude soluble parasite antigens with Bacillus Calmette-Guérin (BCG) as an adjuvant, protection averages around 50% in most mouse strains [17]. Although studies as detailed as those performed in the attenuated vaccine model have not been carried out, available evidence suggests that elimination of subsequent infection may occur at the skin site of challenge [18].

1. Immune Response in the Attenuated-Vaccine Model

In C57Bl/6 mice immunized by a single exposure to irradiated schistosome cercariae, resistance is dependent on T and B lymphocyte function but independent of complement, IgM production, or IgE-mediated immediate-type hypersensitivity [19–21]. In this model, protection requires the presence of CD4+ T cells, but not CD8+ cells [22,23]. Moreover, resistance appears to be dependent on the function of the TH_1, rather than TH_2, subset of CD4+ cells. Thus, treatment of vaccinated mice with antibodies to IFN-γ depletes immunity, while treatment with antibodies against the TH_2 cytokines IL-4 and IL-5 fails to diminish resistance [24,24a]. Interestingly, however, treatment with antibodies against IL-5, which is a powerful colony-stimulating and differentiation factor for eosinophils, completely eliminates circulating and tissue eosinophilia in vaccinated mice [24]. Resistance is also unaffected by whole-body irradiation [25]. These latter two observations suggest that radiosensitive polymorphonuclear cells are not essential in the effector mechanism of resistance.

Several lines of evidence suggest that lymphokine-activated macrophages play a key role in this model. CD4+ T cells isolated from the spleens of vaccinated mice proliferate and produce substantial amounts of IFN-γ upon in vitro stimulation with parasite antigens [7,26]. That such a response also occurs in vivo is indicated by the observation that activated macrophages, displaying cytotoxicity for schistosomula as well as tumor cell targets, can be recovered from the peritoneal cavities of vaccinated animals within 24 hours after intraperitoneal injection of soluble parasite antigens [7,26].

Genetic studies have provided the most direct evidence of a role for IFN-γ–activated macrophages in resistance to *S. mansoni* in vivo. Although most inbred strains of mice respond to vaccination with irradiated cercariae by developing significant levels of resistance to challenge infection, P strain mice vaccinated with irradiated cercariae consistently demonstrate levels of resistance no greater than 20% [27]. In other systems, the immunologic defect of P strain mice has been related to underproduction of a priming signal (presumably IFN-γ) for macrophage activation, as well as decreased responsiveness of macrophages to lymphokine-activating stimuli [28,29]. Vaccinated P strain mice fail to develop delayed-type hypersensitivity to schistosome antigens, they demonstrate impaired ability to produce IFN-γ in response to either antigen or mitogen stimulation, and their peritoneal macrophages are unresponsive to lymphokine activation [30]. Furthermore, intraperitoneal antigen challenge does not recruit activated, larvicidal macrophages in vaccinated P mice [30]. In contrast, humoral immunity appears relatively intact in these animals, and they show no deficit in several other aspects of cellular response, including production of IL-2, IL-3, and macrophage chemotactic factors, blastogenic potential, helper cell activity, and allospecific cytolytic T cell reactivity [31,32].

The availability of high- and low-responder strains of mice offered a unique opportunity to investigate the mechanism of protective immunity induced by vaccination with irradiated cercariae. In genetic crosses between C57Bl/6 and P mice, P-associated defects in resistance and macrophage activation were both found to be inherited in a fully recessive manner, and a highly significant correlation was observed between the level of macrophage larvicidal activity and the level of resistance to challenge infection in individual animals within segregating generations [33]. Such an association is consistent with a cause-and-effect relationship and provides strong in vivo evidence implicating activated macrophages as immune effector cells in this model.

How might these cells function in resistance to challenge infection? In vivo studies in mice vaccinated with irradiated cercariae show a massive increase in number of Ia+, oxidatively reactive, macrophages in the lungs within 2 to 3 weeks [34]. Challenge infection of immunized animals stimulates an anamnestic T lymphocytic response in the lungs, while the macrophage population remains elevated [35]. Lymphocytes recovered from the lungs of vaccinated and challenged mice by bronchoalveolar lavage produce IFN-γ [36], as do cells harvested from the lymph nodes draining the lungs [37]. Investigation of the fate of challenge parasites in vaccinated mice indicates larvae that pass through the lungs become trapped in a mononuclear cell–rich inflammatory reaction histologically reminiscent of delayed hypersensitivity response, a TH_1-associated phenomenon [6]. Autoradiographic tracking of challenge parasites suggests that elimination in the lungs is a protracted process lasting several weeks [38]. One interpretation of the role of cellular effector mechanisms that has been put forward for this model is that of a trapping mechanism that impedes the migration of the challenge larvae in the lungs, where they gradually die [6,38]. However, ultrastructural studies of larvae trapped in inflammatory lesions in the skin, lungs, or liver of vaccinated rodents reveal muscle disruption and internal vacuolation reminiscent of that seen in parasites killed by activated macrophages in vitro [39].

2. *Immune Response in the Nonliving-Vaccine Model*

To further test the role of a macrophage-mediated effector mechanism in resistance, a nonliving-vaccine protocol was designed to induce cell-mediated immunity, and its ability to protect mice against challenge *S. mansoni* infection was examined. The intradermal (ID) immunization route was chosen on the basis of early experiments on induction of delayed hypersensitivity to haptenated splenocytes, in which it was found that the ID route of inoculation enhanced, while the intravenous route suppressed, T cell response [40]. Initial experiments utilized crude schistosome antigen preparations such as killed larvae or soluble adult worm antigens, although subsequently the parasite antigen paramyosin was found to reproduce these effects [41]. In all cases, using different immunization routes and different adjuvants, induction of resistance to challenge infection correlates with stimulation of antigen-specific cell-mediated immunity and macrophage activation [17,42–44]. Further evidence that these responses are integral to resistance was provided by the observation that P strain mice, with defective macrophage function, fail to demonstrate resistance following ID immunization with BCG plus dead larvae [17]. The most definitive proof that T cell–mediated rather than humoral resistance mechanisms are involved in this model is the observation that B cell–deficient (μ-suppressed) mice, but not T cell–deficient (nude) mice, develop significant protective immunity in response to ID immunization with BCG plus dead larvae [45]. Moreover, resistance can be transferred to naive recipients with B cell–depleted spleen cell populations, but not with serum, from ID-immunized animals [45]. Spleen and lymph node cell populations from ID-immunized mice produce high levels of IFN-γ and IL-2 in response to stimulation with schistosome antigens in vitro, but negligible amounts of IL-4 and IL-5, thus displaying a lymphokine production profile characteristic of TH_1 cells. Recently, a schistosome antigen–responsive IFN-γ–producing TH_1 cell clone from these mice has been shown to transfer both resistance to challenge *S. mansoni* infection and the ability to generate activated larvicidal macrophages in response to antigen challenge (S. James et al., unpublished data), again supporting a role for this effector mechanism in protective immunity against this parasite.

C. Relevance to Human Immunity

It is clear that IFN-γ–treated human monocyte-derived macrophages can kill larval schistosomes in vitro [46,47]. There have as yet been no studies in naturally infected human populations that have implicated macrophage activity in resistance. However, one such study did indicate that the T cell blastogenic activity to schistosome antigens observed in infected individuals before chemotherapy correlated indirectly with the level of reinfection [48], suggesting a possible linkage between resistance and cell-mediated immunity in humans. Moreover, earlier studies reported that monocytes from patients with either tuberculosis or light schistosome infection were activated for augmented in vitro larval killing [49]. The requirement for working primarily with peripheral blood monocytes rather than tissue-derived macrophages makes comparison between results in human and murine studies difficult. More important, although it is currently the only system ethically available, information derived from analysis of the immune mechanisms involved in resistance to reinfection after chemotherapy may not be directly extrapolatable to a vaccination situation.

Certainly the evidence obtained in murine models implicating activated macrophages in protective immunity has spurred many investigators to begin examination of the schistosome antigens recognized by IFN-γ–producing TH_1 cells in their search for a defined vaccine against schistosomiasis.

III. GRANULOMA MACROPHAGES AS INFLAMMATORY, ACCESSORY AND EFFECTOR CELLS

In experimental murine schistosomiasis mansoni, egg deposition starts around the fifth week of the infection. During the ensuing weeks soluble egg antigens (SEAs) sensitize the system and induce a strong CD4+ T cell–mediated immune response [50]. This response generates a vigorous granulomatous inflammatory reaction around parasite eggs in the tissues that peaks between 8 and 10 weeks of the infection. Concurrently, splenic and granuloma lymphocytes produce high levels of various lymphokines. With progress to the chronic stage of the infection, the granulomatous response undergoes a spontaneous down-modulation [51,52] in part mediated by T suppressor cells [53]. At the same time, splenic and granuloma lymphokine production diminishes, and specific anti-SEA antibodies attain high levels in the circulation.

The schistosome eggs are large, measuring 150 μm in length. They cannot be phagocytosed, and in the tissues they are surrounded by macrophages, eosinophils, fibroblasts, etc. of the granuloma. Thus, the granuloma effectively isolates the egg from the surrounding tissue parenchyma and prevents damage caused by substances (enzymes) secreted by the embryo within. Over the years, liver and, to a lesser extent, lung granuloma macrophage functions have been characterized after enzymatic dispersal of the isolated granulomas. Judged by Wright stain, macrophages constitute one-third of the total granuloma population [54]. Macrophages appear heterogeneous with respect to esterase staining, size, maturity, intracellullar organelle content, and inclusions [55].

A. Macrophage Recruitment

On the assumption that macrophages of the granuloma are derived from the bone marrow but may enlarge their numbers by focally secreted cytokine(s), the number of macrophage progenitor cells in the bone marrow (GM-CFC) and liver granulomas (M-CFC) was assayed, and production of the cytokine colony-stimulating factor-1 (CSF-1) by isolated liver granuloma explants was measured. Colony counts obtained from murine bone marrow cells cultured with CSF-1 showed that macrophage progenitor cells peaked in numbers between 6 and 12 weeks of the infection, coinciding with the formation of large granulomas, and subsequently declined with the onset of down-regulation of the lesions. In contrast, granuloma macrophages showed a gradual increase in the number of progenitor cells between 6 and 20 weeks of the infection. Indeed, cells of the immunomodulated granuloma stimulated with recombinant CSF-1 produced a significantly higher number of colonies compared with their vigorous granuloma-derived counterparts. During the acute phase of the infection (6–10 weeks), CSF-1 levels in the circulation were elevated. Moreover, CSF-1 production by monolayers of adherent granuloma macrophages derived from the vigorous or the immunomodulated lesions was significantly higher in cultures of the latter. Thus, the bone marrow under the stimulus of CSF-1 produced high numbers of macrophage progenitor cells to sustain the vigorous stage of the granulomatous response. With the down-modulation of the inflammatory response, circulating CSF-1 declined to normal levels, but the amount of CSF-1 produced by adherent granuloma macrophages doubled, indicating that focal production of macrophages may be an important mechanism for the supply of monocytes/macrophages. Because CSF-1 is also capable of macrophage activation and induction of additional cytokine production, the focal production of this cytokine may play an important role in the course of the inflammation or fibrosis [56].

B. Accessory Cell Function

By immunofluorescence, the majority of liver and lung granuloma macrophages were shown to express major histocompatibility complex (MHC) class II, I-A, and I-E antigens [57]. Interestingly, both the vigorous and the immunomodulated granuloma macrophages expressed these antigens to an equal degree. The immune accessory function of granuloma macrophages was examined in both the homologous and the heterologous immune systems. At a lymphocyte-to-macrophage ratio of 30:1, both vigorous and modulated granuloma macrophages reconstituted, albeit not fully, the diminished proliferative responses of adherent cell–depleted mesenteric lymph node cells. Use of a lower cell ratio was inhibitory to blastogenesis. Reconstitution of the response of normal thymocytes to concanavalin A stimulus was similarly achieved. Inhibition studies with specific monoclonal antibodies confirmed that the accessory function in SEA presentation was mediated by I-A and I-E subclass MHC II antigens [58]. Granuloma macrophages at 1% cell concentration could also reconstitute in vitro the specific IgM and IgG plaque-forming antibody responses of burro erythrocyte–primed splenic cells. Similarly, the T cell proliferative responsiveness of accessory cell–depleted lymph node cells primed to GAT (a synthetic terpolymer peptide) was also achieved. In each case the macrophage-to-lymphocyte ratio was crucial, because higher concentrations of macrophages proved to be inhibitory [59]. Macrophages of the vigorous as well as the down-modulated liver granulomas were also capable of inducing a hapten-specific delayed hypersensitivity skin response in mice [60]. Granuloma macrophage culture supernatants were shown to contain a 13- to 21-kDa factor that costimulated mitogen-induced thymocyte proliferation, thus having the characteristics of an IL-1–like substance [61]. Serial determinations also demonstrated that maximal IL-1 secretion by lung granuloma macrophages occurs during the first 2 days after the formation of the lesions. Cells of the vigorous granulomas produced more IL-1 than those of the immunomodulated lesions [62].

The foregoing experiments demonstrated that the inflammatory macrophages of the granuloma also play an accessory role in a variety of immune events. Admixture experiments that used supraoptimal concentrations of macrophages from vigorous granulomas and lymph node cells also indicated a clear macrophage-mediated suppressive effect of lymphoproliferation. Cultures of granuloma cells with a 30% macrophage content derived from either the vigorous or the immunomodulated granulomas failed to respond to mitogenic (concanavalin A) or antigenic stimulation. Because previous experiments clearly demonstrated the presence and functionality of granuloma T cells [63], nonresponsiveness was attributed to the suppressor effect exerted by either the eosinophils or the macrophages of the granulomas. One cycle of plate-adherence removed over 80% of esterase-positive macrophages, leaving behind 15% to 20% lymphocytes, 10% macrophages, and about 70% eosinophils. Whereas such cultures acquired a low level of mitogenic responsiveness, antigen-specific proliferation was not obtained. However, two consecutive cycles of adherence that lowered the number of macrophages to 2% alleviated suppression and allowed good concavavalin A– or SEA-induced lymphoproliferation. Readdition of adherent macrophages to cell cultures of the vigorous granuloma depleted of macrophages suppressed the mitogenic response in a dose-dependent manner, confirming the suppressor effect. Partial relief of the suppression was achieved by the addition to the cultures of indomethacin and anti–IFN-α/β antiserum. It is noteworthy that cell cultures of the immunomodulated granulomas showed a lesser effect of alleviation than cells of the vigorous lesions. Whereas these experiments demonstrated that granuloma T cell proliferation is regulated by macrophage-secreted IFN-α/β and probably prostaglandin, it is noteworthy that the antagonists could not alleviate the suppression observed in cultures that contained the full complement of adherent macrophages. This may indicate that the weakly adherent macrophages removed after the first cycle of adherence exert suppression over the T cells by other as yet unidentified inhibitors or signals [64].

It was recently postulated that granuloma macrophages are capable of inducing anergy in the TH_1 type subset of cloned lymphocytes. This was based on experiments in which pigeon cytochrome C–reactive T cell clones cocultured with granuloma macrophages showed greatly diminished proliferative responses compared with the effect of normal splenic adherent cells. The poor stimulatory capacity of the granuloma macrophage was also demonstrated by showing that 10^3 to 10^4 times higher amounts of antigen fragments were needed to achieve stimulation of the lymphocytes comparable to that of splenic cells. Moreover, cocultures of T clones with granuloma macrophages in the presence of antigen fragment rendered the lymphocytes hyporesponsive to subsequent restimulation with normal splenic cells and the antigen. On the basis of these observations it was proposed that granuloma macrophages participate in the induction of TH_1 subset unresponsiveness during the down-modulation phase of schistosome infection. However, the above data are not incompatible with observations showing a macrophage-secreted interferon–mediated suppressive effect of the lymphoproliferative response [65].

C. Macrophage Activation

The activated state of macrophages of the vigorous liver granuloma has been demonstrated by a variety of observations. Vigorous granuloma macrophages expressed IgG and IgG2b-specific Fc receptors in over 95% of the population, and two-thirds of the adherent cells expressed C3R. Macrophages that displayed receptors also showed strong Fc and C3 receptor–mediated phagocytosis [54]. Fc receptor display is clearly related to the level of granuloma macrophage activation. Macrophages of the foreign-body type, from plastic bead–induced lung granulomas, displayed a significantly lesser percentage of receptor-positive cells compared with those derived from schistosome egg–induced granulomas [66]. In chronically infected mice, macrophages of the smaller immunomodulated granulomas consistently showed lower values in most of the above parameters, indicating a lower lever of cell activation. A further strong corollary for granuloma macrophage activation was obtained from experiments that examined tumoricidal or larvicidal actions. Macrophages of the vigorous granulomas at an effector–target cell ratio of 5 : 1 or 10 : 1 displayed 30% to 40% cytolytic activity to mouse mammary tumor cell lines. The observed cytolytic potential of the macrophages was lost upon prolonged in vitro culture, indicating that they were dependent on exogenous activation signals. Compared with macrophages of the vigorous granulomas, cells from the small immunomodulated lesions at the same effector–target cell ratio displayed significantly lesser cytolytic activity. Thus, diminished cytolytic activity could be directly related to the previously demonstrated lower level of activation of granuloma cells [67]. More relevantly, macrophages of the vigorous liver granulomas cultured for 24 hours with schistosomula showed a low level of larvicidal activity. When contact between cells and larvae was augmented by the addition of serum from infected mice and rabbit complement, larvicidal activity at 48 hours of culture significantly improved. It is interesting that the larvicidal action of such activated macrophages was expressed by the ADCC type of reaction. Demonstration of larvicidal capability of liver granulomas indicates that they may participate in the anthelmintic response during the reinfection of the host [55]. The activated state of the macrophages was further confirmed because macrophages of the vigorous granulomas in the absence of exogenous stimulus produced four times more O_2^- than cells of the immunomodulated lesions. Macrophages of schistosome egg–induced lung granulomas were also examined for prostaglandin E_2 (PGE_2) and hydroxyeicosatetraenoic acid (HETE) production. Cells of the vigorous lesions produced almost twice as much PGE_2 and more HETE than macrophages of the down-modulated granulomas. Production of HETE, a potent proinflammatory agent, and PGE_2, a down-regulator of T lymphocyte function, emphasizes the important potential of the granuloma macrophages to participate in both the inflammatory and the regulatory processes of the granulomatous response [57].

It is plausible that macrophages within the granuloma undergo activation by locally produced cytokines. The growing list of cytokines produced within granulomas includes migration inhibitory factor (MIF) [68], eosinophil stimulation promoter (ESP) [69], IL-2 [63], IL-4, IL-5 [70], IFN-γ [71], IL-1 [61,62], TNF-α [62], and CSF-1 [56]. Many of these cytokines may provide the activating signals by an autocrine or paracrine pathway. Because of the dynamics of lymphocyte activation, cytokines are produced at a higher level during the acute compared with the chronic stage of the granulomatous response. Thus macrophages receive stronger signals for activation within the vigorous as opposed to the down-modulated granulomas.

D. Tissue Fibrosis

Tissue fibrosis is a sequel to granulomatous inflammation and is a major contributory factor in the pathology of schistosomiasis. Explants of vigorous granulomas that contain significant deposits of collagen and other connective tissue matrix [72] cultured in serum-free medium elaborated soluble substances that stimulated fibroblast proliferation [73] and collagen and fibronectin synthesis [74]. Supernatant fluid of cultured, adherent granuloma macrophages also exerted chemotactic and proliferative effects on fibroblasts [75,76]. The large-molecular-weight chemotactic molecule showed antigenic similarity to fibronectin [75], whereas the proliferation-inducer molecule appeared to be distinct from IL-1. Thus granuloma macrophages are capable of triggering fibroblast movement and proliferation and can contribute to the fibrotic process. This process is in part regulated by neutral proteases (collagenase, elastase) that can break down deposited connective tissue and influence the extent of tissue fibrosis. Collagenase and elastase enzymes were shown to be secreted in equal amounts by the macrophages of the vigorous and immunomodulated granulomas. The enzyme-active substances showed the characteristics of trypsin-activatable metalloproteinases [77]. Granuloma macrophages also produce the protease inhibitors α_2-macroglobulin and α_1-protease inhibitor, the former being the predominant product. Macrophages of the vigorous granulomas produced significantly higher amounts of protease inhibitors compared with their counterparts of the immunomodulated lesions. Fractionation of supernatant fluids of granuloma explants showed that the protease inhibitors were mostly found complexed with collagenase and elastase enzymes [78]. Thus macrophages may also influence the tempo of connective tissue breakdown within granulomas.

In sum, macrophages can fulfill a highly protective role also against a large metazoan invader, i.e., schistosome worms and eggs. Within the granuloma, macrophages not only provide protection against parasite secretions, but also perform a variety of biologic functions that influence the duration of inflammation and the outcome of the pathology of the chronic granulomatous process. Whether different macrophage populations can carry out different tasks, or one macrophage upon the receipt of the proper signal can switch biologic functions, is still to be determined.

ACKNOWLEDGMENT

Supported by National Institutes of Health Grant AI-12913.

REFERENCES

1. Boros DL. Immunopathology of *Schistosoma mansoni* infection. Clin Microbiol Rev 1989; 2:250–69.
2. Damian RT. Immunity in schistosomiasis: a holistic view. Contemp Top Immunobiol 1984; 12:359–419.
3. Capron A, Dessain JP. Immunologic aspects of schistosomiasis. Annu Rev Med 1992; 43:209–18.
4. Sher A, Moser G. Immunologic properties of developing schistosomula. Am J Pathol 1981; 102:121–6.
5. Mangold BL, Dean DA. The migration and survival of gamma-irradiated *Schistosoma mansoni* larvae and the duration of host-parasite contact in relation to the induction of resistance in mice. Parasitology 1984; 88:249–66.

6. Wilson RA, Coulson PS. Lung-phase immunity to schistosomes: a new perspective on an old problem. Parasitol Today 1989; 5:274–8.
7. James SL, Natovitz PC, Farrar WL, Leonard EJ. Macrophages as effector cells of protective immunity in murine schistosomiasis: macrophage activation in mice vaccinated with radiation-attenuated cercariae. Infect Immun 1984; 44:569–75.
8. Pearce EJ, James SL. Post lung-stage schistosomula of *Schistosoma mansoni* exhibit transient susceptibility to macrophage-mediated cytotoxicity *in vitro* that may relate to late phase killing *in vivo*. Parasite Immunol 1986; 8:513–27.
9. McLaren DJ, James SL. Ultrastructural studies of the killing of schistosomula of *Schistosoma mansoni* by activated macrophages *in vitro*. Parasite Immunol 1985; 7:315–31.
10. Scott P, James SL, Sher A. The respiratory burst is not required for killing of intracellular and extracellular parasites by a lymphokine-activated macrophage cell line. Eur J Immunol 1985; 15:553–8.
11. James SL, Glaven JA. Effects of inhibitors of tumoricidal activity upon schistosomulum killing by activated macrophages. Infect Immun 1987; 55:3174–80.
12. James SL, Glaven J, Goldenberg S, Meltzer MS, Pearce E. Tumour necrosis factor (TNF) as a mediator of macrophage helminthotoxic activity. Parasite Immunol 1990; 12:1–13.
13. James SL, Glaven J. Macrophage cytotoxicity against schistosomula of *Schistosoma mansoni* involves arginine-dependent production of reactive nitrogen intermediates. J Immunol 1989; 143:4208–12.
13a. Oswald IP, Wynn TA, Sher A, James SL. Interleukin-10 inhibits macrophage microbicidal activity by blocking the endogenous production of tumor necrosis factor α required for interferon γ-induced activation. Proc. Natl. Acad. Sci (USA) 1992; 89:8676–80.
14. James SL, Hibbs JB Jr. The role of nitrogen oxides as effector molecules of parasite killing. Parasitol Today 1990; 6:303–305.
15. Gazzinelli RT, Oswald IP, James SL, Sher A. IL-10 inhibits parasite killing and nitrogen oxide production by IFNγ activated macrophages. J Immunol 1992; 148:1792–6.
15a. Oswald IP, Gazzinelli RI, Sher A, James SL. IL-10 synergizes with IL-4 and transforming growth factor β to inhibit macrophage cytotoxic activity. J Immunol 1992; 148:3578–82.
16. Dean DA. A review: *Schistosoma* and related genera: acquired resistance in mice. Exp Parasitol 1983; 55:1–104.
17. James SL, DeBlois LA. Induction of protective immunity against *Schistosoma mansoni* by a nonliving antigen. II. Response of mouse strains with selective immune defects. J Immunol 1986; 136:3864–71.
18. James SL. Induction of protective immunity against *Schistosoma mansoni* by a nonliving antigen. V. Effects of varying the immunization and infection schedule and site. Parasite Immunol 1987; 9:531–41.
19. Sher A, Hieny S, James SL, Asofsky R. Mechanisms of protective immunity against *Schistosoma mansoni* in mice vaccinated with irradiated cercariae. II. Analysis of immunity in hosts deficient in T lymphocytes, B lymphocytes and complement. J Immunol 1982; 128:1880–4.
20. Correa-Oliveira R, Sher A. Defective immunoglobulin M responses to vaccination or infection with *Schistosoma mansoni* in *xid* mice. Infect Immun 1985; 50:409–14.
21. Sher A, Correa-Oliveira R, Hieny S, Hussain R. Mechanisms of protective immunity against *Schistosoma mansoni* in mice vaccinated with irradiated cercariae. IV. Analysis of the role of IgE antibodies and mast cells. J Immunol 1983; 131:1460–5.
22. Vignali DA, Crocker P, Bickle QD, Cobbold S, Waldmann H, Taylor MG. A role for CD4+ but not CD8+ T cells in immunity to *Schistosoma mansoni* induced by 20 krad–irradiated and Ro 11-3128–terminated infections. Immunology 1989; 67:466–72.
23. Kelly E, Colley DG. *In vivo* effects of monoclonal anti-L3T4 antibody on immune responsiveness of mice infected with *Schistosoma mansoni*: reduction of irradiated cercariae–induced resistance. J Immunol 1988; 140:2737–45.
24. Sher A, Coffman RL, Hieny S, Cheever AW. Ablation of eosinophil and IgE responses with anti–IL-5 or anti–IL-4 antibodies fails to affect immunity against *Schistosoma mansoni* in the mouse. J Immunol 1990; 145:3911–6.
24a. Smythies LE, Coulson PS, Wilson RA. Monoclonal antibody to IFN-γ modifies pulmonary inflammatory responses and abrogates immunity to *Schistosoma mansoni* in mice vaccinated with attenuated cercariae. J Immunol 1992; 149:3654–58.
25. Vignali DA, Bickle QD, Taylor MG. Studies on immunity to *Schistosoma mansoni in vivo*: whole-body irradiation has no effect on vaccine-induced resistance in mice. Parasitology 1988; 96:49–61.

26. Caulada-Benedetti Z, Al-Zamel F, Sher A, James SL. Comparison of Th1 and Th2-associated immune reactivities stimulated by single versus multiple vaccination of mice with irradiated *Schistosoma mansoni* cercariae. J Immunol 1991; 146:1655–60.
27. James SL, Sher A. Mechanisms of protective immunity against *Schistosoma mansoni* in mice vaccinated with irradiated cercariae. III. Identification of a mouse strain, P/N, that fails to respond to vaccination. Parasite Immunol 1983; 5:567–75.
28. Boraschi D, Meltzer MS. Defective tumoricidal capacity of macrophages from P/J mice: characterization of the macrophage defect after *in vivo* and *in vitro* activation stimuli. J Immunol 1980; 125:771–6.
29. Boraschi D, Meltzer MS. Defective tumoricidal capacity of macrophages from P/J mice: tumoricidal defect involves abnormalities in lymphokine derived activation stimuli and in mononuclear phagocyte responsiveness. J Immunol 1980; 125:777–83.
30. James SL, Correa-Oliveira R, Leonard EJ. Defective vaccine-induced immunity to *Schistosoma mansoni* in P strain mice. II. Analysis of cellular responses. J Immunol 1984; 133:1587–93.
31. Correa-Oliveira R, Sher A, James SL. Defective vaccine-induced immunity to *Schistosoma mansoni* in P strain mice. I. Analysis of antibody responses. J Immunol 1984; 133:1581–6.
32. James SL, Deblois LA, Al-Zamel F, Glaven J, Langhorne J. Defective vaccine-induced immunity to *Schistosoma mansoni* in P strain mice. III. Specificity of the associated defect in cell-mediated immunity. J Immunol 1986; 137:3959–67.
33. James SL, Correa-Oliveira R, Sher A, Medvitz L, McCall D. Genetic association of defects in macrophage larvicidal activity and vaccine-induced resistance to *Schistosoma mansoni* in P strain mice. Infect Immun 1987; 55:1884–9.
34. Menson EN, Wilson RA. Lung-phase immunity to *Schistosoma mansoni*: definition of alveolar macrophage phenotypes after vaccination and challenge of mice. Parasite Immunol 1990; 12:353–66.
35. Aitken R, Coulson PS, Wilson RA. Pulmonary leukocytic responses are linked to the acquired immunity of mice vaccinated with irradiated cercariae of *Schistosoma mansoni*. J Immunol 1988; 140:3573–9.
36. Smythies LE, Pemberton RM, Coulson PS, Mountford AP, Wilson RA. T cell–derived cytokines associated with pulmonary immune mechanisms in mice vaccinated with irradiated cercariae of *Schistosoma mansoni*. J Immunol 1992; 148:1512–18.
37. Pemberton RM, Smythies LE, Mountford AP, Wilson RA. Patterns of cytokine production and proliferation by T lymphocytes differ in mice vaccinated or infected with *Schistosoma mansoni*. Immunology 73:327–333.
38. Wilson RA, Coulson PS, Dixon B. Migration of the schistosomula of *Schistosoma mansoni* in mice vaccinated with radiation-attenuated cercariae, and normal mice: an attempt to identify the timing and site of parasite death. Parasitology 1986; 92:101–16.
39. McLaren DJ. Will the real target of immunity to schistosomiasis please stand up. Parasitol Today 1989; 5:279–82.
40. Greene MI, Benacerraf B. Studies on hapten specific T cell immunity and suppression. Immunol Rev 1980; 50:163–86.
41. Pearce EJ, James SL, Hieny S, Lanar D, Sher A. Induction of protective immunity against *Schistosoma mansoni* by vaccination with schistosome paramyosin (Sm97), a non-surface parasite antigen. Proc Natl Acad Sci USA 1988; 85:5678–82.
42. James SL. Induction of protective immunity against *Schistosoma mansoni* is dependent on the method of antigen presentation. J Immunol 1985; 134:1956–60.
43. James SL, Pearce EJ. The influence of adjuvant on induction of protective immunity by a non-living vaccine against schistosomiasis. J Immunol 1988; 140:2753–9.
44. James SL. Induction of protective immunity against *Schistosoma mansoni* by a nonliving antigen. III. Correlation of resistance with induction of activated larvicidal macrophages. J Immunol 1986; 136:3872–7.
45. James SL, Sher A. Cell-mediated immune response to schistosomiasis. Curr Top Microbiol Immunol 1990; 155:21–31.
46. Cottrell B, Pye C, Butterworth A. Cytotoxic effects *in vitro* of human monocytes and macrophages on schistosomula of *Schistosoma mansoni*. Parasite Immunol 1989; 11:91–104.
47. James SL, Woods Cook K, Lazdins JK. Activation of human monocyte-derived macrophages to kill schistosomula of *Schistosoma mansoni in vitro*. J Immunol 1990; 145:2686–90.
48. Colley DG, Barsoum IS, Dahawi HSS, Gamil F, Habib M, El Alamy MA. Immune responses and

immunoregulation in relation to human schistosomiasis in Egypt III. Immunity and longitudinal studies of *in vitro* responsiveness after treatment. Trans Roy Soc Trop Med Hyg 1986; 80:952–7.
49. Ellner JJ, Mahmoud AAF. Cytotoxicity of activated macrophages for the multicellular parasite. *Schistosoma mansoni*. Lymphokines 1981; 3:231–56.
50. Boros DL, Warren KS. Delayed hypersensitivity granuloma formation and dermal reaction induced and elicited by a soluble factor isolated from *Schistosoma mansoni* eggs. J Exp Med 1970; 132:488–507.
51. Boros DL, Pelley RP, Warren KS. Spontaneous modulation of granulomatous hypersensitivity in schistosomiasis mansoni. J Immunol 1975; *144*:1437–43.
52. Colley DG. Immune responses to a soluble schistosomal egg antigen preparation during chronic primary infection with *Schistosoma mansoni*. J Immunol 1975; 115:150–6.
53. Boros DL. Immunoregulation of granuloma formation in murine schistosomiasis mansoni. Ann NY Acad Sci 1986; 465:313–23.
54. Wellhausen SR, Boros DL. Comparison of Fc, C3 receptors and Ia antigens on the inflammatory macrophage isolated from vigorous or immunomodulated liver granulomas of schistosome-infected mice. J Reticuloendothel Soc 1981; 30:191–203.
55. McLaren DJ, Boros DL. *Schistosoma mansoni*: schistosomulicidal activity of macrophages isolated from liver granulomas of infected mice. Exp Parasitol 1983; 56:346–57.
56. Clark CR, Chen BDM, Boros DL. Macrophage progenitor cell and colony-stimulating factor production during granulomatous schistosomiasis mansoni in mice. Infect Immun 1988; 56:2680–5.
57. Chenuse SW, Kunkel SL, Higashi GI, Ward PA, Boros DL. Production of superoxide anion, prostaglandins and hydroxyeicosatetraenoic acids by macrophages from hypersensitivity-type (*Schistosoma mansoni* egg) and foreign body–type granulomas. Infect Immun 1983; 42:1116–25.
58. Elliott DE, Boros DL. Schistosome egg antigen(s) presentation and regulatory activity by macrophages isolated from vigorous or immunomodulated liver granulomas of *Schistosoma mansoni*–infected mice. J Immunol 1984 132:1506–10.
59. Schook LB, Wellhausen SR, Boros DL, Niederhuber JE. Accessory cell function of liver granuloma macrophages of *Schistosoma mansoni*–infected mice. Infect Immun 1983; 42:882–6.
60. Sunday ME, Stadecker MJ, Wright JA, Aoki I, Dorf ME. Induction of immune responses by schistosome granulomas macrophages. J Immunol 1983; 130:2413–7.
61. Elliott EE, Righthand VF, Boros DL. Characterization of regulatory (interferon α/β) and accessory (LAF/IL-1) monokine activities from liver granuloma macrophages of *S. mansoni* infected mice. J Immunol 1987; 138:2653–62.
62. Chensue SW, Otterness IG, Higashi GI, Forsch CS, Kunkel SL. Monokine production by hypersensitivity (*Schistosoma mansoni* egg) and foreign body (Sephadex bead)–type granuloma macrophages. Evidence for sequential production of IL-1 and tumor necrosis factor. J Immunol 1989; 142:1281–6.
63. Ragheb S, Boros DL. Characterization of granuloma T lymphocyte function from *Schistosoma mansoni*–infected mice. J Immunol 1989; 142:3239–46.
64. Elliott DE, Ragheb S, Wellhausen SR, Boros DL. Interactions between adherent mononuclear cells and lymphocytes from granulomas of mice with schistosomiasis mansoni. Infect Immun 1990; 58:1577–83.
65. Stadecker MJ, Kamisato JK, Chikunguwo SM. Induction of T helper cell unresponsiveness to antigen by macrophages from schistosomal egg granulomas. A basis for immunomodulation in schistosomiasis? J Immunol 1990; 145:2696–700.
66. Amsden AF, Boros DL. Fc receptor bearing macrophages isolated from hypersensitivity and foreign body granulomas. Delineation of macrophage dynamics, Fc receptor density/avidity and specificity. Am J Pathol 1979; 96:457–76.
67. Loveless SE, Wellhausen SR, Boros DL, Heppner GH. Tumoricidal macrophages isolated from liver granulomas of *Schistosoma mansoni*–infected mice. J Immunol 1982; 128:284–8.
68. Boros DL, Warren KS, Pelley RP. The secretion of migration inhibitory factor by intact schistosome egg granulomas maintained. Nature 1973; 246:224–6.
69. James SL, Colley DG. Eosinophils and immune mechanisms: production of the lymphokine eosinophil stimulation promoter (ESP) *in vitro* by isolated intact granulomas. J Reticuloendothel Soc 1975; 18:283–93.
70. Grzych J-M, Pearce E, Cheever A, Caulada ZA, Caspar P, Heiny S, Lewis F, Sher A. Egg deposition is the major stimulus for production of Th2 cytokines in murine schistosomiasis mansoni. J Immunol 1991; 146:1322–7.

71. Chensue SW, Terebuh PD, Warmington KS, Hershey SK, Evanoff HL, Kunkel SL, Higashi GI. Role of IL-4 and IFNγ in egg-induced hypersensitivity granuloma formation. Orchestration, relative contribution, and relationship to macrophage function. J Immunol 1992; 148:900–6.
72. Grimaud JA, Boros DL, Takiya C, Mathew RC, Emonard H. Collagen isotypes, laminin and fibronectin in granulomas of the liver and intestines of *Schistosoma mansoni*–infected mice. Am J Trop Med Hyg 1987; 37:335–44.
73. Wyler DJ, Wahl SM, Wahl LM. Hepatic fibrosis in schistosomiasis: egg granulomas secrete fibroblast stimulating factor *in vitro*. Science 1978; 202:438–40.
74. Wyler DJ, Ehrlich HP, Postlethwaite AE, Raghow R, Murphy MM. Fibroblast stimulation in schistosomiasis. VII. Egg granulomas secrete factors that stimulate collagen and fibronectin synthesis. J Immunol 1987; 138:1581–6.
75. Wyler DJ, Postlethwaite AE. Fibroblast stimulation in schistosomiasis. IV. Isolated egg granulomas elaborate a fibroblast chemoattractant *in vitro*. J Immunol 1983; 130:1371–5.
76. Wyler DJ, Stadecker MJ, Dinarello CA, O'Dea JF. Fibroblast stimulation in schistosomiasis. V. Egg granuloma macrophages spontaneously secrete a fibroblast-stimulating factor. J Immunol 1985; 132:3142–8.
77. Truden JL, Boros DL. Collagenase, elastase and nonspecific protease production by vigorous or immunomodulated liver granulomas and granuloma macrophages/eosinophils of *S. mansoni* infected mice. Am J Pathol 1985; 121:166–75.
78. Truden JL, Boros DL. Detection of α_2-macroglobulin, α_1-protease inhibitor and neutral protease-antiprotease complexes within liver granulomas of *S. mansoni* infected mice. Am J Pathol 1988; 130:281–8.

28

Toxoplasma–Macrophage Interactions

Miles H. Beaman,* Carlos S. Subauste, Sin-Yew Wong, and Jack S. Remington

Stanford University School of Medicine and Palo Alto Medical Foundation, Palo Alto, California

I. INTRODUCTION

The coccidian parasite *Toxoplasma gondii* is an important opportunistic pathogen of immunocompromised patients (including those with malignancies, organ transplants, and the acquired immune deficiency syndrome [AIDS]) [1] and newborn infants [2,3]. Cell-mediated immunity has been shown to be critical for defense against this pathogen, and macrophages (Mϕs) have been demonstrated to be major contributors to this defense [4]. The role of Mϕs in *T. gondii* infection has been studied for more than two decades, and in vitro models of Mϕ activation and killing have been widely used because the organism is easily visualized microscopically, multiplies readily within resident murine macrophages, and may be inhibited or killed by activated Mϕs. Initially, we will review the nonspecific effector function of resident and activated Mϕs in *T. gondii* infection in vitro. Because the toxoplasmacidal activity of Mϕs varies in cells from different anatomic compartments and the different animals studied, each will be reviewed in turn. Next, the mechanisms of toxoplasmacidal activity in Mϕs will be reviewed, and finally, the different cytokines that modulate Mϕ function during *T. gondii* infection will be discussed.

II. EFFECT OF MACROPHAGE PHENOTYPE ON RESISTANCE TO *TOXOPLASMA GONDII*

A. Resident, Activated, and Elicited Macrophages in *Toxoplasma gondii* infection

The term "resident" or "normal" Mϕ refers to macrophages harvested from animals without prior infection or use of eliciting agents. Normal Mϕs cultured in vitro and stimulated with soluble lymphocyte products (which contain cytokines such as interferon-γ [IFN-γ]) or lipopolysaccharide (LPS) are functionally altered and exhibit increased microbicidal activity, phagocytic capacity, and tumoricidal activity and are termed activated Mϕs. Macrophages from animals infected with *T. gondii* show similar functional alterations and are also called activated Mϕs. The term "elicited" Mϕ is used for Mϕs harvested from animals stimulated with an agent such as proteose-peptone, thioglycolate, or other inflammatory agents. These elicited Mϕs are also altered functionally and exhibit toxoplasmacidal activity.

B. Murine Peritoneal Macrophages

Resident murine peritoneal Mϕs readily support intracellular multiplication of *T. gondii* in a unrestricted fashion. In contrast, activated murine peritoneal Mϕs can kill and inhibit *T. gondii* [4].

**Current affiliation:* Fremantle Hospital, University of Western Australia, Perth, Australia

Studies from our laboratory have revealed that mice chronically infected with *T. gondii* are resistant to antigenically and phylogenetically diverse organisms including bacteria [5–8], fungi [9], viruses [10,11], and helminths [12,13]. Although the activation process is specific and T cell–mediated, the effector function of the activated Mϕ is nonspecific and confers upon the host nonspecific resistance to infection by other organisms [14]. Activated peritoneal Mϕs from mice chronically infected with *T. gondii* are resistant in vitro to *Listeria monocytogenes* [6], Mengovirus [10], and *Cryptococcus* [9] infection. In addition, activated murine peritoneal Mϕs exhibit nonspecific cytostasis against tumor target cells [15]. Resident murine peritoneal Mϕs kill *T. gondii* tachyzoites if they are coated with specific antibody [16–20].

Of interest is the fact that activated murine peritoneal Mϕs can lose their toxoplasmacidal activity if chronically infected mice are experimentally infected with high titers of murine cytomegalovirus [11], and this correlates with reactivation of *T. gondii* disease.

C. Murine Alveolar Macrophages

Different toxoplasmacidal activity of resident and activated alveolar Mϕs has been obtained when different strains of mice were used. Ryning et al. [21,22] found that alveolar Mϕs from chronically infected Swiss-Webster mice killed or inhibited multiplication of *T. gondii* more readily than alveolar Mϕs from uninfected mice. These activated alveolar Mϕs had greater toxoplasmacidal activity only during the period when there was an intense inflammatory infiltrate in the lungs. Once the inflammatory process resolved, the microbicidal activity of alveolar Mϕs from infected mice was no different from that of uninfected mice [21]. When chronically infected Balb/c mice were used, Pomeroy et al. demonstrated that bronchoalveolar and lung digest macrophages were permissive to *T. gondii* infection and replication even during periods of inflammation in the lung [11], suggesting the possibility of different Mϕ subsets within the pulmonary compartment or differing states of activation in acute and chronic inflammation. In contrast, activated peritoneal Mϕs from both Swiss-Webster and Balb/c mice chronically infected with *T. gondii* continued to have toxoplasmacidal activity throughout infection [11,21].

D. Human Tissue Macrophages

Resident human alveolar [23,24] and pelvic peritoneal [24,25] Mϕs have remarkable ability to kill or inhibit *T. gondii* in vitro. Unlike the situation in murine tissue Mϕs, human tissue Mϕs from individuals with or without prior *T. gondii* infection had similar toxoplasmacidal activity [23,25]. Alveolar Mϕs from AIDS patients displayed similar intracellular antimicrobial activity and responsiveness to in vitro activation by T lymphocyte products and IFN-γ when compared with those from non-AIDS patients. These results suggest that there was no inherent defect in the tissue Mϕs from AIDS patients and that they were capable of toxoplasmacidal activity if appropriately activated [26].

E. Human Monocyte-Derived Macrophages

Monocytes are the precursors of tissue Mϕs [27]. Monocytes from peripheral blood cultured in vitro exhibit the morphologic appearance of tissue Mϕs and are termed human monocyte-derived Mϕs (HMDMs). The role of monocytes and HMDMs in resistance against *T. gondii* has been studied extensively. Disparate observations have been reported by various investigators and most likely result from different durations of in vitro culture before infection. Normal HMDMs cultured in vitro for 7 to 10 days were incapable of significant killing or inhibition of *T. gondii* [28–31]. Studies using seropositive and seronegative donors showed similar results [30]. In contrast, freshly isolated human monocytes cultured for less than 24 hours before challenge degraded and retarded multiplication of intracellular *T. gondii* [23,32]. These results revealed that monocytes cultured in vitro for 7 or more days lose their ability to kill *T. gondii*. Interestingly, other authors have shown that prolonged in vitro culture is required before HMDMs are capable of being activated by some cytokines, such as interferon-β, to inhibit *T. gondii* replication [33]. Since maturation of monocytes in vivo may differ from their maturation in vitro, HMDMs may have limitations as a model for studying the resistance of human tissue Mϕs in *T. gondii* infection.

When HMDMs were cultured in the presence of lymphocytes from infected individuals and toxoplasma lysate antigen (TLA) or streptokinase-streptodornase (SK-SD), they were stimulated to kill *T. gondii* [30,34]. These findings were later found to be due to the elaboration of Mϕ-activating factors by lymphocytes and will be discussed below.

It is of interest that HMDMs readily killed *T. gondii* that were coated with immune sera [17,30].

F. Rat Peritoneal Macrophages

Unlike mice and newborn rats, adult rats are highly resistant to *T. gondii*. Both resident [35] and elicited [36] peritoneal Mϕs from adult rats killed *T. gondii*. McCabe et al. [35] reported two other interesting observations. When Mϕs of adult rats were cultured for a longer period, they appeared to lose some of their toxoplasmacidal activity (90.4% *T. gondii* killed if the duration of culture before challenge was 5 hours compared with 53.2% killing for Mϕs cultured for 168 hours). Macrophages from neonatal rats were infected with and killed *T. gondii* to the same extent as Mϕs from adult rats. This result suggests that the susceptibility of neonatal rats to *T. gondii* infection is not the result of a lack of toxoplasmacidal activity of their Mϕs.

G. Rat Alveolar Macrophages

Similar to our observations with resident human alveolar Mϕs, resident rat alveolar Mϕs cultured in vitro kill *T. gondii* [37].

III. MECHANISMS OF INHIBITION AND KILLING OF *TOXOPLASMA GONDII*

A. Oxidative Mechanisms

Human and murine mononuclear phagocytes utilize oxygen-dependent mechanisms to kill or inhibit intracellular growth of *T. gondii* [17,38–40]. Murray et al. evaluated the susceptibility of extracellular tachyzoites to selected oxygen intermediates and reported that parasites were resistant to hydrogen peroxide (H_2O_2), most likely because of the presence of parasite catalase [38]. Although neither superoxide (O_2^-) nor H_2O_2 alone was toxoplasmacidal, their interaction was required for killing of the parasite. Their results suggested that O_2^- and H_2O_2 are precursor metabolites in the killing of *T. gondii* and that hydroxyl radical and singlet oxygen are directly toxoplasmacidal.

Murray et al. reported a direct correlation between the capacity of Mϕs to generate oxidative metabolites and the influence of these cells on the intracellular fate of *T. gondii* [39]. Whereas normal murine peritoneal Mϕs released little H_2O_2 and allowed unrestricted multiplication of intracellular *T. gondii*, peritoneal Mϕs from mice chronically infected with *T. gondii* released four times more H_2O_2 and inhibited intracellular multiplication of this parasite. These investigators also observed that incubation of activated murine peritoneal Mϕs with scavengers of O_2^-, H_2O_2, $OH\bullet$, and $^1O_2$ resulted in reversal of their anti–*T. gondii* activity [39]. Normal peritoneal Mϕs were found to inhibit intracellular multiplication of *T. gondii* when an exogenous source of oxidative metabolites was present.

Further evidence on the importance of oxygen-dependent mechanisms against *T. gondii* came from studies performed by Wilson et al. [17]. They determined the relationship between intracellular survival of *T. gondii* and production of toxic oxygen metabolites by human and murine mononuclear phagocytes. They demonstrated that there was only a partial correlation between the anti–*T. gondii* activity of different mononuclear phagocytes and their generation of oxygen reactive metabolites after phagocytosis of either *Candida albicans* or latex particles (Table 1). In contrast, when these cells internalize *T. gondii*, a good correlation between anti–*T. gondii* activity and increased generation of oxygen metabolites was observed. These results would suggest that the ability of *T. gondii* to survive within certain mononuclear phagocytes was dependent at least in part on the lack of production of reactive oxygen metabolites by these cells.

1. Inhibition of Phagolysosome Fusion

Following phagocytosis of microorganisms, delivery of lysosomal contents to the phagocytic vacuole is a phenomenon that correlates with death of such microorganisms. As described above, however, *T. gondii* is capable of surviving and multiplying within phagocytic cells. Jones et al. conducted studies to assess the delivery of lysosomal contents to vacuoles containing living or dead tachyzoites and concluded that the inability of peritoneal Mϕs from normal mice to kill in-

TABLE 1 Quantitative Reduction of NBT[a]

	Phagocytic stimulus				
	None	*Candida*	Latex particles	*Toxoplasma*	Δ *Toxoplasma*/Δ *Candida*[d]
Cell preparation	NBT red.[b]	Δ NBT red[c]	Δ NBT red.	Δ NBT red.	
Human mononuclear (11)[e]	2.9 ± 1.0[i]	43.6 ± 4.5	39.5 ± 8.3	11.2 ± 1.4	0.25 ± 0.04
Human macrophage (4)	2.1 ± 1.7	40.0 ± 8.3	72.1 ± 13.6	1.9 ± 0.7[j]	0.04 ± 0.01[j]
Mouse mononuclear					
Normal[f] (16)	1.9 ± 0.2	17.9 ± 1.7[k]	9.5 ± 2.9	1.9 ± 0.5[l]	0.08 ± 0.02[l]
C. parvum[5] (4)	4.3 ± 0.7	20.7 ± 5.2	ND[m]	6.7 ± 2.1[n]	0.31 ± 0.04[n]
Chronic Toxoplasma[h] (5)	3.8 ± 0.5	21.4 ± 1.9	ND[m]	7.1 ± 2.4[o]	0.31 ± 0.09[o]

[a]Reaction mixtures (in duplicate) contained 0.1 ml of NBT (2 mg/ml); 0.25 ml of the suspension of *Candida*; 2-μm latex particles, or *Toxoplasma* (or, in controls, 0.25 ml KRPG); 0.5 ml of cell preparations that contained 1.5×10^6 monocytes or macrophages; and sufficient KRPG to effect a final volume of 1 ml in siliconized glass tubes. The final ratio of organisms or particles to cells was 10 : 1 for *Candida*, ≈600 : 1 for 2-μm latex particles, and 25 : 1–30 : 1 for *Toxoplasma*.
[b]Nanomoles of NBT reduced (red.) per 10^6 monocytes or macrophages per 30 min.
[c]Increase in nanomoles of NBT reduced per 10^6 monocytes or macrophages per 30 min after subtraction of values for resting leukocytes and for organisms alone. *Candida* and latex particles alone did not reduce NBT. NBT reduction by *Toxoplasma* preparations alone was 2.4 ± 0.5, 0.3 ± 0.3, and 1.2 ± 0.5 nmol/30 min for experiments with human mononuclear cell preparations (MCP), human macrophage cell preparations, and mouse MCP, respectively.
[d]Increase in NBT reduction during phagocytosis of *Toxoplasma*/increase in NBT reduction during phagocytosis of *Candida*.
[e]Numbers in parentheses are the number of experiments performed with controls. *Candida*, and *Toxoplasma*: the number of experiments with latex particles was 4, 2, and 2 for human MCP, human macrophage cell preparations, and normal mouse MCP, respectively.
[f]MCP from normal mice.
[g]MCP from mice given 1,400 μg formalin-killed *Corynebacterium parvum* intraperitoneally 7 d previously.
[h]MCP from mice chronically infected with C56 strain of *Toxoplasma*.
[i]Mean ± SEM.
[j]$p < 0.005$ compared with human MCP.
[k]$p < 0.001$ compared with human MCP and human macrophage cell preparations.
[l]$p < 0.001$ compared with human MCP.
[m]ND, not determined.
[n]$p < 0.02$ compared with mouse MCP from age-matched normal controls that were studied in parallel.
[o]$p < 0.05$ compared with mouse MCP from age-matched normal controls that were studied in parallel.
Source: Ref. 17

tracellular *T. gondii* was linked to the lack of fusion of lysosomal granules with vacuoles containing *T. gondii* [41]. In contrast, lysosomal labels were detected in vacuoles that contained dead parasites. It appeared that these differences were due to some property of the parasite rather than of the Mϕ, since frequently vacuoles containing digested parasites and lysosomal markers were located in the same Mϕ near to vacuoles with viable *T. gondii* and no lysosomal markers.

Wilson et al. reported that both fusion of lysosomes with phagosomes that contained *T. gondii* and increased oxidative metabolic activity paralleled destruction of the parasites in human monocytes and in activated murine Mϕ [17]. They suggested that lysosomal contents might act in concert with reactive oxygen metabolites as a first-line mechanism of defense that allows mononuclear phagocytes to kill intracellular *T. gondii*.

Joiner et al. have recently reported that the intracellular fate of *T. gondii* is altered by modifying the route of entry of the parasites with engagement of receptors for the Fc region of immunoglobulins on the surface of the host cell [42]. In Chinese hamster ovary cells transfected with a macrophage-lymphocyte Fc receptor, vacuoles that contained viable *T. gondii* did not fuse with lysosomes. However, when tachyzoites were precoated with anti–*T. gondii* antibodies, fusion of the vacuoles with lysosomes was observed. Unlike Mϕs, this occurred in the absence of a respiratory burst. Other authors have proposed that the ability to avoid fusion with lysosomes is probably due to modifications of the vacuole membrane at the time of infection [43].

2. *Antibody-Dependent Killing of* Toxoplasma gondii

Antibody also plays an important role in resistance against *T. gondii* [44]. Anti–*T. gondii* antibody in the presence of complement results in the killing of extracellular parasites [45]. Whereas treatment of the extracellular tachyzoites with heat-inactivated serum (to remove complement activity) does not affect their viability, it increases the rate of infection of normal Mϕs [20]. This may be due to the presence of Fc receptors on the surface of Mϕs [46]. In contrast, parasites exposed to anti–*T. gondii* antibody were not capable of infecting fibroblasts or HeLa cells [21]. The most important effect of coating parasites with anti–*T. gondii* antibody is that it empowered normal Mϕs to inhibit or kill the organisms once they were intracellular [16,18,21,30,47]. Anderson et al. reported that once *T. gondii* had infected HMDMs, addition of anti–*T. gondii* antibodies did not alter the inability of these Mϕs to kill the parasite [30]. However, when *T. gondii* was incubated with anti–*T. gondii* antibodies before infection of Mϕs, the intracellular organisms were inhibited or killed by these Mϕs (Figure 1). This phenomenon required the incubation of parasites with *T. gondii*–specific antibodies [16]. A similar alteration of the intracellular fate of antibody-coated *T. gondii* has been observed in murine peritoneal and alveolar Mϕs [16,18,21].

The altered intracellular fate of antibody-coated *T. gondii* might be explained by the fact that this treatment resulted in fusion of lysosomes with phagocytic vacuoles [18]. In addition, coating with specific antibodies acts as a trigger for the respiratory burst of HMDMs and normal murine peritoneal Mϕs [17].

Erbe et al. reported that whereas human myeloid cells (monocytes, peritoneal Mϕs, neutrophils, and eosinophils) mediated lysis of the parasites in the absence of *T. gondii* antibodies [48], lysis was significantly increased after opsonization with anti–*T. gondii* antibodies. Monocytes effected this lysis at least partially through nonphagocytic mechanisms. Whereas monocytes treated to inhibit phagocytosis were not able to kill unopsonized *T. gondii*, such treatment only partially inhibited killing of opsonized parasites. The role that the receptors for the Fc region of IgG (FcγRs) played in the killing was studied using bispecific antibodies made by cross-linking anti–*T. gondii* antibodies to antibodies specific for human leukocyte surface antigens [48]. Whereas killing was significantly increased by bispecific antibodies that targeted the parasite to all three FcγRs (FcγRI–FcγRIII), lysis was also enhanced in the presence of bispecific antibodies that targeted *T. gondii* to non-FcγR antigens. These results indicated that engagement of FcγRs on the surface of monocytes was not necessary for increasing antibody-dependent lysis of *T. gondii*.

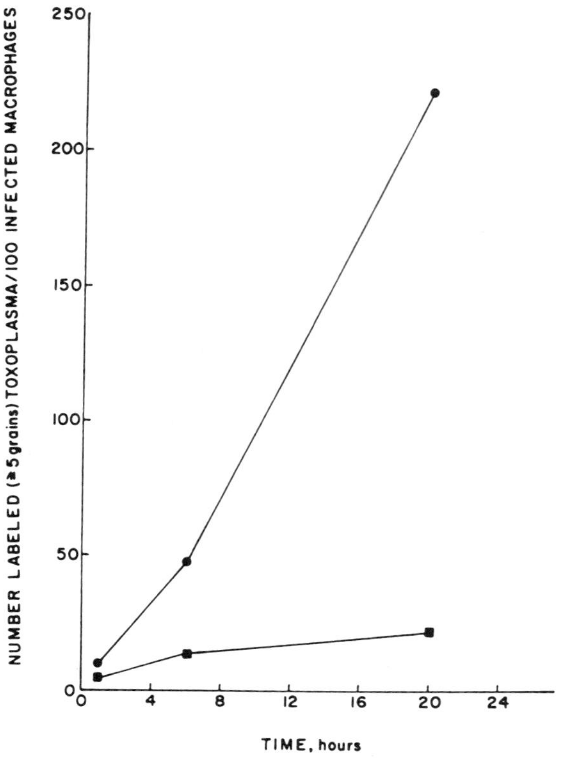

FIGURE 1 Effect of normal human monocyte-derived macrophages on *Toxoplasma gondii* that had been exposed to anti–*T. gondii* antibody before infection. (●) Macrophages infected with *T. gondii* that had not been exposed to antibody; (■) macrophages infected with *T. gondii* that had been exposed to antibody. (From Ref. 30.)

B. Nonoxidative Mechanisms

There is evidence that in addition to oxygen-dependent mechanisms, nonoxidative mechanisms play a significant role in inhibition and killing of *T. gondii* in both phagocytic and "nonprofessional" phagocytic cells [17,24,28,37,49]. When Wilson et al. studied the killing of *T. gondii* by monocytes from a patient with chronic granulomatous disease (CGD) and compared the results with those obtained in monocytes from a normal individual, they found that monocytes from the CGD patient still killed *T. gondii*, although less efficiently than normal monocytes [17] (Figure 2). In addition, nonlysosomal biochemical processes appeared to be involved, since this anti–*T. gondii* activity was not blocked by incubation of the monocytes with inhibitors of either lysosomal cationic proteins, lysosomal neutral proteases, or myeloperoxidase.

Even though the results with monocytes from CGD patients suggest that an oxygen-independent mechanism also participated in killing of *T. gondii*, this was not absolute proof, since these cells can release small levels of oxygen metabolites [17,50]. Studies conducted by Catterall et al. with rat alveolar Mϕs and human alveolar and peritoneal Mϕs provided further evidence for the existence of oxygen-independent mechanisms [24,37]. They demonstrated that these cells kill *T. gondii* in the absence of a demonstrable respiratory burst. Scavengers of toxic oxygen metabolites did not have an effect on the toxoplasmacidal activity of these cells.

To examine whether oxygen-independent mechanisms are needed for optimal anti–*T. gondii* activity in normal mononuclear phagocytes, Murray et al. also studied fresh monocytes from

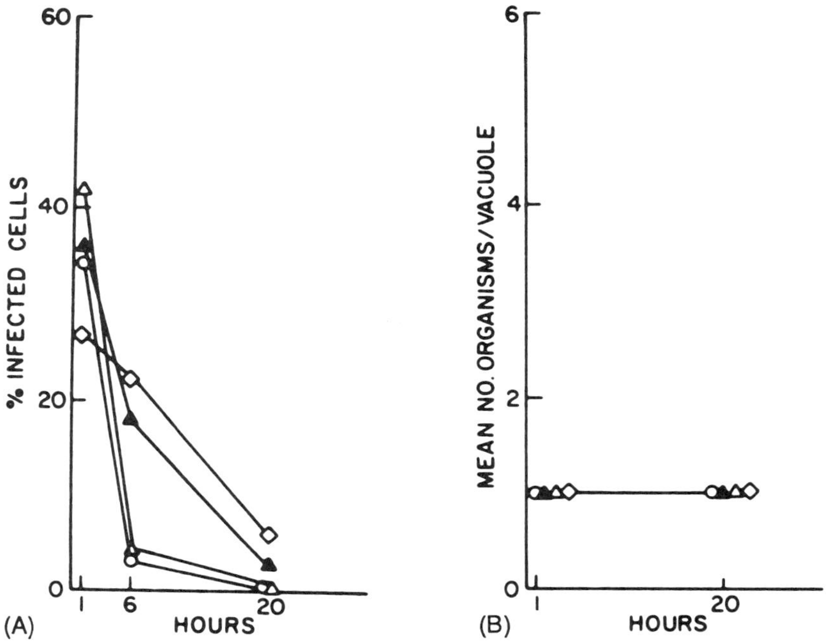

FIGURE 2 Fate of *Toxoplasma gondii* within monocytes from a child with X-linked chronic granulomatous disease (▲), his heterozygote mother (◇), his normal sister (△), and an unrelated normal control (○). The abscissa shows the time after addition of *T. gondii* to monolayers. (A) Percentage of cells that contained *T. gondii*. (B) Replication of *T. gondii*, expressed as mean number of organisms per vacuole. (From Ref. 17.)

normal subjects and CGD patients and infected them with *T. gondii* after cultivation in vitro [28]. The anti–*T. gondii* activity and capacity to generate O_2^- and H_2O_2 of normal monocytes persisted throughout the first 72 hours of in vitro culture. In contrast, CGD monocytes lost their anti–*T. gondii* activity after 48 hours, indicating a waning of oxygen-independent mechanisms. The authors interpreted these results as suggesting that oxygen-independent mechanisms are not necessary for normal mononuclear phagocytes to exert their anti–*T. gondii* activity.

Locksley et al. have reported that infection of resident macrophages results in an alteration of lipoxygenase products released by Mϕs [51]. Incubation of murine resident peritoneal Mϕs with either calcium ionophore or opsonized zymosan resulted in release of leukotrienes (LT) B_4 and C_4, and LTC_4 and LTD_4, respectively. In contrast, when Mϕs were challenged with viable tachyzoites of *T. gondii*, production of hydroxyeicosatetranoic acids and an unidentified compound that appeared to be a lipoxygenase product was observed. However, under conditions in which there is inhibition of intracellular multiplication of the parasites, such as when the tachyzoites were preincubated with anti–*T. gondii* antibodies or when activated Mϕs were used, production of LTD_4 predominated. The authors suggested that the inability of resident Mϕs to produce leukotrienes when infected with *T. gondii* may explain the relative absence of neutrophils and monocytes from the inflammatory response elicited during infection with *T. gondii*.

Pfefferkorn et al. demonstrated that increased activity of indoleamine 2,3-dioxygenase (IDO) in human fibroblasts is one oxygen-independent mechanism of inhibition of intracellular replication of *T. gondii* induced by IFN-γ [52,54]. Induction of this enzyme results in degradation of tryptophan and starvation of *T. gondii* for this amino acid [52–56]. Experiments performed by Murray et al. revealed that tryptophan degradation by IDO also plays a role as an oxygen-independent mechanism of inhibition of *T. gondii* growth in certain human mononuclear phagocytes [57]. Although monocytes, HMDMs, and alveolar Mϕs from normal humans and monocytes and HMDMs from patients with CGD degraded tryptophan upon exposure to IFN-γ, repletion of tryptophan achieved variable effects on anti–*T. gondii* activity (Table 2). Increasing the concentration of tryptophan in the culture medium only partially reversed the antiparasitic activity of IFN-γ–treated HMDMs from normal individuals and CGD patients, but failed to induce any significant change in the activity of IFN-γ–treated human alveolar Mϕs. Furthermore, treatment of unstimulated normal and CGD monocytes with tryptophan did not affect their anti–*T. gondii* activity. The authors interpreted these data as suggesting that tryptophan degradation may play a role in control of *T. gondii* replication only in certain human mononuclear phagocytes. They argued against tryptophan starvation by itself as being sufficient to effectively inhibit intracellular replication of *T. gondii*, basing their argument on the fact that IFN-α and IFN-β also induced

TABLE 2 Degradation of Extracellular Tryptophan and Expression of IDO Activity[a]

Cells and treatment (U/ml)	Tryptophan (μM)	IDO activity (nmol/hr per mg)
T24		
Medium[b]	34 ± 1 (3)	
Cells + medium	34 ± 1 (3)	0 (3)
IFN-γ (300)	<5 (3)	5.3 ± 0.8 (3)
Normal MDM		
Medium[b]	37 ± 1 (13)	
Cells + medium	36 ± 2 (13)	0 (4)
IFN-γ (100)	<5 (13)	9.7 ± 0.9 (4)
IFN-γ (500)	<5 (13)	10.7 ± 1.4 (4)
IFN-α (500)	6 ± 3 (3)	5.4 (2)
IFN-β (500)	<5 (3)	5.3 (2)
Alveolar macrophages		
Medium[b]	36 ± 1 (7)	
Cells + medium	37 ± 1 (7)	NT
IFN-γ (100)	12 ± 4 (3)	NT
IFN-γ (500)	6 ± 1 (7)	NT
Mouse macrophages		
Medium[b]	46 ± 2 (3)	
Cells + medium	45 ± 1 (3)	0 (2)
IFN-γ (100)	42 ± 2 (3)	0 (2)
IFN-γ (500)	41 ± 3 (3)	0 (2)

[a]Adherent cells were treated with the indicated agents for 24 hr (T24 cells, mouse peritoneal macrophages) or 72 hr (MDM, alveolar macrophages). Culture supernatants were then removed for tryptophan assay or lysates were prepared for IDO assay. Results indicate the means ± standard errors of the means of *n* experiments (number in parentheses) each performed in triplicate (tryptophan level) or duplicate (IDO activity), except for IDO activity of MDM treated with IFN-α and IFN-β and IFN-γ-treated mouse macrophages (means of two similar experiments for each).
[b]Culture medium without cells incubated for 24 to 72 h at 37°C.
IDO = indoleamine 2,3-dioxygenase; IFN = interferon; MOM = monocyte-derived macrophages; NT = not tested.
Source: Ref. 57

degradation of tryptophan although, according to their group, Nathan's group, and Pfefferkorn's group, these cytokines do not inhibit the intracellular replication of *T. gondii* [56–58]. However, Schmitz et al. have reported that, when combined with LPS, IFN-β induced remarkable inhibition of *T. gondii* replication in HMDMs [33]. Furthermore, since addition of tryptophan to the culture medium partially reversed the effect of the IFN-β–LPS combination, this activity was considered to be mediated by tryptophan starvation. Other oxygen-independent mechanisms must apply for murine Mϕs, since IFN-γ induces marked activation to inhibit or kill *T. gondii* but does not induce IDO and tryptophan degradation in these cells [57].

Adams et al. have recently demonstrated that activated murine peritoneal Mϕs have a nonoxidative mechanism of inhibition of intracellular replication of *T. gondii* that is dependent on the metabolism of L-arginine [59]. Critical to this mechanism would appear to be the generation of inorganic nitrogen oxides from a terminal guanidino nitrogen atom of L-arginine [59]. These studies were performed using competitive inhibitor of the synthesis of inorganic nitrogen oxides and L-citrulline from arginine. In murine peritoneal Mϕs activated by a combination of IFN-γ and LPS, the ability to inhibit the multiplication of *T. gondii* was blocked when these Mϕs were incubated with N^{G}monomethyl-L-arginine (N^{G}MMLA) (Figure 3). The production of NO_2^-, NO_3^- and citrulline correlated with the inhibition of intracellular replication of *T. gondii*; murine peritoneal Mϕs activated to inhibit replication of *T. gondii* produced high levels of NO_2^-, NO_3^-, and citrulline. Furthermore, addition of N^{G}MMLA not only abrogated inhibition of *T. gondii* multiplication but also reduced production of NO_2^-, NO_3^-, and citrulline. The data were not sufficient to allow for a conclusion as to whether killing of *T. gondii* was also mediated by the L-arginine metabolic pathway.

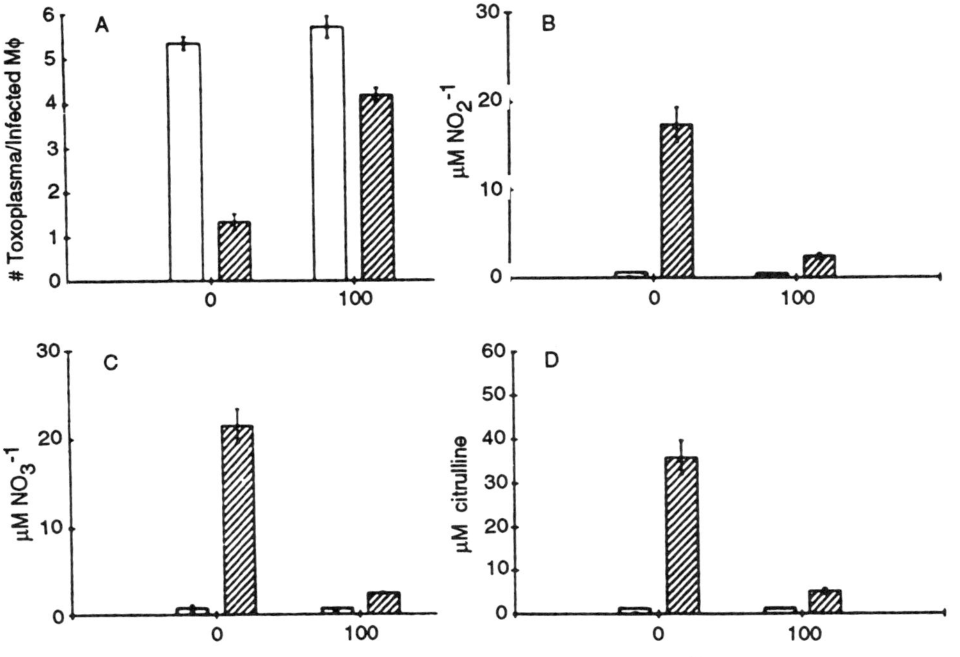

FIGURE 3 Effects of N^{G}monomethyl-L-arginine on microbistatic capacity of activated macrophages for *Toxoplasma gondii* and production of the L-arginine degradation pathway products NO_2^-, NO_3^-, and citrulline. Open bars, control macrophages. Hatched bars, macrophages activated by IFN-γ. Mean +/– SD of triplicate samples. (From Ref. 59.)

Langermans et al. have reported that endogenous TNF-α is involved in the generation of L-arginine–dependent anti–*T. gondii* activity induced by IFN-γ–treated murine peritoneal Mϕs [60]. When these investigators incubated Mϕs with IFN-γ and rabbit anti–mouse TNF-α antiserum for 24 hours before infection with *T. gondii*, the IFN-γ–induced toxoplasmastatic activity was abrogated and a significant inhibition of the release of NO_2^- was observed. These same authors have proposed that reactive nitrogen intermediates may not affect *T. gondii* directly but rather may act via another mechanism to inhibit its intracellular growth [60]. When the IFN-γ–dependent release of NO_2^- was inhibited by incubation of Mϕs with either N^GMMLA or the NADPH-inhibitor diphenylene iodonium (DPI) during the 24-hour period of incubation of the Mϕs with recombinant IFN-γ before infection or during the 24 hours before and throughout the infection, both inhibition of the production of NO_2^- and induction of toxoplasmastatic activity were observed. However, when either N^GMMLA or DPI was present only during the 18-hour infection period of IFN-γ–activated Mϕs, whereas the production of NO_2^- was almost completely inhibited, anti–*T. gondii* activity was only slightly inhibited.

In experiments performed to determine whether reactive nitrogen intermediates play a role in the anti–*T. gondii* activity of activated human Mϕs, Murray et al. reported that in HMDMs activated with IFN-γ, killing or inhibition of the intracellular growth of *T. gondii* was not affected by continuous treatment with N^GMMLA or arginase [61]. Similar lack of effect of incubation with N^GMMLA or arginase was observed when human peripheral blood mononuclear cells (PBMCs) that had been activated by in vivo administration of IFN-γ were used. Furthermore, HMDMs failed to secrete NO_2^- upon stimulation with IFN-γ. The authors concluded that, as opposed to murine Mϕs, L-arginine–dependent production of reactive nitrogen intermediates does not appear to play a role in the in vitro antimicrobial effects of human Mϕs. The authors could not rule out that generation of reactive nitrogen intermediates plays an anti–*T. gondii* role in vivo, since increased production of these intermediates has been reported in cancer patients treated with recombinant interleukin-2 (IL-2) [62].

By restricting the availability of iron, intracellular multiplication of *Legionella pneumophila* and *Histoplasma capsulatum* was inhibited [63,64]. To determine whether such a mechanism is responsible for nonoxidative inhibition of the intracellular growth of *T. gondii*, Murray et al. treated resting HMDMs with deferoxamine (an iron chelator) and treated IFN-γ–activated HMDMs with excess iron [65]. Addition of deferoxamine did not affect the intracellular replication of *T. gondii*, whereas the growth of *L. pneumophila* was completely inhibited. Incubation of IFN-γ–activated HMDMs with iron-saturated transferrin did not affect the replication of *T. gondii*, whereas the inhibitory effect on replication of *L. pneumophila* by IFN-γ was completely reversed. The authors suggested that limiting the availability of iron was probably not an antimicrobial mechanism applicable to a broad range of intracellular organisms. We have confirmed that intracellular iron depletion does not appear to inhibit *T. gondii* replication (J. Huskinson-Mark and J. S. Remington, unpublished data).

C. Role of Macrophages in the Immunosuppression Induced by *Toxoplasma gondii*

Infection with *T. gondii* can result in nonspecific immunosuppression in mice and humans [44,66–73]. Suzuki et al. reported that both primary antibody response and initiation of memory cells are suppressed in mice acutely infected with this parasite [66,67]. This suppression is nonspecific for antigen and immunoglobulin class and appears to be due to the appearance of suppressor Mϕs in the spleens of infected animals [66,67]. These authors reported that spleen cells from infected animals strongly inhibited the in vitro anti–sheep red blood cell (SRBC) antibody response of normal spleen cells [66]. The cells responsible for this suppression were plastic-adherent and irradiation-resistant, and their inhibitory effect was abrogated by treatment with either silica or carrageenan, which are selectively cytotoxic for Mϕs [66,68,69]. The suppressive effect of Mϕs appeared to be due to their inhibitory effect on the proliferation of lymphocytes in primary antibody response, required direct contact with the responder target cell, and was not inhibited by indomethacin or catalase [69].

The induction of suppressive Mϕs in mice is controlled by both H-2–linked and nonlinked genes [70]. Whereas C57BL/6 (B6), C57BL/10 (B10), and SJL mice infected with *T. gondii* showed markedly suppressed antibody responses to keyhole limpet hemocyanin, infected DBA/2 and C3H/He mice showed moderate suppression and infected BALB/c mice showed unabated antibody responses [70]. Both H-2–linked and nonlinked genes appeared to regulate this immunosuppression, since infected B10.D2 (which have B10 background and the same H-2 haplotype as BALB/c mice) exhibited weaker immunosuppression than B10 mice but stronger than BALB/c mice [70]. Using in vitro primary antibody response to SRBC, these authors also reported that activation of suppressive Mϕs was controlled by both H-2–linked and nonlinked genes [70].

The suppressive effect of murine splenic Mϕs induced during infection with *T. gondii* also affected the proliferative responses of normal murine lymphocytes to *T. gondii* antigen in vitro [69] and affected anti–*T. gondii* antibody responses [71]. It has been proposed that this immunosuppression to parasite antigens may be detrimental to the host [70,71].

Luft et al. reported that patients with recently acquired clinically significant *T. gondii* infection develop functional as well as quantitative alterations in their T lymphocyte subpopulations [72]. Cytofluorometric analysis of PBMCs showed that patients with recently acquired symptomatic toxoplasmosis had a significant increase in both absolute and relative numbers of CD8+ T lymphocytes, whereas patients with asymptomatic lymphadenopathy had either an increase in the absolute and relative number of CD8+ T lymphocytes or a decrease in the absolute and relative number of CD4+ T lymphocytes, or both [72]. Lymphocyte proliferation studies showed that the peak lymphocyte response to *T. gondii* antigen on day 8 was significantly depressed in these patients when compared with that in chronically infected control subjects. The kinetics of this depression suggested that a suppressive cell was being induced. Luft et al. conducted studies to determine whether adherent regulatory cells are generated in vitro by *T. gondii* antigen [73]. Adherent mononuclear cells obtained from PBMCs of patients chronically infected with *T. gondii* were incubated with *T. gondii* antigen for 8 days [73]. Addition of these cells resulted in suppression of the proliferative response of autologous PBMCs to this antigen (Figure 4). In order for adherent mononuclear cells to acquire suppressive activity, CD4+ T lymphocytes were required during the incubation with *T. gondii* antigen. Taken together, these results demonstrated the induction of adherent suppressive cells in vitro and the role of CD4+ T lymphocytes in the induction of these cells.

IV. EFFECT OF CYTOKINES ON INHIBITION OF *TOXOPLASMA GONDII*

A. Interferon-γ

Interferon is an important immunomodulating cytokine of T lymphocyte or natural killer (NK) cell origin [74]. It plays a major role in resistance to *T. gondii* infection in animal models in vitro and in vivo [75].

1. *Effect of Interferon-γ on in Vitro* Toxoplasma *infection*

As described above, we reported that HMDMs, when cultured in the presence of immune lymphocytes and TLA, developed the ability to inhibit or kill intracellular *T. gondii* [30]. Subsequently it became apparent from studies performed by us and others that supernatants from TLA-stimulated immune lymphocytes (or nonimmune lymphocytes stimulated with mitogens)

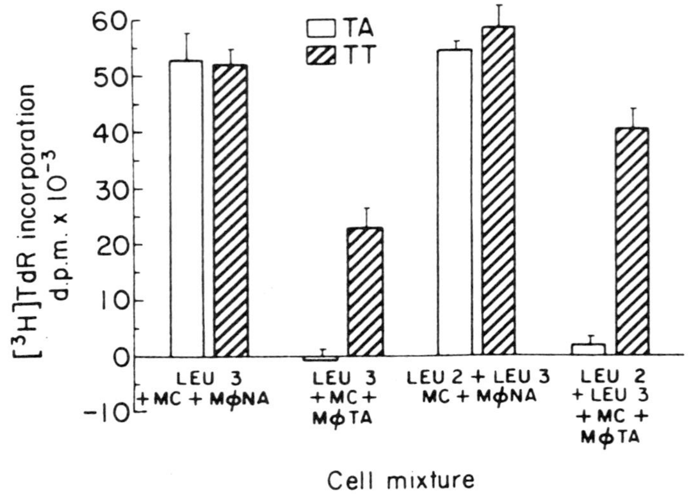

FIGURE 4 Effect of monocytes on *Toxoplasma* antigen–induced T lymphocyte proliferation. Adherent cells were harvested from mononuclear cells that were cultured with no antigen (Mϕ NA) or *T. gondii* antigen (Mϕ TA). After 8 days the proliferative response of CD4+ (Leu 3) and/or CD8+ (Leu 2) T cells cocultured with fresh monocytes (MC) and the pretreated adherent cells (Mϕ NA or Mϕ TA, respectively) to TA and tetanus toxoid (TT) were assessed. (From Ref. 73.)

could induce a similar activity in Mϕs [18,31,32,76,77]. Wilson and Haas identified IFN-γ as the major macrophage-activating factor (MAF) in lymphocyte culture supernatants and made the observation that neonatal monocytes manifested impaired responses to MAF, a fact that may help to explain the susceptibility of neonates to progressive *T. gondii* infection [78]. Nathan et al. confirmed that IFN-γ is the major lymphokine that activates HMDM antitoxoplasma activity under experimental conditions in vitro [79].

Murray et al. reported the ability of alveolar Mϕs from patients with AIDS to develop enhanced toxoplasmacidal activity in response to IFN-γ (whether in the form of conditioned medium or of recombinant protein) and suggested that an important factor in the immunosuppression of AIDS may be impairment of production of IFN-γ (rather than decreased sensitivity of Mϕs to this cytokine) [26]. This group subsequently studied (1987) the ability of HMDMs from AIDS patients to exhibit enhanced *T. gondii* killing in response to recombinant IFN-γ (rIFN-γ) (when infection was initiated 6 days after in vitro activation with cytokine) and found that this was similar to that exhibited by IFN-γ–treated HMDM obtained from normal controls [80]. This study also showed that IFN-γ administered in vivo could induce prolonged activation against *T. gondii* of HMDMs from AIDS patients (if cultured for 5 to 7 days before infection). In contrast, a study by Eales et al., using HMDMs obtained from HIV-infected patients that were cultured for 6 days before stimulation with rIFN-γ and infected 24 hours later, showed impairment of killing of *T. gondii* in comparison with similarly treated HMDMs from non-HIV infected donors [81]. In this study, HMDMs from AIDS patients exhibited a similar degree of toxoplasmacidal activity when compared with normal HMDMs if conditioned medium was used as the source of lymphokine (presumably because of additional factors present in this preparation). This latter study suggests that HMDMs of AIDS patients may have a subtle impairment in response to IFN-γ that becomes apparent only under certain culture conditions.

Similar studies by our group and others have established the ability of recombinant murine IFN-γ (rMuIFN-γ) to activate murine peritoneal macrophages in vivo [82,83], alveolar macrophages in vitro and in vivo and peritoneal macrophages in vitro [83], and rat alveolar macrophages in vitro [84] to kill *T. gondii*. Interestingly, alveolar macrophages are activated by lower doses of IFN-γ than peritoneal macrophages [83], but the degree of activation induced is greater in the latter (suggesting that there may be tissue-specific differences in Mϕ activation induced by IFN-γ). A study by Suzuki et al. noted that treatment of mice with a monoclonal antibody to IFN-γ before infection with *T. gondii* prevented the activation of peritoneal macrophages that occurred in infected control antibody-treated animals, emphasizing the important role this cytokine plays in Mϕ function [75].

*2. Possible Mechanisms for Interferon-γ Anti-*Toxoplasma *Activity*

As discussed above, IFN-γ can induce killing or inhibition of replication of *T. gondii* in activated Mϕs by several different mechanisms including oxidative [41,79] and nonoxidative pathways [22,41,78]. Nonoxidative mechanisms that have been described include induction of IDO [54,56] and generation of reactive nitrogen intermediates derived from L-arginine [59]. Recently, it was noted that induction of heat shock proteins (HSPs) (polypeptides that are produced by cells under stressful conditions to preserve cellular functions) in Mϕs induced by *T. gondii* was associated with resistance of the animal to this infection [85]. Previous studies [86] had shown that IFN-γ treatment of Mϕs can result in induction of HSPs in these cells: these two studies suggest that one additional mechanism of IFN-γ–induced resistance to *T. gondii* infection might include induction of HSPs in activated Mϕs.

3. Role of Other Cytokines in Modulation of Interferon-γ Action

In all of the experiments described above, IFN-γ was used alone to activate Mϕs against *T. gondii*, and the results were interpreted as being due to this cytokine alone. Recent studies have revealed that other cytokines act in synergy with IFN-γ to achieve activation of Mϕs. This was initially noted when the ability of rMuIFN-γ to activate Mϕs in vitro to kill *T. gondii* was shown to be partially inhibited by a polyclonal neutralizing antibody against tumor necrosis factor (TNF), suggesting that TNF may be necessary for the optimal activation of Mϕs [87]. Support for this concept came from recent experiments showing that under stringent endotoxin-free conditions IFN-γ was unable to activate Mϕs to kill *T. gondii* but that this ability was restored by the addition of small quantities of TNF [88]. As will be detailed later, both IL-6 [86] and IL-10 [90] are able to prevent IFN-γ–mediated activation of Mϕs under certain conditions in vitro.

B. Tumor Necrosis Factor-α

Tumor necrosis factor-α is produced predominantly by monocytes/Mϕs and plays a major role in the activation of Mϕs in various infection models [91]. Its importance in *Toxoplasma* infection, however, is still being elucidated. Murine TNF-α did not activate murine Mϕs to inhibit replication of *T. gondii*, although it was capable of activating Mϕs to inhibit intracellular replication of *Trypanosoma cruzi* [92]. A recent study was also unable to demonstrate activation of murine Mϕs against *T. gondii* using TNF-α alone, but the combination of TNF-α and IFN-γ was synergistic when compared with IFN-γ administered alone. Black et al. demonstrated that TNF-α was able to activate murine Mϕs to limit replication of *Mycobacterium avium-intracellulare* (MAI) after successful phagolysosomal fusion, but when individual Mϕs were coinfected with mycobacteria and *T. gondii* the ability of the TNF-α to induce Mϕs to inhibit replication of MAI was lost despite phagolysosome fusion [93] (Figure 5). This important observation is relevant to AIDS, in which infection with multiple organisms is often seen, and it sugests that *T. gondii* may impair the response of Mϕs to some cytokines. The role that TNF-α plays in IFN-γ–induced activation of Mϕs and the association of reactive nitrogen intermediates with this effect have been discussed above [60].

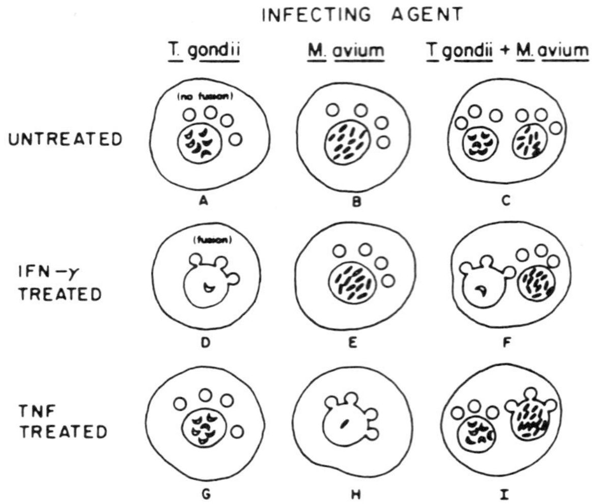

FIGURE 5 Schematic representation of the relationship between intracellular fate and lysosomal fusion with phagosomes containing *Toxoplasma gondii* or *Mycobacterium avium* (MAC) in singly infected and coinfected macrophages (Mϕs) that are treated with cytokines. (*Large circles*) Mϕs; (*small circles*) phagosomes and lysosomes. (*Column on right*) Mϕs coinfected with both agents. (A and B) Both organisms replicate without fusion of phagosomes with lysosomes within untreated Mϕs; (C) these results were the same within coinfected Mϕs; (D) Growth of *T. gondii* is inhibited (successful fusion), while (E) MAC replicates (no fusion) within interferon-γ (IFN-γ)–treated Mϕ; (F) these fates were not changed within coinfected Mϕs; (G) *T. gondii* replicates (no fusion) but (H) growth of MAC is inhibited (successful fusion) within tumor necrosis factor (TNF)-treated Mϕs; (I) in contrast, MAC replicated despite successful fusion in Mϕs coinfected with both agents. The fate of *T. gondii* was not altered in coinfected Mϕs. (From Ref. 93.)

C. Interleukin-1

The major source for IL-1 production is the Mϕ, but NK cells and B lymphocytes can also produce this cytokine [91]. When administered alone, neither recombinant human IL-1a nor IL-1b appears to be able to activate Mϕs to inhibit replication of *T. gondii* in vitro, and coadministration with IFN-γ did not induce synergistic activity [87].

D. Interferon-α

Interferon-α is produced by leukocytes and was first recognized because of its antiviral properties [74]. Wilson and Westall reported slight inhibitory activity of IFN-α on *T. gondii* replication in HMDMs in vitro [94], although Nathan et al. were unable to demonstrate any such effect using a similar experimental protocol [58].

E. Interferon-β

Interferon-β is produced by fibroblasts and, like IFN-α, has antiviral activity [74]. Wilson and Westall [94] noted a weak inhibitory effect of IFN-β on *T. gondii* replication in HMDMs. As outlined above, other authors were unable to demonstrate an effect of IFN-β on *T. gondii* replication in HMDMs in vitro [53–55]. These studies used relatively short-term culture of HMDMs before treatment of the monolayer with cytokine and infection with *T. gondii*. In contrast, Schmitz et al. demonstrated that IFN-β alone could effect a slight inhibition of replication of *T. gondii* in HMDMs that were cultured for 10 days before their treatment and infection [33]. Marked inhibition of *T. gondii* replication resulted, however, when IFN-β was combined with LPS. The effect of IFN-β and LPS is mediated, at least in part, by tryptophan starvation, as these treatments markedly enhanced IDO induction and supplemental amino acid could partially reverse the observed effect. These observations were confirmed by Orellana et al., who showed that the effect of IFN-β was not demonstrable in HMDMs cultured for 5 days only or in murine peritoneal Mϕs [95].

F. Interleukin-6

Interleukin-6 is a potent pleiotropic cytokine with multiple cellular sources including monocytes and Mϕs [91]. Beaman and Remington reported that IL-6 markedly enhanced intracellular replication of *T. gondii* in murine peritoneal Mϕs in vitro [89]. This study also demonstrated impairment by recombinant human IL-6 of the ability of rMuIFN-γ to activate Mϕs to kill *T. gondii* when both cytokines were administered together before infection of the monolayer. This did not occur, however, if IL-6 was administered after IFN-γ had already activated the Mϕs.

G. Interleukin-10

Interleukin-10 (originally known as cytokine synthesis inhibitory factor) is a negative regulatory cytokine that originates predominantly from T and B lymphocytes [96]. Recently, IL-10 was shown to inhibit the IFN-γ–mediated activation of elicited murine peritoneal Mϕs to kill *T. gondii* [90]. This occurred when IL-10 was administered before or at the same time as, but not after, IFN-γ (i.e., analogous to the effect of IL-6 treatment on Mϕs described above). This study also found that the IL-10–mediated inhibition of IFN-γ function was mediated by the suppression of production of reactive nitrogen intermediates (which mediate some of the actions of IFN-γ, as discussed above).

H. Granulocyte-Monocyte Colony-Stimulating Factor

Nathan et al. reported that granulocyte-monocyte colony-stimulating factor (GM-CSF) did not activate HMDMs to inhibit *T. gondii* replication in vitro [58]. In addition, Hughes et al. noted that recombinant murine GM-CSF was unable to activate murine Mϕs or bovine monocytes for toxoplasma killing in vitro [97].

ACKNOWLEDGMENTS

This study was supported in part by a WA and MG Saw Postgraduate Medical Research Fellowship from the University of Western Australia, an HMDP Fellowship from the Ministry of Health of Singapore, Public Health Service Grants AI04717 and AI30320 from the National Institutes of Health, the University of California Universitywide AIDS Research Program, the Center for AIDS Research at Stanford University, and a MacArthur Foundation grant in Molecular Parasitology.

REFERENCES

1. Luft BJ, Remington JS. Toxoplasmic encephalitis. J Infect Dis 1988; 157:1–6.
2. McCabe R, Remington JS. Toxoplasmosis: the time has come. N Engl J Med 1988; 318:313–5.
3. Remington JS, Desmonts G. Toxoplasmosis. In: Remington JS, Klein JO, eds. Infectious diseases of the fetus and the newborn infant. 3rd ed. Philadelphia: WB Saunders, 1990:89–198.
4. Remington JS, Krahenbuhl JL, Mendenhall JW. A role for activated macrophages in resistance to infection. Infect Immun 1972; 6:829–34.
5. Ruskin J, Remington JS. Immunity and intracellular infection: resistance to bacteria in mice infected with a protozoan. Science 1968; 160:72–4.

6. Ruskin J, Remington JS. Role for the macrophage in acquired immunity to phylogenetically unrelated intracellular organisms. Antimicrob Agents Chemother 1968; 8:44–7.
7. Krahenbuhl JL, Levy L, Remington JS. Resistance to *Mycobacterium leprae* in mice infected with *Toxoplasma gondii* and *Besnoitia jellisoni*. Infect Immun 1974; 10:1068–71.
8. Krick JA, Remington JS. Resistance to infection with *Nocardia asteroides*. J Infect Dis 1975; 131:665–72.
9. Gentry LO, Remington JS. Resistance against Cryptococcus conferred by intracellular bacteria and protozoa. J Infect Dis 1971; 123:22–31.
10. Remington JS, Merigan TC. Resistance to virus challenge in mice infected with protozoa or bacteria. Proc Soc Exp Biol Med 1969; 131:1184–8.
11. Pomeroy C, Kline S, Jordan MC, Filice GA. Reactivation of *Toxoplasma gondii* by cytomegalovirus disease in mice: antimicrobial activities of macrophages. J Infect Dis 1989; 160:305–11.
12. Wing EJ, Remington JS. Role of activated macrophage in resistance against *Trichinella spiralis*. Infect Immun 1978; 21:398–404.
13. Mahmoud AAF, Warren KS, Strickland GT. Acquired resistance to infection with *Schistosoma mansoni* induced by *Toxoplasma gondii*. Nature 1976; 263:56–7.
14. McCleod R, Remington JS. Studies on the specificity of killing of intracellular pathogens by macrophages. Cell Immunol 1977; 34:156–174.
15. Krahenbuhl JL, Remington JS. The role of activated macrophages in specific and nonspecific cytostasis of tumor cells. J Immunol 1974; 113:507–16.
16. Anderson SE Jr, Bautista SC, Remington JS. Specific antibody–dependent killing of *Toxoplasma gondii* by normal macrophages. Clin Exp Immunol 1976; 26:375–80.
17. Wilson CB, Tsai V, Remington JS. Failure to trigger the oxidative metabolic burst by normal macrophages. Possible mechanism for survival of intracellular pathogens. J Exp Med 1980; 151:328–46.
18. Jones TC, Len L, Hirsch JG. Assessment in vitro of immunity against *Toxoplasma gondii*. J Exp Med 1975; 141:466–82.
19. Stadtsbaeder S, Nguyen BT, Calvin-Preval MC. Respective role of antibodies and immune macrophages during acquired immunity against toxoplasmosis in mice. Ann Immunol (Paris) 1975; 126c:461–74.
20. Hauser WE, Remington JS. Effect of monoclonal antibodies to phagocytosis and killing of *Toxoplasma gondii* by normal macrophages. Infect Immun 1981; 32:637–40.
21. Ryning FW, Remington JS. Effect of alveolar macrophages on *Toxoplasma gondii*. Infect Immun 1977; 18:746–53.
22. Ryning F, Krahenbuhl JL, Remington JS. Comparison of cytotoxic and microbicidal function of bronchoalveolar and peritoneal macrophages. Immunology 1981; 42:513–19.
23. McLeod R, Estes R, Mack DG, McCleod EG. Effect of human alveolar macrophages and peripheral blood monocytes on *Toxoplasma gondii*. J Infect Dis 1983; 147:957.
24. Catterall JR, Black CM, Leventhal JP, Rizk NW, Wachtel JS, Remington JS. Nonoxidative microbicidal activity in normal human alveolar and peritoneal macrophages. Infect Immun 1987; 55:1635–40.
25. Israelski DM, Araujo FG, Wachtel JS, Heinrichs L, Remington JS. Differences in microbicidal activities of human macrophages against *Toxoplasma gondii* and *Trypanosoma cruzi*. Infect Immun 1990; 58:263–5.
26. Murray HW, Gellene RA, Libby DM, Rothermel CD, Rubin BY. Activation of tissue macrophages from AIDS patients: in vitro respose of AIDS alveolar macrophages to lymphokines and interferon-γ. J Immunol 1985; 135:2374–7.
27. Van Furth R, Diesselhoff-den Dulk MMC, Raeburn JA, van Zweith TL, Crofton RW, Blusse van Oud Albas A, Characteristics, origin and kinetics of human and murine mononuclear phagocytes. In: van Furth R, ed. Mononuclear phagocytes. The Hague, the Netherlands: Martinus Nijhoff, 1980:279–98.
28. Murray HW, Rubin BY, Carriero SM, Harris AM, Jaffee EA. Human mononuclear phagocyte antiprotozoal mechanisms: oxygen-dependent vs oxygen-independent activity against intracellular *Toxoplasma gondii*. J Immunol 1985; 134:1982–8.
29. Stadtsbaeder S. Multiplication des toxoplasmes dans les globules blancs humains mis en culture. Rev Belge Pathol Med Exp 1965; 31:280–4.
30. Anderson SE Jr, Remington JS. Effect of normal and activated human macrophages on *Toxoplasma gondii*. J Exp Med 1974; 139:1154–74.
31. Borges JS, Johnson WD Jr. Inhibition of multiplication of *Toxoplasma gondii* by human monocytes exposed to T-lymphocyte products. J Exp Med 1975; 141:483–96.

32. Wilson CB, Remington JS. Activity of human blood leukocytes against *Toxoplasma gondii*. J Infect Dis 1979; 140:890–5.
33. Schmitz JL, Carlin JM, Borden EC, Byrne GI. Beta interferon inhibits *Toxoplasma gondii* growth in human monocyte-derived macrophages. Infect Immun 1989; 57:3254–6.
34. Anderson SE, Bautista S, Remington JS. Induction of resistance to *Toxoplasma gondii* in human macrophages by soluble lymphocyte products. J Immunol 1976; 117:381–7.
35. McCabe RE, Remington JS. Mechanisms of killing of *Toxoplasma gondii* by rat peritoneal macrophages. Infect Immun 1986; 52:151–5.
36. Chincilla M, Guerrero M, Solano E. Lack of multiplication of Toxoplasma in macrophages in vitro. N Parasitol 1982; 68:952–5.
37. Catterall JR, Sharma SD, Remington JS. Oxygen-independent killing by alveolar macrophages. J Exp Med 1986; 163:1113–31.
38. Murray HW, Cohn ZA. Macrophage oxygen-dependent antimicrobial activity. Susceptibility of *Toxoplasma gondii* to oxygen intermediates. J Exp Med 1979; 150:938–49.
39. Murray HW, Juangbhanich CW, Nathan CF, Cohn ZA. Macrophage oxygen-dependent antimicrobial activity. The role of oxygen intermediates. J Exp Med 1979; 150:950–64.
40. Murray H, Cohn ZA. Macrophage oxygen-dependent antimicrobial activity III. Enhanced oxidative metabolism as an expression of macrophage activation. J Exp Med 1980; 152:1596–1609.
41. Jones TC, Hirsch JG. The interaction between *Toxoplasma gondii* and mammalian cells II: The absence of lysosmal fusion with phagocytic vacuoles containing living parasites. J Exp Med 1972; 136:73–94.
42. Joiner KA, Fuhrman SA, Miettinen HM, Kasper LH, Mellman I. *Toxoplasma gondii*: Fusion competence of parasitophorous vacuole in Fc receptor–transfected fibroblasts. Science 1990; 249:641–6.
43. Sibley LD. Active modification of host cell phagosome by *Toxoplasma gondii*. In: Moulder JW, ed. Intracellular parasitism. Boca Raton, Florida: CRC Press, Inc., 1989:245–57.
44. Krahenbuhl JL, Remington JS. The immunology of Toxoplasma and toxoplasmosis. In: Cohen S, Warren KS, eds. Immunology of parasitic infections. 2nd ed. Oxford: Blackwell Scientific, 1982:356–421.
45. Sabin AB, Feldman HA. Dyes as microchemical indicators of a new immunity phenomenon affecting a protozoan parasite (Toxoplasma). Science 1948;108:660–63.
46. Van de Winkel JGJ, Anderson CL. Biology of human immunoglobulin G Fc receptors. J Leukocyte Biol 1991;49:511–24.
47. Vischer WA, Suter E. Intracellular multiplication of *Toxoplasma gondii* in adult mammalian macrophages cultivated in vitro. Proc Soc Exp Biol Med 1954; 86:413–9.
48. Erbe DV, Pfefferkorn ER, Fanger MW. Functions of the various human IgG Fc receptors in mediating killing of *Toxoplasma gondii*. J Immunol 1991; 146:3145–51.
49. Murray HW, Byrne GI, Rothermel CD, Cartell DM. Lymphokine enhances oxygen-independent activity against intracellular pathogens. J Exp Med 1983; 158:234–9.
50. Nathan CF. Mechanisms of macrophage antimicrobial activity. Trans R Soc Trop Med Hyg 1983; 77:620–30.
51. Locksley RM, Fankhauser J, Henderson WR. Alteration of leucotriene release by macrophages ingesting *Toxoplasma gondii*. Proc Natl Acad Sci USA 1985; 82:6922–6.
52. Pfefferkorn ER. Interferon-γ blocks the growth of *Toxoplasma gondii* in human fibroblasts by inducing the host cells to degrade tryptophan. Proc Natl Acad Sci USA. 1984; 81:908–12.
53. Pfefferkorn ER, Rebhun S, Eckel M. Characterization of an indoleamine 2,3-dioxygenase induced by gamma-interferon in cultured human fibroblasts. J Interferon Res 1986; 6:267–79.
54. Pfefferkorn ER, Eckel M, Rebhun S. Interferon-γ suppresses the growth of *Toxoplasma gondii* in human fibroblasts through starvation for tryptophan. Mol Biochem Parasitol 1986; 20:215–24.
55. Pfefferkorn ER. Interferon gamma and the growth of *Toxoplasma gondii* in fibroblast. Ann Inst Pasteur Microbiol 1986; 137:348–52.
56. Pfefferkorn ER, Guyre PM. Inhibition of growth of *Toxoplasma gondii* in cultured fibroblasts by human recombinant gamma interferon. Infect Immun 1984; 44:211–6.
57. Murray HW, Szuro-Sudol A, Wellner D, Oca MJ, Granger AM, Libby DM, Rothermel CD, Rubin BY. Role of tryptophan degradation in respiratory burst–independent antimicrobial activity of gamma interferon–stimulated human macrophages. Infect Immun 1989; 57:845–9.
58. Nathan CF, Prendergast TJ, Wiebe ME, Stanley ER, Platzer E, Remold HG, Welte K, Rubin BY,

Murray HW. Activation of human macrophages. Comparison of other cytokines with interferon-γ. J Exp Med 1984; 160:600–5.
59. Adams LB, Hibbs JB, Taintor RR, Krahenbuhl JL. Microbiostatic effect of murine activated macrophages for *Toxoplasma gondii*. Role for synthesis of inorganic nitrogen oxides from L-arginine. J Immunol 1990; 144:2725–9.
60. Langermans JAM, van der Hulst MEB, Nibbering PH, Hiemstra PS, Fransen L, van Furth R. IFN-γ–induced L-arginine–dependent toxoplasmastatic activity in murine peritoneal macrophages is mediated by endogenous tumor necrosis factor-α. J Immunol 1992; 148:568–74.
61. Murray HW, Teitelbaum RF. L-arginine–dependent reactive nitrogen intermediates and the antimicrobial effect of activated human mononuclear phagocytes. J Infect Dis 1992; 165:513–7.
62. Hibbs JB, Westfelder C, Taintor R, Vavrin Z, Kablitz C, Baranowski RL, Ward JH, Menlove RL, McMurry MP, Kushner JP, Samlowski WE. Evidence for cytokine inducible nitric oxide synthesis from L-arginine in patients receiving interleukin-2 therapy. J Clin Invest 1992; 89:867–77.
63. Byrd TF, Horwitz MA. Interferon gamma–activated human monocytes downregulate transferrin receptors and inhibit the intracellular multiplication of *Legionella pneumophila* by limiting the availability of iron. J Clin Invest 1989; 83:1457–65.
64. Lane TE, Wu-Hsieh BA, Howard DH. Iron limitation and the gamma interferon–mediated antihistoplasma state of murine macrophages. Infect Immun 1991; 59:2274–8.
65. Murray HW, Granger AM, Teitelbaum RF. Gamma interferon–activated human macrophages and *Toxoplasma gondii*, *Chlamydia psittaci*, and *Leishmania donovani*: antimicrobial role of limiting intracellular iron. Infect Immun 1991; 59:4684–6.
66. Suzuki Y, Watanabe N, Kobayashi A. Nonspecific suppression of primary antibody responses and presence of plastic-adherent suppressor cells in *Toxoplasma gondii*–infected mice. Infect Immun 1981; 34:30–5.
67. Suzuki Y, Watanabe N, Kobayashi A. Nonspecific suppression of initiation of memory cells in *Toxoplasma gondii*–infected mice. Infect Immun 1981; 34:36–42.
68. Suzuki Y, Kobayashi A. Suppression of unprimed T and B cells in antibody responses by irradiation-resistant and plastic-adherent suppressor cells in *Toxoplasma gondii*–infected mice. Infect Immun 1983; 40:1–7.
69. Suzuki Y, Kobayashi A. Macrophage-mediated suppression of immune responses in Toxoplasma-infected mice. I. Inhibition of proliferation of lymphocytes in primary antibody responses. Cell Immun 1984; 85:417–27.
70. Suzuki Y, Kobayashi A. Macrophage-mediated suppression of immune responses in Toxoplasma-infected mice. II. Both H-2–linked and nonlinked control of induction of suppressor macrophages. Cell Immun 1985; 91:375–84.
71. Suzuki Y, Joh K, Kobayashi A. Macrophage-mediated suppression of immune responses in Toxoplasma-infected mice. III. Suppression of antibody responses to parasite itself. Cell Immun 1987; 110:218–25.
72. Luft, BJ, Kansas G, Engleman EG, Remington JS. Functional and quantitative alterations in T-lymphocyte subpopulations in acute toxoplasmosis. J Infect Dis 1984; 150:761–7.
73. Luft BJ, Pedrotti PW, Remington JS. In vitro generation of adherent mononuclear suppressor cells to Toxoplasma antigen. Immunology 1988; 63:643–8.
74. Murray HW. Interferon-gamma, the activated macrophage, and host defense against microbial challenge. Ann Intern Med 1988; 108:595–608.
75. Suzuki Y, Orellana MA, Schreiber RD, Remington JS. Interferon-γ: the major mediator of resistance against *Toxoplasma gondii*. Science 1988; 240:516–8.
76. Sethi KK, Pelster B, Suzuki N, Pierkarski G, Brandis H. Immunity to *Toxoplasma gondii* induced in vitro in non-immune mouse macrophages with specifically immune lymphocytes. J Immunol 1975; 115:1151–8.
77. Shirihata T, Shimizu K, Suzuki N. An in-vitro study on lymphocyte-mediated immunity in mice experimentally infected with *Toxoplasma gondii*. Jpn J Vet Sci 1975; 37:2351–43.
78. Wilson CB, Haas JE. Cellular defenses against *Toxoplasma gondii* in newborns. J Clin Invest 1980; 73:1606–16.
79. Nathan CF, Murray HW, Weibe ME, Rubin BY. Identification of interferon-γ as the lymphokine that activates human macrophage: oxidative metabolism and antimicrobial activity. J Exp Med 1983; 158:670–89.

80. Murray HW, Scavuzzo D, Jacobs JL, Kaplan MH, Libby DM, Schindler J, Roberts RB. In vitro and in vivo activation of human mononuclear phagocytes by interferon-γ. J Immunol 1987; 138:2457–62.
81. Eales L-A, Moshtael O, Pinching AJ. Microbiocidal activity of monocyte derived macrophages in AIDS and related disorders. Clin Exp Immunol 1987; 67:227–35.
82. McCabe RE, Luft BJ, Remington JS. Effect of murine interferon gamma on murine toxoplasmosis. J Infect Dis 1984; 150:961–2.
83. Black CM, Catterall JR, Remington JS. In vivo and in vitro activation of alveolar macrophages by recombinant interferon-γ. J Immunol 1987; 138:491–5.
84. Badger AM, Hutchman JS, Sung C-P, Bugelski PJ. Activation of rat alveolar macrophages by gamma interferon to inhibit *Toxoplasma gondii* in vitro. J Leukocyte Biol 1987; 42:447–84.
85. Nagasawa H, Oka M, Maeda K, Jian-Guo C, Hisaeda H, Ito Y, Good RA, Himeno K. Induction of heat shock protein closely correlates with protection against *Toxoplasma gondii* infection. Proc Natl Acad Sci USA 1992; 89:3155–8.
86. Koga T, Wand-Wurttenburger A, DeBruyn J, Munk ME, Schoel, Kaufmann SHE. T cells against a bacterial heat shock protein recognize stressed macrophages. Science 1989; 245:1112–5.
87. Chang HR, Grau GE, Pechere J-C. Role of TNF and IL-1 in infections with *Toxoplasma gondii*. Immunology 1990; 69:33–7.
88. Sibley LD, Adams L, Fukutomi Y, Krahenbuhl JL. Tumor necrosis factor-α triggers antitoxoplasmal activity of IFN-γ primed macrophages. J Immunol 1991; 147:2340–5.
89. Beaman MH, Remington JS. IL-6 impairs macrophage killing of *Toxoplasma gondii* and IFN-γ function. Clin Res 1991; 39:176Abstr.
90. Gazzinelli RT, Oswald IP, James SL, Sher A. IL-10 inhibits parasite killing and nitrogen oxide production by IFN-γ activated macrophages. J Immunol 1992; 148:1792–6.
91. Bendtzen K. Interleukin 1, interleukin 6 and tumor necrosis factor in infection, inflammation and immunity. Immunol Lett 1988; 19:183–92.
92. De Titto EH, Catterall JR, Remington JS. Activity of recombinant tumor necrosis factor on *Toxoplasma gondii* and *Trypanosoma cruzi*. J Immunol 1986; 137:1342–5.
93. Black CM, Bermudez LEM, Young LS, Remington JS. Co-infection of macrophages modulates interferon-γ and tumor necrosis factor–induced activation against intracellular pathogens. J Immunol 1990; 172:977–80.
94. Wilson CB, Westall J. Activation of neonatal and adult human macrophages by alpha, beta and gamma interferons. Infect Immun 1985; 49:351–6.
95. Orellana MA, Suzuki Y, Araujo F, Remington JS. Role of beta interferon in resistance to *Toxoplasma gondii* infection. Infect Immun 1991; 59:3287–90.
96. Zlotnik A, Moore KW. Interleukin 10. Cytokine 1991; 3:366–71.
97. Hughes HPA, Speer CA, Kyle JE, Dubey JP. Activation of murine macrophages and a bovine monocyte cell line by bovine lymphokines to kill the intracellular pathogens *Eimeria bovis* and *Toxoplasma gondii*. Infect Immun 1987; 55:784–91.

29

Macrophages in Experimental Chagas' Disease

Raymond E. Kuhn

Wake Forest University, Winston-Salem, North Carolina

I. INTRODUCTION

Trypanosoma cruzi is an obligate intracellular protozoan parasite of mammals and is the etiologic agent of Chagas' disease. The parasite is found throughout the tropical, and in parts of temperate, regions of the Western Hemisphere and is thought to infect 20 to 40 million persons. The intermediate hosts are reduviid bugs of several species, and parasites are transmitted to mammals via infected feces that contaminate mucous membranes or abrasions of the host. Infective trypomastigotes enter the mammalian host, invade cells, then transform into the replicative amastigote form, which divide by binary fission. After differentiating back into trypomastigotes, parasites emerge from the dead cell to reinvade other host cells. Chagas' disease is a chronic disease resulting from the infection years to decades after initial contact with the parasite. Pathologic consequences of chronic Chagas' disease include myocardiopathies and enlargement of the hollow viscera. No effective vaccines or treatments are available. There is, therefore, intense interest in the immunobiology of this parasitic infection.

Macrophages play a multiplicity of roles in mammals infected with the protozoan parasite *T. cruzi*. As an obligate intracellular parasite, *T. cruzi* invades and replicates within a variety of host cell types but particularly in muscle cells and macrophages. At the same time, with appropriate cytokine stimulation, macrophages play key roles in phagocytic destruction of the parasite and as antigen-presenting cells in the induction of acquired immune responses. Finally, *T. cruzi* infection in mice and humans is accompanied by profound, long-term immunosuppression of both nonspecific and parasite-specific immune responses. This immunosuppression is mediated, at least in part, by suppressor macrophages.

II. ANTIBODY AND COMPLEMENT

Several investigations have shown that passive transfer of immune serum from immunized or convalescent mice can confer varying degrees of protection in naive mice infected with *T. cruzi* [1]. Although antibodies play a definite role in immunity to the parasite, antibody-dependent, complement-mediated lysis of blood-form trypomastigotes may not be the primary means by which circulating parasites are removed from the blood or interstitial fluids. Under conditions in which parasites should be susceptible to antibody-dependent, complement-mediated lysis in vitro, a large proportion of either blood-form or fibroblast-derived trypomastigotes are resistant to destruction [2]. It appears that the parasite has evolved means by which it can avoid lysis by utilizing membrane components that inactivate or interrupt the complement cascade [3,4]. The principal means by which antibody and complement participate in immune clearance of *T. cruzi* from the blood, therefore, may involve opsonic activities leading to phagocytosis of trypomastigotes and amastigotes.

Treatment of normal mice with immune mouse serum greatly increases immune clearance of parasites injected shortly thereafter [5]. Heat-inactivated immune mouse serum also has been shown to increase the infectivity of trypomastigotes for resident macrophages in culture [6]. The opsonic capacity of antibodies to enhance macrophage interiorization of *T. cruzi* is apparently, as expected, mediated via Fc-receptors [7].

Complement component C3 is also critical, perhaps more so than antibodies, for increased clearance of *T. cruzi* from the blood of infected mice. Studies by Kierszenbaum and co-workers [8] showed that treatment of mice with cobra venom factor greatly increased parasitemia and mortality in infected mice, and these investigators attributed the effect of depleting complement to loss of antibody-dependent, complement-mediated lysis of the parasite. More recently, Mota and Umekita [9] have shown that, whereas treatment of mice with parasite-specific immunoglobulin G (IgG) led to very rapid clearance from the blood, treatment of the mice with cobra venom factor completely reversed the effects of the antibody on immune clearance.

The role of C3 in phagocytosis may be especially significant in macrophage–*T. cruzi* interactions, although complement receptors (CR3) are not preferentially incorporated into phagocytic vesicles containing trypomastigotes, even when phagocytosis occurs in the presence of immune serum [10].

III. CYTOKINES AND PHAGOCYTIC KILLING OF *TRYPANOSOMA CRUZI*

Pioneering studies by Taliaferro and Pizzi [11] and others demonstrated the significance of macrophages in immunity to *T. cruzi* and laid the foundation for future studies of macrophages in experimental Chagas' disease. Extensive studies by Hoff [12] showed that activated macrophages capable of killing *T. cruzi* in vitro were evident in infected mice by day 21 of infection in C3H mice, but that the activation of the macrophages was nonspecific, i.e., the same effects were seen by stimulation of macrophages with Bacillus Calmette-Guérin (BCG), and *T. cruzi*– or BCG-stimulated macrophages were highly effective in vitro in killing *Listeria*. Hoff [12] also showed in these early studies that antigen activation of macrophages in vivo, of itself, did not provide significant protection to *T. cruzi*–infected mice.

Although activated macrophages were shown to have greatly enhanced capabilities in inhibiting intracellular growth of *T. cruzi* amastigotes, it was not clear if the effect was trypanocidal or trypanostatic, nor was it clear how invading parasites, which ultimately would reside free in the cytoplasm of macrophages and replicate, avoided phagolysosomal destruction following engulfment. In a controversial study, Noguiera and Cohn [13] reported that parasites have the ability to escape the phagolysosome by mechanisms not elucidated at the time. In the intervening period, several studies on macrophages in experimental Chagas' disease have focused on understanding the mechanisms for stimulating macrophage–mediated killing of *T. cruzi* and the parasite's ability to avoid these mechanisms.

In recent years numerous cytokines that affect the activities of macrophages have been identified and are available in purified form. Experiments examining the levels of specific cytokines in serum or as products of host leukocytes, as well as the effects of specific cytokines provided exogenously in vitro or in vivo, are now providing new insights into activities of macrophages in control of parasitic infection.

The most thoroughly studied of the cytokines in experimental Chagas' disease is interferon-γ (IFN-γ) [14–18]. Treatment of macrophages or macrophage-like cell lines with IFN-γ in vitro inhibits the intracellular replication of amastigotes of *T. cruzi* by trypanostatic/trypanocidal mechanisms without affecting the initial invasion or interiorization of the parasite [16,17]. In vivo administration of recombinant IFN-γ (rIFN-γ) has multiple effects on both mice that are highly susceptible and mice that are resistant to *T. cruzi* infection [18].

One of the earliest studies by Nathan et al. [19] showed that supernatants of spleen cells stimulated by antigen in vitro contained factors that dramatically enhanced macrophage killing of *T. cruzi*. Enhanced killing was found to be dependent on the continued exposure of infected macrophages to the factors with the supernatant. These investigators also clearly demonstrated that killing of *T. cruzi* by activated macrophages correlated directly with the production of H_2O_2. It is now clear that the major functional cytokine in the supernatant of stimulated spleen cells for activation of trypanostatic/trypanocidal activity of macrophages is IFN-γ.

Wirth et al. [15] examined the effect of IFN-γ and determined that macrophages were stimulated specifically by this lymphokine to kill the parasite in vitro. With the availability of purified lymphokines from recombinant methods, Reed and co-workers [17,20] began a series of experiments to determine the contribution of IFN-γ and other mediator/signal molecules on immunity to *T. cruzi* and, specifically, the trypanostatic/trypanocidal activation of macrophages. These studies revealed that recombinant murine IFN-γ (rMuIFN-γ), as well as recombinant murine granulocyte-macrophage colony-stimulating factor (rMuGM–CSF), could stimulate extraordinary inhibition of *T. cruzi* in treated macrophages [20]. Although both cytokines had significant effects, 1 ng/ml of rMuIFN-γ was effective in inhibiting parasites by approximately 95%, whereas 10 ng/ml of rMuGM-CSF was effective in inhibition by less than 80%. The differences in effectiveness of the two cytokines may be related to the four times lower stimulation of H_2O_2 production of macrophages treated with rMuGM-CSF as compared with rMuIFN-γ [20]. They further demonstrated that neither cytokine was toxic to the parasite and that neither affected initial interaction/interiorization of the parasite.

Notwithstanding the effectiveness of IFN-γ in increasing resistance to *T. cruzi* in infected macrophages, it has not proved possible to obtain 100% inhibition of parasites by macrophages. In an attempt to determine potential additive or synergistic effects of various cytokines on IFN-γ–enhanced stimulation of macrophages, Golden and Tarleton [16] found that lipopolysaccharide (LPS) significantly increased the trypanocidal effect of IFN-γ on infected macrophages and that the effect was not due to LPS stimulation of tumor necrosis factor-α (TNF-α) or TNF-β. It was also shown that a complex mixture of cytokines in the supernatant of PMA-stimulated EL-4 cells was relatively ineffective in stimulating trypanocidal activity in peritoneal macrophages, but that addition of IFN-γ to supernatants of stimulated EL-4 cells was much more effective than IFN-γ alone. Treatment of the supernatant with monoclonal antibody (mAb) against interleukin-4 (IL-4), but not mAb against IL-2 and IL-5, abrogated the additive effectiveness of the mixture of cytokines. However, treatment of macrophages with recombinant IL-4 (rIL-4) alone had no inhibitory effect, and the combination of rIL-4 and rIFN-γ was no more effective than IFN-γ alone. On the basis of these results the authors [16] suggested that the synergism of IFN-γ and the other cytokines in the supernatant is complex and interactive.

There are, however, reports using different experimental systems and designs that provide data differing from the results of Golden and Tarleton [16]. Wirth et al. [21] used amastigotes to infect macrophages and found that rIL-4 alone did stimulate uptake and intracellular killing of parasites, and that this effect was related to IL-4 stimulation of oxygen reduction metabolites, including H_2O_2, in the macrophages. Also, Munoz-Fernandez et al. [22] reported a profound synergistic effect of addition of rMuIFN-γ and rMuTNF-α on trypanocidal activity of infected macrophages. They also found that TNF-α or LPS alone had little effect on macrophage inhibition of *T. cruzi*. Of note is a report by Wirth and Kierszenbaum [23] that rTNF alone is not effective in stimulating macrophage trypanostatic/trypanocidal activity but is highly effective in the presence of LPS and imploring caution in the utilization of potentially LPS-contaminated reagents, with the suggestion that earlier reports of the stimulatory activity of TNF were due to LPS contamination of reagents [24].

Although it is clear that administration of IFN-γ, either in vivo or to cultures of macrophages, inhibits *T. cruzi*, there is not a correlation of IFN-γ production in vivo with susceptibility or resistance in different strains of mice. Silva et al. [18] have shown that concanavalin A–stimulated splenocytes from infected susceptible C57BL/6 (B6) mice and resistant (C57B1/6 × DBA)F_1 (B6D2) mice produced similar levels of IFN-γ, and that B6, but not B6D2, mice produced elevated levels of IFN-γ in the absence of mitogen stimulation. However, these investigators determined that B6 mice produced significantly higher levels of IL-10 than did B6D2 mice and suggested that IL-10 down-regulates IFN-γ production in infected mice and, importantly, that IL-10 could block IFN-γ–mediated inhibition of *T. cruzi* in macrophages. Transforming growth factor-β (TGF-β) may also play a role in the regulation of IFN-γ–mediated effects in infected mice [25]. Susceptible B6 mice were shown to produce significant levels of TGF-β by day 8 of infection, and although the levels of TGF-β in resistant B6D2 mice were not reported, injection of both B6 and B6D2 mice with rTGF-β increased parasitemia and resulted in 50% mortality in the resistant B6D2 mice.

Nabors and Tarleton [26] have examined the differential control of IFN-γ and IL-2 in resistant and susceptible mice and found, as well, that both produce significant levels of IFN-γ. Interleukin-2 deficiency is a well-characterized phenomenon in *T. cruzi*–infected mice and is central to the profound immunosuppression of T-dependent responses observed in infected mice. Examination of mRNA levels for IFN-γ and IL-2 revealed substantial IFN-γ levels but selective down-regulation of IL-2 mRNA. A surprising finding in this study was that cells responsible for production of IFN-γ were Thy-1+, CD4–, CD8–, NK–cells, with little production of IFN-γ by CD4+ and CD8+ cells.

Although it is well documented that IFN-γ will provide protection to susceptible mice, but only when injected early in infection, the role of endogenous IFN-γ in providing protection has only recently been determined. Injection of anti–IFN-γ mAb dramatically increases parasitemia levels, decreases longevity, and increases mortality in both resistant and susceptible mouse strains [18,27]. The studies described above clearly demonstrate a central role of macrophage-mediated inhibition of *T. cruzi* both in vivo and in vitro as well as the significant contributions of cytokines in enhancement and regulation of macrophage activity. However, it should be noted that there seem not to be any deficiencies in monokine production in mice infected with *T. cruzi*. It would appear, as related above, that the principal microbicidal activities of macrophages are regulated basically by lymphokines as stimulators or inhibitors of trypanostatic/trypanocidal activity.

IV. KILLING MECHANISMS

Macrophages' utilization of oxygen-reduction metabolites as a principal antimicrobial mechanisms has been accepted for some time [19]. However, it is becoming increasingly apparent that nitric oxide, inducible in macrophages via nitric oxide synthase production, plays a major, perhaps central, role in diverse microbicidal and other activities [28]. Recent studies have shown that levels of nitric oxide synthase in activated murine macrophages correlate with resistance to *Leishmania major* in inbred strains of mice [29]. Both IFN-γ and TNF-α, in the presence of LPS, induced high levels of nitric oxide in resistant mice (CBA) and much less in susceptible mice (BALB/c).

Likewise, Munoz-Fernandez et al. [22] found similar results when examining nitric oxide production in *T. cruzi*–infected J774 cells or murine peritoneal macrophages. Tumor necrosis factor-α and IFN-γ induced significant levels of nitric oxide, which correlated with trypanocidal activity of the infected host cells. Addition of N-monomethyl-L-arginine, a specific competitive inhibitor for nitric oxide synthase, eliminated the trypanostatic/trypanocidal effect of the cytokine-stimulated macrophages [22].

V. ESCAPE MECHANISMS

Nogueira et al. [13] showed that *T. cruzi* escaped the phagolysosome and then divided in the cytosol of host macrophages. Until recently, little was known regarding the means by which parasites could accomplish this task. However, a series of studies by Andrews and co-workers [30,31] showed that about 70% of interiorized parasites were extravacuolar at 2 hours after infection of macrophages and strongly implicated a parasite-derived hemolytic protein, which is optimally active at pH 5.5, in the escape of the parasite from the phagolysosome. In attempts to provide support of the hypothesis that this acid-active hemolysin aids in escape from the phagolysosome, experiments were performed in which the pH of the phagolysosome of infected macrophages was increased using various chemical manipulations (e.g., chloroquine). Increasing the vacuolar pH substantially increased the percentage of parasites retained within phagolysosomes of infected cells. This 60- to 75-kDa hemolysin is capable of forming channels in membranes and is homologous to complement component C9.

VI. IMMUNOREGULATION

Mice infected with *T. cruzi* demonstrate a profound immunosuppression of parasite-specific and nonspecific responses that affects both T-dependent and T-independent responses [32–34]. This immunosuppression begins early in infection and persists well into the postacute (chronic) phase and is characterized, in part, by deficiency in IL-2 production (see above). Early studies identified macrophage-like adherent cells that could down-regulate normal spleen cell responses to mitogens, heterologous erythrocytes, and other antigens [33]. Although immunosuppression is demonstrable in lymph node and spleen cell populations of infected mice, suppressor cells have been described only in spleen cells [34]. To date, little is known regarding the mechanism or mechanisms by which these suppressor macrophages can down-regulate immune responses.

Recent studies have found, however, that there are at least two subpopulations of splenic suppressor macrophages [34]. Using limiting dilution-partition analysis, it has been shown that there are three dose-zones of immunoregulation when spleen cells of infected mice are added to normal splenocytes. The first of these, representing the most abundant macrophage suppressor cells, down-regulates anti–sheep erythrocyte and mitogenic responses. The second or intermediate dose-zone of regulator cells overcomes the first dose-zone of suppression (i.e., has an immunoenhancing effect) and is mediated by resident T cells or, with T-depleted splenocytes from infected mice, exogenously supplied, antigen-primed T cell lines. The third dose-zone of regulation by spleen cells of infected mice suppresses responses in a dose-dependent manner and is mediated by macrophages.

It has been determined as well that the most abundant (low-dose zone) population of suppressor macrophages does not down-regulate IL-2 production, and that this suppression cannot be overcome by exogenously supplied IL-2 [35]. The least abundant of the suppressor macrophages (third dose-zone) dramatically inhibits PHA-induced IL-2 production of normal spleen cells. The suppression of PHA-induced proliferation by this least abundant population of suppressor macrophages, however, is readily overcome by the addition of IL-2 to cultures. Finally, splenic suppressor cells can directly inhibit the proliferative response of CTLL-2 cells to IL-2.

Immunosuppression during the course of experimental Chagas' disease has an extraordinary effect on the outcome of the disease, as demonstrated by the effects of administration of cytokines to infected mice, particularly IFN-γ and GM-CSF (see above). To date, however, it has not been determined what the conditions or stimuli are for the induction of suppression and suppressor cells, nor have mechanisms been elucidated that define the means by which suppressor macrophages down-regulate host immune cells. Although little information exists on the role of suppressor T cells in experimental Chagas' disease, recent studies have shown that IL-2 production is down-regulated selectively by transcriptional control of IL-2 mRNA and that this down-regulation is dependent on splenic T cells [26,36,37]. It is also known that maintenance of immunosuppression in cultures of splenic cells is dependent on the presence of parasite antigens, suggesting a role for antigen-specific T cells in the maintenance of suppressor cell activities [36,37].

A possible role for nitric oxide in immunosuppression in experimental Chagas' disease has not been examined to date. Numerous studies have determined that IFN-γ stimulation of macrophages induces enhanced nitric oxide production, and that this correlates with suppression of nonspecific lymphoproliferation [38,39]. It will be particularly interesting to determine if one or both of the suppressor macrophage subpopulations described above exert their suppressive activities, at least in part, via nitric oxide production.

VII. SUMMARY AND CONCLUSIONS

The foregoing provides a basis for considering that macrophages are important in basically every aspect of the host-parasite relationship in experimental Chagas' disease. The myriad of activities of macrophages and the diverse responses of these cells to *T. cruzi* and various stimulatory and inhibitory cytokines as measured both in vitro and in vivo are suggestive of the complexity of the cell-cell interactions of the host during the course of infection and the evasive activities of the parasite.

REFERENCES

1. Kierszenbaum F, Howard JG. Mechanisms of resistance against experimental *Trypanosoma cruzi* infection: the importance of antibodies and antibody-forming capacity in the Biozzi high and low responder mice. J Immunol 1976; 116:1208–11.
2. Jacobson KC, Fletcher RC, Kuhn RE. Binding of antibody and resistance to lysis of trypomastigotes of *Trypanosoma cruzi*. Parasite Immunol 1992; 14:1–12.
3. Norris KA, Bradt B, Cooper NR, So M. Characterization of a *Trypanosoma cruzi* C3 binding protein with functional and genetic similarities to the human complement regulatory protein, decay-accelerating factor. J Immunol 1991: 147:2240–7.
4. Joiner K, Sher A, Gaither T, Hammer C. Evasion of alternative complement pathway by *Trypanosoma cruzi* results from inefficient binding of factor B. Proc Natl Acad Sci USA 1986; 83:6593–7.
5. Umekita LF, Mota I. Role of the mononuclear phagocyte system in the immune and nonspecific clearance of *Trypanosoma cruzi* bloodstream trypomastigotes. Brazilian J Med Biol Res 1989; 22:1489–95.
6. Kierszembaum F, Lima MF, Wirth JJ. Effects of antiserum to *Trypanosoma cruzi* on the uptake and rate of killing of vector-borne, metacyclic forms of the parasite by macrophages. Int J Parasitol 1985; 15:409–13.
7. Rodriguez de Cuna C, Kierszenbaum F, Wirth JJ. Binding of the specific ligand to Fc receptors on *Trypanosoma cruzi* increases the infective capacity of the parasite. Immunology 1991; 72:114–20.
8. Budzko DB, Pizzimenti MC, Kierszenbaum F. Effects of complement depletion in experimental Chagas' disease: immune lysis of virulent blood forms of *Trypanosoma cruzi*. Infect Immun 1975; 11:86–91.
9. Mota I, Umekita LF. The effect of C3 depletion on the clearance of *Trypanosoma cruzi* induced by IgG antibodies. Immunol Letters 1989; 21:223–6.
10. Hall BF, Furtado CC, Joiner KA. Characterization of host cell–derived membrane proteins of the vacuole surrounding different intracellular forms of *Trypanosoma cruzi* in J774 cells. J Immunol 1991; 147:4313–21.
11. Taliaferro WH, Pizzi T. Connective tissue reactions in normal and immunized mice to a reticulotropic strain of *Trypanosoma cruzi*. J Infect Dis 1995; 96:199–208.
12. Hoff, R. Killing *in vitro* of *Trypanosoma cruzi* by macrophages from mice immunized with *T. cruzi* of BCG, and absence of cross-immunity on challenge *in vivo*. J Exp Med 1975; 142:299–311.

13. Nogueira N, Cohn Z. *Trypanosoma cruzi*: mechanism of entry and intracellular fate in mammalian cells. J Exp Med 1976; 143:1402–20.
14. Kierszenbaum F, Sonnenfeld G. Characterization of the antiviral activity produced during *Trypanosoma cruzi* infection and protective effects of exogenous interferon against experimental Chagas' disease. J Parasitol 1982; 68:194–8.
15. Wirth JJ, Kierszenbaum F, Sonnenfeld G, Zlotnik A. Enhancing effects of gamma interferon on phagocytic cell association with and killing of *Trypanosoma cruzi*. Infect Immun 1985; 49:61–6.
16. Golden JM, Tarleton RL. *Trypanosoma cruzi*: cytokine effects on macrophage trypanocidal activity. Exp Parasitol 1991; 72:391–402.
17. Reed SG. *In vivo* administration of recombinant IFN-gamma induces macrophage activation, and prevents acute disease, immune suppression, and death in experimental *Trypanosoma cruzi* infection. J Immunol 1988; 140:4342–7.
18. Silva JS, Morrissey PJ, Grabstein KH, Mohler KM, Anderson D, Reed SG. Interleukin 10 and interferon γ regulation of experimental *Trypanosoma cruzi* infection. J Exp Med 1992; 175:169–74.
19. Nathan C., Nogueira N, Juangbhanich C, Ellis J, Cohn Z. Activation of macrophages *in vivo* and *in vitro*. Correlation between hydrogen peroxide release and killing of *Trypanosoma cruzi*. J Exp Med 1979; 149:1056–68.
20. Reed SG, Nathan CF, Pihl DL, Rodricks P, Shanebeck K, Conlon PJ, Grabstein KH. Recombinant granulocyte/macrophage colony-stimulating factor activates macrophages to inhibit *Trypanosoma cruzi* and release hydrogen peroxide. Comparison with interferon γ. J Exp Med 1987; 166:1734–46.
21. Wirth JJ, Kierszenbaum F, Zlotnik A. Effects of IL-4 on macrophage functions: increased uptake and killing of a protozoan parasite (*Trypanosoma cruzi*). Immunol 1989; 66:296–301.
22. Munoz-Fernandez MA, Fernandez MA, Fresno M. Synergism between tumor necrosis factor-α and interferon-γ on macrophage activation for the killing of *Trypanosoma cruzi* through a nitric oxide–dependent mechanism. Eur J Immunol 1992; 22:301–7.
23. Wirth JJ, Kierszenbaum F. Recombinant tumor necrosis factor enhances macrophage destruction of *Trypanosoma cruzi* in the presence of bacterial endotoxin. J Immunol 1988; 141:286–8.
24. De Titto EJ, Catterall R, Remington JS. Activity of recombinant tumor necrosis factor on *Toxoplasma gondii* and *Trypanosoma cruzi*. J Immunol 1986 137:1342–5.
25. Silva JS, Twardzik DR, Reed SG. Regulation of *Trypanosoma cruzi* infections *in vitro* and *in vivo* by transforming growth factor β (TGF-β). J Exp Med 1991; 174:539–45.
26. Nabors GS, Tarleton RL. Differential control of IFN-γ and IL-2 production during *Trypanosoma cruzi* infection. J Immunol 1991; 146:3591–8.
27. Torrico F, Heremans H, Rivera MT, Marck EV, Billiau A, Carlier Y. Endogenous IFN-γ is required for resistance to acute *Trypanosoma cruzi* infection in mice. J Immunol 1991; 146:3626–32.
28. Ding AH, Nathan CF, Stuehr DJ, Release of reactive nitrogen intermediates and reactive oxygen intermediates from mouse peritoneal macrophages. Comparison of activating cytokines and evidence for independent production. J Immunol 1988; 141:2407–12.
29. Liew FY, Li Y, Moss D, Parkinson C, Rogers MV, Moncada S. Resistance to *Leishmania major* infection correlates the induction of nitric oxide synthase in murine macrophages. Eur J Immunol 1991; 21:3009–14.
30. Ley V, Robbins ES, Nussensweig V, Andrews NW. The exit of *Trypanosoma cruzi* from the phagosome is inhibited by raising the pH of acidic compartments. J Exp Med 1990; 171:401–13.
31. Andrews NW. The acid-active hemolysin of *Trypanosoma cruzi*. Exp Parasitol 1990 71:241–4.
32. Tarleton RL. Regulation of immunity in *Trypanosoma cruzi* infection. Exp Parasitol 1991; 73:106–9.
33. Cunningham DS, Kuhn RE. Trypanosoma cruzi–induced suppression of the primary immune response in murine cell cultures to T-cell–dependent and –independent antigens. J Parasitol 1980; 66:16–27.
34. Cerrone MC, Kuhn RE. Macrophage regulation of immune responses of spleen cells from mice infected with *Trypanosoma cruzi*. Cell Immunol 1991; 138:423–36.
35. Cerrone MC, Ritter DM, Kuhn RE. Effect of antigen-specific T helper cells or interleukin-2 on suppressive activity of macrophage subsets detected in spleens of *Trypanosoma cruzi*–infected mice as determined by limiting dilution-partition analysis. Infect Immun 1992; 60:1489–98.
36. Tarleton RL. *Trypanosoma cruzi*–induced suppression of IL-2 production. I. Evidence for the presence of IL-2–producing cells. J Immunol 1988; 140:2763–8.

37. Tarleton RL. *Trypanosoma cruzi*–induced suppression of IL-2 production. II. Evidence for a role for suppressor cells. J Immunol 1988; 140:2769–73.
38. Denham S, Rowland IJ. Inhibition of the reactive proliferation of lymphocytes by activated macrophages: the role of nitric oxide. Clin Exp Immunol 1992; 87:157–62.
39. Tomioka H, Saito H. Characterization of immunosuppressive functions of murine peritoneal macrophages induced by various agents. J Leukocyte Biol 1992; 51:24–31.

IV

MACROPHAGE–FUNGAL INTERACTIONS

30

Macrophage Interactions with *Candida*

Richard Calderone and Joy Sturtevant

Georgetown University School of Medicine, Washington, D.C.

I. INTRODUCTION

One only needs to examine the various forms of candidosis to understand the underlying host defenses critical to protection. Thus, mucosal and/or cutaneous candidosis is associated with a breakdown in one of several aspects of cell-mediated immunity (CMI) [1]. Chronic mucocutaneous candidosis (CMCC) is the prototypical case in point. Patients either lack functional T lymphocytes, are unable to produce mediators of immunity, or are unable to recognize antigen, in this case antigens of *Candida albicans*. While defects may occur in other aspects of immunity, the predominant defect is in CMI. The lack of patient lymphocyte proliferation in the presence of *Candida* antigens is thought to be due to a suppression induced either by the presence of serum inhibitors of proliferation or by some immunosuppressive component of the organism [1]. A similar type of suppression has been observed in patients with chronic vaginal candidosis [1]. T cell defects are also associated with pharyngeal mucosal infections of AIDS patients. However, systemic invasion in CMCC, in vaginitis, or in AIDS patients is rare, probably because of an intact or nearly intact neutrophil response and other host factors [2]. In neutropenic patients, systemic invasion by the organism may occur, an observation that suggests a critical role for the neutrophil in protection.

Mucosal barriers, serum factors such as lactoferrin [3], and the complement system [4] are also critical to the host defense reactions against this organism. Activation of the alternative complement pathway leads to the depostion of C3b on the surface of yeast cells [5]. Presumably, complement receptors (CR1 or CR3) of phagocytes use this ligand in accomplishing phagocytosis.

The role of phagocytic cells such as neutrophils in protection has been mentioned above. This conclusion is based on three lines of evidence. First, the histopathological response to invasion by *Candida*, at least in the early stages of infection and in the absence of immunocompromising conditions, is usually composed of microabscesses containing polymorphonuclear leukocytes (PMNLs) surrounded by mononuclear cells, the latter in relatively small number. As the organism is eliminated from lesions, the number of mononuclear cells increases [6]. Compared with this observation is the compromising condition of neutropenia, which is often associated with disseminated disease. Second, in vitro studies using either mouse, human, or guinea pig PMNLs have documented the candidicidal activities of PMNLs [7–9]. In these studies, killing activity by PMNLs was measured using quantitative plate culture, dye exclusion, or ^{51}chromium release assays. Correlation of techniques for measuring viability was fairly good [7]. It should be noted, however, that the extent of killing of *Candida* by PMNLs in vitro varies considerably among investigators. The reasons for this variability are uncertain but may reflect effector : target ratios, the choice of opsonizing reagent, and the method involved in determining the extent of killing. Killing of blastoconidia by PMNLs occurs through both oxidative and nonoxidative mechanisms [10–13].

While most of the in vitro studies with PMNLs were done with blastoconidia (yeast) of the organism, filamentous forms of the organism appear also to be susceptible to oxidative killing, although attachment and subsequent processing of the organism by the PMNLs remains extracellular (nonphagocytic) for the most part [11,14,15].

Finally, a third observation which points to a role of PMNLs in protection against *C. albicans* is that treatment of animals with agents that selectively deplete PMNLs leads to systemic disease [16]. Thus, induction of neutropenia by treatment of mice with cyclophosphamide (CY) rendered them more susceptible to *Candida* invasion of the kidney and other organs. Recovery by animals from invasion occurred if the time of inoculation was extended after CY treatment. Associated with this recovery was a restoration of PMNL numbers and an unidentified splenic factor.

While the PMNL performs an important function in protection against *Candida* invasion, especially systemic disease, the role of mononuclear phagocytes is uncertain. It is the purpose of this chapter to focus on both in vitro and in vivo studies of monocyte-macrophage interactions with *Candida* and to define a role for this cell type in resistance to infection.

II. *CANDIDA*-NONACTIVATED MONOCYTE/MACROPHAGE INTERACTIONS

Human blood monocytes may possess two candidicidal mechanisms [17]. One of these systems appears to be identical to the myeloperoxidase-H_2O_2-halide system of PMNLs (MPO), while the second, a myeloperoxidase-independent system, in comparison is weakly effective. As an extension of these studies, the ability of rabbit alveolar macrophages (AMs) and resident peritoneal macrophages (PMs) to kill *Candida* was also investigated [18]. While killing by nonstimulated cells was relatively low (15% by PMs and 28% by AMs), enhanced killing was observed when cells were obtained from animals treated 3 weeks before infection with complete Freund's adjuvant. The use of inhibitors of macrophage metabolic function allowed these investigators to determine that the candidicidal mechanism for rabbit AMs and PMs was distinct from the MPO system of human granulocytes and monocytes.

In this regard, a comparison of the killing activity of human PMNLs, monocytes and macrophages against *C. albicans* when cultured both aerobically and anaerobically was recently undertaken [19] (Table 1). Monocytes killed the organism to a similar extent regardless of the culture conditions, while a 2.5-fold increase in killing was observed with macrophages when the assay was performed aerobically. The optimum time for killing by both types of cells was 6 hours. In comparison, optimum killing by PMNLs occurred at 24 hours and under aerobic conditions. Killing by PMNLs and macrophages was compared in the presence of inhibitors of oxidative

TABLE 1 A Comparison of the Candidicidal Activity of Human Monocytes (MO), Macrophages (MA), and Neutrophils (PMNL)

	MO	MA	PMNL
% Killing	20	50	50
Conditions			
Time	6 hr	6 hr	24 hr
Aerobic/anaerobic	Equal	Aerobic	Aerobic
Inhibitors	ND	SOD, catalase	SOD, catalase, thiourea

Source: Ref. 19.
ND = no data; SOD = superoxide dismutase.

metabolism. An inhibition of killing was observed for both cells in the presence of superoxide dismutase (SOD) and catalase; however, methionine or thiourea also inhibited PMNL function, although not to the same degree as with SOD and catalase. Since thiourea is an inhibitor of hydroxyl radical formation, the authors postulate that the lack of this oxidative metabolite may explain why macrophages do not have enhanced killing of *Candida* even though ingestion and respiratory burst by macrophages is greater than with PMNLs. Other observations also suggest a role for oxidative metabolites in the killing of *C. albicans* [20]. Freshly isolated human monocytes could ingest and kill *C. albicans*. When cultured in vitro, the monocytes continued to ingest the organism but lost their ability to kill. Correlated with the loss of candidicidal activity was a decrease in the activity of MPO and the production of hydroxyl radicals. However, the level of O_2^- was similar regardless of the age of the monocytes. These data suggest that the candidicidal activity of monocytes is associated with MPO and hydroxyl levels of the cell and not with superoxide generation.

While the previously discussed data would point toward a role for oxidative metabolites as candidicidal, other investigators have shown little correlation between the candidicidal activity of cells from a monocyte-macrophage lineage and the production of reactive oxygen intermediates (ROIs). Nylon wool–nonadherent, nonparenchymal liver cells [21] and bone marrow [22] cells that phenotypically were identified as macrophage precursors have before their maturation the ability to kill both *Candida* and tumor cells. In the case of the bone marrow–derived precursors, candidicidal activity was lost when these cells were cultured and allowed to mature [22]. The relationship of these precursor cells to candidicidal activity and ROIs was investigated [23]. As stated above, bone marrow, spleen, and liver macrophage precursor cells were effective in killing *C. albicans* yeast cells. If cells were allowed to mature in culture and then were assayed for their effects against the organism, only liver and spleen cells retained their ability to kill *Candida*. Bone marrow and peritoneal macrophages lost over 50% of their candidicidal activity. The levels of ROIs as determined by chemoluminescence (CL) were not correlated with killing activity. For example, peritoneal macrophages displayed the highest level of CL but were the least effective cell with regard to anti-*Candida* activity, in comparison with Kupffer and liver cells, which had 25% to 30% of the CL of peritoneal cells but were strongly candidicidal.

The difference in data on the effects of the oxidative metabolites may be related to the system studied. Thus, the observations suggesting that ROIs may play some role in protection was made with human cells. The mouse system suggests that ROIs may not be of prime importance.

III. THE EFFECT OF ACTIVATORS OF MONOCYTE/MACROPHAGE FUNCTION ON THEIR CANDIDICIDAL ACTIVITY

Of particular interest, as stated above, was the observation that stimulated macrophages displayed enhanced killing of the organism. Similarly, peritoneal macrophages from mice previously immunized with viable blastoconidia [24] or Bacillus Calmette-Guérin (BCG) or macrophages exposed to lymphokines of phytohemagglutinin-treated lymphocytes [25] were able to retard the germination of yeast forms of the organism for several hours. Retardation of germ tube development by PMs was associated with the development of cell-mediated immunity in animals immunized with viable *Candida* cells [24]. Activation of mouse macrophages to a greater killing capacity has also been reported using an immunoreactive fraction from *Escherichia coli* [26]. These examples clearly demonstrate the enhanced killing of *Candida* by mononuclear phagocytes by induction with both specific and nonspecific treatments.

Activation of monocytes/macrophages would appear to be an essential step in a protective response by these cells. Unstimulated cells have often been reported to have little effect on *Candida* and may, in fact, serve as a vehicle of transportation for the organism, as suggested by others [27]. For example, rapid killing of human peripheral blood–enriched monocytes was observed when these cells were incubated with viable blastoconidia of *C. albicans* [28]. Killing of these monocytes was measured using a chromium-release assay. The rapid killing, exhibited only with strains of *C. albicans* and not with *C. krusei* or *C. stellatoidea*, was not associated with germination and outgrowth of the organism from within the monocytes, since release of chromium occurred within 15 to 30 minutes following phagocytosis, while germination generally requires 1 to 2 hours. Killing of monocytes was also verified by differential dye staining.

The mechanism or mechanisms that contribute to the relative insensitivity of *C. albicans* to resident (unstimulated) macrophages have not been studied to any great extent. The organism seems to be sensitive to the products of oxidative metabolism of phagocytic cells. In this regard, the susceptibilities of *C. albicans* and five bacteria to the oxidative products produced by the xanthine oxidase system have been determined [29]. In an in vitro assay, only *C. albicans* and *Mycobacterium tuberculosis* were essentially resistant to the products of this system, in comparison to *Escherichia coli*, *Listeria monocytogenes*, *Salmonella typhimurium*, and *Staphylococcus aureus*. Killing of *C. albicans* occurred at concentrations of H_2O_2, sodium iodide, and ferrous sulfate 10 times greater than the amount needed to kill *E. coli*, *S. aureus*, and *S. typhimurium*. On the other hand, *C. albicans* was second to *S. aureus* in triggering a respiratory burst following phagocytosis by peritoneal macrophages. These observations indicate that *C. albicans* may be less sensitive compared to oxidative products of phagocyte metabolism, even though the organism can stimulate production of such metabolites.

Although it seems unlikely that phagolysosomal fusion (PL-F) is prevented following phagocytosis by mononuclear phagocytes, this process has been studied to some extent [30,31]. Initially, acridine orange (AO) was used to study the intracellular events following phagocytosis of yeast forms of *C. albicans* by resident mouse peritoneal macrophages [30]. Using this staining technique, PL-F was not observed. However, upon further study using Thorotrast, an electron-opaque lysosomal marker, PF-L was observed [31]. It was believed that AO was unable to penetrate viable *C. albicans*, so that what appeared to be a lack of PL-F fusion was really the resistance of the organism to destruction by the macrophage.

A. Activation of Macrophages by Granulocyte-Macrophage Colony-Stimulating Factor

There is agreement among investigators that granulocyte-macrophage colony-stimulating factor (GM-CSF) activates macrophages to inhibit the growth of or kill *C. albicans* [32–36] (Table 2). Recombinant GM-CSF (rGM-CSF) enhanced the phagocytosis and killing of *C. albicans* by

TABLE 2 Effects of Granulocyte-Macrophage Colony-Stimulating Factor (GM-CSF) on the Candidicidal and Phagocytosis Activity of Human and Murine Monocytes/Macrophages

System	U/ml GM-CSF	% Phagocytosis		% Killing		Reference
		No GM-CSF	GM-CSF	No GM-CSF	GM-CSF	
mPEM[a]	6 ng/ml	41	64	10	28	32
mPEM[a]	1000[b]	ND	ND	0	50	33
mGG2EE[c]	3200	35	78	30	29	34
	800	35	70	30	29	
hMo[d]	5–500	ND	ND	9	41	35
hMo[d]	10	ND	ND	—	>90	36

[a]Murine peritoneal exudate macrophages.
[b]Cells were incubated with CSF for 48 hr before the addition of *C. albicans*.
[c]Murine macrophage cell line.
[d]Human monocytes.
ND = no data.

mature murine peritoneal macrophages [32]. Interleukin-3 (IL-3) also enhanced phagocytosis, although not as much as rGM-CSF, while synergy or an additive effect was not observed with macrophages stimulated by both cytokines. Resident, nonstimulated cells minimally killed *Candida* (only 10%), while activation of macrophages increased the killing to approximately 30%.

Resident and proteose-peptone–elicited peritoneal macrophages were compared for their ability to kill *C. albicans* in the presence of GM-CSF [33]. Macrophages were preincubated with CSF-1 for 6, 24, or 48 hours, blastoconidia were added, and the extent of killing was determined by quantitative plate cultures after 10 or 24 hours. Resident macrophages were not affected by CSF-1 at any concentration of CSF-1. However, with exudate macrophages, significant killing of *C. albicans* was observed with macrophages pretreated with CSF-1 for 24 or 48 hours, in comparison with untreated macrophages. For example, macrophages treated with approximately 10,000 U of CSF-1 for 24 hours killed approximately 50% of the organisms compared with untreated macrophages. Longer exposure of macrophage cells to CSF-1 increased the killing capacity of exudate macrophages. Thus, only 1000 U of CSF-1 was needed to obtain similar killing profiles if the macrophages were pretreated for 48 hours. In comparison with the data presented above [32], resident, untreated macrophages were able to kill *Candida* substantially; however, killing was not enhanced in the presence of the cytokine, as stated above.

The effect of CSF-1 on the anti-*Candida* activity of a macrophage cell line has been studied [34]. In contrast to the data presented above, CSF-1 did not increase the killing by this macrophage cell line, although phagocytosis was enhanced significantly at CSF-1 concentrations of 800 to 3200 U. The difference in response of these cells in comparison with the data presented above [32,33] may be related to the type of macrophage, since the concentrations of CSF-1 as well as the length of pretreatment are within the range of the other studies.

Observations on the effect of GM-CSF on human monocytes and their ability to kill *C. albicans* have also been published [35,36]. Purified populations of monocytes obtained by leukophoresis were incubated with radiolabeled *C. albicans* blastoconidia, and the extent of killing of the organism was measured by the release of radiolabeled material [35]. Correlations between the amount of released isotope and viability of the organism had been established, i.e., greater loss of isotope was directly correlated with a decrease in yeast viability. Compared with untreated control monocytes, the extent of killing by monocytes incubated with 500 U of rGM-CSF was enhanced fivefold when the effector:target (E:T) ratio was 2.5:1. Less cytotoxicity was observed at a higher (E:T) ratio (10:1).

In a similar type of study, peripheral blood mononuclear cells were incubated with blastoconidia of *C. albicans* at different E:T ratios overnight. Subsequently, radiolabeled glucose was added and the inhibition of growth of *Candida* by the monocytes was measured as an inhibition of glucose uptake. This assay offers several advantages with regard to measuring the effects of phagocytic cells on the organism, including speed and the avoidance of subjectivity in the case of determining viability with vital stains. On the other hand, it is uncertain whether the organism was killed by the monocytes. Nevertheless, augmented growth inhibition of blastoconidia was observed in the presence of 0.01 to 100 U/ml of GM-CSF at E:T ratios of 30:1 and 10:1. Minimal inhibition of glucose uptake occurred at an E:T ratio of 3:1. The time course of induction of monocyte activity by GM-CSF was determined. The anti-*Candida* activity of monocytes incubated with GM-CSF for 6 hours was similar to that of untreated monocytes. However, GM-CSF was able to prolong the cytotoxic activity of monocytes over a period of 12 to 30 hours. In comparison, the candidicidal activity of untreated monocytes decreased with time in culture.

The activation of mononcytes by GM-CSF results in changes in the number of mannose receptors [33] and is also accompanied by an increase in the level of superoxide anion (O_2^-) [35]. In the former instance, the effect of mannose on the binding of viable *C. albicans* to both resident (PC) and proteose-peptone–elicited peritoneal (PEC) murine macrophages was investigated.

D-mannose (0.001M) inhibited the binding of the organism to PECs treated with GM-CSF but did not inhibit the binding of the organism to untreated PECs nor was any effect of mannose observed with treated or untreated PCs. In addition to D-mannose, mannosamine and *N*-acetylglucosamine (NAGA) also inhibited binding, although NAGA was not as effective. Other sugars that had no effect on the binding of the organism to PECs included L-fucose, D-galactose, and L-mannose. On the basis of these observations, it was suggested that increased binding and uptake of *C. albicans* (and, hence, more growth inhibition) may be associated with an increase in the number of mannose receptors by PECs.

The role of GM-CSF in superoxide production by human monocytes was investigated [35]. In preliminary experiments, it was established that *C. albicans* caused a 10-fold increase in the production of superoxide when cultured alone with human monocytes, suggesting a role for ROIs in the growth inhibition of the organism. In subsequent experiments, either superoxide dismutase or catalase caused a dose-dependent reduction in the ability of monocytes to kill the organism. Heat-inactivation of the enzymes before use overcame this effect. The effect of the ROIs on the organism was also studied by using scavengers of hydroxyl radicals and singlet oxygen scavengers. A substantial reduction in monocyte cytotoxicity was observed with each of the inhibitors used. The combined approach of using both enzymes and inhibitors of ROIs resulted in the conclusion that these products of monocyte metabolism were of importance in the candidicidal activity of these cells. The effect of GM-CSF on human monocyte production of superoxide anion was determined [35]. Monocytes exposed to 500 U/ml of GM-CSF for 14 hours (no organism) produced over a 100% increase in O_2^- upon the addition of cytochrome C. However, if GM-CSF was added at the same time as cytochrome C, no increase in production was observed, showing that the induction of superoxide by monocytes was not caused directly by the cytokine but was probably a result of the cytokine's induction of monocyte differentiation. The specificity of GM-CSF on the generation of superoxide was examined. Granulocyte-macrophage colony-stimulating factor but not granulocyte colony-stimulating factor (G-CSF) caused an increase in superoxide production by monocytes; also, monoclonal antibody to GM-CSF, when added to the culture medium, inhibited the increased superoxide production in GM-CSF–stimulated monocytes. Thus, the data from these studies suggest that rGM-CSF stimulation of respiratory-burst activity coincides with the candidicidal activity of monocytes treated with the cytokine. Since total inhibition of monocyte candidicidal activity was not overcome with any one of the treatments described above (inhibitors of ROIs, intermediates, anti–GM-CSF, or mannose), it is likely that more than one mechanism is involved in the anti-*Candida* activities of mononuclear cells.

B. Activation of Monocytes by Other Cytokines

The role of other cytokines such as interferon-α (IFN-α), IFN-β, or IFN-γ, tumor necrosis factor (TNF), and interleukins (IL-1, IL-3, and IL-4) in the activation of monocytes/macrophages against *Candida* has been investigated [32–34, 36–38] (Table 3). Interleukin-3 has been shown to increase the number of murine peritoneal macrophages that phagocytose *C. albicans* [32]. An increase in the number of organisms per macrophage was also observed, although the amount of killing by treated cells was no greater than by untreated macrophages. On the other hand, IL-3 and IL-4 had no effect on the phagocytosis and killing activity of a murine macrophage cell line, GG2EE [33]. In this same study, it was observed that IL-1–treated cells did inhibit the growth of the organism more than untreated macrophages. With regard to the activation of human monocytes by IL-3, growth inhibition of *C. albicans* was observed when cells were incubated with 0.01 to 100 U/ml IL-3 for 24 hours before the addition of the organism [36]. On the basis of these few studies, it would appear that of the interleukins, IL-3 may have potentiating effects on both human and murine macrophages; IL-1 has a very restrictive capacity to activate a murine macrophage cell line to increased killing of the organism. There is only a single study on the role of TNF in inducing

TABLE 3 The Effect of Cytokines Interferon-α,β and -γ, Interleukins (IL-1, IL-3) and Tumor Necrosis Factor (TNF) on the Anti-*Candida* Activity of Murine and Human Monocytes/Macrophages

Cytokine	Macrophage[a]	Treatment[b]	Effect on *Candida*	Reference
IL-3	mPEC	30 min, 4 ng/ml	Greater uptake Killing unaffected	32
IFN-α,β	mPEC	6–48 hr, 333–1000 U/ml	No effect	33
IFN-α,β	mPC	6–24 hr, 100–1000 U/ml	Candidicidal	33
IL-1, TNF, IFN-γ	GG2EE	Overnight	Growth inhibition	34
IL-3, IL-4	GG2EE	Overnight	No effect	34
IL-3,	hPBC	24 hr, 0.01–100 U/ml	Growth inhibition	36
IFN-γ	hPBC	24 hr, 100 U/ml	No effect	36
IFN-γ	mPC	4–24 hr, 100 U/ml	Increased killing	38
IFN-γ	mPC	Overnight, 200 U/ml	Increased killing	39

[a] mPEC, murine peritoneal exudate cells; mPC, murine resident peritoneal cells; GG2EE, murine macrophage cell line; hPBC, human peripheral blood cells (monocytic).
[b] Conditions for treatment of the macrophage cell with the cytokine.

anti-*Candida* effects, again with the GG2EE macrophage cell line [34]. As with IL-1, TNF also induced these cells to a greater level of killing than with untreated cells.

Several studies on the effect of the interferons on the growth-inhibitory properties of monocytes/macrophages have been reported [33,34,36,38,39]. In the murine system, IFN-α,β induced resident peritoneal macrophages but not peritoneal exudate cells to kill *Candida* to a greater extent than untreated cells [33]. Similarly, murine resident peritoneal cells killed *Candida* to a greater extent than untreated macrophages [38,39]. In each of these studies, the effect of IFN-γ was dose-dependent, and the anti-*Candida* activity of IFN-γ-treated macrophages was inhibited when anti-IFN-γ antibody was included in the assay medium. Spleen cells of normal mice were cultured in the presence of 5 μg of concanavalin A (Con A) [39]. Supernatants from these cultures were also able to activate macrophages to kill the organism to a greater extent than with macrophage cultures incubated only with fresh tissue culture medium. More recently, the mechanisms underlying the augmented activity of IFN-γ and culture supernatant–treated macrophages has been studied [40]. *C. albicans,* in comparison with *Candida parapsilosis*, which was killed by resident and activated macrophages, was killed only by macrophages activated by IFN-γ or Con A–treated spleen culture supernatants. The mechanism of killing induced by either IFN-γ or the culture supernatants was investigated using either inhibitors or enzymes that destroy ROIs. The augmented killing activity of macrophages induced by IFN-γ was inhibited by SOD. However, SOD had less effect on the activation by culture supernatants; instead, activity was inhibited by catalase, dimethyl sulfoxide, or azide. These data suggest that the mechanism resulting in increased killing by macrophages treated with IFN-γ is different from the mechanism induced in Con A–stimulated macrophages.

An effect of IFN-γ on resident macrophages was observed when cells were treated with the cytokine for 24 hours but not after incubation for 48 to 72 hours [39]. By comparison, optimum production of H_2O_2 by similarly treated cells occurred at 48 to 72 hours, suggesting that ROIs may not play a role in the killing of the organism. The level of other types of intermediates was not determined. From both normal and activated macrophages, granule extracts were prepared and tested for their activity against *Candida*. From activated macrophages only, an extract was obtained that was able to kill the organism. The active component(s) was heat-stable and acid-extractable but destroyed by trypsin or by treatment with polyaspartic acid, the latter used to neutralize the killing activity of cationic peptides from neutrophils. The authors believe that this protein(s) is an important candidicidal component that exerts its activity in the acidic milieu of the phagolysosome. Interestingly, treatment of phagocytosing cells with reagents that increased cellular pH inhibited the killing activity of macrophages treated with IFN-γ. Lysosomal extracts from resident alveolar macrophages have also been shown to inhibit *C. albicans* [41,42]. The active material could also be extracted with citric acid and thus appeared to have cationic properties. The uptake of certain amino acids by *Candida* was affected while glucose, uridine, or Rb^+ was unaffected.

A role for IFN-γ in the treatment of patients with chronic obstructive pulmonary disease has been suggested [43]. Both alveolar macrophages and peripheral blood monocytes from these patients were found to be significantly less effective than cells from control patients in the killing of *C. albicans*. However, when cultured in the presence of IFN-γ alone or in combination with lipopolysaccharide, the candidicidal activity of these cells was restored to the level of those from normal individuals. Secretion of TNF and IL-1 was induced in patient's cells following treatment with IFN-γ and lipopolysaccharide.

Recently, the induction of cytokines by a macrophage cell line (ANA-1) in the presence of hyphal (H) and yeast (Y) forms of *C. albicans* was studied [44]. ANA-1 macrophages selectively responded to only (H) forms of the organism. Using Northern blot techniques, the authors showed that TNF activity was up-regulated while lysozyme production was down-regulated. The authors suggest a role for TNF in the killing of the organism, perhaps in a manner similar to that involved in the killing of tumor cells.

IV. MOLECULAR MIMICRY IN *CANDIDA ALBICANS*

A variety of mammalian cells, including phagocytic cells such as PMNs and mononuclear cells, possess receptors for the complement C3 conversion products C3b, iC3b, and C3dg [45]. Additionally, B lymphocytes produce a cell surface receptor for C3d [46]. The complement receptors are designated CR1, CR2, CR3, and CR4 and differ biochemically as well as in ligand specificity. Thus, it was of interest when Heidenreich and Dierich [47] first determined that *C. albicans* possessed CR2- and CR3-type activity in that the organism was able to form rosettes with antibody-sensitized sheep erythrocytes (EA) coated with either iC3b (EAiC3b) or C3d (EAC3d). Other investigators observed similar results [48–52].

The biochemistry of the *Candida* CR2 has been studied [50–52]. The protein is glycosylated, appears to have a molecular mass (SDS-PAGE) of 60 kDa, is found on the surface of hyphae of the organism but is located at the level of the plasma membrane in blastoconidia, and would appear to recognize a broad range of ligands including fibrinogen [53] and laminin [54]. The CR3 activity has not been purified, and there is no clear agreement that the CR2 and CR3 activities are associated with one or different proteins.

Speculation on the role of this protein or proteins in *C. albicans* has focused on a molecular-mimicry type of activity [48,55]. Thus, the organism is seen as able to escape host defense mechanisms by possessing proteins of a nature similar to host cells. A more likely role for this protein or proteins is to provide adherence capabilities for the organism, at least in its recognition of endothelial cells [55].

V. SUMMARY

The monocyte/macrophage, in comparison to the neutrophil, would appear to have a limited role in protection against *C. albicans*. This statement is based on the observations of several investigators who report that these cells have very little killing capacity unless they are activated by cytokines such as IFN-γ and GM-CSF. The mechanisms of killing by these cells appear to include both oxidative and nonoxidative mechanisms, the latter perhaps being more important. The mechanisms of killing may be different for monocytes and macrophages. Granulocyte-macrophage colony-stimulating factor and its effect on monocytes has been studied using *Candida* as a target organism. Two explanations for the enhancement of monocyte killing by this cytokine have been proposed: GM-CSF augments both superoxide anion and the level of mannose receptors on treated monocytes. Both of these changes could be significant in the increased killing capacity of these cells.

ACKNOWLEDGMENTS

The authors wish to thank Dr. Andrea Bailey for reading the manuscript. Part of the research discussed was supported by Public Health Service grant AI25738 to R.C. from the NIH.

REFERENCES

1. Domer JE, Carrow EW. Candidiasis. In: Cox RA, ed. Immunology of fungal diseases. Boca Raton, Florida: CRC Press, 1989:57–92.
2. Greenfield RA. Host defense system interactions with *Candida*. J Med Vet Mycol 1992; 30:89–104.
3. Kirkpatrick CH, Green I, Rich RR, Schade AL. Inhibition of growth of *Candida albicans* by iron-saturated lactoferrin: relation to host defense mechanisms in chronic mucocutaneous candidiasis. J Infect Dis 1971; 124:539–45.
4. Calderone R, Linehan L. The role of complement in host resistance to systemic fungal infection. In: Kurstak E, ed. Immunology of fungal diseases. New York: Marcel Dekker 1989; :225–43.
5. Kozel TR, Brown RR, Pfrommer GS. Activation and binding of C3 by *Candida albicans*. Infect Immun 1987; 55:1890–4.
6. Pearsall N, Lagunoff D. Immunological responses to *Candida albicans*: mouse-thigh lesion as a model for experimental candidiasis. Infect Immun 1974; 9:999–1002.
7. Bistoni F, Baccarini M, Blasi E, Puccetti P, Marconi P. A radiolabel release assay for phagocytic killing of *Candida albicans*. J Immunol Methods 1982; 52:369–74.
8. Schuit K. Phagocytosis and intracellular killing of pathogenic yeast by human monocytes and neutrophils. Infect Immun 1979; 24:932–8.
9. Cutler JE, Thompson BD. A simple and inexpensive method for assessing *in vitro* candidacidal activity of leukocytes. J Immunol Methods 1984; 66:27–33.
10. Lehrer RI. Measurement of candidacidal activity of specific leukocyte types in mixed cell populations. I. Normal, myeloperoxidase-deficient, and chronic granulomatous disease neutrophils. Infect Immun 1970; 2:42–7.
11. Wagner DK, Collins-Leach C, Sohnle PG. Inhibition of neutrophil killing of *Candida albicans* by substances which quench hypochlorous acid and chloramines. Infect Immun 1986; 51:731–5.
12. Lehrer RI, Szklarek D, Ganz T, Selsted ME. Correlation of binding of rabbit granulocyte peptides to *Candida albicans* with candidicidal activity. Infect Immun 1985; 49:207–11.
13. Selsted ME, Szklarek D, Ganz T, Lehrer RI. Activity of rabbit leukocyte peptides against *Candida albicans*. Infect Immun 1985; 49:202–6.
14. Diamond RD, Clark RA, Haudenschild CC. Damage to *Candida albicans* hyphae and pseudohyphae by the myeloperoxidase system and oxidative products of neutrophil metabolism. J Clin Invest 1980; 66:908–17.
15. Diamond RD, Krzesicki R. Mechanism of attachment of neutrophils to *Candida albicans* pseudohyphae in the absence of serum, and of subsequent damage to pseudohyphae by microbial processes of neutrophils *in vitro*. J Clin Invest 1978; 61:360–9.
16. Bistoni F, Baccarini M, Blasi E, Marconi P, Puccetti P, Garaci E. Correlation between *in vivo* and *in vitro* studies of modulation of resistance to experimental *Candida albicans* infection by cyclophosphamide in mice. Infect Immun 1983; 40:46–55.
17. Lehrer RI. The fungicidal mechanisms of human monocytes. I. Evidence for myeloperoxidase-linked and myeloperoxidase-independent candidicidal mechanisms. J Clin Invest 1975; 55:338–46.

18. Lehrer RI, Ferrari LG, Patterson-Delafield J, Sorrell T. Fungicidal activity of rabbit alveolar and peritoneal macrophages against *Candida albicans*. Infect Immun 1980; 28:1001–8.
19. Thompson HL, Wilton JM. Interaction and intracellular killing of *Candida albicans* by human polymorphonuclear leukocytes, monocytes and monocyte-derived macrophages in aerobic and anaerobic conditions. Clin Exp Immunol 1992; 87:316–21.
20. Sasada M, Kubo A, Nishimura T, Kakita T, Moriguchi T, Yamamoto K, Uchino H. Candidacidal activity of monocyte-derived human macrophages: relationship between *Candida* killing and oxygen radical generation by human macrophages. J Leukocyte Biol 1987; 41:289–94.
21. Decker T, Baccarini M, Lohmann-Matthes ML. Liver-associated macrophage precursors as natural cytotoxic effectors against *Candida albicans* and Yac-1 cells. Eur J Immunol 1986; 16:693–9.
22. Baccarini M, Bistoni F, Lohmann-Matthes ML. *In vitro* natural cell-mediated cytotoxicity against *Candida albicans*: macrophage precursors as effector cells. J Immunol 1985; 136:837–43.
23. Decker T, Lohmann-Matthes ML, Baccarini M. Heterogeneous activity of immature and mature cells of the murine monocyte-macrophage lineage derived from different anatomical districts against yeast-phase *Candida albicans*. Infect Immun 1986; 54:477–86.
24. Evron R. *In vitro* phagocytosis of *Candida albicans* by peritoneal mouse macrophages. Infect Immun 1980; 28:963–71.
25. Maiti PK, Kumar R, Mohapatra LN. Candidacidal activity of mouse macrophages *in vitro*. Infect Immun 1980; 29:477–82.
26. Hockertz S. Immunomodulating effect of lysed immunoreactive fractions of selected *Escherichia coli* strains on the macrophage system. An *in vitro* study. Arzneimittelforschung 1990; 40:1068–72.
27. Stanley VC, Hurley R, The growth of *Candida* species in cultures of mouse peritoneal macrophages. J Pathol 1969; 97:357–66.
28. Danley DL, Polakoff J. Rapid killing of monocytes *in vitro* by *Candida albicans* yeast cells. Infect Immun 1986; 51:307–13.
29. Yamada Y, Saito H, Tomioka H, Jidoi J. Relationship between the susceptibility of various bacteria to active oxygen species and to intracellular killing by macrophages. J Gen Microbiol 1987; 133:2015–21.
30. Goren MB. Some paradoxes of macrophage function. Adv Exp Med Biol 1983; 162:31–50.
31. Mor N, Goren MB. Discrepancy in assessment of phagosome lysosome fusion with two lysosomal markers in murine macrophages infected with *Candida albicans*. Infect Immun 1987; 55:1663–7.
32. Riklis I, Kleitter Y, Bleiberg I, Fabian I. Biological properties *in vitro* of a combination of recombinant murine interleukin-3 and granulocyte-macrophage colony-stimulating factor. Eur J Haematol 1989; 43:375–81.
33. Karbassi A, Becker JM, Foster JS, Moore RN. Enhanced killing of *Candida albicans* by murine macrophages treated with macrophage colony stimulating factor: evidence for augmented expression of mannose receptors. J Immunol 1987; 139:417–21.
34. Blasi E, Farinelli S, Varesio L, Bistoni F. Augmentation of GG2EE marcophage cell line–mediated anti-*Candida* activity by gamma interferon, tumor necrosis factor, and interleukin-1. Infect Immun 1990; 58:1073–7.
35. Smith PD, Lamerson CL, Banks SM, Saini SS, Wahl LM, Calderone RA, Wahl SM. Granulocyte-macrophage colony-stimulating factor augments human monocyte fungicidal activity for *Candida albicans*. J Infect Dis 1990; 161:999–1005.
36. Wang M, Friedman H, Djeu JY. Enhancement of human monocyte function against *Candida albicans* by the colony-stimulating factors (CSF), IL-3, granulocyte-macrophage-CSF, and macrophage-CSF. J Immunol 1989; 143:671–7.
37. Fabian I, Kletter Y, Mor S, Geller-Bernstein C, Ben-Yaakov M. Activation of human eosinophil and neutrophil functions by haematopoietic growth factors: comparisons of IL-1, IL-3, IL-5 and GM-CSF. Br J Haematol 1992; 80:137–43.
38. Watanabe K, Kagaya K, Yamada T, Fukazawa Y. Mechanism for candidacidal activity in macrophages activated by recombinant gamma interferon. Infect Immun 1991; 59:521–8.
39. Brummer E, Morrison CJ, Stevens DA. Recombinant and natural gamma-interferon activation of macrophages *in vitro*: different dose requirements for induction of killing activity against phagocytizable and nonphagocytizable fungi. Infect Immun 1985; 49:724–30.
40. Brummer E, Stevens DA. Candidacidal mechanisms of peritoneal macrophages activated with lymphokines or γ-interferon. J Med Microbiol 1989; 28:173–81.

41. Peterson EM, Calderone RA. Inhibition of specific amino acid uptake in *Candida albicans* by lysosomal extracts from rabbit alveolar macrophages. Infect Immun 1978; 21:506–13.
42. Peterson EM, Calderone RA. Growth inhibition of *Candida albicans* by rabbit alveolar macrophages. Infect Immun 1977; 15:910–5.
43. Vechiarelli A, Dottorini M, Puliti M, Todisco T, Cenci E, Bistoni F. Defective candidacidal activity of alveolar macrophages and peripheral blood monocytes from patients with chronic obstructive pulmonary disease. Am Rev Respir Dis 1991; 143:1049–54.
44. Blasi E, Pitzurra L, Puliti M, Lanfrancone L, Bistoni F, Early differential molecular response of a macrophage cell line to yeast and hyphal forms of *Candida albicans*. Infect Immun 1992; 60:832–7.
45. Hynes R. Integrins: a family of cell surface receptors. Cell 1989; 56:549–54.
46. Nemerow GR, Moore M, Cooper NR. Structure and function of the B-lymphocyte Epstein-Barr Virus/C3d receptor. Adv Cancer Res 1990; 54:273–300.
47. Heidenreich F, Dierich MP. *Candida albicans* and *Candida stellatoidea*, in contrast to other *Candida* species, bind iC3b and C3d but not C3b. Infect Immun 1985; 50:598–600.
48. Edwards JE, Gaither TA, O'Shea JJ, Rotrosen D, Lawley TJ, Wright SA, Frank MM, Green I. Expression of specific binding sites on *Candida* with functional and antigenic characteristics of human complement receptors. J Immunol 1986; 137:3577–83.
49. Gilmore BJ, Retsinas EM, Lorenz JS, Hostetter MK. An iC3b receptor on *Candida albicans*: structure, function and correlates for pathogenicity. J Infect Dis 1988; 157:38–46.
50. Calderone RA, Linehan L, Wadsworth E, Sandberg AL. Identification of C3d receptors on *Candida albicans*. Infect Immun 1988; 56:252–8.
51. Linehan L, Wadsworth E, Calderone RA. *Candida albicans* C3d receptor, isolated by using a monoclonal antibody. Infect Immun 1988; 56:1981–6.
52. Saxena A, Calderone RA. Purification and characterization of the extracellular C3d-binding protein of *Candida albicans*. Infect Immun 1990; 58:309–14.
53. Tronchin G, Bouchara JP, Robert R. Dynamic changes of the cell wall surface of *Candida albicans* associated with germination and adherence. Eur J Cell Biol 1989; 50:285–90.
54. Bouchara J, Tronchin G, Annaix V, Robert R, Senet JM. Presence of laminin receptors on *Candida albicans*. Infect Immun 1990; 58:48–54.
55. Gustafson KS, Vercellotti GM, Bendei CM, Hostetter MK. Molecular mimicry in *Candida albicans*: role of an integrin analogue in adhesion of the yeast to human endothelium. J Clin Invest 1991; 87:1896–1902.

31

Interaction of *Histoplasma capsulatum* Yeasts and Conidia with Human and Animal Macrophages

Simon L. Newman and Ward E. Bullock

University of Cincinnati College of Medicine, Cincinnati, Ohio

I. INTRODUCTION

Histoplasma capsulatum (Hc) is a dimorphic fungal pathogen of worldwide importance that causes a broad spectrum of disease activity. In the United States, Hc is endemic to the Ohio and Mississippi river valleys, and it has been estimated that 500,000 new infections occur in the United States each year [1]. Although the course of infection is mild in most immunocompetent individuals, Hc may produce progressive disseminated infections in individuals immunocompromised by hematologic malignancies [2–5] or cytotoxic therapy [6–8], or in individuals with the acquired immunodeficiency syndrome (AIDS) [9–16].

H. capsulatum exists outside the human host as a mycelial form consisting of hyphae that bear both macro- and microconidia [17,18]. Infection with Hc is acquired by inhalation of microconidia or small mycelial fragments (5–8 μm) into the terminal bronchioles and alveoli of the lung. Inhaled microconidia subsequently convert into the yeast form that is responsible for the pathogenesis of histoplasmosis. *H. capsulatum* yeasts are phagocytosed by alveolar macrophages (AMs), within which they multiply [19–22]. Presumably, the dividing yeasts destroy the AMs and subsequently are ingested by other resident AMs and by macrophages (Mϕs) recruited to the loci of infection. Repetition of this cycle results in spread of infection to hilar lymph nodes and transient dispersal of yeasts to other organs during the acute phase of primary histoplasmosis (1–2 weeks). Thereafter, the maturation of specific cell-mediated immunity (CMI) against Hc activates Mϕs to halt yeast proliferation, with gradual resolution of the disease process in most immunocompetent hosts [23–25].

The interaction of Mϕs with Hc is the key event in the pathogenesis of histoplasmosis. Macrophages first provide an environment for fungal replication and dissemination and subsequently act as the final effector cells to remove the organism from the host. Thus, the focus of this chapter will be to review specifically what is known about the interaction of Hc microconidia and yeasts with human and animal Mϕs. The reader is referred to other recent reviews for a more thorough discussion of fungal metabolism and dimorphism [26], clinical manifestations of histoplasmosis [19], mechanisms of pathogenesis [27], and the immunological response of the host to Hc [19,28].

II. RECOGNITION OF *HISTOPLASM CAPSULATUM* YEASTS AND CONIDIA

During the early phase of infection, Mφs recognize, bind, and ingest Hc, thereby providing the organisms with access to a permissive intracellular environment for replication. This initial interaction between Hc and Mφs is crucial in the pathogenesis of histoplasmosis. Initial in vitro studies demonstrated that the binding of Hc yeasts to human cultured monocyte–derived Mφs is temperature- and cation-dependent. Thus yeasts rapidly attach to adherent Mφs at 37° C, but not at 4° C, and only in the presence of both Ca and Mg ions [29].

Macrophage recognition of unopsonized Hc yeasts is mediated via the CD18 family of adhesion-promoting glycoproteins (LFA-1 [CD11a, CR3 [CD11b], p150,95 [CD11c]) [29]. Each receptor molecule of the CD18 family contains a unique α-chain subunit noncovalently linked to a common β-chain subunit [30]. Experiments using α- and β-chain–specific monoclonal antibodies (mAbs) suggest that the yeasts bind independently to each of the three receptors. In these studies, Mφs were adhered to surfaces coated with the mAbs to down-modulate specific receptors from the apical to the basal surface of the Mφ. Removal of either CD11a, CD11b, or CD11c had little effect on the attachment of Hc yeasts. However, down-modulation of all three CD18 receptors with an anti–β-chain mAb resulted in complete inhibition of yeast binding.

In additional studies using anti-CD11/CD18 mAbs, it has been demonstrated that CD18 receptors also mediate the attachment of unopsonized Hc yeasts to human AMs [31] and to human neutrophils (PMNs) [32]. Furthermore, as is found for yeasts, CD18 receptors mediate the attachment of Hc microconidia to human cultured Mφs and AMs [31]. Thus, both stages of the fungus are recognized by Mφ CD18 receptors. No evidence has been obtained to suggest a role for Fc receptors, mannose-fucose receptors, or β-glucan receptors, or for Mφ-derived complement components [29]. Similar studies to identify specific receptors that mediate the attachment of Hc yeasts to nonhuman Mφs have not been performed.

The identity of the ligand on the surface of Hc yeasts and conidia that is recognized by Mφ CD18 receptors is unknown. Preliminary experiments in our laboratory suggest that the ligand is a heat-stable glycoprotein. Binding of Hc yeasts to Mφs is inhibited by preincubation of the yeasts with trypsin, chymotrypsin, pronase, and iodoacetamide. A freeze-thaw extract of the yeasts yields a soluble (100,000 × g) supernatant that completely blocks the attachment of yeasts to Mφs. The inhibitory activity of the crude supernatant is specific for Hc yeast binding to Mφ CD18 receptors, as the supernatant does not block the binding of sheep erythrocytes (E) coated with complement component C3b (EC3b) to complement receptor type 1 (CR1), EC3bi to complement receptor type 3 (CR3), or *Candida glabrata* to mannose/fucose receptors [29]. The inhibitory activity of the supernatant is heat stable (56° C, 30 minutes), and its activity does not decay on prolonged storage at 4° C (S. L. Newman, H. Henderson, and L. Gootee, unpublished data).

III. PHAGOCYTOSIS OF *HISTOPLASMA CAPSULATUM* YEASTS AND CONIDIA

To distinguish between ingested versus bound but uningested organisms, Hc yeasts and conidia were labeled with fluorescein isothiocyanate (FITC), and then were incubated with monolayers of human Mφs for varying periods. At the end of the incubation, the fluorescence of bound but uningested organisms was quenched with trypan blue. In this assay, internalized yeasts or conidia appear bright green, while bound uningested organisms appear either dark green or reddish brown [31].

Using this assay it was demonstrated that phagocytosis of unopsonized yeasts by human cultured Mφs and freshly adherent AMs is rapid, and both cell populations ingest an equivalent number of yeasts. After 1 to 2 hours of incubation, 85% to 95% of Mφs contain ingested yeasts, indicating that yeasts are not phagocytosed by a subpopulation of Mφs. In comparison with Mφs, monocytes ingest significantly fewer yeasts. However, in all three cell populations, 75% to 80% of attached yeasts are ingested at all time points [31]. Thus, the rate of phagocytosis of yeasts by monocytes/Mϕs is determined by the rate of attachment of the yeasts to the phagocyte membrane.

Differences in the rates of phagocytosis of Hc yeasts by human monocytes, cultured Mϕs, and AMs may relate to the different numbers of CD11a/CD18, CD11b/CD18, and CD11c/CD18 receptors on these cells [33]. Alternatively, monocyte and Mϕ CD18 receptors may differ in their binding affinity for yeasts and/or in their ability to trigger a phagocytic signal. A third possibility is that attachment could be mediated by one receptor, and ingestion triggered by another receptor. In this regard, it has been suggested that CD11b/CD18 receptors on U937 Mϕs stimulated by phorbol ester mediate *binding* of EC3bi, whereas CD11c/CD18 triggers *ingestion* of this particle [34]. Currently, we cannot distinguish between these possibilities.

Human cultured Mϕs and AMs also phagocytose unopsonized microconidia to a similar extent, and Mϕs phagocytose microconidia almost as efficiently as yeasts [31]. Peritoneal Mϕs (PMs) and AMs from outbred Swiss albino mice also phagocytose Hc microconidia and small hyphal fragments at rates comparable to yeasts [35]. These observations may be relevant with regard to the transformation of microconidia into yeasts in vivo. Since acute infection of humans with Hc frequently is asymptomatic and self-limited, the sequence of events in the early inflammatory response to inhaled microconidia is unknown [19]. Furthermore, the fate of microconidia in the lungs of experimental animals has not been described [20]. Thus, the events that are initiated once microconidia enter the pulmonary alveoli are unknown.

As microconidia are ingested rapidly by human and murine AMs in vitro, it is likely that upon exposure to Hc in nature, microconidia are inhaled into the pulmonary alveoli and phagocytosed by resident AMs within which an unknown proportion transform into yeasts. This event initiates primary infection, which in turn stimulates the development of Hc-specific CMI, with eventual resolution of the infection. Support for this hypothetical sequence of events is provided by the fact that in regions where histoplasmosis is endemic, up to 90% of sampled adult populations exhibit CMI to Hc as demonstrated by delayed-type skin test reactions or by blastogenic transformation of their peripheral blood lymphocytes to histoplasmin [19]. In addition, it has been reported that following in vitro infection of AMs from naive mice by microconidia and hyphal fragments of Hc, the number of intracellular fungi increased over 48 hours, whereas there was no increase in fungi within AMs from mice that had been immunized with Hc [35].

A. Nature of the Phagocytic Signal and Role of the Cellular Cytoskeleton

Several lines of evidence suggest that the phagocytic signal generated by the binding of Hc yeasts to Mϕ CD18 receptors differs from that initiated by the binding of EC3bi to Mϕ CD11b/CD18 (CR3). First, binding of EC3bi to monocytes, to Mϕs cultured in suspension, or to freshly adherent AMs does not initiate a phagocytic signal [36–39]. That is, EC3bi bind to these phagocytes but are not ingested. In contrast, all three cell populations phagocytose Hc yeasts [31]. Second, Mϕs cultured adherently can phagocytose EC3bi [36,37], whereas Mϕs cultured in suspension cannot [38]. Conversely, Mϕs obtained by either culture method phagocytose equivalent numbers of unopsonized Hc yeasts [31].

Additional evidence that different phagocytic signals are triggered by binding of Hc and EC3bi to Mϕs is their differential requirements for cellular cytoskeleton for attachment and ingestion. Disruption of actin microfilaments with cytochalasin D [40] inhibits *binding*, as well as the ingestion of Hc yeasts, whereas inhibition of microtubule polymerization with nocodazole [41] does not inhibit either yeast binding or ingestion by Mϕs [31]. In contrast, phagocytosis of EC3b and EC3bi by adherently cultured Mϕs via CR1 and CR3, respectively, requires intact actin microfilaments *and* microtubules. However, neither cytoskeletal element is required for binding of

EC3b or EC3bi [42]. Finally, phagocytosis of EC3b or EC3bi by cultured human Mϕs does not stimulate the production of H_2O_2 [43], whereas phagocytosis of Hc yeasts does trigger it [29]. Characterization of the biochemical pathway by which Hc triggers phagocytosis may provide insight into the ability of this microorganism to survive within the hostile environment of the Mϕ phagolysosome.

IV. STIMULATION OF THE RESPIRATORY BURST AND PHAGOLYSOSOMAL FUSION

An immediate consequence of the phagocytosis of microorganisms by Mϕs is the stimulation of the cell's microbicidal armamentarium, which consists of the respiratory burst and the fusion of lysosomes with the phagocytic vacuole. Thus, a microorganism that is confined in the phagocytic vacuole is directly exposed to toxic oxygen metabolites and various lysosomal hydrolases and cationic peptides that destroy the organism. Intracellular parasites have evolved several ways of evading these Mϕ defense mechanisms, including inhibition of phagolysosomal fusion [44–47], escaping from the phagocytic vacuole into the cytoplasm [48], and initiating phagocytosis without stimulating the respiratory burst [49–51].

Several studies have demonstrated that Hc yeasts are killed in vitro by a combination of H_2O_2, Fe^{2+}, and iodide [52], or H_2O_2, horseradish peroxidase, and iodide [53–55; S. L. Newman and L. Gootee, unpublished data). However, respiratory burst products do not appear to mediate Mϕ antifungal activity against Hc yeasts in phagocytic cells of either the human or murine system. First, Hc yeasts activate the respiratory burst of human monocytes/Mϕs and neutrophils upon phagocytosis [29,32,52], but are not killed [52,56,57]. Second, the intracellular growth of Hc yeasts proceeds at a similar rate both in freshly isolated monocytes and in cultured Mϕs [57] that have lost their myeloperoxidase (MPO) and exhibit a decreased ability to produce toxic oxygen metabolites [58–60]. Third, phagocytosis of serum-opsonized Hc yeasts by human neutrophils stimulates a strong respiratory burst, and superoxide anion (O_2^-) becomes trapped within the phagocytic vacuole [32]. Thus, yeasts should be exposed directly to toxic oxygen metabolites. However, all of the fungistatic activity of human neutrophils is mediated by a nonoxidative mechanism or mechanisms [61].

In contrast to human Mϕs, phagocytosis of unopsonized Hc yeasts by resident [62,63] or interferon-γ (IFN-γ)–activated [64] mouse PMs does not stimulate the respiratory burst. Phagocytosis of yeasts opsonized in fresh or heat-inactivated normal mouse serum or Hc immune serum does stimulate the respiratory burst of mouse PMs [63]. However, the intracellular growth of yeasts is not impaired [65].

Although phagocytosis of unopsonized Hc yeasts by mouse PMs does not stimulate the respiratory burst, phagolysosomal fusion does occur upon ingestion of yeasts by the mouse P388D1 Mϕ cell line [66] and resident PMs [67]. Phagolysosomal fusion also is stimulated upon ingestion of Hc yeasts by human Mϕs [68]. Since Hc yeasts multiply within these Mϕs, the yeasts apparently are able to evade the normally destructive effects of Mϕ lysosomal hydrolases.

V. INTRACELLULAR REPLICATION OF *HISTOPLASMA CAPSULATUM* YEASTS

In vitro studies on the intracellular fate of Hc within Mϕs demonstrate that yeasts multiply readily within rabbit, mouse, and human AMs [35,56,57,69], mouse, guinea pig, and human PMs [56,65,70–72], and human monocytes and cultured Mϕs [56,57]. The intracellular generation time of yeasts in mice and guinea pig PMs is 10 to 12 hours [65,72]. *H. capsulatum* yeasts multiply within human monocytes, cultured Mϕs, and AMs with intracellular generation times of 14.2, 18.5, and 19.9 hours, respectively [57]. Although serum clearly is not required for phagocytosis of Hc yeasts by Mϕs, opsonization of the yeasts with normal or Hc immune serum also does not effect the intracellular growth of Hc yeasts in either mouse or guinea pig PMs [65] or in human Mϕs [57].

The strategies used by Hc yeasts to adapt to the intracellular environment of Mϕs are unknown. As discussed above, Hc yeasts survive and multiply within the phagocytic vacuoles of Mϕs despite the stimulation of the respiratory burst and phagolysosomal fusion. In human [68] and mouse Mϕs [70] that have ingested killed Hc yeasts, the organisms are digested completely, suggesting that the yeast's thick cell wall is not sufficient by itself to enable this organism to survive within Mϕ phagolysosomes. In human cultured Mϕs, chloroquine, a drug that raises the intraphagosomal pH in Mϕs [73,74], completely protects killed yeasts from digestion by Mϕs when it is present in the culture medium [68]. Thus, preformed Mϕ lysosomal hydrolases are sufficient to digest killed yeasts.

When human Mϕs infected with viable Hc yeasts are cultured in the presence of cycloheximide, an antibiotic that inhibits protein synthesis in yeasts [75,76] and mammalian cells [77,78], the yeasts are digested [68]. Cycloheximide apparently penetrates the phagolysosomal compartment and inhibits the yeast from making a protein or proteins required for intracellular survival and growth. These experiments also suggest that Mϕs themselves are not required to synthesize protein to digest the yeasts, and that fusion of preformed lysosomal hydrolases with phagocytic vacuoles is sufficient for the digestion of yeasts to proceed.

The antifungal agents amphotericin B, nystatin, and ketoconazole also can enter human and murine Mϕs and compromise the intracellular survival and growth of Hc yeasts. Addition of concentrations of amphotericin B as low as 0.08 μg/ml to the culture medium after Mϕ phagocytosis of viable yeasts promotes complete intracellular digestion of the yeasts by both human cultured Mϕs [68] and mouse PMs [71,79]. Culture of infected human Mϕs with nystatin and ketoconazole results only in inhibition of the growth of yeasts [68]. The fact that amphotericin B promotes the digestion of viable Hc yeasts within human Mϕs undoubtedly is relevant to the high efficacy of amphotericin B in treating histoplasmosis [80].

As Hc yeasts must synthesize proteins to survive within Mϕs, it is possible that the yeasts secrete a product or products mediating a global effect on Mϕ fungicidal mechanisms. This hypothesis was tested by allowing human cultured Mϕs to phagocytose a mixture of viable yeasts and FITC-labeled heat-killed yeasts. After 1 hour of phagocytosis some monolayers were fixed, and the remaining Mϕs were cultured for a further 24 hours. The percentage of Mϕs that contained viable yeasts (visible by phase-contrast microscopy), heat-killed yeasts (visible via fluorescence microscopy), or both then was quantified. After 1 hour of phagocytosis, 25% to 30% of Mϕs had ingested one or more viable and/or heat-killed yeasts. The percentage of phagocytic Mϕs ingesting both viable and heat-killed yeasts was 32%. However, after culture for 24 hours, all of the visible yeasts within Mϕs were viable; the heat-killed yeasts had been digested. In addition, the viable yeasts had multiplied within the Mϕs [68]. This observation suggests that although viable Hc yeasts can survive and multiply within Mϕ phagolysosomes, they do not secrete a factor that affects the activity of other phagolysosomes within the same Mϕ. Thus, the yeasts' survival strategy appears to focus on modifying its phagolysosomal environment.

An as yet unpublished study with the mouse P388D1 Mϕ cell line suggests that one possible mechanism by which Hc yeasts evade the action of lysosomal hydrolases is by preventing acidification of Mϕ phagolysosomes [81]. Fluroescein isothiocyanate–labeled Hc yeasts and zymosan particles were incubated with adherent Mϕs, and the intraphagosomal pH was quantified by the ratio of fluorescence at 490 and 450 nm [73]. The pH of zymosan-containing phagolysosomes was found to be approximately 5.0 over a period of 1 hour. In contrast, both the virulent Hc strain G186AR and the avirulent G186AS strain [82,83] resided in Mϕ phagolysosomes with a pH > 6.0 over the same period. It is speculated that the relatively neutral pH associated with Hc-containing Mϕ phagolysosomes may reflect either active buffering by the yeasts or a malfunction of the proton pump in the organelle's membrane.

VI. MACROPHAGE ACTIVATION

A. Murine Macrophages

The importance of CMI in host defense against *H. capsulatum* was first demonstrated by Salvin [23]. In his early experiments, resistance to a lethal inoculum of Hc yeasts was induced in mice and guinea pigs by prior injection of dead yeasts or by sublethal infection with viable yeasts. Resistance to a subsequent lethal inoculum of yeasts did not correlate with the titer of humoral factors (complement-fixing antibodies), and resistance could not be transferred by homologous immune serum [23].

Early in vitro studies employing the murine system demonstrated CMI to Hc by the capacity of PMs from Hc-immune mice to restrict the intracellular growth of yeasts, whereas PMs from normal mice did not. Inhibition of growth was dependent on the presence of immune lymphocytes [84,85]. Lymphokines generated from cultures of immune splenic T cells stimulated with Hc antigens, or recombinant murine IFN-γ, activate resident mouse PMs to inhibit the intracellular growth of yeasts [86–88]. However, no killing of the yeasts ever has been observed. In addition, Mφs from Hc-immunized mice are better able to restrict the growth of mycelial elements than Mφs from control animals. However, only PMs and AMs from mice treated with Bacillus Calmette-Guérin (BCG) actually reduce the numbers of viable fungi [35].

H. capsulatum yeasts multiply within the mouse P388D1 and IC-21 Mφ cell lines with generation times similar to their rate of growth in mouse PMs. Interestingly, however, IFN-γ is incapable of activating these Mφs to inhibit the intracellular growth of Hc yeasts. P388D1 Mφs also do not mediate fungistatic activity after exposure sequentially or simultaneously to IFN-γ and lipopolysaccharide (LPS) [89]. Likewise, Lane et al. [90] have presented data in abstract form showing that IFN-γ alone does not activate fungistatic activity against Hc yeasts by splenic Mφs from mice. Activation of fungistasis in splenic Mφs required preincubation for 18 hours with 2000 U/ml of IFN-γ in combination with 100 ng/ml LPS or 2000 U/ml of tumor necrosis factor-α (TNF-α). Neither LPS nor TNF-α alone activated splenic Mφs [90]. In addition, anti–TNF-α antibody partially inhibited the ability of IFN-γ and LPS to activate splenic Mφs, suggesting a possible autocrine role for TNF-α in the activation of splenic Mφs [90].

These in vitro experiments correlate with in vivo studies that also suggest a possible role for both IFN-γ and TNF-α in murine host defense against Hc. Treatment of mice with a polyclonal anti–TNF-α antibody [91] or antibody to IFN-γ [92] accelerates mortality from systemic *Histoplasma* infection and increases significantly the yeast colony counts in the liver and spleen, as compared with infected controls. In addition, TNF-α is detected in the supernatant of mouse PMs incubated with Hc yeasts and in bronchoalveolar lavage fluid of mice challenged intranasally with Hc [91].

B. Human Monocytes/Macrophages

The cytokines that activate human monocyte/Mφ fungistatic activity against Hc yeasts have begun to be investigated only recently. Initial studies in our laboratory demonstrated that PHA stimulation of human lymphocytes generated a cytokine supernatant that could rapidly activate cultured human monocyte-derived Mφs to inhibit the intracellular growth of Hc yeasts. Optimal activation took place when the cultured Mφs were infected with Hc yeasts for 1 hour, the monolayers were washed, and the infected Mφs were cultured for 24 hours in the presence of the cytokine supernatant. Preincubation of Møs for 24 hours with the cytokine supernatant before infection of the Mφs resulted in less inhibition of yeast growth. Furthermore, opsonization of Hc yeasts in normal serum or Hc-immune serum did not affect their intracellular growth in either control Mφs or cytokine-activated Mφs [57].

Recent attempts to identify the cytokine or cytokines in the PHA-generated supernatant that are responsible for the activation of Mϕ fungistatic activity have been unsuccessful. Recombinant cytokines that do not rapidly activate Mϕ fungistatic activity include interleukin 1 (IL-1), IL-2, IL-3, IL-6, granulocyte-Mϕ colony-stimulating factor (GM-CSF), Mϕ CSF (M-CSF), granulocyte CSF (G-CSF), IFN-γ, TNF-α, LPS, transforming growth factor-β (TGF-β), and platelet-derived growth factor (PDGF). Combinations of cytokines and growth factors that are without effect include IFN-γ + TNF-α, IFN-γ + IL-3, IFN-γ + GM-CSF, IFN-γ + LPS, IFN-γ + M-CSF, GM-CSF + IL-3, and M-CSF + IL-3 [93].

We do find, however, that in long-term cultures IL-3, GM-CSF, and M-CSF can activate human Mϕs to inhibit the intracellular growth of yeasts [93]. In these experiments, freshly isolated monocytes or 5- to 7-day suspension-cultures monocyte-derived Mϕ are adhered in 96-well tissue culture plates and incubated in medium, or stimulated with varying concentrations of CSFs for varying periods of time. Macrophages then are washed, are incubated with Hc yeasts for 24 hours, and then are pulsed with ^{3}H-leucine for 24 hours to quantify the growth of the yeasts.

Stimulation of Mϕ fungistatic activity by CSFs is concentration-dependent and requires 3 to 5 days of culture with freshly isolated monocytes. Colony-stimulating factors do not stimulate Mϕ fungistatic activity if monocytes first are allowed to differentiate into Mϕs and then further cultured in the presence of the CSFs. Thus, it appears that CSFs must be present during the process of monocyte differentiation into mature Mϕs for the CSFs to induce Mϕ fungistatic activity against Hc yeasts [93].

Maximum antifungal activity is obtained when monocytes are cultured with IL-3, GM-CSF, or M-CSF individually. Culture of monocytes with varying combinations of two CSFs, or the addition of a second CSF during the 24-hour incubation with yeasts, does not enhance Mϕ antifungal activity above that obtained with a single CSF. In addition, neither IFN-γ nor TNF-α augments or inhibits the ability of CSFs to activate monocyte/Mϕ-mediated fungistasis [93]. As has been found in the murine system, only inhibition of Hc yeast growth, but not killing, has been observed in activated human Mϕs [57,93].

The in vitro activation of antifungal activity in human monocytes/Mϕs contrasts with the in vitro and in vivo data reported in the murine system, which suggest that IFN-γ and TNF-α both may play a role in the activation of Mϕ anti-*Histoplasma* activity [88,90–92]. Thus, regardless of whether human monocytes/Mϕs are cultured for 24 hours to 7 days, neither IFN-γ nor TNF-α, nor a combination of the two cytokines, activates Mϕs to inhibit the intracellular growth of Hc yeasts. Furthermore, addition of IFN-γ or TNF-α to CSF-activated Mϕs also does not enhance Mϕ fungistatic activity [93].

The pattern of intracellular multiplication of Hc yeasts in human monocytes/Mϕs, and the ability of cytokines to activate monocytes/Mϕs to inhibit the intracellular growth of Hc, are distinct in several respects from those reported for most other intracellular pathogens.

First, Hc yeasts grow somewhat better in freshly isolated monocytes than in cultured Mϕs or in AMs. In contrast, previous studies have found that viral [94], fungal [60,95], bacterial [96–98], and protozoan [58,99] intracellular pathogens multiply more rapidly in cultured Mϕs than in monocytes. This loss in Mϕ antimicrobial activity coincides with a decline in the ability of cultured Mϕs to be stimulated to produce toxic oxygen metabolites [58–60].

A second difference in the interaction of Hc yeasts with human Mϕs compared with other intracellular parasites is that optimal activation of Mϕs requires either that PHA-generated cytokines of unknown character be added to the Mϕs immediately after infection with yeasts [57] or that monocytes be cultured for 5 to 7 days in the presence of CSFs [93]. In contrast, preincubation of cultured human Mϕs with IFN-γ for 24 to 48 hours before infection activates Mϕs to inhibit the intracellular growth of *Trypanosoma cruzi* [100], *Leishmania donovani* [58], *Toxoplasma gondii* [99], and *Chlamydia psittaci* [98]. Recombinant human IFN-γ does not activate human cultured Mϕs to inhibit the intracellular growth of Hc yeasts [56,93].

Interleukin-3, GM-CSF, and M-CSF also activate human monocyte/Mϕ antimicrobial activity against other intracellular pathogens including *T. cruzi* [101,102], *L. donovani* [103], *L. amazonensis* [102,104], and *Mycobacterium avium* complex [105]. In contrast to Hc, however, IL-3 acts synergistically with GM-CSF and M-CSF in stimulating Mϕ antileishmanial activity [102]. Furthermore, the ability of IL-3 to stimulate Mϕ antileishmanial activity is inhibited by antibodies to TNF-α, and TNF-α alone can activate Mϕs to inhibit the intracellular growth of *L. amazonensis* [102]. In addition, GM-CSF activation of human Mϕs to inhibit the intracellular growth of *M. avium* complex is augmented by TNF-α, but suppressed by IFN-γ [105].

VII. MECHANISMS OF MACROPHAGE-MEDIATED FUNGISTASIS

A. Murine Macrophages

As mentioned previously, the intracellular replication of Hc yeasts in murine PMs is inhibited by IFN-γ, but the yeasts are not killed [86–88]. One interpretation of this observation is that IFN-γ induces PMs to restrict the availability of specific nutrients that Hc yeasts require to replicate. Precedents for nutritional starvation by activated Mϕs have been reported for *Legionella pneumophila* [106] and *T. gondii* [107]; IFN-γ–activated human monocytes inhibit the intracellular growth of *L. pneumophila* by restricting the availability of iron [106], and the growth of *T. gondii* is inhibited in IFN-γ–treated human fibroblasts via tryptophan starvation [107]. *H. capsulatum* yeasts require iron to multiply extracellularly in tissue culture medium [108], and the fungus also is a functional methionine-cysteine auxotroph at 37° C [109–111].

Recent studies by Lane et al. [112] demonstrate that iron also is required for the growth of Hc yeasts within murine PMs. Thus, incubation of infected PMs with the iron chelator deferoxamine suppresses the intracellular growth of Hc yeasts; growth suppression can be reversed with iron-saturated human transferrin (holotransferrin). More important, holotransferrin reverses the ability of IFN-γ–activated PMs to inhibit the intracellular growth of Hc yeasts. In preliminary studies, ferrous sulfate has been found to block the ability of IFN-γ–LPS–activated murine splenic Mϕs to inhibit the intracellular growth of Hc yeasts [90]. Thus, one mechanism by which cytokine-activated murine PMs may mediate fungistatic activity against Hc yeasts is restriction of the availability of iron within the phagolysosome. In other experiments, no evidence was obtained to suggest that the anti-*Histoplasma* effect of INF-γ–activated PMs was mediated by limiting the availability of methionine-cysteine [113].

Nitric oxide (NO) produced from L-arginine metabolism has been reported to be involved in the killing of several intracellular parasites by mouse PMs [114,115]. Recent studies presented in abstract form suggest that NO also may play a role in the anti-*Histoplasma* activity of cytokine-activated murine Mϕs [90,116]. L-arginine was required in the medium for IFN-γ–activated resident PMs, or for PM's for Hc-immune mice, to inhibit the growth of Hc yeasts. Furthermore, N^G-monomethyl-L-arginine (L-NMMA), an inhibitor of nitric oxide synthetase [117], inhibited the fungistatic activity of IFN-γ–activated murine PMs. In addition, L-NMMA has been reported to inhibit the anti-*Histoplasma* activity of murine splenic Mϕs activated with IFN-γ–LPS [90]. Thus, NO also may contribute to the antifungal armamentarium of murine Mϕs against Hc yeasts.

B. Human Monocytes/Macrophages

The mechanism or mechanisms by which CSFs stimulate human Mϕ fungistatic activity against Hc yeasts are unknown. However, as discussed previously, it is unlikely that fungistasis is mediated through an oxygen-dependent mechanism, in contrast to Mϕ antimicrobial activity against *L. amazonensis*, *L. donovani*, *T. cruzi*, and *Candida albicans* [101,118–120].

Nitric oxide also probably does not play a role in the ability of activated human Mϕs to inhibit the intracellular growth of Hc yeasts. Regardless of the cytokines used to stimulate human monocytes/Mφs, nitrite (quantified using the Greiss reagent [121]) is not detected in the culture medium. Moreover, L-NMMA does not affect the intracellular growth of Hc yeasts within human Mφs (S. L. Newman and L. Gootee, unpublished data). These results agree with the reports of others that NO does not appear to play a role in the antimicrobial activity of human Mφs against *T. gondii*, *C. psittaci*, *L. donovani* [122], or *Cryptococcus neoformans* [123]. Further studies will be required to define the mechanism or mechanisms by which activated human Mφs mediate fungistatic activity against Hc yeasts.

VIII. FUNGISTATIC ACTIVITY OF NEUTROPHILS

Because acute pulmonary histoplasmosis is rarely fatal, the early inflammatory response to Hc is poorly understood in humans. Clearly, PMNs are part of the inflammatory response to Hc in animal models of histoplasmosis [20,21]. However, because most individuals possess CMI to the fungus in areas where Hc is endemic, the specific role of PMNs in the immune response to Hc is obscure.

In vitro studies of PMN fungicidal activity against Hc yeasts have generated conflicting results. In an early study, glycogen-induced guinea pig peritoneal neutrophils were thought to kill Hc yeasts after 3 hours of incubation. However, the definition of killing was based entirely on altered stainability of the ingested Hc yeasts [53], a methodology that has not been accepted widely [124]. Moreover, neither ultrastructural nor tinctorial changes have been observed by Schaffner et al. after phagocytosis of Hc yeasts by human PMNs [52]. Furthermore, at a high phagocyte-to-fungus ratio, human PMNs apparently kill the mycelial phase but not the yeast phase of Hc, even though opsonized yeasts induce the PMN respiratory burst [52].

More recent reports demonstrate that human [55] and murine [125] PMNs can inhibit the growth of yeasts by 47%, compared with yeasts cultured in media alone. However, fungistasis was observed only if the PMN : yeast ratio was 500 : 1. We have demonstrated that, in vitro, human PMNs possess potent fungistatic activity against Hc yeasts using the incorporation of ^{3}H-leucine as a sensitive assay to quantify the multiplication of Hc yeasts [61]. Neutrophil-mediated fungistasis was evident as early as 2 hours and was detectable at PMN : yeast ratios as low as 1 : 2. Maximum PMN-mediated fungistasis was achieved at PMN : yeast ratios of 10 : 1 to 50 : 1, and after 24 hours of incubation. Remarkably, neutrophils sequestered yeasts intracellularly for up to 5 days. Only on the sixth day of culture were yeasts observed extracellularly. In contrast, Hc yeasts grew better within freshly isolated monocytes than in tissue culture medium.

Fungistasis mediated by PMNs required the presence of either fresh or heat-inactivated human serum, but ingestion of the yeasts per se was not required. Thus, PMN-mediated fungistasis was equivalent in the presence or absence of cytochalasin D. Clearly, however, the yeasts bound to neutrophils, and binding was sufficient to induce signal transduction and PMN-mediated fungistasis.

Attachment of serum opsonized Hc yeasts to neutrophils was mediated predominantly via CR1 and CR3. However, engagement of the FcRIII also was important, as heat-inactivated serum promoted PMN-mediated fungistasis almost as well as fresh complement-preserved serum.

Coculture of PMNs and Hc yeasts with agents that stimulate the respiratory burst and enhance PMN-mediated cytotoxicity (IFN-γ, IL-8, IL-3, GM-CSF, G-CSF, and PMA) failed to induce PMN fungicidal activity or enhance PMN-mediated fungistasis. Furthermore, endogenous PMN fungistatic activity was not blocked by agents known to inhibit the respiratory burst or by inhibitors of NO synthetase.

Neutrophils from patients with chronic granulomatous disease (CGD) inhibited the growth of Hc yeasts as well as PMNs from normal controls. The CGD PMNs lack the capacity to produce toxic oxygen metabolites because of an NADPH oxidase deficiency [126–128]. Therefore, all of the fungistatic activity of CGD PMNs must be mediated by a nonoxidative mechanism or mechanisms. In addition, raising the intraphagosomal pH with the lysosomotropic agent ammonium chloride [73] almost completely inhibited the fungistatic activity of normal PMNs.

The major components of the neutrophil that mediate nonoxidative antimicrobial activity are the lysosomal hydrolases and cationic peptides that reside in the specific and azurophil granules [129–131]. The fact that Hc yeasts rapidly multiply within human monocytes and Mϕs [56,57] indicates that neutrophils must contain some enzymes or cationic peptides not found in these phagocytes. One obvious candidate is the cationic peptides known as "defensins" [130,131]. Although defensins have been found in Mϕs from other mammalian species, human Mϕs do not contain these antimicrobial agents [130,131].

In pilot experiments, we have determined that lysates of both the specific and azurophil granules mediated fungistatic activity against Hc yeasts. No fungistatic activity was detected in a membrane fraction, the cytosol, or intact specific or azurophil granules (S. L. Newman and L. Gootee, unpublished data). Although preliminary, the data would indicate that defensins alone, which reside in the azurophil granules [130,131], cannot account for all of the fungistatic activity mediated by human neutrophils.

The potent fungistatic activity against Hc yeasts exhibited by human neutrophils in vitro suggests that PMNs may play a more important role in host defense against Hc in vivo than previously recognized. Indeed, it is possible that neutrophils may be capable of slowing the course of the infection, and under certain circumstances they may prevent dissemination of the yeasts from the lung. It also is conceivable that neutrophils may damage the yeasts in such a manner as to make them vulnerable to inflammatory Mϕs. Thus, inflammatory Mϕs may phagocytose yeast-containing neutrophils [132] and subsequently kill the partially damaged yeasts. Either or both of these postulated mechanisms of defense involving neutrophils during the early phase of the inflammatory response to Hc may explain, at least in part, the fact that a substantial percentage of pulmonary infections by this organism are subclinical and self-limiting. It also is worth noting that during the course of disseminated histoplasmosis in immunocompromised patients, neutrophils engorged with Hc yeasts can be observed in peripheral blood smears in 30% to 40% of cases [133].

IX. CONCLUSIONS

Over the past several years considerable progress has been made toward understanding the interaction of Hc with human and murine Mϕs. Clearly Hc yeasts have uniquely adapted themselves to survive and multiply within the normally hostile confines of Mϕ phagolysosomes. The mechanism or mechanisms by which Hc yeasts accomplish this legerdemain are still unknown. Critical questions concerning Mϕ-Hc interaction that remain unanswered include: What are the cytokines that activate human and murine Mϕs to *kill* Hc yeasts? What are the destructive mechanisms of fungicidal Mϕs? Do PMNs play a significant role in vivo in host defense against Hc, or is their potent fungistatic activity merely an in vitro phenomenon? These and other questions should provide interested scientists with significant intellectual stimulation in their attempts to understand this fascinating fungus. Furthermore, insight gained from continuing research on Mϕ-Hc interactions may provide us with the rationale to design new drugs for the treatment of disseminated histoplasmosis in immunocompromised patients.

ACKNOWLEDGMENTS

The authors' research on *H. capsulatum* was supported by NIH grants A123985 and A128392, and a grant from the Western-Southern Foundation.

REFERENCES

1. Ajello L. Distribution of *Histoplasma capsulatum* in the United States. In: Ajello L, Chick EW, Furculow MF, eds. Histoplasmosis. Springfield: CC Thomas, 1971:103–22.
2. Reddy P, Gorelick DF, Brasher CA, Larsh H. Progressive disseminated histoplasmosis as seen in adults. Am J Med 1970; 48:629–36.
3. Smith JW, Utz JP. Progressive disseminated histoplasmosis. A prospective study of 26 patients. Ann Intern Med 1972; 76:557–65.
4. Davies SF, Khan M, Sarosi GA. Disseminated histplasmosis in immunologically suppressed patients. Occurrence in a nonendemic area. Am J Med 1978; 64:94–100.
5. Bryan CS, DiSalvo AF. Overwhelming opportunistic histoplasmosis. Sabouraudia 1979; 17:208.
6. Kaufman CA, Israel KS, Smith JW, White AC, Schwarz J, Brooks GF. Histoplasmosis in immunosuppressed patients. Am J Med 1978; 64:923–32.
7. Wheat LJ, Slama TG, Norton JA, Kohler RB, Eitzen HE, French MLV, Sathapatayavongs B. Risk factors for disseminated or fatal histoplasmosis. Ann Intern Med 1982; 96:159–63.
8. Wheat LJ, Smith EJ, Sathapatyavongs B, Batteiger B, Filo RS, Leapman SB, French MV. Histoplasmosis in renal allograft recipients, two large urban outbreaks. Arch Intern Med 1983; 143:703–7.
9. Jones PG, Cohen RL, Batts DH, Silva Jr J. Disseminated histoplasmosis, invasive pulmonary aspergillosis, and other opportunistic infections in a homosexual patient with acquired immune deficiency syndrome. Sex Transm Dis 1983; 10:202–4.
10. Wheat LJ, Small CB. Disseminated histplasmosis in the acquired immune deficiency syndrome. Arch Intern Med 1984; 144:2147–9.
11. Bonner JR, Alexander J, Dismukes WE, App W, Griffin FM, Little R, Shin MS. Disseminated histoplasmosis in patients with the acquired immune deficiency syndrome. Arch Intern Med 1984; 144:2178–81.
12. Wheat LJ, Slama TG, Zeckel ML. Histoplasmosis in the acquired immune deficiency syndrome. Am J Med 1985; 78:203–10.
13. Mandell W, Goldberg DM, Neu HC. Histoplasmosis in patients with the acquired immune deficiency syndrome. Am J Med 1986; 81:974–8.
14. Huang CT, McGarry T, Cooper S, Saunders R. Disseminated histoplasmosis in the acquired immunodeficiency syndrome—report of 5 cases from a nonendemic area. Arch Intern Med 1987; 147:1181–4.
15. Graybill JR. Histoplasmosis in AIDS. J Infect Dis 1988; 158:623–6.
16. Kurtin PJ, McKinsey DS, Gupta MR, Driks M. Histoplasmosis in patients with acquired immunodeficiency syndrome. Am J Clin Pathol 1990; 93:367–72.
17. DeMonbruen WA. The cultivation and characteristics of Darling's *Histoplasma capsulatum*. Am J Trop Med 1934; 14:93–125.
18. Howell A. Studies on *Histoplasma capsulatum* and similar form species. 1. Morphology and development. Mycologia 1939; 31:191–216.
19. Deepe GS, Bullock WE. Histoplasmosis: A granulomatous inflammatory response. In: Gallin JI, Goldstein IM, Snyderman R, eds. Inflammation: basic principles and clinical correlates. 2nd ed. New York: Raven Press, 1992:943–58.
20. Procknow JJ, Page MI, Loosli CG. Early pathogenesis of experimental histoplasmosis. Arch Pathol 1960; 69:413–26.
21. Baughman RP, Kim CK, Vinegar A, Hendricks DE, Schmidt DJ, Bullock WE. The pathogenesis of experimental pulmonary histoplasmosis. Correlative studies of histopathology, bronchoalveolar lavage, and respiratory function. Am Rev Respir Dis 1986; 134:771–6.
22. Berry CL. The production of disseminated histoplasmosis in the mouse: the effects of changes in reticuloendothelial function. J Pathol 1969; 97:441–57.
23. Salvin SB. Acquired resistance in experimental histoplasmosis. Trans NY Acad Sci 1955/56; 18:462–8.
24. Hill GA, Marcus S. Nature of resistance in mouse histoplasmosis. Bacteriol Proc 1957; 81–2.
25. Newberry WM, Chandler JW, Chin TDY, Kirkpatrick CH. Immunology of the mycoses. I. Depressed lymphocyte transformation in chronic histoplasmosis. J Immunol 1968; 100:436–43.

26. Maresca B, Kobayashi GS. Dimorphism in *Histoplasma capsulatum*: a model for the study of cell differentiation in pathogenic fungi. Microbiol Rev 1989; 53:186–209.
27. Eissenberg LG, Goldman WE. *Histoplasma* variation and adaptive strategies for parasitism: new perspectives on histoplasmosis. Clin Microbiol Rev 1991; 4:411–21.
28. Wu-Hsieh B, Howard DH. Histoplasmosis. In: Cox RA, ed. Immunology of the fungal diseases. Boca Raton, Florida: CRC Press, 189:200–25.
29. Bullock WE, Wright SD. Role of adherence promoting receptors, CR3, LFA-1, and p150,95, in binding of *Histoplasma capsulatum* by human macrophages. J Exp Med 1987; 165:195–210.
30. Sanchez-Madrid F, Nagy J, Robbins E, Simon P, Springer TA. A human leukocyte differentiation antigen family with distinct alpha subunits and a common beta subunit: the lymphocyte-function associated antigen (LFA-1), the C3bi complement receptor (OKM1/Mac-1), and the p150,95 molecule. J Exp Med 1983; 158:1785–1803.
31. Newman SL, Bucher C, Rhodes J, Bullock WE. Phagocytosis of *Histoplasma capsulatum* yeasts and microconidia by human cultured macrophages and alveolar macrophages. Cellular cytoskeleton requirement for attachment and ingestion. J Clin Invest 1990; 85:223–30.
32. Schnur RA, Newman SL. The respiratory burst response to Histoplasma capsulatum by human neutrophils. Evidence for intracellular trapping of superoxide anion. J Immunol 1990; 144:4765–72.
33. Myones BL, Dalzell JG, Hogg N, Ross GD. Neutrophil and monocyte cell surface p150,95 has iC3b-receptor (CR4) activity resembling CR3. J Clin Invest 1988; 82:640–51.
34. Ross GD, Reed W, Dalzell JG, Becker SE, Hogg N. Macrophage cytoskeleton association with CR3 and CR4 regulates receptor mobility and phagocytosis of iC3b-opsonized erythrocytes. J Leukocyte Biol 1992; 51:109–17.
35. Kimberlin CL, Hariri AR, Hempel HO, Goodman NL. Interactions between *Histoplasma capsulatum* and macrophages from normal and treated mice: comparison of the mycelial and yeast phases in alveolar and peritoneal macrophages. Infect Immun 1981; 34:6–10.
36. Newman SL, Musson RA, Henson PM. Development of functional complement receptors during *in vitro* maturation of human monocytes into macrophages. J Immunol 1980; 125:2236–44.
37. Newman SL, Devery-Pocius JE, Ross GD, Henson PM. Phagocytosis by human monocyte-derived macrophages. Independent function of receptors for C3b (CR1) and iC3b (CR3). Complement 1984; 1:213–27.
38. Wright SD, Silverstein SC. Tumor-promoting phorbol esters stimulate C3b and C3b' receptor-mediated phagocytosis in cultured human monocytes. J Exp Med 1982; 156:1149–64.
39. Reynolds HY, Atkinson JP, Newball HH, Frank MM. Receptors for immunoglobulin and complement on human alveolar macrophages. J Immunol 1975; 114:1813–9.
40. Puszkin E, Puszkin S, Lo LW, Tannenbaum SW. Binding of cytochalasin D to platelet and muscle myosin. J Biol Chem 1973; 248:7754–61.
41. De Brabender MJ, Van de Veire RML, Aerts FEM, Borgers M, and Janssen PAJ. The effects of methyl [5-(2-thienylcarbonyl)-1H-benzimidazole-2-yl] carbamate (R17934; NSC 238159), a new synthetic anti-tumor drug interfering with microtubules, on mammalian cells. Cancer Res 1976; 36:905–16.
42. Newman SL, Mikus LK, Tucci MA. Differential requirements for cellular cytoskeleton in human macrophage complement receptor– and Fc receptor–mediated phagocytosis. J Immunol 1991; 146:967–74.
43. Wright SD, Silverstein SC. Receptors for C3b and C3bi promote phagocytosis but not the release of toxic oxygen from human phagocytes. J Exp Med 1983; 158:2016–23.
44. Armstrong JA, Hart PD. Response of cultured macrophages to *Mycobacterium tuberculosis*, with observations on fusion of lysosomes with phagosomes. J Exp Med 1971; 134:713–40.
45. Jones TC, Hirsch JG. The interaction between *Toxoplasma gondii* and mammalian cells. II. The absence of lysosomal fusion with phagocytic vacuoles containing living parasites. J Exp Med 1972; 136:1173–94.
46. Goren MB, Hart PD, Young MR, Armstrong JA. Prevention of phagosome-lysosome fusion in cultured macrophages by sulfatides of *Mycobacterium tuberculosis*. Proc Natl Acad Sci USA 1976; 73:2510–4.
47. Horwitz MA. The Legionnaires' disease bacterium *(Legionella pneumophila)* inhibits phagosome-lysosome fusion in human monocytes. J Exp Med 1983; 158:2108–26.
48. Tilney LG, Portnoy DA. Actin filaments and the growth, movement, and spread of the intracellular bacterial parasite, *Listeria monocytogenes*. J Cell Biol 1989; 109:1597–1608.

49. Anderson SE, Remington JS. Effect of normal and activated human macrophages on *Toxoplasma gondii*. J Exp Med 1974; 139:1154–74.
50. Wilson CB, Tsai V, Remington JS. Failure to trigger the oxidative burst by normal macrophages. Possible mechanism for survival of intracellular pathogens. J Exp Med 1980; 151:328–46.
51. Holzer TJ, Nelson KE, Schauf V, Crispen RG, Anderson BR. *Mycobacterium leprae* fails to stimulate phagocytic cell superoxide anion generation. Infect Immun 1986; 51:514–20.
52. Schaffner A, Davis CE, Schaffner T, Markert M, Douglas H, Braude AI. *In vitro* susceptibility of fungi to killing by neutrophil granulocytes discriminates between primary pathogenicity and opportunism. J Clin Invest 1986; 78:511–24.
53. Howard DH. Fate of *Histoplasma capsulatum* in guinea pig polymorphonuclear leukocytes. Infect Immun 1973; 8:412–19.
54. Howard DH. Comparative sensitivity of *Histoplasma capsulatum* conidiospores and blastospores to oxidative antifungal systems. Infect Immun 1981; 32:381–7.
55. Brummer E, Kurita N, Yoshida S, Nishimura K, Miyaji M. Fungistatic activity of human neutrophils against *Histoplasma capsulatum*: correlation with phagocytosis. J Infect Dis 1991; 164:158–62.
56. Fleischmann J, Wu-Hsieh B, Howard DH. The intracellular fate of *Histoplasma capsulatum* in human macrophages is unaffected by recombinant human interferon-γ. J Infect Dis 1990; 161:143–5.
57. Newman SL, Gootee L, Bucher C, Bullock WE. Inhibition of intracellular growth of *Histoplasma capsulatum* yeasts cells by cytokine-activated human monocytes and macrophages. Infect Immun 1991; 59:737–41.
58. Murray HW, Cartelli DM. Killing of intracellular *Leishmania donovani* by human mononuclear phagocytes. Evidence for oxygen-dependent and -independent leishmanicidal activity. J Clin Invest 1983; 72:32–44.
59. Pabst MJ, Hedegaard HB, Johnston RB. Cultured monocytes require exposure to bacterial products to maintain an optimal oxygen radical response. J Immunol 1983; 128:123–8.
60. Sasada M, Kubo A, Nishimura T, Kakita T, Moriguchi T, Yamamoto K, Uchino H. Candidacidal activity of monocyte-derived human macrophages: relationship between candida killing and oxygen radical generation by human macrophages. J Leukocyte Biol 1987; 41:289–94.
61. Newman SL, Gootee L, Gabay JE. Human neutrophil-mediated fungistasis against *Histoplasma capsulatum*. Localization of fungistatic activity to the azurophil granules. J. Clin. Invest. 1993; (in press).
62. Eissenberg LG, Goldman WE. *Histoplasma capsulatum* fails to trigger release of superoxide from macrophages. Infect Immun 1987; 55:29–34.
63. Wolf JE, Kerchberger V, Kobayashi GS, Little JR. Modulation of the macrophage oxidative burst by *Histoplasma capsulatum*. J Immunol 1987; 138:582–6.
64. Wolf JE, Abegg AL, Travis SJ, Kobayashi GS, Little JR. Effects of *Histoplasma capsulatum* on murine macrophage functions: inhibition of macrophage priming, oxidative burst, and antifungal activities. Infect Immun 1989; 57:513–9.
65. Howard DH. Intracellular growth of *Histoplasma capsulatum*. J Bacteriol 1965; 89:518–23.
66. Eissenberg LP, Schlesinger PH, Goldman WE. Phagosome-lysosome fusion in P388D1 macrophages infected with *Histoplasma capsulatum*. J Leukocyte Biol 1988; 43:483–91.
67. Taylor ML, Espinosa-Schoelly ME, Iturbe R, Rico B, Casasola J, Goodsaid F. Evaluation of phagolysosome fusion in acridine orange stained macrophages infected with *Histoplasma capsulatum*. Clin Exp Immunol 1989; 75:466–70.
68. Newman SL, Gootee L, Morris R, Bullock WE. Digestion of *Histoplasma capsulatum* yeasts by human macrophages. J Immunol 1992; 149:574–80.
69. Sanchez S, Carbonell LM. Immunological studies on *Histoplasma capsulatum*. Infect Immun 1975; 11:387–94.
70. Howard DH. Observation on tissue cultures of mouse peritoneal exudates innoculated with *Histoplasma capsulatum*. J Bacteriol 1959; 78:69–78.
71. Howard DH. Effect of Mycostatin and Fungizone on the growth of *Histoplasma capsulatum* in tissue culture. J Bacteriol 1960; 79:442–9.
72. Howard DH. Intracellular behavior of *Histoplasma capsulatum*. J Bacteriol 1964; 87:33–8.
73. Ohkuma S, Poole B. Fluorescence probe measurement of the intralysosomal pH in living cells and the perturbation of pH by various agents. Proc Natl Acad Sci USA 1978; 75:3327–31.

74. Geisow MJ, D'arcy Hart P, Young MR. Temporal changes of lysosome and phagosome pH during phagolysosome formation in macrophages: studies by fluorescence spectroscopy. J Cell Biol 1981; 89:645–52.
75. Siegel MR, Sisler HD. Site of action of cycloheximide in cells of Saccharomyces pastorianus. I. Effect of the antibiotic on cellular metabolism. Biochim Biophys Acta 1964; 87:70–82.
76. Siegel MR, Sisler HD. Site of action of cycloheximide in cells of Saccharomyces pastorianus. II. The nature of inhibition of protein synthesis in a cell free system. Biochim Biophys Acta 1964; 87:83–9.
77. Ennis HL, Lubin M. Cycloheximide: aspects of inhibition of protein synthesis in mammalian cells. Science 1964; 146:1474–6.
78. Gorski J, Axman MC. Cycloheximide (Actidione) inhibition of protein synthesis and the uterine response to estrogen. Arch Biochem Biophys 1964; 105:517–20.
79. Baum GL, Artis D. Susceptibility to amphotericin B of phagocytosed yeasts of cells of *Histoplasma capsulatum*. Am J Med Sci 1965; 249:113–7.
80. Gallis HA, Drew RH, Pickard WW. Amphotericin B: 30 years of clinical experience. Rev Infect Dis 1990; 12:308–29.
81. Eissenberg LG, Goldman B, Schlesinger PH. Phagolysosomes containing *Histoplasma capsulatum* fail to acidify normally. Abs Am Soc Microbiol 1992; p511.
82. Klimpel KR, Goldman WE. Isolation and characterization of spontaneous avirulent variants of *Histoplasma capsulatum*. Infect Immun 1987; 55:528–33.
83. Klimpel KR, Goldman WE. Cell walls from avirulent strains of *Histoplasma capsulatum* lack α-(1,3)-glucan. Infect Immun 1988; 56:2997–3000.
84. Howard DH, Otto V, Gupta RK. Lymphocyte-mediated cellular immunity in histoplasmosis. Infect Immun 1971; 4:605–10.
85. Howard DH. Further studies on the inhibition of *Histoplasma capsulatum* within macrophages from immunized animals. Infect Immun 1973; 8:577–81.
86. Wu-Hsieh B, Howard DH. Inhibition of growth of *Histoplasma capsulatum* by lymphokine stimulated macrophages. J Immunol 1984; 132:2593–7.
87. Wu-Hsieh B, Zlotnik A, Howard DH. T-cell hybridoma–produced lymphokine that activates macrophages to suppress intracellular growth of *Histoplasma capsulatum*. Infect Immun 1984; 43:380–5.
88. Wu-Hsieh BA, Howard DH. Inhibition of the intracellular growth of *Histoplasma capsulatum* by recombinant murine gamma interferon. Infect Immun 1987; 55:1014–6.
89. Wu-Hsieh B, Howard DH. Macrophage cell lines P388D1 and IC-21 stimulated with gamma interferon fail to inhibit the intracellular growth of *Histoplasma capsulatum*. Infect Immun 1989; 57:2903–5.
90. Lane TE, Wu-Hsieh BA, Howard DH. IFN-gamma cooperates with LPS or TNF-alpha to activate mouse splenic macrophages to an antihistoplasma state. Abs Am Soc Microbiol 1992; p505.
91. Smith JG, Magee DM, Williams DM, Graybill JR. Tumor necrosis factor-α plays a role in host defense against *Histoplasma capsulatum*. J Infect Dis 1990; 162:1349–53.
92. Graybill JR, Ahrens J, Smith J, Williams D. Does interferon gamma play a role in murine histplasmosis? Abs Am Soc Microbiol 1989; p469.
93. Newman SL, Gootee L. Colony stimulating factors (CSFs) activate human macrophages to inhibit the intracellular growth of *Histoplasma capsulatum* yeasts. Infect Immun 1992; 60:4593–7.
94. Daniels CA, Kleinerman ES, Snyderman R. Abortive and productive infections of human mononuclear phagocytes by type 1 herpes simplex. Am J Pathol 1978; 91:119–29.
95. Diamond RD, Bennett JE. Growth of *Cryptococcus neoformans* within human macrophages *in vitro*. Infect Immun 1973; 7:231–6.
96. Czuprynski CJ, Campbell PA, Henson PM. Killing of *Listeria monocytogenes* by human neutrophils and monocytes, but not by monocyte-derived macrophages. J Reticuloendothel Soc 1983; 34:29–44.
97. Horwitz MA, Silverstein SC. Activated human monocytes inhibit the intracellular multiplication of Legionnaires' disease bacteria. J Exp Med 1981; 154:1618–35.
98. Rothermel CD, Rubin BY, Jaffe EA, Murray HW. Oxygen-independent inhibition of intracellular *Chlamydia psittaci* growth by human monocytes and interferon-γ–activated macrophages. J Immunol 1986; 137:689–92.
99. Murray HW, Rubin BY, Carriero SM, Harris AM, Jaffe EA. Human mononuclear phagocyte antiprotozoal mechanisms: oxygen-dependent vs oxygen-independent activity against intracellular *Toxoplasma gondii*. J Immunol 1985; 134:1982–8.

100. Nogueira N, Chaplan S, Reesink M, Tydings J, Cohn ZA. *Trypanosoma cruzi*: induction of microbicidal activity in human mononuclear phagocytes. J Immunol 1982; 128:2142–6.
101. Reed SG, Nathan CF, Pihl DL, Rodricks P, Shanebeck K, Conlon PJ, Grabstein KH. Recombinant granulocyte/macrophage colony-stimulating factor activates macrophages to inhibit *Trypanosoma cruzi* and release hydrogen peroxide. Comparison with interferon γ. J Exp Med 1987; 166:1734–46.
102. Ho JL, Reed SG, Sobel J, Arruda S, He SH, Wick EA, Grabstein KH. Interleukin-3 induces antimicrobial activity against *Leishmania amazonensis* and *Trypanosoma cruzi* and tumorcidal activity in human peripheral blood–derived macrophages. Infect Immun 1992; 60:1984–93.
103. Weiser WY, Van Niel A, Clark SC, David JD, Remold HG. Recombinant human granulocyte/macrophage colony-stimulating factor activates intracellular killing of *Leishmania donovani* by human monocyte-derived macrophages. J Exp Med 1987; 166:1436–46.
104. Ho JL, Reed SG, Wick EA, Giordano M. Granulocyte macrophage and macrophage colony stimulating factors activate intramacrophage killing of *Leishmania amazonensis*. J Infect Dis 1990; 157:925–30.
105. Bermudez LEM, Young LS. Recombinant granulocyte-macrophage colony-stimulating factor activates human macrophages to inhibit growth or kill *Macobacterium avium* complex. J Leukocyte Biol 1990; 48:67–73.
106. Byrd TF, Horwitz MA. Interferon gamma–activated human monocytes downregulate transferrin receptors and inhibit the intracellular multiplication of *Legionella pneumophila* by limiting the availability of iron. J Clin Invest 1989: 83:1457–65.
107. Pfefferkorn ER. Interferon-γ blocks the growth of *Toxoplasma gondii* in human fibroblasts by inducing the host cells to degrade tryptophan. Proc Natl Acad Sci USA 1984; 81:908–12.
108. Sutcliffe MC, Savage AM, Alford RH. Transferrin-dependent growth inhibition of yeast-phase *Histoplasma capsulatum* by human serum and lymph. J Infect Dis 1980; 142:209–19.
109. Boguslawski G, Akagi JM, Ward LG. Possible role for cysteine biosynthesis in conversion from mycelial to yeast form of *Histoplasma capsulatum*. Nature 1976; 261:336–8.
110. Stetler DA, Boguslawski G. Cysteine biosynthesis in a fungus *Histoplasma capsulatum*. Sabouraudia 1979; 17:23–4.
111. Howard DH, Dabrowa N, Otto V, Rhodes J. Cysteine transport and sulfite reductase activity in a germination-defective mutant of *Histoplasma capsulatum*. J Bacteriol 1980; 141:417–21.
112. Lane TE, Wu-Hsieh BA, Howard DH. Iron limitation and the gamma interferon–mediated antihistoplasma state of murine macrophages. Infect Immun 1991; 59:2274–8.
113. Wu-Hsieh BA, Howard DH. Intracellular growth inhibition of *Histoplasma capsulatum* induced in murine macrophages by recombinant gamma interferon is not due to a limitation of the supply of methionine or cycteine to the fungus. Infect Immun 1992; 60:698–700.
114. Green SJ, Mellouk S, Hoffman SL, Meltzer MS, Nacy CA. Cellular mechanisms of nonspecific immunity to intracellular infection: cytokine-induced synthesis of toxic nitrogen oxides from L-arginine by macrophages and hepatocytes. Immunol Letters 1990; 25:15–20.
115. Nathan CF, Hibbs JB. Role of nitrix oxide synthesis in macrophage antimicrobial activity. Curr Opin Immunol 1991; 3:65–70.
116. Wu-Hsieh B, Nakamura LT, Odesa S, Howard DH. L-arginine is required for macrophage activation to inhibit *Histoplasma capsulatum*. Abs Am Soc Microbiol 1991; p416.
117. Palmer RMJ, Rees DD, Ashton DS, Moncada S. L-arginine is the physiologic precursor for the formation of nitric oxide by vascular endothelial cells. Biochem Biophys Res Commun 1988; 153:1251–6.
118. Ho JL, Reed SG, Sobel J, Arruda S, He SH, Wick EA, Grabstein KH. Interleukin-3 induces antimicrobial activity against *Leishmania amazonensis* and *Trypanosoma cruzi* and tumoricidal activity in human peripheral blood–derived macrophages. Infect Immun 1992; 60:1984–93.
119. Smith PD, Lamerson CL, Banks SM, Saini SS, Wahl LM, Calderone RA, Wahl SM. Granulocyte-macrophage colony-stimulating factor augments human monocyte fungicidal activity for *Candida albicans*. J Infect Dis 1990; 161:999–1005.
120. Weiser WY, Van Niel A, Clark SC, David JD, Remold HG. Recombinant human granulocyte/macrophage colony-stimulating factor activates intracellular killing of *Leishmania donovani* by human monocyte-derived macrophages. J Exp Med 1987; 166:1436–46.
121. Green LC, Wagner DA, Glogowski J, Skipper PL, Wishnok JS, Tannenbaum SR. Analysis of nitrate, nitrite, and [^{15}N]nitrate in biological fluids. Anal Biochem 1982; 126:131–8.

122. Cameron ML, Granger DL, Weinberg JB, Kozumbo WJ, Koren HS. Human alveolar and peritoneal macrophages mediate fungistasis independently of L-arginine oxidation to nitrite or nitrate. Am Rev Respir Dis 1990; 142:1313–9.
123. Murray HW, Teitelbaum RF. L-arginine–dependent reactive nitrogen intermediates and the antimicrobial effect of activated human mononuclear phagocytes. J Infect Dis 1992; 165:513–7.
124. Kwon-Chung KJ, Tewari RP, Tom WK. Determination of viability of *Histoplasma capsulatum* yeast cells: comparison between dye exclusion and colony count methods. J Med Vet Mycol 1982; 25:107–14.
125. Kurita N, Brummer E, Yoshida S, Nishimura K, Miyaji M. Antifungal activity of murine polymorphonuclear neutrophils against *Histoplasma capsulatum*. J Med Vet Mycol 1991; 29:133–43.
126. Baehner RL, Nathan DG. Leukocyte oxidase: defective activity in chronic granulomatous disease. Science 1967; 155:835–7.
127. Curnutte JT, Whitten DM, Babior BM. Defective superoxide production by granulocytes from patients with chronic granulomatous disease. N Engl J Med 1974; 290:593–7.
128. Curnutte JT, Kipnes RS, Babior BM. Defect in pyridine nucleotide–dependent superoxide production by a particulate fraction from the granulocytes of patients with chronic granulomatous disease. N Engl J Med 1975; 293:628–32.
129. Henson PM, Henson JE, Fittschen C, Kimani G, Bratton DL, Riches DW. Phagocytic cells: degranulation and secretion. In: Gallin JI, Goldstein IM, Snyderman R, eds. Inflammation: basic principles and clinical correlates. New York: Raven Press, 1988:363–90.
130. Spitznagel JK. Antibiotic proteins of human neutrophils. J Clin Invest 1990; 86:1381–6.
131. Ganz T, Selsted ME, Lehrer RI. Defensins. Eur J Haematol 1990; 44:1–8.
132. Newman SL, Henson JE, Henson PM. Phagocytosis of senescent neutrophils by human monocyte-derived macrophages and rabbit inflammatory macrophages. J Exp Med 1982; 156:430–42.
133. Nightengale SD, Parks JM, Pounders SM, Burns DK, Reynolds J, Hernandez SA. Disseminated histoplasmosis in patients with AIDS. S Med J 1990; 83:624–30.

32

Macrophage–*Cryptococcus* Interactions

Stuart M. Levitz

Boston University School of Medicine, The Boston University Hospital, Boston, Massachusetts

Cryptococcosis, defined as disease caused by the encapsulated yeast–like fungus *Cryptococcus neoformans*, has emerged as a major cause of morbidity and mortality in patients with impaired cell-mediated immunity, especially those with acquired immune deficiency syndrome (AIDS). Rare cases of cryptococcosis also occur among patients without apparent immune defects. Meningitis and, to a lesser extent, pneumonia are the most frequent life-threatening complications of cryptococcosis, although morbidity due to involvement of virtually any organ system can occur. This chapter focuses on the critical role macrophages are presumed to play in host defenses against cryptococcosis.

I. DESCRIPTION OF THE PATHOGEN

C. neoformans is unique among the major fungal pathogens in its possession of a polysaccharide capsule, which is thought to be a major virulence factor. On the basis of antigenic determinants on the capsule, four serotypes (A, B, C, and D) have been recognized [1,2]. Genetic and biochemical studies have established two varieties of *C. neoformans*. *C. neoformans* var. *neoformans* includes serotypes A and D, while *C. neoformans* var. *gattii* includes serotypes B and C [1]. While distinctive features characterize the epidemiology and ecology of the individual varieties and serotypes of *C. neoformans* [1], studies specifically examining macrophage responses as a function of variety or serotype have not been published.

In clinical specimens, the yeast cell of *C. neoformans*, exclusive of capsule, typically has a diameter of about 4 to 6 μm. The size of the surrounding capsule varies greatly in size, ranging in thickness from less than 1 μm to greater than 30 μm [3]. In nature, the yeast cells tend to be smaller and poorly encapsulated [4,5]. Since exposure to *C. neoformans* usually occurs as a consequence of inhalation of airborne organisms, the small size of environmental isolates should favor alveolar deposition. Thus, it is likely that bronchoalveolar macrophages, the resident phagocyte of the lungs, form the first line of host defenses against cryptococcosis. Moreover, the large size that *C. neoformans* can attain in vivo may physically preclude phagocytosis from occurring, although groups of phagocytes can surround and possibly kill such organisms [6].

As opposed to other pathogens that cause disease preferentially in the host with compromised cell-mediated immunity, *C. neoformans* is not an obligate intracellular parasite. Although the organism is apparently capable of intracellular parasitism, it is unclear under what circumstances, if any, *C. neoformans* uses the macrophage as an intracellular sanctuary. Moreover, it is unknown whether some cases of cryptococcosis arise as a result of reactivation from a latent focus.

II. BINDING AND PHAGOCYTOSIS OF *CRYPTOCOCCUS NEOFORMANS* BY MACROPHAGES

For the purposes of this discussion, binding refers to attachment of a particle to a macrophage. Bound particles can be either attached to receptors on the surface of the macrophage or phagocytosed (inside the cell).

A. Binding in the Absence of Opsonins

Since *C. neoformans* does not possess a mechanism to actively invade macrophages, macrophage effector cell activity against cryptococcosis is dependent on the ability of macrophage receptors to recognize ligands on the surface of the organism. Several investigators have examined recognition of unopsonized *C. neoformans* by phagocyte populations including macrophages. For most macrophage populations studied, including mouse peritoneal macrophages, mouse bronchoalveolar macrophages, mouse bone marrow–derived macrophage clones, human monocytes, and human monocyte-derived macrophages, binding of encapsulated yeasts was minimal in the absence of opsonins [7–13].

Bolaños and Mitchell demonstrated that rat bronchoalveolar macrophages were able to bind strains of *C. neoformans* with capsules of small and medium (but not large) size [14]. Binding was mediated, at least in part, by mannose receptors. Recently, my laboratory has found that human bronchoalveolar macrophages bind unopsonized, encapsulated *C. neoformans*. Binding appeared to be independent of mannose receptors and was inhibitable by components in serum of large molecular size [15].

As opposed to the situation with encapsulated organisms, where binding is usually low in the absence of opsonins, macrophages generally bind unopsonized acapsular mutants of *C. neoformans* quite avidly [10,12,16,17]. Kozel and Gotschlich examined the mechanisms by which capsule inhibits binding of unopsonized organisms [16,18]. Cryptococcal capsule confers on the organism a negative surface charge and a hydrophilic surface. However, chemical modification of these physicochemical features did not increase macrophage binding. The authors concluded that capsule does not directly modulate phagocytic function, but rather inhibits binding by presenting a surface that is not recognized by the phagocyte [16].

B. Binding of Opsonized *Cryptococcus neoformans*

Following incubation of encapsulated *C. neoformans* in normal human serum, large amounts of complement, mostly in the form of the cleavage fragment iC3b, are deposited on the surface of the capsule [7,19,20]. Macrophages obtained from several different species and anatomical sites have been shown to bind encapsulated *C. neoformans* in the presence of normal serum, but not serum lacking a functional complement system [7–13,17,21–23]. Moreover, incubation of *C. neoformans* with purified proteins of the alternative complement pathway results in C3 deposition on the capsular surface and binding by human monocyte-derived macrophages. Incubation of normal human serum with *C. neoformans* also results in the generation of chemotactic factors (presumably C5a) for neutrophils and monocytes [24,25]. In general, binding by macrophages of serum-opsonized encapsulated *C. neoformans* is inversely proportional to capsule size [14,21,26,27].

In contrast to C3, immunoglobulin G (IgG) from normal serum binds at sites throughout the capsule and thus, for the most part, is unavailable for recognition by macrophage Fc receptors [7,19]. However, specific anticapsular antibody is deposited on the capsular surface and is a potent opsonin [7,8,11–13,19,28,29]. Similar to the situation seen with other particulate stimuli, binding of *C. neoformans* to macrophages is augmented when the organisms are opsonized with both specific antibody and complement, compared with antibody or complement alone [12].

Human macrophages possess at least three defined complement receptors for fragments of C3; CR1 (CD35), CR3 (CD11b/CD18), and CR4 (CD11c/CD18) [30]. Several lines of evidence, taken together, strongly suggest that all three aforementioned complement receptors participate in binding of serum opsonized *C. neoformans* to monocyte-derived macrophages [11]. First, monoclonal antibodies directed against any of these three complement receptors profoundly inhibited binding. Second, migration of CR to the cryptococcal binding site could be demonstrated by immunofluorescence studies. Third, inactivation of selected complement receptors either by proteolytic cleavage using trypsin or elastase, or chelation of divalent cations with EDTA, also inhibited binding.

The above experimental observations lend strong support to the concept that opsonization is an important aspect of host defenses against cryptococcosis. This concept also gains support from clinical studies showing that certain patients with cryptococcosis have impaired opsonic activity. Serum and cerebrospinal fluid from patients with cryptococcal fungemia and meningitis, respectively, have been shown to be markedly depleted of complement components and incapable of opsonizing *C. neoformans* for phagocytosis [31,32]. Moreover, many patients with cryptococcosis do not have detectable antibody to the fungus, and such patients tend to have a poorer prognosis than patients with antibody [33]. Low levels of opsonins may be found in the bronchoalveolar space, even under normal conditions [34,35].

C. T Cell–Derived Suppressor Factor

During the course of experimental murine cryptococcosis, adherent and nonadherent suppressor cells are generated that function to limit the immune response [36–38]. Blackstock, Hall, and colleagues have characterized an antigen-specific suppressor CD8+ T cell producing a T suppressor factor that inhibits binding of *C. neoformans* by murine peritoneal macrophages [39–41]. The T suppressor factor has a molecular weight of about 70 kDa and appears to be functionally equivalent to hapten-specific T suppressor factors previously described [42,43]. The T suppressor factor inhibits only a small subset (8% to 27%) of the macrophages and will inhibit binding of targets other than *C. neoformans*.

D. Phagocytosis (Internalization) of *Cryptococcus neoformans*

Following binding of a particle to a phagocytic receptor, phagocytosis of the particle may or may not take place, depending on the particular receptor involved and its state of activation [11,44,45]. While complement receptors do not constitutively phagocytose, under certain circumstances they can be up-regulated to do so. Moreover, they can act synergistically with mannose or Fc receptors to internalize particles coated with C3 and mannose or IgG [11,46]. Critical macrophage functions, including antimicrobial activity and antigen processing, may depend on internalization of bound microorganisms.

Consistent with these observations, the majority of serum-opsonized encapsulated *C. neoformans* are not internalized by macrophages in vitro unless specific anticapsular antibody is added (Figure 1) [11–13,15,47]. In contrast, serum-opsonized acapsular strains are internalized, perhaps because the acapsular strains have ligands (e.g., mannans, β-glucans) exposed on the cell wall. Such cell wall ligands are masked from the surface of the encapsulated organisms and thus not available to bind to their corresponding receptors on macrophages [11].

Generally, internalization (but not binding) is thought to be a process requiring energy and cytoskeletal reorganization [46]. However, binding of serum-opsonized encapsulated *C. neoformans* by monocyte-derived macrophages is inhibited by incubation in the cold, and by inhibitors of receptor capping and actin polymerization. In contrast, energy is not required for binding of serum-opsonized acapsular *C. neoformans* [11].

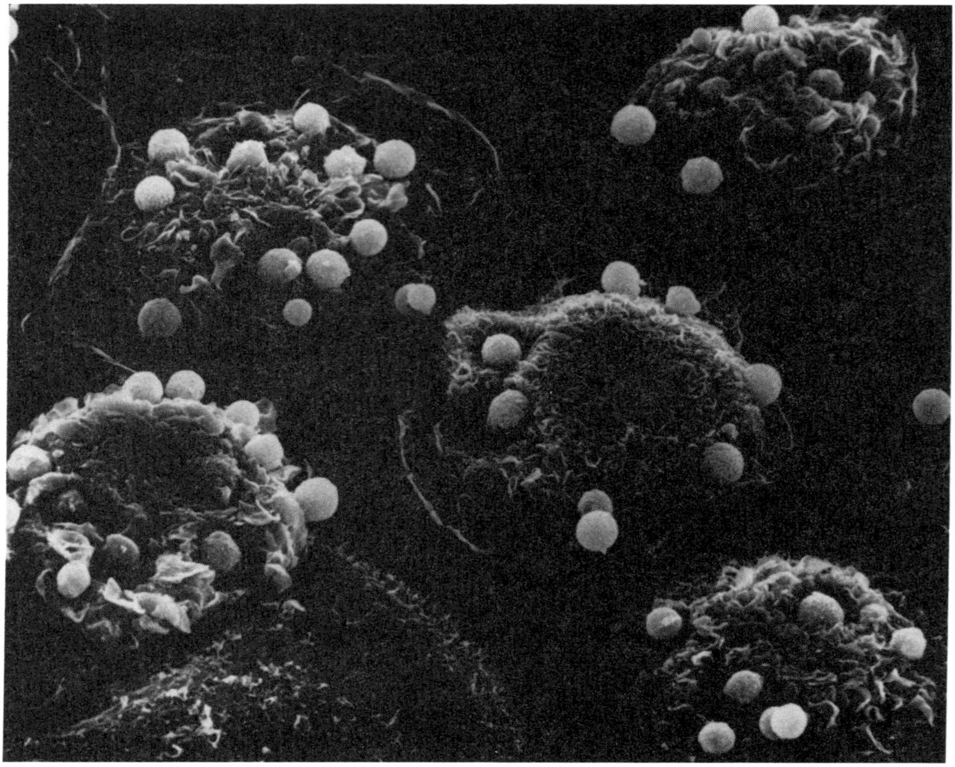

(A)

FIGURE 1 Low-power (A) and high-power (B) scanning electron micrographs of the interaction between human monocyte-derived macrophages and encapsulated *Cryptococcus neoformans*. Macrophages were cultured on a poly-L-lysine–coated plastic surface for 6 days, and then challenged with fungi for 30 minutes in the presence of 10% pooled human serum. Unbound organisms were washed away before fixation. Note most of the organisms appear attached to the surface of the cell but not actually internalized. Many of the fungi appear to be partially engulfed. Scanning electron micrographs by Robert H. Liss based on material prepared by Abdul Tabuni and Stuart Levitz. Copyright 1990 by Sears Associates Incorporated. Used by permission.

III. INHIBITION AND KILLING OF *CRYPTOCOCCUS NEOFORMANS* BY MACROPHAGES

The literature concerning the anticryptococcal activity of macrophages underscores the heterogeneous nature of macrophage populations. As discussed below, inhibition and killing of *C. neoformans* vary depending on the species of origin, anatomical location, and state of activation of the macrophage.

While this review focuses on macrophages, other effector cells (including neutrophils, monocytes, natural killer [NK] cells, and interleukin-2–stimulated lymphocytes) have been shown to inhibit or kill *C. neoformans* under defined conditions [48–51]. Given the complexity and redundancy of the immune system, it is likely that many different types of effector cells are ultimately responsible for inhibition and killing of *C. neoformans* in the host. Moreover, some effectors (e.g., bronchoalveolar macrophages, NK cells, and neutrophils) may be critical early in infection, before an antigen-specific T cell response has developed, while others (e.g., activated macrophages and lymphocytes) may become vital should natural defenses fail and the infection progress.

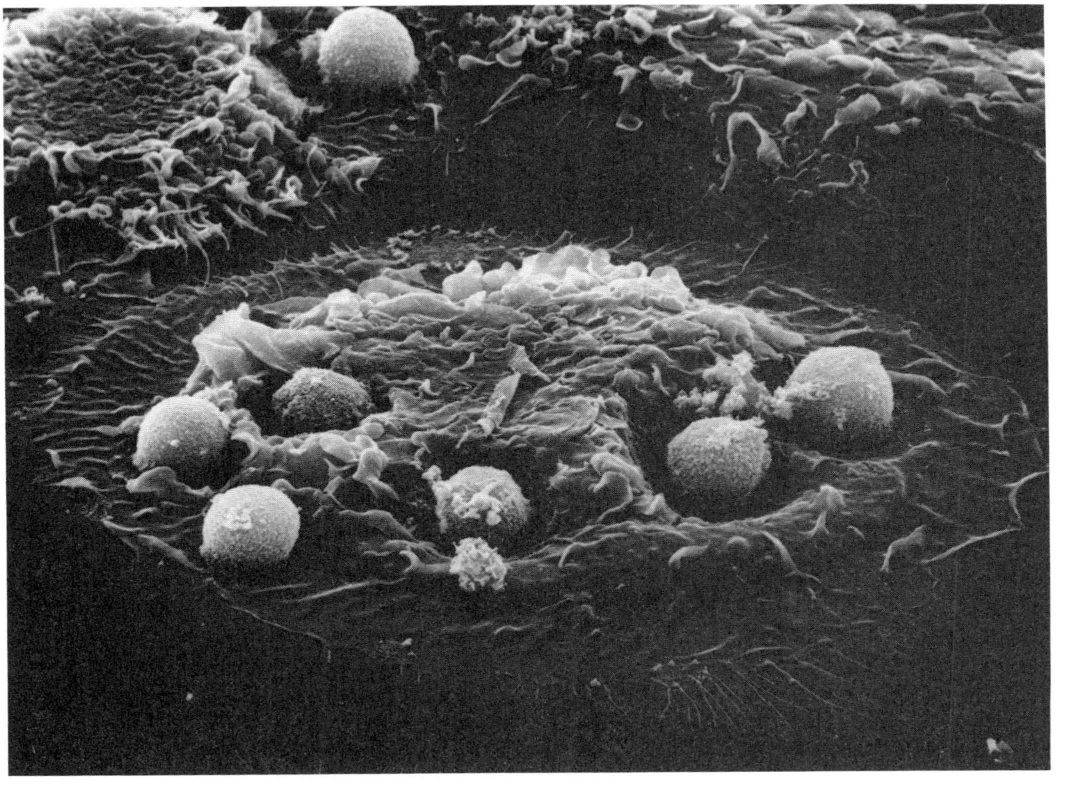

(B)

FIGURE 1 *(continued)*

Most investigators in this field try to distinguish between fungistatic (inhibition of growth) and fungicidal (killing) activity. However, often the distinction is difficult to make. Most assays feature an 18- to 24-hour incubation period with macrophages and *C. neoformans* following which macrophages are lysed and the number of viable organisms determined by colony counts. Killing is said to occur if fewer colonies are found at the end of the incubation than at the beginning. The problem arises because during the incubation period some of the yeast cells may be killed while others proliferate. Thus, if half the inoculum is killed, and the other half replicates once, the colony counts will be the same at the start and finish of the incubation. In fact, alternative assays, such as microcolony assays and assays employing vital stains, have been shown to demonstrate killing in situations where colony counting assays do not [12,52].

A. In Vivo Studies

Administration of silica intravenously to mice to destroy the hosts' macrophages followed by challenge of the mice with an inoculum of *C. neoformans* resulted in decreased survival as well as a greater number of colony counts in spleens, lungs, and brains compared with control mice [53]. However, it is unclear from this study what is the mechanism or mechanisms underlying the detrimental effects of silica. In addition to adversely affecting macrophage-mediating killing of *C. neoformans*, silica's deleterious actions in this model could be secondary to diminished antigen-presentation or monokine production. Moreover, silica treatment may have affected cell types other than macrophages.

Prolonged survival following cryptococcal challenge has been seen if mice were first stimulated with Bacillus Calmette-Guérin (BCG), *Listeria monocytogenes*, *Toxoplasma gondii*, or *Besnoitia jellisoni* [53,54]. However, as with the studies with silica discussed above, the intrinsic mechanisms responsible for this effect are unclear.

In studies performed in my laboratory, mice were given an intraperitoneal injection of interferon-γ (or normal saline for the control group) followed 18 hours later by an intraperitoneal challenge with either an encapsulated or an acapsular strain of *C. neoformans*. After an additional 18 hours, the peritoneal cavity was lavaged and the viability of the recovered organisms determined. The interferon-treated mice, but not the control mice, killed the encapsulated strain. In contrast, both groups of mice killed the acapsular strain, although killing was augmented in the interferon group. Since over 90% of the cells in the peritoneal cavity were macrophages, and in view of the known macrophage-activating effects of interferon-γ, killing was probably mediated by macrophages. However, the contribution of other cell types, acting alone or synergistically with macrophages, cannot be ruled out [12].

Recently, Hill demonstrated that following an intratracheal challenge of mice with *C. neoformans*, multinucleated giant cells (presumably derived from the fusion of macrophages) form that surround the fungus [55]. Interestingly, mice depleted of CD4+ (but not CD8+) T cells did not form multinucleated giant cells. When the mice underwent bronchoalveolar lavage 3 weeks following the cryptococcal challenge, fungi were found in association with macrophages or giant cells in the normal mice, but not in the CD4-depleted mice. Moreover, normal mice, but not CD4-deficient mice, effectively cleared *C. neoformans* from the lungs, as measured by a reduction in colony counts. These results suggest that CD4+ T cells produce cytokines at the site of infection that promote granuloma formation and activate macrophages [55].

B. In Vitro Studies

Because bronchoalveolar macrophages are likely to be the first phagocyte to encounter *C. neoformans* following initial exposure to the organism, many cryptococcologists have focused on this cell type. Even in the absence of specific activation, bronchoalveolar macrophages freshly lavaged from mice and rats kill *C. neoformans* [10,14,56]. Paradoxically, encapsulated organisms are more efficiently killed by mouse bronchoalveolar macrophages than are acapsular mutants [10]. The antifungal activity of rat bronchoalveolar macrophages diminishes after overnight culture but is partially restored by the addition of interferon-γ [57]. Growth inhibition, but not killing, of *C. neoformans* by human bronchoalveolar macrophages has been demonstrated [58–60]. Addition of gp120 (the envelope glycoprotein of human immunodeficiency virus-1, the etiological agent of AIDS) to human bronchoalveolar macrophages significantly reduced the fungistatic activity of human bronchoalveolar macrophages [60]. The deleterious effect of gp120 on this first line of defense may help explain the exquisite vulnerability of patients with AIDS to cryptococcosis.

In the absence of activation, several investigators have been unable to show significant growth inhibition or killing of *C. neoformans* by mouse resident peritoneal macrophages [12,54,61–64]. However, these investigators have defined conditions whereby activators empower the macrophage to inhibit or kill. Peritoneal macrophages obtained from mice infected with *L. monocytogenes* or *B. jellisoni*, but not uninfected mice, were able to partially withstand a subsequent in vitro challenge with *C. neoformans* [54]. Granger et al. were able to obtain complete fungistasis only if the peritoneal macrophages were induced by peptone from BCG-infected mice, and endotoxin plus a nonopsonic serum factor were present in the culture medium [61,62]. In later experiments, fungistasis was obtained when interferon-γ was substituted for BCG [64]. Culture with interferon-γ also activated murine resident peritoneal cells to kill *C. neoformans* [12]. Finally, amphotericin B stimulated growth inhibition of murine peritoneal cells, apparently as a result of amphotericin B's immunoadjuvant properties [63].

C. neoformans has a propensity to infect the central nervous system and cause meningitis.

Perfect et al. found that cerebrospinal fluid macrophages obtained from rabbits with experimental cryptococcal meningitis were incapable of inhibiting growth of *C. neoformans* [65]. Flesch et al. demonstrated that activation with interferon-γ and endotoxin empowered murine bone marrow–derived macrophages to inhibit *C. neoformans* [66].

Several laboratories have demonstrated killing of opsonized *C. neoformans* by human peripheral blood monocytes [48,49,67]. Monocytes obtained from patients with AIDS, cryptococcosis, Hodgkin's disease, or sarcoidosis killed *C. neoformans* as well as control monocytes [48,67]. As was the case with murine bronchoalveolar macrophages [10], human monocytes killed encapsulated strains more efficiently than a capsule-free mutant [49].

Upon in vitro culture, monocytes differentiate and acquire many of the characteristics of tissue macrophages. Diamond and Bennett demonstrated that human macrophages obtained by culturing blood monocytes on plastic surfaces failed to inhibit growth of *C. neoformans* [68]. However, macrophages derived from monocytes cultured on extracellular matrix proteins appear to more closely resemble tissue macrophages [69,70]. Therefore, my laboratory examined the influence of culture surface on the anticryptococcal activity of macrophages obtained by differentiation of monocytes [47]. We found that monocytes cultured on surfaces coated with fibronectin or poly-L-lysine (but not plastic, laminin, or collagen) were able to effect complete fungistasis of a serum-opsonized encapsulated strain. Activity was not improved by incubating the cells with interferon-γ, tumor necrosis factor, granulocyte-macrophage colony stimulating factor, 1,25-dihydroxycholecalciferol (vitamin D_3), or lymphocyte supernatants. In contrast, if an acapsular strain was used, complete growth inhibition was observed regardless of the choice of culture surface, opsonins, or cytokines [47].

In other experiments, culture of human blood monocytes and lymphocytes for 15 days with cryptococcal antigens under conditions resulting in a lymphoproliferative response generated effector cells capable of killing encapsulated *C. neoformans*. Both adherent (mostly macrophages) and nonadherent cells were necessary for killing to be seen. An acapsular strain was killed by adherent cells alone [71].

C. Mediators of Inhibition and Killing

As reviewed elsewhere in this book, a myriad of mechanisms have been described whereby macrophages can inhibit and kill invading microbes. The contribution some of these mechanisms make to anticryptococcal activity has been investigated.

Studies by Diamond et al. established that *C. neoformans* is killed by H_2O_2, an oxidative product of the phagocytic respiratory burst [48]. However, as with most fungi, relatively high concentrations of H_2O_2 (which may not be obtainable in the cell) are required for killing. Neutrophil-mediated killing of *C. neoformans* appears to proceed by oxidative mechanisms, since chronic granulomatous disease neutrophils do not kill and there is a lag in killing of cells treated with the myeloperoxidase inhibitor azide [48]. Comparable studies have not been done with monocytes. Since monocytes, but not macrophages, possess myeloperoxidase, it is tempting to hypothesize that the loss of fungicidal activity seen upon monocyte culture is due to the loss of myeloperoxidase activity.

Encapsulated *C. neoformans* have been shown to be weak stimulators of the respiratory burst in murine resident peritoneal macrophages, unless the organisms are opsonized with both a source of complement and anticapsular antibody. In contrast, the burst was stimulated by a capsule-free mutant, with or without opsonization. However, there was no correlation between killing and stimulation of a burst for either strain [12]. Similarly, murine bronchoalveolar macrophages appear to kill encapsulated *C. neoformans* by nonoxidative mechanisms, as bronchoalveolar macrophages released minimal amounts of H_2O_2 after stimulation with encapsulated *C. neoformans*, and killing was not reduced by respiratory burst inhibitors or scavengers [10]. Flesch et al. demonstrated that killing of *C. neoformans* by interferon-γ–activated bone marrow macrophages is independent of toxic oxygen radicals and mediated by secreted protein(s) of apparent molecular mass of 15 and 30 kDa [66].

Nitric oxide, and other reactive nitrogen intermediaries, are formed when activated macrophages oxidize L-arginine to nitrite, nitrate, and L-citrulline [72]. In a series of elegant experiments, Granger et al. demonstrated that growth inhibition of *C. neoformans* by activated murine peritoneal macrophages is absolutely dependent on this nitrogen oxidation mechanism [73,74]. Moreover, reagent nitric oxide was shown to be fungistatic at low concentrations and fungicidal at high concentrations [72]. In support of the above results, my laboratory correlated nitrite synthesis with killing of *C. neoformans* by murine resident peritoneal cells [12]. However, human macrophages may be incapable of nitric oxide generation [75], and fungistasis mediated by human macrophages appears to be independent of this pathway [22].

IV. CRYPTOCOCCAL ANTIGEN PRESENTATION BY MACROPHAGES

In addition to the effector cell functions discussed above, macrophages play critical immunological roles as antigen-presenting cells, and as cells capable of secreting a variety of monokines such as interleukin-1 and tumor necrosis factor [76–78]. Unfortunately, little research has been published concerning these important macrophage functions as they relate to *C. neoformans*. Collins and Bancroft demonstrated that murine macrophages challenged in vitro with an unopsonized encapsulated strain had a reduced ability to stimulate specific T cell proliferation when compared with an acapsular mutant strain. However, this reduced ability appeared to be the result of capsule's inhibiting of phagocytosis, since macrophages challenged with opsonized, encapsulated *C. neoformans* efficiently processed and presented antigen. Antigen processing and presentation was chloroquine-sensitive and major histocompatibility complex–restricted [79].

V. SUMMARY

Macrophages are a heterogeneous population that vary depending on their species of origin, anatomic location, state of activation, and conditions of culture. Moreover, macrophages normally interact with other cells both within and without the immune system. It is clear from the data reviewed in this chapter that all of these aforementioned variables greatly influence macrophage–*C. neoformans* interactions. While circumstantial evidence strongly supports a major role for the macrophage in host defenses against cryptococcosis, the nature and extent of the contribution macrophages make remain to be defined. One major challenge for researchers in this field will be to design experiments that closely mimic what occurs in human physiological and pathological states.

REFERENCES

1. Levitz SM. The ecology of *Cryptococcus neoformans* and the epidemiology of cryptococcosis. Rev Infect Dis 1991; 13:1163–9.
2. Bhattacharjee AK, Bennett JE, Glaudemans CPJ. Capsular polysaccharides of *Cryptococcus neoformans*. Rev Infect Dis 1984; 6:619–24.
3. Perfect JR. Cryptococcosis. Infect Dis Clin North Am 1989; 3:77–102.
4. Neilson JB, Fromtling RA, Bulmer GS. *Cryptococcus neoformans*: size range of infectious particles from aerosolized soil. Infect Immun 1977; 17:634–8.
5. Powell KE, Dahl BA, Weeks RJ, Tosh FE. Airborne *Cryptococcus neoformans*: particles from pigeon excreta compatible with alveolar deposition. J Infect Dis 1972; 125:412–5.
6. Kalina M, Kletter Y, Aronson M. The interaction of phagocytes and the large-sized parasite *Cryptococcus neoformans*: cytochemical and ultrastructural study. Cell Tiss Res 1974; 152:165–74.
7. Kozel TR, Pfrommer GST, Guerlain AS, Highison BA, Highison GJ. Role of the capsule in phagocytosis of *Cryptococcus neoformans*. Rev Infect Dis 1988; 10:S436–9.

8. Davies SF, Clifford DP, Hoidal JR, Repine JE. Opsonic requirements for the uptake of *Cryptococcus neoformans* by human polymorphonuclear leukocytes and monocytes. J Infect Dis 1982; 145:870–4.
9. Kitz DJ, Johnson CR, Kobayashi GS, Medoff G, Little JR. Growth inhibition of *Cryptococcus neoformans* by cloned cultured murine macrophages. Cell Immunol 1984; 88:489–500.
10. Levitz SM, DiBenedetto DJ. Paradoxical role of capsule in murine bronchoalveolar macrophage–mediated killing of *Cryptococcus neoformans*. J Immunol 1989; 142;659–65.
11. Levitz SM, Tabuni A. Binding of *Cryptococcus neoformans* by human cultured macrophages: requirements for multiple complement receptors and actin. J Clin Invest 1991; 87:528–35.
12. Levitz SM, DiBenedetto DJ. Differential stimulation of murine resident peritoneal cells by selectively opsonized encapsulated and acapsular *Cryptococcus neoformans*. Infect Immun 1988; 56:2544–51.
13. Griffin FM Jr. Roles of macrophage Fc and C3b receptors in phagocytosis of immunologically coated *Cryptococcus neoformans*. Proc Natl Acad Sci USA 1981; 78:3853–7.
14. Bolaños B, Mitchell TG. Phagocytosis and killing of *Cryptococcus neoformans* by rat alveolar macrophages in the absence of serum. J Leukocyte Biol 1989; 46:521–8.
15. Levitz SM, Tabuni A, Wagner R, Kornfeld H, Smail EH. Binding of unopsonized *Cryptococcus neoformans* by human bronchoalveolar macrophages: inhibition by a component in serum of large molecular size. J Infect Dis 1992 (in press).
16. Kozel TR, Gotschlich EC. The capsule of *Cryptococcus neoformans* passively inhibits phagocytosis of the yeast by macrophages. J Immunol 1982; 129:1675–80.
17. Small JM, Mitchell TG. Strain variation in antiphagocytic activity of capsular polysaccharides from *Cryptococcus neoformans* serotype A. Infect Immun 1989; 57:3751–6.
18. Kozel TR. Dissociation of a hydrophobic surface from phagocytosis encapsulated and non-encapsulated *Cryptococcus neoformans*. Infect Immun 1983; 39:1214–9.
19. Kozel TR, Highison B, Stratton CJ. Localization on encapsulated *Cryptococcus neoformans* of serum components opsonic for phagocytosis by macrophages and neutrophils. Infect Immun 1984; 43:574–9.
20. Kozel TR, Pfrommer GST. Activation of the complement system by *Cryptococcus neoformans* leads to binding of iC3b to the yeast. Infect Immun 1986; 52:1–5.
21. Bolaños B, Mitchell TG. Phagocytosis of *Cryptococcus neoformans* by rat alveolar macrophages. J Med Vet Mycol 1989; 27:203–17.
22. Bulmer GS, Tacker JR. Phagocytosis of *Cryptococcus neoformans* by alveolar macrophages. Infect Immun 1975; 11:73–9.
23. Kozel TR, Wilson MA, Pfrommer GST, Schlageter AM. Activation and binding of opsonic fragments of C3 on encapsulated *Cryptococcus neoformans* by using an alternative complement pathway reconstituted from six isolated proteins. Infect Immun 1989; 57:1922–7.
24. Laxalt KA, Kozel TR. Chemotaxigenesis and activation of the alternative complement pathway by encapsulated and non-encapsulated *Cryptococcus neoformans*. Infect Immun 1979; 26:435–40.
25. Diamond RD, Erickson NF III. Chemotaxis of human neutrophils and monocytes induced by *Cryptococcus neoformans*. Infect Immun 1982; 38:380–2.
26. Granger DL, Perfect JR, Durack DT. Virulence of *Cryptococcus neoformans*: regulation of capsule synthesis by carbon dioxide. J Clin Invest 1985; 76:508–16.
27. Mitchell TG, Friedman L. In vitro phagocytosis and intracellular fate of variously encapsulated strains of *Cryptococcus neoformans*. Infect Immun 1972; 5:491–8.
28. Kozel TR, McGaw TG. Opsonization of *Cryptococcus neoformans* by human immunoglobulin G: role of immunoglobulin G in phagocytosis by macrophages. Infect Immun 1979; 25:255–61.
29. Kozel TR, Follette JL. Opsonization of encapsulated *Cryptococcus neoformans* by specific anticapsular antibody. Infect Immun 1981; 31:978–84.
30. Law SKA. C3 receptors on macrophages. J Cell Sci 1988; 9:S67–97.
31. Diamond RD, May JE, Kane MA, Frank MM, Bennett JE. The role of the classical and alternate complement pathways in host defenses against *Cryptococcus neoformans* infection. J Immunol 1974; 112:2260–70.
32. Macher AM, Bennett JE, Gadek JE, Frank MM. Complement depletion in cryptococcal sepsis. J Immunol 1978; 120:1686–90.
33. Diamond RD, Bennett JE. Prognostic factors in cryptococcal meningitis: a study in 111 cases. Ann Intern Med 1974; 80:176–81.
34. Reynolds HY, Newball HH. Analysis of proteins and respiratory cells obtained from human lungs by bronchial lavage. J Lab Clin Med 1974; 84:559–73.

35. Robertson J, Caldwell JR, Castle JR, Waldman RH. Evidence for the presence of components of the alternative (properdin) pathway of complement activation in respiratory secretions. J Immunol 1976; 117:900–3.
36. Murphy JW. Influence of cryptococcal antigens on cell-mediated immunity. Rev Infect Dis 1988; 10:S432–5.
37. Breen JF, Lee IC, Vogel FR, Friedman H. Cryptococcal capsular polysaccharide-induced modulation of murine immune responses. Infect Immun 1982; 36:47–51.
38. Sotomayor CE, Rubinstein HR, Riera CM, Masih DT. Immunosuppression in experimental cryptococcosis in rats. Induction of afferent T suppressor cells to a nonrelated antigen. J Med Vet Mycol 1987; 25:67–75.
39. Morgan MA, Blackstock RA, Bulmer GS, Hall NK. Modification of macrophage phagocytosis in murine cryptococcosis. Infect Immun 1983; 40:493–500.
40. Blackstock R, McCormack JM, Hall NK. Induction of a macrophage-suppressive lymphokine by soluble cryptococcal antigens and its association with models of immunologic tolerance. Infect Immun 1987; 55:233–9.
41. Blackstock R, Hall NK, Hernandez NC. Characterization of a suppressor factor that regulates phagocytosis by macrophages in murine cryptococcosis. Infect Immun 1989; 57:1773–9.
42. Blackstock R, Zembala M, Asherson GL. Functional equivalence of cryptococcal and haptene-specific T suppressor factor (TsF): I.Picryl and oxazolone-specific TsF, which inhibit transfer of contact sensitivity, also inhibit phagocytosis by a subset of macrophages. Cell Immunol 1991; 136:435–47.
43. Blackstock R, Zembala M, Asherson GL. Functional equivalence of cryptococcal and haptene-specific T suppressor factor (TsF): II. Monoclonal anti-cryptococcal TsF inhibits both phagocytosis by a subset of macrophages and transfer of contact sensitivity. Cell Immunol 1991; 136:448–61.
44. Newman SL, Devery-Pocius JE, Ross GD, Henson PM. Phagocytosis of human monocyte-derived macrophages. Independent function of receptors for C3b (CR_1) and iC3b (CR_3). Complement 1984; 1:213–27.
45. Bianco C, Griffin FM Jr, Silverstein SC. Studies of the macrophage complement receptor. Alteration of receptor function upon macrophage activation. J Exp Med 1975; 141:1278–90.
46. Silverstein SC, Greenberg S, DiVirgilio F, Steinberg TH. Phagocytosis. In: Paul WE, ed. Fundamental immunology. 2nd ed. New York: Raven Press, 1989:703–20.
47. Levitz SM, Farrell TP. Growth inhibition of *Cryptococcus neoformans* by cultured human monocytes: role of the capsule, opsonins, the culture surface, and cytokines. Infect Immun 1990; 58:1201–9.
48. Diamond RD, Root RK, Bennett JE. Factors influencing killing of *Cryptococcus neoformans* by human leukocytes in vitro. J Infect Dis 1972; 125:367–76.
49. Miller MF, Mitchell TG. Killing of *Cryptococcus neoformans* strains by human neutrophils and monocytes. Infect Immun 1991; 59:24–8.
50. Murphy JW, McDaniel DO. In vitro reactivity of natural killer (NK) cells against *Cryptococcus neoformans*. J Immunol 1982; 128:1577–83.
51. Levitz SM. Activation of human peripheral blood mononuclear cells by interleukin-2 and granulocyte-macrophage colony-stimulating factor to inhibit *Cryptococcus neoformans*. Infect Immun 1991; 59:3393–7.
52. Hidore MR, Nabavi N, Sonleitner F, Murphy JW. Murine natural killer cells are fungicidal to *Cryptococcus neoformans*. Infect Immun 1991; 59:1747–54.
53. Monga DP. Role of macrophages in resistance of mice to experimental cryptococcosis. Infect Immun 1981; 32:975–8.
54. Gentry LO, Remington JS. Resistance against *Cryptococcus* conferred by intracellular bacteria and protozoa. J Infect Dis 1971; 123:22–31.
55. Hill JO. CD4+ T cells cause multinucleated giant cells to form around *Cryptococcus neoformans* and confine the yeast within the primary site of infection in the respiratory tract. J Exp Med 1992; 175:1685–95.
56. Bolaños B, Mitchell TG. Killing of *Cryptococcus neoformans* by rat alveolar macrophages. J Med Vet Mycol 1989; 27:219–28.
57. Mody CH, Tyler CL, Sitrin RG, Jackson C, Toews GB. Interferon-γ activates rat alveolar macrophages for anticryptococcal activity. Am J Respir Cell Mol Biol 1991; 5:19–26.

58. Weinberg PB, Becker S, Granger DL, Koren HS. Growth inhibition of *Cryptococcus neoformans* by human alveolar macrophages. Am Rev Respir Dis 1987; 136:1242–7.
59. Cameron ML, Granger DL, Weinberg JB, Kozumbo WJ, Koren HS. Human alveolar and peritoneal macrophages mediate fungistasis independently of L-arginine oxidation to nitrite or nitrate. Am Rev Respir Dis 1990; 142:1313–9.
60. Wagner RP, Levitz SM, Tabuni A, Kornfeld H. HIV-1 envelope protein (gp120) inhibits the activity of human bronchoalveolar macrophages against *Cryptococcus neoformans*. Am Rev Resp Dis 1992; 146:1434–8.
61. Granger DL, Perfect JR. Durack DT. Macrophage-mediated fungistasis in vitro: requirements for intracellular and extracellular cytotoxicity. J Immunol 1986; 136:672–80.
62. Granger DL, Perfect JR, Durack DT. Macrophage-mediated fungistasis: requirement for a macromolecular component in serum. J Immunol 1986; 137:693–701.
63. Aslanzadeh J, Mormol JS, Little JR. Anticryptococcal activity of amphotericin B–stimulated macrophages. Immunopharmacol Immunotoxicol 1991; 13:465–83.
64. Perfect JR, Granger DL, Durack DT. Effects of antifungal agents and γ interferon on macrophage cytotoxicity for fungi and tumor cells. J Infect Dis 1987; 156:316–23.
65. Perfect JR, Hobbs MM, Granger DL, Durack DT. Cerebrospinal fluid macrophage response to experimental cryptococcal meningitis: relationship between in vivo and in vitro measurements of cytotoxicity. Infect Immun 1988; 56:849–54.
66. Flesch IEA, Schwamberger G, Kaufmann SHE. Fungicidal activity of IFN-γ–activated macrophages: extracellular killing of *Cryptococcus neoformans*. J Immunol 1989; 142:3219–24.
67. Washburn RG, Tuazon CU, Bennett JE. Phagocytic and fungicidal activity of monocytes from patients with acquired immunodeficiency syndrome. J Infect Dis 1985; 151:565–6.
68. Diamond RD, Bennett JE. Growth of *Cryptococcus neoformans* within human macrophages in vitro. Infect Immun 1973; 7:231–6.
69. Brown EJ. The role of extracellular matrix proteins in the control of phagocytosis. J Leukocyte Biol 1986; 39:579–91.
70. Kaplan G, Gaudernack G. In vitro differentiation of human monocytes. Differences in monocyte phenotypes induced by cultivation on glass or on collagen. J Exp Med 1982; 156:1101–14.
71. Levitz SM, Farrell TP, Maziarz RT. Killing of *Cryptococcus neoformans* by human peripheral blood mononuclear cells stimulated in culture. J Infect Dis 1991; 163:1108–13.
72. Alspaugh JA, Granger DL. Inhibition of *Cryptococcus neoformans* replication by nitrogen oxides supports the role of these molecules as effectors of macrophage-mediated cytostasis. Infect Immun 1991; 59:2291–6.
73. Granger DL, Hibbs JB Jr Perfect JR, Durack DT. Specific amino acid (L-arginine) requirement for the microbiostatic activity of murine macrophages. J Clin Invest 1988; 81:1129–36.
74. Granger DL, Hibbs JB Jr Perfect JR, Durack DT. Metabolic fate of L-arginine in relation to microbiostatic capability of murine macrophages. J Clin Invest 1990; 85:264–73.
75. Murray HW, Teitelbaum RF. L-arginine–dependent reactive nitrogen intermediates and the antimicrobial effect of activated human mononuclear phagocytes. J Infect Dis 1992; 165:513–7.
76. Nathan CF. Secretory products of macrophages. J Clin Invest 1987; 79:319–26.
77. Ezekowitz RAB. Macrophages. In: Nelson DS, ed. Natural immunity. Australia: Academic Press, 1989:15–38.
78. Sorg C, Lohmann-Matthes M-L. Macrophages and accessory cells of the immune system. Immunol Today 1989; 10:S27–9.
79. Collins HL, Bancroft GJ. Encapsulation of Cryptococcus neoformans impairs antigen-specific T-cell responses. Infect Immun 1991; 59:3883–8.

33

Macrophage–*Aspergillus* Interactions

Andreas Schaffner

University Hospital, Zurich, Switzerland

I. INTRODUCTION

The interactions between aspergilli and macrophages have some unique but also more universally applicable aspects. If the role of macrophages is compared with the role of other host defense systems potentially operative against *Aspergillus,* it becomes apparent that mononuclear phagocytes are, together with neutrophil granulocytes, responsible for the solid natural resistance against infection with these ubiquitous fungi, to which most humans and animals are continuously exposed [1]. Phagocytosis in the absence of opsonins, followed by the destruction of conidia, the spores from aspergilli, without instruction by macrophage-activating cytokines, documents the importance of mononuclear phagocytes as an autonomous host defense system that is operative even in the absence of complex antigen-dependent host reactions involving lymphocytes.

However, before we turn to details of the encounter between aspergilli and macrophages it appears appropriate to present the characteristics of aspergilli, and to describe the pathogenic attributes of these fungi.

II. *ASPERGILLUS* AN OPPORTUNISTIC PATHOGEN

Aspergilli are ubiquitous saprobic fungi living under a wide spectrum of conditions. Because aspergilli cannot take ecological advantage either from colonizing or infecting a living host, such events must be viewed as accidents. Nevertheless, some members of the genus *Aspergillus* have properties permitting them to colonize and invade the host consequent to failing host defense systems.

The first pathogenic quality of *Aspergillus* consists in its omnipresence, which is based on its wide spectrum of habitats and its profuse production of conidia, the asexual spores of aspergilli, which become easily airborne and thereby contaminate virtually every environment. Therefore aspergilli, while not as virulent as some of the fungi discussed in previous chapters, are always near at hand to profit from the opportunity of failing immune defenses and to colonize and infect the host. Similarly, the physical properties of conidia from many species of *Aspergillus* are perfect for deposition in terminal airways, deep within the host.

Many aspergilli, in contrast to the majority of other fungi, are thermotolerant or even thermophilic and dispose of enzyme systems that function at 37° to 39° C, a quality mandatory for a pathogen causing deep-seated infection in the warm-blooded host. For their germination, the process of transformation from the spore stage into the tissue-invasive hyphal form, aspergilli have very modest nutritional requirements, e.g., condidia from *A. fumigatus* germinate readily in phosphate buffer supplemented with dextrose [1].

In order to gain access to constituents of organic macromolecules from a large variety of substrates, aspergilli dispose of many digestive exoenzymes, including all four classes of proteases [2]. This property, which probably contributes to the pathogenic potential of aspergilli [3,4], can be viewed as an expression of the striving of these molds to occupy every possible niche consisting of inert organic material. On the other hand, it has been proposed that exoenzymes such as the elastase of *A. fumigatus* have evolved in response to the attack of protozoa feeding on fungal spores [5]. In any event, such enzymes might play an important role for *Aspergillus* attempting to survive phagocytosis by macrophages or for the ability of pathogenic aspergilli to cross anatomical tissue barriers such as blood vessel walls.

Similarly, the production of toxic metabolic products such as gliotoxins [6–9] or phospholipids that inhibit the complement system [10,11] can be viewed as fungal weapons developed against competing microorganisms such as amebae, the archetype of mononuclear phagocytes. It is of note, however, that there is no evidence that any of these factors is a relevant virulence factor in aspergillosis, with the possible exception of elastase [4].

In conclusion, if the interactions between aspergilli and macrophages are to be analyzed, we have to consider that the fungus is a versatile and proficient saprophyte without special skills to overcome host defenses such as macrophages, professional antimicrobial executioners that were perfected during evolution under the constant pressure of offensive microorganisms.

III. INTERACTION BETWEEN CONIDIA AND MONONUCLEAR PHAGOCYTES

A. Recognition, Binding and Ingestion of Conidia

The natural route of infection for aspergilli is the respiratory tract, with a majority of infections starting in the lungs, after deposition of conidia in terminal airways. In this environment macrophages resident to the alveoli form a first host defense system that is effective against fungal spores. Murine alveolar macrophages recognize and attach to conidia from *A. fumigatus* through lectin-like attachment sites in the absence of opsonins [12]. Comparably, human monocytes bind conidia by a β-1, 4-glucoside–inhibitable receptor [13]. These observations explain previous studies finding that phagocytes recognize, bind, and ingest conidia in the absence of antibody, thus permitting phagocytosis of spores by resident alveolar macrophages in an environment where complement and immunoglobulins are not readily available. The rate of phagocytosis can, however, be increased in vitro by preopsonization of spores; nevertheless, a role has yet to be defined for opsonins in natural infection with spores [14,15].

As a result of the attachment and ingestion of conidia, typical phagocytic responses occur that include a respiratory burst with secretion of reactive oxygen intermediates [16], phagolysosomal fusion [1,17], and inhibition of germination, killing, and digestion of spores [18–20]. It has been proposed that conidia from *A. fumigatus* release antiphagocytic material that interferes with migration and spreading of phagocytes [21,22] or with the complement system [10]. However, the circumstances of the encounter between inhaled conidia, which usually reach alveoli one at a time or in small numbers and there meet resident macrophages in an environment deficient in complement, speak against a role of such virulence factors in natural infection. Furthermore, alveolar macrophages in experimental infections efficaciously eliminate conidia even after a challenge with extreme numbers of spores [1,18,19,23,24].

Some of the interactions between fungal spores and macrophages depend on the developmental stage of the conidia. Conidia are not formed to resist exogenous stress in a dormant state, as bacterial spores endure heat and desiccation. Conidia are, however, designed to become readily airborne in order to introduce aspergilli into every possible environment. Apparently, therefore, the surface of mature conidia is exquisitely hydrophobic, to avoid soaking that would preclude aerosolization. It is no surprise, then, that in order to experiment with conidia and macrophages in vitro, spores have to be wetted to become suspended in an aqueous milieu. This is achieved by using a mild detergent or a homogenizing device to forcefully soak conidia. Furthermore, conidia are usually subjected to several washings before they are added to macrophages in the test tube. Such manipulations are inclined to alter the properties of conidia and wash off surface material that might modify the interaction between immune mechanisms and the fungus. Nevertheless, freshly prepared conidia retain specific surface characteristics that affect their interaction with host defenses. Such spores poorly activate the complement system by the classical pathway [25], and freshly prepared spores, even when opsonized, will trigger only a modest respiratory burst in neutrophil granulocytes during phagocytosis [15]. Furthermore, freshly prepared conidia appear also to react poorly with other host proteins such as C-reactive protein [26]. Again, whether these surface properties of conidia are relevant in natural infection remains doubtful, since opsonins are sparse or absent in terminal airways and phagocytosis and killing of conidia occurs readily after inhalation of spores [23].

B. Killing of Ingested Conidia

Freshly prepared conidia from *A. fumigatus*, whether opsonized with complement or with specific antibody, are resistant to killing by neutrophil granulocytes [14,15,23]. Also, phagocytosis of conidia by mononuclear phagocytes from humans and rodents during the first several hours after ingestion does not result in killing of conidia [14,20,23,27,28]. Already, early studies noted that conidia are highly resistant to oxidants [14], and more recent studies have confirmed that freshly prepared conidia, before their swelling and germination, are resistant to oxidative and nonoxidative antimicrobial products of phagocytes [15,28]. It is only after the spores swell and become metabolically active within the phagosome, i.e., 4 to 6 hours after phagocytosis, that killing of spores starts [20]. Accordingly it has been shown that spores activated metabolically in a mycological broth become progressively more susceptible to antimicrobial products of phagocytes during the process of swelling, which includes morphological alterations in the composition of the cell wall, and precedes transformation into the tissue-invasive hyphal phase [15,28].

Macrophages can exert two different modes of anticonidial function: inhibiting germination of spores or killing spores. Metabolic activation of conidia has varying effects on their susceptibility to macrophages. Preswelling of spores in broth before phagocytosis abolishes the inhibitory function of macrophages. Resident mouse peritoneal macrophages that are unable to kill conidia from *A. fumigatus*, *A. flavus*, or *A. niger* can retard germination of freshly prepared, but not of metabolically activated, conidia in vitro and in vivo [20,23,29]. In contrast, rabbit alveolar macrophages, which have the ability to kill conidia, kill preincubated spores more rapidly, an observation that is not surprising, since freshly prepared spores are killed only after their intraphagosomal metabolic activation [20]. Swollen spores, in contrast to "resting" spores, 1) are amenable opsonization by the classical pathway of the complement system [25], 2) trigger a notable respiratory burst [15], and 3) are in addition more susceptible to reactive oxygen intermediates and defensins than resting spores [28]. Accordingly, spores from *Aspergillus* species that fail to swell and to start the transitional process from the conidial to the hyphal stage are not killed by macrophages over prolonged periods in vitro and can persist over many days in the reticulo-histiocytic system after intravenous injection in the dormant state (A. Schaffner, unpublished data).

In experimental infections of rodents through the respiratory route, alveolar macrophages ingest conidia, inhibit their germination, and kill the fungus in this stage [18,19,23,24]. In vitro, alveolar but not resident peritoneal macrophages kill conidia even in the absence of serum or other opsonins. Killing starts after a delay of several hours with a surprisingly slow killing rate of 90% per 24 hours (Figure 1), which corresponds well to the elimination rate of conidia seen in the lungs of mice after a respiratory challenge [20,23]. Killing follows an exponential first-order kinetic, both in the lung of animals infected experimentally and in the test tube, making it clear that alveolar macrophages do not require instruction by cytokines or other mediators of acquired immunity to deal successfully with conidia.

The antimicrobial system or systems of alveolar macrophages responsible for the killing of conidia have not been defined. The competence of macrophages to kill conidia appears to depend on the anatomical source of the phagocyte [20]. Resident peritoneal macrophages from rabbits and mice are unable to kill a relevant fraction of spores from *A. fumigatus*, *A. flavus*, and even *A. niger* in vitro, possibly because they do not possess the necessary killing systems. In contrast, alveolar macrophages from the same animals studied in parallel handle spores efficiently. Blood-derived lapine and human macrophages are as good in handling conidia as are the alveolar cells from the homologous species [20].

The observation that conidia trigger a respiratory burst at the time they are ingested but are killed only hours later raises the question whether reactive oxygen intermediates (ROIs) are still available as antimicrobial system many hours after phagocytosis and initiation of O_2^- and H_2O_2 generation. This question, which applies to many intracellular host-parasite interactions, has not yet been addressed. However, several observations make it clear that other than oxidative killing mechanisms are operative against conidia and responsible for the conidiacidal activity of macrophages: 1) Human blood-derived macrophages cultured in vitro for 10 days, which lose their capacity to produce significant amounts of ROIs, deal with spores as efficiently as fresh monocytes or monocytes maintained in culture for only 2 days, which both are competent in ROI secretion [20]; 2) monocytes from children with chronic granulomatous disease handle spores as well as

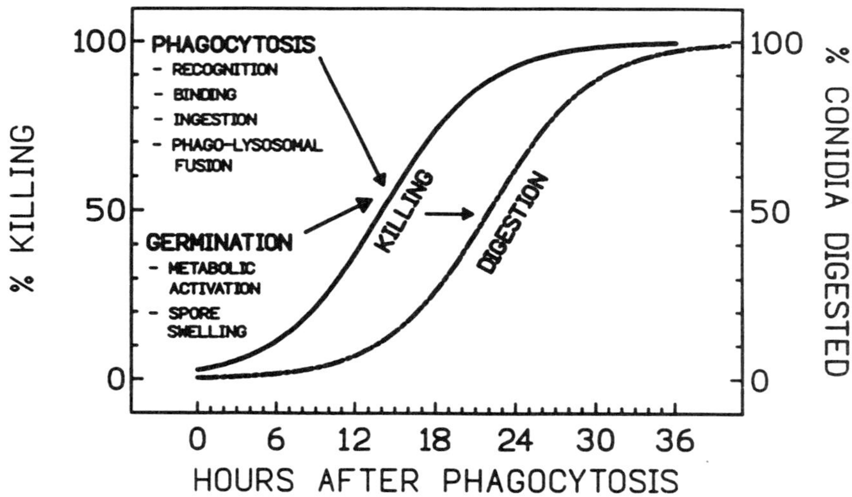

FIGURE 1 Killing and digestion of conidia by alveolar macrophages. Killing of conidia from *Aspergillus fumigatus* starts only following a lag phase of about 6 hours after phagocytosis by alveolar macrophages, at a time when spores begin to swell and initiate germination. Note that after killing has begun the number of living conidia decreases only gradually, with a killing rate of "only" about 90% per 24 hours. Killing is followed by complete digestion of spores with a delay of about another 6 hours. Also, fungal metabolism appears to be essential for digestion of spores, because spores killed before phagocytosis are not readily digested. (Adapted from Ref. 20.)

control phagocytes [30]; and 3) under strictly anaerobic conditions spores are killed by alveolar rabbit macrophages at a rate equivalent to the killing rate under aerobic conditions [20].

Killing of conidia is followed by complete digestion of spores [20], a process that lags about 6 to 12 hours behind killing (Figure 1). In contrast to living spores, heat-killed spores are not readily digested, an observation indicating that fungal enzyme systems are required to break down chitin and other fungus-specific compounds during digestion of spore walls [20; A. Schaffner, unpublished data].

C. Effect of Glucocorticoids on Anticonidial Activity of Macrophages

Macrophages exposed to pharmacological concentrations of glucocorticoids ingest conidia in vitro and in vivo at a normal rate and respond adequately with a respiratory burst and secretion of ROIs but fail to inhibit spore germination and to kill conidia [1,16–19,23,29–32]. This impairment is mediated by the glucocorticoid receptor, because dose-response studies of the effect of different glucocorticoids on the anticonidial action of macrophages follow their relative activity at the hormone receptor, and this activity can be antagonized by glucocorticoid antagonists [30]. Impairment of antimicrobial activity of macrophages is not specific for *Aspergillus*. Extension of the work with conidia has shown that glucocorticoids also impair the antimicrobial activity of human and murine macrophages to *Listeria, Nocardia,* and *Salmonella* [16,17,30,32]. In models of experimental infection with *A. fumigatus* [23] and *Listeria monocytogenes* [33], impairment of the antimicrobial activity of resident macrophages by glucocorticoids was found to be the essential immunosuppressive effect of glucocorticoids.

An unconfirmed study proposed that glucocorticoids affect anticonidial activity by stabilizing lysosomes, thereby preventing phagolysosomal fusion. Evidence provided in support of this hypothesis was indirect and was limited to the observation that phagosomes from glucocorticoid-treated mice that hold conidia from *A. flavus* had an appearance that was ultrastructurally distinct from that of phagosomes of control mice [31]. Ultrastructural studies depending on the lysosomal marker enzyme acid phosphatase or the fluorescent lysosomal stain acridine orange have not confirmed this hypothesis [1,17]. Thus conidia are recognized by glucocorticoid-treated macrophages, are ingested, induce a respiratory burst, and induce phagolysosomal fusion, yet conidia germinate and pierce and escape the handicapped phagocyte unharmed. Taken together these observations indicate that glucocorticoids, through a receptor-mediated effect, impair nonoxidative killing systems that are operative in the phagolysosome of the unimpaired phagocyte. Kinetic observations indicate that functional impairment becomes evident 24 to 36 hours after glucocorticoid exposure is begun. Because of this, and because repetitious short-lived glucocorticoid pulses fail to induce a functional impairment, it appears logical to assume that glucocorticoids, through their catabolic effects on protein synthesis, affect one or several nonoxidative killing systems of macrophages. Nonoxidative killing systems of macrophages are poorly defined. Furthermore nonoxidative killing systems of macrophages appear poorly conserved, in an evolutionary sense, among species. For example, the defensins, the cationic low-molecular-weight peptide antibiotics [34,35] that are abundant in rabbit alveolar macrophages, are not found in human alveolar, peritoneal, or blood-derived macrophages [1]. Similarly, inducible nitric oxide synthase, which generates important amounts of nitric oxide, to which antimicrobial activity has been attributed, appears to belong to the burst-independent armature of phagocytes from rats and mice but not man or rabbit [36].

In summary, the nonoxidative killing system or systems of mononuclear phagocytes that are responsible for the reliable natural resistance of the host against *Aspergillus* spores and possibly many other pathogens remain to be defined. Systematic studies on the effects of glucocorticoids on the anticonidial ativity of mononuclear phagocytes appear therefore to be appropriate to elucidate such nonoxidative antimicrobial systems.

D. Relation of Anticonidial Activity of Macrophages to Activation by Cytokines

Interferon-γ, probably the most important macrophage-activating cytokine, does not affect the anticonidial activity of human blood-derived or alveolar macrophages [16,30]. Similarly, we have found no increase of the anti-conidial activity of macrophages treated with lipopolysaccharide (LPS), granulocyte-macrophage colony-stimulating factor, interleukin-1, or tumor necrosis factor-α (A. Schaffner, unpublished data). Also, the defective anticonidial activity of glucocorticoid-exposed macrophages cannot be restored by IFN-γ or the other aforementioned cytokines or LPS. In contrast to the anticonidial activity, however, listericidal activity can be restored by IFN-γ in glucocorticoid-exposed macrophages. This restoration of listericidal function by the cytokine parallels the activation of ROI-generating systems, which appear to become responsible for the killing of *Listeria* in glucocorticoid-treated phagocytes but cannot contribute to killing of *Aspergillus* spores or other microorganisms that are more resistant to oxidative killing mechanisms [16,17,30].

In agreement with these observations, anticonidial function cannot be boosted in intact animals by prior homologous infection with *Aspergillus* or heterologous macrophage activation by prior infection with intracellular pathogens such as *Listeria* or *Salmonella typhimurium*, which provide a strong homologous as well as heterologous immunity against each other, even in cortisone-treated animals, but not against *Aspergillus*.

IV. INTERACTION BETWEEN HYPHAE AND MONONUCLEAR PHAGOCYTES

In vivo, hyphae from aspergilli ellicit a suppurative, granulocytic host response. Granulocytes attack and kill hyphae in vitro [23,37,38], and granulocytopenia renders animals susceptible to a challenge with the hyphal phase of aspergilli [38]. Monocytes, although by far outnumbered by granulocytes, can attack and kill hyphae from *A. fumigatus* [20,38,39] as well as from other *Aspergillus* species [38]. Oxidative killing systems, primarily the halide-myeloperoxidase–dependent systems, appear responsible for most of the killing, but nonoxidative killing systems appear also to contribute to killing, particularly when monocytes from patients with chronic granulomatous disease are studied [39]. Also, human alveolar macrophages studied at high effector-to-target ratio are capable of killing hyphae, as shown by their capacity to sterilize microcultures (A. Schaffner, unpublished data). In summary, mononuclear phagocytes play a minor role in host defense against *Aspergillus* hyphae. It appears conceivable, however, that in rare instances, in which granulocytic defenses are specifically defective with intact monocyte and macrophage functions, mononuclear phagocytes can substitute for some of the deficient granulocyte function.

V. CONCLUSION

Mononuclear phagocytes play a major role in host defense against aspergilli by reliably killing spores after their inhalation and deposition in the terminal airways. The perpetual nature of this encounter between the accidental intruder and macrophages would make it unpractical to call for complex systems for the recognition of aspergilli, or activation of macrophages for a successful kill of spores. The interaction of the macrophage with spores from aspergilli and other molds therefore provides an excellent illustration of the primeval immune function of mononuclear phagocytes: autonomous recognition and destruction of microbial invaders independent from auxiliary immune systems. This indispensable function of macrophages is neglected by many immunologist who limit their interest in the archetype of phagocytes to their interactions with lymphocytes.

ACKNOWLEDGMENTS

Part of this work was supported by a grant of the Swiss National Science Foundation.

REFERENCES

1. Schaffner A. Host defence in aspergillosis. In: Bennett JE, Hay RJ, Peterson PK, eds. New strategies in fungal disease. Edinburgh: Churchill Livingstone, 1992:98–112.
2. Cohen BJ. The proteases of aspergilli. In: Smith JE, Pateman H, eds. Genetics and physiology of *Aspergillus*. London: Academic Press, 1977:282–92.
3. Miyaji M, Nishimura K. Relationship between proteolytic activity of *Aspergillus fumigatus* and the fungus invasiveness of mouse brain. Mycopathologia 1977; 62:161–6.
4. Kothary MH, Chase T Jr, Macmillan JD. Correlation of elastase production by some strains of *Aspergillus fumigatus* with ability to cause pulmonary invasive aspergillosis in mice. Infect Immun 1984; 43:320–5.
5. Seaton A, Robertson MD. *Aspergillus*, asthma, and amoebae. Lancet 1989; 1:893–4.
6. Eichner RD, Al Salami M, Wood PR, Mullbacher A. The effect of gliotoxin upon macrophage function. Int J Immunopharmacol 1986; 8:789–97.
7. Mullbacher A, Waring P, Eichner RD. Identification of an agent in cultures of *Aspergillus fumigatus* displaying anti-phagocytic and immunomodulating activity in vitro. J Gen Microbiol 1985; 131:1251–8.
8. Mullbacher A, Eichner RD. Immunosuppression in vitro by a metabolite of a human pathogenic fungus. Proc Natl Acad Sci USA 1984; 81:3835–7.
9. Eichner RD, Mullbacher A. Hypothesis: Fungal toxins are involved in aspergillosis and AIDS. Aust J Exp Biol Med Sci 1984; 62:479–84.
10. Washburn RG, DeHart DJ, Agwu DE, Bryant Varela BJ, Julian NC. *Aspergillus fumigatus* complement inhibitor: production, characterization, and purification by hydrophobic interaction and thin-layer chromatography. Infect Immun 1990; 58:3508–15.
11. Washburn RG, Hammer CH, Bennett JE. Inhibition of complement by culture supernatants of *Aspergillus fumigatus*. J Infect Dis 1986; 154:944–51.
12. Kan VL, Bennett JE. Lectin-like attachment sites on murine pulmonary alveolar macrophages bind *Aspergillus fumigatus* conidia. J Infect Dis 1988; 158:407–14.
13. Kan VL, Bennett JE. Beta 1,4-oligoglucosides inhibit the binding of *Aspergillus fumigatus* conidia to human monocytes. J Infect Dis 1991; 163:1154–6.
14. Lehrer RI, Jan RG. Interaction of *Aspergillus fumigatus* spores with human leukocytes and serum. Infect Immun 1970; 1:345–50.
15. Levitz SM, Diamond RD. Mechanisms of resistance of *Aspergillus fumigatus* conidia to killing by neutrophils in vitro. J Infect Dis 1985; 152:33–42.
16. Schaffner A, Rellstab P. Gamma-interferon restores listericidal activity and concurrently enhances release of reactive oxygen metabolites in dexamethasone-treated human monocytes. J Clin Invest 1988; 82:913–9.
17. Schaffner A, Schaffner T. Glucocorticoid-induced impairment of macrophage antimicrobial activity: mechanisms and dependence on the state of activation. Rev Infect Dis 1987; 9(suppl 5):S620–S629.
18. Merkow LP, Epstein SM, Sidransky H, Verney E, Pardo M. The pathogenesis of experimental pulmonary aspergillosis. An ultrastructural study of alveolar macrophages after phagocytosis of *A. flavus* spores in vivo. Am J Pathol 1971; 62:57–74.
19. Epstein SM, Verney E, Miale TD, Sidransky H. Studies on the pathogenesis of experimental pulmonary aspergillosis. Am J Pathol 1967; 51:769–88.
20. Schaffner A, Douglas H, Braude AI, Davis CE. Killing of *Aspergillus* spores depends on the anatomical source of the macrophage. Infect Immun 1983; 42:1109–15.
21. Robertson MD, Seaton A, Raeburn JA, Milne LJ. Inhibition of phagocyte migration and spreading by spore diffusates of *Aspergillus fumigatus*. J Med Vet Mycol 1987; 25:389–96.
22. Robertson MD, Seaton A, Milne LJ, Raeburn JA. Resistance of spores of *Aspergillus fumigatus* to ingestion by phagocytic cells. Thorax 1987; 42:466–72.
23. Schaffner A, Douglas H, Braude A. Selective protection against conidia by mononuclear and against

mycelia by polymorphonuclear phagocytes in resistance to *Aspergillus*. Observations on these two lines of defense in vivo and in vitro with human and mouse phagocytes. J Clin Invest 1982; 69:617–31.
24. Gernez Rieux C, Voisin C, Aerts C, Wattel F, Gosselin B. Experimental aspergillosis in the guinea pig. Dynamic study of the role of alveolar macrophages in the defense of the respiratory tract, after massive inhalation of *Aspergillus fumigatus* spores. Rev Tuberc Pneumol 1967; 31:705–25.
25. Kozel TR, Wilson MA, Farrell TP, Levitz SM. Activation of C3 and binding to *Aspergillus fumigatus* conidia and hyphae. Infect Immun 1989; 57:3412–7.
26. Richardson MD, Shankland GS, Gray CA. Opsonizing activity of C-reactive protein in phagocytosis of *Aspergillus fumigatus* conidia by human neutrophils. Mycoses 1991; 34:141–3.
27. Kurup VP. Interaction of *Aspergillus fumigatus* spores and pulmonary alveolar macrophages of rabbits. Immunobiology 1984; 166:53–61.
28. Levitz SM, Selsted ME, Ganz T, Lehrer RI, Diamond RD. In vitro killing of spores and hyphae of *Aspergillus fumigatus* and *Rhizopus oryzae* by rabbit neutrophil cationic peptides and bronchoalveolar macrophages. J Infect Dis 1986; 154:483–9.
29. Sidransky H, Epstein SM, Verney E, Horowitz C. Experimental visceral aspergillosis. Am J Pathol 1972; 69:55–70.
30. Schaffner A. Therapeutic concentrations of glucocorticoids suppress the antimicrobial activity of human macrophages without impairing their responsiveness to gamma interferon. J Clin Invest 1985; 76:1755–64.
31. Merkow L, Pardo M, Epstein SM, Verney E, Sidransky H. Lysosomal stability during phagocytosis of *Aspergillus flavus* spores by alveolar macrophages of cortisone-treated mice. Science 1968; 160:79–81.
32. Schaffner A. Experimental basis for the clinical epidemiology of fungal infections. A review. Mycoses 1989; 32:499–515.
33. Schaffner A, Douglas H, Davis CE. Models of T cell deficiency in listeriosis: the effects of cortisone and cyclosporin A on normal and nude BALB/c mice. J Immunol 1983; 131:450–3.
34. Ganz T, Selsted ME, Szklarek D, et al. Defensins. Natural peptide antibiotics of human neutrophils. J Clin Invest 1985; 76:1427–35.
35. Selsted ME, Harwig SS, Ganz T, Schilling JW, Lehrer RI. Primary structures of three human neutrophil defensins. J Clin Invest 1985; 76:1436–9.
36. Schneemann M, Schoedon G, Hofer S, Blau N, Guerrero L, Schaffner A. Nitric oxide synthase is not a constituent of the antimicrobial armature of human mononuclear phagocytes. J Infect Dis 1993; 167: (in press).
37. Diamond RD, Krzesicki R, Epstein B, Jao W. Damage to hyphal forms of fungi by human leukocytes in vitro. A possible host defense mechanism in aspergillosis and mucormycosis. Am J Pathol 1978; 91:313–28.
38. Schaffner A, Davis CE, Schaffner T, Markert M, Douglas H, Braude A. In vitro susceptibility of fungi to killing by neutrophil granulocytes discriminates between primary pathogenicity and opportunism. J Clin Invest 1986; 78:511–24.
39. Diamond RD, Huber E, Haudenschild CC. Mechanisms of destruction of *Aspergillus fumigatus* hyphae mediated by human monocytes. J Infect Dis 1983; 147:474–83.

34

Macrophages and *Blastomyces dermatitidis*

Robert W. Bradsher

University of Arkansas for Medical Sciences, Little Rock, Arkansas

I. INTRODUCTION

Blastomycosis is a systemic fungal infection caused by *Blastomyces dermatitidis*. The spectrum of manifestations ranges from self-limited infections, focal cutaneous or pulmonary lesions, and destructive foci in various organs to rapidly progressive and fatal infection. Blastomycosis was first described by Gilchrist in Baltimore in 1894 as a skin infection; the illness was subsequently known as Gilchrist's disease [1] after he isolated the fungus and named it *Blastomyces dermatitidis* [2]. Initially, cases were perceived as involving a primary skin infection, and it was not until Schwartz and Baum [3] described primary pulmonary blastomycosis that the pathophysiology of this infection was more clearly understood. The vast majority of blastomycosis cases are now recognized to begin from a pulmonary portal of entry [4,5], although cases of cutaneous inoculation of the fungus in laboratory workers have been reported [6]. In addition, reports of cases of environmental cutaneous inoculation have been described that may have been pulmonary in origin [7,8].

II. MICROBIOLOGY

B. dermatitidis is a round, budding, thick-walled yeast cell with daughter cells forming from a broad-based bud. Like *Histoplasma capsulatum*, *Coccidioides immitis*, and *Paracoccidioides brasiliensis*, *B. dermatitidis* is a thermally dimorphic yeast that relates to the transition from mycelial forms in nature to yeast in tissue. This also is seen in the properties of growth in cultures as mycelial at 25° C and as yeast at 37° C. Medoff and coworkers [9] described this physiologic change from mycelial to yeast as a heat-related insult with resultant uncoupling of oxidative phosphorylation. The perfect, or sexual, stage is called *Ajellomyces dermatitidis*, with the imperfect, or conidial, stage given the more familiar name *Blastomyces dermatitidis* [4]. The organism in clinical specimens varies from 5 to 15 μm in size, and the number of organisms in tissue can vary greatly.

III. PATHOPHYSIOLOGY

Infection with this organism begins with inhalation of conidia of the mycelial phase of the fungus into the lung, which is followed by clearing of the organism by bronchial pulmonary phagocytes [5]. As the fungus undergoes transition to yeast cells, growth can occur in the lung itself or spread via the bloodstream and lymphatics to distant sites. With development of immunity, inflammatory reactions occur at the initial infection site and at these metastatic foci. The pathologic response is initially suppurative, with polymorphonuclear leukocyte (PMN) infiltration, and is followed by a granulomatous formation with lymphocyte and monocyte-derived macrophages. This pylogranulomatous response is typical of blastomycosis, although necrosis or fibrosis can also be found.

IV. EPIDEMIOLOGY

Generally, the fungus causes infection in the states surrounding the Mississippi and Ohio River valleys. The organism was once thought to be found only on the North American continent, prompting the title North American blastomycosis in contradistinction to another disease termed South American blastomycosis. This terminology has been abandoned because blastomycosis cases have been reported with increased frequency on other continents [10,11]. Secondly, the appropriate name for an infection formerly called South American blastomycosis and due to *P. brasiliensis* is paracoccidioidomycosis.

V. MANIFESTATIONS

Blastomycosis causes constitutional symptoms including weight loss, fever, malaise, fatigue, and other nonspecific symptoms. Most patients are diagnosed with pneumonia, which is either an acute pulmonary infection, mimicking acute bacterial pneumonia, or a more chronic, indolent pneumonia that resembles tuberculosis [4]. Cutaneous involvement is the most common form of extrapulmonary disease. This form manifests as either a verrucous lesion or an ulcer, with both types having yeast-phase organisms present in skin biopsy histology, by culture, or by direct smears. Involvement of the genitourinary tract, bones, nervous system, or other organs may also be found in humans with blastomycosis [4,5].

VI. IMMUNOLOGY

A. Cellular and Humoral Immunity

Cellular immunity is considered to be the major protection factor in preventing progressive disease in blastomycosis, as with other slow-growing but pathogenic fungi such as *H. capsulatum*, *P. brasiliensis*, and *C. immitis* [12,13,14]. These organisms evoke cellular immunity as detected by skin tests or by in vitro markers of antigen-induced lymphocyte reactivity. Although antibodies may be helpful for diagnosis, humoral immunity is not critical for protection, since patients with hypogammaglobulinemia handle fungal infections adequately [15]. Mice with congenital absence of the thymus are more susceptible to these fungi because they lack cellular immunity [16]. With restoration of thymus function, protection from further infection is demonstrated [17]. Cellular immunosuppression with transplantation or with human immunodeficiency virus (HIV) infection leads to infection with these types of fungi [18,19].

The assessment of host response in human blastomycosis has lagged behind investigation of other fungal infections because of lack of suitably active and specific antigens. For example, blastomycin, a crude mycelial-phase filtrate of *B. dermatitidis*, has been used for delayed hypersensitivity skin tests. In two large series [20,21], however, 59% and 100% of patients with positive cultures for *B. dermatitidis* had negative reactions to blastomycin skin tests.

B. In Vivo Markers

Experimental animal models of blastomycosis have suggested that cellular immunity is the essential arm of the immune system for preventing progressive disease. Spencer and Cozad [22] immunized mice with either live or killed yeast cells of *B. dermatitidis*. Delayed hypersensitivity was subsequently detected by antigen-induced footpad swelling; increased thickness as a marker of cellular immunity was noted on day 12 and 20 after immunization. More important, levels of delayed hypersensitivity correlated with resistance of immunized mice to lethal challenge with live *B. dermatitidis* organisms [23]. Subsequent studies of T lymphocyte–depleted mice confirmed that cellular immunity was responsible for the resistance to this fungus [17].

Deepe and colleagues [24] described a reproducible model of disseminated blastomycosis in mice. Tail vein injection of a virulent *B. dermatitidis* strain caused a progressive infection, with a mortality rate of 75% at 8 weeks. The number of organisms increased in the lung and brain by 200-fold from the first week to the fifth week of infection. Initially, acute inflammatory cells were found, but by 3 weeks, multiple granulomas were present. Footpad response to a specific *Blastomyces* antigen was not found, presumably secondary to anergy. A different route of infection (subcutaneous inoculation) with a lower yeast inoculum was associated with clear-cut delayed hypersensitivity as measured by spleen cell proliferation in response to *Blastomyces* antigen. Suppression of immune response with a large antigenic challenge was considered to mimic human disease with widespread dissemination.

C. In Vitro Markers

Markers of cellular immunity have correlated with protection of humans from progression of infection with *H. capsulatum* [12], *P. brasiliensis* [13], and *C. immitis* [14]. Because of inadequate antigen, study of in vitro markers in blastomycosis lagged behind these other fungal infections. A more active and specific *Blastomyces* antigen was developed and reported by Cox and Larsh [25,26] and Deighton et al. [27]. The antigen, called B-ASWS, is an alkali- and water-soluble preparation of yeast cells of *B. dermatitidis*. Sensitized animals responded in a specific fashion by footpad thickness assays following injection of the antigen. Cox [28] used B-ASWS antigen in human lymphocyte proliferative assays using cells from healthy donors. She demonstrated cross-reactivity to the B-ASWS antigen in histoplasmin skin test–positive donors. However, less cross-reaction occurred than with blastomycin as antigen [28].

When patients with previously treated blastomycosis were chosen as lymphocyte donors, rather than healthy donors with no history of infection or those with previous histoplasmosis, clear-cut differentiation was made with B-ASWS–induced lymphocyte reactivity assays [29]. Patients with blastomycosis treated up to 12 years before study had positive responses, with values of ^{3}H-thymidine uptake ranging from 11,000 to over 250,000 counts per minute and stimulation indices from 4 to 110. In subsequent studies, patients with acute pulmonary blastomycosis who had only 2 to 5 weeks of symptoms did not have in vitro markers of specific *Blastomyces* immunity [30]. Within 4 weeks of antifungal therapy, however, retesting of B-ASWS–induced lymphocyte reactivity demonstrated evidence of development of specific immunity [30]. In contrast, patients with extrapulmonary blastomycosis had symptoms for much longer before diagnosis (3 to 9 months) and had significant lymphocyte responses to B-ASWS at the onset of therapy [30]. The time to develop specific cellular immunity in these patients with pulmonary blastomycosis was similar to that for development of immunity in the disseminated blastomycosis murine model of Deepe [24].

VII. PATHOLOGY

The inflammatory response in blastomycosis is pyogranulomatous [4,5]. Initially, PMN infiltration is noted, followed by mononuclear cell infiltration and granuloma formation. The first leukocyte response is thought to be nonspecific, while the formation of a granuloma is considered to be the result of development of specific cellular immunity. In human infections with *B. dermatitidis*, pathological sections frequently demonstrate the fungus either inside or attached to monocytes, macrophages, or giant cells. In one large series of pathologic findings [31], from one to several yeast cells were located in the cytoplasm of giant cells as a routine finding. Therefore, this infection is noted to have epithelioid granulomatous tissue reactions typical of activated macrophage–pathogen interactions. Because both PMNs and macrophages are present in the pathologic picture of blastomycosis, research with both cell types in interaction with *B. dermatitidis* will be reviewed.

VIII. INTERACTION OF *BLASTOMYCES DERMATITIDIS* WITH POLYMORPHONUCLEAR LEUKOCYTES

The role of PMNs in defense against blastomycosis has been controversial in the literature. Despite the almost universal presence of these acute inflammatory cells in histologic sections from patients, whether the cell is a defense or actually an aid to the infection has been debated. The studies will be reviewed according to laboratories of study.

A. Little Rock

One of the initial studies of the interaction of PMNs and *B. dermatitidis* was reported by Sixbey and coworkers at the University of Arkansas for Medical Science [32]. One strain of the fungus was utilized for study of chemotaxis, chemiluminescence, phagocytosis, and killing. The organism was from a clinical isolate and was not characterized further. This point was criticized by subsequent investigators because of the high rates of phagocytosis (95% of yeast organisms after 30 minutes of incubation), which was confirmed by electron microscopy. Since with other strains of *B. dermatitidis*, other investigators have not been able to reproduce phagocytosis of yeast cells by PMNs, it is considered that Sixbey and colleagues studied an unusually small yeast cell strain of this fungus.

In addition to phagocytosis, Sixbey and others described a 29% reduction in colony-forming units with a ratio of 10 PMNs to 1 yeast cell in coculture experiments [32]. The possibility of clumping of the yeast as a reflection of the drop in colony-forming units was not mentioned. Chemiluminescence of granulocytes was noted in response to the yeast, and broth culture supernatants of the organism contained PMN chemotactic activity [32].

B. San Jose

The laboratories of Brummer and Stevens at Santa Clara Valley Medical Center have actively pursued the interactions of *B. dermatitidis* and PMNs. The first report was of human cells with enhancement of growth of the fungus with PMN cocultures [33]. The donors had no history of blastomycosis, and up to 45% enhancement of replication at 24 hours and 68% enhancement by 72 hours was reported. Even after two cycles of freezing and thawing the PMNs to ensure nonviability, enhancement of growth was 40% compared with a 43% enhancement with viable PMNs [33]. Subsequently, Brummer and Stevens [34] reported coculture of *B. dermatitidis* and murine PMNs obtained by thioglycolate broth–induced peritoneal exudate cells with further purification to PMN populations. Enhancement of fungal replication was noted with both living and killed PMNs, but heating the PMN lysate at 80° C for 30 minutes removed the enhancement of growth of *B. dermatitidis*. In additional in vivo experiments, injection of *B. dermatitidis* with autologous PMNs caused 90% enhancement of growth of subcutaneous organisms in comparison with organisms injected alone [34]. It was postulated that PMNs might actually accelerate or exacerbate the infection rather than help control it. Other laboratory examinations of the interactions of PMNs and *B. dermatitidis* have not reproduced this enhancement of fungal growth, but different PMN-to-fungus ratios have been utilized.

Brummer et al. [35] subsequently reported immune activation of murine PMNs with subsequent killing of *B. dermatitidis*; cells were obtained from injection of nonviable *B. dermatitidis* into the peritoneal cavity of immunized mice. A reduction with such PMNs in coculture of 60% from the initial inoculum compared with growth in media alone was noted. The results were interpreted as support for the concept that PMNs elicited by immunologic reaction have enhanced nonspecific fungicidal activity, since both *Candida albicans* and *B. dermatitidis* were killed [35]. Results were similar with peritoneal PMNs elicited from mice with sodium caseinate compared with murine peripheral blood PMNs or thioglycolate-induced peritoneal PMNs [36].

In another set of experiments, Brummer and Stevens [37] used *Blastomyces* antigen–induced culture supernatants of spleen cells from immunized mice to activate murine PMNs; more effective killing of *B. dermatitidis* in coculture was found. Enhancement of fungal growth with nonactivated PMNs was not reported in these studies as compared with earlier studies, but fungicidal activity was increased in a dose-dependent fashion with supernatants that presumably contained lymphokines capable of activating the leukocytes. Further support to this PMN activation was added with the report by Brummer et al. [38] that immunologically activated PMNs had a ten times higher peak of oxidative burst as measured by chemiluminescence. Killing of *B. dermatitidis* and the oxidative burst was eliminated by the presence of catalase in these experiments. This supports the hypothesis that hydrogen peroxide generation and other products of oxidative metabolism by PMN are responsible for killing of the yeast [38]. With further work in this area, Morrison, Isenberg, and Stevens [39] reported that activated peritoneal PMNs from mice produced greater amounts of superoxide anion in response to *B. dermatitidis* than thioglycolate-induced peritoneal murine PMNs. In subsequent studies, Morrison and Stevens [40] reported an inverse correlation of in vivo virulence of *B. dermatitidis* strains to susceptibility to killing by murine PMNs and the capacity to induce superoxide anion production by PMNs. Sugar et al. [41,42] reported susceptibility of *B. dermatitidis* in a cell-free system to hydrogen peroxide and the iron-myeloperoxidase-halide system of microbial killing. The virulence of a strain of the yeast for mice was again shown to correlate to relative resistance to hydrogen peroxide killing. No evidence for hydrogen peroxide production by *B. dermatitidis* was found in this study, but subsequently, Schaffner et al. [43] reported that hydrogen peroxide was produced by *B. dermatitidis*.

Another source of murine PMN harvest for *B. dermatitidis* interactions, peripheral blood obtained from axillary bleeding, was reported by Morrison, Brummer, and Stevens [44]. A reduction in yeast in cocultures ranged from 33% with cells from normal, nonimmune mice to 54% with peripheral PMNs from immunized mice given killed yeast intraperitoneally. Killing was significantly greater than with peripheral blood PMNs either from nonimmune animals given intraperitoneal yeast or from immunized animals without intraperitoneal antigens. A curious observation was that peripheral blood PMNs from nonimmune mice had greater fungicidal activity than similar cells from immunized mice without intraperitoneal challenge [44]. The formation of different populations of peripheral blood PMNs as a result of localization of some PMNs at the resolving subcutaneous abscesses at the *Blastomyces* immunization site was postulated as the reason for this difference [44]. The greater killing with PMNs from stimulated immune animals was hypothesized to be a result of soluble mediators from the local immune reaction that activated the peripheral blood PMNs to have greater fungicidal activity [44]. Morrison and colleagues further confirmed greater killing after lymphokine activation of murine PMNs from either the peritoneal cavity [45] or peripheral blood [46]. With in vitro addition of recombinant interferon-γ to PMNs, subsequent greater killing of *B. dermatitidis* in cocultures was noted.

C. San Antonio

Drutz and Frey [47] reported on interactions of human PMNs and *B. dermatitidis* in experiments performed at the University of Texas at San Antonio. The human volunteers had no evidence of prior blastomycosis. With a methylene blue dye exclusion technique to assess fungicidal activity, approximately 18% of five yeast strains were killed by PMNs using a ratio of 10 PMNs to 1 yeast cell; one strain had 8% killing under the same conditions. Killing of yeast was confirmed by ^{51}chromium release in one strain of *B. dermatitidis* previously grown in the presence of the isotope.

More than 97% of the yeast were surrounded by, but extracellular to, the PMNs by 2 hours of coculture incubation. Electron-dense material presumably from degranulation of PMN granules encircled the yeast cells on electron-microscopic examination. Staining for enzymes confirmed leakage from the PMNs enveloping yeast cells.

Experiments were also performed to examine the effect of PMNs on *B. dermatitidis* conidia [47]. Phagocytosis for four fungal strains was measured at 91% at 3 hours. Killing of the conidia occured with reduction of colony-forming units by 52%. Optimal phagocytosis and killing was noted in the presence of fresh serum plus calcium and magnesium. Reduction of nitroblue tetrazolium dye with PMNs and either yeast or conidia cultures suggested that superoxide anion generation was occurring. Conidia caused a greater degree of oxidative burst as measured by luminol-enhanced chemiluminescence than did yeast cells.

This work by Drutz and Frey supported the concept that human PMNs were effective in killing conidia, which are the infectious form of this fungus in nature. If a small enough number of conidia were inhaled, the PMN response might be adequate to eliminate the fungus before transformation to the yeast form and subsequent progressive infection [47].

D. Zurich, San Diego

Schaffner and colleagues compared the susceptibility of dimorphic fungi and opportunistic fungi to killing by human PMNs [43]. The major conclusions to their studies were that virulent dimorphic fungi (*B. dermatitidis*, *H. capsulatum*, *P. brasiliensis*, and *Sporothrix schenckii*) were more resistant to killing by PMNs than were opportunistic organisms (*Aspergillus*, Mucorales, and *Petriellidium boydii*). Specifically with *B. dermatitidis*, approximately 45% to 55% of conidia were killed by human PMNs, but the yeast-phase organisms were resistant to killing. When comparing these results with the observed killing by PMNs of yeast reported by Drutz and Frey [47], Schaffner et al. questioned the use of methylene blue permeability as a marker for killing [43]. Even with low numbers of yeast and a high concentration of PMNs, killing was not observed [43].

Schaffner et al [43] noted in chemiluminescence assays that *B. dermatitidis* yeast cells triggered a notable oxidative burst in the absence of opsonins and even without PMNs. Production of hydrogen peroxide by the yeast itself was measured. Therefore, chemiluminescence with luminol-amplification with *B. dermatitidis* was called into question [32,38,47]. However, in experiments using lucigenin rather than luminol, unequivocal evidence was obtained of PMN oxidative burst stimulation by *B. dermatitidis*, as with the other dimorphic fungi that were studied.

IX. INTERACTION OF *BLASTOMYCES DERMATITIDIS* WITH MACROPHAGES

As with PMNs, the role of macrophages in controlling growth of *B. dermatitidis* has been the topic of interest of a number of laboratories. Summaries of the studies from various laboratories follow.

A. Norman

Cozad and colleagues at the University of Oklahoma have reported a series of experiments pointing to the importance of cell-mediated immunity in blastomycosis [17,22,23]. Delayed hypersensitivity to *Blastomyces* antigen was demonstrated by transfer of lymphocytes from immunized mice but not by serum [48]. The adoptive transfer of immunity also provided protection from challenge with live organisms [23]. McDaniel and Cozad examined one of the effector mechanisms of protection of this immunization [49]. Peritoneal macrophages from mice immunized with killed *B. dermatitidis* were obtained without exudate-inducing agents. Monolayers of macrophages were challenged with live *B. dermatitidis* yeast, with subsequent examination for phagocytosis. Scanning electron microscopy was superior to light microscopy in confirming the intracellular location of the yeast; the murine cells were observed to have characteristics of activated macrophages. In other experiments, extracellular yeast were removed from the macrophage monolayers after 3 hours of incubation to allow examination of the intracellular fate of the fungus. Macrophages from nonsensitized animals had a 1.5 log increase in yeast numbers over 96 hours, while macrophages from immune mice inhibited the intracellular growth of the yeast [49]. A significant number of nonsensitized macrophages had evidence of yeast cells protruding from the cell surface, and even yeast cell destruction of some nonsensitized macrophages was observed. Even though there was not spectacular killing of intracellular yeast in the sensitized macrophages, these cells did inhibit intracellular proliferation. This supported the hypothesis that macrophage activity correlated with specific cellular immunity to *B. dermatitidis*.

B. San Jose

In addition to studies of *B. dermatitidis* and PMNs, Brummer and Stevens at the Santa Clara Medical Center have performed a series of experiments to examine interactions of this fungus with murine macrophages. Brummer and colleagues [50] reported reductions of colony-forming units with peritoneal macrophages in coculture experiments at a ratio of 1000 macrophages to 1 yeast cell. Untreated mice cells had a 24% reduction of colony-forming units of yeast compared with the growth of the fungus in medium alone. Fetal calf serum–induced peritoneal macrophages had a 49% reduction, while concanavalin A–treated mice macrophages had a 64% reduction in yeast numbers. *Blastomyces*-immunized mice had peritoneal macrophages obtained 4 weeks after fungal challenge. A reduction in yeast cells of 41% was noted with these cells. In another set of experiments, Brummer and others [51] reported that a virulent strain of *B. dermatitidis* could escape inhibition of growth in coculture with peritoneal murine macrophages under conditions similar to the above [50]. With cells from untreated mice, avirulent strains of the fungus were inhibited from growth, while the virulent strain escaped inhibition at 72 hours [51]. When cells from nonlethally challenged mice were examined, replication of both virulent and avirulent strains was inhibited at 24 hours, but the virulent strain escaped inhibition of growth at 72 hours of culture [51]. Similar findings were obtained with macrophages activated by prior concanavalin A treatment of the mice. However, inflammatory peritoneal macrophages obtained by thioglycolate injection intraperitoneally inhibited fungal growth without as much escape from inhibition by the virulent strain. In subsequent work, Brummer, Sugar, and Stephens [52] repeated the work to show that complement and antibodies were not necessary for inhibition of growth of the fungus. In additional studies [53], protein synthesis inhibitors could block killing if added before macrophage activation but not if added after macrophage activation. No effect was caused by corticosteroid or cyclosporine coincubation [53].

A major difference in these studies compared to those of McDaniel and Cozad [49] is the ratio of macrophages to yeast and the fact that extracellular yeast were not removed from cultures as performed by McDaniel and Cozad.

Bummer and Stevens [33] reported interactions of monocytes and macrophages from peripheral blood of healthy human volunteers with no history of blastomycosis. With a ratio of 2000 cells to 1 yeast, colony-forming units of fungus were reduced by monocyte coculture by 65% and 45% of avirulent and virulent strains, respectively, compared with growth of fungus in medium alone. After maturation to macrophages, the reduction of growth of yeast was 85% [33]. Since supernatants from macrophage cultures caused no inhibition of growth, it was considered that soluble factors were not responsible for the depression of replication.

Brummer and colleagues next examined the ability of resident peritoneal murine macrophages to kill *B. dermatitidis* after activation by interferon-γ in vitro [54] and the mechanisms for the fungicidal activity. Using the same methods as in earlier studies, peritoneal macrophages activated by interferon-γ resulted in a 37% reduction in yeast numbers [54]. A 100 times higher concentration of interferon-γ was required for activation of macrophages to kill *B. dermatitidis* than for *Candida albicans*. The investigators were not able to show phagocytosis of the *B. dermatitidis* strain. Spleen cells cultured with concanavalin A had supernatant that could activate murine macrophages to kill the fungus [54]. With similar methods, inhibitors of oxidative metabolism in the cocultures did not inhibit the killing of the yeast [55]. This suggested that killing by macrophages occurred independent of oxidative burst [55].

Sugar, Brummer, and Stevens compared murine alveolar macrophages to peritoneal macrophages for fungicidal activity for *B. dermatitidis* [56]. Alveolar and peritoneal macrophages had similar killing activity. Cells from immune animals had 15% to 24% killing of the yeast compared with cultures in medium alone. Cells from unstimulated animals had killing of 21% to 33%. The equivalent fungicidal activity of normal and immune mice was of note; the investigators speculated that the results might have been altered by a more prolonged incubation with immune cells because of lymphokine alterations of the macrophage [56]. Brummer and colleagues [57,58] furthered these observations of murine alveolar macrophages and *B. dermatitidis* with activation by either interferon-γ or lymphokines. Killing by the cells was increased by 28% after concanavalin A and by 25% after interferon-γ [57]. Without activation, this study could not demonstrate alveolar macrophage killing of this fungus. In another set of experiments, similar results were found, with 37% killing with alveolar macrophages from interferon-γ–treated mice compared with 4% killing without this in vivo activation of macrophages [58].

C. Boston

Sugar and Field [59] reported from the Boston University Medical Center the results of respiratory challenge of mice with *B. dermatitidis*. A 50% mortality was noted after 30 days after an inoculum of only 5 colony-forming units. Lavage studies showed early PMN recruitment with both a virulent and an avirulent strain. However, PMN predominance continued with the virulent strain but was followed by macrophage infiltration with the avirulent strain. Histologically, microabscesses were seen early, followed by granuloma formation of giant cells [59].

Sugar and Picard [60] examined growth of a virulent and an avirulent strain of *B. dermatitidis* conidia upon coculture with murine alveolar macrophages. With these unstimulated murine cells, conidia were blocked from transition from this mycelial phase to the yeast phase. Hydrogen peroxide was also able to block this phase transition. Germination to hyphal forms was undisturbed by alveolar macrophage coculture. This inability to convert to the tissue phase might be another nonspecific host defense, like PMNs [47], that could account for subclinical blastomycosis.

D. San Antonio

In addition to the experiments with human PMNs and *B. dermatitidis* described above, Drutz and Frey also examined monocyte and macrophage interactions with this fungus [47]. Monocytes phagocytosed 40% of conidia and killed 35% of the fungi as detected by the methylene blue dye exclusion technique. Monocyte-derived macrophages from these subjects with no history of blastomycosis were shown to ingest and kill more that 80% of conidia. Monocytes were unable to phagocytose yeast, but macrophages ingested approximately 20% of the yeast. Of interest, 40% of the yeast were killed by the macrophages, apparently by extracellular mechanisms [47].

E. Little Rock

Our laboratory at the University of Arkansas for Medical Sciences performed three sets of experiments with peripheral blood mononuclear cells [61], monocyte-derived macrophages [62], and alveolar macrophages [63] respectively from persons with culture-proven blastomycosis. After specific cellular immunity to *Blastomyces* antigen (B-ASWS) was demonstrated in patients with previously treated blastomycosis [29,30], we compared the responses of peripheral blood mononuclear cells from such patients to those of cells from normal controls with no history of blastomycosis [61]. After peripheral venipuncture and density-gradient separation of cells, 74% to 80% of the cells were lymphocytes and the remainder were nonspecific-esterase staining cells consistent with monocytes. A clinical isolate of *B. dermatitidis* was added at ratios of yeast to human cells ranging from 0.001 to 10. Uptake of tritiated thymidine by lymphocytes was measured after 5 days of incubation of these tumbled suspension cocultures. Microscopic slides were made from the contents of unspun tissue cultures for each of the 5 days of incubation. Only the blastomycosis patients' cells were demonstrated to have monocytes that had engulfed yeast [61]. Counts of yeast increased over the 5 days of coculture of both immune and nonimmune donor cells. After the 5×10^4 initial inoculum of *B. dermatitidis*, nonimmune cell cultures had 1.9×10^5 yeast at 72 hours and 3.2×10^5 at 120 hours. Counts from immune donor cell cultures at the same time points were 2.5×10^5 and 3.6×10^5, respectively. Therefore, growth of the fungus was different with immune cells than with normal cells. The absence of inhibition of replication of yeast by the immune cells was in contradistinction to the report of murine macrophages by McDaniel and Cozad [49]; this difference was considered to be due to extracellular growth of the fungus, since removal was not possible with these tumbled suspension cultures.

Inocula of 10^3 to 10^7 yeast did not cause lymphocytic uptake of tritiated thymidine as a measure of proliferation in cultures from the control group, but the yeast did stimulate uptake in a dose-dependent fashion with cultures from the blastomycosis patient group. Therefore, live yeast caused specific lymphocyte stimulation from blastomycosis donors, like the soluble antigen B-ASWS [29,30].

In another set of experiments with cells from patients with culture-proven blastomycosis or normal controls [62], we examined macrophage monolayers with *B. dermatitidis* challenge. Peripheral blood mononuclear cells were placed in tissue culture plates at a concentration of 5×10^5 monocytes per well. After incubation, nonadherent cells were removed and monocytes matured to macrophages over 72 hours. Yeast challenge, performed at a ratio of 1 yeast to 5 macrophages, was added to monolayers for a 2-hour incubation with subsequent removal of extracellular organisms. Phagocytosis was assessed by light and electron microscopy, and growth of intracellular *B. dermatitidis* was assessed by detergent disruption of monolayers and counting of yeast at the end of the yeast challenge and 72 hours later. Previously infected donors had 35% to 44% of macrophages with ingested yeast, while phagocytosis was observed with macrophages from nonimmune donors in only 12% to 16% of cells. Electron microscopy confirmed the intracellular location of the yeast. The numbers of yeast increased in coculture with nonimmune cells, but growth was inhibited in cultures of immune macrophages, paralleling the experience reported by McDaniel and Cozad with immune murine macrophages [49]. The mean increase for growth of yeast inside the nonimmune cells was 8.8% ± 4%; a 4% ± 8% decrease in yeast numbers was found in cultures of cells from infected patients [62]. Growth of the organism in media alone was similar to that inside normal cells (85% ± 3%).

Because alveolar macrophages are considered to be one of the first lines of defense [63], we studied macrophages obtained by bronchoalveolar lavage performed on patients with recently treated blastomycosis and on healthy control subjects [64]. In addition, comparisons were made with peripheral blood monocyte–derived macrophages from the same donors. Using methods as described for the previous study [62], a greater number of macrophages from blastomycosis patients had ingested *B. dermatitidis* than of the cells from normal controls [64]. The range for blastomycosis patients was 25% to 38% of macrophages with intracellular yeast after challenge, compared with 12% to 18.8% for nonimmune donor cells. No significant differences were found in phagocytosis rates between peripheral macrophages and alveolar macrophages [64].

After removal of extracellular yeast at 2 hours of incubation, parallel monolayers were lysed then and 72 hours later for determination of intracellular growth. Alveolar macrophages from immune donors had a mean of 8.3×10^4 yeast at two hours and 7.1×10^4 yeast at 72 hours [64]. This compares with mean yeast numbers from alveolar macrophage coculture from nonimmune donors of 7.1×10^4 at 2 hours and 1.7×10^5 at 72 hours [64]. Peripheral macrophage cultures followed the same pattern of growth inside nonimmune donor cells and inhibition of intracellular growth with blastomycosis patients' cells. Again, no differences between alveolar and peripheral macrophages were detected for either group.

As with the previous study [62], there were high numbers of *B. dermatitidis* at the initial 2-hour time point in the immune macrophage cocultures than with the nonimmune cells [64]. This is considered to be due to the increased rate of phagocytosis by the immune cells, since extracellular yeast were removed before counts were done. The rate of phagocytosis of *B. dermatitidis* by nonimmune cells of 12% to 18% compares to the rate of 17.6% of macrophages from normal host reported by Drutz and Frey [47]. Murine peritoneal macrophages were reported by Brummer et al [50,54] to be unable to ingest *B. dermatitidis* because of the size of the yeast, but McDaniel and Cozad [49] were able to demonstrate murine macrophage phagocytosis of this yeast.

One reason for increased phagocytosis and intracellular growth inhibition by alveolar [64] and peripheral [62,64] macrophages was thought to be immune activation of the cells in the coculture experiments. In the macrophage monolayers, 2% to 5% of the cells were lymphocytes. These lymphocytes probably were stimulated by the live *B. dermatitidis* organisms to secrete cytokines. Lymphocyte activation with immune cells was clearly shown with the tumbled suspension culture technique in response to live yeast challenge [61]. Brummer and co-workers [57,58] demonstrated interferon-γ enhancement of peritoneal macrophages from nonimmune mice to kill *B. dermatitidis* in coculture. Residual lymphocytes in the macrophage monolayers were considered to be the source of the cytokines in our experiments [63,64]. This hypothesis was examined by treating macrophages from nonimmune donors with supernatants from cultures of B-ASWS–stimulated lymphocytes obtained from an immune donor [64]. Macrophages induced by active supernatant inhibited intracellular *B. dermatitidis* growth and had greater phagocytosis than macrophages treated with supernatants from nonstimulated cultures of lymphocytes [64]. This supports the hypothesis that macrophages from immune donors are activated in vitro by the stimulated lymphocytes present in the monolayer tissue cultures.

X. CONCLUSIONS

B. dermatitidis infections begin with inhalation of conidia. Both PMNs and alveolar macrophages have been shown to ingest and kill these conidia [47,60]. This defense could then keep the human host from developing subsequent clinical infection with this organism. The numbers of persons with subclinical or asymptomatic blastomycosis is not as well known [4,5] as for histoplasmosis or coccidioidomycosis, since skin tests with blastomycin are not considered to provide results as reliable as the other fungal skin tests. Epidemics of blastomycosis include persons who are infected but recover without therapy [65]. Lymphocyte reactivity studies of persons from the endemic area of blastomycosis also confirm that some individuals become infected with this organism but recover spontaneously [66]. The elimination of conidia by macrophages and PMNs is a likely mechanism for these subclinical infections.

Macrophages from humans are able to ingest and inhibit the growth of intracellular *B. dermatitidis* [47,62,64]. Macrophages from immune mice have likewise been reported to ingest yeast and eliminate fungal replication [49]. This supports the concept that cellular immunity is the critical host defense in blastomycosis. The mechanisms of the killing of *B. dermatitidis* by macrophages are not fully understood and should be an area of further research.

REFERENCES

1. Gilchrist TC. Protozoan dermatitis. J Cutan Gen Dis 1894; 12:496–9.
2. Gilchrist TC, Stokes WR. Case of pseudo–lupus vulgaris caused by blastomycosis. J Exp Med 1898; 3:53–78.
3. Schwarz J, Baum GL. Blastomycosis. Am J Clin Pathol 1951; 21:999–1029.
4. Bradsher RW. Blastomycosis. Infect Dis Clin NA 1988; 2:877–98.
5. Sarosi GA, Davies SF. Blastomycosis. State of the art. Am Rev Respir Dis 1979; 120:911–38.
6. Larson DM, Eckman MR, Alber RL, et al. Primary cutaneous (inoculation) blastomycosis: an occupational hazard to pathologists. Am J Clin Pathol 1983; 79:253–5.
7. Butka BJ, Bennett SR, Johnson AC. Disseminated inoculation blastomycosis in a renal transplant recipient. Am Rev Respir Dis 1984; 130:1180–4.
8. Bradsher RW, Abernathy RS. Blastomycosis following cutaneous trauma. Am Rev Respir Dis. 1985; 132:931.
9. Medoff G, Painter A, Kobayashi GS. Mycelial to yeast-phase transition of dimorphic fungi *Blastomyces dermatitidis* and *P. brasiliensis*. J Bacteriol 1987; 169:4055–60.
10. Fragoyannis S, VanWyk G, DeBeer M. North American blastomycosis in South Africa. S Afr Med J 1977; 51:169–71.
11. Baily GG, Robertson VJ, Neill P, Garrido P, Levy LF. Blastomycosis in Africa: clinical features, diagnosis and treatment. Rev Infect Dis 1991; 13:1005–8.
12. Alford RH, Goodwin RA. Variations in lymphocyte reactivity to histoplasmin during chronic pulmonary histoplasmosis. Am Rev Respir Dis 1973; 108:85–92.
13. Musatti CC, Rezkallah MT, Mendes E, Mendes NF. In vivo and in vitro evaluation of cell-mediated immunity in patients with paracoccidioidomycosis. Cell Immunol 1976; 24:365–78.
14. Cox RA, Vivas JR, Gross A, Lecara G, Miller E, Brummer E. In vivo and in vitro cell-mediated responses in coccidioidomycosis. Am Rev Respir Dis 1976; 114:937–43.
15. Biggar WD, Menwissen HJ, Good RA. Successful defense against *Histoplasma capsulatum*. Arch Intern Med 1971; 128:585–7.
16. Williams DM, Graybill JR, Drutz DJ. *Histoplasma capsulatum* infection in nude mice. Infect Immun 1978; 21:973–7.
17. Longley RE, Cozad GC. Thymosin restoration of cellular immunity to *Blastomyces dermatitidis* in T-cell-depleted mice. Infect Immun 1983; 40:733–40.
18. Pappas PG, Pottage JC, Powderly WG, Fraser VJ, Stratton CW, McKenzie S, Tapper ML, Chmel H, Bonebrake FC, Blum R, Shafer RW, King C, Dismukes WE. Blastomycosis in patients with the acquired immunodeficiency syndrome. Ann Intern Med 1991; 116:847–53.
19. Fraser VJ, Keath EJ, Powderly WG. Two cases of blastomycosis from a common source; DNA restriction analysis to identify strains. J Infect Dis 1991; 163:1378–80.
20. Witorsch P, Utz JP. North American blastomycosis: a study of 40 patients. Medicine 1968; 47:169–200.
21. Busey JF. Blastomycosis: a review of 198 collected cases in VA hospitals. Am Rev Respir Dis 1964; 89:659–72.
22. Spencer HD, Cozad GC. Role of delayed hypersensitivity in blastomycosis of mice. Infect Immun 1973; 7:329–34.
23. Cozad GC, Chang CT. Cell-mediated immunoprotection in blastomycosis. Infect Immun 1980; 28:398–403.
24. Deepe GSJr, Taylor CL, Bullock WE. Evaluation of inflammatory response and cellular immune response in a murine model of disseminated blastomycosis. Infect Immun 1985; 50:183–9.
25. Cox RA, Larsh HW. Isolation of skin test-active preparations form yeast-phase cells of *Blastomyces dermatitidis*. Infect Immun 1974; 10:42–7.
26. Cox RA, Larsh HW. Yeast and mycelial-phase antigens of *Blastomyces dermatitidis*: comparison using disc gel electrophoresis. Infect Immun 1974; 10:48–53.
27. Deighton F, Cox RA, Hall NJ, Larsh HW. In vivo and in vitro cell-mediated immune responses to a cell wall antigen of *Blastomyces dermatitidis*. Infect Immun 1977; 15:429–35.
28. Cox RA. Cross-reactivity between antigens of *C. immitis*, *H. capsulatum*, and *B. dermatitidis* in lymphocyte transformation assays. Infect Immun 1979; 25:932–8.

29. Bradsher RW, Alford RH. *Blastomyces dermatitidis* antigen-induced lymphocyte reactivity in human blastomycosis. Infect Immun 1981; 33:485–90.
30. Bradsher RW. Development of specific immunity in patients with pulmonary and extrapulmonary blastomycosis. Am Rev Respir Dis 1984; 129:430–4.
31. Schwarz J, Salfelder K. Blastomycosis: a review of 152 cases. Curr Top Pathol 1977; 65:165–200.
32. Sixbey JW, Fields BT, Sun CN, Clark RA, Nolan CM. Interactions between human granulocytes and *Blastomyces dermatitidis*. Infect Immun 1979; 23:41–4.
33. Brummer E, Stevens DA. Opposite effects of human monocytes, macrophages, and polymorphonuclear neutrophils on replication of *Blastomyces dermatitidis* in vitro. Infect Immun 1982; 36:297–303.
34. Brummer E, Stevens DA. Enhancing effect of murine PMN on the multiplication of *B. dermatitidis* in vitro and in vivo. Clin Exp Immunol 1983; 54:587–94.
35. Brummer E, Sugar AM, Stevens DA. Immunological activation of polymorphonuclear neutrophils for fungal killing: studies with murine cells and *Blastomyces dermatitidis* in vitro. J Leukocyte Biol 1984; 36:505–20.
36. Brummer E, McEwen JG, Stevens DA. Fungicidal activity of murine inflammatory polymorphonuclear neutrophils: comparison with murine peripheral blood PMN. Clin Exp Immunol 1986; 66:681–90.
37. Brummer E, Stevens DA. Activation of murine polymorphonuclear neutrophils for fungicidal activity with supernatants from antigen-stimulated immune spleen cell cultures. Infect Immun 1984; 45:447–52.
38. Brummer E, Sugar AM, Stevens DA. Enhanced oxidative burst in immunologically activated but not elicited PMN leukocytes correlates with fungicidal activity. Infect Immun 1985; 49:396–401.
39. Morrison CJ, Isenberg RA, Stevens DA. Enhanced oxidative mechanisms in immunologically activated versus elicited polymorphonuclear neutrophils: correlations with fungicidal activity. J Med Microbiol 1988; 25:115–21.
40. Morrison CJ, Stevens DA. Mechanisms of fungal pathogenicity: correlation of virulence in vivo, susceptibility to killing by PMN in vitro, and neutrophil superoxide anion induction among *Blastomyces dermatitidis* isolates. Infect Immun 1991; 59:2744–9.
41. Sugar AM, Chahal RS, Brummer E, Stevens DA. Susceptibility of *Blastomyces dermatitidis* strains to products of oxidative metabolism. Infect Immun 1983; 41:908–12.
42. Sugar AM, Chahal RS, Brummer E, Stevens DA. The iron–hydrogen peroxide–iodide system is fungicidal: activity against the yeast phase of *Blastomyces dermatitidus*. J Leukocyte Biol 1984; 36:545–8.
43. Schaffner A, Davis CE, Schaffner T, Market M, Douglas H, Braude A. In vitro susceptibility of fungi to killing by PMN discriminates between primary pathogenicity and opportunism. J Clin Invest 1986; 78:511–24.
44. Morrison CJ, Brummer E, Stevens DA. Effects of a local immune reaction on peripheral blood polymorphonuclear neutrophil microbicidal function: studies with fungal targets. Cell Immunol 1987; 110:176–82.
45. Morrison CJ, Brummer E, Isenberg RA, Stevens DA. Activation of murine polymorphonuclear neutrophils for fungicidal activity by recombinant gamma interferon. J Leukocyte Biol 1987; 41:434–40.
46. Morrison CJ, Stevens DA. Enhanced killing of *Blastomyces dermatitidis* by gamma interferon–activated murine peripheral blood polymorphonuclear neutrophils. Int J Immunopharmacol 1989; 11:855–62.
47. Drutz DJ, Frey CL. Intracellular and extracellular defenses of human phagocytes against *B. dermatitidis* conidia and yeasts. J Lab Clin Med 1985; 105:737–50.
48. Scilliam JJ, Cozad GC, Spencer HD. Passive transfer of delayed hypersensitivity to *Blastomyces dermatitidis* between mice. Infect Immun 1974; 10:705–11.
49. McDaniel LS, Cozad GC. Immunomodulation by *Blastomyces dermatitidis*: functional activity of murine peritoneal macrophages. Infect Immun 1983; 40:733–40.
50. Brummer E, Morozumi PA, Stevens DA. Macrophages and fungi: in vitro effects of method of macrophage induction, activation by different stimuli, and soluble factors on *Blastomyces*. J Reticuloendothel Soc 1980; 28:507–18.
51. Brummer, Morozumi PA, Philpott DE, Stevens DA. Virulence of fungi: correlation of virulence of *B. dermatitidis* in vitro with escape from macrophage inhibition of replication in vitro. Infect Immun 1981; 32:864–71.
52. Brummer E, Sugar AM, Stevens DA. Activation of peritoneal macrophages by concanavalin a or

Mycobacterium bovis BCG for fungicidal activity against *Blastomyces dermatitidis* and effect of specific antibody and complement. Infect Immun 1983; 39:817–22.

53. Brummer E, Hanson LH, Stevens DA. Kinetics and requirements for activation of macrophages for fungicidal activity: effect of protein synthesis inhibitors and immunosuppressants on activation and fungicidal mechanism. Cell Immunol 1991; 132:236–45.
54. Brummer E, Morrison CJ, Stevens DA. Recombinant and natural gamma-interferon activation of macrophages in vitro: different dose requirements for induction of killing activity against phagocytizable and nonphagocytizable fungi. Infect Immun 1985; 49:724–30.
55. Brummer E, Stevens DA. Fungicidal mechanisms of activated macrophages: evidence for nonoxidative mechanisms for killing of *B. dermatitidis*. Infect Immun 1987; 55:3221–4.
56. Sugar AM, Brummer E, Stevens DA. Fungicidal activity of murine broncho-alveolar macrophages against *Blastomyces dermatitidis*. J Med Microbiol 1986; 21:7–11.
57. Brummer E, Stevens DA. Activation of pulmonary macrophages for fungicidal activity by gamma-interferon or lymphokines. Clin Exp Immunol 197; 70:520–8.
58. Brummer E, Hanson LH, Restrepo A, Stevens DA. In vivo and in vitro activation of pulmonary macrophages by IFN-gamma for enhanced killing of *P. brasiliensis* or *B. dermatitidis*. J Immunol 1988; 140:2786–9.
59. Sugar AM, Field KG. Characteristics of the pulmonary cellular immune response to two strains of *Blastomyces dermatitidis* in the mouse. Am Rev Respir Disease 1985; 132:1319–23.
60. Sugar AM, Picard M. Macrophage- and oxidant-mediated inhibition of the ability of live *B. dermatitidis* conidia to transform to the pathogenic yeast phase: implications for the pathogenesis of dimorphic fungal infections. J Infect Dis 1991; 163:371–5.
61. Bradsher RW. Live *Blastomyces dermatitidis* yeast-induced responses of immune and nonimmune human mononuclear cells. Mycopathologia 1984; 87:159–66.
62. Bradsher RW, Ulmer C, Marmer DJ, Townsend JW, Jacobs RF. Intracellular growth and phagocytosis of *Blastomyces dermatitidis* by monocyte-derived macrophages from previously infected and normal subjects. J Infect Dis 1985; 151:57–64.
63. Green GM, Jakab GH, Low RB, Davis GS. Defense mechanisms of the respiratory membrane. Am Rev Respir Dis 1977; 115:479–514.
64. Bradsher RW, Balk RA, Jacobs RF. Growth inhibition of *Blastomyces dermatitidis* in alveolar and peripheral macrophages from patients with blastomycosis. Am Rev Respir Dis 1987; 135:412–7.
65. Klein BS, Vergeront JM, Weeks RJ, et al. Isolation of *B. dermatitidis* in soil associated with a large outbreak of blastomycosis in Wisconsin. N Engl J Med 1986; 314:529–34.
66. Vaaler AK, Bradsher RW, Davies SF. Evidence of subclinical blastomycosis in forestry workers in northern Minnesota and northern Wisconsin. Am J Med 1990; 89:470–5.

V

MACROPHAGE–VIRAL INTERACTIONS

35

The Macrophage and HIV-1

Wen-Zhe Ho, Ravi Cherukuri, and Steven D. Douglas

The Children's Hospital of Philadelphia, University of Pennsylvania School of Medicine, Philadelphia, Pennsylvania

I. INTRODUCTION

Human immunodeficiency virus (HIV-1) infection and acquired immunodeficiency syndrome (AIDS) are characterized by disease manifestations that are linked to immunologic defects caused by the severe depletion of CD4+ lymphocytes [1,2]. These effects are preceded by and/or concomitant with HIV-1 infection of monocytes/macrophages and the subsequent impairment of numerous other important immunologic functions [3–5]. The global immunosuppression that results from the depletion of CD4 lymphocytes or from lymphocyte dysfunction causes HIV-1–infected individuals to be susceptible to a variety of opportunistic infections and/or malignancies [6]. In monocytes/macrophages HIV-1 infection can alter crucial immunoregulatory functions such as cytokine production and accessory cell function [7–10]. The complex interaction between monocytes/macrophages and HIV-1 is of special importance, since these cells are relatively refractory to the cytopathic effects of HIV-1 and probably serve as a major reservoir for the virus, as well as a vehicle for disseminating HIV-1 to various tissues [11,12]. Furthermore, by expressing major histocompatibility complex (MHC) class II antigens, HIV-1 infected monocytes may deliver virus to the CD4+ T lymphocytes during the process of normal antigen presentation, thus acting as a potential reservoir for HIV-1 transmission [13]. In this section we will analyze the interaction of HIV-1 with various cells of the monocyte/macrophage lineage and the role of these cell types in HIV-1 disease progression.

II. MONOCYTES/MACROPHAGES AS A RESERVOIR FOR HIV-1

There are conflicting reports concerning the level of HIV-1 proviral DNA present in the peripheral blood monocytes of HIV-1–seropositive individuals and whether or not this cell population is the first to be infected with HIV-1. Innocenti et al. and others have recently reported that at least 50% of the monocytes from these individuals are positive for HIV-1 provirus by polymerase chain reaction (PCR) analysis using *gag*, *env*, or long terminal repeat (LTR) primer pairs [14,15]. In contrast, other groups have found only low levels of HIV-1 DNA (<14%) in the monocytes of seropositive subjects [16,17]. Furthermore, the available data indicate that patients whose monocytes test positive for HIV-1 provirus also have lower CD4+ T cell counts than patients whose monocytes are negative for HIV-1 (means of 269 CD4/mm^3 and 573 CD4/mm^3, respectively) [14]. This suggests a possible correlation between disease as measured by total CD4 count and integration of HIV-1 DNA into the monocyte cellular genome. Human immunodeficiency virus may enter monocytes early in the course of infection and accumulate in the cytoplasm, allowing it to escape detection by PCR [18]. An alteration in cytokine levels caused by a reduction in the number of circulating lymphocytes may stimulate viral replication [14]; for example, diminished levels of interferon (IFN), a cytokine product of T cells, may result in enhanced HIV-1 activity in macrophages.

In vitro, HIV-infected monocyte cultures show a great disparity between the number of infected cells and the amount of infectious virus that is released as compared with lymphocyte cultures. While HIV-1 mRNA levels reveal that there are three times more infected cells in monocyte cultures than in lymphocyte cultures inoculated with the same titer of virus, assays for reverse transcriptase (RT) activity reveal that the T cells release 10 to 100 times more virions into culture supernatant than monocytes [19]. This basic difference in HIV-1 interaction between T cells and macrophages can in part be explained by electron-microscopic analysis. In T cells active HIV-1 assembly and budding is seen around the plasma membrane, and there is very little intracytoplasmic accumulation of viral particles (Figure 1) [18–20]. In contrast, infected macrophages have most of their HIV-1 sequestered in intracellular vacuoles, and there are few, if any, viral particles associated with the cell membrane (Figure 2) [18,19]. Virion assembly and budding does occur within these vacuoles, which morphologically resemble the Golgi complex, and virions originating from these vacuoles are infectious [21]. Thus, HIV-1–infected monocytes/macrophages may escape detection by the immune system and persist in various tissues.

Human immunodeficiency virus infection in lymphocytes is dependent on DNA synthesis and

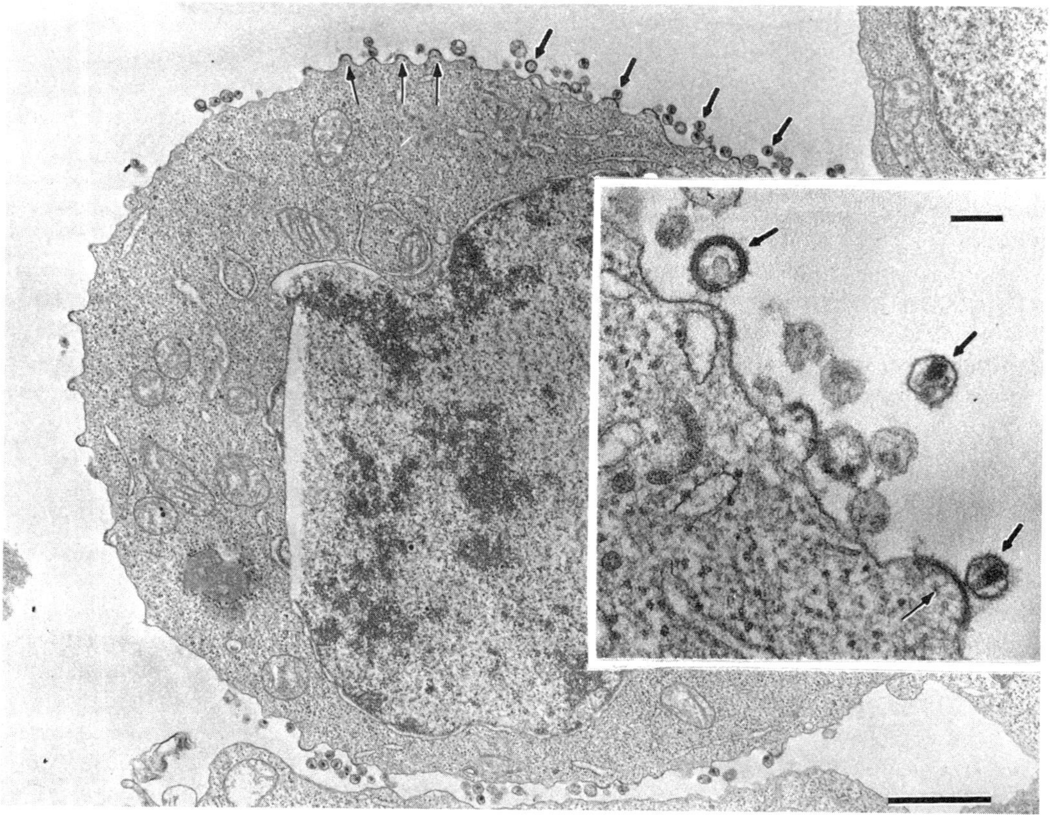

FIGURE 1 Electron photomicrographs of an HIV-1 chronically-infected human T cell line (ACH-2) stimulated by monokine-enriched supernatants from primary human monocyte cultures. Numerous viral particles are budding at the cell membrane. There are no evident intracellular virions (× 30,000, insert × 95,000X).

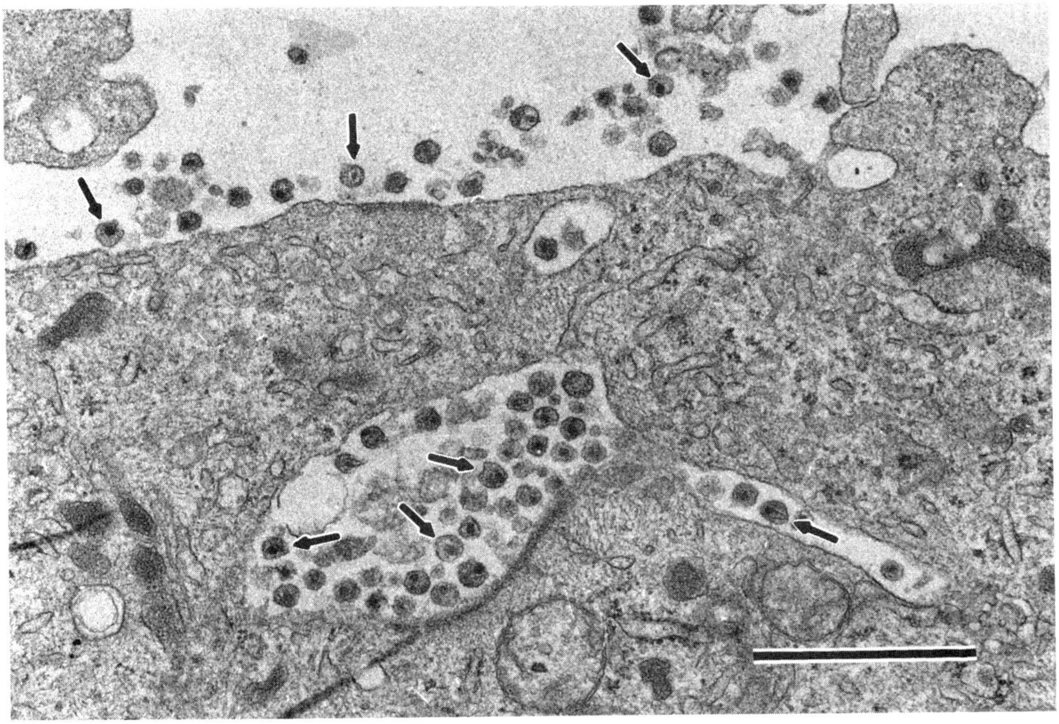

FIGURE 2 Electron photomicrographs of HIV-1 Bal–infected cord monocyte-derived macrophages 25 days after infection. Numerous viral particles sequestered within intracytoplasmic vacuole. There are few or no virions evident at the cell membrane (× 18,750).

cell proliferation. In monocytes/macrophages, however, controversial results have been reported. Schuitemaker et al. reported that the proliferation of monocytes during differentiation into macrophages is a prerequisite for productive infection with HIV [22], while Weinberg et al. showed that productive HIV-1 infection in monocytes is dependent neither on proliferation nor on cellular DNA synthesis. They showed that monocytes that have been irradiated to impair cellular DNA synthesis can still be efficiently infected with HIV-1 [Ref. 23]. These cells are resistant to the cytopathic effects of the virus, providing evidence that monocytes/macrophages may serve as persistent reservoirs for HIV-1. Presumably, an HIV-1 provirus integrated into the cellular genome of a monocyte/macrophage utilizes a promoter or some cellular factor endogenous to the cell that allows it to replicate independently of the rest of the cellular genome. Although the mechanism by which monocytes support productive infection without DNA synthesis and cell proliferation is unknown, these observations support the idea that the replicative process utilized by HIV-1 in monocytes/macrophages differs from the one utilized in T cells and other proliferating cells.

Similar to lymphocytes, HIV-1 entry in most cells of the monocyte/macrophage lineage is mediated by the cell surface receptor CD4. Treatment of most monocyte/macrophage cells in vitro with soluble CD4 or anti-CD4 antibody results in reduced HIV-1 infection [11,20,24–29]. However, entry of an antibody–HIV-1 complex via the Fc receptor is a second mechanism by which HIV-1 can infect macrophages [30–33]. Antibodies from the sera of HIV-1–seropositive individuals have been shown to enhance HIV-1 infectivity of monocytes/macrophages in vitro as well as the U937 promonocytic cell line [34]. Furthermore, monoclonal antibodies directed against a subtype of the Fc receptor (FcRIII) can block this enhancement of HIV-1 infectivity in peripheral blood mononuclear cells and other cell populations [30]. This phemonenon of antibody-mediated enhancement of HIV-1 infection may explain why neutralizing or cytotoxic antibodies to HIV-1 are ineffective against viral spread in vivo [35]. Fc receptor–mediated HIV-1 infection in monocytes/macrophages is not, however, completely independent of the CD4 receptor [36].

III. SUSCEPTIBLE MONOCYTE/MACROPHAGE TISSUE TARGETS FOR HIV-1

Monocytes/macrophages infected with HIV-1 have been detected in various tissues (Table 1) including the brain [25,37–39,185], placenta [79–82], skin [40–44], liver [45–47], lungs [21, 48,49,68], lymph nodes [50,51], and peripheral blood [52–54,183,184]. Myelomonocytic bone marrow stem cells and blood monocyte precursor cells have also been determined to be targets for HIV-1 [55–57]. Viral infection of these cells may be related to the variety of hematologic abnormalities seen in adults and children with AIDS [55]. While the frequency of HIV-1–infected macrophages varies with tissue origin, it can range from 10% to more than 50% in tissues such as the central nervous system (CNS), lungs, and lymph nodes [37,49,50]. Donor-to-donor variation was also seen in the experiments using monocyte-derived macrophages infected with different HIV-1 strains in vitro [52].

Although freshly isolated monocytes do not support productive infection of HIV-1 in vitro [22,23,61], monocyte-derived macrophages are susceptible to infection by the virus [52,58–62]. Surprisingly, monocyte-derived macrophages express fewer CD4 receptors than freshly isolated blood monocytes [27,60,61,63,64]. This apparent paradox may be explained by the fact that differentiation of monocytes induces cellular nuclear factor κB (NF-κB) activity, which in turn enhances HIV-1 gene expression [65–67]. The treatment of monocytes/macrophages with activators such as lipopolysaccharide (LPS) normally induces binding of NF-κB to sites on the host cellular genome that regulate the production of various cellular proteins. However, the HIV-1-LTR also contains two NF-κB binding sites to which cellular NF-κB in HIV-1–infected cells may bind and thereby initiate transcription of HIV-1 RNA. There is constitutive NF-κB activity in human monocytes and macrophages, but this level has not yet been quantified [21,65].

Human immunodeficiency virus also infects alveolar macrophages (AMs) both in vitro and in vivo [21,48,58,68]. Productive infection as quantified by p24gag antigen ELISA in AMs is higher than in freshly isolated monocytes but not as high as in differentiated blood monocytes [21]. Altered cytokine production has been observed in HIV-1–infected AMs [69,70]. The high level of virion production seen in AMs implicates this cell type as a major contributor to the immu-

TABLE 1 Susceptible Tissue Macrophage Targets for HIV-1 In Vitro

Monocytye/macrophage cell types	HIV-1 strain(s)	References
1. Periperal blood MDM	Macrophage-tropic (Bal, ADA-M, SF162,DV)	52–54,183,184
2. Cord blood macrophages	Macrophage-tropic (Bal, ADA-M)	27
3. Alveolar macrophages	Macrophage-tropic (JR-FL)	21,48,49,68
4. Peritoneal macrophages	Macrophage-tropic (Bal)	60
5. Microglia	Macrophage-tropic (AD87, Bal, JR-FL)	25,37–39,185
6. Kupffer cells (liver)	Lymphocyte-tropic (IIIB) Macrophage-tropic (BRU)	45–47
7. Langerhans cells (skin)	Lymphotropic-tropic (IIIB)	40–44
8. Hofbauer cells (placenta)	Macrophage-tropic (BRU)	79–82
9. Hemopoietic progenitor cells	Macrophage-tropic (D117III)	55–57

MDM = primary monocyte-derived macrophages.

nopathogenesis of lung disorders seen in AIDS patients. In addition, peritoneal macrophages (PMs) isolated from normal women by laparoscopy are more easily infected with HIV-1 than blood monocytes or AMs [60]. Peritoneal macrophages from some donors show higher infectibility than those from other donors, indicating that some individuals may be a higher risk for HIV-1 infection than others. As peritoneal macrophages are localized in the endometrium, these cells may play a critical role in the heterosexual transmission of HIV-1 via the seminal fluid.

Cells of the monocyte/macrophage lineage are the primary sites of HIV-1 infection and replication in the CNS [38,71–75]. Microglial cells, which play an important role in immune regulation of the CNS, can be productively infected by monocyte-tropic HIV-1 strains but not lymphocyte-tropic strains [38]. This productive infection, which is CD4-dependent [25], leads to syncitial formation and cell death [38]. Microglial cells express markers of the monocyte/macrophage lineage, such as low-density lipoprotein receptor (LDL-R), Leu M3, Leu M5, EBM11,OKM-1, and OX42 [38,76]. Although microglia originate from the same bone marrow precursor cells as monocytes/macrophages, they interact very differently with HIV-1. Infection of microglial cells by HIV-1 appears to more closely resemble HIV infection of CD4 lymphocytes in that there is rapid viral production and cytotoxicity [38]. Also, virions in microglial cells actively bud from the plasma membrane instead of being sequestered in intracellular vacuoles. Similar to monocytes/macrophages, however, brain microglial cells can proceed with HIV-1 replication without cellular proliferation [38].

Although the interaction of HIV-1 and monocytes/macrophages isolated from the umbilical cord blood of neonates has not been investigated extensively, there is evidence that these cells are more easily infected by HIV-1 monocyte-tropic strains in vitro than their adult counterparts [27]. This fact may be relevant to the enhanced susceptibility of neonates to HIV-1 disease in vivo and supports the concept that monocytes/macrophages may plan an important role in the pathophysiology of congenital AIDS [27]. Our laboratory has observed that cord monocyte-derived macrophages secrete significantly less fibronectin [77] and tumor necrosis factor-α (TNF-α) in vitro than adult monocyte-derived macrophages. Also, the CD4 receptor is a major HIV-1 entry pathway for cord blood macrophages [27]. Among human placental cell populations there is a macrophage subpopulation, called Hofbauer cells, located in the chorionic villous stroma [78]. While the interaction between HIV-1 and Hofbauer cells has been studied very little, these cells can express CD4 as well as Fc receptors and are susceptible to productive HIV-1 infection, suggesting that they may play a role in the vertical transmission of HIV-1 [79–83].

IV. CELL TROPISM

Isolates of HIV-1 have been categorized into three types according to their tropism: monocyte-tropic isolates are those that readily infect monocytes/macrophages [84–86], lymphocyte–tropic isolates are those that efficiently infect T cells [11,19], and amphitropic isolates are those that efficiently infect both monocytes/macrophages and lymphocytes [25,87]. Interestingly, when HIV-1 isolated from monocytes is serially passaged in T cells it has a reduced capacity to reinfect monocytes/macrophages [88]. For example, when a mixed inoculum of monocyte-tropic and lymphocyte-tropic viruses is used to infect a T cell population, the progeny virions of this infection can only efficiently reinfect T-cells [88]. One explanation for these observations is that monocyte-tropic viruses undergo some change in gene expression upon infection of T cells that causes them to lose their tropism for monocytes. Another explanation is that lymphocyte-tropic HIV-1 may have some intrinsic advantage over monocyte-tropic HIV-1 that allows them to replicate more rapidly. T cell–tropic HIV-1 isolates can infect monocytes in the presence of enhancing antibodies, suggesting that a CD4-independent mode of entry is involved [30].

The basis for HIV-1 variants exhibiting particular tropism is unknown, although it is probably associated with the viral glycoprotein gp120 [89]. The production of viral *env* gene products may be associated with the phenomenon of tropism, since HIV-1–infected T cells synthesize and assemble primarily envelope glycoproteins, while infected monocytes produce primarily capsid proteins [19]. In T cells, 35% of the total viral proteins produced are *env* gene products, while these gene products constitute less than 10% of the total viral proteins produced by infected monocytes. Electron microscopy of virus from the culture supernatant of infected T cells reveals the characteristic spikes that are images of gp120 (Figure 1) [11]. These spikes are absent on virus from infected monocytes (Figure 2).

Hwang et al have identified a 20-amino-acid sequence in the envelope V3 loop from the macrophage-tropic Bal isolate that is sufficient to confer macrophage tropism on HTLV-IIIB [90]. Although the exact mechanism by which the V3 loop determines tropism is unclear, it probably does not directly affect CD4-gp120 binding, since HIV-1 variants containing defective or missing V3 loops retain the ability to bind to target cells [91,186]. It may in fact interfere with the membrane fusion process that occurs very soon after binding. Furthermore, as the principal determinant of CD4 neutralization of HIV-1, the V3 loop plays an important role in the interaction between the host immune system and the virus; thus, V3 is pivotal both in HIV-1, tissue tropism and host immune response [92].

V. HIV-1, MONOCYTES/MACROPHAGES, AND CENTRAL NERVOUS SYSTEM DISEASE

Central nervous system disease in AIDS patients is characterized by diffuse encephalopathy, cranial or peripheral neuropathies [93], astrocytosis [94], syncitial formation, demyelination of neural cells, and cerebral vasculitis [95,96]. The human immunodeficiency virus probably does not affect cells of the CNS directly, but it may act on them indirectly in three possible ways. It may cause infected brain macrophages or microglial cells to produce neurotoxins or cytokines that have adverse effects on neural cells [97,98]. There are at least two cytokines that produce these effects of CNS disease in cells in vitro, namely, interleukin-1 (IL-1) and TNF-α [99]. Interleukin-1 has modulating effects on neural cells, such as the induction of astrocyte proliferation [100], a prominent histologic finding in AIDS dementia complex [101], and TNF-α is cytotoxic for glioma cell lines [102]. Although severe dementia in AIDS patients is usually characterized by large numbers of brain cells infected with HIV-1 [103], there have been several reports of AIDS patients who suffer from dementia but show almost no signs of HIV-1 in their CNS tissue [104,105]. Therefore, the persistent secretion of cytotoxic factors by HIV-1–infected macrophages may be the cause of prolonged neural cell dysfunction [98]. Tumor necrosis factor-α is known to have various neurotoxic effects on cells of the CNS [102,106,107] and is enhanced in macrophages by HIV-1 infection. Production of IL-1β and TNF-α is potentially important in the immunopathogenesis of HIV-1 infection, since release of these cytokines is known to activate T and B cells as well as induce fever, cachexia, and demyelination [108,109].

Alternatively, HIV-1 viral antigens may damage monocytes/macrophages in the CNS and cause subsequent neurologic dysfunction by interacting with their cellular membranes. The interaction between HIV-1 gp120 and cell surfaces can cause neuronal cell death in hippocampal cultures of fetal mice [110], induce monocytes to produce harmful physiological metabolites [111], inhibit the function of neuroleukin [112], and cause other metabolic changes that result in cell death [110,111,113]. In many of these cases, mere binding of HIV-1 to the target cell membrane without HIV-1 infection has the capacity to effect biological changes in these cells. This may hold true for CNS monocytes/macrophages and other related cells as well [114]. Finally, Tardieu et al. have recently described a third possible mechanism by which HIV-1–infected monocytes/macrophages may cause CNS dysfunction: HIV-1 infection appears to enhance the capacity of monocytic cells to adhere directly to neural cells and thereby cause subsequent cytopathic effects [115].

While productive infection of primary neural cells has not been very successful in vitro [8], high levels of HIV-1 infection have been documented in the brain tissue and cerebrospinal fluid of AIDS patients [71,116–118]. Intrathecal synthesis of antibody against HIV-1 has been shown and suggests viral replication within the CNS in cells with the morphologic appearance of macrophages and in multinucleated giant cells [119]. Electron-microscopic studies demonstrate HIV-1 particles within multinucleated giant cells in the brain and within astrocytes [120]. In situ hybridization studies indicate that the affected cell in the brain is a microglial-mononuclear phagocyte [121]. Furthermore, the pathogenesis of HIV-1 infectivity in the CNS is characterized by an aggregation of microglial nodules [12]. Cells of the monocyte/macrophage lineage such as microglia and multinucleated giant cells may serve as reservoirs of HIV-1 infection in the brain in the same way as blood monocytes/macrophages do in other organs.

Beyond the various neurological complications seen in AIDS patients and in persons infected with HIV-1, there is increasing evidence to suggest that the psychological symptoms seen in these populations may be related to the progression of HIV-1 disease. As those at risk for AIDS experience a wide range of disruptive "life events," both depression and anxiety have been shown to be highly prevalent in HIV-1–seropositive patients, those with AIDS-related complex (ARC), and AIDS patients [122,123]. While there is little evidence that directly correlates relevant neurochemical changes with immunologic changes, the occurrence of major or chronic stressful life events may have implications for health through their impact on immunologic functions [124]. Stress conditions are usually accompanied by an altered blood level of various CNS-released peptides and neurohormones. At least 25 pharmacologically active neuropeptides are involved in the integrative communication within the CNS, many of which are altered in depressive states [125]. If alterations in circulating levels of neuropeptides can be correlated with changes in monocytes/macrophage function in vitro and with psychological studies, these will become invaluable clinical markers. Also, a distinction between neurogenic phenomena (direct brain involvement by HIV-1 or opportunistic infection, neoplasia or side effects of treatment) and psychological disorders (resultant from knowledge of HIV-1 infection, progression, and social consequences) is heuristically useful [126].

Monocytes/macrophages play a central role in the afferent and efferent limbs of the immune system. Monocytes/macrophages express surface receptors for different CNS-secreted molecules such as somatotropin [127], neurotensin [128], β-endorphin [129], substance P [130], and ACTH [131]. Hartung [132] has recently reviewed the capacity of neuropeptides to up- or down-regulate the state of activation of macrophages. In ARC or AIDS patients, abnormal neuropeptide levels may be related to severe psychological disturbances. However, neuropeptide and neurohormone blood levels in stressed HIV-1–infected patients are unknown. The work of Peck [133] and Koff and Dunegan [131] showed that the exposure of human blood monocyte-derived macrophages in vitro to different neuropeptides released in stress-related conditions resulted in altered ability to kill a tumor cell target or to display an oxidative burst upon activation. They also showed that the effect of IFN-γ on monocyte-derived macrophage activation is modulated in the presence of different neuropeptides or neurohormones [131,133]. In addition, Lotz et al. [134] have recently shown that substance P and substance K induce the release in human blood monocytes of IL-1, TNF-α, and IL-6, which are known to enhance HIV-1 replication in both T cells and monocytic cell lines. Neuropeptides and neurohormones may play a central role in stressed HIV-infected patients by affecting monocyte/macrophage functions, which may further trigger disease progression and immunological deficiency. Therefore, the possible relationship between stress-released neuropeptides and neurohormones and the modulation of monocyte/macrophage functions may play an important role in the immunopathogenesis of HIV-1 in the CNS [135].

VI. HIV-1, CYTOKINES, AND MONOCYTES/MACROPHAGES

Human immunodeficiency virus infection, expression, and replication are regulated in an autocrine loop involving cytokines and other soluble factors. Interleukin-1, IL-2, IL-4, IL-6, TNF-α and -β,

GM-CSF, and endotoxin LPS have each been shown to enhance HIV-1 infection in T cells and/or macrophages (Table 2) [99,136–142]. Furthermore, TNF-α with IL-6 or GM-CSF can synergistically enhance HIV-1 infection, as can GM-CSF in combination with endotoxin LPS [136]. TGF-β and INF-γ, on the other hand, have been shown to inhibit HIV-1 expression in monocyte/macrophages [88,137,145,152–154,179,187]. The fact that increased levels of certain cytokines have been detected in the sera of HIV-1–infected individuals implicates these factors in disease progress [99,142]. Infection of peripheral blood mononuclear phagocytes by HIV-1, as well as infection of promonocytic cell lines, results in the increased production of prostaglandins, IL-1-β, TNF-α, IL-6, IL-8, and GM-CSF [111,143–145]. Blood, cerebrospinal fluid, and macrophage cultures from AIDS patients all have high levels of IL-1, TNF-α, and IL-6 [146,147]. In particular, AMs isolated from HIV-1–infected individuals produce high levels of TNF-α and IL-1 [69]. Secondary factors such as LPS or phorbol myristate acetate (PMA) are sometimes necessary to stimulate production of these cytokines. Infection may be ruled out as a prerequisite for enhanced cytokine production in many cases because inactivated HIV-1 and gp120 can both produce the same results in some macrophages in terms of cytokine production as exposure to normal HIV-1 [111,143,144].

Most cytokines that enhance HIV-1 replication in monocytes/macrophages probably exert their effects by inducing endogenous DNA-binding proteins such as Sp-1, AP1, and, most notably, NF-κB [70,148–150]. For example, TNF-α and phorbol esters induce NF-κB binding activity in U937 cells [19,151]. Transcription of HIV-1 in T cells and monocytes/macrophages treated with TNF-α and IL-1 is also stimulated by means of an NF-κB–like enhancer protein [99]. These cellular transcription factors may act on one or more of several *cis*-acting sequences of the HIV-1 LTR [19]. Other cytokines exert their influences through intracellular factors other than NF-κB; this is the case for GM-CSF, which enhances HIV-1 replication without causing NF-κB binding activity, and for IL-6, which probably enhances HIV-1 expression in U937 at a post-transcriptional level [59]. NF-κB binding activity is present constitutively in monocytes but not in U937 [65,149,70].

TABLE 2 Effect of Cytokines and Other Soluble Factors on HIV-1 Production in Monocytes/Macrophages In Vitro

Cytokine	Cell type	Effect on HIV production	Measure of HIV production	References
1. TNF-α	MDM, U1, U937	↑	p24 ELISA, RT, LTR-CAT assay	99,136,177
2. IL-1-α	MDM	↑	p24 ELISA, RT	178
3. IL-2	THP-1	↑	RT	137
4. IL-3	MDM	↑	p24 ELISA	141
5. IL-4	MDM	↑	p24 ELISA	138
6. IL-6	MDM, U1	↑	p24 ELISA, RT	140,178
7. TGF-β	MDM, U1	↑ ↓	RT, Western blot	145,179
8. IFN-α IFN-β IFN-γ	MDM, U1	↓	p24 ELISA, RT, immunoprecip.	88,137,152–154,187
9. GM-CSF	MDM	↑	p24 ELISA	99,139,141,142
10. M-CSF	MDM	↑	p24 ELISA, RT	59,141
11. LPS	U937 MDM	↑ at low conc. ↓ at high conc.	p24 ELISA, RT, LTR-CAT assay	137,178,181,182

TNF = tumor necrosis factor; MDM = primary monocyte derived macrophages; IL = interleukin; TGF = transforming growth factor; IFN = interferon; GM-CSF = granulocyte-macrophage colony-stimulating factor; M-CSF = macrophage colony-stimulating factor; LPS = lipopolysaccharide.

Interferons (IFNs) protect macrophages from HIV-1 infection in vitro, as well as suppressing viral replication in macrophages that are already infected [140,152–154]. There are conflicting hypotheses as to the mechanisms by which IFNs exert their inhibitory effects on HIV-1. For example, IFNs may disrupt viral assembly, down-regulate the expression of viral genes, or degrade viral RNA. Some investigators have reported that IFN prevents the formation of proviral DNA, suggesting that it inhibits early events in viral production [19]. In contrast, studies with chronically infected cells show that IFNs exert their inhibitory effects by blocking viral release from the cellular membrane [153,155].

Interferon-γ treatment of macrophages dramatically decreases CD4 expression, and there is evidence to suggest that HIV-1 stimulates IFN production in macrophages [152]. Therefore, IFN induced by HIV-1 infection may be an endogenous regulator responsible for the extended asymptomatic phase that follows HIV-1 infection and may even cause viral latency in monocytes/macrophages. In contrast, Gendelman et al. report that HIV-1–infected monocytes have defective IFN-α production [164]. Elevated levels of IFN and IFN antigens have been detected in the sera of HIV-1–infected individuals, although HIV-1 has not been conclusively determined to be the causative agent of this enhanced IFN production in vivo [152]. The cell-to-cell interaction of HIV-1–infected monocytes/macrophages induces IFN-α and IFN-γ [156].

VII. EFFECT OF HIV-1 ON MONOCYTE/MACROPHAGE FUNCTION

A variety of changes have been reported in monocytes from HIV-1–infected adult patients in vivo, and these include differences in cell number [119], phenotypic marker expression [157], chemotaxis [158], microbicidal activity [28,159], and phagocytosis [160]. For example, decreased oxidative burst responses of monocytes and macrophages from asymptomatic HIV-1–seropositive individuals have been reported [161–163]. Interestingly, oxidative burst is reduced in asymptomatic HIV-seropositive individuals almost as much as it is in AIDS patients [161]. Also, monocytes isolated from AIDS patients with Kaposi's sarcoma or opportunistic infections show abnormal migratory responses when compared with those of healthy control patients [165,166]. This is consistent with clinical observations showing that monocytes from AIDS patients have significantly impaired responses to inflammatory stimuli in vivo [167]. Alveolar macrophages isolated from HIV-1 seropositive individuals have enhanced IL-1β and IL-6 production, which may contribute to enhanced accessory cell function in these HIV-1–infected cells [168].

Infection of monocytes/macrophages by HIV-1 in vitro also results in the impairment of numerous immunological functions including chemotactic activity, surface protein expression, phagocytosis, and cytokine production. Monocytes from healthy donors infected with HIV-JRFL, a monocyte-tropic strain, are impaired in antibody-dependent cell-mediated cytotoxicity (ADCC) and intracellular killing of *Candida pseudotropicalis* [9]. This is probably related to the fact that HIV-1–infected monocytes/macrophages exhibit reduced Fc and C3 receptor–mediated phagocytosis [5]. Infection by HIV-1 also increases the capacity of monocytes to adhere to extracellular matrix in vitro [169]. The poor in vitro T cell mitogenic response of HIV-1–infected patients may be explained by a monocyte defect that has been reported in AIDS patients [4]. Furthermore, Locksley et al have reported that HIV-1 infected macrophages produce a so-called "contra IL-1" that blocks IL-1–mediated T cell activation [170]. Another way in which HIV-1 impairs normal macrophage function is by impairment of their in vitro chemotactic activity. This occurs soon after viral exposure and before viral production, as monocytes treated with purified HIV-1 gp120 and those treated with HIV-1 show a similar degree of reduced chemotactic responsiveness. Exposure of monocytes to either HIV-1 or gp120 causes a reduction in the number of the chemotactic receptors for C5a and FMLP [171].

In vitro HIV-1 infection of monocytes has two distinct phases; acute and chronic infection [18]. During the acute phase, which lasts about 2 weeks, infected cells undergo HIV-1–induced cytopathic effects and release relatively high levels of progeny virions through budding from the plasma membrane. After this phase, the monocytes enter a chronic stage where the level of virion release is very low, and the morphology of the cells appears normal. Here, virus is located almost exclusively within intracellular Golgi vacuoles. During both phases, there is active accumulation of HIV-1 in these vacuoles. In vitro studies show that most of the cells (60% to 80%) in an HIV-1–infected population remain stable during the acute phase; however, a small population does undergo morphologic changes in their cytoplasmic structures such as formation of multinucleated giant cells [18]. Interaction with activated B lymphocytes or T lymphocytes may enhance HIV-1 expression in monocytes/macrophages that are in a chronic stage of infection [172,173]. Abnormalities in the monocyte/macrophage population of HIV-1 infected individuals may be due, in part, to the absence of or to reduced levels of T cell–derived lymphokines that modulate their activities.

The expression of class II MHC determinants (HLA-DP, -DQ, -DR) by monocytes and macrophages infected with HIV-1 and changes in cell surface epitopes during infection and maturation are points of some controversy. Evidence from the laboratories of Melendez-Guerrero et al. and others suggests that HIV-1 infection does not directly affect the expression of the class II MHC antigens HLA-DR or HLA-DQ [174,175]. In contrast, Wahl et al. have shown that HIV-1– and gp120-treated monocytes differentiate into macrophages as evidenced by HLA-DR marker expression [171]. Apparently, HIV-1 binding may cause premature activation of monocytes by the induction of NF-κB binding activity. While some studies of phenotypic marker expression on monocytes from HIV-1–infected adults show no difference in comparison with health controls with regard to expression of CD4, CD11, transferrin, and Fc-IgG receptors [175,176], others have demonstrated that HIV-1 infection does indeed down-regulate the expression of surface CD4 and CD14 in monocytes/macrophages [174]. CD4 mRNA and cellular CD4 protein levels are the same in HIV-1–infected monocytes/macrophages as in uninfected cells, indicating that CD4 receptor expression in infected cells may be blocked at a post-transcriptional level. In addition, CD4 marker expression on monocytes is reduced when these cells differentiate into macrophages [75].

VIII. SUMMARY

As described above, monocytes/macrophages from such diverse areas as the CNS, lungs, peripheral blood, cord blood, and liver are susceptible to infection by HIV-1 and may serve as reservoirs for transmitting the virus throughout the body. In contrast to lymphocytes, the chronic and low-level mode of infection that HIV-1 takes on within monocytes/macrophages allows these cells to contribute to persistent viral infection. Infection of these cells by HIV-1 can alter a variety of their immunological functions such as cytokine production, chemotaxis, accessory cell function, and microbicidal activity. Furthermore, macrophages and macrophage-related microglial cells in the brain and CSF are thought to play a critical role in the pathogenesis of CNS disease. Infection with HIV-1 in monocytes/macrophages is regulated in an autocrine manner by a panel of cytokines and other soluble factors. Most HIV-1 isolates exhibit specific tropism for monocytes/macrophages, lymphocytes, or, in some cases, both cell types; this phenomenon of tropism may be linked to structural variabilities between different isolates. While in vitro studies clearly demonstrate that the replicating cycle of HIV-1 in monocytes/macrophages differs from its replicating cycle in lymphocytes, elucidation of the role of monocyte/macrophages in mediating the severe immunosuppression characteristic of AIDS is far more from complete, and further investigation into the role of monocytes/macrophages in the immunopathogenesis of HIV-1 infection is necessary to design therapeutic approaches to control viral infection and disease progression.

ACKNOWLEDGMENTS

The authors are grateful to Natale Tomassini for electron microscopy assistance and to Ann Murray for secretarial assistance. The studies performed in the authors' laboratory were supported in part by W. W. Smith Charitable Trust grant A4089, NIMH-MH grant 47422, NIH-NS grant 27405, and PAF/AmFar grant 500-185-11-PG and 500-270-13-PGR.

REFERENCES

1. DeWolf F, Lang JMA, Houweling JTM. Number of CD4+ cells and the levels of core antigens and antibodies to the human immunodeficiency virus as predictors of AIDS among seropositive homosexual men. J Infect Dis 1988; 158:1360–7.
2. Fauci AS. The human immunodeficiency virus: infectivity and mechanisms of pathogenesis. Science 1988; 239:617–22.
3. Poli G, Botazzi B, Acero R, Bergani L, Rossi V, Introna M, Lazzarin A, Galli M, Mantovani A. Monocyte function in intravenous drug abusers with lymphadenopathy syndrome and in patients with acquired immunodeficiency syndrome: selective impairment of chemotaxis. Clin Exp Immunol 1985; 62:136–42.
4. Prince HE, Moody DJ, Shubin BI, Fahey JL. Defective monocyte function in AIDS: evidence from a monocyte-dependent T cell proliferative system. J Clin Immunol 1985; 5:21–25.
5. Bender BS, Augor FA, Quinn TC, Redfield R, Gold J, Folks TM. Impaired antibody-dependent cell-mediated cytotoxic activity in patients with the acquired immunodeficiency syndrome. Clin Exp Immunol 1986; 64:166–72.
6. Rosenberg ZF, Fauci AS. Immunopathegenesis of HIV infection. FASEB 1991; 5:2382–90.
7. Meltzer MS, Gendelman HE. Mononuclear phagocytes as targets, tissue reservoirs, and immunoregulatory cells in human immunodeficiency virus disease. In Macrophage biology and activation, Russell SW, Gordon S., eds., p. 239–63, Springer-Verlag, New York, 1992.
8. Gendelman HE, Morahan PS. Macrophages in viral infections in the macrophage, Lewis CE, McGee J. O'D., eds., p. 189–91, Oxford University Press. Oxford, 1992.
9. Baldwin GC, Fleischmann J, Chung Y, Koyanagi Y, Chen IS, Golde DW. Human immunodeficiency virus causes mononuclear phagocyte dysfunction. Proc Natl Acad Sci USA 1990; 87:3933–7.
10. Ennen J, Seipp A, Norley SG, Kurth R. Decreased accessory cell function of macrophages after infection with human immunodeficiency virus type 1 in vitro. Eur J Immunol 1990; 20:2451–6.
11. Meltzer MS, Nakamura M, Hansen BD, Turpin JA, Kalter DC, Gendelman HE. Macrophages as susceptible targets for HIV infection, persistent viral reservoirs in tissue, and key immunoregulatory cells that control levels of virus replication and extent of disease. AIDS Res Hum Retroviruses 1990; 6:967–71.
12. Bender BS, Davidson DL, Kline RL, Brown C, Quinn RC. Role of the mononuclear phagocyte system in the immunopathegenesis of human immunodeficiency virus infection and the acquired immunodeficiency syndrome. Rev Infect Dis 1988; 10:1142–54.
13. Mann DL, Gartner S, Le Sane F, Buchow H, Popovic M. HIV-1 transmission and function of virus-infected monocyte/macrophages. J Immunol 1990; 144:2152–8.
14. Innocenti P, Ottmann M, Morand P, Leclercq P, Seigneurin JM. HIV-1 in blood monocytes: frequency of detection of proviral DNA using PCR and comparison with the total CD4 count. AIDS Res Hum Retroviruses 1992; 8:261–8.
15. Hufert FT, v. Laer D, Schramm C, Tarnok A, Schmitz H. Detection of HIV-1 DNA in different subsets of human peripheral blood mononuclear cells using the polymerase chain reaction. Arch Virol 189; 106:341–5.
16. Schnittman SM, Psallidopoulos MC, Lane HC, Thompson L, Baseler M, Massari F, Fox CH, Salzman NP, and Fauci AS. The reservoir of HIV-1 in human peripheral blood is a T cell that maintains expression of CD4. Science 1989; 245:305–8.
17. Spear GT, Ou CY, Kessler HA, Moore JL, Schocheteman G, Landay AL. Analysis of lymphocytes, monocytes, and neutrophils from human immunodeficiency virus (HIV)-infected persons for HIV DNA. J Infect Dis 1990; 162:1239–44.

18. Orenstein JM, Meltzer MS, Phipps T, Gendelman HE. Cytoplasmic assembly and accumulation of human immunodeficiency virus types 1 and 2 in recombinant human colony-stimulating factor-1–treated human monocytes: an ultrastructural study. J Virol 1988; 62:2578–86.
19. Meltzer MS, Skillman DR, Hoover DL, Hanson BD, Turpin JA, Kalter DC, Gendelman HE. HIV and the immune system: macrophages and the human immunodeficiency virus. Immunol Today 1990; 11:217–23.
20. Gomatos PJ, Stamatos NM, Gendelman HE, Fowler A, Hoover DL, Kalter DC, Burke DS, Tramont EC, Meltzer MS. Relative inefficiency of soluble recombinant CD4 for inhibition of infection by monocyte-tropic HIV in monocytes. J Immunol 1990; 144:4183–8.
21. Meltzer MS, Skillman DR, Gomatos PJ, Kalter DC, Gendelman HE. Role of mononuclear phagocytes in the pathogenesis of human immunodeficiency virus infection. Annu Rev Immunol 1990; 8:169–94.
22. Schuitemaker H, Kootstra NA, Koppelman MHGM, Bruisten SM, Huisman HG, Tesmette M, Miedema F. Proliferation-dependent HIV-1 infection of monocytes occurs during differentiation into macrophages. J Clin Invest 1992; 89:1154–60.
23. Weinberg JB, Mathhews TJ, Cullen BR, Malim MH. Productive human immunodeficiency virus type 1 (HIV-1) infection of nonproliferating human monocytes. J Exp Med 1991; 174:1477–82.
24. Kazazi F, Mathijs JM, Foley P, Cunninham AL. Variations in CD4 expression by human monocytes and macrophages and their relationship to infection with the human immunodeficiency virus. J Gen Virol 1989; 70:2661–72.
25. Jordan CA, Watkins BA, Kufta C, Dubois-Dalco M. Infection of brain microglial cells by human immunodeficiency virus type 1 is CD4 dependent. J Virol 1991; 65:736–42.
26. Collman R, Godfrey B, Cutilli JR, Rhodes A, Hassan NF, Sweet R, Douglas SD, Friedman H, Nathanson N, Gonzalez-Scarano F. Macrophage-tropic strains of human immunodeficiency virus type 1 utilize the CD4 receptor. J Virol 1990; 64:4468–76.
27. Ho WZ, Lioy J, Song L, Cutilli JR, Polin RA, Douglas SD. Infection of cord blood monocyte-derived macrophages with human immunodeficiency virus type-1. J Virol 1992; 66:573–9.
28. Mabondzo A, Le Naour R, Raoul H, Clayette P, Lafuma C, Barre-Sinoussi FC, Cayre Y, Dormont D. In vitro infection of macrophages by HIV: correlation with cellular activation, synthesis of tumour necrosis factor alpha and proteolytic activity. Res Virol 1991; 142:205–12.
29. Clapham PR, Weber JN, Whitby D, McIntosh K, Dagleish AG, Maddon PJ, Deen KC, Sweet RW, Weiss RA. Soluble CD4 blocks the infectivity of diverse strains of HIV and SIV for T-cells and monocytes but not for brain and muscle cells. Nature 1989; 337:368–70.
30. Homsy J, Meyer M, Tateno M, Clarkson S, Levy JA. The Fc and not CD4 receptor mediates antibody enhancement of HIV infection in human cells. Science 1989; 244: 1357–60.
31. Takeda A, Ennis FA. FcR-mediated enhancement of HIV-1 infection by antibody. AIDS Res Hum Retroviruses 1990; 6:999–1004.
32. Lenert P, Mehta RL, Zanetti M. Synthetic peptides of human CD4 enhance binding of Ig to monocyte/macrophage cells: characterization and mapping studies. J Immunol 1992; 148:1759–63.
33. Jounalt T, Chapuis F, Bahraoui E, Gluckman JC. Infection of monocytic cells by HIV-1: combined role of FcR and CD4. Res Virol 1991; 142:183–8.
34. Haubrich RH, Takeda A, Koff W, Smith G, Ennis FA. Studies of antibody-dependent enhancement of human immunodeficiency virus (HIV) type 1 infection mediated by Fc receptors using sera from recipients of a recombinant gp 160 experimental HIV-1 vaccine. J Infect Dis 1992; 165:545–8.
35. Robinson WE, Montefiori DC, Mitchell WM. A human immunodeficiency virus type 1 (HIV-1) infection–enhancing factor in seropositive sera. Biochem Biophys Res Commun 1987; 149:693–9.
36. Perno CF, Baseler MW, Broder S, Yarchoan R. Infection of monocytes by human immunodeficiency virus type 1 blocked by inhibitors of CD4-gp120 binding, even in the presence of enhancing antibodies. J Exp Med 1990; 171:1043–56.
37. Koenig S, Gendelman HE, Orenstein JM, Dal Canto MS, Pezeshkpour EM, Yungbluth M, Janotta F, Aksamit A, Martin MA, Fauci AS. Detection of AIDS virus in macrophages in brain tissue from AIDS patients with encephalopathy. Science 1986; 233:1089–93.
38. Watkins BA, Dorn HH, Kelly WB, Armstrong RC, Potts BJ, Michaels F, Kufta CV, Dubois-Dalcq M. Specific tropism of HIV-1 for microglial cells in primary human brain cultures. Science 1990; 249:549–53.

39. Peudenier S, Hery C, Montagnier L, Tardieu M. Human microglial cells: characterization in cerebral tissue and in primary culture, and study of their susceptibility to HIV-1 infection. Ann Neurol 1991; 29:152–161.
40. Tschackler E, Groh V, Popovic M, Mann D, Konrad K, Safai B, Eron L, Veronese F, Wolff K, Stingl G. Epidermal Langerhans cells—a target of HTLV-III/LAV infection. J Invest Dermatol 1987; 88:233–7.
41. Zambruno G, Mori L, Marconi A, Mongiardo N, De Rienzo B, Bertazzani V, Gianetti A. Detection of HIV-1 in epidermal Langerhans cells of HIV-infected patients using the polymerase chain reaction. J Invest Dermatol 1991; 96:979–82.
42. Braathen LR, Ramirez G, Kunze RO, Mork C, Strand O. Latent infection of epidermal Langerhans' cells in HIV-positive individuals. Res Virol 1991; 142:119–21.
43. Rappersberger K, Gartner S, Schenk P, Stingl G, Groh V, Tschachler E, Mann DL, Wolff K, Konrad K, Popovic M. Langerhans cells are an actual site of HIV-1 replication. Intervirol 1988; 29:185–94.
44. Ramirez G, Braathen LR, Lunze RO, Gelderblom H. In vitro infection of human epidermal Langerhans cells with HIV. Adv Exp Med Biol 1988; 237:901–5.
45. Schweitzer C, Keller F, Schmitt MP, Jaeck D, Adloff M, Schmitt C, Royer C, Kirn A, Aubertin AM. Morphine stimulates HIV replication in primary cultures of human Kupffer cells. Res Virol 1991; 142:189–95.
46. Schmitt MP, Gendrault JL, Schweitzer C, Steffan AM, Beyer C, Royer C, Jaeck D, Pasquali JL, Kirn A, Aubertin AM. Permissivity of primary cultures of human Kupffer cells for HIV-1. AIDS Res Hum Retroviruses 1990; 6:987–91.
47. Cao YZ, Dietrich D, Thomas PA, Huang YX, Mirabile M, Ho DD. Identification and quantitation of HIV-1 in the liver of patients with AIDS. AIDS 1992; 6:65–70.
48. Salahuddin SZ, Rose RM, Groopman JE, Markham PD, Gallo RC. Human lymphotropic virus type III infection of human alveolar macrophages. Blood 1987; 68:281–4.
49. Plata F, Autran B, Martins LP, Wain-Hobson S, Raphael M, Mayaud C, Denis M, Guillon JM, Debre P. AIDS virus–specific cytotoxic T lymphocytes in lung disorder. Nature 1987; 328:348–51.
50. Armstrong JA, Horne R. Follicular dendritic cells and virus-like particles in AIDS-related lymphadenopathy. Lancet 1984; 2:370–2.
51. Kaneshima H, Shih CC, Namikawa R, Rabin L, Outzen H, Machado SG, McCune JM. Human immunodeficiency virus infection of human lymph nodes in the SCID-hu mouse. Proc Natl Acad Aci USA 1991; 88:4523–7.
52. Collman R, Hassan NF, Walker R, Godfrey B, Cutilli J, Hastings JC, Friedman H, Douglas SD, Nathanson N. Infection of monocyte-derived macrophages with human immunodeficiency virus type 1 (HIV-1). J Exp Med 1989; 170:1149–63.
53. Novak RM, Holzer TJ, Kennedy MM, Heynen CA, Dawson G. The effect of interleukin-I (BSF-1) on infection of peripheral blood monocyte-derived macrophages with HIV-1. AIDS Res Hum Retroviruses 1990; 6:973–6.
54. Hufert FT, v. Laer D, Schramm C, Tarnok A, Schmitz H. Detection of HIV-1 DNA in different subsets of human peripheral blood mononuclear cells using the polymerase chain reaction. Arch Virol 1989; 106:341–5.
55. Burstein Y, Rashbaum WK, Hatch WC, Calvelli T, Golodner M, Soeiro R, Lyman WD. Alterations in human fetal hematopoiesis are associated with maternal HIV infection. Ped Res 1992; 32:155–9.
56. Folks TM, Kessler SW, Orenstein JM, Justement JS, Jaffe ES, Fauci AS. Infection and replication of HIV-1 in purified progenitor cells of normal human bone marrow. Science 1988; 242:919–22.
57. Kojouharoff G, Ottmann OG, von Briesen H, Geissler G, Rubsamen Waigmann H, Hoelzer D, Ganser A. Infection of granulocyte/monocyte progenitor cells with HIV1. Res Virol 1991; 142:151–7.
58. Rich EA, Chen ISY, Zack JA, Leonard ML, O'Brien WA. Increased susceptibility of differentiated mononuclear phagocytes to productive infection with human immunodeficiency virus-1 (HIV-1). J Clin Invest 1992; 89:176–83.
59. Kalter DC, Nakamura M, Turpin JA, Baca LM, Hoover DL, Dieffenbach C, Ralph P, Gendelman HE, Meltzer MS. Enhanced HIV replication in macrophage colony-stimulating factor–treated monocytes. J Immunol 1991; 146:298–306.

60. Olafsson K, Smith MS, Marshburn P, Carter SG, Haskill S. Variation of HIV infectibility of macrophages as a fuction of donor, stage of differentiation, and site of origin. J AIDS 1991; 4:154–64.
61. Valentin A, Gegerfelt AV, Matsuda S, Nilsson K, Asjo B. In vitro maturation of mononuclear phagocytes and susceptibility to HIV-1 infection. J AIDS 1991; 4:751–9.
62. Pauza CD, Galindo J, Richman DD. Human immunodeficiency virus infection of monoblastoid cells: cellular differentiation determines pattern of virus replication. J Virol 1988; 62:3558–64.
63. Kazazi F, Mathijs JM, Foley P, Cunninham AL. Variations in CD4 expression by human monocytes and macrophages and their relationship to infection with the human immunodeficiency virus. J Gen Virol 1989; 70:2661–72.
64. Valentin A, Matsuda S, Asjo B. Characterization of the in vitro maturation of monocytes and the susceptibility to HIV infection. AIDS Res Hum Retroviruses 1990; 6:977–8.
65. Griffin GE, Leung K, Folks TM, Kunkel S, Nabel GL. Activation of HIV gene expression during monocyte differentiation by induction of NF-κB. Nature 1989; 339:70–3.
66. Kawakami K, Scheiderweit C, Roeder RG. Identification and purification of a human immunoglobulin-enhancer-binding protein (NFκB) that activates transcription from a human immunodeficiency virus type 1 promoter in vitro. Proc Natl Acad Sci USA 1988; 85:4700–4.
67. Neuveut C, Suzan M, Querat G, Sire J. HIV-1 infection of human monocytes and macrophages promotes induction or translocation of NF-κB–related factors. Res Virol 1991; 142:227–31.
68. Plata F, Garcia-Pons F, Ryter A, Lebargy F, Goodenow MM, Quan Dat MH, Autran B, Mayaud C. HIV-1 infection of lung alveolar fibroblasts and macrophages in humans. AIDS Res Hum Retroviruses 1990; 6:979–86.
69. Cox RA, Anders GT, Cappelli PJ, Johnson JE, Blanton HM, Seaworth BJ, Treasure RL. Production of tumor necrosis factor-alpha and interleukin-1 by alveolar macrophages from HIV-1–infected persons. AIDS Res Hum Retroviruses 1990; 6:431–41.
70. Osborn L, Kunkel S, Nabel GJ. Tumor necrosis factor alpha and interleukin 1 stimulate the human immunodeficiency virus enhancer by activation of the nuclear factor kappa β. Proc Natl Acad Sci USA 1989; 86:2336–40.
71. Vazeux R, Brousse N, Jarry A, Henin D, Marche C, Vendrenne C, Mikol J, Wolff M, Michon C, Rozenbaum W, Bureau JF, Montagnier L, Brahic M. AIDS subacute encephalitis. Identification of HIV-infected cells. Am J Pathol 1987; 126:403–10.
72. Stoler MH, Eskin TA, Benn S, Angerer RC, Angerer LM. Human T-cell lymphotropic virus type-III infection of the central nervous system. JAMA 1986; 256:2360–4.
73. Wiley CA, Schrier RD, Melson GA, Lampert PW, Oldstone MBA. Cellular localization of human immunodeficiency virus infection within the brains of acquired immune deficiency syndrome (AIDS) patients. Proc Natl Acad Sci USA 1986; 83:7089–93.
74. Erfle V, Stoeckbauer P, Kleinschmidt A, Kohleisen B, Mellert W, Stavrou D, Brack-Werner R. Target cells for HIV in the central nervous system: macrophages or glial cells? Res Virol 1991; 142:139–44.
75. Peudenier S, Hery C, Heng Ng K, Tardieu M. HIV receptors within the brain: a study of CD4 and MHC-II on human neurons, astrocytes and microglial cells. Res Virol 1991; 142:145–9.
76. Hassan NF, Campbell DE, Rifat S, Douglas SD. Isolation and characterization of human fetal brain derived microglia *in vitro* culture. Neuroscience 1991; 41:149–58.
77. Gerdes JS, Douglas SD, Kolski GB, Yoder MC, Polin RA. Decreased fibronectin biosynthesis by human cord blood mononuclear phagocytes in vitro. J Leukocyte Biol 1984; 35:91–9.
78. Takahashi K, Naito M, Katabuchi H, Higashi K. Development, differentiation, and maturation of macrophages in the chorionic villi of mouse placenta with special reference to the original of Hofbauer cells. J Leukocyte Biol 1991; 50:57–68.
79. Mano H, Chermann JC. Replication of human immunodeficiency virus type 1 in primary cultured placental cells. Res Virol 1991; 142:95–104.
80. Valente P, Main EK. Role of the placenta in perinatal transmission of HIV. Obstet Gynecol Clin North Am 1990; 17:607–16.
81. Lewis SH, Reynolds-Kohler C, Fox HE, Nelson JA. HIV-1 in trophoblastic and villous Hofbauer cells, and haematological precursors in eight-week fetuses. Lancet 1990; 335:565–8.
82. Amirhessami-Aghili N, Spector SA. Human immunodeficiency virus type 1 infection of human placenta: potential route for fetal infection. J Virol 1991; 65:2231–6.

83. Sedmak DD, Davis DH, Singh U, van de Winkel JGJ, Anderson CL. Expression of IgG FC receptor antigens in placenta and on endothelial cells in humans. Am J Pathol 1991; 138:175–80.
84. Gendelman HE, Baca LM, Husayni H, Turpin JA, Skillman D, Kalter DC, Orestein JM, Hoover DL, Meltzer MS. Macrophage-HIV interaction: viral isolation and target cell tropism. J AIDS 1990; 4:221–8.
85. Valentin A, Albert J, Fenyo EM, Asjo B. HIV-1 infection of normal human macrophage cultures: implication for silent infection. Virology 1990; 177:790–4.
86. Schuitemaker H, Kootstra NA, de Goede REY, de Wolf F, Miedema F, Tesmette M. Monocytotropic human immunodeficiency virus type 1 (HIV-1) variants detectable in all stages of HIV-1 infection lack T-cell line tropism and syncytium-inducing ability in primary T-cell culture. J Virol 1991; 65:356–63.
87. Massari FE, Poli G, Schnittman SM, Psallidopoulos MC, Davey V, Fauci AS. In vivo T lymphocyte origin of macrophage-tropic strains of HIV: role of monocytes during in vitro isolation and in vivo infection. J Immunol 1990; 144:4628–32.
88. Gendelman HE, Baca LM, Turpin J, Kalter DC, Hansen B, Orenstein JM, Dieffenbach CW, Freidman RM, Meltzer MS. Regulation of HIV replication in infected monocytes by IFN-α: mechanisms for viral restriction. J Immunol 1990; 145:2669–76.
89. O'Brien WA, Koyanagi Y, Namazie A, Zhao JQ, Diagne A, Idler K, Zack JA, Chen ISY. HIV-1 tropism for mononuclear phagocytes can be determined by regions of gp120 outside the CD4-binding domain. Nature 1990; 348:69–72.
90. Hwang SS, Boyle TJ, Lyerly HK, Cullen BR. Identification of the envelope V3 loop as the primary determinant of cell tropism in HIV-1. Science 1991; 253:71–4.
91. Freed EO, Myers DJ, Risser R. Identification of the principal neutralizing determinant of human immunodeficiency virus type 1 as a fusion domain. J Virol 1991; 65:190–4.
92. Hwang SS, Boyle TJ, Lyerly HK, Cullen BR. Identification of envelope V3 loop as the major determinant of CD4 neutralization sensitivity of HIV-1. Science 1992; 257:535–7.
93. Britton CB, Miller JR. Neurological complications in acquired immunodeficiency syndrome (AIDS). Neurol Clin 1984; 2:315–39.
94. Neilson SL, Petito CK, Urmacher CD, Posner JB. Subacute encephalitis in acquired immune deficiency syndrome: a postmortem study. Am J Clin Pathol 1984; 82:678–82.
95. Faulstich ME. Acquired immune deficiency syndrome: an overview of central nervous system complications and neuropsychological sequelae. Int J Neurosci 1986; 30:249–54.
96. Gabuza DH, Ho DD, de la Monte SM, Hirsch MS, Rota TR, Sobel RA. Immunohistochemical identification of HTLV-III antigen in brains of patients with AIDS. Ann Neurol 1986; 20:289–95.
97. Pulliam L, Herndier BG, Tang NM, McGrath MS. Human immunodeficiency virus–infected macrophages produce soluble factors that cause histological neurochemical alterations in cultured human brains. J Clin Invest 1991; 87:503–12.
98. Giulian D, Vaca K, Noonan CA. Secretion of neurotoxins by mononuclear phagocytes infected with HIV-1. Science 1990; 250:1593–6.
99. Merrill JE. Cytokines and retroviruses. Clin Immunol Immunopathol 1992; 64:23–7.
100. Giulian D, Lachman LB. Interleukin-1 stimulation of astroglial proliferation after brain injury. Science 1985; 228:497–9.
101. Navia BA, Cho ES, Petitio CK, Price RW. The AIDS dementia complex: II. Neuropathology. Ann Neurol 1986; 19:525–35.
102. Rutka JT, Giblin JRT, Berens ME, Bar-shiva E, Tokuda K, McCulloch JR, Rosenblum ML, Bessalu TE, Aggarwal BB, Bodell WJ. The effects of human recombinant tumor necrosis factor on glioma-derived cell lines: cellular proliferation, cytotoxicity, morphological and radioreceptor studies. Int J Cancer 1988; 41:573–82.
103. Vazeux R. AIDS encephalopathy and tropism of HIV for brain monocytes/macrophages and microglial cells. Pathobiology 1991; 59:214–8.
104. Navia BA, Jordan BD, Price RW. The AIDS complex dementia. I. Clinical features. Ann Neurol 1986; 19:517–24.
105. McArthur JC, Becker PS, Parisi JE, Trapp B, Selnes OA, Cornbluth DR, Balakrishnan J, Griffin JW, Price D. Neuropathological changes in early HIV-1 dementia. Ann Neurol 1989; 26:681–4.
106. Selmaj KW, Raine CS. Tumor necrosis factor mediates myelin and oligodendrocyte damage *in vitro*. Ann Neurol 1988; 23:339–46.

107. Mastroianni CM, Paoletti F, Valenti C, Vullo V, Jirillo E, Delia S. Tumour necrosis factor (TNF-α) and neurological disorder in HIV infection. J Neurol Neurosurg Psych 1992; 55:219–21.
108. Beutler B, Cerami A. Cachectin: more than a tumor necrosis factor. N Engl J Med 1987; 316:379–85.
109. Dinarello CA. Interleukin-1 and its biologically related cytokines. Adv Immunol 1989; 44:153–84.
110. Brenneman DE, Westbrook GL, Fitzgerald SP, Ennist DL, Elkins KL, Ruff DL, Pert CB. Neuronal cell killing by the envelope protein of HIV and its prevention by vasoactive intestinal peptide. Nature 1988; 335:639–42.
111. Wahl LM, Corcoran ML, Pyle SW, Arthur LO, Harel-Bellan A, Farrar WL. Human immunodeficiency virus glycoprotein (gp120) induction of monocyte arachidonic acid metabolites and interleukin 1. Proc Natl Acad Sci USA 1989; 86:621–5.
112. Gurney ME, Heinrich SP, Lee MR, Yin HS. Molecular cloning and expression of neuroleukin, a neurotropic factor for spinal and sensory neurons. Science 1986; 23:566–74.
113. Dreyer EB, Kaiser PK, Offermann JT, Lipton SA. HIV-1 coat protein neurotoxicity prevented by calcium channel antagonists. Science 1990; 248:364–7.
114. Lee M, Ho DD, Gurney ME. Functional interaction and partial homology between human immunodeficiency virus and neuroleukin. Science 1986; 237:9254–8.
115. Tardieu M, Hery C, Peudenier S, Boespflug O, Montagnier L. Human immunodeficiency virus type 1–infected monocytic cells can destroy human neural cells after cell-to-cell adhesion. Ann Neurol 1992; 32:11–7.
116. Shaw GM, Harper ME, Hahn BH, Epstein LG, Gajdusek DC, Price RW, Navia BA, Petito CK, O'Hara CJ, Groopman JE, Cho ES, Leske JM, Wong-Staal F, Gallo RC. HTLV-III infection in brains of children and adults with AIDS encephalopathy. Science 1985; 227:177–80.
117. Levy JA, Shimabukuro J. Hollander H, Mills J, Kaminsky L. Isolation of AIDS-associated retroviruses from cerebrospinal fluid and brain of patients with neurological symptoms. Lancet 1985; 2:586–8.
118. Ho DD, Tota TR, Schooley RT, Kaplan JC, Allan JD, Groopman JE, Resnick L, Felsenstein D. Isolation of HTLV-III from cerebrospinal fluid and neural tissues of patients with neurologic syndromes related to the acquired immunodeficiency syndrome. N Engl J Med 1985; 313:1493–7.
119. Sei Y, Petrella RJ, Tsang P, Bekesi JG, Yokoyama MM. Monocytes in AIDS (letter). N Engl J Med 1986; 315:1611–2.
120. De la Monte SM, Ho DD, Schooley RT, Hirsch MS, Richardson EP, Jr. Subactute encephalomyelitis of AIDS and its relation to HTLV-III infection. Neurology 1987; 37:562–9.
121. Gendelman HE, Orenstein JM, Baca LM, Weiser B, Burger H, Kalter DC, Meltzer MS. The macrophage in the persistence and pathogenesis of HIV infection. AIDS 1989; 3:475–95.
122. Ostrow DG, Joseph J, Monjan A, Kessler R, Emmons C, Phair J, Fox R, Kingsley L, Dudley J, Chmiel JS. Psychosocial aspects of AIDS risk. Psychopharmacol Bull 1986; 22:678–83.
123. Perry SW, Tross S. Psychiatric problems in AIDS inpatients at the New York Hospital: preliminary report. Public Health Rep 1984; 99:200–5.
124. Kiecolt-Glaser JK, Glaser R. Major life changes, chronic stress, and immunity. In: Bridge TP, Mirsky AF, Goodwin FK, eds. Psychological, neuropsychiatric, and substance abuse aspects of AIDS. New York: Raven Press, 1988:217–24.
125. Janowsky DS, Golden RN, Rapaport M, Cain JJ, Gallin JC. Neurochemistry of depression and mania. In: Georgatas A, Cancro R, eds. Depression and mania. New York: Elsevier, 1988:244–64.
126. Solomon GF, Kemeny ME, Temoshok L. Psychoneuroimmunologic aspects of human immunodeficiency virus infection. In Psychoneuroimmunology. 2nd ed., Ader R, Felton DL, Cohen N., eds., p. 1081–1106, Academic Press, San Diego, 1991.
127. Edwards CK III, Ghiasuddin SM, Schepper JM, Yunger LM, Kelley KW. A newly defined property of somatotropin: priming of macrophages for production of superoxide anion. Science 1988; 239:769–71.
128. Lemaire I. Neurotensin enhances IL-1 production by activated alveolar macrophages. J Immunol 1988; 140:2983–8.
129. Peterson PK, Sharp B, Gekker G, Brummitt C, Keane WF. Opioid-mediated suppression of interferon gamma production by cultured peripheral blood mononuclear cells. J Clin Invest 1987; 80:824–31.
130. Ruff MR, Wahl SM, Pert CB. Substance P receptor–mediated chemotaxis of human monocytes. Peptides 1985; 6:107–11.

131. Koff WC, Dunegan MA. Modulation of macrophage-mediated tumoricidal activity by neuropeptides and neurohormones. J Immunol 1985; 135:350–4.
132. Hartung HP. Activation of macrophages by neuropeptides. Brain Behav Immun 1988; 2:275–81.
133. Peck R. Neuropeptides modulating macrophage function. Ann NY Acad Sci 1987; 496:264–70.
134. Lotz M, Vaughan JH, Carson DA. Effect of neuropeptides on production of inflammatory cytokines by human monocytes. Science 1988; 241:1218–21.
135. Hassan NF, Douglas SD. Stress-related neuroimmunomodulation of monocyte-macrophage function in HIV-1 infection. Clin Immunol Immunopathol 1990; 54:220–7.
136. Latham PS, Lewis AM, Varesio L, Pavlakis GN, Felber BK, Ruscetti FW, Young HA. Expression of human immunodeficiency virus long terminal repeat in the human promonocytic cell line U937: effect of endotoxin and cytokines. Cell Immunol 1990; 129:513–8.
137. Tsunetsugu-Yokata Y, Honda M. Effect of cytokines on HIV release and IL-2 receptor α expression in monocytic cell lines. J AIDS 1990; 3:511–6.
138. Novak RM, Holzer TJ, Kennedy MM, Heynen CA, Dawson G. The effect of interleukin 4 (BSF-1) on infection of peripheral blood monocyte-derived macrophages with HIV-1. AIDS Res Hum Retroviruses 1990; 6:973–6.
139. Folks TM, Justement JS, Kinter A, Dinarello CA, Fauci AS. Cytokine-induced expression of HIV-1 in a chronically infected promonocyte cell line. Science 1987; 238:800–2.
140. Poli GT, Bressler P, Kinter A, Duhe, Timmer WC, Rabson A, Justement JS, Stanley S, Fauci AS. Interleukin-6 induces human immunodeficiency virus expression in infected monocytic cells alone and in synergy with tumor necrosis factor-α by transcriptional and post-transcriptional mechanisms. J Exp Med 1990; 172:151–158.
141. Koyanagi Y, O'Brien WA, Zhao JQ, Golde DW, Gasson JC, Chen ISY. Cytokines alter production of HIV-1 from primary mononuclear phagocytes. Science 1988; 241:1673–5.
142. Matsuyan T, Kobayashi N, Yamamoto N. Cytokines and HIV infection: is AIDS a tumor necrosis factor disease? AIDS 1991; 5:1405–1417.
143. Nakajima K, Martinez-Maza O, Hirano T, Breen EC, Nishanian PG, Salazar-Gonzales JF, Fahey JL, Kishimoto T. Induction of IL-6 (B-cell stimulatory factor-2/IFN-beta 2) production by HIV. J Immunol 1989; 142:531–6.
144. Merrill JE, Koyanagi Y, Chen ISY. Interleukin-1 and tumor necrosis factor alpha can be induced from mononuclear phagocytes by human immunodeficiency virus type 1 binding to the CD4 receptor. J Virol 1989; 63;4404–8.
145. Poli G, Kinter AL, Justement JS, Bressler P, Kehrl JH, Fauci AS. TGF-β suppresses HIV expression and replication in infected cells of the monocyte/macrophage lineage. J Exp Med 1991; 173:589–97.
146. Roux-Lombard P, Modoux C, Cruchaud A, Dayer JM. Purified blood monocytes from HIV-infected patients produce high levels of TNF-α and IL-1. Clin Immunol Immunopathol 1989; 50:374–84.
147. Merril JE, Chen IS. HIV-1, macrophages, glial cells, and cytokines in AIDS nervous system disease. FASEB 1991; 5:2391–7.
148. Jones KA, Kadonaga JT, Luciw PA, Tjian R. Activation of the AIDS retrovirus promoter by the cellular transcription factor, Sp1. Science 1986; 232:755–7.
149. Nabel G, Baltimore D. An inducible transcription factor activates expression of human immunodeficiency virus in T cells. Nature 1987; 326:711–3.
150. Siekevitz M, Josephs SF, Dukovich M, Peffer N, Wong-Staal F, Greene WC. Activation of the HIV-1 LTR by T cell mitogens and the transactivator protein of HTLV-1. Science 1987; 238:1575–8.
151. Griffin GE, Leung K, Folks TM, Kunkel S, Nabel GJ. Induction of NF-κB during monocyte differentiation is associated with activation of HIV-gene expression. Res Virol 1991; 142:233–8.
152. Kornbluth RS, Oh PS, Munis JR, Cleveland PH, Richman DD. The role of interferons in the control of HIV replication in macrophages. Clin Immunol Immunopathol 1990; 54:200–19.
153. Poli G, Orenstein JM, Kinter A, Folks TM, Fauci AS. Interferon-α but not AZT suppresses HIV expression in chronically infected cell lines. Science 1989; 244:575–7.
154. Mace K, Gazzolo L. Interferon-regulated viral replication in chronically HIV-1–infected promonocytic U937 cells. Res Virol 1991; 142:213–20.
155. Pitha PM, Wivel NA, Fernie BF, Harper HP. Effect of interferon on murine leukemia virus infection. IV. Formation of non-infectious virus in chronically infected cells. J Gen Virol 1979; 42:467–80.

156. Rokos K, Pauli G. Interferon induction by cell-cell interaction of HIV-infected monocytes/macrophages with lymphocytes. Res Virol 1991; 142:221–5.
157. Heagy W, Kelley VE, Strom TB, Mayer K, Shapiro HM, Mandel R, Finberg R. Decreased expression of human class II antigens on monocytes from patients with acquired immunodeficiency syndrome: increased expression with interferon-γ. J Clin Invest 1984; 74:2089–96.
158. Smith PD, Ohura K, Masur H, Lane HC, Fauci AS, Wahl SM. Monocyte function in the acquired immune deficiency syndrome. J Clin Invest 1984; 74:2121–8.
159. Eales LJ, Moshtael O, Pinching AJ. Microbicidal activity of monocyte derived macrophages in AIDS and related disorders. Clin Exp Immunol 1987; 67:227–35.
160. Estevez ME, Ballart IJ, Diez RA, Planes N, Scaglione C, Sen L. Early defect of phagocytic cell function in subjects at risk for acquired immunodeficiency syndrome. Scand J Immunol 1986; 24:215–21.
161. Muller F, Rollag H, Froland SS. Reduced oxidative burst responses in monocytes and monocyte-derived macrophages from HIV-infected subjects. Clin Exp Immunol 1990; 82:10–5.
162. Spear GT, Kessler HA, Rothberg L, Phair J, Landay AL. Decreased oxidative burst activity of monocytes from asymptomatic HIV-infected individuals. Clin Immunol Immunopathol 1990; 54:184–91.
163. Muller F, Rollag H, Froland SS. Effects of alpha- and beta-interferon on oxidative burst responses of monocytes and monocyte-derived macrophages from patients with HIV infection. J Biol Regul Agents 1990; 4:93–7.
164. Gendelman HE, Freidman RM, Joe S, Baca LM, Turpin JA, Dveksler G, Meltzer MS, Dieffenbach C. A selective defect of IFN-α production in HIV-infected monocytes. J Exp Med 1990; 172:1433–42.
165. Pauza CD, Galindo J. Persistent human immunodeficiency virus type 1 infection of monoblastoid cells lead to accumulation of self-integrated viral DNA and to production of defective virions. J Virol 1989; 63:3700–7.
166. Poli G, Botazzi B, Acero R, Bersoni L, Rossi V, Introna M, Lazzarin A, Galli M, Mantovani A. Monocyte function in intravenous drug users with lymphadenopathy syndrome and in patients with acquired immunodeficiency syndrome: selective impairment of chemotaxis. Clin Exp Immunol 1985; 62:136–42.
167. Macher AM, Kovacs JA, Gill V, Roberts GD, Ames J, Park CH, Straus S, Lane HC, Parrillo JE, Fauci AS, Masur H. Bacteremia due to *Mycobacterium avium-intracellulare* in the acquired immunodeficiency syndrome. Ann Intern Med 1983; 99:782–5.
168. Twigg HL, Iwamoto GK, Soliman DM. Role of cytokines in alveolar macrophage accessory cell function in HIV-infected individuals. J Immunol 1992; 149:1462–9.
169. Dhawan S, Vargo M, Meltzer MS. Interactions between HIV-infected monocytes and the extracellular matrix: increased capacity of HIV-infected monocytes to adhere to and spread on extracellular matrix associated with changes in extent of virus replication and cytopathic effects in effected cells. J Leukocyte Biol 1992; 52:62–9.
170. Locksley RM, Crowe S, Sadick MD, Heinzel FP, Gardner KD, McGrath MS, Mills J. Release of interleukin 1 inhibitory activity (contra-IL-1) by human monocyte-derived macrophages infected with human immunodeficiency virus *in vitro* and *in vivo*. J Clin Invest 1988; 82:2097–105.
171. Wahl SM, Allen JB, Gartner S, Orenstein JM, Popovic M, Chenoweth DE, Arthur LO, Farrar WL, Wahl LM. HIV-1 and its envelope glycoprotein down-regulate chemotactic ligand receptors and chemotactic function of peripheral blood monocytes. J Immunol 1989; 142:3553–9.
172. Schrier RD, McCutchan JA, Venable JC, Nelson JA, Wiley CA. T-cell–induced expression of human immunodeficiency virus in macrophages. J Virol 1990; 64:3280–8.
173. Rieckmann P, Poli G, Kehrl JH, Fauci AS. Activated B lymphocytes from human immunodeficiency virus–infected individuals induce virus expression in infected T cells and a promonocytic cell line, U1. J Exp Med 1991; 173:1–5.
174. Melendez-Guerro LM, Nicholson JKA, McDougal JS. In vitro infection of monocytes with HIVbal. Effect on cell surface expression of CD4, CD14, HLA-DR, and HLA-DQ. AIDS Res Hum Retroviruses 1990; 6:731–41.
175. Haas JS. Riethmuller G, Ziegler-Heitbrock HWL. Monocyte phenotype and function in patients with the acquired immunodeficiency syndrome and AIDS-related disorders. Scand J Immunol 1987; 26:371–9.

176. Davidson BL, Kline RL, Rowland J, Quinn TC. Surface markers of monocyte function and activation in AIDS. J Infect Dis 1988; 185:483–6.
177. Folks TM, Clouse KA, Justement J, Rabson A, Duh E, Kehrl JH, Fauci AS. Tumor necrosis factor alpha expression of human immunodeficiency virus in a chronically infected T cell clone. Proc Natl Acad Sci USA 1989; 86:2365–8.
178. Von Briesen H, von Mallinckrodt C, Esser R, Muller S, Becker K, Rubsamen-Waigmann H, Andreesen R. Effect of cytokines and lipopolysaccharides on HIV infection of human macrophages. Res Virol 1991; 142:197–204.
179. Lazdins JK, Klimkait T, Alteri M, Walker M, Woods-Cook K, Cox D, Bilbe G, Shipman R, Cerletti N, McMaster G. TGF-β: upregulator of HIV replication in macrophages. Res Virol 1991; 142:239–42.
180. Hammer SM, Gillis JM, Pinkston P, Rose RM. Effect of zidovudine and granulocyte-macrophage colony-stimulating factor on human immunodeficiency virus replication in alveolar macrophages. Blood 1990; 75:1215–9.
181. Bernstein MS, Tong-Starksen SE, Locksley RM. Activation of human monocyte-derived macrophages with lipopolysaccharide decreases human immunodeficiency virus replication in vitro at the level of gene expression. J Clin Invest 1991; 88:540–5.
182. Latham PS, Lewis AM, Varesio L, Pavlakis GN, Felber BK, Ruscetti FW, Young HA. Expression of human immunodeficiency virus long terminal repeat in the human promonocyte cell line U937: effect of endotoxin and cytokines. Cell Immunol 1990; 129:513–8.
183. Potts BJ, Maury W, Martin MA. Replication of HIV-1 in primary monocyte cultures. Virology 1990; 175:465–76.
184. Gendelman HE, Orenstein JM, Martin MA, Ferrua C, Mitra R, Phipps T, Wahl LA, Lane CH, Fauci AS, Burke DS, Skillman D, Meltzer MS. Efficient isolation and propagation of human immunodeficiency virus on recombinant colony-stimulating factor 1–treated monocytes. J Exp Med 1988; 167:1428–41.
185. Watkins BA, Dorn HH, Kelly WB, Armstrong RC, Potts BJ, Michaels F, Kufta CV, Dubois-Dalcq M. Specific tropism of HIV-1 for microglial cells in primary human brain cultures. Science 1990; 249:549–53.
186. Skinner MA, Langlois AJ, McDanal CB, McDougal JS, Bolognesi DP, Matthews TJ. Neutralizing antibodies to an immunodominant envelope sequence do not prevent gp120 binding to CD4. J Virol 1988; 62:4195–200.
187. Francis ML, Meltzer MS, Gendelman HE. Interferons in the persistence, pathogenesis, and treatment of HIV infection. AIDS Res Hum Retroviruses 1992; 8:199–207.

36

Lentivirus Infection of Macrophages

Janice E. Clements, M. Christine Zink, Opendra Narayan

The Johns Hopkins University School of Medicine, Baltimore, Maryland

Dana H. Gabuzda

Dana Farber Cancer Institute, Boston, Massachusetts

I. INTRODUCTION

Lentiviruses are a family of nononcogenic retroviruses that cause slowly progressing chronic diseases with unusually long periods between infection of the host and the manifestation of clinical disease [1]. A lentivirus of sheep, visna virus, was the prototype for the description of "slow viruses" by Sigurdsson in the 1950s [2,3]. Lentiviral diseases involve multiple organ systems of the host, and the major target cells for infection by these viruses are monocytes and macrophages [4–6]. Lentiviruses that cause immunodeficiency disease infect lymphocytes in addition to macrophages. Lentiviruses have been isolated from sheep (maedi-visna virus or ovine progressive pneumonia virus [OPPV]), goats (caprine arthritis encephalitis virus [CAEV]), horses (equine infectious anemia virus [EIAV]), cats (feline immunodeficiency virus [FIV]), nonhuman primates (simian immunodeficiency virus [SIV]), and humans (human immunodeficiency virus [HIV]) [7]. The recognition that acquired immunodeficiency syndrome (AIDS) is caused by a lentivirus (HIV) has focused a great deal of scientific interest on the lentiviruses.

Lentiviruses, like the other retroviruses, are RNA viruses that encode and carry in the virion an RNA-dependent DNA polymerase or reverse transcriptase. When the virus infects a cell, this enzyme transcribes the genomic RNA into a double-stranded DNA copy that integrates into the host-cell DNA. Thus, the viral genome becomes a part of the cellular DNA and is duplicated during cell division. Retroviruses (with the exception of lentiviruses) productively infect only actively dividing cells, presumably because this cellular environment provides optimal conditions for synthesis and integration of viral DNA. In contrast, lentiviruses replicate efficiently in nondividing, terminally differentiated cells both in vivo and in cell culture. Rather than depend on cell division, the lentiviruses require activation and/or differentiation of the host cell for productive replication [8–10].

Lentivirus gene expression is under the control of viral sequences in the long terminal repeats (LTRs) (Figure 1) [11,12]. Enhancer and promoter sequences are located in the U3 region of the LTR, and viral RNA transcription is accomplished by the cellular RNA polymerase II. Cellular transcription factors regulate viral RNA transcription, and this cellular regulation is dependent on the particular cell the virus replicates in and the activation and/or differentiation state of the cell [13,14]. When the cellular factors are present viral mRNA is transcribed and viral genes are

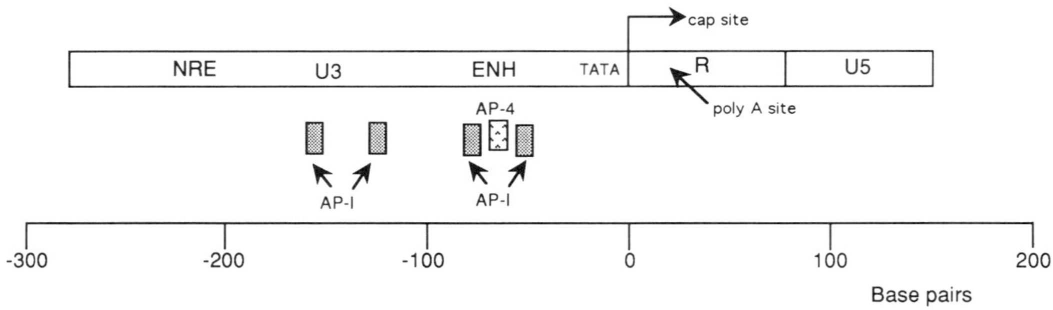

FIGURE 1 The visna virus long terminal repeat. The U3, R, and U5 regions are indicated. The core enhancer consists of the AP-1 and AP-4 sites most proximal to the TATA box. The cap site and polyadenylation signal are indicated.

synthesized. Lentiviruses encode structural and enzymatic proteins that constitute the viral particle [15]. In addition, lentiviruses encode regulatory proteins [16]. One of these proteins, the transactivating or *tat* gene, increases gene expression from the virus. A second protein, the *rev* gene, regulates the level of unspliced and partially spliced viral mRNAs in the cytoplasm of the infected cell. Synthesis of both structural and regulatory proteins is required for productive replication of virus. This type of replication may progress at a steady low rate and cause no cytopathicity or at a high rate with attendant cytopathic effects characterized by multinucleated giant cell formation.

Lentiviruses are transmitted by infected cells and free virus in body fluids; the sheep and goat lentiviruses are disseminated through colostrum, milk, and respiratory exudates, and those of humans are transmitted by blood, sexual contact, and probably also by breast milk [17–19]. The lentiviruses of animals and humans generally cause multiorgan disease. The pathologic manifestations of lentiviral disease can be broken down into two broad classifications. There is primary disease related to viral replication in macrophages, including nonsuppurative giant cell encephalitis seen frequently in HIV-infected humans and SIV-infected primates [20,21], the macrophage- and lymphocyte-rich encephalitis seen in sheep and young goats [5,22], the lymphocytic interstitial pneumonia seen in people with HIV [23] and lentivirus-infected sheep and goats [5,24], and the giant cell pneumonia seen in SIV-infected macaques [20]. In addition, species infected with lentivirus that replicate in CD4+ lymphocytes, such as HIV, SIV, and FIV, frequently develop secondary opportunistic infections associated with the immunosuppression that results from the loss of CD4 lymphocytes [20,25,26].

The lentiviruses of sheep, goats, horses, and primates infect cells of the monocyte/macrophage lineage in vivo, and the pathogenesis of the disease process can be correlated with infection in these cells [4,5,27,28]. There is a complex interaction between the lentivirus and the macrophage. Cellular factors are required for viral gene expression, and in turn viral genes affect the expression of cellular genes. In the bone marrow of lentivirus-infected animals, monocytes and monocyte precursors contain integrated provirus. However, viral gene expression is activated only when the monocytes mature into macrophages. Studies on a monocytic cell line as well as primary monocytes and macrophages have shown that specific cellular factors are required for the activation of the viral promoter [13,14]. These factors are only present when the monocytes mature into macrophages (see section III below). Once viral gene expression is activated, viral gene products affect expression of the macrophage-specific markers that play a direct role in the lymphoproliferative reaction elicited in infected organs and tissues. This is a review of the current understanding of the role of the macrophage in the pathogenesis of lentiviral disease, cellular tropism of lentiviruses, and the molecular events that occur during the infection of macrophages.

II. VIRUS–HOST INTERACTIONS

The ovine and caprine lentiviruses cause widespread infection in nature and persist throughout the life of the infected animal. The nature of these lentiviral diseases, with their long incubation period, gradual onset, and chronic progressive course, requires controlled but continuous virus replication in the host [1,7]. The ovine and caprine lentiviruses use a number of strategies to avoid the immune system of the host. Antigenic variation of the envelope glycoprotein of visna virus and the blocking of neutralizing antibodies by the heavily glycosylated envelope glycoprotein of CAEV are two mechanisms that facilitate a continuous replication of these viruses in immunocompetent hosts [29]. Another strategy is infection of cells of the immune system. Infection in the monocytes results in a latent infection in which no viral gene products are expressed in the cell [4,9,10,30]. This provides a reservoir of infected cells that are not subject to immune elimination. Virus gene expression is dependent on maturation of the monocyte into a macrophage [3,13,14,30]. Macrophage development leads to productive viral replication and the consequences in the disease process discussed below.

Chronic pneumonia is the most common pathologic manifestation of ovine lentiviral infections [24,31]. Animals with pneumonia usually become cathectic and frequently have mastitis and/or arthritis. Occasionally, infected animals also develop neurological disease characterized by progressive paralysis and wasting. In the case of goats, arthritis is the main pathologic manifestation of infection in adult animals [32]. In young goats encephalitis is a more common disease manifestation. In all cases, the lesions in affected organs are associated with the infiltration of macrophages, many of which are infected. In addition, there is a lymphoproliferative response.

Studies using in situ hybridization to detect viral RNA as well as immunocytochemistry to identify cell phenotype have indicated that many of the macrophages at the site of a pathologic lesion are infected and expressing lentiviral RNA transcripts [32,33]. Similar studies done on blood-derived monocytes indicate that few express viral genes. However, when these cells are placed in culture they mature into macrophages in 4 to 5 days and develop macrophage cell markers. At this time, viral gene expression can be detected in about 15% of the cells.

The reservoir of infected cells appears to be in the bone marrow [4]. Two types of study have been used to identify infected cells. First, in situ hybridization and combined immunocytochemistry on sections of bone marrow from infected animals showed foci of macrophage precursor cells expressing low levels of viral mRNA. Another approach has been to aspirate the bone marrow cells and separate them on the basis of density by countercurrent centrifugal elutriation. This study again indicated that the infected cells included monocyte precursors (monoblasts) and large precursor blast cells containing viral RNA by in situ hybridization [34].

Infection in the macrophage and activation of viral gene products alters the normal function of the macrophage [35]. Major histocompatibility complex (MHC) class II genes are up-regulated, and this results in the recruiting of lymphocytes to the site. Whether other changes occur on the surface of the macrophage remains to be established. The presence of infected macrophages and lymphocytes elicits from the lymphocyte the expression of a specific lentivirus-induced lymphokine called LV-IFN in the ovine/caprine system [36,37]. In sheep, LV-IFN causes further up-regulation of MHC class II and prevents maturation of the virus. Retardation of virus replication is mediated in part by a direct inhibitory effect on the viral promoter. One potential effect of LV-IFN in the infected host would be to maintain continual expression of MHC II antigen and viral antigen in the infected macrophage while virion assembly is inhibited. This sets the stage for a severe inflammatory reaction; this was shown to occur by infusion of LV-IFN into the brain of infected sheep via ALZET pumps, whereupon these treated animals developed severe encephalitis [38].

III. ROLE OF DIFFERENT LENTIVIRAL STRAINS IN INFECTION OF DIFFERENT POPULATIONS OF TISSUE MACROPHAGES

Ovine lentiviruses cause synovitis-arthritis, interstitial pneumonia, mastitis, and encephalitis. All of these syndromes are associated with lentivirus replication in macrophages in affected tissues. In the United States, arthritic and pneumonic diseases are the main clinical signs of ovine lentiviral infection, whereas central nervous system disease is rare. In order to determine whether arthrotropic virus could also cause infection and neuropathological changes in brain, we cultivated virus from macrophages aspirated from an arthritic joint of a sheep and inoculated four sheep intracerebrally. All four animals became persistently infected, as evidenced by virus recovery from peripheral blood monocytes, but none of the animals developed infection in the brain [39]. The virus was then passaged intracerebrally seven times in new sheep, and the seventh-passage virus was used for evaluation of neurovirulence. Inoculation of four animals intracerebrally with this agent caused infection in the brain and onset of inflammatory lesions in all four animals. Macrophages cultured from the brain tissue produced infectious virus. This study thus illustrated that neurotropic ovine lentivirus was not only macrophage-tropic but had a specific tropism for brain macrophages. Further, macrophage-tropic virus that was associated with lesions in nonneural tissues was not necessarily neurotropic.

Studies on neuropathogenesis of SIV infection of rhesus macaques confirmed that virus causing infection in brain had to be macrophage-tropic. Lymphocyte-tropic virus $SIV_{mac}239$ did not infect macrophages in the animal, and the virus failed to infect cells in the brain after intracerebral inoculation in two macaques [40]. Virus infection in these animals was confined to lymphoid tissues and this resulted in immunosuppression and onset of AIDS. Neuroadaptation of this virus was performed by inoculating infectious bone marrow cells intracerebrally into new macaques. This resulted in development of a virus that now replicated in the brain and caused neurological lesions. Furthermore, similar to the neurotropic ovine lentivirus, cultivation of macrophages from the brain of macaques with encephalitis resulted in production of infectious macrophage-tropic virus in supernatant fluids. Another finding, reminiscent of the ovine neurotropic viral properties, was that although macrophages from the brain of encephalitic macaques produced virus, splenic macrophages from the same animals failed to produce virus despite the finding that these cells had viral DNA. This reinforced the finding from infected sheep that neurotropic virus had a general tropism for macrophages but an apparently much higher affinity for the neural subset of tissue macrophages.

A major conclusion that can be drawn from the studies of viral neuroadaptation in both sheep and macaques is that the molecular basis for neurological disease may lie in those viral sequences that confer on the virus the ability not only to replicate in macrophages in general but to replicate in brain macrophages in particular. A virus strain with a specific tissue tropism is most probably selected in specific tissues from among the many viruses that evolve during lentivirus infection. The fact that neurological disease in sheep is relatively rare in nature suggests that these variants are rarely selected. In contrast, the high frequency of neurological disease in HIV infections suggests that this virus may be naturally neuroinvasive and/or that repliction of neurovirulent virus may be potentiated by immunosuppression, a condition that does not occur to any significant extent in ovine lentivirus–infected sheep.

IV. MOLECULAR MECHANISM OF LATENCY AND ACTIVATION OF LENTIVIRUSES IN MONOCYTES AND MACROPHAGES

Visna virus, as well as other lentiviruses, infects monocyte precursors and monocytes in vivo. Gene expression of lentiviruses is highly restricted in these cells, but transcriptional activation occurs when monocytes differentiate into macrophages [13,14]. Macrophage maturation induces transcriptional activation from the visna virus LTR. Recent studies have identified the sequences in the visna virus promoter (contained in the U3 region of the viral LTR) that are involved in the differential regulation of the virus in monocytes and macrophages. These studies have been done both in U937 cells, a human monocytic cell line that differentiates into macrophages after treatment with the phorbol ester 12-0-tetrad-canoyl phorbol 13-acetate (TPA), and in primary human monocytes and macrophages [13].

To examine the sequences responsible for transcriptional regulation, the visna virus LTR was placed upstream of the bacterial gene chloramphenicol acetyltransferase (CAT). Activity from the promoter could be measured by transfection of the plasmid DNA into cells and enzymatic assay of the CAT protein produced in the cell. The differential expression of the visna virus promoter could be reproduced in the U937 cells. Low levels of CAT activity were measured in the U937 monocytes, and the CAT activity was induced up to fourfold in the TPA-treated U937 macrophages. Quantitation of the CAT-specific mRNA indicated that the increase in the steady-state level of RNA accounted for the induced activity. This supported the idea that induction of gene expression was wholly at the transcriptional level [13].

To identify the precise sequences in the promoter that were involved in this transcriptional activation, 5' deletions were made in the U3 sequences and these deleted promoters were placed upstream of the CAT gene [14]. When these deletion mutants were transfected into U937 cells and treated with TPA, it was found that the basal level of transcription decreased substantially when sequences were deleted to –111 (Figure 2B). However, there was still a difference in the level of CAT activity when the cells were treated with TPA in the –111, –91, and –81 deletion mutants. When sequences were deleted to –67, then TPA inducible activity was lost. A very similar result was obtained when the deletion mutants were transfected into primary human monocytes and the cells treated with TPA (Figure 2A). Examination of the nucleotide sequences in the region of the viral promoter showed that the –67 mutant removed an AP-1 binding site and disrupted an AP-4 binding site [42–44].

Cellular factors that activate transcription by interacting with AP-1 sites include AP-1/JUN, FOS, and FOS-related antigens; these cellular factors are induced by TPA treatment of monocytic cells [45]. AP-1/JUN and FOS form a complex that activates transcription via AP-1 binding sites. JUN protein forms homodimers as well as heterodimers with the FOS protein, which associates with the AP-1 site; the heterodimers of JUN and FOS bind with a higher affinity than the homodimer to the AP-1 sequence [46,47]. AP-4 is another transcription factor induced by phorbol ester that activates transcription by acting in concert with AP-1.

To further investigate the activation of the visna virus LTR in U937 monocytes and macrophages, nuclear extracts from these cells were used to delineate the sequences recognized by cellular DNA binding proteins [48]. DNase I footprinting of the LTR sequences from –140 to +22 revealed that there were three regions recognized by cellular proteins. Two of the regions showed identical DNase I protection patterns using nuclear extracts from either monocytic U937 cells or TPA-induced U937 cells. One DNase I-protected region (from –87 to –81) contains the sequence TTTTTTG, which is similar to sequences protected from DNase I digestion in the HTLV-I LTR (between –75 and –69) [49] and in the HIV LTR between (–14 and –9) [50]. This sequence has not been identified as the recognition site for a known transcription factor. However, its presence in footprinted regions of several retroviral LTRs implicates its potential significance in regulating transcription of these viruses.

The region (–57 to –47) containing the AP-1 recognition site most proximal to the TATA box showed inducible binding of cellular factors. There was only partial protection of this site with nuclear extracts from monocyte U937 cells, but the protection increased with the time of induction of the cells with TPA [13]. It has previously been demonstrated that multiple AP-1 binding sites are necessary to confer maximum TPA-inducibility on a promoter [42]. We found that while all three AP-1 sites appear to be involved in the regulation of gene expression, the AP-1 site most proximal

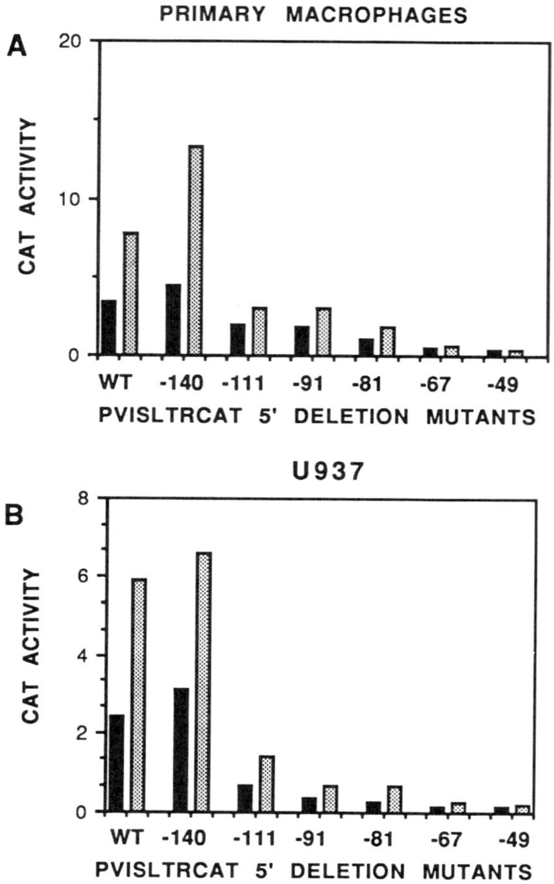

FIGURE 2 Activity of pVISLTRCAT 5' deletion mutants in primary macrophages (Figure 2A) and U937 cells (Figure 2B) (12). The black bars represent CAT activity in uninduced cells and gray bars represent CAT activity in TPA-induced cells. CAT activity is expressed as the percentage conversion of chloramphenicol to its acetylated form per 30 μg of protein per 20-hour assay (Figure 2A) or per 250 μg of protein per 6-hour assay (Figure 2B) (12).

to the TATA box is clearly the most critical. In addition, this AP-1 site was DNase I–protected while the other two were not. Multiple binding sites for the same transcription factor are frequently of unequal importance, probably as a result of other variables such as proximity to the TATA box and flanking sequences.

Using DNase I footprinting, two regions of the visna virus LTR have been identified that are important for the regulation of gene expression in primary macrophages (Figure 3A,B,C). One region located between –41 and –55 (–41 to –60 in Figure 3C), contains sequences homologous to the AP-1 binding site. This region is critical for basal activity and cellular activation of the visna virus LTR in primary macrophages and U937 cells as well as viral transactivation in sheep cells. The second region maps between –71 and –108 and includes part of one 43–base pair repeat and another potential AP-1 binding site. The same sequences were DNase I-protected using U937 nuclear extracts, except that only one AP-1 site was protected (–81 to –108).

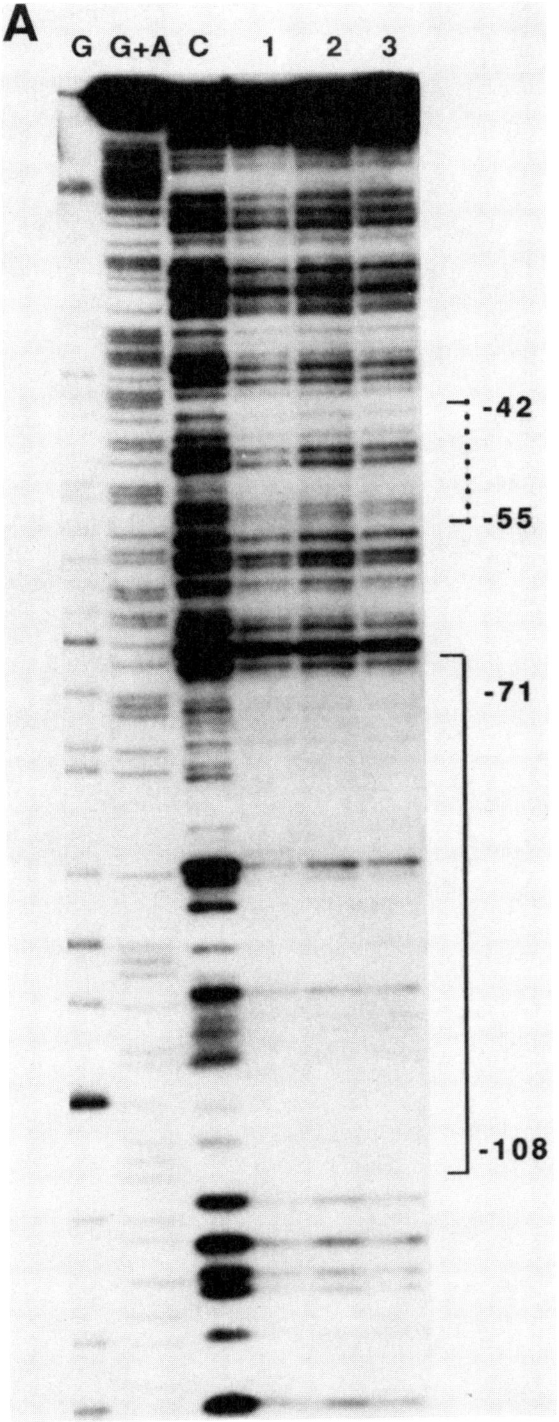

FIGURE 3 DNase I footprinting of the –140 to +27 region of the visna virus long terminal repeat (LTR) with primary monocyte/macrophage nuclear extracts. Lanes G and G+A are Maxam and Gilbert sequencing reactions; C, control lanes with 50 μg of bovine serum albumin and no extract; 1, freshly isolated day 1 monocytes; 2, 10- to 14-day macrophages; 3, 10- to 14-day macrophages treated with TPA for 24 hours. (A) Sense strand with 50 μg of nuclear extracts. (B) Anti-sense strand with 50 μg of nuclear extracts. (C) Anti-sense strand with 100 μg of nuclear extract.

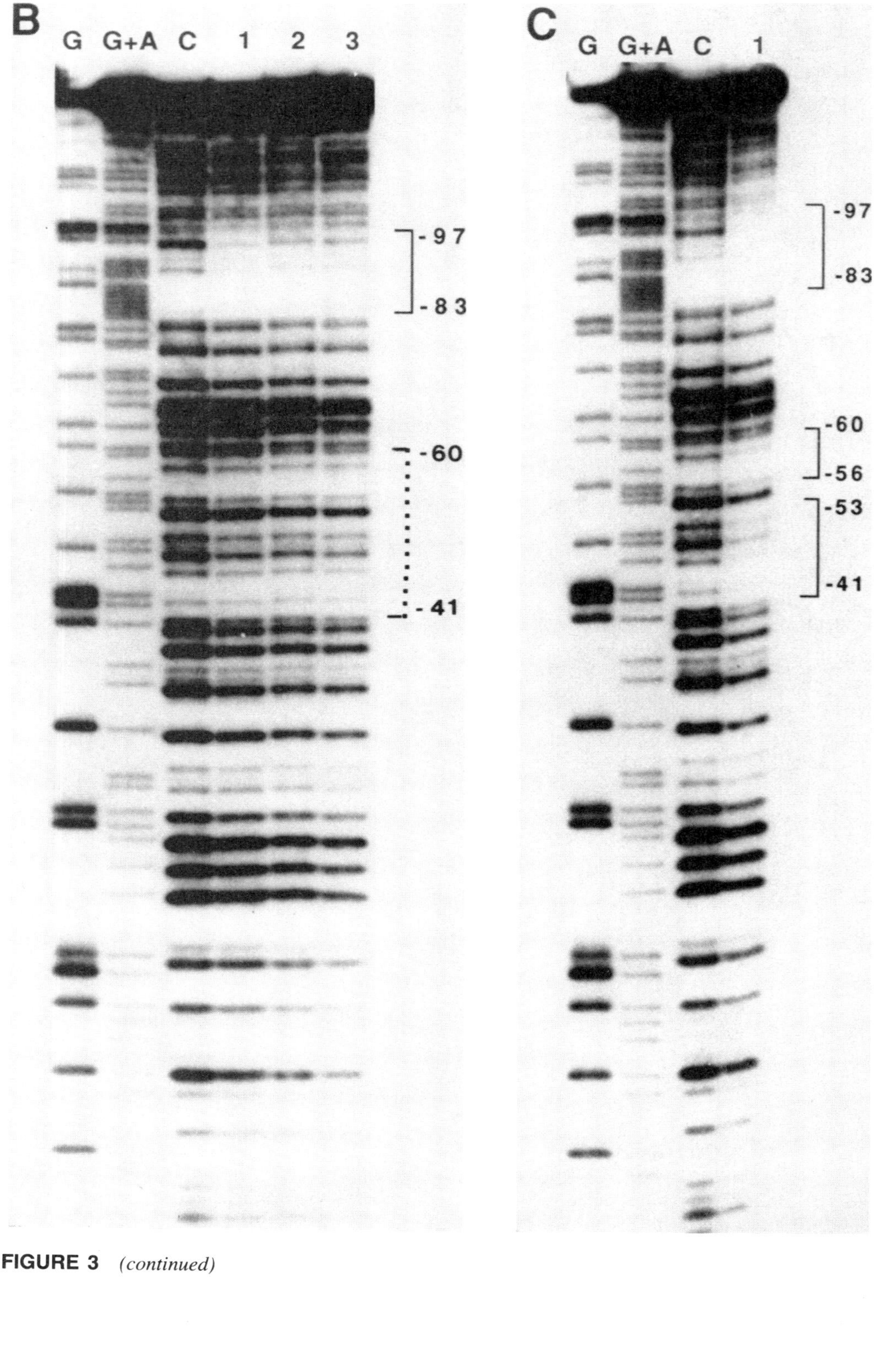

FIGURE 3 *(continued)*

It was now of interest to specifically identify the cellular proteins that bound to the visna virus promoter and specifically the region –49 to –59 that contained the AP-1 binding site. Since the cellular proteins JUN and FOS have been shown to bind to this sequence, two approaches were used to show that these were indeed the cellular proteins involved in transcriptional activation. To examine whether these cellular proteins bound to the AP-1 site in the visna virus LTR, a synthetic oligonucleotide containing the visna virus AP-1 sequence (–49 to –59) was used in gel-mobility shift assays with both TPA-induced U937 nuclear extracts and in vitro transcribed and translated JUN and FOS proteins [14]. Gel shifts using the U937 nuclear extracts revealed the presence of a specific DNA-protein complex in TPA-treated cells, whose intensity appeared to be greatest at 48 hours after TPA stimulation. The intensity of this complex could be substantially reduced in extracts that had been preincubated with anti-JUN and anti-FOS antibodies, but not in the presence of an irrelevant antibody. Using in vitro synthesized proteins, JUN was found to bind to the AP-1 oligo, but as previously demonstrated, FOS alone did not bind [46,47]. An enhancement of binding was detected when both JUN and FOS were used for the binding studies [46,47]. These results indicate that JUN and FOS are an integral part of the complex that binds and activates the visna virus LTR [14]. It is also likely that other cellular components exist in this complex, as antibody interference assays did not completely inhibit binding to the AP-1 oligo.

The level of the mRNAs of *c-jun* and *c-fos* was examined in the U937 cells. Little or no mRNA for either of these genes was present before TPA treatment, but the RNA increased rapidly (by 4 hours). The concentration of JUN and FOS proteins in the cells were also found to be increased with TPA treatment. Other studies have also shown that FOS mRNAs are up-regulated by TPA treatment of U937 cells [51,52]. Thus, it appears that the cellular proteins JUN and FOS are important in the activation of viral gene expression in the monocytic cell line U937 and are likely to be involved in vivo in the activation of viral gene expression. This provides another example of how lentiviruses use host cell factors to activate expression of their viral genomes. It has been previously shown that the human immunodeficiency virus (HIV) uses the NF-κB family of factors to up-regulate transcription from its LTR [reviewed in 53].

It is interesting to note that the viral gene *tat*, which activates the viral promoter, acts through the same AP-1 site in the visna virus promoter as JUN and FOS [41,54]. Thus, an important aspect of visna virus transcriptional regulation will be to determine if viral factors can up-regulate JUN and/or FOS production in the cell, thereby creating a feedback mechanism for viral gene expression. This type of regulation may potentially exist through the action of the visna Tat protein, although the mechanisms is not yet clear. Certainly other retroviral transactivating proteins, such as HTLV-I Tax, have been shown to be involved in cellular signal transduction pathways for regulation of viral gene expression [55].

V. SUMMARY

The ovine and caprine lentiviruses infect monocytes, and the viral DNA is integrated into the cellular DNA. The provirus remains silent until the monocyte matures into a macrophage. Intrinsic to this maturation is the induction of a class of immediate early genes in the monocyte that includes the transcription factors JUN and FOS. These transcription factors are thought to couple short-term signals in the cell to long-term cellular differentiation by regulation of specific cellular genes. Thus, JUN and FOS bind to the AP-1 site in the promoters of cellular genes and activate their transcription, resulting in maturation of the monocyte into a macrophage. In addition, these cellular factors activate the same AP-1 sequence in the visna virus LTR, leading to transcriptional activation, full viral gene expression, and production of progeny virus. The expression of viral antigens in the context of MHC class II on the macrophage leads to the production of cytokines and a lymphoproliferative response that causes the lesions in specific target organs in an infected animal. We still understand only the framework of these events. The specific mechanisms by which viral genes alter macrophage gene expression and the molecular basis of different viral tropism for specific tissue macrophages, i.e. microglia, remain to be determined.

REFERENCES

1. Narayan O, Clements JE. Lentiviruses. In: Fields BN, Knipe DM, eds. Virology. 2nd ed. New York: Raven Press, 1990:1571–89.
2. Sigurdsson B. Maedi, a slow progressive pneumonia of sheep: an epizootiological and pathologic study. Br Vet J 1954; 110:255–70.
3. Sigurdsson B, Palsson PA, Grimsson H. Visna, a demyelinating transmissible disease of sheep. J Neuropathol Exp Neurol 1957; 16:389–403.
4. Gendelman HE, Narayan O, Molineaux S, Clements JE, Ghotbi Z. Slow, persistent replication of lentiviruses: role of tissue macrophages and macrophage precursors in bone marrow. Proc Natl Acad Sci USA 1985; 82:7086–90.
5. Zink MC, Yager JA, Myers JD. Pathogenesis of caprine arthritis encephalitis virus. Cellular localization of viral transcripts in tissues of infected goats. Am J Pathol 1990; 136:843–54.
6. Narayan O. Role of macrophages in the immunopathogenesis of visna-maedi of sheep. Prog Brain Res 1983; 59:233–5.
7. Narayan O, Clements JE. Biology and pathogenesis of lentiviruses. J Gen Virol 1989; 70:1617–39.
8. Zack IA, Arrigo SJ, Weitsman SR, Go AS, Haislip A, Chen IS. HIV-1 entry into quiescent primary lymphocytes: molecular analysis reveals a labile, latent viral structure. Cell 1990; 61:213–22.
9. Narayan O, Wolinsky JS, Clements JE, Strandberg JD, Griffin DE, Cork LC. Slow virus replication: the role of macrophages in the persistence and expression of visna viruses of sheep and goats. J Gen Virol 1982; 59:345–56.
10. Narayan O, Kennedy Stoskopf S, Sheffer D, Griffin DE, Clements JE. Activation of caprine arthritis-encephalitis virus expression during maturation of monocytes to macrophages. Infect Immun 1983; 41:67–73.
11. Braun MJ, Clements JE, Gonda MA. The visna virus genome: evidence for a hypervariable site in the env gene and sequence homology among lentivirus envelope proteins. J Virol 1987; 61:4046–54.
12. Saltarelli M, Querat G, Konings DA, Vigne R, Clements JE. Nucleotide sequence and transcriptional analysis of molecular clones of CAEV which generate infectious virus. Virology 1990; 179:347–64.
13. Gabuzda DH, Hess JL, Small JA, Clements JE. Regulation of the visna virus long terminal repeat in macrophages involves cellular factors that bind sequences containing AP-1 sites. Mol Cell Biol 1989; 9:2728–33.
14. Shih DS, Carruth LM, Anderson M, Clements JE. Involvement of FOS and JUN in the activation of visna virus gene expression in macrophages through an AP-1 site in the viral LTR. Virology 1992 (in press).
15. Coffin JM. Retroviridae and their replication. Fields BN, Knipe DM, eds. Virology 1990:1437–1500.
16. Cullen BR, Greene WC. Functions of the auxiliary gene products of the human immunodeficiency virus type 1. Virology 1990; 178:1–5.
17. Chu SY, Peterman TA, Doll LS, Buehler JW, Curran JW. AIDS in bisexual men in the United States: epidemiology and transmission to women. Am J Public Health 1992; 82:220–4.
18. Lifson AR, Curran JE. Epidemiology of AIDS: current trends and prevention. In: Gottlieb MS, et al., eds. Current topics in AIDS. New York: John Wiley & Sons, 1987:7–30.
19. Davis MK. Human milk and HIV infection: epidemiologic and laboratory data. In: Mestechy J, ed. Immunology of milk and the neonate. New York: Plenum Press, 1991:271–80.
20. Letvin NL, King NW. Immunologic and pathologic manifestations of the infection of rhesus monkeys with simian immunodeficiency virus of macaques. J AIDS 1990; 3:1023–40.
21. Shaw GM, Harper ME, Hahn BH, et al. HTLV-III infection in brains of children and adults with AIDS encephalopathy. Science 1985; 227:177–82.
22. Georgsson G, Nathanson N, Palsson PA, Petursson G. The pathology of visna and maedi in sheep. Front Biol 1976; 44:61–96.
23. Travis WD, Fox CH, Devaney KO, et al. Lymphoid pneumonitis in 50 adult patients infected with the human immunodeficiency virus: lymphocytic interstitial pneumonitis versus nonspecific interstitial pneumonitis. Hum Pathol 1992; 23:529–41.

24. Oliver RE, Gorham JR, Parish SF, Hadlow WJ, Narayan O. Ovine progressive pneumonia: pathologic and virologic studies on the naturally occurring disease. Am J Vet Res 1981; 42:1554–9.
25. Baskerville A, Ramsay A, Cranage MP, et al. Histopathological changes in simian immunodeficiency virus infection. J Pathol 1990; 162:67–75.
26. Pedersen NC, Barlough JE. Clinical overview of feline immunodeficiency virus. J Am Vet Med Assoc 1991; 199:1298–1305.
27. Narayan O, Wolinsky JS, Clements JE, Strandberg JD, Griffin DE, Cork LC. Slow virus replication: the role of macrophages in the persistence and expression of visna viruses of sheep and goats. J Gen Virol 1982; 59:345–56.
28. Meltzer MS, Skillman DR, Gomatos PJ, Kalter DC, Gendelman HE. Role of mononuclear phagocytes in the pathogenesis of human immunodeficiency virus infection. Annu Rev Immunol 1990; 8:169–94.
29. Clements JE, Gdovin SL, Montelaro RC, Narayan O. Antigenic variation in lentiviral diseases. Annu Rev Immunol 1988; 6:139–59.
30. Gendelman HE, Narayan O, Kennedy-Stoskopf S, et al. Tropism of sheep lentiviruses for monocytes: susceptibility to infection and virus gene expression increase during maturation of monocytes to macrophages. J Virol 1986; 58:67–74.
31. Narayan O, Cork LC. Lentiviral diseases of sheep and goats: chronic pneumonia leukoencephalomyelitis and arthritis. Rev Infect Dis 1985; 7:89–98.
32. Zink MC, Yager JA, Myers JD. Pathogenesis of lentiviruses: cellular localization of viral transcripts in tissues of goats infected with caprine arthritis-encephalitis virus. Am J Pathol 1990; 136:843–54.
33. Zink MC, Gorrell MD, Narayan O. The neuropathogenesis of visna virus infection in sheep. Semin Neurosciences 1991; 3:125–30.
34. Narayan O, Zink MC, Gorrell MD, McEntee MF, Sharma DP, Adams RJ. Retrovirus-induced arthritis in animals. J Rheumatol 1992 (in press).
35. Kennedy PG, Narayan O, Ghotbi Z, Hopkins J, Gendelman HE, Clements, JE. Persistent expression of Ia antigen and viral genome in visna-maedi virus–induced inflammatory cells. Possible role of lentivirus-induced interferon. J Exp Med 1985; 162:1970–82.
36. Narayan O, Sheffer D, Clements JE, Tennekoon G. Restricted replication of lentiviruses. Visna viruses induce a unique interferon during interaction between lymphocytes and infected macrophages. J Exp Med 1985; 162:1954–69.
37. Zink MC, Narayan O. Lentivirus-induced interferon inhibits maturation and proliferation of monocytes and restricts the replication of caprine arthritis-encephalitis virus. J Virol 1989; 63:2578–84.
38. Zink MC, Adams RJ, Narayan O. Augmentation of encephalitis in lentivirus-infected goats by intraventricular infusion of the lentiviral cytokines, LV-IFN. Lab Invest 1992 (submitted):
39. Zink MC, Sheffer D, Meyer AL, et al. Pathogenesis of encephalitis caused by a neurovirulent strain of ovine lentivirus adapted from a field isolate lacking neurotropism. Am J Pathol 1992 (submitted):
40. Sharma DP, Zink MC, Anderson MG, et al. Derivation of neurotropic SIV from exclusively lymphocyte-tropic parental virus: pathogenesis of infection in macaques. J Virol 1992 (in press):
41. Hess JL, Small JA, Clements JE. Sequences in the visna virus long terminal repeat that control transcriptional activity and respond to viral *trans*-activation: involvement of AP-1 sites in basal activity and *trans*-activation. J Virol 1989; 63:3001–15.
42. Angel P, Imagawa M, Chiu R, et al. Phorbol ester-inducible genes contain a common cis-element recognized by a TPA-modulated *trans*-acting factor. Cell 1987; 49:729–39.
43. Bohmann D, Bos TJ, Admon A, Nishimura T, Vogt PK, Tjian R. Human proto-oncogene *c-jun* encodes a DNA binding protein with structural and functional properties of transcription factor AP-1. Science 1987; 238:1386–92.
44. Mermod N, Williams TJ, Tjian R. Enhancer binding factors AP-4 and AP-1 act in concert to activate SV40 late transcription in vitro. Nature 1988; 332:557–61.
45. Curran T, Franza BR. Fos and Jun: the AP-1 connection. Cell 1988; 55:395–7.
46. Nakabeppu Y, Tyder K, Nathans D. DNA binding activities of three murine JUN proteins: stimulation by FOS. Cell 1988; 55:907–15.
47. Rauscher FJ, Voulalas PJ, Franza BR, Curran T. FOS and JUN bind cooperatively to the AP-1 site: reconstitution in vitro. Genes Dev 1988; 2:1687–99.
48. Dignam JD, Lebowitz RM, Roeder RG. Accurate transcription initiation by RNA polymerase II in a soluble extract from isolated mammalian nuclei. Nucl Acids Res 1983; 11:1475–89.

49. Nyborg JK, Dynan WS, Chen ISY, Wachsman W. Binding of host-cell factors to DNA sequences in the long terminal repeat of human T-cell leukemia virus type I: implications for viral gene expression. Proc Natl Acad Sci USA 1988; 85:1457–61.
50. Garcia JA, Wu FK, Mitsuyasu R, Gaynor RB. Interactions of cellular proteins involved in the transcriptional regulation of the human immunodeficiency virus. EMBO 1987; 6:3761–70.
51. Auwerx J, Staels B, Sassone-Corsi P. Coupled and uncoupled induction of fos and jun transcription by different second messengers in cells of hematopoietic origin. Nucl Acids Res 1990; 18:221–8.
52. Sherman ML, Stone RM, Datta R, Bernstein SH, Kufe DW. Transcriptional and post-transcriptional regulation of *c-jun* expression during monocytic differentiation of human myeloid leukemic cells. J Biol Chem 1990; 265:3320–3.
53. Jones KA. HIV *trans*-activation and transcription control mechanisms. New Biol 1989; 1:127–35.
54. Gdovin SL, Clements JE. Molecular mechanisms of visna virus tat: identification of the targets for transcriptional activation and evidence for a post-transcriptional effect. Virology 1992; 188:438–50.
55. Arima N, Molitor J, Smith M, Kim J, Daitoku Y, Greene W. Human T-cell leukemia virus type I tax induces expression of the rel-related family of κB enhancer-binding proteins: evidence for a pre-translational component of regulation. J Virol 1991; 65:6892–9.

37

The Role of Macrophages and Macrophage Cytokines in Host Resistance to Herpes Simplex Virus

Jill A. Hendrzak and Page S. Morahan

The Medical College of Pennsylvania, Philadelphia, Pennsylvania

Macrophages (Mϕs) are widely distributed in major organs of the body, and by their initial encounter with virus particles they may determine the subsequent course of infection [1]. Two types of Mϕ-mediated resistance to virus infections have been described. The ability of Mϕs to phagocytize and degrade viruses or inhibit virus growth inside the Mϕ itself is defined as intrinsic resistance. The ability of Mϕs, alone or with other cells and factors, to inhibit virus replication in surrounding cells that are permissive to virus infection, or to destroy infected cells, is defined as extrinsic resistance [2].

The critical role of Mϕs in natural resistance to herpes simplex virus (HSV) infection has long been recognized, since the seminal work of Johnson [3], and Mϕs have been shown to exhibit both intrinsic and extrinsic antiviral activities against HSV. The interaction of Mϕs and HSV at the molecular level has recently been reviewed [4]. The considerable number of studies with HSV makes this viral infection a useful prototype in which to consider the role of Mϕs in resistance to viral infections [1–4]. This review will focus on recent work since 1988, and particularly on the role of cytokines and the early inflammatory response to HSV infection.

I. INTRINSIC RESISTANCE OF MACROPHAGES TO HERPES SIMPLEX VIRUS INFECTION

A. Potential Mechanisms

Macrophages of all species studied are quite nonpermissive to HSV infection as compared with completely permissive cells. However, factors such as age, differentiation, activation, culture conditions, tissue source, and genetic background influence the degree of permissivity [2,5–8]. There are two general mechanisms by which Mϕs mediate their intrinsic antiviral activity: 1) phagocytosis and degradation of viral particles; and 2) inhibition of viral replication. Macrophage intrinsic resistance to HSV probably involves a combination of both mechanisms, but the relative importance of each has not yet been defined.

B. In Vitro Studies on the Phagocytic and Degradative Function of Macrophages

One of the principal functions of Mϕs is their phagocytic capability of ingesting, digesting, and disposing of foreign material, such as bacteria and virus particles [1]. Surprisingly little attention has been paid to this function of Mϕs as it relates to viruses, but Van Strijp and colleagues have recently readdressed this issue. They have reported that human monocytes can rapidly degrade the DNA and membrane proteins of phagocytosed herpes simplex virus type 1 (HSV-1) [9]. As with other microorganisms, the addition of opsonizing antibody plus complement markedly increased uptake [10]. Recently, they demonstrated that surfactant protein A can also act as an opsonin in the phagocytosis of HSV-1 by alveolar Mϕs [11]. The identification of additional normal and inflammatory host proteins that enhance phagocytosis has important consequences for better understanding the early interaction between HSV and Mϕs at the initial site of infection.

C. In Vitro Studies on the Molecular Basis for the Block in Herpes Simplex Virus Type 1 Replication in Mouse and Human Macrophages

In addition to the phagocytic and degradative function, Mϕs can inhibit virus replication by interfering with any stage of the viral replication cycle. The block in HSV-1 replication in resident mouse peritoneal Mϕs has recently been reviewed [4,12], and it appears to occur in the expression of the HSV early genes after initiation of transcription. Dysfunction of an immediate early gene product may also play a role [13].

Freshly isolated human blood monocytes, like mouse Mϕs, are also generally resistant to HSV infection [14–17]. The block in HSV replication in human monocytes apparently occurs even earlier than in mouse peritoneal Mϕs, since no viral DNA was associated with the nuclear fraction of the monocytes [16]. A 30-kDa cellular protein shown to be a form of tumor necrosis factor, which is induced upon infection of monocytes with HSV, has been hypothesized to be involved in the inhibition of viral replication [17].

D. Correlation of Intrinsic Resistance of Macrophages in Vitro with Pathogenesis in Vivo

A variety of early studies correlated the ability of Mϕs to restrict HSV replication in vitro with the resistance of an animal to infection in vivo [3,8,18]. Recent molecular pathogenesis studies have focused on determining which regions of the HSV genome are required for HSV replication in Mϕs and the correlation with virulence in mice. The immediate early gene sequence that was responsible for intraperitoneal virulence of HSV-1 was found to be required for transcription of viral genes, synthesis of viral DNA, and induction of cytopathic effect in Mϕs [19]. Studies have been initiated to characterize the viral proteins encoded by this gene sequence [20].

It is not yet clear whether Mϕ intrinsic resistance to HSV during infection in vivo can be correlated with enhanced host resistance to infection. If Mϕs are in an activated state as a result of the infection, they may be more permissive to HSV replication, since activated or differentiated Mϕs generally show increased permissiveness to HSV in vitro [4,7,12]. It should be noted, however, that the replication of HSV in activated Mϕs is still orders of magnitude less than in permissive cells. Techniques such as fluorescent antibody staining or in situ hybridization will be useful to determine the expression of specific viral genes and therefore the extent of viral phagocytosis and replication in Mϕs, as compared with other cells, during infection with HSV in vivo.

II. ROLE OF CYTOKINES IN MACROPHAGE INTRINSIC RESISTANCE TO HERPES SIMPLEX VIRUS INFECTION

While Mϕs are clearly able to phagocytose HSV and are quite nonpermissive for HSV replication, the precise intracellular antiviral mechanisms remain to be defined. Since Mϕs produce a variety of cytokines, there has been considerable interest in the antiviral role of cytokines.

A. Interferon and Macrophage Intrinsic Antiviral Activity

1. Effects of Endogenous and Early-Induced Interferon in Macrophages on Herpes Simplex Virus Replication

Mogenson and Virelizier (21) have described an interferon (IFN)-Mϕ "alliance" that operates before and early after virus infection to provide antiviral resistance to infection. Interferon can be induced in Mϕs in vivo by continuous stimulation with endogenous IFN inducers such as endotoxin (Figure 1) [21–23]; these low levels of endogenous IFN could maintain the Mϕ in an antiviral state. After infection, IFN can also be rapidly induced in the virus-infected Mϕ, resulting in an antiviral state if the IFN is induced sufficiently early (Figure 1). The time of induction is critical. Otherwise, IFN production may be a coincidental event rather than a requirement for intrinsic resistance.

An endogenous IFN-mediated antiviral state before infection does not appear to play a major role in the intrinsic resistance of mouse peritoneal Mϕs to HSV, since treatment of CD-1 mice with anti-IFN antibody did not reverse nonpermissiveness [7]. However, early production of IFN in response to HSV infection of Mϕs in vitro has been shown to be likely to be involved [4,7,24].

The role of previously induced endogenous IFN and early HSV-induced endogenous IFN in the intrinsic resistance of human monocytes to HSV has also been studied. Interferon-α, induced in monocytes by infection with high multiplicities of HSV, prevented the productive replication of HSV [25]. Furthermore, HSV-exposed monocytes of normal healthy volunteers produced significantly higher amounts of IFN-α than did such monocytes from patients with recurrent herpes lesions [26]. Factors in addition to IFN are also probably involved, since the addition of anti-IFN serum shortened, but did not completely eliminate, the initial lag period of virus production in monocytes [25]. Studies at the molecular level have confirmed that IFN induction does not completely explain the intrinsic resistance [16,17]. Although high titers of IFN-α/β were detected early after HSV infection of monocytes in vitro, HSV replication was found to be inhibited before viral DNA entry into the nucleus, before IFN had time to be produced [16].

Therefore, there is clear evidence for a role of early HSV-induced IFN in the intrinsic antiviral resistance of mouse Mϕs and human monocytes to infection with HSV, and other factors must also contribute. It is possible that if the earlier non-IFN-mediated antiviral mechanisms are insufficient, HSV-induced IFN functions to inhibit viral replication. Further investigation, using molecular techniques and anti-IFN antibodies, is needed to define more precisely the relative role of HSV-induced IFN and other mechanisms in Mϕ antiviral resistance.

2. Effect of Exogenous Interferon on Herpes Simplex Virus Replication in Macrophages

In addition to 1) endogenous IFN produced in vivo, which leads to a Mϕ IFN-induced antiviral state, and 2) early IFN induced by HSV infection of Mϕs, 3) exogenous IFN-α and IFN-β produced by other Mϕs, or IFN-γ produced by lymphocytes, can affect Mϕ intrinsic resistance (Figure 1). To measure the effect of exogenous IFN, it has generally been necessary to use activated Mϕs that are semipermissive for HSV replication.

One particular semipermissive system that has been studied is that of Mϕs differentiated by culture in vitro. Treatment of cultured mouse splenic Mϕs with IFN-α or IFN-β inhibited HSV-1 replication [27]. IFN-α/β was later shown to inhibit HSV immediate early and early gene expression at the transcriptional level [28]. Treatment of cultured human monocyte-derived Mϕs with IFN-α also enhanced resistance to the cytopathic effects of HSV-1 [29], and IFN-α and IFN-β were shown to restore the impaired intrinsic antiviral activity in Mϕs from patients with neoplasia [30]. As in mouse Mϕs, IFN-β was shown to block HSV replication by inhibiting the synthesis of immediate early and early proteins [31]. However, inhibition of cellular differentiation of monocytes to Mϕs by IFN also appears to be involved [25].

The primary source of exogenous IFN-α produced during HSV infection may be monocytes or Mϕs [25], or a recently characterized subset of peripheral blood mononuclear cells known as

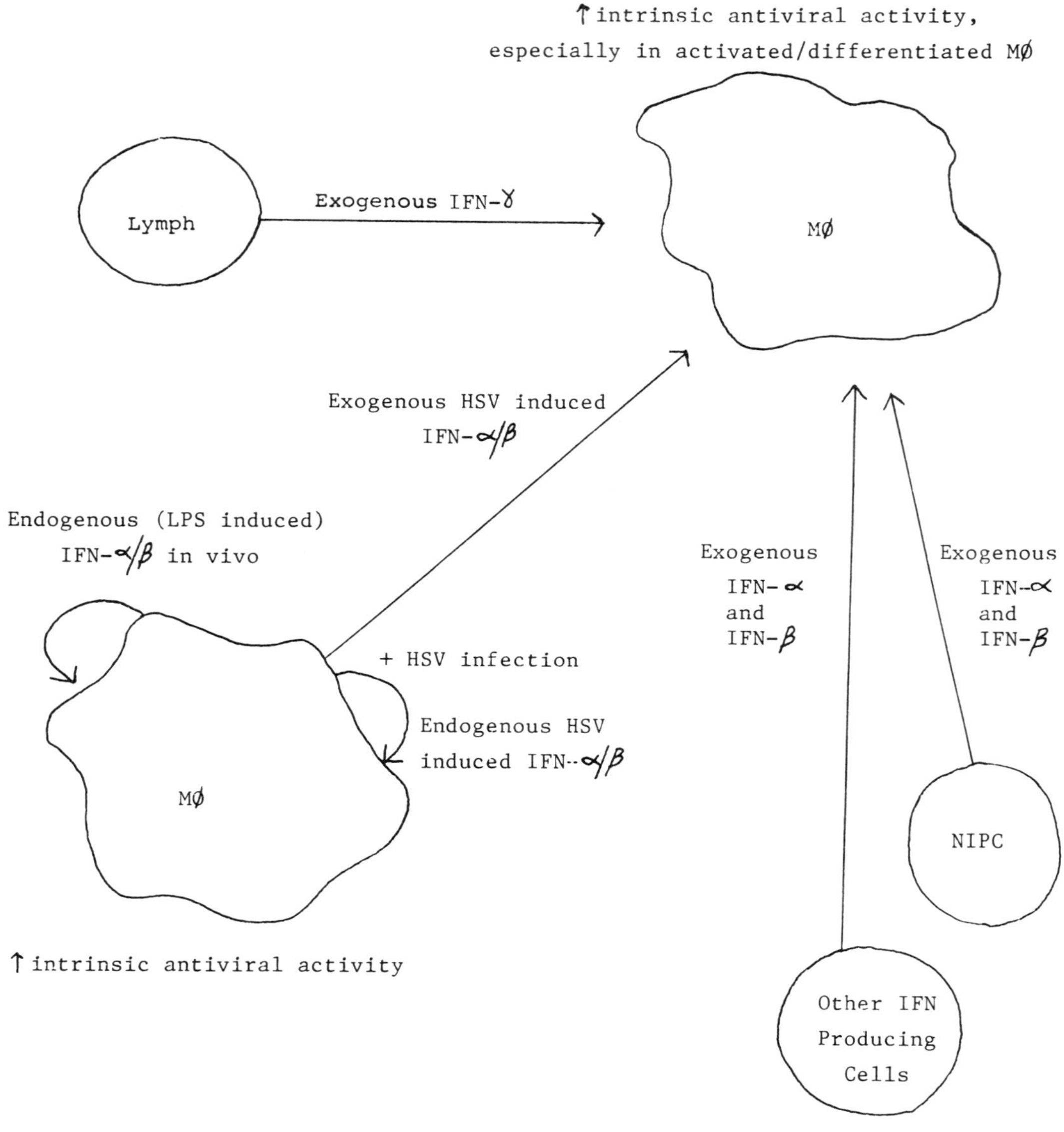

FIGURE 1 Both endogenous and exogenous sources of interferon (IFN) may enhance macrophage (Mϕ) intrinsic antiviral activity against herpes simplex virus (HSV). Endogenous lipopolysaccharide (LPS)- or HSV-induced IFN-α/β can act on Mϕs in an autocrine fashion to enhance intrinsic antiviral activity. Interferon-α/β induced in Mϕs upon infection with HSV is also a source of exogenous IFN, which can enhance intrinsic antiviral activity in other Mϕs. Other sources of exogenous IFN include IFN-γ produced by lymphocytes and IFN-α or IFN-β produced by natural IFN-α–producing cells (NIPC) or other IFN-producing cells.

natural IFN-α–producing cells (NIPCs), or as yet unidentified cells. Natural IFN-α–producing cells were first thought to represent either a small subpopulation in the monocyte/Mφ lineage or null lymphocytes [32], but it now appears that they are not of the myeloid or lymphoid lineage and are also distinct from natural killer (NK) and dendritic cells [33].

In addition to IFN-α and IFN-β, crude IFN-γ inhibited HSV replication in mouse peritoneal Mφs [34,35]. Using cultured mouse splenic Mφs and recombinant murine IFN-γ, the inhibition was shown to be at the translational level rather than the transcriptional level of HSV immediate early gene expression [36]. In contrast to its effect on mouse Mφ, IFN-γ treatment had no effect on HSV replication in differentiated human Mφs [30,31]. The investigators attributed the higher sensitivities of the human Mφs to IFN-α and IFN-β as compared with IFN-γ to differences in the binding of these IFNs to receptors [31].

In summary, both endogenous and exogenous IFN can inhibit HSV replication in Mφs in vitro. Further investigation is required to establish the most effective type of IFN for immunotherapy in HSV infections, as well as the molecular mechanisms by which IFN inhibits cellular differentiation and HSV replication.

B. Effect of Colony-Stimulating Factors on Herpes Simplex Virus Replication in Macrophages

As shown in the IFN studies, the interaction between Mφs and viruses is influenced by their state of differentiation. Monocyte-derived Mφs are generally more permissive to HSV than are freshly isolated monocytes [25]. The colony-stimulating factors (CSFs), which promote the proliferation and differentiation of hematopoietic progenitor cells, can also affect some functions of mature Mφs [37] and are currently being tested for their effects on Mφ antiviral activity.

Bone marrow–derived Mφs differentiated in the presence of granulocyte-macrophage CSF (GM-CSF) showed increased permissiveness for HSV as compared with resident peritoneal Mφs [38]. Bone marrow–derived Mφs that were differentiated with macrophage colony-stimulating factor (M-CSF) were more resistant to HSV replication than were GM-CSF–treated bone marrow–derived Mφs; this is probably due to the induction of endogenous IFN-α by M-CSF, which has an antiproliferative effect [39]. In this study, however, the permissiveness of CSF-treated bone marrow–derived Mφs was not compared with that of resident peritoneal Mφs.

The effect of M-CSF on the intrinsic resistance of mature Mφs to HSV is not yet clear. In an early study, there was no significant difference in HSV-1 yield when thioglycollate-elicited peritoneal Mφs incubated in crude M-CSF were compared with controls [34]. Resident peritoneal Mφs incubated for 24, 48, or 72 hours with 10-1000 U of recombinant human M-CSF restricted HSV-1 replication to the same extent as did control Mφs (J.A. Hendrzak and P.S. Morahan, unpublished data). In contrast, Ben-Hur et al. [19] have reported that incubation of resident mouse peritoneal Mφs with crude M-CSF enhanced HSV-1 DNA synthesis and progeny virus production. Mouse splenic Mφs cultured in crude M-CSF for 1 week also are reported to have increased permissiveness for HSV-1 [27]. Different sources of Mφs, state of Mφ differentiation due to culture conditions, and doses and purity of the M-CSF used may explain the different results obtained in these studies. More investigation on the in vitro effects of the CSFs, as well as other cytokines, on Mφ intrinsic antiviral activity are needed before they can be used optimally in immunotherapy.

III. MACROPHAGE EXTRINSIC ANTIVIRAL RESISTANCE TO HERPES SIMPLEX VIRUS

A. Potential Mechanisms

There are three general mechanisms by which Mφs, alone or in concert with other cells and factors, may mediate extrinsic antiviral resistance to HSV: 1) inhibition of the spread of HSV between cells [40,41]; 2) inhibition of HSV replication in permissive cells [42]; and 3) selective lysing of HSV-infected cells [43,44].

B. Role of Cytokines in Macrophage Extrinsic Antiviral Activity

Cytokines may activate Mϕs for enhanced extrinsic antiviral activity, similar to the way cytokines such as IFN can enhance Mϕs intrinsic antiviral activity (Figure 1, 2). Interferon-α and IFN-γ, as well as Mϕ-activating factor (MAF), can activate monocytes to lyse HSV-infected cells [45–47]. Not all extrinsic antiviral activity appears to be IFN-mediated; Mϕ-mediated inhibition of HSV replication in permissive cells appears to be an IFN-independent mechanism [48].

Macrophages can also secrete cytokines that then mediate the extrinsic antiviral activity (Figure 2). The role of Mϕ-secreted monokines in addition to IFN has begun to receive considerable attention. Tumor necrosis factor (TNF), a cytokine produced mainly by cells of the monocyte/Mϕ lineage, was shown to act directly to lyse HSV-infected cells without harming uninfected cells [49], and to inhibit HSV replication in permissive cells [50,51]. The antiviral effect of TNF has been reported to be partially mediated by IFN-β, but not by IFN-α or IFN-γ [50,51]. However, TNF could synergize with the exogenous addition of IFN-γ to inhibit HSV-2 replication [51,52]. In a recent study involving HSV-induced atherosclerosis, TNF and interleukin-1 (IL-1) were shown to inhibit HSV replication in arterial smooth muscle cells, which was correlated with the prevention of cholesterol ester accumulation [53]. Not all studies have shown, however, that exogenous TNF alone has inhibitory effects on HSV-infected cells [52,54].

Besides acting directly on permissive cells, monokines secreted by Mϕs may act on other immune cells, which then mediate antiviral activity. Recombinant human TNF-α, but not GM-CSF, was recently reported to prime human polymorphonuclear leukocytes (PMNs) to internalize complement-coated HSV particles [55]. Both Mϕs and the monokines they produce were found to be required for HSV-induced IL-2 production by T cells from HSV-1 seropositive individuals [56].

In summary, cytokines that activate Mϕs or are secreted by Mϕs are definitely involved in extrinsic antiviral mechanisms. Their actions may be either direct or through effects on other immune cells.

IV. EFFECT OF HERPES SIMPLEX VIRUS ON MARCOPHAGE FUNCTION

Because the initial interaction between Mϕs and HSV may clearly affect the course of disease, the effect of HSV on various Mϕ functions has been studied. Enhancement, suppression, and no change in Mϕ functions have been reported. For example, HSV infection of mouse bone marrow–derived Mϕs has been reported to induce expression of Ia antigen on infected cells, but no changes in phagocytic ability or Fc receptor expression [57]. Infection of mouse peritoneal Mϕs by HSV has been reported to inhibit chemotaxis [58], Mϕ-mediated tumoricidal activity [59], and antibody-dependent cellular cytotoxicity and phagocytosis [60].

There are also conflicting reports on the effect of HSV on the intracellular oxidative responses of Mϕs. Chemiluminescence induced by zymosan was reported to be impaired in HSV-1–infected mouse peritoneal Mϕs [61], while rabbit alveolar Mϕs and mouse peritoneal Mϕs could be primed by HSV-2 for an increased respiratory burst [62,63].

Recently, three studies have shown that the synthesis of cytokines by Mϕs may also be altered by HSV infection. Activation of mouse peritoneal Mϕs with the immunomodulators *Corynebacterium parvum* or *Salmonella typhimurium* increased their permissiveness to HSV-1 infection in vitro, and upon infection, IL-1β mRNA levels were markedly reduced [64]. Interleukin-1β mRNA transcripts could not be detected in either uninfected or HSV-infected resident peritoneal Mϕs. Early after peripheral blood human mononuclear cells were infected in vitro with HSV-1, however, TNF-α, IL-1β, and IL-6 mRNA transcription were found to be weakly enhanced, with the greatest

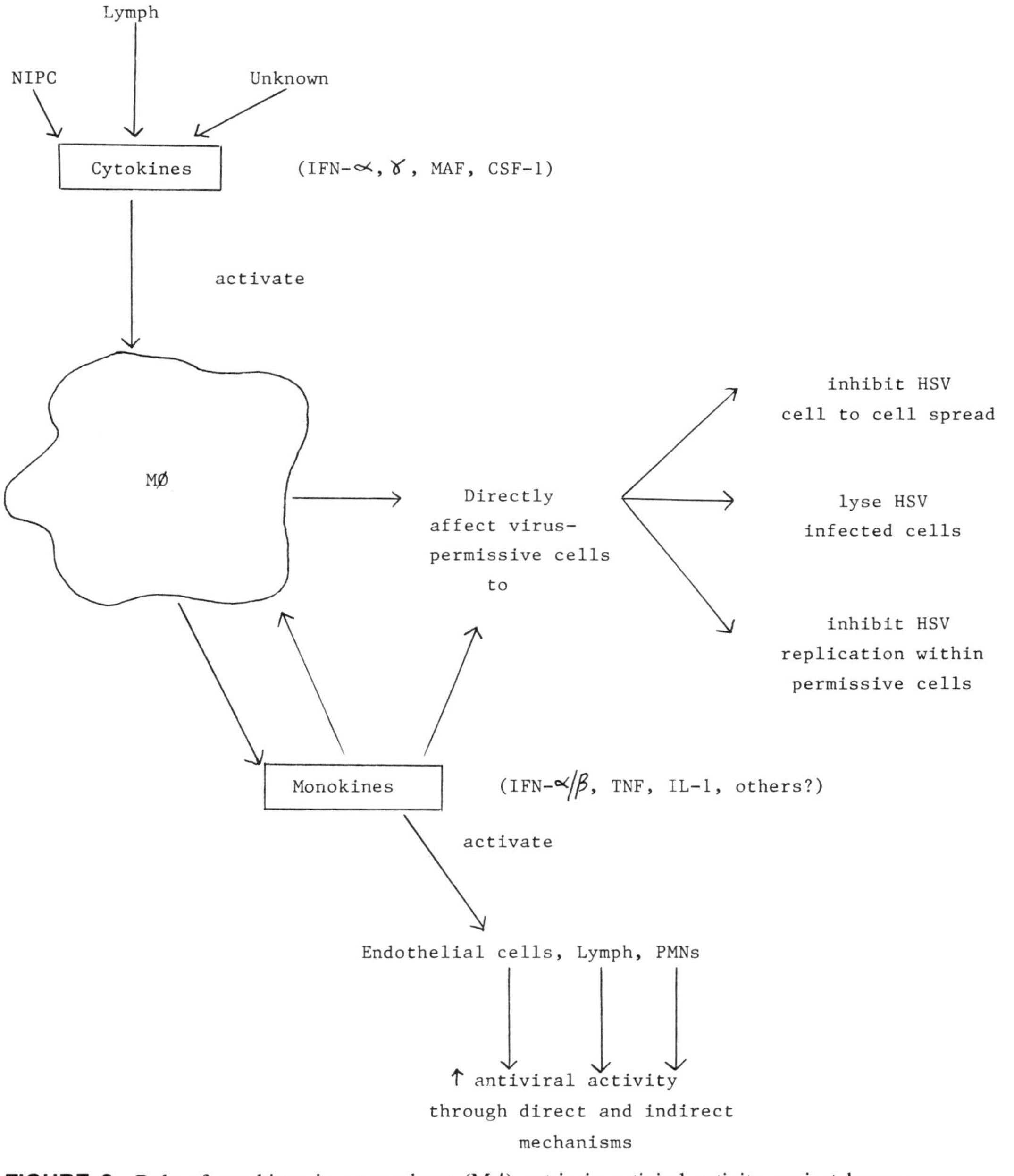

FIGURE 2 Role of cytokines in macrophage (Mϕ) extrinsic antiviral activity against herpes simplex virus (HSV). Cytokines produced by various cells can activate Mϕs, which then directly exhibit extrinsic antiviral activity against HSV. Macrophages also produce monokines, which then can mediate the extrinsic antiviral activity directly, or indirectly through effects on other immune cells. CSF-colony-stimulating factor; IFN-interferon; IL-interleukin; MAF-macrophage-activating factor; NIPC-natural IFN-α–producing cells; PMN-polymorphonuclear leukocyte; TNF-tumor necrosis factor.

effect on TNF-α gene transcription [65]. In vivo, HSV-1 infection of mice was shown to induce IL-1α and IL-1β gene transcription in peritoneal exudate cells, with a peak at 6 hours after infection [66]. The investigators did not examine whether infected or uninfected inflammatory cells were responsible for this early IL-1 RNA synthesis.

Because different sources of Mϕs and experimental systems were used in these studies, no definitive conclusions can be reached. Clearly, HSV infection has the potential to induce alterations in Mϕ functions. The mechanisms by which they occur, and their relationship to the outcome of disease in vivo, merit further study.

V. ACTIVATION OF MACROPHAGES IN VIVO FOR ENHANCED ANTIVIRAL ACTIVITY

A. Antiviral Effects of Cytokines In Vivo

Because of the demonstrated antiviral potential of cytokines in vitro, a number of recent investigations have focused on treatment with cytokines for the immunotherapy of experimental HSV infections. As with the in vitro studies previously described, most of the research on the effects of cytokines in vivo has dealt with the IFNs.

1. *Antiviral Efficacy of Interferon in Herpes Simplex Virus Infections*

Studies in mice infected with HSV have shown that IFN is produced early after infection [67], and it has been concluded that virus-induced IFN plays a role in natural antiviral resistance. Direct proof of this concept has come from depletion studies. Depletion of either IFN-α/β or IFN-γ with anti-IFN antibodies in vivo, early after infection, clearly increases the mortality of infected mice [68,69].

Given the importance of IFN in natural resistance to HSV, it is not surprising that all three types of IFN administered to mice are protective against HSV infection. Prophylactic and/or early therapeutic treatment with a variety of IFNs has significantly increased the resistance of mice to HSV-1 and HSV-2 infections [70–72]. In a comparative study, all three IFN types (α,β,γ) provided strong protection against systemic HSV-2 infection [73].

The IFNs, therefore, show potential for immunotherapy for the treatment of HSV infections in humans. In clinical trials, IFN-α was shown to be effective against both HSV-1 and HSV-2 infections [74]. The exact antiviral mechanisms remain to be determined. Although IFNs have direct antiviral effects [75], they may also act indirectly. For example, passive transfer of adherent murine spleen cells incubated in IFN-γ has been shown to mediate protection of newborn mice against HSV [76]. Administration of IFN-β to mice increased the cytotoxic activity of their peritoneal exudate cells against HSV-1–infected cells in vitro, as compared with cells from untreated mice [77]. Further investigation is required to elucidate the relative roles of direct IFN-mediated antiviral action, IFN-mediated activation of Mϕs, IFN-mediated activation of NK cells, and other indirect IFN effects in immunotherapy (Figures 1,2).

2. *Effect of the Colony-Stimulating Factors in Herpes Simplex Virus Infections*

The CSFs have been shown to have a protective effect against HSV in both normal and immunosuppressed mice. Multiple injections of recombinant human M-CSF protected guinea pigs against HSV-2, and this was correlated with increases in both the number of blood monocytes and their cytotoxic activity against HSV-infected cells [78]. Multiple injections of recombinant murine GM-CSF protected cyclophosphamide-treated mice from systemic HSV-1 infection [79]. However, combined repeated prophylactic and therapeutic treatment with 2×10^6 U recombinant human M-CSF or 1×10^4 U recombinant murine GM-CSF failed to protect immunocompetent mice against HSV-2 infection (D. Stewart and P.S. Morahan, unpublished data). Furthermore, after M-CSF treatment of mice according to this schedule, there was no change in the intrinsic antiviral resistance of either the peritoneal or liver Mϕs to HSV-1 infection in vitro [80]. Because the CSF source, route, schedule, and dose and the infection models used differed substantially in these four studies, it is not possible to draw any general conclusions. It is encouraging that certain CSF regimens can have a protective effect against HSV infection in vivo, since CSFs have been shown to increase Mϕ permissiveness to HSV in vitro (Section II.B.).

3. Effect of Other Cytokines in Herpes Simplex Virus Infections

Very few studies have investigated the effect of other cytokines on antiviral resistance and Mϕ antiviral activity in vivo. In a model of therapy for HSV infections in neonatal mice, which involved transferring murine Mϕs from adult mice in combination with repeated administration of human recombinant IL-2, protection was reported to be mediated by IFN-γ and was associated with increased antibody-dependent cellular cytotoxicity [81]. Because of the presence of IL-2 receptors on activated monocytes [82], a direct effect of IL-2 on Mϕs cannot be excluded. Chong et al. [83] have also reported that human recombinant IL-2 could reduce the severity of HSV infection in mice, and they showed that enhanced resistance was correlated with an increase in Mϕ intrinsic antiviral activity.

Tumor necrosis factor-α has recently been shown to exert a protective effect in mice infected with HSV-1 [84]. Whether the effect is direct or mediated through IFN remains to be established; IFN could not be detected in the serum. The persistence or loss of activity in mice treated with anti-IFN antibodies needs to be determined. Combined repeated prophylactic and therapeutic treatment of mice with several doses around 10^5 U recombinant human IL-6, another Mϕ-secreted monokine, did not protect mice against HSV-2 infection (L. Wu and P.S. Morahan, unpublished data).

In summary, treatment of animals with several cytokines including IFN-α, IFN-β, IFN-γ, M-CSF, GM-CSF, IL-2, and TNF can protect against HSV infections in experimental animal models. This protection has been correlated in some investigations with increased Mϕ antiviral activity. Some cytokines appear to act through IFN production, while others may act independently of IFN. Establishment of the optimal cytokine and treatment schedule remains an important effort, since the improper administration of cytokines can disturb the cytokine network in vivo. For example, the uncontrolled production of IFN-γ caused abnormal responses of other cytokines in a patient infected with HSV [85]. Also, TNF has been shown to reactivate a persistent HSV infection in murine splenic Mϕs [86].

B. Antiviral Effects in Vivo of Immunomodulators Other Than Cytokines

It has been well established that immunomodulators such as microbially derived materials and synthetic compounds can protect mice against HSV infections [2]. The protection has often been correlated with activation of Mϕ antiviral resistance mechanisms and/or NK cells early in the course of infection [2,87]. Some immunomodulators cause an influx of Mϕs with intrinsic antiviral activity to the site of infection, which may be crucial for recovery [80,88,89]. Macrophage depletion studies have indicated that Mϕs appear to be a critical determinant in the protective effect of at least one immunomodulator [90], but not others [91–93]. These data suggest that circulating monocytes and certain populations of tissue Mϕs may not be required for the protective effects of these immunomodulators.

Some immunomodulators appear to exert their antiviral effects by inducing the production of IFN-α/β. Whether the IFN acts directly or through activating Mϕs or other cells remains to be established. Using anti-IFN serum administered to mice, the protective effect of Poly ICLC and CL246,738 against HSV-2 infection was shown to be mediated by IFN [94]. A Mϕ-IFN "alliance" [21] might be involved; Poly IC–induced IFN-β has been shown to increase Mϕ intrinsic antiviral activity [95]. The protective effect of heat-killed *Lactobacillus casei*, an IFN inducer, was also correlated with increased Mϕ intrinsic activity [96]. Macrophage intrinsic resistance was additionally enhanced by IFN-β treatment in vitro or the in vivo administration of UV-inactivated HSV-1 antigen, which also induced IFN. It has not yet been established whether IFN production is a common mechanism of action among all immunomodulators protective against HSV infections.

Macrophage-activating immunomodulators have also been shown to protect mice immunosuppressed with cyclophosphamide treatment, and the authors speculated that the restoration of Mϕs and stimulation of their functional activities may be due to the induction of CSF production by these immunomodulators [89,97]. The antiviral efficacy of immunomodulators in immunosuppressed mice gives support to their usefulness in viral infections in immunocompromised patients.

VI. ROLE OF MACROPHAGES IN THE INFLAMMATORY RESPONSE TO HERPES SIMPLEX VIRUS INFECTION

There has been recent interest in better characterizing the protective and pathogenic inflammatory responses induced by HSV infection. Inflammatory infiltrates have been examined both at the initial site of virus replication and in other organs after viral dissemination had occurred.

A. Early Inflammatory Events at the Site of Virus Inoculation

The mouse peritoneum, after an intraperitoneal infection, serves as a convenient model to study the early cellular response of the host at the site of inoculation. It has been well established that peritoneal Mϕs are the primary cells that take up the virus during the early hours after infection [3,98–100] and therefore may serve as a barrier to prevent the further spread of infection. The type of peritoneal inflammatory response differs depending on the strain of infecting virus. After intraperitoneal inoculation of mice with 2×10^7 PFU of HSV-1, KOS strain, there was an influx of PMNs into the peritoneum beginning on days 1 to 3 after infection, and a monocytosis, lymphocytosis, and persisting PMN influx by day 5 [101] (Table 1). This viral dose did not cause lethal infection. In contrast, inoculation of mice with an LD_{90} dose (7×10^3 PFU) of HSV-2 MS strain, resulted in a quite different inflammatory response. A lymphocytosis was apparent days 3 and 5 after infection, with no apparent change in the total number of Mϕs or PMNs [101] (Table 1), but a decrease by day 5 in the number of recoverable resident Mϕs, which appeared activated both phenotypically and morphologically [100]. Very different inflammatory responses have also been reported recently with HSV-1 and HSV-2 infections of the anterior chamber of the eye [102]. These studies emphasize that although these two herpesviruses are very similar structurally, they are biologically quite distinct.

As with the above infections, skin lesions induced by HSV-1 or HSV-2 represent a combination of direct cell death due to viral infection and an associated inflammatory response [103]. In some HSV-induced vesicles, T cells were predominant early after infection (12 to 24 hours). Monocytes/Mϕs, prominent on day 2 after infection, were presumably activated by IFN-γ, since a substance copurifying with it could be detected in the Mϕ cytoplasm [104]. In one HSV-2–induced genital lesion biopsied for electron microscopy, PMNs were observed that had phagocytosed the virus, while monocytes showed limited viral replication and cytopathic effects [105]. The authors interpreted these results as indicating monocytes may function as vehicles for viral dissemination, but there was no direct investigation of this point. Only Mϕs from suckling mice have actually been shown to be involved in the dissemination of HSV infection [3].

B. Inflammation After Virus Dissemination

Macrophages may also be involved in inflammatory processes that occur after the spread of HSV from the initial site of infection. In herpes simplex encephalitis, T cell–mediated mechanisms appear to contribute to tissue injury in the brain [106] and may interact with Mϕs to extend central

TABLE 1 Comparison of the Early Host Inflammatory Response After Intraperitoneal HSV-1 or HSV-2 Infection[a]

Virus	Day post-infection	Group	Total PCs[b] (× 10^6)	Mϕ (× 10^6)	Lymphocytes (× 10^6)	PMNs (× 10^6)
HSV-1	1	Control	8.7[c] (2.4)	6.4 (1.4)	1.3 (0.4)	0.7 (0.5)
		HSV-1	8.5 (1.4)	2.6 (2.2)	1.0 (1.0)	4.3 (3.2)[d]
	3	Control	9.2 (1.3)	6.3 (1.0)	2.1 (0.5)	0.4 (0.4)
		HSV-1	11.0 (6.3)	5.8 (2.7)	1.8 (1.0)	3.1 (4.1)[d]
	5	Control	8.7 (2.1)	6.5 (1.3)	1.9 (0.9)	0.1 (0.4)
		HSV-1	48.1 (28.3)[d]	28.3 (16.3)[d]	16.5 (10.7)[d]	2.3 (1.4)[d]
HSV-2	1	Control	9.7 (3.0)	5.3 (1.6)	3.7 (1.1)	0.5 (0.3)
		HSV-2	8.1 (1.6)	4.1 (0.6)	2.7 (0.8)	0.4 (0.1)
	3	Control	8.9 (1.3)	5.8 (0.9)	2.8 (1.4)	0.2 (0.2)
		HSV-2	13.0 (1.4)[d]	5.7 (1.1)	6.9 (1.4)[d]	0.1 (0.1)
	5	Control	9.2 (1.9)	5.7 (1.3)	3.2 (0.7)	0.1 (0.1)
		HSV-2	12.1 (3.4)	5.7 (1.4)	6.2 (2.1)[d]	0.1 (0.04)

[a]Mice were inoculated intraperitoneally with 2 × 10^7 PFU of HSV-1 or 7 × 10^3 PFU of HSV-2, and PCs were harvested on the days indicated.
[b]Total number of peritoneal cells per mouse.
[c]Data are the mean values in representative experiments of differential cell counts of PCs obtained from 4–5 mice per group. Numbers in parentheses represent the standard error.
[d]$p < .05$ for the virus infected group in comparison with the control group.
HSV-1 = herpes simplex virus type I; HSV-2 = herpes simplex virus type 2; PC = peritoneal cell; PMN = polymorphonuclear leukocyte; PFU = plaque forming unit.

nervous system demyelination [107]. This immunopathogenic role of Mϕs has been observed in other viral infections in the central nervous system [108].

In sites other than the central nervous system, Mϕs appear to play a protective role during inflammatory responses to HSV infection. In an experimental model of murine hepatitis [109], HSV-1 infection resulted in an early influx of PMNs, Mϕs, and lymphocytes into necrotic lesions in the liver by day 3 after infection, while an inflammatory reaction was almost completely absent in HSV-2–infected animals [110]. Similar to the studies involving HSV-1 and HSV-2 infection in the mouse peritoneum [100,101], a more intensive inflammatory infiltrate was associated with increased survival. Furthermore, depletion of Mϕs by pretreatment of mice with silica increased viral replication and necrotic lesions in the liver after infection with either virus type. In HSV esophagitis in humans, which occurs more frequently in immunocompromised patients, aggregates of Mϕs were shown to be characteristics of the associated inflammatory response and have been proposed to serve as a diagnostic marker for the disease [111].

There has been recent interest in the role of adhesion molecules in inflammatory responses to HSV infections. Adhesion molecules have been identified that are reported to be required for Mϕ adherence to HSV-infected endothelium during inflammation [112–114]. We have found, however, that treatment of mice with an antibody to the murine type 3 complement receptor, which inhibits inflammatory cell recruitment into the peritoneum [115], did not affect natural or MVE-2 immunomodulator-enhanced resistance to peritoneal HSV-2 infection (J.A. Hendrzak, L. Wu, and P.S. Morahan, unpublished data).

Further investigation is necessary to determine when Mϕs are critical during HSV-induced inflammation, and what adhesion events are required. Delineation of the molecules and kinetics of cellular events may be useful in immunopharmacological intervention to modify the host inflammatory response for increased antiviral efficacy.

VII. CONCLUSIONS AND FUTURE PERSPECTIVES

The spectrum of Mϕ-mediated antiviral defense mechanisms against HSV is becoming more clearly defined. In vitro studies have shown that a number of cytokines are involved in both Mϕ intrinsic and extrinsic antiviral activities against HSV. Also, Mϕs have recently been shown to play an important role during early host inflammatory responses to HSV.

The knowledge gained from studies of Mϕ-HSV interactions in vitro must continue to be applied to investigations of the more complex interactions in vivo. The ability of Mϕs at the initial site of infection to phagocytose and degrade HSV as compared with other inflammatory cells, as well as their permissiveness for viral growth, must be assessed and correlated with the outcome of the disease. Further investigation of the role of cytokines in Mϕ antiviral resistance mechanisms in vitro should lead to the development of more optimal treatment schedules in experimental HSV infections. Treatment of HSV infections with cytokines that act on Mϕs or are secreted by Mϕs may be useful, especially in immunocompromised individuals. Macrophages continue to be an important component in the development of immunotherapy for HSV infections.

REFERENCES

1. Mims CA. Aspects of the pathogenesis of virus diseases. Bacteriol Rev 1964; 28:30–71.
2. Morahan PS, Connor JR, Leary KR. Viruses and the versatile macrophage. Br Med Bull 1985; 41:15–21.
3. Johnson RT. The pathogenesis of herpes virus encephalitis. II. A cellular basis for the development of resistance with age. J Exp Med 1964; 120:359–74.
4. Wu L, Morhan PS. Macrophages and other nonspecific defenses: Role in modulating resistance against herpes simplex virus. In: Rouse BT, ed. Current topics in microbiology and immunology. Immunobiology of herpes simplex virus infection. New York: Springer-Verlag, 1992:89–110.
5. Kirchner H, Hirt HM, Rosenstreich DL, Mergenhagen SE. Resistance of C3H/HeJ mice to lethal challenge with herpes simplex virus. Proc Soc Exp Biol Med 1978; 157:29–32.
6. Sarmiento M. Intrinsic resistance to viral infection. Mouse macrophage restriction of herpes simplex virus replication. J Immunol 1988; 141:2740–8.
7. Sit MF, Tenney DJ, Rothstein JL, Morahan PS. Effect of macrophage activation on resistance of mouse peritoneal macrophages to infection with herpes simplex virus types 1 and 2. J Gen Virol 1988; 69:1999–2010.
8. Mogensen SC. Role of macrophages in natural resistance to virus infections. Microbiol Rev 1979; 43:1–26.
9. Van Strijp JAG, Miltenburg LAM, Van Der Tol ME, Van Kessel KPM, Fluit AC, Verhoef J. Degradation of herpes simplex virions by human polymorphonuclear leukocytes and monocytes. J Gen Virol 1990; 71:1205–9.
10. Van Strijp JAG, Van Kessel KPM, Van Der Tol ME, Fluit AC, Snippe H, Verhoef J. Phagocytosis of herpes simplex virus by human granulocytes and monocytes. Arch Virol 1989; 104:287–98.
11. Van Iwaarden JF, Van Strijp JAG, Ebskamp MJM, Welmers AC, Verhoef J, Van Golde LMG. Surfactant protein A is opsonin in phagocytosis of herpes simplex virus type 1 by rat alveolar macrophages. Am J Physiol 1991; 261:L204–L209.
12. Wu L, Morahan PS, Leary K. Mechanisms of intrinsic macrophage-virus interactions *in vitro*. Microbial Pathogenesis 1990; 9:293–301.
13. Wu L, Morahan PS, Leary K. Regulation of herpes simplex virus type 1 gene expression in nonpermissive murine resident peritoneal macrophages. J Leukocyte Biol 1993; 53:61–65.
14. Daniels CA, Kleinerman ES, Snyderman R. Abortive and productive infections of human mononuclear phagocytes by type 1 herpes simplex virus. Am J Pathol 1978; 91:119–29.
15. Plaeger-Marshall S, Ank BJ, Altenburger KM, Pizer LI, Johnston RB Jr, Stiehm ER. Replication of herpes simplex virus in blood monocytes and placental macrophages from human neonates. Pediatr Res 1989; 26:135–139.
16. Albers I, Kirchner H, Domke-Opitz I. Resistance of human blood monocytes to infection with herpes simplex virus. Virology 1989; 169:466–9.

17. Sarmiento M, Kleinerman ES. Innate resistance to herpes simplex virus infection. Human lymphocyte and monocyte inhibition of viral replication. J Immunol 1990; 144:1942–53.
18. Stevens JG, Cook ML. Restriction of herpes simplex virus by macrophages. An analysis of the cell-virus interaction. J Exp Med. 1971; 133:19–38.
19. Ben-Hur T, Rosen-Wolff A, Lamade W, Darai G, Becker Y. HSV-1 DNA sequence determining intraperitoneal pathogenicity in mice is required for transcription of viral immediate-early genes in macrophages. Virology 1988; 163:397–404.
20. Ben-Hur T, Moyal M, Rosen-Wolff A, Darai G, Becker Y. Characterization of RNA transcripts from herpes simplex virus-1 DNA fragment *Bam*HI-B. Virology 1989; 169:1–8.
21. Mogensen SC, Virelizier JL. The interferon-macrophage alliance. Interferon 1987; 8:55–84.
22. Gessani S, Belardelli F, Borghi P, Boraschi D, Gresser I. Correlation between the lipopolysaccharide response of mice and the capacity of mouse peritoneal cells to transfer an antiviral state. Role of endogenous interferon. J Immunol 1987; 139:1991–8.
23. Vogel SN, Fertsch D. Macrophages from endotoxin-hyporesponsive (Lps^d) C3H/HeJ mice are permissive for vesicular stomatitis virus because of reduced levels of endogenous interferon: possible mechanism for natural resistance to virus infection. J Virol 1987; 61:812–8.
24. Ellermann-Eriksen S, Sommerlund M, Mogensen SC. Differential sensitivity of macrophages from herpes simplex virus–resistant and –susceptible mice to respiratory burst priming by interferon-α/β. J Gen Virol 1989; 70:2139–47.
25. Linnavuori KH, Hovi T. Restricted replication of herpes simplex virus in human monocyte cultures: role of interferon. Virology 1983; 130:1–9.
26. Linnavuori KH. History of recurrent mucocutaneous herpes correlates with relatively low interferon production by herpes simplex virus–exposed cultured monocytes. J Med Virol 1988; 25:61–8.
27. Domke, I, Straub P, Jacobsen H, Kirchner H, Panet A. Inhibition of replication of herpes simplex virus in mouse macrophages by interferons. J Gen Virol 1985; 66:2231–6.
28. Mittnacht S, Straub P, Kirchner H, Jacobsen H. Interferon treatment inhibits onset of herpes simplex virus immediate-early transcription. Virology 1988; 164:201–10.
29. Lazdins J, Alteri E, Cook KW, Burgin C, Gangemi JD. Use of human monocytes in the evaluation of antiviral drugs: quantitation of HSV-1 cytopathic effects. Antiviral Res 1990; 13:175–86.
30. Merendino RA, Arena A, Liberto MC, Mesiti M, Chillemi S, Iannello D, Bonina L. Role of exogenous interferons on intrinsic antiviral activity of macrophages from patients affected by neoplasia. J Chemother 1990; 2:116–122.
31. Domke-Opitz I, Straub P, Kirchner H. Effect of interferon on replication of herpes simplex virus types 1 and 2 in human macrophages. J Virol 1986; 60:37–42.
32. Sandberg K, Gobl AE, Funa K, Alm GV. Characterization of the blood mononuclear leucocytes producing alpha interferon after stimulation with herpes simplex virus in vitro, by means of combined immunohistochemical staining and *in situ* RNA-RNA hybridization. Scand J Immunol 1989; 29:651–8.
33. Sandberg K, Eloranta ML, Johannisson A, Alm GV. Flow cytometric analysis of natural interferon-α producing cells. Scand J Immunol 1991; 34:565–76.
34. Sethi KK, Brandis H. *In vitro* acquisition of resistance against herpes simplex virus by permissive murine macrophages. Arch Virol 1979; 59:157–72.
35. Seid JM, Liberto M, Bonina L, Leung KN, Nash AA. T cell–macrophage interactions in the immune response to herpes simplex virus: the significance of interferon-γ. J Gen Virol 1986; 67:2799–802.
36. Klotzbucher A, Mittnacht S, Kirchner H, Jacobsen H. Different effects of IFNγ and IFNα/β on "immediate early" gene expression of HSV-1. Virology 1990; 179:487–91.
37. Lee MT, Warren MK. CSF-1 induced resistance to viral infection in murine macrophages. J Immunol 1987; 138:3019–22.
38. Leary K, Connor JR, Morahan PS. Comparison of herpes simplex virus type 1 DNA replication and virus production in murine bone marrow–derived and resident peritoneal macrophages. J Gen Virol 1985; 66:1123–9.
39. Kruse A, Kirchner H, Zawatsky R, Domke-Opitz I. In vitro development of bone-marrow-derived macrophages. Influence of mouse genotype on response to colony-stimulating factors and autocrine interferon induction. Scand J Immunol 1989; 30:731–40.

40. Lodmell DL, Niwa A, Hayashi K, Notkins AL. Prevention of cell-to-cell spread of herpes simplex virus by leukocytes. J Exp Med 1973; 137:706–20.
41. Hayashi K, Kurata T, Morishima T, Nassery T. Analysis of the inhibitory effect of peritoneal macrophages on the spread of herpes simplex virus. Infect Immun 1980; 28:350–8.
42. Morahan PS, Morse SS, McGeorge MB. Macrophage extrinsic antiviral activity during herpes simplex virus infection. J Gen Virol 1980; 46:291–300.
43. Chapes SK, Tompkins WAF. Cytotoxic macrophages induced in hamsters by vaccinia virus: selective cytotoxicity for virus-infected targets by macrophages collected late after immunization. J Immunol 1979; 123:303–310.
44. Stanwick TL, Campbell DE, Nahmias AJ. Cytotoxic properties of human monocyte-macrophages for human fibroblasts infected with herpes simplex virus: interferon production and augmentation. Cell Immunol 1982; 70:132–47.
45. Koff WC, Fidler IJ, Showalter SD, Chakrabarty MK, Hampar B, Ceccorulli LM, Kleinerman ES. Human monocytes activated by immunomodulators in liposomes lyse herpes-virus-infected but not normal cells. Science 1984; 224:1007–9.
46. Koff WC, Showalter SD, Chakrabarty MK, Hampar B, Ceccorulli LM, Kleinerman ES. Human monocyte-mediated cytotoxicity against herpes simplex virus–infected cells: activation of cytotoxic monocytes by free and liposome-encapsulated lymphokines. J Leukocyte Biol 1985; 37:461–72.
47. Visser MR, Verhoef J. Human nonspecific effector leukocytes: possible function in herpes simplex virus infection. Path Immunopathol Res 1986; 5:140–55.
48. Morse SS, Morahan PS. Activated macrophages mediate interferon-independent inhibition of herpes simplex virus. Cell Immunol 1981; 58:72–84.
49. Koff WC, Fann AV. Human tumor necrosis factor-alpha kills herpesvirus-infected but not normal cells. Lymphokine Res 1986; 5:215–21.
50. Mestan J, Digel W, Mittnacht S, Hillen H, Blohm D, Moller A, Jacobsen H, Kirchner H. Antiviral effects of recombinant tumour necrosis factor *in vitro*. Nature 1986; 323:816–9.
51. Wong GHW, Goeddel DV. Tumour necrosis factors α and β inhibit virus replication and synergize with interferons. Nature 1986; 323:819–22.
52. Feduchi E, Alonso MA, Carrasco L. Human gamma interferon and tumor necrosis factor exert a synergistic blockade on the replication of herpes simplex virus. J Virol 1989; 63:1354–9.
53. Etingin OR, Hjjar DP. Evidence for cytokine regulation of cholesterol metabolism in herpesvirus-infected arterial cells by the lipoxygenase pathway. J Lipid Res 1990; 31:299–305.
54. Oh SH, Trinchieri G, Bandyopadhyay S, Starr SE. Natural killer cell-mediated lysis of herpes simplex virus–infected fibroblasts: inability to detect soluble factors that contribute to lysis. Cell Immunol 1990; 127:221–9.
55. Van Strijp JAG, Van Der Tol ME, Miltenburg LAM, Van Kessel KPM, Verhoef J. Tumour necrosis factor triggers granulocytes to internalize complement-coated virus particles. Immunology 1991; 73:77–82.
56. Clouse KA, Orosz CG, Sheridan JF. Influence of macrophages on HSV-1 induced IL-2 production by human lymphocytes. Fed Proc 1986; 45:862.
57. Howie S, Norval M, Maingay J, McBride WH. Interactions between herpes simplex virus and murine bone marrow macrophages. Arch Virol 1986; 87:229–39.
58. Kleinerman ES, Synderman R, Daniels CA. Depression of human monocyte chemotaxis by herpes simplex and influenza viruses. J Immunol 1974; 113:1562–7.
59. Koff WC, Dunegan MA, Chakrabarty MK, Hampar B, Showalter SD. Herpes simplex virus–induced suppression of macrophage-mediated tumoricidal activity in murine macrophages. Cancer Res 1987; 47:1534–7.
60. Plaeger-Marshall S, Wilson LA, Smith JW. Alteration of rabbit alveolar and peritoneal macrophage function by herpes simplex virus. Infect Immun 1983; 41:1376–9.
61. Sipka S, Boldogh I, Abel G, Csongor J, Fachet J, Szegedi G. Effects of herpes virus infections on the chemiluminescence induced by zymosan phagocytosis in mouse peritoneal macrophages. Acta Microb Hungarica 1987; 34:115–20.
62. Giridhar G, Hayakawa H, Kucera LS, Myrvik QN. Priming of rabbit alveolar macrophages for enhanced oxidative responses by herpes simplex virus type 2 infection. J Leukocyte Biol 1991; 49: 442–8.

63. Mogensen SC, Ellermann-Eriksen S, Sommerlund M. Herpes simplex virus type 2 primes mouse macrophages for an early and genetically determined respiratory burst mediated by interferon-α/β. J Gen Virol 1989; 70:1371–9.
64. Wu L, Eisenstein TK, Morahan PS. Effect of herpes simplex virus type 1 infection on cytokine gene expression in activated murine peritoneal macrophages. In: Walsh L, Block T, Crowell RL, Jungkind DL, eds. Innovations in antiviral development and the detection of virus infection. New York: Plenum Press, 1992:167–72.
65. Gosselin J, Flamand L, D'Addario M, Hiscott J, Stefanescu I, Ablashi DV, Gallo RC, Menezes J. Modulatory effects of Epstein-Barr, herpes simplex, and human herpes-6 viral infections and coinfections of cytokine synthesis. J Immunol 1992; 149:181–7.
66. Sprecher E, Becker Y. Induction of interleukin-1 α and β gene transcription in mouse peritoneal exudate cells after intraperitoneal infection with herpes simplex virus-1. Arch Virol 1990; 110:259–69.
67. Zawatsky R, Hilfenhaus J, Marcucci F, Kirchner H. Experimental infection of inbred mice with herpes simplex virus type 1. I. Investigation of humoral and cellular immunity and of interferon induction. J Gen Virol 1981; 53:31–8.
68. Gresser I, Tovey MG, Maury C, Bandu MT. Role of interferon in the pathogenesis of virus disease in mice as demonstrated by the use of anti-interferon serum. II. Studies with herpes simplex, Moloney sarcoma, vesicular stomatitis, Newcastle disease, and influenza viruses. J Exp Med 1976; 144:1316–23.
69. Stanton GJ, Jordan C, Hart A, Heard H, Langford MP, Baron S. Nondetectable levels of interferon gamma is a critical host defense during the first day of herpes simplex virus infection. Microbial Pathogenesis 1987; 3:179–83.
70. Zawatsky R, Gresser I, DeMaeyer E, Kirchner H. The role of interferon in the resistance of C57BL/6 mice to various doses of herpes simplex virus type 1. J Infect Dis 1982; 146:405–10.
71. Gangemi JD, Lazdins J, Dietrich FM, Matter A, Poncioni B, Hochkeppel HK. Antiviral activity of a novel recombinant human interferon-αB/D hybrid. J Interferon Res 1989; 9:227–37.
72. Kunder SC, Wu L, Morahan PS. Role of NK cells in immunomodulator-mediated resistance to herpesvirus infection. Antiviral Res (in press).
73. Pinto AJ, Morahan PS, Brinton M, Stewart D, Gavin E. Comparative therapeutic efficacy of recombinant interferons-α, -β, and -γ against alphatogavirus, bunyavirus, flavivirus, and herpesvirus infections. J Interferon Res 1990; 10:293–8.
74. Ho M. Interferon as an agent against herpes simplex virus. J Invest Dermat 1990; 95:158S–160S.
75. Hartshorn KL, Hirsch MS. Interferons. In: Peterson PK, Verhoef J, eds. The antimicrobial agents annual 3. New York: Elsevier, 1988:371–83.
76. Kohl S, Loo LS, Drath DB, Cox P. Interleukin-2 protects neonatal mice from lethal herpes simplex virus infection: a macrophage-mediated, γ interferon–induced mechanism. J Infect Dis 1989; 159:239–47.
77. Watanabe T, Yamori T, Yatani R. Primary barrier and amplifier in herpes simplex virus infection in BALB/c mice. Mie Med J 1989; 39:297–306.
78. Ho RJY, Chong KY, Merigan TC. Antiviral activity and dose optimum of recombinant macrophage colony-stimulating factor on herpes simplex genitalis in guinea pigs. J Immunol 1991; 146:3578–82.
79. Iida J, Saiki I, Ishihara C, Azuma I. Protective activity of recombinant cytokines against Sendai virus and herpes simplex virus (HSV) infections in mice. Vaccine 1989; 7:229–33.
80. Hendrzak JA, Pinto AJ, Morahan PS. Intrinsic resistance to herpes simplex virus type 1 infection in liver Kupffer cells and peritoneal macrophages from normal and immunomodulator-treated mice. Nat Immun 1992; 11:26–33.
81. Kohl S, Loo LS. Use of interleukin-2 and macrophages to treat herpes simplex virus infection in neonatal mice. J Infect Dis 1988; 157:1187–91.
82. Wahl SM, McCartney-Francis N, Hunt DA, Smith PD, Wahl LM, Katona IM. Monocyte interleukin 2 receptor gene expression and interleukin 2 augmentation of microbicidal activity. J Immunol 1987; 139:1342–7.
83. Chong KT. IL-2 induced resistance to viral infection in murine macrophages. J Leukocyte Biol 1987; 42:351.
84. Rossol-Voth R, Rossol S, Schutt KH, Corridori S, De Cian W, Falke D. *In vivo* protective effect of tumour necrosis factor α against experimental infection with herpes simplex virus type 1. J Gen Virol 1991; 72:143–7.

85. Imashuku S, Hibi S. Cytokines in haemophagocytic syndrome. Br J Haematol 1991; 77:438–40.
86. Domke-Opitz I, Kirchner H. Stimulation of macrophages by endotoxin results in the reactiviation of a persistent herpes simplex virus infection. Scand J Immunol 1990; 32:69–75.
87. Morahan PS, Morse SS. Macrophage-virus interactions. In: Proffitt MR, ed. Virus-lymphocyte interactions: implications for disease. New York: Elsevier-North Holland, 1979:17–35.
88. Ikeda S, Sai K, Nichimura C, Yamamoto A. Antiherpes activity of the immunomodulator OK-432, a streptococcal preparation, in immunosuppressed mice. Antiviral Res 1988; 10:299–304.
89. Ikeda S, Tominaga T, Nishimura C, Homma JY, Kiso M, Hasegawa A. Antiherpes activity of chemically synthesized lipid A–subunit analogue GLA-60 in immunosuppressed mice. Antiviral Res 1989; 11:173–80.
90. Harada Y, Kurokawa E, Endo H, Kimura T, Saito M, Sugawara Y, Irie H, Ito K, Fujii M, Shimizu Y, Kumagai K. Protection of OK-432, a *Streptococcus pyogenes* preparation, against lethal infection of mice with herpes simplex virus. Microbiol Immunol 1989; 33:467–77.
91. Morahan PS, Kern ER, Glasgow LA. Immunomodulator-induced resistance against herpes simplex virus. Proc Soc Exp Biol Med 1977; 154:615–20.
92. Pinto AJ, Stewart D, Van Rooijen N, Morahan PS. Selective depletion of liver and splenic macrophages using liposomes encapsulating the drug dichloromethylene diphosphonate: effects on antimicrobial resistance. J Leukocyte Biol 1991; 49:579–86.
93. Morahan PS, Dempsey WL, Volkman A, Connor J. Antimicrobial activity of various immunomodulators: independence from normal levels of circulating monocytes and natural killer cells. Infect Immun 1986; 51:87–93.
94. Kunder SC, Kelly KM, Morahan PS. Biological response modifier mediated resistance to herpesvirus infections requires induction of α/β interferon. Antiviral Res (in press).
95. Pyo S, Gangemi JD, Ghaffar A, Mayer EP. Poly I:C-induced anti–herpes simplex virus type 1 activity in inflammatory macrophages is mediated by induction of interferon-β. J Leukocyte Biol 1991; 50:479–87.
96. Watanabe T, Yamori T. Primary resistance induced in mice by *Lactobacillus casei* following infection with herpes simplex virus. J Jap Assoc Infect Dis 1989; 63:182–8.
97. Iida J, Nishi N, Saiki I, Mizukoshi N, Ishihara C, Tokura S, Azuma I. Macrophage activation and host augmentation against Sendai or herpes simplex virus (HSV) infections with synthetic polypeptides in mice. Int J Immunopharmacol 1989; 11:249–58.
98. Armerding D, Mayer P, Scriba M, Hren A, Rossiter H, In-Vivo modulation of macrophage functions by herpes simplex virus type 2 in resistant and sensitive inbred mouse strains. Immunobiology 1981; 160:217–27.
99. Yamada M, Arao Y, Uno F, Nu S. Mechanism of differences in pathogenicity between two variants of a laboratory strain of herpes simplex virus type 1. Microbiol Immunol 1986; 30:1259–70.
100. Hendrzak JA, Morahan PS. Alterations in resident peritoneal macrophages in the early inflammatory response to herpes simplex virus type 2 infection. Manuscript in preparation.
101. Hendrzak JA, Morahan PS. Early inflammatory response to herpes simplex virus infections. Eleventh Annual Philadelphia Immunology Conference, Philadelphia, 1990.
102. Dix RD, Streilein JW, Cousins S, Atherton SS. Histopathologic characteristics of two forms of experimental herpes simplex virus. Curr Eye Res 1987; 6:47–52.
103. Whitley RJ. Herpes simplex viruses. In: Fields BN, Knipe DM, eds. Fields Virology. 2nd ed. Vol. 2. New York: Raven Press, 1990:1843–87.
104. Cunningham AL, Turner RR, Miller AC, Para MF, Merigan TC. Evolution of recurrent herpes simplex lesions. An immunohistologic study. J Clin Invest 1985; 75:226–33.
105. Boddingius J, Dijkman H, Hendriksen E, Schift R, Stolz E. HSV-2 replication sites, monocyte and lymphocytic cell infection and virion phagocytosis by neutrophils, in vesicular lesions on penile skin. J Cutan Pathol 1987; 14:165–75.
106. Sobel RA, Collins AB, Colvin RB, Bhan AK. The *in situ* cellular immune response in acute herpes simplex encephalitis. Am J Pathol 1986; 125:332–8.
107. Townsend JJ. Macrophage response to herpes simplex encephalitis in immune competent and T cell-deficient mice. J Neuroimmunol 1985; 7:195–206.
108. McFarland HF, Johnson RT. The role of the inflammatory response in viral infections. In: Notkins AL, ed. Viral immunology and immunopathology. New York: Academic Press, 1975:137–48.

109. Mogensen SC, Teisner B, Andersen KH. Focal necrotic hepatitis as a biological marker for differentiation of herpesvirus hominis type 1 and type 2. J Gen Virol 1974; 25:151–5.
110. Schirmacher P, Worsdorfer M, Lubbe K, Falke D, Thoenes W, Dienes HP. HSV hepatitis in the mouse: a light and electron microscopic study with immunohistology and in situ hybridization. Virchows Arch B Cell Pathol 1989; 56:351–61.
111. Greenson JK, Beschorner WE, Botnott JK, Yardley JH. Prominent mononuclear cell infiltrate is characteristic of herpes esophagitis. Hum Pathol 1991; 22:541–9.
112. Kohl S, Loo LS, Schmalstieg FS, Anderson DC. The genetic deficiency of leukocyte surface glycoprotein Mac-1, LFA-1, p150,95 in humans is associated with defective antibody-dependent cellular cytotoxicity in vitro and defective protection against herpes simplex virus infection in vivo. J Immunol 1986; 137:1688–94.
113. Etingin OR, Silverstein RL, Hajjar DP. Identification of a monocyte receptor on herpesvirus-infected endothelial cels. Proc Natl Acad Sci USA 1991; 88:7200–3.
114. Span AHM, Van Dam-Mieras MCE, Mullers W, Endert J, Muller AD, Bruggeman CA. The effect of virus infection on the adherence of leukocytes or platelets to endothelial cells. Eur J Clin Invest 1991; 21:331–8.
115. Rosen H, Gordon S. Monoclonal antibody to the murine type 3 complement receptor inhibits adhesion of myelomonocytic cells in vitro and inflammatory cell recruitment in vivo. J Exp Med 1987; 166:1685–1701.

Index